# THE ROUTLEDGE INTERNATIONAL HANDBOOK OF HUMAN-ANIMAL INTERACTIONS AND ANTHROZOOLOGY

This diverse, global, and interdisciplinary volume explores the existing research, practice, and ethical issues pertinent to the field of human-animal interactions (HAIs), interventions, and anthrozoology, focusing on the perceived physical and mental health benefits to humans and the challenges derived from these relationships.

The book begins by exploring the basic theoretical principles of anthrozoology and HAI, such as the evolution and history of the field, the importance of language, the economic costs and current perspectives to physical and mental wellbeing, the origins of domestication of animals, anthropomorphism, and how animals fit into human societies. Chapters then move onto practice, covering topics such as how animals help childhood and adulthood development, pet ownership, disability, the roles of pets for people with psychiatric disorders, the links between animal and domestic abuse, and then more widely into the therapeutic roles of animals, animal-assisted therapies, interactions outside the home, working animals, animals in popular culture, and animals in research, for leisure, and food. Including chapters on a wide range of animals, from domesticated pets to wildlife, this collection examines the benefits yet also reveals the complexity, and often dark side, of human-animal relations.

Interweaving accessible commentaries with revealing chapters throughout the text, this collection would be of interest to students and practitioners in the fields of mental health, psychology, veterinary medicine, zoology, biology, social work, history, and sociology.

**Dr. Aubrey H. Fine** is a Professor Emeritus at California State Polytechnic University. He is the author and editor of numerous books on human-animal interactions, animal-assisted interventions, children, and parenting.

**Dr. Megan K. Mueller** is an Associate Professor of Human-Animal Interaction at the Cummings School of Veterinary Medicine at Tufts University who studies the role of human-animal interaction in youth and families.

**Dr. Zenithson Y. Ng** is a Clinical Associate Professor of small animal primary care at the University of Tennessee College of Veterinary Medicine with interests spanning all aspects of the human-animal bond.

**Dr. Alan M. Beck** is a Professor who explores the relationship between people, animals, and their environment. He emphasizes the evolutionary bases for the interactions.

**Dr. Jose M. Peralta** is a Veterinary School Professor of animal welfare who is interested in the impact that the interaction between animals and humans has on the animals and their wellbeing.

# ROUTLEDGE INTERNATIONAL HANDBOOKS

**THE ROUTLEDGE INTERNATIONAL HANDBOOK OF NEW CRITICAL RACE AND WHITENESS STUDIES**
Edited by: Rikke Andreassen, Catrin Lundström, Suvi Keskinen, Shirley Anne Tate.

**ROUTLEDGE HANDBOOK OF ASIAN PARLIAMENTS**
Edited by: Po Jen Yap and Rehan Abeyratne

**THE ROUTLEDGE INTERNATIONAL HANDBOOK OF CHILDREN'S RIGHTS AND DISABILITY**
Edited by: Angharad E. Beckett and Anne-Marie Callus

**ROUTLEDGE HANDBOOK OF MACROECONOMIC METHODOLOGY**
Edited by: Jesper Jespersen, Victoria Chick and Bert Tieben

**THE ROUTLEDGE HANDBOOK OF DIGITAL SOCIAL WORK**
Edited by: Antonio López Peláez and Gloria Kirwan

**THE ROUTLEDGE INTERNATIONAL HANDBOOK OF ECONOMIC SOCIOLOGY**
Edited by: Milan Zafirovski

**THE ROUTLEDGE INTERNATIONAL HANDBOOK OF INTERSECTIONALITY STUDIES**
Edited by: Kathy Davis and Helma Lutz

**THE ROUTLEDGE INTERNATIONAL HANDBOOK OF THE PSYCHOLOGY OF MORALITY**
Edited by: Naomi Ellemers, Stefano Pagliaro and Félice van Nunspeet

**THE ROUTLEDGE INTERNATIONAL HANDBOOK ON HUMAN-ANIMAL INTERACTIONS AND ANTHROZOOLOGY**
Aubrey H. Fine, Megan K. Mueller, Zenithson Y. Ng, Alan M. Beck, and Jose M. Peralta

# THE ROUTLEDGE INTERNATIONAL HANDBOOK OF HUMAN-ANIMAL INTERACTIONS AND ANTHROZOOLOGY

*Edited by*
*Aubrey H. Fine, Megan K. Mueller,*
*Zenithson Y. Ng, Alan M. Beck, and Jose M. Peralta*

Routledge
Taylor & Francis Group
NEW YORK AND LONDON

Designed cover image: © Getty Images

First published 2024
by Routledge
605 Third Avenue, New York, NY 10158

and by Routledge
4 Park Square, Milton Park, Abingdon, Oxon, OX14 4RN

*Routledge is an imprint of the Taylor & Francis Group, an informa business*

© 2024 selection and editorial matter, Aubrey H. Fine, Megan K. Mueller, Zenithson Y. Ng, Alan M. Beck, and Jose M. Peralta; individual chapters, the contributors

The right of Aubrey H. Fine, Megan K. Mueller, Zenithson Y. Ng, Alan M. Beck, and Jose M. Peralta to be identified as the authors of the editorial material, and of the authors for their individual chapters, has been asserted in accordance with sections 77 and 78 of the Copyright, Designs and Patents Act 1988.

All rights reserved. No part of this book may be reprinted or reproduced or utilised in any form or by any electronic, mechanical, or other means, now known or hereafter invented, including photocopying and recording, or in any information storage or retrieval system, without permission in writing from the publishers.

*Trademark notice*: Product or corporate names may be trademarks or registered trademarks, and are used only for identification and explanation without intent to infringe.

*Library of Congress Cataloging-in-Publication Data*
Names: Fine, Aubrey H., editor. | Mueller, Megan K., editor. | Ng, Zenithson Y., editor. | Beck, Alan M., editor. | Peralta, José Manuel, editor.
Title: The Routledge international handbook of human-animal interactions and anthrozoology / edited by Aubrey H. Fine, Megan K. Mueller, Zenithson Y. Ng, Alan M. Beck, Jose M. Peralta.
Description: New York, NY: Routledge, 2024. | Includes bibliographical references and index.
Identifiers: LCCN 2023007637 (print) | LCCN 2023007638 (ebook) | ISBN 9781032153322 (hardback) | ISBN 9781032153339 (paperback) | ISBN 9781032153346 (ebook)
Subjects: LCSH: Human-animal relationships. | Animal welfare. | Animals and civilization.
Classification: LCC QL85 .R685 2024 (print) | LCC QL85 (ebook) | DDC 591.5—dc23/eng/20230412
LC record available at https://lccn.loc.gov/2023007637
LC ebook record available at https://lccn.loc.gov/2023007638

ISBN: 9781032153322 (hbk)
ISBN: 9781032153339 (pbk)
ISBN: 9781032153346 (ebk)

DOI: 10.4324/9781032153346

Typeset in Bembo Std
by codeMantra

Aubrey H. Fine EdD

I would like to acknowledge all the scholars, researchers, and practitioners who, over the past decades, have made an impact on the growing field of anthrozoology. Personally, I would like to dedicate this book to my grandchildren Leora, Matteo, and Alina who have been blessed with the opportunity of growing up with animals. They are learning with experience how rich their lives have become with the animals they share it with. I also want to dedicate the book to all the animals in my family members' lives today. They include Hunter, Colby, Von, Amber, Daniella, Metropolis, Eclipse, Fey, Ketzy, and Mystic. Finally, to Sasha, my little gerbil, who over five decades ago opened my eyes to the significance of the human–animal bond.

Megan K. Mueller PhD

To the people and animals who graciously share their lives with us in our research and practice, and who remind us of why this work matters. And to Jett, the dog who started it all.

Zenithson Y. Ng DVM

Dedicated to my parents, sister, brother, and grandpa, whose unconditional support allowed me to dream big. Dedicated to the countless mentors and colleagues who continue to teach me all the things I don't know every single day. To all the animals who have touched my life, and of course, Grace.

Alan M. Beck, ScD

Dedicated to the many authors of this book, whose selfless efforts have allowed all to better understand and appreciate our relationship with animals and nature. I also thank Gail Beck who shared advice and continuously encouraged me to read, edit, and learn.

Jose M. Peralta DVM, PhD

To Amaya and Ana for their continued love and support. To my parents for always showing me the right path with their example. To all animals for making our lives so much more interesting and complete.

# CONTENTS

*About the Editors* xiv
*About the Contributors* xvii
*Foreword* xxiv
Temple Grandin

1  An Introduction to HAI and Anthrozoology 1
   *Aubrey H. Fine, Alan M. Beck, and Megan K. Mueller*

2  Valuational (Ethical) Inquiry, Animals, and Us: Contemporary
   Framings and Future Imaginaries 17
   *Raymond Anthony and Andreia De Paula Vieira*

3  A Legal Opinion: Pets and Housing in the United States 37
   *Dianne Prado*

4  All for One Health and One Health for All 49
   *Zenithson Y. Ng and Isabella Pfeiffer*

5  Current Perspectives on the Health Effects (Mental and Physical)
   of Human-Animal Interaction (HAI) 65
   *Erika Friedmann and Lincy Koodaly*

6  Challenges and Opportunities in Establishing the Evidence Base
   for the Impact of Human-Animal Interactions on Human Health
   and Well-Being 79
   *Kerri E. Rodriguez, Erin Flynn, Jaci Gandenberger,
   Marisa Motiff, and Kevin N. Morris*

## Contents

7  Anthropomorphism and Anthropocentrism in the Anthropocene    93
*Randall Lockwood*

8  Pets and Housing: A One Health One Welfare Issue    109
*Brinda Jegatheesan, Elizabeth Ormerod, Taryn M Graham,
Wendy Stone, Emma R. Power, Debbie Rook, and Sandra McCune*

9  Diversity, Equity, Inclusion, and Belonging in Human-Animal
Interactions: Reflections on Practice    123
*Sloane Hawes, Lizett Gutierrez, Lindsay Rojas, Rebecca Cohen,
Bridget Camacho, Ashley Taeckens, Tess Hupe,
and Nina Ekholm Fry*

10  Across Worlds: Cultural, Religious, and Societal Factors Influencing
Interactions with Animals    139
*Georgitta Valiyamattam*

11  Animal Cognition    153
*Lena Provoost*

12  Animal Sentience, Animal Emotions, and Human-Animal Interactions    169
*Annika Bremhorst, Catia Caeiro, Anna Zamansky, and Sabrina Karl*

13  Animal Domestication in West Eurasia: What the Histories of Dogs,
Cattle, and Horses Tell Us about the Domestication Process    183
*Benjamin S. Arbuckle*

14  Why Do We Love the Pets We Have? The Role of Animals and Family    200
*Alan M. Beck, Aubrey H. Fine, Stanley Coren, Erin King,
Steve Feldman, and Lindsey Braun*

15  Animals in Development: Childhood and Adolescence    214
*Erin King, Eli D. Halbreich, and Megan K. Mueller*

16  Animals in Development: Aging Adults    228
*Jessica Bibbo and Ann M. Toohey*

17  LGBTQIA+ Populations and Human-Animal Interactions    245
*Angela Matijczak and Shelby McDonald*

18  Pet Ownership and Persons with Developmental and Physical
Disabilities: Impact on Mental and Physical Health    259
*Gretchen K. Carlisle and Steph M. Craven*

## Contents

19 Homelessness and Pet Ownership — 272
*Leslie Irvine*

20 Caregiver Burden in Companion Animal Owners — 286
*Mary Beth Spitznagel, Mark D. Carlson, and Christopher M. Fulkerson*

21 Human – Animal Interactions and the Impact of Pet Loss — 299
*Susan P. Cohen*

22 Animal Hoarding: The Human-Animal Bond Gone Awry — 314
*Phil Arkow*

23 Co-Occurrence of Animal Abuse and Intimate Partner Violence — 331
*Maya Gupta and Shelby McDonald*

24 Introducing Dogs and Cats into Families — 347
*Zarah Hedge, Amanda Kowalski, and Madelaine Steevens*

25 Veterinary Social Work — 360
*Pamela Linden*

26 Zoonoses — 373
*Jason W. Stull*

27 Working with Each Other: Animal Training to Coexist — 388
*Seana Dowling-Guyer*

28 Safe Interactions with Animals — 404
*Carri Westgarth, Sara Owczarczak-Garstecka, and John Tulloch*

29 The Case for Conservation in the Age of the Anthropocene — 421
*Eric G. Strauss and Carolyn A. Ristau*

30 Ethics, Wellbeing and Wild Lives — 438
*William S. Lynn, Liv Baker, Francisco Santiago-Ávila, and Kristin L. Stewart*

31 Tourism and Its Impacts on Wildlife — 453
*Neil Carr and Paul Tully*

32 Human-Wildlife Conflict and Interaction — 467
*Brian J. MacGowan*

33  Oh, the Places I've Seen and the Places We Will Go: A Commentary on the Evolution of AAI  477
    *Aubrey H. Fine*

34  The Practice of Animal-Assisted Interventions: Exploring the Value of AAT/AAA  494
    *Taylor Chastain Griffin, Jen VonLintel, Erin Trella, and Annie Kate Hudson*

35  Animals in Education  509
    *Nancy R. Gee*

36  Around the Arena: Equine-Assisted Services' Foundation in Understanding Human-Horse Interactions  522
    *Aviva Vincent and Robin Peth-Pierce*

37  Assistance and Emotional Support Animals  537
    *Marguerite E. O'Haire, Leanne O. Nieforth, Clare L. Jensen, and Sarah C. Leighton*

38  Handler-Dog Interaction: Role of Operational Working Dogs from Past to Present  553
    *Melissa A. Singletary, Sarah Krichbaum, and Lucia Lazarowski*

39  Working Animals and Human-Animal Interactions  568
    *Tamara Tadich and João B. Rodrigues*

40  Human-Nonhuman Interactions in the Animal Laboratory  582
    *Lesley Sharp, Garet Lahvis, Amy Robinson-Junker, and Larry Carbone*

41  Companion Animals in Times of Crisis: Contemporary Evidence from the COVID-19 Pandemic  596
    *Jaume Fatjó and Jonathan Bowen*

42  Human-Animal Interactions in an Agricultural and Production Setting  613
    *Jose M. Peralta and Taylor Haefs*

43  Farm-Based Interventions Considering One Health – One Welfare  626
    *Lena Lidfors, Bente Berget, and Karen Thodberg*

44  Public Perceptions of Farm Animal Welfare: The Case of Cow-Calf Separation  644
    *Lara V. Sirovica and Marina A.G. von Keyserlingk*

| | | |
|---|---|---|
| 45 | Human-Animal Interactions in Western Art<br>*Laura D. Gelfand* | 658 |
| 46 | Sinister Canines in the Media: Animals in Popular Culture<br>*Margo DeMello* | 677 |
| 47 | I'll Get You… and Your Little Dog Too! Animal Cruelty and Kindness in the Cinema over the Last Century<br>*Randall Lockwood* | 690 |
| 48 | Animals in Our Leisure and Pleasure: Research for a More Species Equitable Future<br>*Carmel Nottle, Janette Young, Torben Nielsen, Kate Dashper, and Zoei Sutton* | 704 |
| 49 | Hunting and the Consumptive Use of Wildlife<br>*Jacob L. Berl and Joshua T. Wise* | 719 |
| 50 | Human-Animal Interactions with Non-Domestic Animals in Captive Settings<br>*Petty Grieve and Marcy Souza* | 729 |
| 51 | Epilogue: A Roadmap for the Future of Anthrozoology<br>*Aubrey H. Fine, Megan K. Mueller, Zenithson Y. Ng, Alan M. Beck, and Jose M. Peralta* | 746 |
| | Appendix 1: Glossary of Terms<br>*Stefanie Malzyner, Nicole Lorig, Kylie Bell, and Morgan Starkweather* | 753 |
| *Index* | | 765 |

# ABOUT THE EDITORS

**Dr. Aubrey H. Fine** is a native of Montreal, Canada. He received his graduate degree from University of Cincinnati in 1982. Dr. Fine has been on the faculty at California State Polytechnic University since 1981 and is presently a Professor Emeritus and a licensed psychologist. In 2001, Dr. Fine was presented the Wang Award given to distinguished professor within the California State University system (23 Universities).

Aubrey has been recognized by numerous organizations for his service and dedication to children, animals, and the community. In July 2016, he received the William McCulloch Award for Excellence in HAI Education and Practice from the International Association of Human Animal Interaction Organizations, in Paris, France. Additionally, he was awarded the Educator of the Year, from the Learning Disability Association of CA and receiving the 2006 CA Poly Faculty Award for Community Engagement.

Aubrey's primary research interests relate to the psycho-social impact of human-animal interactions and animal-assisted interventions, social skills training and children with ADHD, and resilience in children. He is the author of several books including *Our Faithful Companions, Parent Child Dance, Therapist's Guide to Learning and Attention Disorders, Fathers and Sons, The Total Sports Experience for Children, Give a Dog Your Heart, The Handbook on Animal Assisted Therapy, Afternoons with Puppy,* and *The Welfare of Animals in Animal Assisted Therapy.* Dr. Fine is a board member of Pet Partners, past chair of the Human Animal Bond Advisory Committee of Pet Partners, as well as the past chair of the steering committee on Human Animal Interactions for the American Veterinary Medical Association.

*About the Editors*

**Dr. Megan K. Mueller,** PhD, is an Associate Professor of Human-Animal Interaction at the Cummings School of Veterinary Medicine at Tufts University within the Center for Animals and Public Policy. Dr. Mueller is Co-Director of the Tufts Initiative for Human-Animal Interaction and is a senior fellow at the Jonathan M. Tisch College of Civic Life at Tufts University. Dr. Mueller is a developmental psychologist, and her research program focuses on assessing the dynamic relationships between people and animals in families and communities. In particular, her work focuses on the psychology of the human-animal bond, and how human-animal interaction can promote thriving for adolescents and their families. Her research has been published in numerous scientific journals and media outlets. Dr. Mueller is a board member of the International Society for Anthrozoology and serves on the Pet Partners Human-Animal Bond Advisory Board.

**Dr. Zenithson Y. Ng** is a Clinical Associate Professor at the University of Tennessee. He earned his undergraduate degree in Animal Science from Rutgers University and his veterinary degree from Cornell University; then completed a small animal rotating internship at the ASPCA in NYC, followed by a canine/feline primary care residency combined with a master's degree in Human-Animal Bond Studies at Virginia Tech. His clinical interests include small animal behavior, dentistry, preventive medicine, and management of chronic disease. His research and teaching interests span all aspects of the human-animal bond including the effect of human-animal interaction on both humans and animals, the veterinary-client relationship, and stress reduction in both veterinary and animal-assisted intervention settings.

**Dr. Alan M. Beck** received his Baccalaureate from Brooklyn College and Master's degree from California State University at Los Angeles. He received his Doctor of Science in Animal Ecology from The Johns Hopkins University School of Public Health. His 1973 book, *The Ecology of Stray Dogs: A Study of Free-Ranging Urban Dogs*, is considered a classic in the field of urban ecology and was republished by Purdue University Press in 2002.

He directed the animal programs for the New York City Department of Health for five years, and then was the first Director of the Center for the Interaction of Animals and Society at the School of Veterinary Medicine, the University of Pennsylvania, for ten years.

Together with Dr. Aaron Katcher, he edited the book, *New Perspectives on Our Lives with Companion Animals* (1983) and co-authored *Between Pets and People: The Importance of Animal Companionship*, first published in 1983 then revised in 1996. In 2011, he edited *The Health Benefits of Dog Walking for Pets and People* (with Rebecca Johnson and Sandra McCune).

In 1990, Dr. Beck became the "Dorothy N. McAllister Professor of Animal Ecology" and Director of the Center for the Human-Animal Bond at the College of Veterinary Medicine, Purdue University, West Lafayette, Indiana. The Center was established to develop a comprehensive understanding of the relationship between people and their companion animals.

**Dr. Jose M. Peralta,** DVM PhD, is a Professor of Animal Welfare at the College of Veterinary Medicine at Western University of Health Sciences. He is a Diplomate of the American and the European Colleges of Animal Welfare and has received awards from the American Veterinary Medical Association and the World Veterinary Association for his contributions to the education and advancement in the scientific field of animal welfare. He is interested in all aspects of animal welfare and has worked on projects in different species, having developed over the last several years a particular interest in the wellbeing of animals used in animal-assisted therapy and in other forms of human-animal interactions.

# ABOUT THE CONTRIBUTORS

**Raymond Anthony** is in the Department of Philosophy, University of Alaska, Anchorage.

**Benjamin S. Arbuckle** is in the Department of Anthropology, University of North Carolina at Chapel Hill.

**Phil Arkow** is at the National Link Coalition, Etowah, NC.

**Liv Baker** is in the Animal Behaviour and Conservation Program, Hunter College, CUNY, NY.

**Alan M. Beck** is at the Center for Human Animal Bond, Purdue University College of Veterinary Medicine, West Lafayette, Indiana.

**Bente Berget** is at the Norwegian Research Center AS, Helse og samfunn, Kristiansand, Norway.

**Jacob L. Berl** is in the Idaho Department of Fish and Game, Coeur d'Alene, ID.

**Jessica Bibbo** is at the Center for Research and Education, Benjamin Rose Institute on Aging, Cleveland, OH.

**Jonathan Bowen** is at the Queen Mother Hospital for Small Animals, Royal Veterinary College, Hertfordshire, United Kingdom.

**Lindsey Braun** is at the Human Animal Bond Research Institute (HABRI), Washington, DC.

**Annika Bremhorst** is at the Institute for Canine Science and Applied Cynology, Uitikon, Switzerland.

**Catia Caeiro** is at the Primate Research Institute, Kyoto University, Inuyama, Japan.

## About the Contributors

**Bridget Camacho** is at the Institute for Human-Animal Connection, University of Denver, Denver, CO.

**Larry Carbone,** independent, San Francisco, CA.

**Gretchen K. Carlisle** is at the Research Center for Human-Animal Interaction, College of Veterinary Medicine, University of Missouri, Columbia, Missouri.

**Mark D. Carlson** is at Stow Kent Animal Hospital, Stow, OH.

**Neil Carr,** PhD, in Tourism, is at the University of Otago, Dunedin, New Zealand.

**Rebecca Cohen** is at Stable Grounds Therapy, New Haven, CT.

**Susan P. Cohen** is an independent consultant, New York, New York.

**Stanley Coren** is in the Department of Psychology, University of British Columbia, Vancouver, Canada.

**Steph M. Craven** is in the College of Veterinary Medicine, University of Missouri, Columbia, Missouri.

**Kate Dashper** is in the School of Events, Tourism and Hospitality Management, Leeds Beckett University, Leeds, United Kingdom.

**Margo DeMello** is in the Department of Anthrozoology at Carroll College, Helena, Montana.

**Seana Dowling-Guyer** is at the Center for Animals & Public Policy, Department of Clinical Sciences/Cummings School of Veterinary Medicine, North Grafton, MA.

**Jaume Fatjó** is Affinity Foundation Chair for Animals and Health in the Department of Psychiatry and Forensic Medicine, School of Medicine, Autonomous University of Barcelona, Bellaterra (Barcelona), Spain.

**Steve Feldman** is at the Human Animal Bond Research Institute (HABRI), Washington, DC.

**Aubrey H. Fine** is in the Department of Education, La Verne, California State Polytechnic University, Pomona, California.

**Erin Flynn** is at the Institute for Human-Animal Connection (IHAC), Graduate School of Social Work, University of Denver, Denver, CO.

**Erika Friedmann** is Professor and Associate Dean for Research at the University of Maryland, Baltimore, MD.

**Nina Ekholm Fry** is at the Institute for Human-Animal Connection, University of Denver, Denver, CO.

## About the Contributors

**Christopher M. Fulkerson** is in the Department of Veterinary Clinical Sciences, Purdue University College of Veterinary Medicine, West Lafayette, IN.

**Jaci Gandenberger** is at the Institute for Human-Animal Connection (IHAC), Graduate School of Social Work, University of Denver, Denver, CO.

**Nancy R. Gee,** PhD, is at the Center for Human-Animal Interaction, School of Medicine, Virginia Commonwealth University, Richmond, VA.

**Laura D. Gelfand** is in the Department of Art + Design, Utah State University, Logan, Utah.

**Taryn M Graham** is at PAWSitive Leadership, Toronto, Canada.

**Petty Grieve** is at Zoo Knoxville, Knoxville, Tennessee.

**Taylor Chastain Griffin** is at Pet Partners, Lavonia, GA.

**Maya Gupta** is in the Department of Strategy & Research, ASPCA, Woodstock, GA.

**Lizett Gutierrez** is at the Institute for Human-Animal Connection, University of Denver, Denver, CO.

**Taylor Haefs** is in the College of Veterinary Medicine, Western University of Health Sciences, Pomona, CA.

**Eli Halbreich** is in the School of Arts & Sciences, Tufts University, Medford, MA.

**Sloane Hawes** is at the Institute for Human-Animal Connection, University of Denver, Denver, CO.

**Zarah Hedge** is the Chief Medical Officer at the San Diego Humane Society, San Diego, CA.

**Annie Kate Hudson** is an Occupational Therapist, Rehab without Walls, Neurosolutions, in Nashville, Tennessee.

**Tess Hupe** is at the Institute for Human-Animal Connection, University of Denver, Denver, CO.

**Leslie Irvine** is in the Department of Sociology, University of Colorado Boulder, Boulder, CO.

**Brinda Jegatheesan** is in the Faculty of Educational Psychology, Learning Sciences & Human Development, University of Washington, Seattle; College of Education, Miller Hall, Seattle, WA.

**Clare L. Jensen** is at the College of Veterinary Medicine, University of Arizona, Oro Valley, AZ.

*About the Contributors*

**Sabrina Karl** is at Messerli Research Institute, University of Veterinary Medicine Vienna, Vienna, Austria.

**Marina A.G. von Keyserlingk** is with the Animal Welfare Program, Faculty of Land and Food Systems, The University of British Columbia, Vancouver, BC, Canada.

**Erin King** is at the Cummings School of Veterinary Medicine and Jonathan M. Tisch College of Civic Life at Tufts University, North Grafton, MA.

**Lincy Koodaly** is a PhD student at the University of Maryland, Baltimore, MD.

**Amanda Kowalski** is at the San Diego Humane Society, San Diego, CA.

**Sarah Krichbaum** is at the Canine Performance Sciences, Auburn University, Alabama.

**Garet Lahvis,** independent, Portland, OR.

**Sarah C. Leighton** is at the College of Veterinary Medicine, University of Arizona, Oro Valley, AZ.

**Lena Lidfors** is in the Department of Animal Environment and Health, Swedish University of Agricultural Sciences, Skara, Sweden.

**Pamela Linden** is at the University of Tennessee Knoxville Colleges of Veterinary Medicine and Social Work, Knoxville, TN.

**Randall Lockwood** is at the American Society for the Prevention of Cruelty to Animals, Falls Church, VA.

**William S. Lynn** is at the Marsh Institute, Worcester, MA.

**Brian J. MacGowan** is in the Department of Forestry and Natural Resources, Purdue University, West Lafayette, IN.

**Angela Matijczak** is in the School of Social Work, Virginia Commonwealth University, Richmond, VA.

**Sandra McCune** is Visiting Professor in Human-Animal Interaction in the School of Psychology and School of Life Sciences, Trustee of the Society for Companion Animal Studies (SCAS), University of Lincoln, United Kingdom.

**Shelby McDonald** is at Children, Families, and Animals Research (CFAR) Group, LLC, Richmond, VA.

**Kevin Morris** is at the Institute for Human-Animal Connection (IHAC), Graduate School of Social Work, University of Denver, Denver, CO.

*About the Contributors*

**Marisa Motiff** is at the Institute for Human-Animal Connection (IHAC), Graduate School of Social Work, University of Denver, Denver, CO.

**Megan K. Mueller** is at the Cummings School of Veterinary Medicine and Jonathan M. Tisch College of Civic Life at Tufts University, North Grafton, MA.

**Zenithson Y. Ng** is in the Department of Small Animal Clinical Sciences, University of Tennessee, 2407 River Dr, Knoxville, TN.

**Leanne O. Nieforth** is at the College of Veterinary Medicine, University of Arizona, Oro Valley, AZ.

**Torben Nielsen** is in the School of Animal and Veterinary Sciences, The University of Adelaide, University of Adelaide, Roseworthy, South Australia.

**Carmel Nottle** is at Allied Health and Human Performance, University of South Australia, Adelaide, South Australia.

**Marguerite E. O'Haire** is at the College of Veterinary Medicine, University of Arizona, Oro Valley, AZ.

**Elizabeth Ormerod** is in the Society for Companion Animal Studies, Cambridgeshire, United Kingdom.

**Sara Owczarczak-Garstecka** is at Dogs Trust, London.

**Jose M. Peralta** is at the College of Veterinary Medicine, Western University of Health Sciences, Pomona, CA.

**Robin Peth-Pierce** is at Cleveland State University.

**Isabella Pfeiffer** is at Thompson Cancer Survival Center Parkwest Medical/Covenant Health, Knoxville, TN.

**Emma Power** is at the Institute for Culture and Society, Western Sydney University, Sydney, Australia.

**Dianne Prado,** Executive Director, Housing Equality, Advocacy, Resource Team, Los Angles, California.

**Lena Provoost** is with Small Animal Behavior Medicine, University of Pennsylvania, Philadelphia, PA.

**Carolyn A. Ristau** is at Columbia University Seminars, New York.

**Amy Robinson-Junker,** independent, Oakland, CA.

*About the Contributors*

**João B. Rodrigues** is in Research and Operational Support Department, The Donkey Sanctuary, Devon, United Kingdom.

**Kerri Rodriguez** is at Human-Animal Bond in Colorado, School of Social Work, Colorado State University, Fort Collins, CO.

**Lindsay Rojas** is at the Institute for Human-Animal Connection, University of Denver, Denver, CO.

**Debbie Rook** is at Northumbria Law School, Northumbria University, UK.

**Francisco Santiago-Ávila** is in the School of the Environment, University of Wisconsin, Madison, WI.

**Lesley Sharp** is in the Department of Anthropology, Barnard College, New York City, New York.

**Melissa A. Singletary** is at the College of Veterinary Medicine, Auburn University, Auburn, AL.

**Lara V. Sirovica** is with the Animal Welfare Program, Faculty of Land and Food Systems, The University of British Columbia, Vancouver, BC, Canada.

**Marcy Souza** is at the University of Tennessee College of Veterinary, Knoxville, TN.

**Mary Beth Spitznagel** is at Psychological Sciences, Kent State University, Kent, OH.

**Madelaine Steevens** is at the College of Veterinary Medicine, Western University of Health Sciences, Pomona, CA.

**Kristin L. Stewart** teaches Anthrozoology at Canisius College, Buffalo, NY.

**Wendy Stone** is at the Center for Urban Transitions, Swinburne University of Technology, Swinburne, Australia.

**Eric G. Strauss** is at Loyola Marymount University, Los Angeles, California.

**Jason W. Stull** is in the Department of Health Management, University of Prince Edward Island, Atlantic Veterinary College; Veterinary Preventive Medicine, The Ohio State University, College of Veterinary Medicine, Stratford, PE, Canada.

**Zoei Sutton** is at the College of Business, Government and Law, Flinders University, Adelaide, South Australia.

**Tamara Tadich** is at Instituto de Ciencia Animal, Facultad de Ciencias Veterinarias. Universidad Austral de Chile.

*About the Contributors*

**Ashley Taeckens-Seabaugh** is at the Institute for Human-Animal Connection, University of Denver, Denver, CO.

**Karen Thodberg** is in the Department of Animal Science – ANIS Welfare, Aarhus University, Tjele, Denmark.

**Ann M. Toohey** is in the Department of Community Health Sciences, Cumming School of Medicine, University of Calgary, Calgary, AB, Canada.

**Erin Trella** is an Occupational Therapist Student at MedStar Washington Hospital Center, Washington, DC.

**John Tulloch** is in the Department of Livestock and One Health, Institute of Infection, Veterinary and Ecological Sciences, University of Liverpool, Neston, Cheshire.

**Paul Tully** is independent researcher, University of Otago, Dunedin, New Zealand.

**Georgitta Valiyamattam** is in the Department of Psychology, Gitam University, India.

**Andreia De Paula Vieira** is an independent researcher and veterinarian, Faculty of Veterinary Medicine, Universidade Positivo (UP), CPUP, a multidisciplinary research center, and the Veterinary School, Curitiba, Brazil.

**Aviva Vincent** is at the University of Tennessee, Veterinary Social Work Certificate Program, Knoxville, Tennessee.

**Jen VonLintel** is a school counsellor at the Thompson School District, Longmont, Colorado.

**Carri Westgarth** is in the Department of Livestock and One Health, Institute of Infection, Veterinary and Ecological Sciences, University of Liverpool, Neston, Cheshire.

**Joshua T. Wise** is in the Department of Fisheries, Wildlife, and Conservation Science, Oregon State University, Corvallis, OR.

**Janette Young** is at Allied Health and Human Performance, University of South Australia, Adelaide, South Australia.

**Anna Zamansky** is at Tech4Animals Lab, Information Systems Department, University of Haifa, Israel.

# FOREWORD

*Temple Grandin*

DEPARTMENT OF ANIMAL SCIENCE COLORADO STATE UNIVERSITY

Photo credit: Rosalie Winard.

The human brain is wired to react to seeing animals. People with epilepsy who were being monitored by implanted electrodes in their amygdala were shown pictures of animals, people, landmarks, or objects. Recordings from 1,443 single neurons indicated that pictures of animals evoked more responses than the other object categories (Mormann et al., 2011). Both cute and aversive animal pictures triggered the same response. The amygdala is an important center in the brain for processing emotional information. Other studies with fMRI imaging have discovered that dangerous animal images, such as crocodiles, provoked a bigger response than pictures of negative objects such as machine guns (Fang et al., 2018; Cao et al., 2014).

Humans have brains that are naturally oriented to pay attention to animals. This may be one of the reasons why people form strong emotional bonds with their pets, farm animals, and wildlife they interact with. *The Handbook on Human Animal Interactions, Interventions, and Anthrozoology* is essential reading for all people who are interested in how animals can improve people's lives. Multiple researchers and practitioners from all over the world have contributed chapters on the beneficial effects of interacting with dogs, cats, horses, and other species of animals as well as the challenges that exist in our relationships. Equine therapy has helped many children with autism. Having pets is also beneficial for mental health. Also covered are situations where interactions with animals are negative such as animal cruelty and hoarding.

Anthrozoology is a wide field which covers all interactions with animals. More people need to know about it and understand its relevance. When I typed "anthrozoology" into Google, it kept auto-correcting for anthropology. When it finally decided to search for anthrozoology, the first ten hits were mostly about companion animals. The book contains several chapters about animal sentience and wildlife. An insightful chapter about wildlife issues was written by William Lynn. He asks an important question about animal conservation policies. What is more important, preserving the species as a whole or caring for individual animals (Lynn and Hadidin, 2019)? The trend toward compassionate conservation puts a high value on individual animal's lives.

Researchers have now learned that on an emotional level, animals are more like us than some people would like to admit. In my early writings, I was not allowed to use emotional words in my papers such as "fear" (Grandin, 1993). The reviewers forced me to call fearful cattle behavior – agitation. Finally in 1997, I published a paper in *Journal of Animal Science* where I used the word "fear" and discussed neuroscience research (Grandin, 1997). The word fear was used for years in neuroscience literature before it started appearing in animal science and veterinary papers.

The work of Jaak Panksepp clearly shows that animals have emotions (Panksepp, 2011; Davis and Montag, 2019). Another indicator that animals have emotions with similarities to people is that drugs such as Prozac (Fluoxetine) are effective in dogs (Sherman et al., 2007). If a dog's brain was totally different, Prozac probably would not work to reduce separation anxiety. There is also good scientific evidence that animals are conscious (Birch, 2020). For readers who are interested in human interactions with pets and therapy animals, this book will provide many practical answers. For people who want answers to complex wildlife or agricultural animal issues, this book provides a starting point for discussion. There is diversity in the chapters integrated into this book including chapters on philosophical concepts and moral reasoning which can help get the scholarly conversation started. It is my belief that the field of Anthrozoology is growing and gaining respect from the scientific community. Reading this book will be a definite asset in expanding your knowledge of the field and help disseminate the significance of human-animal interactions.

# References

Birch, J. 2020. Dimension of animal consciousness, *Trends in Cognitive Science* 24, pp. 789–801.
Cao, Z., Zhao, Y., Tan, T., Chen, G., Ning, X., Zhan, L., and Yang, J. 2014. Distinct brain activity in processing negative pictures of animals and objects – Role of human contexts, *Neuroimage* 84 doi:10.1016/j.neuroimage.2013.09.064.
Davis, K.L. and Montag, C. 2019. Selected principles of Pankseppian affective neuroscience, *Frontiers in Neuroscience*, https://doi.org/10.3389/fnins.2018.01025.
Fang, Z., Li, H., Chen, G. and Yang, J.J. 2018. Unconscious processing of negative animals and objects: Role of the amygdala revealed by fMRI, *Frontiers in Humana Neuroscience*, doi:10/3389/fnhum.2016.00146.
Grandin, T. 1993. Behavioral agitation during handling of cattle is persistent over time, *Applied Animal Behavior Science*, 36, pp. 1–9.
Grandin, T. 1997.Assessment of steers during handling and transport, *Journal of Animal Science*, 72, pp. 249–257.
Lynn, W. and Hadidin, J. 2019. Bridging the gap between animal protection and traditional conservation biology and well-being of individuals, Well-being International, wellbeingint.org.

Mormann, F., Dubois, J., Kornbluth, S., Milosavljevic, M., Cerf, M. et al. 2011. A category specific responses to animals in the right human amygdala, *Nature Neuroscience* 14, pp. 1247–1249.
Panksepp, J. 2011. The basic emotional circuits of mammalian brains: Do animals have affective lives? *Neuroscience and Biobehavioral Reviews* 35, pp. 1791–1804.
Sherman, B., Landsberg, G.M., Reissner, I.R., Ciribassi, J.J., Horowitz, D. et al. 2007. Effects of reconcile (Fluoxetine) chewable tablets plus behavior management for canine separation anxiety, *Veterinary Therapeutics* 8, pp. 18–23.

# 1
# AN INTRODUCTION TO HAI AND ANTHROZOOLOGY

*Aubrey H. Fine, Alan M. Beck, and Megan K. Mueller*

## Introduction

Our commitment to the natural environment and our interactions with nonhuman animals has always been considered an important ingredient in the history of humankind. From a scholarly standpoint, there have been many who have been interested in our bond and connection with animals and the significance of these interactions. However, it wasn't until the 1980s that a new interdisciplinary field emerged called Anthrozoology, defined as the study of the interactions between humans and other animals. John Bradshaw, from the United Kingdom, is considered the individual who coined the term (Bradshaw, 2010).

Over the past several decades, the discipline has grown and includes a wide array of topics and subfields, including conceptualizing human-animal relationships, One Health, animal assisted interventions/assistance animals, conservation, animal abuse, animal welfare, ecotourism and others. This volume is a testimony to many of the subjects and topics that are integrated into the field of anthrozoology. Furthermore, at the end of chapter (Appendix 1), the readers will find a limited glossary of terms that are used in various disciplines under the umbrella of Anthrozoology.

Siddiq and Habib (2017) explain that the field has grown into examining, conceptualizing and critically evaluating the complex relationships between humans and nonhuman animals. As an interdisciplinary effort, the field's purpose is to investigate these unique relationships to better understand why these encounters occur and find better ways to enrich safe and respectful interactions. As noted, anthrozoology is interdisciplinary and includes academics and practitioners from diverse disciplines, including psychology, psychiatry, ethology, anthropology, human medicine, child development, social work, education, occupational therapy, nutrition, kinesiology, gerontology, zoology, animal sciences and veterinary medicine.

Hines (2003) provides an insightful historical overview of the earlier years of the field and discusses many of the milestones that occurred in the field's evolution. Readers are encouraged to review that article for many of the historical highlights, as well as some of the concerns in the growth of the field. The scientific interest in human-animal interactions (HAI) appeared to grow incrementally as a discipline when early research began to suggest health and psychological benefits of HAI. Consequently, numerous academic centers throughout the globe were established, as well as the first journal that was eventually called Anthrozoös, which had its launch in 1987. Hines (2003) points out that there were numerous international activities that were generated at that time that helped create a global community that fostered intellectual engagement. For example, in 1974, the Joint Advisory Committee on Pets in Society was established in the United Kingdom. The first center in the United States focusing on HAI was a Center on Interaction of Animals and Society, which opened at the University of Pennsylvania in 1977. In 1980, Alan Beck was brought in as the Center's permanent director. In October 1981, the Center hosted the country's first national conference on human-animal interaction. Also in 1981, the American Veterinary Medical Association convened a task force on the human-animal bond. Hines (2003) also identifies many of the key early publications on HAI that helped to launch the field, including *Pet Animals in Society* by RS Anderson in 1975 and the book by Katcher and Beck (1983) entitled *New Perspectives on Our Lives with Companion Animals*.

Resulting from these initial efforts, the field of anthrozoology has grown in size and scope, the infrastructure has similarly continued to expand. The International Society for Anthrozoology (ISAZ) was founded in 1991 and continues to thrive, with over 300 members from nearly 30 countries. *Anthrozoös*, now the flagship journal of ISAZ, continues to be a successful outlet for anthrozoology research. The International Association of Human-Animal Interaction Organizations (IAHAIO) was founded in 1992, fostering a global venue for HAI researchers and practitioners. In more recent years, the Human-Animal Interaction section of the American Psychological Association has become a robust platform for psychology-focused HAI work, supporting the *Human-Animal Interaction Bulletin* journal. In 2021, the Center for the Human-Animal Bond hosted its recurring conference for academic centers and institutes related to human-animal interaction, with participation from 16 centers from a wide range of universities, encompassing focal areas including veterinary medicine, arts and sciences, public health, medicine, social work, nursing and many others. Anthrozoology continues to be an interdisciplinary field with regard to the scientific literature, with a range of discipline specific journals (*Anthrozoös, Human-Animal Interaction Bulletin, Society & Animals, Animals*) in addition to much of the research appearing in journals of other allied disciplines, allowing for broad visibility of research within the scope of anthrozoology. The field is interdisciplinary not only because of the scholars involved, but also due to the areas of study. Some may argue that there is an overlap between the fields of animal welfare and anthrozoology; however, there is a key difference between the two fields. The field of animal welfare promotes the importance of respecting and providing for the well-being for nonhuman animals by humans themselves. The field of anthrozoology, on the other hand, focuses on the importance of these interactions or relationships between humans and nonhumans (Siddiq & Habib, 2016).

Let's read about some of these early year milestones by hearing from three of the pioneers who made their mark on the field's evolution. Each of these individuals has made a significant contribution to the field with their leadership in establishing the field's first organization (ISAZ), the first journal (Anthrozoos) and the pioneer years of instituting a research agenda.

# Spotlight 1

## The Early Years of the Delta Society and Anthrozoology: Events Leading to the Establishment of the International Society for Anthrozoology (ISAZ)

*Lynette Hart*

*Figure 1.1* Lynette Hart with husband Ben Hart

The Center on Interaction of Animals and Society at the University of Pennsylvania launched a new field for the United States in 1977 (Figure 1.1). Funding from the Geraldine Dodge Foundation for this Center made it possible to bring together a strong scholarly team for a solid start that set the stage for Erika Friedmann and colleagues (1980) to do her pivotal research while she was a graduate student, showing that pet owners had better survival one year after heart attack than non-owners. An important 1981 conference held in Philadelphia then led to Aaron Katcher MD and Alan Beck ScD (1983) editing *New Perspectives on Our Lives with Companion Animals*: a book that became an essential classic resource.

Related efforts were also underway for establishing the Delta Society, led mainly by veterinarians in the western US. Leo Bustad DVM and Calvin Schwabe MS DVM MPH ScD were both at UC Davis in the early 1970s, sharing overlapping interests and exchanging ideas on the importance of pets, leading Schwabe (1984) to include a long chapter, "Animals and Mental Health," in his epic book on One Medicine, *Veterinary Medicine and Human Health*, 3rd ed. Leo Bustad became the dean of Washington State University (WSU) College of Veterinary Medicine in 1973, over time devoting most of his energy to furthering local and international activities related to "the human-animal bond." In 1977, the term "human-animal bond" was coined by three veterinarians—Robert K Anderson, and two of his students, Stanley Diesch and Bill McCulloch—and psychiatrist Michael McCulloch (D Anderson, 2004). Also in 1977, psychiatrist Michael McCulloch, Bill McCulloch and Bustad organized the Delta Foundation. (Tragically, Michael McCulloch was murdered in 1985.) Bustad launched his WSU People-Pet Partnership in 1979, sponsoring local programs for equine-assisted therapy, pets in nursing homes and dog training by inmates in a prison. These model programs provided content for his lectures that inspired people worldwide to provide contact with companion animals.

As described by RK Anderson in 2011 interviews, the next step was a reconfigured Delta Society, established in 1981 as a follow-up to the Delta Foundation. Delta Society founders were RK Anderson DVM MPH DACVPM DACVB from University of Minnesota and three of his graduate students, Joe Quigley, DVM, Stan Diesch, DVM MPH DACVPM, and Bill McCulloch DVM, as well as Leo Bustad, who became the founding first president of the Delta Society (now known as Pet Partners). RK Anderson (2011) mentions in his published interview that these five names were inscribed on the wall of the first Delta Society offices in Renton, Washington. These founding veterinarians' interests in the Delta Society bridged into the emerging area of veterinary public health, and complemented the work of CENSHARE (Center to Study Human-Animal Relationships and Environments), which was established in 1981 at the university of Minnesota. Over time, Bustad built a Delta Society board of directors that within a few years included involvement of Earl Strimple, DVM; Alan Beck; Bill McCulloch; Andrew Rowan, PhD; Ben Hart, DVM, PhD, DACVB; Lynette Hart, MA, PhD; and others. Ben Hart had been at the UC Davis School of Veterinary Medicine since 1964, where he was teaching veterinary behavior, and Lynette Hart moved to UC Davis in 1982, where she focused on animal behavior and teaching human-animal interactions. Some modest support was provided the UCD center by KalKan, a Waltham affiliate and by Bill Balaban.

Leo's peripatetic lecturing that inspired people around the world was primarily addressed to the veterinary profession. Leo's special relationship with Hill's Pet Nutrition facilitated funding some of his projects, including the publication and distribution of a K-12 Humane Animal Care Curriculum in 1986. Not surprisingly, veterinary schools became the home of the emerging discipline of human-animal interactions, and those initially conceiving the Delta Society were largely veterinarians. However, the faculty conducting research over the subsequent years invariably had PhD degrees and were not veterinarians.

In 1983, veterinarians Bill Winchester of UC Irvine, Anderson, Bustad and Diesch co-hosted two back-to-back similar conferences at University of Minnesota and UC Irvine. The proceedings were published: *The Pet Connection: Its Influence on Our Health and Quality of Life*, edited by Anderson, Ben Hart and Lynette Hart (1984). Research was a goal of the Delta Society and the *Journal of the Delta Society* was launched in 1984–1985 with Alan Beck as editor, continuing for about a year. Soon after, the quarterly scholarly journal *Anthrozoos* was established in 1987 with Andrew Rowan of Tufts University as editor.

The momentum of the field of human-animal interactions kept growing, and in August, 1986, the Delta Society and IAHAIO held a joint international conference in Boston that revealed the extreme worldwide interest in human-animal interactions. Conferees came representing many different organizations that provided people supportive contact with animals. The conference emphasis for Delta Society and IAHAIO focused on delivering therapeutic exposure to animals, by offering training and furthering organizations with that emphasis. For those relatively few persons doing research pertaining to human-animal interactions, the conference was exciting but was somewhat narrowly focused on therapeutic uses of animals, and it did not provide optimal opportunities to meet others conducting research on the broad field of anthrozoology and have relaxed conversations with them.

Throughout this time, finding published work on human-animal interactions was extremely challenging due to the interdisciplinary scope of the field and the poor indexing of relevant keywords and terms. In a study, Librarian David Anderson at UC Davis documented that most publications on human-animal interactions at that time appeared in veterinary journals, with some others in nursing and psychology journals. He went on to select keywords and develop and publish *The Interactions Bibliography* using a variety of titles (Anderson, 1988–1999), a quarterly periodical that ran for ten years, providing relevant citations and including abstracts of published papers and theses in the field, offering crucial support for scholars searching for the field's research literature. Essentially, he did much to define the scope of the field and make the research easy to access.

At this time, only a handful of people had faculty positions at universities and were engaged in studying and publishing on human-animal interactions. The field of anthrozoology was just emerging; the people involved had completed their graduate studies in other disciplines, most in animal behavior or ecology. It was, and to some extent remains, a bottom-up field driven by the keen interest of students. In Europe, Waltham was providing research funding to a few scientists working on the animal behavior side: James Serpell at Cambridge University, John Bradshaw at Southampton University and Dennis Turner at University of Zurich. Coincidentally, Lynette and Ben Hart spent a sabbatical leave in Cambridge that provided additional opportunities for several discussions at Waltham and with the European scientists involved. A sentiment on the need to establish a scholarly society broadly oriented on human-animal interactions was building. Ivan Burger at Waltham facilitated the Waltham Symposium in August 1990 that was held in Harrogate in conjunction with the First European Congress of the British Small Animal Veterinary Association; *Pets, Benefits & Practice* (Burger, 1990) was a publication based on the proceedings of that symposium. Then in the following year, on April 12, 1991, ISAZ was formally established in Cambridge, England, during a dog behavior workshop hosted by James Serpell that also led to Serpell's (1996) classic edited book: *The Domestic Dog*. The International Society for Anthrozoology (ISAZ) was launched with Erica Friedmann as the first President and John Bradshaw as Secretary. Ivan Burger, Benjamin Hart, Lynette Hart, James Serpell and Dennis Turner were additional founders.

## References

Anderson, D. C. (1988–1999). *Humans & Other Species/Interactions of Man and Animals/The Interactions Bibliography: The Quarterly Journal of Resources on Human-Animal Relationships*. Penryn, CA: Rockydell Resources.

Anderson, D. C. (2004). The human-companion animal bond. *The Reference Librarian*, 41(86), 7–23. https://doi.org/10.1300/J120v41n86_02

Anderson, R. K., Hart, B. L., & Hart, L. A. (Eds.) (1984). *The Pet Connection: Its Influence on Our Health and Quality of Life*. Minneapolis: CENSHARE, University of Minnesota.

Anderson, R. K., & Tobbell, D. A. (May 31, 2011). *Interviews with Doctor Robert K. Anderson*. June 13, 2011. https://ahc-ohp.lib.umn.edu/wp-content/uploads/2013/01/RKAnderson.pdf

Burger, E. H. (Ed.) (1990). *Pets, Benefits and Practice*. Waltham Symposium 20. London: BVA Publications.

Friedmann, E., Katcher, A. H., Lynch, J. J., & Thomas, S. A. (1980). Animal companions and one-year survival of patients after discharge from a coronary care unit. *Public Health Reports*, 95(4), 307.

Katcher, A. H., & Beck, A. M. (Eds.) (1983). *New Perspectives on Our Lives with Companion Animals*. Philadelphia: University of Pennsylvania Press.

Schwabe, C. W. (1984). Animals and mental health. *Veterinary Medicine and Human Health*. Philadelphia, PA: Williams & Wilkins, 3rd ed., pp. 609–644.

Serpell, J. (Ed.) (1996). *The Domestic Dog: Its Evolution, Behaviour, and Interactions with People*. Cambridge: Cambridge University Press.

## Spotlight 2

## The Origins and First Ten Years of Anthrozoos

*Andrew N. Rowan*
President, WellBeing International, Potomac, Maryland

*Figure 1.2* Andrew Rowan

After I had led the organization of an international human-animal bond conference in Boston in 1986, Leo Bustad approached me about launching an academic journal for the Delta Society (Figure 1.2). I was initially reluctant to take on the challenge, but Leo would not relent and I eventually agreed to look into the project. This new journal would be my third adventure in academic publishing – the first two being the FRAME journal, ATLA (Alternatives to Laboratory Animals – still going strong in its 50th volume) and the International Journal for the Study of Animal Problems (published by The Humane Society of the United States for the four years from 1980 to 1983 – those volumes are now available on the Wellbeing International Studies Repository – see https://www.wellbeingintlstudiesrepository.org/ijsap/all_issues.html).

I found out that Tufts University (where I worked at the time) was part of a university publishing consortium – the University Press of New England (UPNE). Even more opportune, UPNE had started a journal division and was looking to add new journals to the division. Our timing was impeccable! The first task that Linda Hines (the Executive Director of the Delta Society) and I had to address was the choice of a suitable name for the new journal. We wanted to avoid names that began with "International" or "Journal" or both and eventually settled on Anthrozoös. There was some discussion about the etymological appropriateness of the name because "anthro" is derived from Latin while "zoos" has Greek origins. However, we felt that "Anthropozoos" (the etymologically correct form) did not flow off the tongue as smoothly as "Anthrozoös" and so exercised editorial discretion and went with "Anthrozoös." The other challenge involved the cover design. While it looks like it might be a copy of some actual cave art, it was actually produced by a member of the UPNE design team.

There were no journals at the time that dealt specifically with human-animal issues and filling the first volume was not a challenge. When launching a new journal, one can usually count on a "pent up" supply of unpublished manuscripts in the first volume. In addition, we were able to include papers derived from the plenary talks at the 1986 Boston international conference. However, finding sufficient quality manuscripts for the second and third volumes is often a more difficult task. It seems that authors are not keen to have their papers published in volume two or three of a new journal. However, by volume four or five, the concern about being in a "new" publication is fading (Anthrozoös is currently publishing Volume 35 and expanded the number of issues per volume from four to six for Volume 31 (a sure sign of the journal's success).

In the early days of Anthrozoös, I wanted to create a journal that brought researchers in the field together. To make sure I was on top of what was being published, I began a subscription to Current Contents (remember when researchers used to scan those weekly publications from the Institute for Scientific Information in order to try to keep up with literature). I would then write to the authors of papers that, from their titles, appeared to be relevant to the human-animal interactions field and ask for a reprint. Abstracts of those articles were then included at the back of each issue of Anthrozoös. I have no idea if Anthrozoös readers found those abstracts useful, but it certainly helped me to broaden my grasp of the field and to come to appreciate the importance of multi-disciplinary scholarship for the human-animal bond.

Today, people searching for relevant literature simply use Google Scholar or some other internet search engine. In addition, human-animal interactions scholarship is no longer a marginal sub-specialty. There are now multiple academic journals in the field (Society and Animals, being the main "competitor" to Anthrozoös). I would also note that the day when an editor of a human-animal interactions journal could, with just a little effort, keep up with all the books being published in the field is long behind us. Reaktion alone has published 100 books in their "animals" series in the past two decades while a number of other publishers also have specialty series on the topic.

I was also committed to the idea that Anthrozoös would encourage and publish case studies of interesting clinical cases. However, in the ten years that I was Editor, and exhorting researchers to send me case studies, I received only two manuscripts that could be considered to be case studies. To this day, I regard the lack of interesting (and disciplined) case studies in Anthrozoös as one of my failures as editor. There were certainly plenty of interesting anecdotes that came my way, and I wrote about a number of them in either my quarterly editorial or in the News and Analysis section, but only one or two authors produced a publishable and systematic case study.

Being an editor of Anthrozoös was a most rewarding activity and I am glad that Leo Bustad did not accept my excuses. There are now many other academic periodicals in the thriving field of human-animal interactions.

# Spotlight 3

# Sparking Scholarly Interest in HAI

*Erika Friedmann*

*Figure 1.3* Erika Friedmann

In the late 1970s, faculty from veterinary medicine, psychology, psychiatry, biology, public health, anthropology and other fields coalesced into several centers focusing on human-animal interactions (Figure 1.3). Many of the members of the centers were practitioners in some aspect of the bond. Some concentrated on animal behavior and behavior problems; others focused on human behavior and physical or mental health. Those on the animal side had an interest in how the humans in the relationship contributed to animal behavior problems and could be used to address them and the role of humans in animal health care and health; those on the human side had an interest in how the animals contributed to human behavior and human health. It was a diverse group trying to understand norms, benefits and problems in the human-animal bond. I became a member of the new University of Pennsylvania Center for the Interaction of Animals and Society as it was forming and became aware of other centers and interest groups springing up across the United States and around the world. In the United Kingdom, The Society for Companion Animal Studies (SCAS) was established in 1979 to promote the study of human-companion animal interactions and raise awareness of the importance of pets in society. The Delta Foundation was founded in the United States in 1977 to study the human-animal bond and subsequently changed to the Delta Society (1991) and to Pet Partners (2012). A series of international conferences for researchers in this fledgling field began in the late 1970s. The first Canadian Symposium on Pets and Society was held in 1976, and while I was still in graduate school and was not invited, I presented at the second conference in 1979.

The founding of the ISAZ in 1991 was a turning point for scholarship in the field. With the gradual shift of Delta to focusing on animal-related interventions and practice, we needed a venue with a broader scholar-based society. I suddenly had a conference to attend every year and colleagues with similar scholarly interests were gathered in one place. I remember being at the early conferences and the interest and excitement of researchers who had finally found someone else to talk with. I would come back totally energized and thinking of new questions and potential collaborations. At the time, human-animal interaction or anthrozoology, a term coined by John Bradshaw to describe his institute at the University of Southampton, did not feel like a bona fide research field.

Few academics were able to focus their scholarship exclusively on human-animal interactions. For many, it was tangential to their parent fields; it would not advance their careers, which largely depended on publications and grant funding. Little funding was available to support it and opportunities to publish and disseminate findings were sparse. Consequently, many researchers focused on other fields and used the research there to also obtain data about human-animal interaction while completing their other more mainstream studies. Most of my early research was unfunded or funded with internal funds that fit that description. My colleague Sue Thomas and I obtained NIH funding for studies of psychosocial aspects of cardiovascular health. Several of my studies and papers were only possible because I was able to add questions about pet ownership into them. Then, in addition to the papers we published about anxiety, depression and social support, we were able to publish our findings about human-animal interactions as related to coronary heart disease.

Along the way, other interactions within the biomedical research community implicitly devalued human-animal interaction research. One of my colleagues presented a seminar at a prestigious hospital/medical school. He described the experiment we had done and the impact of the experimental intervention on blood pressure and blood pressure responses to stressors. The cardiology faculty attending were riveted. They couldn't believe how good the results were. They kept asking questions about randomization methods, side effects and data analysis techniques. They were all in and ready to prescribe the intervention right away. Finally, when he could hold out no longer, he revealed that the intervention was having a dog present. There was silence in the room. The attendees lost interest, thanked him for his time and left the room shaking their heads and talking disbelievingly among themselves. It was clear they thought the seminar was a waste of their time. The problem was dog visits weren't billable! No drug company reps were going to bring lunches to their offices, a common practice at the time.

We live in a strange world where the popular perception of human-animal interaction as a panacea collides with the science of what we know. I knew my research was well known, when completing a Washington Post crossword puzzle (July 12, 2013) I saw the clue "**They're said to lower one's blood pressure**" and the answer was "pets"!

The public perception of pet's being good for people also had a downside.

Imagine my frustration at being told in the mid-1990s that my study of the differences in behavioral social responses of nursing home residents with dementia to visits that included a pet and visits by the same people without a pet wouldn't be funded because "everyone knows" that visits from pets are good for the residents. Why do this research? Lots of nursing homes have volunteers who bring pets to visit. I wish all our questions had been answered then, but what we have learned so far leads us to ask more nuanced questions.

The movement of anthrozoology from the fringe to a mainstream area of scholarship was a gradual process with many important milestones along the way. The three NIH related conferences – the Consensus Development Workshop of 1987, NIH/WALTHAM Invited Symposium: Directions in HAI Research: Child Development, Health, and Therapeutic Interventions in 2008 and the NIH/Waltham Workshop: Animal Assisted Interventions in Special Populations in 2015 validated the developing stature of the field. I'm grateful to have been a part of all these opportunities and to see our field now recognized as a bona fide scholarly discipline. There are now several journals, including the stalwart of the field – Anthrozoos, and more recent additions, Society and Animals, Human-Animal Interactions Bulletin, Animals and others. Papers regularly appear in broader higher impact journals too. NIH now funds research on human-animal interaction. NIH funding is a hallmark of scientific recognition for studies involving human health. The institution of Requests for Applications in 2009 through a public-private partnership with Waltham, with dedicated funds and separate review panels to encourage research in a fledgling field, was another important step. Subsequent movement to Program Announcement Reviewed where studies compete with other foci for funding within more interdisciplinary review panels reflects the progression of our science.

The field is burgeoning. I sometimes look back at our early conferences with nostalgia. I could attend every talk and often was inspired by presentations only tangentially related to human health, my specific area of research. Now when I attend, I am constantly torn with conflicting priorities at multi-track scientific sessions and somewhat overwhelmed in my desire to talk with many attendees. It's very important to me that the field continues to grow and keeps its roots in the broad context of all human-animal interactions – from interactions with wild animals, to zoos, to anthropology and archaeology, to history and films, to the welfare of the animals and to impact on human health. I look back and think of my old and new colleagues and am grateful to be part of this field, almost from the beginning. The "old guard" of the first generation is retiring. I am amazed at how far we've come together and, of course, look forward in anticipation of the accomplishments in human-animal interaction scholarship yet to come.

## Challenges

During its early decades, the field of Anthrozoology and HAI has faced ridicule and criticism from other professions. For instance, Dr. Boris Levinson encountered disapproving remarks from colleagues when presenting his ideas on the value of animal-assisted therapy at various psychology meetings, including his meeting at the American Psychological Association in the 1960s (Hines, 2003). This forum is discussed more in Chapter 33. Much of the early criticism regarding the evolving field was attributed to the lack of top-quality gold standard research that would include randomized controlled studies. Most of the early research studies were flawed and had small samples. In 1984, the Delta Society held a "Conference for Research on the Interactions of Animals and People." This conference was instrumental in specifying research questions for studying the human-animal bond and addressing strategies for best practices within the field (Zeglen, Lee, & Brudvik, 1984). As previously noted, the first scientific journal publishing peer-reviewed studies in human-animal interactions was Anthrozoös, which evolved from the Journal of the Delta Society and was founded by Dr. Andrew Rowan.

Historically, funding for the advancement of HAI/HAB has been minimal. Pet food companies such as Waltham/Mars, Hill's Pet Nutrition, Ralston Purina and The Petfood Institute (USA) have long sponsored and supported research in the field of human-animal bond (HAB) (Hines, 2003). Although the field has made profound advancements in enriching the lives of animals and people, research, funding and recognition are still necessary to develop the full potential of HAB. Over the past 15 years, there has been more funding available (e.g., NIH partnership with Waltham and funding opportunities from non-profit organizations and foundations such as the Human Animal Bond Research Institute [HABRI], the Society for Companion Animal Studies and others).

## Definition of Human-Animal Bond

So much has been written over the past few decades exploring the value of human-animal interactions (both direct and indirect) on human well-being. Many scholars over the past few decades have also argued that our understanding of these relationships must go beyond our discussion of the benefits only for humans, but rather must encompass the implications of these relationships on animals as well. The term "bond" was borrowed from the common language used that described the significance of the relationship between parents and their infants. The first "official" use of the term "human-animal bond" appeared in the *Proceedings of the Meeting of the Group for the Study of Human–Companion Animal Bond* in Dundee, Scotland, March 23–25, 1979.

Perhaps the most widely used definition of the human-animal bond was formulated by the American Veterinary Medical Association. The organization defined the bond as

> A mutually beneficial and dynamic relationship between people and other animals that is influenced by behaviors that are essential to the health and well-being of both. This includes, but is not limited to, emotional, psychological, and physical interactions with people, other animals, and the environment.
>
> *(JAVMA, 1998, p. 1675)*

Spotlight 4 highlights one of the early pioneers in the field and his commitment to spreading the value of HAI globally. Meet Dr. William McCulloch.

## Spotlight 4

## The Legacy of an HAB Pioneer: Some Call Him William McCulloch. I Just Call Him My Friend

*Aubrey Fine*

*Figure 1.4*  Bill (right) and Michael McCulloch (left)

I have known Bill McCulloch for a couple of decades (Figure 1.4). I treasure not only his wisdom and insights, but also his friendship. Perhaps one of the greatest honors bestowed on my professional career was when I was presented the William McCulloch Award by the International Association of Human Animal Interactions Organizations in Paris in 2016. Truly, Bill was one of the early pioneers in defining the human-animal bond, and he was one of the leaders that established the Delta Foundation, now known as Pet Partners.

In January and February of 2022, Bill and I had a couple of conversations about the early years of the field. (I'm taking the liberty of editing some of our comments). When we began our conversation, he highlighted that the term the bond was shamelessly borrowed from our perception of the infant-mother bond. When the term was first used to describe the relationship with pets and their humans (at a meeting in Dundee Scotland in the 1979), those that were present appreciated how the concept captured the significance that many express. Our conversations were rich and fruitful, and we jumped into talking about many aspects of our unique relationships with nonhuman animals.

Initially, we began talking about his earliest experiences in appreciating the importance of the bond. Bill quickly shared a story about the passing of his 11-year-old dog, Nibs, when he was a senior in high school. Unfortunately, he was out of town when Nibs died. His father buried Nibs on their property, but when Bill returned home, he was angry that he didn't have the opportunity. He wanted to dig up Nibs and bury his four-legged friend himself. Bill disclosed that he needed closure and the opportunity to say good-bye to his four-legged buddy. It was that moment that Bill believes he began appreciating the importance of the bond: way before he became a veterinarian.

Many years later, a serendipitous event occurred while Bill was a practicing veterinarian. He had worked with a couple who had to recently euthanize their Beagle. A few weeks later, the couple visited Bill (in 1959) in his office to thank him for his sensitivity. They presented him with a small gift: a wallet. They wanted Bill to recognize how appreciative they were of his efforts in being compassionate about their loss. In a note given to Bill, they shared,

> You know that my husband and I could not have children and our beagle was like our child and a member of our family to us. We appreciated that you understood this attachment and the pain that we were going through. You know, Dr. McCulloch, we love you for that.

To this day, Bill has that wallet in his drawer at home.

After talking for a while, Bill went on to describe his early involvement in establishing the Delta Foundation in 1977. He identified many of the early leaders, including Dr. Leo Bustad, and Dr. R.K. Andersen. He passionately explained that the organization's initial mission was to emphasize the human health benefits of pets. The name Delta originated from their beliefs about the intersection between People, Pets and the Environment. Proudly, Bill shared that his brother Michael was the first president of the Foundation. His presidency was followed by Dr. Leo Bustad. Their first board meeting was at Michael's beach home on the Washington coast and the board meetings usually lasted several days.

When I questioned Bill about how he met Dr. Bustad, he went on to tell me about his early military service in the mid-1950s. He noted he met Dr. Bustad at a Sunday religious service where Bustad was leading the sermon that day. Afterward, he introduced himself and their friendship began at that moment.

You could see the twinkle in Bill's eyes as he went over the history of the HAI movement. Bill noted that after the initial National Institutes of Health meeting in 1981 focusing on the scientific evidence of the health benefits of the bond, Bill took this information over to the AVMA. He presented at their two-day meeting with the outcome being the Board's approval of establishing the organizations first committee on HAI. Bill was the initial chair of the committee. Over the past several decades, the AVMA has continued its significant interest in fostering and highlighting the importance of the bond.

Those of you who have ever met Dr. McCulloch will easily recognize how important he perceives the human-animal bond in our lives today. He not only advocates its value but lives the talk. Bill always closes his emails by signing "In the bond." It was the work of some of our early pioneers, like William McCulloch, who led the way to the evolution of our modern-day field of anthrozoology.

In addition to defining the human-animal bond, much work over the past several decades has focused on elucidating the theoretical frameworks that are relevant to understanding the human-animal bond. Many of these theories relate to human affinity for animals (e.g., the biophilia hypothesis), the relational components of human-animal interactions (e.g., attachment theory, social support theory) and how reciprocal relationships with animals inform health and well-being (e.g., biopsychosocial model). These models, and others, are discussed in later chapters in this volume (see Chapters 5 and 14 in particular).

## HAI/HAB and Its Impact

The scholarly interest in human-animal interactions and the bond has evolved over the past several decades due to more reliable research, funding and the platforms available to promote the research. Society's interest in the field has gained momentum not only due to the current research available, but also through media coverage of the value of various HAI, including animal-assisted interventions, pet companionship and animals in the wild (Fine & Andersen, 2021; Hines, 2003). Hines (2003) stresses that in the early years, the progress that the field attained was mainly attributed to advances in veterinary medicine, conferences promoting the value of HAI, and the establishment of the various university centers, and publications. Thanks to the media, public awareness of human-animal activities and research grew. More attention was given not only to the importance of domestic animals in our lives but also to our curiosity of wildlife. Assistance animal programs for people with disabilities in the 1970s and 1980s were also influential in educating the mainstream on the value of HAI and its contribution to human well-being. For example, organizations such as The Seeing Eye (founded in 1929) and Guide Dogs for the Blind (founded in 1942) set the tone for the potential of human-animal interactions (Hines, 2003).

## Challenges for the Future of the Field

As the field continues to grow, it faces a number of key challenges. The breadth and quality of research in anthrozoology has increased exponentially, with now a focus on understanding the processes and mechanisms underlying human-animal relationships. As the breadth of the field expands, it will be critical to continue to provide opportunities for interdisciplinary researchers and practitioners to collaborate in a cohesive manner.

The field also continues to grapple with the often-unexpected complexity of how human-animal relationships do (or do not) contribute to health and well-being for both the people and animals involved. Often, the underlying assumption in human-animal interaction is that these relationships produce positive outcomes, and often personal experiences with animals are a motivating factor for scholars to enter the field. However, as the body of evidence increases, it is becoming clear that human-animal relationships are not uniformly positive, or, in fact, uniform at all. Human-animal interactions are embedded within complex contexts, at the individual, family, community and global levels, and therefore there is an inherently high degree of complexity. Often, our research methods run the risk of being reductive for reasons of feasibility, and so therefore there are many open questions about our relationships with animals and to what extent the existing research sheds light on the nuances. These open questions present an opportunity for a new and evolving generation of researchers and practitioners to continue to move the field forward into the future.

## Concluding Remarks

There is indeed a great deal of excitement in the field of anthrozoology as it continues to evolve as a robust, multidisciplinary field. It is evident, as the field continues to germinate, more attention is being placed on the credible research being published as well as the establishment of more Centers (from a variety of colleges/schools in their university), international organizations and meetings, and evolutions in standards of practice. In essence, the multidisciplinary nature of anthrozoology continues to flourish, providing new career pathways and opportunities for many interested in applying their knowledge.

Peter Drucker once said, "the best way to predict the future is to create it." Over the past 40 years, an interdisciplinary group of professionals has made an imprint on the evolving field of anthrozoology and shepherded it from its infancy as a marginal subfield, to a place of maturity as a robust discipline. Although it is difficult to make a prediction regarding a trajectory, one can make reasonable prognostications. The primary areas of change are influenced by scholarship and research, funding, policy, and training and education. It is evident that more quality research has been initiated over the past 40 years in numerous areas that the field covers. The increase in scholarship has been impacted by more funding from both private and public agencies, which has allowed for the design and execution of high quality research studies in diverse areas of the field. Furthermore, with the growth of numerous academic centers globally, and national and international meetings, we are seeing greater opportunities for a new generation of scientists and practitioners to become more engaged in studying and influencing HAI. Through the increased access to educational programs, there are more opportunities than ever for careers in anthrozoology and HAI across the research/practice continuum. There continues to be significant interest in the field from the general public, making effective science communication and translation a top priority moving forward.

Jacque Fresco once said "we cannot predict the future, but we will most surely live it." Every action and decision we make or don't, ripples into the future. In essence, we have the capability and the knowledge to direct those ripples. It behooves the new leadership, that the field continues to evolve, and they must take charge in moving toward the future. The collection of papers that follow this chapter provide additional insights into the numerous topics that are covered within this discipline. One must wonder if the early founders of the field would be pleased to see the fruits of their efforts continuing to flourish and blossom. It is the intention of this book to shed more insight into the intricacies that are now being researched and debated within the field and provide direction and suggestions for its future.

### Discussion Questions

1. What were some of the early challenges faced by the founders of the field, and how does that inform the work still being conducted today?
2. What are some strengths and challenges to the interdisciplinary nature of anthrozoology and human-animal interactions?
3. What theories resonate with your own academic or practical interests in anthropology and human-animal interactions, and how might you apply them to your work?

## References

Anderson, R. S. (Ed.) (1975). *Pet Animals and Society: A BSVMA Symposium*. London: Bailliere Tindall.

Bradshaw J. (2010). Anthrozoology. In Mills, D. (Ed.), *The Encyclopedia of Applied Animal Behaviour and Welfare*. Wallingford: CABI, pp. 28–30.

Fine, A. H., & Andersen, S. J. (2021). A commentary on the contemporary issues confronting animal assisted and equine assisted interactions. *Journal of Equine Veterinary Science*, 100, 103–436. https://doi.org/10.1016/j.jevs.2021.103436

Hines, L. M. (2003). Historical perspectives on the human-animal bond. *American Behavioral Scientist*, 47(1), 7–15. https://doi.org/10.1177/0002764203255206

Journal of the American Veterinary Medical Association. (1998). Statement from the committee on the human-animal bond. *Journal of the American Veterinary Medical Association*, 212(11), 1675.

Katcher, A. H., & Beck, A. M. (1983). *New Perspectives on Our Lives with Companion Animals*. Philadelphia: University of Pennsylvania Press.

Siddiq, Abu, & Habib, Ahsan. (2017). Anthrozoology – An emerging robust multidisciplinary subfield of anthropological science. *Green University Review of Social Sciences*, 3, 45–67. https://doi.org/10.6084/m9.figshare.13159736

Zeglen, M. E., Lee, R. G., & Brudvik, A. (1984). The delta society invitational conference for research on the interactions of animals and people: An overview. *Journal of the Delta Society*, 1(1), 6–20.

# 2
# VALUATIONAL (ETHICAL) INQUIRY, ANIMALS, AND US

## Contemporary Framings and Future Imaginaries

*Raymond Anthony and Andreia De Paula Vieira*

### Introduction

An important aspect of valuational inquiry[1] into the shape of human-animal relationships is to make visible both the values underpinning various interspecies entanglements and the normative ideals reflected in our activities involving animals. This type of inquiry must be done in collaboration with applied ethology and animal welfare science (AWS). Applied ethology is the science that seeks to discover the behavioral processes underpinning the various contexts in which humans utilize and relate to animals (Jensen 2017), including animal domestication and co-evolution. AWS, which emerged out of ethical concern for animals, is the rigorous use of scientific methods to study the quality of life of animals, including companion animals, wildlife, research animals and those farmed for food (Fraser et al. 1997).

Modern-day animal domestication and interspecies co-evolution reflect deep interconnections between humans and animals. While humans have tended to shape the behavior and activities of animals, animals have also significantly affected human behavior, structures and institutions (Haraway 2003). The SARS-CoV-2 pandemic has brought our positive and negative relationships with animals into acute focus. For example, many of us looked to companion animals or pets for comfort during the lockdowns.[2] However, when animals (e.g., mink and hamsters) were depicted as disease reservoirs or "epidemic villains" (Lynteris 2019, p. 1), public authorities quickly approved the emergency killing of healthy animals to prevent or 'stamp out' the spread of disease and minimize transmission to humans (Splitter 2020). Critical investigation and analysis of emergencies such as a pandemic enables a concentrated focus on relationality beyond species boundaries through examining our moral convictions toward animals and understanding the attitudes and species-specific factors that shape our moral obligations to individual and groups of animals during normal times (De Paula Vieira and Anthony 2021; Irvine 2009). Emergencies also accentuate the importance of integrating philosophical and ethical inquiry with applied ethology and AWS to enable productive rethinking of key issues in animal care, and to narrow the divide between policy and other decision makers and the lay public so that tangible steps can be taken to understand the perspectives and needs of animals. An integrated approach between ethical inquiry and these sciences can also advance a more accurate appraisal of the ethical, economic, spiritual, cultural, aesthetic and emotional values that animals bring to humans.

The current era of the Anthropocene (Crutzen and Stroemer 2000), when humans influence Earth processes more than any other time in history, challenges humanity to reexamine some of its views toward animals. While a binary framing of animals as either friend or foe (equal or lesser) may be easy to conceive for policy purposes during a disaster like the COVID-19 pandemic, in reality our experiences of animals are complex and multifaceted (Korthals 2016). Our lives and fates are inextricably intertwined (e.g., Höglin et al. 2021; Kirk and Worboys 2011), including with Earth systems (e.g., https://www.ipcc.ch/report/ar6/wg2/). Jointly, ethical inquiry, applied ethology and AWS can reveal the complexities embodied within the various institutionalized structures of human use of animals and redefine how to strengthen more equitable, humane and just human-animal interactions and relationships.

In this chapter, we explore some influential arguments about the moral status of animals in philosophical animal ethics (AE) that favor the abovementioned binary framing (of equal to or lesser than human). Next, we consider how the relationship between ethics, AWS and applied ethology is central to imagining improvements in our practices concerning animals by understanding their interests and perspectives through more scientific approaches, advancing the welfare of animals, advocating professionally for them when under our care, and strengthening human-animal relationships. We close by highlighting three possible frontiers for imaginative work for joint valuational inquiry, AWS and applied ethology exploration.

We use the following interlocking terms to structure our ethical inquiry and discussion: *Conceptual or philosophical framings* are akin to mental filters that influence how we make sense of our world (Goffman 1974). These filters can influence what we currently (and should) highlight in our relationships with animals, thus, the shape of our experiences with animals. Animal extremist groups (e.g., Animal Liberation Front), for example, use slanted messaging, well-cropped videos and savvy media tactics to convey a narrow version of human-animal relations. It is important that ethicists, scientists, veterinarians and other professionals working with animals be vigilant against inappropriate uses of their scholarship and research to advance a political or monetary agenda and call out mischaracterizations and misrepresentations in how we should relate to animals as part of faulty framing.

*Imaginaries* are ideas to help us think past our current frames. They are a form of 'deep seeing' that can help us reexamine and explore the values, institutions, symbols, practices and customs (Berry 2012) that govern our collective lives. Imaginaries can lead to transformation and growth by motivating new horizons or trajectories to envision the nature of moral problems through open-ended inquiry (Dewey 1982/1920). Collective imaginaries are powerful ideas that unlock nascent ethical concern, innovate thinking and motivate action toward what could be. Proponents of animals' issues may use imaginaries to advance social change by pointing to shortcomings in existing practices and alternative ways of relating to, and improving the lives of animals.

## The Ethics of Human-Animal Relationships: A Primer

In the 1960s, a shift in the philosophical literature sparked an increased interest in the ethics of human-animal relationships. Rachel Carson's (1962) *Silent Spring* and Ruth Harrison's (1964) *Animal Machines* spurred public concern and academic interest in commercial animal use and the environment. In 1975, Australian philosopher, Peter Singer, published *Animal Liberation: A New Ethics for Our Treatment of Animals*, calling for greater ethical scrutiny of animal exploitation. From the mid-1970s to 1990s, philosophers working in AE applied traditional ethical approaches (e.g., utilitarianism, contractarianism, rights-based theory, virtue ethics, various strands of feminist ethics) to frame animals' significance outside of humans (e.g., Anderson 2004; Hursthouse 2011; McMahon 2016; McPherson 2016). A number of influential philosophical and ethical expositions on animals and human-animal interactions have since followed, which critique those who benefit from animal use for neglecting their interests (e.g., Bovenkerk and Keulartz 2016; DeGrazia 2011; Rollin 1981, 1998). While some criticism of animal industry practices is warranted, animal caretakers, including farmers and researchers, have been committed to making improvements based on the best available science on animals' welfare, but lately, they have been constrained by institutional structures and systems (Anthony 2017; Bennett and Thompson 2018; Mench, Sumner, and Rosen-Molina 2011; Norwood and Lusk 2011). For example, industrialized production practices and technologies and the economics of agriculture have resulted in a loss of control by and less autonomy of farmers, curtailing their ability to actualize their animal husbandry commitments (Rollin 1995b). While animal farmers participate in animal agriculture, categorizing them as "just part of the industry" risks failing to take account of changing agricultural ownership and the landlord-tenant relationship (Petrzelka 2014) and the true value of farmers in providing animal care and husbandry. Indeed, farmers make wide contributions to others' welfare (animals included), sustainability and food security (Thompson 2010). It has been suggested that those who benefit from nature, including animals should conform to a "social license to operate" (e.g., Moffat et al. 2016) to curtail their over-exploitative tendencies toward nature and animals. However, framing human-animal relationships predominantly in terms of a 'social license' may minimize the rich and complex co-mingled realities involved in: (i) living *from* animals, which coincides with the ability of animals to provide nourishment, companionship, knowledge, therapies and protection; (ii) living *with* animals, which recognizes the interdependence between species including spiritual, social, emotional and communal aspects; (iii) living *through* animals, which refers to the identity and cultural forming activities as a result of human-animal-environmental interactions; and (iv) living *as* animals, which acknowledges our own animal natures and vulnerabilities as essential to being human.

Valuational (ethical) inquiry, together with the current scientific comprehension of animals' preferences and needs, should inform policymaking of whose and which values regarding animals should be considered (and why) and how to fairly distribute the benefits and burdens of these policies/decisions. In animal agriculture, the burden to guarantee animal welfare falls on everyone in the food value chain, including citizen-consumers and regulators who are physically, intellectually and emotionally distant to animal agriculture (Thompson 2021). When collective habits break down, new imaginaries are needed. Ethics is central to integrating and discussing a broad swath of values, interests, and ways of knowing and to making lucid how normative ideas, beliefs or values[3] are included in our decision-making and how they are realized through our decisions, and (together with science) what ought to be pursued to address infelicities and shortcomings.

While both science and ethics are about truth-seeking, science focuses on methods of discovery through observation and hypothesis testing and answers to "what is…" or "how do…" questions: "What is the impact of a particular pain management treatment on castrated dogs?" "How do calves separated from the cow soon after birth, fed higher milk volumes or provided social partners fair health-wise, socially and emotionally upon weaning throughout different life stages?" (De Paula Vieira et al. 2008). Ethics focuses on right action and the best way to treat others, seeking answers to "how ought we to…" questions: "How ought we to decide which animal management option(s) to adopt for animals under our care?"

Ethics is not merely the expression of opinion, but must bend to justification in the form of reasoned deliberation of the relevant evidence (including an understanding of the pertinent scientific matters and particularities of each context) and the strength of the inferences made and consideration of alternative assumptions and scientific approaches. While ethics is not subjective, it should not be understood to indicate a single or dominant philosophical vision. People ought to be encouraged to reflect and investigate the moral prescriptions or constraints placed upon them from time to time instead of "relying on an illusion, as if trying to give the moral 'ought' a magic force" (Foot 1972).

Ethics does not occur in isolation. It involves relationality, dialogue and cooperation to solve difficult moral and social issues. Ethics is a collective activity in pursuit of how we ought to live, and includes both debate and deliberation over differences in our norms and scrutiny of scientific evidence, goals and assumptions. These activities are characteristic of a community's ethical life (Dewey 1982/1920). Hence, moral agents are not only conferred rights as members of a community, but they also have duties toward each other, including animals as members of our communities in agriculture, biomedical, breeding and teaching facilities, zoos and in our homes.

Those concerned about AE issues may explore:

- Which animals should be ascribed moral standing and on what basis?
- What kind of responsibilities do human beings have toward (which) animals?
- How ought humans justify using animals in varying contexts?
- Should our responsibilities to animals include only individuals or also be expanded to populations of animals?

The ethics of human-animal relationships is complex since our relationship with animals is diverse and our entanglements with different individuals and species run deep and broad. While the utilization of animals is part of our co-evolving human-animal entanglements, our interactions with animals is not limited to their instrumentalization. Our ethical views about animals must acknowledge the multiple ways people relate through, as, in and from animals and to nature and each other. These views are informed by co-evolution, domestication, historical and sociological factors, such as custom, family and peers, religion, political ideology, level of education, law and regulations, economics, health, work, sports, husbandry, climate and our relationship to technology (Fraser 2008; Guerrini 2016; Kendall, Lobao and Sharp 2006; Rollin 1995a). Thus, understanding and strengthening our ethical relationships to animals is an ongoing project, which should be sensitive to the preferences of animals and their direct caretakers (e.g., farmers, breeders, researchers, zoological managers) and the diverse contexts in which animals and peoples have co-evolved historically and continue to do so socially and morally to promote healthy relationships.

## The Philosophical Significance of Framings – Where We Come From

Those working in AE tend to use one of the following philosophical *frames*: animals are passive victims of human exploitation and must be protected through moral or legal protections (Regan 1983; Rowlands 1998; Wise 2001) or by undertaking reforms that improve their quality of life or welfare (Fraser 1999; Singer 1975/1990); some animals are (near) moral persons (Aaltola 2011; Varner 2012; Waldau 2010), carrying strict implications for how they should be treated, including an end to their exploitation; animals are partners or fellow travelers, whose narratives and lives are intimately intertwined with ours (Midgley 1983). They are also "active agents of symbolization, history, learning and the breaking down and reconstruction of established world perspectives" (Korthals 2016, p. 73). Taking in the perspectives of animals can teach us about ourselves, including instilling ethical and epistemic knowledge and humility "to become receptive to listening to nature's story" (Burton and Brady 2016, p. 89; Dawkin 1993; Mench 1998).

The first two framings (moral and legal rights and welfare interests) explore the philosophical and ethical grounds that distinguish humans from other animals and provide critiques of the basis for sameness and difference. They explore the basis on which animals have moral standing in their own right (i.e., what attributes should be morally considerable, a.k.a. the ethics of personhood studies), and the responsibilities humans have toward them. Here, if animals deserve direct consideration as morally considerable beings in their own right, then human beings are responsible for ensuring that their rights or interests are not violated – in essence, humans should minimize negative experiences and advance positive experiences for these animals. Philosophers Tom Regan and Peter Singer have been two of the leading voices in this animal rights vs animal welfare utilitarianism space.

The instrumentalizing of animals (regarding animals as mere utility objects) as the main modality of human-animal relationships leads proponents of inclusion such as Regan and Singer to frame animals as passive beings in need of protection from exploitation. Proponents tend to emphasize non-interference claims that certain animals can make against human beings and that the grounds for human uses of animals should be ethically questioned. A strong rights-based approach, for example, seeks to either eliminate uses of animals or has strong proscriptions against use, overuse and misuse of animals (Norcross 2004). A consequence of this framing is constant negotiation of (perceived) human-animal conflicts.

Early arguments in this first wave of AE centered on the inner lives of animals – that is, if and the extent to which animals possess morally significant attributes such as consciousness and mental capacities like emotions, pain and suffering and specific kinds of cognition (Allen and Bekoff 1999; Andrews 2020). Animals with the requisite mental or social attributes should be bumped up into the moral community (at least to the status of moral patients or subjects). Some, such as dolphins, chimpanzees and other great apes, conceivably embody the salient attributes for moral personhood (Bernstein 1998; DeGrazia 1996; Midgley 2003; Sapontzis 1987; Warren 1986). Much of the initial concern for animals also focused on the moral status of individuals, an approach termed 'moral individualism' (Rachels 1990) in contrast with 'moral relationalism' (May 2014). The atomist nature of the first wave of AE has been at odds with holistic views that give primacy to species interests (Rolston III 1989; Sagoff 1984) or broader common or public good concerns (De Paula Vieira and Anthony 2021). It has also resulted in schisms involving human obligations to wild versus captive or domesticated animals (Anderson 2004).

Both Singer and Regan seek to overcome perceived institutionalized abuses of animals or *speciesism* (Ryder 1971 in Waldau 2010) – the notion that humans deserve greater moral priority over animals simply because they are human. Singer's and Regan's arguments are linked to analytic philosophy, a form of reasoning that concludes that the interests or rights of animals should be protected just in case we also do so for 'marginal' or 'non-paradigm' human beings who many possess the same or fewer capacities as animals (Warren 1987). Their arguments seek to unseat historical antecedents offered by philosophers such as Rene Descartes (1637/1988) and Immanuel Kant (1963), who excluded animals from direct moral consideration because animals did not bear moral-making qualities such as human language and reason. Exclusivist sentiments (which typically play up human linguistic and reason-giving capabilities) can still be found in contemporary variants of contractarianism (Carruthers 1992; Narveson 1983) and neo-Kantianism (Korsgaard 2015, 1996).

First wave AE weakened the traditional anthropocentric frame that membership in the moral community should be restricted to humans. Singer's and Regan's arguments have been used to support legal arguments and judicial activism in favor of legal animal rights. However, the scientific basis of these rights is ambiguous and when absent or contested can injure animals and their human caretakers alike (especially if the basis of rights ignores animals' species-specific qualities). However, Singer and Regan have been criticized on several fronts, including perpetuating a form of human-centrism (Birke 1994), where possession of human capacities (i.e., animals must be like humans) continue to be the measure and price of admission into the moral community (Vance 1995). But perhaps there is not just one archytype of moral considerability (Crist 1999; Plumwood 1991)? Arguably, we should refrain from understanding the perspectives of animals through primarily human lenses and accept the value of animals based on scientific evidence provided by applied ethology and AWS (by respecting differences without overemphasizing them), without requiring sameness with humans (Baker 2000).

Lilly-Marlene Russow (1988) argued that as a moral category, Regan's definition of subjects-of-a-life lacks specificity in its characterization of harm and opportunities and thus is not useful for guiding action, especially in the face of interspecies conflicts. Further, while Regan and Singer were careful to argue that humans and animals are not equal in every way,[4] their respective approaches tend to lump animals into an amorphous group with a single moral identity and fail to appreciate species uniqueness and diversity of capacities that ground differences in moral treatment (Anthony 2009) and scientifically informed management and animal-care practices.

Further, a rights-based notion of animal agency can conflict with our everyday interactions with animals and both Singer and Regan minimize the role of daily interactions and dependencies in shaping our moral responsibilities to animals (Warren 1987). Relations to humans (e.g., pets to their owners; patients to their veterinarians) create a set of practical (acquired) moral responsibilities to animals (Burgess-Jackson 1998). Bringing animals into human households or communities results in dependencies, including a responsibility to care well for them (Midgley 1983). Welfare in this instance can be expressed as working toward establishing evidence-based (e.g., science and values investigation and deliberation) standards governing the humane handling, (veterinary and routine) care, treatment, transportation and humane endpoints (slaughter or euthanasia) during the life of animals under human charge. The content of 'welfare' needs filling in through rigorous scientific understanding of animals' needs.

## The Philosophical Significance of Imaginaries – Where We Might Be/Are Headed

A central challenge for AE in the Anthropocene has been to strike the right balance between responsibilities toward animals and duties to people and the planet, and to be responsive to those working directly with animals (e.g., in food production, conservation, breeding and pet industries, teaching and biomedical research), who have been alienated by the animal welfare utilitarian-rights-based or the ethics of personhood framings. Practical issues, including existential ones, permeate human-animal-environment relationships and act as a social force on the imagination. In turn, collective imagination can encourage more creative non-idealized approaches in AE (contra Singer or Regan) in the face of material innovation and normative habits that are fractured and no longer advance social goods. For example, contemporary concerns for the ethical treatment of animals have led to practitioners' greater interest in the ethics of human-animal interconnections and animal welfare issues (e.g., Hobson-West and Jutel 2020; Lord et al. 2017). Laypersons, scientists, farmers, policy makers, breeders and ethicists, ethologists, and veterinarians in direct proximity to animals have called for moral and practical approaches that focus not on *whether* animals have moral status (Fraser 1999), but *how to* provide them with the appropriate care based on their species-specific needs and adaptations (Rollin 1995a; Russow 1999).

A revitalization and re-envisioning of husbandry ethics or animal care seems urgent (Thompson 2021) in the Anthropocene. In the debates moving forward, relevant scientific facts about the natures or adaptations of animals and their contexts are increasingly pivotal in human-animal relationships (Fraser 1999; Grandin 2020). Also, narratives regarding the roles animals play in society are instructive (Anthony and Varner 2019; Barnhardt and Kawagley 2005; Linzey 1995; Preece and Fraser 2000; Rollin 1995a; Reo and Powys Whyte 2012; Thompson 1993). While training in animal welfare is important ethically, an organization's institutional culture of husbandry, responsible management or animal care is increasingly more important as a way to inculcate science-informed animal welfare practices and care as core values[5]. Animal welfare scientist David Fraser (2008) has challenged scientists to show greater awareness and self-reflection of and engagement with the norms and values that guide their research on ethology and welfare. He encourages ethicists and philosophers to be informed about the daily circumstances of animal use and to grapple with the on-the-ground constraints of scientists working with animals and veterinarians regarding animal husbandry so that any ethical recommendations they issue can be actionable and lead to practicable solutions for animals and those who depend on them.

In the decades to come, humans will need responsive ethical imaginaries for how to live well with animals. These imaginaries may take at least three forms. First, climate change, including natural disasters, zoonoses and conflicts related to diminishing natural resources will likely shape future human-animal-environment relationships and will require *novel* imaginaries and solutions[6] grounded in human and interspecies relational solidarity, buttressed by compassion and the ethics of care reflected into practical animal-management strategies.

Compassion is a moral sentiment beyond just sympathizing with the plight of others or feeling as they do (i.e., empathy). It involves acknowledging the plight of others (i.e., "feeling for") and taking steps to advance their well-being (Singer and Klimecki 2014, p. r875). Compassion stretches our moral imagination by inviting active concern (caring about) and involvement in (caring for) others. Compassion is *doing* relational solidarity by acting with evidence-based knowledge of how we should relate to the animal. It can also encompass an interdependency with other species, founded on a realization that 'we're all in this together.' Compassion at the interspecies level can turn attention to common interests and vulnerabilities and minimize the false choice that we must give primacy to either animals or humans.

An ethics of care mindset encourages "attentive companionship with and for another, and attentive commitment to another" (Jennings 2018). It anchors moral responsibility toward animals by serving as the normative basis that underpins caregiving or caretaking (Held 2006; Midgley 1983) through scientifically informed practices. Instead of abstract notions, such as rights or harms and benefits, ethical responsibility is defined in concrete terms through our actual (and diverse) entanglements with and responsiveness to others (Gilligan 1993).

A proper understanding of interspecies relational solidarity involves cultivating acts of compassion, understanding how animals experience their realities and reimagining the ethical basis of science, technology and professionalism. This is especially urgent at a time when the impacts of climate change and human encroachment and scarce resources are becoming more pronounced and conservation concerns and zoonotic diseases are becoming more frequent. Interspecies solidarity highlights our interdependency with other creatures and human beings' interrelatedness with others (in relation to their local environments) in mutually constitutive communities (List and Vermeule 2014; Sharma 2016; Thompson 2016). Instead of the binary framing of equal to or lesser than, it suggests a 'one-for-all and all-for-one' approach – welfare is an universal good that transcends species and their contexts (Degeling et al. 2016).

Epistemically, relational solidarity reveals porous interspecies boundaries and a strong entanglement of fates. Ethically, it can suggest policy or interventions that advance the welfare of both animals and humans. A commitment to relational solidarity encourages investigation into the scientific-socio-ecological-political-economical-legal factors underpinning interspecies health and welfare. It provides the grounds for understanding animal interests in broader relational terms to and simultaneously with the environment, people, and the economy as a dynamic system. A practical example of implementing relational solidarity includes antimicrobial stewardship strategies that acknowledge interspecies co-vulnerabilities and benefit both animals and humans through careful consideration of how to minimize risk factors through interspecies and comparative microbiological and immunological research and better environmental management.

A relational account wedded to an ethics of care approach can bridge the gap between abstruse theory-heavy approaches and the practical, adaptive responsiveness required to match the lived experiences and broader social and material environments in which human-animal relationships are embedded. For one thing, it tends to focus on a plurality of views and relationships at the personal level (i.e., I have acquired responsibilities of care because this is/these are *my* animal(s) or patient(s)). Further, it brings into view what 'ought implies can' entails, which can be empowering for owners, caretakers and professionals. Ideal care should be guided by the best scientific and professional practices contextualized by short-, medium- and long-term plans, and engagement with a variety of worldviews and knowledge systems. It should also be guided by a commitment to understand animals as they are (not through anthropomorphism).

The relational account highlighted above aims to minimize friction (e.g., humans vs. animals/individual vs. group interests) by creating safe discourse spaces, recognizing (inter- and intraspecies) vulnerabilities and constraints, and promoting transparent values and norms that shape our co-evolving relationships with animals. It advances solutions informed by our knowledge of each species' capacities through attentiveness to ongoing relationships (not bounded by species) in different human-animal-environmental contexts; not by punishing or forbidding existing relationships with animals.

Since our ethical responses will have to be more nuanced and involve multifaceted accounts of how we relate to animals, a comprehensive account of human-animal welfare and human-animal behavior will be vital in providing the baseline and flexibility for how to ensure animals of all species fare well in the Anthropocene. The One Health and One Welfare multidisciplinary frameworks provide promising examples of imaginaries at the intersection of human-animal-environment health. One Health, which emerged at the beginning of the 21st century in response to the persistence of emerging infectious diseases, champions the view that the health of humans and other animals is contiguous (Kelly et al. 2017; Meijboom and Nieuwland 2018; OIE 2016). It can also be a useful tool for promoting animal welfare and considering animals' experiences during a disaster (Anthony and De Paula 2022).

Novel imaginaries should look beyond narrow framings of animals as victims of pain and suffering to be rescued from human exploitation. While there is a case to be made that we ought to be vigilant in our use of and respect for animals, we also ought to strengthen our bonds with them. Looking to Indigenous relationships with animals as co-agents, partners or fellow travelers can inform broader and more sustainable interspecies relationships (Kawagley 1995). Further, by looking to how animals have evolved to meet their challenges, we can learn new adaptive strategies necessary to navigate our changing environmental circumstances (Bodenhorn 1990; Reo and Pows Whyte 2012).

Stronger interdisciplinary collaborations between AE (including the philosophies of mind and science), AWS and applied ethology, especially to understand what matters to animals, are increasingly necessary. These sciences investigate animals' perspectives or animal-based measures and can inform ethical deliberations about practices involving animals with a view toward improving their quality of life. AWS integrates ethology, neuroscience, psychology, physiology, environmental and health factors to identify how well animals are thriving in their different contexts (De Paula Vieira et al. 2008; Fraser 2008).

Despite some conceptual ambiguity regarding the scope of animal welfare and its assessment criteria (Weary and Robbins 2019), AWS and applied ethology can broaden how scientists, veterinarians, ethicists and other professionals consider 'good quality of life' by incorporating the perspectives of animals and their vulnerabilities in relation to human activities. They will also be essential in informing the development of evidence-based assessments in concert with ethical objectives to minimize harm to the fewest animals.

AWS and applied ethology can advance good outcomes for animals through scientific, technical, systematic and species-specific guidance to manage animals as well as strategies to minimize suffering and loss of life by improving animal care and management. Training in both AWS and applied ethology is vital to recognize when animals are distressed and what constitutes poor and good animal welfare. These sciences can measurably improve our understanding of how animals and humans actually relate.

Applied ethology and AWS have been instrumental in helping humans realize that our ways of understanding animals has been too narrow, and have permitted interspecies conflict-oriented framing of our discourse regarding animals. The relational view, together with the sciences discussed above, serve as the evidential grounds for engaging with owners, caretakers and decision makers to modify the ways animals and human-animal relationships are understood and treated. These sciences jointly play an important role in detailing the actual relationship between species-specific behaviors and how animals' interests are frustrated or rewarded when they are actualized by their owners or caretakers. For example, applied ethology and AWS have already provided evidence supporting higher cognitive capacities in animals, including planning and execution and emotions (Panksepp 2005). Birds, fish and other non-mammals have the capacity to learn from past experiences (Allen and Fortin 2013; Braithwaite 2010). Evidence also points to pain, suffering, and positive affective states, as well as higher order thought, episodic memory, self-consciousness, future-oriented beliefs and empathy as being widespread throughout the animal kingdom (Bekoff 2007; Gennaro 2016; Varner 2012). Also, there is evidence showing that it is doubtful that animals are harmed morally or have their "rights" violated in every instance in which their natural proclivities are frustrated. Animals demonstrate behavioral flexibility and may rely on a number of adaptive abilities to compensate for a momentary frustration or discomfort (Fraser et al. 1997).

Applied ethology and AWS will feature significantly in enriching our understanding of animal minds across species (Andrews and Beck 2017; Beauchamp and Frey 2011; Lurz 2011; Varner 2012). Scientists such as Dawkins (1993), Harding, Paul and Mendl (2004), Duncan (2006) and Weary et al. (2017) have provided vital evidence regarding the affective states of and emotions in animals. Knowing about consciousness and capacities for pain and suffering dovetail with best practices for performing euthanasia (AVMA 2020) and minimizing unpleasant effects on research animals (NAP 2011). This knowledge will inform welfare assessment frameworks like Fraser's Three Domains model (Fraser et al. 1997), Mellor's Five Domains model (Mellor et al. 2020) and Triangulation model (Paul et al. 2022). New frontiers in AE, involving under-studied species (e.g., marine life and insects) (Meijboom and Bovenkerk 2013), and the impacts of emerging (bio or neuro) technological manipulations on animals (Takala and Häyry 2014) also rely on the latest science regarding animal minds and capacities. The interconnections between ethics and these sciences point to the importance of interdisciplinary collaboration in comparative human-animal welfare studies and as a guard against untutored anthropomorphism (Daston and Mitman 2005).

While it is difficult to predict the exact shape of human-animal relationships in the decades to come, if past is prologue, we can expect that there will be imaginaries that are *extensions* of existing frameworks. It is likely, for better or worse, that some version of the welfare utilitarianism vs rights-based framing will still have currency among academic circles (at least). For example, animal-rights proponents will likely continue to press the significance of animal personhood and dignity arguments to underpin future animal law and regulatory initiatives (Francione 2008; Wise 2001). If this materializes, the legal (and political) system will be used more frequently to enforce narrower forms of "acceptable" relationships between humans and animals. However, the rise of animal protectionism as a political movement (e.g., Brazil, Spain) could harm animals and their owners, especially if it is lacking in scientific credibility (e.g., cherry picking or incomplete representations of welfare science data and the complexity of human-animal relationships) and a comprehensive understanding of the structure, processes, historical and cultural significance of human-animal relations.

We predict a greater convergence of disciplines in that AWS, applied ethology and the social sciences and humanities (Davies et al. 2020; Desmond 2022) will play significant roles in exploring future questions about the moral status of animals and human-animal relationships. This second type of imaginary will likely animate animal-centric capacities in concert with social sciences and humanities research on how to strengthen human-animal relationships and improve welfare across the animal kingdom. The pairing of valuational (ethical) inquiry with AWS and applied ethology is relevant and vital since: (i) it will unite diverse voices to identify and remedy possible harms to animals; and (ii) it will acknowledge the importance of comparative accounts in leading to fruitful and practicable solutions for those who benefit from animal use. This pairing (iii) is essential for pinpointing knowledge gaps regarding the impacts of policies and technologies on animals, as a corrective to untutored approaches of what is in the best interest of animals and serving as a basis for enhanced modes of animal-centered care, management and husbandry training. Furthermore, (iv) it can unearth valuable stakeholder feedback and advance effective multi-stakeholder communication strategies, both critical for identifying biases and implementing practices that accord with societal values and expectations regarding what is owed to animals. Careful attention should be paid, however, to distinguishing between social science research that highlights what people want for animals and what AWS and applied ethology tell us is important from animals' points of view. Additionally, more attention should be given to unlocking pleasant experiences and maximizing positive animal welfare, while minimizing pain and suffering (Rault et al. 2020).

A third type of imaginary will be *technologically motivated*. Genetic engineering may advance animal welfare (over traditional selective breeding techniques) by breeding animals that adapt better to their environment (Croney et al. 2018). In some instances, proposed modifications to animals to fit their circumstances are motivated by good intentions, i.e., to extinguish welfare problems thought irredeemable by engineering animals' housing conditions. Through extreme genetic make-overs, animals can be created that are numb to or dis-enhanced relative the conditions under which they are raised, e.g., "blind hens" or totally insentient (i.e., resulting in animals no longer having conscious experiences) or decerebrate organisms (Rollin 1995b; Sandøe et al. 2014). Genetic engineering to modify the natures or capacities of animals to limit their suffering or negative experiences so they fit better within their environments, while a noble goal (e.g., eliminating the poll gene in dairy cattle), may result in public resistance or backlash if extrinsic and intrinsic aspects of the technology result in unintended consequences for animals and human-animal relationships. Ethical concern regarding how genetic manipulation may affect animal welfare takes four general forms (Anthony and Thompson 2003):

1   Animals may suffer directly as a result of the effects of modification and manipulation.
2   Animals may suffer indirectly as a result of the effects of modification and manipulation.
3   Animals may suffer from consumption of or treatment with genetically modified products.
4   The natures of animals are changed in substantial ways to benefit humans.

If we can enhance an individual animal's welfare by transcending its "nature" through genetic modification or other technological enhancements, ought we not do it? Rollin (1995b) has addressed this issue by employing the conservation of welfare principle: genetic alteration is morally problematic if it would worsen an animal's well-being relative to unaltered fellow creatures or if the individual in question would have enjoyed a better life without the intervention. Rollin's view focuses on meeting the evolutionarily inherited species endowments or adaptations of farmed animals, i.e., their *teli* or distinctive natures. *Telos* is "a nature, a function, a set of activities intrinsic to [an individual of a particular species], evolutionarily determined and genetically imprinted." Those of us who interact with animals know that

> [They] have natures – the pigness of the pig, the cowness of the cow, 'fish gotta swim, birds gotta fly' – which are essential to their well-being. Common sense tells us that animals who are built to move need to move to feel good; there is no point in trying to prove that they are fine if kept immobile.

Animals should be valued for their evolutionarily established species typical "natures" (Rollin 1995a, p.159).

In other instances, use of genetic engineering technologies reflects the pinnacle of hubris since humans have the power to make animals into anything we want. Here, the status of animals or their welfare seems to be of secondary importance, subservient to a 'because we can' mantra. Gene editing and other conventional biotechnologies pave the way for breeding animals to have the traits that we fancy (Davis 2022; Palmer 2011), genetically engineering and cloning pets and livestock (Rollin 1995b) or resurrecting extinct species like the woolly mammoth or saber tooth tiger. The motivations behind 'de-extinction' are varied, including righting the wrong of being responsible for anthropogenic species extermination (Jebari 2016). It can be argued that humanity should not willfully produce chimeras with poor welfare or disrupt the welfare of on-target species and ecosystems in the process of recreating extinct animals (see for example, Rohwer and Marris 2018; Sandler 2014).

Lately, there is growing excitement for synthetic, cultured or clean "meat and milk" (Kadim et al. 2015; Shapiro 2018). These highly processed products are touted as more efficient ways of producing protein that also side-step environmental and animal welfare issues. Also, using biotechnological approaches to culture meat in laboratories is suggested as a viable alternative for meat consumers who desire to be more 'ethical' but who do not wish to drastically modify their diet. The prospect of consuming animal-based products not directly resulting from sentient beings aims to use considerably fewer animals than conventional livestock farming. However, some animals will still have to be reared to harvest the cells to produce synthetic protein (and there is a veritable fog around the lives of donor animals cultivated for the use of bovine fetal serum as cell culture system). Long-term studies on the health and environmental consequences of this protein and on the human-animal relationship are lacking. Also, aside from overcoming regulatory hurdles, convincing skeptical publics and ensuring equity of access, more immediately, scale-up looks to be an insurmountable challenge since large amounts of energy and capital will be needed to grow the "meat" in more and bigger bioreactors and produce the growth media. Further, removing animals from agriculture ignores centuries of human-animal co-evolution. By reducing food from animal sources to their protein, this technology once again sees food animals as passive subjects in need of rescue and limits human-animal relationships considerably by commodifying their bits and pieces. Further, it is an extreme instrumentalization of what we put inside our bodies (Thompson 2014). Synthetic protein consumption does not benefit animals per se, since *no* animals that come into existence can benefit from this technology (Palmer 2011). It is a fair question to ask if consumers of synthetic protein are benefitting only themselves, displacing farmers and farming communities and landscapes in the process and feeling good about not supporting the perceived suffering of farm animals. Paul Thompson (2021) is critical of those who are seemingly only concerned about the ethical significance of their own consumption habits by embracing synthetic protein production over ethical husbandry. He observes that this "complete instrumentalization of the animal" is a form of "structural narcissism" and perpetuates "a pattern of disengagement," since no attempt is made to either "actually consider [how] the condition in which animals live reproduces a pattern of normative practice" or interest shown to engage substantively with the animals' experiences.

The emergence of precision automated and monitoring technologies, through heightened computer processing power and sensor technologies (e.g., 'Big Data,' artificial intelligence and robotics) pose new ways of thinking about the moral status of animals, animal care and human-animal relationships. These technologies have huge potential to develop knowledge-informed solutions to manage and care for animals and improve our animal-based systems. Networked 'smart' technologies that continuously monitor and manage individual animals automatically via algorithms can be approached fruitfully when ethical theorizing and philosophical reflection about the moral status of these technologies is paired with AWS and applied ethology.

These technologies are only as good as their technical inputs and those minding their development, design and implementation. Hence, the devices must consider the animals themselves and human as final users (i.e., veterinarians, farm managers or slaughter operators or technicians, caretakers or owners). Precision livestock farming, when designed by ethologists and animal welfare scientists with animals and caretakers as end users in mind, will allow for welfare improvements and stronger human-animal bonds. It is necessary that animal welfare scientists, applied ethologists and veterinarians direct the technological requisites of the interfaces, software and hardware systems in order to promote technological developments attentive to animals' welfare and needs. Through ethograms, indicators of behavior, affective states and health, video technology, imagery, physiological and pathological parameters, social networks, geolocation and so on, together with animal and human factors, end-user experiences can be enhanced. Valuational inquiry by social scientists and humanities scholars can provide insights on how these technologies mediate or hinder healthful and respectful interspecies interactions. Valuational inquiry can also accentuate awareness about the complexity of human-animal-technology relationships and stress relationality and interconnectivity with animals. Once again, interdisciplinarity will be pivotal in developing a holistic view of shared ends, one that considers human-animal co-evolution against the backdrop of environmental, historical, cultural and social factors.

Much of our current understanding about the welfare and behavior of animals has been possible because of technological advancements. Other ethical opportunities or challenges include:

a   Ensuring technologies enable animals to communicate their preferences.
b   Promoting technologies that enhance human-animal interaction and do not alienate caretakers from their husbandry/stewardship responsibilities.
c   Encouraging technologies that involve (i) self-reflection on values and assumptions regarding large-scale animal use (e.g., agriculture, breeding) and sustainability-related 'profit-planet-people' trade-offs, and (ii) deliberation of how to advance animal welfare concepts and frameworks (Wathes et al. 2008).
d   Being transparent about the animal welfare assumptions driving a particular technological design.
e   Being honest about the limits of what the technology can reveal about animals' minds, bodies and natures.
f   Ensuring that technological devices do not replace animal care or displace veterinary care (especially when brought to scale) but are thoughtfully integrated into a management system that advances positive welfare outcomes for animals and human-animal interactions.
g   The development of animal-robot chimeras and trans-animalism, including linking animals' brains to communicate with computers and AI (see, e.g., https://neuralink.com/approach/).

## Conclusion

The moral status of animals within different societies is determined by larger culturally specific social valuations and by institutional and political structures that support animal ownership and the roles owners, caretakers and professionals play in advancing the lived experiences of animals. First-wave AE sought to address the mere instrumentalization and commodification of animals (and associated harms), correlated to their invisibility in our communities and limited exposure to them. However, the ethical concern for animals should not eclipse the diverse ways in which we interact with them. Human-animal species co-evolution (e.g., Haraway 2003; Sundman et al. 2019), such as interspecies physical and social proximity and historical and cultural co-mingling, form part of the social, ethical and epistemic world we share with animals. Ethical framings that narrowly portray animals as victims or at odds with human interests ought to be re-contextualized in the Anthropocene. Moving forward, those working in AE should be open to interdisciplinarity, including collaboration with applied ethologists and animal welfare scientists. Valuational inquiry explores ethical, cultural, esthetic, legal, scientific, veterinary and technological elements that inform how animals ought to be treated. Coupled with applied ethology and AWS, the complicated calculus of how to minimize harm or accentuate benefits to animals based on their capacities can be operationalized. This kind of interdisciplinarity is essential when exploring what matters to animals and being responsive to problems faced by those who interact with and care for animals daily, including ensuring that animal welfare oriented legislation will actually benefit animals.

Promising novel imaginaries center on relationality with animals and seek to position the perspectives and interests of animals as serious bases of normativity within inter- and intraspecies entanglements. The relational view upholds compassion and care as pivotal virtues and practices in our relationship with animals. It also underlies alternative ways of seeing and being with animals, including better ways to situate animals' lived experiences within the human world. Knowledge from AWS and applied ethology reflects the view that healthy human-animal relationships and interactions should be grounded on actual measures of animal health, feelings, behaviors and natural adaptations. Those working to advance interspecies relationality will need to develop new tools to advance positive human-animal relations and cultivate productive ways of integrating ethics and the sciences discussed here to enhance both human and animal lives.

## Guided Questions

1. What has been the impact of first-wave AE on our attitudes toward animals?
2. How might imaginaries that consider the perspective of animals in different contexts help us to develop a more comprehensive view of the animal-human relationship?
3. How might animal welfare science, applied ethology and a relational ethics based on compassion and care inform animal care cum management strategies?
4. What role might technology play in the future of caring for animals and managing animal-human relationships and what are some of the ethical caveats, inherent and indirect, in applying technologies to animal welfare and husbandry?
5. How should science and ethics be employed to cultivate knowledge and understanding of human-animal futures grounded in a commitment to equity, inclusion, diversity and belonging?

## Notes

1 Valuational inquiry in this context involves applying the humanities (philosophy, ethics, art, Indigenous ways of knowing) and social sciences (sociology, anthropology, history) to human-animal studies and the practices, social structures and meanings and institutions involving animals. It is dedicated to understanding and addressing human-animal interactions and relationships through discursive, narrative and descriptive investigation and analysis and philosophical conceptualizing. Our focus in this chapter involves primarily philosophical ethics.

2   This close relationship with our pets during a crisis is playing out in Ukraine at the time of writing, as owners are reluctant to abandon them despite threats to their own security and life (https://www.bbc.com/news/newsbeat-60593791). One Ukrainian refugee stated: "My dog is 12 and a half and she struggled to walk and fell down every kilometre or so and couldn't stand up again. I stopped cars and asked for help but everyone refused; they advised us to leave the dogs. But our dogs are part of our family. My dog has experienced all the happy and sad moments with us. Mum's dog is all she has left of her former life. So my husband, at times, carried our dog on his shoulders" (https://www.theguardian.com/world/2022/mar/08/ukraine-russia-crisis-border).

3   Values also inform formal ethical standards, such as professional codes of conduct (OIE Animal Welfare principles, the leading intergovernmental organization responsible for monitoring animal health globally; OLAW) and governance processes (e.g., animal care and use committees). They also influence animal protection directives (e.g., Article 13 of Title II, of The Lisbon Treaty) (2009) (https://ec.europa.eu/food/animals/welfare_en) and the UK government's 2021 Action Plan for Animal Welfare (https://www.gov.uk/government/publications/action-plan-for-animal-welfare).

4   Regan contends that "we are not saying that dogs and cats can do calculus, or that pigs and cows enjoy poetry." However, since animals are "psychological beings, with an experiential welfare of their own," this makes humans and animals equal in terms of being rights-bearers (Regan 1983). Singer's focus is on animal interests a la utility considerations, anthropocentric claims to superiority are quashed by "this hard fact: in suffering, the animals are our equals" (Singer 1975).

5   Examples include commitments/practice by animal industry groups to improve animal care and welfare (e.g., https://www.porkcares.org/ethical-principles/animal-well-being/) and veterinary organizations that have made animal welfare a centerpiece of their advocacy of animals (e.g., https://www.avma.org/javma-news/2011-01-01/veterinarians-oath-revised-emphasize-animal-welfare-commitment) and through the force of animal welfare legislation (https://food.ec.europa.eu/animals/animal-welfare/animal-welfare-practice_en).

6   Imaginaries that go beyond egalitarian and impartial approaches linked to the traditional moral theories used by, for example, Singer (utilitarianism) and Regan (deontology) and which embrace non-ideal philosophical approaches based on relationality and connectivity are echoed in two important edited volumes by Bovenkerk and Keulartz: (2016) *Animal Ethics in the Age of Humans: Blurring Boundaries in Human-animal Relationships* and (2021) *Animals in Our Midst: The Challenges of Co-existing with Animals in the Anthropocene*.

# References

Aaltola, E. (2011). The philosophy behind the movement: Animal studies Vs. Animal rights. *Society and Animals*, 19(4): 393–406.

Allen, C., and Bekoff, M. (1999). *Species of Mind: The Philosophy and Biology of Cognitive Ethology*. Cambridge: MIT Press.

Allen, T. A., and Fortin, N. (2013). The evolution of episodic memory. *Proceedings of the National Academy of Science*, 110: 10379–10386.

American Veterinary Medical Association (AVMA) (2020). *Guidelines for the Euthanasia of Animals: 2020 Interim* Edition*. Retrieved from www.avma.org/KB/Policies/Documents/euthanasia.pdf.

Anderson, E. (2004). Animals and the values of non-human life. In C. Sunstein and M. Nussbaum (Eds.), *Animal Rights* (pp. 277–297). New York: Oxford University Press.

Andrews, K. (2020). *The Animal Mind: An Introduction to the Philosophy of Animal Cognition* (2nd ed.). New York: Routledge.

Andrews, K., and Beck, J. (Eds.) (2017). *The Routledge Handbook of Philosophy of Animal Minds*. London: Routledge.

Anthony, R. (2009). Farming animals and the capabilities approach: Understanding roles and responsibilities through narrative. *Ethics, Society and Animals*, 17(3): 259–280.

——— (2017). Sustainable animal agriculture and environmental virtue ethics. In Philip Brey and David M. Kaplan (Eds.), *Philosophy, Technology and the Environment* (pp. 213–228). Cambridge: The MIT Press.

Anthony, R., and De Paula Vieira, A. (2022). One health animal disaster management: An ethics of care approach. *Journal of Applied Animal Welfare Science*, 25(2): 180–194 https://doi.org/10.1080/10

888705.2022.2040360; Open Access https://www.tandfonline.com/doi/full/10.1080/10888705.2022.2040360.

Anthony, R., and Thompson, P. B. (2004). Biosafety, ethics, and regulation of transgenic animals. In S. Parekh (Ed.), *Handbook of Genetically Modified Organisms—Mammalian, Microbial, and Plant Cells: Technologies, Economic Impact and Challenges* (pp. 183–205). Totowa, NJ: Humana Press.

Anthony, R., and Varner, G. (2019). *Subsistence Hunting*, edited by Bob Fischer, Routledge Animal Ethics Encyclopedia. New York: Routledge.

Baker, S. (2000). *The Postmodern Animal*. London: Reaktion Books, p. 160.

Barnhardt, R., and Kawagley, A. O. (2005). Indigenous knowledge systems and Alaska native ways of knowing. *Anthropology and Education Quarterly*, 36(1): 8–23.

Beauchamp, T., and Frey, R. (2011). *The Oxford Handbook of Animal Ethic*. New York: Oxford University Press.

Bekoff, M. (2007). *The Emotional Lives of Animals*. Novato, CA: New World Library.

Bennett, R. M., and Thompson, P. (2018). Economics. In M. C. Appleby, I. A. S. Olsson and F. Galindo (Eds.), *Animal Welfare* (3rd ed., pp. 335–348). Wallingford, Oxfordshire: CAB International.

Berry, W. (2012). *It All Turns on Affection: The Jefferson Lecture and Other Essays*. Retrieved from https://www.neh.gov/about/awards/jefferson-lecture/wendell-e-berry-biography.

Bernstein, M. H. (1998). *On Moral Considerability: An Essay on Who Morally Matters*. Oxford: Oxford University Press.

Birke, L. (1994). *Feminism, Animals and Science: The Naming of the Shrew*. Philadelphia, PA: Open University Press.

Bodenhorn, B. (1990). "I'm not the great hunter, my wife is": Inupiat and anthropological models of gender. *Etud/Inuit Stud*, 14: 55–74.

Bovenkerk, B., and Keulartz, J. (Eds.) (2016). *Animal Ethics in the Age of Humans Blurring Boundaries in Human-animal Relationships*. Cham: Springer International Publisher.

Braithwaite, V. (2010). *Do Fish Feel Pain?* Oxford: Oxford University Press.

Burgess-Jackson, K. (1998). Doing right by our animal companions. *Journal of Ethics*, 2(2): 159–185.

Burton, S., and Brady, E. (2016). What is it like to be a bird? Epistemic humility and human-animal relations. In B. Bovenkerk and J. Keulartz (Eds.), *Animal Ethics in the Age of Humans: Blurring Boundaries in Human-Animal Relationships* (pp. 89–101). Cham: Springer Publishing.

Carruthers, P. (1992). *The Animals Issue*. Cambridge: Cambridge University Press.

Carson, R. (1962). *Silent Spring*. Boston, MA: Houghton Mifflin Company.

Crist, E. (1999). *Images of Animals: Anthropocentrism and Animal Mind*. Philadelphia, PA: Temple University Press.

Croney, C., Muir, W., Ni, J. Q. et al. (2018). An overview of engineering approaches to improving agricultural animal welfare. *Journal of Agricultural and Environmental Ethics*, 31: 143–159.

Crutzen, P. J., and Stroemer, E. F. (2000). The "anthropocene". *IGBP Newsletter*, 41: 17–18.

Daston, L., and Mitman, G. (Eds.) (2005). *Thinking with Animals: New Perspectives on Anthropomorphism*. New York: Columbia University Press.

Davies, G., Gorman, R., Greenhough, B., et al. (2020). Animal research nexus: A new approach to the connections between science, health and animal welfare. *Medical Humanities*, 46: 499–511.

Davis, N. (2022). *The Guardian*. https://www.theguardian.com/lifeandstyle/2022/jan/26/first-uk-hairless-french-bulldog-litter-prompts-extreme-breeding-concerns.

Dawkins, M. S. (1993). *Through Our Eyes Only? The Search for Animal Consciousness*. Oxford: W.H. Freeman.

Degeling, C., Lederman, Z., and Rock, M. (2016). Culling and the common good: Re-evaluating harms and benefits under the one health paradigm. *Public Health Ethics*, 9(3): 244–254.

DeGrazia, D. (1996). *Taking Animals Seriously: Mental Life and Moral Status*. Cambridge: Cambridge University Press.

——— (2011). The ethics of confining animals: From farms to zoos to human homes. In T. L. Beauchamp and R. G. Frey (Eds.), *The Oxford Handbook of Animal Ethics* (pp. 738–768). New York: Oxford University Press.

De Paula Vieira, A., and Anthony, R. (2021). Reimagining human responsibilities towards animals for disaster management in the anthropocene. In B. Bovenkerk and J. Keulartz (Eds.), *Animals in Our Midst: The Challenges of Co-Existing with Animals in the Anthropocene* (pp. 223–254). Cham: Springer Open Access. https://link.springer.com/chapter/.

De Paula Vieira, A., Guesdon, V., de Passillé, A. M., von Keyserlingk, M. A. G., and Weary, D. M. (2008). Behavioural indicators of hunger in dairy calves. *Applied Animal Behaviour Science*, 109: 180–189.

Descartes, R. (1637/1988). Discourse on method. In *Descartes: Selected Philosophical Writings*. J. Cottingham et al., Trans. Cambridge: Cambridge University Press.

Desmond, J. (2022). Medicine, value, and knowledge in the veterinary clinic: Questions for and from medical anthropology and the medical humanities. *Frontiers Veterinary Science*, 9: 780482. https://doi.org/10.3389/fvets.2022.780482.

Dewey, J. (1982[1920]). *The Middle Works of John Dewey, Volume 12, 1899–1924*, edited by J. Boydston. Carbondale, IL: Southern Illinois University Press.

Duncan, I. J. H. (2006). The changing concept of animal science. *Applied Animal Behaviour Science*, 100(1–2): 11–19.

Foot, P. (1972). Morality as a system of hypothetical imperatives. *The Philosophical Review*, 81(3): 305–316.

Francione, G. (2008). *Animals as Persons: Essays on the Abolition of Animal Exploitation*. New York: Columbia University Press.

Fraser, D. (1999). Animal ethics and animal welfare science: Bridging the two cultures. *Applied Animal Behaviour Science*, 65(3): 171–189.

——— (2008). *Understanding Animal Welfare: The Science in Its Cultural Context*. Oxford, UK: Wiley-Blackwell.

Fraser, D., Weary, D. M., Pajor, E. A., and Milligan, B. N. (1997). A scientific conception of animal welfare that reflects ethical concerns. *Animal Welfare*, 6: 187–205.

Gennaro, R. (2016). Animal consciousness and higher-order thoughts. In Jacob Beck & Kristin Andrews (Eds.), *The Routledge Handbook of Philosophy of Animal Minds*. London: Routledge Publishers.

Gilligan, C. (1993). *In a Different Voice: Psychological Theory and Women's Development*. Cambridge, MA: Harvard University Press.

Goffman, E. (1974). *Frame Analysis: An Essay on the Organization of Experience*. Cambridge, MA: Harvard University Press.

Grandin, T. (Ed.) (2020). *Improving Animal Welfare: A Practical Approach* (3rd ed.). Wallingford: CABI.

Guerrini, A. (2016). Deep history, evolutionary history, and animals in the anthropocene. In Bernice Bovernkerk and Jozef Keulartz (Eds.), *Animal Ethics in the Age of Humans Blurring Boundaries in Human-Animal Relationships* (pp. 25–38). Cham: Springer International Publishers.

Haraway, D. J. (2003). *The Companion Species Manifesto Dogs, People, and Significant Otherness*. Chicago, IL: Prickly Paradigm Press.

Harding, E. J., Paul, E. S., and Mendl, M. (2004). Cognitive bias and affective state. *Nature*, 427(6972): 312. Retrieved from https://www.nature.com/articles/427312a.

Harrison, R. (1964). *Animal Machines*. London: Vincent Stuart Publishers.

Held, V. (2006). *The Ethics of Care: Personal, Political and Global*. New York: Oxford University Press.

Hobson-West, P., and Jutel, A. (2020). Animals, veterinarians and the sociology of diagnosis. *Sociology of Health & Illness*, 42(2): 393–406.

Höglin, A., Van Poucke, E., Katajamaa, R. et al. (2021). Long-term stress in dogs is related to the human–dog relationship and personality traits. *Scientific Reports*, 11: 8612. https://doi.org/10.1038/s41598-021-88201-y.

Hursthouse, R. (2011). Virtue ethics and the treatment of animals. In T. L. Beauchamp and R. G. Frey (Eds.), *The Oxford Handbook of Animal Ethics* (pp. 119–143). New York: Oxford University Press.

Irvine, L. (2009). *Filling the Ark: Animal Welfare in Disasters*. Philadelphia, PA: Temple University Press.

Jebari, K. (2016). Should extinction be forever? *Philosophy of Technology*, 29: 211–222.

Jennings, B. (2018). Solidarity and care as relational practices. *Bioethics*, 32(9): 553–561.

Jensen, P. (Ed.) (2017). *The Ethology of Domestic Animals: An Introductory Text* (3rd ed.). Wallingford, UK: CABI.

Kadim, I. T., Mahgoub, O., Baqir, S., and Faye, B. (2015). Purchas R. Cultured meat from muscle stem cells: A review of challenges and prospects. *Journal Integrative Agriculture*, 14: 222–233. https://doi.org/10.1016/S2095-3119(14)60881-9.

Kant, I. (1963). Duties to animals and spirits. In L. Infield (ed.) *Lectures on Ethics* (pp. 239–241). New York: Harper and Row.

Kawagley, A. O. (1995). *A Yupiaq World View: A Pathway to Ecology and Spirit*. Prospect Heights, IL: Waveland Press.

Kelly, T. R., Karesh, W. B., Kreuder Johnson, C., Gilardi, K. V. K., Anthony, S. J., Goldstein, T., Olson, S. H., Machalaba, C., PREDICT, Consortium, Mazet, J. A. et al. (2017). One Health proof of concept: Bringing a transdisciplinary approach to surveillance for zoonotic viruses at the human–wild animal interface. *Preventive Veterinary Medicine*, 137: 112–118.

Kendall, H., Lobao, L., and Sharp, J. (2006). Public concern with animal well-being: Place, social structural location, and individual experience. *Rural Sociology*, 71(3): 399–428.

Kirk, R. G. W., and Worboys, M. (2011). Medicine and species: One medicine, one history? In Mark Jackson (Ed.), *The Oxford Handbook of the History of Medicine* (pp. 561–577). Oxford: Oxford University Press.

Korsgaard, C. M. (1996). *The Sources of Normativity*. Cambridge: Cambridge University Press.

——— (2015). A Kantian case for animal rights. In T. Visak and R. Garner (Eds.), *The Ethics of Killing Animals* (pp. 154–177). New York: Oxford University Press.

Korthals, M. (2016). Human-animal interfaces from a pragmatist perspective. In Bernice Bovernkerk and Jozef Keulartz (Eds.), *Animal Ethics in the Age of Humans Blurring Boundaries in Human-Animal Relationships* (pp. 73–88). Cham: Springer International Publishers.

Linzey, A. (1995). *Animal Theology*. Urbana, Chicago: University of Illinois Press.

List, C., and Vermeule, A. (2014). Independence and interdependence: Lessons from the hive. *Rationality and Society*, 26(2): 172.

Lord, L.K., Millman, S.T., et al. (2017). A model curriculum for the study of animal welfare in colleges and schools of veterinary medicine. *JAVMA-Journal of the American Veterinary Medical Association*, 250(6): 632–640. doi:10.2460/javma.250.6.632

Lurz, R. (2011). *Mindreading Animals: The Debate over What Animals Know about Other Minds*. Cambridge, MA: MIT Press.

Lynteris, C. (Ed.) (2019). *Framing Animals as Epidemic Villains. Histories of Non-human Disease Vectors*. Cham: Palgrave Macmillan.

May, T. (2014). Moral individualism, moral relationalism, and obligations to non-human. *Animals Journal of Applied Philosophy*, 31(2): 155–168. https://doi.org./10.1111/japp.12055.

McMahon, J. (2016). The comparative badness of animal suffering and death. In T. Višak and R. Garner (Eds.), *The Ethics of Killing Animals* (pp. 65–85). New York: Oxford University Press.

McPherson, T. (2016). Why i am a vegan (and you should be one, too). In A. Chignell, T. Cuneo and M. C. Halteman (Eds.), *Philosophy Comes to Dinner: Arguments about the Ethics of Eating* (pp. 73–91). New York: Routledge.

Meijboom, F. L. B., and Bovenkerk, B. (2013). Beneath the surface: Killing of fish as a moral problem. In H. Röcklinsberg and P. Sandin (Eds.), *Proceedings 11th Congress of the European Society for Agricultural and Food Ethics* (pp. 245–250). Wageningen: Wageningen Academic Publishers.

Meijboom, F. L. B., and Nieuwland, J. (2018). Manifold health: The need to specify one health and the importance of cooperation in (bio)ethics. In S. Springer and H. Grimm (Eds.), *Professionals in Food Chains* (pp. 266–271). Wageningen: Wageningen Academic Publishers.

Mellor, D. J., Beausoleil, N. J., Littlewood, K. E., McLean, A. N., McGreevy, P. D., Jones B., and Wilkins, C. (2020). The five domains model: Including human–animal interactions in assessments of animal welfare. *Animals*, 10(10): 1870. https://doi.org/10.3390/ani10101870.

Mench, J. (1998). Why it is important to understand animal behavior. *ILAR Journal*, 39: 20–26.

Mench, J. A., Sumner, D. A., and Rosen-Molina, J. T. (2011). Sustainability of egg production in the United States—The policy and market context. *Poultry Science*, 90: 229–240.

Midgley, M. (1983). *Animals and Why They Matter*. Athens, GA: University of Georgia Press.

——— (2003). Is a dolphin a person? In Susan Armstrong & Richard Botzler (Eds.), *The Animal Ethics Reader* (pp. 166–172). New York: Routledge.

Moffat, K., Lacey, J., Zhang, A., and Leipold, S. (2016). The social licence to operate: A critical review. *Forestry: An International Journal of Forest Research*, 89(5): 477–488. https://doi.org/10.1093/forestry/cpv044.

Narveson, J. (1983). Animal rights revisited. In H. B. Miller and W. H. Williams (Eds.), *Ethics and Animals* (pp. 45–59). Clifton, NJ: Humana Press.

Norcross, A. (2004). Puppies, pigs and people: Eating meat and marginal cases. *Philosophical Perspectives*, 18: 229–245.

Norwood, F. B., and Lusk, J. (2011). *Compassion by the Pound: The Economics of Farm Animal Welfare*. New York: Oxford University Press.

OIE. (2016). *Guidelines on Disaster Management and Risk Reduction in Relation to Animal Health and Welfare and Veterinary Public Health*. Retrieved from https://www.oie.int/fileadmin/Home/eng/Animal_Welfare/docs/pdf/Others/Distermanagement-ANG.pdf.

Palmer, C. (2011). Animal disenhancement and the non-identity problem: A response to Thompson. *NanoEthics*, 5: 43–48.

Panksepp, J. (2005). *Affective Neuroscience: The Foundations of Human and Animal Emotions*. New York: Oxford University Press.

Paul, E. S., Browne, W., Mendl, M. T. et al. (2022). Assessing animal welfare: A triangulation of preference, judgement bias and other candidate welfare indicators. *Animal Behaviour*, 186: 151–177.

Petrzelka, P. (2014). Absentee landlords and agriculture. In P. B. Thompson and D. M. Kaplan (Eds.), *Encyclopedia of Food and Agricultural Ethics*. Dordrecht: Springer. https://doi.org/10.1007/978-94-007-0929-4_56.

Plumwood, V. (1991). Nature, self and gender: Feminism, environmental philosophy, and the critique of rationalism. *Hypatia*, 6(1): 3–27.

Preece, R., and Fraser, D. (2000). The status of animals in biblical and christian thought: A study in colliding values. *Society & Animals*, 8(3): 245–263.

Rachels, J. (1990). *Created from Animals: The Moral Implications of Darwinism*. Oxford: Oxford University Press, p. 173.

Rault, J. L., Hintze, S., Camerlink, I., and Yee, J. R. (2020). Positive welfare and the like: Distinct views and a proposed framework. *Frontiers Veterinary Science*, 2(7): 370. https://doi.org/10.3389/fvets.2020.00370.

Regan, T. (1983). *The Case for Animal Rights*. Berkeley, CA: University of California Press.

Reo, N. J., and Whyte, K. P. (2012). Hunting and morality as elements of traditional ecological knowledge. *Human Ecology*, 40: 15–27.

Rohwer, Y., and Marris, E. (2018). An analysis of potential ethical justifications for mammoth de-extinction and a call for empirical research. *Ethics, Policy & Environment*, 21: 1, 127–142.

Rollin, B. E. (1981). *Animal Rights and Human Morality*. Buffalo, NY: Prometheus Press.

——— (1995a). *Farm Animal Welfare: Social, Bioethical and Research Issues*. Ames: Iowa State University Press.

——— (1995b). *The Frankenstein Syndrome: Ethical and Social Issues in the Genetic Engineering of Animals*. New York: Cambridge University Press.

——— (1998). *The Unheeded Cry: Animal Consciousness, Animal Pain, and Science*. Ames: Iowa State University Press.

Rolston III, H. (1989). The value of species. In Tom Regan and Peter Singer (Eds.), *Animal Rights and Human Obligations* (pp. 252–255). Prentice Hall.

Rowlands, M. (1998). *Animal Rights. A Philosophical Defense*. London: Macmillan Press Ltd.

Russow, L.-M. (1988). Regan on inherent value. *Between the Species*, 4: 46–54.

——— (1999). Bioethics, animal research, and ethical theory. *ILAR Journal*, 40: 15–21.

Sagoff, M. (1984). Animal liberation and environmental ethics: Bad marriage, quick divorce. *Osgoode Hall Law Journal*, 22(2): 297–307.

Sandler, R. (2014). The ethics of reviving long extinct species. *Conservation Biology*, 28(2): 354–360.

Sandøe, P., Hocking, P. M., Förkman, B., Haldane, K., Kristensen, H. H., and Palmer, C. (2014). The blind hens' challenge: Does it undermine the view that only welfare matters in our dealings with animals? *Environmental Values*, 23(6): 727–742.

Sapontzis, S. F. (1987). *Morals, Reason and Animals*. Philadelphia, PA: Temple University Press.

Shapiro P. (2018). Clean meat: How growing meat without animals will revolutionize dinner and the world. *Science*, 359: 399. https://doi.org/10.1126/science.aas8716.

Sharma, K. (2016). *Interdependence: Biology and Beyond*. New York: Fordham University Press.

Singer, P. (1975). *Animal Liberation*. New York: Avon Books.

——— (1990). *Animal Liberation Second Edition*. New York: Random House.

Singer, T., and Klimecki, O. M. (2014). Empathy and compassion. *Current Biology*, 24(18): R875–R878.

Splitter, J. (2020). Farmers face their worst-case scenario: 'Depopulating' chickens, euthanizing pigs and dumping milk. *Forbes*. Available at: https://www.forbes.com/sites/jennysplitter/2020/04/28/

farmers-face-their-worst-case-scenarios-depopulating-chickens-euthanizing-pigs-and-dumping-milk/#571aee873003. Accessed 28 April 2020.

Sundman, A. S., Van Poucke, E., Svensson Holm, A. C. et al. (2019). Long-term stress levels are synchronized in dogs and their owners. *Scientific Reports*, 9: 7391. https://doi.org/10.1038/s41598-019-43851-x.

Takala, T., and Häyry, M. (2014). Neuroethics and animals: Methods and philosophy. *Cambridge Quarterly of Healthcare Ethics*, 23(2): 182–187.

The National Academies Press (NAP) (2011). *Guide for the Care and Use of Laboratory Animals* (8th ed.). Retrieved from https://grants.nih.gov/grants/olaw/guide-for-the-care-and-use-of-laboratory-animals.pdf.

Thompson, N. (2016). Metaphysical Interdependence. In Mark Jago (Ed.), *Reality Making*, Mind Association Occasional Series. Oxford: Oxford Academic. https://doi.org/10.1093/acprof:oso/9780198755722.003.0003, accessed 13 May 2023..

Thompson, P. B. (1993). Animals in the agrarian ideal. *Journal of Agricultural and Environmental Ethics*, 6(special supplement 1): 36–49.

——— (2014). Artificial meat. In: Sandler, R.L. (Eds.), *Ethics and Emerging Technologies* (pp. 516–530). London: Palgrave Macmillan. https://doi.org/10.1057/9781137349088_340.

——— (2010). *The Agrarian Vision: Sustainability and Environmental Ethics*. Lexington: University Press of Kentucky.

——— (2021). The vanishing ethics of husbandry. In Bernice Bovenkerk and Jozef Keulartz (Eds.), *Animals in Our Midst: The Challenges of Co-existing with Animals in the Anthropocene*, Cham: Springer Open Access, pp. 203–221. https://link.springer.com/chapter/10.1007/978-3-030-63523-7_13.

Vance, L. (1995). Beyond just-so stories: Narrative, animals, and ethics. In Carol Adams & Josephine Donovan (Eds.), *Animals and Women: Feminist Theoretical Explorations* (pp. 163–191). NC: Duke University Press.

Varner, G. (2012). *Personhood, Ethics, and Animal Cognition: Situating Animals in Hare's Two-Level Utilitarianism*. New York: Oxford University Press.

Waldau, P. (2010). *Animal Rights: What Everyone Needs to Know*. New York: Oxford University Press.

Warren, M. A. 1987. Difficulties with the strong animal rights position. *Between the Species*, 2(4): 433–441.

Wathes, C. M., Kristensen, H. H., Aerts, J. M., and Berckmans, D. (2008). Is precision livestock farming an engineer's daydream or nightmare, an animal's friend or foe, and a farmer's panacea or pitfall? *Computers and Electronics in Agriculture*, 64(1): 2–10.

Weary, D. M., Droege, P., and Braithwaite, V. A. (2017). Behavioural evidence of felt emotions: Approaches, inferences and refinements. *Advances in the Study If Behavior*, 49: 27–48.

Weary, D., and Robbins, J. A. (2019). Understanding the multiple conceptions of animal welfare. *Animal Welfare*, 28(1): 33–40.

Wise, S. (2001). *Rattling the Cage: Toward Legal Rights for Animals*. Cambridge, MA: Perseus Books.

# 3
# A LEGAL OPINION: PETS AND HOUSING IN THE UNITED STATES

*Dianne Prado*

## Introduction

According to the 2020–2021 National Pet Owners Survey, about 85 million people in the United States (70%) are pet owners.[1] Over 70% of a person's income goes toward paying either rent or their mortgage.[2] This doesn't include additional fees that are added on for having pet (i.e. pet rent, pet deposit). In LA County, 70% of renters with pets are behind in rent and will be facing eviction if the current emergency protections are eliminated.[3] Housing and pet ownership are interrelated as our pets are part of the family – to house ourselves is to house our pets. The definition of a pet requires a human owner and as such we can't fail to consider the barriers a pet owner faces when finding housing. This article will discuss the implications of housing and pets as a commodity, the barriers pet owners face, including the eviction process, defenses, and rights of pet owners, and suggest policy measures that can be taken to ensure that people and their pets remain together and housed.

## Commodity of Housing and Pets

In 1823, in Johnson v. McIntosh,[4] Supreme Court Chief Justice John Marshall upheld the McIntosh family's ownership of land purchased from the federal government which was taken from the Native American people already on the land. It was held Native Americans only had a right to occupancy and held no title to the land. This was based on the "Discovery Doctrine," which goes back five centuries and was created to grant Europeans authority to invade, capture, and take over the territories and people they found. While Justice Marshall stated Native Americans were the "rightful occupants of the soil," they also had no rights simply because it was their homeland. Justice Marshall stated that Native Americans "power to dispose of the soil at their own will to whomsoever they pleased was denied by the original fundamental principle that discovery gave exclusive title to those who made it." Ironically, in a nation that now wants to exclude immigrants, the entire notion of the discovery doctrine is based on those immigrating to the United States and claiming title by way of "discovery." This was further formalized in two more cases in 1831, Cherokee Nation v. Georgia,[5] and in 1832, Worcester v. Georgia.[6] The cases are known as the "Marshall Trilogy." In 2005, the Supreme Court cited Marshall's Discovery Doctrine in New York v. Oneida Indian Nation of New York, relying upon it as the basis of the federal government's dominion over land once controlled by Native Americans.[7] And with this we have private property rights inherently founded upon racist principles disregarding the existence of the people already living and tending to the land.

To date, we continue this model. Making a profit over people, forcing them to pay to have a roof over their heads. And to add insult to injury, if you dare own an animal, you pay extra to have that animal under the same roof. Creating the model of pets as a commodity. Just as we treat housing as a privilege, we also treat ownership of an animal as a privilege. Let's continue down the path as housing as a privilege before we dive deeper down into the world of pet privilege.

If you want to own, first you needed to be of white descent. The history of redlining – government created boundaries identifying certain land as valuable and safe, and other land, those where people of color lived, not valuable and unsafe – created racist bank lending practices which only provided loans for homes to white people. Further the concept of single-family homes further eradicated the communal living created by the Native Americans and instilled the mind-set of everyone for themselves – individual versus community. These racist practices have left a lasting impact with current demographics showing that 71.99% of white people are homeowners, in comparison to 46% Latinx, and 41.6% Black.[8] Additionally, 75% of redlined areas are still the most economically distressed today, 66% have mostly minority residents, 91% of the areas marked as 'best' are middle to upper class, and 85% of areas marked as 'best' are still predominantly White. Not only is homeownership impacted by the foundational racism of property rights, but this also has a disproportionate effect on renters of color, pushed out of homeownership, and forced to rent.

It is important to understand the foundation, the history, and basis of property rights to understand how housing became a privilege versus a right. Such as the ownership of an animal has been formed to be a privilege versus a right. If you want to own a pet and get it from an 'adoption agency' or 'animal rescue organization', then you are required to meet their qualifications on who is deserving of this pet. And the qualifications disproportionately disqualify low-income, people of color. Surprised? We have redlined the animal industry without any regard to the consequences – or more accurately with complete regard to the consequences because this industry, this country as a whole is functioning exactly how it was created to function – those in power with money, get more power and money, and keep everyone else as far away from the power and money by giving just enough power and money to feed the delusion the people actually have power and money. Capitalism in the name of democracy.

The commodification of housing and pets has allowed for the racist foundations of these industries to continue to permeate. We must recognize and acknowledge this first – we must come to the table with the clear understanding that housing is a human right and we have domesticated animals to the point that we have now created an industry in owning them, versus creating a value based in love and compassion. Instead of asking how much love this person can give this animal, we ask how much room do they have in their home, do they have an enclosed back yard, do they have a sufficient salary to love this animal, and we all carry bias that affects how we interact with one another and yet want to ignore this, want to shame ourselves for this, instead of asking our bias, our racism, to come sit at the table with us so we can uncover our biases and hold them, understand them, where they came from in order to let them go. We can't truly come to the table if we don't first recognize our implicit bias and admit that the industries we are working within, were not created to benefit low-income, people of color. That these industries are here to capitalize on profit and that does not benefit the community. It divides, it destroys, and as you will continue to read, it tears apart families and causes unnecessary evictions and unnecessary surrendering of pets to animal shelters.

## Renting With Pets

The average American renter pays about $1,326 a month.[9] The real median personal income for a worker in the United States in 2020 is $35,805.[10] Over 40% of salary goes toward keeping a roof over one's head. Another report shows that one in four renters spend over 70% of their income on housing costs, one in four renters eligible for federal assistance received funding, Black and Latinx renters constitute approximately 80% of the people facing eviction, and Black households are twice as likely of being evicted than white households.[11] These are the basic uphill battles faced disproportionately by people of color, not including the added factor of renting with a pet. My Pit Bull is Family has created a national database for pet-inclusive housing[12] – when I inserted the zip code of my office space in South LA county and searched for pet-inclusive listings in the 50-mile radius, the search came up with only 13 properties, only two of which had no pet fees.

The lack of housing, let alone, pet-inclusive housing, results in the unnecessary surrender of pets to shelters. A report conducted by Michelson Found Animals[13] states that 14% of pet owners have surrendered their pet to the shelter due to a housing issue, with 22% stating that reason was fear of eviction. The study further shows that over 50% of the population are pet owners and 72% of renters reported that 'pet-friendly' housing is hard to find. Pet-friendly and pet-inclusive housing is further defined with the main difference being pet-friendly housing merely being a housing provider that allows pets, yet has restrictions, while pet-inclusive housing is a housing provider who doesn't place restrictions on the breed, size or number of pets. Additionally, the report shows that renters with pets, on average, stay 21% longer than those in non-pet-friendly housing, landlord/housing operators also report that pet-friendly vacancies are 83% easier to fill than non-pet-friendly vacancies. Finally, pet damage is reported being less than $250.

Given these statistics, it would seem to be inevitable to ensure that all housing would allow for pets in order to keep families together and housed; however, given the commodification of housing and pets, it becomes instead a way to market to landlords to gain profit – that is to capitalize on the fact that over 50% of the population are pet owners and will need a home for both human and the pet. The same report touts that pet-friendly vacancies are good for the bottom line with an added revenue of up to $1.5 billion in pet fees deposits that owner operators could be collecting. Despite the fact the same report shows that only 9% of pets reported caused damage and 2% of pets that caused damage required a security deposit deduction, it still calls for owner operators to charge pet fees in the name of profit.

When it comes to pet fees, including additional pet security deposit and monthly pet rent, each state has their own regulations.[14] For example, in California, a security deposit is restricted to twice the amount of rent (for an unfurnished unit)[15] and an additional pet security deposit can't be required. Additionally in CA, there is state-wide rent control[16] for certain properties and each locality can have their own local rent control laws. On the other end of the spectrum, we have Texas, which has no law governing security deposits or additional pet fees.

While it is true, we continue to have private property which is driven by the market, the same truth holds that these same property rights are based on racist principles. Therefore, to continue the conversation about increasing profit for landlords and maximizing potential, is furthering the disproportionate effect it will have on people of color, and particularly low-income people of color, and is in fact perpetuating racism. We must stop the verbiage that to have pet inclusivity we need to play well with landlords and market to landlords. Instead, we need to all band together and demand the local city, county, and state governments to ban additional pet fees, including pet rent and deposits. If people can barely find jobs to that will pay the rent, how do we expect their pets to find jobs to pay pet rent?

## Evictions and Pets

What is an eviction? This may sound ridiculous but growing up I was privileged to never face housing insecurity. I only came to know about evictions, or what some courts call unlawful detainers, when I was an attorney in Los Angeles (LA County). I had no idea that owners could file paperwork in the courts to render a family homeless. Given the discussion of private property rights earlier, owners have a right to reclaim their property in an expedited fashion. Each state, city, and/or county has created their own laws on how the eviction process works. Some places have strong eviction protections for tenants, while others do not.[17] Much like the pet fees.

Due to the rising costs of rent and stagnancy of wages, this leaves renters, and specifically renters with pets vulnerable to eviction. Out of 3,295,198 households in LA County, about 1,723,389 have pets.[18] The City of Los Angeles also has the highest rates of renters in the nation, with 64% of renters as opposed to homeowners.[19] An estimated amount of 50% of renters own pets.[20] A recent report showed that 49,000 families with pets in Los Angeles County will be facing eviction for failure to pay rent in 2021.[21]

A typical tenant story being threatened with eviction looks like the following: Tenant has lived in their apartment for years; Tenant has paid rent and overall "kept their head down" and "minded their own business." Tenant gets a dog, owner is aware, and even likes the dog. Owner decides to sell the building. New owner comes in. Next month new owner sends tenant a three-day notice to cure or quit, requiring the dog be removed or the tenant move out. The new owner attaches to the three-day notice the original lease the tenant signed with the old owner.

So what now? Will the tenant be evicted in three days if they don't comply? Is this landlord able to do this? It is important to understand that regardless of how quick and unjust the legal process is to evict someone, there is in a fact a **legal process** that must be followed. People can't just be thrown out of their home. Do illegal evictions exist? Meaning, do owners take matters into their own hands and change locks, yes. We call this illegal self-help eviction. The process to get the tenant back into their home is lengthy and ironically requires a legal process. Sadly, since these evictions are done illegally, there isn't data to capture how often this occurs.

Illegal evictions aside, the legal process typically requires first a written notice to be sent to the tenant. This notice initiates the eviction process and gives a certain number of days a tenant will need to either do something and/or move/vacate their home. In the case of tenants with pets, this written notice is a X-day notice to remove the pet or quit, or a three-day notice to quit alleging the dog is a nuisance. If the tenant does not remove the pet or leave their home, then the owner can begin the court process by filing what is called the unlawful detainer/eviction action in their local court. Normally owners need to pay for this filling unless they qualify for a fee waiver. After an eviction is filed, the tenant must be provided a copy of the documents filed in the courthouse – these documents are called a Summons and Complaint. The tenant then has a certain number of days to respond to the eviction, normally called filing an Answer. If the tenant does not file an Answer in the time frame allotted, the tenant automatically loses, and a default judgment is entered against the tenant. This judgment has a lasting detrimental effect on the tenant as it is not only reported to credit agencies, but even worse, the tenant is added to a "blacklist" of tenants who have been evicted. This list is then used to screen out tenants when they are looking for new housing. On the other hand, if the tenant does timely answer, then they have an opportunity to defend themselves in the eviction action. Typically, owners are represented by attorneys 90% of the time, while tenants go unrepresented 90% of the time.[22] Given the lack of attorney representation by tenants, most tenants lose their cases or enter into one-sided agreements mostly benefitting the landlord. In recent years, a movement advocating for a guaranteed right to counsel in eviction actions has created laws in numerous states and cities, including, New York, Washington, Philadelphia, San Francisco, ensuring low-income tenants facing eviction are represented by an attorney.[23]

Ultimately, if a tenant loses his/her case, either by failing to answer or taking the case to trial and losing, then a writ is issued by the court and soon after a sheriff shows up at the tenant's home with the court order and forces the tenant to leave their home with whatever they can carry in their arms, if the tenant hasn't left already. In most states, a lock out notice is placed on the door about 5 days prior to the lockout date.

As animal advocates, it is crucial to determine where the tenant is in the eviction process. The graph on the following page shows the process of the eviction process in California. When a person comes distraught saying they are being kicked out, the question is, do you have a sheriff's notice posted on your door or is there a sheriff that came to lock you out? If the answer to that is no, then the next question is: have you received any written notice from your landlord or received any documents from court entitled Summons & Complaint? If the tenant hasn't received any documents from court, we are still in what housing advocates call the prevention stage. This means that an eviction has yet to be filed and can hopefully be prevented from being filed. If court documents have been received, we now know the tenant has a certain number of days to answer the lawsuit and should seek representation as soon as possible (Figure 3.1).

## Eviction Defenses for Renters with Pets

When a person is at a shelter with their pet and saying the owner is forcing them to get rid of their pet, it begs to ask a couple more questions. None of them starting with why would you be doing this? As discussed earlier, did they receive anything in writing? If they did not and the owner is verbally harassing them, the first thing you can do is provide assurance and discuss that they have rights and the owner must go through the legal eviction process with the courts. Instinctually, the tenant will likely shudder at the idea of any legal process, as no one typically wants to be involved in legal issues. That is why how we talk to the tenant surrendering the pet to the shelter is important to determine how to best help them. The following are some suggestions to navigate this conversation:

Ask them the pet's name, how long has the pet been with them, who in the family loves the pet the most? You might get the response that the pet actually belongs to the minor child. Does the child perform better in school, sleep better, eat their vegetables easier because the pet is around? They child has ADHD they say, so yes, Billy, their four-year-old, 45 lb mix, does help their child, and could be classified by a mental health provider as an emotional support animal. Just like that, the person has helped us uncover one of the main defenses available – a fair housing right. The person has a child with a disability whose animal alleviates symptoms of their disability and thereby they can ask for their dog to remain in

*Dianne Prado*

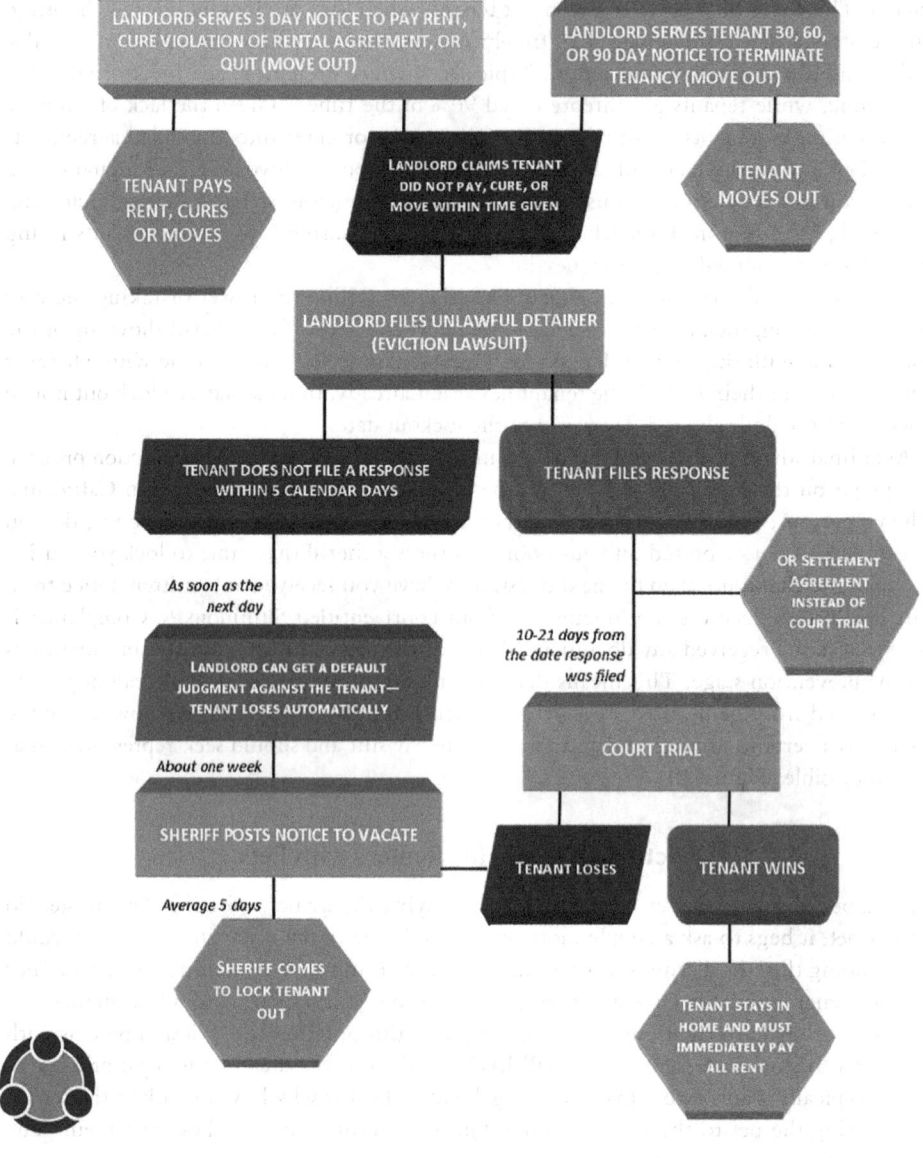

*Figure 3.1* Tenants together. https://www.tenantstogether.org/resources

the home as an assistance animal. We will get into further discussion regarding fair housing rights further below.

Let's say the questioning doesn't lead down the path of discovering either the person themselves or anyone in their family is a person with a disability. Then with a couple more questions, we can uncover various other common defenses: How long they have had the dog? How long has the owner or management been aware you've had the dog? Have you been paying rent after they knew you had the dog? How did they know you had the dog? Because they came into the apartment when the bathroom was leaking and the dog greeted them at the door, they looked at the dog as you moved your creature out of the way and it followed you both, wagging their tail to the bathroom.

Acceptance of rent with knowledge of the "breach" (here having the dog) is a defense called waiver.[24] The other form of this defense is called estoppel,[25] which is a defense in equity that is based on the actions of the owner to which the tenant relies upon.

Or you could ask what happened that led to the owner saying the dog had to go after all these years? You could possibly get an answer that it all started with the bathroom leak, that went unaddressed for a month, and then became a bigger leak that caused the owner to have to change the entire piping and redo the bathroom floors. The owner was very upset and told them they either had to get rid of the dog or pay an additional deposit and pet rent for the dog. And here we have a retaliation[26] defense. The tenant was asserting their rights to have safe and secure housing and the owner got upset and retaliated against the tenant and threatened them with having to remove their dog.

One final line of questioning can lead to the very common scenario of a new owner or a new management company who wants to come in and change all the rules. The new owner steps into the same shoes of the prior owner. If the prior owner allowed pets, the new owner must also allow pets.[27]

It's not about asking the person the right legal question; it's about asking the questions that naturally you would want to ask if you weren't busy judging or making your own assumptions or creating your own solution of what the person should do. The renter with the pet is the person that needs to come up with the solution. We are here to offer a different perspective, to offer resources and information for the person to come to the best solution for them, their family, which includes their companion animal.

## Fair Housing Rights

While the Civil Rights Act in the United States (also known as the Fair Housing Act) was passed in 1968, which prohibited discrimination in housing, it has not worked as racial discrimination in housing still exists and class segregation has worsened.[28] Our focus with fair housing rights is not to discuss the racial disparities that still exist despite its enactment,[29] but rather discuss fair housing laws and how it specifically applies to humans, their pets, and housing.

Animals are defined assistance animals in the fair housing context.[30] They are separated into two different categories: service animals and support animals. Service animals are trained to help the person perform an activity. Support animals do not have a special skill, their mere existence as an animal provides comfort and support to the human. Service animals have additional protections to be in public spaces, while support animals do not. Neither service nor support animal has a certificate. A service animal normally has proof of training and can also show the task they are performing for their human. A support animal on the other hand does not require any specialized training. A support animal must be a domesticated animal, and in the fair housing context, their presence alleviates symptoms of the person's disability. Breed and size restrictions are not applicable for support animals.

Federal fair housing laws provide the floor of protections,[31] and each state has their own fair housing laws which provide the same or stronger protections. A rental owner has an obligation to always abide by fair housing laws and engage in the interactive process when the tenant with a disability requests a change or modification to a rule or policy. This is called a reasonable accommodation. No special form, letter, or language is necessary to request a reasonable accommodation. Nor does it need to be in writing. It is recommended to have it in writing to prove a request was made; fair housing laws do not require a request in writing. To request an accommodation, a tenant must be a person with a disability and the request being made alleviates symptoms of the person's disability which allows the person to have the equal use and enjoyment of their housing. To provide proof of disability, a tenant can self-verify or also provide a letter from a reliable third party that can verify the letter. To self-verify a tenant could provide proof of disability benefits. A reliable third party can be any person that can verify by personal knowledge that the tenant is a person with a disability, as defined by the relevant federal or state law. Examples of reliable third parties are: health service providers, therapists, social worker, group and guidance counselors.[32] After a tenant provides a letter from a reliable third party that can verify the letter, the owner must engage in the interactive process. Engaging in the interactive process does not mean rejecting the request. Engaging in the interactive process doesn't mean harassing the tenant for private medical information. It can mean calling the person who wrote the letter to verify what is stated in the letter. If the owner has concerns regarding the animal posing a threat, fair housing laws require the owner prove that it is a direct threat and that the threat cannot be mitigated. A determination that an individual poses a direct threat must rely on an individualized assessment that is based on reliable objective evidence. The assessment must consider: (1) the nature, duration, and severity of the risk of injury; (2) the probability that injury will actually occur; and (3) whether there are any reasonable accommodations that will eliminate the direct threat.[33]

In practice fair housing law plays out in the following hypothetical scenario. A family has lived in their two-bedroom rent-controlled apartment for 22 years. Throughout the years, their requests for repairs have gone unanswered. They were forced to make repairs themselves. During the pandemic, the family faced eviction because the landlord alleged the apartment was cluttered. The eviction action was eventually dismissed. Around midnight, a couple of days later, the son takes out the family's two dogs outside for a quick walk. As he is coming back to the apartment, before reaching the corner to his door, he unleashes the dogs. At that same time, the manager and her daughter were turning the corner, and everyone is startled. The manager and daughter scream and the younger dog bites the manager's leg. The ambulance and police officers are called. The manager is taken to the hospital and the bite cleaned, no stitches needed, bruising occurs thereafter. Soon thereafter, the family receives a three-day notice to quit claiming the dog is a nuisance.

What do we do here? A landlord would claim, this is an insurance issue, once a dog bites it needs to be removed or confined. Ok, so how do we determine if the confining is an option. We turn to the assessment of whether the animal poses a direct threat and the varied factors: severity, duration, and whether another or additional reasonable accommodation mitigates or eliminates the direct threat. In this case, the family has owned the dogs for over two years without any incident. This event happened at midnight, and because of the failure to keep the dogs leashed, the dogs were placed in an unfair situation of being screamed at in front of the home. To date, the dogs are still in the unit and there has been no further incident. An assessment from a dog behaviorist further shows that the dogs don't exhibit any signs of aggression and rather would benefit from a training course. In this case, it appears the threat has and can be mitigated and the family and their companion animals can remain together and housed.

Before ending our discussion on assistance animals, let's touch quickly on the constant push and pull regarding the "abuse" of assistance animals and that anyone can make their animal an assistance animal to get into housing. The bigger picture question here is not the abuse of assistance animals, rather the lack of pet-inclusive housing that causes desperation and people that perhaps do not have a disability to state they do have one to remain housed with their pet. Our focus should not be to create more stringent laws cracking down on assistance animal fraud, rather creating policies that call for housing to include pets, without additional fees. The war on support animals needs to end here and the fight to ensure people and their pets remain together and housed should be shouted from the rooftops.

## Homeownership and Pets

It is the author's opinion that some insurance companies are one of the drug cartels of the homeownership world. The insurance industry is a racket – truly a racketeering industry. To start, insurance companies were part of redlining and labeling neighborhoods of color as "high risk" – therefore part of racist foundations of private property. Once you finally cross the red tape of making enough money and having worthy 'credit', stable salary, likely dual income, pass the racist test, then you need to find an insurance company that you pay money into each month to be able to get that money back should something occur to your house or apartment. These companies then compete with one another and create rules and regulations of what events can and can't happen that they will cover. The more events covered, the more expensive the insurance. One of the more expensive events is that of an incident with an animal, and particularly with a certain breed of dogs.[34] And with that disqualifications begin for who can obtain homeowner's and/or renter's insurance.

## Fair Housing Rights and Insurance Companies

Because insurance companies are private, just as property is a private right, it can have the right to deny or reject a person, under certain restrictions. Fair housing rights come into play in order to put a bandaid on the racist foundation of property rights; otherwise, it was too overtly racist and we can't actually come to grips with the truth our entire principles need to be thrown out the window. Federal fair housing regulations prevent discrimination against people in their homes. People can't be discriminated against based on natural original, race, color, and gender, whether they are a person with a disability. Insurance companies also need to abide by fair housing laws as they are providing insurance on a home/rental unit.

Now this comes hand in hand with insurance companies charging more for certain incidents or disqualifying a person from insurance entirely because they own a dog of a certain breed. If the dog is the person's support animal, the insurance company must evaluate that reasonable accommodation request. If the request does not impose an undue financial and administrative burden, then it must granted.[35] Meaning an insurance company must change their breed restrictive policies and insure the homeowner, including their dog of any breed. Extensive work has been done to combat insurance company's discriminatory practices by organizations like the Animal Farm Foundation[36] and My Pit Bull Is Family,[37] and both organizations list insurance companies who do not discriminate based on breed of dog. The list is small, but hopefully with pressure from the people, these racist practices are banned outright.

## Pets as Property and Unhoused People with Pets: A Personal Case Study

According to most federal and state laws,[38] pets are considered property, with little to no legal rights. While there are laws that protect against animal abuse and cruelty, like the 1966 Animal Welfare Act, the biggest hurdle we have yet to overcome is that of pets being treated like a piece of furniture.

*Dianne Prado*

I became a pet owner the spring of 2009, when I was driving to work one morning on the 110-north freeway in Los Angeles. I was in the slow lane, that is when there is not traffic which in LA is extremely rare and in the morning in LA, unheard of. As I am moving inches at the fast pace of 5 mph, to my amazement I see a white and gray spotted dog, with a gray patch on the eye, ears cropped, stocky, looking around 60 lbs, just walking down the edge of the freeway lane. It appeared he had come down the off ramp. I pull over to the side and get out of the car and grab an extra leash I had in the back of my car. I walk toward this creature, who has a huge smile walking toward me and as I put out the back of my hand, he lays down and rolls on his back to give me his belly. I slowly approach him and rub his belly, put the leash over his head, and slowly turn him back so he can get up and walk to the car with me. He turns back and as I stand up and pull on the leash, he lays there, with his entire body pressed on the ground. He wasn't going anywhere. So I did what I had to do. I picked him up from behind and he happily let me carry him to my car. He then sat in the front seat, with a grin from cropped ear to cropped ear, tongue hanging out, drool everywhere. And it was at that moment that I acquired property which I would then need to figure out what I was going to do with "it" and the "it" not being an "it" at all, already had his name, Falcor. Yes, Falcor, my flying dream dragon.

The first few days and weeks we searched to see if Falcor had a family. I put out flyers at the local animal shelters, used social media, and even asked to see if anyone wanted to foster Falcor since we already had two dogs who could be quite a handful. Falcor being Falcor calmed the other two creatures, his love was big enough for all of us plus some, so it always helped the energy in the room. And so it was, Falcor after about two months officially became my first dog I owned. I grew up with golden retrievers, but having a dog as a child is very different from having a pet as an adult and more importantly making the choice to own this pet.

Falcor became a part of who I was. A part of my being. He lived my joy, my pain, my everything. I was fortunate to take Falcor on an impromptu road trip across the United States with me in my Toyota Prius. It started as a five-day journey to Zion National Park and trip miraculously became find the best and most glorious swimming holes, lakes, rivers, oceans for Falcor to swim in. He was a swimmer you see, very much a champion in his past life. I learned during this road trip that very little is needed to take care of my dog and myself in terms of what society has told us is acceptable or the being a good owner. That my love for Falcor and his love for me, and our responsibility to keep each other safe, healthy, and happy, mattered very little to the material things we had; our love was founded upon the space and time we provided one another. We both needed a secure space to sleep, and we found shelter in my car, a tent, a hotel, a room in a home. We both needed food, so Falcor ate eggs in the morning with his kibble. I ate eggs sans kibble. If I munched on an apple or any other fruit, Falcor would partake. And our journey for the best swimming holes led to the most epic walks through forests, mountains, and flower trails. All of this journeying had nothing to do with whether my dog was allowed inside or not, if he ate vegan, or where he would sleep. It had everything to do with the time and love we shared. Falcor was literally my road dog.

The stigma that unhoused people with pets are underserving and unworthy of owning a pet needs to end and I believe animal welfare advocates are slowly but surely coming around to understanding this. Unhoused people will feed their animals before themselves, will remain unhoused and in their tent because housing that can be provided won't allow their pet. Our focus should not be on judging the unhoused person, rather discussing how we can help both that person and their pet. The readers are encouraged to read Leslie Irving's chapter on Homelessness and Pet Ownership within this text.

## Where Do We Go From Here?

As animal advocates, we need to engage with housing advocates and truly understand that fighting for animals includes fighting for the human and the home they are both living in. We need to take active steps to engage in not only animal, but also housing legislation. Get involved and informed with how evictions are happening in your city or state,[39] advocate for a right to counsel, advocate for pet-inclusive housing laws which bans pet fees. The fight for housing as a human right is a fight for keeping people and their companion animals together. These are not separate issues. As we look ahead, our causes are intertwined, and we need to ensure they are aligned.

## Discussion Questions

1. How does commodification of housing and pets impact people with pets?
2. Do renters with pets have rights?
3. What can we do to ensure people and their pets remain together and housed?

## Notes

1. https://www.iii.org/fact-statistic/facts-statistics-pet-ownership-and-insurance.
2. https://www.aspeninstitute.org/blog-posts/the-covid-19-eviction-crisis-an-estimated-30-40-million-people-in-america-are-at-risk/.
3. https://www.census.gov/quickfacts/fact/table/losangelescountycalifornia#.
4. https://tile.loc.gov/storage-services/service/ll/usrep/usrep021/usrep021543/usrep021543.pdf.
5. https://supreme.justia.com/cases/federal/us/30/1/.
6. https://supreme.justia.com/cases/federal/us/31/515/.
7. Indigenous Values Initiative, "Sherrill v. Oneida Opinion of the Court," *Doctrine of Discovery Project* (1 August 2018), https://doctrineofdiscovery.org/sherrill-v-oneida-opinion-of-the-court/.
8. US Census Bureau.
9. https://worldpopulationreview.com/state-rankings/average-rent-by-state.
10. https://fred.stlouisfed.org/series/MEPAINUSA672N.
11. https://www.aspeninstitute.org/blog-posts/the-covid-19-eviction-crisis-an-estimated-30-40-million-people-in-america-are-at-risk/.
12. https://www.mypitbullisfamily.org/housing.
13. https://www.foundanimals.org/pets-and-housing/2021-pet-inclusive-housing-report/.
14. http://www.landlord.com/security-deposit-law-guide.htm.
15. https://codes.findlaw.com/ca/civil-code/civ-sect-1950-5.html.
16. https://tenantprotections.org.
17. https://nationalevictions.com/home/welcome/states-eviction-process/.
18. https://www.laalmanac.com/environment/ev21d.php.
19. https://www.zillow.com/research/cities-gaining-renters-20915/.
20. https://ebusiness.avma.org/Files/ProductDownloads/2019%20ECO-PetDemoUpdateErrata-FINAL-20190501.pdf.
21. https://evictionlab.org/eviction-tracking/.
22. https://static1.squarespace.com/static/52b7d7a6e4b0b3e376ac8ea2/t/5b1273ca0e2e72e-c53ab0655/1527935949227/CA_Evictions_are_Fast_and_Frequent.pdf.
23. http://www.civilrighttocounsel.org/map.
24. https://www.justia.com/trials-litigation/docs/caci/300/336/.
25. Cornell Law School. https://www.law.cornell.edu/wex/estoppelRetrieved July 3, 2023.
26. CA civil Code 1940, 1941.1.
27. New owners steps into old owner shoes: Burris, S. Syringe Distribution Laws Map. Temple University Center for Public Health Law Research, Policy Surveillance Program. Accessed July 1, 2015.

28 https://shelterforce.org/2018/09/05/the-most-important-housing-law-passed-in-1968-wasnt-the-fair-housing-act/.
29 https://shelterforce.org/2019/02/21/long-before-redlining-part-2/.
30 https://www.hud.gov/program_offices/fair_housing_equal_opp/assistance_animals.
31 https://www.hud.gov/sites/dfiles/PA/documents/HUDAsstAnimalNC1-28-2020.pdf.
32 Id.
33 https://www.hud.gov/sites/documents/huddojstatement.pdf.
34 https://www.iii.org/article/spotlight-on-dog-bite-liability.
35 http://www.fhcwm.org/uploads/files/reports/HUD_Memo_-_RA_Breed_Restrictions_Insurance.pdf.
36 https://animalfarmfoundation.org/endhousinginsurancebreedrestrictions/.
37 https://www.mypitbullisfamily.org/insurance.
38 https://betterpet.com/pet-ownership-laws/.
39 https://evictionlab.org.

# 4
# ALL FOR ONE HEALTH AND ONE HEALTH FOR ALL

*Zenithson Y. Ng and Isabella Pfeiffer*

## Introduction

Altruistic people often identify their life's purpose as improving the health and well-being of living things. Some focus on humans, some focus on animals, and others focus on the living earth. When we strive to achieve improved wellness for all, we must consider a few important constantly changing and challenging issues. The population reached 7.7 billion in 2019 and is estimated to reach 10.9 billion by the year 2100 (Desa, 2019). The rising human population forces intrusion into previously undisturbed ecosystems and animals living within them. More people live now in close contact with wild and domestic animals, both livestock and pets. International travel has made it easier for possible zoonoses to be spread across borders and around the globe through movement of people, animals, and animal products. The unnatural overcrowding of humans and animals in these ever-changing environments increases the risk of infectious disease transmission and health concerns for all living things. In fact, three out of every four emerging infectious diseases in humans originate from animal species (Jones et al., 2008). The ongoing COVID-19 pandemic has clearly emphasized the connection between humans and animals and the need to identify methods of mitigating death and disease (King, 2021).

Therein exists a need to maintain the health and welfare of life at the intersection of humans, animals, and the environment, all of which are inextricably linked. We can fulfill this need within the context of One Health, which is defined as, "an integrated, unifying approach that aims to sustainably balance and optimize the health of people, animals, and the environment" (United Nations Environment Programme, 2021). It describes how different experts, whether they advocate for the health and welfare of people, animals, or the environment, view problems and can collaborate to come up with solutions in an interdisciplinary manner. From a myopic standpoint, medical doctors only treat humans; veterinarians only treat animals; and environmentalists only treat the environment. The One Health approach challenges all experts to strategically work together and share unique perspectives that would not be considered independently. Collaboration and communication between all parties ensures that the solution that benefits one arm ideally also benefits all others, and at minimum, does not negatively impact another. Doing so creates innovative solutions and greater impact with the potential to help protect and save untold millions of lives in our present and future generations (onehealthinitiative.com).

Traditionally, One Health has focused on areas such as food safety, zoonotic disease, environmental health, antimicrobial resistance, and neglected tropical disease (Mackenzie & Jeggo, 2019). However, it can be used as a framework to shape the approach to virtually any problem affecting humans, animals, or the environment. This chapter will provide an overview of the history, impact, and current and future trends in One Health.

## History of One Health

Historically, priests and others applied their knowledge gained through taking care of animals or even animal sacrifices and then used that knowledge for the benefit of human individuals. Priests in ancient Egypt would study cattle anatomy and physiology through domestication and ritual bull sacrifices and then translate these insights to help their human flock (Driesch, 2003; Gordon & Schwabe, 2004; Schwabe, 1984). This is the most rudimentary concept of One Health, and these interactions led to the beginning of human medicine as a discipline. Unfortunately, despite being so closely intertwined in the beginning, human and veterinary medicine have been developed as distinct disciplines after that and are still struggling to find their way back to the One Medicine paradigm. Veterinary medicine has been described as a distinct discipline in the Zhou dynasty in China (11th through 13th century) (Driesch, 2003) The first veterinary school in Western society was founded in Lyon, France in 1762. Founder Claude Bourgelat was heavily criticized when he recommended clinical human medicine training for veterinarians (Driesch, 2003). It was not until advances in cell pathology led the German physician Rudolf Virchow, also known as "the father of modern pathology" and the "Pope of medicine" to show interest in comparative medicine spiked by the observation of similar disease processes in humans and animals in the 19th century (Saunders, 2000; Silver, 1987; "Rudolph Virchow," 2022). One of his students, Sir William Osler, is credited with having coined the term One Medicine (Dukes, 2000). Osler was also considered the first physician to, not only study, but also teach veterinary pathology to students ("William Osler," 2022). Almost another decade later in 1976, Calvin Schwabe re-thought the concept of One Medicine (Schwabe, 1984). His work in African communities and the close relationship between priests in charge of animal and human well-being has led to a new understanding of the term One Medicine. These traditional African pastoral societies can still be found today (Majok & Schwabe, 1996). The One Medicine concept now fully recognizes the systemic interactions between animals and humans in regards to nutrition, medicine/health and livelihood. Unfortunately, despite all these efforts, human and veterinary medicine are slow to fully integrate the concept of One Medicine, even though both sciences share a common ground of knowledge in anatomy, physiology, pathology, and the origins of diseases in all species (Schwabe, 1984). Traditionally termed comparative or translational medicine topics, such as cancer and zoonoses research, have been explored more readily to the mutual benefit for both humans and animals. However, true collaboration under the holistic assumption that human, animal (domesticated and wildlife), and environment, including climate and well-being, are all dependent on each other is harder to achieve on a wide spread level (Forget & Lebel, 2001; Lebel, 2003; Rapport et al., 1998, 1999).

Out of the One Medicine concept, focusing initially on zoonotic diseases, One Health has emerged, now also incorporating ecosystem health (Rapport et al., 1998, 1999). Zoonoses are infectious diseases that can be transferred from animals to humans. Current research estimates that about 60% of human infectious diseases have an animal origin, but even though likely, we do not know the current number of the opposite scenario (WHO,

2022, "Neglected Zoonotic diseases"). For example, TseTse fly control in Africa did not originate as an effort to minimize human sleeping disease, but instead, control of this species originated to treat and control Trypanosomiasis in cattle (Dukes, 2000).

In October 2008, a strategic plan to reduce the risks of infectious diseases within the human-animal-ecosystem was first made public during the 6th International Ministerial Conference on Avian and Pandemic Influenza (FAO, OIE, WHO, UNSIC, UNICEF, WB, 2008). This plan has been further developed under the trademark-protected term "One World One Health™" in 2009 (Public Health Agency of Canada, 2009). In addition, functional genomics of human and animal genes are pulled together by a large International Knockout Mouse Consortium (**IKMC) (www.knockoutmouse.org**).

The overarching concept of One Health not only focuses on human and animal health, but it also includes the whole ecosystem and states that all parts are inextricably linked (Zinsstag et al., 2011). Researching the connections and links between those systems can and should benefit not just one part of the system but all partners. Trying to elevate the well-being and health of any one of the components can also benefit others when keeping the One Health approach in mind. Any research or solution that harms one of those components does not align with the One Health concept.

## Organizational One Health

The One Health approach can be applied at all levels, whether it be at the community, subnational, national, regional, or global level. One Health requires effective governance to direct communication, collaboration, and coordination amongst all stakeholders (Figure 4.1) (UNEP FAO, WHO, OIE, 2021). This subsequently leads to capacity building, defined as the process of developing and strengthening the skills, instincts, abilities, processes, and resources that organizations and communities need to survive, adapt, and thrive in a fast-changing world (Eade, 2007). Various groups and organizations work on a national and global level to advocate issues that fall under the One Health umbrella.

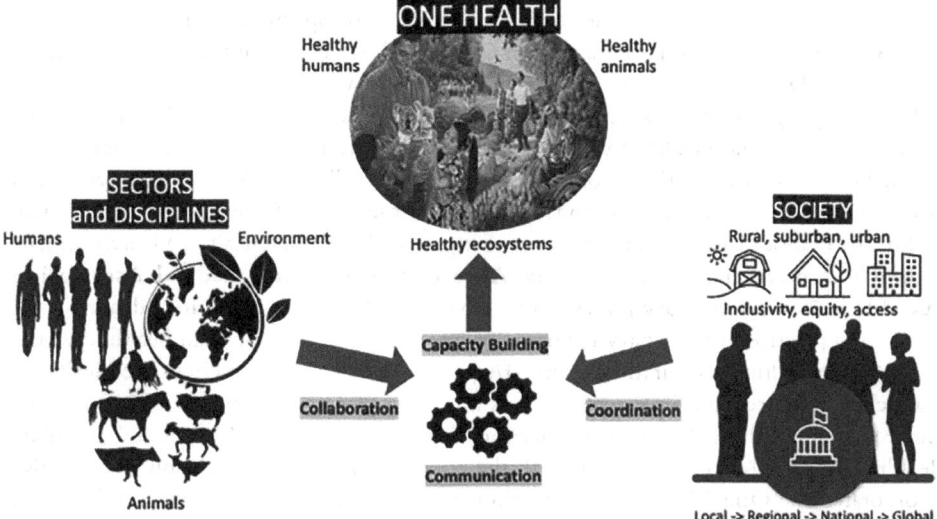

*Figure 4.1* Definition of one health (UNEP FAO, WHO, OIE, 2021)

On a global level, the **World Health Organization (WHO)** works closely with the **Food and Agriculture Organization (FAO)** of the United Nations and the **World Organization for Animal Health (OIE),** to promote multisectoral responses to public health risks (WHO, 2022, "One Health"). Each of these organizations serves a specific mission. The WHO is a specialized agency of the United Nations responsible for the international health of all people and leading global health responses (https://www.who.int/). The FAO is a specialized agency of the United Nations that leads international efforts to defeat hunger and improve nutrition and food security (https://www.fao.org/home/en).

The World Organization for Animal Health was formerly called the Office International des Epizooties (OIE), and the organization still maintains the historical acronym today. The OIE is an intergovernmental organization that coordinates, supports, and promotes animal disease control at a global level. In addition, the OIE collects and disseminates information on the distribution and occurrence of animal diseases to govern trade in animals and animal products (https://www.oie.int/en/home/). While the WHO, FAO, and OIE cover areas of human and animal health, the **United Nations Environment Programme (UNEP)** is a global authority for the environment with programs focusing on climate, nature, pollution, and sustainable development (UNEP, 2022, "About UN Environment Programme"). Together, the WHO, FAO, OIE, and UNEP established the definition and mission of One Health (UNEP FAO, WHO, OIE, 2021), which is stated at the beginning of this chapter.

On a national level, the **Centers for Disease Control and Prevention** (CDC) in the United States is a federal agency dedicated to "saving lives/protecting people" (www.cdc.gov). They demonstrate their efforts and priorities in One Health through education and action (CDC, 2022, "One Health in Action"). One Health in action is an important step toward educating the public but also furthering projects to achieve the best health for everyone. Specifically, the CDED provides continuing education in terms of Zoonoses & One Health Updates (ZOHU) that provides timely education on zoonoses and infectious disease (CDC, 2022, "ZOHU Calls/Webinars").

Another example of an organization operating on a national level is the **Animal and Plant Health Inspection Service** (APHIS) of the US Department of Agriculture (USDA). APHIS maintains a mission to protect the health and value of America's agricultural and natural resources. They exercise the concept of One Health, specifically through the Veterinary Services (VS) program unit, although One Health is a common theme throughout all programs (USDA, 2022, "Animal Health Program Review").

Some may argue that veterinary medicine, out of all the professions in this arena, is the driving force for One Health. Every graduating veterinarian takes the veterinarian's oath to benefit society through the "promotion of public health" (AVMA, 2022 "Veterinarian's Oath). Veterinarians are well positioned to connect the dots between animal and human health, as zoonotic disease is a primary area of expertise that human health and environmental health advocates may be lacking. The **American Veterinary Medical Association (AVMA)** leads the veterinary profession by advocating for its members and advancing the science and practice of veterinary medicine to improve animal and human health (AVMA, 2022, "One Health"). Naturally, the AVMA plays a key role in advancing One Health by channeling resources into initiatives, such as the AVMA's Antimicrobial Use Data for Antimicrobial Stewardship, Food Safety Advisory Committee, and AVMA Council on Public Health. The AVMA, in conjunction with the American Medical Association (AMA), also incorporated the **One Health Commission,** which is a 501(c)(3) organization dedicated to driving teamwork and raising awareness and education about One Health (One Health Commission, 2022, "What is One Health"). Their slogan is to "Connect" One Health

Advocates and Stakeholders to "Create" networks and teams that work together across disciplines to "Educate" about One Health.

These are just a few examples of major organizations that exist specifically to communicate, collaborate, and coordinate all issues regarding One Health. Organizations are gaining traction everyday as we identify more and more issues that can be managed through a One Health perspective. The next section will describe some of the issues that are prioritized in One Health.

## One Health Issues

### *Issues Prioritizing Human Health and Welfare*

#### *Zoonoses*

Studying every aspect of zoonotic diseases is a priority in One Health to improve human and animal health. Zoonoses are infectious diseases that can be transmitted from animals to humans and it is currently estimated that about 60% of infectious diseases are zoonotic (WHO, 2022, "Neglected Zoonotic diseases"). Effective management and prevention of zoonotic diseases includes studying modes of transmission, life cycles of parasites, potential reservoirs in the wild, and possible modes of interactions between humans and animal hosts, and developing and deploying effective treatments or preventative vaccines for both humans and animals. Environmental changes, changes in living conditions, and increase in populations are all factors that need to be considered when trying to successfully manage and control zoonotic diseases, especially those with wildlife as hosts or reservoirs. Along with brucellosis and MRSA (multiresistant *staphylococcus aureus*), which are both caused by bacteria, rabies, and avian influenza are important examples of zoonotic diseases. Trypanosomiasis (human sleeping disease) and hookworm infections, which are both caused by parasites, are also important examples of zoonotic diseases. The canary in the coal mine, i.e. monitoring of animal sentinel species for zoonotic diseases, may become an integral part of bio-surveillance and may potentially be bidirectional, with humans acting as sentinels for animals. For example, clusters of human Brucellosis cases may be able to indicate outbreaks in cattle. Brucellosis in cattle can lead to large economic losses and animal suffering if undetected (Rabinowitz et al., 2009).

#### *Safeguarding Food Supply*

Veterinarians are not only responsible for the livestock's health before and right up to slaughter, but they are also responsible for human and animal consumption food safety. According to the American Veterinary Medicine Association (AVMA), veterinarians are "central to ensuring safe, wholesome, and correctly labeled and packaged meat and poultry products. Federal inspection is provided by the U.S. Department of Agriculture (USDA) Food Safety and Inspection Service (FSIS)" (AVMA, 2022, "Food safety"). Veterinarians involved in Food Supply Veterinary Medicine (FSVM) help to protect the world's food supply from farm to table (AVMA, 2022, "What is FSVM?"). They are ensuring the animals' well-being and welfare as well as safety of their meat, egg, and dairy products. Food-borne illnesses such as salmonella can be detrimental and life-threatening for humans and animals alike. Raw pet food diets have been implicated in containing potentially harmful and zoonotic organisms, such as *E. coli*, *Salmonella*, and *Listeria* species, and parasites such as *Toxoplasma gondii* (van Bree et al., 2018). Those pathogens can potentially be transmitted to susceptible human owners from infected pets or by handling foods without proper precautions and hygiene (Cammack et al., 2021).

## *Antibiotic Resistance and Hormones*

Antimicrobial resistance due to antibiotic overuse in human and veterinary medicine is a historical, but still pressing, problem. The use of continuous low dose administration of antibiotics especially in crowded livestock conditions to improve production gain and to minimize possible infections is a well-known issue and has been blamed, in part, for the increase in antibiotic resistance. Another issue of antimicrobial resistance is the use of hormone products and their effects on the environment when released into the water cycle. Fish farming uses hormones to increase production. When used incorrectly, hormones may enter the water cycle and can have potentially negative effects on humans. Exposure to hormones can cause endocrine disorders, such as early puberty in children and advances in bone age, along with negative effects on growth, modification of sexual characteristics, and cancer development (Cox et al., 2020).Treatment of chickens with anabolic steroids to increase meat production may lead to positive doping tests in athletes if the meat is consumed before the withdrawal period (Kicman et al., 1994).These are examples of how treating one species can affect the whole ecosystem. Guidelines for antibiotic and hormone use and waste management are developed and need to be strictly implemented (Aidara-Kane et al., 2018).

## *Medical Advances*

Animals and humans share the same diseases and are researched using similar methodology and clinical trials. Animal models are typically used to test newly developed drugs. Dogs, for example, are a near perfect animal model for comparative cancer research. Besides developing similar cancers as their human counterparts, dogs are outbred, have a comparably short life span, live in the same environment as their humans, and even drink the same water as there is no standard of care for most cancers. When their human caretakers are not able to afford the standard of care, they are willing to try experimental therapies, such as new surgical techniques, radiation treatments, or drug therapies. Many spontaneous canine cancers share commonalities in regards to histology and biologic behavior, common genetic mutations, and tumor suppressors, or oncogenes, which makes dogs ideal subjects for novel therapies that have the potential to benefit humans and dogs (Rodriguez, 2014; Simpson et al., 2017). Humans and dogs also share about 85% of their gene pool. For example, osteosarcoma (OSA), a malignant cancer of bones, has very strong similarities, clinically and genetically, between humans and dogs. This means dogs can be used as an excellent model for new therapies trying to improve survival. OSA incidence rates in dogs are seven to ten times higher than in people, and the prognosis in both species is relatively poor, with a five-year survival rate of about 60% in people. This prognosis has not improved in decades (Simpson et al., 2017). The one-year survival rate in dogs is about 45%, and most deaths occur from metastatic disease (tumor spread), as local control can generally be achieved by radical surgical excision, i.e. amputation of the affected limb in most cases. Improved and novel treatment regimens are urgently required to improve survival in both humans and dogs with OSA (Rodriguez, 2014; Simpson et al., 2017). Any treatments that are successful on dogs and subsequently applied to humans will obviously benefit human OSA patients, but also future canine patients.

## Issues Prioritizing Animal Health and Welfare

### Animal-Assisted Interventions

Traditionally, we only consider the human and animal as components of the human-animal bond, even though the application of the environment and nature may be important as well. The incorporation of nature into the human-animal bond employs the highly regarded biophilia hypothesis, which describes the innate human instinct to connect with nature and other living things (Wilson, 1984). Most research settings studying the benefit of or the risk associated with animal assisted interventions (AAIs) focus mainly on the human side in this setting. The risk of possible zoonotic disease transmission, allergies, injuries through scratches or bites, and fall risk in elderly patients are all things that have been a major focus in assessing the outcome and benefit of AAIs. All this is focused on human well-being. Very few studies actually describe the health, welfare, living conditions, and the possible benefit for the animals used in those interventions (Bert et al., 2016; Friedman & Krause-Parello, 2018). Most AAIs involve companion animals, but with the rising use of horses and farm animals in AAIs, there is a rising need to consider their living conditions and environment in addition to their well-being and their possible benefit from those interactions with us humans. A recent literature review exploring AAIs in the Green Care Framework found that all evaluated publications primarily dealt with farm animals, and in 36 out of 44 records, there was no reference to animal health or welfare (Galardi et al., 2021). Most studies focused solely on the human benefit of the interaction. Keeping the One Health approach in mind, any research or solution that harms one of those components does not align with the One Health concept. Future studies should not only focus on the human benefit but also on the environmental impact as well as the animal benefit from AAIs. For example, Green Care involving people caring for animals and the landscape may be investigated in an interdisciplinary manner, with mental health researchers measuring the impact of the intervention on mental health outcomes, animal health researchers measuring animal welfare outcomes, and ecologists measuring plant health and species diversity in the environment.

### Wildlife Conservation

Tourism has been heavily relied on as a means to support endangered species. In some cases, >50% of park funding stems from tourism (Buckley et al., 2012). Research has shown that fluctuation or downturn in tourism can be detrimental to the conservation of a species. In addition, some of the heavily desired species are not necessarily in the most need of conservation. Another problem can be the resulting crowding and pollution caused by tourists when not managed properly with a possible negative impact on the environment and other wildlife in addition to the endangered species (Buckley et al., 2012; Sorice et al., 2006).

### Trap-Neuter-Release (TNR) Programs for Feral Cats

The issue of managing free roaming cats is a highly debated controversy within One Health. Controlling the stray cat population becomes necessary for conservation of native species and for the stray cats themselves, as they are at risk for road trauma, injuries from cat fights, and infectious diseases. Some of these infectious diseases have zoonotic potential, which increases the risk for their human neighbors. One method of management involves trap and removal (euthanasia) of cats, which weighs heavily from a social and ethical perspective (Barrows,

2004). A common method of trying to control those stray cat populations are TNR programs. Feral cats are trapped, neutered, potentially medicated/vaccinated, and then they are released back in the "wild," usually at the original point of capture. Even though TNR is advocated as a humane way to minimize euthanasia and is an effective way for population control amongst the stray cat population, TNR programs come with their own issues when considering a One Health approach. TNR programs theoretically improve feral feline welfare by preventing euthanasia and reducing the burden on shelters/rescue groups and the emotional burden of euthanasia for people involved (Barrows, 2004). TNR advocates also claim to reduce the overall size of the stray cat population and prevent the vacuum effect, i.e. new cats migrating in and taking over the vacant space when a colony has been removed either via euthanasia or adoption. Most programs fail to show a reduction in colony size, and evidence exists that colonies are not closed populations with new stray cats entering monitored populations. TNR programs that successfully reduce population size usually have large numbers of adoptions, which creates exactly the vacuum effect they are claiming to avoid (Loyd & DeVore, 2010). Although adopted cats may have positive outcomes, the void from their removal in a colony invites the migration of intact, unvaccinated, and diseased cats into the colony. Therefore, TNR programs that maintain feral cat populations may also increase the public health risk. Cats are known to carry potentially zoonotic diseases, such as gastrointestinal parasites, toxoplasmosis, and rabies. Human exposure to rabies is more commonly caused by cats than any other domestic animal. Throughout 1993–2002, cats accounted for one-third of cases of human exposure to rabies in New York State (Eidson & Bingman, 2010). In addition, the health and well-being of feral cat colonies have been found to be a reflection of socioeconomic status, as lower-income communities are associated with lower neuter and higher pregnancy rates, further highlighting the One Health connection (Finkler, Hatna & Terkel, 2011).

## *Responsible pet ownership*

Under the title "Healthy Pets, Healthy People," the CDC offers information and resources on responsible pet ownership, facts on benefits and risks of having pets, zoonotic and animal-related diseases and more (CDC, 2022, "Healthy Pets, Healthy People"). They especially focus on diseases that animals, including common companion animals, exotic pets, domesticated farm animals, and wildlife can spread and explain measurements one can take to help people and their pets stay healthy (CDC, 2022, "Pets and Other Animals"). Uninformed, but still well intended pet ownership, feeding, or other actions taken toward wildlife carry the risk of spreading and furthering diseases and may also increase the risk of injury or damage not only to us humans, but also animals and the environment.

## **Issues Prioritizing Environmental Health and Welfare**

### *Climate Change*

As mentioned in other sections, the environment is part of the One Health concept, albeit often overlooked or only treated as a bystander when looking into zoonotic diseases. Advocates of EcoHealth position it as a successor to One Health by emphasizing the importance of issues such as climate change and sustainability for health, branching out from medicine to environmental and social sciences (Zinsstag et al., 2011). Climate change, particularly global warming, has a profound impact on the environment and, of course, the animals and humans in it (Cahill et al., 2013; Schleuning et al., 2016). Climate change alters the availability of resources necessary for the health of all life, from microorganisms in the environment to the animals that survive on them. This progressive change will result in the extinction of a significant number of an estimated 15–37% of species by 2050 (Thomas et al., 2004). These sobering statistics compel us to utilize a One Health approach that focuses on technologies to decrease greenhouse gas emissions and strategies for carbon sequestration to preserve the planet and minimize negative health consequences.

## *Pollution*

It is impossible to deny that pollution is harmful to the environment and the humans and animals living within it (Khan & Ghouri, 2011). Human death from pollution is estimated to be more than 16% worldwide (Landrigan et al., 2018). Moreover, animal deaths have even been utilized as biomarkers of pollutants in the environment (Noyes & Lema, 2015). The One Health conundrum in pollution is that humans are the source of pollution, whether it be to air, water, or land. However, humans, along with animals and the environment, are the recipients of the negative health consequences of pollution. One Health solutions should focus on educating the human culprits of pollution to minimize waste and the impact that it has on all life forms.

## *Urbanization*

One of the other greatest displays of humans' negative effect on the environment and animals is the process of urbanization (McKinney, 2008). Population concentration and migration are important factors in disease transmission for both animals and humans, as well as decreasing plant and animal species richness. Brucellosis, for example, is now a highly endemic zoonotic disease in bison herds confined in national parks after human settlement in the Great Plains of the United States restricted bison movements. This high infection rate within those bison herds acts as a reservoir that puts livestock and even humans, such as ranch workers and packing house employees at an increased infection risk (Altizer et al., 2011). On the other side of the globe, livestock exports from Mongolia are banned because of endemic brucellosis, increasing the national herd, which in turn exerts heavy pressure on increasingly fragile pastures (Zinsstag et al., 2011).

Ecosystem approaches to health or "Eco-Health" consider inextricable ties between ecosystems, society, and health of animals and humans (Rapport et al., 1998; Rapport et al., 1999). The in-depth analysis of those eco-health links led to the conclusion that mercury poisoning of fish and impending health risks for humans in the Amazon basin were not due as initially thought to upstream gold mining, but they were due to soil erosion following widespread deforestation in the area (Forget & Lebel, 2001).

To increase awareness for the inclusivity of the concept of One Health and EcoHealth, Zinsstag et al., are proposing to provisionally use the term "health in social-ecological systems" (HSES). As he writes: "In this way, we explicitly include the health of humans and animals as a quantitative and qualitative interaction and outcome process in social-ecological systems" (Ostrom, 2007; Zinsstag et al., 2011).

## One Health Action

Strong advocacy, leadership, and funding are necessary for conducting research and executing action to solve One Health issues. We must reemphasize that stakeholders from each facet of One Health communicate, collaborate, and coordinate to create effective change. The significant barrier to creating this change is not considering all voices. When public policy and funding decisions are made regarding human health, veterinarians and environmentalists are typically not at the table. An interdisciplinary approach that is inclusive of all perspectives fosters the capacity building that One Health needs to impact the issues that arise at the intersection of humans, animals, and the environment. Education and awareness need to transcend the scientific or academic level; it needs to speak at the governmental level to make a difference.

Specifically, the One Health Act of 2016 was introduced to the US House and Senate to create a comprehensive strategy to address zoonotic disease such as Zika and Ebola (One Health Commission, 2022, "One Health Act Reintroduced in US House and Senate"). It tasked the CDC, Department of Homeland Security, and USDA to work together and respond to these zoonotic disease outbreaks. Although that bill was not passed, One Health maintained momentum in government legislation. The "Advancing Emergency Preparedness through the One Health Act" was introduced in 2019 to federally mandate that the Department of Health and Human Services, Department of Agriculture, and other federal agencies communicate and implement a federal One Health framework for emergency preparedness (H.R. Rep. No. 2061, 2021). Although unsuccessful, Congress passed a massive 2.3 trillion-dollar Omnibus Appropriations bill to include language directing federal agencies to develop a One Health Framework in 2020; the "Advancing Emergency Preparedness through the One Health Act" was reintroduced again in 2021. At the time of writing, advocates are still striving to rally cosponsors and pass this legislation in Congress. Passing of this bill will be a milestone that paves a new path for the future of One Health.

## Unanswered Questions and Future Research Opportunities

The field of One Health is ripe with opportunity for researchers and practitioners from every discipline to collaborate on the infinite questions that integrate the study of human, animal, and environmental systems. It is a challenge to deliberately and thoroughly address all three arms of the One Health equation in a research or policy proposal; therefore, experts from each discipline, not just one or two, should be fully invested in a project to truly embrace the One Health approach. The discipline that is often overlooked in discussions of One Health is the environment. The One Health approach has mainly focused on emerging zoonoses from domestic or wildlife without encompassing the role of the inclusive ecosystems. Human and animal populations alike will be more and more difficult to maintain when the ecosystem deteriorates (Destoumieux-Garzón et al., 2018). Therefore, research should integrate the specific role of the environment when investigating questions regarding disease transmission and animal extinction (Estrada-Peña et al., 2014).

Another opportunity for strengthening One Health comes from the human medical sector. Although the AMA demonstrated initial enthusiasm for One Health through the joint incorporation of the One Health Commission with the AVMA in 2007, the AMA's commitment to this area has since subsided. In fact, the AMA does not have a One Health committee, nor does it reference One Health on their website at this time (Wolfe, 2015). One likely reason for this is that the AMA's president, Dr. Ronald Davis, who was a strong champion for the incorporation of the One Health Commission in 2007, passed away shortly thereafter. Since his passing, no other leader within the AMA has revived the incorporation of the One Health Commission as a priority. The human medical field's absence from One Health may be largely a function of the extreme specialization of human physicians versus the general practitioner mindset of veterinarians, who work with both animals and people on a daily basis.

To impart the importance of One Health in human medicine and foster careers of future physicians in the area, these concepts should be taught in the medical school curriculum just as they are taught in the veterinary curriculum (Larsen, 2021; NAVMEC Board of Directors, 2011; P. M. Rabinowitz et al., 2017). Interestingly, a human cardiologist, Dr. Barbara Natterson-Horowitz, along with Kathryn Bowers introduced the concept of **Zoobiquity,** which is defined as, "the approach to finding innovative solutions to the most challenging issues in our human lives, from physical and mental health to social change" (Natterson-Horowitz & Bowers, 2013). Zoobiquity is different from One Health in that it focuses on evolutionary medicine, comparing the shared similarities in disease and pathophysiology between humans and animals.

> We could improve the health of all species by learning how animals live, die, get sick and heal in their *natural* settings. [...] I wanted to break down the wall between physicians, veterinarians, and evolutionary biologists because together we are uniquely situated to explore the animal-human overlap where it matters most urgently - in the effort to heal our patients.
>
> *(Natterson-Horowitz & Bowers, 2013)*

They contend that answers in human health can be answered by exploring the same questions that arise in animal health. Zoobiquity maintains public awareness through collaborations and conferences, actively using this approach to focus on our understanding and remedying cardiovascular diseases and infertility (Karolinska Institutet, 2022, "Zoobiquity Event at Karolinska Institutet").

Although it can be argued that One Health encompasses all aspects of health, it has typically focused on physical or clinical health, without specific attention to welfare and well-being. In an effort to ensure that welfare and well-being are not overlooked, the concept of One Welfare has been developed to complement One Health. **One Welfare** serves to highlight the interconnections between animal welfare, human well-being and the environment (Colonius & Earley, 2013). One Welfare helps to promote key global objectives such as supporting food security, sustainability, reducing human suffering and improving productivity within the farming sector through a better understanding of the value of high welfare standards (www.onewelfareworld.org). Integrating One Health and One Welfare together facilitates a holistic approach that maximizes the synergistic benefits between people, animals, and the environment. The term One Welfare is a notable addition to the research literature so any publications pertinent to these types of outcomes can be found under this search term (Pinillos et al., 2016).

The way the One Health approach can create change is through public education and awareness. Those invested in the One Health mission should strive for building mechanisms to better share and communicate advances in research between humans, animals, and the environment. Research typically occurs within a vacuum of a single discipline, and often a very specific and specialized area of that discipline. It may be rare for a researcher to explore studies outside of their area of focus, let alone consult with experts outside their profession, to seek answers to their questions. It is clear that answers or different approaches to finding answers can be discovered through the lens of other disciplines. Therefore, researchers should share their discoveries in scientific meetings and journals different from their usual audiences to achieve a broader impact. This knowledge sharing can foster discussion and collaboration in the interdisciplinary nature of One Health.

Fortunately, the knowledge transfer of One Health is promoted in numerous ways, including conferences at regional, national, and international levels. Specifically, the **One Health Initiative** was created as a movement to forge collaboration between numerous professionals and professional organizations as a hub for communications and progress regarding One Health (https://onehealthinitiative.com/). Furthermore, public awareness for One Health has been celebrated globally every November 3rd since 2016, known as "One Health Day" (https://onehealthday.com/). Raising awareness enables the field to create more interdisciplinary programs in education, training, research, and established policy; more information sharing related to disease detection, diagnosis, education, and research; more prevention of both infectious and chronic diseases; and development of new therapies and approaches to treatment (www.Onehealthcommision.org).

## Conclusions

The broad and deeply complex field of One Health continues to evolve and change the way in which we live. It accelerates research discoveries, promotes the power of public health, and improves healthcare, welfare, and wellness for all humans, animals, and the environment. However, this elevated state cannot be achieved without education and cross collaboration. We hope that bringing awareness to the challenges and opportunities at the intersection of humans, animals, and the environment recruits the next generation of One Health advocates. One Health begins with one individual from any discipline who invites the conversation to make a difference. At the end of the day, no one person or field of study can effectively address the problems that arise at the human-animal-ecosystem interface. Whether they be human and animal health professionals, wildlife specialists, anthropologists, economists, environmentalists, behavioral scientists, or sociologists, all have the potential to play a significant role in practicing and advancing One Health. Together, we use this transdisciplinary and holistic approach to broaden our focus from the health of one into the health of all.

### Discussion Questions

1. Identify a problem that affects people, animals, or the environment. How do you employ a One Health approach to solve it?
2. What careers or professions integrate or advocate for One Health?
3. What is the economic impact of a One Health approach?
4. How do we improve communication and collaboration between disciplines?

## References

Aidara-Kane, A., Angulo, F. J., Conly, J. M., Minato, Y., Silbergeld, E. K., McEwen, S. A., & Collignon, P. J. (2018). World health organization (WHO) guidelines on use of medically important antimicrobials in food-producing animals. *Antimicrobial Resistance & Infection Control*, 7(1), 7. https://doi.org/10.1186/s13756-017-0294-9

Altizer, S., Bartel, R., & Han, B. A. (2011). Animal migration and infectious disease risk. *Science*, 331(6015), 296–302. https://doi.org/10.1126/science.1194694

American Veterinary Medical Association (AVMA). (2022, March 28). *Food Safety*. https://www.avma.org/resources-tools/avma-policies/food-safety.

American Veterinary Medical Association (AVMA). (2022, March 28). *One Health.* https://www.avma.org/resources-tools/one-health

American Veterinary Medical Association (AVMA). (2022, March 28). *Veterinarian's Oath.* https://www.avma.org/resources-tools/avma-policies/veterinarians-oath

American Veterinary Medical Association (AVMA). (2022, March 28). *What Is FSVM?* https://www.avma.org/resources-tools/food-supply-veterinary-medicine/what-is-fsvm

Barrows, P. L. (2004). Professional, ethical, and legal dilemmas of trap-neuter-release. *Journal-American Veterinary Medical Association, 225*, 1365–1368.

Bert, F., Gualano, M. R., Camussi, E., Pieve, G., Voglino, G., & Siliquini, R. (2016). Animal assisted intervention: A systematic review of benefits and risks. *European Journal of Integrative Medicine, 8*(5), 695–706. https://doi.org/10.1016/j.eujim.2016.05.005

Buckley, R. C., Castley, J. G., Pegas, F. de V., Mossaz, A. C., & Steven, R. (2012). A population accounting approach to assess tourism contributions to conservation of IUCN-Redlisted mammal species. *PLoS ONE, 7*(9), e44134. https://doi.org/10.1371/journal.pone.0044134

Cahill, A. E., Aiello-Lammens, M. E., Fisher-Reid, M. C., Hua, X., Karanewsky, C. J., Yeong Ryu, H., Sbeglia, G. C., Spagnolo, F., Waldron, J. B., Warsi, O., & Wiens, J. J. (2013). How does climate change cause extinction? *Proceedings of the Royal Society B: Biological Sciences, 280*(1750), 20121890. https://doi.org/10.1098/rspb.2012.1890

Cammack, N. R., Yamka, R. M., & Adams, V. J. (2021). Low number of owner-reported suspected transmission of foodborne pathogens from raw meat-based diets fed to dogs and/or cats. *Frontiers in Veterinary Science, 8*, 741575. https://doi.org/10.3389/fvets.2021.741575

Centers for Disease Control and Prevention (CDC). (2022, March 28). *Keeping Pets Healthy Keeps People Healthy Too!* https://www.cdc.gov/healthypets/index.html

Centers for Disease Control and Prevention (CDC). (2022, March 28). *One Health in Action.* https://www.cdc.gov/onehealth/in-action/index.html

Centers for Disease Control and Prevention (CDC). (2022, March 28). *Pets and Other Animals.* https://www.cdc.gov/healthypets/pets/index.html

Centers for Disease Control and Prevention (CDC). (2022, March 28). *ZOHU Calls/Webinars.* https://www.cdc.gov/onehealth/zohu/index.html

Colonius, T. J., & Earley, R. W. (2013). One welfare: A call to develop a broader framework of thought and action. *Journal of the American Veterinary Medical Association, 242*(3), 309–310. https://doi.org/10.2460/javma.242.3.309

Cox, S., Sommardahl, C., Fortner, C., Davis, R., Bergman, J., & Doherty, T. (2020). Determination of grapiprant plasma and urine concentrations in horses. *Veterinary Anaesthesia and Analgesia, 47*(5), 705–709. https://doi.org/10.1016/j.vaa.2020.04.006

Desa, U. N. (2019). World population prospects 2019: Highlights. *New York (US): United Nations Department for Economic and Social Affairs, 11*(1), 125.

Destoumieux-Garzón, D., Mavingui, P., Boetsch, G., Boissier, J., Darriet, F., Duboz, P., Fritsch, C., Giraudoux, P., le Roux, F., Morand, S., Paillard, C., Pontier, D., Sueur, C., & Voituron, Y. (2018). The one health concept: 10 years old and a long road ahead. *Frontiers in Veterinary Science, 5*, 14. https://doi.org/10.3389/fvets.2018.00014

Driesch, A. (2003). *Geschichte der Tiermedizin. 5000 Jahre Tierheilkunde.* Schattauer.

Dukes, T. W. (2000). That other branch of medicine: An historiography of veterinary medicine from a canadian perspective. *Canadian Bulletin of Medical History, 17*(1), 229–243. https://doi.org/10.3138/cbmh.17.1.229

Eade, D. (2007). Capacity building: Who builds whose capacity? *Development in Practice, 17*(4–5), 630–639. https://doi.org/10.1080/09614520701469807

Eidson, M., & Bingman, A. K. (2010). Terrestrial rabies and human postexposure prophylaxis, New York, USA. *Emerging Infectious Diseases, 16*(3), 527–529. doi: 10.3201/eid1603.090298. PMID: 20202438; PMCID: PMC3322005.

Estrada-Peña, A., Ostfeld, R. S., Peterson, A. T., Poulin, R., & de la Fuente, J. (2014). Effects of environmental change on zoonotic disease risk: An ecological primer. *Trends in Parasitology, 30*(4), 205–214. https://doi.org/10.1016/j.pt.2014.02.003

Finkler, H., Hatna, E., & Terkel, J. (2011). The impact of anthropogenic factors on the behavior, reproduction, management and welfare of urban, free-roaming cat populations. *Anthrozoös, 24*(1), 31–49.

Food and Agriculture Organization (FAO), World Organisation for Animal Health (OIE), World Health Organization (WHO), United Nations System Influenza Coordination (UNSIC), United

Nations Children's Fund (UNICEF), World Bank (WO). (2008). *Contributing to One World, One Health: A Strategic Framework for Reducing Risks of Infectious Diseases at the Animal-Human-Ecosystems Interface*. https://www.preventionweb.net/files/8627_OWOH14Oct08.pdf

Forget, G., & Lebel, J. (2001). An ecosystem approach to human health. *International Journal of Occupational and Environmental Health*, 7(2 Suppl), S3–S8.

Friedman, E., & Krause-Parello, C. A. (2018). Companion animals and human health: Benefits, challenges, and the road ahead for human–animal interaction. *Revue Scientifique et Technique de l'OIE*, 37(1), 71–82. https://doi.org/10.20506/rst.37.1.2741

Galardi, M., de Santis, M., Moruzzo, R., Mutinelli, F., & Contalbrigo, L. (2021). Animal assisted interventions in the green care framework: A literature review. *International Journal of Environmental Research and Public Health*, 18(18), 9431. https://doi.org/10.3390/ijerph18189431

Gordon, A., & Schwabe, C. (2004). *The Quick and the Dead*. Styx/Koninklijke Brill NV.

H.R.2061–117th Congress (2021–2022). *Advancing Emergency Preparedness Through One Health Act of 2021*. (2021, May 26). https://www.congress.gov/bill/117th-congress/house-bill/2061

Jones, K., Patel, N., Levy, M., Storeygard, A., Balk, D., Gittleman, J., & Daszak, P. (2008). Global trends in emerging infectious diseases. *Nature*, 451, 990–993.

Karolinska Institutet. (2022, March 28). *Zoobiquity Event at Karolinska Institutet*. https://news.ki.se/zoobiquity-event-at-ki

Khan, M., & Ghouri, A. (2011). Environmental pollution: Its effects on life and its remedies. *Researcher World: Journal of Arts, Science & Commerce*, 2(2), 276–285.

Kicman, A. T., Cowan, D. A., Myhre, L., Nilsson, S., Tomten, S., & Oftebro, H. (1994). Effect on sports drug tests of ingesting meat from steroid (methenolone)-treated livestock. *Clinical Chemistry*, 40(11), 2084–2087. https://doi.org/10.1093/clinchem/40.11.2084

King, T. A. (2021). The one medicine concept: Its emergence from history as a systematic approach to re-integrate human and veterinary medicine. *Emerging Topics in Life Sciences*, 5(5), 643–654. https://doi.org/10.1042/ETLS20200353

Landrigan, P. J., Fuller, R., Acosta, N. J. R., Adeyi, O., Arnold, R., Basu, N. (Nil), Baldé, A. B., Bertollini, R., Bose-O'Reilly, S., Boufford, J. I., Breysse, P. N., Chiles, T., Mahidol, C., Coll-Seck, A. M., Cropper, M. L., Fobil, J., Fuster, V., Greenstone, M., Haines, A., & … Zhong, M. (2018). The lancet commission on pollution and health. *The Lancet*, 391(10119), 462–512. https://doi.org/10.1016/S0140-6736(17)32345-0

Larsen, R. J. (2021). Shared curricula and competencies in one health and health professions education. *Medical Science Educator*, 31(1), 249–252. https://doi.org/10.1007/s40670-020-01140-7

Lebel, J. (2003). *Health: An Ecosystem Approach*. International Development Research Centre.

Loyd, K., & DeVore, J. (2010). An evaluation of feral cat management options using a decision analysis network. *Ecology and Society*, 15(4), 10. http://www.ecologyandsociety.org/vol15/iss4/art10/

Mackenzie, J. S., & Jeggo, M. (2019). The one health approach—Why is it so important? *Tropical Medicine and Infectious Disease*, 4(2), 88. https://doi.org/10.3390/tropicalmed4020088

Majok, A., & Schwabe, C. (1996). *Development Among Africa's Migratory Pastoralists*. Greenwood Publishing.

McKinney, M. L. (2008). Effects of urbanization on species richness: A review of plants and animals. *Urban Ecosystems*, 11(2), 161–176. https://doi.org/10.1007/s11252-007-0045-4

Natterson-Horowitz, B., & Bowers, K. (2013). *Zoobiquity: What Animals Can Teach Us About Health and the Science of Healing* (1st ed.). Knopf.

NAVMEC Board of Directors. (2011). The North American veterinary medical education consortium (NAVMEC) looks to veterinary medical education for the future: "Roadmap for veterinary medical education in the 21st century: Responsive, collaborative, flexible." *Journal of Veterinary Medical Education*, 38(4), 320–327. https://doi.org/10.3138/jvme.38.4.320

Noyes, P. D., & Lema, S. C. (2015). Forecasting the impacts of chemical pollution and climate change interactions on the health of wildlife. *Current Zoology*, 61(4), 669–689. https://doi.org/10.1093/czoolo/61.4.669

One Health Commission. (2022, March 28). *One Health Act Reintroduced in US House and Senate*. https://www.onehealthcommission.org/index.cfm/38050/48310/one_health_act_reintroduced_in_us_house_and_senate

One Health Commission. (2022, March 28). *What Is One Health?* https://www.onehealthcommission.org/en/why_one_health/what_is_one_health/

Ostrom, E. (2007). A diagnostic approach for going beyond panaceas. *Proceedings of the National Academy of Sciences*, 104(39), 15181–15187. https://doi.org/10.1073/pnas.0702288104

Pinillos, R. G., Appleby, M. C., Manteca, X., Scott-Park, F., Smith, C., & Velarde, A. (2016). One Welfare - a platform for improving human and animal welfare. *Veterinary Record, 179*(16), 412–413. https://doi.org/10.1136/vr.i5470

Public Health Agency of Canada. (2009). *One World One Health: From Ideas to Action*. Public Health of Canada.

Rabinowitz, P. M., Natterson-Horowitz, B. J., Kahn, L. H., Kock, R., & Pappaioanou, M. (2017). Incorporating one health into medical education. *BMC Medical Education, 17*(1), 45. https://doi.org/10.1186/s12909-017-0883-6

Rabinowitz, P., Scotch, M., & Conti, L. (2009). Human and animal sentinels for shared health risks. *Veterinaria Italiana, 45*(1), 23–24.

Rapport, D. J., Böhm, G., Buckingham, D., Cairns, J., Costanza, R., Karr, J. R., de Kruijf, H. A. M., Levins, R., McMichael, A. J., Nielsen, N. O., & Whitford, W. G. (1999). Ecosystem health: The concept, the ISEH, and the important tasks ahead. *Ecosystem Health, 5*(2), 82–90. https://doi.org/10.1046/j.1526-0992.1999.09913.x

Rapport, D., Costanza, R., Epstein, P., Gaudet, C., & Levins, R. (Eds.) (1998). *Ecosystem Health*. Blackwell Sciences.

Rodriguez, C. O. (2014). *Using Canine Osteosarcoma as a Model to Assess Efficacy of Novel Therapies: Can Old Dogs Teach Us New Tricks?* (pp. 237–256). https://doi.org/10.1007/978-3-319-04843-7_13

Rudolph Virchow. (2022, February 9). In *Wikipedia*. https://en.wikipedia.org/wiki/Rudolf_Virchow

Saunders, L. Z. (2000). Virchow's contributions to veterinary medicine: Celebrated then, forgotten now. *Veterinary Pathology, 37*(3), 199–207. https://doi.org/10.1354/vp.37-3-199

Schleuning, M., Fründ, J., Schweiger, O., Welk, E., Albrecht, J., Albrecht, M., Beil, M., Benadi, G., Blüthgen, N., Bruelheide, H., Böhning-Gaese, K., Dehling, D. M., Dormann, C. F., Exeler, N., Farwig, N., Harpke, A., Hickler, T., Kratochwil, A., Kuhlmann, M., … & Hof, C. (2016). Ecological networks are more sensitive to plant than to animal extinction under climate change. *Nature Communications, 7*(1), 13965. https://doi.org/10.1038/ncomms13965

Schwabe, C. (1984). *Veterinary Medicine and Human Health* (3rd ed.). Williams and Wilkins.

Silver, G. A. (1987). Virchow, the heroic model in medicine: Health policy by accolade. *American Journal of Public Health, 77*(1), 82–88. https://doi.org/10.2105/AJPH.77.1.82

Simpson, S., Dunning, M. D., de Brot, S., Grau-Roma, L., Mongan, N. P., & Rutland, C. S. (2017). Comparative review of human and canine osteosarcoma: Morphology, epidemiology, prognosis, treatment and genetics. *Acta Veterinaria Scandinavica, 59*(1), 71. https://doi.org/10.1186/s13028-017-0341-9

Sorice, M. G., Shafer, C. S., & Ditton, R. B. (2006). Managing endangered species within the use–preservation paradox: The Florida Manatee (Trichechus manatus latirostris) as a tourism attraction. *Environmental Management, 37*(1), 69–83. https://doi.org/10.1007/s00267-004-0125-7

Thomas, C. D., Cameron, A., Green, R. E., Bakkenes, M., Beaumont, L. J., Collingham, Y. C., Erasmus, B. F. N., de Siqueira, M. F., Grainger, A., Hannah, L., Hughes, L., Huntley, B., van Jaarsveld, A. S., Midgley, G. F., Miles, L., Ortega-Huerta, M. A., Townsend Peterson, A., Phillips, O. L., & Williams, S. E. (2004). Extinction risk from climate change. *Nature, 427*(6970), 145–148. https://doi.org/10.1038/nature02121

UN Environment Programme (UNEP). (2022, March 28). *About UN Environment Programme*. UNEP. https://www.unep.org/about-un-environment

United Nations Environment Programme. (2021). *Joint Tripartite (FAO, OIE, WHO) and UNEP Statement Tripartite and UNEP Support OHHLEP's Definition of "One Health."* Https://Wedocs.Unep.Org/20.500.11822/37600

United States Department of Agriculture (USDA). (2022, March 28). *Animal Health Program Review*. https://www.aphis.usda.gov/aphis/ourfocus/animalhealth/program-overview

van Bree, F. P. J., Bokken, G. C. A. M., Mineur, R., Franssen, F., Opsteegh, M., van der Giessen, J. W. B., Lipman, L. J. A., & Overgaauw, P. A. M. (2018). Zoonotic bacteria and parasites found in raw meat-based diets for cats and dogs. *Veterinary Record, 182*(2), 50–50. https://doi.org/10.1136/vr.104535

William Osler. (2022, 15 March). Wikipedia. https://en.wikipedia.org/w/index.php?title=William_Osler&action=history

Wilson, E. (1984). *Biophilia*. Harvard University Press.

Wolfe, L. (2015, June 10). *Why the Human Side Lags Behind in One Health*. News.

World Health Organization (WHO). (2022, March 28). *One Health*. World Health Organization. https://www.euro.who.int/en/health-topics/health-policy/one-health

World Health Organization (WHO). (2022, March 28). *Neglected Zoonotic Diseases*. World Health Organization. https://www.who.int/teams/control-of-neglected-tropical-diseases/neglected-zoonotic-diseases

Zinsstag, J., Schelling, E., Waltner-Toews, D., & Tanner, M. (2011). From "one medicine" to "one health" and systemic approaches to health and well-being. *Preventive Veterinary Medicine, 101*(3–4), 148–156. https://doi.org/10.1016/j.prevetmed.2010.07.003

# 5
# CURRENT PERSPECTIVES ON THE HEALTH EFFECTS (MENTAL AND PHYSICAL) OF HUMAN-ANIMAL INTERACTION (HAI)

*Erika Friedmann and Lincy Koodaly*

## Historical Background

Humans are demonstrated to live in association with animals from the beginning of recorded history and the human-animal bond is an ancient concept. Paleolithic cave art often includes hybrid figures that interweave human and animal components (Comba, 2010). Anthropologists argue that humans and animals are similar and differ from animals only in their culture and diversity (Ingold, 2018). In Greek mythology, the dog is commonly represented as an intermediary between this world and the next. Ancient Egyptian relics frequently document the importance of cats, dogs, and snakes where they were treated as gods or spirits (Serpell, 2010). In the early years of Christianity in Europe, saints and holy men of Britain and Ireland such as St. Francis Assissi, St. Christopher, St. Bernard, St. Gertrude of Nievelles, and St. Philip Neri among others had strong associations with animals (Serpell, 2010). Human-animal connections in other parts of Northern and Eastern Europe were demonstrated through Shamanism in which a shaman had unique powers to transform self and heal humans using their animal connections (Serpell, 2010). Shamanism as documented as early as the late 17th century suggests no clear distinction between humans and animals at the beginning of the time and that shamans reestablish the relationship with animals and become symbiotic, performing unnatural tasks such as healing through spiritual connections between animals and people (Francfort et al., 2001).

In the late 17th century, John Locke recommended pets for children to help develop their social skills and maturation (Locke & Axtell, 1968). The 18th century saw the emergence of a sympathetic attitude toward animals and pet keeping expanded to middle-class families that migrated to urban areas. The first documented use of animal companionship for the treatment of the mentally ill was in England in the late 18th century and later the idea slowly seeped into prisons and long-term care institutions (Serpell, 2015). Animals were a part of the moral therapeutic environment for the treatment of the mentally ill, developed by William Tuke and instituted at 'The Retreat' in York in 1792. This model gradually replaced the punitive treatments previously employed (Serpell, 2015). Florence Nightingale's documentation notes reference small pets as the best companions for the sick and chronically ill.

Despite the values attributed to these practices, in the early 20th-century awareness of infection and the need to control it led to the elimination of animals from hospitals. HAI continued to be used by practitioners in other settings to help select groups of mental health patients. For example, Levinson documents the incorporation of animals into psychotherapy as transitional objects to encourage patients to interact with their therapists (Levinson, 1972).

A gradual move away from the biomedical model of health to include environmental, social, psychological, and behavioral contributors to physical health that developed after the middle of the 20th century primed examination of these types of factors as contributors to health outcomes that previously were considered as the pure domain of mechanistic medicine. It was in this context that the contributions of HAI to people's physical health were noticed again. A study based on the idea of the integration of mental and physical aspects of health established the potential role of HAI on human health outcomes. In this remarkable study of 92 patients hospitalized with coronary heart disease, patients who owned pets showed better survival than non-owners (Friedmann et al., 1980). These findings inspired more research into the mechanisms for the physiological benefits of pet ownership and into other ways HAI can influence human health.

The 1980 study specifically investigated pet ownership. The mechanisms for the health benefits from pet ownership that were theorized based on this research included reducing anxiety and stress by making people feel safer and more comfortable in their environment (social and physical); helping people feel more supported, connected, and needed, and less lonely by providing a reason for living, an opportunity for nurturing and caring for others and focus outside of the self; and finally being more active by providing an impetus for exercise and physical function.

Once the possibilities were considered, it became evident that many of these benefits could occur from a variety of forms of HAI, not just from pet ownership. This led to a series of experimental and quasi-experimental studies, mostly examining the effect of HAI on stress and stress responses (Allen et al., 2001, 2002; Alonso, 1999; Eddy, 1995; Friedmann et al., 1983, 1993, 2007; Tsai et al., 2010; Wilson, 1991). Research evidence indicated that interacting with animals by being in the presence of animals, focusing on the animals without physically interacting with them, and interacting with animals including touching them all can provide short-term benefits in some of the ways that were hypothesized (Friedmann et al., 2010).

This evidence also supported using animals as an intervention to help people who do not own them on the basis that people can benefit from visits with animals they might enjoy. These visits can be focused on specific therapeutic goals and incorporated into therapeutic modalities or can provide general support as parts of more informal unstructured interactions. More recently, trained therapy dogs are used in animal assisted interventions, but untrained "friendly dogs" or pets of investigators were used for this purpose frequently in the 20th century. The general support and companionship function is particularly important for individuals who are not able to meet the demands of responsible pet ownership.

Additional evidence supports that interaction with animals provides models for interacting with other people. HAI supports other aspects of overall well-being including promoting child development, developing moral understanding, responsibility for others, healthy relationships with others, and developing respect for other beings and nature (Esposito et al., 2011).

The examples above include caring for animals, keeping animals as pets, visits with animals, and incorporation of animals into therapeutic regimens. Viewing animals in natural or artificial environments may also provide some of these benefits to the people viewing or interacting with them. Some of these mechanisms might also be used in other contexts. Many advertisers use pet animals in photos or video productions to provide a positive environment for their products. Animals can also serve as a focus of attention or study in their manmade environments such as zoos and natural environments. Menageries began in Egypt more than 4500 years ago (2500 BC) and expanded as the collections of "exotic" animals for religion, display, and to advertise the status of their owners (Tribe, 2004). They expanded in Europe during missions of exploration and then systematic collections for exhibitions at private homes or in public displays (Baratay & Hardouin-Fugier, 2004). Modern zoos began in the late 18th century to exhibit, conduct education about, and develop scientific research on animals (Tribe, 2004). Zoos became a focus as major tourist attractions in the later half of the 20th century (Nekolný & Fialová, 2018). Cageless designs starting in the beginning of the 20th century followed by environmental enrichment and more natural environments in the mid to late 20th century, increased opportunities for people to connect with nature and the animals they observed in it (Ash, 2018).

Animals can also be used as service/or assistance animals, to provide support services to their handlers. Dogs have served as guides for individuals with diminished sight for at least 20 centuries. A Roman fresco from the 1st century A.D. depicts a dog leading a blind person. Dogs trained for this task came into common use in Germany and Switzerland to assist blinded veterans of World War I; and the Seeing Eye was founded in the United States in 1929. Guide Dogs for the Blind Association was founded in the United Kingdom in 1934. The Americans with disabilities Act of 1990 was the first legal recognition of service animals in the United States. The use of service animals to meet specific needs of individuals with disabilities also has been expanded to include support for additional physical and mental challenges. The current US definition of a service animal includes only animals individually trained "to perform tasks for the benefit of an individual with a disability including a physical, sensory, psychiatric, intellectual, or other mental disability" (Brown, 2019).

## Biopsychosocial Model and HAI

The biopsychosocial model is an excellent framework to explain potential mental and physical health benefits from HAI (Figure 5.1). In this model, biological, psychological, and social factors are integrated into health outcomes. Both enhancements and diminishments in characteristics composing any of these factors impact all other factors and contribute to health outcomes. Within the biopsychosocial model, HAI can act in multiple ways to support mental and physical health and well-being (Compitus, 2021; Gee et al., 2021). In the biological realm, pets like dogs encourage physical activity and have been linked to greater cardiac health. The presence of dogs has been shown to reduce blood pressure and heart rate and cortisol secretion. Dog interactions are also known to be associated with oxytocin secretion, which is a buffer for stress, and a promoter of trust and bonding. These physiological changes are conducive to better physical health. Some of the proposed psychological enhancement mechanisms involve decreasing anxiety and stress, and providing feelings of safety. The non-judgmental social support, and bonding provided by the dogs promote a sense of well-being to humans in addition to the comfort and feeling of safety that come with dog ownership. The social aspect includes providing a source of companionship and an opportunity to interact with others enhancing nurturing and giving a purpose to live. The clear connections between the realms are demonstrated through multiple examples, including the physiological and psychological components of the stress response. Feelings of stress and anxiety are accompanied by physiological changes including increased heart rate, blood pressure, stress hormones, and other stress-induced body responses. Loneliness and lack of companionship are often accompanied by depression and depression can have negative consequences such as feelings of helplessness, sleep disturbances, lack of appetite, low self-esteem, and poor concentration.

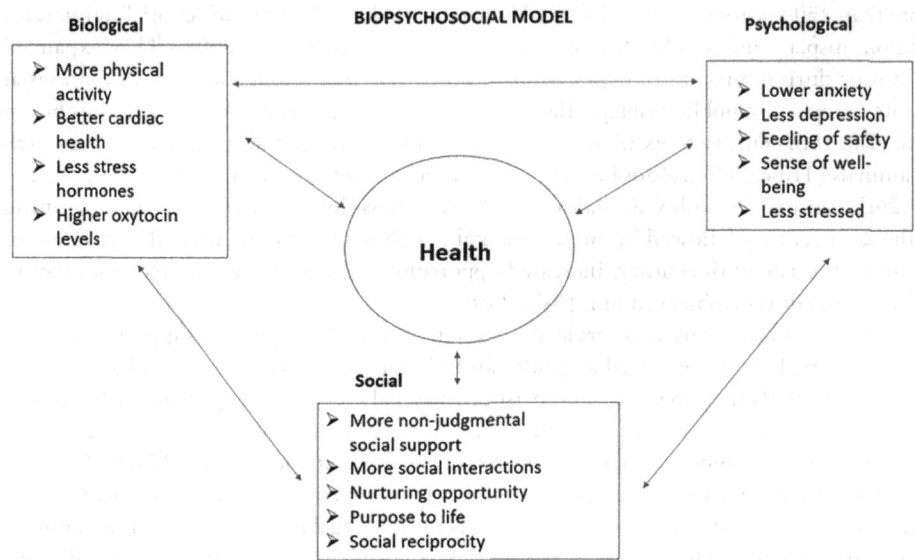

*Figure 5.1* Illustration of the biopsychosocial model for the interrelationships of biological, social, and psychological contributors to health

The early research employed experimental methods to document the decreases in stress in the presence of an animal and show that the impact of the presence of an animal on stress responses was independent of the contribution of support from other people. Much of this research used cardiovascular stress indicators as outcomes, including blood pressure and heart rate. The focus of possible positive health and well-being derived outcomes from HAI has expanded immensely in the past three decades. Researchers now address outcomes that include post-traumatic stress disorder (PTSD), cognition, physical function, loneliness, depression, general well-being, and pain, to name a few.

In addition to the experimental studies, a stream of epidemiological studies used population-based data to evaluate the relationship of pet ownership to cardiovascular health and health behaviors. National surveys document a variety of associations of cardiovascular health measures related to pet ownership. And several surveys tout the physical activity differences between dog owners and owners of other pets or non-owners. The combination of the experimental and epidemiological studies of dog ownership and cardiovascular health led to the American Heart Association's (2013) scientific statement.

> There are a number of methodologically sound studies, and there is a substantial body of data that suggests that pet ownership is associated with a reduction in CVD [cardiovascular disease] risk factors and increased survival in individuals with established CVD.
>
> *(Levine et al., 2013)*

Several theories have been suggested to explain the basis for the impact of HAI that is demonstrated in the biopsychosocial model. The biophilia hypothesis posits that connection with nature and the natural world leads to people's enjoyment and feeling of connectedness with animals (Beetz, 2017). Both attachment theory and the social support theory also indicate that HAI enhances feelings of connectedness with animals and helps buffer negative or inadequate feelings or negative perceptions and enrich the social realm (Collis & McNicholas, 1998; Meehan et al., 2017). Major components of attachment theory include an affectionate bond, being a safe haven in times of distress, seeking physical proximity, experiencing separation anxiety, and viewing the other as a safe base for explorations of new experiences or situations. HAI relationships can demonstrate several aspects of attachment theory. Social support theory focuses on provision of psychological and material resources from individual's interpersonal relationships generally types of resources include instrumental, emotional, and informational support. HAI is thought to be a contributor to emotional support (Hill et al., 2020; Meehan et al., 2017; Rockett & Carr, 2014).

## HAI and Influence on Physical and Mental Outcomes

Animal interactions have become popular although the research related to their health benefits is still in its adolescence. In 2013, nearly three-quarters of US households owned pets and the numbers are on the rise (Bureau & Henderson, 2013). HAI is related to physical fitness, both directly and indirectly. Pet owners report greater physical activity and fewer limitations in mobility than non-owners (Curl et al., 2017). Dogs provided social and emotional support and motivation for adherence to an exercise enhancing weight reduction program (Kushner et al., 2006).

One reason HAI is thought to benefit people by reducing stress is the non-judgmental nature of the animal involved in the interaction. Mood and well-being improved more interacting with an animal only compared with interacting with an animal and handler (Grajfoner et al., 2017). A person experienced lower stress from a standard stressor with a pet present than with a friend selected for being supportive present (Allen et al., 2002). Lowering psychological stress levels promotes physical well-being by reducing the stress biomarkers, including blood pressure, heart rate, cortisol, and other stress hormones in the body.

Among psychological outcomes other than stress and anxiety, a large body of research suggests that HAI is associated with alleviating depression (Souter & Miller, 2007; Virues-Ortega et al., 2012). A study of patients with coronary heart disease suggests that the greatest impact of HAI on patient survival might be among people who are depressed (Friedmann et al., 2011).

As expected within the biopsychosocial model, HAI also promotes social support. The support provided by HAI appears to be independent of the support from other people (Friedmann et al., 2010). The contributions of HAI to psychological, social, and physical health appear to be most important for people who are socially isolated (Fine & Friedmann, 2018). Thus, in the COVID-19 pandemic, support from HAI became important as people were forced to socially isolate themselves.

HAI research performed in humans of various ages, and stages of life indicate that HAI has the potential to influence the wellness of humans across the life span. While much of the research has been conducted on individuals with specific needs or health conditions, or in vulnerable populations, it may be generalizable to more healthy members of the population. For example, pet ownership or visits with pets in shelters or owned by others may decrease the declines in cognitive and physical function associated with aging.

We review several of the key groups. More in-depth coverage is included in individual chapters of this book addressing these populations or outcomes.

## HAI and Children

Child development is greatly affected by the surroundings and interactions with their families and environment. Studies indicate that child-animal interactions can alleviate trauma and help support children's coping and resilience (Gee, Griffin et al., 2017; McCardle et al., 2011). Counseling sessions that include small pocket pets such as hamsters, gerbils, and guinea pigs help build rapport, enhance the counseling relationship, and facilitate the counseling results in elementary school children (Flom, 2005). Reading to an animal is a popular intervention to foster reading in educational settings, despite limited evidence for its efficacy (Hall et al., 2016). Young children completed cognitive tasks with fewer errors when a dog was involved than without one (Gee et al., 2010). HAI can cause direct positive effects on children's motivation, engagement, self-regulation, and social interaction and indirect effects on social-emotional development and learning (Grajfoner et al., 2017). See also Chapter 17 of this book.

## HAI and College Students

College students experience high levels of stress related to their education, new roles, and responsibilities of their independent lives. Several studies suggest that participating in animal visitation programs has the potential to improve college students' mood states. A study involving an animal-assisted campus visitation program in a college showed that a ten-minute HAI improved the positive emotion and decreased levels of stress and negative emotions before examinations (Pendry et al., 2018). In another study, group-administered canine visits improved the well-being of university students by lowering perceived stress. The mere presence of an animal can reduce stress in college students. A ten-minute animal visitation program reduced the salivary cortisol levels of college students significantly (Pendry et al., 2021) Another study provides evidence that interacting with therapy dogs and their handlers, improved executive functions more than a formalized stress management program for university students who have been classified at high risk for academic failure (Pendry et al., 2021). The topic of animals in education is addressed more thoroughly in Chapter 35.

## HAI and Older Adults

Evidence indicates that older adults living in the community or care facilities can benefit from HAIs. Evidence from a variety of studies supports the therapeutic psychological and physical health benefits of animals in the lives of older adults for reducing loneliness, depression, and improving cardiovascular health and physical activity, although there are associated challenges to be considered (Krause-Parello et al., 2019).

Community surveys and health-care utilization data associate health benefits with pet ownership among older adults(Gee & Mueller, 2019; Gee, Mueller et al., 2017; Siegel, 1990; Thorpe et al., 2006). The presence of their pet dogs or cats was associated with lower home ambulatory blood pressure in older adults with hypertension (Friedmann et al., 2015). During the COVID pandemic, dog walking moderated the impact of social isolation on the development of loneliness among older adults (Carr et al., 2021). Interaction with dogs has been shown to reduce depression in some studies, although the results have been inconsistent (Gee & Mueller, 2019). In a care setting such as a nursing home, mental hospital, or rehabilitation facility, those most impaired may benefit the most because of their limited interactions with the external world. Among older adults with dementia residing in a nursing home, social behavior was enhanced by visits from a person accompanied by a dog or computerized dog (AIBO) more than visits from the person alone (Kramer et al., 2009). A 12-week animal-assisted intervention for cognitively impaired assisted living residents resulted in less depression and better physical function compared with a reminiscing intervention (Friedmann, et al., 2015). A more complete discussion of human-animal interaction in older adults is included in Chapter 16.

## HAI and Vulnerable Populations

HAI has also been investigated with health outcomes in vulnerable populations, including individuals with autism or other developmental disorders, those who experienced trauma, psychiatric patients, and prison inmates.

### *Individuals with Autism and Developmental Disorders*

Small mammals (hamsters) may act as social buffers and cause improvements in multiple areas of functioning known to be impaired in autism spectrum disorders, namely increased social interaction and communication as well as decreased problem behaviors, autistic severity, and stress (O'Haire et al., 2015). Hippotherapy or equine-assisted therapy has been effective for improving autism symptoms, social cognition, and social communication and reducing hyperactivity and irritability (Gabriels et al., 2015; Harris & Williams, 2017). The introduction of dogs to their household also led to improvements in stress levels of children with autism that reverted once the dog was removed from the residence (Viau et al., 2010). Chapter 18 addresses pet ownership for individuals with developmental disabilities and Chapter 36 addresses the use of horses in therapeutic regimens.

### *Individuals Who Experienced Trauma*

HAI is used in situations of trauma to assist both direct victims of trauma and first responders. The presence of a therapy dog was associated with decreased stress indicators such as salivary immunoglobulin, and heart rate during their forensic interviews for child sexual abuse (Krause-Parello & Friedmann, 2014). A case study of two students who had suffered sexual abuse in childhood provided evidence that extensive hippotherapy, 20-, 30-minute horseback riding sessions, improved coordination, balance, socialization, and self-esteem (Guerino et al., 2015). Psychiatric service dogs assist individuals with post-traumatic stress disorder (PTSD) by responding to their owners' specific symptoms. Evidence supports that these dogs reduce veterans' PTSD symptoms and suicidal behaviors and ideations over up to 18 months (VA, 2019). There is less robust evidence for the use of psychiatric service dogs to improve quality of life, stress, anxiety, depression, sleep, and behavior or other outcomes among individuals with PTSD. Chapter 37 addresses assistance animals in more detail.

### *Psychiatric Patients*

HAI for psychiatric patients is usually investigated in the form of animal-assisted activities (AAAs). A 15-minute animal-assisted therapeutic session with a dog before the procedure was effective at reducing fear, and depression, but not anxiety in psychiatric patients awaiting electroconvulsive therapy (Barker et al., 2003). Animal-assisted therapy using dogs reduced anxiety in hospitalized adults with a variety of psychiatric diagnoses, including psychotic and mood disorders compared to those who underwent recreational therapy sessions

(Barker & Dawson, 1998). Hippotherapy also benefitted women with mental health disorders by increasing their confidence and self-concept and providing social stimulation; these effects led to transferable skills in the rider's lives (Burgon, 2003). Raising farm animals was associated with improved self-efficacy and coping ability in patients with schizophrenia, affective disorders, anxiety, and personality disorders (Berget et al., 2008). Equine-assisted therapy, which did not include horseback riding, was more effective than canine-assisted therapy and conventional therapy in reducing violence-related incidents and aggression-related measures but not non-violence-related incidents in psychiatric inpatients (Nurenberg et al., 2015).

## Prisoners

Many institutions have used animal raising or training or AAA programs as part of their rehabilitation programs without much evidence for efficacy (Mulcahy & McLaughlin, 2013); however, the staff perceives that the prisoners benefit (Mercer et al., 2015). A Florida dog-raising program resulted in decreasing the likelihood of re-arrest for any reason and re-arrest for a new crime within one year of release (Hill, 2016) and a larger study that controlled for several other possible confounders supported decreases in recidivism. A systematic review suggested that prison-based animal-assisted activities improved anxiety, stress, recidivism, and other social variables in male and female inmates (Villafaina-Domínguez et al., 2020). However, most of the studies reviewed demonstrated associations but not causation. In contrast, a more recent study with a comparison group found no differences overall between participants in an eight-week biweekly dog training program in official recidivism outcomes and self-reported criminal behavior, but some differences in outcomes and efficacy in subgroups of inmates (Duindam et al., 2021).

## Possible Negative Effects of HAI

HAI can be physically harmful or stressful, leading to negative physical, psychological, and social health. Among the many health hazards associated with HAI, bites and scratches are the most common and can cause localized infections (Plaut et al., 1996). Safe interactions with animals are the topic of Chapter 28. Animal allergies also are common. Allergists frequently recommend rehoming pets when family members have allergies. Many owners persist in keeping them, leading to increased health burdens and costs. Current research suggests that immune responses may be lessened through early exposure (Ownby & Johnson, 2016; Plaut et al., 1996). Older adults caring for pets, especially cats and dogs, are at risk for falls and these falls can cause fractures, contusions, and abrasions, with severe health consequences (Stevens et al., 2010). Even for younger individuals, HAI can have risks. An untrained dog can pull too hard on the leash causing the walker to trip and fall or have joint injuries. Pets can be carriers of zoonotic diseases that are transmitted to humans; and immunocompromised people are at a higher risk (Esch & Petersen, 2013). COVID-19 raised concerns in pet owners. Pets could get infected with COVID-19 but not necessarily be symptomatic, risking transmission to humans in contact with, although the chances of transmission from pets to humans are low (CDC). Chapter 28 addresses safe interactions with animals and 26 addresses zoonotic diseases in detail.

Dog walkers derive physical benefit from walking their dogs, but in walking their dogs, they can also put themselves in danger. Higher outdoor activity is linked to the perceived crime-related safety of neighborhoods (Foster & Giles-Corti, 2008). Dog owners may put themselves at risk by walking in less safe neighborhoods and at less well-traveled times.

Attachment to pets, while providing a source of social support, can have negative consequences for those who keep them. Attachment to pets can lead owners to behave in ways that endanger the owners' safety and health. During natural calamities and disasters, people refused to evacuate without their pets and, if forced to evacuate, return prematurely to disaster areas to search for their pets (Chadwin, 2017). The US, Congress passed the Pets Evacuation and Transportation Standards Act of 2006 (PETS Act; Pub L No. 109–308, HR 3858) requiring planning for dogs, cats, birds, rabbits, rodents, and turtles. It has not been entirely successful in mitigating owners' inability to evacuate with their pets (Babcock & Smith, 2020). Owners may sacrifice their medical needs for concern for their pets and even put their pet's needs ahead of their own (Friedmann & Son, 2009). Over 20% of pet owners declined to visit family or friends or take a trip out of concern for their pet's welfare (Friedmann et al., 2020). Pet keeping entails an economic and physical burden and the death of a pet is a traumatic emotional experience in many cases equivalent to losing a close family member. Chapters 20 and 21 address these disadvantages associated with pet ownership, respectively. Thus, although humans can benefit from HAI, their adverse effects should be kept in mind as well.

## Gaps in Knowledge and New Directions

Most studies of HAI study dogs, although there are a few studies that include fish, goats, cats, and other domestic animals and even more exotic species, including non-human primates, snakes, and crickets. Studies of animal-assisted interventions largely use therapy dogs or horses as the intervention. Other animals have been successful in isolated instances. For example, goats, hamsters, and cricket raising have been efficacious in the few studies that include them. Most of these studies focus on the health or functional benefits to the people. Much less attention is paid to what actually happens during the interaction and to the stress and behavior of the animals involved. Additional research is required to develop an understanding of the complex interactions, how animal and human characteristics, and the relationship between the different players in the interaction impact the health effects of the interaction.

Studies of the relationship of pet ownership to health also used population-based data, but largely rely on cross-sectional data that relate current pet ownership to current health outcomes. These studies are hampered by the role of health and other factors related to health in the decision to keep a pet, and in the history of the person before the data collection. Longitudinal studies are necessary to learn more about how pet ownership is related to changes in health status over time. Application of advanced statistical techniques, such as propensity score models that incorporate the predictors of pet ownership or linear mixed models that allow time-varying pet ownership and individual starting points (intercepts) for outcomes would strengthen the designs and enhance interpretation of results as evidence of causality. Studies of populations allow examination of the context for the association of pet ownership with health. Pet ownership could promote health in some circumstances and inhibit it in others. The role of pet ownership as a moderator or mediator of other health-related outcomes is an important area of research. For example, if social isolation is related to the use of health services, does pet ownership moderate or mediate that relationship? Does attachment to the pet make a difference in the relationship? Does the type of pet make a relationship? We also must be able to conduct multivariate analyses that allow investigation of whether the relationship differs according to other characteristics of the individual, for example, socioeconomic status and gender.

Most animal-assisted intervention programs involve domestic pets, especially dogs as interventions in individuals' homes or institutional settings. Interactions with other species might be effective for some groups or to address specific needs. For example, caring for insects, was a cost-effective and safe intervention that lowered depression and improved cognitive function in Korean community-dwelling older adults with limited mobility (Ko et al., 2016). In many Western countries, crickets would be considered pests, not pets.

Animal-assisted intervention programs sometimes involve other species, but the efficacy of the different species has not been compared. Typical types of interventions with horses target physical function such as coordination and socialization skills for individuals with disabilities or poor function. Interventions with dogs often target more varied outcomes, including physical function, cognitive function, wellness, anxiety, social interaction, depression, pain, stress, and general well-being. Interventions with small mammals targeted social interaction in children. These distinctions are important in that they identify specific types of targets that can be effectively addressed within each species. Different people have different needs. Specifying desired outcomes and the target populations, as well as the different species and even characteristics within the species, are all crucial to the development and application of animal-assisted interventions. An intervention designed to help depressed or anxious older adults might not be efficacious for children with the need for assistance with social skill development. While there is general agreement that animal-assisted interventions enhance a wide variety of outcomes for many groups of people, additional research is needed and even in the process to learn more about tailoring interventions.

In this same vein, very little information is available on the specifics of animal-assisted therapy or other animal-assisted interventions. There is general agreement that teams that deliver these interventions are trained for the safety of all involved. There is a little standard beyond that. There is tremendous variability in the operationalization of HAI as a therapeutic modality. It's not just the species, but the way the animal and intervention are introduced can matter. For example, programs using reading to animals to improve children's reading might use a dog, but results may differ when using a dog lying beside them that is trained to look at the book rather than when using a dog that is lying on their lap and constantly moving and nuzzling the child seeking attention and interrupting concentration.

Optimal dosing has not been addressed systematically. Of prime importance for therapeutic interventions are the number, frequency, and duration of each session, and the length of the treatment series. Most experimental studies examined only one HAI session and compared it to no intervention or one session of another intervention. Studies of HAI intervention programs include several sessions administered in a series and may compare pre-post session changes for HAI sessions compared with changes with no activity or with another intervention, compare pre-post intervention series changes for HAI intervention series with no activity or with another intervention, or compare a combination of the two. Unfortunately, research has not begun to systematically establish intervention characteristics or dose requirements for outcomes. For example, one program might have five minute long visits to each individual twice a week, another might have ten-minute individual visits weekly, and another may include one hour-long visits with groups of an indeterminate number of individuals weekly. Much of the literature doesn't provide well-defined descriptions of the activities and how they are presented. The introduction of online appendices to research journal articles could be used to better describe interventions and their implementation. There are sparse data on the duration of effects once the end of the treatment. These issues need to be addressed to enhance understanding of the cost and benefit of animal-assisted interventions.

Since animal-assisted interventions are neither a panacea nor feasible to provide for everyone, it is important to understand where they will be most effective and develop guidance for appropriate targets for intervention.

While the research focuses on the health benefits of assistance animals for the people targeted, additional attention to the "other end of the leash" and the other people in the family of the recipients of the family is warranted. More needs to be known about the stress and burden on the animals involved in HAI and how that can be mitigated. Very little is known about the positive or negative impact of service animals on the other people in the handler's home. As the use of service animals proliferates this issue will become more pressing.

## Conclusion

Research and practice have contributed greatly to our understanding of the contributions of HAI to human health and well-being in the past 40–50 years. There's still much we have to learn to maximize the benefits, minimize the risks and support the well-being of people and the animals with whom they interact.

## References

Allen, K., Blascovich, J., & Mendes, W. B. (2002). Cardiovascular reactivity and the presence of pets, friends, and spouses: The truth about cats and dogs. *Psychosomatic Medicine*, *64*(5), 727–739.

Allen, K., Shykoff, B. E., & Izzo Jr, J. L. (2001). Pet ownership, but not ACE inhibitor therapy, blunts home blood pressure responses to mental stress. *Hypertension*, *38*(4), 815–820.

Alonso, Y. (1999). Cardiovascular responses to a pet snake. *The Journal of Nervous and Mental Disease*, *187*(5), 311–313.

Ash, M. G. (2018). Zoological Gardens. In H. A. Curry, N. Jardine, J. A. Secord & E. C. Spary (Eds.), *Worlds of natural history* (pp. 418–432). Cambridge University Press.

Babcock, S. A., & Smith, D. G. (2020). Pets in comprehensive disaster planning: The post–hurricane Katrina experience. *American Journal of Public Health*, *110*(10), 1500–1501.

Baratay, E., & Hardouin-Fugier, E. (2004). *Zoo: A history of zoological gardens in the West*. Reaktion Books.

Barker, S. B., & Dawson, K. S. (1998). The effects of animal-assisted therapy on anxiety ratings of hospitalized psychiatric patients. *Psychiatric Services*, *49*(6), 797–801. https://doi.org/10.1176/PS.49.6.797/ASSET/IMAGES/LARGE/B313T1.JPEG

Barker, S. B., Pandurangi, A. K., & Best, A. M. (2003). Effects of animal-assisted therapy on patients' anxiety, fear, and depression before ECT. *The Journal of Ect*, *19*(1), 38–44.

Beetz, A. M. (2017). Theories and possible processes of action in animal assisted interventions. *Applied Developmental Science*, *21*(2), 139–149.

Berget, B., Ekeberg, Ø., & Braastad, B. O. (2008). Animal-assisted therapy with farm animals for persons with psychiatric disorders: effects on self-efficacy, coping ability and quality of life, a randomized controlled trial. *Clinical Practice in and Epidemiology in Mental Health*, 4 https://doi.org/10.1186/1745-0179-4

Brown, S. E. (2019). *Individuals with disabilities and their assistance animals: A brief history and definitions*. ADA Knowledge Translation Center. Retrieved 03/27 from https://adata.org/legal_brief/individuals-disabilities-and-their-assistance-animals-brief-history-and-definitions

US Bureau of Labor Statistics, & Henderson, S. (2013). *Spending on pets: "Tails" from the consumer expenditure survey*. Beyond the Numbers 2(16) US Bureau of Labor Statistics. https://ecommons.cornell.edu/handle/1813/79421

Burgon, H. (2003). Case studies of adults receiving horse-riding therapy. *Anthrozoös*, *16*(3), 263–276.

Carr, D., Friedmann, E., Gee, N. R., Gilchrist, C., Sachs-Ericsson, N., & Koodaly, L. (2021). Dog walking and the social impact of the COVID-19 pandemic on loneliness in older adults. *Animals 2021*, *11*(7), 1852–1852. https://doi.org/10.3390/ANI11071852

CDC (2020, 11/15/2021). *What you should know about COVID-19 and pets*. Retrieved 03/27 from https://www.cdc.gov/healthypets/covid-19/pets.html

Chadwin, R. (2017). Evacuation of pets during disasters: A public health intervention to increase resilience. *American Journal of Public Health, 107*(9), 1413–1417.

Collis, G.M., & McNicholas, J. (1998). A theoretical basis for health benefits of pet ownership: Attachment versus social support. In C. Wilson, & D. Turner (Eds.), *Companion Animals in Human Health* (pp. 105–122). Sage Publications.

Comba, E. (2010). *Mixed human-animal representations in Palaeolithic art: An anthropological perspective.* IFRAO International Congress-Ariège 2010.

Compitus, K. (2021). *The* biopsychosocial model as a framework for understanding the human–animal bond. In K. Compitus (Ed.), *The Human-Animal Bond in Clinical Social Work Practice* (pp. 15–25). Springer Publishing. https://doi.org/10.1007/978-3-030-87783-5_3

Curl, A. L., Bibbo, J., & Johnson, R. A. (2017). Dog walking, the human–animal bond and older adults' physical health. *The Gerontologist, 57*(5), 930–939.

Duindam, H. M., Creemers, H. E., Hoeve, M., & Asscher, J. J. (2021). Breaking the chains? The effects of training a shelter dog in prison on criminal behavior and recidivism. *Applied Developmental Science, 26*(4), 813–826.

Eddy, T. J. (1995). Human cardiac responses to familiar young chimpanzees. *Anthrozoös, 8*(4), 235–243.

Esch, K. J., & Petersen, C. A. (2013). Transmission and epidemiology of zoonotic protozoal diseases of companion animals. *Clinical Microbiology Reviews, 26*(1), 58–85. https://doi.org/10.1128/CMR.00067-12/ASSET/B575CC88-9B93-40EA-9A66-047CD94CCBB6/ASSETS/GRAPHIC/ZCM9990924070013.JPEG

Esposito, L., McCune, S., Griffin, J. A., & Maholmes, V. (2011). Directions in human–animal interaction research: Child development, health, and therapeutic interventions. *Child Development Perspectives, 5*(3), 205–211.

Fine, A. H., & Friedmann, E. (2018). Involving our pets in relationship building–pets and elder well-being. In L. W. Kaye & C. M. Singer (Eds.), *Social isolation of older adults: Strategies to bolster health and well-being* (pp. 136–148). Springer Publishing. doi: 10.1891/9780826146991.0012

Flom, B. L. (2005). Counseling with pocket pets: Using small animals in elementary counseling programs. *Professional School Counseling,* 469–471.

Foster, S., & Giles-Corti, B. (2008). The built environment, neighborhood crime and constrained physical activity: An exploration of inconsistent findings. *Preventive Medicine, 47*(3), 241–251. https://doi.org/10.1016/J.YPMED.2008.03.017

Francfort, H.-P., Hamayon, R. N., & Bahn, P. G. (2001). *The concept of shamanism: Uses and abuses* (Vol. 10). Akadémiai Kiadó Budapest.

Friedmann, E., Galik, E., Thomas, S.A., Hall, P.S., Chung, S-Y, McCune, S. (2015). Evaluation of a Pet Assisted Living (PAL) intervention for improving functional status in assisted living residents with mild to moderate cognitive impairment: *A pilot study. American Journal of Alzheimer's Disease and Other Dementias,* 30(3):276–89.

Friedmann, E., Gee, N. R., Simonsick, E. M., Studenski, S., Resnick, B., Barr, E., ... & Hackney, A. (2020). Pet ownership patterns and successful aging outcomes in community dwelling older adults. *Frontiers in Veterinary Science, 7,* 293.

Friedmann, E., Katcher, A. H., Lynch, J. J., & Thomas, S. A. (1980). Animal companions and one-year survival of patients after discharge from a coronary care unit. *Public Health Reports, 95*(4), 307–312. http://www.ncbi.nlm.nih.gov/pubmed/6999524

Friedmann, E., Katcher, A. H., Thomas, S. A., Lynch, J. J., & Messent, P. R. (1983). Social interaction and blood pressure: influence of animal companions. *The Journal of Nervous and Mental Disease, 171*(8), 461–465.

Friedmann, E., Locker, B. Z., & Lockwood, R. (1993). Perception of animals and cardiovascular responses during verbalization with an animal present. *Anthrozoös, 6*(2), 115–134.

Friedmann, E., & Son, H. (2009). The human–companion animal bond: How humans benefit. *Veterinary Clinics of North America: Small Animal Practice, 39*(2), 293–326. https://doi.org/10.1016/J.CVSM.2008.10.015

Friedmann, E., Son, H., & Tsai, C. (2010). The animal-human bond: Health and wellness. In A.H. Fine (Ed.), *Handbook on Animal Assisted Therapy, Theoretical Founcations and Guidelines for Practice* (3rd ed., pp. 85–107). Elsevier.

Friedmann, E., Thomas, S. A., Cook, L. K., Tsai, C.-C., & Picot, S. J. (2007). A friendly dog as potential moderator of cardiovascular response to speech in older hypertensives. *Anthrozoös, 20*(1), 51–63.

Friedmann, E., Thomas, S. A., & Son, H. (2011). Pets, depression and long-term survival in community living patients following myocardial infarction. *Anthrozoös, 24*(3), 273–285.

Friedmann, E., Thomas, S. A., Son, H., Chapa, D., & McCune, S. (2015). Pet's presence and owner's blood pressures during the daily lives of pet owners with pre- to mild hypertension, *Anthrozoös, 26*(4), 535–550. https://doi.org/10.2752/175303713X13795775536138

Gabriels, R. L., Pan, Z., Dechant, B., Agnew, J. A., Brim, N., & Mesibov, G. (2015). Randomized controlled trial of therapeutic horseback riding in children and adolescents with autism spectrum disorder. *Journal of the American Academy of Child & Adolescent Psychiatry, 54*(7), 541–549.

Gee, N. R., Crist, E. N., & Carr, D. N. (2010). Preschool children require fewer instructional prompts to perform a memory task in the presence of a dog. *Anthrozoös, 23*(2), 173–184.

Gee, N. R., Griffin, J. A., & McCardle, P. (2017). Human–animal interaction research in school settings: Current knowledge and future directions. *AREA Open 3*(3), 233285841772434–233285841772434. https://doi.org/10.1177/2332858417724346

Gee, N. R., Mueller, M. K., & Curl, A. L. (2017). Human–animal interaction and older adults: An overview. *Frontiers in Psychology, 0*(AUG), 1416–1416. https://doi.org/10.3389/FPSYG.2017.01416

Gee, N. R., & Mueller, M. K. (2019). A systematic review of research on pet ownership and animal interactions among older adults. *Anthrozoös, 32*(2), 183–207. https://doi.org/10.1080/08927936.2019.1569903

Gee, N. R., Rodriguez, K. E., Fine, A. H., & Trammell, J. P. (2021). Dogs supporting human health and well-being: A biopsychosocial approach. *Frontiers in Veterinary Science, 8,* 630465.

Grajfoner, D., Harte, E., Potter, L. M., & McGuigan, N. (2017). The effect of dog-assisted intervention on student well-being, mood, and anxiety. *International Journal of Environmental Research and Public Health, 14*(5), 483.

Guerino, M. R., Briel, A. F., & Araújo, M. d. G. R. (2015). Hippotherapy as a treatment for socialization after sexual abuse and emotional stress. *Journal of Physical Therapy Science, 27*(3), 959–962. https://doi.org/10.1589/JPTS.27.959

Hall, S. S., Gee, N. R., & Mills, D. S. (2016). Children reading to dogs: A systematic review of the literature. *PLoS One, 11*(2), e0149759.

Harris, A., & Williams, J. M. (2017). The impact of a horse riding intervention on the social functioning of children with autism spectrum disorder. *International Journal of Environmental Research and Public Health, 14*(7), 776.

Hill, L. B. (2016). *Becoming the person your dog thinks you are: An assessment of Florida prison-based dog training programs on prison misconduct, post-release employment and recidivism.* The Florida State University.

Hill, L., Winefield, H., & Bennett, P. (2020). Are stronger bonds better? Examining the relationship between the human–animal bond and human social support, and its impact on resilience. *Australian Psychologist, 55*(6), 729–738.

Ingold, T. (2018). One world anthropology. *Journal of Ethnographic Theory, 8*(1–2), 158–171. https://doi.org/10.1086/698315

Ko, H. J., Youn, C. H., Kim, S. H., & Kim, S. Y. (2016). Effect of pet insects on the psychological health of community-dwelling elderly people: A single-blinded, randomized, controlled trial. *Gerontology, 62*(2), 200–209. https://doi.org/10.1159/000439129

Kramer, S. C., Friedmann, E., & Bernstein, P. L. (2009). Comparison of the effect of human interaction, animal-assisted therapy, and AIBO-assisted therapy on long-term care residents with dementia. *Anthrozoös, 22*(1), 43–57.

Krause-Parello, C. A., & Friedmann, E. (2014). The effects of an animal-assisted intervention on salivary alpha-amylase, salivary immunoglobulin a, and heart rate during forensic interviews in child sexual abuse cases. *Anthrozoös, 27*(4), 581–590.

Krause-Parello, C. A., Gulick, E. E., & Basin, B. (2019). Loneliness, depression, and physical activity in older adults: The therapeutic role of human–animal interactions. *Anthrozoös, 32*(2), 239–254.

Kushner, R. F., Blatner, D. J., Jewell, D. E., & Rudloff, K. (2006). The PPET study: People and pets exercising together. *Obesity, 14*(10), 1762–1770.

Levine, G. N., Allen, K., Braun, L. T., Christian, H. E., Friedmann, E., Taubert, K. A., ... & Lange, R. A. (2013). Pet ownership and cardiovascular risk: A scientific statement from the American Heart Association. *Circulation, 127*(23), 2353–2363.

Levinson, B. M. (1972). *Pets and Human Development*. Charles C. Thomas.

Locke, J., & Axtell, J. L. (1968). *The educational writings of John Locke*. CUP Archive.

McCardle, P. E., McCune, S. E., Griffin, J. A., & Maholmes, V. E. (2011). *How animals affect us: Examining the influences of human–animal interaction on child development and human health. Directions in Human–Animal Interaction Research: Child Development, Health and Therapeutic Interventions,* American Psychological Association.

Meehan, M., Massavelli, B., & Pachana, N. (2017). Using attachment theory and social support theory to examine and measure pets as sources of social support and attachment figures. *Anthrozoös, 30*(2), 273–289.

Mercer, J., Gibson, K., & Clayton, D. (2015). The therapeutic potential of a prison-based animal programme in the UK. *Journal of Forensic Practice, 17*(1), 43–54. https://doi.org/10.1108/JFP-09-2014-0031

Mulcahy, C., & McLaughlin, D. (2013). Is the tail wagging the dog? A review of the evidence for prison animal programs. *Australian Psychologist, 48*(5), 370–378.

Nekolný, L., & Fialová, D. (2018). Zoo tourism: What actually is a zoo? *Czech Journal of Tourism, 7*(2), 153–166.

Nurenberg, J. R., Schleifer, S. J., Shaffer, T. M., Yellin, M., Desai, P. J., Amin, R., ... & Montalvo, C. (2015). Animal-assisted therapy with chronic psychiatric inpatients: Equine-assisted psychotherapy and aggressive behavior. *Psychiatric Services, 66*(1), 80–86. https://doi.org/10.1176/APPI.PS.201300524/ASSET/IMAGES/LARGE/APPI.PS.201300524F1.JPEG

O'Haire, M. E., McKenzie, S. J., Beck, A. M., & Slaughter, V. (2015). Animals may act as social buffers: Skin conductance arousal in children with autism spectrum disorder in a social context. *Developmental Psychobiology, 57*(5), 584–595.

Ownby, D., & Johnson, C. C. (2016). Recent understandings of pet allergies. *F1000Research, 5*. doi: 10.12688/f1000research.7044.1

Pendry, P., Carr, A., Roeter, S. M., & Vandagriff, J. L. (2018). Experimental trial demonstrates effects of animal-assisted stress prevention program on college students' positive and negative emotion. *Human-Animal Interaction Bulletin*. https://doi.org/10.1079/hai.2018.0004

Pendry, P., Carr, A. M., Vandagriff, J. L., & Gee, N. R. (2021). Incorporating human–animal interaction into academic stress management programs: Effects on typical and at-risk college students' executive function. *AREA Open 7*. https://doi.org/10.1177/23328584211011612

Plaut, M., Zimmerman, E. M., & Goldstein, R. A. (1996). Health hazards to humans associated with domestic pets. *Annual Review of Public Health, 17*, 221–245. https://doi.org/10.1146/ANNUREV.PU.17.050196.001253

Rockett, B., & Carr, S. (2014). Animals and attachment theory. *Society & Animals, 22*(4), 415–433.

Serpell, J. A. (2010). Animal-assisted interventions in historical perspective. In A.H. Fine (Ed.) *Handbook on Animal-Assisted Therapy* (3rd edition, pp. 17–32). Elsevier. https://doi.org/10.1016/B978-0-12-381453-1.10002-9

Serpell, J. A. (2015). Animal-assisted interventions in historical perspective. A. H. Fine (Ed), *Handbook on Animal-Assisted Therapy* (4th Edition, pp. 11–19) Elsevier.—. https://doi.org/10.1016/B978-0-12-801292-5.00002-X

Siegel, J. M. (1990). Stressful life events and use of physician services among the elderly: The moderating role of pet ownership. *Journal of Personality and Social Psychology, 58*(6), 1081.

Souter, M. A., & Miller, M. D. (2007). Do animal-assisted activities effectively treat depression? A meta-analysis. *Anthrozoös, 20*(2), 167–180.

Stevens, J. A., the, S., & Haileyesus, T. (2010). Dogs and cats as environmental fall hazards. *Journal of Safety Research, 41*(1), 69–73.

Thorpe, R. J., Kreisle, R. A., Glickman, L. T., Simonsick, E. M., Newman, A. B., & Kritchevsky, S. (2006). Physical activity and pet ownership in year 3 of the Health ABC study. *Journal of Aging and Physical Activity, 14*(2), 154–168.

Tribe, A. (2004). Zoo tourism. In K. Higgenbottom (Ed.) *Wildlife Tourism: Impacts, Management and Planning* (pp. 35–56). Common Ground Publishing.

Tsai, C.-C., Friedmann, E., & Thomas, S. A. (2010). The effect of animal-assisted therapy on stress responses in hospitalized children. *Anthrozoös, 23*(3), 245–258.

VA. (2019). *Can service dogs improve activity and quality of life in veterans with PTSD?* Retrieved 10/29 from https://clinicaltrials.gov/ct/show/NCT02039843?order=1

Viau, R., Arsenault-Lapierre, G., Fecteau, S., Champagne, N., Walker, C.-D., & Lupien, S. (2010). Effect of service dogs on salivary cortisol secretion in autistic children. *Psychoneuroendocrinology, 35*(8), 1187–1193.

Villafaina-Domínguez, B., Collado-Mateo, D., Merellano-Navarro, E., & Villafaina, S. (2020). Effects of dog-based animal-assisted interventions in prison population: A systematic review. *Animals 2020, 10*(11), 2129–2129. https://doi.org/10.3390/ANI10112129

Virues-Ortega, J., Pastor-Barriuso, R., Castellote, J. M., Poblacion, A., & de Pedro-Cuesta, J. (2012). Effect of animal-assisted therapy on the psychological and functional status of elderly populations and patients with psychiatric disorders: A meta-analysis. *Health Psychology Review, 6*(2), 197–221.

Wilson, C. C. (1991). The pet as an anxiolytic intervention. *Journal of Nervous and Mental Disease, 179*(8), 482–489.

# 6
# CHALLENGES AND OPPORTUNITIES IN ESTABLISHING THE EVIDENCE BASE FOR THE IMPACT OF HUMAN-ANIMAL INTERACTIONS ON HUMAN HEALTH AND WELL-BEING

*Kerri E. Rodriguez, Erin Flynn, Jaci Gandenberger, Marisa Motiff, and Kevin N. Morris*

The field of human-animal interaction (HAI), including research on pet ownership and animal-assisted interventions (AAIs), has seen tremendous growth in the past few decades. With an increase in funding and educational opportunities, as well as support from health providers, practitioners, and policymakers, research describing the effects of HAI on both human and animal health and wellbeing continues to expand and improve in quality (Griffin et al., 2019; O'Haire, 2010). Although HAI research portrayed in the media is highly skewed toward the positive (Herzog, 2011), increasing attention from the media and the public has also contributed to the increasing popularity of the field. As the field of HAI continues to grow, many have argued that the field is in critical need of more evidence-based research in order to compare efficacy of different approaches, inform best practice standards, and guide present and future policy (Fine et al., 2019; McCune et al., 2014; Wilson & Barker, 2003). By doing so, the field can move toward establishing an evidence-based practice (EBP) in which there is consistent and reliable scientific evidence showing for whom and in what contexts HAI can improve human health and wellbeing outcomes while also maintaining the wellbeing of animals involved. However, challenges including methodological heterogeneity related to research design, measurement, and accounting for the inherent variability in HAIs will need to be addressed if the field of HAI is to realize its potential of being established as an EBP. Below, we outline each of these considerations to discuss both challenges and opportunities in conducting HAI research to contribute to the evidence base of the field.

## Study Designs in HAI Research

One of the most fundamental research questions in establishing the evidence base of HAI research is quantifying the efficacy of animal interactions in improving human health outcomes. A majority of early positive research studies in the field of HAI have aimed to evaluate this preliminary efficacy by conducting single-group, pre-post design studies. While "groundbreaking in hypothesis generation" (Schuck et al., 2018), these correlational studies were unable to provide causational, conclusive findings on the efficacy of HAI due to lack of control groups and randomization. To continue to build on these foundational studies and improve the evidence base of HAI, more rigorous studies have incorporated randomized designs and carefully selected control groups to establish greater causality in research findings (e.g., Friedmann & Gee, 2019; Kazdin, 2019).

One study design that has greatly strengthened the evidence base in the field is the randomized controlled trial (RCT) design. Randomization balances participant characteristics between control and treatment groups, which minimizes the effect of confounding variables on outcomes by isolating the effects of the intervention. Another hallmark of an ideal RCT is the blinding of participants, providers, and outcome assessors to which participants are allocated to the treatment and control groups, which reduces biases in outcomes. One important consideration for HAI research is that blinding whether a participant is receiving an AAI or not is often impossible or impractical. A 2014 review of 11 RCTs evaluating the efficacy of animal-assisted therapies (AATs) found that only 4/11 studies blinded participants to treatment groups, and only 3/11 studies blinded providers and outcome assessors to treatment groups (Kamioka et al., 2014). Another consideration for RCTs in the field of HAI is that in many cases, it is not always feasible or ethical to randomize a participant to receive animal interaction. For example, it is extremely difficult to truly randomize a large group of individuals to owning or not owning a pet. For these reasons, RCTs have been less widespread in HAI than in other intervention sciences.

While some have referred to RCTs as the "gold-standard" of intervention research, others have pointed out that RCTs are not always superior to other types of evidence (Grossman & Mackenzie, 2005). Rather, different study designs may be more appropriate for different research questions and situations. For example, AAIs in practice settings often involve highly individualized, complex, and dynamic relationships between people, animals, and the environment. Although AAIs can be designed for group participation, they are often built for the unique needs and goals of individuals. Study designs that focus on the individual accurately reflect the realistic context and personalization utilized in AAI implementation. In addition, for research questions that are interested in understanding the role of HAI in human development, person-centered inquiry can allow for understanding the effects at individual levels that can inform group level findings (Geldhof et al., 2021). Therefore, while RCTs have been steadily increasing in the study of AAIs (Maujean et al., 2015), other types of research designs including quasi-experimental, observational, population-based, and qualitative methodologies are just as important and necessary in building the field's evidence base.

## *Control Groups*

Regardless of the study design, an important consideration in designing a research study in HAI is careful consideration of the control or comparison group. In pet-ownership research, limitations in control group options have posed a challenge to establishing an evidence base. Most studies have used cross-sectional designs to compare health outcomes among those who own pets to those who do not (Scoresby et al., 2021). However, these cross-sectional designs can only offer correlational associations, often fail to measure and account for confounding variables, and rely on convenience samples with a high degree of selection bias (Friedmann & Gee, 2019; Herzog, 2011). To establish a more rigorous evidence base, cross-sectional designs are using more sophisticated control groups (e.g., comparing pet owners to those who plan on getting a pet; Martin et al., 2021), including more appropriate demographic and socioecological covariates, and using propensity score matching to account for confounders (e.g., Miles et al., 2017).

In AAI research, studies have largely been dominated by having an inactive control group, in which the comparison condition does not receive any intervention or treatment. For example, a study may compare participants who receive AAI to participants who do not receive AAI at all, or who are on the waitlist to participate in AAI in the future (e.g., Pedersen et al., 2012). This waitlist control design provides an ethical advantage and improves ease of recruitment as all participants are allowed access to AAI, even if delayed. In studies in which participants are actively applying to engage in an AAI program (e.g., equine-assisted therapy or receiving a service dog), this design also ensures that both the treatment and control groups are equally motivated to engage in the animal program before randomization.

However, the use of inactive control designs has been criticized in reviews of AAI research as they are unable to account for nonspecific treatment effects, or effects that are not specific to the intended AAI treatment (Marino, 2012). Specifically, an inactive control group design cannot determine if the treatment effects observed are due to the actual animal itself, or due to other components of the intervention such as novelty, excitement, or social interaction with peers or adults (such as the animal handler). Secondly, inactive controls are unable to account for treatment expectations. Thus, these studies cannot determine if participants receiving AAI improve because of the intervention itself, or because they are aware that they are receiving an intervention and have expectations of improvement. For example, a participant who knows that they are in the AAI group may intentionally or unintentionally exhibit or report better outcomes because they believe they should, or because they want the study to have positive results because of their own affinity or inherent belief that AAIs can be beneficial. Some strategies have been used to combat this bias, including incorporating objective physiological outcomes that are not as subject to bias, or even deceiving participants about the "real" reason for the animal's presence to limit potential expectation effects (e.g., Wagner et al., 2021). However, this is not possible to do in all research designs or ethical protocols.

To overcome the limitations of inactive control groups, HAI researchers are increasingly using active control group designs, in which the comparison condition receives a different variation of the AAI intervention or a different kind of intervention altogether. In contrast to inactive control group designs, active control groups can account for both treatment expectation biases and aid in understanding the active components of the intervention. For example, researchers may compare participants randomized to receive individual or group therapy incorporating animals to participants randomized to receive the same therapy without animals (e.g., Schuck et al., 2018). In this context, the experimental design attempts to control for as many factors as possible (e.g., social interaction, physical environment) to isolate precise treatment effects due to the animal's presence. Other studies may compare outcomes of those receiving AAI to those receiving another intervention such as music therapy or art therapy (e.g., Gebhart et al., 2020). With these designs, researchers can determine the efficacy of AAI as a complementary or additive component of an already-established intervention.

Studies using active control groups are also important in examining the mechanisms and active components of AAI to help answer *why* animals are beneficial for human health and wellbeing and under what conditions. For example, researchers may compare participants who receive AAI sessions with a live animal to control participants who receive similar sessions with a stuffed or robotic version of the animal (e.g., Gabriels et al., 2015). In this case, novelty, social peer interaction, and activity can be controlled for, which allows researchers to isolate the effect of interacting with a live animal on outcomes. In an even more complex dismantling design, a 2015 study compared participants receiving equine-assisted psychotherapy to three control conditions: human-only psychotherapy (to isolate the effect of an animal), canine-assisted psychotherapy (to isolate the effect of the horse versus the dog), and no psychotherapy at all (to isolate the absolute effect of any treatment; Nurenberg et al., 2015). Studies using these kinds of thoughtful and theoretically driven active control groups are crucial in advancing the understanding of the mechanisms and potential therapeutic components of AAIs (Kazdin, 2019).

Controlled observational designs with active control groups have also been instrumental in understanding the underlying theoretical mechanisms of how HAI contributes to positive human health outcomes. Many of these studies have assessed how animal interaction may confer physiological benefits (e.g., influencing neurochemical or cardiovascular activity) during times of stress. For example, a 2019 study aimed to isolate the distinct contributions of touching an animal on physiological reactivity by assessing salivary cortisol levels of participants randomly assigned to hands-on petting of animals, waiting in line while observing others petting animals, viewing images of animals, or waiting silently (Pendry & Vandagriff, 2019). In this active control group design, researchers can directly quantify the indirect contributions of socializing with peers, viewing animals' exposure, or neither on physiological outcomes. Other studies have contributed to understanding how an animal's presence may serve to buffer subjective and objective measures of stress and arousal in a laboratory setting (e.g., Crossman et al., 2020). A benefit of these studies is that the experimental setting can be more highly controlled by the researcher to isolate the effect of the live animal interaction. However, these designs may lack ecological validity as most HAIs don't occur in a lab setting; HAIs often happen in complex settings where multiple variables are interacting. Nonetheless, studies examining these mechanistic underpinnings of HAIs are crucial for the continued development of the evidence base.

## Measurement in HAI Research

In addition to rigorous research designs and control groups, another important consideration for establishing the evidence base of HAI lies in measurement and assessment. No matter how strong the research design, results will ultimately be reliant on the reliability (i.e., consistency of measurement) and validity (i.e., accuracy of measurement) of the outcome measures used. In attempting to quantify the role that animals play in human health and wellbeing, a large question that presents both a challenge and opportunity is defining what is meant by "health and wellbeing." Whether it be on a personal or community level or on a psychological, social, or physiological level, HAI research conceptualizes health and wellbeing in a variety of ways. In addition, there is a growing focus in the field on assessing the mutual health benefits of HAIs on both the humans and animals involved (McCune et al., 2014). Therefore, researchers must face many decisions in determining *what* to measure and *how* to measure it when building an evidence base of HAI research.

When choosing *what* to measure, the choice of outcomes is a critical first step. Outcomes should be chosen based on the researcher's expectations of what human or animal aspects may be sensitive to change from HAI, to what degree or magnitude they are expected to change, and over what period of time the change may occur (Coster, 2013). When possible, these decisions should be guided largely by theory, which provides meaningful parameters to guide research by identifying concepts to describe what we observe and explaining relationships between concepts (Haardörfer, 2019). This allows researchers to systematically test assumptions to examine phenomena, identify solutions to observed problems, or predict outcomes. Anchoring conceptualization and measurement of outcomes in theory also strengthens the study design and increases the transparency of the research (Kazdin, 2019).

Once it has been determined *what* to measure, researchers must then determine *how* to measure their outcomes. It is critical that any measures selected are valid representations of the construct of interest (Marino, 2012) and have established reliability and validity (Rodriguez et al., 2018). Many have argued that continued standardization of assessment is critically needed to advance the rigor of HAI research (Fine et al., 2015; Herzog, 2015). Wherever possible, use of common, reliable, and valid measures will allow the field of HAI to compare and replicate studies to advance its empirical evidence base (Asendorpf et al., 2016). When researchers need to adapt or create a new assessment tool, they should rigorously test and report its psychometric properties and provide detailed protocols for its use. It is also beneficial to use multiple forms of assessment to allow for triangulation of the data and to account for the limitations inherent to any one mode of measurement. Use of such multimodal assessment will increase the validity of study findings and yield a more comprehensive understanding of the construct under study (Eid & Diener, 2006).

Finally, researchers must determine the timing of outcome measurement, which may include short-term outcomes, long-term outcomes, or both (Wilson & Barker, 2003). Quantifying the short-term effects of HAI is important for establishing initial efficacy and feasibility, testing mechanisms of action, and informing future directions of research. However, quantifying the long-term effects of HAI is also essential to assess whether HAI supports progress toward broader health or developmental goals while establishing the relevance of the field to policymakers, leaders, and stakeholders. Therefore, both short-term and long-term measurement of outcomes are beneficial to examining health and wellbeing outcomes of HAI.

## *Questionnaires*

The most frequently used form of subjective assessment in the field of HAI is questionnaires (O'Haire et al., 2018). Questionnaires can yield useful information about a wide range of personal experiences including individual attitudes, values, emotions, and perceived health. In pet ownership research, questionnaires are a valuable tool to learn about animals through the perspectives of people who interact with them, including perceptions of an animal's health, wellbeing, temperament, or behavioral traits. In AAI research, questionnaires are important tools to measure self-reported or proxy-reported health and wellbeing outcomes such as anxiety/depression, self-esteem, perceived pain, or positive and negative affect.

When using questionnaires, researchers should prioritize using standardized instruments with established psychometric properties to reduce systematic error and bias (Wilson & Barker, 2003). Consistent use of valid and reliable forms of measurement across studies is critical to the field's ability to conduct meta-analyses that allow for direct comparisons to be made across studies involving different populations and interventions (Serpell et al., 2017). Wherever possible, investigators should use existing standardized assessments developed in other fields because it allows us to understand the efficacy of HAI in relation to established interventions and will increase the generalizability and clinical utility of outcomes. Where development of new measurement tools or adaptation of existing tools is needed, investigators should be guided by a strong theoretical basis and a clear research question and should prioritize psychometric evaluation of the assessment tool prior to use. Finally, researchers using questionnaires should be cognizant of the risk of bias that can arise from subjective measures and take steps to limit this potential bias. Some questionnaires have versions validated for multiple response groups, such as child, parent, or teacher-report. Using multiple versions allows for the assessment of a phenomenon from multiple perspectives, which yields complementary data and can strengthen conclusions drawn from the analysis.

## *Qualitative Methods*

Another form of subjective measurement used often in the field of HAI is qualitative methods. Integrating qualitative methodology into a study can provide rich descriptive data that hold a high level of ecological validity beyond standardized measures. Qualitative questionnaires, interviews, or open-ended questions may allow researchers to more fully investigate the complexity of human phenomena, especially those centering around real-world social and ethical issues (Gough & Lyons, 2016). Indeed, HAI studies have used qualitative methodology to capture rich details in thoughts and experiences surrounding topics such as how pets may contribute to suicidal protection (Young et al., 2020) or the role of pets for sexual and gender minority adults (McDonald et al., 2022). Qualitative studies are also useful in describing *why* HAIs make a difference in a participant's life by gathering nuanced information about participants' connections to animals and/or their pets. Qualitative work allows researchers to identify themes of experience mentioned by multiple participants, which can provide valuable information to build on hypothesized theories of HAI such as social support and attachment (e.g., Hartsell & Zolnikov, 2023).

There are many types of qualitative research designs that can be used depending on the goal of the study. For example, if a researcher is attempting to establish a new theory in the field to describe why animals are impactful during an AAI, a grounded theory design could be used (e.g., Kuzara et al., 2019). If the researcher aims to understand the core experience of individuals newly placed with a service dog and how they may bond with and benefit from their service dogs, a phenomenological approach could be used (Krause-Parello & Morales, 2018). In these designs, researchers should be cognizant of their own biases as they create narratives from the data collected. Methodological rigor is improved by engaging in processes of reflexive assessment of the researcher's own positionality throughout the study and, in some cases, the use of multi-rater coding and inter-rater reliability (Morse, 2015). Other studies have used rich creative processes as data, such as drawings, storytelling, and photovoice to capture personal thoughts and experiences with animals (e.g., KoÃ & Arslan, 2016). This kind of measurement is particularly valuable in HAI research involving populations that have limited verbal and abstract thinking capabilities, such as young children or individuals with intellectual disabilities. Taken together, qualitative measurement should be considered just as crucial to adding to the evidence base of HAI as standardized self-report measurement.

## *Behavioral Observation*

When measuring more subjective constructs that are prone to bias, assessment can be strengthened by combining questionnaires or qualitative data with more objective forms of assessment such as direct behavioral observation. Although behavioral coding can be a complex and resource-intensive process and has thus been less widely used in HAI research (O'Haire et al., 2018), behavioral observation is advantageous when the research question centers on some form of overt human behavior or when behaviors cannot be reliably assessed through self-report (Chorney et al., 2015). Behavioral observation can also strengthen study findings by capturing multiple facets of a construct, including the often unmeasured "I" in HAI (Griffin et al., 2019). Behavioral observation can either be conducted live, in which the observer is physically present to observe the interaction, or via a recording, where the observer watches the interaction on a videotape. While a live observer has the advantage of viewing the entire environment and adjusting their viewpoint as necessary to observe interactions, disadvantages include limitations to conducting reliability analyses, inability to code multiple individuals separately in a group setting, and a potential to miss behavioral subtleties. On the other hand, recorded behavioral observations can be beneficial for allowing several observers to code the same videotapes to ensure reliability and allows for capturing of subtle behaviors (i.e., by slowing down or replaying footage), but can be limited by poor audio or visual resolution and inflexible viewpoints.

Observation is also beneficial to measure both human- and animal-related aspects of AAI, including assessing animal welfare, quantifying frequencies and durations of certain HAIs, and as a measure of treatment fidelity. Where research questions focus on constructs related solely to human behavior, researchers should choose a standardized behavior coding tool that has established validity and inter-rater reliability (Cone, 1999). Where the behaviors of interest are specific to animals, investigators should use ethograms developed in the field of animal behavior. Often, measurement of HAI focuses on interactions between humans and animals and requires tools beyond those that are species-specific. In these cases, researchers have developed their own coding systems (e.g., Grandgeorge et al., 2015). As with questionnaires, adapting standardized behavioral assessments developed in other fields for HAI research may bolster theoretical understanding of HAI.

## *Physiology*

Another objective form of measurement that can be combined with subjective measures to greatly strengthen the robustness of an HAI study is physiological assessments. Physiological assessments can not only quantify internal states of humans and animals to complement self-report or behavioral observation, but can also be used to assess the potential biopsychosocial mechanisms of HAIs (Gee et al., 2021). Recent innovations have made the measurement of physiological activity easier and less invasive (Dreschel & Granger, 2016) contributing to the increasing popularity of these measures in HAI studies (O'Haire et al., 2018). Given the improvement in affordability, accessibility, and reliability of non-invasive physiological assessments and the significant recent focus on the underlying mechanisms of how animal interaction may benefit human health and wellbeing, HAI researchers should continue to incorporate more physiological assessments in studies (Griffin et al., 2019; Rodriguez et al., 2018). However, researchers should be aware of the methodological and logistical intricacies involved in physiological data collection. Research involving physiological assessment requires careful planning, knowledge of biomarker-specific best practices in measurement, and appropriate statistical analyses to account for between-individual physiological variability.

## **Accounting for Variability in HAI**

Beyond research design and measurement, a final consideration in establishing the evidence base in HAI research is the high degree of heterogeneity across the field. First, research is conducted with a wide variety of human and animal populations. On the animal side, while dogs and horses are the most common animals incorporated into AAIs, studies also evaluate outcomes from interactions with a variety of other species from guinea pigs to farm animals (Charry-Sánchez et al., 2018; Nieforth et al., 2021). Even in research involving only one species such as dogs, there are differences in animals' temperaments, rearing and training histories, and physical characteristics. Currently, many studies do not report on these specific variables, despite research suggesting these characteristics may influence AAI outcomes (Rodriguez et al., 2021).

On the human side, the effects of HAI have been studied across the human lifespan and for a myriad of health challenges. For example, HAI studies feature populations ranging from typically developing children, to nursing home residents with dementia, to military veterans with post-traumatic stress disorder, and more. Variability in human subjects in HAI research has rarely been taken sufficiently into account. While demographics such as gender identity and age are often controlled for in HAI research, other characteristics such as cultural/racial identity, socioeconomic status, familial structure, other forms of social support, attitudes toward animals, and current/past animal experiences are less often measured or controlled for. In fact, several studies on pet ownership have found that controlling for these confounding variables can significantly impact findings (Mueller et al., 2021b). Measuring and quantifying the role of these individual differences is important in understanding the question of *for whom* HAI is most beneficial, which has not yet been well-studied (Geldhof et al., 2021). In addition, studies on outcomes of HAI have been extremely limited in diversity, with most participants being White, cis-gendered, and female-identifying which can limit the validity of findings (Herzog, 2021). When feasible, HAI studies should aim to recruit a diverse and representative population and be careful to measure and report detailed characteristics of participants that may be confounding to outcomes.

Another source of variability in the field is the nature of the "interaction" in HAI. In AAI research, there are significant differences across interventions being studied including the activities engaged in, duration and dosage of animal contact, and personnel involved. For example, in a systematic review of AAIs for children with autism, intervention duration ranged from one session to 80 sessions, with total animal contact time ranging from 3 to 100 hours (Nieforth et al., 2021). Further, even within a single program, AAIs can vary by clients' individual needs and are not often manualized or standardized. In some cases, that is seen as part of the appeal of AAIs; they can be individualized to each participant (Gandenberger et al., 2023). It is also, to an extent, unavoidable with AAIs; unlike a pill, the timing or dose of an interaction with another living being cannot be perfectly controlled, particularly within AAIs in which the animal's welfare and preferences must be prioritized. In research on the impacts of pet ownership for human health and wellbeing, there is significant variability in the quality and quantity of the interactions between people and their pets. Research rarely accounts for considerations such as the pet "parenting" styles, frequency and type of training that owners engage in with their pet, or the attachment styles between the owner and their pet (McNicholas et al., 2005; Rodriguez et al., 2021). In addition, self-selection biases and other unaccounted for confounding variables limit conclusions drawn from pet owner vs non-pet owner comparisons (Chur-Hansen et al., 2010). Pet owners and non-pet owners may be systematically different in sociodemographic and contextual factors (Mueller et al., 2021b), leading to inherent population variability that should be carefully accounted for in research design and participant recruitment.

## Future Directions

Thus far, we have described several methodological considerations that HAI researchers face when aiming to conduct evidence-based research in the field. As the field continues to grow and improve in methodological rigor, below we outline three specific future directives that will be necessary to continue to strengthen the evidence base of HAI: synthesis of research findings, manualization and treatment fidelity, and facilitating transparent reporting across studies.

## *Synthesis of Research Findings*

Continued, periodic reviews of findings in the field are incredibly important to continue to strengthen the evidence base of HAI research. While narrative reviews have been instrumental in exploring initial patterns of findings across studies, meta-analytic reviews are necessary to estimate pooled effects from several studies and improve the precision of estimates from the effects of HAI. Meta-analyses are also essential to reaching more definitive conclusions about the processes, mediators, and moderators of HAI, including participant, animal, and intervention characteristics (e.g., Nimer & Lundahl, 2007). Significant moderators identified in meta-analyses are critical in facilitating the continued translation of research to practice for clinicians and mental health professionals (Esposito et al., 2011).

While beneficial, meta-analyses have been difficult to conduct in the field due to the significant heterogeneity observed across study interventions, outcomes and assessment tools, and assessment time points. In a 2021 systematic review of 54 studies measuring the impact of pets on mental health, 75 different scales were used to measure the outcome of "mental health," which prevented authors from making any cross-study comparisons (Scoresby et al., 2021). Therefore, when possible, researchers should attempt to replicate measures that have been previously promising in other studies to facilitate meta-analytic efforts. In addition to differences in measurement and outcomes, meta-analyses must also consider variation in research designs when pooling effects. For example, pooling both active and inactive control group designs may inaccurately estimate effects and lead to incorrect conclusions. Therefore, researchers must carefully select comparable studies to facilitate meaningful cross-study comparisons that can aid in estimating reliable effects from HAIs.

## *AAI Manualization and Treatment Integrity*

Although high-quality studies are increasingly adding to the evidence base of understanding the efficacy of AAIs, studies in the field have greatly lacked in reporting the intervention details surrounding manualization and treatment integrity, both of which are essential in establishing an EBP. Manualizing the detailed protocol of a given AAI not only allows for the intervention to be replicated across research and practice, but also allows for the measurement of treatment integrity, or the degree to which an intervention was delivered as intended. Carefully measuring and reporting treatment integrity is critical for determining whether null or variable effects from an AAI study are due to the lack of effects of the intervention itself or due to the inadequate delivery of the intervention (Kazdin, 2019; Schoenwald et al., 2011). Yet, treatment integrity is rarely measured or reported in AAI studies. For example, in a 2021 systematic review of RCTs assessing outcomes from AAIs for students in higher education, none of the 11 RCTs assessed reported any sort of monitoring or measuring of the intervention's integrity (Parbery-Clark et al., 2021). The reasons for this omission may certainly be due to the flexible nature of AAIs, but may also include barriers such as a lack of time and resources, a lack of knowledge about treatment fidelity and manualization importance, or the absence of reporting requirements for these details (Perepletchikova et al., 2009). Future steps for the field thus may include allocating funding and resources for manualization and treatment integrity procedures, implementing specific guidelines surrounding treatment integrity in AAI research, and encouraging reporting of this information in funding applications or manuscript submissions.

## *Transparent Reporting*

A final future direction in the field of HAI research is increasing the transparency of reporting findings across studies. Both in the field of HAI and across many other fields in the social sciences, there is a tendency that positive findings are more likely to be published than null or negative findings, a phenomenon commonly referred to as the "file drawer" effect (Franco et al., 2014; Herzog, 2015). In HAI research, there can be a bias toward not sharing findings in which animal interaction is associated with no effect (or even a negative effect) on human health outcomes. This publication bias presents a significant challenge to the growth of the field for many reasons. First, understanding when HAI does and does *not* lead to benefits is both equally important for establishing the evidence base of the field and understanding therapeutic mechanisms of action (Rodriguez et al., 2021). Second, in situations in which null or negative findings are omitted from systematic reviews or meta-analyses, this can lead to inflated effect estimates that inhibit the field's ability to make valid and translatable conclusions. Lastly, HAI is a field characterized by significant public and media interest in findings that confirm perceptions that animals, especially pets, are beneficial for our health and wellbeing. Facilitating transparent reporting and intentionally publicizing null findings can help establish a more accurate public perception of the areas in which animals may, and may not, influence health outcomes. Unfortunately, null findings both in HAI and across other fields are more difficult to get published which further contributes to the problem (Stern & Simes, 1997). Although recent years have seen an increasing willingness from journals and researchers to publish null findings in the field (e.g., Mueller et al., 2021a) and pre-print services are decreasing barriers to disseminating findings, continued effort is necessary to address publication bias in the field of HAI.

In addition to publication bias, outcome reporting bias also poses a significant threat to the growth of the evidence base in the field of HAI. This bias occurs when authors measure many outcomes or timepoints in a study, but selectively pick and choose which results to publish on the basis of which were positive. This may occur due to a hesitancy to publish null findings as referenced above, or potentially due to pressure from funding agencies or organizations. To increase transparent reporting, preregistration of study outcomes, a-priori hypotheses, and planned analyses are increasingly becoming the norm across many fields of research (Nosek et al., 2018). Although clinical trials have long been required to preregister study details due to funding and regulatory reporting processes (e.g., clinicaltrials.gov), databases such as the Open Science Framework are increasingly allowing for preregistration for non-RCT studies that HAI researchers can take part in. Outcome reporting bias can also occur in the form of an abstract reporting bias, in which authors highlight only the positive outcomes in the manuscript's abstract or title without specifying the null or negative outcomes This can lead to skewed conclusions made about the article from readers and members of the media, which may overinflate the benefits of HAIs for human health (Herzog, 2015).

## Conclusion

While the field of HAI has faced obstacles, the quality and quantity of research studies in the field are increasing. Opportunities for unique studies that explore the interactions between humans and animals and their impact on human health and wellness continue to expand with the development of new technologies and incorporation of expertise from other fields. We have discussed how establishing the evidence base of HAI requires close attention to methodological details, including research design, control groups, and assessment, as well as strategies that reduce bias and improve generalizability of findings. To continue expanding the reputability of the field and create EBPs, we suggest continued cross-study syntheses of findings, a more widespread use of manualization and treatment integrity in interventions research, and improvements to facilitating transparent reporting.

## Discussion Questions

1. What are some of the key differences between an active control group and an inactive control group in HAI research?
2. What might qualitative and quantitative methods uniquely contribute to advancing the evidence base of the HAI field?
3. When choosing how to measure outcomes from HAI, what factors should researchers consider when choosing to use questionnaires, behavioral observation, qualitative methods, or physiological methods?
4. How are RCTs contributive to the field of HAI and how might the limitations of this study approach be addressed through other types of designs?
5. Why is transparent reporting of outcomes especially important in the field of HAI?

## References

Asendorpf, J. B., Conner, M., De Fruyt, F., De Houwer, J., Denissen, J. J., Fiedler, K., ... & Wicherts, J. M. (2013). Recommendations for increasing replicability in psychology. *European Journal of Personality, 27*(2), 108–119.

Charry-Sánchez, J. D., Pradilla, I., & Talero-Gutiérrez, C. (2018). Animal-assisted therapy in adults: A systematic review. *Complementary Therapies in Clinical Practice, 32*, 169–180.

Chorney, J. M., McMurtry, C. M., Chambers, C. T., & Bakeman, R. (2015). Developing and modifying behavioral coding schemes in pediatric psychology: A practical guide. *Journal of Pediatric Psychology, 40*(1), 154–164.

Chur-Hansen, A., Stern, C., & Winefield, H. (2010). Commentary: Gaps in the evidence about companion animals and human health: Some suggestions for progress. *International Journal of Evidence-Based Healthcare, 8*(3), 140–146.

Cilekciler, N., K. & Arslan, R. (2016). The concept of "pets" as perceived by 4-8 year-old children, Contemporary Educational Researches Journal. *6*(4), 154–160.

Cone, J. D. (1999). Observational assessment: Measure development and research issues. In P. C. Kendall, J. N. Butcher, & G. N. Holmbeck (Eds.), *Handbook of research methods in clinical psychology* (2nd ed., pp. 183–223). John Wiley & Sons Inc.

Coster, W. J. (2013). Making the best match: Selecting outcome measures for clinical trials and outcome studies. *American Journal of Occupational Therapy, 67*(2), 162–170.

Crossman, M. & Herzog, H. (2019). The research challenge: Threats to the validity of animal-assisted therapy studies and suggestions for improvement. In Fine, A. (Ed.), *Handbook on animal-assisted therapy: Foundations and guidelines for animal-assisted interventions* (5th ed., pp. 474–485). Elsevier.

Crossman, M. K., Kazdin, A. E., Matijczak, A., Kitt, E. R., & Santos, L. R. (2020). The influence of interactions with dogs on affect, anxiety, and arousal in children. *Journal of Clinical Child & Adolescent Psychology, 49*(4), 535–548.

Dreschel, N. A., & Granger, D. A. (2016). Advancing the social neuroscience of human-animal interaction: The role of salivary bioscience. In L. S. Freund, S. McCune, L. Esposito, N. R. Gee &

P. McCardle (Eds.), *The social neuroscience of human–animal interaction* (pp. 195–216). American Psychological Association.

Eid, M., & Diener, E. (2006). Introduction: The Need for Multimethod Measurement in Psychology. In M. Eid & E. Diener (Eds.), *Handbook of multimethod measurement in psychology* (pp. 3–8). American Psychological Association. https://doi.org/10.1037/11383-001.

Esposito, L., McCune, S., Griffin, J. A., & Maholmes, V. (2011). Directions in human–animal interaction research: Child development, health, and therapeutic interventions. *Child Development Perspectives, 5*(3), 205–211.

Fine, A. H., Beck, A. M., & Ng, Z. (2019). The state of animal-assisted interventions: Addressing the contemporary issues that will shape the future. *International Journal of Environmental Research and Public Health, 16*(20), 3997.

Fine, A. H., Tedeschi, P., & Elvolve, E. (2015). Forward thinking: The evolving field of human–animal interactions. In A. H. Fine (Ed.), *Handbook on animal-assisted therapy: Foundations and guidelines for animal-assisted interventions* (4th ed., pp. 21–36). Elsevier Inc.

Franco, A., Malhotra, N., & Simonovits, G. (2014). Publication bias in the social sciences: Unlocking the file drawer. *Science, 345*(6203), 1502–1505.

Friedmann, E., & Gee, N. R. (2019). Critical review of research methods used to consider the impact of human–animal interaction on older adults' health. *The Gerontologist, 59*(5), 964–972.

Gabriels, R. L., Zhaoxing, P., DeChant, B., Agnew, J. A., Brim, N., & Mesibov, G. (2015). Randomized controlled trial of therapeutic horseback riding in children and adolescents with autism spectrum disorder. *Journal of the American Academy of Child & Adolescent Psychiatry, 55*(7), 541–549.

Gandenberger, J., Motiff, M., Flynn, E., & Morris, K. N. (2023). Staff perspectives on the targeted incorporation of nature-based interventions for children and youth at a residential treatment facility. *Residential Treatment for Children & Youth, 40*(1), 67–86.

Gebhart, V., Buchberger, W., Klotz, I., Neururer, S., Rungg, C., Tucek, G., Zenzmaier, C., & Perkhofer, S. (2020). Distraction-focused interventions on examination stress in nursing students: Effects on psychological stress and biomarker levels. A randomized controlled trial. *International Journal of Nursing Practice, 26*(1), e12788.

Gee, N. R., Rodriguez, K. E., Fine, A. H., & Trammell, J. P. (2021). Dogs supporting human health and well-being: A biopsychosocial approach. *Frontiers in Veterinary Science, 8*(246), 630465.

Geldhof, G. J., Flynn, E., Olsen, S. G., Mueller, M. K., Gandenberger, J., Witzel, D. D., & Morris, K. N. (2021). Emotion regulation and specificity: The impact of animal-assisted interventions on classroom behavior. *Journal of Applied Developmental Psychology, 73*, 101253.

Gough, B., & Lyons, A. (2016). The future of qualitative research in psychology: Accentuating the positive. *Integrative Psychological and Behavioral Science, 50*(2), 234–243.

Grandgeorge, M., Bourreau, Y., Alavi, Z., Lemonnier, E., Tordjman, S., Deleau, M., & Hausberger, M. (2015). Interest towards human, animal and object in children with autism spectrum disorders: An ethological approach at home. *European Child & Adolescent Psychiatry, 24*(1), 83–93.

Griffin, J. A., Hurley, K., & McCune, S. (2019). Opinion: Human–animal interaction research: Progress and possibilities. *Frontiers in Psychology, 10*, 2803.

Grossman, J., & Mackenzie, F. J. (2005). The randomized controlled trial: Gold standard, or merely standard? *Perspectives in Biology and Medicine, 48*(4), 516–534.

Haardörfer, R. (2019). Taking quantitative data analysis out of the positivist era: Calling for theory-driven data-informed analysis. *Health Education & Behavior, 46*(4), 537–540.

Hartsell, M., & Zolnikov, T. R. (2023). Understanding attachment in homeless adolescents and emerging adults with pets. *Journal of Social Distress and Homelessness, 32*(1), 114–122.

Herzog, H. (2011). The impact of pets on human health and psychological well-being: Fact, fiction, or hypothesis? *Current Directions in Psychological Science, 20*(4), 236–239.

Herzog, H. (2021, May 24). Women dominate research on the human–animal bond. *Psychology Today*. https://www.psychologytoday.com/us/blog/animals-and-us/202105/women-dominate-research-the-human-animal-bond

Kamioka, H., Okada, S., Tsutani, K., Park, H., Okuizumi, H., Handa, S., Oshio, T., Park, S.-J., Kitayuguchi, J., Abe, T., Honda, T., & Mutoh, Y. (2014). Effectiveness of animal-assisted therapy: A systematic review of randomized controlled trials. *Complementary Therapies in Medicine, 22*(2), 371–390.

Kazdin, A. E. (2019). Methodological standards and strategies for establishing the evidence base of animal-assisted therapies. In A. H. Fine (Ed.), *Handbook on animal-assisted therapy: Foundations and guidelines for animal-assisted interventions* (5th ed., pp. 451–463). Elsevier.

Krause-Parello, C. A., & Morales, K. A. (2018). Military veterans and service dogs: A qualitative inquiry using interpretive phenomenological analysis. *Anthrozoös, 31*(1), 61–75.

Kuzara, S., Pendry, P., & Gee, N. R. (2019). Exploring the handler-dog connection within a university-based animal-assisted activity. *Animals, 9*(7), 402.

Marino, L. (2012). Construct validity of animal-assisted therapy and activities: How important is the animal in AAT? *Anthrozoös, 25*(3), 139–151.

Martin, F., Bachert, K. E., Snow, L., Tu, H.-W., Belahbib, J., & Lyn, S. A. (2021). Depression, anxiety, and happiness in dog owners and potential dog owners during the covid-19 pandemic in the united states. *PloS One, 16*(12), e0260676.

Maujean, A., Pepping, C. A., & Kendall, E. (2015). A systematic review of randomized controlled trials of animal-assisted therapy on psychosocial outcomes. *Anthrozoos, 28*(1), 23–36.

McCune, S., Kruger, K. A., Griffin, J. A., Esposito, L., Freund, L. S., Hurley, K. J., & Bures, R. (2014). Evolution of research into the mutual benefits of human–animal interaction. *Animal Frontiers, 4*(3), 49–58.

McDonald, S. E., Matijczak, A., Nicotera, N., Applebaum, J. W., Kremer, L., Natoli, G., O'Ryan, R., Booth, L. J., Murphy, J. L., & Tomlinson, C. A. (2022). "He was like, my ride or die": Sexual and gender minority emerging adults' perspectives on living with pets during the transition to adulthood. *Emerging Adulthood, 10*(4), 1008–1025.

McNicholas, J., Gilbey, A., Rennie, A., Ahmedzai, S., Dono, J.-A., & Ormerod, E. (2005). Pet ownership and human health: A brief review of evidence and issues. *BMJ, 331*(7527), 1252–1254.

Miles, J. N., Parast, L., Babey, S. H., Griffin, B. A., & Saunders, J. M. (2017). A propensity-score-weighted population-based study of the health benefits of dogs and cats for children. *Anthrozoös, 30*(3), 429–440.

Morse, J. M. (2015). Critical analysis of strategies for determining rigor in qualitative inquiry. *Qualitative Health Research, 25*(9), 1212–1222.

Mueller, M. K., Anderson, E. C., King, E. K., & Urry, H. L. (2021a). Null effects of therapy dog interaction on adolescent anxiety during a laboratory-based social evaluative stressor. *Anxiety, Stress, & Coping, 34*(4), 365–380.

Mueller, M. K., King, E. K., Callina, K., Dowling-Guyer, S., & McCobb, E. (2021b). Demographic and contextual factors as moderators of the relationship between pet ownership and health. *Health Psychology and Behavioral Medicine, 9*(1), 701–723.

Nieforth, L. O., Schwichtenberg, A. J., & O'Haire, M. E. (2021). Animal-assisted interventions for autism spectrum disorder: A systematic review of the literature from 2016 to 2020. *Review Journal of Autism and Developmental Disorders*, 1–26.

Nimer, J., & Lundahl, B. (2007). Animal-assisted therapy: A meta-analysis. *Anthrozoös, 20*(3), 225–238.

Nosek, B. A., Ebersole, C. R., DeHaven, A. C., & Mellor, D. T. (2018). The preregistration revolution. *Proceedings of the National Academy of Sciences, 115*(11), 2600–2606.

Nurenberg, J. R., Schleifer, S. J., Shaffer, T. M., Yellin, M., Desai, P. J., Amin, R., Bouchard, A., & Montalvo, C. (2015). Animal-assisted therapy with chronic psychiatric inpatients: Equine-assisted psychotherapy and aggressive behavior. *Psychiatric Services, 66*(1), 80–86.

O'Haire, M. E. (2010). Companion animals and human health: Benefits, challenges, and the road ahead. *Journal of Veterinary Behavior: Clinical Applications and Research, 5*(5), 226–234.

O'Haire, M. E., Bibbo, J., Mueller, M. K., Ng, Z. Y., Buechner-Maxwell, V. A., & Hoffman, C. F. (2018). Overview of centers and institutes for human-animal interaction in the United States. *Human-Animal Interaction Bulletin, 6*, 3–22.

Parbery-Clark, C., Lubamba, M., Tanner, L., & McColl, E. (2021). Animal-assisted interventions for the improvement of mental health outcomes in higher education students: A systematic review of randomised controlled trials. *International Journal of Environmental Research and Public Health, 18*(20), 10768.

Pedersen, I., Martinsen, E. W., Berget, B., & Braastad, B. O. (2012). Farm animal-assisted intervention for people with clinical depression: A randomized controlled trial. *Anthrozoös, 25*(2), 149–160.

Pendry, P., & Vandagriff, J. L. (2019). Animal visitation program (AVP) reduces cortisol levels of university students: A randomized controlled trial. *Aera Open, 5*(2), 2332858419852592.

Perepletchikova, F., Hilt, L. M., Chereji, E., & Kazdin, A. E. (2009). Barriers to implementing treatment integrity procedures: Survey of treatment outcome researchers. *Journal of Consulting and Clinical Psychology, 77*(2), 212.

Rodriguez, K. E., Guérin, N. A., Gabriels, R. L., Serpell, J. A., Schreiner, P. J., & O'Haire, M. E. (2018). The state of assessment in human-animal interaction research. *Human Animal Interaction Bulletin, 6*, 63–81.

Rodriguez, K. E., Herzog, H., & Gee, N. R. (2021). Variability in human-animal interaction research. *Frontiers in Veterinary Science, 7*, 619600.

Schoenwald, S. K., Garland, A. F., Chapman, J. E., Frazier, S. L., Sheidow, A. J., & Southam-Gerow, M. A. (2011). Toward the effective and efficient measurement of implementation fidelity. *Administration and Policy in Mental Health and Mental Health Services Research, 38*(1), 32–43.

Schuck, S. E., Johnson, H. L., Abdullah, M. M., Stehli, A., Fine, A. H., & Lakes, K. D. (2018). The role of animal assisted intervention on improving self-esteem in children with attention deficit/hyperactivity disorder. *Frontiers in Pediatrics, 6*, 300.

Scoresby, K. J., Strand, E. B., Ng, Z., Brown, K. C., Stilz, C. R., Strobel, K., Barroso, C. S., & Souza, M. (2021). Pet ownership and quality of life: A systematic review of the literature. *Veterinary Sciences, 8*(12), 332.

Serpell, J., McCune, S., Gee, N., & Griffin, J. A. (2017). Current challenges to research on animal-assisted interventions. *Applied Developmental Science, 21*(3), 223–233.

Stern, J. M., & Simes, R. J. (1997). Publication bias: Evidence of delayed publication in a cohort study of clinical research projects. *BMJ, 315*(7109), 640–645.

Wagner, C., Gaab, J., Locher, C., & Hediger, K. (2021). Lack of effects of the presence of a dog on pain perception in healthy participants: A randomized controlled trial. *Frontiers in Pain Research, 2*, 714469.

Wilson, C. C., & Barker, S. B. (2003). Challenges in designing human-animal interaction research. *American Behavioral Scientist, 47*(1), 16–28.

Young, J., Bowen-Salter, H., O'Dwyer, L., Stevens, K., Nottle, C., & Baker, A. (2020). A qualitative analysis of pets as suicide protection for older people. *Anthrozoös, 33*(2), 191–205.

# 7
# ANTHROPOMORPHISM AND ANTHROPOCENTRISM IN THE ANTHROPOCENE

*Randall Lockwood*

## Introduction

Humankind is already well into the epoch termed the Anthropocene (Waters et al., 2016) in which our actions appear immutably in the lithostratigraphic record. We now live in a world where human-made materials exceed all living biomass (Elhacham et al., 2020), despite the fact that humans make up only .01% of the living biomass of Earth (Bar-On et al., 2018). Our treatment of and attitudes to the natural world have altered and will continue to impact the planet in ways that are likely to persist well past the age of human domination and perhaps even human existence.

Whether we date the onset of the Anthropocene to the start of the Industrial Revolution or to far earlier beginnings, such as at the start of agriculture, the common factor is that it represents an era in which humanity began to see itself as superior to and apart from, rather than a part of, the natural world. A core feature of much of Western philosophy that developed alongside the growth of civilization marking the Anthropocene was the conceptualization of the human condition as a middle station between animality and divinity, where humans were the closest to the gods, who were usually drawn as likenesses to human beings Such anthropocentric thinking fostered the belief that human-like god or gods created animals expressly for the sake of human beings – a view carried through from Aristotle to Saint Augustine, Saint Thomas Aquinas, Descartes and Kant (Steiner, 2010). As Steiner notes, such anthropocentric conceptions of animals and their moral status undercut the force of arguments advanced on behalf of animals. In addition, I would argue, it clouded efforts to fully understand and explain the behavior of many creatures. Anthropocentrism leads us to ignore the needs and interests of other creatures. It also allows us to ignore the long history of the existence of a world before humans, or to contemplate the likelihood of a world without them.

One of the earliest critiques of seeing the gods as entities in human form who attend to human interests came from Xenophenes (6th century BCE) who used the term *anthropomorphism* to describe the assertion that "gods are born, wear their own clothes and have a voice and body". He further noted "if horses and oxen had hands and could draw pictures, their gods would look remarkably like horses and oxen".[1]

The use of the term anthropomorphism has evolved over the centuries into the classic definition described by Guthrie (1997, p. 51) as "the attribution of human characteristics to non-human things or events". Such characteristics can include physical appearance (e.g. believing a god to look human), emotional states perceived to be uniquely human or internal mental states and motivations. They may be applied to anything that acts or is believed to act independently, including non-human animals, natural forces, religious agents or inanimate objects (Epley et al., 2008).

## The Rise and Fall and Rise of Anthropomorphism

Early accounts of natural history were both anthropomorphic and anthropocentric. Descriptions of animals were often intended as parables (e.g. Aesop's Fables) or to provide lessons of appropriate behavior or show the potential power or wrath of God or gods.

In the 19th century, the compendia of anthropomorphic anecdotes eventually gave way to improved descriptions as in Fabre's writings on insects (Fabre, 1937), but the frequency of errors or misinterpretations made it difficult to equate the study of animal behavior with the quantitative sciences such as physics, chemistry and astronomy that were having a growing impact.

Charles Darwin stood at the intersection of these two traditions. He made extensive but careful use of anecdotal reports in many of his greatest works. His descriptions of behavior were also highly anthropomorphic. He wrote, for example:

> Dogs exhibit their affection by desiring to rub against their masters…I have also seen dogs licking cats with whom they were friends. This habit probably originated in the females' carefully licking their puppies- the dearest object of their love- for the sake of cleansing them.
>
> *(Darwin, 1872, p. 118)*

Darwin's anthropomorphism was partly based on literary style. We must remember that he was a best-selling author, with the first edition of *Origin of Species* selling out in a single afternoon. But Darwin's choice of language also reflected his firm belief in the continuity of mental experiences across human and non-human species.

In 1882, Darwin's protégé, George Romanes, offered an approach to assessing animal feelings by suggesting his principal of "ejection" or "double inference", the idea that we can infer the mental state of animals having some similarities to ourselves by reflecting on our own mental states under similar conditions (Romanes, 1885). This is similar to the suggested approach of applied anthropomorphism (Lockwood, 1989), Servais's category of "Inferred anthropomorphism" and Burghardt's "critical anthropomorphism" (Burghardt, 2016) discussed below.

In 1894, Romanes' student, C. Lloyd Morgan, released his *Introduction to Comparative Psychology*. The book echoed many of Romanes' ideas and stressed the importance of *both* subjective and objective approaches to understanding the mentality of animals. However, Morgan's book is most remembered for a simple statement on method which, taken out of context, became the guiding light for American biologists and psychologists. This idea, now immortalized as "Morgan's Canon," was brief and deceptively simple. He wrote (p. 53):

> In no case may we interpret an action as the outcome of the exercise of higher psychical faculty if it can be interpreted as the outcome of the exercise of one that stands lower in the psychological scale.

Morgan's Canon had the effect of both freeing psychology from some of its pseudoscientific trappings while, at the same time, closing the door on some of the most interesting questions possible. Most of the major figures in the study of animal behavior from 1890 to the 1950s embraced Morgan's idea. Pavlov (1927) wrote that animals should be studied as "physiological facts, without any need to resort to fantastic speculations as to the existence of any possible subjective states... which may be conjectured as an analogy with ourselves." (p. 16). Wells (1956) notes that Pavolv was particularly critical of primatologists such as Wolfgang Kohler and Robert Yerkes, commenting that:

> Pavolv's sharpest polemics were reserved for, not the open and avowed theologians, but for those who under the cover of science in effect preached the doctrine of the mysterious, disembodied psyche.
>
> *(p. 61)*

In America, John B. Watson set the tone for American animal research for the next 40 years when he proclaimed the official position of Behaviorism. He maintained that states of consciousness are not objectively verifiable and for that reason can *never* become data for science (Watson, 1924). A direct outgrowth of this approach was the proliferation of mechanical models of animal and human behavior that attempted to avoid any taint of anthropomorphism. The research of Skinner(1938), and his ideological descendants thus often took the form of precisely measured studies of trivial events without the context of evolutionary significance.

In the 1940s and 1950s, while American psychologists followed a largely Behaviorist tradition, European ethologists such as Niko Tinbergen, Konrad Lorenz and Karl von Frisch were rediscovering many of the Darwinian traditions. Tinbergen's methods were the result of a desire to understand the entire pattern of an animal's adaptations to its natural environment, as well as the desire to study mechanisms of behavior without damaging the animal in any way. He described his procedures as "physiology without breaking the skin" (Tinbergen, 1958). Lorenz, Tinbergen, von Frisch, and others performed naturalistic observations and natural experiments, all in accordance with accepted scientific method. However, they felt free to discuss their results, as Darwin had, with reference to animal minds, intentions, and feelings.

As with Darwin, part of this was for literary effect. Like Darwin's *Origin of Species*, Lorenz's books *King Solomon's Ring* (1952) and *On Aggression* (1966) and Tinbergen's *Curious Naturalists* (1958) became best-sellers. But the anthropomorphic tone of these books reflected a very different philosophical viewpoint from that found in the writings of most American comparative psychologists. It was one that had been neglected and even opposed for some time in this country. Lorenz made it clear that his anthropomorphism was not to be confused with softheaded sentimentality. He wrote:

> Believe me, I am not mistakenly assigning human properties to animals; on the contrary, I am showing you what an enormous inheritance remains in man to this day.
>
> *(1952, p. 152)*

For those of us being trained in animal behavior in the 1960s and 1970s, there was clearly a potential clash between the growing prestige of the Nobel Prize-winning ethologists and the entrenched American behaviorist tradition. For much of the last century, the prevailing view had been that the use anthropomorphism in describing or explaining non-human behavior undermined the objective, rational basis of animal studies, in an era when many scholars in comparative psychology were striving to put the field on an equal footing to physics. The tendency to use, or avoid, anthropomorphism was often presented as the primary distinction between "serious" scientists and the rest of the research world. This certainly was made clear in many of the texts first studied by my generation of students of animal behavior.

Breland and Breland (1966, p. 3) warned:

> Sometimes we fall into the dangerous pit of anthropomorphism… or the tendency to think of animals as if they were human.

Fortunately, many ethologists and psychologists sought a balance between rejecting any notion of the continuity of mental experiences between humans and animals and recognizing the parallels, within limits, of these experiences to ask meaningful questions about behavior. The "ejection" approach of Romanes was given a more systematic definition in an early influential text on animal behavior by Margaret Floy Washburn (1917). She advised:

> When a being whose structure resembles ours receives the same stimulus that affects us and moves in the same way as a result, he has an inner experience which resembles our own. We may extend this inference to the lower animals, *with proper safeguards* [emphasis added], just as far as they present resemblances in structure or behavior to ourselves.

More recently, a compromise approach began to emerge establishing possible guidelines and methods for "critical anthropomorphism" to more clearly delineate the promise and danger of applying anthropomorphism to understanding animal and human behavior.

Dawkins (1980), Markowitz (1982), Burghardt (1991, 2016) and others have been very influential in establishing research protocols for applying "critical anthropomorphism" to evaluate admittedly subjective ideas about what animals may like or dislike, and to assign relative weights to the importance animals place on their needs.

## Taxonomies of Anthropomorphism

As the potential value of anthropomorphism as a research construct has been realized, a variety of approaches have been used to create a taxonomy of different forms of anthropomorphism. Epley et al. (2008) distinguish between "strong" and "weak" anthropomorphism. Strong anthropomorphism entails behaving toward an agent as if it possessed human-like traits along with the belief that the agent actually possesses these traits. That would include many religious beliefs as well as the interpretation of much animal behavior as reflecting higher forms of cognition or emotion. Weak anthropomorphism may only involve an "as-if" component, such as the belief that the malfunctioning of a car or computer is the result of some conscious motive. Some anthropomorphic attribution may bridge these categories depending on the interpreter. For example, seeing "Mother Nature" or "Karma" as a sentient force acting on the world may be allegorical to some observers, while being seen as an actual process by others.

Servais (2018) provides a simpler taxonomy, recognizing three types of anthropomorphism:

1. Inferred without evidence
2. Hypothesized and tested
3. Recognized or directly perceived

Servais sees anthropomorphism as applied to animals as a process that does not only depend on the characteristics of the animal, but also on the nature of the relationship and interaction between the person and the animal. This becomes particularly important in assessing the behavior of animals we frequently interact with, such as companion animals and some livestock.

A generation ago, I suggested distinguishing five different forms of anthropomorphism (Lockwood, 1989):

1.  Allegorical – descriptions of animal or other non-human behavior that are not intended to be interpreted as fact, such as fables, cartoons and literary or political allegories such as *Alice in Wonderland, Animal Farm* or *Watership Down*.
2.  Personification – Directly treating animals as if they were human, e.g. dressing them in human clothes or having "weddings" or birthday celebrations for companion animals
3.  Superficial – Interpreting the underlying basis of an animal's behavior only on the simplistic external appearance, e.g. describing the fear grimace of a chimpanzee as "smiling" or the aggressive "kissing" behavior of gourami fish as affection.
4.  Explanatory – Using circular reasoning to assume that a behavior has been explained because it has been given a name – such as describing a dog's destruction of a favorite shoe when left alone as "spiteful", thus assuming that the dog has the ability for spiteful motivation.
5.  Applied – The use of our personal perspective on what it's like to be a living being to hypothesize what it is like to be another living being of our own or some other species and then develop ways to gather information to test that idea. This approach mirrors what has been referred to in cognitive psychology as having a "theory of mind" (Baron-Cohen, 1991; Leudar et al., 2004).

Fisher (1996) offered a taxonomy that focused on the types of errors that can be produced by anthropomorphism.

1.  Imaginative – Representing animals or inanimate objects as similar to us – analogous to the above category of allegorical anthropomorphism
2.  Interpretive – Like explanatory anthropomorphism, interpreting an animal's behavior as being caused by mechanisms or constituted in ways similar to human traits. This category of errors is further subdivided into Categorical errors, in which traits are ascribed to creatures in which the traits don't ever apply. This is further divided into species and predicate type errors. Species errors include unwarranted attribution of human traits that might be warranted in species more closely related to humans but which are unlikely to be found in the animal under discussion, including invertebrates. Predicate errors are those in which the behavior may not be appropriately attributed to the animal in question. For example, description of "rape" may be appropriate for ducks as well as primates, but "embarrassment" might not.
3.  Situational – These errors involve misinterpretation of a behavior that might be analogous to human behaviors in situations other than that being described, as in the above superficial error of misinterpreting a chimpanzee fear grimace as a smile.

## The Value of Anthropomorphism

Regardless of how anthropomorphism is defined or measured, a central question has always been "is it a good thing" or, as Breland and Breland (1966) warned "a dangerous pit"? Does such a widespread approach to viewing the world in human terms help or hinder the understanding of the way the world works? More narrowly, the question has usually been framed within the context of understanding, explaining and accurately predicting the behavior of animals.

Although many, if not most, researchers in animal behavior have embraced the value of using "critical anthropomorphism" and its analogs in generating testable hypotheses about the inner workings of animals, there are still those who question the process (e.g. Blumberg & Wasserman, 1995). Bruni et al. (2018) refer to such opposition as "anti-anthropomorphism", admitting that anthropomorphism can fail if we do not have enough information about the biology, ecology and evolutionary history of the animal of interest, adding (p. 3):

> ... In the spirit of Darwin's writings ... if you are aware of the ethological constraints of what we are saying, you can feel free to use our empathy to make a behavioral prediction and then evaluate it.

Sheets-Johnstone (1992) applies the term "reverse anthropocentrism" – to the dogmatic refusal to attribute human traits to non-human animals and thus see humanity as separate from and superior to all other species. Griffin (1992) calls the dogmatic refusal to attribute human traits to non-human animals an "inverse Clever Hans error", in reference to the discredited reports of a horse that appeared to do mathematical calculations by responding to the subtle cues of his handler. De Waal (1999) terms the blindness to the human-like characteristics of other animals, or the animal-like characteristics of ourselves "anthropodenial".

The growing recognition of the potential benefits of applying anthropomorphic thinking to the understanding of animal and human behavior has led to intriguing advances in six distinct areas – cognitive ethology, applied animal behavior, human mental health, animal welfare, animal legislation/litigation and human-machine interaction. Examples of work in each of these areas is provided below.

## Anthropomorphism and Cognitive Ethology

As the scientific community has come to terms with the challenges and opportunities of anthropomorphism, it has fostered the emergence of animal models that examine non-human thinking and feeling in their proper context in the form of the growing discipline of cognitive ethology (Keeley, 2004).

Donald Griffin is widely recognized as the founding father of the field of cognitive ethology, building on the foundation laid by Darwin and the European ethologists, with several influential works setting the tone for a rich area of research (Griffin, 1976, 1984, 1992). He fully embraced the value of critical anthropomorphism noting (1984, p. 23): "Behaviorism should be abandoned not so much because it belittles the value of living animals, but because it leads us to a seriously incomplete and hence misleading picture of reality."

This Renaissance in the way of thinking has given rise to research in a wide array of fields, including animal communication, animal learning, comparative psychopathology, the study of the human/companion animal bond and animal welfare (Lockwood, 1989) and has spawned dozens of academic courses, symposia, books and several journals, including the *Journal of Animal Sentience*, the *Journal of Animal Cognition*, the *Journal of Animal Behavior and Cognition* and the *Journal of Experimental Psychology: Animal Learning and Cognition*.

## Anthropomorphism and Applied Animal Behavior

Although Darwin (1872) based much of his writing about animal communication on his observations of dogs, they were largely ignored as subjects of serious behavioral research for decades, in part because of the stigma of anthropomorphism attached to the study of a species so familiar to us. However, the last decade has seen a proliferation of writings by respected ethologists and comparative psychologists willing to entertain ideas of canine sentience in their analysis of behavior, including Bradshaw (2011), Miklosi (2014), Reid (2012) and Serpell (2017), among many others.

A good example of using anthropomorphism to generate testable ideas is the proliferation of studies of the possibility of jealousy in dogs. This is a question that clearly marks the difference between "strong" and "weak" anthropomorphism, since, as Epley et al. (2008) note (p. 152): "Thinking that one's dog is jealous is one thing, but thinking that one's god is jealous is quite another."

Several researchers have attempted to validate the presence of the specific emotional response of jealousy in dogs. Harris and Prouvost (2014) adapted a paradigm from human infant studies to examine jealousy in domestic dogs. They found that dogs exhibited significantly more jealous behaviors (e.g., snapping, getting between the owner and object, pushing/touching the object/owner) when their owners displayed affectionate behaviors toward what appeared to be another dog as compared to nonsocial objects. They concluded that the results lend support to the hypothesis that jealousy has some "primordial" form that exists in human infants and in at least one other social species besides humans.

Cook et al. (2018) used the C-BARQ scale to estimate dogs' aggressiveness and used noninvasive brain imaging (fMRI) to measure activity in their amygdala (an area involved in aggression). They reported that more aggressive dogs had more amygdala activation data while watching their caregiver give food to a realistic fake dog than when they put the food in a bucket. They suggested that this may have some similarity to human jealousy, adding to a growing body of evidence that differences in specific brain activities correlate with differences in canine temperament.

Other studies of human-dog interactions have suggested the importance of viewing the attribution of anthropomorphism in terms of the relationship between the human and animal. Desmet et al. (2017) conducted an fMRI study of humans to investigate whether brain regions involved in mentalizing about human behavior are also engaged when observing dog behavior. They showed that these brain regions are more engaged when observing dog behavior that is difficult to interpret compared to dog behavior that is easy to interpret. These results were not only obtained when participants were instructed to infer reasons for the behavior but also when they passively viewed the behavior, indicating that these brain regions are activated by spontaneous mentalizing processes. This suggests that observers of canine behavior are actively invoking a canine theory of mind when watching dogs in much the same way we do when observing fellow humans.

## Anthropomorphism and Human Mental Health

While most research on anthropomorphism has focused on the perception of and attitudes toward animals, there has been growing interest in the tendency of some people to develop anthropomorphic attachments to objects as well, particularly among those diagnosed with Hoarding Disorder (American Psychiatric Association, 2013). Hoarding disorder involves a greater emotional attachment to possessions and a higher level of emotional comfort derived from objects. This may be due to a greater tendency in hoarders to view their possessions as having human-like characteristics or mental states (Frost & Hartl, 1996; Frost et al., 1995). This is particularly true of animal hoarders, in whom the inclination to anthropomorphize objects is augmented by the common inclination to view companion animals in human terms (Arluke et al., 2017; Lockwood, 2018). Because hoarders are usually socially isolated, it has been suggested that they may anthropomorphize to fulfill a need for social affiliation (Epley et al., 2007).

In examining the role of anthropomorphism in hoarding, Timpano and Shaw (2013) found that greater hoarding symptoms were associated with the tendency to anthropomorphize objects, even after taking into account general mood and anxiety symptoms, but anthropomorphism was not significantly linked with clutter.

Anthropomorphism may also play a role in improving the social skills of children with Autism Spectrum Disorder (ASD). Carlisle (2015) surveyed 70 parents of children with ASD. Children in homes owning dogs had greater mean scores for social skills, using the Social Skills Improvement System Rating Scale. Parents described their children as attached to their dogs. Children owning dogs completed the Companion Animal Bonding Scale, and reported strong bonding with dogs. Atherton and Cross (2018) noted that individuals with ASD, show an increased interest related to anthropomorphism and may even show improved Theory of Mind (ToM) abilities when judging the mental states of anthropomorphic characters. Although individuals with ASD traditionally show deficits on a wide range of ToM tests, such as recognizing facial emotions, these deficits may be reduced if the stimuli presented are cartoon or animal-like rather than in human form. Individuals with ASD also self-report that they may identify more closely with animals than other humans. That has been confirmed in detail in the writings of Temple Grandin (Grandin, 2006; Grandin & Johnson, 2006), a noted animal behaviorist with ASD whose work with innovative livestock handling methods has been shaped by her unique ability to empathize with the emotional lives of the animals with which she works.

## Anthropomorphism and Animal Welfare

It is important to recognize that much opposition to thinking of animal behavior, cognition and emotion in human terms lies not in the fact that it produces errors of fact or prediction – but that it presents one of the greatest moral challenges to an anthropocentric status quo built on the utilization and exploitation of animals for the presumed benefit of humans. Such conflict has a long history.

The notion of the continuity of mental experience between animals and humans was clear to some philosophers even before the Darwinian Revolution. Voltaire, in his *Philosophical Dictionary* (1763) entry on animals, offered a critique of the Cartesians. He wrote:

> There are barbarians who seize this dog…and nail him down to a table and dissect him alive, to show you the mesaraic veins! You discover in him all the same organs of feeling as in yourself. Answer me, mechanist, has Nature arranged all the springs of feeling in this animal to the end that he might not feel?

Here Voltaire is offering a surprisingly modern argument from physiological homology. As we gain an increasing understanding of the anatomical, physiological and pharmacological events that underlie the experience of emotional states in animals and man, we are forced to realize the close parallels that exist.

Humphry Primatt, a Christian Doctor of Divinity whose 1776 work *The Duty of Mercy* inspired the Reverend Arthur Broome to help found the Royal Society for the Prevention of Cruelty to Animals, stated his comparative view powerfully and succinctly (p. 21):

> Pain is pain, whether it be inflicted on man or beast; and the creature that suffers it, whether man or beast, being sensible of the misery of it while it lasts, suffers evil.

Darwin was among the first scientists to recognize the implications that the continuity of mental experience had for the ethical treatment of animals. His ideas played an important role in shaping the animal welfare movement of the Victorian era. The notion of continuity between human and non-human animal mental/emotional experience raised new moral and ethical issues concerning our responsibility for preventing unnecessary animal suffering.

Darwin avoided becoming embroiled in the growing antivivisection movement. He was reluctant to speak out against his scientific colleagues and was quite adamant in his faith in the contributions of much of their work to human welfare. However, he eloquently attacked, in anthropomorphic terms, practices that he perceived as inhumane. In an essay entitled "Trapping Agony" (Darwin, 1863) he wrote:

> If we attempt to realize the sufferings of a cat, or other animal when caught, we must fancy what it would be to have a limb crushed during a whole long night, between the iron teeth of a trap, and with the agony increased by constant attempts to escape.

According to Jamieson and Bekoff (1996), the fusion of concerns about animal welfare and interest in animal sentience and cognition began to develop in the period between 1974 and 1976, driven in part by Singer's (1975) *Animal Liberation*, Nagel's (1974) influential paper "What is it like to be a bat?", Wilson's (1975) *Sociobiology: The New Synthesis*, and Griffin's (1976) *The Question of Animal Awareness: Evolutionary Continuity of Mental Experience*. Together these works brought increased attention and focus to the value of considering the mental states of animals in developing laws and regulations regarding their treatment, a process previously considered to be tainted by anthropomorphism.

## Anthropomorphism and Legislation and Litigation

Laws addressing cruelty to animals have been progressively strengthened throughout the last several decades (Arkow & Lockwood, 2022). Despite concerns about the effects of cruelty on the animals themselves, today, as in the 18th century, animals are legally viewed as property and animal cruelty is legally considered to be a crime against the property owner. In 2016, the FBI began tracking crimes against animals as part of the National Incident Based Reporting System (NIBRS). The system tracks four distinct categories of crimes – neglect, intentional abuse, organized abuse (dogfighting, cockfighting) and animal sexual abuse. However, for classification purposes the "victim" in each of these cases is coded as "society", rather than individual animals (DeSousa, 2016).

One of the most comprehensive attempts to address animal suffering through regulations began in 1965, when the United Kingdom government commissioned an investigation, led by Professor Roger Brambell, into the welfare of intensively farmed animals. This was driven in part, by concerns raised in the book, *Animal Machines* (Harrison, 1964). The goal of the Brambell Commission was not to offer a critique of the agricultural industry, but rather to define practices that would maximize the quality of life for animals in confinement. As a result of the Commission's initial report, the Farm Animal Welfare Advisory Committee was created to monitor the livestock production sector. In 1979, this was replaced by the Farm Animal Welfare Council, and that year, they codified what has come to be known as **the Five Freedoms:**

1. **Freedom from hunger and thirst,** by ready access to water and a diet to maintain health and vigor;
2. **Freedom from discomfort,** by providing an appropriate environment;
3. **Freedom from pain, injury, and disease,** by prevention or rapid diagnosis and treatment;
4. **Freedom to express normal behavior,** by providing sufficient space, proper facilities, appropriate company of the animal's own kind; and
5. **Freedom from fear and distress,** by ensuring conditions and treatment which avoid mental suffering.

The Five Freedoms clearly accept the reality of mental experiences in animals that earlier would have been considered overly anthropomorphic, yet they have achieved worldwide recognition and are a component of national legislation, marketing and farm assurance schemes. They have withstood much criticism and, within Great Britain, are the cornerstone of government and industry policy and the Codes of Recommendations for the Welfare of Livestock (Farm Animal Welfare Council, 2009). Wathes (2010), a member of the Farm Animal Welfare Council, notes that the Council's most recent analysis should move beyond the current test of whether the animal suffers to a new standard of whether the animal has a life worth living, from the animal's point of view.

Recently, the role of sentience in assessing farm animal welfare has attracted greater attention. Nawroth et al. (2019) review the welfare implications of current farm animal cognition research, particularly focusing on cattle, horses, pigs and small ruminants. They identify a variety of cognitive capacities they argue should be taken into consideration when assessing welfare or the absence of welfare. These include physical cognition: categorization, numerical ability, object permanence, reasoning, tool use; social cognition: individual discrimination and recognition, communication with humans, social learning, attribution of attention, prosociality and fairness. In the past, such concerns focused on species that were obviously cognitively complex such as great apes, elephants and cetaceans. In the case of companion animal cruelty, McMillan (2003) has expressed concern that the enforcement of animal cruelty laws focuses on documenting physical harm to the animal, ignoring the reality of the potential for the induction of psychological suffering in sentient victims.

The status of animals as a special category of property that is capable of pain, suffering and death, and to which we can form unique attachments, is rarely recognized in legislation or case law, but has increasingly attracted the attention of legal scholars who argue for the recognition of a new category of "living property" or "sentient property" (Bernays, 2018; Favre, 2009; Hankin, 2006; Matlack, 2006).

This trend has advanced even further with the efforts of the Nonhuman Rights Project (NhRP) founded by Steven Wise in 2007 as a project of the Center for the Expansion of Fundamental Rights and renamed the NhRP in 2012.[2] The strategy of the NhRP is to argue at the state legal level that animals scientifically demonstrated to be self-aware, autonomous beings (including great apes, elephants, dolphins and whales) should be recognized as legal persons protected under U.S. common law, including the *writ of habeas corpus,* which prevents unlawful imprisonment and *the writ of de homin replegiano*, which offers legal "things" the opportunity to challenge their legal "thinghood" and establish rights to bodily liberty (Mills & Wise, 2015). The first efforts of the NhRP were to attempt to secure the release of four chimpanzees from various facilities in New York. These petitions were denied on the grounds that chimpanzees were not persons. An appellate court concurred suggesting it should be up to the legislatures to set government policy (Wise, 2017).

In 2021, the New York Court of Appeals made legal history when it agreed to hear the *habeas corpus* case of Happy, an Asian elephant kept at the Bronx Zoo for over 40 years, often in isolation from other elephants. NhRP wanted Happy transferred to an elephant sanctuary with a better opportunity to live with other elephants. One basis for the argument was that Happy was the first of her species to demonstrate self-recognition in a mirror test. The petition also gave detailed accounts of many aspects of the species' demonstration of empathy and the capacity for boredom and depression (Nonhuman Rights Project, 2018). On June 14, 2022, New York State Court of Appeals ruled 5-2 that elephants have no constitutional rights, with Chief Judge Janet DiFiore writing in the majority decision, "nothing in our precedent or, in fact, that of any other state or federal court, provides support for the notion that the writ of habeas corpus is or should be applicable to nonhuman animals" (State of NY Court of Appeals, 2022, p. 9). However, the majority did note (p. 4): "it is essentially undisputed—that elephants are intelligent beings, who have the capacity for self-awareness, long-term memory, intentional communication, learning and problem-solving skills, empathy, and significant emotional response".

The dissenting opinion from Justice Rivera gave more weight to the admittedly anthropomorphic perception of the plight of the elephant (p. 108):

> Captivity is anathema to Happy ... She is held in an environment that is unnatural to her and that does not allow her to live her life as she was meant to: as a self-determinative, autonomous elephant in the wild. Her captivity is inherently unjust and inhumane. It is an affront to a civilized society, and every day she remains a captive—a spectacle for humans—we, too, are diminished.

Within the European Union the *Treaty on the Functioning of the European Union* recognized the concept of animal sentience as relevant to EU policies, stating that States "shall, since animals are sentient beings, pay full regard to the welfare requirements of animals when they formulate EU policies".[3] With the advent of Brexit, there was concern that removal from this provision could impact animal welfare in the United Kingdom. Thus in 2022 Parliament enacted the Animal Welfare (Sentience) Act (c.22).[4] The Act makes provision for an Animal Sentience Committee with functions relating to the effect of government policy on the welfare of animals as sentient beings. It defines "animal" to mean (a) any vertebrate other than *homo sapiens,* (b) any cephalopod mollusk, and (c) any decapod crustacean. In response to concerns that the Act might overlook the potential for sentience in invertebrates, it includes a provision that "The Secretary of State may by regulations amend this section so as to bring invertebrates of any description within the meaning of 'animal' for the purposes of this Act, where they are not already within that meaning."

These many changes in the last decade demonstrate that perceptions of animals that would have been dismissed as sentimental and "anthropomorphic" in earlier times have been strongly influenced by cognitive ethology and a public willingness to view many animals as having human-like mental and emotional experiences.

## Anthropomorphism and Human-Machine Interactions

Our meaningful and successful relationships with other people or other creatures depend on our natural tendency to infer, observe or project our own emotional and intellectual experiences on those others – the essence of anthropomorphism. Increasingly, our world will involve adapting that skill to our interactions with non-living entities – including robots – as workmates, companions, caretakers and providers of other human needs. This demand has given rise the emerging field of Social Robotics (SR) (Damiano and Dumouchel, 2018). Researchers in this field view anthropomorphism not as a cognitive error or sign of immature or emotional thinking, but as a near universal human tendency that evolved to facilitate interactions with others. One objective of SR is to use applied anthropomorphism to engineer robots that activate anthropomorphism in users to create a comfortable "social presence" that can support long-lasting relationships.

Traditionally, interaction with inanimate toys such as dolls and stuffed animals involves the cognitive activity of ascribing (or imagining) human traits – a characteristic of play, particularly in children. Applied anthropomorphism suggests that a much stronger social presence is produced when people, including adults, can infer the human-like traits through observation of the robot's appearance, autonomous movement and appropriate response to human movements. Damiano and Dumouchel suggest that refining these techniques to make better social robots may also improve our knowledge of how we build Theories of Mind in dealing with other animals and people.

Objections to SR echo the backlash against cognitive ethology, arguing that the field blurs the cognitive boundaries between humans and animals or humans and robots. Critics fear that this destabilizes the anthropocentric view of human uniqueness. The fears in the case of social robots are even greater than simply narrowing the gap between human and animal since it is likely that social robots will surpass humans in many abilities (Ferrari et al., 2016). Other critics (e.g. Turkle, 2010) see SR as a cheat, an illusion of relationship that presses our Darwinian buttons and turns people away from social relationships. Likewise Sharkey and Sharkey (2010) express concern that SR's exploit and even amplify children's natural anthropomorphism – focusing on the old notion of anthropomorphism as a system of false beliefs rather than as a tool for interaction – a tool that if properly used can build skills in forming other social interactions. They fear that anthropomorphic views of SR's serve as a form of compensation for poor relationships or an impairment of ability to form such relationships. Melson et al. (2009) raise the concern that increasing interaction with animals primarily through robotics, virtual reality and other media could come at the expense of direct engagement with living animals, potentially resulting in what Louv (2005) has termed "Nature deficit disorder". Ultimately, the goal, as put forward by Damiano and Dumouchel, is to build social robots that can work as social connectors and reinforce, rather than weaken, human-human bonds.

Several projects highlight this potential application. Tamura et al. (2004) demonstrated AIBO, a robotic dog, to severely demented elderly people living in a geriatric home and observed their reactions. The presentation of AIBO resulted in positive outcomes for the severe dementia patients, including increased communication between the patients and AIBO. Similar results were reported by Kramer et al. (2009). Shibata and Wada (2011) report on several healthcare studies with elderly using a soft seal robot (Paro), documenting improvements in behavior, stress hormone levels and cortical neuron activity. Bharatharaj et al. (2017) conducted a robot-assisted therapeutic study over seven weeks to investigate the changes in stress levels of children with ASD using a parrot-inspired therapeutic robot (KiliRo). They investigated urinary and salivary samples of participating children to report changes in stress levels before and after interacting with the robot. The results showed that the bio-inspired robot-assisted therapy could significantly help reduce the stress levels of children with ASD.

As noted by Melson et al. (2009) better understanding of human-robot interactions will help pinpoint interventions where such interactions may be optimal. More generally "robotic pets may ultimately have a place within the complex relationship that humans have with animals" (p. 564).

## Conclusion

Until recently, anthropomorphism and anthropocentrism could be considered to be parallel rather than opposing views of human relationships to animals and nature in that both approaches ignore or deny the basic intrinsic nature of the animal in question or the inorganic/insentient nature of an object. Anthropomorphism without validation involves acting on a view of the animal without regard to its true nature and interests while anthropocentrism starts with the assumption that those interests, if they exist, are irrelevant to human needs. The acceptance of the validity of anthropomorphism as an intrinsic human tool for the establishment of connections to people, animals and nature, and the refinement of critical anthropomorphism as a technique for generating complex and important questions, has opened the door to array of undertakings in cognitive ethology, human-animal relations, mental health, animal law and robotics. All of these inquiries demonstrate that the antithesis of anthropomorphism is not simply anthropocentrism – but dehumanization – seeing anyone and anything as less than human or less deserving of moral or ethical concern. Embracing the benefits of anthropomorphism could provide a much needed tool for countering anthropocentric views of the world and resolving human conflicts with other humans, and with nature itself.

## Notes

1 from plato.stanford.edu/entries/Xenophanes - entries (B14) and (B17) accessed 3/14/2022
2 https://www.nonhumanrights.org/our-story/.
3 https://*eur-lex.europa.eu*. Retrieved 8/3/2021.
4 https://www.legislation.gov.uk/ukpga/2022/22/enacted Animal Welfare (Sentience) Act 2022 retrieved 7/7/2022.

## References

American Psychiatric Association (2013). *Diagnostic and Statistical Manual of Mental Disorders 5.* Arlington, VA: American Psychiatric Publishing.

Arkow, P., & Lockwood, R. (2022). Defining animal cruelty. In. C. L. Reyes & M. Brewster (Eds.), *Animal Cruelty: A Multidisciplinary Approach to Understanding 3rd Edition* (pp. 3–33). Durham, NC: Carolina Academic Press

Arluke, A., Patronek, G., Lockwood, R., & Cardona, A. (2017). Animal hoarding. In A. Linzey & C. Linzey (Eds.), *The Palgrave International Handbook of Animal Abuse Studies* (pp. 107–129). London: Palgrave Macmillan.

Atherton, G., & Cross, L. (2018). Seeing more than human: Autism and anthropomorphic theory of mind. *Frontiers in Psychology, 9*, 528.

Baron-Cohen, S. (1991). Precursors to a theory of mind: Understanding attention in others. In A..Whiten (Ed.), *Natural Theories of Mind: Evolution, Development, and Simulation of Everyday Mindreading* (pp. 233–251). Cambridge, MA: Blackwell.

Bar-On, Y. M., Phillips, R., & Milo, R. (2018). The biomass distribution on Earth. *Proceedings of the National Academy of Sciences, 115*(25), 6506–6511.

Bernays, K. A. (2018). We've still got feelings: Re-presenting pets as sentient property. *Arizona Law Review, 60*, 485.

Bharatharaj, J., Huang, L., Al-Jumaily, A., Elara, M. R., & Krägeloh, C. (2017, September). Investigating the effects of robot-assisted therapy among children with autism spectrum disorder using bio-markers. *IOP Conference Series: Materials Science and Engineering, 234*(1), 012017.

Blumberg, M. S., & Wasserman, E. A. (1995). Animal mind and the argument from design. *American Psychology, 50*, 133–144.

Bradshaw, J. (2011). *Dog Sense*. New York: Basic Books.

Breland, K., & Breland, M. (1966). *Animal Behavior*. Toronto: Macmillan

Bruni, D., Perconti, P., & Plebe, A. (2018). Anti-anthropomorphism and its limits. *Frontiers in Psychology, 9*, 2205.

Burghardt, G. M. (1991). Cognitive ethology and critical anthromorphism: A snake with two heads and hognose snakes that play dead. In C. A. Ristau (Ed.), *Cognitive Ethology: The Minds of Other Animals* (pp. 53–90). San Francisco, CA: Erlbaum.

Burghardt, G. (2016). Mediating claims through critical anthropomorphism. *Animal Sentience, 1*(3), 17.

Carlisle, G. K. (2015). The social skills and attachment to dogs of children with autism spectrum disorder. *Journal of Autism and Developmental Disorders, 45*(5), 1137–1145.

Cook, P., Prichard, A., Spivak, M., & Berns, G. S. (2018). Jealousy in dogs? Evidence from brain imaging. *Animal Sentience, 3*(22), 1.

Damiano, L., & Dumouchel, P. (2018). Anthropomorphism in human–robot co-evolution. *Frontiers in Psychology, 9*, 468.

Darwin, C. (1863). Trapping agony. *Gardeners' Chronicle and Agricultural Gazette*. August.

Darwin, C. (1872). *The Expression of the Emotions in Man and Other Animals*. Chicago: University of Chicago Press.

Dawkins, M. S. (1980). *Animal Suffering: The Science of Animal Welfare*. New York: Chapman and Hall.

Desmet, C., van der Wiel, A., & Brass, M. (2017). Brain regions involved in observing and trying to interpret dog behaviour. *PLoS One, 12*(9), e0182721.

DeSousa, D. (2016). *NIBRS User Manual for Animal Control Officers and Humane Law Enforcement*. http://www.nacanet.org/page/NIBRS_Manual

De Waal, F. B. M. (1999). Anthropomorphism and anthropodenial: Consistency in our thinking about humans and other animals. *Philosophical Topics, 27*, 255–280.

Elhacham, Emily, & Ben-Uri, Liad, et al. (2020). Global human-made mass exceeds all living biomass. *Nature, 588*(7838), 442–444.

Epley, N., Waytz, A., Akalis, S., & Cacioppo, J. T. (2007). On seeing human: A three factor theory of anthropomorphism. *Psychological Review, 114*, 864–886.

Epley, N., Waytz, A., Akalis, S., & Cacioppo, J. T. (2008). When we need a human: Motivational determinants of anthropomorphism. *Social Cognition, 26*(2), 143–155.

Fabre, J. H. (1937). *Social Life in the Insect World*. Translated by B. Miall. London: Penguin Books.

Farm Animal Welfare Council. (2009). *Farm Animal Welfare in Great Britain: Past, Present and Future*. https://www.gov.uk/government/publications/fawc-report-on-farm-animal-welfare-in-great-britain-past-present-and-future. Accessed 9/15/2021

Favre, D. (2009). Living property: A new status for animals within the legal system. *Marquette Law Review, 93*, 1021.

Ferrari, F., Paladino, M. P., & Jetten, J. (2016). Blurring human–machine distinctions: Anthropomorphic appearance in social robots as a threat to human distinctiveness. *International Journal of Social Robotics, 8*(2), 287–302.

Fisher J. A. (1996). The myth of anthropomorphism. In M. Bekoff & D. Jamieson (Eds.), *Readings in Animal Cognition* (pp. 3–16). Cambridge, MA: MIT Press.

Frost, R. O., Hartl, T. L., Christian, R., & Williams, N. (1995). The value of possessions in compulsive hoarding (patterns of use and attachment). *Behaviour Research and Therapy, 33*, 897–902.

Frost, R. O., & Hartl, T. L. (1996). A cognitive-behavioural model of compulsive hoarding. *Behaviour Research and Therapy, 34*, 341–350.

Grandin, T. (2006). *Thinking in Pictures: And Other Reports from My Life with Autism*. New York: Vintage.

Grandin, T., & Johnson, C. (2006). *Animals in Translation: Using the Mysteries of Autism to Decode Animal Behavior*. Boston, MA: Houghton Mifflin Harcourt.

Griffin. (1976). *The Question of Animal Awareness: Evolutionary Continuity of Mental Experience*. New York: Rockefeller University Press.

Griffin, D. R. (1984). *Animal Thinking*. Cambridge, MA: Harvard University Press.
Griffin, D. R. (1992). *Animal Minds*. Chicago, IL: University of Chicago Press.
Guthrie, S. E. (1997). Anthropomorphism: A definition and a theory. In R. W. Mitchell, N. S. Thompson & H. L. Miles (Eds.), *Anthropomorphism, Anecdotes, and Animals* (pp. 50–58). Albany: State University of New York Press.
Hankin, S. J. (2006). Not a living room sofa: Changing the legal status of companion animals. *Rutgers Journal of Law and Public Policy, 4*, 314.
Harris, C. R., & Prouvost, C. (2014). Jealousy in dogs. *PLoS One, 9*(7), e94597.
Harrison, R. (1964). *Animal Machines: The Factory Farming Industry*. London: Vincent Stuart.
Jamieson, D., & Bekoff, M. (1996). Afterword: Ethics and the study of animal cognition. In M. Bekoff & D. Jamieson (Eds.), *Readings in Animal Cognition Readings in Animal Cognition* (pp. 359–371). Cambridge: MIT Press.
Keeley, B. L. (2004). Anthropomorphism, primatomorphism, mammalomorphism: Understanding cross-species comparisons. *Biology and Philosophy, 19*(4), 521–540.
Kramer, S. C., Friedmann, E., & Bernstein, P. L. (2009). Comparison of the effect of human interaction, animal-assisted therapy, and AIBO-assisted therapy on long-term care residents with dementia. *Anthrozoös, 22*(1), 43–57.
Leudar, I., Costall, A., & Francis, D. (2004). Theory of mind: A critical assessment. *Theory & Psychology*, 14, 571–578.
Lockwood, R. (1989a). Anthropomorphism is not a four-letter word. In R. J. Hoage (Ed.), *Perceptions of Animals in American Culture* (pp. 41–57). Washington, DC: Smithsonian Institution Press.
Lockwood, R. (1989b). Anthropomorphism and the assessment of animal welfare: How golden is the Golden Rule. *Animal Behavior Society*, Northern Kentucky University, June 1989.
Lockwood, R. (2018). Animal hoarding: The challenge for mental health, law enforcement, and animal welfare professionals. *Behavioral Sciences & the Law, 36*(6), 698-716.
Lorenz, K. Z. (1952). *King Solomon's Ring*. London: Methuen.
Lorenz, K. Z. (1966). *On Aggression*. London: Methuen.
Lucretius (1715 translation). *On the Nature of Things*. London: J. Matthews. https://www.google.com/books/edition/T_Lucretius_Carus_Of_the_Nature_of_Things, accessed 7/3/2022.
Markowitz, H. (1982). *Behavioral Enrichment in the Zoo*. New York: van Nostrand Reinhold.
Matlack, C. B. (2006). *We've Got Feelings Too!: Presenting the Sentient Property Solution*. Davidson, NC: Log Cabin Press.
McMillan, F. D. (2003). A world of hurts—is pain special?. *Journal of the American Veterinary Medical Association, 223*(2), 183–186.
Melson, Gail F., Kahn, P. H. Jr, Beck, A., & Friedman, B. (2009). Robotic pets in human lives: Implications for the human–animal bond and for human relationships with personified technologies. *Journal of Social Issues, 65*(3), 545–567.
Miklósi, Á. (2014). *Dog Behaviour, Evolution, and Cognition*. Oxford: Oxford University Press.
Mills, B. M., & Wise, S. M. (2015). The writ de homine replegiando: A common law path to nonhuman animal rights. *George. Mason University Civil Rights Law Journal, 25*, 159.
Morgan, C. L. (1894). *An Introduction to Comparative Psychology: With Diagrams* (Vol. 27). New York: W. Scott.
Nagel, T. (1974). What is it like to be a bat? *Philosophical Review, 83*, 435–450.
Nawroth, C., Langbein, J., Coulon, M., Gabor, V., Oesterwind, S., Benz-Schwarzburg, J., & von Borell, E. (2019). Farm animal cognition—linking behavior, welfare and ethics. *Frontiers in Veterinary Science, 6*, 24.
Nonhuman Rights Project (2018). *Petition on Behalf of Happy v. James J. Breheny*. (Filed October 21, 2018, index 18-45164).
Pavlov, I.P. (1927). *Conditioned Reflexes*. (1960 reprint). New York: Dover Books.
Primatt, H. (1776). *The Duty of Mercy and the Sin of Cruelty to Brute Animals* (1992 reprint). Sussex: Centaur Press.
Reid, P. (2012) *Dog Insight*. Wenatchee, WA: Dogwise Publishing.
Romanes, G. J. (1885). *Mental Evolution in Animals*. London: Kegan Paul, Trench & Co.
Serpell, J. (Ed.) (2017). *The Domestic Dog: Its Evolution, Behavior and Interactions with People*. Cambridge: Cambridge University Press.
Servais, V. (2018). Anthropomorphism in human–animal interactions: A pragmatist view. *Frontiers in Psychology, 9*, 2590.

Sharkey, N., & Sharkey, A. (2010). The crying shame of robot nannies: An ethical appraisal. *Interaction Studies, 11*(2), 161–190.

Sheets-Johnstone, M. (1992). Taking evolution seriously. *American Philosophical Quarterly, 29*, 343–352.

Shibata, T., & Wada, K. (2011). Robot therapy: A new approach for mental healthcare of the elderly— A mini-review. *Gerontology, 57*(4), 378–386.

Singer, P. (1977). *Animal Liberation*. New York: Avon Books.

Skinner, B. F. (1938). *The Behavior of Organisms: An Experimental Analysis*. New York: Appleton-Century-Crofts.

State of New York Court of Appeals. (2022). No. 52, In the Matter of Nonhuman Rights Project, Inc. &c., Appelant, v. James J. Brehency &c., et al., Respondents (June 14, 2022).

Steiner, G. (2010). *Anthropocentrism and its Discontents: The Moral Status of Animals in the History of Western philosophy*. Pittsburgh, PA: University of Pittsburgh Press.

Tamura, T., Yonemitsu, S., Itoh, A., Oikawa, D., Kawakami, A., Higashi, Y., ...& Nakajima, K. (2004). Is an entertainment robot useful in the care of elderly people with severe dementia? *The Journals of Gerontology Series A: Biological Sciences and Medical Sciences, 59*(1), M83–M85.

Timpano, K. R., & Shaw, A. M. (2013). Conferring humanness: The role of anthropomorphism in hoarding. *Personality and Individual Differences, 54*(3), 383–388.

Tinbergen, N. (1958). *Curious Naturalists*. London: Country Life.

Turkle, S. (2010). In good company?: On the threshold of robotic companions. In Y. Wilks (Ed.) *Close Engagements With Artificial Companions* (pp. 3–10). Philadelphia, PA: John Benjamins.

Waters, C. N., et al. (2016). The Anthropocene is functionally and stratigraphically distinct from the Holocene. *Science, 351*, 6269.

Wathes, C. (2010). Lives worth living?. *Veterinary Record, 166*(15), 468–469.

Watson, J. B. (1919). *Psychology: From the Standpoint of a Behaviorist*. Philadelphia, PA: Lippincott.

Wells, H. K. (1956). *Ivan P. Pavlov: Toward a Scientific Psychology and Psychiatry*. New York: International Publishers.

Wilson, E. O. (1975). *Sociobiology: The New Synthesis*. Cambridge, MA: Harvard University Press.

# 8
# PETS AND HOUSING
## A One Health One Welfare Issue

*Brinda Jegatheesan, Elizabeth Ormerod, Taryn M Graham,*
*Wendy Stone, Emma R. Power, Debbie Rook, and Sandra McCune*

### Introduction

Many countries are experiencing a paradigm shift in understanding of the role of companion animals in society. A growing body of evidence demonstrates that having a bonded relationship with an animal can positively influence human health, with benefits across the lifespan (Hodgson & Darling, 2011) and may create ripple effects that extend beyond the individual to non-pet owners and the broader community (Wood et al., 2005). Increasingly, people are considering pets as important members of their families. Demographic and cultural shifts are partly responsible for the popularity of pets, including smaller family size; child free status; geographical separation; higher levels of divorce; and an aging population (Franklin & Tranter, 2011; Graham et al., 2018; Ormerod, 2008a). More people now live alone, and many suffer from stress, loneliness, and depression (Cacioppo & Hawkley, 2009; Franklin & Tranter, 2011), which have become especially prevalent during the COVID-19 pandemic (Salari et al., 2020). People may be turning to animal companionship to help prevent or mitigate these effects (Wells et al., 2022). However, a recent review (Herzog, 2022) found that in many studies, the mental health of pet owners was not measurably better than that of people without a companion animal during the pandemic lockdowns. Nonetheless, several surveys show pet owners are convinced they have personally experienced the mental health benefits of pet ownership (Merkouri et al., 2022).

Despite growing social recognition of the potential value and importance of companion animals within households (Power, 2008), lack of pet-friendly housing options can influence who gets to live with pets, where and how. Renters in many jurisdictions face challenges relative to homeowners (Carlisle-Frank et al., 2005; Power, 2017; Saunders et al., 2017; Stone et al., 2021), and such restrictions on pets affect households across the lifespan. For example, single-parent families that rent, experience elevated risks (Graham & Rock, 2019), while younger adults with pets face challenges trying to find a place to rent when leaving home (Graham et al., 2018). Older adults also face challenges whether wishing to age-in-place or needing to move into more supportive or assisted-type housing scenarios (Fox & Ray, 2019; Toohey & Rock, 2019).

Notably, for many individuals, home is within an institution. Although animals are increasingly being introduced to health and social care facilities such as long-term or retirement homes (Kogan, 2001), prisons and psychiatric hospitals (Ormerod, 2008b), for example, as part of animal-assisted interventions, many older adults may wish to keep pets themselves but are all too often denied this choice, due to housing providers operating no-pet policies (Matsuoka et al., 2020). The need to nurture animals is strong and should not be denied (Kellert & Wilson, 1993). Allowing people to continue living with their pets, should they wish, and to remain together, should they move, can prevent an avoidable grief and perhaps even help mitigate relocation stress syndrome (Ormerod, 2012).

A One Health framework recognizes the interdependence of human health and animal health against the backdrop of a shared environment, yet tends to focus on threats, such as zoonotic diseases, food safety, or antimicrobial resistance. In response, a One Welfare framework recognizes the social interconnectedness between human well-being, animal welfare, and the integrity of the environment and how improvements to one influences others (Pinillos et al., 2016). Pet ownership in housing is a One Health/One Welfare issue, as previous studies report various health and social benefits to owning a companion animal; however, opportunity to experience these benefits depends heavily on the types of housing and neighborhoods in which people can afford to live (Graham et al., 2018; Power, 2017). In this chapter, we overview connections between companion animal ownership and housing pathways. We overview how housing opportunities are shaped by pet-connected housing policies and explore the implications for pet-inclusive households and their companion animals, including in times of crisis such as homelessness and natural disaster. We also overview the opportunities provided by pet-friendly housing, alongside mitigating potential risks.

## Companion Animals within a Housing Pathways Perspective

Housing pathways is an important framework in housing studies to conceptualize the experiences that households have within the housing system. The approach is valuable in bringing analysis of resident experiences together with knowledge of institutional settings in housing systems, including everyday practices of governance and administration. Focus is on the ways that households navigate housing systems, including accessing, securing, living in, and moving housing. Research shows that companion animals play a significant role across all elements of housing pathways.

First, in many households, companion animals are considered key family and household members (Power, 2008). More than simply possessions found inside a house, pets are an important part of how many households imagine and make a home. Consequently, in many households, pets are actively considered within housing choices and decisions. This can include decisions relating to moving house, for example, prioritizing housing that is in proximity to parks for dog walking or away from risks like major roads that could cause injury to roaming pets. They can also influence preferences for particular features such as the presence of a yard or secure fencing. Pets can also shape decisions internal to a house, e.g. internal furnishing and decoration (such as choosing flooring types that are easy to clean), or reorganizing and furnishing housing in ways that meet pets' unique needs. Veterinary specialists Harvey and Malik (2020), for instance, advocate for pet-friendly features such as the need for cats to have vertical space as well as horizontal space inside a house, 'multiple points of safety and seclusion', access to warmth via the sun or a heater, and secure window finishings. In Japan in 2020, reasonably priced studios were fitted with cat-friendly and safe amenities to encourage cat-loving people to keep their pets living indoors by educating them on what keeps their feline companions satisfied. The model apartments were built after the animal welfare center began receiving an increasing number of complaints from neighbors about more strays due to people feeding them and flower beds being destroyed as a result of cats using them as litter boxes (Hamamoto, 2022).

In addition to household preferences, housing pathways with companion animals are also shaped by the institutional context, particularly cross-tenure housing policy. When pets are restricted or banned from housing, families that include companion animals can face substantial challenges accessing and securing a place to live. In the private rental sector in Australia, for instance, Power (2017) identifies pets as a pathway to housing insecurity. The limited availability of pet-friendly rentals made it difficult for families with pets to identify and secure housing, and led many to report making financial and geographical compromises.

A lack of housing that allows pets can lead to residents taking the risk of not declaring their companion animal to the landlord or property manager (Power, 2017). Renters that did not declare their pets described the ongoing stresses of this housing decision, underpinned by the risk of eviction or poor landlord references if the illegally kept pet was identified by the landlord. In addition to shaping where households live, this shows how pets can also shape experiences within housing – shaped in this case by pet-restrictive housing policy.

Pets can also shape housing pathways and opportunities at times of crisis, including in the context of financial or other household-scale crises such as homelessness and domestic partner violence. People may delay leaving abusive domestic relationships, putting themselves and their animals in prolonged danger, if they cannot find emergency sheltering options that allow pets (Wuerch et al., 2020) or are unaware of them, due to language barriers, dependency on their partner, and/or social isolation. Animal welfare organizations should advertise widely about any emergency sheltering options they have, to help spread awareness and reduce barriers to help-seeking (Jegatheesan et al., 2020).

People with lower incomes may be forced to choose homelessness rather than be separated from their companion animal. In one study, 18% of homeless respondents reported refusing housing that was not pet-inclusive (Cronley et al., 2009). Pet-inclusive housing options are required to support such households. This includes a combination of temporary shelter solutions and permanent housing options, wherein restrictions on size of pets should not apply (Graham et al., 2018).

Families that are already disadvantaged within the housing system are more likely to be affected, with pet ownership exacerbating existing challenges in accessing secure housing. Low-income households experience particular risks. Housing advertised as pet-friendly tends to receive more applicants than housing that does not allow pets (Graham & Rock, 2018; Power, 2017), which allows housing providers to be more selective (Toohey & Krahn, 2018). During periods of housing under-supply or high demand for pet-friendly properties, landlords may also charge extra for pet ownership (Graham & Rock, 2018), which puts pet owners with greater financial resources at an advantage. By contrast, pet owners with fewer resources may be forced to accept lower-quality housing, in less-desirable neighborhoods, in order to keep their pets (Graham et al., 2018; Koohsari et al., 2022; Power, 2017). There is some evidence that pet owners may stay longer in their rentals as compared to tenants without pets (Graham & Rock, 2018; Power, 2017), which is often promoted as a reason for housing providers to open their properties up to pet owners, given the potential to keep turnover costs low (Carlisle-Frank et al., 2005). However, research shows that tenants with pets report staying put, not because they feel happy with their housing *per se*, but because of how difficult it is to find housing that allows pets and because of extra pet charges they will never recoup (Graham et al., 2019).

Restrictions and bans on pets in housing may lead some pet owners to limit seeking necessary health support. For example, there is evidence that pet care responsibilities can impact decisions to seek medical care (Applebaum et al., 2020; Polick et al., 2021). Healthcare practitioners could prevent this by including animal companions in care plans. People in the helping professions, and especially in housing, are frequently unaware that such bonds should be respected and supported (Matsuoka et al., 2020) and further policies are needed to protect such bonds.

Pets can also shape housing pathways during natural disasters. Pet owners may risk their lives and that of their pets, if they cannot take their animals with them (Travers et al., 2017). This was seen following Hurricane Katrina in the United States, for example, when many experienced animal loss (Hunt et al., 2008). The Federal Pets Evacuation and Transportation Standards (PETS) was enacted as a result of Hurricane Katrina, requiring pets to be saved during emergencies (Mike et al., 2011). In 1995, after the Great Hanshin earthquake in Japan, elderly pet owners in Kobe spent the winter outdoors because new housing, built as emergency centers, prohibited pets. City officials were perplexed when the newly built housing offered was rejected by those with pets. The properties had 'no pet' policies. It was only when housing officials met the older people and their pets that they fully appreciated the strength of their bond. Local animal welfare groups introduced the Pet Committee concept to the city's housing committee and helped them to understand the value of well-managed, positive pet policies (Ormerod, 2022).

Re-housing challenges also emerge after natural disasters. In a flood case study in Canada (Graham & Rock, 2019), demand for pet-friendly housing outweighed available supply in the year following the flood. With flood-diminished housing supply, landlord restrictions on the species, size or breed of pet created further challenges for households with pets, with dog inclusive households particularly affected. Tenants with pets frequently had to make compromises on where they lived (Graham & Rock, 2019). Restrictions and bans on pets in housing intensify existing challenges as a consequence of overall housing supply reducing (i.e. with flood affected properties removed from the market), while the number of households competing for available housing increases (including tenants and homeowners forced to relocate from flood affected properties).

## Benefits and Risks of Allowing Companion Animals in Housing

So far, we have set out some ways that companion animals can shape housing pathways, particularly in the context of changing relationships between people and pets, pet-restrictive housing policy and in times of crisis. In this section, we turn to consider the benefits and risks of allowing companion animals in housing, and how these are mediated through housing policy and related programming and support.

### *Benefits of Allowing Companion Animals in Housing*

With the potential benefits of pets achieving growing awareness, and pets assuming increasingly significant positions within households, there is growing recognition of the potential benefits of allowing pets within housing. Benefits can accrue at both individual and community levels.

Individual benefits of living with companion animals include positive health and well-being outcomes. In New Zealand, the attitudes and emotional well-being of older women in subsidized housing where cats were allowed or banned were compared (Mahalski et al., 1988); the positive effects of allowing cats outweighed the negative effects, and the 'no-cat' policy was reversed in the buildings where cats had been previously banned. Similarly,

Headey et al. (2007) reported on the effects of allowing pets in urban areas in China following a ten-year ban. Thereafter, dogs have become popular pets and presented the opportunity for a 'natural' experiment comparing dog owners and non-owners. Participants were all women, and those with dogs reported taking more exercise, improved sleep, better fitness and health. Their work attendance was better, and they made fewer doctor appointments.

Marginalized and disadvantaged communities can gain particular individual benefits from animal companionship. A Canadian study found the odds of being depressed were three times greater for homeless youth who did not own pets (Lem et al., 2016). They found that despite pet ownership among homeless youth having many liabilities, including impairing their ability to access shelter, services, and housing and employment opportunities, companion animals may offer both physical and psychosocial benefits that youth have difficulty attaining.

Benefits can also accrue within communities. In the United States, California housing managers who expressed concerns prior to the introduction of a state law allowing pets in social assisted housing for older people were found to be favorable in follow-up (Hart & Mader, 1986). They found the animals improved mental attitude, increased regular exercise, provided security, and created a sense of community and, while only a minority adopted pets after the fact, many others were sharing in pet care in ways that encouraged friendship between residents.

Companion animals can also be creators of social capital, facilitating the development of safer, friendlier communities by bringing neighbors together (Wood et al., 2015). Pet owners are more likely to report social benefits such as helpfulness, friendliness and trust between neighbors, supporting claims that pet ownership is a valuable and positive feature in community and neighborhood life (Wood et al., 2017). Dog ownership, in particular, is associated with higher social capital with the strongest associations found for owners who walked their dogs and may be in part related to dog walking as a mechanism for improved natural surveillance and increased perceptions of neighborhood safety (Christian et al., 2016). Pinchak and colleagues (2022), testing whether dog ownership is a deterrent to offenders by providing more 'eyes on the street', found that higher concentrations of dog owning households were related to lower criminality in neighborhoods.

National health surveys from both Germany and Australia have shown that pet owners may use less medication and make about 15% fewer visits to doctors per year than non-owners, and that those who have owned pets the longest are the healthiest (Headey & Grabka, 2007). These reductions in healthcare usage mean that pet ownership potentially saves billions of dollars/Euros in government and private healthcare expenditure annually (Headey et al., 2002).

Pets can facilitate positive interactions with housing providers too. Some housing providers found that renting to dog owners provided companionship for their own pets, for instance, if a landlord rented out a basement unit of their home, so long as the pets met beforehand and sharing of yard space had been discussed (Graham et al., 2018). Important to the success of this scheme – and for any other housing providers opening their properties up to pet owners – is careful planning and access to good advice and support services to help keep families together and promote benefits as well as mitigate perceived risks.

## *Risks Connected with Companion Animals in Housing*

The potential for noise from pets is often identified as a risk, particularly in denser urban neighborhoods or apartment/condominium settings, where noise can more readily flow between housing units (Carlisle-Frank et al., 2005). There are also potential health risks, including injury if animals bite or are aggressive to household or non-household members in the broader community (Rock et al., 2014). Companion animals can host zoonoses and there is risk of some infections transferring from animals to people, such as ringworm in cats (Haverkos et al., 2011). The preference of some people to not live near animals, including due to fear or allergy, can also drive challenges.

However, risks brought by companion animals are often exaggerated, may not be rooted in actual experience, and can frequently be mitigated through education and responsible care (Carlisle-Frank et al., 2005). Pet-inclusive housing policy and housing design that recognizes the needs of companion animals, the households and communities that they are part of, can play an important role in actively mediating potential risks and supporting positive relationships between people and companion animals. For example, design that minimizes sound transfer between apartments/condominiums can reduce noise transfer from both people and companion animals, avoiding a common source of complaint within dense urban neighborhoods (Koohsari et al., 2022; Power, 2015, 2018). Similarly, housing policy can provide opportunities for neighbors to identify and manage persistent problems, as discussed later.

One of the key risks that needs to be managed via coherent policy settings is the real or perceived risks associated with pet related damage to property. In the Australian context, for example, there has been policy debate about how to mitigate financial risks. Approaches range from pet-restricted housing (banning pets), to payment of pet-bonds where animals are present within residential property leases, to increased insurances required for property owners (Stone et al., 2021). Despite the perception of risk, however, there is scant international evidence in support of a pet-property damage association. Indeed, US research in 2005 indicates that every-day property damage in the form of wear and tear, is equivalent among pet and non-pet inclusive households that rent (Carlisle-Frank et al., 2005).

### *Risks from Not Allowing Companion Animals in Housing*

Restrictions and bans on pets in housing engender risks, both to people, their pets, and community well-being, as described earlier with regard to housing pathways. Further risks can come from poorly designed legislation, and again, lower income households can face elevated risks. An example of this is a policy that allows landlords to charge extra for pet ownership, which is common in some jurisdictions (e.g. Western Australia, see Stone et al., 2021; Alberta, Canada, see Graham & Rock, 2018; parts of the United States; see Applebaum et al., 2021). Arguments for additional pet charges include that this can help landlords to manage property damage that may occur as a result of pets being kept within a rental property. However, such policy can place undue burden on low-income households (Applebaum et al., 2021). On the other hand, in contexts where pet-restrictive policy is the norm, it is possible that additional pet charges may support the supply of pet-friendly housing. These tensions and emerging consequences were seen in the context of England's Tenancy Fees Act 2019. Designed to reduce financial burdens on renters, this act had the effect of removing the right of private landlords to require tenants to pay a reasonable deposit against possible pet damage. Consequently, more landlords prohibited pets and many of those that continued to allow pets charged a non-refundable 'pet rent' (Berezai, 2021). However, in 2022, the government published its white paper on A Fairer Private Rented Sector, which includes amending the Tenancy Fees Act to permit landlords to require pet damage insurance (Department for Levelling up, Housing and Communities, 2022). A recent UK survey found the majority (62%) of pet owners were willing to pay pet damage insurance and/or a pet deposit (YouGov 2021).

Restrictions and bans on pets in housing also have animal welfare consequences. Housing issues are a leading cause of pet surrender to animal shelters (Coe et al., 2014; New et al., 1999) and vulnerable populations are especially at risk (Eagan et al., 2022; Ly et al., 2021). The anguish felt by families who have relinquished animals may be worse than other forms of pet bereavement, given there is no closure and such loss carries much guilt. For example, John Chadwick in the United Kingdom was lost to suicide ten days after giving up his pet dogs and cat in order to secure social housing and avoid homelessness (Bell, 2017). Maidstone Borough Council, the local authority that had a duty to help John as a homeless person, subsequently introduced the John Chadwick Pet Policy (https://www.scas.org.uk/give-up-your-pets-or-your-home-in-loving-memory-of-john-chadwick/), to help ensure people with pets experiencing homelessness are offered pet friendly accommodation. This policy provides a useful precedent and could be implemented elsewhere.

## Global Legislation and Future Directions

Globally, the combination of contextual, cultural and local housing system conditions shapes the extent to which pets are able to be included within housing and living arrangements as well as how such inclusion can be practiced. Institutional settings reflect country and cultural contexts and in turn shape housing and shelter outcomes and the lives of residents, in a process of mutually reinforcing norms and practices. Path dependency thinking (Malpass, 2011) asserts that policies, programs and practices that are built around established norms, become enmeshed and in turn reinforced via resourcing, everyday practices, ongoing processes of policy decision-making and regulatory implementation. Policy-making built upon norms, regulatory regimes and practices, can be more or less pet and animal tolerant and accepting, or exclusive and risk-averse, or may be some combination of approaches – and will be more or less rigid according to the ways policy is constructed and reformed within housing regimes (Soaita et al., 2021). The combination and sum of 'what has always been' can act as a barrier to regulatory and practice changes within housing or other fields of social policy.

Yet, points of innovation and practice change-making within laws, and policy and practice realms that support these, can also be identified. With regard to pets and housing, key change-points range from multi-national influences of organizations such as the United Nations or World Health Organization, to local level 'disruptors' such as city leaders or even citizen-led court-based appeals for law reform. The International Association of Human Animal Interaction Organizations (IAHAIO) has issued a series of Declarations regarding companion animals in society that it recommends should be adopted by international bodies and national and local governments. The IAHAIO Geneva Declaration (1995) acknowledges the need for *"the universal non-discriminatory right to pet ownership in all places and reasonable circumstances, if the pet is properly cared for and does not contravene the rights of non-pet owners"*. The IAHAIO Tokyo Declaration (2007) goes further to state: *"It is a universal, natural and basic human right to benefit from the presence of animals"*, highlighting the need for housing policy, in particular, to *"allow the keeping of companion animals if they can be housed properly and cared for adequately, while respecting the interest of people not desiring direct contact with such animals"*.

Pet-inclusive housing policy must be made holistically. Rather than trying to promote pet-friendly housing as an 'economic opportunity' for landlords, advocates should work toward fair housing policies that make it illegal to discriminate based on family or household status (including pet ownership) and toward supporting landlords, including making sure pet owners comply with housing rules concerning any nuisance concerns and the cleanliness of their units. Support should be provided to pet-owning tenants as well, to avoid or resolve any behavioral problems that could cause disturbance in rental housing (e.g., barking) or damage to the housing unit (e.g., by making sure to not leave pets alone for long periods of time, ensuring a pet receives enough stimulation and exercise, or by taking training classes) (Graham et al., 2018). The creation of a Pet Committee can greatly enhance the smooth running of pet-keeping in communities and is especially recommended for condominiums and flats. Members should include pet owners, residents without pets, a veterinarian, a knowledgeable representative from a local humane group, among others. The Committee can draft pet-keeping rules for approval by management; orient new residents to the rules and provide them with pet care materials; act as first-line for complaints related to pets, strays or wildlife and seek to resolve these. A Pet Committee only refers situations to management if these cannot be resolved, thus alleviating management of much of the responsibility for pet-related issues. This successful concept was developed in the United States and has been introduced to other countries. Published Pet Committee rules can be considered and adapted to suit any housing type (ASPCA, n.d.).

National registers of pet-friendly landlords, enabling pet owners to find accommodation, have been established. These include Lets with Pets (Dogs Trust) and the Purrfect Landlord (Cats Protection) in the United Kingdom and several in the United States (e.g. Abodo, My Pitbull is my Family, People with Pets).

Policies such as extra pet charges require additional consideration because they disproportionately harm populations that are already economically and socially disadvantaged (low-income and non-White renters as well as older adults), thereby contributing further to their housing insecurity (Matsuoka et al., 2020; Toohey & Krahn, 2018). Rather than band-aid responses (such as permitting pet-bonds) action should be taken to ensure that housing policy is pet-inclusive across tenure and sector. However, when this is not possible and if pet charges must be made, particular regard should be given to the economic and housing security of lower-income households. For example, the amount should be a percentage of total monthly rent, be capped, and be refundable, to incentivize good tenancy (Graham et al., 2018) and so as to not disadvantage those who may already be struggling to afford housing.

Pet-owning tenants often have little bargaining strength in negotiations with the landlord. The landlord is able to dictate the terms of the lease regarding pets and can readily find another tenant if a prospective or current tenant is unhappy with a 'no pet' covenant in the lease. To redress this imbalance, and promote fairness between the parties, some countries and states have passed legislation regulating the keeping of pets in rental housing. Examples include Ontario, Canada, and more recently, Victoria, Australia. In Ontario, a blanket ban on pets in leases is void:

> A provision in a tenancy agreement prohibiting the presence of animals in or about the residential complex is void.
>
> *(section 14, Residential Tenancies Act 2006)*

The tenant's right to keep companion animals is not unfettered and provisions exist to balance the conflicting interests of the parties (Rook, 2018). For example, the landlord can apply to the Landlord and Tenant Board to evict a tenant with a companion animal where the pet causes a nuisance to other tenants, causes them to suffer a serious allergic reaction, or threatens the safety of the landlord or other tenants. Failure to abide by animal control bylaws applicable in the local jurisdiction, for example, any licensing or micro-chipping requirements or the need to pick up dog feces, constitutes an illegal act that is a ground for terminating the tenancy. The Ontario legislation provides a valuable example of how to regulate the use of 'no pet' covenants in a way that balances the competing interests of the respective parties, permitting responsible pet owners to live with well-behaved pets while also addressing the needs of landlords and other tenants to protect their property or neighborhood. That said, there are loopholes around the law, especially in Toronto, Ontario's largest city, where housing stock is low and demand high, so it is not uncommon to see rental ads stipulating 'no pets allowed' (Toohey & Krahn, 2018). Once enacted, recommended policies must therefore be actively enforced and legal aid may also be needed, to help make tenants aware of their rights.

In 2020, the state of Victoria in Australia gave tenants the right to request to keep companion animals in their rental housing. Tenants must first obtain the landlord's consent which can only be refused by order of the Victorian Civil and Administrative Tribunal. The legislation includes similar exceptions to those in Ontario, Canada that protect the interests of landlords and neighboring tenants. In 2022, the English government announced proposals to pass legislation as part of its Renters Reform laws to give tenants a right to request to have a pet in their rental housing (Department for Levelling up, Housing and Communities, 2022). Private landlords will not be able to unreasonably refuse consent and tenants will have the right to challenge a landlord's refusal. To balance the interests of landlords and meet their concerns about potential property damage, landlords will be able to request that tenants have insurance to cover the cost of any damage to the property caused by their pet.

Even within one country context, pet-inclusive or permissive regulatory complexities across housing manifest in sometimes incoherent and inconsistent ways. Pet permissive laws and practices may vary across, for example, housing tenures (such as ownership and rental or other living arrangements), types of housing stock (such as free-standing detached housing compared with apartments/condominiums, for example) and location of housing (for example within high- or low-density neighborhoods, or across provincial boundaries). In some country contexts, the tenure, stock and locational differences may additionally be variable for population cohorts, where rights vary according to citizenship status and administration of such categories.

The federated context of Australia is one example of a national context in which approaches to pets and housing vary in multiple and complex ways. In their review of national policy and practice of housing and pets, Stone, Power et al. (2021) reviewed policies relating to pets and housing within housing tenures (rental, ownership and other forms of living arrangements such as crisis accommodation or residential van parks); types of housing stock (with a particular focus on restrictions associated with strata title affecting high-density housing design); across state and territory (provincial) internal country borders. In addition to finding inconsistent and highly variable definitions of 'animal', 'pets', 'companion animal', and so on, the extent to which households in this context are afforded rights to include animals in their housing varies primarily along two lines across Australian states and territories. Most notably, residents without property ownership (private tenants and tenants of social/government owned housing) generally face more restricted access to pet housing than homeowners, as did those living in higher density housing complexes (including shared 'strata' title). In the absence of a coordinated national approach to housing policy, the authors argue, divergent and often exclusive and risk averse, rather than inclusive and progressive, pet policies will remain a feature of at least some parts of the country, the impacts of which have most restrictive outcomes for renters and people with limited tenure security.

In Europe, France was the first country to implement legislation to regulate the use of 'no pet' covenants. It enacted legislation in 1970 stipulating that any prohibition of pets in residential tenancies is deemed to be void (Article 10 of the Law of 9 July 1970). The fact that legislation has existed in France for over 50 years without the need to revoke it, and the fact that the scope of the law was extended in 2011 to include holiday rentals *(Association Union fédérale des consommateurs de l'Isère – Que Choisir v. Association Clévacances Isère)*, demonstrates that legal governance of 'no pet' covenants can successfully work in practice. Understanding the component parts of the success of this enactment, and how these can be modified and adapted for other international contexts is critical.

The constitution of India 1949 recognizes and grants every citizen the right to choose to live with a companion animal and that pet owners are to be free from intimidation to give up their pets. Housing associations and landlords are prohibited from banning pets regardless of the breed type and size of dogs. In Singapore's public housing of more than a million flats, 63 dog breeds are approved by the country's Housing and Development Board. Cats are not permitted due to concerns about difficulties in keeping them solely indoors, wandering around the housing estate, shedding, urinating and defecating in public spaces (e.g., corridors, stairways) and neighbors being disturbed due to caterwauling sounds. In predominantly Islamic countries such as Malaysia and Indonesia, religious views of dogs (Jegatheesan, 2019) and subsequent restrictions on dogs in apartments and homes (owned or rented) are strictly enforced (Howell et al., 2022).

Utilizing human rights' arguments to protect the human-companion animal relationship within housing policy and law is one avenue that requires further consideration. There is a strong argument that the human-companion animal relationship falls within 'private life and family' under Article 8, European Convention on Human Rights, an argument that has succeeded in a housing case in Belgium (Civ. Liege 21 October 1986, J.L.M.B., 1987).

Article 8 states:

*"Everyone has the right to respect for his private and family life, his home and his correspondence."*

Defining family membership is infused with complexity and controversy. In recent years, the concept of family has been extended to recognize the Cultural Construction of Kinship (Charles & Davies, 2008) and embrace the idea of 'families we choose' (Weston, 1991). In many places across the world, it is illegal to discriminate against family status in access to housing and pets should be protected and recognized as such within housing policy (Graham, 2019).

## Conclusion

Existing evidence indicates that the range of perceived costs and risks to landlords and property owners/managers associated with more pet-permissive housing, pose distinct challenges for the ability of households and communities to achieve the relative benefits of wider pet ownership. Current practices privilege those who can afford extra pet charges, leaving a wide range of pet owners at risk of relinquishment or eviction (Applebaum et al., 2021). Further, these risks and costs are shaped by country and locational legislative frameworks and cultural norms and practices in reinforcing ways. In order to more confidently open the way for a more pet-inclusive housing future, landlord investors, property owners and managers, as well as investors in larger-scale rental industry and property groups, require evidence-based education about the benefits of pet ownership (health, wellbeing, community safety, long-tenancies and maintenance of properties) as well as education to off-set perceived risks (property damage and associated costs, anti-social behaviors). Part of such education is establishing a more solid global evidence-base, building on existing research that indicates that pet owning tenants do not do any more damage than other types of tenants.

From an urban planning perspective, greater attention paid toward housing and neighborhood design could better support pets and their people, to prevent any damages or nuisance. Greater collaboration between animal services and human social services could also help reduce the risk of relinquishments due to behavior concerns or housing issues. For example, traditional safety net programs in shelters which focus on connecting with owners of companion animals at risk of relinquishment and if possible, to address issues that may result in the animal being kept in the home (Weiss, 2015). A more universal approach is to recast the role of landlords and property investors as housing providers, ensuring that these property owners and managers have a holistic and thorough understanding of their responsibilities to provide housing for all household types, with pets considered family members, as well as the insurances and finances in place to support the provision of quality housing, whether pets are included in households or not (Stone et al., 2021). Lastly, housing policy should reflect the significant roles that companion animals play as considered members of the family.

## Questions for Discussion

1. Why is the topic of pets and housing an important focus for research and policy development?
2. How do restrictions on pets in housing impact human health and development and animal welfare?
3. What policies/practices/programs exist that can protect pets in housing? What else can be done to support people to keep pets in their housing?
4. Climate change is predicted to increase the frequency and intensity of natural disasters. What housing challenges do households with pets face during and in the period after a natural disaster? What policies, practices or programs might help to address these challenges?
5. How do restrictions on pets, especially dogs, in housing and temporary shelters impact on pet owners living with homelessness?

## References

Abodo. (n.d.) *Pet-Friendly Apartment Rentals Allowing Dogs and Cats.* https://www.rentable.co/pet-friendly

Anchor Housing Trust. (n.d.) *Pets Policy: Anchor Retirement Home Pet Policy* https://www.anchor.org.uk/guides-and-support/pet-friendly

Animal and Veterinary Service. (n.d.) https://www.nparks.gov.sg/avs

Animal Welfare Board of India. (2019) *Laws for Pets in India & Rights of Pet Owners — Know the Guidelines by Animal Welfare Board,* May 23, 2019. https://www.lawnn.com/laws-for-pets-in-india/#:~:text=Furthermore%2C%20the%20Indian%20Constitution%20has%20also%20incorporated%20provisions, to%20live%20with%20or%20without%20a%20companion%20animal

Applebaum, J. W., Adams, B. L., Eliasson, M. N., Zsembik, B. A., & McDonald, S. E. (2020) How pets factor into healthcare decisions for COVID-19: A one health perspective. *One Health, 11,* 100176.

Applebaum, J. W., Horecka, K., Loney, L., & Graham, T. M. (2021) Pet-friendly for whom? An analysis of pet fees in Texas rental housing. *Frontiers in Veterinary Science 8,* 767149.

Bell, G. (2017) John Chadwick took his own life when told he could not live with pets in Maidstone. *KentOnLine.* https://www.kentonline.co.uk/maidstone/news/man-took-his-life-after-125128/

Berezai, J. (2021) *Heads for Tails! How Amending the Tenant Fees Act Could Be the Answer to More Pets in Rented Accommodation.* https://www.advocatseastmids.org.uk/wp-content/uploads/2021/09/Heads-for-Tails-September-2021.pdf

Cacioppo, J. T., & Hawkley, L. C. (2009) Perceived social isolation and cognition. *Trends Cognitive Sciences, 13*(10), 447–454.

Carlisle-Frank, P., Frank, J. M., & Nielsen, L. (2005) Companion animal renters and pet-friendly housing in the US. *Anthrozoös, 18*(1), 59–77.

Cats Protection. (n.d.) *Purrfect Landlords.* https://www.cats.org.uk/what-we-do/campaigning/purrfectlandlords

Charles, N. & Davies, C.A. (2008) My family and other animals: Pets as Kin. *Sociological Research Online, 13*(5), 1–14.

Christian, H., Wood, L., Nathan, A., Kawachi, I., Houghton, S., Martin, K., & McCune, S. (2016) The association between dog walking, physical activity and owner's perceptions of safety: Cross-sectional evidence from the US and Australia. *BMC Public Health, 16*(1), 1010.

Coe, J. B., Young, I., Lambert, K., Dysart, L., Nogueira Borden, L., & Rajić, A. (2014) A scoping review of published research on the relinquishment of companion animals. *Journal of Applied Animal Welfare Science, 17*(3), 253–273.

Cronley, C., Strand, E. B., & Patterson, D. A., et al. (2009) Homeless people who are animal caretakers: A comparative study. *Psychological Reports, 105*(2), 481–499.

Department for Levelling Up, Housing and Communities. (2022) *A Fairer Private Rented Sector.* https://www.gov.uk/government/publications/a-fairer-private-rented-sector

Dogs Trust. (n.d) *Lets with Pets.* https://www.letswithpets.org.uk/

Eagan, B. H., Gordon, E., & Protopopova, A. (2022) Reasons for guardian-relinquishment of dogs to shelters: Animal and regional predictors in British Columbia, Canada. *Frontiers in Veterinary Science, 9,* 857634.

Fox, M., & Ray, M. (2019) No pets allowed? Companion animals, older people and residential care. *Medical Humanities, 45,* 211–222.

Franklin, A., & Tranter, B. (2011) *Housing, Loneliness and Health,* AHURI Final Report No. 164, Australian Housing and Urban Research Institute Limited, Melbourne, https://www.ahuri.edu.au/research/final-reports/164.

Graham, T. M. (2019) *Enhancing Access to Quality Rental Housing for People with Pets as Healthy Public Policy.* [Doctoral dissertation, University of Calgary]. http://hdl.handle.net/1880/109854

Graham, T. M., Lucyk, K., Diep, L., & Rock, M. J. (2019) Discrimination towards people partnered with assistance dogs in Canada: Implications for policy and practice. *Society & Animals, 30*(2), 210–245.

Graham, T. M., Milaney, K. J., Adams, C. L., & Rock, M. J. (2018) Pets negotiable: How do the perspectives of landlords and property managers compare with those of younger tenants with dogs? *Animals, 8*(3), 32.

Graham, T. M., & Rock, M. J. (2019) The spillover effect of a flood on pets and their people: Implications for rental housing. *Journal of Applied Animal Welfare Science, 22*(3), 229–239.

Hamamoto, Toshihiro. (2022, June 12) Japan pref.'s model rooms teach cat owners how to keep feline buddies happy as indoor pets. *The Mainichi.* https://mainichi.jp/english/articles/20220610/p2a/00m/0li/024000c

Hart, L. A., & Mader, B. (1986) The successful introduction of pets into California public housing for the elderly. *California Veterinarian, 40*(5), 17–21.

Harvey, A., & Malik, R. (2020) Don't let them out: 15 ways to keep your indoor cat happy. *The Conversation.* https://theconversation.com/dont-let-them-out-15-ways-to-keep-your-indoor-cat-happy-138716.

Haverkos, L., Hurley, K. J., McCune, S., & McCardle, P. (2011) Public health implications of pets: Our own animals and those of others. In McCardle, P., McCune, S., Griffin, J. A., Esposito, L. & Freund, L. (Eds) *Animals in Our Lives: Human-Animal Interaction in Family, Community & Therapeutic Settings.* Chapter 4, pp. 55–82. Brookes Publishing Company.

Headey, B., Grabka, M., Kelley, J., Reddy, P., & Tseng, Y. -P. (2002) Pet ownership is good for your health and saves public expenditure too: Australian and German longitudinal evidence. *Australian Social Monitor, 5*(4), 93.

Headey, B., & Grabka, M. M. (2007) Pets and human health in Germany and Australia: National longitudinal results. *Social Indicators Research, 80*(2), 297–311.

Headey, B., Na, F., & Zheng, R. (2007) Pet dogs benefit owners' health in a "Natural" experiment in China. *Social Indicators Research,* 2008, 87, 481–493.

Hodgson, K., & Darling, M. (2011) Pets in the family: Practical approaches. *Journal of the American Animal Hospital Association, 47*(5), 299–305.

Howell, T. J., Nieforth, L., Thomas-Pino, C., Samet, L., Agbonika, S., Cuevas-Pavincich, F., & Fry, N. E., et al. (2022) Defining terms used for animals working in support roles for people with support needs. *Animals, 12*(15), 1975.

Hunt, M., Al-Awadi, H., & Johnson, M. (2008) Psychological sequelae of pet loss following Hurricane Katrina. *Anthrozoös, 21*(2), 109–121.

IAHAIO Geneva Declaration. (1995) https://iahaio.org/geneva-declaration/

IAHAIO. (2007) *Tokyo Declaration.* https://iahaio.org/tokyo-declaration/

Jegatheesan, B. (2019) Influence of cultural and religious factors on attitudes towards animals. In Fine, A. H. (Ed) *Handbook on Animal-Assisted Therapy: Theoretical Foundations and Guidelines for Practice,* 5th edition, pp. 43–49. Academic Press.

Jegatheesan, B., Ormerod, E., Enders-Slegers, M. J., & Boyden, P. (2020) Understanding the link between animal cruelty and family violence: The bioecological systems model. *International Journal of Environmental Research and Public Health, 17,* 3116.

Kellert, S. R., & Wilson, E. O. (1993) *The Biophilia Hypothesis.* Island Press.

Kogan, L. R. (2001) Effective animal-intervention for long term care residents. *Activities, Adaptation & Aging, 25*(1), 31–45.

Koohsari, M. J., Yasunaga, A., & McCormack, G. R., et al. (2022) The design challenges for dog ownership and dog walking in dense urban areas: The case of Japan. *Frontiers in Public Health, 10,* 904122.

Lem, M., Coe, J. B., Haley, D. B., Stone, E., & O'Grady, W. (2016) The protective association between pet ownership and depression among street-involved youth: A cross-sectional study. *Anthrozoös, 29*(1), 123–136.

Ly, L. H., Gordon, E., & Protopopova, A. (2021) Exploring the relationship between human social deprivation and animal surrender to shelters in British Columbia, Canada. *Frontiers in Veterinary Science, 8,* 656597.

Mahalski, P. A., Jones, R., & Maxwell, G. M. (1988) The value of cat ownership to elderly women living alone. *The International Journal of Aging and Human Development, 27*(4), 249–260.

Malpass, P. (2011) Path dependence and the measurement of change in housing policy. *Housing, Theory and Society, 28,* 305–319.

Matsuoka, A., Sorenson, J., Graham, T. M., & Ferreira, J. (2020) No pets allowed: A trans-species social justice perspective to address housing issues for older adults and companion animals. *Aotearoa New Zealand Social Work Review, 32*(4), 55–68.

Merkouri, A., Graham, T. M., O'Haire, M. E., Purewal, R., & Westgarth, C. (2022) Dogs and the good life: A cross sectional study of the association between the dog-owner relationship and owner mental wellbeing. *Frontiers in Psychology, 13,* 903647.

Mike, M., Mike, R., & Lee, C. J. (2011) Katrina's animal legacy: The pets act. *Journal of Animal Law and Ethics, 4*(1), 133–160.

My Pitbull is Family. (n.d.) https://www.mypitbullisfamily.org/

New, J. C. J., Salman, M. D., Scarlett, J. M., Kass, P. H., Vaughn, J. A., Scherr, S., & Kelch, W. J. (1999) Moving: Characteristics of dogs and cats and those relinquishing them to 12 U.S. animal shelters. *Journal of Applied Animal Welfare Science, 2*(2), 83–96.

Ormerod, E. (2008a) Bond-centered practice: Lessons for Veterinary Faculty and Students. *Journal of Veterinary Medical Education, 35*(4), winter 2008, 545–552.

Ormerod, E. (2008b) Companion animals and offender rehabilitation - experiences from a prison therapeutic community in Scotland. *Therapeutic Communities, 29,* 3.

Ormerod, E. (2012) Supporting older people with pets in sheltered housing. *In Practice, 34,* 170–173.

Ormerod, E. (2022) The challenges of no pets accommodation. *Veterinary Times, 52*(22), 12–14.

People with Pets. (n.d.) www.Peoplewithpets.com

Pinchak, N. P., Browning, C. R., Boettner, B., Calder, C. A., & Tarrence, J. (2022) Paws on the street: Neighborhood-level concentration of households with dogs and urban crime. *Social Forces, 101*(4), 1888–1917.

Pinillos, R. G., Appleby, M., Manteca, X., Scott-Park, F., Smith, C., & Velarde, A. (2016) One welfare: A platform for improving human and animal welfare. *Veterinary Record, 22,* 412–413.

Polick, C. S., Applebaum, J. W., Hanna, C., Jackson, D., Tsaras-Schumacher, S., Hawkins, R., Conceicao, A., O'Brien, L. M., Chervin, R. D., & Braley, T. J. (2021) The impact of pet care needs on medical decision-making among hospitalized patients: A cross-sectional analysis of patient experience. *Journal of Patient Experience, 8,* 23743735211046089.

Power, E. (2008) Furry families: Making a human-dog family through home. *Social and Cultural Geography, 9*(5), 535–555.

Power, E. R. (2015) Placing community self-governance: Building materialities, nuisance noise and neighbouring in self-governing communities. *Urban Studies, 52*(2), 245–260.

Power, E. R. (2017) Renting with pets: a pathway to housing insecurity? *Housing Studies, 32*(3), 336–360.

Power, E. R. (2018) Restrictions on pets in multi-owned properties. In Altmann, E. & Gabriel, M. (Eds) *Multi-Owned Properties in the Asia-Pacific Region,* pp. 155–173. Palgrave Macmillan.

Rock, M. J., Adams, C. L., Degeling, C., Massolo, A., & McCormack, G. (2014) Policies on pets for healthy cities: a conceptual framework. *Health Promotion International, 30*(4), 976–986.

Rook, D. (2018) For the love of Darcie – Recognising the human-companion animal relationship in housing law and policy. *Liverpool Law Review, 39,* 29–46.

Salari, N., Hosseinian-Far, A., Jalali, R., Vaisi-Raygani, A., Rasoulpoor, S., Mohammadi, M., Rasoulpoor, & Khaledi-Paveh, B. (2020) Prevalence of stress, anxiety, depression among the general population during the COVID-19 pandemic: A systematic review and meta-analysis. *Globalization and Health, 16*(1), 1–11.

Saunders, J., Parast, L., Babey, S. H., & Miles, J. V. (2017) Exploring the differences between pet and non-pet owners: Implications for human-animal interaction research and policy. *PLoS One, 12*(6), e0179494.

Soaita, A. M., Marsh, A., & Gibb, K. (2022). Policy movement in housing research: a critical interpretative synthesis. *Housing Studies, 38*(1), 107–127.

Stone, W., Power, E., Tually, S., James, A., Faulkner, D., Goodall, Z., & Buckle, C. (2021) *Housing and Housing Assistance Pathways with Companion Animals: Risks, Costs, Benefits and Opportunities,* AHURI Final Report. No. 350, Australian Housing and Urban Research Institute Limited, Melbourne https://www.ahuri.edu.au/research/final-reports/350

The Constitution of India. (1950) https://www.constitutionofindia.net/constitution_of_india

The Housing and Development Board. (n.d.) *Keeping Pets.* https://www.hdb.gov.sg/residential/living-in-an-hdb-flat/keeping-pets

Toohey, A. M., & Krahn, T. M. (2018) 'Simply to be let in': Opening the doors to lower-income older adults and their companion animals. *Journal of Public Health, 40*(3), 661–665.

Toohey, A. M., & Rock, M. J. (2019) Disruptive solidarity or solidarity disrupted? A dialogical narrative analysis of economically vulnerable older adults' efforts to age in place with pets. *Public Health Ethics, 12*(1), 15–29.

Travers, C., Degeling, C., & Rock, M. (2017) Companion animals in natural disasters: A scoping review of scholarly sources. *Journal of Applied Animal Welfare Science, 20*(4), 324–343.

Weiss, E. (2015) Safety nets and support for pets at risk of entering the sheltering system. In Weiss, E. Mohan-Gibbons, H. & Zawistowski, S. (Eds) *Animal Behavior for Shelter Veterinarians and Staff,* Chapter 15, pp. 286–291. Wiley Blackwell.

Wells, D., Clements, M., Elliott, L., Meehan, E., Montgomery, & Williams, G. (2022) Quality of the human-animal bond and mental wellbeing during a COVID-19 Lockdown. *Anthrozoos.* Published online ahead of print. DOI: 10.1080/08927936.2022.2051935

Weston, K. (1991) *Families we Choose.* Columbia University Press.

Wood, L., Giles-Corti, B., & Bulsara, M. (2005) The pet connection: Pets as a conduit for social capital? *Social Science & Medicine, 61*(6), 1159–1173.

Wood, L., Martin, K., Christian, H., Nathan, A., Lauritsen, C., Houghton, S., & McCune, S. (2015) The pet factor – Companion animals as a conduit for getting to know people, friendship formation and social support. *PloS One, 10*(4), e0122085.

Wood, L., Martin, K., Christian, H., Houghton, S., Kawachi, I., Vallesi, S., & McCune, S. (2017) Social capital and pet ownership—a tale of four cities. *SSM-Population Health, 3,* 442–447.

World Health Organisation. (1981) *WHO/WSAVA Guidelines to Reduce Human Health Risks Associated With Animals in Urban Areas.* Published by WHO.

Wuerch, M. A., Giesbrecht, C. J., Price, J. A., Knutson, T., & Wach, F. (2020) Examining the relationship between intimate partner violence and concern for animal care and safekeeping. *Journal of Interpersonal Violence, 35*(9–10), 1866–1887.

# 9
# DIVERSITY, EQUITY, INCLUSION, AND BELONGING IN HUMAN-ANIMAL INTERACTIONS
## Reflections on Practice

*Sloane Hawes, Lizett Gutierrez, Lindsay Rojas, Rebecca Cohen, Bridget Camacho, Ashley Taeckens, Tess Hupe, and Nina Ekholm Fry*

**Preamble**

At the Institute for Human-Animal Connection, our mission is to elevate the relationships between people, other animals, and the environment to improve the health and wellbeing of all. We demonstrate our commitment to diversity, equity, inclusion, and belonging (DEIB) by recognizing how identities, contexts, and systems intersect to disparately affect humans and, by extension, other animals, and the environment. We are dedicated to understanding, confronting, and dismantling the ongoing historic, systemic marginalization and oppression in human-animal interactions. While this chapter includes perspectives often left out of research and education in human-animal interactions, we acknowledge that each of the authors of this chapter holds both privileged and marginalized identities that inform our experiences in the field. The perspectives we share in this chapter are reflections of our lived experiences. We recognize that due to systemic oppression, some voices continue to be excluded. We also acknowledge that the University of Denver was built on the land and labor of Black, Indigenous, and People of Color (BIPOC) in ways that violate the fundamental principles of equity. In the spirit of healing, the authors acknowledge and honor the Cheyenne and Arapaho Tribes, the indigenous people of the land upon which the university stands. We offer this chapter with the hope that we will all critically examine our positionality in systems of marginalization and take concrete steps toward reparations.

The fields of human-animal interactions and anthrozoology have evolved substantially over the last decade. While these fields share a commitment to promoting the health and wellbeing of humans, other animals, and the natural environment, some of our systems and practices may be intentionally or unintentionally creating or perpetuating harm. We wrote this chapter as an introductory guide to thinking critically about diversity, equity, inclusion, and belonging (DEIB) within your work as a student or professional in these fields. The purpose of this chapter is to highlight the importance of DEIB within human-animal interactions, particularly for those who may not have previously considered its significance. We hope this chapter will contribute to an increased focus on equity in both study and practice by providing tangible takeaways for reflection and action within your work. In this chapter, we use *human-animal interactions* as an umbrella term to encompass a number of disciplines and areas of study that include relationships between people and animals.

## Why Is Diversity, Equity, Inclusion, and Belonging Important in Human-Animal Interactions?

You may arrive at this chapter and wonder what DEIB have to do with the field of human-animal interactions. Those working within the field generally share a commitment to improving the health and wellbeing of people, other animals, and the natural environment. DEIB efforts are integral to the **collective liberation** needed to achieve this goal.

Understanding issues surrounding DEIB is necessary when our social and political systems have created and continue to perpetuate **oppression** and inequity. In these systems, constructs of **identity** have been used to maintain **power** and **privilege** for some, while marginalizing others. For example, race is a human social construct, separate from ethnicity or culture, that was created to separate a superior, white, human race and subjugate all others (Kang et al., 2017). In the United States, the construct of race has been used to justify the exploitation and enslavement of African, Indigenous, and Asian peoples. Scholars and practitioners in antiracism and animal liberation have explored how constructions of identity have also been used to perpetuate harm and oppression of other animals.

In the book, *Racism and Zoological Witchcraft: A Guide to Getting Out*, Ko (2019) examines the ways in which white supremacy "relies upon zoological ideas to bolster its power" (p. 35). Ko (2019) discusses how words used to describe animals have been weaponized to dehumanize and animalize marginalized communities, particularly BIPOC. For example, in 1783, George Washington rationalized the extermination of American Indians by describing them as wolves who deserved "total ruin" (Dunbar-Ortiz, 2015). In systems where **speciesism** and **white supremacy** are not addressed, the following hierarchy of oppression is created and maintained: humans are seen as better and more important than animals; white humans are seen as better and more important than BIPOC; BIPOC are seen as less than animals (Everett et al., 2019; Horta, 2010; Ko & Ko, 2017) [see *Supremacy and Speciesism in Present Day* box]. This hierarchy of oppression underpins the policies and practices in the present-day animal protection movement. The early animal protectionist movement was initially led by wealthy, white, Protestant males, who established anti-cruelty laws targeting BIPOC and immigrant communities working in industries involving working animals. This included the criminalization of certain culture's animal practices, including kosher slaughter and cockfighting (Davis, 2015). Later contributors to the animal protectionist movement were associated with the child protection and women's suffrage movement, yet disregarded racial justice issues. If practitioners working in human-animal interactions are to support the health and wellbeing of the people and animals living in the communities we serve, we must engage in self-reflection around how systems of oppression and marginalization have played out in our own lived experience and seek to change the systems that create and perpetuate this hierarchy of oppression.

If this is your first time encountering concepts such as power and privilege, it is understandable that discussing issues related to inequity may feel overwhelming. Engaging with topics related to DEIB requires us to think critically about all the areas where oppression and marginalization exist due to racism, sexism, homophobia, xenophobia, classism, ageism, and ableism. We acknowledge that these conversations can be particularly difficult given the lack of shared knowledge and history around how these systems originated and persist in society today. If you or your ancestors have experienced marginalization, these conversations can be (re)traumatizing. If you or your ancestors have participated in oppression and marginalization (either intentionally or unintentionally), these conversations can be uncomfortable and result in feelings of shame and/or blame [see *Trauma-Informed Diversity, Equity, and Inclusion* box]. We encourage you to engage in self-reflection and personal/professional development around DEIB topics with mentorship and support. If done with intention to acknowledge and shift the systems of power that affect us all, those working with human-animal interactions have a unique opportunity to create spaces for healing and understanding across species boundaries.

There are a number of ways an individual or organization can engage in anti-oppressive practice to promote DEIB in the human-animal interactions field. For example, in October of 2021, the American Psychological Association (APA) released a statement as an organization titled *Apology to People of Color for APA's Role in Promoting, Perpetuating, and Failing to Challenge Racism, Racial Discrimination, and Human Hierarchy in U.S.* where they list the actions they are taking to address how the history of the organization is steeped in colonialism and white supremacy. While this form of acknowledgment is an important first step for systems change, the foundation for sustained improvements in systems of oppression is behavior change (e.g., **person-first language**, addressing **microaggressions**). Throughout this chapter, we include several vignettes from HAI practitioners to provide examples of how DEIB manifests in their work.

---

**Trauma-Informed Diversity, Equity, Inclusion, and Belonging**

When topics related to inequity are discussed in workplaces and academic settings without explicit practices in place to support those with oppressed identities during the conversation, it can be traumatizing for those involved. This includes experiences of being tokenized, spoken over, spoken for, invalidated, and having your voice excluded in favor of centering and prioritizing those with privileged identities. Furthermore, people with privileged identities may be labeled as "fragile" (i.e., "white fragility" and "male fragility") because they experience stress and reactivity when confronting the ways they may have caused harm. It is important to note that the trauma experienced by a person in an oppressed position is not the same as a reaction that a person in a privileged position may have. The former is a perpetuation of harm. The latter is part of a growth process that should take place with attention to reducing harm. In the book *My Grandmother's Hands: Racialized Trauma and the Pathway to Mending our Hearts and Bodies*, Menakem (2017) speaks specifically to Black people, white people, and people in law enforcement and centers the importance of understanding the impact of racialized trauma on our bodies, how it is perpetuated, and how it can be healed. Menakem's work is of critical importance to the meaningful implementation of anti-oppressive work and provides specific tools to facilitate and support this growth and development.

---

**Supremacy and Speciesism in Present Day**

When caring about and taking action for animals, we should be careful to dismantle the white supremacist trope that only white people are truly human and that animals associated with white people should have priority over BIPOC. There have been several examples of this narrative in recent events. In 2021, U.S. Federal Border Patrol Agents in Texas used horses to chase and whip people from Haiti who were seeking asylum at the border between the United States and Mexico. Images and videos in the media captured the interactions between the predominately white border control agents and the Black asylum seekers. The pain and distress exhibited by the horses while steered by the border control agents in interactions with asylum seekers became a topic of discussion in the U.S. horse community. However, the stress and pain of the Black asylum seekers and the human rights violations taking place was only acknowledged secondarily, if at all. In 2022, at the time of writing this chapter, the war in Ukraine began. Media coverage showed white people and their pets fleeing Ukraine to nearby countries for safety, while individuals of North African and Indian descent were blocked or made to wait. These scenarios highlight the global, historical context of white supremacy and human supremacy, which explicitly values white bodies and other animals over Black and Brown bodies. It is important to recognize the systems that displace others from sentience and species categories, such as the notion that the lives of other-than-human animals should suffer or be deprioritized.

## How Can I Integrate Considerations for Diversity, Equity, Inclusion, and Belonging in my Human-Animal Interactions Work?

Human services practitioners trained in human-animal interactions, such as therapists and educators, commonly incorporate direct interactions between people and animals in their work. The following vignettes highlight the work of three practitioners who work with diverse client populations in mental health-related services. The first vignette is written by a clinical social worker who shares about assessing family systems beliefs around human-animal interactions, how interactions with animals can be integrated as a technique within the treatment plan, and how human-animal interactions can help BIPOC and non-English language dominant clients in the United States make therapeutic gains.

### Vignette #1 – Written by: Lizett Justa Gutierrez, LCSW, MA, RVT

I am a bilingual South Amerindian descendant of parents from rural regions who view animals as guardians for our livestock, pest control, carriers of germs or disease, and sometimes for consumption, but not as pets. My lived experience has not always aligned with the dominant culture's appreciation of human-animal interactions in the United States.

I bring this lived experience into my therapy practice of serving Latinx families of predominantly Mexican and Central American descent by recognizing that the Latinx community is a diaspora that encompasses rich cultural differences, such as Spanish dialects, regional differences based on a client's rural or urban origin from their home country, and socio-political differences. These differences follow a Latinx individual throughout their lifetime and can either stay static or change once immersed with the U.S. culture's norms.

To effectively serve clients in the United States from diverse backgrounds, I have come to understand how engaging with **cultural humility** can affect my therapeutic rapport with the families I serve. When I begin my work with a new family, I complete a biopsychosocial assessment with each participating family member to gain insight into whether human-animal interactions would be appropriate in individual or family therapy sessions. This culturally responsive assessment process includes collecting the intergenerational perspectives on the topic of animals as pets. In my experience conducting these assessments, I have observed how the first generation is more likely to acculturate and embrace the U.S. appreciation of pets and human-animal interactions, while their immigrant parents or grandparents are more likely to refer to pets as a privilege. By asking members of the family system to share their stories about interacting with animals, I build trust and refine how I approach the treatment process.

If interactions with animals are part of my proposed treatment plan for a child, I will provide the parent or guardian(s) with a thorough overview of how these interactions can benefit the child's treatment goals. The parent or guardian(s) can then make an informed decision on whether or not they want interactions with animals to be included in our family system work or within their child's individual sessions.

When I incorporate the family's pet into the session, I am utilizing the triangle model where the clinician, the pet, and the client are engaged and working toward clinical goals (Jones et al., 2019). Working with the family pet present is useful for my therapeutic relationship with the client, especially when the parental presence causes undue stress to the child. When the family pet is present in our sessions, the pet provides emotional safety for the child, thereby allowing the child to be vulnerable, share insights around attachment, and practice self-awareness, self-efficacy, and security when the client is processing trauma.

The other interventions I might employ with the family pet present are play therapy, role play, and education around welfare and general care of the pet. When working with families who do not have pets or are not attuned to human-animal interactions in general, I engage child clients through videos, art collage, and building on their connection to nature and animals by having sessions in green spaces and with puppets or stuffed animals.

---

The second vignette is written by a clinical social worker in child welfare who shares details on how barriers related to DEIB have impacted their ability to provide services for diverse clients.

---

## Vignette #2 – Written by: Lindsay Rojas, LCSW, MSc, AASW

I am the daughter of immigrants, with heritage of South American, Indigenous, and mixed White descent. As a bilingual and bicultural clinician, I have worked in a variety of mental health settings, including in homes, offices, agencies, non-profits, private practice, and court-ordered child welfare. Through my experiences, I have seen how the clinical benefits of incorporating animals into interventions are generalizable across multiple settings. I have observed firsthand how clients learn to attach to, depend on, and feel supported by my therapy dog, Dante, and in turn, by me as the therapist. Yet, despite the profound benefits of human-animal interactions that I have observed as a mental health professional, I have navigated a number of systemic barriers that often prohibit my clients from low-income and marginalized backgrounds from being able to experience these advantages.

In general, integrating interactions with animals into my work has made me more approachable. This has been crucial when serving communities and populations who have been stigmatized due to mental illness and who have experienced high rates of therapist turnover. My clients in the child welfare system were initially hesitant to engage with me, largely due to the negative associations they have had with previous therapists who had become burned out by the intensity and demands of the work. However, I have witnessed clients who would initially not talk or respond at the mention of the word therapy proceed to talk at length on topics, discussions, and activities involving my therapy dog, Dante.

In sessions with my therapy dog, I learn important information about my clients much sooner than I would without him present. For example, clients in the foster care system were able to identify with Dante, a dog who was rescued and rehomed. Talks involving Dante would often lead to discussions about previous pet ownership and animal keeping habits within the home. One such discussion resulted in a disclosure that their animals did not have access to adequate resources in the home, such as food, shelter, and veterinary care, which also reflected and directly correlated to my client's own living conditions.

My desire to act on my social justice values has often been directly challenged or undermined by barriers in the U.S. healthcare system such as low reimbursement rates, tedious documentation requirements, lengthy insurance-related credentialing processes, and the availability of in-person versus telehealth services. The fear of recoupment of payment by health insurance companies years after services were provided has led to several of my therapist colleagues deciding that they will no longer accept health insurance and only accept direct payments from clients to protect their livelihood.

To increase my ability to provide therapy to clients who are under resourced or forced into poverty due to discrimination and marginalization, my colleagues and I often opt to use a sliding scale fee structure or provide services on a pro-bono basis to a limited number of clients on the caseload. It is the larger systems that create these healthcare inequities, which, in turn, impact therapist income, and impact which clients are served by therapists who have sought out specialized training, such as in human-animal interactions.

---

Accessibility issues are also present for those interested in integrating human-animal interactions into their professional practice. Specifically, having a career in the field of human-animal interactions generally requires higher education, where enrollment and graduation rates are significantly higher for white students (American Council on Education, 2019). Further, there are a limited number of graduate programs or professional development certificates where one can receive training in human-animal interactions as a treatment technique and even fewer opportunities to explore ways to integrate human-animal interactions into community or policy practice settings. Many human-animal interactions practitioners complete graduate training in a mental health or nonprofit management degree program and then pursue additional education or experiences in human-animal interactions after graduating. This extensive, expensive education and training process generates barriers to incorporating human-animal interactions in clinical practice and likely contributes to why this area, and allied fields like veterinary medicine and animal sheltering, also lack racial/ethnic diversity in the workforce (AAVMC, 2020; Brown, 2005; Data USA, n.d.).

To balance this issue of representation in the field, research by Risley-Curtiss et al. (2006b) brought to the forefront the voices of clinicians-in-training from diverse heritages and their experiences with the human-animal bond. They acknowledge the importance of this research and its potential to engage and recruit more ethnically diverse clinicians to enhance their skill set by pursuing education, training, and supervision in human-animal interactions with the caveat that the curriculum addresses the voices and experiences of culturally diverse populations. In discussions of DEIB, it is important to consider factors that impact who can enter the fields of human-animal interactions. Patricia E. Kelly, Founder, President, and CEO of the Ebony Horsewomen, Incorporated (EHI) in Hartford, CT has experienced various gatekeepers that deter or prevent more BIPOC learners and professionals from pursuing work in her area of practice of psychotherapy that incorporates equine (P.E. Kelly, personal communication, May 9, 2022). These barriers include: high cost, training locations in suburban and rural areas, a lack of focused marketing for BIPOC communities, disproportionately negative or traumatic lived experiences with mental health systems, as well as training curriculum that centers the experience of white people and further marginalizes BIPOC experiences. She also noted a lack of adequate training in equine psychology and behavior in existing programs. To address these systemic issues, Mrs. Kelly developed the Gallop Cultural Competency Equine-Assisted Psychotherapy Training and Certification Program. Through a grant from Hartford Foundation for Public Giving, in 2022 EHI has been able to offer fully funded scholarships to Connecticut-based BIPOC, including 20 mental health providers and students and 15 aspiring equine specialists to obtain Gallop certification. She noted, "it has been more trying, actually, to find horse specialists because the criteria include that you have to come in with horse experience. The only BIPOC people in Connecticut that we have run across with that experience have been my alumni, and that's it. I think our next proposal will be to try to extend it beyond Connecticut and bring in others. Emerging efforts to increase diversity amongst professionals of human-animal interactions include scholarship opportunities for educational programs (e.g., Jodie G. Blackwell scholarship supporting African American veterinarians, Institute for Human-Animal Connection's Diversity in Human-Animal Interactions scholarship), but more work is needed to expand these opportunities and ensure the programming promotes inclusivity and belonging.

---

The third vignette is written by a clinical psychologist who addresses dynamics of identities within helping professions and their work with equine interactions.

---

## Vignette #3 – Written by Rebecca Cohen, Psy.D.

I am writing this section through the lens of the identities I hold. The identities that are most relevant to this discussion are my race, my ethnocultural religious background, my class, and my gender: I am a white-bodied Ashkenazi Jewish woman primarily of Eastern European descent who grew up with a mostly middle-class and cisgender experience. These identities are significant to this discussion because they played a major role in how I was raised to position myself as a helper. As a young girl, I was taught to explicitly and implicitly care for others; I was taught to be empathic. As my interests developed in my teen years, through experiences with horses and noticing racial disparities in access to horses in my community, the path I was encouraged to pursue was to "help solve the problem." This mindset was steeped in language urging me to "save" people who were "less fortunate than me," and these ideas became important parts of my identity and were central to my decision to become a psychologist.

Early on, I sought experiences that supported this dynamic of "me helping people in need," and I didn't see any problem with this, and I was rewarded for it. It wasn't until I was deep into my graduate studies in psychology that peers and professors challenged me to interrogate my language use and the power dynamics inherent in how I thought about and discussed my work. Then, I began to learn about the history and context of my position as a white woman wanting to "help" or "save" "others." This language is part of the legacy of benevolent oppression and, more specifically, white saviorism. During my graduate education, the professors challenged us to turn the lens on ourselves, understand our positionality, consider our own identities, learn about them, and examine the roles of power, privilege, and oppression from our experiences.

Specific to my fields of practice, the mental health and child welfare fields, the Child Savers Movement of the late 19th and early 20th centuries played a significant role in the development of the juvenile justice and child welfare systems that we have today (Platt,

1969). The "child savers" were white women, often wives of influential white men, who entered politics through their mission to save children of poor and immigrant identities. Understanding this history and my role in its current form required me to make significant changes to my approach.

Extracting these values of saviorism from my identity was among my first lessons in cultural humility and required me to ask questions that were difficult and often painful to answer. At that point, I was in the planning phases of my dissertation and this learning caused me to make some serious changes. I pivoted from a study looking at the impacts of incorporating horses into psychotherapy for "at-risk" youth to an examination of how practitioners trained in using equine interactions as a technique in therapy think about the role of culture in their work with youth labeled as "at-risk" (Cohen, 2011).

I've spent the last ten years continuing on this path toward deepening my understanding of what it means to be an ally to people who are marginalized in ways that I am not. I've learned about my own marginalized identities and the interconnectedness of my oppression to that of others. This changes how I think about positioning myself in my work (and the inherent privilege of having some choice in this positioning). Rather than sourcing my sense of identity from helping those I perceive as being "in need," I find meaning in understanding my own cultural history, the positions it places me in, and in practicing the exploration of how to deepen authentic connections with the humans and other animals I interact with.

I also try to be mindful of how and where I do my own personal development work. I acknowledge that when I'm in a group with people who do not share my privileged identities, my learning and unlearning related to power, privilege, and oppression in their presence can (at the very least) place upon them undue emotional and mental labor, or worse, inflict or reinforce trauma. As such, practices of explicit consent have become central in identifying people who want to play a supportive role in this accountability work. Looping back to the concept of cultural humility: regardless of the effort I believe I'm making to minimize harm as I do this work, I'm aware that I will sometimes cause harm. Because of this, I remain receptive to feedback and am willing to integrate that feedback in a meaningful way.

In clinical practice, this includes implementing ethnographic skills and an ecological model to examine how social systems impact us, the clients we serve, and the therapeutic relationship between us. I apply this thinking to my case formulations and treatment. I seek out training, consultation, and dialogue in pursuit of exploration of how I can invest in ongoing movements of creative transformation.

## Future Directions for Research on Human-Animal Interactions and Diversity, Equity, Inclusion, and Belonging

There is an acknowledged lack of human-animal interactions research focused on diverse populations in the United States. Further, the researchers and practitioners conducting this research are also not diverse, which presents a tremendous challenge for providing training and mentorship to students and young professionals in the field (Faver & Cavazos Jr., 2008; Risley-Curtiss, 2010; Risley-Curtiss et al., 2006a, 2006b). The limited literature that exists on engaging marginalized populations in human-animal interactions indicates that improving engagement with historically and systemically excluded populations requires an emphasis on cultural responsiveness, with particular attention to relationships and trust-building that honor individuals' intersectional identities and lived experiences, while also addressing structural barriers to accessing care (Decker Sparks et al., 2018; Gandenberger et al., 2021; Hawes et al., 2021; Jenkins & Rudd, 2021). This initial research on best practices in diverse communities demonstrates awareness of the importance of shifting the way we evaluate services and programs from an individualistic lens that stigmatizes individuals for being vulnerable to an equity lens that recognizes the structural conditions and power dynamics that create vulnerability (AMA & AAMC, 2021). However, more research is needed if DEIB issues are to be addressed in the human-animal interactions field.

Most studies have focused on documenting DEIB issues in the human-animal interactions field. For example, some studies have begun to examine the spatial and social disparities in how pet safety net programs offered through animal shelters and humane law enforcement are applied and utilized (Dyer & Milot, 2019; Hawes et al., 2020a, 2020b; Ly et al., 2021; Marceau, 2019; Thompson et al., 2019). However, a majority of the instrumentation being utilized in the human-animal interactions field fails to recognize how structural conditions, such as **social determinants of health,** that are in focus for DEIB efforts may impact individual responses in ways that are inaccurate reflections of their attachment to their pet. One scale to measure the human-animal bond includes questions like, "Nothing would ever convince me to give up my pet" and "If my pet needed extensive veterinary care, I would pay for whatever it takes," (Human-Animal Bond Research Institute, 2022). Other sources of data are using deficit-focused language to describe individuals who experience structural barriers to access to veterinary care as individuals who "speak English less than well," "have no access to a vehicle" or are "experiencing" poverty (The Veterinary Care Accessibility Project, 2022). By focusing on individual characteristics and situations, this approach blames individuals for their lack of access to veterinary care rather than identifying social and structural barriers to accessing pet care. This deficit-focused approach to conceptualizing the human-animal bond has served to perpetuate the historical legacy of the animal welfare movement to limit or criminalize pet ownership for low-income and BIPOC communities, (Applebaum, MacLean, & McDonald, 2021; Hawes, Hupe, & Morris, 2020b). This individualistic and U.S. centric lens of the human-animal bond is particularly troubling given the lack of evidence to support claims of racial or ethnic differences in caring for pets and the growing body of research demonstrating how structural factors such as resource deserts, lack of affordable or acceptable access to pet support services, and housing security are the primary drivers of poor health and welfare outcomes for both people and their pets (Applebaum et al., 2021; Guenther, 2020; LaVallee et al., 2017; Mueller et al., 2018; Poss & Everett, 2006; Rauktis et al., 2020; Rose et al., 2020).

In this way, there is a need to rethink our existing instrumentation in the human-animal interactions field and expand the current operationalizations of key constructs in the field such as health, welfare, wellbeing, and mutually beneficial, to include consideration of key DEIB issues, including racial equity, economic and housing security, gender and sexual diversity, disability rights, and ecological justice (Jenkins & Rudd, 2021). This next generation of instrumentation should be developed to address the critical gap in availability of culturally responsive instrumentation to measure the human-animal bond and other key constructs in the human-animal interactions field. While there are many validated measures of the human-animal bond, each of these instruments has been validated with predominantly white populations (Brown, 2002). One instrument developed to measure human emotional attachment to companion animals was initially developed and validated in English and took over 20 years to be translated into Spanish with Spanish speakers from Mexico (Ramirez et al., 2014). Given the multitude of Spanish dialects, colloquial meanings, and ways of communicating among Latinx populations, a need remains to include the perspective of Spanish speakers with various educational and socioeconomic backgrounds in human-animal interactions instrumentation. Instrumentation can be developed and validated with diverse populations to measure key constructs in the human-animal interactions field through the use of rigorous methods, such as utilizing a bilingual instrumentation development committee, or by conducting cognitive interviews to avoid mistranslation and ensure cultural appropriateness of items (Salgado-Santamaria et al., 2023).

Community-based research methodologies are critical for understanding and removing the social, cultural, and logistical barriers that curtail services and programs that incorporate human-animal interactions. However, BIPOC and individuals with low socioeconomic status are underrepresented in public health research (Habibi et al., 2015). Researchers have considered historically marginalized communities "hard to reach" in research because of their challenges in accessing, engaging, and retaining participants from these communities (Bonevski et al., 2014). Common barriers to conducting research with marginalized groups include mistrust of researchers, fear of authority, scheduling difficulties, or a lack of comprehensive informed consent procedures (Bonevski et al., 2014). However, there are numerous best practices identified for involving marginalized populations in research, including building trusting relationships with community leaders, speaking the language of the focus community, using (a) familiar and accessible language(s), respecting participants' schedules and cultural norms, and offering compensation for participation (Stonewall et al., 2017). These strategies for engagement with historically marginalized populations are typically resource-intensive for research institutions (Bonevski et al., 2014). However, when BIPOC are effectively recruited and engaged, they are equally if not more likely to participate in research than non-Hispanic white individuals (Rogers & Lange, 2013; Wendler et al., 2006).

Community-based and participatory research methodologies in the field of HAI can promote inclusivity by providing opportunities for historically excluded voices to be heard. For example, qualitative research can help capture community perspectives, understand lived experience, and provide nuanced detail to complex phenomena (Sofaer, 1999). Other approaches, such as transformative mixed-methods research and participatory action research, involve collaboration with the population affected by the issue being studied. Community-based research approaches have the potential to address health inequities and social justice issues by democratizing the production of knowledge and accelerating the translation of research into practice (George et al., 1998). These research processes encourage power-sharing between the research group and participants, promote capacity-building, and improve the sustainability of programs and partnerships (Cargo & Mercer, 2008). Overall, human-animal interaction researchers should practice due diligence to ethically implement research methodologies that partner with marginalized communities as participants as opposed to subjects. Using methodologies that follow a social justice framework can mitigate the historically harmful, and sometimes violent nature, in which knowledge has been unethically extracted from marginalized communities (Jenkins & Rudd, 2021).

## Glossary

Common language is needed to achieve DEIB outcomes in our field. Without the adoption of universal terms, practitioners and researchers are at risk of conceptualizing HAI in different ways, resulting in varying methodological applications (Santaniello et al., 2020). For your references, we've included basic definitions of the terms used throughout this chapter.

**Belonging** a fundamental human need and motivation guided by a desire to be affiliated with and respected by others that we seek to acquire and maintain throughout the life course. If inclusion is the process of creating a culture/environment that allows individuals to feel valued, then belonging is the feeling of connectedness and respect one may experience when inclusivity efforts are successful (Smith et al., 2021; Witwer, 2021).

**Collective liberation** acknowledges that multiple oppressions exist, and that we must work in solidarity to undo oppression in ourselves, our families, our communities, and our institutions, in order to achieve a world that is truly free; requires that we center the voices and lived experiences of those who have been most marginalized; depends on our communities to build shared power and accountability that foster a just and transformed world (Center for Racial Justice in Education, 2022).

**Cultural humility** the commitment to self-evaluation and critique, to redressing the power imbalances in the provider-client dynamic, and to developing mutually beneficial and non-paternalistic partnerships with communities on behalf of individuals and defined populations (AMA &AAMC, 2021).

**Diversity** refers to the identities we carry, including race, gender, sexual orientation, class, age, country of origin, education, religion, geography, physical or cognitive abilities, or other characteristics; valuing diversity means recognizing differences between people, acknowledging that these differences are a valued asset, and striving for diverse representation as a critical step toward equity (AMA & AAMC, 2021; Race Forward, 2015).

**Equity** refers to fairness and justice and is distinguished from equality; while equality means providing the same to all, equity requires recognizing that we do not all start from the same place because power is unevenly distributed; the process of achieving equity is ongoing, requiring us to identify and overcome uneven distribution of power as well as intentional and unintentional barriers arising from bias or structural root causes (AMA & AAMC, 2021).

**Identity** an individual's sense of self, which involves continuity, or the feeling that one is the same person today that one was yesterday or last year (despite physical or other changes); such a sense is derived from one's body sensations; one's body image; and the feeling that one's memories, goals, values, expectations, and beliefs belong to the self (APA, 2021b).

**Inclusion** a state of being valued, respected, and supported; the process of creating a culture and environment that allows individuals to feel valued for their unique qualities and experience a sense of belonging (AMA & AAMC, 2021).

**Justice** a future state where the root causes of inequity (e.g., racism, sexism, class oppression) have been dismantled and barriers have been removed; an achievable goal that requires the sustained focus, investment, and energy of leaders and communities working together holding each other accountable to redesign our structures, policies, and practices to deliver the high quality and safest possible conditions that allows for everyone to reach their highest potential (AMA & AAMC, 2021).

**Microaggression** Everyday verbal, nonverbal and environmental slights, snubs or insults, whether intentional or unintentional, which communicate hostile, derogatory or negative messages to persons targeted solely for their membership in historically marginalized groups (AMA & AAMC, 2021).

**Person-first language** language that emphasizes the individual over the person's condition, e.g., "woman with diabetes," rather than "diabetic woman." Most fields of practice encourage the use of person-first language whenever possible. However, keep in mind that not all individuals prefer person-first language; some, like the deaf community, prefer to emphasize their shared group identity as "people who are deaf." It is important to acknowledge that different communities and individuals have different standards and preferences (AMA & AAMC, 2021).

**Positionality** refers to our individual identities and the intersection of those identities and statuses with systems of privilege and oppression; shapes our psychological experiences, worldview, perceptions others have of us, social relationships, and access to resources;

includes how an individual actively understands and negotiates the ways that our identities affect our relationships because of power dynamics related to privilege and oppression (APA, 2021b).

**Power** the ability or authority to influence and make decisions that impact other people; power is often held by those who have privilege, which is a set of advantages systemically conferred on a particular person or group of people (AMA & AAMC, 2021).

**Oppression** when one group has more access to power and privilege than another group, and when that power and privilege is used to maintain the status quo (i.e., domination of one group over another). Thus, oppression is both a state and a process, with the state of oppression being unequal group access to power and privilege, and the process of oppression being the ways in which inequality between groups is maintained (APA, 2021b).

**Speciesism** the belief that the human species is inherently superior to other species and so has rights or privileges that are denied to other sentient animals (Ryder, 1975).

**Social determinants of health** the underlying community-wide social, economic, and physical conditions in which people are born, grow, live, work and age that affect a wide range of health, functioning, and quality-of-life outcomes and risks; these determinants and their unequal distribution according to social position, result in differences in health status between population groups that are avoidable and unfair (AMA & AAMC, 2021).

**White supremacy** the ideological belief that biological and cultural whiteness is superior, as well as normal and healthy; this is a pervasive ideology that continues to polarize the U.S. and undergird racism (APA, 2021b).

## Reflection Questions

Self-awareness and vulnerability are important steps in advocating for change and becoming a part of the solution. We spark change by asking ourselves and processing difficult questions, posing these difficult questions to the agencies and communities we are part of, and allowing the answers to guide our interdisciplinary work to uphold best practices within the human-animal interactions field. To help you get started, we have compiled a list of questions to help strengthen your self-awareness and assess the agencies and communities you serve.

### Individual/Personal and Agency

1. How does your identity and/or beliefs inform your work with individuals and/or communities?
2. How do you build rapport with the individuals you work with? How do you take their cultural backgrounds into consideration when building rapport?
3. How do you integrate individual and/or community voice(s) into your interventions and/or services?
4. Are there areas in your practice that – either intentionally or unintentionally – reinforce oppressive ideologies, policies, and/or practices?
5. The central part of anti-oppression is behavior change. List at least two short-term and two long-term action steps you can take to advance DEIB in the human-animal interactions field?

6   What are your agency's values regarding diversity? How does your agency incorporate these values into practice with individuals and/or communities?
7   What policies and/or procedures does your agency have that support DEIB? What policies and/or procedures are barriers to DEIB?
8   What populations are excluded from the services provided by your organization?
9   What barriers exist that impact the accessibility of your organization's services?
10  Are there staff members at the agency, at each level of leadership, who reflect the cultural backgrounds and values of the clients served?

## Summary/Conclusions

By bringing attention to DEIB issues in our work, students and practitioners in human-animal interactions are better positioned to support the health and wellbeing of people, other animals, and the environment. Future directions of the field include: deepening practices of self-awareness related to dynamics of power, privilege, and oppression among providers; conducting research on culturally appropriate and responsive interventions that incorporate recognition of structural barriers experienced by marginalized communities and other systems of oppression; and implementation of specific strategies to increase the diversity of the field.

## References

American Council on Education. (2019). *Race and ethnicity in higher education*. https://www.acenet.edu/Research-Insights/Pages/Race-and-Ethnicity-in-Higher-Education.aspx

American Medical Association and Association of American Medical Colleges' Center for Health Justice. (2021). *Advancing health equity: A guide to language, narratives and concepts*. https://www.ama-assn.org/system/files/ama-aamc-equity-guide.pdf

American Psychological Association. (2021a). *Apology to people of color for APA's role in promoting, perpetuating, and failing to challenge racism, racial discrimination, and human hierarchy in U.S*. https://www.apa.org/about/policy/racism-apology

American Psychological Association. (2021b). *Equity, diversity, and inclusion framework*. https://www.apa.org/about/apa/equity-diversity-inclusion/framework

Applebaum, J. W., Horecka, K., Loney, L., & Graham, T. M. (2021). Pet-friendly for whom? An analysis of pet fees in Texas rental housing. *Frontiers in Veterinary Science, 8*, 1–13. Doi: 10.3389/fvets.2021.767149

Applebaum, MacLean, E. L., & McDonald, S. E. (2021). Love, fear, and the human-animal bond: On adversity and multispecies relationships. *Comprehensive Psychoneuroendocrinology* (Online), 7, 100071. Doi: 10.1016/j.cpnec.2021.100071

Association of American Veterinary Medical Colleges. (2020). *AAVMC admissions: Analysis of 2019 student survey analysis*. https://www.aavmc.org/wp-content/uploads/2021/03/2019-Admissions-Analysis-Monograph.pdf

Bonevski, B., Randell, M., Paul, C., Chapman, K., Twyman, L., Bryant, J., Brozek, I., & Hughes, C. (2014). Reaching the hard-to-reach: A systematic review of strategies for improving health and medical research with socially disadvantaged groups. *BMC Medical Research Methodology, 14*(1), 42. http://www.biomedcentral.com/1471-2288/14/42

Brown, S. E. (2002). Ethnic variations in pet attachment among students at an American school of veterinary medicine. *Society & Animals, 10*(3), 249–266. Doi: 10.1163/156853002320770065

Brown, S. E. (2005). The under-representation of African American employees in animal welfare organizations in the United States. *Society & Animals, 13*(2), 153–162. https://www.animalsandsociety.org/wp-content/uploads/2016/01/brown.pdf

Cargo, M., & Mercer, S. L. (2008). The value and challenges of participatory research: Strengthening its practice. *Annual Review of Public Health, 29,* 325–350. Doi: 10.1146/annurev.publhealth.29.091307.083824

Center for Racial Justice in Education. (n.d.). *Mission, vision, values.* https://centerracialjustice.org/mission-vision-values/

Cohen, R. A. (2011). *Exploring multicultural considerations in equine facilitated psychotherapy.* (Publication No. 3516479) [Doctoral Dissertation, John F. Kennedy University]. ProQuest Dissertations Publishing.

Data USA. (n.d.). *DATA USA: Veterinarians.* https://datausa.io/profile/soc/veterinarians#demographics

Davis, J. M. (2015). The history of animal protection in the United States. *The Organization of American Historians.* https://www.oah.org/tah/issues/2015/november/the-history-of-animal-protection-in-the-united-states/

Decker Sparks, J. L., Camacho, B., Tedeschi, P., & Morris, K. N. (2018). Race and ethnicity are not primary determinants in utilizing veterinary services in underserved communities in the United States. *Journal of Applied Animal Welfare Science, 21*(2), 120–129. Doi: 10.1080/10888705.2017.1378578

Dunbar-Ortiz, R. (2015). *An indigenous peoples' history of the United States.* Beacon Press.

Dyer, J. L., & Milot, L. (2019). Social vulnerability assessment of dog intake location data as a planning tool for community health program development: A case study in Athens-Clarke County, GA, 2014–2016. *PLoS ONE, 14*(12), e0225282. Doi: 10.1371/journal.pone.0225282

Everett, J. A. C., Caviola, L., Savulescu, J., & Faber, N. S. (2019). Speciesism, generalized prejudice, and perceptions of prejudiced others. *Group Processes & Intergroup Relations, 22*(6), 785–803. Doi: 10.1177/1368430218816962

Faver, C. A., & Cavazos, Jr., A. M. (2008). Love, safety, and companionship: The human-animal bond and Latino families. *Journal of Family Social Work, 11*(3), 254–271. Doi: 10.1080/10522150802292350

Gandenberger, J., Hawes, S. M., Wheatall, E., Pappas, A., & Morris, K. N. (2021). Development and initial validation of the animal welfare cultural competence inventory (AWCCI) to assess cultural competence in animal welfare. *Journal of Applied Animal Welfare Science, 13,* 1–12. Doi: 10.1080/10888705.2021.2008934

George, M. A., Daniel, M., & Green, L. W. (1998). Appraising and funding participatory research in health promotion. *International Quarterly of Community Health Education, 18*(2), 181–197. https://doi-org.du.idm.oclc.org/10.2190/C1B5-7PPE-7TYL-7YN8

Guenther, K. M. (2020). *The lives and deaths of shelter animals.* Stanford University Press.

Habibi, A., Der Sarkissian, A., Gomez, M., & Ilari, B. (2015). Developmental brain research with participants from underprivileged communities: Strategies for recruitment, participation, and retention. *Brain, Mind, and Education, 9*(3), 179–186. Doi: 10.1111/mbe.12087

Hawes, S. M., Hupe, T., & Morris, K. N. (2020b). Punishment to support: The need to align animal control enforcement with the human social justice movement. *Animals, 10*(10), 1902. Doi: 10.3390/ani10101902

Hawes, S. M., Hupe, T. M., Winczewski, J., Elting, K., Arrington, A., Newbury, S., & Morris, K. N. (2021). Measuring changes in perceptions of access to pet support care in underserved communities. *Frontiers in Veterinary Science, 8*(745345). Doi: 10.3389/fvets.2021.745345

Hawes, S. M., Ikizler, D., Loughney, K., Temple Barnes, A., Marceau, J., Tedeschi, P., & Morris, K. N. (2020a). A quantitative study of Denver's breed-specific legislation. *Animal Law Review, 26,* 195–270. https://heinonline.org/HOL/P?h=hein.journals/anim26&i=209

Horta, O. (2010). What is speciesism? *Journal of Agricultural and Environmental Ethics, 23,* 243–266. Doi:10.1007/s10806-009-9205-2

Human-Animal Bond Research Institute. (n.d.). *International survey of pet owners and veterinarians.* https://habri.org/international-hab-survey

Jenkins, J. L., & Rudd, M. (2021). Decolonizing animal welfare through a social justice framework. *Frontiers in Veterinary Science, 8*(787555). Doi: 10.3389/fvets.2021.787555

Jones, M. G., Rice, S. M., & Cotton, S. M. (2019). Incorporating animal-assisted therapy in mental health treatments for adolescents: A systematic review of canine assisted psychotherapy. *PloS One, 14*(1), e0210761–e0210761. Doi: 10.1371/journal.pone.0210761

Kang, M., Lessard, D., Heston, L., & Nordmarken, S. (2017). *Introduction to women, gender, sexuality studies: Grounding theoretical frameworks and concepts*. University of Massachusetts Amherst Libraries. https://openbooks.library.umass.edu/introwgss/chapter/social-constructionism/

Ko, A. (2019). *Racism and zoological witchcraft: A guide to getting out*. Lantern Books.

Ko, A., & Ko, A. (2017). *Aphro-ism: Essays on pop culture, feminism, and black veganism from two sisters*. Lantern Books.

LaVallee, E., Mueller, M. K., & McCobb, E. (2017). A systematic review of the literature addressing veterinary care for underserved communities. *Journal of Applied Animal Welfare Science, 20*(4), 381–394. Doi: 10.1080/10888705.2017.1337515

Ly, L. H., Gordon, E., & Protopopova, A. (2021). Inequitable flow of animals in and out of shelters: Comparison of community-level vulnerability for owner-surrendered and subsequently adopted animals. *Frontiers in Veterinary Science, 8*, 784389. Doi: 10.3389/fvets.2021.784389

Marceau, J. (2019). *Beyond cages: Animal law and criminal punishment*. Cambridge University Press. Doi: 10.1017/9781108277877

Menakem, R. (2017). *My grandmother's hands: Racialized trauma and the pathway to mending our hearts and bodies*. Central Recovery Press. Doi: 10.1080/02615479.2021.1976760

Mueller, M. K., Gee, N. R., & Bures, R. M. (2018). Human-animal interaction as a social determinant of health: Descriptive findings from the health and retirement study. *BMC Public Health, 18*(1): 305. Doi: 10.1186/s12889-018-5188-0

Platt, A. M. (1969). *The child savers: The invention of delinquency*. University of Chicago Press.

Poss, J. E., & Everett, M. (2006). Impact of a bilingual mobile spay/neuter clinic in a U.S./Mexico border city. *Journal of Applied Animal Welfare Science, 9*(1), 71–77. Doi: 10.1207/s15327604jaws0901_7

Race Forward. (2015). *Race reporting guide*. Race Forward. https://www.raceforward.org/reporting-guide

Ramírez, M. T. G., Quezada Berumen, L. D. C., & Hernández, R. L. (2014). Psychometric properties of the lexington attachment to pets scale: Mexican version (LAPS-M). *Anthrozoös, 27*(3), 351–359. Doi: 10.2752/175303714X13903827487926

Rauktis, M. E., Lee, H., Bickel, L., Giovengo, H., Nagel, M., & Cahalane, H. (2020). Food security challenges and health opportunities of companion animal ownership for low-income adults. *Journal of Evidence-Based Social Work, 17*(6), 662–676. Doi: 10.1080/26408066.2020.1781726

Risley-Curtiss, C. (2010). Social work practitioners and the human—Companion animal bond: A national study. *Social Work, 55*(1), 38–46. Doi: 10.1093/sw/55.1.38

Risley-Curtiss, C., Holley, L. C., & Wolf, S. (2006a). The animal-human bond and ethnic diversity. *Social Work, 51*(3), 257–268. Doi: 10.1093/sw/51.3.257

Risley-Curtiss, C., Holley, L. C., Cruickshank, T., Porcelli, J., Rhoads, C., Bacchus, D. N. A., Nyakoe, S., & Murphy, S. B. (2006b). "She was family" women of color and animal-human connections. *Affilia, 21*(4), 433–447. Doi: 10.1177/0886109906292314

Rogers, W., & Lange, M. (2013). Rethinking the vulnerability of minority populations in research. *American Journal of Public Health, 103*(12), 2141–2146. Doi:10.2105/AJPH.2012.301200

Rose, D., McMillian, C., & Carter, O. (2020). Pet-friendly rental housing: Racial and spatial inequalities. *Space and Culture, 26*(1), 1–14. Doi: 10.1177/1206331220956539

Ryder, R. D. (1975). *Victims of science: The use of animals in research*. Davis-Poynter.

Salgado-Santamaria, X., Hawes, S. M., Dazzio, R., Trainor, M., Flynn, E., & Morris, K. N. (2023). Using cognitive interviews to develop a Spanish version of the one health community assessment to measure one health in Spanish-speaking populations. *Human Animal Interactions* [Manuscript submitted for publication].

Santaniello, A., Dicé, F., Claudia Carratú, R., Amato, A., Fioretti, A., & Menna, L. F. (2020). Methodological and terminological issues in animal-assisted interventions: An umbrella review of systematic reviews. *Animals, 10*(759), 1–20. Doi: 10.3390/ani10050759

Smith, S. G., Johnson, K. R., Davis, C., & Banks, P. B. (2021). Belonging as a pathway to diversity, equity, and inclusion. *eJACD, 88*(2), 25–33. https://www.acd.org/wp-content/uploads/Final_eJACD_vol88-2_2021-07-22_f.pdf#page=25

Sofaer, S. (1999). Qualitative methods: What are they and why use them? *Health Services Research, 34*(5 Pt 2), 1101–1118. https://www.ncbi.nlm.nih.gov/pmc/articles/PMC1089055/pdf/hsresearch00022-0025.pdf

Stonewall, J., Fjelstad, K., Dorneich, M., Shenk, L., Krejci, C., & Passe, U. (2017). Best practices for engaging underserved populations. *Proceedings of the Human Factors and Ergonomics Society Annual Meeting, 61*(1), 130–134. Doi: 10.1177/1541931213601516

The Veterinary Care Accessibility Project. (2022). *Veterinary care accessibility score.* https://www.access-tovetcare.org

Thompson, A. J., Pickett, J. T., & Intravia, J. (2019). Racial stereotypes, extended criminalization, and support for breed-specific legislation: Experimental and observational evidence. *Race and Justice, 12*(2), 303–321. Doi: 10.1177/2153368719876332

Wendler, D., Kingston, R., & Madans, J. (2006). Are racial and ethnic minorities less willing to participate in health research? *PLoS Med, 3*(2), e19. Doi: 10.1371/journal.pmed.0030019

Witwer, R. F. (2021). *DEI and belonging: Changing the narrative and creating a culture of belonging in nonprofit organization* [Master's capstone research report]. University of San Francisco. https://cpb-us-w2.wpmucdn.com/usfblogs.usfca.e0du/dist/9/244/files/2021/05/witwerrakiya_6199833_68188178_Rakiya-Witwer-622-Capstone-Report.pdf

# 10
# ACROSS WORLDS
## Cultural, Religious, and Societal Factors Influencing Interactions with Animals

*Georgitta Valiyamattam*

It is widely accepted that our beliefs and practices regarding animals, companion animals or others, are heavily influenced by our identities. These identities may emerge from multiple sources including but not limited to religion, ethnicity, socioeconomic and political influences. Further, such identities may be modulated by the processes of cultural continuity, discontinuity and acculturation. Our identities and the attitudes that we consequently hold, hugely modulate our relationships with our non-human counterparts. These relationships are often '*intense, complex and paradoxical*' determining which species we love, fear, revere, consume, or are simply repelled by (Herzog, 1988). Most importantly, these relationships are also as unique as our identities.

Our interactions with animals are not by any means infrequent. As Herzog (1988) writes, "*It would be difficult to overestimate the significance of animals in the social and psychological life of our species. Images of animals are everywhere: in our language, religions, dreams, television programs, and folklore*". Animal-assisted intervention (AAI) presents one such channel (and an extremely significant one) where our lives intertwine with those of animals. Hence, the '*cultural and symbolic baggage*' (Serpell, 2004, p. 148) that animals carry would invariably influence our interactions with them, within the context of AAI.

Vital neurological evidence suggests that our brain may be hardwired to respond to animals, perhaps, as a consequence of our evolution (Mormann et al., 2011). O.E Wilson's biophilia hypothesis also puts forward a similar perspective emphasizing on the innate human need to connect with non-animals among other elements of nature (Wilson, 1984). AAIs essentially harness these responses to promote human health and wellness. This would also imply that AAIs can have potential benefits across cultural boundaries. What are the potential cultural considerations that can emerge in such a scenario? How can we navigate these to ensure that the benefits of AAI are not limited/lost due to cultural variations?

As the field of AAI grows rapidly with reports of increasingly widespread practice, it becomes imperative to understand the presence of cultural variables and their dynamism. Smith (2019) observes that it can be tempting for many practitioners across disciplines to fantasize reaching a level of '*permanent cultural enlightenment*' (p. 478), after putting in a large enough amount of effort in exploring human diversity. AAI practitioners may also be tempted likewise or sometimes even carry such illusive notions. A greater exposure to and a discussion of cultural variables can be crucial in fostering cultural humility.

This chapter is divided into two parts: In the first part, we discuss case examples from practice in the Indian setting. The diversity of races, ethnicities and religions in India along with acculturation to the West makes it in parts, a melting pot, a salad bowl and even a 'thali' where cultures thrive separately yet together (Biswas, 2021). A *'thali'* in Hindi, refers to the traditional Indian plating, where food of different flavours-sweet, sour, salty, bitter and spicy are all served together in separate portions. Our goal in presenting these examples is to emphasize on the multiplicity of cultural factors that may influence AAI in multicultural settings. We discuss how contradictory cultural values may coexist, rendering the reliable decoding of cultural variables a tricky exercise. As Smith (2019) puts it, *'thinking with'* case examples from practice may serve as a way of reflecting on cultural variables and developing cultural humility. In the second part of this chapter, we focus briefly on how cultural relevance may necessitate a rethinking of the popular conceptualizations of AAI.

## AAI in Multicultural Settings: Case Examples from Practice in India

The first two case examples in this section emerge from situations involving a close interaction with animals and can be applicable to AAI scenarios. The other three examples emerge from AAI practice.

### How Do We Love Our Animals?

Cultural processes can greatly influence how emotions are displayed and understood including bodily expressions and physical proximity. While such influences have been largely discussed in the realm of human-human interactions, they may also hold true for human-animal interactions. The case example below illustrates this point.

*Case example:* Mr. C, had been visiting my clinic for weekly counselling sessions for nearly five months following the death of his wife. With the outbreak of the pandemic, most of my clients had to be shifted to an online platform. Mr. C was the only geriatric client I had at the time (he was 72). Since he was finding it difficult to use the online platform, I offered to continue bi-weekly sessions for him at my house. I also informed him that I had two dogs at home and they were good with guests. Mr. C had pets in the past and he assured me that he was fine with the dogs as long as they did not 'jump/climb on to him'. During our later conversations, Mr. C mentioned of how he strongly disapproved of his son allowing his pet dog on to the bed or his lap and feeding him at the dinner table.

> When I visited him this time, (his son stayed in the US for work) I was very upset by this. From my 30s till date, I have had 4 dogs and the last one just passed away this summer, (shows photos from his cell phone where garlanded portraits of four dogs adorn the entrance to his flat). When they died, I followed all the rituals, the way we would do for a family member. All of them were taken care of till the very end...They had their own beds in the verandah, and I made sure they were protected from the weather. Rani and Browny (the last two dogs he had) would be satisfied only if I gave them some rice from my plate, in addition to their food. Each night after dinner I used to roll out a ball of rice from my plate and put it in their food bowls. They would play in the courtyard and rest in the verandah or drawing room. They would not jump on me, and when I sat in the verandah, they would sit near me, so I could pat them. None of them ever entered the kitchen, or the bedroom. This (referring to his son's behaviour) is all a Western concept. I don't think this is the way to show affection... (shrugging).

During the few continued sessions, before the second wave of the pandemic hit, Mr. C enjoyed seeing my dogs play in the courtyard. *"I am just happy to see them"*, he mentioned on one occasion *"I feel more relaxed"*. On few other occasions he would also gently stroke their foreheads when they passed by or sat near him.

Edmund Leach (1964) uses social distance to describe our relationships with animals with the ones being most familiar/closest to us either subject to utmost affection or considered taboo. Similarly, those farther away from home/less personally familiar, e.g., animals in the farms/fields, may not be seen as pets and hence also consumed. Finally, those that are socially and physically farthest in space such as wild animals may be seen as beyond consumption and perhaps other utilitarian benefits. While this model may hold true across cultures, with companion animals being relatively closest in social and physical proximity, what comprises physical closeness with an animal/species may vary across cultures. Similar sentiments have also been captured by Brown (2002) in her examination of how pet attachment with ethnic variations influenced the level to which an animal was anthropomorphized and the ways in which attachment was expressed.

Research on human social interactions reveals how ingroup-outgroup effects exist in our perception of emotions and their expressions (examples include Chiao et al., 2008 and Derntl et al., 2012). The same may hold true in how we perceive human-animal interactions with our own/mainstream cultural experiences unconsciously presumed as the normative standard. This can lead to erroneous assumptions, particularly in the context of AAI, when a client's receptiveness to animals is judged from the AAI practitioner's definition of what receptiveness should constitute. As Mr. C remarked *"Seeing my discomfort, my son and his friends who were visiting him there (in the USA) remarked – 'You do not like dogs'… It is ridiculous to hear this"*.

## Religion, Animals and AAI

Across the major religions, attitudes towards animals are marked by both commonalities and differences. Commonalities include first, the belief of an interdependence among all creatures, which is also hierarchical, since humans occupy the highest space in the material realm. A second point of convergence is the dimension of divinity, which points to a 'common source and a common goal to all life', beyond the human-animal divide (Caruana, 2020). In addition to this, sacred scriptures of the major religions and their continued teachings are also seen to espouse kindness towards animals, with those helping an animal being praised. Animals in communion with humans are also seen as important elements in God's plan (Caruana, 2020).

However, beyond these commonalities, religious positions related to animals also differ at several levels. For instance, while the hierarchy of living beings and the superior status that humans occupy is unbroken in Abrahamic religions such as Judaism, Christianity and Islam, a similar hierarchy in Eastern religions such as Hinduism is offset also by the reincarnation of deities as animals. Similarly, while the hierarchical conceptualization is also maintained to an extent in Buddhism, both in Buddhism and more so in Jainism, the idea of ahimsa or non-violence emerges as a driving force emphasizing the intrinsic worth of all living beings (Caruana, 2020). Differences in religious teachings and their interpretations are also seen to influence other aspects of our contact with animals. Examples include which species or species specifics we refrain from interacting with (an instance is the taboo associated with certain species such as pigs and dogs in several sects of Islam (Harris, 2012; Berglund, 2014), which species are meant to be consumed or forbidden as food (e.g., prohibitions on the consumption of the meat of pigs in Islam and of camels, hares, badgers and pigs in Judaism) (Silber, 1994), which species receive a protected existence (e.g., the revered and protected status of cows in Hinduism, leading to the ban on the sale and consumption of beef in several Indian states (Krishna, 2010; Biswas, 2015; The Indian Express, 2015) and even whether or not we choose to have pets (e.g., keeping pets is discouraged in Jainism for several reasons, foremost being the desire to not intrude into the lives of animals (Berry, 1998). Several such examples have also been stated in other related research (e.g., Szucs et al., 2012; Jegatheesan, 2015).

Religion has been seen as a formidable factor in implicitly or explicitly shaping our attitudes towards animals. However, religious influences even within a religion are not by any means a uniform canvas; they can differ according to the levels of religiosity (Heleski et al., 2006). and the religious denomination (Hand & Liere, 1984). For instance, within a group of Christian survey respondents in Ohio, USA, those from the Catholic denomination (84%) were seen to be significantly more supportive of animal rights, than those from the Protestant denomination (77%) (Nibert, 1994). Further, as societies grow economically and move towards a largely individualistic ideology (Santos et al., 2017), differences in religious interpretations (earlier expected at the level of denomination), may also emerge at an individual level.

***Case example:*** Ms. F a 19-year-old student and a practicing Muslim, had requested for an urgent appointment at a very short notice. We were then conducting sessions in a special school, where a few rooms were allotted to us. She requested to visit me there for a brief session. The presence of our therapy dog also brought with it the fear of hurting religious sentiments, since dogs are usually a taboo for Ms. F's community. We discussed removing D, our therapy dog from the scene, and a team member offered to drop her back home, since therapy sessions for the kids were over for the day. However, after some thought, we decided to have the dog in the adjoining room, and informed Ms. F of the same. After discussing D's presence with her, I told her that, she could choose if she wanted to have D around or not. I assured her that, some clients chose not to have D around and some did. I also offered to move the session to an entirely new space (my office space), if needed. Ms. F admitted that dogs were unwelcome at home and that she had a pet cat. *"But I absolutely love being with dogs"*, she said, *"I chose to visit you here not just because its closer to my college; it's also because I knew you would have a dog here and I wanted to play with it (smiling)"*. She was deeply religious herself, but did not see this as problematic to her personal beliefs.

> Initially my parents used to get angry about this, because I used to play with the dogs in the neighborhood. But I used to argue back a lot, countering them with how I interpreted the scriptures. Eventually they relented. I have seen my parents arrange food for a few stray animals including dogs near our home…they are very kind…but the dog cannot enter our house, that physical distance is defined… My parents know that I will adopt a dog when I move out for work. But of course, I will not do it in my parent's house…that will be too much for them to handle, both personally and socially…It's a kind of compromise there (smiling).

In one of our later conversations on the topic, she told me,

> Many people including my parents often assume this…but I am very clear, it was not a western influence or something I started fancying due to social media. It may be true for some others but for me, it was just that I liked dogs from the very beginning… even

as a very young kid, I liked them the same way that I liked the cats in my house or the rabbits my cousin had. I did not see a reason why we should differentiate between dogs and other animals, and I used to search for resources to argue and prove my point. Many people wouldn't do that.

Similar sentiments can be found in other instances too (e.g., see Berglund, 2014). According to R, a graduate student and aspiring HAI researcher, unlike their parents and grandparents, she and several of her cousins and friends no longer believed that dogs were taboo. This was mostly driven by cultural discontinuity and sometimes by acculturation processes. A common thread was the tendency to question and seek information beyond what was handed down intergenerationally. As R put in, *"For instance there are verses in the Quran that I wish to go by…they tell us that dogs can be kept for hunting and guarding and that you can eat what the dog has hunted for you"*. *"This does not mean a huge shift in opinion"*, she cautioned further, *"we had difficulties in navigating popular sentiments when we wished to start an NGO for street dogs in our Muslim dominated area* (Malappuram, Kerala). *But yes, there is definitely some gradual shift in attitudes"*.

Farida's example illustrates how contradictory beliefs may coexist within the same religious and familial boundaries. Over the next seven months that Farida visited me, D become a very important link facilitating our bond. I was absolutely grateful that we had not gone by any presumptions.

## Social Identities, Influences and AAI

Historically, animals and the practices surrounding them have shaped our social identities. For instance, pet-keeping was often seen as a symbol of luxury reserved for the elite and ruling classes both in the Asian and European societies. These often involved exotic, specifically bred or pure-bred species (Walsh, 2009; Fujimura, & Nommensen, 2017). Apart from pets, the possession of certain animals such as horses or cattle has and continues to be seen as an indicator of wealth (Blazina et al., 2011; The Cattle Economy of the Maasai, n.d.). Similarly, certain practices involving animals such as bull fighting are associated with a sense of ethnic pride. For instance, Jallikattu in the southern Indian state of Tamil Nadu, the ban on which had to be revoked due to the mass protests centring around the Tamil identity (Tilak, 2017). The social identity attached to animals can also influence AAIs often in combination with several other factors.

***Case example:*** One of the special education schools that had signed up for our AAI program was situated within an urban and largely affluent community. Mr. K, the parent of an autistic child, attending this school was very excited about his child being a part of the Animal-assisted activity (AAA) sessions. He came up to us and said, *"I love dogs and I'm so happy to see that you are doing this the right way"*. As we chatted further, we realized that he was referring not to the intervention process per se, but specifically to the breed of our therapy dog. The dog accompanying us to this particular school happened to be a golden-retriever. Mr. K explained how a group of practitioners in the city had engaged in AAA a few years back and had been horribly wrong with their practice. *"They knew nothing! They used stray dogs…how could I let my child participate?"* He also shared how he had seen dogs of a *'good breed'* helping children on social media. *"I also wish to buy a dog of a good breed"*, he added. The AAA instance that Mr. K was referring to, was a noble effort by a local NGO, and had incorporated a trained Indian dog. Unfortunately, native Indian dog breeds are usually referred to as street/stray dogs, and seen as inferior to the foreign pure-bred dogs.

Ideas like those expressed by Mr. K can emerge from different sources, sometimes reflective of cultural hegemony. A key way in which this hegemony works across several scenarios is through the media, which constructs what is acceptable or the norm, including breeds that are most desirable or acceptable as pets (e.g., Ghirlanda et al., 2013, 2014). For a majority of the upper middle-class population in India with disposable incomes permitting the rearing of a companion animal, having a pure-bred, foreign breed pet is a status symbol (Dutta, 2015). This perhaps reflects economic colonialism. Further, it is often combined with a negative attitude towards stray dogs, mostly revolving around the fear of dog bites and the risk of disease.

Navigating through Mr. K's perception of a companion animal or a therapy animal required different levels of consideration and several discussions, which moved gradually from the concept of breeds and their origins, to what made a good companion animal and a good therapy dog. Towards the close of the sessions, Mr. K agreed that therapy with an Indian dog would probably have the same benefits for his daughter. He also admitted that, though Indian dogs make good companion animals, he would probably still buy a foreign breed first and then think of adopting an Indian dog eventually. "*All my relatives and friends will be very judgmental, if I have an Indian dog in the house right away*", he explained.

Mr. K's concerns regarding the breed of the therapy dog, illustrate how considerations regarding the choice of a therapy animal may go beyond the usually highlighted species preferences. In our conversations, we did not expect to nor did we bring about a complete change in his viewpoints. Scenarios such as these would require a patient navigation through deep-seated ideas and beliefs, going back and forth, without expecting immediate breakthroughs.

## The Intertwining of Cultural Factors and the Utilitarian Value of Animals in AAI

The last two case examples in this chapter try to reflect on the multiplicity of factors framing our attitudes towards animals. We also examine how these attitudes can both influence and be influenced by the therapeutic setting.

**Case Example**: While seeking consent from caregivers before we started AAA sessions for a new batch of kids in a special education school, a mother, Ms. N, shared how her child was extremely scared of dogs and hence would not be a good fit. Interestingly, we had observed her son AK start whimpering at a distance, when he saw the therapy dog D entering the facility, a few days before for a different session. Even though D was a considerable distance away, in a bid to scare her, he bent down and gestured as though he was picking up a stone to throw at her. I shared this instance with his mother, and enquired if AK had possible negative experiences with dogs. Surprisingly, there were none. Ms. N, however, shared how she had been chased by stray dogs twice as a young child.

> When I was around four, there was a newspaper report of a dog bite, and then a lot of people in my family were narrating stories of dog bites and the multiple injections that follow to ensure rabies prevention. They were also talking of rabies-related deaths. I was so scared that, I would scream and run whenever I saw a dog approach or bark. Twice the dogs chased me, and I cried for hours after that. Later, one of my uncles taught me how to scare dogs away by bending down and acting as if I was picking up a stone to throw. I came to know only in my 20s that, perhaps, my reaction (screaming and running) was the very reason the dogs chased me. But my fear has never really reduced, so I still use my uncle's trick many times,

she said. Her son's fear and the accompanying reactions were learnt vicariously from his social space.

"*It's not that I don't like animals*", she clarified,

> I feed doves and whenever I see cows at the temple or on the street, I feed them. Even AK puts his palm in mine as I feed. My mother practiced this for several years…she used to feed the cows and a few used to wait outside our gate patiently for her in the morning. 'Cows are gentle just like mothers' my mother used to say.

We assured her that her feelings were understood, along with some brief conversations revolving around animals, their temperaments and what triggers certain animal behaviours.

Ms. N's relationships with animals were clearly a mix of multiple factors – society, religion and personal experiences – a tricky brew that was not necessarily negative. However, it fostered relationships with some species and hampered those with certain others. Her childhood encounters were a classic example of the complex human-dog interaction in several communities in India, thanks to an increasing number of stray dogs, the struggle for limited food sources, the fear of rabies and the lack of widespread public knowledge about how to tackle this scenario/interact with the animals. Measures to tackle these have ranged between an implementation of the animal birth control (ABC) program (Fitzpatrick et al., 2016) to extra-legal killing of dogs fuelled by periodic public uproar, political inclinations and irresponsible media reporting (Karlekar, 2008; HT, 2016). Recent studies have called for measures that conceptualize multispecies habitats rather than singularly viewing animals as disease agents requiring management (Srinivasan et al., 2019).

Few days later, when she came to pick her son up from the school, Ms. N stood for a while watching other kids playing with the therapy dog and remarked, "*Your dog is different…gentle like a cow. The kids are very happy, and other parents were also saying the same. Do you think AK can get over his fear and join this program?*"

As for Ms. N's case, we had to explain how we would need to partner with her, in helping her child get over his fear of dogs. This would mean tackling her own fears first. While this would be a useful exercise, we also pointed out that, there were other interventions (not involving animals) that AK could benefit from. Navigating scenarios such as these can often become relatively protracted exercises, as we experienced (with Ms. N opting to partner with us), but rich in terms of the knowledge exchanged and the lessons learnt mutually.

Since ancient times, the significance of animals and their welfare have been heavily influenced by their utilitarian value, both economic and social. For instance, in the ancient religion of Zoroastrianism, the classification of non-human animals as 'good' and 'evil' was probably rooted to a large extent in their perceived utilitarian value. The value of the dog for herding cattle in their earlier nomadic past and the benefits of the cow as a milch animal were perhaps instrumental in these animals being viewed as beneficent, whereas the creatures that 'pricked, bit, or stung, and seemed hideous and repulsive to human beings' were seen as vile or 'khrafstars' (Foltz, 2010). Within the therapeutic setting, animals assume a unique utilitarian value as healers, and this can influence how they are perceived. This was demonstrated in Ms. N's case, where despite her continued fear, the utilitarian value of the animal within the therapeutic context had acted as a motivating factor in seeking out AAI.

*Case Example*: In another instance, where we worked with a school as part of our research with autistic children, a mother approached me. She asked us to include her child, who was diagnosed with cerebral palsy, in the AAA sessions. I explained to her that, since the work we were doing was part of our research, we had limited resources to do that immediately.

We assured her that we would conduct extra sessions to include her child and a few other interested kids in the coming month. She told us, *"I saw the kids laughing and playing with your rats; so my friend and I wanted to have our kids in the session too"*. It took us a few moments to recognize that, from a distance, she had mistaken our guinea pig co-therapists for rats.

What was really surprising was that the mothers were approving of their kids playing with creatures commonly seen as unclean household pests. When I explained to her that we worked with guinea pigs and not rats, she said, *"Oh, I am sorry...but I would be fine even if they are rats, as long as my child is active and happy"*. The fact that the utilitarian value of a therapy animal can potentially transcend societal and cultural barriers to such an extent was definitely important and also slightly unsettling. With the power that this gave to the therapist by default, there could also emerge chances of misuse. In a similar example, in the ultra-orthodox Israeli Jewish community where problems exist regarding contact with certain animals, rulings of high-ranking rabbis in some communities permitted animal-assisted psychotherapy (AAP) based on a principle in Judaism which allowed certain religious laws to be broken for the sake of 'saving a soul' (N. Parish-Plass, personal communication, October 25, 2022).

## AAI in Multicultural Settings: Is a Different Conceptualization of AAI Possible?

AAIs have found acceptance in several communities. However, it is also important to note that, sometimes, certain communities as a whole or some members of a community may not share the same enthusiasm or favourable attitudes. The reluctance to engage in AAI may emerge from different sources.

In several societies, interaction with animals may be viewed as associated with a risk of disease. Experiences from practice in India and Japan have noted several barriers in bringing a therapy animal into a facility, for instance, a hospital or a school, as there is a disproportionate focus on the risks involved as compared to potential benefits. In an AAI case example from Japan, an elderly man felt it was '*dirty*' to bring a dog into a facility and cited how dogs used to be kept 'outside' when he was young (M. Yamamoto, personal communication, September 28, 2022). While Japan has seen a pet boom in recent years (Hoffman, 2011), with the emergence of the concept of the 'katie dobutsu' or the family animal (Fujimura, & Nommensen, 2017), this has been a gradual transition from the predominantly utilitarian view of animals. Dogs, for instance, were valued as hunting companions or as guard dogs. Closer interactions with animals in physical space, as in the case of 'lapdogs', were reserved for certain exotic species and for the rich and elite classes (Fujimura, & Nommensen, 2017). Further, recent studies suggest that an increase in the number and variety of companion animals has also been accompanied by an increased risk for zoonotic diseases. This is due to the high population density in Japan and the relatively lower percentages of companion animal registration and vaccination (Takahashi-Omoe & Omoe, 2012). A related scenario is seen in the Nigerian context where dogs belonging to the native breeds were long seen as hunting companions. They blended harmoniously into the family structure, efficiently executing their part in hunting without any form of restraint or control. However, in the recent decades fuelled by several socioeconomic changes, foreign and exotic dog breeds have replaced the native breeds as companion animals in Nigerian families. This is with little regard to whether the particular breed fits into the family dynamics and lifestyle and is accompanied by a lack of awareness among the owners about the standards of optimal care for these breeds (for e.g., vaccination, feeding and training). The resultant cases of canine aggression including bites (sometimes fatal) have enhanced the risk perceptions surrounding dogs thereby limiting the potential for positive interactions (S. Agbonika, personal communication, October 24, 2022). Thus, multiple factors may operate cumulatively, enhancing perceptions of the risk surrounding animals and the spaces they occupy.

While several studies and anecdotal accounts have popularly linked AAI with positive affect, this may not always be the case. Personal and community level experiences with animals in general or with certain species may differ. The potential negative perceptions related to certain species or species specifics emerging from religious and related beliefs, have already been discussed earlier in this chapter. In addition, species that have coexisted with communities/individuals may also evoke unpleasant memories due to the nature of their interactions. Certain animal species for instance, may symbolize power and discrimination. An example is the role of the dog in enforcing the discrimination against African Americans. As Stewart (2019) writes,

> There has been a perpetual image of a person in a position of power brandishing a dog against African Americans. The slaveowner viciously releasing his bloodhounds to hunt enslaved people turned into the officer charging his K-9 dog to attack unarmed Black men.
>
> (p. 184)

The fear and suspicion may also be directed more specifically to certain breeds like the German Shepherds, who were associated with the brutal policing and control of Black people during the Apartheid regime in South Africa (Shear, 2008, L. Tucker, personal communication, October 25, 2022). In a different case example from Japan which also echoes the importance of personal/community level experiences, an elderly man stated that, he did not wish to interact with a horse, as it reminded him of the hard times in his life when he used to work with horses (M. Yamamoto, personal communication, September 28, 2022). Further, the role that an animal is commonly assigned within a community (even when it is of a high utilitarian value), may interfere with its role within a therapeutic setting. Dogs for instance continue to be valued in several communities (e.g., South Africa and India) as watchdogs and guard dogs acting as deterrents to criminal activity. However, this also implies that the dog can be 'dangerous' or 'aggressive' which then evokes fear and interferes with its image as a therapy animal.

As discussed earlier, the socio-political influences on our attitudes towards animals can also translate to the AAI setting. South Africa, for instance, is a nation characterized by both cultural diversity (and hence often referred to as the Rainbow Nation') and socioeconomic disparity. These variations are also reflected in the attitudes towards companion animals. Across the socioeconomic and demographic divide, companion animals (particularly dogs and cats) are treated with affection as family members by a substantial majority. However, this is also offset by a politicization of human-animal relationships in several quarters. Animal advocacy efforts in South Africa for instance are met with suspicion as *"promoting racist rhetoric and maintaining White power by drawing on discourses of primitiveness surrounding Black people's ability to sufficiently care for their animals"* (L. Tucker, personal communication, October 25, 2022). This has been accompanied by recent onslaughts by Black politicians and media personalities who criticize how White people prioritize the needs of their pets over their domestic workers such as childminders. An instance would be paying for medical aid for pets but being unwilling to extend a similar support to domestic workers. While there are a number of practitioners (psychologists, counsellors, social workers, occupational therapists) who engage in AAT, across South Africa, this practice is largely limited to private healthcare or private social services. Further, a majority of the volunteers and administrative staff identify as White females from middle to upper-income families (as is the case in *Pets as Therapy*-one of the longest running organizations providing AAI in several South African provinces). The separations in terms of where AAI is provided (in private rather than public healthcare settings) and who provides it, serve to further reinforce the inequities around AAI and the perception that it is a '*more privileged form of support*' meant for the higher socioeconomic strata or select demographic groups (L. Tucker, personal communication, October 25, 2022). In circumstances such as these, AAIs and their accessibility may involuntarily serve as a yardstick for socioeconomic disparity, thereby drawing mixed responses.

AAI, as it is popularly conceptualized, involves trained animals. However, this can be an unfamiliar concept in many communities, who have interacted largely with farm animals, free ranging animals within the community or untrained companion animals. The relative foreignness of the concept may also make AAIs unrelatable to such communities. Many Japanese feel sorry for the 'strict training' that assistance dogs undergo and feel that animals are '*used*' and '*exploited*' in the context of AAIs (Miura et al., 2002; Takahashi & Yamamoto, 2017; M. Yamamoto, personal communication, September 28, 2022). To counter this, the Japan Guide Dog Association, which is the largest guide dog training organization, even made a TV commercial to educate people not to feel sorry for guide dogs as they were also enjoying their work (M. Yamamoto, personal communication, October 8, 2022). In certain other instances, an individual may have had little or no experience of a companion animal, and very little exposure to animals at a personal level.

In such scenarios, it can be both easy and erroneous to conclude, that interventions involving animals may not work well with certain individuals or communities. Culturally responsive approaches to AAI would require us to think beyond whether or how a particular individual/community can thrive within the popular conceptualization of AAI. It would also demand a rethinking of how AAIs are conceived. The Israeli model of AAP presents one such example (Parish-Plass, 2013). It is unique in several ways; first, instead of the presence of a single animal/species as is commonly the case with AAI, AAP distinguishes itself with the therapy zoo emerging from its traditions of the kibbutz petting zoos. Just as in the petting zoos, the therapy setting also comprises of a variety of animals that the therapist is trained in handling. Second, while it is ensured that the animal is socialized and does not pose a threat to the client, the animal is also not specifically 'trained' to show the behaviours that are generally expected of a therapy animal, for instance, following commands, being well-behaved, not showing negative emotions or being able to perform certain tricks. The focus here is to tap on the spontaneity and 'aliveness' of the animal as a medium within the therapeutic process, thus creating 'natural interactions' (Parish-Plass, 2013) It is important to note, that this may not always involve positive affect, but allows clients to reflect on their own feelings. For instance, Parish-Plass (2013) talks of the example of a young client, who was able to break the ice and delve into his own feelings of abandonment (by his mother at a very early age), due to a spontaneous reaction from the therapy dog Mushu. Mushu barked and jumped at the door, when he was left behind in the therapy room (a behaviour that would be considered improper for a trained therapy dog). On seeing this, the child said "*Oh, no! She thinks we are leaving her behind. Poor Mushu! We can't abandon her. She needs me!*" (p. 433).

It may be argued that AAP represents what is most suited to psychotherapy. However, this approach can also work well with other interventional and therapeutic activities involving animals. For instance, an animal displaying natural behaviour can be more relatable to communities, who have not had much experience with trained animals and can lessen notions of the animal being 'used'. While this may not always be possible, the flexibility of a therapist in working with different species can be useful in navigating potential taboos associated with a specific species. In the case of clients who have had little experience with animals at close quarters (also accompanied at times by a fear of animal presence), opportunities that allow interactions with several species and at a pace suitable to oneself, can be useful in enabling an understanding of their own needs and preferences. Spaces that allow flexible levels of contact can also be valuable. Nancy Parish-Plass shares the instance of an Ethiopian child from an emergency shelter, who was scared of animals, as he had little experience of close interactions with animals. However, as he slowly explored the therapist's room and the species variety that it offered, at his own pace, fear gave way to an expression of several repressed feelings (N. Parish-Plass, personal communication, September 27, 2022). What if this child was seen as unsuitable for AAI, due to the initial fear that he displayed? Spaces that allow choices can be very valuable. This may include first, the choice of interacting with animals or not, and second, the choice to interact in multiple ways. For instance, multiple species or flexible levels of contact (an example is the client in case example one, who experienced a sense of relaxation watching dogs play from a distance).

Rooting AAI practice within the framework of native philosophies can also be helpful. An example is the placement of AAI within the Maori perspective of health and wellbeing. The Maori model – Te Whare Tapa Wha described by Sir Mason Durie sees wellbeing as a wharenui/meeting house comprising four walls – Taha wairua (spiritual wellbeing), Taha tinana (physical wellbeing), Taha hinengaro (mental wellbeing) and Taha whānau (social wellbeing). The foundation of this house is the Whenua (environmental wellbeing). Whenua/land includes the soil and landforms, animals, plants and people and forms the link that connects us to us both to our ancestors and the generations to come. A balance between these five dimensions is considered essential to health (Durie, 1998). AAI practice within this framework in New Zealand involves an initial grounding body scan which allows the client to acknowledge being in the present with feet touching the ground, noticing the connection between one's body and the earth and to all beings and the surrounding nature. This then becomes symbolic of the interconnectedness of all things and prepares the client for a complete engagement with the animal (the horse in this context), by remaining in the present and letting go of any baggage of the past (K. Howieson, personal communication, October 25, 2022). Similar models that echo the biophilia hypothesis can also be found in other cultures, an example being the ancient Indian philosophical perspective of 'Vasudhaiva Kutumbakam' that sees the world and all its beings as one interconnected family.

A rethinking of how AAIs are delivered would also benefit from broader community level estimates – these may include among others, data about the settings where AAI is delivered within a community, the demographics of who delivers AAI and the populations served. This can then enable a better understanding of how socio-political influences around AAI may operate and who may be intentionally or unintentionally excluded – an example being the racial politics surrounding AAI in South Africa (L. Tucker, personal communication, October 25, 2022). Also, while the practice of AAI spreads to newer communities, thereby aiding greater awareness, this has not yet translated into a significant body of cultural knowledge. A primary reason is the comparatively limited local research, which even when significant, rarely translates into peer-reviewed academic literature (L. Tucker, personal communication, October 25, 2022, N. Parish-Plass, personal communication,

September 27, 2022). A systematic recording of experiences emerging from multicultural AAI practice can be pivotal in making both AAI and the cultural considerations involved a more visible and less esoteric subject both within and outside the community. It can also reveal the unique benefits provided by the inclusion of animals within traditional treatment paradigms. An example is of the Arab and the ultra-orthodox Jewish communities in Israel where attending psychotherapy sessions is seen as a social embarrassment. The Muslim Arab and ultra-orthodox Jewish communities also have reservations regarding contact with certain animals, yet AAP sessions are seen as more acceptable. AAP sessions are viewed more as activities thereby buffering the inhibitions otherwise traditionally associated with psychotherapy (N. Parish-Plass, personal communication, September 27, 2022).

## Conclusion

With societies becoming increasingly multicultural, many AAI practitioners seek out ways to best engage with cultural differences. Jegatheesan (2015) and Smith (2019) in their articles, focus on the principles that can promote cultural competence and cultural humility in interactions involving animals. Key among these is a continuing self-awareness that one's own beliefs emerge from a unique sociocultural backdrop, and that one may be instinctively and incorrectly predisposed to consider one's beliefs normative or universal. Equally vital is an understanding of the client's sociocultural background, and also an appreciation of the multiple layers and intersections uniquely framing an individual's identity. While these would enable the development of cultural competence, cultural humility, by its very definition, would also demand change at a systemic level. Culturally humble AAI would require going beyond 'choosing' clients for AAI, based on whether they fit existing/popular AAI frameworks. Further, it would involve examining if AAI approaches can be reframed to reach diverse clients.

Above all, as societies evolve rapidly, AAI practitioners may benefit from continually educating themselves. Such education can range from humble and open conversations with their clients to questioning possible presumptions that they themselves may carry. These presumptions may pertain as to how we (AAI practitioners) see ourselves, how we see our clients, how we see the process of AAI and finally as to how we conceptualize cultural differences. For instance, how do we define cultures as *ours* versus *theirs*? What are the demarcations of cultural difference: religion, geography ethnicity, gender, sexual identity, skin colour, wealth or others? Where and when do we expect to encounter multicultural considerations in our practice? As Gottlieb (2021) writes, offering a truly inclusive definition of culture, "...*every relationship becomes cross-cultural in some way. Even a client who seems so much 'like us' has identities that we do not hold, and every client we perceive as different will ultimately have some identities we share*". This also implies that cultural considerations in AAI are perhaps more frequent than we may expect or realize. The worlds beyond ours then, may not be that far away.

## Acknowledgements

I sincerely thank Nancy Parish-Plass (Chairperson, Israeli Association of Animal-Assisted Psychotherapy), Sunday Agbonika (Dogalov Human Support Initiative, Nigeria), Leigh Tucker (University of the Western Cape and Pets as Therapy, South Africa), Karen Howieson (Sentient Clan Services, New Zealand), Mariko Yamamoto (Teikyo University of Science, Japan) and Ayisha Rushda (Team Incubation, Kerala, India) for their valuable inputs.

## Discussion Questions

1 What constitutes multicultural practice? Have you come across any such scenarios, particularly in the AAI setting?
2 Many practitioners face a double issue cross-culturally namely the attitudes towards psychotherapy/counselling and the attitudes towards animals. How can these be best navigated?
3 The article reveals philosophical perspectives such as the Te Whare Tapa Wha and Vasudhaiva Kutumbakam that echo the biophilia hypotheses. Can you think of similar examples?
4 Can you think of ways in which AAIs can be made more flexible in catering to the needs of multicultural populations?

## References

Berglund, J. (2014). Princely companion or object of offense? The dog's ambiguous status in Islam. *Society & Animals, 22*(6), 545–559.
Berry, R. (1998). *Food for the gods: Vegetarianism & the world's religions.* U.S: Pythagorean Publishers.
Biswas, S. (2015, October 15). Why the humble cow is India's most polarising animal. *BBC News*. https://www.bbc.com/news/world-asia-india-34513185
Biswas, S. (2021, July 8). Pew survey: India is neither a melting pot nor a salad bowl. *BBC News*. https://www.bbc.com/news/world-asia-india-57723926
Blazina, C., Boyra, G., & Shen-Miller, D. S. (2011). *The psychology of the human-animal bond.* Berlin: Springer.
Brown, S. E. (2002). Ethnic variations in pet attachment among students at an American school of veterinary medicine. *Society & Animals, 10*(3), 249–266.
Caruana, S. J. L. (2020). Different religions, different animal ethics? *Animal Frontiers, 10*(1), 8–14.
Chiao, J. Y., Iidaka, T., Gordon, H. L., Nogawa, J., Bar, M., Aminoff, E., et al. (2008). Cultural specificity in amygdala response to fear faces. *Journal of Cognitive Neuroscience, 20*(12), 2167–2174.
Derntl, B., Habel, U., Robinson, S., Windischberger, C., Kryspin-Exner, I., Gur, R. C., et al. (2012). Culture but not gender modulates amygdala activation during explicit emotion recognition. *BMC Neuroscience, 13*, 54.
Durie, M. (1998). *Whaiora: Maōri health development.* Oxford University Press.
Dutta, S. (2015, February 20). Dear Indian dog owners, Saint Bernards and Huskies do not belong in Delhi or Mumbai. *Quartz*. https://qz.com/india/343312/dear-indian-dog-owners-saint-bernards-and-huskies-do-not-belong-in-delhi-or-mumbai
Fitzpatrick, M. C., Shah, H. A., Pandey, A., Bilinski, A. M., Kakkar, M., Clark, A. D., … & Galvani, A. P. (2016). One Health approach to cost-effective rabies control in India. *Proceedings of the National Academy of Sciences, 113*(51), 14574–14581.
Foltz, R. (2010). Zoroastrian attitudes toward animals. *Society & Animals, 18*(4), 367–378.
Fujimura, C. K., & Nommensen, S. (2017). *Cultural dimensions of well-being: Therapy animals as healers.* Lexington Books.
Ghirlanda, S., Acerbi, A., Herzog, H., & Serpell, J. A. (2013). Fashion vs. function in cultural evolution: The case of dog breed popularity. *PLoS ONE, 8*(9), e74770. doi:10.1371/journal.pone.0074770
Ghirlanda, S., Acerbi, A., & Herzog, H. (2014). Dog movie stars and dog Breed popularity: A case study in media influence on choice. *PLoS ONE, 9*(9), e106565. doi:10.1371/journal.pone.0106565
Gottlieb, M. (2021). *What is cultural humility? 3 Principles for social workers.* The New Social Worker. https://www.socialworker.com/feature-articles/practice/what-is-cultural-humility-3-principles-for-social-workers/
Hand, C. M., & Van Liere, K. D. (1984). Religion, mastery-over-nature, and environmental concern. *Social Forces, 63*(2), 555–570.
Harris, M. (2012). The abominable pig. In C. Counihan & P. V. Esterik (Eds.), *Food and Culture* (pp. 73–85). Routledge.

Heleski, C. R., Mertig, A. G., & Zanella, A. J. (2006). Stakeholder attitudes toward farm animal welfare. *Anthrozoös, 19*(4), 290–307.

Herzog Jr, H. A., & Burghardt, G. M. (1988). Attitudes toward animals: Origins and diversity. *Anthrozoös, 1*(4), 214–222.

Hoffman, M. (2011, July 17). It seems Japan has literally gone to the dogs. *The Japan Times.* https://www.japantimes.co.jp/news/2011/07/17/national/media-national/it-seems-japan-has-literally-gone-to-the-dogs/

HT. (2016). Stop killing stray dogs in Kerala: Supreme Court tells vigilante groups. *Hindustan Times,* 18th November. http://www.hindustantimes.com/indianews/stop-killing-stray-dogs-in-kerala-supreme-court-tells-vigilante-groups/ story-wnPjOLRR3vFAlX5ToHD5KI.html

Jegatheesan, B. (2015). Influence of cultural and religious factors on attitudes toward animals. In A. Fine (Ed.) *Handbook on animal-assisted therapy* (pp. 37–41). Academic Press.

Karlekar, H. (2008). *Savage humans and stray dogs: A study in aggression.* Sage Publications India.

Krishna, N. (2010). *Sacred animals of India.* Penguin Books India.

Leach, E. R. (1964). *Anthropological aspects of language: Animal categories and verbal abuse.* MIT Press.

Miura, A., Bradshaw, J. W., & Tanida, H. (2002). Attitudes towards assistance dogs in Japan and the UK: A comparison of college students studying animal care. *Anthrozoös, 15*(3), 227–242.

Mormann, F., Dubois, J., Kornblith, S., Milosavljevic, M., Cerf, M., Ison, M., ... & Koch, C. (2011). A category-specific response to animals in the right human amygdala. *Nature Neuroscience, 14*(10), 1247–1249.

National Geographic. (n.d.). *The cattle economy of the maasai.* https://education.nationalgeographic.org/resource/cattle-economy-maasai

Nibert, D. A. (1994). Animal rights and human social issues. *Society & Animals, 2*(2), 115–124.

Parish-Plass, N. (Ed.) (2013). *Animal-assisted psychotherapy: Theory, issues, and practice.* Purdue University Press.

Santos, H. C., Varnum, M. E., & Grossmann, I. (2017). Global increases in individualism. *Psychological Science, 28*(9), 1228–1239.

Serpell, J. A. (2004). Factors influencing human attitudes to animals and their welfare. *Animal Welfare-Potters Bar Then Wheathampstead, 13,* S145–S152.

Shear, K. (2008). Police dogs and state rationality in early twentieth-century South Africa. In L.V. Sittert & S.S. Swart (Eds.) *Canis Africanis* (pp. 193–216). Brill.

Silber, D. S. (1994). *The Jewish Dietary laws and their foundation.* Retrieved from https://dash.harvard.edu/handle/1/8889478

Smith, Y. (2019). Pets and human diversity: Toward culturally competent, culturally humble psychotherapy. In In L. Kogan & C. Blazina (Eds.), *Clinician's Guide to Treating Companion Animal Issues* (pp. 477–496). Academic Press.

Srinivasan, K., Kurz, T., Kuttuva, P., & Pearson, C. (2019). Reorienting rabies research and practice: Lessons from India. *Palgrave Communications, 5*(1), 1–11.

Stewart, S. (2020). Man's best friend? How dogs have been used to oppress African Americans. Mich. J. Race & L., 25(2), 183–208.

Szucs, E., Geers, R., Jezierski, T., Sossidou, E. N., & Broom, D. M. (2012). Animal welfare in different human cultures, traditions and religious faiths. *Asian-Australasian Journal of Animal Sciences, 25*(11), 1499–1506.

Takahashi, R., & Yamamoto, M. (2017, June 22). *Attitudes toward guide dogs in Japan: Descriptive analysis of comments on Twitter.* ISAZ Conference, UC Davis, California.

Takahashi-Omoe, H., & Omoe, K. (2012). Social environment and control status of companion animal-borne zoonoses in Japan. *Animals, 2*(1), 38–54.

*The states where cow slaughter is legal in India* (2015, October 8). The Indian Express, August 18, 2022. https://indianexpress.com/article/explained/explained-no-beef-nation/

Tilak, S. G. (2017, January 20). Jallikattu matters: It's a symbol of self-assertion, people power in Tamil Nadu. *Hindustan Times.*

Walsh, F. (2009). Human-animal bonds I: The relational significance of companion animals. *Family Process, 48*(4), 462–480.

Wilson, E. O. (1984). *Biophilia: the human bond with other species.* Cambridge, MA: Harvard University Press., USA

# 11
# ANIMAL COGNITION
*Lena Provoost*

## Introduction to Animal Cognition

Cognition is defined as the mental processes utilized to gain knowledge and understanding through sense, experience, and thought. The fascinating field of comparative psychology investigates similarities and differences between human and animal behavior, while comparative cognition focuses on the biological mechanisms driving behavior. Studies in animal cognition ask what type of knowledge a species learns, how, and why. The answers to these questions and their driving factors can contribute to improved welfare in various aspects of animal management systems, therapeutics, and training. Cognition studies continue to focus on non-human primates (NHP), rats, pigeons, other bird types, and dogs (Beran et al., 2014). However, there are studies, though not as in-depth, on insects (Perry et al., 2017), cats (Vitale et al., 2015), horses (Matsuzawa et al., 2017), dolphins (Herman et al., 1989), and sea lions (Kastak et al., 1994).

Cognition is classified as either social or non-social. Non-social cognition investigates perceptual abilities, attention, learning, memory, and executive function (Tulsky et al., 2003). While social cognition explores the way in which individuals adapt and interact with conspecifics, as well as other species, some consider this emotional intelligence (Brabec et al., 2012). Not surprising, current literature suggests overlap in social and non-social cognition due to shared neural circuits (Rodriguez-Rajo et al., 2020).

## Historical Perspectives

Charles Darwin spurred the research of the parallels between human, animal emotion, and cognitive abilities (1872). In 1882, George Romanes, a friend of Darwin and naturalist, began reporting on cognitive abilities in animals, from scorpions to elephants; however, he was known to anthropomorphize his findings. In 1913, the psychologist John Watson ushered in the age of behaviorism. Watson pressed for more rigorous scientific methods to study animal behavior. It was during this time that animal studies were conducted in laboratories where observations could be measured in a systematic way.

At the time, psychologist Charles Henry Turner embraced behaviorism. While working as a high school educator, he contributed a plethora of laboratory-based data on the behavior of insects, including bees, wasps, and cockroaches (Turner, 1913, 1912). In 1912, he found that wasp nest building was not entirely driven by instinct after observing that the removal of the queen resulted in subpar nests (Turner). Margaret Floy Washburn, the first woman awarded a PhD in psychology, investigated the theory of awareness, the connection between motor movement and social consciousness (1903). In 1908, Washburn published *The Animal Mind*, a text that included empirical data for many species, marking the beginning of animal cognition as a field. Lastly, the work of Ivan Pavlov and Edward Thorndike on classical conditioning and learning theory, respectively, have provided the foundational methods for animal cognition studies. Please refer to the following for detailed reviews: Dewsbury (1984) and Wasserman (1997).

## The Brain and Learning

To appreciate the complexity of cognition, a brief review of the brain and learning may be helpful. Fundamentally, learning improves odds of survival. So, one might not find it surprising to discover that the cellular process of associative learning was first elucidated in the sea slug, among other invertebrates (Hawkins & Byrne, 2015). In comparison to invertebrates, the brains of mammals, reptiles, and avians are more complex. The mammalian forebrain plays a major role in cognition. Reptiles have a forebrain, however, the extent of its layers varies and avians have structures comparable to a forebrain. The forebrain is comprised of three major areas: cerebrum, thalamus, and hypothalamus. The cerebral cortex, the outer layer of the cerebrum, is covered with folds called gyri where neuronal cell bodies and dendrites are found. The four lobes of the cortex are: frontal, parietal, occipital, and temporal. Each play a role in cognition, memory, learning, reasoning, emotion, sensory processing, and movement. However, the frontal lobe, particularly the prefrontal cortex, has a role in behavior regulation. While the dorsolateral prefrontal cortex is involved with executive function (Boland et al., 2007; Diamond, 2013), which includes cognitive flexibility, impulse control, and attention (Mars & Grol, 2007). The ventromedial prefrontal cortex is involved with emotional processing and has a role in reward, risk analysis, and social cognition (Hiser & Koenigs, 2018).

Below the cerebral cortex is the limbic system and basal ganglia that regulate emotion, learning, memory, reward, and pleasure (Sabatinelli et al., 2007). The limbic system is comprised of the amygdala, hippocampus, and hypothalamus (Bear et al., 2007). The hypothalamus plays a key role in organizing behavior (Carlson, 2010; Rodan, 2010) and the hippocampus commits events to long-term memory (Ressler, 2010).

There are many ways to categorize learning. This section will focus on four types: perceptual, stimulus-response, motor, and relational. On a molecular level, learning results in observable neuronal changes that aid in strengthening synapses, the communication between neurons. Keep in mind that the hippocampus within the limbic system plays a key role in learning and memory.

Perceptual learning simply involves recognition of a stimulus or object through sight, smell, or sound. Of course, to notice any stimulus, the visual, olfactory, or auditory systems must be intact. Hence, the neural pathway will depend on the sensory association cortex activated. For instance, visual recognition requires the lateral geniculate nucleus of the thalamus to send information about an object to the visual association cortex. From this point, neurons synapse on the extrastriate cortex where the information is split into two streams, dorsal and ventral. The dorsal stream synapses on the parietal lobe and is involved with stimulus perception. While the ventral stream synapses on the temporal lobe and is involved with stimulus recognition (Carlson, 2010).

Stimulus-response learning, as the name implies, is the association of a stimulus with another stimulus (classical conditioning) or an action (operant conditioning). Classical conditioning, or emotional learning, involves the amygdala. The most common example is Pavlov's dog experiment in which after repeatedly pairing a ringing bell with food, the dogs eventually salivated at hearing the bell ring without the food. Classical conditioning occurs by pairing a neutral stimulus with an unconditioned stimulus (US), eliciting an unconditioned response (UR). When repeatedly paired, the neutral stimulus becomes a conditioned stimulus (CS), eliciting a conditioned response (CR). In classical conditioning, the amygdala receives information of the CS from a sensory association cortex, while the somatosensory cortex receives information of the US. Neurons then project that information to the hypothalamus, midbrain, pons, and medulla (Carlson, 2010). The more a CS and US are paired, the more responsive the neurons within each synapse will become; forming a neural network that will either increase or decrease in strength, as explained by Hebbian rule.

Operant conditioning is more complex because motor learning is involved. Consider teaching a dog to sit for a treat. The verbal or visual cue you provide to the dog is the stimulus required for the dog to respond with a behavior, such as sitting. The behavior is then reinforced with a treat, making the dog more likely to sit in the future. Because body movement is involved and associated with a cue, the motor neurons associated with the movement strengthen and build muscle memory. The pathway of operant conditioning begins with a perception of the stimulus via a sensory association cortex and ends at the frontal lobe and motor association cortex. It can follow one of two pathways, either a direct transcortical neural circuit to the motor association cortex or the basal ganglia and thalamus that synapse at the supplemental motor area. The motor cortex is responsible for learning and executing complex movements, while the supplementary motor area controls behavioral sequences (Carlson, 2010).

Lastly, relational learning involves associations between behaviors, environments, and stimuli with outcomes (Carlson, 2010). Relational learning is considered an executive function because it draws on experience to be used for decision-making or planning. For instance, dogs and horses have been shown to have excellent relational and spatial learning abilities. If you have ever ridden a trail horse, you likely have no reason to direct the horse as it traverses a common trail. This type of learning is also observed in dogs that lose their vision unbeknown to their owner as they navigate their home without any concern, but their deficits become much more evident when furniture is moved, much like it happens to humans that have a mental map of a familiar area.

## Sensory Capabilities

Since sensory capabilities play such a large role in cognition, a brief review of a select number of animals is provided below.

### *Horse*

The eyes of a horse are laterally placed with a total field of vision of ~350°; they are dichromatic and have difficulty discriminating red (Hangii & Ingersoll, 2012). Little research has investigated the olfactory ability in horses, but they can distinguish conspecifics based on odor (Hothersall et al., 2010). Horses can taste sweet, salt, bitter, and sour (Jankunis & Whishaw, 2013). There are two areas on their body which are particularly sensitive to touch, their muzzles which have vibrissae (Chang et al., 2016) and the area in which a rider's leg meets their abdomen (Saslow, 2002).

## Cow

Cows have a 330° field of vision with limited depth perception (Adamczyk et al., 2015). They are also dichromatic and can see red, orange, and yellow better than blue and green (Marino & Allen, 2017). They have a hearing range of frequencies from 23 to 35,000 Hz (Heffner & Heffner, 1983) and approximately 20,000 taste buds that can sense sweet, salt, bitter, and sour (Adamczyk et al., 2015; Marino & Allen, 2017). Like horses, cattle have vibrissae surrounding their muzzle, making this area sensitive to touch (Chang et al., 2016).

## Cats

The sensory abilities of cats are very similar to dogs, both being dichromatic with a field of vision of ~200–250°. Cats have 5–10 times more olfactory epithelium than humans (Heath, 2007), and approximately 470 taste buds that can sense salt, sour, bitter, and umami (Li et al., 2006; Pekel et al., 2020). Their whiskers are sensitive to touch, and they have rapidly discharged specialized epidermal cell units scattered all over their bodies that make them highly sensitive to touch (Overall, 2013) (Table 11.1).

*Table 11.1* Sensory abilities of select animals used in comparative cognition studies

|  | Human | NHP | Pigeon | Rat | Dog |
|---|---|---|---|---|---|
| Vision | Trichromatic; 120–180° | Apes and old-world monkeys: trichromatic (Carvalho et al., 2017) Prosimians: dichromatic (Carvalho et al., 2017) | Tetrachromatic with UV sensitive cones (Emery, 2006) | 76° field of vision; dichromatic with UV sensitive cones (Jacobs et al., 2001) | Dichromatic (blues/greens); 200–250° field of vision (Neitz et al., 1989) |
| Auditory (Hz) | 64–23,000 ("How Well," 2017) | Varies depending on species Lowest frequency in *Macaca spp.* at 28–34,000 (Coleman, 2009) Highest frequency in *Galago spp.* ranging from 70 to 65,000 (Coleman, 2009) | 1–2,000 (Smolders et al., 1999) | 200–76,000 ("How Well," 2017) | 64–45,000 ("How Well," 2017) |
| Olfaction | 400 olfactory receptors (Morrison, 2014) | 500–1,000 olfactory receptors (Rouquier et al., 2000) | Poorly developed (Wiltschko, 1996) | 1,4000 olfactory receptors (Freeman et al., 2020) | 125–300 million olfactory receptors (Miklosi, 2016) |

*Table 11.1* (Continued)

|  | Human | NHP | Pigeon | Rat | Dog |
|---|---|---|---|---|---|
| Taste | 2,000–10,000 taste buds Sweet, salt, bitter, sour, and umami | ~2,300 taste buds (Hevezi et al., 2009) Sweet, salt, bitter, sour, and umami (Hevezi et al., 2009) | Small number of taste buds, estimated 300–400 (Kare & Mason, 1986) | 227 taste buds (Miller and Spangler, 1982) Sweet, salt, bitter, sour, and umami (Miller & Spangler, 1982) | 1,700 taste buds (Pekel et al., 2020) Sweet, salt, bitter, sour, and umami (Miklosi, 2016) |
| Tactile sensitive structures | Forehead and palm (Ackerley et al., 2014) | Fingers (Verendeev et al., 2015) | Many types of mechanoreceptors in beaks and skin (Necker, 2000) | Whiskers on snout (Chang et al., 2016) | Vibrissae around muzzle (Chang et al., 2016) |

## Cognitive Domains

To conceptualize cognition, domains are classified by the specific brain region from which it arises or as a lower- versus higher-order function (Harvey, 2019). For example, the higher order domain of executive function orchestrates lower-order domains of learning and memory to produce an integrated response to the environment. Many cognitive domains and subdomains are cited in literature and can be confusing due to a lack of consensus (Spreng et al., 2010; Sachdev et al., 2014).

Major domains include executive function, learning, memory, attention, language, and visuospatial memory (Boland et al., 2007; Harvey, 2019). In humans, cognitive deficits and behavioral changes can be associated with pathologies of brain regions that moderate cognitive domains. This information is collected from human neuropsychological tests, the effects of known brain lesions in humans and animals (Graham et al., 1997; Meeter et al., 2004), and comparative studies (Miklosi et al., 2004; Gomez, 2005). Before delving further into what these domains represent, remember that they are not independent of each other. Ultimately, the nature of the brain and thus cognitive domains are all intricately part of the whole, connected by the vast neural circuits found throughout.

### *Perception*

Perception is dependent on sensory capability (see above), previous experience, attention, and relevance to the individual. For instance, it may not be of much value for nocturnal animals to have trichromatic vision when they are most active at night. Some cognitive functions may also decline with age as observed in humans, NHP, and dogs (Tapp et al., 2003). Studies in NHP and pigeons have investigated if animals perceive global and local stimulus attributes. In this sense, global attributes can be thought of as a painting while local attributes are the images that comprise the painting. The concept is called global precedence hypothesis (Navon, 1977) and humans have been shown to focus on a global attribute (whole painting) before focusing on local attributes (images) of a stimulus. Humans perceive global attributes before switching their attention to the local attribute (Navon, 1977). Studies in baboons and pigeons have shown that they have a precedence to perceive the local attributes of the stimulus (Fagot & Deruelle, 1997; Cook, 2001).

## Attention

With survival in mind, the ability to attend to the environment for potential risk and benefit is essential. This process is categorized as attention and considered an executive function. It has two major subdomains, selective- and sustained-attention (Harvey, 2019). The executive function aspect of attention is mediated by the medial prefrontal cortex (Birrell & Brown, 2000). Selective attention is comparable to divided attention as it relies on the ability to filter information that may help reach a goal. Complete, or sustained attention is a function of vigilance requiring constant attention on a specific stimulus. Human mental health disorders, such as patients with schizophrenia (Smid et al., 2013) and attention-deficit/hyperactivity disorder (Bailey et al., 2009), have difficulty with selective and sustained attention (Smid et al., 2013), which can impair quality of life.

## Memory

There are many types of memory, with most following a process of information encoding, storing, and retrieving; the exception being working memory, where information is consciously held for a purpose and not stored (Carlson, 2010; Harvey, 2019). Working memory is considered an executive function. Other types of memory include explicit/episodic, procedural, semantic, and prospective. Explicit/episodic memory is one of two forms of long-term memory and is the ability to recall events or sequences of events. In humans, evaluating explicit memory involves language and recital of events. As you can imagine, explicit memory is difficult to investigate in animals without language and so it has been strongly argued that animals do not have explicit memory (Suddendorf & Corballis, 1997). However, there is a plethora of compelling evidence to suggest that many animals, from birds (Clayton et al., 2003), to rats (Ego-Stengal et al., 2010), and NHP (Vincent et al., 2007) do exhibit forms of explicit memory (Pastalkova et al., 2008; Corballis, 2013).

Implicit memory is unconscious and used for everyday tasks; one type of implicit memory is procedural memory. Procedural memory refers to learned motor actions or skills, such as swimming or riding a bike. For animals, this can be operantly learned behaviors, such as sitting and targeting. Semantic memory is the long-term storage of information, which can be reconsolidated as more information about a stimulus or event is learned (Carlson, 2010). The ability to plan and execute future behaviors is prospective memory (Harvey, 2019). There is no debate that animals have procedural, semantic, and prospective memory (McLean, 2004; Adachi et al., 2007; Nagasawa et al., 2011; Smith et al., 2016).

## Pattern Learning and Visuospatial Memory

Pattern learning has been well investigated and observed in humans and animals. It is the equivalent of stimulus response learning on a larger scale. It is how an individual perceives a relation (pattern) among events in the world (classical conditioning), or among events in the world and behaviors (operant conditioning). It is realistic to consider that animals utilize spatial and temporal relationships to know when and where to forage for food and how to efficiently reach that goal (Brown et al., 2001; Wasserman & Zentall, 2006). Despite pattern learning utilizing spatial and temporal information, it is not affected by the presence or absence of a visual landmark and considered its own domain (Brown et al., 2001; Wasserman & Zentall, 2006).

Unlike pattern learning, visuospatial learning and memory, do require a landmark to maintain and manipulate information that is no longer present in the environment (Nikolova & Macken, 2016). This is sometimes referred to as the ability to make a mental or cognitive map in your mind about your surroundings and is necessary for problem solving. Visuospatial memory helps individuals to navigate and find places or things. It is an executive function requiring working memory and perception.

## *Evaluating Non-Social Cognition*

Non-social cognition is measured in a variety of ways using neuropsychological tests modified for animals. A few tests include match to sample, non-match to position (NMP) with or without a delay, sensory discrimination, visual search, attention oddity, and reversal learning (Bensky et al., 2013). The modality of testing depends on the neuropsychological task and what species are being investigated. Equipment can range from electronic touch screens, Wisconsin and Toronto General Testing apparatuses, T- and Y-mazes, and modifications thereof. Specific training and testing procedures are designed, and animals trained, with a designated number of trials and criterion to determine success or failure of the task.

## *Delayed Non-Match to Position*

A delayed non-match to position (DNMP) task evaluates visuospatial learning and memory. It is administered in two phases, a sample position phase followed by a non-match to position phase. The subject is required to navigate a specified space for a target location/position using a marker (i.e., landmark). During the sample phase, the subject is shown the negative position identified by the marker; then the test field is blinded to the subject for a time delay. This is followed by the non-match to position phase in which the subject is shown one marker in the negative position and one identical marker in the positive position (not marked in the sample phase). The subject is only provided a reward if the marker identifying the positive position is selected. The subject learn that a reward is earned when the positive target location is selected. As the time delay increases, the demand on working memory increase as well. There are several variations of this task, including variable or progressive time delays (Adams et al., 2000; Zanghi et al., 2015) and variable or match to position (Zhangi et al., 2015). In dogs, the DNMP task consistently finds that as age increases so do errors (Adams et al., 2000; Studzinski et al., 2006).

## *Discrimination*

Sensory discrimination and visual search tasks evaluate perception and attention. The simplest is a two-choice discrimination task, using vision, olfaction, or auditory capabilities. In visual discrimination, the subject is required to differentiate between two differently shaped, sized, or colored items (Head et al., 1998). This is considered a lower order cognitive domain and requires an intact sensory system, such as vision (Saalmann & Kastner, 2009), olfaction (Kjelvik et al., 2014), or audition (Ruan et al., 2014). Interestingly, in dogs, auditory recognition is not impaired after hippocampal lesions like it is in NHP, indicating that auditory association cortex may be organized differently in this species (Kowalska, 2000).

## Visual Search/Attention Oddity

Visual search tasks evaluate perception, selective attention, working memory, and inhibitory control. In these tasks, the subject is required to attend to a target stimulus while ignoring distractors; this is sometimes referred to as an attention oddity (AO) task. Subjects are trained to select a target stimulus and then simultaneously shown an increasing number of similar or dissimilar distractors. Distractors that appear similar in shape, size, or color require a higher level of selective attention as the subject must stay on task by ignoring the distractors. In senior dogs, the number of errors on an AO task increase with increasing number of distractors (Snigdha et al., 2012) regardless of high-, moderate-, or low-ranking memory performance (Zhangi et al., 2015). Indicating that DNMP is not predictive of selective attention (Zhangi et al., 2015) and highlighting their separate domains. Attention is frequently measured in reaction time, which should reflect the duration of processes required to elicit a response (Wasserman & Zentall, 2006). Reaction time will decrease with repeated or increased stimulus presentation in humans, pigeons, and NHP (Wasserman & Zentall, 2006).

## Task Switching

Task-switching paradigms require less training and can be used to measure selective attention. In one study, dogs were asked to switch between finding food on the floor and making eye contact with the experimenter; the time to make eye contact and the time to find food on the floor was measured (Wallis et al., 2014; Chapagain et al., 2017). They found that middle-aged dogs (3–6 years of age) performed the best. Attention capture and sustained attention have also been investigated in dogs. Attention capture is measured by the time it takes the dog to make eye contact with a relevant stimulus, and sustained attention is measured as the time the dog stays focused on the stimulus (Wallis et al., 2014; Chapagain et al., 2017). In humans, poor performance is associated with increased age and attributed to loss of inhibition.

## Reversal Learning

Executive function, in particular inhibition and learning flexibility, can be assessed with a reversal-learning task (Collie et al., 2000). Such tasks are comprised of two stages, an initial object reward stage and a reversal object reward stage. During the initial object-reward contingency, the subject is consistently rewarded for selecting object B versus object A. After reaching criterion, the reversal object reward stage is started, and the subject is only rewarded when selecting object A while ignoring the previously rewarded object B. Reversal learning tasks can use discrimination contingencies or time to navigate a maze. Reversal learning requires subjects to suppress or inhibit previously learned responses. Response inhibition and reversal learning is mediated by the orbital frontal cortex (McAlonan & Brown, 2003). When compared with young and middle-aged beagles, old and senior dogs require more trials to reach criterion for both initial object-reward and reversal object-reward (Tapp et al., 2003).

## Social Cognition

The three major social cognitive processes involved with interactions include social perception, understanding, and decision-making (Green & Horan, 2010). Since it is difficult to ascertain how animals feel or process emotion, comparative studies investigate how an animal interacts (Miklosi et al., 2016), how their presence impacts the social situation and how they solve social problems within a group (Miklosi et al., 2004; Rooney & Bradshaw, 2006; Bensky et al., 2013). For example, cats within a large colony will perceive members from scent and allow them to pass into the territory without dispute. However, if the cat belonging to the colony is alerted to a non-colony member, the cat may hiss, growl, swat, and chase after the non-colony member. Such body postures and vocalizations are understood as offensive territorial behaviors to which the non-colony member cat must respond by deciding to flee or fight. In this scenario, both cats must be able to perceive each other as affiliates or not, understand the body language and emotional state of each other, and decide how to interact or not.

## Limitations

A major limitation of cognitive studies in animals is the validity of analysis and interpretation of cognitive domains in animals (Miklosi & Abdai, 2021). This is likely impacted by the lack of in-depth comparative data and ecological relevance (Stevens, 2010). For instance, early comparative studies found that chimpanzees form abstract generalizations about stimuli after learning a same/different paradigm via a match-to-sample task; however, it was reported that pigeons were not able to make such abstract generalizations (Young et al., 1999). Years later, it was found that pigeons can make abstract generalizations but require more stimuli when presented with the task (Katz et al., 2008).

Another example comes from Tolman (1948) investigating how quickly rats navigated a maze. With two groups of rats individually placed into a maze with food or a maze without food (Tolman, 1948). The rats in the maze with food always navigated more quickly and with less errors versus the rats without food. To determine if rats in the maze without food had learned anything, he placed them in a maze with food, and found that they were able to navigate through the maze quickly and with less errors. He concluded that the rats without food in the maze were less motivated, not less intelligent. These examples highlight some of the constraints that cognitive testing methods inherently have regarding motivation and construct.

Despite the way in which results are often reported in cognition studies (i.e., errors), it is the authors' belief that cognition should not be viewed as a direct measure of intelligence. Rather cognition should be viewed as an indirect measure of the way in which sensory input, experience, knowledge, and recall is integrated to provide an appropriate or inappropriate response given the circumstances and relevance to the individual.

## Impact on People

Comparative cognition allows researchers to better understand how cognitive processes work, including the age and life factors that may strengthen or weaken abilities. Some animals are used as model species to better understand the processes and progression associated with primary disease states; models also help develop and evaluate treatments for humans and animals. This knowledge may be useful to predict cognitive ability, and refine methods to identify, intervene, and prevent cognitive impairment associated with primary and secondary disease states. Many animals are affected by age associated cognitive decline like humans, while some animals suffer from cognitive dysfunction due to underlying disease processes.

Both inherited and acquired factors can affect cognition in humans and animals (Kim & Park, 2017). Natural occurring disease models provide several advantages such as a shared environment, lifestyle, and exposure to similar stressors and chemicals. In fact, aging dogs are used as disease models for Alzheimer's disease. Several studies in dogs report that by the age of eight years, beta-amyloid accumulation without neuritic plaques (Cummings et al., 1996; Head et al., 2002; Tapp et al., 2004), oxidative damage (Head et al., 2002) and frontal lobe atrophy recognizable on MRI (Studzinski et al., 2006) are present. There are also common neurological diseases, such as mucopolysaccharidosis and ceroid lipofuscinosis that cause progressive degeneration of the central nervous system and impair cognition (Sanders et al., 2011). Naturally occurring disease of animals not only allows us to study disease processes between human and animal, it is an opportunity to evaluate the impact of our shared environments on health.

## Impact on Welfare

There is a relationship between cognition and welfare. It is well documented that high states of arousal due to fear or stress can impact learning, resulting in repercussions on cognition. For example, companion pet owners may report that their pet does not respond to verbal cues while on walks; handlers of working animals may be unable to redirect their working horse's attention when highly aroused; and extremely fearful animals that are newly introduced into zoos or shelters, may lack any motivation to investigate their environment. Thus, cognition may be a good measure of welfare.

Welfare principles state that animals be free of discomfort and allowed to practice normal and species-typical behaviors. Studies have shown that welfare can be improved or enhanced by providing enriched environments and mental stimulation. This may look different depending on the type of animal you work with, but a common form of enrichment used is food foraging puzzles, which can be quite motivating.

## Practical Perspectives

Caring for the cognitive health of animals in our world can have many advantages for them and us. Being aware that most animals are affected by age associated cognitive decline, with some even suffering from forms of dementia, can help to identify affected or at-risk animals. This can allow for the initiation of timely interventions or, at the very least, expectations of caretakers.

Adapted training procedures based on cognitive milestones may be advantageous for animals as more knowledge is gained about cognitive development. Such information may be useful in identifying and selecting animals that are better suited for specific types of work and help those that work with animals to keep track of progress. Again, being able to make cognitive or behavioral interventions to ensure the success and welfare of animals utilized in this manner may be worthwhile.

Improved management strategies, from the way animals are sheltered and fed, should be implemented based on their perception and sensory abilities. Ideally, housing should mimic natural environments and allow normal behaviors to reduce distress. Likewise, gaining a better understanding of how and why animals make the choices they do and what they can readily discriminate can potentially help avoid negative interactions between people and wildlife (i.e., wolves and livestock producers).

## Ethical Considerations

Management practices and the environments that sentient beings are subjected to should be routinely scrutinized to propel growth and improvement. One can argue the humaneness of keeping animals in unnatural conditions, but change is often a slow process, and though thought should be given to alternative options, a focus should be on how we can improve the lives of the animals we have in our care, in our fields, on our farms, in our sanctuary, and homes, now.

## Future Research Demands

The procedural training of animals to perform cognitive tasks can be extremely time consuming and citizen science has been a way to propel cognition research. However, additional investigation is warranted to discover how the results of neuropsychological tasks performed in the laboratory setting compare to similar tasks performed in a home setting and if such methods can be improved. Future research should also focus on cognitive abilities of zoo and farm animals with the aim of improving their types of enrichment and environment.

There are studies in humans demonstrating that diet, mental stimulation, and exercise all help to reduce the risk of cognitive dysfunction. However, studies on preventative and maintenance measures for cognitive health in animals are lacking. There is limited data on how disease processes (inflammatory mediators, cytokines, etcetera), anesthesia protocols, and medications affect effect cognition in animals. These aspects warrant further investigation and development of novel therapeutics.

## Questions

1. What are the two classifications of cognition?
2. What are the four types of learning discussed in this chapter?
3. Do animals have areas that are more sensitive to touch than others, and if so, how might you approach that animal with such knowledge?
4. Name three neuropsychological tests used to evaluate non-social cognition in animals, and which one is most sensitive to age (i.e., the older the animal, the more errors).

## Answers

1. Social cognition and non-social cognition.
2. Perceptual, stimulus-response, motor, and relational learning.
3. Absolutely animals have areas that are more sensitive to touch than others. One should be cognizant of these areas when working with them by avoiding overstimulation of these areas and being aware of how these areas are utilized for non-social and social cognitive tasks.
4. Delayed non-match to position*, discrimination testing, task switching, reversal learning.

## References

Ackerley, R., Carlsson, I., Wester, H., Olausson, H., & Backlund Wasling, H. (2014). Touch perceptions across skin sites: Differences between sensitivity, direction discrimination and pleasantness. *Frontiers in Behavioral Neuroscience, 8*, 54.

Adachi, I., Kuwahata, H., & Fujita, K. (2007). Dogs recall their owner's face upon hearing the owner's voice. *Animal Cognition, 10*(1), 17–21.

Adamczyk, K., Górecka-Bruzda, A., Nowicki, J., Gumulka, M., Molik, E., Schwarz, T.,... & Klocek, C. (2015). Perception of environment in farm animals – A review. *Annals of Animal Science, 15*(3), 565.

Adams, B., Chan, A., Callahan, H., Siwak, C., Tapp, D., Ikeda-Douglas, C.,... & Milgram, N. W. (2000). Use of a delayed non-matching to position task to model age-dependent cognitive decline in the dog. *Behavioural Brain Research, 108*(1), 47–56.

Bailey, U. L., Lorch, E. P., Millich, R., & Chamigo, R. (2009). Developmental changes in attention and comprehension among children with attention deficit hyperactivity disorder. *Child Development, 80*(6), 1842–1855.

Bear, M. F., Connors, B. W., & Paradiso, M. A. (2007). *Neuroscience; exploring the brain*, 3rd edn. Baltimore, MD: Lippincott Williams & Wilkins.

Bensky, M. K., Gosling, S. D., & Sinn, D. L. (2013). The world from a dog's point of view: Are view and synthesis of dog cognition research. *Advances in the Study of Behavior, 45*, 209–406.

Beran, M. J., Parrish, A. E., Perdue, B. M., & Washburn, D. A. (2014). Comparative cognition: Past, present, and future. *International Journal of Comparative Psychology, 27*(1), 3–30.

Birrell, J. M., & Brown, V. J. (2000). Medial frontal cortex mediates perceptual attentional set shifting in the rat. *Journal of Neuroscience, 20*(11), 4320–4324.

Boland, L., Saul, R. E., & Purisch, A. D. (2007). *Neuropsychology for psychologists, health care professionals, and attorneys*. Boca Raton, FL: CRC Press.

Brabec, C. M., Gfeller, J. D., & Ross, M. J. (2012). An exploration of relationships among measures of social cognition, decision making, and emotional intelligence. *Journal of Clinical and Experimental Neuropsychology, 34*(8), 887–894.

Brown, M. F., Zeiler, C., & John, A. (2001). Spatial pattern learning in rats: Control by an iterative pattern. *Journal of Experimental Psychology: Animal Behavior Processes, 27*(4), 407.f

Carlson, N. R. (2010). *Physiology of behavior*, 10th edn. Boston, MA: Pearson Education Inc.

Carvalho, L. S., Pessoa, D. M. A., Mountford, J. K., Davies Wayne, I. L., & Hunt David M. (2017). The Genetic and evolutionary drives behind primate color vision. *Frontiers in Ecology and Evolution*, 1–12. https://www.frontiersin.org/article/10.3389/fevo.2017.00034 https://doi.org/10.3389/fevo.2017.00034. ISSN 2296-701X

Chang, W., Kanda, H., Ikeda, R., Ling, J., DeBerry, J. J., & Gu, J. G. (2016). Merkel disc is a serotonergic synapse in the epidermis for transmitting tactile signals in mammals. *Proceedings of the National Academy of Sciences, 113*(37), E5491–E5500.

Chapagain, D., Viranyi, Z., Wallis, L. J., Huber, L., Serra, J., & Range, F. (2017). Aging of attentiveness in border collies and other pet dog breeds: The protective benefits of lifelong training. *Front Aging Neuroscience, 9*(100), 1–14.

Clayton, N. S., Bussey, T. J., & Dickinson, A. (2003). Can animals recall the past and plan the future? *Nature Review Neuroscience, 4*, 685–690. https://doi.org/10.1038/nrn1180

Coleman, M. N. (2009). What do primates hear? A meta-analysis of all known nonhuman primate behavioral audiograms. *International Journal of Primatology, 30*, 55–91. https://doi.org/10.1007/s10764-008-9330-1

Collie, A., Maruff, P., Yucel, M., Danckert, J., & Currie, J. (2000). Spatiotemporal distribution of facilitation and inhibition of return arising from the reflexive orienting of covert attention. *Journal of Experimental Psychology: Human Perception and Performance, 26*(6), 1733.

Cook, R. G. (2001). *Avian visual cognition*. Retrieved January 8, 2022 from http://pigeon.psy.tufts.edu/avc/cook/default.htm

Corballis, M. C. (2013). Wandering tales: Evolutionary origins of mental time travel and language. *Frontiers in Psychology, 4*, 485.

Cummings, B. J., Head, E., Ruehl, W., Milgram, N. W., & Cotman, C. W. (1996). The canine as an animal model of human aging and dementia. *Neurobiology of Aging, 17*, 259–268.

Darwin, C. R. (1872). *The expression of the emotions in man and animals*, 1st edn. London: John Murray.

Dewsbury, D. A. (1984). *Comparative psychology in the twentieth century*. Stroudsburg, PA: Hutchinson Ross Publishing Company.

Diamond, A. (2013). Executive functions. *Annual Review, 64*, 135–168.

Ego-Stengel, V., & Wilson, M. A. (2010). Disruption of ripple-associated hippocampal activity during rest impairs spatial learning in the rat. *Hippocampus, 20*(1), 1–10. https://doi.org/10.1002/hipo.20707. PMID: 19816984; PMCID: PMC2801761.

Emery, N. J. (2006). Cognitive ornithology: The evolution of avian intelligence. *Philosophical Transactions Royal Society B, 361*, 23–43.

Fagot, J., & Deruelle, C. (1997). Processing of global and local visual information and hemispheric specialization in humans (*Homo sapiens*) and baboons (*Papio papio*). *Journal of Experimental Psychology: Human Perception and Performance, 23*(2), 429.

Freeman, A. R., Ophir, A. G., & Sheehan, M. J. (2020). The giant pouched rat (Cricetomys ansorgei) olfactory receptor repertoire. *PloS One, 15*(4), e0221981.

Gomez, J. C. (2005). Species comparative studies and cognitive development. *Trends in Cognitive Sciences, 9*(3), 118–125.

Graham, K. S., & Hodges, J. R. (1997). Differentiating the roles of the hippocampal complex and the neocortex in long term memory storage: Evidence from the study of semantic dementia and Alzheimer's disease. *Neuropsychology, 2*(1), 77–89.

Green, M. F., & Horan, W. P. (2010). Social cognition in schizophrenia. *Current Directions in Psychological Science, 19*(4), 243–248.

Hangii, E. B., & Ingersoll, J. F. (2012). Lateral vision in horses: A behavioral investigation. *Behavioural Processes, 91*, 70–76.

Harvey, P. D. (2019). Domains of cognition and their assessment. *Dialogues in Clinical Neuroscience, 21*(3), 227–237. https://doi.org/10.31887/DCNS.2019.21.3/pharvey

Hawkins, R. D., & Byrne, J. H. (2015). Associative learning in invertebrates. *Cold Spring Harbor Perspectives in Biology, 7*(5), a021709.

Head, E., Callahan, H., Muggenburg, B. A., Cotman, C. W., & Milgram, N. W. (1998). Visual-discrimination learning ability and beta-amyloid accumulation in the dog. *Neurobiology of Aging, 19*, 415–425.

Head, E., Liu, J., Hagen, T. M., Muggenburg, B. A., Milgram, N. W., Ames, B. N., & Cotman, C. W. (2002). Oxidative damage increases with age in a canine model of human brain aging. *Journal of Neurochemistry, 82*(2), 375–381.

Heath, S. E. (2007). Behavior problems and welfare. In: Rochlitz, I. (Eds.), *The welfare of cats* (pp. 91–118). Dordrecht: Springer.

Heffner, R. S., & Heffner, H. E. (1983). Hearing in large mammals: Horses (*Equus caballus*) and cattle (Bos taurus). *Behavioural Neuroscience, 97*(2), 299–309.

Herman, L. M., Hovancik, J. R., Gory, J. D., & Bradshaw, G. L. (1989). Generalization of visual matching by a Bottlenosed Dolphin (Tursiops truncates): Evidence for invariance of cognitive performance with visual and auditory materials. *Journal of Experimental Psychology: Animal Behavior Processes, 15*, 124–136.

Hevezi, P., Moyer, B. D., Lu, M., Gao, N., White, E., Echeverri, F.,... & Zlotnik, A. (2009). Genome-wide analysis of gene expression in primate taste buds reveals links to diverse processes. *PloS One, 4*(7), e6395.

Hiser, J., & Koenigs, M. (2018). The multifaceted role of ventromedial prefrontal cortex in emotion, decision making, social cognition, and psychopathology. *Biological Psychiatry, 83*(8), 638–647.

Hothersall, B., Harris, P., Sörtoft, L., & Nicol, C. J. (2010). Discrimination between conspecific odor samples in the horse (*Equus caballus*). *Applied Animal Behavior Science, 126*(1–2), 37–44.

How well do dogs and other animals hear? (2017, April 10). Retrieved from https://lsu.edu/deafness/HearingRange.html

Jacobs, G. H., Fenwick, J. A., & Williams, G. A. (2001). Cone-based vision of rats for ultraviolet and visible lights. *Journal of Experimental Biology, 204*(14), 2439–2446.

Jankunis, E. S., & Whishaw, I. Q. (2013). Sucrose bobs and quinine gapes: Horse (*Equus caballus*) responses to taste support phylogenetic similarity in taste reactivity. *Behavioral Brain Research 256*(1), 284–290.

Kare, M. R., & Mason, J. R. (1986). The chemical senses in birds. In Sturkie, P. D. (ed.), *Avian physiology* (pp. 59–73). New York: Springer-Verlag.

Kastak, D., & Schusterman, R. J. (1994). Transfer of visual identity matching-to-sample in two California sea lions (*Zalophus californians*). *Animal Learning & Behavior, 22,* 427–435.

Katz, J. S., Bodily, K. D., & Wright, A. A. (2008). Learning strategies in matching to sample: If-then and configural learning by pigeons. *Behavioural Processes, 77*(2), 223–230. https://doi.org/10.1016/j.beproc.2007.10.011

Kim, M., & Park, J. M. (2017). Factors affecting cognitive function according to gender in community-dwelling elderly individuals. *Epidemiology and Health, 39,* e2017054. https://doi.org/10.4178/epih.e2017054

Kjelvik, G., Saltvedt, I., White, L. R., Stenumgård, P., Sletvold, O., Engedal, K., ... & Håberg, A. K. (2014). The brain structural and cognitive basis of odor identification deficits in mild cognitive impairment and Alzheimer's disease. *BMC Neurology, 14*(1), 1–10.

Kowalska, D. M. (2000). Cognitive functions of the temporal lobe in the dog: A review. *Progress in Neuro-Psychopharmacology and Biological Psychiatry, 24*(5), 855–880.

Li, X., Li, W., Wang, H., Bayley, D. L., Cao, J., Reed, D. R., ... & Brand, J. G. (2006). Cats lack a sweet taste receptor. *The Journal of Nutrition, 136*(7), 1932S–1934S.

Marino, L., & Allen, K. (2017). The psychology of cows. *Animal Behavior and Cognition, 4*(4), 474–498.

Mars, R. B., & Grol, M. J. (2007). Dorsolateral prefrontal cortex, working memory, and prospective coding for action. *Journal of Neuroscience, 27*(8), 1801–1802.

Matsuzawa, T. (2017). Horse cognition and behavior from the perspective of primatology. *Primates, 58*(4), 473–477.

McAlonan, K., & Brown, V. J. (2003). Orbital prefrontal cortex mediates reversal learning and not attentional set shifting in the rat. *Behavioural Brain Research, 146*(1–2), 97–103.

McLean, A. N. (2004). Short-term spatial memory in the domestic horse. *Applied Animal Behaviour Science, 85*(1–2), 93–105.

Meeter, M., & Murre, J. M. J. (2004). Consolidation of long-term memory: Evidence and alternatives. *Psychological Bulletin, 130*(6), 843–857.

Miklósi, Á. (2016). *Dog behaviour, evolution, and cognition.* Oxford: Oxford University Press.

Miklósi, Á., & Abdai, J. (2021). The lack of validity hinders research in animal cognition. *Learning Behaviour, 49,* 259–260. https://doi.org/10.3758/s13420-020-00460-3

Miklosi, A., Topal, J., & Csanyi, V. (2004). Comparative social cognition: What can dogs teach us? *Animal Behaviour, 67,* 995–1004.

Miller Jr, I. J., & Spangler, K. M. (1982). Taste bud distribution and innervation on the palate of the rat. *Chemical Senses, 7*(1), 99–108.

Morrison, J. (2014). Human nose can detect 1 trillion odors. *Nature News,* 1–2. doi:10.1038/nature.2014.14904.

Nagasawa, M., Murai, K., Mogi, K., & Kikusui, T. (2011). Dogs can discriminate human smiling faces from blank expressions. *Animal Cognition, 14*(4), 525–533.

Navon, D. (1977). Forest before trees: The precedence of global features in visual perception. *Cognitive Psychology, 9*(3), 353–383.

Necker, R. (2000). The somatosensory system. In Whittow, G. C. (ed.), *Sturkie's avian physiology* (pp. 57–69). San Diego: Academic Press.

Neitz, J., Geist, T., & Jacobs, G. H. (1989). Color vision in the dog. *Visual Neuroscience, 3,* 119–125. https://doi.org/10.1017/s0952523800004430

Nikolova, A., & Macken, B. (2016). The objects of visuospatial short-term memory: Perceptual organization and change detection. *Quarterly Journal of Experimental Psychology, 69*(7), 1426–1437.

Overall, K. L. (2013). *Clinical behavioral medicine for small animals,* 2nd edn. (No. Sirsi) i9780801668203). St. Louis, MO.

Pastalkova, E., Itskov, V., Amarasingham, A., & Buzsáki, G. (2008). Internally generated cell assembly sequences in the rat hippocampus. *Science, 321,* 1322–1327. https://doi.org/10.1126/science.1159775

Pekel, A. Y., Mülazımoğlu, S. B., & Acar, N. (2020). Taste preferences and diet palatability in cats. *Journal of Applied Animal Research, 48*(1), 281–292.

Perry, C. J., Barron, A. B., & Chittka, L. (2017). The frontiers of insect cognition. *Current Opinion in Behavioral Sciences, 16,* 111–118.

Ressler, K. J. (2010). Amygdala activity, fear, and anxiety: Modulation by stress. *Biological Psychiatry, 67*(12), 1117–1119.

Rodan, I. (2010). Understanding feline behavior and application for appropriate handling and management. *Topics Companion Animal Medicine, 25*(4), 178–188.

Rodríguez-Rajo, R., García-Rudolph, A., Sánchez-Carrión, R., Aparicio-López, C., Enseñat-Cantallops, A., & García-Molina, A. (2020). Social and nonsocial cognition: Are they linked? A study on patients with moderate-to-severe traumatic brain injury. *Applied Neuropsychology: Adult, 5,* 1–10. https://doi.org/10.1080/23279095.2020.1845171

Romanes, G. J. (1882). *Animal intelligence.* New York: D. Appleton.

Rooney, N. J., & Bradshaw, J. W. S. (2006). Social cognition in the domestic dog: Behavior of spectators towards participants in interspecific games. *Animal Behavior, 72,* 343–352.

Rouquier, S., Blancher, A., & Giorgi, D. (2000). The olfactory receptor gene repertoire in primates and mouse: Evidence for reduction of the functional fraction in primates. *Proceedings of the National Academy of Sciences, 97*(6), 2870–2874.

Ruan, Q., Ma, C., Zhang, R., & Yu, Z. (2014). Current status of auditory aging and anti-aging research. *Geriatrics & Gerontology International, 14,* 40–53.

Saalmann, Y. B., & Kastner, S. (2009). Gain control in the visual thalamus during perception and cognition. *Current Opinion in Neurobiology, 19*(4), 408–414.

Sabatinelli, D., Bradley, M. M., Lang, P. J., Costa, V. D., & Versace, F. (2007). Pleasure rather than salience activates human nucleus accumbens and medial prefrontal cortex. *Journal of Neurophysiology, 98,* 1374–1379.

Sachdev, P. S., Blacker, D., Blazer, D. G., Ganguli, M., Jeste, D. V., Paulsen, J. S., & Petersen, R. C. (2014). Classifying neurocognitive disorders: The DSM-5 approach. *Nature Reviews Neurology, 10*(11), 634–642.

Sanders, D. N., Kanazono, S., Wininger, F. A., Whiting, R. E., Flournoy, C. A., Coates, J. R.,... & Katz, M. L. (2011). A reversal learning task detects cognitive deficits in a Dachshund model of late-infantile neuronal ceroid lipofuscinosis. *Genes, Brain and Behavior, 10*(7), 798–804.

Saslow, C. A. (2002). Understanding the perceptual world of horses. *Applied Animal Behaviour Science, 78,* 209–224.

Smid, H. G., Martens, S., de Witte, M. R., & Bruggeman, R. (2013). Inflexible minds: Impaired attention switching in recent-onset schizophrenia. *PloS One, 8*(10), e78062.

Smith, A. V., Proops, L., Grounds, K., Wathan, J., & McComb, K. (2016). Functionally relevant responses to human facial expressions of emotion in the domestic horse (*Equus caballus*). *Biology Letters, 12*(2), 20150907.

Smolders, J. W. T., Mueller, M., & Klinke, R. (1999). The upper frequency range in pigeon hearing: Response properties in auditory nerve and cohlear nucleus—is tuning different in birds? *The Journal of the Acoustical Society of America, 105*(2), 1110–1110.

Snigdha, S., Christie, L-A., DeRivera, C., Araujo, J.A., Milgram, N. W., & Cotman, C. W. (2012). Age and distraction are determinants of performance on a novel visual search task in aged beagle dogs. *Age, 34,* 67–73.

Spreng, R. N., Wojtowicz, M., & Grady, C. L. (2010). Reliable differences in brain activity between young and old adults: A quantitative meta-analysis across multiple cognitive domains. *Neuroscience & Biobehavioral Reviews, 34*(8), 1178–1194.

Stevens J. R. (2010). The challenges of understanding animal minds. *Frontiers in Psychology, 1,* 203. https://doi.org/10.3389/fpsyg.2010.00203

Studzinski, C. M., Christie, L. A., Araujo, J. A., Burnham, W. M., Head, E., Cotman, C. W., & Milgram, N. W. (2006). Visuospatial function in the beagle dog: An early marker of cognitive decline in a model of human aging and dementia. *Neurobiology of Learning and Memory, 86*(2), 197–204.

Suddendorf, T., & Corballis, M. C. (1997). Mental time travel and the evolution of the human mind. *Genetic Social and General Psychology Monographs, 123,* 133–167.

Tapp, P. D., Siwak, C. T., Estrada, J., Head, E., Muggenburg, B. A., Cotman, C. W., & Milgram, N. W. (2003). Size and reversal learning in the beagle dog as a measure of executive function and inhibitory control in aging. *Learning & Memory, 10*(1), 64–73.

Tapp, P. D., Siwak, C. T., Gao, F. Q., Chiou, J. Y., Black, S. E., Head, E., ... & Su, M. Y. (2004). Frontal lobe volume, function, and β-amyloid pathology in a canine model of aging. *Journal of Neuroscience, 24*(38), 8205–8213.

Tolman, E. C. (1948). Cognitive maps in rats and men. *Psychological Review, 55*(4), 189.

Tulsky, D. S., & Price, L. R. (2003). The joint WAIS-III and WMS-III factor structure: Development and cross-validation of a six-factor model of cognitive functioning. *Psychological Assessment, 15*(2), 149.

Turner, C. H. (1912). An orphan colony of Polistes pallipes Lepel. *Psyche, 19*, 184–190. https://doi.org/10.1155/1912/67292

Turner, C. H. (1913). Behavior of the common roach (Periplaneta orientalis L.) on an open maze. *Biological Bulletin, 25*, 348–365. https://doi.org/10.2307/1536129

Verendeev, A., Thomas, C., McFarlin, S. C., Hopkins, W. D., Phillips, K. A., & Sherwood, C. C. (2015). Comparative analysis of Meissner's corpuscles in the fingertips of primates. *Journal of Anatomy, 227*(1), 72–80.

Vincent, J. L., Patel, G. H., Fox, M. D., Snyder, A. Z., Baker, J. T., Van Essen, D. C.,... & Raichle, M. E. (2007). Intrinsic functional architecture in the anaesthetized monkey brain. *Nature, 447*(7140), 83–86.

Vitale Shreve, K. R., & Udell, M. A. (2015). What's inside your cat's head? A review of cat (Felis silvestris catus) cognition research past, present and future. *Animal Cognition, 18*(6), 1195–1206.

Wallis, L. J., Range, F., Müller, C. A., Serisier, S., Huber, L., & Virányi, Z. (2014). Lifespan development of attentiveness in domestic dogs: Drawing parallels with humans. *Frontiers in Psychology, 5*, 71.

Washburn, M. F. (1903). The genetic function of movement and organic sensations for social consciousness. *The American Journal of Psychology, 14*, 337–342. https://doi.org/10.2307/1412306

Washburn, M. F. (1908). *The animal mind; a textbook of comparative psychology.* New York: The Macmillan Company.

Wasserman, E. A. (1997). The science of animal cognition: Past, present, and future. *Journal of Experimental Psychology: Animal Behavior Processes, 23*, 123–135. https://doi.org/10.1037/0097-7403.23.2.123

Wasserman, E. A., Wasserman, E. A., & Zentall, T. R. (eds.) (2006). *Comparative cognition: Experimental explorations of animal intelligence.* Oxford: Oxford University Press.

Watson, J. B. (1913). Psychology as the behaviorist views it. *Psychological Review, 20*, 158–177. https://doi.org/10.1037/h0074428

Wiltschko, R. (1996). The function of olfactory input in pigeon orientation: Does it provide navigational information or play another role? *The Journal of Experimental Biology, 199*(1), 113–119.

Young, M. E., Wasserman, E. A., Hilfers, M. A., & Dalrymple, R. M. (1999). An examination of the pigeon's variability discrimination using lists of successively presented stimuli. *Journal of Experimental Psychology: Animal Behavior Processes, 25*, 475–490.

Zanghi, B. M., Araujo, J., & Milgram, N. W. (2015). Cognitive domains in the dog: Independence of working memory from object learning, selective attention, and motor learning. *Animal Cognition, 18*(3), 789–800.

# 12
# ANIMAL SENTIENCE, ANIMAL EMOTIONS, AND HUMAN-ANIMAL INTERACTIONS

*Annika Bremhorst, Catia Caeiro, Anna Zamansky, and Sabrina Karl*

## Animal Sentience and Animal Emotions

Animal sentience can be understood as the ability of non-human animals (hereinafter referred to as animals) to experience feelings (Duncan, 2006; Proctor et al., 2013). Acknowledging animals (certainly mammals) as sentient beings seems widely accepted today, but this particularly has not always been the case. Feelings are subjective states available only to the individual experiencing them (Duncan, 2006). This fact poses challenges for scientific research to this day, since feelings cannot be measured directly and objectively (Mendl et al., 2010). Feelings remain the foremost concern when humans think of animals and their welfare state (see Mendl et al., 2010). However, the current scientific approach largely bypasses animal feelings and instead focuses more on the study of animal pain and emotions, which include feelings, in addition to other components that can be objectively measured (Mendl et al., 2010).

## What Are Emotions?

There are different views on what emotions are (Paul & Mendl, 2018). A common approach is to consider them as intense, short-term affective states triggered by events associated with rewards or punishers (Mendl et al., 2010; Rolls, 2013). When rewards are present/expected or punishments are removed, positive emotions are likely to occur, while negative emotions are potentially triggered when punishments are present/expected or rewards are removed (Mendl et al., 2010). Emotions can be classified by using different approaches, the most prominent ones being the *categorical* and the *dimensional* emotion approach. The *categorical* emotion approach suggests a certain number of fundamental emotion systems based on neuronal brain structures homologous across mammals (Panksepp, 2005). Thus, at least some specific emotions can be distinguished, whereby this number can vary between authors (e.g. Jaak Panksepp suggested seven basic emotion systems: FEAR; RAGE; PANIC/GRIEF; SEEKING; LUST; CARE; PLAY (Panksepp, 2005); Paul Ekman only considered six distinct emotions: fear; sadness; disgust; anger; happiness; surprise (Ekman, 1992)). According to the *dimensional* approach, emotions are differentiated by a few dimensions, usually including valence (i.e. positive/negative emotions) and often the animals' arousal state (high/low)

(Mendl et al., 2010). However, the categorical and dimensional approaches are not mutually exclusive but can be integrated, e.g., by classifying a distinct emotion as positive or negative (Dawkins 2006a; Mendl et al., 2010).

## How Can Emotions Be Objectively Measured in Animals?

Identifying emotions in animals is a challenging endeavour (Paul et al., 2005). Since animals cannot verbally communicate, we must resort to objectively measurable ways for correctly inferring their emotions. Emotions are accompanied by different types of changes, including alterations in feelings, heart rate, cognitive processes, vocalisations, facial expressions, or body postures (Mendl et al., 2010; Scherer, 2005). Therefore, emotions are considered as multi-component states that include a subjective (feeling), neuro-physiological, cognitive, and behavioural component (Kremer et al., 2020; Scherer, 2005). While the subjective component is inaccessible to direct measurement in animals, there are various methods to objectively measure changes in the other components during emotions (Mendl et al., 2010; Paul et al., 2005). Hence, physiological, cognitive, or behavioural changes that occur when an animal is experiencing an emotion can potentially serve as indicators (Mendl et al., 2010). Behavioural changes that accompany emotions are immediate sources of information that can be assessed non-invasively (Dawkins, 2006b) and possibly even automatically (see later in this chapter), and in almost any situation by mere observation. Especially in HAI, assessing such emotional expressions is possibly the most practical and therefore a very common approach to identifying animal emotions and so we will focus on the behavioural component of emotions in the remaining chapter.

## Why Is It Important to Study Animal Emotional Expressions (for Research and More Applied Contexts Including HAI)?

Charles Darwin (1872) suggested that emotional expressions (e.g., facial expressions) are broadly similar between different species in terms of both appearance and meaning. Since then, researchers have been interested in studying how emotional and communicative expressions have evolved (Bliss-Moreau et al., 2021; Parr & Waller, 2006; Waller et al., 2020). However, recent research using more objective and detailed tools has challenged Darwin's assumption (see, e.g., Caeiro et al., 2017). While anatomical homologies between species suggest a common ancestor of potential emotional facial cues, behaviourally (in terms of appearance and function) there appears to be less continuity. This suggests that emotional expressions are developed for each species and are thus species-specific.

Humans and other animals have different production and perception systems related to emotional expressions. Therefore, the question may arise about how interactions between species work. For example, researchers interested in this question want to understand how animals learn to read human emotional expressions – either through social learning (i.e., learning by associating external events with motivational states, such as "Human makes a happy face after I followed instructions") or mental simulation (i.e., learning by mimicking observed responses to external events, such as "If people withdraw when confronted with a scary object, I will withdraw too"). While animals learn some behaviours only through imitation, social learning seems more likely when it comes to facial expressions. Recent studies on the perception of human facial expressions in dogs (Correia-Caeiro et al., 2020) support social learning processes and suggest that this occurs through the memory of emotional cues from past interactions (Proops et al., 2018).

From an applied point of view, the correct assessment of animal emotions is a cornerstone particularly of animal welfare practices, which increasingly include not only the reduction of negative emotions but also the attempt to promote positive states. However, what poses a major challenge is that emotional expressions in animals are not only subtle and complex, but also species-specific (and thus different from those of humans). In particular, misunderstanding emotional expressions has implications for animal welfare. For example, the same facial expression (e.g., a bared-teeth display which looks similar to the human "smile") can have different meanings across species (Waller & Dunbar, 2005) and even within a species, depending on the context in which it is displayed (Clark et al., 2020). In dogs, it has been shown that particularly children mistake the bared teeth for a friendly facial expression because it morphologically resembles a human smile (Meints & De Keuster, 2009).

These examples illustrate the important role of emotional expressions in animals, whether it is for better understanding its origins and evolutionary processes in different species (including humans) or for direct application within HAI in a range of contexts in which emotional cues are relevant (e.g., owner–companion animal relationship improvement, training). Hence, there has been growing interest in considering and including animal emotional expressions in many fields of research and HAI. However, identifying these animal emotional expressions remains a current ongoing challenge.

## Animal Emotional Expressions: Research Approaches, Insights, and Challenges

The most common approaches to identifying emotional expressions are observing animals in naturalistic situations in which the respective emotion occurs (e.g., Caeiro et al., 2017) or experimentally inducing the emotion, by creating a situation that is likely to evoke the emotion of interest (e.g., Bremhorst et al., 2019; Bremhorst et al., 2021; Karl et al., 2022). Behavioural changes that are reliably shown across different situations in which the particular emotion is likely to be experienced represent potential indicators that can help infer the specific emotion at hand (Mendl et al., 2010). However, there are different types of challenges that this kind of research faces.

Our knowledge about which expressions animals produce when experiencing different emotions is still very scarce and many more studies are needed to systematically identify and fully understand emotional expressions even of species very close to us, such as non-human primates and domestic dogs. However, research aimed at identifying such emotional expressions seems to have steadily increased in recent years. In humans, decades of research have focused on examining facial expressions associated with emotions (e.g., Ekman & Rosenberg, 2012; Keltner & Ekman, 2003). In animal studies, facial expressions as a possible indicator of emotions have historically been investigated less often (Descovich et al., 2017). Recently, however, there has been a growing number of dog-focused studies aiming to link expressions in the eye, ear, and mouth area to canine emotions (e.g., Bremhorst et al., 2019; Bremhorst et al., 2021; Caeiro et al., 2017; Gähwiler et al., 2020).

In dogs, possible expressions of emotion in the eye area are, for example, blinks. Eye blinks have been linked to negative emotions such as fear in dogs (Mills, 2005). Empirical data supported this assumption: one study (Gähwiler et al., 2020) showed that dogs blinked more frequently during fireworks, a situation eliciting negative emotions in many dogs (McPeake et al., 2017; Storengen & Lingaas, 2015). However, individual differences in blink frequency were considerable, so it is unlikely that blinking can be considered a reliable potential indicator at least of fear in dogs (Gähwiler et al., 2020). Blinking has also been associated with emotions in cats. Like dogs, cats also increased their blinking rate in fearful situations (Bennett et al., 2017). Interestingly, Humphrey et al. (2020) showed that the speed of blinking is important to consider in cats since slow blinking has been associated with positive cat-human communication. Other features of the eye area that have also been explored in animal emotion studies are, e.g., the exposure of eye white in emotional situations in cows (e.g., Sandem et al., 2002) and the increase of eye wrinkles in horses during negative situations (Hintze et al., 2016).

Ear movements are considered important for communication in different animal species, such as dogs (McGreevy et al., 2012), with several existing studies having linked ear movements to canine emotions. For example, upward ear movements were observed during a positive emotion when dogs anticipated the arrival of a reward (Bremhorst et al., 2019; Bremhorst et al., 2021; Caeiro et al., 2017). In a negative condition, dogs increasingly moved their ears backwards and flattened them against the head, e.g., during fear (Gähwiler et al., 2020) or frustration (Bremhorst et al., 2019; Bremhorst et al., 2021). Ear movements have also been linked to emotions in other species. Specifically, flattened ears were observed also in pigs (Reimert et al., 2013) or horses (as reviewed by Stomp et al., 2018) during negative emotions. However, it should not be concluded that backwards-directed ears generally indicate negative emotions – emotional expressions can be species-specific. For example, in cows backwards-rotated ears were observed in a presumably positive situation when being stroked (De Oliveira & Keeling, 2018). Furthermore, within a species, it may not be sufficient to associate an emotional expression generally with positive or negative emotions – different positive or negative emotions may trigger different behaviours. For example, sheep point their ears backwards during a negative situation, but only if it is unfamiliar and uncontrollable, likely eliciting fear (Boissy et al., 2011). If the situation is controllable, potentially triggering anger, their ears point up (Boissy et al., 2011).

The above observations exemplify key points to consider when analysing and interpreting animal (emotional) behaviour, including that a direct translation of behaviour to emotion between species should not be constructed in the absence of an empirical basis. Furthermore, it is important that, when evaluating behaviour, the behavioural context must also be considered. For example, one study observed that sheep in a positive situation (feeding on fresh hay) had backwards-oriented ears (Reefmann et al., 2009). However, considering the context, it must be borne in mind that to eat the hay, the animals elevated their heads, causing their loosely hanging ears to turn backward (Reefmann et al., 2009). This observation shows that the animals' ear position may be an artefact of the context rather than a potential emotional expression. In general, it is always preferable not to interpret behaviour in isolation – instead, the triangulation of behaviours with other sources of information, including the situational context, the animals' arousal level, and the goal of its actions, can help to more systematically assess and interpret animal emotions (Mills, 2017).

Also mouth movements may represent emotions (see for a review, Descovich et al., 2017). A typical mouth movement that has been implicated as a potential emotional expression (among other functions) in dogs is lip-licking. Several studies reported that dogs frequently exhibited this behaviour in different, mainly negative or stressful situations (e.g., Beerda et al., 1997). Interestingly, dogs do not increase their rate of lip-licking when the stress or aversiveness of a situation increases. In fact, in cases of extreme threat, lip-licking is rarely produced (Firnkes et al., 2017). The latter may seem counterintuitive. However, it has been suggested that lip-licking in dogs has a de-arousing effect (Mariti et al., 2017; Part et al., 2014). In situations where dogs are severely threatened, an effort to reduce arousal is unlikely to be an adaptive response, which could be one reason why lip-licking is rarely shown in very negative situations. As suggested by Shepherd's Ladder of Aggression (Shepherd, 2009), in the event of an extreme threat, dogs' behavioural responses will rather escalate gradually; starting with more subtle signals, such as the described lip-licking, through several progressive stages culminating in biting as the most extreme form of escalation. Although the Ladder of Aggression has not yet been fully empirically investigated, some studies provided data supporting the escalation of some of these signals (Mariti et al., 2017; Owczarczak-Garstecka et al., 2018).

In addition to facial expressions, body cues have been considered possible indicators of animal emotions (e.g., reviewed by De Oliveira & Keeling, 2018). Particularly the tail has been examined in studies of animal emotions in pigs (Reimert et al., 2013), goats (Briefer et al., 2015), or dogs (McGowan et al., 2014). Thereby, different aspects of the tail movement have been considered, namely, tail position and tail wagging, including the laterality and amplitude of the wagging tail (Quaranta et al., 2007). For example, a high tail position was associated with positive emotions in goats (Briefer et al., 2015). A tail down position was more commonly observed in pigs during negative emotions (Reimert et al., 2013), and this tail position has also been linked to fear in dogs (Barrera et al., 2010). Tail wagging has been observed in positive emotions in dogs (McGowan et al., 2014) and pigs (Reimert et al., 2013). Interestingly, at least in dogs, the direction of the tail-wagging movement also seems to be important; in presumably positive states (when dogs are looking at their owner), it has been observed that dogs' tails tend to wag more to the right (Quaranta et al., 2007). The same directional bias was also observed when the dogs saw an unfamiliar human or a cat – although the tail wagging amplitude varied (owner: strong wagging, stranger: medium wagging, cat: very little wagging). A left-sided tendency to wag the tail occurred when dogs were confronted with an unfamiliar dog exhibiting agonistic behaviour (Quaranta et al., 2007).

Although many studies focus on a single element of an animal's behavioural expression, e.g., facial expressions, it is important to remember that the emotional display is multimodal, orchestrating vocalisations, body posture, and facial movements that can all be essential aspects of the emotionally expressive behaviour (as discussed by Caeiro et al., 2017). Future research should therefore increasingly investigate this multimodality of animal emotional expressions, e.g., by analysing facial and body expressions in combination, as was done for instance in a study with cows (De Oliveira & Keeling, 2018).

There are several challenges that research aiming to identify animal emotional expressions is confronted with, for example, the assumption that when facing emotional triggers, behaviours are always displayed or that all displayed behaviours are necessarily emotional expressions. Internal states can generate behavioural cues that can be observed and measured. However, this is not necessarily the case. Emotional responses may not always generate noticeable cues, or some cues may even be suppressed. In addition, some of the external responses observed may not be strictly correlated with internal emotions; e.g., some facial expressions have been associated with emotions, but do all facial expressions depict an emotion? Or can they also transmit other communicative signals? In humans, we know that not all facial expressions are associated with emotions. The classical example to illustrate this subtle but essential difference is the Duchenne smile and the non-Duchenne smile, commonly displayed by people in different situations (Figure 12.1). The Duchenne smile (which is felt, with true pleasure) differs from the non-Duchenne smile (which is not felt, usually shown during greeting) by a single muscular contraction (Ekman et al., 1990).

To separate emotions from communicative or other behaviours, it might be too simple to classify these expressions holistically or generically (e.g., classifying a face as "happy" or

*Figure 12.1* Left – Non-Duchenne smile, Right – Duchenne smile. Unlike the individual on the left, the individual on the right displays wrinkles in the eye corners, which are usually only present in a Duchenne (felt) smile.
*Images adapted from Pixabay.com/PublicDomainPictures and Pixabay.com/RAEng_Publications.*

"smiling"). Instead, a fine-grained and objective analysis is required to avoid *a priori* assumptions about what emotional expressions will look like in a given species. One way to successfully do this for facial expressions is to use a tool called the Facial Action Coding System (FACS). FACS has been the gold standard for objective and anatomical measurement of discrete facial movements (i.e. Action units: AUs) in humans for several decades (Ekman & Friesen, 1978; Ekman et al., 2002). One of the immediate benefits of this system is that it relies on the facial anatomy (i.e., individual muscle contraction) rather than on the assumed meaning of a composed (holistic) facial expression. This is particularly relevant when using a comparative approach between different species, since, as discussed earlier, expressions can be species-specific. FACS has been adapted for over 11 non-human species (all freely available at www.animalFACS.com), including primates and domestic animals, like dogs (Waller et al., 2013), and cats (Caeiro et al., 2017). This tool cannot only be used for an objective measurement of facial expressions within a species, but also for comparisons between different species and, importantly, for HAI. For example, using the species-specific AnimalFACS, different facial expressions shown during emotional responses were identified in chimpanzees (Parr et al., 2007), dogs (Bremhorst et al., 2019; Caeiro et al., 2017), and cats (Bennett et al., 2017). This suggests that emotional expressions are commonly produced on faces of various species, but also that they can vary widely between species and contexts. These studies are fairly recent, but they demonstrated the importance of taking a bottom-up approach with a deeper and more detailed analysis of emotional expressions for each species separately. They also show the need to avoid holistic evaluations, which are likely to be highly biased toward anthropocentrism and thus less informative on other animals' emotional experiences and expressions.

## Perceiving and Recognising Emotional Expressions

The multimodal expressions that correlate with emotions can be perceived in social interactions by other individuals who can extract meaning from these cues. HAI are particularly complex interactions because they involve the perception of emotional expressions between different species that can vary considerably in their behaviour (different expressions are used to communicate) and morphology (faces/bodies are different). Humans need to look at animals and interpret what they are trying to tell us, just as animals need to look at people and attempt to decipher what we want from them or how we are feeling to achieve successful interactions. However, there is a wide discrepancy in how each species does this. For humans, social interactions tend to be primarily based on sight, and they focus on the face to extract a variety of information, including emotional cues (Bruce & Young, 2013). For example, humans concentrate almost exclusively on faces of dogs that are responding to a variety of emotional triggers and that produce various facial and body cues (Correia-Caeiro et al., 2021).

The way humans perceive and process emotional expressions in other animals tends to be holistic (i.e., looking at the individual as a whole and not the distinct signals), and as if they were processing human emotional expressions. For example, the same brain regions (Spunt et al., 2017) and gaze patterns (Correia-Caeiro et al., 2021, 2020) are activated to process human and dog emotional expressions. Consequently, humans may not recognise important emotional expressions in animals that humans cannot produce themselves (e.g., ear or tail movements). In general, the interpretation of dog emotion cues seems to be a difficult task for humans to do (Correia-Caeiro et al., 2021, 2020; Demirbas et al., 2016; Owczarczak-Garstecka et al., 2018; Tami & Gallagher, 2009), at least without extensive training beforehand, and a range of factors might influence this ability, including cultural background (Amici et al., 2019), experience (Kujala et al., 2012; Wan et al., 2012), and even empathy and personality traits (Kujala et al., 2017). Additionally, people (including dog owners) often have unrealistic expectations about dog behaviour and cognitive abilities. This has implications not only in the field of HAI but also in the human public health field, where subtle emotional expressions are often lost in the early warning stages of fear/discomfort/pain in dogs, potentially leading to physical aggression, like dog bites (Owczarczak-Garstecka et al., 2018).

Not only may humans overlook important subtle emotional cues from animals, but they also tend to overestimate their ability to interpret the cues they observe correctly. This usually involves making many assumptions and interpretations about animal emotions, whereby anthropomorphic projections are made and seem to be common. For example, humans believe they are intuitively able to recognise many emotions in dogs (Konok et al., 2015) and tend to represent the affective space of animals similarly to that of other humans (Konok et al., 2015; Morris et al., 2008). However, empirical studies mostly indicate a human lack of proficiency in understanding emotional expressions in dogs (Hecht et al., 2012; Horowitz, 2009; Meints & De Keuster, 2009; Ostojić et al., 2015). Crossing this anthropomorphic barrier is not easy; combining empirical studies focused on understanding both the production and perception of emotional expressions in HAI, with expert training and experience with these cues, may improve people's ability to understand animals. To be able to provide this expert training though, much more research in this area is needed.

A particular challenge with this type of research is over-reliance on studying negative emotions in animals, as well as stereotypical human emotions that may or may not exist in other species. For example, dog owners usually interpret the "guilty look" (i.e., cowering, ears lowered and tail between legs) in dogs that transgressed a social rule as the animals *feeling* guilt. Guilt is not only a secondary, complex emotion, but also studies have shown this display is more likely to be a set of cues consistent with submission or fear due to the scolding of owners than guilt (Horowitz, 2009). This focus on negative or "human emotions", that understandably tries to address the more immediate emotional needs of individuals (by e.g., decreasing/removing fear or pain), or that uses more automatic or empathy-based assumptions to try to identify other species' emotions, may make human observers miss relevant emotion cues that are not stereotypically present in human behaviour, e.g., positive anticipation cues that have been identified in dogs (Bremhorst et al., 2019; Bremhorst et al., 2021; Caeiro et al., 2017). Additionally, there may be emotions with associated cues that only exist in environmental and behavioural niches that are species-specific, such as in species that live in widely different habitats (e.g., ocean), or that have sensory modalities we do not have (e.g., echolocation), which allows for the emergence of discrete emotions humans cannot experience (Bliss-Moreau, 2017), and potential for associated cues that humans cannot naturally perceive.

While humans focus more on emotional faces (Correia-Caeiro et al., 2021), animals seem to extract emotional expressions differently. For example, dogs (Correia-Caeiro et al., 2021) and monkeys (Eliza Bliss-Moreau, 2017) focus more on emotional bodies and look much less at faces. Although this is a relatively new area of research, there has been a steadily growing interest in understanding how animals perceive human facial expressions over the past decade. Most of these studies deal with the corresponding processes in domestic animals such as dogs (Barber et al., 2016; Müller et al., 2015), horses (Proops et al., 2018; Smith et al., 2016), and goats (Nawroth et al., 2018). Regarding dogs, for instance, there is accumulating evidence that they are sensitive to human facial expressions of emotion and can use these facial expressions to guide their actions (Hare & Tomasello, 2005).

## Automated Recognition of Animal Emotional Expressions

Many of the challenges in identifying and recognising emotional expressions in animals (e.g., anthropomorphic biases, neglect of non-human behaviours such as ear or tail movements, need for extensive observer training) can be addressed by using automated approaches. The data used in such approaches varies from different physiological signals to body and facial expression analysis through computer vision techniques, which are gaining more popularity due to their non-invasive nature (Dzedzickis et al., 2020).

In humans, a large body of work on automation is for instance available for facial expression analysis, including commercial professional software (e.g., Lewinski et al., 2014). Some of these tools employ descriptive techniques that offer segmentation of facial expressions into their individual components, such as FACS Action Units. These methods save time required to learn and use the FACS tool, and some researchers even consider it more objective and reliable than the latter, eliminating subjectivity and bias (Bartlett et al., 1999; Cohn & Ekman, 2005).

In animals, research about automated emotion and pain recognition, and particularly automated facial expression analysis, is only just beginning and seems to face additional challenges to the human research. Hummel et al. (2020) highlighted some of these challenges in addressing the automation of pain recognition in animals, which can also be generalised to emotion recognition. First, much less data is available for animals compared to humans. Second, the outer appearance of the heads and faces of many animal species is influenced by the large variance in textures and shapes, more than in humans. Finally, and perhaps most crucially – while self-reporting of emotional or pain state is the norm in the human domain, there is no verbal basis for establishing ground truth in animals.

It is therefore not surprising that automation has so far only been addressed in a few species and for a few behaviours. Rather than the identification of animal emotion, the focus has been mainly on pain assessment in animals, like rodents (Sotocina et al., 2011), sheep (Mahmoud et al., 2018), and cats (Feighelstein et al., 2022). Automating facial movement recognition in emotional contexts has so far only been addressed for non-human primates.

For instance, Morozov et al. developed a prototype system for the automated coding of six FACS Action Units or AUs (i.e., facial movements caused by individual facial muscles) for macaques. Another study developed an approach to automatically analyse a number of basic primate facial expressions, including a neutral expression, lip-smacking, chewing, and opening of the mouth (Blumrosen et al., 2017).

In addition to the studies mentioned above, there are approaches to automatically evaluate body expressions in animals. Ferres et al. (2022) focused on automated pose estimation to classify emotions in dogs, including anger, fear, happiness and relaxation. The dataset used for this development included images selected from the internet based on text describing the circumstances under which the pictures were taken. While this approach allows for the collection of a low-effort and large image dataset, it should be used cautiously, as it also may introduce human bias during image selection (e.g., selecting only overt/exaggerated behaviours) or by trusting potentially inaccurate descriptions that were made by people (see the previous section for more explanations on why people generally are not proficient at identifying dog emotion).

In sum, automated emotion recognition in animals is only just beginning to emerge. Its development is strongly challenged by the discussed difficulties in data collection, selection, and validation. The ground truth must therefore be provided by using complex and well-controlled experimental designs, e.g., controlled induction of emotions (e.g., Bremhorst et al., 2019) or pain (e.g., Broomé et al., 2019). Regardless of how the ground truth is built, consideration must be given to the potential for introducing bias and error into the automated system for animals, which is a more challenging problem than for humans. The development of this area is envisioned to advance animal emotion research, and have far-reaching applications for HAI, e.g., in the form of mobile apps that can help humans to correctly identify animal emotions for better communication and understanding their needs and welfare.

## Summary

In this chapter, we reflected on the critical role of animal emotions, particularly in the context of HAI as well as animal welfare. We reviewed fundamental theoretical concepts in the field of animal emotions, highlighting what animal emotions are, when they can be observed, and how they can be measured objectively, considering challenges such as animal's inability to self-report internal states. It is important to remember that behavioural expressions are species-specific, and correctly inferring animal emotions requires empirical evidence for potential emotion indicators, which is relatively fragmented even in species close or familiar to us. Additionally, we discussed known human biases and challenges in interpreting animal behaviour and deducing animal emotions, which generally seems to be a challenging task for humans, at least for those without in-depth expert training. To systematically assess animal emotions, it is highly recommended to consider information from different sources, including the context the animal is in. Computer-vision techniques have high potential to be a game changer in animal emotion recognition, and we briefly mentioned the technical challenges that need to be overcome in this domain. The scientific community has embarked on a remarkable journey in studying animal emotions, and we are only just beginning to understand their complex nature. We hope that the aspects presented and discussed in this chapter can be used to advance public discourse on the societal challenges of understanding animal emotions in the context of animal welfare and HAI.

## Discussion Questions

1. Emotional expressions are often species-specific, but some are universal in that they accompany an emotion across species (e.g., fear is usually accompanied by withdrawal from the threatening stimulus) – can you think of other examples of potentially universal expressions?
2. Recognising emotional expressions in other species is generally a challenging task for humans – what might help in improving this ability?
3. How could using FACS or automated technologies to identify animals' emotions improve daily HAI in different fields, e.g., at home with pets, with animals in zoos etc., or professional activities like AAI, e.g., by recording and analysing sessions regarding the animals' welfare?
4. What are the possible implications of automating emotion detection in animals on HAI in the future? How do you envision the future role of technology in HAI?

## References

Amici, F., Waterman, J., Kellermann, C. M., Karimullah, K., & Bräuer, J. (2019). The ability to recognize dog emotions depends on the cultural milieu in which we grow up. *Scientific Reports, 9*(1), 1–9. https://doi.org/10.1038/s41598-019-52938-4

Barber, A. L. A., Randi, D., Müller, C. A., & Huber, L. (2016). The processing of human emotional faces by pet and lab dogs: Evidence for lateralization and experience effects. *PLoS One, 11*(4), e0152393. https://doi.org/10.1371/journal.pone.0152393

Barrera, G., Jakovcevic, A., Elgier, A. M., Mustaca, A., & Bentosela, M. (2010). Responses of shelter and pet dogs to an unknown human. *Journal of Veterinary Behavior: Clinical Applications and Research, 5*(6), 339–344. https://doi.org/10.1016/j.jveb.2010.08.012

Bartlett, M. S., Hager, J. C., Ekman, P., & Sejnowski, T. J. (1999). Measuring facial expressions by computer image analysis. *Psychophysiology, 36*(2), 253–263. https://doi.org/10.1017/S0048577299971664

Beerda, B., Schilder, M. B. H., Van Hooff, J. A. R. A. M., & De Vries, H. W. (1997). Manifestations of chronic and acute stress in dogs. *Applied Animal Behaviour Science, 52*(3–4), 307–319. https://doi.org/10.1016/S0168-1591(96)01131-8

Bennett, V., Gourkow, N., & Mills, D. S. (2017). Facial correlates of emotional behaviour in the domestic cat (*Felis catus*). *Behavioural Processes, 141*, 342–350. https://doi.org/10.1016/j.beproc.2017.03.011

Bliss-Moreau, Eliza. (2017, October 1). Constructing nonhuman animal emotion. *Current Opinion in Psychology, 17*, 184–188. https://doi.org/10.1016/j.copsyc.2017.07.011

Bliss-Moreau, E., Williams, L. A., & Karaskiewicz, C. L. (2021). Evolution of emotion in social context. In T. Shackelford & V. Weekes-Shackelford (Eds.), *Encyclopedia of Evolutionary Psychological Science* (pp. 2487–2499). Cham: Springer.

Blumrosen, G., Hawellek, D., & Pesaran, B. (2017). Towards automated recognition of facial expressions in animal models. *Proceedings of the 2017 IEEE International Conference on Computer Vision Workshops*, 2810–2819. doi: 10.1109/ICCVW.2017.332.

Boissy, A., Aubert, A., Désiré, L., Greiveldinger, L., Delval, E., & Veissier, I. (2011). Cognitive sciences to relate ear postures to emotions in sheep. *Animal Welfare, 20*(1), 47–56.

Bremhorst, A., Mills, D. S., Würbel, H., & Riemer, S. (2021). Evaluating the accuracy of facial expressions as emotion indicators across contexts in dogs. *Animal Cognition, 25*(1), 121–136.

Bremhorst, A., Sutter, N. A., Würbel, H., Mills, D. S., & Riemer, S. (2019). Differences in facial expressions during positive anticipation and frustration in dogs awaiting a reward. *Scientific Reports, 9*(1), 19312. https://doi.org/10.1038/s41598-019-55714-6

Briefer, E. F., Tettamanti, F., & McElligott, A. G. (2015). Emotions in goats: Mapping physiological, behavioural and vocal profiles. *Animal Behaviour, 99*, 131–143. https://doi.org/10.1016/j.anbehav.2014.11.002

Broomé, S., Gleerup, K. B., Andersen, P. H., & Kjellström, H. (2019). Dynamics are important for the recognition of equine pain in video. *Proceedings of the IEEE/CVF Conference on Computer Vision and Pattern Recognition*, 12667–12676.

Bruce, V., & Young, A. (2013). *Face perception*. London: Psychology Press.

Caeiro, C. C., Burrows, A., & Waller, B. M. (2017). Development and application of cat FACS: Are human cat adopters influenced by cat facial expressions? *Applied Animal Behaviour Science, 189*, 66–78. https://doi.org/10.1016/j.applanim.2017.01.005

Caeiro, C. C., Guo, K., & Mills, D. S. (2017). Dogs and humans respond to emotionally competent stimuli by producing different facial actions. *Scientific Reports, 7*(1), 15525. https://doi.org/10.1038/s41598-017-15091-4

Clark, E. A., Kessinger, J. N., Duncan, S. E., Bell, M. A., Lahne, J., Gallagher, D. L., & O'Keefe, S. F. (2020). The facial action coding system for characterization of human affective response to consumer product-based stimuli: A systematic review. *Frontiers in Psychology, 11*, 920.

Cohn, J. F., & Ekman, P. (2005). Measuring facial action. In J. A. Harrigan, R. Rosenthal & K. R. Scherer (Eds.), *The new handbook of methods in nonverbal behavior research* (pp. 9–64). Oxford: Oxford University Press.

Correia-Caeiro, C., Guo, K., & Mills, D. S. (2020). Perception of dynamic facial expressions of emotion between dogs and humans. *Animal Cognition, 1*, 1–12. https://doi.org/10.1007/s10071-020-01348-5

Correia-Caeiro, C., Guo, K., & Mills, D. (2021). Bodily emotional expressions are a primary source of information for dogs, but not for humans. *Animal Cognition, 24*(2), 267–279. https://doi.org/10.1007/S10071-021-01471-X/FIGURES/4

Darwin, C. (1872). *The expression of the emotions in man and animals*. London: John Murray.

Dawkins, M. S. (2006a). Through animal eyes: What behaviour tells us. *Applied Animal Behaviour Science, 100*(1–2), 4–10. https://doi.org/10.1016/j.applanim.2006.04.010

Dawkins, M. S. (2006b). A user's guide to animal welfare science. *Trends in Ecology & Evolution, 21*(2), 77–82.

Demirbas, Y. S., Ozturk, H., Emre, B., Kockaya, M., Ozvardar, T., & Scott, A. (2016). Adults' ability to interpret canine body language during a dog–child interaction. *Anthrozoös, 29*(4), 581–596. https://doi.org/10.1080/08927936.2016.1228750

De Oliveira, D., & Keeling, L. J. (2018). Routine activities and emotion in the life of dairy cows: Integrating body language into an affective state framework. *PLoS One, 13*(5), e0195674. https://doi.org/10.1371/journal.pone.0195674

Descovich, K., Wathan, J., Leach, M. C., Buchanan-Smith, H. M., Flecknell, P., Farningham, D., & Vick, S.-J. (2017). Facial expression: An under-utilised tool for the assessment of welfare in mammals. *ALTEX, 34*(3), 409–429. https://doi.org/10.14573/altex.1607161

Duncan, I. J. H. (2006). The changing concept of animal sentience. *Applied Animal Behaviour Science, 100*(1–2), 11–19. https://doi.org/10.1016/j.applanim.2006.04.011

Dzedzickis, A., Kaklauskas, A., & Bucinskas, V. (2020). Human emotion recognition: Review of sensors and methods. *Sensors, 20*(3), 592. https://doi.org/10.3390/S20030592

Ekman, Paul. (1992). An argument for basic emotions. *Cognition and Emotion, 6*(3/4), 169–200.

Ekman, Paul, Davidson, R. J., & Friesen, W. V. (1990). The Duchenne smile: Emotional expression and brain physiology II. *Journal of Personality and Social Psychology, 58*(2), 342–353. https://doi.org/10.1037/0022-3514.58.2.342

Ekman, P., & Friesen, W. V. (1978). *Facial coding action system (FACS): A technique for the measurement of facial actions*. California, CA: Consulting Psychologists Press.

Ekman, P., Friesen, W. V., & Hager, J. (2002). *Facial action coding system (FACS): The manual & The investigator's guide*. Salt Lake City: A Human Face, Research Nexus.

Ekman, Paul, & Rosenberg, E. L. (2012). *What the face reveals: Basic and applied studies of spontaneous expression using the facial action coding system (FACS)*. Oxford: Oxford University Press.

Feighelstein, M., Shimshoni, I., Finka, L. R., Luna, S. P., Mills, D. S., & Zamansky, A. (2022). Automated recognition of pain in cats. *Scientific Reports, 12*(1), 1–10.

Ferres, K., Schloesser, T., & Gloor, P. A. (2022). Predicting dog emotions based on posture analysis using deeplabcut. *Future Internet, 14*(4), 97. https://doi.org/10.3390/FI14040097

Firnkes, A., Bartels, A., Bidoli, E., & Erhard, M. (2017). Appeasement signals used by dogs during dog–human communication. *Journal of Veterinary Behavior, 19*, 35–44. https://doi.org/10.1016/J.JVEB.2016.12.012

Gähwiler, S., Bremhorst, A., Tóth, K., & Riemer, S. (2020). Fear expressions of dogs during new year fireworks: A video analysis. *Scientific Reports, 10*, 16035. https://doi.org/10.1038/s41598-020-72841-7

Hare, B., & Tomasello, M. (2005). Human-like social skills in dogs? *Trends in Cognitive Sciences, 9*(9), 439–444. https://doi.org/10.1016/j.tics.2005.07.003

Hecht, J., Miklósi, Á., & Gácsi, M. (2012). Behavioral assessment and owner perceptions of behaviors associated with guilt in dogs. *Applied Animal Behaviour Science, 139*(1–2), 134–142. https://doi.org/10.1016/J.APPLANIM.2012.02.015

Hintze, S., Smith, S., Patt, A., Bachmann, I., & Würbel, H. (2016). Are eyes a mirror of the soul? What eye wrinkles reveal about a horse's emotional state. *PLoS One, 11*(10), e0164017. https://doi.org/10.1371/journal.pone.0164017

Horowitz, A. (2009). Disambiguating the "guilty look": Salient prompts to a familiar dog behaviour. *Behavioural Processes, 81*(3), 447–452. https://doi.org/10.1016/j.beproc.2009.03.014

Hummel, H. I., Pessanha, F., Salah, A. A., Van Loon, T. J. P. A. M., & Veltkamp, R. C. (2020). Automatic pain detection on horse and donkey faces. *2020 15th IEEE International Conference on Automatic Face and Gesture Recognition, FG 2020*, 793–800. https://doi.org/10.1109/FG47880.2020.00114

Humphrey, T., Proops, L., Forman, J., Spooner, R., & McComb, K. (2020). The role of cat eye narrowing movements in cat-human communication. *Scientific Reports, 10*, 16503. https://doi.org/10.1038/s41598-020-73426-0

Karl, S., Anderle, K., Völter, C. J., & Virányi, Z. (2022). Pet dogs' behavioural reaction to their caregiver's interactions with a third party: Join in or interrupt? *Animals, 12*(12), 1574.

Keltner, D., & Ekman, P. (2003). Facial expressions of emotion. In R. J. Davidson, K. R. Scherer & H. H. Goldsmith (Eds.), *Series in affective science. Handbook of affective sciences* (pp. 415–432). Oxford: Oxford University Press.

Konok, V., Nagy, K., & Miklósi, Á. (2015). How do humans represent the emotions of dogs? The resemblance between the human representation of the canine and the human affective space. *Applied Animal Behaviour Science, 162*, 37–46. https://doi.org/10.1016/j.applanim.2014.11.003

Kremer, L., Holkenborg, S. K., Reimert, I., Bolhuis, J. E., & Webb, L. E. (2020). The nuts and bolts of animal emotion. *Neuroscience & Biobehavioral Reviews, 113*, 273–286.

Kujala, M. V., Kujala, J., Carlson, S., & Hari, R. (2012). Dog experts' brains distinguish socially relevant body postures similarly in dogs and humans. *PLoS ONE, 7*(6), e39145.

Kujala, M. V, Somppi, S., Jokela, M., Vainio, O., & Parkkonen, L. (2017). Human empathy, personality and experience affect the emotion ratings of dog and human facial expressions. *PLoS One, 12*(1), e0170730. https://doi.org/10.1371/journal.pone.0170730

Lewinski, P., den Uyl, T. M., & Butler, C. (2014). Automated facial coding: Validation of basic emotions and FACS AUs in FaceReader. *Journal of Neuroscience, Psychology, and Economics, 7*(4), 227–236.

Mahmoud, M., Lu, Y., Hou, X., McLennan, K., & Robinson, P. (2018). Estimation of pain in sheep using computer vision. In R. J. Moore (Ed.), *Handbook of Pain and Palliative Care* (pp. 145–157). Cham: Springer. https://doi.org/10.1007/978-3-319-95369-4_9

Mariti, C., Falaschi, C., Zilocchi, M., Fatjó, J., Sighieri, C., Ogi, A., & Gazzano, A. (2017). Analysis of the intraspecific visual communication in the domestic dog (Canis familiaris): A pilot study on the case of calming signals. *Journal of Veterinary Behavior: Clinical Applications and Research, 18*, 49–55. https://doi.org/10.1016/j.jveb.2016.12.009

McGowan, R. T. S., Rehn, T., Norling, Y., & Keeling, L. J. (2014). Positive affect and learning: Exploring the "Eureka Effect" in dogs. *Animal Cognition, 17*(3), 577–587. https://doi.org/10.1007/s10071-013-0688-x

McGreevy, P. D., Starling, M., Branson, N. J., Cobb, M. L., & Calnon, D. (2012). An overview of the dog-human dyad and ethograms within it. *Journal of Veterinary Behavior: Clinical Applications and Research, 7*(2), 103–117. https://doi.org/10.1016/j.jveb.2011.06.001

McPeake, K., Affenzeller, N., & Mills, D. (2017). Noise sensitivities in dogs: A new licensed treatment option. *Veterinary Record, 180*(14), 353–355.

Meints, K., & De Keuster, T. (2009). Brief report: Don't kiss a sleeping dog: The first assessment of "The Blue Dog" bite prevention program. *Journal of Pediatric Psychology, 34*(10), 1084–1090. https://doi.org/10.1093/JPEPSY/JSP053

Mendl, M., Burman, O. H. P., & Paul, E. S. (2010). An integrative and functional framework for the study of animal emotion and mood. *Proceedings. Biological Sciences / The Royal Society, 277*(1696), 2895–2904. https://doi.org/10.1098/rspb.2010.0303

Mills, D. (2005). Management of noise fears and phobias in pets. *In Practice, 27*, 248–255.

Mills, D. (2017). Perspectives on assessing the emotional behavior of animals with behavior problems. *Current Opinion in Behavioral Sciences*, *16*, 66–72. https://doi.org/10.1016/j.cobeha.2017.04.002

Morris, P., Doe, C., & Godsell, E. (2008). Secondary emotions in non-primate species? Behavioural reports and subjective claims by animal owners. *Cognition and Emotion*, *22*(1), 3–20. https://doi.org/10.1080/02699930701273716

Müller, C. A., Schmitt, K., Barber, A. L. A., & Huber, L. (2015). Dogs can discriminate emotional expressions of human faces. *Current Biology*, *25*(5), 601–605. https://doi.org/10.1016/j.cub.2014.12.055

Nawroth, C., Albuquerque, N., Savalli, C., Single, M. S., & McElligott, A. G. (2018). Goats prefer positive human emotional facial expressions. *Royal Society Open Science*, *5*(8), 180491. https://doi.org/10.1098/RSOS.180491

Ostojić, L., Tkalčić, M., & Clayton, N. S. (2015). Are owners' reports of their dogs' 'guilty look' influenced by the dogs' action and evidence of the misdeed? *Behavioural Processes*, *111*, 97–100. https://doi.org/10.1016/J.BEPROC.2014.12.010

Owczarczak-Garstecka, S. C., Watkins, F., Christley, R., & Westgarth, C. (2018). Online videos indicate human and dog behaviour preceding dog bites and the context in which bites occur. *Scientific Reports*, *8*(1), 1–11. https://doi.org/10.1038/s41598-018-25671-7

Panksepp, J. (2005). Affective consciousness: Core emotional feelings in animals and humans. *Consciousness and Cognition*, *14*, 30–80. https://doi.org/10.1016/j.concog.2004.10.004

Parr, L. A., & Waller, B. M. (2006). Understanding chimpanzee facial expression: Insights into the evolution of communication. *Social Cognitive and Affective Neuroscience*, *1*(3), 221–228. https://doi.org/10.1093/scan/nsl031

Parr, L. A., Waller, B. M., Vick, S. J., & Bard, K. A. (2007). Classifying chimpanzee facial expressions using muscle action. *Emotion*, *7*(1), 172–181. https://doi.org/10.1037/1528-3542.7.1.172.

Part, C. E., Kiddie, J. L., Hayes, W. A. A., Mills, D. S., Neville, R. F., Morton, D. B., & Collins, L. M. (2014). Physiological, physical and behavioural changes in dogs (Canis familiaris) when kennelled: Testing the validity of stress parameters. *Physiology and Behavior*, *133*, 260–271. https://doi.org/10.1016/j.physbeh.2014.05.018

Paul, E. S., Harding, E. J., & Mendl, M. (2005). Measuring emotional processes in animals: The utility of a cognitive approach. *Neuroscience and Biobehavioral Reviews*, *29*(3), 469–491. https://doi.org/10.1016/j.neubiorev.2005.01.002

Paul, E. S., & Mendl, M. T. (2018). Animal emotion: Descriptive and prescriptive definitions and their implications for a comparative perspective. *Applied Animal Behaviour Science*, *205*, 202–209. https://doi.org/10.1016/j.applanim.2018.01.008

Proctor, H. S., Carder, G., & Cornish, A. R. (2013). Searching for animal sentience: A systematic review of the scientific literature. *Animals*, *3*(3), 882–906. https://doi.org/10.3390/ani3030882

Proops, L., Grounds, K., Smith, A. V., & McComb, K. (2018). Animals remember previous facial expressions that specific humans have exhibited. *Current Biology*, *28*(9), 1428–1432. https://doi.org/10.1016/J.CUB.2018.03.035

Quaranta, A., Siniscalchi, M., & Vallortigara, G. (2007). Asymmetric tail-wagging responses by dogs to different emotive stimuli. *Current Biology*, *17*(6), 199–201. https://doi.org/10.1016/j.cub.2007.02.008

Reefmann, N., Bütikofer Kaszàs, F., Wechsler, B., & Gygax, L. (2009). Ear and tail postures as indicators of emotional valence in sheep. *Applied Animal Behaviour Science*, *118*(3–4), 199–207. https://doi.org/10.1016/j.applanim.2009.02.013

Reimert, I., Bolhuis, J. E., Kemp, B., & Rodenburg, T. B. (2013). Indicators of positive and negative emotions and emotional contagion in pigs. *Physiology and Behavior*, *109*(1), 42–50. https://doi.org/10.1016/j.physbeh.2012.11.002

Rolls, E. T. (2013). What are emotional states, and why do we have them?. *Emotion Review*, *5*(3), 241–247.

Sandem, A. I., Braastad, B. O., & Bøe, K. E. (2002). Eye white may indicate emotional state on a frustration-contentedness axis in dairy cows. *Applied Animal Behaviour Science*, *79*(1), 1–10. https://doi.org/10.1016/S0168-1591(02)00029-1

Scherer, K. R. (2005). What are emotions? And how can they be measured? *Social Science Information*, *44*(4), 695–729. https://doi.org/10.1177/0539018405058216

Shepherd, K. (2009). Behavioural medicine as an integral part of veterinary practice. In D. F. Horwitz & D. S. Mills (Eds.), *BSAVA Manual of Canine and Feline Behavioural Medicine*, (2nd Edition, pp. 10–23). Gloucester: BSAVA Library.

Smith, A. V., Proops, L., Grounds, K., Wathan, J., & McComb, K. (2016). Functionally relevant responses to human facial expressions of emotion in the domestic horse (Equus caballus). *Biology Letters, 12*(2), 20150907. https://doi.org/10.1098/rsbl.2015.0907

Sotocina, S. G., Sorge, R. E., Zaloum, A., Tuttle, A. H., Martin, L. J., Wieskopf, J. S., ... & Mogil, J. S. (2011). The rat grimace scale: A partially automated method for quantifying pain in the laboratory rat via facial expressions. *Molecular Pain, 7*(1), 1744–8069. https://doi.org/10.1186/1744-8069-7-55

Spunt, R. P., Ellsworth, E., & Adolphs, R. (2017). The neural basis of understanding the expression of the emotions in man and animals. *Social Cognitive and Affective Neuroscience, 12*(1), 95–105. https://doi.org/10.1093/SCAN/NSW161

Stomp, M., Leroux, M., Cellier, M., Henry, S., Lemasson, A., & Hausberger, M. (2018). An unexpected acoustic indicator of positive emotions in horses. *PLoS One, 13*(7), e0197898. https://doi.org/10.1371/JOURNAL.PONE.0197898

Storengen, L. M., & Lingaas, F. (2015). Noise sensitivity in 17 dog breeds: Prevalence, breed risk and correlation with fear in other situations. *Applied Animal Behaviour Science, 171*, 152–160. https://doi.org/10.1016/J.APPLANIM.2015.08.020

Tami, G., & Gallagher, A. (2009). Description of the behaviour of domestic dog (Canis familiaris) by experienced and inexperienced people. *Applied Animal Behaviour Science, 120*(3–4), 159–169. https://doi.org/10.1016/j.applanim.2009.06.009

Waller, B. M., & Dunbar, R. I. M. (2005). Differential behavioural effects of silent bared teeth display and relaxed open mouth display in chimpanzees (Pan troglodytes). *Ethology, 111*, 129–142. https://doi.org/10.1111/J.1439-0310.2004.01045.X.

Waller, B. M., Julle-Daniere, E., & Micheletta, J. (2020). Measuring the evolution of facial 'expression' using multi-species FACS. *Neuroscience and Biobehavioral Reviews, 113*, 1–11. https://doi.org/10.1016/J.NEUBIOREV.2020.02.031

Waller, B. M., Peirce, K., Caeiro, C. C., Scheider, L., Burrows, A. M., McCune, S., & Kaminski, J. (2013). Paedomorphic facial expressions give dogs a selective advantage. *PLoS One, 8*(12), e82686. https://doi.org/10.1371/journal.pone.0082686

Wan, M., Bolger, N., & Champagne, F. A. (2012). Human perception of fear in dogs varies according to experience with dogs. *PLoS One, 7*(12), e51775. https://doi.org/10.1371/JOURNAL.PONE.0051775

# 13
# ANIMAL DOMESTICATION IN WEST EURASIA

## What the Histories of Dogs, Cattle, and Horses Tell Us about the Domestication Process

*Benjamin S. Arbuckle*

Animal domestication is often linked to the so-called Neolithic Revolution, that period beginning around 10,000 years ago in SW Asia when some communities shifted from hunting and gathering wild plants and animals to farming and herding domesticates (Braidwood and Reed 1957; Childe 1936). These economic and technological innovations are seen as a major step toward the development of 'civilization'. As a result of this framing, scholarly interest in the histories of domestic animals have tended to focus on those animals of central importance to the rise of Eurasian states including the most ancient livestock domesticates including cattle, sheep, goats, and pigs; animals such as horses and donkeys which historically played central roles in transportation; as well as globally distributed pet species including dogs and cats.

In order to reconstruct these animal histories, scholars have developed a suite of 'traditional' methods focused on these economically and culturally significant domestic species. Unsurprisingly, these traditional approaches are centered within western conceptual frameworks including the Linnaean classification system, Darwinian evolutionary theory, and the Modern Synthesis as well as notions of social and technological evolution (Larson and Fuller 2014; Zeder 2012). This work at the intersection of biology, social sciences, and history has been a productive area of research resulting in an enormous volume of high-profile multi-disciplinary scholarly work (e.g., Frantz et al. 2019).

In this chapter, I discuss the history of research on animal domestication. My focus is on zoological, genomic, and especially archaeological approaches focusing on west Eurasian mammalian livestock. I begin by describing what I call 'traditional' approaches to domestication situated within what anthropologist Descola (2013) refers to as a perspective of 'naturalism'. These approaches take as their starting point a division between nature and culture, the wild and the domestic, and seek to follow these bifurcated paths back to their shared point of origin (i.e., domestication events) deep in the past. I then move to recent scholarship based on the accumulation of new lines of evidence, including paleogenomics, which is reformulating and expanding some of the narratives surrounding animal domestication. Rather than presenting brief summaries of the domestication histories of dozens of taxa, I instead focus on three well-known case studies targeting dogs, cattle, and horses. These examples facilitate the exploration of new evidence for the nature and diversity of human-animal relationships in the past. In some cases these results support traditional views of animal domestication, but in other cases they make us re-evaluate how domestic processes unfold, which taxa they are allowed to include, and how the boundaries of the wild and domestic are conceived.

## Traditional Approaches to Domestication

As Purugganan (2022) has recently pointed out, there are many definitions of domestication, none of which is satisfactory to everyone. Most definitions of animal domestication focus on its role as a long term co-evolutionary process between two species, a domesticator and a domesticant. The process is generally seen as instigated by the former who uses the latter for specific purposes—usually as a food source. Although co-evolutionary and mutualistic relationships have parallels in many other biological systems (Purugganan 2022), research on domestication often takes an anthrocentric perspective viewing this process as uniquely linked to human agency (although see Rindos 1983). The overwhelming emphasis of most domestication studies is on documenting the biological differences between wild and domestic animals, the features domesticates share in common, and the human role in 'engineering' these new forms which are often viewed as economic responses to scarcity (Zeder 2012).

At its core, the concept of animal domestication reflects a worldview in which 'culture' and the anthropogenic environments created by humans, are perceived as beyond, or otherwise separated from, 'nature' (Ingold 1994). This perspective has been described by Descola (2013) as representing an ontology of 'naturalism', a point of view in which a unique human interiority is contrasted with the shared biological/physical properties (exteriorities) of all things. Domestication narratives engage with this point of view by exploring ways in which humans transport other species across the nature-culture boundary from a state of being in the wild to becoming domestic.

In addition to representing a biological process, Meadow (1989) emphasizes that domestication also represents an important change in human-animal relationships, with implications for human social organization. This emphasis on the social implications of animal domestication has a much deeper history, going back to 18th- and 19th-century scholars including Turgot (1793) and Morgan (1877) and reflects a remarkably resilient lens through which research on the history of animal domestication is still viewed.

Of relevance here is an influential publication by Turgot (1793) in which he suggests that cattle domestication led to the origins of farming which in turn led to private property, inequality, and the rise of 'civilization' itself. Turgot placed the origins of cattle herding vaguely among the 'Asiatic nations' (Turgot 1793: 32; also see de Mortillet 1879) which Peake and Fleure (1927) later narrowed down to the upper reaches of the Euphrates river, a region archaeologist Braidwood (Braidwood and Reed 1957) famously called the Hilly Flanks of the Fertile Crescent. Although the direction of the relationship between livestock domestication and agriculture is now assumed to be reversed, this historical narrative placing animal domestication at the center of narratives of the origins of the modern world has had an immense impact on the structure of subsequent research.

As a result, the exploration of the histories of domestic animal became a major focus of newly professionalized archaeological research in the late 19th and early 20th centuries. Zoologists began studying the animal remains recovered from prehistoric archaeological sites and documenting differences between purported wild and domestic varieties. Of particular importance was Rütimeyer's (1861) analysis of the animal remains from Neolithic lake dwellings in Switzerland where he was among the first to describe skeletal differences between wild and domestic animals particularly noting the smaller size of the latter—a method still widely in use today.

Following Rütimeyer's work, Darwin (1868) was among those who documented a range of phenotypic traits that often distinguish domestic animals from their wild counterparts including a decrease in body size, changes in horn shape, and craniofacial morphology as well as changes in behavior, pelage, and reproduction. Documenting these changes, known collectively as the 'domestication syndrome' (DS) (Hammer 1984) has become the central pillar of the methods used to identify early domestic livestock in the archaeological record (Meadow 1989). These methods have been utilized to document the presence of domestic cattle, sheep, goats, and pigs in the Fertile Crescent region of southwest Asia during the Neolithic period (c. 10,000 years ago) (e.g., Zeder 2012; but see Lord et al. 2020).

The broader history of domestication processes has often been understood as progressing in terms of stages and pathways. Vigne (2011: 173) describes domestication as 'an ultimate phase' of intensification in human-animal relationships which include a gradient from anthropophily to wildlife management to intensive breeding and pet keeping. The impression presented by 'stage' models of domestication is that domestication is the logical, linear, and inevitable result of intensive human-animal interactions and that relationships of husbandry of different species (e.g., sheep, reindeer, ferrets) are to be conceived of as comparable entities/systems. Tsing (2018: 234) argues that in this approach, domestication is a 'progress concept', a notion of 'collective advancement' deeply embedded in Enlightenment thought and traceable from Turgot through a multigenerational daisy chain to scholarly approaches of the current century.

Rather than representing stages, Zeder (2012) has presented perhaps the most influential recent model for domestication, describing three distinctive 'pathways' leading to domestic animal populations within different contexts of human interaction. The commensal pathway emphasizes the agency of nonhuman taxa able to colonize a variety of anthropogenic niches. Selection within these environments may result in changes including some features of the DS associated with reduced stress response and alterations in diet. At some point, the commensal relationship may shift toward increased human management of feeding, breeding, and death. Dog domestication is often described as developing through this pathway.

The prey pathway, by contrast, is initiated by human hunting behaviors. In this model, millennia of hunting game animals shifts toward wildlife management in the face of resource depression, which in turn evolves into husbandry including penning, selective culling, and control over breeding. These changes are intentionally 'designed to increase prey availability' (Zeder 2012: 242). Domestic ruminants including cattle, sheep, and goats are considered to have taken the prey pathway to domestication.

Finally, the directed pathway represents a 'fast track' to domestication after the domestication of prey animals provided a clear behavioral precedent. Zeder (2012) suggests horses and donkeys took this route based on a perceived rapid domestication process targeting specific uses such as traction (but see below). More recent examples of hamster, rat, and mouse domestication for use as pets and lab animals are also examples of the directed pathways which occur rapidly and as a result of intentional efforts to 'domesticate' (Hulme-Beaman et al. 2021).

Traditional approaches to domestication emphasize wild and domestic as binary, essentially static categories which can be identified in the archaeological record through the identification of phenotypic changes associated with the DS or, more recently, genetic evidence linking an ancient individual to a lineage of modern domesticates (Frantz et al. 2020). Traditional approaches tend to highlight unilinear staged models or emphasize Zeder's pathways,

> **Case Study 1: Dog (*Canis familiaris* Linnaeus 1758)**
>
> Dogs have long been considered the earliest domestic animal and their presence among pre-farming societies has been a matter of intense discussion since the 19th century (e.g., de Mortillet 1879: 233). Although Darwin (1859: 23) assumed dogs derived from multiple wild canid species, current scholarship sees dogs as derived from one or more extinct populations of Eurasian gray wolf (*Canis lupus*). Recent ancient DNA research suggests that dog ancestors diverged from wolves c.28,000 years ago perhaps from an east Eurasian source (Bergstrom et al. 2022). However, west Eurasian and SW Asian wolf ancestry is also evident in many dog populations and both the timing and geography of dog origins have proven difficult to address. Despite the unknowns, it is clear that dogs represent a uniquely early domestication event that took place within the context of mobile hunting and gathering economies.

which center economic paradigms of resource scarcity and intentional human agency. Often domestication is referred to as being achieved, accomplished, or even 'designed' linguistically placing humans in an active position (Bogaard et al. 2021). I now turn to three case studies focused on the domestication of dogs, cattle, and horses. In these case studies, I examine both the traditional narratives and the complications that have arisen in recent scholarship into the histories of these domestic animals in the deep past.

Explanations of dog domestication generally fall into one of two camps. In one model, dog domestication proceeded along a commensal pathway instigated primarily by wolves themselves (Larsen and Fuller 2014). Here wolves are imagined as colonizing the margins of human settlements cooperating and/or scavenging alongside human hunters rather than competing with or predating upon them (Larson et al. 2012). Selection within the anthropogenic environment favored less aggressive wolves making them more likely to be tolerated by humans, opening up access to dietary niches associated with human settlements, and perhaps leading to a cascade effect of biological changes in phenotype, development, and behavior (Larsen et al. 2012).

A second, older model favors human initiated pet keeping as the primary mechanism for dog domestication (Galton 1865; Germonpré et al. 2021; Serpell 2021). Recent support for this model argues that ancient humans were unlikely to tolerate large dangerous predators near settlements, but that a uniquely human predilection for pet keeping (Shipman's [2010] 'animal connection' or Serpell's [2021] alloparenting) provides a better explanation for the establishment of an apex predator within human communities. In this view, repeated episodes of capturing and raising wolf pups created a population of habituated and tame animals socialized within a multispecies community. Although these pups likely suffered high mortality and some surviving adults returned to free-living lifeways (see Koungoulos 2021), others remained attached to human communities. Reproduction within this group of docile and anthropophillic proto-dogs eventually led to persistent dog populations well-adapted to operating within or on the outskirts of human communities where they were tolerated and in some cases cared for by humans.

Although recent genetic results suggest dog origins extend back to the Last Glacial Maximum, recognizing the skeletal remains of dogs from these early periods has proven problematic. Despite attempts to identify 'incipient' or proto-dogs based largely on cranial morphology from sites dating prior to c.15,000 years ago (e.g., Germonpré et al.

2021), it is exceedingly difficult to evaluate these arguments based on morphological features alone. Moreover, early skeletal remains often represent disarticulated finds derived from low-density cave deposits providing no evidence of the connection between humans and canids. Thus, it is difficult to know if Pleistocene foragers viewed 'proto-dogs' as something later generations would recognize as 'dogs' or if these animals and relationships had no modern analogs. The possibility that some forager communities may have continued to raise tame wolf cubs, while others cohabitated with early dogs, and that early dogs themselves may have lived within human settlements under intensive care, on the outskirts of human settlements with no care, or even free ranging lives (Koungoulos 2021) emphasizes the difficulties of interpreting genomics, phenotype, and lived experience in the deep past.

The lack of clear evidence for human-dog interaction for much of the Pleistocene makes it difficult to understand the processes that led to dogs diverging from wolves. This situation changes toward the end of the Pleistocene when dog-human co-burials become apparent in the archaeological record. Particularly important examples come from the site of Bonn-Oberkassel in Germany where a double human burial dated to the Magdalenian culture (c. 14,000 years ago) includes a complete canid, identified through its mitochondrial DNA sequence as belonging to a dog lineage (Figure 13.1) (Janssens et al. 2018). The dog is a juvenile animal of modest size exhibiting evidence for severe illness suggesting that it was under human care. It was also covered in red ochre, a treatment often associated with human burials. At approximately the same time in the southern Levant, several dog-human co-burials appear in the Natufian culture. In one particularly famous example from the site of 'Ain Mallaha, a juvenile dog was found placed partially under the head of a deceased adult human (Tchernov and Valla 1997). Although the precise relationships between co-buried humans and dogs is unclear, it does suggest an intimate ontological connection between the two species.

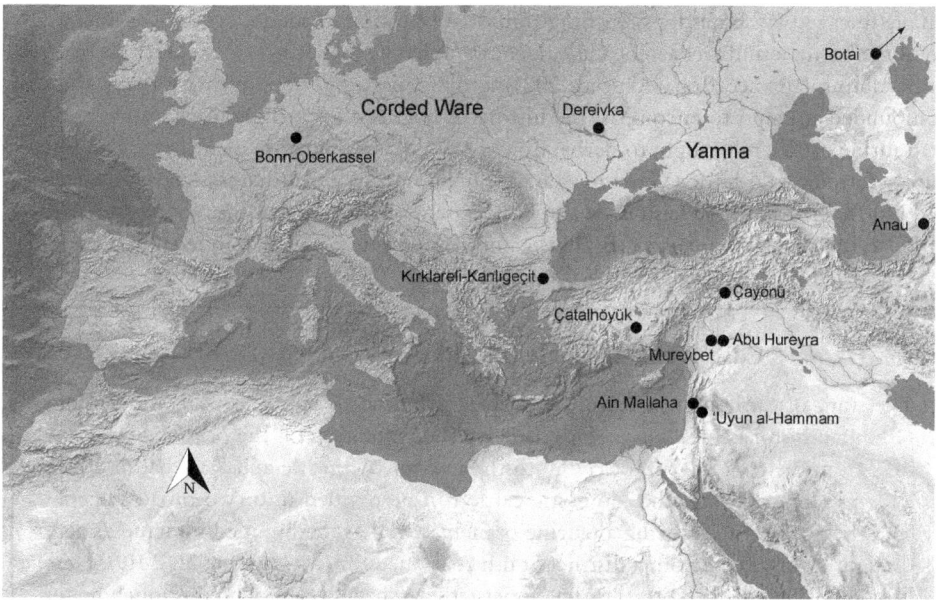

*Figure 13.1* Map showing archaeological sites mentioned in the text

Domestication narratives are often anchored to notions of the utility of the domesticant. For dogs, hunting aids and companionship are often listed as the primary motivations, while food source is generally rejected despite widespread archaeological evidence that both wolves and dogs were consumed (Price et al. 2021). Serpell (2021) points out that untrained village dogs rarely make effective hunting aids and suggests that early dogs may not have had any practical uses at all. This statement is a departure from most domestication paradigms, which emphasize economic utility. Is it possible to imagine the events leading to dog domestication as serving no particular purpose, not playing a clear role in a human centered progress narrative? If so, dogs could be seen as an unpredictable accident. Not the goal of a forward-thinking hunter, but rather the fortuitous result of interaction between a network of wolves, humans, prey taxa, and Eurasian environments toward the end of the Pleistocene.

Serpell (2021) goes on to argue that dog domestication via pet-keeping 'makes sense' as a result of intense selection of humans for alloparenting, therefore centering human agency. However, human agency is often over-emphasized in dog domestication narratives. Following Tsing (2018) dogs can be seen as 'unruly' and agentive in creating their own histories. For example, although recent studies have emphasized the shared histories of some dog and human lineages (e.g., Perri et al. 2021) dogs can also be decoupled from humans; they do not stay at home. Dogs become widespread across Eurasia by the end of the Pleistocene suggesting they spread across many communities independent of human action. Genetic evidence also shows that dogs frequently moved back into the wild introducing dog genes back into wolf populations, while also incorporating genomes of west Eurasian and SW Asian wolves into regional dog populations through admixture (Bergstrom et al. 2022; Perri et al. 2021).

This new picture of dog domestication strains against the expectations of tradition approaches to domestication linked to narratives of economic progress, assumptions about when dogs became dogs (based on phenotype or ancestry), and static categories of domestic and wild. Instead, dogs extend back in time where the boundaries of dog and wolf are porous and expectations based on modern analogs lose their utility. Rather than telling a simple linear story about dogs reaching 'full domestication' through a commensal pathway, combined genetic and archaeological evidence emphasize that dogs are in a perpetual process of becoming (also see Bogaard et al. 2021), a dynamism and heterogeneity that has so far confounded attempts to encompass the history of dogs within a single simple story.

Taurine cattle are thought to derive from a SW Asian population of aurochs (*Bos primigenius*), a form of wild cattle (Verdugo et al. 2019). Subspecies of aurochs ranged across Eurasia and North Africa in the Pleistocene and Holocene with the last aurochs reportedly expiring in Poland in the 17th century CE. This game animal was widely exploited across its range

---

### Case Study 2: Cattle (*Bos taurus* Linnaeus 1758)

The story of cattle domestication has played a central role in European conceptions of human history and prehistorians have lavished attention on the origins of this animal from the time of Turgot (1793) to the present (Arbuckle and Kassebaum 2021). This research indicates that domestic cattle can be divided into two broad varieties (and their hybrids) including a taurine branch as well as the humped indicine variety thought to originate from a distinct South Asia ancestor (McHugo et al. 2019). Here I focus on the history of the taurine variety, the archaeology of which is much better documented than its indicine cousin.

and is thought to have become domesticated through Zeder's prey pathway. In the early 20th century, prehistorians imagined cattle domestication as the first step in the rise of western civilization and associated this innovation with the so-called 'Aryan' peoples of central Asia (Mucke 1898; Turgot 1793). Duerst's (1908: 440) work on the archaeofaunal remains from the site of Anau, modern Turkmenistan, was among the first to link cattle domestication with early agricultural communities.

As the 20th century progressed, the perceived geographic center of cattle domestication shifted south to the Fertile Crescent region of SW Asia (Braidwood and Reed 1957; Childe 1936) where scholars have identified a transformation in the skeletal remains of cattle recovered from early Neolithic farming communities particularly centered along the Upper Euphrates river. Here, beginning in the mid-ninth millennium BCE, Helmer (Helmer et al. 2005) argues for the earliest changes in phenotype associated with the transformation of aurochs into domestic cattle. These changes initially include a disruption in sexual dimorphism with smaller male body size hypothesized to reflect human control over cattle mating. Over the following centuries, small sized cattle, well below the 'normal' size of local aurochs, with increasingly variable horn morphologies emerge in the archaeological record from Iraqi Kurdistan to southeastern Turkey and the Damascus Basin (Arbuckle and Kassebaum 2021).

This pattern is often interpreted as reflecting the domestication of a single Euphrates aurochs population, which then spread from its initial hearth to other Neolithic communities across SW Asia; and from there into adjacent regions including NE Africa, Arabia, Central Asia, and Europe. This notion of a single hearth of domestication has been supported by an analysis of ancient and modern cattle mitochondrial DNA sequences (Bollongino et al. 2012).

However, addition evidence suggests the story may be more complex. For example, only a few centuries after the earliest proposed cattle domestication in the Euphrates basin, early farmers transported cattle to the island of Cyprus. These animals were clearly under some form of human management (Vigne 2011) but displayed little evidence for a domestic phenotype. Furthermore, Hongo (Hongo et al. 2009) describes the slow local transformation of aurochs into domestic cattle at the Neolithic site of Çayönü in southeastern Turkey. In the southern Levant, Munro (Munro et al. 2018) has argued that increases in the frequency of aurochs remains in the early ninth millennium BCE reflect the development of management relationships with local aurochs populations prior to their domestication—evident a millennium later.

In the Euphrates region, Ducos (1978) argued that an early system of *proto-elevage* was practiced at the site of Mureybet on the Syrian Euphrates reflecting the management of phenotyptically wild aurochs over many generations. Arbuckle and Kassebaum (2021) have expanded this model suggesting that a long period of pre-domestic aurochs management may have preceded the appearance of domestic cattle in many regions of SW Asia. Systems of management that did not lead toward a domestic phenotype are also known from the later Neolithic of central Anatolia, where at the site of Çatalhöyük, robust aurochs-like cattle were utilized for 700 years before changes in body size and horn shape are evident (summarized in Arbuckle and Kassebaum 2021).

These findings from diverse regions of the Fertile Crescent and Cyprus suggest that the appearance of a domestic cattle phenotype does not represent the initiation of close relationships between aurochs and humans but rather these relationships, perhaps with parallels to reindeer herding, emerged much earlier (Zeder 2012). Unfortunately, these pre-domestic forms of cattle management are challenging to recognize since they diverge from the expectations of tradition approaches which rely on documenting features of the DS. This situation is beginning to change with the application of new techniques to the earliest sedentary communities in SW Asia. For example, a recent study discovered evidence for the widespread use of ruminant dung in the Epipaleolithic levels of the site Abu Hureyra, located on the Syrian Euphrates, prior to the advent of farming (Smith et al. 2022). This suggests that animals were 'tended' within the early pre-agricultural village, and, although the species that produced the dung is not certain, aurochs is one possible candidate.

The notion that multiple regions within SW Asia were involved in the emergence of domestic taurine cattle is also supported by ancient DNA results. Verdugo (Verdugo et al. 2019) found three distinctive lineages represented in Neolithic cattle from SW Asia. These results indicate that early domestic cattle were strongly linked to local aurochs populations and emphasize that cattle management was a spatially complex process. Moreover, as domestic cattle moved across Europe in the Neolithic period, Verdugo show that they admixed with European aurochs in a process Larsen (Larsen and Fuller 2014) refers to as 'introgressive capture'; evidence that permeability between wild and domestic populations continued for millennia after initial cattle domestication in the Fertile Crescent.

Finally, Verdugo also finds evidence for the admixture, or hybridization, between the distinctive taurine and indicine lines of domestic cattle. Beginning in the third millennium BCE, evidence for the admixture of zebu bulls into taurine herds is evident on the Iranian plateau and then more widely in SW Asia and beyond (McHugo et al. 2019). This highlights the fact that domestic cattle are not a static entity but rather are engaged in a continuous process of becoming. This emphasis on change is reflected in genomic data but also in traditional archaeological approaches documenting body size. Although a dramatic decrease is body size is evident in SW Asian cattle in the eighth millennium BCE associated with the proliferation of domestic phenotypes, cattle populations from archaeological sites from the Neolithic (8000 BCE) to Bronze Age (2000 BCE) and beyond are always heterogenous and continue to exhibit changes in size in every period continuing to the present (Figure 13.2).

Overall, the example of cattle domestication highlights several features of the domestication process. These include the limits of traditional methods focusing on features of the DS to 'see' pre-domestic relationships between co-residence and management. These

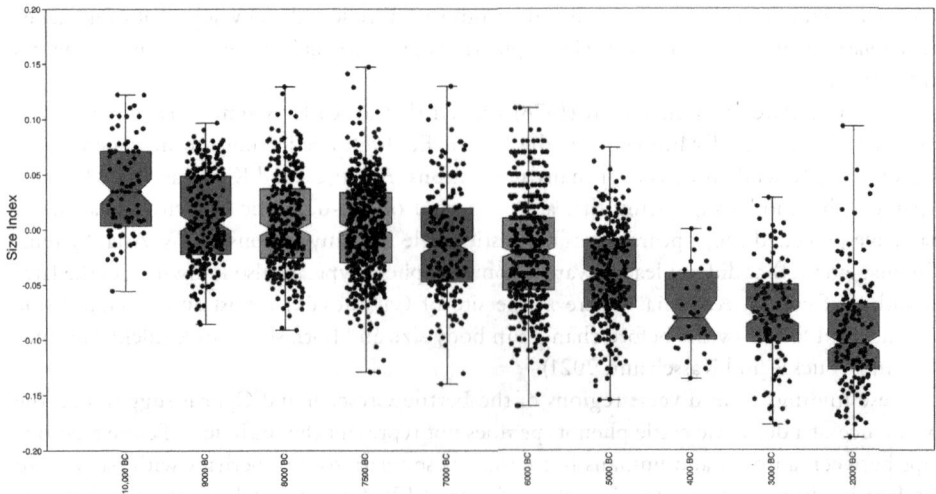

*Figure 13.2* Box and jitters plot showing changes in cattle body size through time from sites in SW Asia. Size Index value 0.0 represents a Mesolithic cow aurochs from Ullerslev, Denmark

> **Case Study 3: Horse (*Equus caballus* Linnaeus 1758)**
>
> Horse domestication has long been a thorny and intractable topic. Horses exhibit few or none of the morphological changes such as decreased body size and disrupted sexual dimorphism associated with the domestication of ruminants including sheep, goat, and cattle (Helmer et al. 2005). This has made the archaeological methods effective for studying the domestication histories of other livestock ineffective in distinguishing the remains of wild and domestic horses. Moreover, horse remains are often rare in the archaeological record and the geographic center of prehistoric horse exploitation, the central Eurasian steppe, is located far from the Fertile Crescent where most research on animal domestication has taken place. These challenges, combined with the extinction of most wild horses, have slowed attempts to understand the process, timing, and geography of horse domestication.

relationships clearly preceded the appearance of domestic phenotypes by centuries or even millennia (see McHugo et al. 2019: Figure 13.1). In addition, genomic and archaeological evidence suggests that, at least in the Neolithic period, the boundaries between domestic cattle and wild aurochs were permeable with local aurochs incorporated into domestic populations in multiple regions. Finally, it is clear that the cattle domestication process represents a continuous process of change extending both far backward into prehistory as well as forward into the future with the 'nature' of domestic cattle refusing to remain static as reflected in taurine-zebu hybrids, feral Texas Longhorns, primitive Heck varieties, and modern dairy breeds.

Although most agree that early horse management originated somewhere on the Eurasian steppe (Warmuth et al. 2012), the processes leading to horse domestication have been difficult to reconstruct. Zeder (2012) suggested horses were domesticated via the directed domestication pathway. Once the technologies of livestock management were transferred from Neolithic SW Asia onto the steppe, the idea goes, herders decided to direct horses—a local game animal—into a domestic relationship in order to access secondary products, specifically traction and perhaps riding. This is based largely on the perceived late domestication of horses compared to the SW Asian barnyard complex of cattle, sheep, goats, and pigs. However, since farming economies arrived on the steppe some four millennia after their appearance in the Fertile Crescent (Anthony 2007), the perception that horse domestication is 'late' is a mirage centered on the chronology of the SW Asian Neolithic. Instead, long traditions of horse hunting on the Eurasian steppe suggest the prey pathway is a more likely model for this process. Based on high frequencies of horse bones in archaeofaunal assemblages from steppic Sredni Stog culture sites such as Dereivka (Ukraine) and the appearance of horse remains in ritual contexts containing domestic ruminants, Anthony (2007) and others (Olsen 2006) have argued horse domestication was under way on the Pontic-Caspian steppe by the fifth millennium BC. Because of the lack of clear domestic phenotypes, as well as the lack of clear horse-riding equipment until the late third millennium BCE (Anthony 2007), arguments for 'early' domestication were controversial until new methodologies allowed researchers to explore details of horse-human relationships indicative of cohabitation and management.

These innovations were first applied at the site of Botai in Kazakhstan, a site representing the eponymous archaeological culture of the fourth millennium BCE and characterized by an unusual animal economy focused almost entirely on horses (Outram et al. 2009). Research at Botai has uncovered evidence for 'horse corrals', while residue analysis from Botai ceramics identified the presence of mare's milk (Outram et al. 2009). Together these provide the earliest convincing evidence for horse domestication interpreted as part of an economic system in which managed horses were used to aid the hunting and acquisition of wild horses (Olsen 2006).

In addition, the recent availability of ancient genomic evidence for Eurasian horses, including those from Botai, has further clarified the nature of horse domestication as well as raising new questions. Genomic work has shown that virtually all domestic horses from historic periods belong to a single lineage known as DOM2 (Librado et al. 2021). Botai horses are not included in this lineage suggesting that they represent a 'failed' or abandoned domestication event and were replaced by DOM2 horses some time after the fourth millennium BCE (Gaunitz et al. 2018). Botai horses are found to be closely related to the only extant wild horse, the Mongolian Przewalski horse and Gaunitz (Gaunitz et al. 2018: 1) argues that Przewalski horses therefore 'are the feral descendants of horses herded at Botai and not truly wild horses'.

Furthermore, the history of DOM2 horses has recently been brought into focus. Analyzing almost 300 ancient horse genomes Librado (Librado et al. 2021) identifies the lower Volga-Don region of the Pontic Caspian steppe as the geographic homeland of DOM2 horses and locates ancient horses as potential ancestors for this lineage dating to c. 3000 BCE associated with the Yamna archaeological culture. A high frequency of horse remains at Yamna burial sites has long been used by scholars as evidence that the Yamna pastoral economy included horse management (Olsen 2006). However, Librado shows that the horses of the Corded-Ware culture of northern Europe, understood to represent a large-scale western migration of the Yamna population, were not part of the DOM2 lineage but instead reflect continuity with preceding late Neolithic wild horses of northern Europe. It is therefore concluded that domestic horses were not part of the expansionary Yamna cultural package (Librado et al. 2021).

Clear evidence for the domestication of DOM2 horses is not apparent until the late third millennium BCE, when this lineage expands geographically into eastern Europe, Anatolia and then across Eurasia in the second millennium BCE where they replace all other horse populations and are closely associated with riding and pulling wheeled vehicles (Librado et al. 2021). Analysis of genetic changes in DOM2 horses indicates selection of genes related to endurance, weight bearing, back stiffness, and increased docility linked to intensive human management and participation in intense exercise through riding and traction (Librado et al. 2021).

Despite the common assumption that the origins of riding and traction in the late third millennium BCE represent a logical beginning of horse domestication (Gibbons 2021), ancient genomic results raise new questions. Although genetic results shine light on the history of the dominant domestic lineage, they are less effective at assessing the history of horse management itself. It is possible that in addition to Botai, other non DOM2 horse populations were kept under various forms of management in the third, fourth, or even fifth millennia BCE. That non-DOM2 horses were managed outside of Botai is supported by remains from the site of Kırklareli-Kanlıgeçit in Turkish Thrace. Here, large numbers of horses have been described in a region with no earlier history of wild horse exploitation, some possessing a rare coat color mutation unlikely to persist in a wild population, dating to the early third millennium BCE (Ludwig et al. 2009). This combination of evidence suggests that the horse population at Kırklareli-Kanlıgeçit and perhaps numerous others outside of the DOM2 lineage were under human management in the third and fourth millennia BCE.

There are limitations inherent to conceiving of the boundary between wild and domestic horses as rigid and impermeable. For example, despite convincing evidence for centuries of horse management at Botai, managed horses do not display phenotypic differences from their hunted counterparts. Here a reindeer model of early animal management may represent a reasonable analog (Stépanoff et al. 2017). In a herding system characterized by frequent recruitment from free ranging herds, it is possible that the Botai horses represent one meta-population including both managed and free-living individuals. In this scenario, rather than representing feral, escaped domesticates (Gaunitz et al. 2018), Przewalski horses may be seen as an eastern expansion of the horse population endemic to the central Asian steppe which included both herded and hunted animals in the fourth millennium BCE. Moreover, the tendency of unruly domesticates to expand outside of their categories is also shown by the genetic history of the tarpan, a free ranging west Eurasian horse which went extinct in the early 20th century CE. Analysis of tarpan's genome shows it to be the result of admixture between local European wild horses and a DOM2-like ancestor perhaps reflecting the incorporation of early domestic horses into wild herds (Librado et al. 2021).

Although recent genomic work has contributed much to our understanding of horse domestication processes, the nature of human and horse relationships on the Eurasian steppe prior to the appearance and spread of the DOM2 lineages remains an open question. What we do now know about the histories of horses, however, provides added richness to our understanding of domestication processes more broadly. This includes an emphasis on domestication relationships that operated in specific times and places but did not develop into persistent domestic species exhibiting distinctive morphological features (e.g., Botai and Kanlıgeçit horses). Although often referred to as 'failed' domestication processes, this vocabulary belies an expectation that domestication processes have an ultimate goal to create a stable and static new species to be used by humans in perpetuity.

The history of horses also emphasizes the capability of specific domestication relationships and domestic lineages to 'erase' earlier ones as the riding and traction of DOM2 horses did to Botai, Kanlıgeçit, and likely numerous other horse-human relationships and populations. This requires thinking beyond well-established aspects of human-animal relationships (e.g., horses and riding) to other possibilities (horses as sources of meat, skins, and milk). Finally, it highlights the porous boundaries of domestication itself, boundaries which in cases like Botai may have included open circulation between horses living in varying degrees of proximity to humans comparable to the intermittent coexistence (Stépanoff et al. 2017) of reindeer herding.

## Discussion

Evidence from these three case studies emphasizes a range of observations on domestication processes not captured in traditional approaches. One of these is that categories of wild and domestic, which often seem clear and stable in the modern world, were not always so in the past.

These categories appear quite leaky in the ancient world particularly in the periods leading up to the emergence of domestic phenotypes when population isolation was often limited or nonexistent (Frantz et al. 2020). However, even after the emergence of phenotypically divergent populations, domesticates are unruly and have a way of escaping boundaries. This is most clearly reflected in processes of feralization and hybridization, in which domesticates do not 'stay in their lane' but rather rejoin free-living populations, colonize entirely new niches with limited human interaction, or incorporate new wild or domestic lineages through admixture. For example, genomics tells us that ancient dogs admixed with wolf populations, and early managed horses (including Botai and perhaps a pre-DOM2 lineage) admixed with wild horses to form the tarpan and Przewalski horses. In addition, feral hogs, 'wild' mustangs, and Longhorn cattle in North America and dingoes in Australian are examples of domestic populations which have expanded beyond human management, colonizing entirely new ecologies on their own.

In terms of hybridization, genomic studies show that European, southern Levantine, and Anatolian aurochs were incorporated into domestic taurine cattle populations while domestic indicine and taurine lineages became intertwined beginning in the Bronze Age. Hybridizations and adaptive introgressions between wild boar and domestic pigs are also well documented across Europe from prehistory to the present (Frantz et al. 2019). Mary (Mary et al. 2022) reports adaptive introgressions from domestic pigs into wild boar populations in modern France. This dynamic feedback between domestic and wild populations is rarely considered in studies of early animal domestication, which often assume the direction of interaction is primarily from wild to domestic and limited temporally only to an early stage in the process of domestication. Thus, rather than reflecting static categories or statuses of being, domestic animals are constantly on the move, both in terms of where they go and what they are.

A second feature highlighted by the case studies is the lack of efficacy of morphological features of the DS in identifying domestication processes. Recent work showing that management of livestock precedes the appearance of phenotypic changes (which in some domestic species never emerge at all) has forced scholars to decouple phenotypic changes associated with the DS from lived experiences of animal populations (e.g., Zeder and Hesse 2000). This represents a difficult shift for archaeozoologists who, since the time of Rütimeyer (1861), have relied on changes in size and shape to differentiate domestic from wild (Germonpré et al. 2021). Despite persistent calls to explore early domestication relationships through means other than changes in phenotype more, many scholars continue to structure domestication research as a hunt for the earliest emergence of domestic phenotypes. This is related to the persistence of the idea that domestication is a homogenous and stable 'status' of being, a common part of traditional staged views of human history and progress.

A third feature of animal domestication relates to its position embedded in narratives of the rise of 'civilization'. As a result, studies of animal domestication (including this chapter) tend to start from present domestication relationships (e.g., dogs, cattle, horses) and work backward to find their origins. This necessarily represents a fraction of the diversity in human-animal relationships that existed in the past and results in a blind spot for domestication relationships that were discontinuous, don't have modern analogs, or which involve taxa that are not important in modern global economies (Bogaard et al. 2021).

As Tsing (2018) has argued, domestication frameworks tend to limit our view of 'other' ways of being in the past. The archaeological record is potentially full of possible examples that have largely remained unexplored. For example, fox-human burials have been described at the site of 'Uyun al-Hammam, Jordan, dating to 15,000 BCE (Maher et al. 2011: 8). The authors describe the foxes (*Vulpes vulpes*) as 'companions' to the deceased humans, summarize extensive evidence that foxes are abundant in many archaeofaunal assemblages in Epipaleolithic and Neolithic sites in SW Asia, and go as far to highlight the parallels between foxes and dogs stating that foxes 'could have been considered as potential domesticates to prehistoric people' (Maher et al. 2011: 8).

However, the authors do not go beyond the past conditional tense when describing the relationship between foxes and prehistoric humans, which has rarely been taken seriously

*Animal Domestication in West Eurasia*

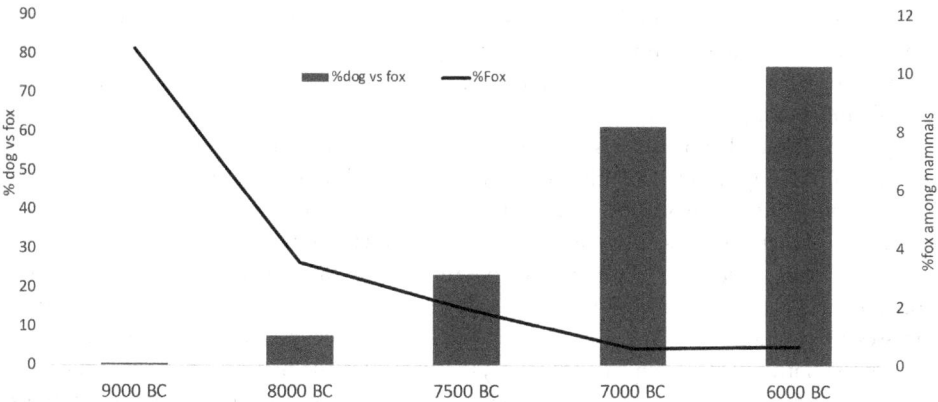

*Figure 13.3* Changes in fox and dog frequencies through the Neolithic period in SW Asia. Bars represent the frequency of dog vs fox remains in archaeofaunal assemblages; line represents the frequency of fox remains among all mammals in archaeofaunal assemblages

or assessed for evidence of management, co-residence, and domestication. The fact that foxes were likely a regular food source in Epipaleolithic and Neolithic communities in SW Asia and that their frequencies decline as that of dogs increases through the Neolithic period suggests the possibility that a short-lived domestication-like relationship emerged in late Epipaleolithic communities (Figure 13.3). These unique human-fox relationships may have been disrupted and replaced by the incorporation of a larger domestic canid, dogs. In addition, there are hints that 'other' taxa including North African Barbary sheep, the extinct Balearic cave goat (*Myotragus balearicus*), and even European Red deer may have been herded, penned, and managed in short-lived systems with limited modern analogs (Higgs and Jarman 1969; Ramis and Bover 2001; Rotunno et al. 2019). These examples suggest that by expanding the field of view past traditional domesticates, scholars may rediscover some of the diversity of domestication relationships that existed, however briefly, in the past.

Many of the features of animal domestication observed here relate to instability, movement, and heterogeneity and do not comfortably sit within traditional approaches. These perspectives, however, are centered in a process philosophy approach which has recently been applied to domestication (Bogaard et al. 2021; Dupre and Nicholson 2018). In a process oriented approach, inspired by the process theory of Alfred North Whitehead and others, species are reimagined as active processes, constantly in motion, and interconnected rather than static, bounded entities (Bogaard et al. 2021). Moreover, the emphasis on connectivity and encouragement to pivot between scales of analysis leads to an emphasis on more complex multispecies communities in contrast to the traditional focus on domesticator-domesticant dyads (Allaby et al. 2021). In this view, there is no homogenous domestic 'status' to move toward but rather a process of co-creating through multispecies interaction constantly redefining what domestic (and wild) relationships and organisms are. This perspective shifts focus to the persistence of domestication relationships and specific lineages, a feature often assumed to be self-evident, not subject to analysis, and deeply embedded within narratives of progress. For example, under traditional approaches, the restricted temporality of Botai horse management invites explanation, whereas the persistence of DOM2 horse management for more than four millennia is taken as self-evident. Similarly, once the dog lineage appears in the paleogenomic record, 'dogs' are perceived as a bounded, effective, and eternal entity, the persistence of which is expected rather than examined. Under the process approach, this persistence is not assumed but rather itself becomes a topic of exploration requiring careful genomic, anatomical, and ecological study.

A process oriented approach also raises questions about the utility of the influential domestication pathways models. Within the pathways model, animals are described as 'completing a journey to domestication' reflecting an emphasis on static categories and a final destination (Zeder 2012: 241). By emphasizing movement and heterogeneity, the process approach accommodates current genomic and archaeological evidence suggesting that domestication unfolds in complex, unruly, and unpredictable ways. In this view, domestication processes are reimagined as endless cycles of interaction or co-created 'paths' (emphasizing the journey), rather than engineered 'pathways'. It also suggests that we can begin replacing the ubiquitous unilinear, staged models of animal domestication with alternate visualizations emphasizing the unruly nature of domesticated animals and their histories (e.g., Allaby et al. 2021; Frantz et al. 2020; McHugo et al. 2019).

## Conclusions

Following anthropologist Levi-Strauss (1962), animal domestication has been 'good to think with'. Traditional approaches have made remarkable progress mapping out the broad histories of many of the world's most economically important domestic species including the dogs, cattle, and horses described here. These approaches tend to reflect a model in which domestication is a homogenous and static status reached through the work of human managers and incorporated into grand narratives of economic progress. Although traditional staged models of animal domestication acknowledge the continuum of human-animal relationships, they tend to focus on the end result of the journey to domestication and on extant domestic lineages at the expense of other possibilities.

Recent research including ancient animal genomes and new archaeological discoveries, however, has produced new results which highlight alternate perspectives on animal domestication emphasizing heterogeneity, instability, unruliness, and encourage the re-incorporation of domestication process into larger multi-species ecological contexts. These broader perspectives on domestication particularly those viewing domestication as a continuous process of becoming rather than a status, and emphasizing the leaky nature of wild and domestic categories, have the potential to significantly alter our understanding of the nature of human-animal relationships in the past and to see beyond the position of naturalism to other ways of imagining human and animal histories.

### Discussion Questions

1. What is the nature of the boundaries between domestic and wild animals? When does a dog become and dog and is no longer a wolf?
2. What problems emerge when trying to identify the earliest domestic animals in the archaeological record?
3. How do genotype, phenotype, and lived experience reflect domestication relationships in different ways?

# References

Allaby, Robin G., Chris J. Stevens, Logan Kistler, and Dorian Q. Fuller. 2021. Emerging evidence of plant domestication as a landscape-level process. *Trends in Ecology & Evolution* 31: 268–279.

Anthony, David W. 2007. *The horse, the wheel, and language: How Bronze-Age riders from the Eurasian steppes shaped the modern world*. Princeton: Princeton University Press.

Arbuckle, Benjamin S., and Theo M. Kassebaum. 2021. Management and domestication of cattle (*Bos taurus*) in Neolithic Southwest Asia. *Animal Frontiers* 11: 10–19.

Bergström, Anders, Laurent Frantz, Ryan Schmidt, Erik Ersmark, Ophelie Lebrasseur, Linus Girdland-Flink, Audrey T. Lin, Jan Storå, Karl-Göran Sjögren, and David Anthony. 2020. Origins and genetic legacy of prehistoric dogs. *Science* 370: 557–564.

Bergström, Anders, David W. G. Stanton, Ulrike H. Taron, … Pontus Skoglund. 2022. Grey wolf genomic history reveals a dual ancestry of dogs. *Nature* 607: 313–320.

Bogaard, A., R. Allaby, B. S. Arbuckle, R. Bendrey, S. Crowley, T. Cucchi,… G. Larson. 2021. Reconsidering domestication from a process archaeology perspective. *World Archaeology* 53(1): 56–77.

Bollongino, Ruth, Joachim Burger, Adam Powell, Marjan Mashkour, Jean-Denis Vigne, and Mark G. Thomas. 2012. Modern taurine cattle descended from small number of near-eastern founders. *Molecular Biology and Evolution* 29: 2101–2104.

Braidwood, Robert, and Charles A. Reed. 1957. The achievement and early consequences of food-production: a consideration of the archaeological and natural-historical evidence. *Cold Spring Harbor Symposium on Quantitative Biology* 22: 19–31.

Childe, V. G. 1936. *Man makes himself*. London: Watts and Company.

Claude, Lévi-Strauss. 1962. *Le totémisme aujourd'hui*. Paris: Presses Universitaires de France.

Darwin, Charles. 1859. *The origin of species by means of natural selection*. London: EA Weeks.

Darwin, Charles. 1868. *The variation of animals and plants under domestication*. London: John Murray.

de Mortillet, G. 1879. Su l'origine des animaux domestiques. *Bulletins de la Société D'anthropologie de Paris* 2: 232–252.

Descola, Philippe. 2013. *Beyond nature and culture*. Chicago, IL: University of Chicago Press.

Ducos, P. 1978. *Tell-Mureybet (Syrie, IXe–VIIe millénaires). Etude archeozoologique et précis d'écologie humaine*. Lyon: CNRS Editions.

Duerst, J. U. 1908. Animal remains from the excavations at Anau, and the horse of Anau and its relation to the races of dometic horses. In Raphael Pumpelly (ed.), *Explorations in Turkestan, expedition of 1904*. Washington, DC: Carnegie Institute of Washington Publication 73 volume II, 341–494.

Dupré, J., and D. Nicholson. 2018. A manifesto for a processual philosophy of biology. In D. J. Nicholson and J. Dupré (eds), *Everything flows: Towards a processual philosophy of biology*. Oxford: Oxford University Press. Doi:10.1093/oso/9780198779636.003.0001.

Evin, Allowen, Joseph Owen, Greger Larson, Mélanie Debiais-Thibaud, Thomas Cucchi, Una Strand Vidarsdottir, and Keith Dobney. 2017. A test for paedomorphism in domestic pig cranial morphology. *Biology Letters* 13: 20170321.

Frantz, Laurent A. F., Daniel G. Bradley, Greger Larson, and Ludovic Orlando. 2020. Animal domestication in the era of ancient genomics. *Nature Reviews Genetics* 21: 449–460.

Frantz, Laurent A. F., James Haile, Audrey T. Lin, Amelie Scheu, Christina Geörg, Norbert Benecke, Michelle Alexander, Anna Linderholm, Victoria E. Mullin, and Kevin G. Daly. 2019. Ancient pigs reveal a near-complete genomic turnover following their introduction to Europe. *Proceedings of the National Academy of Sciences* 116: 17231–17238.

Galton, F. 1865. The first steps towards the domestication of animals. *Transactions of the Ethnological Society of London* 3: 122–138.

Gaunitz, Charleen, Antoine Fages, Kristian Hanghøj, Anders Albrechtsen, Naveed Khan, Mikkel Schubert, Andaine Seguin-Orlando, Ivy J. Owens, Sabine Felkel, and Olivier Bignon-Lau. 2018. Ancient genomes revisit the ancestry of domestic and Przewalski's horses. *Science* 360: 111–114.

Germonpré, Mietje, Martine Van den Broeck, Martina Lázni-ková-Galetová, Mikhail V. Sablin, and Hervé Bocherens. 2021. Mothering the orphaned pup: The beginning of a domestication process in the Upper Palaeolithic. *Human Ecology* 49: 677–689.

Gibbons, A. 2021. Ancient DNA reveals long-sought homeland of modern horses. *Science* 374(6566): 384–385.

Hammer, K. 1984. Das Domestikationssyndrom. *Kulturpflanze* 32: 11–34.

Helmer, D., L. Gourichon, H. Monchot, J. Peters, and M. Sana Segui. 2005. Identifying early domestic cattle from Pre-Pottery Neolithic sites on the Euphrates using sexual dimorphism. In J-. D. Vigne, J. Peters and D. Helmer (eds), *The first steps of animal domestication*. Oxford: Oxbow, 86–95.

Higgs, E., and M. R. Jarman. 1969. The origins of agriculture: A reconsideration. *Antiquity* 43: 31–41.

Hongo, Hitomi, Jessica Pearson, Banu Oksuz, and Gulçin Ilgezdi. 2009. The process of ungulate domestication at Cayonu, southeastern Turkey: A multidisciplinary approach focusing on Bos sp. and Cervus elephus. *Anthropozoologica* 44: 63–78.

Hulme-Beaman, A., D. Orton, and T. Cucchi. 2021. The origins of the domesticate brown rat (Rattus norvegicus) and its pathways to domestication. *Animal Frontiers* 11(3): 78–86.

Ingold, Tim. 1994. From trust to domination: An alternative history of human-animal relations. In Aubrey Manning, and James Serpell (eds), *Animals and human society: Changing perspectives*. London: Routldege, 1–22.

Janssens, Luc, Liane Giemsch, Ralf Schmitz, Martin Street, Stefan Van Dongen, and Philippe Crombé. 2018. A new look at an old dog: Bonn-Oberkassel reconsidered. *Journal of Archaeological Science* 92: 126–138.

Koungoulos, Loukas. 2021. Domestication through dingo eyes: An Australian perspective on human-canid interactions leading to the earliest dogs. *Human Ecology* 49: 691–705.

Larson, Greger, and Dorian Q. Fuller. 2014. The evolution of animal domestication. *Annual Review of Ecology, Evolution, and Systematics* 45: 115–136.

Larson, Greger, Elinor K. Karlsson, Angela Perri, Matthew T. Webster, Simon Y. W. Ho, Joris Peters, Peter W. Stahl, Philip J. Piper, Frode Lingaas, and Merete Fredholm. 2012. Rethinking dog domestication by integrating genetics, archeology, and biogeography. *Proceedings of the National Academy of Sciences* 109: 8878–8883.

Librado, Pablo, Naveed Khan, Antoine Fages, Mariya A Kusliy, Tomasz Suchan, Laure Tonasso-Calvière, Stéphanie Schiavinato, Duha Alioglu, Aurore Fromentier, and Aude Perdereau. 2021. The origins and spread of domestic horses from the Western Eurasian steppes. *Nature* 598: 634–640.

Lord, Kathryn A., Greger Larson, Raymond P. Coppinger, and Elinor K. Karlsson. 2020. The history of farm foxes undermines the animal domestication syndrome. *Trends in Ecology & Evolution* 35: 125–136.

Ludwig, A., M. Pruvost, M. Reissmann, N. Benecke, G. A. Brockmann, P. Castaños, ... M. Hofreiter. 2009. Coat color variation at the beginning of horse domestication. *Science* 324(5926): 485–485.

Maher, L. A., J. T. Stock, S. Finney, J. J. Heywood, P. T. Miracle, and E. B. Banning. 2011. A unique human-fox burial from a pre-natufian cemetery in the Levant (Jordan). *PLoS One* 6(1): e15815.

Mary, Nicolas, Nathalie Iannuccelli, Geoffrey Petit, Nathalie Bonnet, Alain Pinton, Harmonie Barasc, Amélie Faure, Anne Calgaro, Vladimir Grosbois, and Bertrand Servin. 2022. Genome-wide analysis of hybridization in wild boar populations reveals adaptive introgression from domestic pig. *Evolutionary Applications* 15: 1115–1128.

McHugo, Gillian P., Michael J. Dover, and David E. MacHugh. 2019. Unlocking the origins and biology of domestic animals using ancient DNA and paleogenomics. *BMC Biology* 17: 1–20.

Meadow, Richard H. 1989. Osteological evidence for the process of animal domestication. In Juliet Clutton-Brock (ed.), *The walking larder: Patterns of domestication, pastoralism, and predation*. London: Unwin Hyman, 80–96.

Morgan, Lewis Henry. 1877. *Ancient society; or, researches in the lines of human progress from savagery, through barbarism to civilization*. London: H. Holt.

Mucke, Johann Richard. 1898. *Urgeschichte des Ackerbaues und der Viehzucht*. Greifswald: J. Abel.

Munro, Natalie D., Guy Bar-Oz, Jacqueline S. Meier, Lidar Sapir-Hen, Mary C. Stiner, and Reuven Yeshurun. 2018. The emergence of animal management in the Southern Levant. *Scientific Reports* 8: 9279.

Olsen, Sandra L. 2006. Early horse domestication on the Eurasian steppe. In *Documenting Domestication: New Genetic and Archaeological Paradigms*. Berkeley: University of California Press, 245–269.

Outram, Alan K., Natalie A. Stear, Robin Bendrey, Sandra Olsen, Alexei Kasparov, Victor Zaibert, Nick Thorpe, and Richard P. Evershed. 2009. The earliest horse harnessing and milking yes have it. *Science* 323: 1332–1335.

Peake, Harold J., and Herbert John Fleure. 1927. *Peasants and potters*. London: Oxford University Press.

Perri, Angela R., Tatiana R. Feuerborn, Laurent A. F. Frantz, Greger Larson, Ripan S. Malhi, David J. Meltzer, and Kelsey E. Witt 2021. Dog domestication and the dual dispersal of people and dogs into the Americas. *Proceedings of the National Academy of Sciences* 118: e2010083118.

Price, Max, Jacqueline Meier, and Benjamin Arbuckle. 2021. Canine economies of the ancient Near East and Eastern Mediterranean. *Journal of Field Archaeology* 46: 81–92.

Purugganan, Michael D. 2022. What is domestication? *Trends in Ecology & Evolution* 37: 663–671.

Ramis, D., and P. Bover. 2001. A review of the evidence for domestication of Myotragus balearicus Bate 1909 (Artiodactyla, Caprinae) in the Balearic Islands. *Journal of Archaeological Science* 28(3): 265–282.

Rindos, David. 1983. *The origins of agriculture: an evolutionary perspective*. New York: Academic Press.

Rotunno, R., A. M. Mercuri, A. Florenzano, A. Zerboni, and S. di Lernia. 2019. Coprolites from rock shelters: Hunter-gatherers "herding" Barbary sheep in the Early Holocene Sahara. *Journal of African Archaeology* 17(1): 76–94.

Rütimeyer, L. 1861. *Die fauna der Pfahlbauten in der Schweiz: Untersuchungen über die Geschichte der wilden und der Haus-Säugethiere von Mittel-Europa*. Basel: Bahnmaier Buchhandlung.

Serpell, James A. 2021. Commensalism or cross-species adoption? A critical review of theories of wolf domestication. *Frontiers in Veterinary Science* 8: 662370.

Shipman, Pat. 2010. The animal connection and human evolution. *Current Anthropology* 51: 519–538.

Smith, Alexia, Amy Oechsner, Peter Rowley-Conwy, and Andrew M. T. Moore. 2022. Epipalaeolithic animal tending to Neolithic herding at Abu Hureyra, Syria (12,800–7,800 calBP): Deciphering dung spherulites. *Plos One* 17: e0272947.

Stépanoff, Charles, Charlotte Marchina, Camille Fossier, and Nicolas Bureau. 2017. Animal autonomy and intermittent coexistences: North Asian modes of herding. *Current Anthropology* 58: 57–81.

Tchernov, Eitan, and François F. Valla. 1997. Two new dogs, and other Natufian dogs, from the southern Levant. *Journal of Archaeological Science* 24: 65–95.

Tsing, Anna Lowenhaupt. 2018. Provocation. nine provocations for the study of domestication. In *Domestication Gone Wild*. Durham: Duke University Press, 231–251.

Turgot, A. R. J. 1793. *Reflections on the Formation and Distribution of Wealth*. London: E. Spragg.

Uerpmann, H. P. 1995. Domestication of the horse. When, where and why? In *colloque d'histoire des connaissance zoologiques (Belgium)*. Liege: Universite de Liege. Inst. De Zoologie (6), 15–29.

Verdugo, Marta Pereira, Victoria E. Mullin, Amelie Scheu, Valeria Mattiangeli, Kevin G. Daly, Pierpaolo Maisano Delser, Andrew J. Hare, Joachim Burger, Matthew J. Collins, and Ron Kehati. 2019. Ancient cattle genomics, origins, and rapid turnover in the Fertile Crescent. *Science* 365: 173–176.

Vigne, Jean-Denis. 2011. The origins of animal domestication and husbandry: A major change in the history of humanity and the biosphere. *Comptes Rendus Biologies* 334: 171–181.

Warmuth, V., A. Eriksson, M. A. Bower, G. Barker, E. Barrett, B. K. Hanks,… Manica, A. 2012. Reconstructing the origin and spread of horse domestication in the Eurasian steppe. *Proceedings of the National Academy of Sciences* 109(21): 8202–8206.

Zeder, M. A. 2012. The domestication of animals. *Journal of Anthropological Research* 68(2): 161–190.

Zeder, Melinda, and Brian Hesse. 2000. The initial domestication of goats (Capra hircus) in the Zagros mountains 10,000 years ago. *Science* 287: 2254–2257.

# 14
# WHY DO WE LOVE THE PETS WE HAVE?
## The Role of Animals and Family

*Alan M. Beck, Aubrey H. Fine, Stanley Coren, Erin King, Steve Feldman, and Lindsey Braun*

### Introduction

Evidence in the archeological record suggests that animals have been an integral part of human culture and society for thousands of years (Benecke, 1987; Morey, 1992, 2006). Through food production, service provision, and companionship, nonhuman animals play a multifaceted role in today's society. When looking specifically at animals within the family system, there are many different models and forms of human-animal interaction (HAI) around the world. This diversity prompts us to focus on the complexity of the human-animal bond and consider how we define the word "*pet*."

Many scholars have concluded that companion animals provide stability in family homes. Triebenbacher (2000) suggested that the more emotionally involved a pet owner is with their companion animal, the more likely that the individual will develop a more significant social bond with that animal over time. Furthermore, (Nagasawa et al., 2009) suggested that early meaningful interactions with animals may promote healthy companionship with others including humans later in life.

Stability seems to be one of the major reasons why people have pets in their lives. Bennett (2000) emphasizes this point and identifies how pets within a family are critical during significant life periods such as widowhood, during a divorce, or becoming an empty nester. Fine (2014) suggests that pets often offer a sense of continuity within the family life cycle. They can serve numerous functions during momentous life events such as the passing of a family member. Within this chapter, we will discuss some of the reasons why people have become so connected to their animals and the benefits and challenges that they experience.

### Pet Ownership

Many definitions of pet ownership stem from the idea that pets are animals kept for no apparent utilitarian purpose, and/or provide long-lasting companionship or pleasure (Gray & Young, 2011; Herzog, 2014). However, determining what qualifies an animal as a "pet" within the family context becomes challenging when looking at the variety of culturally specific human-animal relationships around the world. For example, in countries like the United States and United Kingdom, dogs and cats are often kept in the home and viewed as "out of place" if found roaming the streets (Philo& Wilber, 2000). However, in other cultures, free roaming animals are common and frequently play the important role of "companion animal" in people's lives (Miele & Bock, 2007). To add to this complexity, within a culture, differences surrounding animals are extremely common, fluid, and often change rapidly. It is also important to acknowledge internal and systemic biases that impact perceptions of the human-animal bond and pet ownership. Much of the research in the field of HAI stems from Eurocentric ideas of pet ownership that often neglect the variability of human-animal relationships found around the world. Therefore, it is important to recognize that human-animal relationships are diverse in nature as are the families and communities that they exist within.

Additionally, animals often play multiple roles in a family at one time depending on the perspective of the relationship, which further clouds the meaning of the term "pet." For example, an owned horse could be considered a working animal but also considered to be a pet depending on the individual family member's relationship with the horse. Similarly, an animal raised for slaughter may be seen as a pet while alive, and family members may have a bond with the animal that is typically associated with companion animal ownership (Rusk et al., 2003). It is important to view multiple kinds of human-animal relationships within the purview of the family structure. To that end, we suggest that the term "pet" is not mutually exclusive with other categorizations of animals (working, livestock, etc.,), and that more diverse research is needed to understand the nuance of the human-animal bond and how pets relate to individual, family, and community health and wellness outcomes.

## Demographic Info About Pets and Families – What Do the Statistics Tell Us?

In the United States, pet ownership is not only increasing, but the commitment to our pets as members of the family has also grown stronger. Families spend billions of dollars annually on pet supplies, which include toys, medication, food, and actual veterinary care (APPA, 2021). In essence, families are committed to and spend a great deal of their discretionary funds on the pets that live with them. According to the 2021–2022 American Pet Products Association (APPA) National Pet Owners Survey, 70% of U.S. households own a pet, which equates to 90.5 million homes.

By pet species, APPA says 9.9 million households have birds, 45.3 million have cats, 69.0 million have dogs, 3.5 million have horses, 11.8 million have freshwater fish, 2.9 million have saltwater fish, 5.7 million have reptiles, and 6.2 million have other small animals. Generationally, Millennials (aged 26–40) are the largest cohort of pet owners at 32%, followed by Boomers (aged 56–74) at 27%, Gen X (aged 41–55) at 24%, Gen Z (aged 18–25) at 14%, and Builders (75+) at 3%. The most common combination of pets to own is a dog and a cat, noted by 30% of pet owners. The average number of cats per cat owning household is 1.74, and the average number of dogs per dog owning households is 1.56 dogs (APPA, 2021–2022).

The pandemic has played a role in accelerating new pet ownership, particularly among the younger generations. In fact, 26% of Gen Z pet owners reported getting a new pet during the pandemic. Despite reported economic challenges and uncertainty, pet spending has continued its year-over-year upward trajectory. Most recently, the 2021–2022 APPA survey found that 35% of pet owners spent more money on their pet and pet supplies over the past 12 months than in the preceding year.

Considering that a growing majority of households have pets, has the strength of the human-animal bond also grown? The question "Do you consider your pet to be a member of your family?" used to be the must-ask question to gauge the strength of the human-animal bond, but when the answer is consistently "yes" for well more than 90% of all pet owners in survey after survey, more sensitive measures of the bond are necessary. In a recent, large survey of 18,000 dog and cat owners across nine countries on four continents (Zoetis-HABRI, 2022), 94% said their pet was "an important member of the family" and they were willing to act upon that strong feeling with 76% saying they would "make a major life change to accommodate my pet." Few differences across dog and cat owners in the nine countries surveyed were noted regarding motivations for getting a pet, how pet owners think about their pets, and regarding agreement to statements about their pet that speak to the strength of the human-animal bond. The majority (68%) of pet owners across the nine countries surveyed reported that companionship was the main motivation for getting their dog or cat. Ninety-one percent of pet owners reported spending time playing or interacting with their pet every day, and 91% of pet owners agreed that "nothing would ever convince me to give up my pet." The survey also found that owning a pet during childhood was associated with current pet ownership, with 79% of cat owners and 77% of dog owners reporting having pets as children. No correlation was found between how long someone had had their pet, and the strength of the human-animal bond, suggesting that the human-animal bond can be instant, much like that of a parent-child relationship.

Further information from the international survey indicates that dog and cat owners across countries and cultures experience similar health benefits resulting from owning a pet. Eighty-seven percent of pet owners said that they had personally experienced positive health benefits from having a pet. Happiness was most frequently articulated as specific benefits from pet ownership. More than just a feeling, 58% of pet owners globally had heard of scientific research demonstrating that pets improve mental or physical health. As pet owners learn more about the HAI science, they indicate an increased likelihood to maintain their pet's veterinary health; spend more time with their pets; and spend more money on their pet's care overall.

Looking ahead, the continued demographic growth of pet ownership, the strong commitment of the next generation of pet owners, the growing, universal nature of the human-animal bond across countries and cultures, and increased knowledge of human-animal bond science will drive a strengthening of the human-animal bond as a powerful force with the potential to positively impact people, pets, and many aspects of our society.

## Theories of the Human-Animal Bond

When explaining why we all become so connected with nonhuman animals, it is important to turn to the existing theories that explain our bond with them. Several theories have been used to conceptualize various aspects of the human-animal bond in different environmental contexts. Many would argue that each of the theories on its own may not capture the complete essence of these relationships, but collectively they provide a framework for interpreting these complex relationships in various settings. These theories will also be expounded upon in numerous other chapters in this book.

### *The Biophilia Hypothesis*

People were always aware of the natural world around them. This awareness was needed for daily survival; people had to know what they can eat or not, which animals they could eat or might eat them or simply to navigate the natural world. We can assume there was a natural selection in favor of people with such awareness. This attraction has been named "the

Biophilia Hypothesis" which is defined as "the innate tendency to focus on life and lifelike processes" (Wilson, 1984 p. 1). Not only are people attracted to nature, but it appears that that focus has positive health effects (Ulrich, 1993). Surgical patients who could view a garden had a shorter length of stay in the hospital and required fewer analgesics than those who had a view of a brick wall (Ulrich, 1984). People interacting with a dog have a drop in blood pressure compared to interacting with a person, or even just resting (Friedmann et al.,1983). Conscious neurosurgical patients observed dozens of photographs of animals, famous people, or recognizable places while their amygdala was being monitored for activity. The amygdala's primary role is in the processing of memory and emotional reactions. Photographs of animals triggered greater activation of the amygdala than views of famous people, landmarks, or common objects, indicating a category-specific recognition that animals are important to people (Mormann et al., 2011).

Kellert (1993) asserted that humans are dependent on the environment for survival. He believes that humans tend to bond and relate with other forms of natural life because of our propensity to fulfill this need and because of our reliance on the environment for survival (Kellert, 1993). The hypothesis has been used to explain why we feel emotional connections and compassion toward docile animals or fear a threatening animal. Both Kellert and Wilson identified different values that individuals seek to fulfill in living with nature. These values will be elaborated upon in later chapters.

## *Attachment Theory*

John Bowlby was the first to bring attention to humans' evolutionary reaction to being caregivers and protectors (Bowlby (1969). Humans often explain their relationships with their companion animals in ways that parallel relationships with children, especially because our pets depend on us for their care and protection (Barba, 1995). Biologically, the hormone oxytocin that plays a significant role in parent/infant bonding also impacts the attachment between owners and pets. While studies about the role of oxytocin in human-dog relationships have mixed findings, research by Odendaal and Meintjes (2003) suggests that when humans and dogs interact, oxytocin levels increases in both species. Furthermore, Nagasawa in (2009) found that dog owners showed higher levels of oxytocin when gazing into the eyes of their pet dogs. The same finding was also identified with humans and their children. However, Marshall-Pescini et al. (2019) did not find a relationship between oxytocin and human-dog interaction and suggested many methodological challenges associated with oxytocin research. Beyond oxytocin, Carlisle (2014) has noted an increase in social behavior among children with pets and strong levels of attachment to their animals. Finally, Fine (2014) pointed out that in both boys and girls, taking care of their pets allows them to act as a nurturer in a socially acceptable manner.

## *Animals Acting as Social Supports*

There has been a great deal of commentary written explaining the value of animals acting as social supports to humans. In the early literature, Cassel (1976) pointed out that the most critical dimension of social support is the meaningfulness of the social contact to either party in the relationship. He goes on to explain that social contact and the frequency of the contact are the prime ingredients to social support acting as a foundation to any relationship. Cobb (1976), on the other hand, focuses more on the meaningfulness of the relationship as the key aspect of social support. He believes that the dimension of feeling cared for and loved by others has a strong impact on feeling support. Cobb identified four key aspects of social support: emotional support, informational support, providing esteem support and finally practical support. McNicholas and Collis (2006) highlight that the belief in feeling supported by a companion animal is extremely crucial in most relationships and can provide a sense of protection from anxiety and other adverse behaviors.

Wells (2019) suggested that being around animals can decrease social isolation and enhance opportunities for social interaction. McNicholas and Collis (2006) have described numerous incidents where animals have acted as support during times of adversity. For example, when people are grieving, they seem to be able to express their intense emotions more easily in the presence of an animal rather than with another human (McNicholas & Collis 2006). On the other hand, McNicholas and Collis (2001), Fine (2014) and Melson (2003) reported that children identified that their pets served as critical outlets for social support in their homes in emotionally difficult times, even over their human family members. They report feeling safer in their homes with the presence of their animal companions.

Strand (2004) adds to this position, pointing out that children seem to use their pet interaction as a "buffer" to help self-calm during a time of adversity. The literature is full of examples of how pets provide an opportunity to express emotions that often cannot be expressed easily, especially with humans. It seems that some people may feel uncomfortable talking about specific fears and difficulties, but they become more comfortable talking in the privacy of their homes surrounded by their pets. Fine (2014) adds that children and single-parent households often felt safer and more secure knowing that their family dog was there with them and would alert them if an intruder entered. In an additional study, McNicholas and Collis (2001) interviewed 22 elementary school children who discussed how they rated the social support from various family members including their pets. The findings highlighted that children chose their pets over family members in difficult situations. Similar findings were reported by Melson (2003) who found that 79% of German fourth graders selected their pets for support when they were feeling sad.

This point of view has been documented during the global lockdown due to COVID-19. Many of the studies emphasized that meaningful companionship with animals supported an increase in resilience in combating isolation. Several studies noted that the animals served as a critical source of emotional support during COVID-19 and provided an outlet for relaxation and a source of normality and routine (Olivia &Johnston, 2020; Ratschen et al., 2020). Young et al. (2020) in Australia identified the significance of the physical contact that our pets provided when being isolated from other humans. People reported that being able to pet and snuggle with their animals dramatically decreased their stress.

## *The Biopsychosocial Model*

Engel (1977, 1980) was the first to develop and discuss the Biopsychosocial Model. The Biopsychosocial Model highlights the reciprocal influences of each biological, psychological, and social factor as they relate to human well-being. Gee et al. (2021) as well as Chapter 5 in this volume discuss how the model can be applied to explain how our relationships between various species of animals and nature may impact each of the three domains of human and animal health and well-being. Applying the Biopsychosocial Model can be helpful in explaining some of the outcomes that are witnessed within HAIs. For example, regarding the psycho-social domains, Krause-Parello (2012) has identified that having a pet dog may serve as a tremendous resource for older women suffering from depression and a sense of loneliness. On the other hand, biologically, there is a wealth of research identifying how interactions with animals can have an impact on heart rate, blood pressure, as well as our neurotransmitters including oxytocin, dopamine, prolactin, and beta endorphins (Powell et al., 2019). Esposito et al. (2011) suggested that the biopsychosocial model can be used to provide a framework for deeper scientific investigation of the impact of HAIs. For example, Esposito et al. (2011) discuss the importance of examining the neuroendocrine responses in both animals and humans and how they are impacted during HAIs.

## *The Cuteness Effect*

Obvious to many is that people are attracted to most of the animals enjoyed as pets because they are attractive, indeed cute; they often have conspicuously large eyes and somewhat small noses—juvenile features. This is not a coincidence—there is a natural selection to find those traits endearing. Lorenz (1943) noticed how infant mammals and birds all had juvenile features when they were born. He postulated that the configuration was genetically determined to encourage caring. Domestic animals retain their juvenile features (neoteny) as part of the domestication process and are unconsciously accepted as being cute and in need of care. Many believe humans are self-domesticated and return our juvenile features (Beck, 2014; Benítez-Burraco, 2021). Chimpanzees and people have many similarities when very young, juvenile facial features, relatively hairless bodies and can stand, while chimpanzees develop mature features, humans just get bigger. "These and many other features define the anatomical relationship between ape and man as the latter's *neoteny*" (Bednarik, 2012, p. 22). Not surprisingly, with the retention of juvenile physical features comes the retention of behavioral features, like being more social and playful. Humans are the only great ape that plays throughout their lives.

## *The IKEA Effect*

We suggest that there is yet another reason for our love of domestic animals, especially dogs. The widely explored cognitive bias known as "The IKEA Effect." Basically, people put greater value on things they constructed themselves compared to comparable products they did not contribute to the construction (Norton et al., 2012). It is named for the IKEA furniture that require some personal assembly before use. There is reward for our effort, competence, and simple ownership (Marsh et al., 2018).

It is a robust phenomenon that appears in people from age five and beyond (Marsh et al., 2018). While this is commonly thought of as a marketing phenomenon, we suggest it may be just another reason for our love of dogs. Dogs were developed by selecting the calmer wolves and eventually consciously breeding them together to develop even tamer wolves. By continually selecting for, at first, calm behavior and perhaps size, but eventually for shape and specific behaviors, we developed breeds. Like consumer products, we created not only dogs, but dog breeds and they are recognized as human achievements. We even have dog shows and awards for breeding achievements. Cats, horses, rabbits, and some livestock are also more valued because people are thought to be part of creating the final product—The IKEA effect.

## **Why we Love the Pets We Love: An Alternate Explanation**

We have been considering a number of reasons for our fondness for certain types of animals. However, watching a baby or a toddler encounter a dog or a cat for the first time gives us a new set of clues. In the vast majority of cases, the response of the child to the animal is not fear, but rather a desire to approach and touch. Since there is not yet much of an opportunity for learning to have encroached, one is left wondering if our affiliative feelings for certain animal species might not be genetically wired. Let us use dogs as our exemplar, since more data and speculation has been generated about the canine-human bond than for other animals.

It has been well established that genetics plays a major role in the friendliness of dogs toward people. The Russian geneticist, Dmitry K. Belyaev, demonstrated that by selective breeding for friendliness and approachability, one could domesticate a naturally hostile and aggressive breed of canids (silver foxes) and turn them into animals which approached, and sought social contact, with humans (Trut, 1999). In effect, he had turned foxes into friendly pet "dogs."

According to some research, the friendliness of our domestic dogs borders on the realm of being pathological. In human beings, there is a genetic abnormality known as Williams–Beuren syndrome. It is quite rare, affecting one out of every 10,000–20,000 people. Individuals with this condition seem to lack any kind of social inhibition. They tend to seek the company of others quite vigorously, and are usually described as being kind-spirited, empathetic, caring, and forgiving. They don't seem to show the normal fear of strangers and tend to run up to and hug completely unfamiliar people. The technical term for the behaviors they are showing is *hypersociability*. It appears to be due to disruptions on the gene that is associated with a protein called GTF21. Von Holdt, Shuldiner, Koch et al. (2017) were able to show that this rare genetic disruption in humans is common in dogs and absent in wolves, which they interpreted to mean that dogs are genetically tuned to be affectionate to people.

Could people be genetically programmed to bond with dogs as well? Several researchers have speculated that dogs and humans are an example of co-evolution (Newby, 1997; Paxton, 2011). Co-evolution occurs when two species reciprocally affect each other's evolution through the process of natural selection. The natural selection involved in this case is demonstrated by the fact that dogs seem to have played a role in why the Cro-Magnon branch of early primates managed to win out over the more powerful Neanderthal primates (Coren, 2008). Early dogs served as an important hunting aid which definitely improved the survival of the Cro-Magnons, but not the Neanderthals who never formed a cooperative relationship with canines (Shipman, 2014). Dogs also served an important warning function in that era where different groups of hominids were known to ambush and kill one another (Coren, 2008).

Evolution works by altering the genetic makeup of animals, thus allowing changes to be passed on to offspring. Evolution also uses an interesting trick. At a behavioral level, it makes anything that will improve survival of the species pleasant for individuals, such as eating or sex. It is not hard to then extrapolate and suggest that since cooperation with canines was serving a survival function, then perhaps the evolutionary process would have encoded our affection for dogs into our genetic makeup. Like most genetic processes, there should be some variability; however, the overall trend would be toward affection and bonding with dogs.

A team of Swedish researchers (Fall et al., 2019) had some incredibly useful tools which allowed them to study this question. First, they had access to the Swedish Twin Registry, which was founded in the late 1950s and contains information on all twins born in Sweden. Secondly, information on dog ownership has been available since January 1, 2001, when two dog registries were established, one at the Swedish Board of Agriculture and the other at the Swedish Kennel Club. There is a legal requirement that all dogs in Sweden must be registered.

The reason why the twin registry was so important is because the most powerful way to answer the question, "Is this particular factor influenced by genetics?" is to use twins.

There are two types of twins. Monozygotic (MZ) or identical twins arise when a single egg is fertilized by a single sperm cell. Usually within the first or second cell division, it splits apart to form two separate individuals which have an identical genetic makeup.

Contrast these identical twins to dizygotic (DZ) or fraternal twins. Fraternal twins come about when two egg cells are fertilized by two different sperm cells and develop in the same womb. In this case the zygotes only share the same amount of genetic material that any pair of brothers or sisters might.

Regardless of whether we are dealing with identical or fraternal twins, both kinds of twins have very similar personal histories since they grow up in the same family, share the same environment, culture, and parenting influences. So, if we compare the similarities in the behaviors of the two types of twins, and we find greater similarities between the identical twins (with their matching genetic makeup), than between the fraternal twins (who only share about half of their genes), this would confirm that we are looking at something which has a genetic component.

These Swedish researchers looked at 35,035 pairs of twins and determined whether they were identical (MZ) or fraternal (DZ) and then determined whether each of the twins had gone on to become dog owners. Obviously, those individuals who become dog owners are more likely to have greater affection for canines. If there is a genetic factor, then the "concordance rate" should be higher for the monozygotic twins. Concordance would be shown when both twins owned dogs, or both twins did not own dogs while a lack of concordance would be shown by the fact that one twin owned a dog, and the other did not. The researchers conclude that "MZ twins had higher concordance rates and tetrachoric correlations (0.58 for females and 0.52 for males) than DZ twins (0.35 for females and 0.30 for males), in line with presence of genetic effects." The size of the inherited component influencing whether we are impelled to own a dog or not, is quite large, 57% for females and 51% for males. In other words, approximately half of the psychological pressure that we feel to own or to not own a dog could be influenced by our DNA with the other half influenced by environmental factors such as our personal histories or culture.

It may appear to be a bit strange and perhaps unbelievable, to suggest that our attitudes toward specific animal species could have a genetic component, however there are other examples. Take the fact that among humans there is an almost universal negative and fearful response to spiders and snakes (Landova et al., 2021; Murray & Foote, 1979). This is clearly genetically determined (Kendler et al., 2001; Landova et al., 2021; Loken et al., 2014). While this kind of twin study can't tell us exactly which genes are involved, they do suggest that genetics and environment could play an important role in determining our predisposition to love and bond with dogs, and similar effects could be working to shape our relationship to other companion animals.

## Global Perspectives of Pet Keeping

Pet-keeping has become increasingly popular around the world (Serpell & Paul, 2011), and is common in many human cultures in some form (Herzog 2014; Shipman, 2015). Globally, the range of animal species kept by humans is vast; however, dogs, cats, and birds are among the most common (Gray & Young, 2011). While there is not an extensive amount of literature on global pet keeping, Gray and Young (2011) performed a systematic review and found 60 societies with research on animals kept for "pet" like purposes. Dogs were present in 53 societies and were reportedly kept for multiple working roles (hunting, guarding, herding, etc.), whereas birds were most commonly reported as "playthings" (Gray & Young, 2011). Other commonly kept species included horses and other hoofed mammals (cattle, sheep, water buffalo), rodents (guinea pigs, prairie dogs, squirrels), primates (monkeys and gibbons), and pigs.

Early anthropologic research suggests that a variety of animal species have functioned as symbols with their own cultural narrative and identity. For example, the Bororo people of central Brazil kept birds, such as Macaws, because they believed the souls of their ancestors to inhabit the bird and it was common for women to keep many birds if they had lost children (Crocker, 1977). Bird keeping is still an extremely common practice in Brazil with strong cultural importance, as well as economic value (Alves et al., 2010; Oliveira et al., 2018). Pet birds are common in both rural and urban areas, with caged birds present not only in family homes, but also public places like grocery stores, bars, and more (Alves et al., 2010). This example emphasizes how human-pet relationships are steeped in cultural identity and more research is needed to accurately describe and measure these complex relationships.

As previously mentioned, what defines an animal as a "pet" is not universally defined, and moreover, even within a culture, social norms about pets are fluid and often change rapidly. In China, pet ownership was banned until 1992 (Headey et al., 2008), yet since then it has grown rapidly and one study found that 62.2% of participants regarded their pets as family member (Wu et al., 2018). Additionally, free -roaming animals and definitions of animal "ownership" also differ cross-culturally and change quickly. For example, some cultures often have large populations of free-ranging dogs in urban and/or rural areas that provide both human benefits (companionship, protection, vermin-control) as well as challenges (zoonotic disease, harassment, etc.,). The World Health Organization (1990) offered four classifications for dogs that are useful in thinking about different types of ownership around the world:

- *Restricted dogs*, fully supervised and/or restricted and entirely dependent on man for food and other resources;
- *Family dogs*, semi-restricted (often able to roam freely for part of the time) and fully dependent on one or more families for food and shelter (Beck, 2002);
- *Neighborhood dogs*, entirely free to roam (sometimes semi-restricted) and only semi-dependent on one or more families for food/shelter;
- *Feral dogs*, wholly unrestricted and not dependent on food deliberately given by any person or group (Reese, 2005).

Pet relationships with varying levels of restriction/supervision are found in many different cultures, and often differ by species. For example, in the United States, "owned" pet dogs are usually thought of as restricted/fully supervised animals, yet the same culture has different attitudes toward cats, which are commonly semi-restricted and live both in and outside of the home. A study in rural and central Chile found that 85.9% of households had dogs, of which 37% of those households had semi-restricted dogs, and 41% of dogs were free-ranging dogs (Astorga et al., 2022). In the Bahamas, 73% of households that own dogs keep them outside (Fielding &Plumridge, 2005) and across the globe in Kenya, about 69% of dogs are never restricted (Kitala et al., 2001). This illustrates how pet-keeping practices may vary by culture and more research into the human-animal bond in diverse settings is needed to understand the full nature of these relationships.

Within cultures with large free-ranging dog populations, pet ownership practices can be complex and diverse. For example, in Thailand, the well-established practice of caring for free-roaming dogs and cats as community members, coincides with new and increasing numbers of people keeping cats and dogs inside the home (Savvides, 2013). A strong "classist" element in dog-keeping practices can be seen in Bangkok and other wealthier parts of the country, in which dogs are used to "create social boundaries" and often accessorized (Kasempimolporn, 2008; McCarthy, 2017). Historically, people of all socioeconomic status would let their dog roam free and scavenge for at least part of the day. However, McCarthy (2017) suggests the recent and rapid change in Thailand from community-based care of animals to individualized ownership in the home has not included all types of cats and dogs – purebred cats and dogs with pedigree have been "invited into many homes as family members, but animals that have lived on the street have remain there" (McCarthy, 2017 p. 98). Nuanced human-pet relationships in Thailand showcase the multi-faceted nature of pet ownership around the globe and why it is important to expand research questions in a culturally specific and comprehensive way.

Thailand's wealthy class adopting pets into child-like relationships is not an uncommon practice in countries with growing consumerism of pet goods. Some researchers have hypothesized that cultural shifts in pet ownership styles are prompted by increased wealth/status, growing consumerism, declining fertility, and/or second demographic transitions (Marinelli et al., 2007; Lesthaeghe, 2014; Volsche et al., 2021). India is a growing market for companion animal-related spending and has varying types of relationships with pets. A recent study of 190 households in Mangalore city, India, found that 56.8% of houses owned pets (62% of those households owning dogs, 14.8% cats, and 23.3% both dogs and cats) and pet-attachment scores were significantly higher among unmarried/divorced/widow owners than married owners (Joseph et al., 2019). Another study compared Indian pet owners to United States owners and found that Indian pet owners were more attached to their pets (LAPS scores) and engaged in more affiliative behaviors such as cuddling and considering their pet preferences (Volsche et al., 2021). However, while these studies help us understand pet ownership in India, most all research on domestic dogs in India has focused on free-ranging dog behavior and disease transmission (Home et al., 2018; Menezes, 2008, and Sarkar et al., 2019) and few studies explore changing and dynamic relationships with animals. Volsche et al. (2021) point out that there is a lack of attention given to varying sociocultural norms, and suggest it is due to "outdated [and] reductionist understandings" of India's pet-relationships. A call for more nuanced approaches to studying cultural attitudes surrounding the human-animal bond is echoed by the sentiment for more diverse perspectives more broadly in the field of HAI.

## Concluding Remarks

Throughout this chapter, we have identified many of the benefits, challenges, and various reasons of why people live with animals as part of their families. We have provided this information from a global perspective describing these unique relationships. Pets are embedded in a complex global web of culturally specific practices. While there is value in past studies, rapidly changing relationships with pets and animals in families call for new studies that have both qualitative and quantitative perspectives. To truly understand the impact of the human-animal bond, we need to understand what aspects of the relationship might be unique to a given individual, community, or culture. Similarly, when discussing social norms of pet ownership, it is important to not overgeneralize and always work to recognize how one's own bias may impact views on cross-cultural animal interactions. Critical reflection of one's own viewpoints and attitudes toward HAIs, especially in how we define "responsible" pet ownership, is crucial to address inequities in research. It is also important to remember that even within cultures, there are sub-cultures and individual differences in pet-keeping practices and values. As the field of HAI grows, understanding differing pet-relationships around the world will be integral to an in-depth understanding of the human-animal bond. It is Incumbent for future scientist and practitioners to continue to unravel the mysteries of the importance of animals in the lives of families. For now, we continue to be the benevolent benefactors of these relationships, appreciating the significance of animals in our lives.

## Questions

1. How can the various theories of HAIs explain why families seem so committed to having animals in their lives?
2. It is important for all people to appreciate the unique ways people interact cross culturally with animals. What are some of the similarities and differences?
3. How does the IKEA Effect explain our bond and connection with animals?

## References

Alves, R. R. D. N., Nogueira, E. E., Araujo, H. F., & Brooks, S. E. (2010). Bird-keeping in the Caatinga, NE Brazil. *Human Ecology, 38*(1), 147–156. https://doi.org/10.1007/s10745-009-9295-5

American Pet Products Association. (2021). *2021–2022 APPA national pet owners survey*. Stamford, CT: American Pet Products Association.

Association, J. P. F. Ownership of Dogs and Cats Survey Results. (2018). https://petfood.or.jp/topics/img/181225.pdf (Accessed on 3 March 2022).

Astorga, F., Poo-Muñoz, D. A., Organ, J., & Medina-Vogel, G. (2022). Why let the dogs out? Exploring variables associated with dog confinement and general characteristics of the free-ranging owned-dog population in a peri-urban area. *Journal of Applied Animal Welfare Science, 25*(4), 311–325.

Barba, B. E. (1995). A critical review of research on the human/companion animal relationship: 1988 to 1993. *Anthrozoös, 8*(1), 9–19. https://doi.org/10.2752/089279395787156509

Beck, A. M. (2002). *The ecology of stray dogs: A study of free-ranging urban animals*. West Lafayette, IN: Purdue University Press.

Beck, A. M. (2014). The biology of the human–animal bond. *Animal Frontiers, 4*(3), 32–36.

Bednarik, R. G. (2012). The origins of human modernity. *Humanities, 1*, 1–53.

Benecke, N. (1987). Studies on early dog remains from Northern Europe. *Journal of Archaeological Science, 14*(1), 31–49.

Benítez-Burraco, A., Ferretti, F., & Progovac, L. (2021). Human self-domestication and the evolution of pragmatics. *Cognitive Science, 45*, e12987. https://doi.org/10.1111/cogs.12987

Bennett, J. R. B. (2000). *Relational experiences between children and pets during parental divorce* (Doctoral dissertation, John F. Kennedy University).

Bowlby, J. (1969). *Attachment and loss* (Vol. 1). London: Hogarth Press.

Carlisle, G. K. (2014). The social skills and attachment to dogs of children with autism spectrum disorder. *Journal of Autism and Developmental Disorders, 45*(5), 1137–1145. https://doi.org/10.1007/s10803-014-2267-7

Cassel, J. (1976). The contribution of the social environment to host resistance. *American Journal of Epidemiology, 104*(2), 107–123. https://doi.org/10.1093/oxfordjournals.aje.a112281

Cobb, S. (1976). Social support as a moderator of life stress. *Psychosomatic Medicine, 38*(5), 300–314. https://doi.org/10.1097/00006842-197609000-00003

Coren, S. (2008). *The modern dog*. New York: Free Press.

Crocker, J.C. (1977). My brother the parrot. In J. D. Sapir & J. C. Crocker (Eds.), *The social uses of metaphor: Essays on the anthropology of rhetoric* (pp. 164–192). Philadelphia: University of Pennsylvania Press.

de Oliveira, W. S. L., de Faria Lopes, S., & Alves, R. R. N. (2018). Understanding the motivations for keeping wild birds in the semi-arid region of Brazil. *Journal of Ethnobiology and Ethnomedicine, 14*(1), 1–14. https://doi.org/10.1186/s13002-018-02436

Engel, G. L. (1977). The need for a new medical model: A challenge for biomedicine. *Science, 196*(4286), 129–136. https://doi.org/10.1126/science.847460

Engel, G. L. (1980). The clinical application of the biopsychosocial model. *American Journal of Psychiatry, 137*, 535–544.

Esposito, L., McCune, S., Griffin, J. A., & Maholmes, V. (2011). Directions in human-animal interaction research: Child development, health, and therapeutic interventions. *Child Development Perspectives, 5*(3), 205–211. https://doi.org/10.1111/j.1750-8606.2011.00175.x

Fall, T., Kuja-Halkola, R., Dobney, K., Westgarth, C., Magnusson, P. K. E. (2019). Evidence of large genetic influences on dog ownership in the Swedish twin registry has implications for understanding domestication and health associations. *Scientific Reports, 9*(1):1-7. Doi: 10.1038/s41598-019-44083-9

Fielding, W. J., & Plumridge, S. J. (2005). Characteristics of owned dogs on the island of new providence, The Bahamas. *Journal of Applied Animal Welfare Science, 8*(4), 245–260.

Fine, A. H. (2014). *Our faithful companions: Exploring the essence of our kinship with animals.* Crawford, CO: Alpine Publications Inc.

Friedmann, E., Katcher, A. H., Thomas, S. A., Lynch, J. J., & Messent, P. R. (1983). Social interaction and blood pressure: Influence of animal companions. *Journal of Nervous and Mental Disease, 17*(18), 461–465.

Gee, N., Rodriguez, K., Fine, A., & Trammell, J. (2021). Dogs supporting human health and well-being: A biopsychosocial approach. *Frontiers in Veterinary Science, 8*, 1–11. https://doi.org/10.3389/fvets.2021.630465

Gray, P. B., & Young, S. M. (2011). Human–pet dynamics in cross-cultural perspective. *Anthrozoös, 24*(1), 17–30. https://doi.org/10.2752/175303711X12923300467285

Headey, B., Na, F., & Zheng, R. (2008). Pet dogs benefit owners' health: A 'natural experiment' in China. *Social Indicators Research, 87*(3), 481–493.

Herzog, H. (2014). Biology, culture, and the origins of pet-keeping. *Animal Behavior and Cognition, 1*(3), 296–308. https://doi.org/10.12966/abc.08.06.2014

Home, C., Bhatnagar, Y., &Vanak, A. (2018). Canine conundrum: Domestic dogs as invasive species and their impacts on wildlife in India. *Animal Conservation, 21*(4), 275–282. https://doi.org/10.1111/acv.12389

Joseph, N., Chandramohan, A. K., D'souza, A. L., Hariram, S., & Nayak, A. H. (2019). Assessment of pet attachment and its relationship with stress and social support among residents in Mangalore city of south India. *Journal of Veterinary Behavior, 34*, 1–6. https://doi.org/10.1016/j.jveb.2019.06.009

Kasempimolporn, S., Jitapunkul, S., & Sitprija, V. (2008). Moving towards the elimination of rabies in Thailand. *Journal of the Medical Association of Thailand, 91*(3), 433–437.

Kellert, S. (1993). Kinship to mastery: Biophilia in human evolution and development. *Choice Reviews Online, 35*(04). https://doi.org/10.5860/choice.35-2067

Kendler, K. S., Myers, J., Prescott, C. A., & Neale, M. C. (2001). The genetic epidemiology of irrational fears and phobias in men. *Archives of General Psychiatry, 58*(3), 257–265.

Kitala, P., McDermott, J., Kyule, M., Gathuma, J., Perry, B., & Wandeler, A. (2001). Dog ecology and demography information to support the planning of rabies control in Machakos district, Kenya. *Acta Tropica, 78*(3), 217–230.

Krause-Parello, C. A. (2012). Pet ownership and older women: The relationships among loneliness, pet attachment support, human social support, and depressed mood. *Geriatric Nursing, 33*(3), 194–203. https://doi.org/10.1016/j.gerinurse.2011.12.005

Landová, E., Janovcová, M., Štolhoferová, I., Rádlová, S., Frýdlová, P., & Sedláčková, K., et al. (2021). Specificity of spiders among fear- and disgust-eliciting arthropods: Spiders are special, but phobics not so much. *PLoS One, 16*(9), e0257726. https://doi.org/10.1371/journal.pone.0257726

Lesthaeghe, R. (2014). The second demographic transition: A concise overview of its development. *Proceedings of the National Academy of Sciences, 111*(51), 18112–18115. https://doi.org/10.1073/pnas.1420441111

Loken, E., Hettema, J., Aggen, S., & Kendler, K. (2014). The structure of genetic and environmental risk factors for fears and phobias. *Psychological Medicine, 44*(11), 2375–2384.

Lorenz, K. (1943). Die angeborenenFormenmöglicherErfahrung. *Zeitschrift fur Tierpsychologie, 5*, 94–125.
Marinelli, L., Adamelli, S., Normando, S., & Bono, G. (2007). Quality of life of the pet dog: Influence of owner and dog's characteristics. *Applied Animal Behaviour Science, 108*(1–2), 143–156.
Marsh, L. E., Kanngiesser, P., & Hood, B. (2018). When and how does labour lead to love? The ontogeny and mechanisms of the IKEA effect. *Cognition, 170*, 245–253.
Marshall-Pescini, S., Schaebs, F. S., Gaugg, A., Meinert, A., Deschner, T., & Range, F. (2019). The role of oxytocin in the dog-owner relationship. *Animals: An Open Access Journal from MDPI, 9*(10), 792. https://doi.org/10.3390/ani9100792
McCarthy, M. (2017). *From street dweller to family member: The dynamic relationship between people, and cats and dogs in Thailand* (Doctoral dissertation, Monash University).
McNicholas, J., & Collis, G. M. (2006). Animals as social supports: Insights for understanding animal-assisted therapy. *Handbook on Animal-Assisted Therapy: Theoretical Foundations and Guidelines for Practice, 2*, 49–72.
McNicholas, G. M., Collis, C., Kent, C., & Rogers, M. (2001). *The role of pets in the support networks of people recovering from breast cancer*. Paper presented at the 9th international conference on human-animal interactions: people and animals, a global perspective for the 21st century, Rio de Janeiro, Brazil.
Melson, G. F. (2003). Child development and the human-companion animal bond. *American Behavioral Scientist, 47*(1), 31–39. https://doi.org/10.1177/0002764203255210
Menezes, R. (2008). Rabies in India. *Canadian Medical Association Journal, 178*(5), 564–566. https://doi.org/10.1503/cmaj.071488
Miele, M., & Bock, B. (2007). Competing discourses of farm animal welfare and agri-food restructuring. *International Journal of Sociology of Agriculture and Food, 15*(3), 1–7.
Morey, D. F. (1992). Size, shape, and development in the evolution of the domestic dog. *Journal of Archaeological Science, 19*(2), 181–204.
Morey, D. F. (2006). Burying key evidence: The social bond between dogs and people. *Journal of Archaeological Science, 33*(2), 158–175.
Mormann, F., Dubois, J., Kornblith, S., Milosavljevic, M., Cerf, M., Ison, M., & Koch, C. (2011). A category specific response to animals in the right human amygdala. *Nature Neuroscience, 14*, 1247–1249. https://doi.org/10.1038/nn.2899
Murray, E. J., & Foote, F. (1979). The origins of fear of snakes. *Behaviour Research and Therapy, 17*(5), 489–493.
Nagasawa, M., Kikusui, T., Onaka, T., & Ohta, M. (2009). Dog's gaze at its owner increases owner's urinary oxytocin during social interaction. *Hormones and Behavior, 55*(3), 434–441. https://doi.org/10.1016/j.yhbeh.2008.12.002
Newby, J. (1997). *Pact for survival: Humans and their animal companions*. Australia & New Zealand: ABC Enterprises.
Norton, M. I., Mochon, D., & Ariely, D. (2012). The IKEA effect: When labor leads to love. *Journal of Consumer Psychology, 22*(3), 453–460.
Odendaal, J. S. J., & Meintjes, R. A. (2003). Neurophysiological correlates of affiliative behaviour between humans and dogs. *The Veterinary Journal, 165*(3), 296–301. https://doi.org/10.1016/s1090-0233(02)00237-x
Oliva, J. L., & Johnston, K. L. (2020). Puppy love in the time of Corona: Dog ownership protects against loneliness for those living alone during the COVID-19 lockdown. *International Journal of Social Psychiatry, 67*(3), 232–242. https://doi.org/10.1177/0020764020944195
Paxton, D. (2011). *Why it's okay to talk to your dog: Co-evolution of people and dogs*. Tingalpa Qld, Australia: Boolarong Press.
Philo, C., & Wilbert, C. (2000). *Animal spaces, beastly places: New geographies of human-animal relations*. London: Routledge.
Powell, L., Edwards, K., Bowan, A., Guastella, A., Drayton, B., Stomatakis, E., & McGreevy, P. (2019). Canine endogenous oxytocin responses to dog-walking and affiliative human-dog interactions. *Animals, 9*, 51. https://doi.org/10.3390/ani9020051
Ratschen, E., Shoesmith, E., Shahab, L., Silva, K., Kale, D., Toner, P., Reeve, C., & Mills, D. S. (2020). Human-animal relationships and interactions during the COVID-19 lockdown phase in the UK: Investigating links with mental health and loneliness. https://doi.org/10.31235/osf.io/6ju9m
Reese, J. F. (2005). Dogs and dog control in developing countries. In D. J. Salem & A. N. Rowan (Eds.), *The state of the animals III: 2005* (pp. 55–64). Washington, DC: Humane Society Press.

Rusk, C. P., Summerlot-Early, J. M., Machtmes, K. L., Talbert, B. A., & Balschweid, M. A. (2003). The impact of raising and exhibiting selected 4-H livestock projects on the development of life and project skills. *Journal of Agricultural Education, 44*(3), 1–11.

Sarkar, R., Sau, S., &Bhadra, A. (2019). Scavengers can be choosers: A study on food preference in free-ranging dogs. *Applied Animal Behaviour Science, 216*, 38–44. https://doi.org/10.1016/j.applanim.2019.04.012

Savvides, N. (2013). Living with dogs: Alternative animal practices in Bangkok, Thailand. *Animal Studies Journal, 2*(2), 28–50.

Serpell, J. A., & Paul, E. S. (2011). Pets in the family: An evolutionary perspective. In C. A. Salmon & T. K.Shackelford (Eds.), *The Oxford handbook of evolutionary family psychology* (pp. 298–309). New York: Oxford University Press.

Shipman, P. (2015). How do you kill 86 mammoths? Taphonomic investigations of mammoth megasites, *Quaternary International. 359*, 38–46. https://doi.org/10.1016/j.quaint.2014.04.048

Strand, E. B. (2004). Interparental conflict and youth maladjustment: The buffering effects of pets. *Stress, Trauma, and Crisis, 7*(3), 151–168. https://doi.org/10.1080/15434610490500071

Triebenbacher, S. L. (2000). The companion animal within the family system: The manner in which animals enhance life within the home. In A. H. Fine (Ed.), *Handbook on animal-assisted therapy: Theoretical foundations and guidelines for practice* (pp. 357–374). New York: Academic Press.

Trut, L. N. (1999). Early canid domestication: The farm fox experiment. *American Scientist, 87*, 160–169.

Ulrich, R. S. (1984). View through a window may influence recovery from surgery. *Science, 2*(24), 420–421.

Ulrich, R. S. (1993). Biophilia, biophobia, & natural landscapes. In S. R. Kellert & E. O. Wilson (Eds.), *The biophilia hypothesis* (73–137). Washington, DC: Island Press.

Volsche, S., Mukherjee, R., & Rangaswamy, M. (2022). The difference is in the details: attachment and cross-species parenting in the United States and India. *Anthrozoös, 35*(3), 393–408. https://doi.org/10.1080/08927936.2021.1996026

Von Holdt, B. M., Shuldiner, E., Koch, I. J., Kartzinel, R. Y., Hogan, A., Brubaker, L., Wanser, S., Stahler, D., Wynne, C. D. L., Ostrander, E. A., Sinsheimer, J. S., & Udell, M. A. R. (2017). Structural variants in genes associated with human Williams-Beuren syndrome underlie stereotypical hypersociability in domestic dogs. *Science Advances, 3*(7), e1700398. https://doi.org/10.1126/sciadv.1700398

Wells, D. L. (2019). The state of research on human–animal relations: Implications for human health. *Anthrozoös, 32*(2), 169–181. https://doi.org/10.1080/08927936.2019.1569902

Wilson, E. O. (1984). *Biophilia*. Cambridge, MA: Harvard University Press.

Wu, C. S. T., Wong, R. S. M., & Chu, W. H. (2018). The association of pet ownership and attachment with perceived stress among Chinese adults. *Anthrozoös, 31*(5), 577–586. https://doi.org/10.1080/08927936.2018.1505269

Young, J., Pritchard, R., Nottle, C. & Banwell, H. (2020). Pets, touch, and Covid-19. *Journal of Behavioral Economics, 4*, 25–33.

Zoetis-HABRI International Human-Animal Bond Survey. (2022). https://habri.org/international-hab-survey/

# 15
# ANIMALS IN DEVELOPMENT
## Childhood and Adolescence

*Erin King, Eli D. Halbreich, and Megan K. Mueller*

While the importance of forming stable attachments in childhood has been discussed since the early 1940s (Bowlby, 1944), child-pet relationships are now increasingly recognized as an influential component of youth development. Companion animals are a key feature of family life, with between 57% and 70% of households owning a pet in the United States (American Pet Products Association, 2021; American Veterinary Medical Association, 2018). Similarly, in the United Kingdom, about 53% of households with children own at least one pet (Pet Food Manufacturers' Association, 2019). Given high rates of pet ownership in households with children, it is important to understand what roles pets play in youth development. Pets can provide a unique and stable relationship for children and adolescents, perhaps because animals are inherently non-judgmental in nature and are often seen as a safe haven (Archer, 1997; Kurdek, 2009). For example, both children and young adults often report turning to dogs, more so than some other humans, as a source of support when they are upset (Kurdek, 2009; McNicholas & Collis, 2001). Similarly, children have reported less conflict and more satisfaction with their pets than with their siblings (Cassels et al., 2017), which supports the idea that pets are seen as a part of the family (Albert & Bulcroft, 1988; Cain, 1985; Melson, 2005). In addition to emotional and social support, pets also give children the chance to develop skills such as responsibility, caretaking, and empathy (Cassels et al., 2017; Legg et al., 1974; Poresky, 1990; Robin & ten Bensel, 1985). Given the prevalence of pet ownership and these developmental aspects of child-pet relationships, it is important to understand how and in what context positive relationships flourish.

Research on the child-animal bond has blossomed since the 1980s, as part of the growing scientific movement around the human-animal bond (Hines, 2003). In 1987, the National Institutes of Health (NIH) in the United States held a conference focused on the health benefits of pets in which a panel of experts highlighted a need for a stronger scientific foundation for research on the role of pets in child development (Esposito et al., 2011; NIH, 1987). Since then, the human-animal interaction (HAI) field has grown rapidly, with numerous institutes, organizations, and journals focusing on research related to HAI (McCune et al., 2020). The field has also expanded with the aid of increased research funding opportunities through the NIH and private foundations (e.g., WALTHAM/Mars, the Human Animal Bond Research Institute, the Society for Companion Animal Studies, the Horses and Humans in Research Foundation), exploring topics relevant to child development such as social and emotional support (Carr & Rockett, 2017; Crossman et al., 2020; Doyle et al., 2010; Meehan et al., 2017), emotional regulation (Ein et al., 2018; Kertes et al., 2017; Polheber & Matchock, 2014), cognitive development (Hurley & Oakes, 2018; Hurley et al., 2010; Hurley & Oakes, 2015; Kovack-Lesh et al., 2014), and more.

## Cross-Cultural Considerations for Research, Practice, and Policy

Although HAI research focused on youth is international in scope, there are a number of cross-cultural questions that remain underexplored. For example, Gustafsson et al. (2020) found that attitudes toward animals are influenced by culture. However, youth attitudes toward animals have not been studied in many countries, leaving gaps in the literature and limiting what we know about large parts of the world. Even within certain geographic regions, there is still considerable variability in how sociodemographic characteristics can predict companion animal relationships. For example, Westgarth et al. (2013) found that pet ownership was associated with gender, ethnicity, and socioeconomic status among 9–10-year-old children in the United Kingdom, and other studies have found similar patterns in other areas of the world. These studies indicate that there are meaningful differences between cultures, and even within cultures, that could change the way people feel about and interact with pets.

## The Role of Pets in Child and Adolescent Development

### Development of Attachment and Identity and the Role of the Pet

In early childhood, children develop an internal working model of social relationships based on their attachment relationships with caregivers (Bowlby, 1958, 1982). Although there is significant variation in developmental outcomes and a number of factors involved in the process, attachment classifications (i.e., secure, insecure avoidant, insecure resistant, disorganized) have been shown to last from infancy to adolescence (Hamilton, 2000), as well as predict emotion recognition (Steele et al., 2008). The internal working model developed during infancy also informs social relationships later in life, and likely affects secondary attachment relationships, such as with friends or pets. Given that pets are often members of a child's immediate environment, and that environmental factors contribute to development (Bronfenbrenner, 1979, 1986; Bronfenbrenner & Morris, 2006), it follows that a pet's presence in the home may be a key component of the developmental system.

Young children often develop relationships with their pets based on their previous experiences with social interactions as well as the specific pet. Hawkins et al. (2017) found that the level of attachment to pets varied by pet ownership, pet type, and gender, but that 76% of children reported that their pet was their best friend, and half of the children in the study reported that their pet provided them with comfort. However, high levels of attachment to companion animals are not uniformly beneficial, as some adolescents with extreme attachment to their pets reported higher levels of loneliness and lower social support scores (Hartwig & Signal, 2020). Pet attachment has been shown to be stable over time for children (Endenburg et al., 2014), but can decrease in adolescence (Muldoon, Williams, & Currie, 2019), indicating that there may be developmental changes in these relationships.

Consistent with other areas of HAI research, dog-human relationships have received a large proportion of focus in youth development. While the specific relationships that children have with pets other than dogs are relatively underexplored, there has been some evidence that dogs may be primed to serve as secondary attachment figures for children. One study found that a dog's ability to follow a child's pointing gestures, the level of perceived social support that the dog provided, and the amount of petting that occurred during a sociability task were associated with the strength of the attachment relationship between the dog and child (Hall et al., 2016). This behavioral connection may explain why attachment to dogs is more pronounced because dogs often demonstrate greater behavioral interactivity compared to other pets (Jalongo, 2015). Dogs are also sensitive to the behavior of humans, and adjust their behavior accordingly (Duranton & Gaunet, 2015). As a result, attachment to dogs in children and adolescents is often reported as stronger than other pets (Muldoon, Williams, Lawrence et al., 2019)—although there are individual differences for specific youth—possibly related to age of the child/adolescent (Hirschenhauser et al., 2017; Zilcha-Mano et al., 2011). For example, one study found that younger children (aged 6–10) reported higher attachment with animals such as dogs and cats compared to animals such as turtles and fish, while older children/adolescents (aged 11–14) reported similar levels of attachment regardless of pet species (Hirschenhauser et al., 2017).

## *Emotional and Social Support*

Existing research has also suggested that pets can be an important source of emotional support for children. Children and early adolescents report turning to their pets for emotional support, often in the form of talking to their pets with the belief that their pet can understand and comfort them (Melson, 2003; Muldoon, Williams, Lawrence et al., 2019). Additionally, children in foster care reported turning to their companion animals during times of distress, and noted the benefit of not needing to verbally communicate with their animals, which could alleviate some pressure from interactions that are already imbued with emotion (Carr & Rockett, 2017). Interactions between adolescents and dogs has also been found to reduce anxiety and raise positive affect (Crossman et al., 2020) as well as alleviate perceived stress before college exams (Barker et al., 2016). Additionally, involving therapy animals in academic skills programs has shown some degree of promise, with certain at-risk students responding positively to both a HAI/program combination, and HAI alone (Pendry et al., 2020).

Pets are often viewed as a source of social support for children and adolescents. As children age into adolescence and develop more complex relationships and communication skills, their interactions with their pets as members of social circle in the home change (Hirschenhauser et al., 2017). For example, children and adolescents report perceiving their pets as important sources of social support (Doyle et al., 2010; Meehan et al., 2017). There have also been possible relationships found between attachment to pets and support networks, such as in rural adolescents (Black, 2012) and adolescents engaging in online social connections (Charmaraman et al., 2020). For some children, dogs may also offer a safe haven during times of social stress (Beetz et al., 2012), and for older adolescents going through posttraumatic growth, spending time with pets was associated with relating to other humans (Dominick et al., 2020). As the research shows, pets can act as a source of social support for children and adolescents, but the relationship, like other social relationships, is not universally positive. For example, if a child exclusively relies on their pet for social interaction, this could potentially contribute to unhealthy social withdrawal. Further research needs to be done to clarify links between these relationship dynamics within the context of child development, and what specific individual and family-level factors support adaptive HAI in the context of pets and social support.

## Social Skills

Companion animals may also provide an opportunity for youth to practice positive social skills and prosocial behavior. Mueller (2014) found that attachment to an animal was associated with the positive development markers of connection with others, social competence, and empathy for older adolescents. Similarly, higher levels of emotional attachment to dogs have been associated with increased communication with adolescents' parents and best friends (with cats, only best friends), which could be explained by the caring activities related to pet ownership (Marsa-Sambola et al., 2017). Interactions with animals in structured settings can also help facilitate social interactions between children in general and youngsters with various developmental disabilities. For example, when used within the context of classroom-based activities, interactions with guinea pigs were associated with increases in social functioning and decreases in problem behaviors for youth on the autism spectrum (O'Haire et al., 2013). Animals could be integrated into interventions to help facilitate relationships between youth, allow for non-judgmental practice of social communication, and foster caring behaviors.

## Cognitive and Language Development

In addition to an emotional attachment relationship between children and their pets, living with pets has been shown to impact young children in domains related to cognition and communication. An early study of 2–5-year-olds found that children initiate communicative contacts with pet dogs twice as often as the dogs initiate communicative contacts with the children (Millot et al., 1988). The ways in which children engage with their dogs compared to interactions with other children are behaviorally distinct; that is, children touch, stroke, and hug their dogs more often than they would other children, which could indicate relationship differences (Millot et al., 1988). Exposure to pets is also associated with infants' cognitive processing of animal faces, indicating that young children's cognitive development may be affected by the pets in their immediate environment (Hurley & Oakes, 2018; Hurley et al., 2010; Hurley & Oakes, 2015; Kovack-Lesh et al., 2014). Research with therapy dogs has demonstrated that preschool children performed better on cognitive tasks when in the presence of a therapy dog, suggesting that interaction with an animal may facilitate cognitive performance (Gee et al., 2012; Gee, Church et al., 2010; Gee, Crist et al., 2010). Additionally, recent population-level research in Japan found positive associations between dog-owning families and five areas of infant development; communication, gross motor, fine motor, problem-solving, and personal-social developmental domains (Minatoya et al., 2020), however, these findings should be interpreted with caution due to the likely presence of confounds. The role of companion animals in shaping cognitive development and facilitating language and communication both in the home and the classroom is an area ripe for additional research.

## Stress and Anxiety

Contact with companion animals may impact psychophysiological responses to stress for children and adolescents. There are several hypothesized theories for how companion animals can impact psychophysiological responses to stress, such as the Human-Animal Interaction Hypothalamic-Pituitary-Adrenal (HAI-HPA) Transactional Model (Pendry & Vandagriff, 2020). The role of animals in contributing to the reduction of anxiety and stress has been predominantly explored context of animal-assisted interventions. For example,

Ein et al. (2018) found in a meta-analysis that exposure to animal-assisted therapy was associated with a decrease in heart rate. Additionally, in a laboratory study, adolescents perceived reduced stress in the presence of pet dogs, and there was an association between a lower cortisol response and increased child-solicited petting as well as less spontaneous proximity-seeking behavior by the dog (Kertes et al., 2017). Furthermore, Polheber and Matchock (2014) found that among university students, the presence of a dog reduced cortisol levels throughout the experiment and reduced heart rate during a stress test.

However, Kerns et al. (2018) found that while the presence of a dog was associated with increased positive affect in children, the presence of the dog did not affect negative affect or heart rate. Additionally, Mueller, Anderson et al. (2021) found no evidence that the presence of a real dog reduced anxiety or autonomic reactivity, or increased cognitive performance compared to the control condition with a stuffed dog. It may be the case that in some studies (particularly those that use laboratory-based stressors), the presence of an animal may not be sufficient to overcome an acutely anxiety-provoking event. There may also be differential effects for a child's own animal as compared to an unfamiliar therapy animal. The relationship between companion animals and anxiety reduction is still underexplored, so further research should assess if HAI is effective in these settings.

## *Adaptive Coping Behaviors*

In addition to supporting youth emotionally, companion animals may contribute positively to child and adolescent adaptive coping behaviors and strategies for supporting resilience. During times of stress, companion animals may help encourage functional activities that are associated with adaptive coping, such as outdoor physical activity, regular walking, social adaptive coping, and other healthy behaviors (Mueller & Callina, 2014; Mueller et al., 2022). Pets in particular have the opportunity to foster a sense of routine and normalcy, even in the face of stressors. However, additional research is needed to disentangle these complex behaviors and to understand the circumstances where pets can encourage adaptive coping.

## *Physical Activity and Health*

In addition to supporting socio/emotional well-being, pets may play an important role in encouraging physical activity and childhood immune development. A scoping review of the literature on dog ownership and youth physical activity levels found that in 9 of the 18 studies which compared physical activity levels of children with and without dogs, an association between dog ownership and increases in physical activity levels was reported (Chase et al., 2022). Numerous studies have found that those in dog-owning households had increased levels of physical activity compared to non-dog-owning families (Engelberg et al., 2016; McMinn et al., 2011; Mueller, King et al., 2021; Owen et al., 2010). Attachment to a pet can also play a role in supporting being active; Gadomski et al. (2017) found that children with higher attachment to their dogs had higher levels of physical activity. Additionally, exposure to pets in early childhood has shown reduced risk of allergy development and reductions in wheezing and atopy (Bufford et al., 2008; Ownby et al., 2002).

However, some research has found mixed results regarding physical activity and pets. For example, Christian et al. (2014) found among Australian adolescents that dog walking (compared to non-dog walking adolescents) was associated with a higher independent mobility index score, and greater odds for walking in the neighborhood, playing in the street, and playing in the yard; however, dog walking was not associated with a change in overall physical activity, walking, or pedometer steps. While Timperio et al. (2008) found that the odds of children being overweight or obese are lower among those who own a dog, multiple studies have not found an association between obesity and dog ownership (Westgarth et al., 2017; Westgarth et al., 2012), suggesting complexity in these relationships given the myriad factors contributing to obesity.

## Barriers to Positive Child-Pet Relationships

While child-pet relationships are ideally mutually beneficial to both the pet and child, there are also potential negative consequences associated with these relationships. First, children (especially young children) do not always understand animal behavior, which can lead to potential physical harm for the child (e.g., bites, scratches) as well as negative animal welfare for the pet. Young children are more likely to incorrectly identify dog behavior (Meints et al., 2010) and approach fearful dogs due to this misunderstanding (Aldridge & Rose, 2019) which can lead to injury and dog bites. Dixon et al. (2012) found that nearly half of a sample of 300 children (5–15 years old) failed a prevention knowledge test about dog bites, and in the United States, young children are disproportionately at risk for dog bites and often need medical care from their injuries (Daniels et al., 2009; Gilchrist et al., 2008). Additionally, during the initial phases of the COVID-19 pandemic, the occurrence of dog bites rose nearly three-fold (Dixon & Mistry, 2020; Tulloch et al., 2021). Given the inherent risk in HAIs, and that most bites occur during normal daily activities (American Veterinary Medical Association, n.d.) it is especially important for young children to develop safe ways of interacting with their pets that incorporate their animals' behavioral needs. Adult supervision and child safety training are critical for the longevity and safety of child-pet relationships—especially as children are spending an increased amount of time at home with their pets. There are a number of safety training programs which aim to improve upon the abilities of families to recognize signals of pet stress and thus reduce dog bite incidence rates, such as the Blue Dog program (Meints & de Keuster, 2009; Schwebel et al., 2012). For comprehensive resources related to the management and prevention of dog bites, see Jakeman et al. (2020), Meints et al. (2018), and *Dog bites: A multidisciplinary perspective* (2017).

Not only is human safety a priority in child-pet relationships, but animal welfare is vital to any relationship. Children have a range of understanding when it comes to animal care and safety, therefore it is important to have adult supervision of child-animal interactions, routines, and care patterns in order to foster positive, healthy relationships. Not every pet is compatible with every child (Hart et al., 2018), therefore it is essential to understand how the needs of a pet, as well as expectations of the child, fit within the family system. In addition, child-pet relationships are not always positive, and some youth have reported acting aggressively or harming their pets (Robin & ten Bensel, 1985). As Chapter 23 highlights, there is often a link between domestic violence, elder abuse, child abuse, and animal cruelty. Child-pet relationships can be complex and embedded within a cycle of abuse that could potentially lead to negative outcomes.

In addition to health and safety concerns, the ability to own and care for a pet is impacted by a web of systemic and historical barriers. Most notably, inequitable access to and the high cost of veterinary care are major issues, especially in historically-marginalized communities (Arrington & Markarian, 2017; Decker Sparks et al., 2018). Jarolmen (1998) found that children and adolescents intensely grieve for their pets, even more so than adults. Therefore, when a pet becomes sick, the ability to provide medical care for the animal is not only important to control for potential zoonotic disease within the child-pet relationship, but also because of the potential emotional toll. In addition to veterinary care, limited access to pet-friendly housing and other resources for pet ownership (pet supply stores, grooming, behavior training, dog friendly parks, etc.) are challenges that impact communities, families, and children with pets.

## Practical and Policy Perspectives

Companion animals contribute to childhood and adolescent development and are an important part of the family system. Removing structural barriers that limit families from mutually-beneficial relationships with animals could potentially increase the number of positive developmental outcomes associated with pet ownership. For example, certain housing policies have historically held discriminatory practices embedded in their pet fee/restriction policies (Applebaum et al., 2021), and by addressing these inequities, more children and adolescents will be able to experience the human-animal bond (for further reading see Chapter 9). Additionally, in order to maximize the potential benefits of child-pet relationships, more accessible education and training resources on healthy and safe dog interaction that focuses on animal behavior is needed. Equity issues around pet ownership have policy implications, especially with regard to accessible veterinary care, safe space, and affordable housing.

## Unanswered Questions and Future Research

Many studies have found that there may be a positive relationship between pet ownership and child/adolescent developmental outcomes, however much of the research is hampered by methodological weaknesses (Purewal et al., 2017). The existing evidence suggests that pets should not be viewed as universally beneficial or detrimental to development, but instead should be conceptualized as individual relationships within a child's life and immediate environment. As such, it is important to keep in mind that some pets may be beneficial for some adolescents and children, but not all. For example, when controlling for demographic variables, pet ownership alone does not appear to be associated with socioemotional outcomes; but Jacobson and Chang (2018) found that attitudes toward animals were positively associated with lower delinquency, higher empathy, less-depressed mood, and prosocial behavior. Additionally, Cassels et al. (2017) found gendered differences in relationships with pets, and dog owners reported more relationship satisfaction compared to non-dog-owners. Mueller et al. (2022) found that some of the adaptive coping benefits associated with dog ownership—such as exercise, healthy eating and sleep habits, and making time to relax—however, the relationship was not significant when controlling for demographic variables. Similarly, it is possible that the health benefits of pet ownership are due more to demographic factors other than pet ownership (Miles et al., 2017), so studies going forward should account for those differences in their analyses.

There is a clear need for research addressing individuality and specificity in human-animal relationships. For example, the preponderance of existing research focuses on dogs (and cats, although often to a lesser extent), with less known about other pets such as hamsters, guinea pigs, rabbits, gerbils, fish, birds, horses, and others. Not all relationships between children and companion animals are the same, and these idiographic variations may produce different outcomes for youth and animals. As noted in prior research, attachment and relationship quality are often more predictive of positive health and well-being outcomes than simply whether there is a pet in the home. Although attachment has been well-studied in childhood and adolescence, existing pet attachment measures often produce high ceiling effects that don't adequately capture variation, and are geared toward dogs and cats. There are other specific behavioral and emotional features of child-animal relationships that should be studied in more detail to help contextualize interactions. For example, a recent paper suggested that integrating pet care routines with medical routines could improve medication adherence in children (Gupta et al., 2018). It is important to explore the interaction between emotional support, social facilitation, responsibility, emotional disclosure, stages of development, and caretaking as intertwined, but distinct features of companion animal interactions.

Attachment, emotional connection, and behavioral aspects of youth-pet relationships can also change across development. For example, prior research has shown that attachment to pets can decrease across adolescence (Muldoon, Williams, & Currie, 2019), perhaps in part reflecting the diversification of adolescents' social resources and relationships over time. As adolescents develop autonomy, their relationships often shift outside of the family context to focus more on peer relationships, and this transition could, in some cases, translate to changes in relationships with companion animals. There may also be changes in *how* youth interact with companion animals depending on their developmental needs at any one time, including changing emphasis on physical comfort, emotional support, and companionship. The role of developmental changes in companion animal relationships is still largely an open question and an area that would benefit from further exploration.

A youth-pet relationship is also part of a whole family system, and studying these relationships in the context of other relationships and interactions within the immediate and broader family system is critical to fully understanding the role of a companion animal in youth development. For example, exploring the role of parents and caregivers in facilitating and structuring relationships between pets and children may vary widely based on a number of factors such as individual animal personality/characteristics, individual child characteristics, age, and parental connection to the pet. All of these aspects are critical in fully understanding the role of companion animals in development.

Furthermore, there have been recent calls for increased diversity in HAI research broadly, as one consistent limitation of existing research has been the homogeneity of research subjects. It is clear that the experiences of families can vary significantly based on sociodemographic characteristics such as residential type and location, socioeconomic status, racial and ethnic background, religion, and family composition. However, this vast diversity of experiences has not adequately been captured with existing research approaches. Future work should prioritize diversity in sampling, as well as the use of different types of research methods to capture the nuances of HAI in diverse families and allow a broad range of stakeholders to share their experiences (Griffin et al., 2019; Rodriguez et al., 2021).

Finally, there is a need for continued evolution in the methodological approaches used for studying youth relationships with companion animals. It is not yet clear if existing measures of relationship quality are sensitive to developmental changes. Research studies should be designed to capture idiographic changes in youth-animal relationships as well as the diversity of outcomes associated with these relationships from proximal, short-term impacts to global, long-term outcomes. The outcomes used in developmental HAI research should be specific to the developmental tasks of the age period being studied, and ideally will involve longitudinal data collection to capture change. These outcomes should also include a range of data sources, including physiological markers, self-report data, behavioral observations, qualitative data, and also include relevant measures of animal health and well-being. Including a breadth of measurement approaches in developmental HAI research designs will allow the field to obtain a more comprehensive understanding of how companion animals can contribute to youth development.

Companion animals clearly play a critical role in childhood and adolescence, as demonstrated by the prevalence of pets in families as well as the extensive body of research in this area. However, there are still many unanswered questions about the specific nuances of youth-animal relationships that warrant further exploration.

> **Discussion Questions**
>
> 1 What is the potential role of relationship quality in supporting both positive and negative relationships between youth and companion animals?
> 2 What features of the family and community systems are important to consider when assessing youth-animal interactions?
> 3 What are the key points that healthcare and education practitioners should know about animals in development?

## References

Albert, A., & Bulcroft, K. (1988). Pets, families, and the life course. *Journal of Marriage and Family, 50*(2), 543–552. https://doi.org/10.2307/352019

Aldridge, G. L., & Rose, S. E. (2019). Young children's interpretation of dogs' emotions and their intentions to approach happy, angry, and frightened dogs. *Anthrozoös, 32*(3), 361–374. https://doi.org/10.1080/08927936.2019.1598656

American Pet Products Association. (2021, June 22). *American pet products association releases newest edition of National Pet Owners Survey* [Press release]. https://www.americanpetproducts.org/press_releasedetail.asp?id=1242

American Veterinary Medical Association. (2018). *AVMA pet ownership and demographics sourcebook* (2017–2018 ed.). Schaumburg, IL: American Veterinary Medical Association.

American Veterinary Medical Association. (n.d.). *Dog bite prevention.* https://www.avma.org/resources-tools/pet-owners/dog-bite-prevention

Applebaum, J. W., Horecka, K., Loney, L., & Graham, T. M. (2021). Pet-friendly for whom? An analysis of pet fees in Texas rental housing. *Frontiers in Veterinary Science, 8*, 767149. https://doi.org/10.3389/fvets.2021.767149

Archer, J. (1997). Why do people love their pets? *Evolution and Human Behavior, 18*(4), 237–259. https://doi.org/10.1016/S0162-3095(99)80001-4

Arrington, A., & Markarian, M. (2017). Serving pets in poverty: A new frontier for the animal welfare movement. *Sustainable Development Law & Policy, 18*(1), 40–43. http://digitalcommons.wcl.american.edu/sdlp/vol18/iss1/11

Barker, S. B., Barker, R. T., McCain, N. L., & Schubert, C. M. (2016). A randomized cross-over exploratory study of the effect of visiting therapy dogs on college student stress before final exams. *Anthrozoös, 29*(1), 35–46. https://doi.org/10.1080/08927936.2015.1069988

Beetz, A., Julius, H., Turner, D., & Kotrschal, K. (2012). Effects of social support by a dog on stress modulation in male children with insecure attachment. *Frontiers in Psychology, 3*, 352. https://doi.org/10.3389/fpsyg.2012.00352

Black, K. (2012). The relationship between companion animals and loneliness among rural adolescents. *Journal of Pediatric Nursing, 27*(2), 103–112. https://doi.org/10.1016/j.pedn.2010.11.009

Bowlby, J. (1944). Forty-four juvenile thieves: Their characters and home-life. *The International Journal of Psychoanalysis, 25*, 19–53.

Bowlby, J. (1958). The nature of the child's tie to his mother. *The International Journal of Psychoanalysis, 39*, 350–373.

Bowlby, J. (1982). Attachment and loss: Retrospect and prospect. *American Journal of Orthopsychiatry, 52*(4), 664–678. https://doi.org/10.1111/j.1939-0025.1982.tb01456.x

Bronfenbrenner, U. (1979). *The ecology of human development: Experiments by nature and design*. Cambridge, MA: Harvard University Press.

Bronfenbrenner, U. (1986). Ecology of the family as a context for human development: Research perspectives. *Developmental Psychology, 22*(6), 723–742. https://doi.org/10.1037/0012-1649.22.6.723

Bronfenbrenner, U., & Morris, P. A. (2006). The bioecological model of human development. In R. M. Lerner & W. Damon (Eds.), *Handbook of child psychology* (6th ed., Vol. 1, pp. 793–828). Hoboken, NJ: John Wiley & Sons.

Bufford, J. D., Reardon, C. L., Li, Z., Roberg, K. A., DaSilva, D., Eggleston, P. A., Liu, A. H., Milton, D., Alwis, U., Gangnon, R. Lemanske, R. F., & Gern, J. E. (2008). Effects of dog ownership in early childhood on immune development and atopic diseases. *Clinical & Experimental Allergy, 38*, 1635–1643. https://doi.org/10.1111/j.1365-2222.2008.03018.x

Cain, A. O. (1985). Pets as family members. *Marriage & Family Review, 8*(3–4), 5–10. https://doi.org/10.1300/J002v08n03_02

Carr, S., & Rockett, B. (2017). Fostering secure attachment: Experiences of animal companions in the foster home. *Attachment & Human Development, 19*(3), 259–277. https://doi.org/10.1080/14616734.2017.1280517

Cassels, M. T., White, N., Gee, N., & Hughes, C. (2017). One of the family? Measuring young adolescents' relationships with pets and siblings. *Journal of Applied Developmental Psychology, 49*, 12–20. https://doi.org/10.1016/j.appdev.2017.01.003

Charmaraman, L., Mueller, M. K., & Richer, A. M. (2020). The role of pet companionship in online and offline social interactions in adolescence. *Child & Adolescent Social Work Journal, 37*, 589–599. https://doi.org/10.1007/s10560-020-00707-y

Chase, C. J., Mueller, M. K., Garvey, C., & Potter, K. (2022). Family dog ownership and youth physical activity levels: A scoping review. *Current Sports Medicine Reports, 21*(1), 18–27. https://doi.org/10.1249/JSR.0000000000000927

Christian, H., Trapp, G., Villanueva, K., Zubrick, S. R., Koekemoer, R., & Giles-Corti, B. (2014). Dog walking is associated with more outdoor play and independent mobility for children. *Preventive Medicine, 67*, 259–263. https://doi.org/10.1016/j.ypmed.2014.08.002

Crossman, M. K., Kazdin, A. E., Matijczak, A., Kitt, E. R., & Santos, L. R. (2020). The influence of interactions with dogs on affect, anxiety, and arousal in children. *Journal of Clinical Child and Adolescent Psychology, 49*(4), 535–548. https://doi.org/10.1080/15374416.2018.1520119

Daniels, D. M., Ritzi, R. B., O'Neil, J., & Scherer, L. R. (2009). Analysis of nonfatal dog bites in children. *Journal of Trauma and Acute Care Surgery, 66*(3), S17–S22. https://doi.org/10.1097/TA.0b013e3181937925

Decker Sparks, J. L., Camacho, B., Tedeschi, P., & Morris, K. N. (2018). Race and ethnicity are not primary determinants in utilizing veterinary services in underserved communities in the United States. *Journal of Applied Animal Welfare Science, 21*(2), 120–129. https://doi.org/10.1080/10888705.2017.1378578

Dixon, C. A., Mahabee-Gittens, E. M., Hart, K. W., & Lindsell, C. J. (2012). Dog bite prevention: An assessment of child knowledge. *The Journal of Pediatrics, 160*(2), 337–341. https://doi.org/10.1016/j.jpeds.2011.07.016

Dixon, C. A., & Mistry, R. D. (2020). Dog bites in children surge during coronavirus disease-2019: A case for enhanced prevention. *The Journal of Pediatrics, 225*, 231–232. https://doi.org/10.1016/j.jpeds.2020.06.071

*Dog bites: A multidisciplinary perspective*. (2017). (D. S. Mills & C. Westgarth, Eds.). Sheffield: 5M Publishing.

Dominick, W., Walenski-Geml, A., & Taku, K. (2020). Associations between pet ownership, posttraumatic growth, and stress symptoms in adolescents. *Anthrozoös, 33*(4), 547–560. https://doi.org/10.1080/08927936.2020.1771059

Doyle, C., Timms, C. D., & Sheehan, E. (2010). Potential sources of support for children who have been emotionally abused by parents. *Vulnerable Children and Youth Studies, 5*(3), 230–243. https://doi.org/10.1080/17450128.2010.487122

Duranton, C., & Gaunet, F. (2015). Canis sensitivus: Affiliation and dogs' sensitivity to others' behavior as the basis for synchronization with humans? *Journal of Veterinary Behavior, 10*(6), 513–524. https://doi.org/10.1016/j.jveb.2015.08.008

Ein, N., Li, L., & Vickers, K. (2018). The effect of pet therapy on the physiological and subjective stress response: A meta-analysis. *Stress & Health, 34*(4), 477–489. https://doi.org/10.1002/smi.2812

Engelberg, J. K., Carlson, J. A., Conway, T. L., Cain, K. L., Saelens, B. E., Glanz, K., Frank, L. D., & Sallis, J. F. (2016). Dog walking among adolescents: Correlates and contribution to physical activity. *Preventive Medicine, 82,* 65–72. https://doi.org/10.1016/j.ypmed.2015.11.011

Endenburg, N., van Lith, H. A., & Kirpensteijn, J. (2014). Longitudinal study of Dutch children's attachment to companion animals. *Society & Animals, 22*(4), 390–414. https://doi.org/10.1163/15685306-12341344

Esposito, L., McCune, S., Griffin, J. A., & Maholmes, V. (2011). Directions in human–animal interaction research: Child development, health, and therapeutic interventions. *Child Development Perspectives, 5*(3), 205–211. https://doi.org/10.1111/j.1750-8606.2011.00175.x

Gadomski, A. M., Scribani, M. B., Krupa, N., & Jenkins, P. (2017). Pet dogs and child physical activity: The role of child–dog attachment. *Pediatric Obesity, 12*(5), e37–e40. https://doi.org/10.1111/ijpo.12156

Gee, N. R., Belcher, J. M., Grabski, J. L., DeJesus, M., & Riley, W. (2012). The presence of a therapy dog results in improved object recognition performance in preschool children. *Anthrozoös, 25*(3), 289–300. https://doi.org/10.2752/175303712X13403555186172

Gee, N. R., Church, M. T., & Altobelli, C. L. (2010). Preschoolers make fewer errors on an object categorization task in the presence of a dog. *Anthrozoös, 23*(3), 223–230. https://doi.org/10.2752/175303710X12750451258896

Gee, N. R., Crist, E. N., & Carr, D. N. (2010). Preschool children require fewer instructional prompts to perform a memory task in the presence of a dog. *Anthrozoös, 23*(2), 173–184. https://doi.org/10.2752/175303710X12682332910051

Gilchrist, J., Sacks, J. J., White, D., & Kresnow, M. J. (2008). Dog bites: Still a problem? *Injury Prevention, 14*(5), 296–301. https://doi.org/10.1136/ip.2007.016220

Griffin, J. A., Hurley, K., & McCune, S. (2019). Human-animal interaction research: Progress and possibilities. *Frontiers in Psychology, 10,* 2803. https://doi.org/10.3389/fpsyg.2019.02803

Gupta, O. T., Wiebe, D. J., Pyatak, E. A., & Beck, A. M. (2018). Improving medication adherence in the pediatric population using integrated care of companion animals. *Patient Education and Counseling, 101*(10), 1876–1878. https://doi.org/10.1016/j.pec.2018.05.015

Gustafsson, E., Alawi, N., & Andersen, P. N. (2020). Companion animals and religion: A survey of attitudes among Omani students. *Society & Animals, 29*(2), 132–152. https://doi.org/10.1163/15685306-bja10005

Hall, N. J., Liu, J., Kertes, D., & Wynne, C. D. L. (2016). Behavioral and self-report measures influencing children's reported attachment to their dog. *Anthrozoös, 29*(1), 137–150. https://doi.org/10.1080/08927936.2015.1088683

Hamilton, C. E. (2000). Continuity and discontinuity of attachment from infancy through adolescence. *Child Development, 71*(3), 690–694. https://doi.org/10.1111/1467-8624.00177

Hart, L. A., Hart, B. L., Thigpen, A. P., Willits, N. H., Lyons, L. A., & Hundenski, S. (2018). Compatibility of cats with children in the family. *Frontiers in Veterinary Science, 5,* 278. https://doi.org/10.3389/fvets.2018.00278

Hartwig, E., & Signal, T. (2020). Attachment to companion animals and loneliness in Australian adolescents. *Australian Journal of Psychology, 72*(4), 337–346. https://doi.org/10.1111/ajpy.12293

Hawkins, R. D., Williams, J. M., & Scottish Society for the Prevention of Cruelty to Animals. (2017). Childhood attachment to pets: Associations between pet attachment, attitudes to animals, compassion, and humane behaviour. *International Journal of Environmental Research and Public Health, 14*(5), 490. https://doi.org/10.3390/ijerph14050490

Hines, L. M. (2003). Historical perspectives on the human-animal bond. *American Behavioral Scientist, 47*(1), 7–15. https://doi.org/10.1177/0002764203255206

Hirschenhauser, K., Meichel, Y., Schmalzer, S., & Beetz, A. M. (2017). Children love their pets: Do relationships between children and pets co-vary with taxonomic order, gender, and age? *Anthrozoös, 30*(3), 441–456. https://doi.org/10.1080/08927936.2017.1357882

Hurley, K. B., Kovack-Lesh, K. A., & Oakes, L. M. (2010). The influence of pets on infants' processing of cat and dog images. *Infant Behavior & Development, 33*(4), 619–628. https://doi.org/10.1016/j.infbeh.2010.07.015

Hurley, K. B., & Oakes, L. M. (2015). Experience and distribution of attention: Pet exposure and infants' scanning of animal images. *Journal of Cognition and Development, 16*(1), 11–30. https://doi.org/10.1080/15248372.2013.833922

Hurley, K., & Oakes, L. M. (2018). Infants' daily experience with pets and their scanning of animal faces. *Frontiers in Veterinary Science, 5,* 152. https://doi.org/10.3389/fvets.2018.00152

Jacobson, K. C., & Chang, L. (2018). Associations between pet ownership and attitudes toward pets with youth socioemotional outcomes. *Frontiers in Psychology, 9*, 2304. https://doi.org/10.3389/fpsyg.2018.02304

Jakeman, M., Oxley, J. A., Owczarczak-Garstecka, S. C., & Westgarth, C. (2020). Pet dog bites in children: Management and prevention. *BMJ Paediatrics Open, 4*(1), e000726. https://doi.org/10.1136/bmjpo-2020-000726

Jalongo, M. R. (2015). An attachment perspective on the child–dog bond: Interdisciplinary and international research findings. *Early Childhood Education Journal, 43*(5), 395–405. https://doi.org/10.1007/s10643-015-0687-4

Jarolmen, J. (1998). A comparison of the grief reaction of children and adults: Focusing on pet loss and bereavement. *OMEGA - Journal of Death and Dying, 37*(2), 133–150. https://doi.org/10.2190/h937-u230-x7d9-cvkh

Kerns, K. A., Stuart-Parrigon, K. L., Coifman, K. G., van Dulmen, M. H. M., & Koehn, A. (2018). Pet dogs: Does their presence influence preadolescents' emotional responses to a social stressor? *Social Development, 27*(1), 34–44. https://doi.org/10.1111/sode.12246

Kertes, D. A., Liu, J., Hall, N. J., Hadad, N. A., Wynne, C. D. L., & Bhatt, S. S. (2017). Effect of pet dogs on children's perceived stress and cortisol stress response. *Social Development, 26*(2), 382–401. https://doi.org/10.1111/sode.12203

Kovack-Lesh, K. A., McMurray, B., & Oakes, L. M. (2014). Four-month-old infants' visual investigation of cats and dogs: Relations with pet experience and attentional strategy. *Developmental Psychology, 50*(2), 402–413. https://doi.org/10.1037/a0033195

Kurdek, L. A. (2009). Young adults' attachment to pet dogs: Findings from open-ended methods. *Anthrozoös, 22*(4), 359–369. https://doi.org/10.2752/089279309X12538695316149

Legg, C., Sherick, I., & Wadland, W. (1974). Reaction of preschool children to the birth of a sibling. *Child Psychiatry & Human Development, 5*(1), 3–39. https://doi.org/10.1007/bf01441311

Marsa-Sambola, F., Williams, J., Muldoon, J., Lawrence, A., Connor, M., & Currie, C. (2017). Quality of life and adolescents' communication with their significant others (mother, father, and best friend): The mediating effect of attachment to pets. *Attachment & Human Development, 19*(3), 278–297. https://doi.org/10.1080/14616734.2017.1293702

McCune, S., McCardle, P., Griffin, J. A., Esposito, L., Hurley, K., Bures, R., & Kruger, K. A. (2020). Editorial: Human-animal interaction (HAI) research: A decade of progress. *Frontiers in Veterinary Science, 7*, 44. https://doi.org/10.3389/fvets.2020.00044

McMinn, A. M., van Sluijs, E. M. F., Nightingale, C. M., Griffin, S. J., Cook, D. G., Owen, C. G., Rudnicka, A. R., & Whincup, P. H. (2011). Family and home correlates of children's physical activity in a multi-ethnic population: The cross-sectional child heart and health study in England (CHASE). *International Journal of Behavioral Nutrition and Physical Activity, 8*(11). https://doi.org/10.1186/1479-5868-8-11

McNicholas, J., & Collis, G. M. (2001). Children's representations of pets in their social networks. *Child: Care, Health and Development, 27*(3), 279–294. https://doi.org/10.1046/j.1365-2214.2001.00202.x

Meehan, M., Massavelli, B., & Pachana, N. (2017). Using attachment theory and social support theory to examine and measure pets as sources of social support and attachment figures. *Anthrozoös, 30*(2), 273–289. https://doi.org/10.1080/08927936.2017.1311050

Meints, K., Brelsford, V., & De Keuster, T. (2018). Teaching children and parents to understand dog signaling. *Frontiers in Veterinary Science, 5*, 257. https://doi.org/10.3389/fvets.2018.00257

Meints, K., & de Keuster, T. (2009). Brief report: Don't kiss a sleeping dog: The first assessment of "The Blue Dog" bite prevention program. *Journal of Pediatric Psychology, 34*(10), 1084–1090. https://doi.org/10.1093/jpepsy/jsp053

Meints, K., Racca, A., & Hickey, N. (2010). How to prevent dog bite injuries? Children misinterpret dogs facial expressions. *Injury Prevention, 16*(Suppl 1), A68. https://doi.org/10.1136/ip.2010.029215.246

Melson, G. F. (2003). Child development and the human-companion animal bond. *American Behavioral Scientist, 47*(1), 31–39. https://doi.org/10.1177/0002764203255210

Melson, G. F. (2005). *Why the wild things are: Animals in the lives of children*. Cambridge, MA: Harvard University Press.

Miles, J. N. V., Parast, L., Babey, S. H., Griffin, B. A., & Saunders, J. M. (2017). A propensity-score-weighted population-based study of the health benefits of dogs and cats for children. *Anthrozoös, 30*(3), 429–440. https://doi.org/10.1080/08927936.2017.1335103

Millot, J. L., Filiatre, J. C., Gagnon, A. C., Eckerlin, A., & Montagner, H. (1988). Children and their pet dogs: How they communicate. *Behavioural Processes*, *17*(1), 1–15. https://doi.org/10.1016/0376-6357(88)90046-0

Minatoya, M., Araki, A., Miyashita, C., Itoh, S., Kobayashi, S., Yamazaki, K., Ait Bamai, Y., Saijyo, Y., Ito, Y., Kishi, R., & The Japan Environment and Children's Study Group. (2020). Cat and dog ownership in early life and infant development: A prospective birth cohort study of Japan Environment and Children's Study. *International Journal of Environmental Research and Public Health*, *17*(1), 205. https://doi.org/10.3390/ijerph17010205

Mueller, M. K. (2014). Is human–animal interaction (HAI) linked to positive youth development? Initial answers. *Applied Developmental Science*, *18*(1), 5–16. https://doi.org/10.1080/10888691.2014.864205

Mueller, M. K., Anderson, E. C., King, E. K., & Urry, H. L. (2021). Null effects of therapy dog interaction on adolescent anxiety during a laboratory-based social evaluative stressor. *Anxiety, Stress & Coping*, *34*(4), 365–380. https://doi.org/10.1080/10615806.2021.1892084

Mueller, M. K., & Callina, K. S. (2014). Human–animal interaction as a context for thriving and coping in military-connected youth: The role of pets during deployment. *Applied Developmental Science*, *18*(4), 214–223. https://doi.org/10.1080/10888691.2014.955612

Mueller, M. K., King, E. K., Callina, K., Dowling-Guyer, S., & McCobb, E. (2021). Demographic and contextual factors as moderators of the relationship between pet ownership and health. *Health Psychology and Behavioral Medicine*, *9*(1), 701–723. https://doi.org/10.1080/21642850.2021.1963254

Mueller, M. K., King, E. K., Halbreich, E. D., & Callina, K. S. (2022). Companion animals and adolescent stress and adaptive coping during the COVID-19 pandemic. *Anthrozoös*, *35*(5), 693–712. https://doi.org/10.1080/08927936.2022.2027093

Muldoon, J. C., Williams, J. M., & Currie, C. (2019). Differences in boys' and girls' attachment to pets in early-mid adolescence. *Journal of Applied Developmental Psychology*, *62*, 50–58. https://doi.org/10.1016/j.appdev.2018.12.002

Muldoon, J. C., Williams, J. M., Lawrence, A., & Currie, C. (2019). The nature and psychological impact of child/adolescent attachment to dogs compared with other companion animals. *Society & Animals*, *27*(1), 55–74. https://doi.org/10.1163/15685306-12341579

NIH. (1987). *Health benefits of pets: Summary of working group*. Washington, DC: NIH Technology Assessment Workshop, US Dept. of Health and Human Services.

O'Haire, M. E., McKenzie, S. J., McCune, S., & Slaughter, V. (2013). Effects of animal-assisted activities with guinea pigs in the primary school classroom. *Anthrozoös*, *26*(3), 445–458. https://doi.org/10.2752/175303713X13697429463835

Owen, C. G., Nightingale, C. M., Rudnicka, A. R., Ekelund, U., McMinn, A. M., van Sluijs, E. M. F., Griffin, S. J., Cook, D. G., & Whincup, P. H. (2010). Family dog ownership and levels of physical activity in childhood: Findings from the child heart and health study in England. *American Journal of Public Health*, *100*(9), 1669–1671. https://doi.org/10.2105/AJPH.2009.188193

Ownby, D. R., Johnson, C. C., & Peterson, E. L. (2002). Exposure to dogs and cats in the first year of life and risk of allergic sensitization at 6 to 7 years of age. *JAMA*, *288*(8), 963–972. https://doi.org/10.1001/jama.288.8.963

Pendry, P., Carr, A. M., Gee, N. R., & Vandagriff, J. L. (2020). Randomized trial examining effects of animal assisted intervention and stress related symptoms on college students' learning and study skills. *International Journal of Environmental Research and Public Health*, *17*(6), 1909. https://www.mdpi.com/1660-4601/17/6/1909

Pendry, P., & Vandagriff, J. L. (2020). Salivary studies of the social neuroscience of human–animal interaction. In D. A. Granger & M. K. Taylor (Eds.), *Salivary bioscience* (pp. 555–581). Cham: Springer. https://doi.org/10.1007/978-3-030-35784-9_23

Pet Food Manufacturers' Association. (2019). *Families and pets*. https://www.pfma.org.uk/households-with-pets-and-children

Polheber, J. P., & Matchock, R. L. (2014). The presence of a dog attenuates cortisol and heart rate in the Trier Social Stress Test compared to human friends. *Journal of Behavioral Medicine*, *37*(5), 860–867. https://doi.org/10.1007/s10865-013-9546-1

Poresky, R. H. (1990). The young children's empathy measure: Reliability, validity and effects of companion animal bonding. *Psychological Reports*, *66*(3), 931–936. https://doi.org/10.2466/pr0.1990.66.3.931

Purewal, R., Christley, R., Kordas, K., Joinson, C., Meints, K., Gee, N., & Westgarth, C. (2017). Companion animals and child/adolescent development: A systematic review of the evidence. *International Journal of Environmental Research and Public Health*, *14*(3), 234. https://doi.org/10.3390/ijerph14030234

Robin, M., & ten Bensel, R. (1985). Pets and the socialization of children. *Marriage & Family Review*, *8*(3–4), 63–78. https://doi.org/10.1300/J002v08n03_06

Rodriguez, K. E., Herzog, H., & Gee, N. R. (2021). Variability in human-animal interaction research. *Frontiers in Veterinary Science*, *7*, 619600. https://doi.org/10.3389/fvets.2020.619600

Schwebel, D. C., Morrongiello, B. A., Davis, A. L., Stewart, J., & Bell, M. (2012). The Blue Dog: Evaluation of an interactive software program to teach young children how to interact safely with dogs. *Journal of Pediatric Psychology*, *37*(3), 272–281. https://doi.org/10.1093/jpepsy/jsr102

Steele, H., Steele, M., & Croft, C. (2008). Early attachment predicts emotion recognition at 6 and 11 years old. *Attachment & Human Development*, *10*(4), 379–393. https://doi.org/10.1080/14616730802461409

Timperio, A., Salmon, J., Chu, B., & Andrianopoulos, N. (2008). Is dog ownership or dog walking associated with weight status in children and their parents? *Health Promotion Journal of Australia*, *19*(1), 60–63. https://doi.org/10.1071/HE08060

Tulloch, J. S. P., Minford, S., Pimblett, V., Rotheram, M., Christley, R. M., & Westgarth, C. (2021). Paediatric emergency department dog bite attendance during the COVID-19 pandemic: An audit at a tertiary children's hospital. *BMJ Paediatrics Open*, *5*(1), e001040. https://doi.org/10.1136/bmjpo-2021-001040

Westgarth, C., Boddy, L. M., Stratton, G., German, A. J., Gaskell, R. M., Coyne, K. P., Bundred, P., McCune, S., & Dawson, S. (2013). Pet ownership, dog types and attachment to pets in 9–10 year old children in Liverpool, UK. *BMC Veterinary Research*, *9*(1), 102. https://doi.org/10.1186/1746-6148-9-102

Westgarth, C., Boddy, L. M., Stratton, G., German, A. J., Gaskell, R. M., Coyne, K. P., Bundred, P., McCune, S., & Dawson, S. (2017). The association between dog ownership or dog walking and fitness or weight status in childhood. *Pediatric Obesity*, *12*(6), e51–e56. https://doi.org/10.1111/ijpo.12176

Westgarth, C., Heron, J., Ness, A. R., Bundred, P., Gaskell, R. M., Coyne, K., German, A. J., McCune, S., & Dawson, S. (2012). Is childhood obesity influenced by dog ownership? No cross-sectional or longitudinal evidence. *Obesity Facts*, *5*(6), 833–844. https://doi.org/10.1159/000345963

Zilcha-Mano, S., Mikulincer, M., & Shaver, P. R. (2011). An attachment perspective on human–pet relationships: Conceptualization and assessment of pet attachment orientations. *Journal of Research in Personality*, *45*(4), 345–357. https://doi.org/10.1016/j.jrp.2011.04.001

# 16
# ANIMALS IN DEVELOPMENT
## Aging Adults

*Jessica Bibbo and Ann M. Toohey*

## Introduction to the Issue

Aging is a lifelong process. Older age does not begin at a milestone birthday such as 50 or 65. However, discussions on aging mainly focus on the second half of the lifespan. Likewise, this chapter will address pets in the lives of people in older adulthood, a period of change, and continuing development.

### Definitions and Conceptualizations of Aging

#### Successful Aging

Successful aging reflects the intersection of avoiding disease and disability, maintaining active engagement, and retaining high cognitive and physical functioning (Rowe & Kahn, 1997). While one of the first aging concepts to refute the inevitability of age-related decline, gerontologists have criticized this definition for overlooking essential subjective aspects of well-being. Successful aging excludes many older people from the opportunity to age "successfully" by not recognizing the multitude of factors which shape the three components of the model (Martinson & Berridge, 2015). The concepts of healthy aging and aging well address some of these limitations.

#### Healthy Aging

Healthy aging focuses on how physical, psychological, social, and environmental factors shape individuals' functional abilities, which in turn enable individuals to maintain well-being and live their values (World Health Organization [WHO], 2020). Functional abilities can be maximized within age-related physical and cognitive limitations through creating "age-friendly" environments (discussed later in this chapter) and empowering older adults' decision-making.

#### Aging Well

Aging well is an inclusive, descriptive, and asset- (rather than deficit-) focused conceptualization of later life (Chapman, 2005). It is aligned with a critical or social constructivist ontological view, which hinges on individuals' own perceptions of their aging trajectory. For instance, those who are living in relative social isolation or with chronic illness or disability may not, by definition, be aging successfully. Yet such older persons may be aging well through a personalized sense of having a meaningful existence, even as they adapt to changes precipitated by life transitions.

## *Ageism*

The WHO Baseline Report on the Decade of Healthy Aging (WHO, 2020) concludes negative views of older adults are persistent. All countries need to adopt programs to redress ageism, defined as age-based discrimination, prejudice, and stereotyping. A common form of ageism within the domain of human-animal interactions is benevolent ageism, where certain rules or conventions are imposed in relation to "paternalistic prejudice" (Cary et al., 2017, p. e28). An example of this is implementing 'no pets' rules in subsidized housing facilities to prevent older adults from making what are considered impractical decisions around pet-related responsibilities (Toohey & Krahn, 2018).

## **Pet Ownership/Guardianship Later in Life**

### *Prevalence*

Mueller and colleagues (2018) found that 51.5% of Americans 50 years and older had a companion animal. In Canada, approximately one third of older adults (i.e., 65 years or older) report living with a companion animal (Toohey et al., 2018). While pets are commonplace in Western countries like the United States, Canada, the United Kingdom, and Australia, pet animals are also increasing in non-Western countries like China and Japan (Headey et al., 2007; Taniguchi et al., 2018). Prevalence of pets for older adults in different countries is difficult to compare, however, since it is rare to see national estimates that consider comparable age categories.

### *Reasons for Living With and Without Pets*

While pet ownership continues throughout the lifespan, rates tend to decline as age increases (Mueller et al., 2018; Toohey et al., 2018). Limitations in physical or cognitive functioning may be reasons for this change. Older adults living with others are more likely to have a pet, as are those in good overall health (Friedmann et al., 2020). Older adults may be less likely to bring in a new pet due to concerns over being outlived by a pet and thereby being unable to fulfill their commitment to their companion (Anderson et al., 2015; Chur-Hansen et al., 2008). While the likelihood of having a pet declines with older age, the strength of the bond does not (Bibbo et al., 2019). Companionship and enjoying the company of animals remain the most common reasons for seeking pets later in life.

## **Aging in Place**

Most older adults would prefer to have choices about where to age (Wiles et al., 2011) and many wish to live independently in their homes for as long as it is possible to safely do so. Efforts to support these desires have resulted in a movement to create "age-friendly" communities (Menec et al., 2011; WHO, 2007), discussed in more detail below. Ideally, age-friendly communities enable independence and social inclusion of older adults. Both individual and systemic factors shape opportunities for older adults to maintain meaningful lives and social connections as they grow older. Similar factors also shape relationships with pets later in life and can influence whether having a companion animal is even possible (Toohey, 2023).

## Aging in Care Facilities

The transition from the home to a facility is a stressful period involving multiple domains. Adults facing this transition need to be actively involved in the decision-making process to facilitate a quicker adjustment to and happiness in the new home (Johnson & Bibbo, 2014; Miller et al., 2016). Pets can be a barrier to making this transition when facilities do not allow residents to bring their pet with them.

While the proportion of long-term care facilities (i.e., assisted living and nursing homes) allowing pets appears to be growing, the policies surrounding pet ownership and visitation remain highly variable (Stull et al., 2018). Older adults living with personal pets credit them with facilitating new interpersonal relationships and creating continuity with their lives prior to the transition, in addition to experiencing the inherent benefits of animal companionship discussed below (Freedman et al., 2021).

## Cultural/Religious and/or Historical Perspectives

### Theoretical Frameworks

Individual-level theories presented in the first chapter of this volume (e.g., attachment, social support, and the bio psychosocial model) have been used to explore how pets influence physical and psychological outcomes in older adulthood (e.g., Friedmann et al., 2020; Garrity et al., 1989; Krause-Parello, 2012). Such studies have offered essential empirical evidence and are presented or referenced throughout this chapter. Additional frameworks present opportunities to capture ways that pets might influence healthy aging or aging well.

Gerontological models (e.g., the convoy model which accounts for how relationships change with age, [Antonucci & Akiyama, 1987]) and theories (e.g., socioemotional selectivity theory which conceptualizes how relationships are prioritized differently as perceived lifetime remaining decreases [Carstensen, 1992]) offer underutilized opportunities for understanding people's relationships to their pets in older adulthood (Bibbo et al., 2019). The principles of the life course perspective (i.e., considering development as a lifelong process, human agency, timing in lives, historical time and place, and linked lives, [Elder et al., 2003]) can provide a more comprehensive understanding of pets in our lives, while acknowledging the factors involved in healthy aging (Bibbo et al., 2019).

Drawing from public health, the socio-ecological framework recognizes multiple layers of influence on people's health, including broad social and cultural factors (Richard et al., 2011). Most research on the prospective influence of companion animals on health and well-being occurs at intra-personal level (i.e., attachments [Crawford et al., 2006] or individual ecologies [Putney, 2013]). Other levels of influence may be equally poignant, for instance interpersonal factors (e.g., supportive social networks); community efforts (e.g., organized pet care support or pet meal programs); and policy-level supports (e.g., pet-friendly housing rules and subsidized veterinary care). A "relational" socio-ecological framework (Toohey et al., 2017) also engages with principles of relational public health ethics, which offer a relational view of autonomy, solidarity and personhood as shaped by social standing (Baylis et al., 2008). Applying this framework recognizes the socio-economic and other forms of disadvantage that may put older adults with pets in precarious situations (Toohey et al., 2017; Toohey & Rock, 2019) and thus may disrupt healthy aging and aging well.

## Impact on People

### Physical Health

The evidence on the impact of pets on older adults' physical health and functioning is mixed. Two systematic reviews (i.e., Gee & Mueller, 2019; Obradović et al., 2020) provide excellent overviews on this topic. The most consistent findings include the positive impact of dog walking on physical and mental health. Dog walking is recognized as simultaneously providing adults an opportunity for physical activity and social engagement (Wood et al., 2005; Wood et al., 2015). The COVID-19 pandemic provided an exceptional opportunity to investigate these associations. Older adults who walked their pet dogs daily experienced less loneliness resulting from the decreased social interactions in the first seven months of the pandemic (Carr et al., 2021). While not all people who live with dogs walk those dogs, those who do may reap the benefits of regular physical activity (Curl et al., 2017; Toohey et al., 2013). The strength of an older person's bond with their companion dog may increase dog-walking (Curl et al., 2017) and also provide motivation to engage in other habits and behaviors that support physical health and well-being.

The human-animal bond with any species can be a motivator for older adults to engage in habits and behaviors that support physical health and well-being For instance, strong bonds with pets have been noted to influence older adults' recovery from acute conditions (e.g., stroke [Johansson et al., 2014]), living with chronic conditions including diabetes (Peel et al., 2010) and chronic pain (Janevic et al., 2019), and to help older adults maintain activities of daily living (Raina et al., 1999). Strong bonds with pets and felt responsibilities for their care can also impede health behaviors, such as delays in seeking needed health care or leaving a hospital against medical advice (Applebaum et al., 2021a; Polick et al., 2021).

### Psychological and Social Health

Studies have found pet ownership associated with both lower and higher rates of depression and loneliness in older adulthood (Gee & Mueller, 2019). Factors that may impact these associations, particularly with loneliness, include interpersonal social support, social isolation, type of companion animal, strength of attachment, and gender (Antonacopoulos & Pychyl, 2010; Branson et al., 2017; Ikeuchi et al., 2021; Krause-Parello, 2012; Pikhartova et al., 2014; Stanley et al., 2014).

Though direct statistical associations between pets and psychological outcomes are inconsistent, the qualitative literature provides robust evidence for the meaning of pets in later life. Caring for a pet creates a meaningful role, which in turn fosters purpose and structure within one's day (Gan et al., 2019; Obradović et al., 2020). The "need to be needed" can be particularly important for older people following a loss, or for those who live alone

(Mueller & Hunter, 2019; Reyes Uribe, 2020). Quantitative studies reinforce the buffering effect pets have on the psychological impacts of loss (Carr et al., 2020; Raina et al., 1999).

Considering aging-in-place, adults living with pets, and especially dogs, report higher social capital within their neighborhoods than those without (Wood et al., 2017). Others have also credited both cats and dogs with facilitating engagement with people (Knight & Edwards, 2008; Mahalski et al., 1988). Dog walking in particular catalyzes social interactions and fosters sense of community, in addition to its physical benefits (Mueller & Hunter, 2019; Toohey et al., 2013). Having a pet may also help offset negative experiences of social isolation experienced by some older adults (Toohey, 2023; Toohey et al., 2018).

## *Health Promotion*

As discussed above, pets support physical, mental, and emotional health and help older adults cope with loss and other adverse events later in life. These prospective impacts combined with the popularity of pets have sparked interest in pets for their relevance to public health, specifically to health promotion. However, when access to companion animals and the prospective benefits they offer is disrupted across socio-economic lines, this becomes a matter of social justice. Social isolation may also increase the likelihood that an older adult may be unable to access the benefits of having a pet. This may be especially true for older adults facing physical challenges or cognitive declines that impede their ability to care for a pet (McLennan et al., 2022; Toohey et al., 2017). Ways that social conditions shape experiences of people and pets merits researchers' consideration (e.g., Applebaum et al., 2021b; Power, 2017; Toohey & Rock, 2019). For instance, older adults who can afford private housing may keep their pets, yet those who must rely on affordable housing are often expected to relinquish them (Toohey & Krahn, 2018). Disrupting relationships with pets for older adults also disrupts equitable opportunities for enhanced mental, physical, and social well-being throughout later life.

## *Living with Dementia*

Dementia is not a normal part of aging, though aging is the greatest risk factor for developing dementia (WHO, 2021). People living with dementia have a better quality of life when they are able to engage in relationships and activities that they find valuable (Hennelly et al., 2021; Martyr et al., 2018). Relationships with pets and pet care activities may offer opportunities to do both. For instance, people living with Alzheimer's disease found pets to be judgment-free companions whose non-verbal communication and touch were both comfortable and comforting (Shell, 2015). Active involvement in pet care may also contribute to a better quality of life for people with mild-to-moderate dementia. In a cross-sectional study surveying people living with dementia, Opdebeeck and colleagues (2020) found being involved in pet care activities was associated with lower depression and a higher quality of life compared to people living with a pet who were not involved in the pet's care. A longitudinal study found that people living with pets had fewer activities of daily living (ADL) limitations, slower disease progression, and less-severe symptoms over a five-year period than those without pets, although these statistical differences may not translate into clinical significance (Rusanen et al., 2021). However, the welfare and well-being of a pet must also be considered as a person's dementia progresses and functioning changes. People living with dementia are likely to require more assistance with pet care as the disease progresses, while at the same time be less likely than people without dementia to ask for or recognize the need for assistance (Bibbo et al., 2022).

## Impact on Animals

Older adults who are past the age of retirement may have more time to spend with their pets, which has positive effects on mental health (Bennett et al., 2015) and also offers benefits for pets. However, inordinate attachments to pets may lead to negative health outcomes for older adults under certain conditions (Chur-Hansen et al., 2009). Some older adults may prioritize pet care over their own needs to avoid the possibility of separation (McNicholas et al., 2005; Polick et al., 2021; Toohey & Rock, 2019). While this strategy may support a pet's welfare in the short term, additional supports for human-animal relationships may be needed in the long term (Toohey, 2023).

### *Risks to Animal Welfare*

The following three issues illustrate the connections between the health and welfare of a person and their pet. The next sections in this chapter suggest ways to address, mitigate, and ideally prevent these situations.

#### *Relinquishment*

Multiple studies confirm that housing issues are among the most significant human-factor predictors of animal relinquishments (Coe et al., 2014; Ly et al., 2021). Transitions into senior housing or assisted living often leave older adults in situations where they must choose between their pet and a home (McNicholas, 2014; Ormerod, 2012; Toohey et al., 2017). The impact on a pet relinquished to a shelter will likely be detrimental, given that older animals are at a higher risk of euthanasia (Hawes et al., 2018).

#### *Neglect*

There is some indication that loneliness, in addition to socio-economic factors such as low income and education, may predict animal neglect (Monsalve et al., 2018). Unintentional neglect may also result from changes in cognitive and physical functioning. In addition to dementia (discussed above), self-neglect is an issue that may be encountered in late life and that can lead to companion animal neglect. Individuals who self-neglect almost certainly create living conditions that are sub-optimal for their companion animals as well (Boat & Knight, 2000). Further, self-neglect has been associated with animal hoarding (Nathanson, 2009).

#### *Hoarding*

A comprehensive systematic review found that many animal hoarders are females in their 50s and 60s; likely to be single, divorced or widowed; and are often living alone (Emmett et al., 2021). Cognitive and/or executive deficits may also be contributing factors (Emmett et al., 2021). The hoarding of animals, which is distinct from the hoarding of objects, is discussed in detail in Chapter 22 of this volume.

## Practical Perspectives

### *The Need for Planning*

The health transitions that both people and their pets experience in later adulthood are more easily managed when pre-emptive plans are made (Anderson et al., 2015; Bibbo et al., 2022; McNicholas, 2014). Changes in health, physical, and cognitive functioning may impair an older person's ability to perform pet care tasks. Similar changes in a pet will likely require more of an older adults' resources, including time dedicated to pet care and finances. Open conversations with family and friends can create strategies to actively manage the shifting health and well-being of people and pets. As already discussed, moving from independent to facility-based living can be a difficult transition. Many facilities do not allow pets, adding a significant stressor to this difficult change (Anderson et al., 2015; McNicholas, 2014). Planning for this transition must include discussions about the pet, which is often cited as the primary delay to a potentially critical move.

Loss of family and friends becomes more common as age increases. As a result, older people develop more effective strategies to manage and adapt to loss. Older adults are often better able to effectively cope with the death of a pet than younger adults (Adams et al., 2000; Kimura et al., 2014), although many will also experience disenfranchised (unacknowledged) grief (Wilson et al., 2021). Unlike younger adults, older people may grieve both the loss of their companion and the loss of their role as pet owner/guardian. Older people living toward the end of the lifespan are more likely than their younger counterparts to consider anticipated changes in abilities (e.g., physical, cognitive) and resources (e.g., finances, housing) that impact their ability to provide for a pet's health and well-being (Anderson et al., 2015; Chur-Hansen et al., 2008). The future needs of an animal companion may be prioritized over the desire to share one's life with a companion animal, resulting in the loss of a meaningful part of everyday life. Arranging for alternatives to pet ownership (e.g., fostering and volunteering, which are discussed below) may ease this transition. Additionally, the death of a person has a direct impact on pets (Anderson et al., 2015). Without a plan identifying someone who is able and willing to take on the responsibility of caring for a pet, the pet may be relinquished to a shelter.

### *Implications for Caregivers*

Older individuals who rely on caregivers for activities of daily living are likely to need assistance in caring for their pets (Bibbo & Proulx, 2018). Pet assistance centers on daily needs (e.g., providing food and water, managing waste) and buying pet food and supplies. Family or friend caregivers may also assist with veterinary care, including scheduling and transportation for appointments and managing medications. However, they also may play and cuddle with the pet. Overall, the care recipient's pet may impact burden, satisfaction, and mastery of care delivery (Bibbo & Proulx, 2019).

The number of caregivers and people requiring care will continue to increase as the population ages. The relationship with a pet may change in positive or negative ways for people on each side of care provision (Connell et al., 2007). Paid caregivers are often not required or even allowed to include pet care tasks as part of the services they provide for their clients (Toohey et al., 2017). Further, clients' pets can present a stressor (e.g., unsanitary conditions, smell) and even danger (e.g., aggressive behavior, bite risk) for professional caregivers and other practitioners who enter their clients' homes (Muramatsu et al., 2019;

Owczarczak-Garstecka et al., 2019). Severe conditions may result in reluctance or refusal to enter a home (Boat & Knight, 2000). Most current in-home services overlook or are unable to address pet care needs within a home (Toohey et al., 2017). As discussed below, community-based volunteer programs may help fill this gap.

## *Implications for Practitioners*

For service providers, including questions about current and past pet ownership at intake will provide a more complete understanding of an adult's family and support network, along with potential resources, stressors, and risks. Professionals who enter clients' homes may be more likely to raise issues related to pet ownership (e.g., basic pet care needs, planning for the pet due to the older adult having to move) with their clients than those who do not enter their clients' homes (Bibbo et al., 2022). Stigma or embarrassment may make adults and caregivers reluctant to bring up the topic of pets during health care conversations despite their being important family members that can impact health care behaviors and decisions (Ryan & Ziebland, 2015). Including pets in conversations and care planning can help build rapport, promote open dialogue, and encourage compliance (Boat & Knight, 2000; Hodgson et al., 2017, 2020).

## *Cross-Sector Collaboration*

A substantial challenge to overcome is the historic separation of organized efforts that address human needs and those that address animal welfare (Cryer et al., 2021; Obradović et al., 2020; Rauktis et al., 2017; Toohey et al., 2017). Programs built on a foundation of cross-sectoral partnerships are necessary to face this challenge. Such collaborations hold promise for preventing older adults and their companion animals from falling through the cracks. Volunteer-based pet care assistance programs have a two-pronged impact, as they both redress social isolation and protect animal welfare (Cryer et al., 2021; McLennan et al., 2022; Toohey et al., 2017). These programs largely rely on community members to perform basic pet care tasks in community or assisted living settings. Such programs also help to redress the root causes of benevolent ageism that may disrupt pet ownership, by supporting older adults with pet care as challenges arise. Lastly, such programs may also be beneficial in short-term caregiving scenarios (e.g., acute conditions or diseases such as cancer [Trigg, 2021]).

## *Common Constraints to Animal Companionship*

Lower income levels and increased social isolation may put older adults at higher risk of experiencing adverse life conditions. Older adults in these circumstances often face numerous barriers to maintaining relationships with pets (Applebaum, et al., 2021b; Toohey et al., 2017; Toohey & Rock, 2019). For instance, appropriate housing choices may be scarce for older adults who do not own their homes when a companion animal is involved, given the typically low supply of affordable, pet-friendly rental housing available (Huss, 2014; Matsuoka et al., 2020; Toohey & Krahn, 2018; Toohey & Rock, 2019). Affordability of appropriate veterinary care may also become challenging, and short-term health or long-term mobility challenges may disrupt older adults' abilities to take care of their pets (Anderson et al., 2015; Toohey et al., 2017). Transportation challenges also pose a major barrier for older adults, who may be unable to access veterinary services and other needs, including grooming (McLennan et al., 2022).

## Current Policy and Legislative Perspectives

### *Age Friendly Framework*

The WHO age-friendly movement was launched in 2007 to create physical and social environments that support an aging population (WHO, 2007). While several domains linked to age-friendliness are relevant to having pets (e.g., housing, social inclusion, outdoor spaces, and others), consideration of companion animals is absent from these guidelines (Toohey et al., 2017). More recently, however, the WHO declared 2021–2030 as the Decade of Healthy Aging. Within this framework, animals are noted, alongside other people, for their role in providing fundamental emotional support for older adults (WHO, 2020, p. 13). This recognition repositions pets to become a recognized part of age-inclusive policies and planning efforts.

### *Elder Maltreatment/Abuse*

Elder maltreatment (i.e., physical, psychological, sexual, financial abuse and/or neglect) is experienced by approximately one in six older adults (Yon et al., 2017). Links between the abuse of older people and pets have received less attention than abuse of younger adults and children (see Chapter 23), though some evidence affirms a similar link (Boat & Knight, 2000). Adult Protective Service (APS) professionals advise that considerations of pets should be included in the intake of clients (Peak et al., 2012). The result has been calls for cross-reporting between APS and humane societies, including legislation to assign mandated reporter status of elder abuse to animal professionals (e.g., veterinarians, humane officers) (Vincent et al., 2020).

## Ethical Considerations

### *The Right vs. Privilege of Pet Ownership*

Having a pet is increasingly viewed as a basic civil right for the responsible pet owner (Huss, 2014). Practically speaking, however, pet ownership/guardianship remains a privilege available only to those who have the resources available to take proper care of their pets. Older adults are particularly vulnerable to restrictions on the rights to have companion animals (Toohey & Krahn, 2018), as discussed throughout this chapter. Our societies face the ethical imperative of ensuring that companion animal relationships are recognized within a civil rights context, and that appropriate supports are in place that both protect this right while also protecting animal welfare. Creating accessible supports that reflect collective responsibility for supporting older people and pets will catalyze this shift. This thinking also reflects the increasing sentiment that pets are family members (Applebaum et al., 2021b; Toohey & Krahn, 2018).

## Unanswered Questions and Future Research Demands

### *Transformation of the Bond Across the Life Course*

Longitudinal studies are needed to advance the field of human-animal interaction (see, for example, Gee & Mueller, 2019). Current quantitative studies are largely cross-sectional, while qualitative work generally occurs within a single stage of life. Existing literature allows us to compare the impact of the human-animal bond between people of different ages.

However, we cannot disentangle the influence of historical time and place and other potential cohort effects that may factor into differences between people of different ages at the same point in time. While the strength of the human-animal bond does not appear to change with age, it is unclear whether this is also the case for the meaning of pets in peoples' lives. It is likely that the relational quality of pet ownership transforms due to major life transitions (e.g., parenthood, marital status, retirement), changes in functioning, and stage of life. Longitudinal qualitative work will help provide a more thorough understanding of whether and how relationships with pets change over the life course.

### *Diversity and Companion Animal Relationships Later in Life*

The effects of social and economic factors on health and well-being are increasingly recognized within and outside our field (e.g., Applebaum et al., 2021b). Future research addressing ethnicity, socio-economic status, gender, and sexual identities, along with both physical and intellectual abilities, is imperative for a more accurate understanding of the relationships with and the salience of pets for an increasingly diverse aging population.

### *Robotic Pets – An Emerging Phenomenon*

The use of robotic pets in care settings for people living with dementia is not new (Koh et al., 2021; Mendes & Palmer, 2016; Minmin et al., 2019). In 2015, Ageless Innovation LLC introduced affordable Joy For All© animatronic pets for families whose older loved ones were lonely or had cognitive impairments. Social service programs have now begun to distribute these devices to socially isolated older adults living in the community as well as in congregate care settings. Several studies have reported positive outcomes of interactions with robotic companion animals, both for community-dwelling older adults (Hudson et al., 2020; Tkatch et al., 2021) and those living in facilities (Koh et al., 2021; Minmin et al., 2019). Use of robotic pets may protect live animals from potential harms in situations where reduced cognitive faculties lead to neglect or mistreatment of animals. In scholarly circles, however, ethical debates around dignity and deception in relation to robotic pets for people living with dementia continue (Chiberska, 2018; Lazar et al., 2016). Future research in this area will help to advance the complexities of this timely phenomenon, by exploring how to balance the benefits that a robotic pet may bring with the prevailing stigma aimed toward older adults who form strong bonds with inanimate, if interactive, objects (Hudson et al., 2020). The social implications of introducing robotic pets into people's lives, whether they live independently or in congregate care facilities, also merits deeper consideration.

### *Promising Practices*

There may be situations where it is not practical for older adults to take on or maintain responsibility for a companion animal (Chur-Hansen et al., 2008). Alternatives to pet-ownership may enable older adults to continue to benefit from human-animal relationships. For instance, multi-generational homes lend themselves to sharing pet care tasks and costs of pet care. The experience of a communal pet differs from having one's own (Freedman et al., 2021) yet still may be preferable to living without any companion animal at all. Joining family, friends or neighbors on dog walks is another way to leverage the benefits, even when it is not possible to have a pet (Toohey & Rock, 2011).

Communal pets may also be introduced into subsidized housing or congregate care facilities. Novel arrangements involve fostering animals for local animal shelters, for instance cats that are awaiting adoption. While pet care assistance programs, as described above, can help keep older adults and their pets together longer, assistance with fostering could also be explored (McLennan et al., 2022). Supported fostering would proffer benefits of voluntarism (Morrow-Howell et al., 2003) while also affording older adults access to companion animals without some of the challenges of sole responsibility. Fostering is just one type of volunteer activity related to companion animals. Encouraging and facilitating older adults' participation in other forms of volunteer work with animal welfare organizations may be particularly beneficial for people who are not able to live with a pet due to constraints such as housing policies or financial resources.

Other promising supports that help older adults to keep their pets include volunteer-based pet care assistance programs, low-cost veterinary services or charitable programs that offset veterinary costs, transportation programs that help older adults with pet care-related needs, temporary fostering arrangements for when older adults require hospitalization and rehabilitation, and pet-friendly shelters for older adults fleeing abusive situations (McLennan et al., 2022).

Animal-assisted interventions (AAI) are another opportunity for people to interact with companion animals. Section 4 in this volume provides information on the impact of AAI on older adults and people living with dementia. The older adults in these studies were the recipients of an activity or intervention, and were not participating with their own companion animal. When appropriate, becoming certified and volunteering with an organization that provides human-animal team visitation may provide an opportunity for older adults living with a pet to engage in their community. Volunteering with a pet is a unique activity integrating the human-animal bond into volunteerism.

## Potential Questions for Discussion

The following questions are presented to begin conversations and critical thinking.

### Disentangling or Tying Knots?

Why are older adults' relationships with pets often viewed as dispensable, even as pets are increasingly experienced as family and proffer an array of health-supportive benefits? How can the interdependencies of human and companion animal health and well-being be recognized and leveraged at a policy level? Is the view that individuals are solely responsible for pets still relevant? Is there a duty to extend social safety nets to protect human-animal bonds?

### How Do We Keep Animal Welfare in the Conversation?

What are some specific types of cross-sectoral collaborations that could benefit aging adults and their pets? In what ways could practitioners also benefit from these collaborations?

### How Do We Preserve Both Safety and Autonomy?

How can we determine when pet ownership is no longer appropriate? What factors make intervention necessary? When it is necessary, how can family, friends, and practitioners intervene to best maintain an adult's right to self-determination while ensuring the health and welfare of the person and their pets? What other alternatives to pet ownership can we promote and support?

# References

Adams, C. L., Bonnett, B. N., & Meek, A. H. (2000). Predictors of owner response to companion animal death in 177 clients from 14 practices in Ontario. *Journal of the American Veterinary Medical Association, 217*(9), 1303–1309. https://doi.org/10.2460/javma.2000.217.1303

Anderson, K. A., Lord, L. K., Hill, L. N., & McCune, S. (2015). Fostering the human-animal bond for older adults: Challenges and opportunities. *Activities, Adaptation & Aging, 39,* 32–42.

Antonacopoulos, N. M. D., & Pychyl, T. A. (2010). An examination of the potential role of pet ownership, human social support and pet attachment in the psychological health of individuals living alone. *Anthrozoös, 23*(1), 37–54. https://doi.org/10.2752/175303710X12627079939143

Antonucci, T. C., & Akiyama, H. (1987). Social networks in adult life and a preliminary examination of the convoy model. *Journal of Gerontology, 42*(5), 519–527. https://doi.org/10.1093/geronj/42.5.519

Applebaum, J. W., Ellison, C., Struckmeyer, L., Zsembik, B. A., & McDonald, S. E. (2021a). The impact of pets on everyday life for older adults during the COVID-19 pandemic. *Frontiers in Public Health, 9,* 652610. https://doi.org/10.3389/fpubh.2021.652610

Applebaum, J. W., MacLean, E. L., & McDonald, S. E. (2021b). Love, fear, and the human-animal bond: On adversity and multispecies relationships. *Comprehensive Psychoneuroendocrinology, 7,* 100071. https://doi.org/10.1016/j.cpnec.2021.100071

Baylis, F., Kenny, N. P., & Sherwin, S. (2008). A relational account of public health ethics. *Public Health Ethics, 1*(3), 196–209. https://doi.org/10.1093/phe/phn025

Bennett, P. C., Trigg, J. L., Godber, T., & Brown, C. (2015). An experience sampling approach to investigating associations between pet presence and indicators of psychological wellbeing and mood in older Australians. *Anthrozoös, 28*(3), 403–420. https://doi.org/10.1080/08927936.2015.1052266

Bibbo, J., Curl, A. C., & Johnson, R. A. (2019). Pets in the lives of older adults: A Life course perspective. *Anthrozoös, 32*(4), 541–554. https://doi.org/10.1080/08927936.2019.1621541

Bibbo, J., Johnson, J., Drost, J. C., Sanders, M., & Nicolay, S. (2022). Pet ownership issues encountered by geriatric professionals: Preliminary findings from an interdiciplinary sample. *Frontiers in Psychology,* 5212. https://doi.org/10.3389/fpsyg.2022.920559

Bibbo, J., & Proulx, C. M. (2018). The impact of a care recipient's pet on the instrumental caregiving experience. *Journal of Gerontological Social Work, 61*(6), 675–684. https://doi.org/10.1080/01634372.2018.1494659

Bibbo, J., & Proulx, C. M. (2019). The impact of a care recipient's pet on caregiving burden, satisfaction, and mastery: A pilot investigation. *Human-Animal Interaction Bulletin, 79*(2), 81–102.

Boat, B. W., & Knight, J. C. (2000). Experiences and needs of adult protective services case managers when assisting clients who have companion animals. *Journal of Elder Abuse & Neglect, 12*(3–4), 145–155. https://doi.org/10.1300/J084v12n03_07

Branson, S., Boss, L., Cron, S., & Turner, D. (2017). Depression, loneliness, and pet attachment in homebound older adult cat and dog owners. *Journal of Mind and Medical Sciences, 4*(1), 38–48. https://doi.org/10.22543/7674.41.P3848

Carr, D., Friedmann, E., Gee, N. R., Gilchrist, C., Sachs-Ericsson, N., & Koodaly, L. (2021). Dog walking and the social impact of the COVID-19 pandemic on loneliness in older adults. *Animals, 11*(7), 1852. https://doi.org/10.3390/ani11071852

Carr, D. C., Taylor, M. G., Gee, N. R., & Sachs-Ericsson, N. (2020). Psychological health benefits of companion animals following a social loss. *The Gerontologist, 60*(3), 428–438. https://doi.org/10.1093/geront/gnz109

Carstensen, L. L. (1992). Social and emotional patterns in adulthood: Support for socioemotional selectivity theory. *Psychology and Aging, 7*(3), 331–338. https://doi.org/10.1037/0882-7974.7.3.331

Cary, L. A., Chasteen, A. L., & Remedios, J. (2017). The ambivalent ageism scale: Developing and validating a scale to measure benevolent and hostile ageism. *The Gerontologist, 57*(2), e27–e36. https://doi.org/10.1093/geront/gnw118

Chapman, S. A. (2005). Theorizing about aging well: Constructing a narrative. *Canadian Journal on Aging/La Revue Canadienne Du Vieillissement, 24*(01), 9–18. https://doi.org/10.1353/cja.2005.0004

Chiberska, D. (2018). The use of robotic animals in dementia care: Challenges and ethical dilemmas. *Mental Health Practice, 21*(10), 23–28. https://doi.org/10.7748/mhp.2018.e1342

Chur-Hansen, A., Winefield, H., & Beckwith, M. (2008). Reasons given by elderly men and women for not owning a pet, and the implications for clinical practice and research. *Journal of Health Psychology, 13*(8), 988–995. https://doi.org/10.1177/1359105308097961

Chur-Hansen, A., Winefield, H. R., & Beckwith, M. (2009). Companion animals for elderly women: The importance of attachment. *Qualitative Research in Psychology*, *6*(4), 281–293. https://doi.org/10.1080/14780880802314288

Coe, J. B., Young, I., Lambert, K., Dysart, L., Nogueira Borden, L., & Rajić, A. (2014). A scoping review of published research on the relinquishment of companion animals. *Journal of Applied Animal Welfare Science*, *17*(3), 253–273. https://doi.org/10.1080/10888705.2014.899910

Connell, C. M., Janevic, M. R., Solway, E., & McLaughlin, S. J. (2007). Are pets a source of support or added burden for married couples facing dementia? *Journal of Applied Gerontology*, *26*(5), 472–485. https://doi.org/10.1177/0733464807305180

Crawford, E. K., Worsham, N. L., & Swinehart, E. R. (2006). Benefits derived from companion animals, and the use of the term "attachment." *Anthrozoös*, *19*(2), 98–112. https://doi.org/10.2752/089279306785593757

Cryer, S., Henderson-Wilson, C., & Lawson, J. (2021). Pawsitive connections: The role of pet support programs and pets on the elderly. *Complementary Therapies in Clinical Practice*, *42*, 101298. https://doi.org/10.1016/j.ctcp.2020.101298

Curl, A. L., Bibbo, J., & Johnson, R. A. (2017). Dog walking, the human–animal bond and older adults' physical health. *The Gerontologist*, *57*(5), 930–939. https://doi.org/10.1093/geront/gnw051

Elder, G. H., Johnson, M. K., & Crosnoe, R. (2003). The emergence and development of life course theory. In J. T. Mortimer & M. J. Shanahan (Eds.), *Handbook of the Life Course* (pp. 3–19). US: Springer. https://doi.org/10.1007/978-0-306-48247-2_1

Emmett, L., Kasacek, N., & Stetina, B. U. (2021). Demographic characteristics of individuals who abuse animals: A systematic review. *People and Animals: The International Journal of Research and Practice*, *4*(1), 3. https://docs.lib.purdue.edu/paij/vol4/iss1/3

Freedman, S., Paramova, P., & Senior, V. (2021). 'It gives you more to life, it's something new every day': An interpretative phenomenological analysis of wellbeing in older care home residents who keep a personal pet. *Ageing & Society*, *41*(9), 1961–1983. https://doi.org/10.1017/S0144686X19001880

Friedmann, E., Gee, N. R., Simonsick, E. M., Studenski, S., Resnick, B., Barr, E., Kitner-Triolo, M., & Hackney, A. (2020). Pet ownership patterns and successful aging outcomes in community dwelling older adults. *Frontiers in Veterinary Science*, *7*, 293. https://doi.org/10.3389/fvets.2020.00293

Gan, G., Hill, A.-M., Yeung, P., Keesing, S., & Netto, J. (2019). Pet ownership and its influence on mental health in older adults. *Aging & Mental Health*, *24*, 1–8. https://doi.org/10.1080/13607863.2019.1633620

Garrity, T. F., Stallones, L., Marx, M. B., & Johnson, T. P. (1989). Pet ownership and attachment as supportive factors in the health of the elderly. *Anthrozoös*, *3*(1), 35–44. https://doi.org/10.2752/089279390787057829

Gee, N. R., & Mueller, M. K. (2019). A systematic review of research on pet ownership and animal interactions among older adults. *Anthrozoös*, *32*(2), 183–207. https://doi.org/10.1080/08927936.2019.1569903

Hawes, S., Kerrigan, J., & Morris, K. (2018). Factors informing outcomes for older cats and dogs in animal shelters. *Animals*, *8*(3), 36. https://doi.org/10.3390/ani8030036

Headey, B., Na, F., & Zheng, R. (2007). Pet dogs benefit owners' health: A 'natural experiment' in China. *Social Indicators Research*, *87*(3), 481–493. https://doi.org/10.1007/s11205-007-9142-2

Hennelly, N., Cooney, A., Houghton, C., & O'Shea, E. (2021). Personhood and dementia care: A qualitative evidence synthesis of the perspectives of people with dementia. *The Gerontologist*, *61*(3), e85–e100. https://doi.org/10.1093/geront/gnz159

Hodgson, K., Darling, M., Freeman, D., & Monavvari, A. (2017). Asking about pets enhances patient communication and care: A pilot study. *Inquiry: The Journal of Health Care Organization, Provision, and Financing*, *54*, 0046958017734030. https://doi.org/10.1177/0046958017734030

Hodgson, K., Darling, M., Monavvari, A., & Freeman, D. (2020). Patient education tools: Using pets to empower patients' self-care—a pilot study. *Journal of Patient Experience*, *7*(1), 105–109. https://doi.org/10.1177/2374373518809008

Hudson, J., Ungar, R., Albright, L., Tkatch, R., Schaeffer, J., & Wicker, E. R. (2020). Robotic pet use among community-dwelling older adults. *The Journals of Gerontology: Series B*, *75*(9), 2018–2028. https://doi.org/10.1093/geronb/gbaa119

Huss, R. J. (2014). Reevaluating the role of companion animals in the era of the aging boomer. *Akron Law Review*, *497*, Valparaiso University Legal Studies Research Paper No 13–4.

Ikeuchi, T., Taniguchi, Y., Abe, T., Seino, S., Shimada, C., Kitamura, A., & Shinkai, S. (2021). Association between experience of pet ownership and psychological health among socially isolated and non-isolated older adults. *Animals, 11*(3), 595. https://doi.org/10.3390/ani11030595

Janevic, M. R., Shute, V., Connell, C. M., Piette, J. D., Goesling, J., & Fynke, J. (2019). The role of pets in supporting cognitive-behavioral chronic pain self-management: Perspectives of older adults. *Journal of Applied Gerontology, 39*(10), 1088–1096. https://doi.org/10.1177/0733464819856270

Johansson, M., Ahlström, G., & Jönsson, A.-C. (2014). Living with companion animals after stroke: Experiences of older people in community and primary care nursing. *British Journal of Community Nursing, 19*(12), 578–584. https://doi.org/10.12968/bjcn.2014.19.12.578

Johnson, R. A., & Bibbo, J. (2014). Relocation decisions and constructing the meaning of home: A phenomenological study of the transition into a nursing home. *Journal of Aging Studies, 30*, 56–63. https://doi.org/10.1016/j.jaging.2014.03.005

Kimura, Y., Kawabata, H., & Maezawa, M. (2014). Frequency of neurotic symptoms shortly after the death of a pet. *Journal of Veterinary Medical Science, 76*(4), 499–502. https://doi.org/10.1292/jvms.13-0231

Knight, S., & Edwards, V. (2008). In the company of wolves: The physical, social, and psychological benefits of dog ownership. *Journal of Aging and Health, 20*(4), 437–455. https://doi.org/10.1177/0898264308315875

Koh, W. Q., Ang, F. X. H., & Casey, D. (2021). Impacts of low-cost robotic pets for older adults and people with dementia: Scoping review. *JMIR Rehabilitation and Assistive Technologies, 8*(1), e25340. https://doi.org/10.2196/25340

Krause-Parello, C. A. (2012). Pet ownership and older women: The relationships among loneliness, pet attachment support, human social support, and depressed mood. *Geriatric Nursing, 33*(3), 194–203. https://doi.org/10.1016/j.gerinurse.2011.12.005

Lazar, A., Thompson, H. J., Piper, A. M., & Demiris, G. (2016). Rethinking the design of robotic pets for older adults. *Proceedings of the 2016 ACM Conference on Designing Interactive Systems*, 1034–1046. https://doi.org/10.1145/2901790.2901811

Ly, L. H., Gordon, E., & Protopopova, A. (2021). Exploring the relationship between human social deprivation and animal surrender to shelters in British Columbia, Canada. *Frontiers in Veterinary Science, 8*, 213. https://doi.org/10.3389/fvets.2021.656597

Mahalski, P. A., Jones, R., & Maxwell, G. M. (1988). The value of cat ownership to elderly women living alone. *International Journal of Aging and Human Development, 27*(4), 249–260. https://doi.org/10.2190/N40Y-68JW-38TD-AT9R

Martinson, M., & Berridge, C. (2015). Successful aging and its discontents: A systematic review of the social gerontology literature. *The Gerontologist, 55*(1), 58–69. https://doi.org/10.1093/geront/gnu037

Martyr, A., Nelis, S. M., Quinn, C., Wu, Y.-T., Lamont, R. A., Henderson, C., Clarke, R., Hindle, J. V., Thom, J. M., Jones, I. R., Morris, R. G., Rusted, J. M., Victor, C. R., & Clare, L. (2018). Living well with dementia: A systematic review and correlational meta-analysis of factors associated with quality of life, well-being and life satisfaction in people with dementia. *Psychological Medicine, 48*(13), 2130–2139.. https://doi.org/10.1017/S0033291718000405

Matsuoka, A., Sorenson, J., Graham, T. M., & Ferreira, J. (2020). No pets allowed: A trans-species social justice perspective to address housing issues for older adults and companion animals. *Aotearoa New Zealand Social Work, 32*(4), 55–68. https://doi.org/10.11157/anzswj-vol32iss4id793

McLennan, K, Rock, M. J., Mattos, E., & Toohey, A. M. (2022). Leashes, litterboxes, and lifelines: Exploring volunteer-based pet care assistance programs for older adults. *Frontiers in Psychology: Positive Psychology, 13*, 873372. https://doi.org/10.3389/fpsyg.2022.873372

McNicholas, J. (2014). The role of pets in the lives of older people: A review. *Working with Older People, 18*(3), 128–133. https://doi.org/10.1108/WWOP-06-2014-0014

McNicholas, J., Gilbey, A., Rennie, A., Ahmedzai, S., Dono, J.-A., & Ormerod, E. (2005). Pet ownership and human health: A brief review of evidence and issues. *BMJ, 331*(7527), 1252–1254. https://doi.org/10.1136/bmj.331.7527.1252

Mendes, A., & Palmer, S. (2016). Robotic animals: Leaps and bounds in dementia care. *Nursing and Residential Care, 18*(8), 423–426. https://doi.org/10.12968/nrec.2016.18.8.423

Menec, V. H., Means, R., Keating, N., Parkhurst, G., & Eales, J. (2011). Conceptualizing age-friendly communities. *Canadian Journal on Aging/Revue Canadienne Du Vieillissement, 30*(03), 479–493. https://doi.org/10.1017/S0714980811000237

Miller, L. M., Whitlatch, C. J., & Lyons, K. S. (2016). Shared decision-making in dementia: A review of patient and family carer involvement. *Dementia, 15*(5), 1141–1157. https://doi.org/10.1177/1471301214555542

Minmin, L., Peng, L., Ping, Z., Mingyue, H., Haiyan, Z., Guichen, L., Huiru, Y., & Li, C. (2019). Pet robot intervention for people with dementia: A systematic review and meta-analysis of randomized controlled trials. *Psychiatry Research, 271*, 516–525. https://doi.org/10.1016/j.psychres.2018.12.032

Monsalve, S., Hammerschmidt, J., Izar, M. L., Marconcin, S., Rizzato, F., Polo, G., & Garcia, R. (2018). Associated factors of companion animal neglect in the family environment in Pinhais, Brazil. *Preventive Veterinary Medicine, 157*, 19–25. https://doi.org/10.1016/j.prevetmed.2018.05.017

Morrow-Howell, N., Hinterlong, J., Rozario, P. A., & Tang, F. (2003). Effects of volunteering on the well-being of older adults. *The Journals of Gerontology: Series B, 58*(3), S137–S145. https://doi.org/10.1093/geronb/58.3.S137

Mueller, M. K., Gee, N. R., & Bures, R. M. (2018). Human-animal interaction as a social determinant of health: Descriptive findings from the health and retirement study. *BMC Public Health, 18*(1), 305. https://doi.org/10.1186/s12889-018-5188-0

Mueller, R. L., & Hunter, E. G. (2019). The intersection of aging and pet guardianship: Influences of health and social support, *People and Animals: The International Journal of Research and Practice, 2*(1), 3.

Muramatsu, N., Sokas, R. K., Lukyanova, V. V., & Zanoni, J. (2019). Perceived stress and health among home care aides: Caring for older clients in a medicaid-funded home care program. *Journal of Health Care for the Poor and Underserved, 30*(2), 721–738. https://doi.org/10.1353/hpu.2019.0052

Nathanson, J. (2009). Animal hoarding: Slipping into the darkness of comorbid animal and self-neglect. *Journal of Elder Abuse & Neglect, 21*(4), 307–324. https://doi.org/10.1080/08946560903004839

Obradović, N., Lagueux, É., Michaud, F., & Provencher, V. (2020). Pros and cons of pet ownership in sustaining independence in community-dwelling older adults: A scoping review. *Ageing & Society, 40*(9), 2061–2076. https://doi.org/10.1017/S0144686X19000382

Opdebeeck, C., Katsaris, M. A., Martyr, A., Lamont, R. A., Pickett, J. A., Rippon, I., Thom, J. M., Victor, C., & Clare, L. (2020). What are the benefits of pet ownership and care among people with mild-to-moderate dementia? Findings from the IDEAL programme. *Journal of Applied Gerontology, 40*(11), 1559–1567. https://doi.org/10.1177/0733464820962619

Ormerod, E. (2012). Supporting older people with pets in sheltered housing. *In Practice, 34*(3), 170–173. https://doi.org/10.1136/inp.e1041

Owczarczak-Garstecka, S. C., Christley, R., Watkins, F., Yang, H., Bishop, B., & Westgarth, C. (2019). Dog bite safety at work: An injury prevention perspective on reported occupational dog bites in the UK. *Safety Science, 118*, 595–606. https://doi.org/10.1016/j.ssci.2019.05.034

Peak, T., Ascione, F., & Doney, J. (2012). Adult protective services and animal welfare: Should animal abuse and neglect be assessed during Adult Protective Services screening? *Journal of Elder Abuse & Neglect, 24*(1), 37–49. https://doi.org/10.1080/08946566.2011.608047

Peel, E., Douglas, M., Parry, O., & Lawton, J. (2010). Type 2 diabetes and dog walking: Patients' longitudinal perspectives about implementing and sustaining physical activity. *The British Journal of General Practice, 60*(577), 570–577. https://doi.org/10.3399/bjgp10X515061

Pikhartova, J., Bowling, A., & Victor, C. (2014). Does owning a pet protect older people against loneliness? *BMC Geriatrics, 14*(1), 1–10. https://doi.org/10.1186/1471-2318-14-106

Polick, C. S., Applebaum, J. W., Hanna, C., Jackson, D., Tsaras-Schumacher, S., Hawkins, R., Conceicao, A., O'Brien, L. M., Chervin, R. D., & Braley, T. J. (2021). The impact of pet care needs on medical decision-making among hospitalized patients: A cross-sectional analysis of patient experience. *Journal of Patient Experience, 8*, 23743735211046090. https://doi.org/10.1177/23743735211046089

Power, E. R. (2017). Renting with pets: A pathway to housing insecurity? *Housing Studies, 32*(3), 336–360. https://doi.org/10.1080/02673037.2016.1210095

Putney, J. M. (2013). Relational ecology: A theoretical framework for understanding the human-animal bond. *Journal of Sociology and Social Welfare, 40*(4), 57–80.

Raina, P. S., Waltner-Toews, D., Bonnett, B., Woodward, C., & Abernathy, T. (1999). Influence of companion animals on the physical and psychological health of older people: An analysis of a one-year longitudinal study. *Journal of the American Geriatrics Society, 47*(3), 323–329. https://doi.org/10.1111/j.1532-5415.1999.tb02996.x

Rauktis, M. E., Rose, L., Chen, Q., Martone, R., & Martello, A. (2017). "Their pets are loved members of their family": Animal ownership, food insecurity, and the value of having pet food available in food banks. *Anthrozoös, 30*(4), 581–593. https://doi.org/10.1080/08927936.2017.1370225

Reyes Uribe, A. C. (2020). Why did I share my life with Glucio? A life course approach to explaining pet ownership motivations in late adulthood. *Anthrozoös, 33*(1), 89–102. https://doi.org/10.1080/08927936.2020.1694314

Richard, L., Gauvin, L., & Raine, K. (2011). Ecological models revisited: Their uses and evolution in health promotion over two decades. *Annual Review of Public Health, 32*, 307–326. https://doi.org/10.1146/annurev-publhealth-031210-101141

Rowe, J. W., & Kahn, R. L. (1997). Successful aging. *The Gerontologist, 37*(4), 433–440. https://doi.org/10.1093/geront/37.4.433

Rusanen, M., Selander, T., Kärkkäinen, V., & Koivisto, A. (2021). The positive effects of pet ownership on Alzheimer's disease. *Journal of Alzheimer's Disease, 84*(4), 1669–1675. https://doi.org/10.3233/JAD-210557

Ryan, S., & Ziebland, S. (2015). On interviewing people with pets: Reflections from qualitative research on people with long-term conditions. *Sociology of Health & Illness, 37*(1), 67–80. https://doi.org/10.1111/1467-9566.12176

Shell, L. (2015). The picture of happiness in Alzheimer's disease: Living a life congruent with personal values. *Geriatric Nursing, 36*(2, supplement), S26–S32. https://doi.org/10.1016/j.gerinurse.2015.02.021

Stanley, I. H., Conwell, Y., Bowen, C., & Van Orden, K. A. (2014). Pet ownership may attenuate loneliness among older adult primary care patients who live alone. *Aging & Mental Health, 18*(3), 394–399. https://doi.org/10.1080/13607863.2013.837147

Stull, J. W., Hoffman, C. C., & Landers, T. (2018). Health benefits and risks of pets in nursing homes: A survey of facilities in Ohio. *Journal of Gerontological Nursing, 44*(5), 39–45. https://doi.org/10.3928/00989134-20180322-02

Taniguchi, Y., Seino, S., Nishi, M., Tomine, Y., Tanaka, I., Yokoyama, Y., Amano, H., Kitamura, A., & Shinkai, S. (2018). Physical, social, and psychological characteristics of community-dwelling elderly Japanese dog and cat owners. *PloS One, 13*(11), e0206399. https://doi.org/10.1371/journal.pone.0206399

Tkatch, R., Wu, L., MacLeod, S., Ungar, R., Albright, L., Russell, D., Murphy, J., Schaeffer, J., & Yeh, C. S. (2021). Reducing loneliness and improving well-being among older adults with animatronic pets. *Aging & Mental Health, 25*(7), 1239–1245. https://doi.org/10.1080/13607863.2020.1758906

Toohey, A. M. (2023). Considering Cats, Dogs, and Contradictions: Pets and Their Relational Influence on Experiences of Aging in Place. *Canadian Journal on Aging/La Revue canadienne du vieillissement*, 1–10. https://doi.org/10.1017/S0714980823000168.

Toohey, A. M., Hewson, J. A., Adams, C. L., & Rock, M. J. (2017). When "places" include pets: Broadening the scope of relational approaches to promoting aging-in-place. *Journal of Sociology & Social Welfare, 44*(3), 119–146.

Toohey, A. M., Hewson, J. A., Adams, C. L., & Rock, M. J. (2018). Is pet ownership relevant to social participation and life satisfaction for older adults who are aging-in-place in Canada? Findings from the Canadian longitudinal study on aging (CLSA). *Canadian Journal on Aging, 37*(2), 200–217.

Toohey, A. M., & Krahn, T. M. (2018). 'Simply to be let in': Opening the doors to lower-income older adults and their companion animals. *Journal of Public Health, 40*(3), 661–665. https://doi.org/10.1093/pubmed/fdx111

Toohey, A. M., McCormack, G. R., Doyle-Baker, P. K., Adams, C. L., & Rock, M. J. (2013). Dog-walking and sense of community in neighborhoods: Implications for promoting regular physical activity in adults 50 years and older. *Health & Place, 22*, 75–81. https://doi.org/10.1016/j.healthplace.2013.03.007

Toohey, A. M., & Rock, M. J. (2011). Unleashing their potential: A critical realist scoping review of the influence of dogs on physical activity for dog-owners and non-owners. *International Journal of Behavioral Nutrition and Physical Activity, 8*(1), 46–54.

Toohey, A. M., & Rock, M. J. (2019). Disruptive solidarity or solidarity disrupted? A dialogical narrative analysis of economically vulnerable older adults' efforts to age in place with pets. *Public Health Ethics, 12*(1), 15–29. https://doi.org/10.1093/phe/phy009

Trigg, J. (2021). Examining the role of pets in cancer survivors' physical and mental wellbeing. *Journal of Psychosocial Oncology, 40*(6), 834–853. https://doi.org/10.1080/07347332.2021.1936337

Vincent, A., Mcdonald, S., Poe, B., & Deisner, V. (2020). The link between interpersonal violence and animal abuse. *Society Register, 3*(3), 83–101. https://doi.org/10.14746/sr.2019.3.3.05

Wiles, J. L., Leibing, A., Guberman, N., Reeve, J., & Allen, R. E. S. (2011). The meaning of "ageing in place" to older people. *The Gerontologist, 52*(3), 357–366. https://doi.org/10.1093/geront/gnr098

Wilson, D. M., Underwood, L., Carr, E., Gross, D. P., Kane, M., Miciak, M., Wallace, J. E., & Brown, C. A. (2021). Older women's experiences of companion animal death: Impacts on well-being and aging-in-place. *BMC Geriatrics, 21*(1), 470. https://doi.org/10.1186/s12877-021-02410-8

Wood, L., Frank, L. D., & Giles-Corti, B. (2010). Sense of community and its relationship with walking and neighborhood design. *Social Science & Medicine, 70*(9), 1381–1390. https://doi.org/10.1016/j.socscimed.2010.01.021

Wood, L., Martin, K., Christian, H., Nathan, A., Lauritsen, C., Houghton, S.,... & McCune, S. (2015). The pet factor-companion animals as a conduit for getting to know people, friendship formation and social support. *PLOS One, 10*(4), e0122085. https://doi.org/10.1371/journal.pone.0122085

Wood, L., Martin, K., Christian, H., Houghton, S., Kawachi, I., Vallesi, S., & McCune, S. (2017). Social capital and pet ownership – A tale of four cities. *SSM - Population Health, 3*, 442–447. https://doi.org/10.1016/j.ssmph.2017.05.002

World Health Organization. (2007). *Global age-friendly cities: A guide*. WHO. https://www.who.int/ageing/publications/Global_age_friendly_cities_Guide_English.pdf

World Health Organization. (2020). *Decade of Healthy Ageing Baseline Report* (Licence: CC BY-NC-SA 3.0 IGO). WHO. https://www.who.int/publications/i/item/9789240017900

World Health Organization. (2021). *Dementia*. WHO. https://www.who.int/news-room/fact-sheets/detail/dementia

Yon, Y., Mikton, C. R., Gassoumis, Z. D., & Wilber, K. H. (2017). Elder abuse prevalence in community settings: A systematic review and meta-analysis. *The Lancet Global Health, 5*(2), e147–e156. https://doi.org/10.1016/S2214-109X(17)30006-2

# 17
# LGBTQIA+ POPULATIONS AND HUMAN-ANIMAL INTERACTIONS

*Angela Matijczak and Shelby McDonald*

## Introduction to LGBTQIA+ Identities and Labels

Individuals who identify as LGBTQIA+ (lesbian, gay, bisexual, transgender, queer, intersex, asexual, and other sexual and gender minority identities) are a diverse group of people. Evidence from nationally representative studies suggests that more than 11 million adults (4.5%), and nearly 2 million youth (9.5%), living in the United States (US) identify as LGBTQIA+ (Conron, 2020; Conron & Goldberg, 2019). Identifying as LGBTQIA+ has become increasingly more common particularly among younger generations, with approximately 30% of the adult LGBTQIA+ population aged between 18 and 24 years (Williams Institute, 2019).

Although attitudes toward LGBTQIA+ people have become more positive in the past decade, LGBTQIA+ people still face high rates of discrimination and other identity-based stressors that contribute to the disproportionately higher rates of physical and mental health problems seen in LGBTQIA+ populations, compared with their heterosexual and cisgender counterparts (King et al., 2008; Kosciw et al., 2020; Meyer et al., 2021; Russell & Fish, 2016; Sterzing et al., 2017; Su et al., 2016). The minority stress model is a theoretical model that is commonly used to understand the pathways through which exposure to minority stressors (e.g., identity-based discrimination, victimization, rejection) leads to health disparities among LGBTQIA+ populations (Meyer, 2003, 2015). In this model, Meyer (2003, 2015) distinguishes between distal stressors, which are objective events or conditions (e.g., victimization), and proximal stressors, which are subjective or personal appraisals (e.g., internalized homophobia and/or transphobia). The model also considers the way that certain factors may moderate the relationship between exposure to minority stress and health outcomes. These moderators may be protective factors (e.g., coping mechanisms, social support) that buffer against the impact of minority stress on health outcomes, or they may be risk factors (e.g., low socioeconomic status) that exacerbate the impact of minority stress on health outcomes. The minority stress model has been supported by evidence from several studies (Kosciw et al., 2018 Meyer et al., 2021).

There is substantial evidence to suggest that social support is an important protective factor for LGBTQIA+ individuals (Matijczak et al., 2020; Simons et al., 2013; Snapp et al., 2015). In particular, support from family members has been identified as a key source of social support for LGBTQIA+ people (Snapp et al., 2015). Additionally, many LGBTQIA+ people choose to form "chosen families," which are diverse social support networks constructed by choice (rather than biological or legal ties) that provide resources and support in the face of stigma and discrimination (Moore & Stambolis-Ruhstorfer, 2013; Weston, 1997). Relationships with chosen family members are characterized by strong bonds that provide psychological benefits and promote resilience among the LGBTQIA+ community (Hailey et al., 2020; Jackson Levin et al., 2020). However, the role of companion animals as members of the family has been largely underexplored in research surrounding chosen families or important relationships in the lives of LGBTQIA+ people. A study conducted by Hull and Ortyl (2019) found that, when asked to identify individuals who were a part of one's chosen family, companion animals were the third most frequently mentioned category, even though the focus of this study was related to human relationships. This is not surprising given evidence that rates of pet ownership are high among LGBTQIA+ populations (67% –71%) and, generally, LGBTQIA+ people report experiencing strong bonds with their companion animals (Community Marketing & Insights, 2019; Harris Interactive, 2010).

## Impact of Human-Animal Interactions (HAI) on the Health and Wellbeing of LGBTQIA+ People

The impact of companion animals on the lives of LGBTQIA+ people is an emerging area of research that has gained attention only within the past decade. This section will detail existing studies that explore the benefits and stressors LGBTQIA+ people may derive from their relationships with their companion animals.

### *Potential Benefits of HAI*

Several studies have found that, generally, pet owners derive a sense of social support and comfort from their interactions with their companion animals (Kurdek, 2009; Meehan et al., 2017; Tomlinson et al., 2020; Zilcha-Mano et al., 2012). The support that companion animals provide may be particularly important to LGBTQIA+ individuals, who may have limited social relationships that are affirming of their identities (Frost et al., 2016; Moore et al., 2021). For example, several qualitative studies found that LGBTQIA+ older adults perceived their pets to be vital sources of social support and that these relationships contributed to their sense of belongingness within their community (MacNamara, 2019; Muraco et al., 2018; Putney, 2014). Indeed, Muraco et al. (2018) found that LGBTQIA+ older adult pet owners reported significantly higher self-reported social support compared to LGBTQIA+ older adults that did not live with a pet. These findings were also mirrored in a qualitative study conducted with LGBTQIA+ emerging adults, in which participants discussed their pets as important confidants that provided unconditional love throughout their journey as an LGBTQIA+ person (McDonald, Matijczak et al., 2021).

It is possible that the emotional support provided by companion animals may be particularly salient for LGBTQIA+ people following experiences of adversity. For example, a theme that emerged from interviews with LGBTQIA+ emerging adults related to the companion animals' support specifically in the context of exposure to minority stressors, such as experiencing rejection from other social relationships (e.g., peers, family members; McDonald, Matijczak et al., 2021). Participants spoke about their pets as non-judgmental and affirming confidants, which was particularly supportive in the context of the coming out process (McDonald, Matijczak et al., 2021). Some studies have found that the routine required to care for a companion animal may contribute to a sense of stability and responsibility that can promote mental health and wellbeing among LGBTQIA+ pet owners struggling with mental health symptoms (McDonald, Matijczak et al., 2021; Schmitz, Carlisle, & Tabler, 2021; Schmitz, Carlisle, Tabler, & Almy, 2021). Furthermore, evidence from a qualitative study with LGBTQIA+ people experiencing homelessness found that bonds with companion animals were perceived to offset some of the stressors uniquely associated with homelessness by providing emotional support, security, and stability (Schmitz, Carlisle, & Tabler, 2021). Additionally, a study conducted with LGBTQIA+ individuals that have experienced, or were currently experiencing, domestic violence found that, not only were companion animals important sources of emotional support and safety, but some participants also cited a desire to protect their companion animals from abuse as the trigger for seeking help (Taylor et al., 2019).

However, few studies have quantitatively investigated relations between exposure to minority stressors, HAI, and psychosocial outcomes. One study found that HAI was a significant mediator of the relationship between exposure to minority stress (e.g., victimization, interpersonal microaggressions) and personal hardiness, an indicator of resilience (McDonald, Murphy et al., 2021). In particular, they found that increases in exposure to minority stress was associated with higher levels of HAI (measured via four domains of the human-pet attachment bond), which in turn was positively associated with self-reported personal hardiness (McDonald, Murphy et al., 2021). This highlights the mechanisms through which companion animals may provide benefits to LGBTQIA+ populations. Another study found that, among a sample of transgender and gender nonconforming adults, living with a companion animal may act as a buffer against the association between exposure to family violence and psychological distress (Riggs et al., 2018). Additionally, a study conducted with LGBTQIA+ emerging adults found that emotional comfort from pets significantly moderated the relation between exposure to victimization and self-esteem; in particular, the negative association between victimization and self-esteem was only significant at low levels of emotional comfort from companion animals (McDonald, O'Connor et al., 2021). These studies highlight the many ways in which living with companion animals and/or having strong bonds with companion animals may be beneficial for the mental health and wellbeing of LGBTQIA+ individuals.

## *Potential Stressors Associated with HAI*

It is important to note that the stressors associated with pet ownership may be pronounced for LGBTQIA+ pet owners. For example, LGBTQIA+ emerging adults noted economic stressors associated with caring for their companion animals, such as concerns about being able to afford resources and challenges related to finding pet-friendly housing (McDonald, Matijczak et al., 2021). These pet-related stressors are particularly important to consider in the context of LGBTQIA+ populations, as research suggests that LGBTQIA+ people are more likely to experience poverty (Badgett et al., 2019; Carpenter et al., 2020; Schneebaum & Badgett, 2019). Furthermore, in the context of domestic violence and/or family violence, LGBTQIA+ people may experience psychological distress when their companion animal becomes the target of abuse (McDonald, Matijczak et al., 2021; Schmitz, Carlisle, & Tabler, 2021). As some domestic violence shelters do not allow companion animals within their facilities (Krienert et al., 2012; Stevenson et al., 2018), this may be a significant barrier to help-seeking and accessing domestic violence-related services (Schmitz, Carlisle, & Tabler, 2021). Among LGBTQIA+ people experiencing homelessness, companion animals may be a barrier to entering shelters or finding affordable, secure housing due to restrictive pet policies (Schmitz, Carlisle, Tabler, & Almy, 2021).

There is also some evidence from quantitative studies that suggests that relationships with pets may actually exacerbate the relationship between minority stressors and mental health symptoms. For example, a study conducted by Matijczak et al. (2020) found that the positive association between exposure to interpersonal microaggressions and depressive symptoms was significant when participants reported high levels of comfort from their companion animals; however, the authors were not able to establish the directions of these relationships due to the cross-sectional design of the study. One possible interpretation is that using a companion animal as a primary source of comfort may actually amplify the harmful relationship between experiencing microaggressions and depressive symptoms (Matijczak et al., 2020). Pets may be viewed as a safer source of social support than humans, particularly in the context of exposure to interpersonal forms of discrimination such as microaggressions. However, it is widely evidenced that human social support is an important protective factor for LGBTQIA+ individuals (Matijczak et al., 2020; Simons et al., 2013; Snapp et al., 2015); thus, relying on one's companion animal for social support and being reluctant to seek out human social support may result in the exacerbation of the association between microaggressions and depressive symptoms (Antonacopoulos & Pychyl, 2010). Indeed, when including social support as a double moderator in these models, evidence suggests that the positive relationship between minority stress and mental health symptoms is strongest for those who report high levels of emotional comfort derived from their companion animal and low levels of social support from human relationships in their lives (Matijczak et al., 2020). This suggests that perhaps a lack of social support and resources that can aid in caring for pets may be a salient stressor for LGBTQIA+ populations that may negatively affect mental health.

## The Intersection of Human and Animal Social Determinants of Health

The social determinants of health that impact human health and wellbeing (e.g., transportation, neighborhood characteristics, income, education, discrimination) likely have direct and indirect effects on pet owners' ability to care for their pets, particularly among marginalized communities. To our knowledge, there is currently no literature examining the welfare of the companion animals of LGBTQIA+ individuals. Given the paucity of research on this topic, in this chapter we discuss how known social and environmental factors that systemically marginalize, oppress, and disadvantage LGBTQIA+ populations may impact the welfare of pets cared for by LGBTQIA+ individuals. We also discuss how bonds between LGBTQIA+ people and their pets may promote animal welfare in the face of adversity, and how individuals' dedication to their pet(s) and related decision-making may intersect with risk and resilience in LGBTQIA+ individuals and communities.

Access to veterinary care (AVC) typically refers to the belief that companion animals should equitably receive holistic care that is compassionate, respectful, and considerate; moreover, this care should improve animal welfare, decrease suffering, and be responsive to the needs of pets and their human caregiver's circumstances (LaVallee et al., 2017; McDonald et al., 2022). Given high rates of pet ownership in the United States, lack of AVC has been identified as a social determinant of animal health and this issue has received increasing attention as a social justice issue among people living in poverty. Nationally representative data suggest that approximately 11.2 million dogs and 8.3 million cats in the United States live in under-resourced homes below the poverty line (ASPCA, 2020). Although few studies have centered or explored the experiences of LGBTQIA+ individuals in AVC, a recent report suggests that, collectively, LGBTQIA+ people in the United States have a poverty rate of 21.6%. Notably, transgender people have particularly high rates of poverty at nearly

30% (Badgett et al., 2019). Rates of poverty among LGBTQIA+ people are notably higher than the rate for cisgender/heterosexual people, which is estimated at 15.7% (Williams Institute, 2019). Given potentially higher rates of both pet ownership and poverty among the LGBTQIA+ population, AVC, or the lack thereof, is one example of how social determinants of human health can impact determinants of animal health (Badgett et al., 2019). For low-resourced LGBTQIA+ individuals experiencing adverse economic conditions, pet-related veterinary costs could lead to considerable financial strain and/or may not be a priority compared to providing the animal, themself, and/or other family members with basic needs such as food, water, and shelter.

In addition, people who live in poverty, especially individuals from racialized minority groups, are more likely to live in neighborhoods with compounded resource scarcity; these areas are frequently referred to as "resource deserts" and have little to no animal welfare infrastructure and lack access to basic veterinary care and pet supply stores (Arrington & Markarian, 2018). LGBTQIA+ pet owners experiencing poverty and living in resource deserts may not be able to provide basic care due to their inability to access veterinary clinics where they live and/or due to the inability to access pet-friendly transportation and related travel supplies (e.g., carrier). Considering that LGBTQIA+ individuals have increased risk for job insecurity, broken family ties, and lower levels of social support, navigating these financial and environmental obstacles to care for pets may be an insurmountable burden, and as previous research indicates, the inability to meet pets' meets could lead to increased stress and reduced feelings of self-worth (Appelbaum et al., 2021; McDonald et al., 2021).

In addition to these barriers to care, it is well known that LGBTQIA+ people face discrimination in health care services (Ruben et al., 2019). This type of discrimination likely extends to animal welfare and health services and is another example of the intersection of human and animal social determinants of health. Although data and empirical evidence are limited, it is estimated that LGBTQIA+ individuals are underrepresented among veterinarians (Parker et al., 2016). Indeed, only 6% of veterinary professionals are LGBTQIA+ (Witte et al., 2020). Therefore, LGBTQIA+ pet owners may struggle to feel represented among those who care for their animals; and many animal health providers may be unfamiliar with navigating human-animal bonds and relationships in LGBTQIA+ families. This may lead to lived experiences of discrimination in accessing or receiving veterinary services, which may be compounded by past or ongoing negative experiences with human health-care providers. Additionally, such experiences may compound with other barriers to care (cost, access) and negatively impact the welfare of LGBTQIA+ people and compromise their ability to care for animals. Due to the limited research in this area, the hypotheses proposed must be tested. Still, it is important that animal welfare and veterinary professionals aim to identify ways to mitigate this potential barrier to the welfare of LGBTQIA+ people's pets. As outlined by San Miguel et al. (2014), steps to improve access could be as simple as clinics displaying a sign to indicate an alliance with LGBTQIA+ people in their community (either physically on display in the clinic and/or online via a website). Furthermore, veterinary clinics, hospitals, and community services can provide relevant training to veterinarians on how to center the experiences and lives of LGBTQIA+ clients, affirm LGBTQIA+ identities (e.g., pronouns), and avoid cisheteronormative assumptions about pet owners in their practice (Topper, 2018).

In addition to considering poverty and environmental barriers to animal care, the intersection of LGBTQIA+ health disparities and pet care warrants attention in relation to animal welfare. Prior research suggests that the physical, mental, and cognitive health of pet owners and characteristics of their social relationships have implications for the quality of care that animals receive (Lockwood, 2002; McDonald et al., 2021). As previously addressed in this chapter, LGBTQIA+ populations, compared to their cisgender and heterosexual counterparts, experience disproportionate risk for a variety of mental health problems, such as anxiety, depression, and suicidal ideation, due to their minoritized status and associated experiences of stress. In addition to the increased prevalence of mental health problems in this population group, recent data suggest that, nearly one in three LGB adults and two in five transgender adults in the United States have a disability (Movement Advancement Project, 2019). The rate of intimate partner violence (also known as domestic violence) is as high or higher among LGBTQIA+ couples as well (Brown & Herman, 2015). Prior studies suggest these individual (mental health, disability) and family-level (intimate partner violence) contextual factors may have implications for animal health and welfare (Collins et al., 2018; McDonald et al., 2021). For example, depression, trauma, cognitive dysfunction, and memory loss are associated with poor self-hygiene behaviors and neglect of child and adult dependents among adults (Mulder et al., 2018; Stewart et al., 2021; Storey, 2020); it has been suggested that these associations likely extend to neglect or omissions of care of companion animals (McDonald et al., 2022). Further, there is some evidence that failure to seek veterinary care, groom pets, and other forms of animal neglect are prevalent among households experiencing intimate partner violence and child abuse (McDonald et al., 2019; Tiplady, 2018). Although these risk factors for compromised animal care and health are not unique to LGBTQIA+ individuals, this population is disproportionately impacted by these social problems and associated outcomes. For LGBTQIA+ individuals, the duality of finding value in one's bond with a pet, while also facing greater risk for the accumulation of barriers to providing care and maintaining that bond, likely makes the experience of pet ownership among LGBTQIA+ individuals, and other groups with similar minoritized and marginalized status, unique in contrast to the more privileged samples that are typically reflected in HAI research.

We draw on the experience of LGBTQIA+ older adults to illustrate this hypothesis. In general, older adults are more likely to have physical limitations or disabilities (CDC, 2018; Okoro et al., 2018) and may require assistance with traveling to veterinary or grooming appointments and/or purchasing supplies and performing animal care activities, such as bathing and nail trimming. Indeed, accessibility disadvantages (e.g., transportation challenges) are a well-known barrier that impacts the welfare of older adults (Assi et al., 2022; Remillard et al., 2022). Compared to their heterosexual and/or cisgender counterparts, LGBTQIA+ older adults are disproportionately more likely to live alone, be childless, and be estranged from family members who could, potentially, provide support for pet care, financial support, transportation to veterinary care, etc., in times of difficulty (Boggs et al., 2017; de Vries, 2014). Combined with the previously addressed impacts of systemic heterosexism and/or cisgenderism, which place LGBTQIA+ older adults at higher risk for poverty, social isolation, financial disparity, mental health problems, disability, housing insecurity/deficiencies, reduced access to health care, and premature institutionalization (Boggs et al., 2017 de Vries, 2014; Espinoza, 2011; Fredriksen-Goldsen et al., 2011; Witten, 2014; Zelle & Arms, 2015), these human social determinants of health have important ties to determinants of animal health and welfare, such as access to veterinary care, access to hygiene services (grooming), basic needs (adequate food, water), and pet-friendly housing. We argue that these systemic barriers and stressors place the pets of LGBTQIA+ older adults at potential risk for omissions of care and relinquishment. Further, even when unintentional or beyond the pet owner's control, these omissions of care may meet legal definitions of animal neglect and, thereby, have very serious consequences for the individual and their pet (McDonald et al., 2022; Norris, 2020; Tiplady,

2013). For instance, if animal neglect is reported to law enforcement, the pet owner could face criminal charges.

Although systemic oppression of LGBTQIA+ older adults likely places pets at risk for compromised health and welfare, it is well known that many people consider and prioritize their pets when making important life decisions, including divisions that impact their own health and welfare (Ascione et al., 2007; Collins et al., 2018; Henwood et al., 2021; Peacock et al., 2012; Polick et al., 2021; Rhoades et al., 2015). For example, prior studies suggest that bonds and care for pets could negatively impact human health-related testing and medical care, such as situations when individuals delay or avoid testing (e.g., COVID-19) or hospitalization because they lack adequate social support or access to services that can provide temporary pet care (Canady & Sansone, 2019; Matijczak et al., 2021; Polick et al., 2021). People may also prioritize meeting pets' basic needs over their own. Studies of individuals experiencing homelessness demonstrate that many housing insecure people place pets' food needs above their own (Irvine et al., 2012; Scanlon et al., 2021). Such prioritization of pets also extends to physical and psychological safety. Studies indicate survivors of intimate partner violence will delay seeking shelter when they are forced to leave animals behind in risky or abusive situations or even live in their car until adequate pet care can be found (Collins et al., 2018; McDonald et al., 2019). Thus, although human social determinants of health that disproportionately impact the LGBTQIA+ population likely have direct and indirect effects on pet health and wellbeing, this may be mitigated or circumvented by the human-animal bond, and the love, respect, and care that LGBTQIA+ people have for their pets, albeit at the detriment of the pet owner's wellbeing.

## Pet-Inclusive Policy and Practices

LGBTQIA+ individuals and communities may be especially likely to benefit from advancements in pet-inclusive policies and practices, particularly in the areas of housing, emergency situations, and physical and mental health services and care. In the context of psychological and social services, Schmitz et al. (2022) suggest that validating and affirming LGBTQIA+ people's lived experiences and subjective interpretations of coping, resilience, and mental health processes in more expansive and diverse ways that include non-human–animal relationships is one step to providing more LGBTQIA+ affirming care. Further, programs that provide direct human services to LGBTQIA+ populations would benefit from developing collaborative relationships with animal welfare organizations that can connect pet-owners with supplies (e.g., pet pantries), resources (e.g., pet carrier), and services (e.g., subsidized veterinary and grooming care) that help foster their ability to care for pets and maintain this important relationship (Arluke, 2021; Hawes et al., 2022; McDonald et al., 2021, 2022; Rauktis et al., 2020). Further, we advocate that pet ownership should be included as a protected status against housing discrimination, particularly in the context of affordable rental housing (Applebaum et al., 2021). Moreover, there is a great need for temporary foster and boarding services for pets of individuals who need to receive in-patient healthcare or are temporarily unavailable to care for pets due to health concerns and related treatment, such as hospitalization (Applebaum et al., 2021). There is also a great need for pet- and LGBTQIA-affirming domestic violence and emergency shelters to keep all members of multispecies families safe (Fraga-Rizo et al., 2022; McDonald et al., 2015, 2019; Sechrist et al., 2022). Such policies and practices have the potential to benefit all multispecies households, but may be critical and essential steps toward ensuring equity for LGBTQIA+ people and communities.

## Unanswered Questions and Future Research Demands

Given that exploring HAI among LGBTQIA+ populations is still an emerging field of research, there are many opportunities and questions remaining for future research to explore. First, it is imperative that studies conducted within HAI science are intentional about the use of inclusive research designs that account for LGBTQIA+ participants. In particular, many studies that investigate the role of HAI on health outcomes measure gender using binary sex categories (male/female) that do not appropriately account for the inclusion of transgender and/or gender expansive participants in their studies (Rudd & Jenkins, 2022). It is also important to note that many funding agencies like the National Institute of Health, a primary funding source for HAI research, only include binary gender categories within inclusion enrollment reports and other funding documents. There are many existing resources that provide suggestions for how to phrase questions related to gender and sexuality in an inclusive manner (e.g., Bauer et al., 2017; Cameron & Stinson, 2019; Henrickson et al., 2020; Suen et al., 2020); it is vital that both researchers and funding agencies explore these resources to ensure that they are able to accurately collect data on participants' identities. Further, most studies do not collect data on sexual orientation (Rudd & Jenkins, 2022); this invisibilizes the LGBTQIA+ population within HAI science and prevents researchers from understanding how the benefits and stressors associated with HAI may be similar or different between heterosexual and cisgender individuals and individuals within the LGBTQIA+ community.

Additionally, to our knowledge, there have not been any studies that have examined the role of companion animals in the lives of LGBTQIA+ children or adolescents. All studies investigating the bond between LGBTQIA+ individuals and their companion animals have been conducted with adult populations. This is important to note as LGBTQIA+ youth can experience minority stressors in various settings (e.g., school, home) and rates of mental health symptoms are high within youth populations (Kosciw et al., 2020; Taylor, 2019; Vance & Rosenthal, 2018). Finally, we believe there is an opportunity for future researchers to explore the role of companion animals in supporting healthy family dynamics between LGBTQIA+ individuals and their family members. There is evidence to suggest that companion animals may serve as a source of support and comedic relief amidst family conflict (Applebaum & Zsembik, 2020; McDonald, Matijczak et al., 2021; Walsh, 2009); however, these there have not been any studies that have specifically investigated the role

---

**Questions for Discussion**

1. How may companion animals provide unique support to LGBTQIA+ individuals? Conversely, how may living with companion animals exacerbate stressors among LGBTQIA+ individuals?
2. What services and policies are needed to better support LGBTQIA+ pet owners and the welfare of their companion animals?
3. What is the relationship between human health equity and animal welfare? How can the social and health disparities faced by LGBTQIA+ individuals impact the welfare of their companion animals?
4. What are important future directions for research to better understand the role of HAI in the lives of LGBTQIA+ individuals?

that companion animals may play in alleviating tension between LGBTQIA+ individuals and their family members during or after identity disclosure. This is an important area of research to consider, as companion animals may be a promising target for the development of interventions that can improve family communication and promote LGBTQIA-affirming interactions between family members.

# References

American Society for the Prevention of Cruelty to Animals. (2020, August 18). *ASPCA estimates number of pets living in poverty with their owners could exceed 24.4 million due to COVID-19 crisis*. https://www.aspca.org/about-us/press-releases/aspca-estimates-number-pets-living-poverty-their-owners-could-exceed-244#:~:text=With%20the%20potential%20for%20a, compared%20to%20pre%2DCOVID%20estimates%20(

Antonacopoulos, N. M. D., & Pychyl, T. A. (2010). An examination of the potential role of pet ownership, human social support and pet attachment in the psychological health of individuals living alone. *Anthrozoös, 23*(1), 37–54. https://doi.org/10.2752/175303710X12627079939143

Applebaum, J. W., MacLean, E. L., & McDonald, S. E. (2021). Love, fear, and the human-animal bond: On adversity and multispecies relationships. *Comprehensive Psychoneuroendocrinology, 7*, 100071. https://doi.org/10.1016/j.cpnec.2021.10007138

Applebaum, J. W., & Zsembik, B. A. (2020). Pet attachment in the context of family conflict. *Anthrozoös, 33*(3), 361–370. https://doi.org/10.1080/08927936.2020.1746524

Arluke, A. (2021). Coping with pet food insecurity in low-income communities. *Anthrozoös, 34*(3), 339–358. https://doi.org/10.1080/08927936.2021.1898215

Arrington, A., & Markarian, M. (2018). Serving pets in poverty: A new Frontier for the animal welfare movement. *Sustainable Development Law & Policy, 18*(1), 11. http://digitalcommons.wcl.american.edu/sdlp/vol18/iss1/11?utm_source=digitalcommons.wcl.american.edu%2Fsdlp%2Fvol18%2Fiss1%2F11&utm_medium=PDF&utm_campaign=PDFCoverPages

Ascione, F. R., Weber, C. V., Thompson, T. M., Heath, J., Maruyama, M., & Hayashi, K. (2007). Battered pets and domestic violence: Animal abuse reported by women experiencing intimate violence and by nonabused women. *Violence against Women, 13*(4), 354–373. https://doi.org/10.1177/1077801207299201

Assi, L., Deal, J. A., Samuel, L., Reed, N. S., Ehrlich, J. R., & Swenor, B. (2022). Access to food and health care during the COVID-19 pandemic by disability status in the United States. *Disability and Health Journal, 15*(3), 101271. https://doi.org/10.1016/j.dhjo.2022.101271

Badgett, M. L., Choi, S. K., & Wilson, B. D. (2019). *LGBT poverty in the United States*. The Williams Institute. https://williamsinstitute.law.ucla.edu/publications/lgbt-poverty-us/

Bauer, G. R., Braimoh, J., Scheim, A. I., & Dharma, C. (2017). Transgender-inclusive measures of sex/gender for population surveys: Mixed-methods evaluation and recommendations. *PloS One, 12*(5), e0178043. https://doi.org/10.1371/journal.pone.0178043

Boggs, J. M., Dickman Portz, J., King, D. K., Wright, L. A., Helander, K., Retrum, J. H., & Gozansky, W. S. (2017). Perspectives of LGBTQ older adults on aging in place: A qualitative investigation. *Journal of Homosexuality, 64*(11), 1539–1560. https://doi.org/10.1080/00918369.2016.1247539

Brown, T. N. T. & Herman, J. L. (2015). Intimate partner violence and sexual abuse among LGBT people. *The Williams Institute*. https://williamsinstitute.law.ucla.edu/wp-content/uploads/IPV-Sexual-Abuse-Among-LGBT-Nov-2015.pdf

Cameron, J. J., & Stinson, D. A. (2019). Gender (mis) measurement: Guidelines for respecting gender diversity in psychological research. *Social and Personality Psychology Compass, 13*(11), e12506. https://doi.org/10.1111/spc3.12506

Canady, B., & Sansone, A. (2019). Health care decisions and delay of treatment in companion animal owners. *Journal of Clinical Psychology in Medical Settings, 26*(3), 313–320. https://doi.org/10.1007/s10880-018-9593-4

Carpenter, C. S., Eppink, S. T., & Gonzales, G. (2020). Transgender status, gender identity, and socioeconomic outcomes in the United States. *ILR Review, 73*(3), 573–599. https://doi.org/10.1177%2F0019793920902776

Centers for Disease Control and Prevention. (2018). *Disability and Health Data System (DHDS)*. http://dhds.cdc.gov

Collins E. A., Cody A. M., McDonald S. E., Nicotera N., Ascione F. R., & Williams J. H. (2018). A template analysis of intimate partner violence survivors' experiences of animal maltreatment: Implications for safety planning and intervention. *Violence Against Women, 24*(4), 452–76. https://doi.org/10.117/107780121769726643

Community Marketing & Insights. (2019). *13th Annual LGBTQ community survey*. https://www.communitymarketinginc.com/documents/temp/CMI-13th_LGBTQ_Community_Survey_US_Profile.pdf

Conron, K. J. (2020). *LGBT youth population in the United States*. The Williams Institute. https://escholarship.org/content/qt18t179cz/qt18t179cz.pdf

Conron, K. J., & Goldberg, S. K. (2019). *Adult LGBT population in the United States*. The Williams Institute. https://escholarship.org/content/qt1zz685gh/qt1zz685gh.pdf

De Vries, B. (2014). LG(BT) persons in the second half of life: The intersectional influences of stigma and cohort. *LGBT Health, 1*(1) 18–23. https://doi.org/10.1089/lgbt.2013.0005

Espinoza, R. (2011). The diverse elders coalition and LGBT aging: Connecting communities, issues, and resources in a historic moment. *Public Policy & Aging Report, 21*(3), 8–12. https://www.lgbtagingcenter.org/resources/pdfs/PPAR%20Summer2011.pdf#page=8

Fraga Rizo, C., Klein, L. B., Chesworth, B., Macy, R. J., & Dooley, R. (2022). Intimate partner violence survivors' housing needs and preferences: a brief report. *Journal of Interpersonal Violence, 37*(1–2), 958–972. https://doi.org/10.1177%2F0886260519897330

Fredriksen-Goldsen, K. I., Kim, H. J., Emlet, C. A., Muraco, A., Erosheva, E. A., Hoy-Ellis, J. & Petry, H. (2011). The aging and health report: Disparities and resilience among lesbian, gay, bisexual, and transgender older adults. *The Public Policy and Aging Report, 21*(3), 3. https://doi.org/10.1093/ppar/21.3.3

Frost, D. M., Meyer, I. H., & Schwartz, S. (2016). Social support networks among diverse sexual minority populations. *American Journal of Orthopsychiatry, 86*(1), 91. https://doi.apa.org/doi/10.1037/ort0000117

Hailey, J., Burton, W., & Arscott, J. (2020). We are family: Chosen and created families as a protective factor against racialized trauma and anti-LGBTQ oppression among African American sexual and gender minority youth. *Journal of GLBT Family Studies, 16*(2), 176–191. https://doi.org/10.1080/1550428X.2020.1724133

Harris Interactive. (2010). *The lesbian, gay, bisexual and transgender (LGBT) population at-a-glance*. http://www.witeck.com/wp-content/uploads/2013/03/HI_LGBT_SHEET_WCC_AtAGlance.pdf

Hawes, S. M., Flynn, E., Tedeschi, P., & Morris, K. N. (2022). Humane communities: Social change through policies promoting collective welfare. *Journal of Urban Affairs, 44*(2), 259–271. https://doi.org/10.1080/07352166.2019.1680244

Henrickson, M., Giwa, S., Hafford-Letchfield, T., Cocker, C., Mulé, N. J., Schaub, J., & Baril, A. (2020). Research ethics with gender and sexually diverse persons. *International Journal of Environmental Research and Public Health, 17*(18), 6615. https://doi.org/10.3390/ijerph17186615

Henwood, B., Dzubur, E., Rhoades, H., St. Clair, P., & Cox, R. (2021). Pet ownership in the unsheltered homeless population in Los Angeles. *Journal of Social Distress and Homelessness, 30*(2), 191–194. https://doi.org/10.1080/10530789.2020.1795791

Hull, K. E., & Ortyl, T. A. (2019). Conventional and cutting-edge: Definitions of family in LGBT communities. *Sexuality Research and Social Policy, 16*(1), 31–43. https://doi.org/10.1007/s13178-018-0324-2

Irvine, L., Kahl, K. N., & Smith, J. M. (2012). Confrontations and donations: Encounters between homeless pet owners and the public. *The Sociological Quarterly, 53*(1), 25–43. https://doi.org/10.1111/j.1533-8525.2011.01224.x

Jackson Levin, N., Kattari, S. K., Piellusch, E. K., & Watson, E. (2020). "We just take care of each other": Navigating 'chosen family' in the context of health, illness, and the mutual provision of care amongst queer and transgender young adults. *International Journal of Environmental Research and Public Health, 17*(19), 7346. https://doi.org/10.3390/ijerph17197346

King, M., Semlyen, J., Tai, S. S., Killaspy, H., Osborn, D., Popelyuk, D., & Nazareth, I. (2008). A systematic review of mental disorder, suicide, and deliberate self harm in lesbian, gay and bisexual people. *BMC Psychiatry, 8*(1), 1–17. https://doi.org/10.1186/1471-244X-8-70

Kosciw, J. G., Clark, C. M., Truong, N. L., & Zongrone, A. D. (2020). The 2019 National school climate survey: The experiences of lesbian, gay, bisexual, transgender, and queer youth in our nation's schools. A Report from GLSEN. *Gay, Lesbian and Straight Education Network (GLSEN)*. https://www.glsen.org/research/2019-national-school-climate-survey

Kosciw, J. G., Greytak, E. A., Zongrone, A. D., Clark, C. M., & Truong, N. L. (2018). *The 2017 National School Climate Survey: The Experiences of Lesbian, Gay, Bisexual, Transgender, and Queer Youth in Our Nation's Schools. Gay, Lesbian and Straight Education Network (GLSEN)*. New York. https://www.glsen.org/sites/default/files/2019-10/GLSEN-2017-National-School-Climate-Survey-NSCS-Full-Report.pdf

Krienert, J. L., Walsh, J. A., Matthews, K., McConkey, K. (2012). Examining the nexus between domestic violence and animal abuse in a national sample of service providers. *Violence and Victims, 27*, 280–295. https://doi.org/10.1891/0886-6708.27.2.280

Kurdek, L. A. (2009). Young adults' attachment to pet dogs: Findings from open-ended methods. *Anthrozoös, 22*(4), 359–369. https://doi.org/10.2752/089279309X12538695316149

Laurent-Simpson, A. (2017, September). Considering alternate sources of role identity: Childless parents and their animal "kids". *Sociological Forum, 32*(3), 610–634. https://doi.org/10.1111/socf.12351

LaVallee, E., Mueller, M. K., & McCobb, E. (2017). A systematic review of the literature addressing veterinary care for underserved communities. *Journal of Applied animal Welfare Science, 20*(4), 381–394. https://doi.org/10.1080/10888705.2017.1337515

Lockwood, R. (2002). Making the connection between animal cruelty and abuse and neglect of vulnerable adults. *The Latham Letter, 23*(1), 10–11. https://www.researchgate.net/publication/308028594_Making_the_connection_between_animal_cruelty_and_abuse_and_neglect_of_vulnerable_adults

MacNamara, M. (2019). LGBT older adults and their pets in southern Appalachia: Results of qualitative research and future research directions. *Innovation in Aging, 3*(Suppl 1), S201. https://doi.org/10.1093/geroni/igz038.729

Matijczak, A., Applebaum, J. W., Kattari, S. K., & McDonald, S. E. (2021). Social support and attachment to pets moderate the association between sexual and gender minority status and the likelihood of delaying or avoiding COVID-19 testing. *Social Sciences, 10*(8), 301. https://doi.org/10.3390/socsci10080301

Matijczak, A., McDonald, S. E., Tomlinson, C. A., Murphy, J. L., & O'Connor, K. (2020). The moderating effect of comfort from companion animals and social support on the relationship between microaggressions and mental health in LGBTQ+ emerging adults. *Behavioral Sciences, 11*(1), 1. https://doi.org/10.3390/bs11010001

McDonald, S. E., Collins, E. A., Nicotera, N., Hageman, T. O., Ascione, F. R., Williams, J. H., & Graham-Bermann, S. A. (2015). Children's experiences of companion animal maltreatment in households characterized by intimate partner violence. *Child Abuse & Neglect, 50*, 116–127. https://doi.org/10.1016/j.chiabu.2015.10.00544

McDonald, S. E., Collins, E. A., Maternick, A., Nicotera, N., Graham-Bermann, S., Ascione, F. R., & Williams, J. H. (2019). Intimate partner violence survivors' reports of their children's exposure to companion animal maltreatment: A qualitative study. *Journal of Interpersonal Violence, 34*(13), 2627–2652. https://doi.org/10.1177/0886260516689977545

McDonald, S. E., Matijczak, A., Nicotera, N., Applebaum, J. W., Kremer, L., Natoli, G.,... & Kattari, S. K. (2021). "He was like, my ride or die": Sexual and gender minority emerging adults' perspectives on living with pets during the transition to adulthood. *Emerging Adulthood, 10*(4), 21676968211025340. https://doi.org/10.1177/21676968211025340

McDonald, S. E., Murphy, J. L., Tomlinson, C. A., Matijczak, A., Applebaum, J. W., Wike, T. L., & Kattari, S. K. (2021). Relations between sexual and gender minority stress, personal hardiness, and psychological stress in emerging adulthood: Examining indirect effects via human-animal interaction. *Youth & Society, 54*(2), 0044118X21990044. https://doi.org/10.1177/0044118X21990044

McDonald, S., O'Connor, K., Matijczak, A., Murphy, J., Applebaum, J., Tomlinson, C., Wike, T., & Kattari, S. (2021). Victimization and psychological wellbeing among sexual and gender minority emerging adults: Testing the moderating role of emotional comfort from companion animals. *Journal of the Society for Social Work and Research, 13*(4). https://doi.org/10.1086/713889

McDonald, S. E., Sweeney, J., Niestat, L., & Doherty, C. (2022). Grooming-related concerns among companion animals: preliminary data on an overlooked topic and considerations for animals' access to health-related services. *Frontiers in Veterinary Science, 86*. https://doi.org/10.3389/fvets.2022.827348

Meehan, M., Massavelli, B., & Pachana, N. (2017). Using attachment theory and social support theory to examine and measure pets as sources of social support and attachment figures. *Anthrozoös, 30*(2), 273–289. https://doi.org/10.1080/08927936.2017.1311050

Meyer, I. H. (2003). Prejudice, social stress, and mental health in lesbian, gay, and bisexual populations: Conceptual issues and research evidence. *Psychological Bulletin, 129*(5), 674–697. https://doi.org/10.1037/0033-2909.129.5.674

Meyer, I. H. (2015). Resilience in the study of minority stress and health of sexual and gender minorities. *Psychology of Sexual Orientation and Gender Diversity, 2*(3), 209. https://doi.org/10.1037/sgd0000132

Meyer, I. H., Russell, S. T., Hammack, P. L., Frost, D. M., & Wilson, B. D. (2021). Minority stress, distress, and suicide attempts in three cohorts of sexual minority adults: A US probability sample. *PLoS One, 16*(3), e0246827. https://doi.org/10.1371/journal.pone.0246827

Moore, M. R., & Stambolis-Ruhstorfer, M. (2013). LGBT sexuality and families at the start of the twenty-first century. *Annual Review of Sociology, 39*, 491–507. https://doi.org/10.1146/annurev-soc-071312-145643

Moore, S. E., Wierenga, K. L., Prince, D. M., Gillani, B., & Mintz, L. J. (2021). Disproportionate impact of the COVID-19 pandemic on perceived social support, mental health and somatic symptoms in sexual and gender minority populations. *Journal of Homosexuality, 68*(4), 577–591. https://doi.org/10.1080/00918369.2020.1868184

Movement Advancement Project. (2019, July). *LGBT people with disabilities.* https://www.lgbtmap.org/lgbt-people-disabilities

Mulder, T. M., Kuiper, K. C., van der Put, C. E., Stams, G. J. J., & Assink, M. (2018). Risk factors for child neglect: A meta-analytic review. *Child Abuse & Neglect, 77*, 198–210. https://doi.org/10.1016/j.chiabu.2018.01.00637

Muraco, A., Putney, J., Shiu, C., & Fredriksen-Goldsen, K. I. (2018). Lifesaving in every way: The role of companion animals in the lives of older lesbian, gay, bisexual, and transgender adults age 50 and over. *Research on Aging, 40*(9), 859–882. https://doi.org/10.1177/0164027517752149

Norris, P. (2020). Animal neglect and abuse. In J. H. Byrd, P. Norris, & N. Bradley-Siemens (Eds.), *Veterinary forensic medicine and forensic sciences* (pp. 307–327). CRC Press.

Okoro, C. A., Hollis, N. D., Cyrus, A. C., & Griffin-Blake, S. (2018). Prevalence of disabilities and health care access by disability status and type among adults—United States, 2016. *Morbidity and Mortality Weekly Report, 67*(32), 882. https://dx.doi.org/10.15585%2Fmmwr.mm6732a3

Parker, J., & Novi, M. (2016). Further thoughts on possible discrimination against LGBT veterinarians. *Journal of the American Veterinary Medical Association, 250*(3), 848–849. https://doi.org/10.2460/javma.250.3.271

Peacock, J., Chur-Hansen, A., & Winefield, H. (2012). Mental health implications of human attachment to companion animals. *Journal of Clinical Psychology, 68*(3), 292–303. https://doi.org/10.1002/jclp.20866

Polick, C. S., Applebaum, J. W., Hanna, C., Jackson, D., Tsaras-Schumacher, S., Hawkins, R., Chervin, R. D., & Braley, T. J. (2021). The impact of pet care needs on medical decision-making among hospitalized patients: a cross-sectional analysis of patient experience. *Journal of Patient Experience, 8*, 23743735211046089. https://doi.org/10.1177/23743735211046089

Putney, J. M. (2014). Older lesbian adults' psychological well-being: The significance of pets. *Journal of Gay & Lesbian Social Services, 26*(1), 1–17. https://doi.org/10.1080/10538720.2013.866064

Rauktis, M. E., Lee, H., Bickel, L., Giovengo, H., Nagel, M., & Cahalane, H. (2020). Food security challenges and health opportunities of companion animal ownership for low-income adults. *Journal of Evidence-Based Social Work, 17*(6), 662–676. https://doi.org/10.1080/26408066.2020.1781726

Remillard, E. T., Campbell, M. L., Koon, L. M., & Rogers, W. A. (2022). Transportation challenges for persons aging with mobility disability: qualitative insights and policy implications. *Disability and Health Journal, 15*(1), 101209. https://doi.org/10.1016/j.dhjo.2021.101209

Rhoades, H., Winetrobe, H., & Rice, E. (2015). Pet ownership among homeless youth: Associations with mental health, service utilization and housing status. *Child Psychiatry & Human Development, 46*, 237–244. https://doi.org/10.1007/s10578-014-0463-5

Riggs, D. W., Taylor, N., Signal, T., Fraser, H., & Donovan, C. (2018). People of diverse genders and/or sexualities and their animal companions: Experiences of family violence in a binational sample. *Journal of Family Issues, 39*(18), 4226–4247. https://doi.org/10.1177/0192513X18811164

Ruben, M. A., Livingston, N. A., Berke, D. S., Matza, A. R., & Shipherd, J. C. (2019). Lesbian, gay, bisexual, and transgender veterans' experiences of discrimination in health care and their relation to health outcomes: A pilot study examining the moderating role of provider communication. *Health Equity, 3*(1), 480–488. https://doi.org/10.1089/heq.2019.0069

Rudd, M. L., & Jenkins, J. L. (2022). Decolonizing animal welfare through a social justice framework. *Frontiers in Veterinary Science, 8*, 787555. https://doi.org/10.3389/fvets.2021.787555

Russell, S. T., & Fish, J. N. (2016). Mental health in lesbian, gay, bisexual, and transgender (LGBT) youth. *Annual Review of Clinical Psychology, 12*, 465–487. https://doi.org/10.1146/annurev-clinpsy-021815-093153

San Miguel, S. F., Reed, W. M., Davis, K. C., Greenhill, L. M., & Sabin, E. A. (2014). Human-centered veterinary medicine. *Journal of the American Veterinary Medical Association, 245*(4), 374–375. https://doi.org/10.2460/javma.245.4.374

Scanlon, L., Hobson-West, P., Cobb, K., McBride, A., & Stavisky, J. (2021). homeless people and their dogs: Exploring the nature and impact of the human–companion animal bond. *Anthrozoös, 34*(1), 77–92. https://doi.org/10.1080/23251042.2022.2031513

Schmitz, R. M., Carlisle, Z. T., & Tabler, J. (2021). "Companion, friend, four-legged fluff ball": The power of pets in the lives of LGBTQ+ young people experiencing homelessness. *Sexualities, 25*(5–6), 1363460720986908. https://doi.org/10.1177/1363460720986908

Schmitz, R. M., Tabler, J., Carlisle, Z. T., & Almy, L. (2021). LGBTQ+ People's Mental Health and Pets: Novel Strategies of Coping and Resilience. *Archives of Sexual Behavior, 50*(7), 3065–3077. https://doi.org/10.1007/s10508-021-02105-6

Schmitz, R. M., Carlisle, Z. T., & Tabler, J. (2022). "Companion, friend, four-legged fluff ball": The power of pets in the lives of LGBTQ+ young people experiencing homelessness. *Sexualities, 25*(5–6), 694–716. https://doi.org/10.1177/1363460720986908

Schneebaum, A., & Badgett, M. L. (2019). Poverty in US lesbian and gay couple households. *Feminist Economics, 25*(1), 1–30. https://doi.org/10.1080/13545701.2018.1441533

Sechrist, S. M., Laplace, D. T., & Smith, P. H. (2022). North Carolina LGBTQ domestic violence response initiative: building capacity to provide safe, affirming services. *Health Education & Behavior, 49*(6), 10901981211067167. https://doi.org/10.1177%2F10901981211067167

Simons, L., Schrager, S. M., Clark, L. F., Belzer, M., & Olson, J. (2013). Parental support and mental health among transgender adolescents. *Journal of Adolescent Health, 53*(6), 791–793. https://doi.org/10.1016/j.jadohealth.2013.07.019

Snapp, S. D., Watson, R. J., Russell, S. T., Diaz, R. M., & Ryan, C. (2015). Social support networks for LGBT young adults: Low cost strategies for positive adjustment. *Family Relations, 64*(3), 420–430. https://doi.org/10.1111/fare.12124

Sterzing, P. R., Ratliff, G. A., Gartner, R. E., McGeough, B. L., & Johnson, K. C. (2017). Social ecological correlates of polyvictimization among a national sample of transgender, genderqueer, and cisgender sexual minority adolescents. *Child Abuse & Neglect, 67*, 1–12. https://doi.org/10.1016/j.chiabu.2017.02.017

Stevenson, R., Fitzgerald, A., & Barrett, B. J. (2018). Keeping pets safe in the context of intimate partner violence: Insights from domestic violence shelter staff in Canada. *Affilia, 33*(2), 236–252. https://doi.org/10.1177%2F0886109917747613

Storey, J. E. (2020). Risk factors for elder abuse and neglect: A review of the literature. *Aggression and Violent Behavior, 50*, 101339. https://doi.org/10.1016/j.avb.2019.10133935

Stewart, V., Judd, C., & Wheeler, A. J. (2021). Practitioners' experiences of deteriorating personal hygiene standards in people living with depression in Australia: A qualitative study. *Health & Social Care in the Community, 30*(4), 589–1598. https://doi.org/10.1111/hsc.134913

Su, D., Irwin, J. A., Fisher, C., Ramos, A., Kelley, M., Mendoza, D. A. R., & Coleman, J. D. (2016). Mental health disparities within the LGBT population: A comparison between transgender and nontransgender individuals. *Transgender Health, 1*(1), 12–20. https://doi.org/10.1089/trgh.2015.0001

Suen, L. W., Lunn, M. R., Katuzny, K., Finn, S., Duncan, L., Sevelius, J.,... & Obedin-Maliver, J. (2020). What sexual and gender minority people want researchers to know about sexual orientation and gender identity questions: a qualitative study. *Archives of Sexual Behavior, 49*(7), 2301–2318. https://doi.org/10.1007/s10508-020-01810-y

Taylor, J. (2019). Mental health in LGBTQ youth: Review of research and outcomes. *Communique, 48*(3), 4–6.

Taylor, N., Fraser, H., & Riggs, D. W. (2019). Domestic violence and companion animals in the context of LGBT people's relationships. *Sexualities, 22*(5–6), 821–836. https://doi.org/10.1177/1363460716681476

Topper, M. J. (2018). Building strong and sustainable partnerships to protect, promote, and advance your interests. *Journal of the American Veterinary Medical Association, 252*(9), 1034–1034. https://doi.org/10.2460/javma.252.9.1034

Tiplady, C. (2013). *Animal abuse: helping animals and people.* CABI.

Tiplady C. M., Walsh D. B., & Phillips C. J. (2018). "The animals are all I have": Domestic violence, companion animals, and veterinarians. *Society & Animals, 26*(5), 490–514. https://doi.org/10.1163/15685306-12341464

Tomlinson, C. A., Matijczak, A., McDonald, S. E., & Gee, N. R. (2020). The role of human-animal interaction in child and adolescent health and development. In *The encyclopedia of child and adolescent health,* Halpern-Felsher, B., Ed. Elsevier.

Vance, S. R., & Rosenthal, S. M. (2018). A closer look at the psychosocial realities of LGBTQ youth. *Pediatrics, 141*(5). https://doi.org/10.1542/peds.2018-0361

Walsh, F. (2009). Human-animal bonds II: The role of pets in family systems and family therapy. *Family Process, 48*(4), 481–499. https://doi.org/10.1111/j.1545-5300.2009.01297.x

Weston, K. (1997). *Families we choose: Lesbians, gays, kinship.* Columbia University Press.

Williams Institute. (2019). *LGBT data & demographics.* https://williamsinstitute.law.ucla.edu/visualization/lgbt-stats/?topic=LGBT#density

Witte, T. K., Kramper, S., Carmichael, K. P., Chaddock, M., & Gorczyca, K. (2020). A survey of negative mental health outcomes, workplace and school climate, and identity disclosure for lesbian, gay, bisexual, transgender, queer, questioning, and asexual veterinary professionals and students in the United States and United Kingdom. *Journal of the American Veterinary Medical Association, 257*(4), 417–431. https://doi.org/10.2460/javma.257.4.417

Witten T. M. (2014). Its not all darkness: Robust, resilience, and successful transgender aging. *LGBT Health, 1*(1), 24–33. https://doi.org/10.1089/lgbt.2013.0017

Zelle, A., & Arms, T. (2015). Psychosocial effects of health disparities on lesbian, gay, bisexual, and transgender older adults. *Journal of Psychosocial Nursing, 53*(7), 25–30. https://doi.org/10.3928/02793695-20150623-04

Zilcha-Mano, S., Mikulincer, M., & Shaver, P. R. (2012). Pets as safe havens and secure bases: The moderating role of pet attachment orientations. *Journal of Research in Personality, 46*(5), 571–580. https://doi.org/10.1016/j.jrp.2012.06.005

# 18

# PET OWNERSHIP AND PERSONS WITH DEVELOPMENTAL AND PHYSICAL DISABILITIES

## Impact on Mental and Physical Health

*Gretchen K. Carlisle and Steph M. Craven*

### Introduction

For people living in the United States, it is common for households to include an extra family member that has four paws, whiskers, and a wet nose. According to the American Pet Products Association National Pet Owners Survey (2021), 70% of U.S. households include a pet, or approximately 90 million families. At the same time, 4% of U.S. children have a disability (Young, 2021), and according to the most recent U.S. Census Bureau report, 26% of adults, or 61 million, adults have a disability (Okoro et al., 2018). Disabilities can be defined through any of the following criteria: (1) use of a device to aid ambulation, (2) vision or hearing deficits, (3) challenges with functional activities including: use of a pen to write, speaking or walking moderate distances, (4) inability to perform Activities of Daily Living such as taking a bath, dressing and using the toilet, and/or (5) mental deficits which interfere with daily life (Taylor, 2018). Increasing evidence over the past decades supports the benefits of the human–animal bond; however, additional research is needed to identify which types of human–animal interaction (HAI) and which species are most beneficial for people with specific conditions or disabilities, which individuals are most likely to benefit, and what burdens of HAI should be considered. This chapter discusses the impact of pets on the mental and physical health of children and adults with different types of developmental and physical disabilities.

### Historical Perspectives on Pet Ownership for People with Developmental and Physical Disabilities

The type of benefits related to companion animals have changed greatly over time, evolving from their duties as hunters, protectors, and herders to the current identification of pets as family members who provide companionship and an increased sense of well-being (Hussein et al., 2021). Pet ownership has a long history, dating back thousands of years. The Egyptians kept pets, mummified them, and were buried with their pets (Eldamaty, 2006). In the late 1700s, William Tuke advocated for the inclusion of companion animals as therapeutic for individuals institutionalized with psychiatric disorders (Tuke, 1964). By the mid-1800s, Florence Nightingale, who is acknowledged as establishing modern nursing, wrote, "A small pet animal is often an excellent companion for the sick, for long chronic cases especially. A pet bird in a cage is sometimes the only pleasure of an invalid confined for years to the same room" (Nightingale, 1969. P. 103). More recent documentation includes research on the beneficial support of companion animals with specific training such as service dogs and/or pets for individuals with disabilities such as emotional support animals.

Service animals are defined as dogs that are individually trained to perform certain tasks or provide support or assistance for a person with a disability (see Chapter 37). The earliest evidence of humans utilizing some form of assistance dog dates back to the 1st century, where murals were found among ancient ruins from a Roman city which revealed a painting of a dog leading a person with a visual impairment (Cohen, 2018). In the late 1700s, a man with a visual impairment trained his own dog to guide him, and soon after in the early 1800s, a manual for guide dogs was published, with recommendations for specific breeds best suited to work as guide dogs including Standard Poodles and German Shepherds, as well as other guidelines (Cohen, 2018). Although the guide dog provides an example of how highly trained animals have assisted people with disabilities for many decades, there are also ways in which pets with no specific training can provide support for people with disabilities, including the more recent introduction of emotional support animals (see Chapter 37), or companion animals as pets in the home.

## Human-Animal Interaction for People with Developmental Disabilities

### Benefits of Pets in Families of Children with Developmental Disabilities

A common developmental disability, autism spectrum disorder (ASD), has an incidence of one in 44 children (Maenner et al., 2021). ASD is characterized by deficits with social communication, presence of restrictive/repetitive behaviors, and sensory processing challenges (DSM-5, APA, 2013). Common comorbid conditions in ASD include anxiety and sleep disturbances (Vasa et al., 2020). HAI has emerged as a beneficial support for children with ASD and their families. Parents of children with ASD reported a reason for having a pet was for social companionship and to encourage their child's ability to learn responsibility through pet care-taking tasks (Carlisle, 2014). In a study echoing the importance of this responsibility, adolescents with ASD who provided more care for their pets had fewer symptoms of depression (Ward et al., 2017). Pets may also provide support for children with social and communication deficits. In families who acquired a new pet, young children with ASD were found to have an increase in prosocial behaviors (Grandgeorge et al., 2012), while Carlisle (2014) found that children already having a pet in the home had greater prosocial skills, compared with children having no pet. Further, parents of children with ASD reported their pet was, associated with their child's increased social interaction with others outside the family, and with companionship for their child (Bystrom et al., 2015; Carlisle, 2014; Harwood et al., 2019).

Many studies on pets include dogs exclusively; however, an exploratory study on the impact of cat adoption on children with ASD found increased empathy, less separation anxiety and fewer problem behaviors after adoption of a cat, versus children living with no cat (Carlisle et al., 2021). Similar findings were identified in a survey of cat owning parents of children with ASD. The parents described their cats as providing a positive and calming influence on their children with ASD (Hart et al., 2018). A noticeable difference was in cases of children with severe symptoms of ASD in which cats did not demonstrate the same companionship as with children having fewer symptoms (Hart et al., 2018). Consistently, both these studies found increased child-cat interactions with younger cats. Interestingly, in exploring eye gaze, children with ASD held greater visual contact with their pet cats versus their pet dog (Grandgeorge et al., 2020). The less-direct eye gaze of cats may decrease the inhibition of those with ASD in making eye contact. These benefits of pets in the home may also extend to parents of children with ASD, and the family as a whole.

Pet ownership in families of children with ASD has been associated with decreased stress for parents (Carlisle et al., 2020; Hall et al., 2016) and improved family functioning (Hall et al., 2016). One study found dog-owning parents of children with ASD having less stress than parents with no dog (Wright et al., 2015). Given the known high stress for parents of children with ASD (Cohrs & Leslie, 2017), it is important to note that dog ownership has been associated with parent and child ratings of improved family functioning, when compared with families having no pet dog (Hall et al., 2016; White, 2020; Wright et al., 2015). Importantly, a higher rating of family functioning was associated with parents having a greater attachment to their dogs (White, 2020). Not all studies include the important variable of pet attachment, and this may be key to understanding the impact of pets within families. In addition to attachment, family income, pet species, and number of pets in the home may also play a role. Parents from lower-income brackets rated themselves as perceiving greater benefits from their pets, while parents in families with both cats and dogs rated themselves as receiving more benefits than parents with only a cat or only a dog (Carlisle et al., 2020). Most of these studies have been small and exploratory in nature, thus limiting generalizability. For parents of children with ASD considering adding a pet to their family, what may be most important is considering the characteristics of their child, themselves, and their family as part of their process in identifying the type of pet which is most likely to be a good match for their child and family. Parents have reported more benefits when there is a good match between their pet, child, and family (Carlisle, 2014). Suggested steps in identifying a pet that would be a good match include considering the following: family lifestyle, available time for pet care, sensory issues (e.g. a quiet pet if loud sounds are a sensory challenge), and inclusion of a child with ASD in selection process if possible. In additional to this, veterinary behaviorists may be able to assist with specific recommendations for an individual family to improve or enhance child-pet interactions.

Less is known about pets in the homes of families of children with other developmental disabilities such as Down syndrome and Cerebral Palsy. There are a limited number of studies on animal-assisted activities (AAA) and these may provide insight on possibilities for interactions with pets in the home. AAA are loosely structured activities involving HAI that promote health and well-being by encouraging socialization, and the interactions are characterized as recreational in nature (Pet Partners, 2022, Chapter 34). Esteves and Stokes (2008) found the presence of a dog in the classroom of children with developmental disabilities was associated with an increase in social communication and positive behaviors. Likewise, when a dog was present in a class of children with Down syndrome, children had increased social engagement (Limond et al., 1997). Additional research in needed to aid in understanding the experience and role of HAI with pets in homes of children with developmental disabilities.

## Benefits of Pets for Adults with Developmental and Intellectual Disabilities

Despite the high prevalence of pet ownership in the United States, the specific benefits of pet ownership for adults with intellectual and developmental disabilities has received relatively little research attention. One study of adults with intellectual disabilities found caring for pets as a preferred leisure activity in the home (Badia et al., 2012). Most of the research related to companion animals assisting adults with disabilities in society is concentrated on service animals or trained therapy animals, rather than on pets in the home. Research on the impact of animals in the lives of adults living with disabilities referencing service dogs, will be discussed in Chapter 37. As with children, there are studies on AAA with adults with disabilities which may be applicable to pets in the home. These AAA have found interacting with companion animals has been associated with improved mood, increased social awareness, and increased attentiveness in adults with developmental disabilities (LeRoux & Kemp, 2009). More benefits of AAA will be highlighted in Chapter 34.

## Benefits of Pets for People with Mental and Physical Disabilities

In the United States, 4% of the population ages 0–17 years has some type of disability (Young et al., 2021), while almost 42% of older adults report a disability (Okoro et al., 2018). In the population of older adults, including those over 50 years, 51% report having pets, and approximately 20% of people 85 years and older own pets (Mueller et al., 2018). Given the high percentage of older adults with a disability and the high prevalence of pet ownership, research on pet ownership in this age group might be especially important. Mobility, in particular, can be a challenge and may be accompanied by aches and pains. For older adults with chronic pain, pets have been described as playing a role with decreased pain levels by increasing the pet owner's ability to manage mood, relax, be joyful, laugh, and engage in increased physical activity and social interaction (Janevic et al., 2020). In the general population of older adults, companionship has been identified as the reason for keeping a pet (Bibbo et al., 2019). During the isolating experience of the Covid-19 pandemic, pets may have filled an important role through a moderating effect on loneliness among older adults (Applebaum et al., 2021). With an increasingly aging population (U.S. Census Bureau, 2019), a greater understanding of HAI with pets among this group of people will be valuable.

Another growing group of persons in the United States is those with obesity. Obesity on its own may be considered a disability, if the person is unable to ambulate normally or has other comorbidities associated with obesity. The age-adjusted prevalence of obesity in the United States is 42%, according to the CDC (Hales et al., 2020). Children and adults with mobility limitations and intellectual or learning disabilities are at the greatest risk for obesity (Chen et al., 2010). People who suffer from medication side effects, pain, chronic fatigue, or a lack of accessible resources for exercise are also at an increased risk of obesity ("Disability and Obesity," 2019). For children with developmental disabilities, activity with pet dogs may be an aid to obesity prevention/treatment (Must et al., 2021). Pet dogs may also benefit adults with disabilities through an increase in social interactions when exercising through dog walking (Bauld et al., 2022). Results on the possible benefits of dog-walking have been conflicting. A meta-analysis of cross-sectional studies on obesity and pets revealed no significant difference in pet ownership versus no pet in adults with obesity (Miyake et al., 2020). Another review by Potter (2021) found dog walking to be potentially beneficial as a motivation for physical activity; however, the research did not identify dog-walking as causative in relation to weight loss. Further research is needed including studies with a longitudinal design, and the accounting of variables such as attachment to pet. This evidence might aid in elucidating the relationship between individuals with a disability who struggle with obesity, and the impact of pet ownership.

Paired exercise practices may benefit humans and may also improve the health of pet dogs. According to an Association for Pet Obesity Prevention survey from 2018, 59% of cats and 56% of dogs were classified as overweight or obese in the United States (Ward, 2017). By observing the excitement of the dog, as well as having companionship during the walk, people may experience a dog as a motivating factor for continuing exercising (Bartges et al., 2017). In this way, the exercise regimen may benefit the human, and improve the health of the dog. Health-care providers for children and adults might consider motivating their human patients to lose weight by encouraging them to partner with their pet in a joint weight loss or exercise program, when medically appropriate.

People with debilitating physical ailments such as chronic disease or cancer are more likely to suffer from mental illness including anxiety and/or depression (Xie et al., 2022) (see Chapter 5). Pets may be beneficial for those with mental disabilities such as schizophrenia, bipolar disorder, or affective psychoses by acting as a family, through unconditional acceptance and by increasing social engagement (Wisdom et al., 2009). A systematic review revealed that pets were associated with support in crisis, reduced feelings of isolation, and as a source of comfort in grief for adults with mental health disabilities (Brooks et al., 2018). Furthermore, pet dogs have been associated with reduced feelings of loneliness, increased calmness, and elevated mood for military veterans with PTSD (Stern et al., 2013). In severe situations, among adults with ASD experiencing depression, those persons who had close relationships with their dogs said their dogs were the reason they did not commit suicide (Barcelos et al., 2021). With 17.2 veteran deaths by suicide every day in the United States (Department of Veteran Affairs, 2021), alternative and integrative therapies and support, including pet ownership, might be important. For those with mental health challenges, even everyday living can be a challenge. Pets may provide support by increasing a person's ability to complete self-care tasks, exercise, keep house, and take medications (Brooks et al., 2018).

Many individuals with disabilities require daily medication to manage their condition and medication compliance is a challenge (Vacek et al., 2013). Pets may help provide a performance cue for individuals to take medication. In a study of adolescents with Type 1 diabetes, the treatment group were given a beta fish and instructed in diabetic self-care paired with fish care, while a control group of similar adolescents were provided with diabetes self-care instructions as usual (Miranda et al., 2015). The adolescents with the fish had a significant improvement in health markers for their diabetes (Maranda et al., 2015). Pets have also been associated with improved medication adherence for chronically ill adults such as those with HIV (Muldoon et al., 2017). Additional research would aid in establishing evidence-based practice for including pet ownership as a component of patient education.

Another benefit of pets for those who are disabled may be related to chronic pain. Pet owners with a positive relationship with their pet, who described themselves as actively attempting to manage their pain through HAI, rated this strategy as helpful and had lower pain scores, compared with those who had a negative pet relationship, or did not actively seek the support of their pet for pain relief (Bradley & Bennett, 2015). In another study evaluating chronic pain in patients who underwent spinal cord stimulation surgery, preliminary evidence suggested pet ownership was associated with less pain and greater procedure satisfaction (Williams et al., 2021). Pain management is not a trivial endeavor. The U.S.

government has declared the opioid epidemic a public health emergency (Jones et al., 2018). Efforts to identify evidence-based integrative and complementary methods to alleviate pain, such as interacting with pets, are indicated.

In summary, pet ownership has been associated with improved social, mental and health, and may aid in promoting a healthy lifestyle overall. These benefits could be very supportive for people living with a disability. Unfortunately, pets are seldom incorporated into the discussion between health-care providers and patients. A One Health approach to care would include pet ownership as part of a medical history (Hodgson et al., 2019). Incorporating consideration of pet ownership as part of a One Health approach might be especially important for children and adults with disabilities.

## Burdens of Pet Ownership for People with Disabilities

Humans often keep pets for the experience of joy, entertainment, support, and other positive benefits discussed in this text. However, pets may also represent a burden to their owners (see Chapter 20). Owning a pet creates a demand for time, energy, and money and these difficulties may be exacerbated for individuals with disabilities. Parents of children with ASD name all these items as challenging aspects of pet ownership (Carlisle, 2014). Other specific challenges may be related to an individual's disability, for instance a person's mobility limitations may make it more difficult to walk a dog, clean a litter box, or be responsible for other pet care-taking activities. Without the physical ability to care for a pet, additional costs for contracting pet care may be required, and without adequate finances, pets may suffer.

Selection of a pet that is a good match for the whole family is especially important. For children or adults with ASD, sensory processing disorders may create a unique challenge in pet ownership, potentially leading to rehoming the animal, which could cause distress for a child with ASD. Loud or sharp sounds, such as a dog barking or a bird shrieking can create sensory overload for an individual and lead to a poor relationship with a pet (Carlisle, 2014). Also, while a pet might be a challenge for a child with sensory deficits, other members of the family might be very attached to the same animal. Siblings of children with ASD have higher levels of depressive symptoms compared with siblings of neurotypical children (Lovell & Wehterell, 2016; Vermaes et al., 2012), and greater anxiety (O'Neill & Murray, 2016). A pet may provide companionship and support for these siblings. Parent direction of joint play activities among children with ASD and their siblings can help strengthen family unity (Tsao et al., 2012). By carefully identifying a pet which is a good match for the whole family, parent may direct sibling interactions with the family pet, thus benefiting the children with ASD and their neurotypical siblings.

For pets in the family, supervision of child-pet interactions are critically important. A child or adult with an intellectual disability may not correctly interpret an animal's behavior, leading to a bite or scratch. In one study, dog bite prevalence was highest among young males and there was an association between the child having a behavioral disorder diagnosis and the risk of a dog bite (Holzer et al., 2019). Parents of children with ASD have described the need to closely monitor child-pet interactions to ensure safety (Carlisle et al., 2018). Observation of child-pet interactions cannot be overemphasized, and medical and veterinary providers can aid in providing educational materials for parents on practices for safe child-pet interactions, as well as educational material to help children interpret pet behavior. One resource available is "Pet Meets Baby" an educational brochure produced by the American Humane Association (2014).

Allergies to pets may also be a burden with the prevalence of pet allergies ranging between 10% and 20% of the global population (Chan & Leung, 2018). Similar to the typical population, individuals with disabilities having allergies should discuss acquiring a pet with their health-care provider. Although "hypoallergenic" pets are becoming a popular trend, there is no evidence to date supporting the validity of truly hypoallergenic pets (Chan & Leung, 2018). In fact, the levels of the canine allergen Can f 1 in dogs, considered to be hypoallergenic, were found to be comparable in non-hypoallergenic dog breeds (Vredegoor et al., 2012).

Zoonotic diseases are a specific concern for individuals with who are immunocompromised, and when compared with the general population, people with disabilities are more likely to have diseases or conditions causing them to be immunocompromised (CDC, 2021). Zoonotic diseases such as salmonellosis, toxoplasmosis, brucellosis, and toxocariasis can be introduced by pets, and therefore place immunocompromised persons at risk (Overgaauw, 2020). Despite the vast impacts of the COVID-19 pandemic, there is no evidence yet identifying pets as transmitting COVID-19 to humans, although humans may be able to transmit COVID-19 to their pets (Korath et al., 2022). One avoidable zoonotic concern is a pet owner's choice of diet for their pet. Feeding raw pet diets offers an increased risk of disease transmission to immunocompromised pet owners, with raw pet food shown to commonly exceed thresholds for counts of Enterobacteriaceae, as well as hosting other infectious organisms including Listeria, shiga toxigenic *Escherichia coli*, *Toxoplasma gondii*, and *Brucella suis* (Davies et al., 2019). Lastly, a stressed animal is more likely to contract a disease (Corsetti et al., 2018), and therefore increases the risk for an immunocompromised person to contract a zoonotic disease. Observing pets for signs of stress and addressing potential behavior concerns with a veterinarian or animal behaviorist may contribute to a safer environment for the pet, and ultimately an immunocompromised owner. Working collaboratively with their human medical provider and veterinarian will help prevent zoonotic diseases in immunocompromised pet owners. Handwashing after interaction with pets is key avoiding the spread of infection.

An unexpected burden of pet ownership may include stress, sadness or grief associated with medical concerns of a pet or pet death (see Chapters 20 and 21). These negative emotions may contribute to more stress or pain for a person with a disability. While parents of children with ASD name learning about loss through death as a benefit of pets, they also report fear and concern that the death of a pet will be a problem for their children (Carlisle et al., 2018). Disenfranchised loss can occur when the loss of a pet is not acknowledged as important, and this type of loss can also occur for persons with developmental or intellectual disabilities due to their grief not being recognized (Lenhardt, 1997). Persons with intellectual disabilities grieve in similar ways to those in the general population and acknowledgement of their right to grieve is important to the person's ability to come to terms with loss (McRitchie et al., 2014). Acknowledging the impact of pet loss and grief is warranted to assist individuals with processing the experience.

Financial concerns are one of the biggest obstacles faced in pet ownership. A survey conducted in 2020 reported that 47% of dog owners were spending $1201/year and described this expense as unexpected (TD Ameritrade, 2020). Financial burden can complicate pet ownership for people with disabilities, given the increased likelihood of their economic insecurity compared with people not having a disability. According to the ADA Participatory Action Research Consortium (2016), the median earning for individuals with a disability was $21,509/year in 2016 compared with those having no disability who earned $31,865/year. Furthermore, although pets are not inexpensive, the cost of a service dog can be much higher. According to the National Service Animal Registry (2019), the average cost of a service dog was estimated at $15,000–$30,000, and as high as $50,000, depending on the extent of a dog's training. Some organizations provide service dogs at no cost to the recipient; however, the shortage of available dogs may lead some with disabilities to seek a dog with a large price tag. One study found a service dog may represent a cost savings over time, compared with an untrained pet, for people with functional impairments or certain chronic diseases (Lundqvist, 2019). The U.S. Department of Veterans Affairs (2021) now accepts applications from veterans, and for those who qualify, the Veterans Administration will reimburse veterinary costs for service dogs. For individuals with disabilities who are not veterans, do not desire or need a service dog, research exploring the cost-benefit analysis of pet ownership would be helpful not only to individuals, but also to potential funding sources helping to cover the cost of veterinary care.

## Animal Welfare in Homes of People with Developmental and Physical Disabilities

In addition to any potential burdens of pet ownership for people with disabilities, there may be welfare concerns for the pet. For example, the owner may not be able to properly complete all the tasks related to the animal's care. Some communities have pet support services available and one such service Meals on Wheels (Meals on Wheels, Central Texas, 2022). If the person has a disability which limits their ability to ambulate, their pet may not receive an appropriate level of exercise and mental stimulation. In the opposite extreme, service dogs may be at risk of too much exercise in some cases (Lane et al., 1998). While no significant evidence of concern for canine welfare was found in this study of veterans and service dogs (Lane et al., 1998), educational information on physical limitations of service dogs should be provided to handlers to prevent overuse injuries or health concerns. Similarly, pet owners should seek the counsel of their veterinarian regarding exercise for their pet. Providing pets with a safe amount of exercise can aid the pet's health and may help avoid the dangers of being overweight. Supportive serves are available to aid persons in keeping their pet healthy. The American Veterinary Medical Association provides an Animal Obesity Toolkit to assist veterinary providers with educating and supporting pet owners in addressing the need for weight loss (American Veterinary Medical Association, 2018). Another resource is the Association for Pet Obesity Prevention, which is a non-profit organization providing education resources to assist those seeking to keep or assist their pet in preventing or addressing obesity (Pet Obesity Prevention, 2022).

As with all pets, regular veterinary care is key to good health for parasite preventatives, vaccinations, and examination. For individuals with a disability, transportation may be a challenge. With a growing trend of mobile veterinary practices, this might provide a valuable option for pet owners who find travel to a veterinary clinic difficult. In addition to transportation, ensuring communication appropriate for the client is important to ensure clients understand instructions and care-taking advise from their veterinary provider. Veterinarians can assist by supporting pet owners and providing information in a format that acknowledges a wide range of differences from a non-verbal child with a best friend guinea pig to an older adult with a visual deficit whose aging cat has developed diabetes. The term "disability" can encompass many conditions. For some types of disabilities, there may be no difficulty for the pet owner in providing adequate care for their pet, whereas for other disabilities, it may not be possible for the person to safely care for their pet without assistance. Like all potential pet owners, it is important for individuals with disabilities seeking a pet to consider what animal might be the best fit for them, their family and their home. Consideration of species, breed, temperament, age, health status, and cost are all considerations for potential pet owners, to be proactive in keeping the welfare of the animal in mind.

## Gaps in the Literature and Need for Research in the Specific Area of Children and Adults with Developmental and Physical Disabilities

A small but gradual increase in funding through non-profit foundations and other sources has provided the resources for a growing number of research studies in HAI over the past few decades. To date, most studies have focused on Animal-Assisted Interventions or Service Dogs, rather than pets in the home. This could be due to the difficult nature of collecting data within the community for a study of pets in the home, and the challenge of confounding variables. Additionally, the opportunity to conduct randomized controlled trials can be limited due to recruitment challenges, in part due to difficulty recruiting participants who are interested in acquiring a pet, yet do not already have a pet.

Specifically, there is a lack of research among many groups of individuals such as people with ADHD, blindness, deafness, cancer, mobility disorders, diabetes, paralysis, multiple sclerosis, cerebral palsy, epilepsy, addiction, and many other disabilities. It is important to note that a research study comparing the experience of service dogs versus pets for individuals with disabilities found that while pets were associated with helping people to "thrive," service dogs were rated as more supportive (Gravrok, 2019). This leads to the question of who might be better matched with a service dog verses a pet. There is a shortage of service dogs, and the wait lists can be years. If some individuals would be better matched with a pet, this would be valuable information. Without more research on the benefits of pets for individuals with disabilities, this question cannot be answered.

Despite the obstacles within the HAI research process on pet ownership, the growing public acceptance of the importance of the bond is encouraging. An increase in educational certificates and degree programs related to anthrozoology will hopefully lead to additional research in the future through educating new professionals in the field, leading to study of the knowledge gaps existing now. Finally, through medical and veterinary providers partnering together, a One Health approach to research and practice will be more comprehensive.

### Potential Questions for Discussion

1. While there is a growing body of research on pets in families of children with ASD, what differences might be found for adults with ASD?
2. What benefits and challenges of pets in the home might be experienced by the parents of children with physical disabilities?
3. Which zoonotic concerns are important for pets in homes of individuals who are immunocompromised?
4. What patient education strategies might be employed by medical providers using a One Health approach?
5. What considerations might a parent of a child with a disability or an adult with a disability consider before acquiring a new pet?
6. What animal welfare concerns might there be for pets in homes of children or adults with disabilities?

## References

American Humane Association. (2014). *Pet meets baby: A guide for bringing children home to pets.* Brochure.

American Pet Products Association. (2021). *2021–2022 American pet products associations national pet owners survey.* Business/Finance Fact Sheet. https://americanpetproducts.org/Uploads/NPOS/21-22_BusinessandFinance.pdf

American Psychiatric Association, DSM-5 Task Force. (2013). *Diagnostic and statistical manual of mental disorders: DSM-5* (5th ed.). American Psychiatric Publishing, Inc. https://doi.org/10.1176/appi.books.9780890425596

American Veterinary Medical Association. (2018, July 23). *Animal obesity toolkit.* https://www.avma.org/blog/new-toolkit-provides-obesity-materials-and-education

Americans with Disabilities Act Participatory Action Research Consortium. (2016). *Median earnings in the past 12 months (in 2011 inflation-adjusted dollars) by disability status by sex for the civilian noninstitutionalized population 16 years and over with earnings.* http://centerondisability.org/ada_parc/utils/indicators.php?id=30

Applebaum, J. W., Ellison, C., Struckmeyer, L., Zsembik, B. A., & McDonald, S. E. (2021). The impact of pets on everyday life for older adults during the COVID-19 pandemic. *Frontiers in Public Health, 9*, 292.

Association for Pet Obesity Prevention. (2022). https://petobesityprevention.org/about

Badia, M., Orgaz, M. B., Verdugo, M. A., & Ullan, A. M. (2012). Patterns and determinants of leisure participation of youth and adults with developmental disabilities. *Journal of Intellectual Disability Research, 57*(4), 319–332.

Barcelos, A. M., Kargas, N., Packham, C., & Mills, D. S. (2021). Understanding the impact of dog ownership on autistic adults: Implications for mental health and suicide prevention. *Scientific Reports, 11*, 23655. https://doi.org/10.1038/s41598-021-02504-8

Bartges, J., Kushner, R. F., Michel, K. E., Sallis, R., & Day, M. J. (2017). One health solutions to obesity in people and their pets. *Journal of Comparative Pathology, 156*(4), 326–333.

Bauld, E., Callaway, L., Warren, N., Lalor, A., & Burke, J. (2023). Pilot of a dog-walking program to foster and support community inclusion for people with cognitive disabilities. *Disability and Rehabilitation, 45*(3), 469–482.

Bibbo, J., Curl, A. L., & Johnson, R. A. (2019). Pets in the lives of older adults: A life course perspective. *Anthrozoös, 32*(4), 541–554.

Bradley, L., & Bennett, P. C. (2015). Companion-animals' effectiveness in managing chronic pain in adult community members. *Anthrozoös, 28*(4), 635–647. https://doi.org/10.1080/08927936.2015.1070006.

Brooks, H. L., Rushton, K., Lovell, K., Bee, P., Walker, L., Grant, L., & Rogers, A. (2018). The power of support from companion animals for people living with mental health problems: A systematic review and narrative synthesis of the evidence. *BMC Psychiatry, 18*(1), 1–12.

Bystrom, K. M., & Lundqvist Persson, C. A. (2015). The meaning of companion animals for children and adolescents with autism: The parents' perspective. *Anthrozoös, 28*(2), 263–275. https://doi.org/10.1080/08927936.2015.11435401

Carlisle, G. K. (2014). Pet dog ownership decisions for parents of children with autism spectrum disorder. *Journal of Pediatric Nursing, 29*(2), 114–123. https://doi.org/10.1016/j.pedn.2013.09.005

Carlisle, G. K., Johnson, R. A., Mazurek, M., Bibbo, J. L., Tocco, F., & Cameron, G. T. (2018). Companion animals in families of children with autism spectrum disorder: Lessons learned from caregivers. *Journal of Family Social Work, 21*(4–5), 294–312. http://doi.org/10.1080/10522158.2017.1394413

Carlisle, G. K., Johnson, R. A., Wang, Z., Brosi, T. C., Rife, E. M., & Hutchison, A. (2020). Exploring human-companion animal interaction in families of children with autism. *Journal of Autism and Developmental Disorders, 50*, 2793–2805. https://doi.org/10.1007/s10803-020-04390-x

Carlisle, G. K., Johnson, R. A., Wang, Z., Bibbo, J., Cheak-Zamora, N., & Lyons, L. A. (2021). Exploratory study of cat adoption in families of children with autism: Impact on children's social skills and anxiety. *Journal of Pediatric Nursing, 58*, 28–35.

CDC. (2021). COVID-19 *Information for people with disabilities.* Centers for Disease Control and Prevention. Published June 21, 2021. https://www.cdc.gov/ncbddd/humandevelopment/covid-19/people-with-disabilities.html

Chan, S. K., Leung, D. Y. M. (2018). Dog and cat allergies: Current state of diagnostic approaches and challenges. *Allergy Asthma and Immunoogical Research, 10*(2), 97–105. https://doi.org/10.4168/aair.2018.10.2.97

Chen, A. Y., Kim, S. E., Houtrow, A. J., & Newacheck, P. W. (2010). Prevalence of obesity among children with chronic conditions. *Obesity*, *18*(1), 210–213.

Cohen J. (2018). *Assistance dogs: Learning new tricks for centuries.* HISTORY. Published August 22, 2018. https://www.history.com/news/assistance-dogs-learning-new-tricks-for-centuries

Cohrs, A. C. & Leslie, D. L. (2017). Depression in parents of children diagnosed with autism spectrum disorder: A claims-based analysis. *Journal of Autism and Developmental Disorders*, *47*, 1416–1422.

Corsetti, S., Borruso, S., Di Traglia, M., Lai, O., Alfieri, L., Villavecchia, A., Cariola, G., Spaziani, A., & Natoli, E. (2018). Bold personality makes domestic dogs entering a shelter less vulnerable to diseases. *PLoS One*, *13*(3), e0193794.

Davies, R. H., Lawes, J. R., & Wales, A. D. (2019). Raw diets for dogs and cats: A review, with particular reference to microbiological hazards. *Journal of Small Animal Practice*, *60*(6), 329–339.

Department of Veterans Affairs. (2021). 2021 national veteran suicide prevention annual report. *Department of Veterans Affairs, Office of Mental Health and Suicide Prevention.* https://www.mentalhealth.va.gov/docs/data-sheets/2021/2021-National-Veteran-Suicide-Prevention-Annual-Report-FINAL-9-8-21.pdf

Disability and Obesity. (2019). *Centers for disease control and prevention.* Published September 6, 2019. https://www.cdc.gov/ncbddd/disabilityandhealth/obesity.html#ref

Eldamaty, M. (2006). *Beloved beasts: Animal mummies from ancient Egypt.* American Univ in Cairo Press.

Esteves, S., & Stokes, T. (2008). Social effects of a dog's presence on children with disabilities. *Antrhozoos*, *21*(1), 5–15.

Grandgeorge, M., Gautier, Y., Bourreau, Y., Mossu, H., & Hausberger, M. (2020). Visual attention patterns differ in dog vs. cat interaction with children with typical development or autism spectrum disorders. *Frontiers in Psychology*, *11*, 2047. https://doi.org/10.3389/fpsyg.2020.02047.

Grandgeorge, M., Tordjman, S., Lazartigues, A., Lemonnier, E., Deleau, M., & Hausberger, M. (2012). Does pet arrival triffer prosocial behaviors in individuals with autism? *PLoS One*, *7*(8), e41739 http://dx.doi.org/10.1371/journal.pone.0041739

Gravrok, J., Howell, T., Bendrups, D., & Bennett, P. (2019). Thriving through relationships: Assistance dogs' and companion dogs' perceived ability to contribute to thriving in individuals with and without a disability. *Disability Rehabilitation: Assistive Technology*, *15*(1), 45–53. https://doi.org/10.1080/17483107.2018.1513574.

Hales, C., & Fryar, C. D. (2020). *QuickStats: Prevalence of obesity and severe obesity among persons ages 2–19 years – National Health and Nutrition Examination Survey, 1999–2000 through 2017–2018.* Morbidity and Mortality Weekly Report, National Center for Health Statistics. https://www.ded.gov/nchs/nhanes.htm

Hall, S. S., Wright, H. F., Mills, D. S., & Schmitz, C. (2016). What factors are associated with positive effects of dog ownership in families with children with autism spectrum disorder? The development of the Lincoln Autism Pet Dog Impact Scale. *PLoS One*, *11*(2), e0149736. https://doi.org/10.1371/journal.pone.0149736

Hart, L. A., Thigpen, A. P., Willits, N. H., Lyons, L. A., Hertz-Picciotto, I., & Hart, B. L. (2018). Affectionate interactions of cats with children with autism spectrum disorder. *Frontiers in Veterinary Science*, *5*, 39. https://doi.org/10.3389/fvets.2018.00039.

Harwood, C., Kaaczmarek, E., & Drake, D. (2019). Parental perceptions of the nature of the relationship children with autism spectrum disorders share with their canine companion. *Journal of Autism and Developmental Disorders*, *49*(1), 248–259.

Hodgson, K., Darling, M., Freeman, D., & Monavvari, A. (2019). Engaging family physicians in one health. *Journal of the American Veterinary Medical Association*, *254*(11), 1267–1269.

Holzer, K. J., Vaughn, M. G., & Murugan, V. (2019). Dog bite injuries in the USA: Prevalence, correlates and recent trends. *Injury Prevention*, *25*(3), 187–190.

Hussein, S. M., Soliman, W. S., & Khalifa, A. A. (2021). Benefits of pets' ownership, a review based on health perspectives. *Journal of Internal Medicine and Emergency Research*, *2*(1), 1–9. https://doi.org/10.37191/Mapsci-2582-7367-2(1)-020

Janevic, M. R., Shute, V., Connell, C. M., Piette, J. D., Goesling, J., & Fynke, J. (2020). The role of pets in supporting cognitive-behavioral chronic pain self-management: Perspectives of older adults. *Journal of Applied Gerontology*, *39*(10), 1088–1096. https://doi.org/10.1177/0733464819856270

Jones, M. R., Viswanath, O., Peck, J., Kaye, A. D., Gill, J. S., & Simopoulos, T. T. (2018). A brief history of the opioid epidemic and strategies for pain medicine. *Pain and Therapy*, *7*(1), 13–21.

Korath, A. D., Janda, J., Untersmayr, E., Sokolowska, M., Feleszko, W., Agache, I., Adel seida, A. A., Hartmann, K., Jensen-Jarolim, E., & Pali-Schöll, I. (2022). One health: EAACI position paper on coronaviruses at the human-animal interface, with a specific focus on comparative and zoonotic aspects of SARS-Cov-2. *Allergy*, 77(1), 55–71.

Lane, D. R., McNicholas, J., & Colli, G. M. (1998). Dogs for the disabled: Benefits to recipients and welfare of the dog, *Applied Animal Behaviour Science*, 59(1–3), 49–60, ISSN 0168–1591, https://doi.org/10.1016/S0168-1591(98)00120-8.

Lenhardt, A. M. (1997). Grieving disenfranchised losses: Background and strategies for counselors. *Journal of Humanistic Education and Development*, 35, 208–216.

Le Roux, M. C., & Kemp, R. (2009). Effect of a companion dog on depression and anxiety levels of elderly residents in a long-term facility. *Psychogeriatrics*, 9, 23–26.

Limond, J. A., Bradshaw, J. W. S., & Cormack, M. K. F. (1997). Behavior of children with learning disabilities interacting with a therapy dog. *Anthrozoos*, 10(2–3), 84–89. https://doi.org/10.2752/089279397787001139

Lovell, B., & Wetherell, M. A. (2016). The psychophysiological impact of childhood autism spectrum disorder on siblings. *Research in Developmental Disabilities*, 49–50, 226–234.

Lundqvist, M., Alwin, J., & Levin, L. Å. (2019). Certified service dogs - A cost-effectiveness analysis appraisal. *PLoS One*, 14(9), e0219911. https://doi.org/10.1371/journal.pone.0219911

Maenner, M. J., Shaw, K. A., Bakian, A. V., Bilder, D. A., Durkin, M. S., Esler, A., Furnier, S. M., Hallas, L., Hall-Lande, J., Hudson, A., Hughes, M. M., Patrick, M., Pierce, K., Poynter, J. N., Salinas, A., Shenounda, J., Vehorn, A., Warren, Z., Constantino, J. N., ... Cogswell, M. E. (2021). Prevalence and characteristics of autism spectrum disorder among children aged 8 years-autism and developmental disabilities monitoring network, 11 sites, United States, 2018. *MMWR Surveillance Summaries*, 70(SS–11), 1–16. http://dx.doi.org/10.15585/mmwr.ss7011a1

McRitchie, R., McKenzie, K., Quayle, E., Harlin, M., & Neumann, K. (2014). How adults with an intellectual disability experience bereavement and grief: A qualitative exploration. *Death Studies*, 38(3), 179–185.

Meals on Wheels of Central Texas. (2022). *Pets assisting the lives of seniors*. https://www.mealsonwheelscentraltexas.org/programs/pals

Miranda, L., Lau, M., Steward, S. M., & Gupta, O. T. (2015). A novel behavioral intervention in adolescents with Type 1 diabetes mellitus improves glycemic control: Preliminary results from a pilot randomized control trial. *The Science of Diabetes Self-Management and Care*, 41(2), 224–230. https://doi.org/10.1177/0145721714567235

Miyake, K., Kito, K., Kotemori, A., Sasaki, K., Yamamoto, J., Otagiri, Nagasawa, M., Kuze-Arata, S., Mogi, K., Kikusui, T., & Ishihara, J. (2020). Association between pet ownership and obesity: A systematic review and meta-analysis. *International Journal of Environmental Research and Public Health*, 17(10), 3498. https://doi.org/10.3390/ijerph17103498

Mueller, M. K., Gee, N. R., & Bures, R. M. (2018). Human-animal interaction as a social determinant of health: Descriptive findings from the health and retirement study. *BMC Public Health*, 18(1), 305. https://doi.org/10.1186/s12889-018-5188-0.

Muldoon, A. L., Kuhns, L. M., Supple, J., Jacobson, K. C., & Garofalo, R. (2017). A web-based study of dog ownership and depression among people living with HIV. *JMIR Mental Health*, 4(4), e53. https://doi.org/10.2196/mental.8180

Must, A., Mule, C. M., Linder, D. E., Cash, S. B., & Folta, S. C. (2021). Animal-assisted intervention: A promising approach to obesity prevention for youth with autism spectrum disorder. *Frontiers in Veterinary Science*, 8, 646081. https://doi.org/10.3389/fvets.2021.646081

National Service Dog Registry. (2019). *How much does a service dog cost: A Buyer's guide for your service dog - national service*. https://www.nsarco.com/blog/service-dog-buyers-guide.html

Nightingale, F. (1969). *Notes on nursing: What it is, and what it is not*. New York: Dover. P. 103.

Okoro, C. A., Hollis, N. D., Cyrus, A. C., & Griffin-Blake, S. (2018). Prevalence of disabilities and health care access by disability status and type among adults-United States, 2016. *MMWR Morbity and Mortality Weekly Report*, 67, 882–887. http://dx.doi.org/10.15585/mmwr.mm6732a3

O'Neill, L. P., & Murray, L. E. (2016). Anxiety and depression symptomatology in adult siblings of individuals with different developmental disability diagnoses. *Research in Developmental Disabilities*, 51–52, 116–125.

Overgaauw, P. A., Vinke, C. M., van Hagen, M. A., & Lipman, L. J. (2020). A one health perspective on the human-companion animal relationship with emphasis on zoonotic aspects. *International Journal of Environmental Research and Public Health*, 17(11), 3789.

Pet Partners. (2022). *Terminology.* https://petpartners.org/learn/terminology/

Potter, K., Marcotte, R. T., Petrucci, G. J., Rajala, C., Linder, D. E., & Balzer, L. B. (2021). Examining the contribution of dog walking to total daily physical activity among dogs and their owners. *Journal for the Measurement of Physical Behaviour, 4*(2), 97–101.

Stern, S. L., Donahue, D. A., Allison, S., Hatch, J. P., Lancaster, C. L., Benson, T. A., Johnson, A. L., Jeffreys, M. D., Pride, D., Moreno, C., & Peterson, A. L. (2013). Potential benefits of canine companionship for military veterans with posttraumatic stress disorder (PTSD). *Society & Animals, 21*(6), 568–581.

Taylor, D. (2018). *Americans with Disabilities: 2014.* Census.gov, U.S. Census Bureau, https://www.census.gov/library/publications/2018/demo/p70-152.html.

TD Ameritrade Inc. (2020). *Examining Americans' financial attitudes on pet ownership.* https://s2.q4cdn.com/437609071/files/doc_news/research/2020/pets-and-finances-survey.pdf

Tsao, L. L., Davenport, R., & Schmiege, C. (2012). Supporting siblings of children with autism spectrum disorders. *Early Childhood Education Journal, 40*(1), 47–54.

Tuke, S. (1964). *Description of the retreat.* London: Dawsons of Pall Mall.

United States (U.S.) Census Bureau. (2019). *National population totals and components of change: 2010–2019.* https://www.census.gov/data/tables/time-series/demo/popest/2010s-national-total.html

United States Department of Veteran Affairs. (2021). *Guide and service dogs.* https://www.va.gov/MS/Veterans/benefits/Guide_and_Service_Dogs.asp

Vacek, J. L., Hunt, S. L., & Shireman, T. (2013). Hypertenstion medication use and adherence among adults with developmental disability. *Disability and Health Journal, 6*(4), 297–302.

Vasa, R. A., Keefer, A., McDonald, R. G., Hunsche, M. C., & Kerns, C. M. (2020). A scoping review of anxiety in young children with autism spectrum disorder. *Autism Research, 13*(12), 2038–2057.

Vermaes, I. P. R., van Susante, A. M. J., & van Bakel, H. J. A. (2012). Psychological functioning of siblings in families of children with chronic health conditions: A meta-analysis. *Journal of Pediatric Psychology, 37*(2), 166–184

Vredegoor, D. W., Willemse, T., Chapman, M. D., Heederik, D. J., & Krop, E. J. (2012). Can f1 levels in hair and homes of different dog breeds: Lack of evidence to describe any dog breed as hypoallergenic. *Journal of Allergy and Clinical Immunology, 130*(4), 904–907.

Ward, A., Arola, N., Bohnert, A., & Lieb, R. (2017). Social-emotional adjustment and pet ownership among adolescents with autism spectrum disorder. *Journal of Communication Disorders, 65*, 35–42. https://doi.org/10.1016/j.jcomdis.2017.01.002

White, L. (2020). *Man's best friend: What is the difference in outcomes (family functioning, quality of life, parental stress and child social communication) in families that have a dog present with children with Autism Spectrum Disorder (ASD): A control comparison study.* [Doctoral dissertation, University of Edinburgh, Scotland]. Edinburgh Archive Reserve. https://doi.org/10.7488/era/1398

Williams, M., Varelas, E. N., Olmsted, Z. T., Sheldon, B. L., Khazen, O., DiMarzio, M., & Pilitsis, J. G. (2021). Can dogs and cats really help our spinal cord stimulation patients? *Clinical Neurology and Neurosurgery, 208*, 106831. h ttps://doi.org/10.1016/j.clineuro.2021.106831.

Wisdom, J. P., Saedi, G. A., & Green, C. A. (2009). Another breed of "service" animals: STARS study findings about pet ownership and recovery from serious mental illness. *American Journal of Orthospsychiatry, 79*(3), 430–436.

Wright, H., Hall, S., Hames, A., Hardiman, J., Mills, R., & Mills, D. (2015). Pet dogs improve family functioning and reduce anxiety in children with Autism Spectrum Disorders. *Anthrozoös, 28*, 611–624.

Xie, Z., Tanner, R., Striley, C. L., & Marlow, N. M. (2022). Association of functional disability with mental health services use and perceived unmet needs for mental health care among adults with serious mental illness. *Journal of Affective Disorders, 15*, 299, 449–455. https://doi.org/10.1016/j.jad.2021.12.040.

Young, N. A. E. (2021). *Childhood disability in the United States: 2019.* American Community Briefs prepared for U.S. Department of Commerce. U.S. Census Bureau.

# 19
# HOMELESSNESS AND PET OWNERSHIP

*Leslie Irvine*

## Introduction to the Issue

Anyone writing about homelessness, with or without pets in the picture, must eventually grapple with the question of definition. There are many ways that people can experience homelessness, and scholars, advocates, and policy makers have long held differing ideas about how to characterize its diverse forms. Scholar and advocate Kim Hopper writes that "the harrowing simplicity of the term *homeless* may have outlived its utility" (1991, p. 38). Nevertheless, a basic distinction exists between the "literal" homeless, who sleep in shelters, abandoned buildings, or vehicles, and the "precariously housed," which includes those "doubling-up" with friends or family (Toro, 2007; Toro & Warren, 1999). Another distinction exists between those experiencing transitional homelessness, which includes spending a few nights in a shelter or doubled up with family during a single period, and those who are episodically or chronically homeless (Benjaminsen & Andrade, 2015). The episodically homeless circulate in and out of shelters, or between shelters and jail. The chronically homeless experience homelessness for over six months, and often permanently. The majority of adults experiencing chronic homelessness have no shelter at all.

As with other ambiguous situations, the search for clarity often requires following the money. In the United States, the McKinney-Vento Homeless Assistance Act (Public Law 100–07), the federal law that provides funding for the majority of programs, defines homelessness as "lack[ing] a fixed, regular, and adequate nighttime residence." This encompasses those sleeping in emergency shelters and the unsheltered, or those sleeping in places "not meant for human habitation" such as empty buildings, vehicles, and sidewalks. Of course, one reason for defining homelessness is to establish its prevalence. Each January, regional or local agencies that coordinate funding for services to homeless families and individuals conduct a point-in-time count on a single night. In January 2020, the annual point-in-time census of the sheltered and unsheltered homeless, and the latest data available at the time of this writing, revealed 580,466 people experiencing homelessness in the United States (U.S. Department of Housing and Urban Development, 2021). The number is likely higher, however, because point-in-time methods do not count the "hidden" homeless, such as those doubled-up with families or friends or housed in hotels. Unlike counts taken over time, point-in-time counts under-represent those episodically or intermittently without permanent shelter (Gould & Williams, 2010; Phelan & Link, 1999). Moreover, the 2020 count occurred just weeks before COVID-19 was declared a national emergency and drove the numbers higher for reasons that include unemployment and eviction. Thus, the 2020 count represents a low estimate of people living without shelter in the United States. International research has found that the United States has the highest homelessness rates among developed nations, with the United Kingdom, France, Australia, and Canada also having "serious problems" (Toro et al., 2007).

According to the National Alliance to End Homelessness, individuals make up 70% of those experiencing homelessness in the United States, with the remaining 30% constituting families with children. Unaccompanied youth (between ages 13 and 24) and veterans each account for 6% of the total number of people experiencing homelessness. Gender, race, and ethnicity are significant predictors of homelessness. The majority of those experiencing homelessness in the United States are white men. Transgender and non-binary individuals account for less than 1%. Among youth, risk factors for homelessness include family conflict, aging out of foster care, and identifying as LGBT (Toro, Dworsky, & Fowler, 2007).

The number of people who have pets while experiencing homelessness remains unknown. Three factors largely account for this lacuna. First, as Kim and Newton (2014) point out, "definitions of homelessness used by policy makers and programme developers have yet to include animals in their language" (p. 50). This has most likely occurred because accounting for pets among the unhoused would add further challenges to already over-burdened service providers. Nevertheless, this lack of recognition means that the numbers of homeless people who have pets, or what Kim and Newton call "inter-species families" can only be estimated, at best. Second, although a growing number of studies examine the practice of pet ownership among people experiencing homelessness, these typically analyze a single location and draw on small samples (Cronley et al., 2009; Donley & Wright, 2012; Henwood et al., 2021; Howe & Easterbrook, 2018; Irvine, 2013; Irvine et al., 2012; Kidd & Kidd, 1994; Labrecque & Walsh, 2011; Lem et al., 2013, 2016; Rew, 2000; Rhoades et al., 2015; Slatter et al., 2012; Thompson et al., 2006). With some exceptions (e.g., Cronley et al., 2009; Rhoades et al., 2015), these studies tend to focus on pet owners and their experiences, rather than attempting to estimate the number of pet owners within a given setting. And third, because safety and health concerns prevent most homeless shelters from co-sheltering people and animals, pet owners are often unsheltered (Scanlon et al., 2021. To keep their animals with them, they sleep in vehicles, in tents, on the street, or in other places not intended for habitation. Because pet owners constitute a hidden population, no definitive, national or international-level data exist on the number of people who have pets while experiencing homelessness. Whether accounting for animals would constitute an added burden to homeless service providers or would not provide a systematic solution for homelessness, no formal mechanism currently exists to collect such data.

Despite the lack of an official means to determine the numbers of pet owners experiencing homelessness, various local service providers report that pet owners account for a substantial minority of the homeless population they reach. According to these reports, between 10 and 25% of clients experiencing homelessness do so with companion, service, or support animals in the United States (Kerman et al., 2019) and the United Kingdom (Scanlon et al., 2021). In the United States, these percentages vary by region. For example, one empirical study based in Knoxville, Tennessee, found that only 5.5% of people surveyed were experiencing homelessness while caring for pets (Cronley et al., 2009), while 12% of the unsheltered homeless in Los Angeles had pets (Henwood et al., 2021). The rate of pet ownership among homeless youth in Los Angeles was 23% (Rhoades et al., 2015). Using the January

2020 U.S. point-in-time census as a basis, an estimate places the number of homeless pet owners between 58,000 and 145,000. Studies conclude that the majority have dogs, although some have cats (and other small animals) despite the challenges posed by life on the street (Squirrell, 2016). People living in abandoned buildings, junkyards, and other permanent sites often care for cat colonies that exist there. Most people experiencing transitional or episodic homelessness had obtained their pets before losing their housing (Irvine, 2013; Williams & Hogg, 2016). Some adopted their pets from shelters, often using a friend's address. They also obtained them from people who could no longer keep them. Homeless youth sometimes keep their dogs intact to breed them and sell or trade the puppies.

## Historical Background

The question of how and when homelessness became a social problem, rather than a problem of individual character, depends on definitions of homelessness. Historically, most skilled occupations required a "journeyman" period while mastering a trade designation. In Western Europe, this meant that a young person traveled widely, or "tramped," in search of jobs with master artisans, for up to five years. Journeymen were technically homeless. However, trade guilds formally organized the activity of tramping, arranging for accommodations or providing funds to reach the next village. As urbanization concentrated jobs in cities, skilled workers settled in one place. Of course, some people still took to the roads in search of work, but "tramping ceased to be a typical feature of artisan life, and the status of the tramp declined. Once perfectly respectable, the tramp came to denote social marginality and vagrancy" (Adler, 1985, p. 341).

Unlike Europe, the United States had no organized system of tramping as a part of trade apprenticeship obligations. However, the activity was nevertheless a part of American history. After the Civil War, unemployed men frequently hopped rail lines looking for employment across the country. Consequently, the term "tramp" came to denote a transient person who lacked a home and a visible means of support. During the Great Depression, impoverished adolescents often struck out on their own in hopes of making a living. An estimated quarter million young people took to the road, seeking work or adventure (Adler, 1985; Minehan, 1934; Urys, 1999). Along with the growing ranks of adults on the road, they were known equally as "tramps" and "hoboes." Their numbers grew as the Depression wore on. They became a social problem that generated social scientific research, fueled public policy debates, and produced responses such as the Federal Transient Relief Service.

After World War II, the meaning of "homelessness" differed from the term's use today. It referred to people, mostly men, who lived "outside normal family life," rather than those who lacked a physical home (Shlay & Rossi, 1992, p. 131). From the 1950s until the 1970s, these were mostly single men, past middle age, who drank heavily and slept in the inexpensive, single room occupancy hotels once abundant in the urban neighborhoods known as "skid rows" (Bahr, 1973; Bahr & Caplow, 1974). The "homeless" men of skid row "had addresses and places in which to sleep" (Shlay & Rossi, 1992, p. 131). They were poor, but many held jobs, albeit menial ones, depending on the availability of work (Bogue, 1963). The "old homelessness" of the skid row era was regarded less as a social problem than an individual one. The separation of skid row from the "better" urban areas meant that many city dwellers seldom if ever saw it. Consequently, it raised no public outcry. And because the men of skid row did not land there because of poverty or a shortage of low-cost housing, most people regarded them as responsible for their circumstances. Individual-level problems labeled "disaffiliation" (Bahr, 1973) and "social maladjustment" (Bogue, 1963) had purportedly led them to the streets.

Homelessness increased dramatically in the United States during the Reagan era (roughly the 1980s) as many relief programs for the poor were eliminated in an attempt to reduce the size of government. Urban renewal programs eliminated affordable housing, and gentrification replaced skid rows with luxury condominiums. The resulting "new homelessness" contrasted dramatically from the "old homelessness" of the skid row era (Bogard, 2001; Hopper, 1991; Rossi, 1989, 1990). The new homeless included families, single women, and minorities. With affordable housing increasingly scarce, people experiencing homelessness slept in doorways, on park benches, and in bus stations. Unlike the alcoholic men of skid row, they were highly visible in public spaces. The numbers of homeless individuals and families in the United States increased significantly following the Great Recession, officially dated from December 2007 to June 2009. Although some critics continue to blame the growing problem on individual character, scholars, and policy makers more typically see the "new homelessness" as a social problem stemming from inequality and a lack of affordable housing.

Along with the growing number of families, single women, and minorities, the new homeless also included increasing numbers of unaccompanied youth. Young people have long had various reasons for leaving home, including poverty, foster-care transitions, and physical or sexual abuse. Although large numbers of young people left home in periods such as the Great Depression, youth homelessness, as a distinct issue, was largely ignored because so many adults were also homeless (Moore, 2005, p. 2; Smollar, 1999). During the 1960s and 1970s, however, the numbers of homeless young people became increasingly visible, especially in cities. What prevailing knowledge had disregarded or attributed to disobedience or delinquency gradually gained recognition as a social problem with heterogeneous predisposing factors. Some homeless youth had run away, but growing numbers had been "thrown away," or forced to leave home by their parents. Many escaped homes "plagued with substance abuse, violence, and other family conflict" (Moore, 2005, p. 2). Because a substantial proportion of the population of homeless youth remains hidden, outside of the shelter system, their numbers are unknown.

## Impact on People

Regardless of whether they have pets, people experiencing homeless people confront many diverse challenges in their everyday lives. Finding food and safe shelter ranks high among the daily challenges. Homeless people's needs may include obtaining mental health counseling, medical services, housing subsidies, and educational and employment assistance. They may also need treatment for substance abuse and addiction. Homeless youth face considerable risk for substance abuse, sexual exploitation, and prostitution.

Homelessness poses psychosocial challenges, too. High among these are the challenges posed by negative perceptions. Homelessness constitutes what the sociologist Everett C. Hughes (1945) calls a "master status," a social position that dominates all other positions and affects all areas of experience. While a master status can be positive or negative, the status of homelessness is always negative. It marks or stigmatizes individuals, discrediting their identity and disqualifying them from social acceptance (Goffman, 1963). Individuals who possess stigmatizing attributes are "reduced in our minds from a whole and usual person to a tainted, discounted one" (Goffman, 1963, p. 3). Although poverty has been stigmatized in the Western world since the Middle Ages, homelessness is stigmatized even more severely (Phelan et al., 1997). Members of the domiciled public often treat homeless people as what

Goffman (1963) calls "nonpersons." Passersby generally deprive them of attention, choosing to ignore or avoid them if possible. The attention homeless people do receive is often negative, taking the form of insults and criticism for bringing on their condition (Snow & Anderson, 1993). Coping with stigma and negative attention adds to the burdens and hardships homeless people already bear.

## *The Challenges of Experiencing Homelessness with Pets*

People who experience homelessness with pets routinely endure stigmatization. This is particularly true for dog owners, who appear in public with their pets. To many observers, homelessness constitutes evidence of failure. It suggests a *moral* failure, in particular, an inability to "take care" of oneself attributed to a lack of integrity, responsibility, work ethic, or some other characteristic. Through a reverse halo effect whereby a perceived negative characteristic colors impression of the entire person, homeless people are regarded as morally unfit to be pet owners. In this construal, they cannot take care of pets because they obviously cannot take care of themselves. Thus, members of the domiciled public openly and often aggressively question homeless people about their ability to care for their animals and even their right to animal companionship (Irvine, 2013; Rhoades et al., 2015). Because dogs serve as "social facilitators," inviting casual exchanges with strangers in public places (Messent, 1983), dog owners experience the worst and the best of those interactions. A large part of the stigmatization of homeless pet owners stems from their perceived inability to feed their animals. Their seeming failure to meet their own basic needs often leads to the perception of pet ownership as selfish (Brewbaker, 2012). Thus, while homeless people are often cast into demeaning and discredited social identities, having a pet can make one especially vulnerable to stigmatization and negative attention.

How members of the homeless population respond to stigma and salvage a positive sense of identity has long been a topic of study (e.g., Snow & Anderson, 1987, 1993). Irvine (2013; Irvine et al., 2012) was the first to examine how homeless pet owners respond to the stigmatizing perceptions of others. Whereas many simply ignored negative comments, attributing them to ignorance and misperception, others became angry and responded with profanity and insults. The latter strategy provided temporary emotional release and claimed a modicum of social power, but ultimately perpetuated negative stereotypes about the homeless. The most common strategy was redefining "good" pet ownership to downplay the need for four walls and emphasize positive aspects unique to their situation, such as the ability to provide ample exercise for their dogs and around-the-clock attention. Redefining pet ownership in this way allows homeless pet owners to construct a positive moral identity that deflected the negative impact of stigma.

One factor that has a significant role in both stigmatizing and redefining pet ownership is the question of how homeless people feed and care for their animals. In the first study of pet ownership among the homeless, Kidd and Kidd (1994) interviewed homeless pet owners in soup kitchens, parks, and on the streets in the San Francisco area. More than half of their sample reported difficulty feeding their pets and agreed that it would be helpful if service providers made pet food available, too. Years later, Irvine et al. (2012) and Irvine (2013) found that homeless pet owners had ample food for their animals. Many reported that members of the public frequently donated pet food. They also received it from animal control officers or veterinarians, or from soup kitchens and food banks. Moreover, they budgeted their limited funds to keep their pets fed. Similarly, in a survey of homeless pet owners in the United Kingdom, Baker (2001) found that they easily obtained food for their animals.

Irvine et al. (2012) and Irvine (2013) noted that although interactions between homeless dog owners and passersby often result in the confrontations described above, they can also result in gestures of good will, as when a passerby donates pet food or treats. Homeless pet owners consistently ensured that their pets ate before they did.

Along with marginalization and negative perceptions, pet ownership brings additional hardships. Homeless individuals consistently report that having a pet limits their access to housing, as most would understandably not live where they could not keep their pets (Cronley et al., 2009; Singer, Hart, & Zasloff, 1995; Slatter et al., 2012). The lack of housing, and thus the lack of a place for a pet to stay, also limits opportunities for employment. For dog owners, in particular, the constant presence of the animal limits access to service providers, such as drop-in centers and soup kitchens (Howe & Easterbrook, 2018). Public transportation often prohibits pets unless they are small enough to fit in a carrier. In a survey of public transportation policies in the 50 largest U.S. cities, only three allow pets to accompany their owners on buses, subways, and trains (Geller, 2022).

Despite the hardships of having a pet while experiencing homelessness, homeless pet owners consistently report that their pets provide countless benefits. These include decreased feelings of loneliness (Rhoades et al., 2015) and increased social interactions with others (Irvine, 2013). Studies of homeless youth and their pets, in particular, have noted the benefits of companionship and the reduction of loneliness (Hartsell & Zolnikov, 2021; Lem et al., 2013, 2016; Rew, 2000). Moreover, because homeless youth often sleep in places that place them at risk for physical and sexual assault, many claim that their dogs provide protection (Irvine, 2013). Brewbaker (2012) and Irvine et al. (2012) found that companion animals decreased stress and enhanced perceptions of self-worth.

In some cases, the sense of responsibility associated with pet ownership prevents people from engaging in destructive activities such as drinking and drug use. An example from Irvine's (2013) work illustrates the power pets have to change lives. During research in San Francisco, Irvine met "Donna," a 53-year-old woman who lived with her mother after decades of homelessness. Donna started drinking and using heroin at age 15. She left home and began using crack cocaine. She turned to prostitution to buy drugs and food. She hitchhiked cross-country numerous times, turning tricks at truck stops along the way. Eventually, Donna contracted HIV, either from "the sex or the needles." By the time Irvine met her, Donna was clean. She said, "I would never, ever shove a needle in my arm anymore" (2013, p. 136). When Irvine asked her how she quit using drugs, Donna said she owed it all to "Athena," the German shepherd/Labrador retriever mix whom she described as "the love of my life" (2013, p. 136).

Donna became Athena's guardian through a formerly homeless woman named Pali, who rescued dogs from death row in a shelter. Pali and Donna knew each other from the street. At the time, Donna and her abusive boyfriend were camping in Donna's mother's backyard. Pali told Donna, "You need a dog in your life" (Irvine, 2013, p. 137). Although a homeless drug addict in an abusive relationship might not seem to be an ideal guardian for a dog, the match worked wonders. As Donna recalled, Athena got her out of the abusive relationship, then got her off drugs. When Irvine asked how Athena had done this, Donna said,

> It was either the dog or him, and I chose the dog. He used to take my money. My shoes. Everything. The guy used to beat me up, and Pali told me it was either the man or the dog, so I chose Athena. I got the dog. Got rid of the man.
>
> *(2013, p. 137)*

Donna then moved into her mother's house. But her mother insisted that she had to be clean to keep Athena and stay in her house. Donna faced a decision. "I realized Athena meant everything to me," she told Irvine.

> I said to myself, 'My dog comes first in my life. Would I rather use drugs, or feed my dog?' And I fell in love with Athena, so I gave up the needle. Gave up the pipe. I gave up liquor. Everything.
>
> *(2013, p. 138)*

As Donna recalled,

> I went through withdrawal, and from there, I went to the methadone clinic. Got on methadone. Athena went with me, and everybody loved her, too. Athena was everything, OK? Athena was everything. Everywhere I went, Athena followed. She knew the pet stores. She sat down at the coffee shop. Everything.
>
> *(2013, p. 138)*

Donna also credited Athena with improving her HIV status. After becoming clean and sober, she felt better and began taking care of herself. As she told Irvine, "When Athena came into my life, everything was beautiful" (2013, p. 138).

Athena died over a year before Irvine and Donna met, at age 13. A pet supply store organized a memorial service and the veterinarian arranged for cremation. Donna kept Athena's ashes to be mixed with her own after she dies. Buddy, the new dog Pali found for her, assuaged some of Donna's grief. Donna had started HIV treatment because, as she told Irvine, "I want to live for my dog" (2013, p. 139). Donna believed Athena had chosen Buddy for her, and had sent her a cat, too. During a hospitalization, Donna recalled, "Every day, I asked the doctor, 'When can I go home so I can be with my pets?' Just to have them in my life means everything" (2013, p. 139).

Donna's story reveals how the human-animal relationship can change lives. Research has found that the relationship can save lives. The benefits of pets for the mental health of their homeless owners included a reduction in suicidality (Irvine, 2013; Rew, 2000) and symptoms of depression (Lem et al., 2016; Rhoades et al., 2015). For instance, Irvine met Denise, a middle-aged woman who had lived in her car with her cat, Ivy, for over eight months. Denise had been a self-employed graphic artist, but severe depression caused her to miss deadlines. She lost some major clients, and the other accounts gradually dwindled. She fell behind on rent, and her landlord evicted her. She tried to find another place before she had to move out of her apartment, but when time ran out, she put her belongings in storage and moved into the car with Ivy. The eviction made it hard for Denise to find another apartment, and having Ivy made it even more difficult. When Irvine asked if she had considered finding Ivy another home, Denise explained:

> I have a history with depression up to suicide ideation, and Ivy, I refer to her as my suicide barrier. And I don't say that in any light way. I would say most days, she's the reason why I keep going, because I made a commitment to take care of her when I adopted her. So, she needs me, and I need her. She is the only source of daily, steady affection and companionship that I have. The only one. I can't imagine being without her, wanting to go on at all, without her.
>
> *(Irvine, 2013, p. 147)*

These accounts demonstrate the powerful emotional connections homeless people can feel with their pets. In Donna's case, a dog turned her life around. In Denise's case, a cat saved her life. To put these narrative examples into concrete empirical terms, people experiencing homelessness report extraordinarily strong levels of attachments to their pets. In studies using the Lexington Attachment to Pets Scale, both men and women who owned pets had higher mean scores on attachment to their pets than those found among the scale's standardized (or housed) population (Singer et al., 1995; Yang et al., 2021).

## Impact on the Animal's Welfare

The criticism faced by homeless pet owners stems mostly from the belief that homelessness has a detrimental impact on the welfare of the animals. This belief is held not only by many members of the public, but also by veterinarians (Williams & Hogg, 2015). However, in research comparing 50 dogs whose owners were homeless with 50 dogs of domiciled owners, Williams and Hogg (2016) found that the dogs of homeless were not significantly less healthy than those belonging to domiciled individuals. Consistent with aspects of redefining good pet ownership discussed above, homeless pet owners reported that their dogs enjoyed ample exercise, frequently off leash. In contrast, most domiciled pet owners did not walk their dogs regularly (Irvine & Cilia, 2016; Williams & Hogg, 2016). Williams and Hogg (2016) also found that dogs of the homeless displayed "very significantly" lower levels of aggression during health checks than did dogs with domiciled owners, and were significantly less likely to be obese (2016, p. 26). They found no significant differences between the two groups of dogs in other areas of physical health, such as skin and ear problems, wounds, infections, and arthritis. "Contrary to much opinion," Williams and Hogg write, "it does not appear from this study that dogs belonging to homeless people are less healthy than those belonging to settled owners" (2016, p. 27).

In addition, homeless pet owners may find access to affordable veterinary care a challenge. In Kidd & Kidd's 1994 study, over half of the pet owners interviewed described obtaining veterinary care as a "serious problem" (p. 718). In addition to the barrier of affordability, pet restrictions on public transportation also impede homeless pet owners' access to veterinary services. More recently, veterinarians and schools of veterinary medicine have established mobile outreach or pop-up clinics to provide services to the pets of the homeless in many places. In San Francisco, one such clinic began after a veterinarian noticed that veterinary emergencies made up the majority of cases involving the pets of the homeless. Providing preventive services reduced those numbers (Irvine, 2013). While most of these clinics provide all services pro bono, some offer certain procedures on a sliding scale.

Along with protecting animal health through routine examinations, vaccinations, and sterilization, veterinary visits also make pet ownership licensing possible. This increases the likelihood that, should animals be confiscated for any reason, such as jail or hospitalization, they would be reunited with their owners. Veterinarians also ensure that owners have sufficient food for their pets. Clinic visits result not only in the provision of veterinary services, but also in conveying a social identity for the homeless person. Over time, the veterinarian and staff learn clients' names, along with those of their animals, and they interact with the clients as equals. Veterinarians and veterinary students gain increased knowledge about the circumstances that lead to homelessness and report reduced bias (Jordan & Lem, 2014). Although access to veterinary services for homeless pet owners has improved since the 1990s, availability still varies by location. Whereas some cities, such as San Francisco, Sacramento, and Seattle on the West Coast and Fort Collins in Colorado, have regularly scheduled clinics, veterinary services for the pets of the homeless remain scarce in many regions. Indeed, access to veterinary care among underserved populations, regardless of housing status, constitutes the veterinary profession's "foremost crisis today" (Arluke & Rowan, 2020, p. 2).

## Practical Perspectives

As veterinarian and founder of The Street Dog Coalition Jon Geller writes, "Providing veterinary care to the pets of people living on the street cannot occur in a silo" (2022, p. 181). Geller and others who practice "street medicine" advocate a "One Health" approach that encompasses both animal and human health and well-being. Although the field of comparative medicine has recognized this interdependency since the 19th century, the "One Health" concept, as such, first gained currency in the 1980s. In 2021, the World Health Organization, together with the Food and Agriculture Organization of the United Nations, the World Organization for Animal Health, and the United Nations Environment Programme, defined One Health as:

> an integrated, unifying approach that aims to sustainably balance and optimize the health of people, animals and ecosystems. It recognizes the health of humans, domestic and wild animals, plants, and the wider environment (including ecosystems) are closely linked and inter-dependent. The approach mobilizes multiple sectors, disciplines and communities at varying levels of society to work together to foster well-being and tackle threats to health and ecosystems, while addressing the collective need for clean water, energy and air, safe and nutritious food, taking action on climate change, and contributing to sustainable development.
>
> *(World Health Organization, 2021)*

Using a One Health approach, a small but growing number of veterinary outreach programs involve a collaboration between veterinary professionals and providers of all aspects of human health care and social services. Through these provider partnerships, homeless clients can receive health care and information. A client who mentions an illness, injury, mental health concern, or other matter during the course of a veterinary examination, or who needs a flu shot or dental procedure, can meet with or obtain a referral to an appropriate provider (Panning, Lem, & Bateman, 2016; Slatter et al., 2012).

The valuable but largely unrecognized role veterinarians play in serving socioeconomically disenfranchised pet owners comes with considerations for the veterinarians themselves (Access to Veterinary Care Coalition, 2018). Veterinary ethics prioritize empathetic care over profit. At the same time, veterinary practices must remain financially viable. To serve clients who cannot pay for private veterinary care, veterinarians in non-profit clinics volunteer their time. In light of the significant student debt load graduates from colleges of veterinary medicine currently bear, pro bono work is not always feasible. The need for accessible veterinary care in communities exists alongside complex financial and logistical realities (Lem, 2019).

## Current Policy Implications

When homeless pet owners discuss their most pressing problems, housing appears high on the list. Irvine (2013) found that with only a few exceptions, most people on the street looked forward to living in a house or apartment again—if the situations would also accommodate their pets. Two issues combine to put housing out of the reach of most pet owners experiencing homelessness. First, the United States (and many other countries) faces a serious shortage of affordable places to live. This shortage is most severe at the lowest income levels because many units are occupied by higher-income tenants who become long-term renters as the traditional next step into home ownership is less attainable (Joint Center for Housing Studies of Harvard University, 2013). Research indicates that rising housing costs, stagnant incomes, and growing student loan debt will continue to delay home ownership for many or put it out of reach altogether (Goodman, Pendall, Zhu, 2015).

Second, even when rental housing is available, landlord concerns about property damage, noise, and liability have generated the practice of prohibiting pets in rental housing. This practice disproportionately affects not only the homeless, but all lower-income individuals and households (O'Reilly-Jones, 2019; Power, 2017). No-pets policies originated in Victorian-era London, when landlords of public housing projects prohibited tenants from keeping dogs because they regarded their owners as unable to care for them (Jones, 1971). Moreover, the dogs of the English lower classes were considered more vulnerable to rabies than were "well-bred" dogs (Ritvo, 1987, p. 179). To this day, the pets of the poor continue to serve as convenient scapegoats for class tensions (Derges et al., 2012). No-pets policies have endured despite high rates of pet ownership and the increasing status of pets as family members (Franklin, 2006; Irvine & Cilia, 2016; Power, 2008). In a nationwide survey of U.S. landlords, only half allowed renters to have pets (Carlisle-Frank, Frank, & Nielsen, 2005). Of these, only 9% imposed no restrictions on size or type of pet. These policies have implications for animal sheltering, as pet prohibitions were the most common additional reason given for relinquishment (New et al., 1999; Shore et al., 2003).

## Ethical Considerations

The topic of homelessness by itself raises ethical considerations about social justice and inequality, housing, employment, and health-care provision, among other issues. Of course, the issue most directly related to human-animal relationships concerns the ethics of pet ownership on the part of homeless people. This begs the more basic question about the ethics of animal "ownership." The practice of keeping a class of animals for human companionship raises moral and ethical questions regardless of the socioeconomic status of the person involved (Irvine, 2004; Pallotta, 2019; Pierce, 2016). As knowledge about the intellectual and emotional capacities of animals increasingly challenges the once-distinct human-animal boundary, our treatment of animals in every domain requires simultaneous reassessment. This includes the domain of companionship. However, pets rarely appear in discussions of animal ethics, which typically involve animals raised for food, confined in laboratories, or kept in captivity in zoos (Pierce, 2016; Spencer et al., 2006). By comparison, most pets live in comfort, even luxury. Yet, as Irvine (2004) asked, are pets pampered—or enslaved? Although the word seems harsh, it highlights the human *use* of pets—for companionship, emotional fulfillment, and other "relatively benign form[s] of exploitation" (Pierce, 2016, p. 27). Yi-Fu Tuan argued that petkeeping constitutes a form of dominance cloaked in affection (1984). Although many would find life without pets difficult to imagine, the question of whether human pleasure justifies exploiting other animals remains open to debate.

Putting aside the idea of phasing out the practice of pet ownership altogether, probing the ethics of pet ownership specifically among the homeless raises questions about who and what kind of person "deserves" animal companionship. In modern Western societies, barriers to pet ownership, whether in the form of no-pets policies, limited access to veterinary care, or food insecurity are highest for those at the lowest income levels (Arluke, 2021; Arluke & Rowan, 2020; Rauktis et al., 2017). If the company of animals conveys benefits to human health and well-being, then denying low-income people the right to have pets deprives those who stand to gain most from those benefits. Of course, the human-pet relationship should be mutually beneficial, and owners must provide for the welfare of their animals, regardless of housing status. Four walls and a roof offer no guarantee that an animal's health and welfare needs will be met.

## Unanswered Questions and Future Research Demands

Future research on homeless pet owners will need to engage with the research on homelessness, in general. A chief concern within this research involves learning how to reduce and prevent homelessness. As Evans, Phillips, and Ruffini write, "the human toll of homelessness is high, and understanding how best to serve this vulnerable population is crucial" (2021, p. 915). The same can be said of the animals who accompany those experiencing homelessness. The welfare needs of animals and the importance of the human-animal bond must have a place in the research to assure that any resulting recommendations are inclusive of all those experiencing homelessness. Ample research has established the supportive role of pets for the homeless. The key now is to incorporate that knowledge into policy and practice.

### Potential Questions for Discussion

- A sizeable proportion of the homeless population struggles with substance abuse and mental illness.
  - What are the risks to anitmal welfare in such situations?
  - What is the ethical and humane response?
- What role do local animal welfare agencies have vis-à-vis homeless pets and their owners?
- Does a dog or cat need four walls to have a good life?
- What is a "good" life for a dog or cat?

## References

Access to Veterinary Care Coalition. (2018). *Access to veterinary care: Barriers, current practices, and public policy*. University of Tennessee Knoxville.
Adler, J. (1985). Youth on the road: Reflections on the history of tramping. *Annals of Tourism Research*, 12(3), 335–354.
Arluke, A. (2021). Coping with pet food insecurity in low-income communities. *Anthrozoös*, 34(3), 339–358.
Arluke, A., & Rowan, A. (2020). *Underdogs: Pets, people, and poverty*. University of Georgia Press.
Bahr, H. M. (1973). *Skid row: An introduction to disaffiliation*. Oxford University Press.
Bahr, H. M., & Caplow, T. (1974). *Old men drunk and sober*. New York University Press.
Baker, O. (2001). *A dog's life: Homeless people and their pets*. Blue Cross.
Benjaminsen, L., & Andrade, S. B. (2015). Testing a typology of homelessness across welfare regimes: Shelter use in Denmark and the USA. *Housing Studies*, 30(6), 858–876.
Bogard, C. J. (2001). Claimsmakers and contexts in early constructions of homelessness: A comparison of New York City and Washington, DC. *Symbolic Interaction*, 24(4), 425–454.
Bogue, D. (1963). *Skid Row in American cities*. University of Chicago Press.
Brewbaker, Erin J. (2012). The experience of homelessness and the human-companion animal bond: a project based upon an investigation at San Francisco Community Clinic Consortium/Veterinary

Street Outreach Services, San Francisco, California. Master's Thesis, Smith College, Northampton, MA. https://scholarworks.smith.edu/theses/869

Carlisle-Frank, P., Frank, J. M., & Nielsen, L. (2005). Companion animal renters and pet-friendly housing in the US. *Anthrozoös, 18*(1), 59–77.

Cronley, C., Strand, E. B., Patterson, D. A., & Gwaltney, S. (2009). Homeless people who are animal caretakers: A comparative study. *Psychological Reports, 105*(2), 481–499.

Derges, J., Lynch, R., Clow, A., Petticrew, M., & Draper, A. (2012). Complaints about dog faeces as a symbolic representation of incivility in London, UK: A qualitative study. *Critical Public Health, 22*(4), 419–425.

Donley, A. M., & Wright, J. D. (2012). Safer outside: A qualitative exploration of homeless people's resistance to homeless shelters. *Journal of Forensic Psychology Practice, 12*(4), 288–306.

Evans, W. N., Phillips, D. C., & Ruffini, K. (2021). Policies to reduce and prevent homelessness: What we know and gaps in the research. *Journal of Policy Analysis and Management, 40*(3), 914–963.

Franklin, A. (2006). 'Be[a]ware of the dog': A post-humanist approach to housing. *Housing, Theory and Society, 23*(3), 137–156.

Geller, J. M. (2022). Street medicine: Caring for the pets of the homeless. *Journal of the American Veterinary Medical Association, 260*(2), 181–185.

Goffman, E. (1963). *Stigma: Notes on the management of spoiled identity*. Prentice Hall.

Goodman, L., Pendall, R., & Zhu, J. (2015). *Headship and homeownership*. Urban Institute.

Gould, T. E., & Williams, A. R. (2010). Family homelessness: An investigation of structural effects. *Journal of Human Behavior in the Social Environment, 20*(2), 170–192.

Hartsell, M., & Zolnikov, T. R. (2021). Understanding attachment in homeless adolescents and emerging adults with pets. *Journal of Social Distress and Homelessness, 32*(1), 1–9.

Henwood, B., Dzubur, E., Rhoades, H., St. Clair, P., & Cox, R. (2021). Pet ownership in the unsheltered homeless population in Los Angeles. *Journal of Social Distress and Homelessness, 30*(2), 191–194.

Hopper, K. (1991). Homelessness old and new: The matter of definition. In D. Culhane & S. Hornburg (Eds.), *Understanding homelessness: New policy and research perspectives* (pp. 9–69). Fannie Mae Foundation.

Howe, L., & Easterbrook, M. J. (2018). The perceived costs and benefits of pet ownership for homeless people in the UK: Practical costs, psychological benefits and vulnerability. *Journal of Poverty, 22*(6), 486–499.

Hughes, E. C. (1945). Dilemmas and contradictions of status. *American Journal of Sociology, 50*(5), 353–359.

Irvine, L. (2004). Pampered or enslaved? The moral dilemmas of pets. *International Journal of Sociology and Social Policy, 24*(9), 5–17.

Irvine, L. (2013). *My dog always eats first: Homeless people and their animals*. Lynne Rienner.

Irvine, L., & Cilia, L. (2016). *The value of pets to public and private health and well-being*. Schaumburg, IL: Report prepared for the American Veterinary Medical Association.

Irvine, L., Kahl, K. N., & Smith, J. M. (2012). Confrontations and donations: Encounters between homeless pet owners and the public. *Sociological Quarterly, 53*(1), 25–43.

Joint Center for Housing Studies of Harvard University. (2013). *America's rental housing: Evolving markets and needs*. Harvard University. http://www.jchs.harvard.edu

Jones, G. S. (1971). *Outcast London: A study in the relationship between classes in Victorian society*. Oxford University Press.

Jordan, T., & Lem, M. (2014). One health, one welfare: Education in practice veterinary students' experiences with community veterinary outreach. *The Canadian Veterinary Journal, 55*(12), 1203–1206.

Kerman, N., Gran-Ruaz, S., & Lem, M. (2019). Pet ownership and homelessness: A scoping review. *Journal of Social Distress and the Homeless, 28*(2), 106–114,

Kidd, A. H., & Kidd, R. M. (1994). Benefits and liabilities of pets for the homeless. *Psychological Reports, 74*(3), 715–722.

Kim, C. H., & Newton, E. K. (2014). My dog is my home: Increasing awareness of inter-species homelessness in theory and practice. In T. Ryan (Ed.), *Animals in social work* (pp. 48–63). Palgrave Macmillan.

Labrecque, J., & Walsh, C. A. (2011). Homeless women's voices on incorporating companion animals into shelter services. *Anthrozoos, 24*(1), 79–95.

Lem, M. (2019). Barriers to accessible veterinary care. *The Canadian Veterinary Journal, 60*(8), 891.

Lem, M., Coe, J. B., Haley, D. B., Stone, E., & O'Grady, W. (2013). Effects of companion animal ownership among Canadian street involved youth: A qualitative analysis. *Journal of Sociology and Social Welfare, 40*(4), 285–304.

Lem, M., Coe, J. B., Haley, D. B., Stone, E., & O'Grady, W. (2016). The protective association between pet ownership and depression among street-involved youth: A cross-sectional study. *Anthrozoös, 29*(1), 123–136.

Messent, P. (1983). Social facilitation of contact with other people by pet dogs. In A. Katcher & A. Beck (Eds.), *New perspectives on our lives with companion animals* (pp. 37–46). University of Pennsylvania Press.

Minehan, T. (1934). *Boy and girl tramps of America*. Grosset and Dunlap.

Moore, J. (2005). *Unaccompanied and homeless youth review of literature (1995–2005)*. National Center for Homeless Education.

New, J. C. J., Salman, M. D., Scarlett, J. M., Kass, P. H., Vaughn, J. A., Scherr, S. & Kelch, W. J. (1999). Moving: Characteristics of dogs and cats and those relinquishing them to 12 U.S. animal shelters. *Journal of Applied Animal Welfare Science, 2*(2), 83–96.

O'Reilly-Jones, K. (2019). When Fido is family: How landlord-imposed pet bans restrict access to housing. *Columbia Journal of Law and Social Problems, 52*(3), 427–472.

Pallotta, N. R. (2019). Chattel or child: The liminal status of companion animals in society and law. *Social Sciences, 8*(5), 158.

Panning, C., Lem, M., & Bateman, S. (2016). Profiling a one-health model for priority populations. *Canadian Journal of Public Health, 107*(3), 222–223.

Phelan, J. C., & Link, B. G. (1999). Who are" the homeless"? Reconsidering the stability and composition of the homeless population. *American Journal of Public Health, 89*(9), 1334–1338.

Phelan, J., Link, B. G., Moore, R. E., & Stueve, A. (1997). The stigma of homelessness: The impact of the label "homeless" on attitudes toward poor persons. *Social Psychology Quarterly, 60*(4), 323–337.

Pierce, J. (2016). *Run, spot, run*. University of Chicago Press.

Power, E. R. (2008). Furry families: Making a human-dog family through home. *Social and Cultural Geography, 9*(5), 535–555.

Power, E. R. (2017). Renting with pets: A pathway to housing insecurity? *Housing Studies, 32*(3), 336–360.

Rauktis, M. E., Rose, L., Chen, Q., Martone, R., & Martello, A. (2017). "Their pets are loved members of their family": Animal ownership, food insecurity, and the value of having pet food available in food banks. *Anthrozoös, 30*(4), 581–593.

Rew, L. (2000). Friends and pets as companions: Strategies for coping with loneliness among homeless youth. *Journal of Child and Adolescent Psychiatric Nursing, 13*(3), 125–132.

Rhoades, H., Winetrobe, H., & Rice, E. (2015). Pet ownership among homeless youth: Associations with mental health, service utilization and housing status. *Child Psychiatry and Human Development, 46*(2), 237–244.

Ritvo, H. (1987). *The animal estate: The English and other creatures in Victorian England*. Harvard University Press.

Rossi, P. H. (1989). *Down and out in America: The origins of homelessness*. University of Chicago Press.

Rossi, P. H. (1990). The old homeless and the new homelessness in historical perspective. *American Psychologist, 45*(8), 954.

Scanlon, L., McBride, A., & Stavisky, J. (2021). Prevalence of pet provision and reasons for including or excluding animals by homelessness accommodation services, *Journal of Social Distress and Homelessness, 30*(1), 1, 77–83,

Shlay, A. B., & Rossi, P. H. (1992). Social science research and contemporary studies of homelessness. *Annual Review of Sociology, 18*(1), 129–160.

Shore, E. R., Petersen, C. L., & Douglas, D. K. (2003). Moving as a reason for pet relinquishment: A closer look. *Journal of Applied Animal Welfare Science, 6*(1), 39–52.

Singer, R. S., Hart, L. A., & Zasloff, R. L. (1995). Dilemmas associated with rehousing homeless people who have companion animals. *Psychological Reports, 77*(3), 851–857.

Slatter, J., Lloyd, C., & King, R. (2012). Homelessness and companion animals: More than just a pet? *The British Journal of Occupational Therapy, 75*(8), 377–383.

Smollar, J. (1999). Homeless youth in the United States: description and developmental issues. *New Directions for Child and Adolescent Development, 85*, 47–58.

Snow, D. A., & Anderson, L. (1987). Identity work among the homeless: The verbal construction and avowal of personal identities. *American Journal of Sociology, 92*(6), 1336–1371.

Snow, D. A., & Anderson, L. (1993). *Down on their luck: A study of homeless street people.* University of California Press.

Spencer, S., Decuypere, E., Aerts, S., & De Tavernier, J. (2006). History and ethics of keeping pets: Comparison with farm animals. *Journal of Agricultural and Environmental Ethics, 19*(1), 17–25.

Squirrell, G. (2016). Homeless People and Their Animal Companions. In A. Höing & A Bennett (Eds.), *Humans and animals: Intersecting lives and worlds* (pp. 147–157). Brill.

Thompson, S. J., McManus, H., Lantry, J., Windsor, L., & Flynn, P. (2006). Insights from the street: Perceptions of services and providers by homeless young adults. *Evaluation and Program Planning, 29*(1), 34–43.

Toro, P. A. (2007). Toward an international understanding of homelessness. *Journal of Social Issues, 63*(3), 461–481.

Toro, P. A., Dworsky, A., & Fowler, P. J. (2007, March). Homeless youth in the United States: Recent research findings and intervention approaches. In D. Dennis, G. Locke & J. Khadduri (Eds.), *Toward understanding homelessness: The 2007 national symposium* (pp. 6–33). Department of Health and Human Services and U.S. Department of Housing and Urban Development.

Toro, P. A., Tompsett, C. J., Lombardo, S., Philippot, P., Nachtergael, H., Galand, B., Schlienz, N., Stammel, N. Yabar, Y., Blume, M., MacKay, L., & Harvey, K. (2007). Homelessness in Europe and the United States: A comparison of prevalence and public opinion. *Journal of Social Issues, 63*(3), 505–524.

Toro, P. A., & Warren, M. G. (1999). Homelessness in the United States: Policy considerations. *Journal of Community Psychology, 27*(2), 119–136.

Tuan, Y-F. (1984). *Dominance and affection: The making of pets.* Yale University Press.

Urys, E. L. (1999). *Riding the rails: Teenagers on the move during the great depression.* Routledge.

U.S. Department of Housing and Urban Development. (2021). *2007–2020 Point-in-Time estimates by state.* https://www.hudexchange.info/resource/3031/pit-and-hic-data-since-2007/

Williams, D. L., & Hogg, S. (2015). Opinions regarding the health and welfare of dogs owned by homeless people. *BSAVA congress proceedings* (pp. 478–478). British Small Animal Veterinary Association Library.

Williams, D. L., & Hogg, S. (2016). The health and welfare of dogs belonging to homeless people. *Pet Behaviour Science, 2016*(1), 23–30.

World Health Organization. (2021, December 1). *Tripartite and UNEP support OHHLEP's definition of "One Health."* https://www.who.int/news/item/01-12-2021-tripartite-and-unep-support-ohhleps-definition-of-one-health

Yang, H., Howarth, A., Hansen, S. R., Harrell, L., & Thatcher, C. D. (2021). Understanding the attachment dimension of human-animal bond within a homeless population: A one-health initiative in the Student Health Outreach for Wellness (SHOW) clinic. *Journal of Applied Animal Welfare Science, 24*(4), 357–371.

# 20
# CAREGIVER BURDEN IN COMPANION ANIMAL OWNERS

*Mary Beth Spitznagel, Mark D. Carlson, and Christopher M. Fulkerson*

## Introduction

Research suggests several health and social benefits of companion animal ownership, but debate exists regarding the notion that companion animal ownership is uniformly beneficial (Herzog, 2011). One rarely considered issue in human-animal interaction that may negatively affect an owner's well-being is the impact of caring for a sick companion animal. Serious illness of the animal will enter into the life of nearly every owner, and management of protracted disease can be complex and time consuming (Christiansen et al., 2013). Recent decades have seen a rise in companion animal owners choosing caregiving over euthanasia, demonstrated through the growth in veterinary palliative care (Shearer, 2019) and specialty veterinary services like oncology (Kiselow, 2019). Medical options previously considered implausible for animals have become standard veterinary practice (Gore et al., 2019). While euthanasia is an available and humane alternative, many companion animal owners do not euthanize until reduced quality of life for the animal necessitates it (Spitznagel et al., 2020). With advanced treatment for serious illness comes caregiving.

"Caregiver burden," or strain encountered while providing care for a loved one with an illness (Zarit et al., 1980), may occur in any protracted caregiving situation. It is a response to the many challenges and problems encountered while providing care; the caregiver often sacrifices their own needs, "giving until it hurts" (Tremont, 2011). Burden is usually multi-faceted, with implications for physical and mental health, social, and economic domains (Paradise et al., 2014), and represents a reaction to both objective and subjective components of providing care (Hébert et al., 2000). These may include demands on time, physical abilities, and financial strain, as well as perceptions and emotional responses to caregiving (Tremont, 2011). Caregivers frequently meet daily instrumental needs for their care recipient, such as medication administration, arranging healthcare appointments, and managing personal care. Burden thus typically increases with time (Adelman et al., 2014) as illness worsens, care demands increase, with the caregiver at times forgoing self-care to provide for the loved one.

Recent work demonstrates that caregiver burden is present in samples of owners caring for a companion animal with a serious illness compared to a healthy companion animal (Spitznagel et al., 2017; Spitznagel et al., 2019a), with a variety of implications for the owner, the companion animal, and even the veterinary healthcare team, all of which will be

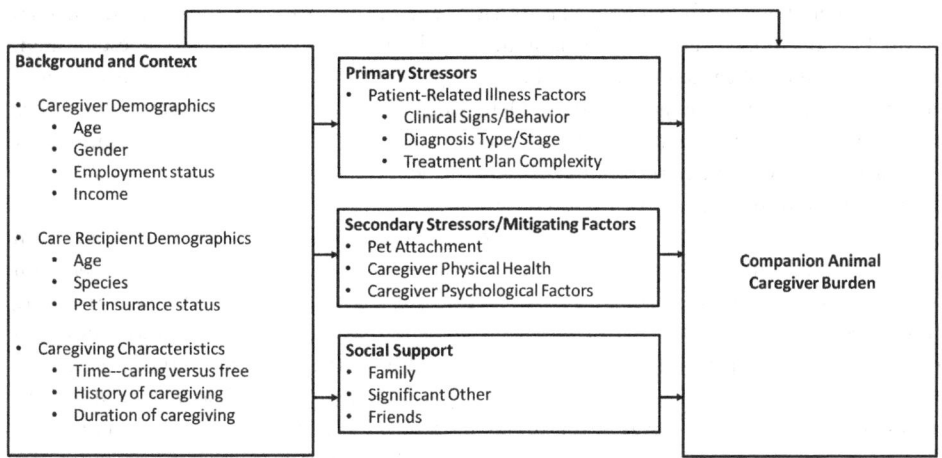

*Figure 20.1* Proposed model of burden on the companion animal caregiver

discussed in this chapter. Many decades of research in human caregiving provide a basis for our understanding of companion animal caregiving, through studies establishing pathways to burden. Pearlin's well-known basic caregiver stress process (Pearlin et al., 1990) conceptualizes burden in caregiving relationships as a product of multiple factors, including (1) the Background and Context of caregiving; (2) Primary Stressors directly related to the care recipient's illness; (3) Secondary Stressors related to the caregiver's own health, as well as the relationship between the care recipient and caregiver; and (4) Social Support, both within and external to the family. In this manner, caregiver burden may result directly from background and contextual influences but may also interact with mitigating primary and secondary stressors, as well as social support factors. Importantly, although companion animal caregiver burden exhibits similarity to burden in human relationships, it differs in several ways from burden in human caregiving (Britton et al., 2018), necessitating adaptation of such models.

Combining knowledge of human caregiving with our research in companion animal caregiving, the authors posit the model in Figure 20.1. A number of risk factors for companion animal caregiver burden have been identified, including background factors, such as the caregiver's age (Spitznagel et al., 2018a), primary stressors including frequency of a companion animal's clinical signs, illness-related behavior problems, and complexity of treatment predict caregiver burden (Spitznagel et al., 2019a; Spitznagel et al., 2018a), as well as secondary stressors including the caregiver's reactivity to problems in the companion animal and coping styles (Spitznagel et al., 2018a), and social support factors (Spitznagel et al., 2018a). All of these potential contributors to caregiver burden have shown relationships with burden, and are included in the proposed model, with the caveat that these factors have been examined only cross-sectionally and in relative isolation to date. Longitudinal investigation of a broad model incorporating these and other factors is needed.

## Historical and Diverse Perspectives on Companion Animal Caregiver Burden

The construct of caregiver burden associated with illness in the companion animal is relatively new – until the past decade, it received little attention. Companion animals have long been considered as members of the family (Berryman, 1985), but whether burden occurs within human-animal relationships as it does in human caregiving was not established. Christiansen and colleagues (2013) conducted a qualitative study to examine this question, interviewing 12 Danish owners of ill or elderly dogs. This work laid a foundation for future studies in companion animal caregiver burden, noting caregiving issues like providing for additional care needs due to the animal's illness, difficulties balancing responsibilities, impact on social life and finances, and emotional strain.

Since that time, multiple quantitative studies have been conducted examining companion animal caregiver burden, its predictors, and correlates. At present, most work in this domain has been conducted in Western cultures, with particular representation of English-speaking samples. Although studies aimed at understanding strain associated with serious illness in a companion animal have been conducted in other cultures, they have not targeted caregiver burden in particular. For example, one study conducted in Japan (Nakano et al., 2019) suggested the presence of elevated symptoms of depression and anxiety in owners following a diagnosis of cancer in the companion animal. While it is tempting to attribute these symptoms to caregiver burden, they might be more reflective of anticipatory grief (Planchon et al., 2002). Because caregiver burden is a separate and distinct construct from other owner experiences in the context of serious companion animal illness, such as quality of life and anticipatory grief (Spitznagel et al., 2021), it is important not to confuse them.

To date, caregiver burden research samples have been comprised primarily of white women with relatively high levels of education and income (Britton et al., 2018; Spitznagel et al., 2020; Spitznagel et al., 2017; Spitznagel et al., 2019a; Spitznagel et al., 2018a; Spitznagel et al., 2021a). This is not necessarily surprising, as many of these factors are known to increase the odds of companion animal ownership, at least in the United States (Tower & Nokota, 2006) where most of the research has been conducted. Varied sampling methods including social media companion animal disease support groups and hospital-based recruitment have been attempted in an effort to broaden sample demographics. Greater diversity, particularly in education and income, has been found when examining samples recruited from general veterinary practices rather than specialty referral clinics (Spitznagel et al., 2018b), and work using social media (Spitznagel et al., 2021b) and consumer panel data collection (Spitznagel et al., 2022) has yielded more representative samples in terms of race/ethnicity and gender. Such factors are critical to consider in companion animal caregiving, as financial strain and differences in race and ethnicity are known to affect the burden in human caregiving (Willert & Minnotte, 2021). Importantly, sample differences to date have not elicited substantially different results, suggesting some robustness to findings in this domain; however, greater diversity in sample representation is needed in the field. In short, the nascent field of companion animal caregiver burden has much room for growth in understanding individual and cultural differences in the caregiving experience.

## The Impact of Caregiver Burden on Humans

Perhaps not surprisingly, the presence of companion animal caregiver burden substantially raises the risk of other problems in psychosocial functioning in the owner. Research also suggests that frequent interactions of veterinary personnel with burdened owners may contribute to reduced psychosocial functioning in individuals working in the field of veterinary medicine; specifically, distressed owners have a greater likelihood of engaging in behaviors that are strongly linked to stress and burnout in veterinary personnel.

## *Correlates of Burden in the Companion Animal Owner*

Several quantitative studies have been undertaken in recent years to better understand the impact of companion animal caregiving on the owner. These studies have relied on the gold standard measure of caregiver burden, the Zarit Burden Interview (ZBI; Zarit et al., 1980), adapted for use in companion animal caregiving (Spitznagel et al., 2017). Of note, although the care recipient population differs from that for which the measure was originally created, this measure has proven robust across a wide variety of populations (Hébert et al., 2000). The construct of caregiver burden, as assessed by this measure, focuses on the subjective and objective challenges experienced by the caregiver, rather than the specifics of the caregiving situation. For example, items address feeling like there is not enough time in the day or financial resources to provide care, and emotional responses like guilt, anger, and fear in the context of caregiving. The lack of reference to specific problems in the care recipient makes this measure easily generalizable to many populations.

The first quantitative study (Spitznagel et al., 2017) of companion animal caregiver burden of which the authors are aware used the ZBI to assess caregiver burden. Participants of this study also completed validated measures of stress, anxiety, depression, and quality of life. Data from a total of 238 owners, all recruited from social media, were analyzed. Half were owners of a dog or cat with a serious illness, such as cancer, diabetes, renal failure, or other disease requiring daily management, and half were demographically (owner age and gender, animal species) matched healthy controls. Results demonstrated that owners of a sick dog or cat endorsed elevated caregiver burden, which was on average, nearly twice that of those with a healthy dog or cat. Although four items were removed from the ZBI due to irrelevance to companion animal caregiving, the average amount of caregiver burden in those with a sick companion animal was still above a suggested cutoff for clinically significant burden in human caregiving (Zarit et al., 1980). Poorer psychosocial functioning was also observed in owners of a sick companion animal, with greater stress, symptoms of depression and anxiety, and lower quality of life relative to those with a healthy companion animal. Of note, the average level of depressive symptoms in individuals with a sick companion animal was also above the threshold for clinically significant symptoms (Lewinsohn et al., 1997). As expected, degree of caregiver burden showed medium to strong correlations with psychosocial dysfunction across the board ($r_s$ ranging from 0.41 through 0.54).

That first study of associations between caregiver burden and broader psychosocial functioning in the context of serious companion animal illness showed intriguing relationships and group differences. However, it was conducted using a social media sample, which might not reflect the average companion animal owner. For this reason, the study design was repeated in a sample of small animal general veterinary hospital clients (for the purpose of this chapter, "clients" are differentiated from "owners" to signify samples of individuals recruited from veterinary facilities; Spitznagel et al., 2019a). The participants were 124 clients, including 62 with a dog or cat with a serious illness and 62 demographically matched clients with a healthy companion animal. The same measures were completed, and a similar pattern of results was observed. Clients with a seriously ill companion animal demonstrated higher levels of caregiver burden, more stress, greater anxiety and depressive symptomology, and lower quality of life. Caregiver burden was once again strongly correlated with measures of psychosocial distress ($r_s$ ranging from 0.50 through 0.59). Once again, average levels of caregiver burden and depressive symptoms were within ranges indicating clinically meaningful elevation.

Combined, these studies are a clear demonstration that in the presence of serious illness in the companion animal, an owner is more likely to experience a caregiver burden. Moreover, robust associations exist between that burden and lower psychosocial functioning. However, the implications of caregiver burden in the companion animal owner may extend beyond the basic owner-companion animal relationship, to the veterinary healthcare team.

## *Correlates of Owner Burden in the Veterinary Care Provider*

Given the context in which caregiver burden occurs (i.e., serious illness in the companion animal), it is important to ask whether frequent interactions with burdened owners impact those working in veterinary medicine. In human medicine, the triadic caregiver-patient-physician relationship can become conflictual, particularly if a caregiver is viewed as demanding (Kravitz et al., 2002) or adding to workload (Boise et al., 1999). Caregiver distress plays a role in healthcare utilization for the care recipient in human medicine (Maust et al., 2017), and similar relationships among these variables may exist in veterinary medicine, as well. Examination of whether owner caregiver burden impacts veterinary personnel is of particular importance in light of the high levels of distress observed in the field. Many occupational hazards have been identified as contributors to stress and burnout in veterinary medicine, such as compassion fatigue, moral distress, overwork, fear of complaints, and frustration with unrealistic client expectations and non-compliance (Moir & Van den Brink, 2019; Moses et al., 2018; Vande Griek et al., 2018).

It has been posited that core to many of these occupational stressors could be the difficult interactions that veterinary personnel have with clients who are experiencing caregiver burden (Spitznagel et al., 2019b). Specifically, presence of caregiver burden in owners of a sick companion animal have been linked to certain behaviors or interactions with veterinary personnel that are in turn related to distress in the field, a phenomenon termed "burden transfer" (Spitznagel et al., 2019b). The theory of burden transfer suggests that caregiver burden in the client underlies stressful encounters with providers, essentially transferring the client burden to veterinary personnel.

Five domains of difficult veterinarian-client interactions have been identified through factor analysis, referred to, in combination, as the "burden transfer DANCE" (Spitznagel et al., 2019b; See Table 20.1): **D**aily hassles (e.g., clients requesting impossible predictions, struggling to make decisions, following the advice of others about the companion animal's health, comparing costs), **A**ffect (e.g., clients needing euthanasia counseling, exhibiting anxiety or sadness), **N**on-adherent/Inconsiderate behaviors (e.g., clients declining work-up or treatment that has been recommended, not showing for appointments), **C**onfrontations (e.g., clients blaming the veterinarian for poor outcomes, payment refusal, lodging a complaint), and **E**xcess Communications (e.g., phone or email contact that is viewed as unnecessary). Highlighting the domain of "Excess Communication," the earlier described study conducted in 124 general veterinary clients extended beyond examining caregiver burden for the owner to also examine the owners' utilization of veterinary services (Spitznagel et al., 2019a). In a comparison of clients with elevated levels of caregiver burden to those without, no significant difference was observed in billable services (i.e., how often the companion animal was brought in for services such as examinations, bloodwork, or procedures), but those with high burden required approximately twice as much non-billable communication (i.e., phone and email responses). The lack of difference in billable contacts suggests that groups were similar in disease severity, and that caregiver burden may drive those higher non-billable contacts. These burden transfer DANCE behaviors in the client are strongly correlated with the level of caregiver burden reported by the client, and in turn, significantly predict stress and burnout for veterinarians (Spitznagel et al., 2019b).

*Table 20.1* Burden transfer DANCE factors, definitions, and examples

| Dance Factor | Definition | Examples |
| --- | --- | --- |
| Daily hassles | Everyday frustrations | Client wants impossible predictions; has difficult making decisions; compares costs |
| Affect | Difficult client emotions | Euthanasia counseling needed; client shows anxiety, sadness |
| Non-adherent/inconsiderate | Non-compliant or thoughtless behaviors | Client declines recommended treatment or work-up, no-shows for appointment |
| Confrontations | Conflict with client | Client becomes upset, blames, refuses to pay for services, makes a complaint |
| Excess communication | Client needs more support than seems warranted | Client makes too frequent contact (e.g., phone, email) |

In summary, research suggests there could be far-reaching consequences of caregiver burden in the companion animal owner, both to the owner and the veterinary team with whom that owner is interacting. Caregiver burden is, on average, elevated in those providing care for a companion animal with serious illness, and the presence of burden is associated with poorer psychosocial functioning including stress, symptoms of anxiety, clinically elevated depressive symptoms, and reduced quality of life. Moreover, higher levels of burden in the companion animal owner are associated with certain behaviors and interactions with veterinary personnel that are in turn related to stress and burnout for those working in the field of veterinary medicine – a field that already suffers from high rates of stress and depression (Nett et al, 2015) as well as suicide (Tomasi et al., 2019).

The cross-sectional designs of the above-described studies cannot shed light on directionality of these associations – it is not fully clear whether the burden of caregiving is the root cause of these psychosocial problems and challenging interactions with veterinary personnel, or if individuals who demonstrate broader psychosocial problems simply experience caregiving as more burdensome and are more likely to engage in behaviors that the veterinary healthcare team experiences as stressful. Regardless, group differences in owners of healthy versus seriously ill companion animals underscore the importance of ongoing research efforts to determine how and why caregiver burden develops in the companion animal owner, and if it might be prevented or reduced.

## The Impact of Caregiver Burden on the Companion Animal

Presence of caregiver burden in the owner of a seriously ill companion animal may also have bearing on the animal. Is a burdened owner more likely to prematurely discontinue treatment, or to consider euthanasia before the companion animal's quality of life indicates it is necessary? Past work has investigated how variables related to the companion animal (e.g., diagnosis, clinical signs) correlate with an owner's decision to euthanize (Marchitelli et al., 2020), and qualitative research has highlighted the importance of owner perspectives in treatment and euthanasia decision-making on the animal's behalf (Christiansen et al., 2016). Recent quantitative work suggests that the owner's burden may affect decisions in this context.

To examine this issue, a "Consideration of Euthanasia" measure was developed to ask the owner about their own actions reflecting stages of the euthanasia decision (e.g., "I have considered euthanasia for this pet," "I have discussed euthanasia of this pet with family or friends," "I have discussed euthanasia of this pet with a veterinarian," "I have taken steps toward euthanizing this pet, e.g., scheduled an appointment"; Spitznagel et al., 2020). The consideration of euthanasia measure correlated strongly ($r=.070$) with actual euthanasia within 30 days of completing this measure. To then investigate the potential influence of caregiver burden and other factors on the euthanasia decision, 345 owners of an elderly or seriously ill cat or dog completed the consideration of euthanasia measure, as well as assessments of owner psychosocial variables (e.g., depression, stress, quality of life), human-animal interaction variables (e.g., caregiver burden, anticipatory grief, companion animal attachment), animal severity of illness (i.e., quality of life), and demographic variables of the owner and animal (e.g., income, age, human gender, animal species). Several factors showed significant correlation with consideration of euthanasia, including companion animal quality of life, owner income, caregiver burden, anticipatory grief, depression, and stress. However, in a regression model controlling for the animal's quality of life, only caregiver burden and income emerged as significant predictors of consideration of euthanasia.

More work is needed to fully understand the role of caregiver burden in decisions that the owner makes on the animal's behalf. However, it appears likely that distress associated with caregiving could exert influence on the consideration of euthanasia. It will be important for future research to determine if this finding is replicated, and to examine the role of burden in other important owner behaviors and decisions.

## Practical Perspectives

The potentially far-reaching consequences of caregiver burden in the companion animal owner raise the question of how to best ameliorate it. Decades of human caregiving research might be used to inform interventions for companion animal caregiver burden. However, efforts to translate work from the literature in human medicine must be informed by knowledge of companion animal caregiving, due to research demonstrating that the presentation of burden differs in these populations.

### *Comparing Human and Companion Animal Caregiving*

Work to better understand differences between caregiving for a human family member compared to a companion animal is limited; however, many similarities as well as important discrepancies in key caregiving experiences, including burden and "positive aspects of caregiving" (e.g., emotional satisfaction, personal growth, experiencing a sense of competency, spiritual growth, and sense of fulfillment of duty; Lloyd et al., 2016; Cohen et al., 2002) have been shown. In a study involving cross-sectional comparison of burden and positive aspects of caregiving for human family and companion animal caregivers (Britton et al., 2018), a total of 369 caregivers (117 companion animal caregivers and 252 family caregivers) were recruited through social media caregiving groups. Many demographic differences were present in these two groups; as such, analyses were conducted first in the full sample, then repeated in a subset ($n=75$ per group) of blindly matched caregivers with similar caregiver age, education, gender, race, income, living situation (all caregivers lived with the care recipient), and duration of caregiving experience (matched for greater versus less than one year). Burden was elevated at a clinically meaningful level in both human and companion animal caregiver groups, though higher overall for human caregivers. Comparable levels of guilt, fear of what the future holds, and financial strain were seen across samples. Although burden was higher in human family caregivers, positive aspects of caregiving, including feeling useful, important, confident, and more positive toward life, were greater in companion animal caregivers.

Thus, while some similarities were seen across groups, the overall caregiving experience was not equivalent. The lower burden and more positive appraisal of caregiving observed in companion animal owners might be expected for several reasons, including the availability of euthanasia. Whereas taking on the responsibility of caregiving may not be an option in the case of human caregiving due to lack of support, including financial limitations (Quinn et al., 2015), euthanasia is an available alternative for the companion animal owner who does not wish to continue care. Having made the decision to provide care rather than euthanize may predispose the companion animal caregiver to a more positive caregiving experience Britton et al., 2018.

Research suggests that a more positive appraisal of caregiving can serve a protective role in the context of caregiving, making a caregiver less likely to experience burden, depression, and even health concerns; for this reason, efforts to increase positive aspects of caregiving have been posited as a potential intervention for burden in human caregiving (Hilgeman et al., 2007). The high levels of positive experiences reported by companion animal caregivers may create a "ceiling" for such effects (Britton et al., 2018).

Differences between human and companion animal caregiving underscore the importance of understanding companion animal caregiving in its own right when considering interventions for burden in the companion animal owner. Greater cognizance of which factors show strong correlations with (or predictive power for) caregiver burden may be of benefit to this end.

## *Understanding Determinants of Caregiver Burden*

An incidental but crucial finding of early caregiver burden group comparisons (i.e., owners of a seriously ill versus healthy companion animal) is that within the group of owners of seriously ill animals, only about half showed meaningfully elevated levels of burden. This finding highlights that although burden is present for many, not everyone in this situation experiences high levels of burden. In fact, a recent cluster analytic study of 345 owners of elderly or seriously ill companion animals demonstrated four distinct owner profiles in regard to levels of distress (Spitznagel et al., 2021a). Some owners exhibited a "Resilient" presentation, characterized by compartmentalized distress about the companion animal (i.e., elevated caregiver burden, but average general quality of life). Others demonstrated a "Distressed" presentation, with pervasive distress, including caregiver burden, anticipatory grief, and reduced quality of life. A third, "Non-Distressed" profile was also present, in which owners reported low levels of distress, appearing relatively unaffected by the companion animal's problems. The final group of owners demonstrated distress due to "Other Influences," with emotional difficulty on general measures that did not appear to be related to the companion animal's condition. Caregiver burden is thus present in some, but not all, owners in this context.

Research of the determinants of caregiver burden may provide insight into who develops caregiver burden and why. Is it possible that certain factors elevate an individual's risk for developing caregiver burden when the companion animal is ill? In the first quantitative effort to understand predictors of companion animal caregiver burden, Spitznagel and colleagues (2019a) examined a group of 62 owners of a seriously ill cat or dog to determine whether challenges associated with elements of the treatment plan were related to higher levels of burden. Is the likelihood of burden increased in the context of difficulty administering prescribed medications, or following other aspects of a treatment plan? Results suggested that changes in daily routine and difficulty following new routines were associated with higher caregiver burden. Although these relationships were highly significant, the magnitude of the correlations was moderate ($r_s = 0.43$ for both), suggesting that development of caregiver burden is more complex than difficulty managing a care routine.

Additional potential risk factors for caregiver burden were investigated in a subsequent study (Spitznagel et al., 2018a). This work investigated a broad cross-sectional model including demographic characteristics (e.g., client income), companion animal clinical signs and behavior problems, and client personal resources including coping skills, problem-solving, sense of personal control, tangible support (e.g., someone to provide instrumental assistance), and emotional support (e.g., someone talk to) in 95 owners of a seriously ill companion animal recruited through a small animal general veterinary hospital. Younger age of the owner, frequency of the animal's clinical problems, the owner's reaction to those problems, and sense of personal control all significantly predicted caregiver burden. Of interest, significant zero-order correlations between greater coping skills and social support were also observed with burden; however, these did not rise to a level of significant prediction in the larger model.

The current research only scratches the surface in our current understanding of the determinants of companion animal caregiver burden. At present, findings suggest that factors beyond a difficult treatment plan or the companion animal's specific problems contribute to the burden. Future research is needed to better understand interactions between the treatment plan, the animal's problems, and the owner's individual characteristics, such as personality characteristics (e.g., conscientiousness, neuroticism), stress reactivity, and self-efficacy.

## Ethical Considerations

The current stage of the literature in companion animal caregiver burden raises an ethical issue that warrants consideration. The potential impact of owner caregiver burden on the owner, the companion animal, and the veterinary healthcare team suggest that when present in a client, efforts to reduce caregiver burden should be made. At present, best practices for alleviating companion animal caregiver burden are unclear. As research unfolds and approaches for reducing burden become clear, who bears this responsibility?

As already established, the burdened client (particularly if presenting with the "Distressed" profile), has clear problems with psychosocial functioning. Clinically meaningful levels of depressive symptoms may be present, as well as high stress and anxiety, and low quality of life (Spitznagel et al., 2017; Spitznagel et al., 2019a). While the owner may realize the distress they are experiencing, they may not be aware that burden in the context of caretaking for a seriously ill companion animal exists, and is very common. Indeed, the veterinary healthcare team may have greater awareness of these issues, as caregiver burden is something encountered in the veterinary clinic setting on a daily basis.

However, veterinary professionals, comprised of individuals with education and training in companion animal medicine, are focused on treating the animal. They are neither responsible for nor competent to treat problems of human psychosocial functioning. Prior work suggests that nearly three-quarters of veterinarians do not refer clients to mental health professionals for grief issues, in part due to lack of awareness or resources, but also because such referrals are viewed as beyond the role or competence of the veterinarian (Dow et al., 2019). As burden is less well-studied and understood in veterinary medicine compared to grief, this number would likely be even lower.

Identification of caregiver burden in the companion animal owner may help a veterinary healthcare team understand and communicate with the owner, but solutions for caregiver burden must be found elsewhere. The question thus arises: how does the veterinary healthcare team, well-equipped to notice burden but not to provide services to address it, handle burden in the client?

The answer to this question may soon be answered. Allied mental health professionals, such as veterinary social workers, are becoming increasingly involved in veterinary settings (Holcombe et al., 2015). Although caregiver burden is not a formally recognized pillar of veterinary social work (*Veterinary social work,* 2022), the closely linked experiences of animal-related grief and bereavement are. Veterinary social workers would, in many cases, be able to provide support to owners experiencing caregiver burden. With the integration of a veterinary social worker into the veterinary healthcare team, answering the question of "who bears the responsibility" of treating caregiver burden becomes clear. In the absence of an integrated veterinary social worker within the veterinary healthcare team, awareness, and recognition of burden in companion animal owners by the veterinary healthcare team could prompt referral of the client to other local mental health professionals. Including training on caregiver burden in veterinary curricula, and importantly when and how to refer clients for additional support, might be considered.

## Unanswered Questions and Future Research Demands

The recent attention to caregiver burden in the companion animal owner is exciting, but there are many avenues of research that have yet to be pursued. Determinants must be delineated through longitudinal research. A more complete understanding of how and why companion animal caregiver burden develops in some (but not all) could help identify owners most at risk for negative outcomes. The model proposed earlier in this chapter suggests that caregiver burden may directly result from background and contextual influences, but may also interact with mitigating primary and secondary stressors, as well as social support factors. To date, elements of the proposed model have been examined only cross-sectionally and in relative isolation. Longitudinal investigation of a broad model is necessary in order to identify pathways to burden. Such work could, in turn, pinpoint intervention targets, reducing negative mental health outcomes like depressive symptoms for the companion animal owner.

Additionally, as previously described, the research to date suffers from relative homogeneity of demographic factors. Although findings have been robust when research has been conducted in samples of greater gender, racial, and socioeconomic diversity, there is an absence of work designed to specifically detect possible group differences. Perhaps just as important as demographic differences in the makeup of human samples in this research is the diversity of the companion animals involved. The research cited in this chapter generally focuses specifically on cats and/or dogs in effort to capture similar companion animal owner experiences. However, the world of human-animal interaction offers a far broader population than the samples that have been examined thus far. Increased sample diversity in both the humans and animals involved in this research is an important future direction.

Finally, much research, including the caregiver burden work described in this chapter, suggests that negative impact of companion animal ownership is possible in some situations. However, research calls within the field of human animal interaction almost uniformly lean toward funding opportunities aimed at determining the benefits of animals to humans. For example, recent requests for applications through both the National Institutes of Health (U.S. Department of Health and Human Services, 2022) and the Human Animal Bond Research Institute (Funding opportunities, 2022) explicitly requested research proposals to examine the impact of human-animal interactions on childhood development, mental and physical health, wellness, and healthy aging. Focusing primarily on benefits of animals to humans, with little attention paid to negative consequences like caregiver burden, will contribute to a biased literature. Future work in the field will benefit from broad calls for research that will help to develop a more complete understanding of the human-animal interaction, identify key targets for intervention, and allow for a better appreciation of the complexities of human-animal relationships.

**Potential Questions for Discussion**

1. What differences between caregiving in companion animals and humans might explain lower levels of caregiver burden in owners of sick companion animals?
2. What are the potential impacts of high caregiver burden in companion pet owners on the veterinary healthcare team?
3. What resources might be available to members of the veterinary healthcare team to support clients experiencing high levels of caregiver burden in the absence of an integrated veterinary social worker?
4. How does the permissibility of euthanasia for companion animals alter the context of burden in caregivers of companion animals versus people?

# References

Adelman, R. D., Tmanova, L. L., Delgado, D., Dion, S., & Lachs, M. S. (2014). Caregiver burden. *JAMA*, *311*(10), 1052. https://doi.org/10.1001/jama.2014.304

Berryman, J., Howells, K., & Lloyd-Evans, M. (1985). Pet owner attitudes to pets and people: A psychological study. *Veterinary Record*, *117*(25–26), 659–661. https://doi.org/10.1136/vr.117.25-26.659

Boise, L., Camicioli, R., Morgan, D. L., Rose, J. H., & Congleton, L. (1999). Diagnosing dementia: Perspectives of primary care physicians. *The Gerontologist*, *39*(4), 457–464. https://doi.org/10.1093/geront/39.4.457

Britton, K., Galioto, R., Tremont, G., Chapman, K., Hogue, O., Carlson, M. D., & Spitznagel, M. B. (2018). Caregiving for a companion animal compared to a family member: Burden and positive experiences in caregivers. *Frontiers in Veterinary Science*. https://doi.org/10.3389/fvets.2018.00325

Christiansen, S. B., Kristensen, A. T., Sandøe, P., & Lassen, J. (2013). Looking after chronically Ill dogs: Impacts on the caregiver's life. *Anthrozoös*, *26*(4), 519–533. https://doi.org/10.2752/175303713x13795775536174

Christiansen, S. B., Kristensen, A. T., Lassen, J., & Sandøe, P. (2016). Veterinarians' role in clients' decision-making regarding seriously ill companion animal patients. *Acta Veterinaria Scandinavica*, *58*(1). https://doi.org/10.1186/s13028-016-0211-x

Cohen, C. A., Colantonio, A., & Vernich, L. (2002). Positive aspects of caregiving: Rounding out the caregiver experience. *International Journal of Geriatric Psychiatry*, *17*(2), 184–188. https://doi.org/10.1002/gps.561

Dow, M. Q., Chur-Hansen, A., Hamood, W., & Edwards, S. (2019). Impact of dealing with bereaved clients on the psychological wellbeing of veterinarians. *Australian Veterinary Journal*, *97*(10), 382–389. https://doi.org/10.1111/avj.12842

*Funding opportunities*. HABRI. (2022, January 3). Retrieved January 23, 2022, from https://habri.org/grants/funding-opportunities/

Gore, M., Lana, S. E., & Bishop, G. A. (2019). Colorado state university, pet hospice program. *Veterinary Clinics of North America: Small Animal Practice*, *49*(3), 339–349. https://doi.org/10.1016/j.cvsm.2019.01.002

Hébert, R., Bravo, G., & Préville, M. (2000). Reliability, validity and reference values of the Zarit burden interview for assessing informal caregivers of community-dwelling older persons with dementia. *Canadian Journal on Aging / La Revue Canadienne Du Vieillissement*, *19*(4), 494–507. https://doi.org/10.1017/s0714980800012484

Herzog, H. (2011). The impact of pets on human health and psychological well-being: Fact, fiction, or hypothesis? *Current Directions in Psychological Science*, *20*(4), 236–239. https://doi.org/10.1177/0963721411415220

Hilgeman, M. M., Allen, R. S., DeCoster, J., & Burgio, L. D. (2007). Positive aspects of caregiving as a moderator of treatment outcome over 12 months. *Psychology and Aging*, *22*(2), 361–371. https://doi.org/10.1037/0882-7974.22.2.361

Holcombe, T. M., Strand, E. B., Nugent, W. R., & Ng, Z. Y. (2015). Veterinary social work: Practice within Veterinary Settings. *Journal of Human Behavior in the Social Environment*, *26*(1), 69–80. https://doi.org/10.1080/10911359.2015.1059170

Kiselow, M. (2019). Private practice oncology. *Veterinary Clinics of North America: Small Animal Practice*, *49*(3), 519–527. https://doi.org/10.1016/j.cvsm.2019.01.010

Kravitz, R. L., Bell, R. A., Franz, C. E., Elliott, M. N., Amsterdam, E., Willis, C., & Silverio, L. (2002). Characterizing patient requests and physician responses in office practice. *Health Services Research*, *37*(1), 217–238.

Lewinsohn, P. M., Seeley, J. R., Roberts, R. E., & Allen, N. B. (1997). Center for Epidemiologic Studies Depression Scale (CES-D) as a screening instrument for depression among community-residing older adults. *Psychology and Aging*, *12*(2), 277–287. https://doi.org/10.1037/0882-7974.12.2.277

Lloyd, J., Patterson, T., & Muers, J. (2016). The positive aspects of caregiving in dementia: A critical review of the qualitative literature. *Dementia*, *15*(6), 1534–1561. https://doi.org/10.1177/1471301214564792

Marchitelli, B., Shearer, T., & Cook, N. (2020). Factors contributing to the decision to euthanize: Diagnosis, clinical signs, and triggers. *The Veterinary Clinics of North America. Small Animal Practice*, *50*(3), 573–589. https://doi-org.proxy.library.kent.edu/10.1016/j.cvsm.2019.12.007

Maust, D. T., Kales, H. C., McCammon, R. J., Blow, F. C., Leggett, A., & Langa, K. M. (2017). Distress associated with dementia-related psychosis and agitation in relation to healthcare utilization and costs. *The American Journal of Geriatric Psychiatry*, *25*(10), 1074–1082. https://doi.org/10.1016/j.jagp.2017.02.025

Moir, F. M., & Van den Brink, A. R. K. (2019). Current insights in veterinarians' psychological well-being. *New Zealand Veterinary Journal*, *68*(1), 3–12. https://doi.org/10.1080/00480169.2019.1669504

Moses, L., Malowney, M. J., & Wesley Boyd, J. (2018). Ethical conflict and moral distress in veterinary practice: A survey of North American veterinarians. *Journal of Veterinary Internal Medicine*, *32*(6), 2115–2122. https://doi.org/10.1111/jvim.15315

Nakano, Y., Matsushima, M., Nakamori, A., Hiroma, J., Matsuo, E., Wakabayashi, H., Yoshida, S., Ichikawa, H., Kaneko, M., Mutai, R., Sugiyama, Y., Yoshida, E., & Kobayashi, T. (2019). Depression and anxiety in pet owners after a diagnosis of cancer in their pets: A cross-sectional study in Japan. *BMJ Open*, *9*(2). https://doi.org/10.1136/bmjopen-2018-024512

Nett, R. J., Witte, T. K., Holzbauer, S. M., Elchos, B. L., Campagnolo, E. R., Musgrave, K. J., Carter, K. K., Kurkjian, K. M., Vanicek, C. F., O'Leary, D. R., Pride, K. R., & Funk, R. H. (2015). Risk factors for suicide, attitudes toward mental illness, and practice-related stressors among us veterinarians. *Journal of the American Veterinary Medical Association*, *247*(8), 945–955. https://doi.org/10.2460/javma.247.8.945

Paradise, M., McCade, D., Hickie, I. B., Diamond, K., Lewis, S. J. G., & Naismith, S. L. (2014). Caregiver burden in mild cognitive impairment. *Aging & Mental Health*, *19*(1), 72–78. https://doi.org/10.1080/13607863.2014.915922

Pearlin, L. I., Mullan, J. T., Semple, S. J., & Skaff, M. M. (1990). Caregiving and the stress process: An overview of concepts and their measures. *The Gerontologist*, *30*(5), 583–594. https://doi.org/10.1093/geront/30.5.583

Quinn, C., Clare, L., & Woods, R. T. (2015). Balancing needs: the role of motivations, meanings and relationship dynamics in the experience of informal caregivers of people with dementia. *Dementia (London, England)*, *14*(2), 220–237. https://doi-org.proxy.library.kent.edu/10.1177/1471301213495863

Shearer, T. (2019). Where have we been, where are we going. *Veterinary Clinics of North America: Small Animal Practice, 49*(3), 325–338. https://doi.org/10.1016/j.cvsm.2019.01.001

Spitznagel, M. B., Anderson, J. R., Marchitelli, B., Sislak, M. D., Bibbo, J., & Carlson, M. D. (2021a). Owner quality of life, caregiver burden and anticipatory grief: How they differ, why it matters. *Veterinary Record, 188*(9). https://doi.org/10.1002/vetr.74

Spitznagel, M. B., Ben-Porath, Y. S., Rishniw, M., Kogan, L. R., & Carlson, M. D. (2019b). Development and validation of a burden transfer inventory for predicting veterinarian stress related to client behavior. *Journal of the American Veterinary Medical Association, 254*(1), 133–144. https://doi.org/10.2460/javma.254.1.133

Spitznagel, M. B., Cox, M. D., Jacobson, D. M., Albers, A. L., & Carlson, M. D. (2019a). Assessment of caregiver burden and associations with psychosocial function, veterinary service use, and factors related to treatment plan adherence among owners of dogs and cats. *Journal of the American Veterinary Medical Association, 254*(1), 124–132. https://doi.org/10.2460/javma.254.1.124

Spitznagel, M. B., Hillier, A., Gober, M., & Carlson, M. D. (2021b). Treatment complexity and caregiver burden are linked in owners of dogs with allergic/atopic dermatitis. *Veterinary Dermatology, 32*(2), 192–e50. https://doi-org.proxy.library.kent.edu/10.1111/vde.12938

Spitznagel, M. B., Jacobson, D. M., Cox, M. D., & Carlson, M. D. (2017). Caregiver burden in owners of a sick companion animal: A cross-sectional observational study. *Veterinary Record, 181*(12), 321. https://doi.org/10.1136/vr.104295

Spitznagel, M. B., Jacobson, D. M., Cox, M. D., & Carlson, M. D. (2018a). Predicting caregiver burden in general veterinary clients: Contribution of companion animal clinical signs and problem behaviors. *The Veterinary Journal, 236*, 23–30. https://doi.org/10.1016/j.tvjl.2018.04.007

Spitznagel, M. B., Marchitelli, B., Gardner, M., & Carlson, M. D. (2020). Euthanasia from the veterinary client's perspective. *Veterinary Clinics of North America: Small Animal Practice, 50*(3), 591–605. https://doi.org/10.1016/j.cvsm.2019.12.008

Spitznagel, M.B., Patrick, K., Hillier, A., Gober, M., Carlson, M. D. (2022). Caregiver burden, treatment complexity, and the veterinarian-client relationship in owners of dogs with skin disease. *Veterinary Dermatology, 33*, 208–e59. https://doi.org/10.1111/vde.13065

Spitznagel, M. B., Solc, M., Chapman, K. R., Updegraff, J., Albers, A. L., & Carlson, M. D. (2018b). Caregiver burden in the veterinary dermatology client: Comparison to healthy controls and relationship to quality of life. *Veterinary Dermatology, 30*(1), 3–e2. https://doi.org/10.1111/vde.12696

Stokes, S., Planchon, L., Templer, D., & Keller, J. (2002). Death of a companion cat or dog and human bereavement: Psychosocial variables. *Society & Animals, 10*(1), 93–105. https://doi.org/10.1163/156853002760030897

Tomasi, S. E., Fechter-Leggett, E. D., Edwards, N. T., Reddish, A. D., Crosby, A. E., & Nett, R. J. (2019). Suicide among veterinarians in the United States from 1979 through 2015. *Journal of the American Veterinary Medical Association, 254*(1), 104–112. https://doi.org/10.2460/javma.254.1.104

Tower, R. B., & Nokota, M. (2006). Pet companionship and depression: Results from a united states internet sample. *Anthrozoös, 19*(1), 50–64. https://doi.org/10.2752/089279306785593874

Tremont, G. (2011). Family caregiving in dementia. *Med Health RI, 94*(2), 36–38.

U.S. Department of Health and Human Services. (2022). *PAR-20–033: Human-animal interaction (HAI) research (R21 clinical trial optional)*. National Institutes of Health. Retrieved January 23, 2022, from https://grants.nih.gov/grants/guide/pa-files/PAR-20-033.html

Vande Griek, O. H., Clark, M. A., Witte, T. K., Nett, R. J., Moeller, A. N., & Stabler, M. E. (2018). Development of a taxonomy of practice-related stressors experienced by veterinarians in the United States. *Journal of the American Veterinary Medical Association, 252*(2), 227–233. https://doi.org/10.2460/javma.252.2.227

*Veterinary social work*. Veterinary Social Work. (2022, January 10). Retrieved January 23, 2022, from https://vetsocialwork.utk.edu/

Willert, B., & Minnotte, K. L. (2021). Informal caregiving and strains: Exploring the impacts of gender, race, and income. *Applied Research in Quality of Life, 16*(3), 943–964.

Zarit, S. H., Reever, K. E., & Bach-Peterson, J. (1980). Relatives of the impaired elderly: Correlates of feelings of burden. *The Gerontologist, 20*(6), 649–655. https://doi.org/10.1093/geront/20.6.649

# 21
# HUMAN – ANIMAL INTERACTIONS AND THE IMPACT OF PET LOSS

*Susan P. Cohen*

## Introduction

Whenever people become fond of something, its loss will cause distress. This is especially true of people who feel bonded to a special animal. Not everyone has a happy childhood and a satisfying partnership as an adult. For some individuals, interactions with pets may be the safest and most meaningful of their lives. For various reasons, they may live alone, isolated from other people, so that their pet is their best friend and only companion. Other pet lovers have spouses, children, jobs, hobbies, health—everything society imagines will make them happy—and still feel a strong affinity to pets and more broadly, to the animal world. Grief at the death of a special animal may cause more anguish than that for a human family member. Recovery from that level of pain can take months or years and never be completely in the past. This chapter will explore affection for animals and how this in the context of how the loss occurred, together with personal characteristics of the mourner illuminates grief.

## Pet Attachment and Loss

Understanding pet loss begins with the attachment people feel for other animals. No one knows when humans began to see other living creatures as more than a danger or food source, but somewhere along the line, the ancestors of dogs started hanging around human campfires and proved themselves useful. As humans discovered the value of planting food and keeping it nearby, early versions of our domestic cats hunted for mice among the stores of grain. Someone realized that riding was faster than walking and tamed a horse. Perhaps humans of thousands of years ago relied on the superior senses of the creatures in their vicinity to warn them of danger, because today, there is ample evidence that in the presence of calm animals, our blood pressure goes down and our oxytocin levels go up (Friedmann, 1983; Nagasawa, 2009).

As the idea of agriculture spread, humans domesticated other animals. Humans began to breed for the characteristics they wanted, which led to animals who were calmer, more willing to live beside people. In the case of dogs, this led to animals with more juvenile characteristics, like smaller snouts and bigger eyes, characteristics that made them even more appealing to humans (Beck, 2014).

Fast forward to contemporary North American Culture: People express their affection for the animals in their lives, as well as ones they have never met who live thousands of miles away by speaking warmly about them, providing good food and medical care, and donating to causes that support them. In the United States, 70% of households have a pet and depending on the survey, 76–95% of respondents consider their dog or cat a member of the family (Cohen, S. P., 2002; Gray, 2022; Spin, 2022).

What do people with pets mean when they say their pet is family? As part of her dissertation research, the author interviewed a subgroup of pet lovers at length, usually in their homes. All were comfortable with the term "member of the family". Many of these participants viewed themselves as parents. They used conventional family terms to talk about their relationships: One woman with a service dog had a wish, "I wish she would speak to me and say, 'Gee, Mom, I'm having a great time. I feel like a real human being'". A man with a dog wondered if his pup felt like Pinocchio and wanted to be "a real boy" (Cohen, 1998, p.154).

None of these participants literally thought they gave birth to their pets. Their responses reflected the depth of responsibility and affection they felt toward an innocent, dependent being who lived in their house and could not take care of themselves. These respondents were willing to make significant sacrifices to keep their pets. Several had serious allergies but refused to part with their cats and dogs. One man kept his dog company while she sat under a full-spectrum lamp for two hours every day because she did not like to go outside. A woman declared that she would "give up a kidney" for anyone in her inner circle, including her cat (Cohen, 1998. P. 154).

Some people have a hierarchy of attachment. They believe this level of bonding and therefore of grief applies just to dogs, or dogs and cats, but not to a tortoise or a chicken. This kind of dismissal forces people to hide their sorrow, a condition known as disenfranchised grief. People are reluctant to admit how much a relationship meant, because they don't want to be laughed at or shunned for "carrying on" about their loss.

Thinking about how people live with animal companions sheds light on the difference in reaction between human and animal death. Particularly in urban and suburban life, pets are an intimate part of everyday life. They take over the couch. They sit under the table hoping someone will slip them a delicious morsel of dinner. They invade the bathroom and stare while their favorite person showers. And at night, they sleep on the bed, maybe under someone's arm. Many people are more relaxed, more themselves with a pet than with any human being.

If this heartache happens to pet lovers who would give up a kidney for their dog or cat, what about other animals? Can they be family too? Are there limits to family feeling? Those with other kinds of house pets, such as parrots, bunnies, bearded dragons, as well as those that live outdoors, like horses and donkeys, also consider them family.

> The relationship [with horses] is deeper, in some ways, than those with our dogs or cats because we communicate with them extremely closely and develop a strong working relationship ... because [her 22-year-old horse] is family to me, we continued to keep him in our show barn- full training and care.
>
> *(Bree Montana, personal communication, 4/10/2022)*

> When [her daughter] was quite little, she would bring the bearded dragons to sit with her in the kitchen ... and she would read to them, or play games... I don't relate to our aquarium fish as family, however.
>
> *(Sharyn Marks, personal communication, 4/15/2022)*

Nature lovers may feel kinship with animals who live on their property. Some say that as soon as one takes responsibility for an animal or names it, it becomes family.

> When there is a critter, that you/spouse/child: feeds, pets, houses, when you spend money to see to its comfort and well-being, when you pay attention to how it is reacting, looking at its behavior—not only is it family, but you will mourn the loss.

Although she feels honor-bound to provide for the animals in her life, family membership is not granted to all of them. "As we run a ranch, we are responsible for hundreds of lives. If we considered them all family, we'd not be able to run our business" (Maria Rivas, personal communication, 4/2022).

Others who get close to animals over time may also feel a family relationship. One group of animal welfare workers the author has gotten to know over several years explained it this way.

> We often have animals with us for months after they have been surrendered or removed as part of legal proceedings. We take care of them, give them behavioral enrichment, treat their medical problems—give them the best life we can while they are with us. They have no one else, so we become their family.
>
> *(personal communication, name withheld)*

Family status is not the only measure of attachment. Dog walkers, people with sunny back porches the neighbor's cat likes to visit, fellow animal lovers at the dog park, the store owner who hands out treats to their favorite canine shoppers, and others who might seem peripheral to a pet may still mourn the loss of a companion animal that is not theirs. In the author's dissertation research, a subgroup of pet lovers who were interviewed at length often included other people's cats and dogs in their social network (Cohen, S. P., 1998).

People often develop a fondness for animals that are not pets at all. On the internet, animal lovers tune into wildlife cams to follow the development of animals, often ones that someone has named, such as a pair of eagles, one of whom was abandoned by his first mate and now has another (National Arboretum Bald Eagle Cam|HDOnTap.com). In addition to familiar wildlife, animal lovers have favorite residents of the local zoo. (Full disclosure, as a child, the author used to visit the original Smokey Bear at the National Zoo and sing him his theme song so that he would feel at home.)

## History of Pet Loss Observance, Counseling, Veterinary Education, and Practice

### *History of Grief for Animals*

Burials of humans with animals date back 14,000 years ago. A sick puppy that had been cared for was buried with an older dog, a man in his 40s, and a woman in her 20s (Geggel, 2018). Ancient Egyptians venerated cats and other special animals, mummifying their remains and carving their images in stone. Archeologists have excavated a 2000-year-old pet cemetery containing cats, dogs, and monkeys, a few with collars and ornaments, suggesting they may have been treated as companions (Osypinska, 2016).

## Acceptance of Pet Loss in Modern Culture

While some human-animal relationships have been memorialized since ancient days, wide acceptance of people's sorrow over the loss of special animals has grown in modern times. A special animal is often treated with the same respect as if it were human by being individually buried or cremated. There may be a memorial service where family and friends gather to pay their respects. Whether they are placed in the backyard or a pet cemetery, pets are often interred with important objects—a favorite toy or blanket. There are greeting cards for pet loss, online discussion groups, and members of the clergy who can provide spiritual guidance to the bereaved and religious observance at the graveside.

Contemporary American culture has often been uncomfortable with death. Older people no longer die at home but in hospitals and nursing homes. As family members live far away from each other, they can't see their grandparents' slow decline or help when Mom breaks her hip. Schools do not teach how to speak to a grieving person, so people make awkward comments that leave the bereaved feeling worse.

Because to outsiders, feelings for non-humans are often seen as less significant, pet lovers are on the receiving end of remarks that devalue the pain they feel, such as "Thank heavens it wasn't one of the children. You can always get another dog". These remarks are another example of disenfranchising the connections humans have with the animals in their lives. Mourners may feel there is something wrong with them or that they must keep quiet to prevent humiliation and perhaps the loss of friends and family who disapprove of their closeness to someone those friends feel is inappropriate.

## *Pet Loss in Fiction*

Although pet loss has been minimized in everyday life, fiction about animal loss has become more common. *Old Yeller* by Fred Gipson, *The Yearling* by Marjorie Kinnan Rawlings, and *Charlotte's Web* by E. B. White center on the death of a pet, a tame deer, and a spider, plus the threat of death to a special pig. The central humans in these stories are children, perhaps because being that attached to an animal seems more acceptable in youngsters who have fewer adult responsibilities. This kind of fiction is designed to tug at the reader's heartstrings, but mature people are expected to outgrow such sentiments. And yet, this is often children's first exposure to loss. In the author's experience, parents often regard these deaths and those of family pets as "practice" for the real thing. This view fails to acknowledge the depth of pain children feel when someone they care for, someone who makes them feel safe is taken away.

One exception to child-oriented pet loss literature is the 13th Century Welsh folktale of Gelert, the brave dog. Like many tales, this story is darker than modern stories. Prince Llywelyn the Great has a dog, Gelert, his favorite hunting companion. One day, Gelert fails to respond to the Prince's call. When the Prince returns home, he finds the cradle of his infant son overturned and Gelert with blood on his muzzle. Seeing no sign of the child, the Prince assumes that Gelert has killed his son and slays him on the spot. As Gelert breathes his last, the Prince hears his son and lifting the cradle, finds the baby unhurt and a huge wolf, whom Gelert has killed, dead. Realizing he has slaughtered his faithful hound and child's defender, the Prince is overcome with grief and buries brave Gelert where all the townspeople will pass by and be inspired (Radeska, 2016).

## Causes of Death and Their Effect on Pet Lovers

### *Euthanasia*

In addition to attachment to companion animals that are as close as other family members, other factors shape how someone will react to pet loss. In the United States, many animals die of euthanasia. The word comes from the Greek and means "good death". By current medical standards, pets are injected by a lethal drug and die very rapidly. These days, it is common for at least one family member to attend the event. Veterinary practices often provide tissues and a comfortable chair for the client. House call practices will send a veterinarian to one's home so that the pet is in familiar surroundings and does not have to be moved. These days, many people have never witnessed a death or seen a family member afterward.

Euthanasia is not an easy decision. Many see pets as children, but even more innocent and helpless (Bulcroft, 1987). When humans get sick, the people around them look for an explanation, perhaps one that proves the other person's bad luck will spare them. "I told Harry he needed to get out of that lounge chair and eat a salad now and then, but would he listen?" In contrast, pets don't smoke and they eat what the people they live with give them. They are not responsible for their own bad health.

The family may balk at "playing God" or feel they are betraying an innocent creature who does not know what is coming. Sometimes there is conflict among family members about whether it is time to intervene. In these moments, it can be helpful to recognize that disease or injury is taking the pet's life. Euthanasia just ensures that the death is gentle and pain-free.

### *Unassisted Death*

Some pet lovers say that they prefer that the dog, cat, or other animal have a "natural" death at home. It can be a relatively peaceful experience for the pet and the family. Unfortunately, some reasons for choosing to have the pet die without assistance are problematic for the patient. Some pet guardians either do not recognize that the pet is in pain or has difficulty breathing. If they do know, they may have no way to bring their pet to a doctor at that moment. Still others are so attached that they cannot bear to separate one minute earlier than is absolutely necessary, even if they know the pet is suffering.

### *Accident*

A third major cause of pet loss is accidental death. Such a death can be traumatic for anyone who sees it happen or feels responsible for it. Dogs run into traffic, cats fall from high-rise buildings, drugs have unexpected side effects. A family's anger at anyone they feel is responsible may either redirect some of their own guilt or keep them stuck in their grief (Cohen & Clark, 2019).

### *Behavior Issues*

While most animal lovers understand death from injury or illness, losing a pet due to behavior issues presents different challenges. As with medical problems, what one family can manage, another cannot. Young animals chew on things, scratch furniture, and wake people who are just trying to get a full night's sleep. Old dogs and cats urinate in the wrong places, get picky about their food, and shed on the carpet. Those are deal-breakers for those who see these animals as unsanitary, suffering, or taking more time and care than the humans in charge can accommodate.

Some animal behavior is more than a nuisance; it is dangerous. The person with an aggressive dog may see it as protection. Others assess the behavior as "not that bad', because "he only bites little dogs and the neighbors". Then, when the dog attacks someone at home, such families excuse it. Like cases of domestic violence, aggressive animals often behave with calm and affection at some times and around certain people. Even if people with problem pets recognize that the animal is unsafe, they may expect rehome it to a place where caretakers love and can handle dangerous animals. They are shocked that no such paradise exists. Many others are heartbroken and feel like failures because they cannot keep their pet from hurting others. "I love him and yet despise the person I feel that I'm turning into (because) of my experience with him" (Buller, 2020).

## *Life Changes, Disappearance, Deliberate Harm*

Behavior problems are not the only non-medical cause for the loss of favored animals. Allergies, financial downturns, loss of housing, and other life changes can force people to surrender animal companions in the hope that someone else will care for them.

As painful as it is to have a loved animal die or be surrendered, at least the human companion knows where it was and perhaps was able to say goodbye. Those who do not know what became of a pet may feel tremendous anxiety and fear for the animal's well-being, as well as guilt for not protecting it. A dog is frightened by thunderstorms or fireworks and bolts from the yard. Cats and birds escape because home repair workers or visiting relatives leave a door open. Pet kidnappings have become more common. In one famous case, two of entertainer Lady Gaga's dogs were stolen in a violent robbery. Her friend and dog walker was shot trying to prevent their abduction. (Dillon, 2022). If the animals are not recovered, pet lovers have no closure and may worry about the pet and search for it for years.

A treasured animal may also be lost through deliberate harm. A neighbor who objects to a cat digging in their yard or to a dog who barks too much may lay out poison to eliminate the problem. (Blum, 2013). Even more common, pets are threatened and killed by domestic abusers as a way of punishing or controlling people in the household. This topic is covered in other chapters, but for now, consider being trapped in frightening relationship because even if those being abused can find somewhere to hide, pets are not allowed there. The intention of pet loss from domestic abuse is not so much to torture the animal but to hurt and control the humans who are attached to it.

## *Lack of Resources*

Underlying many animal deaths is a lack of resources to treat illness and accidents. Financially secure animal lovers in big cities of developed countries have treatment options for sick pets that are unavailable to others. Specialty practices offer chemotherapy for cats with cancer, total hip replacements for big dogs, and advanced diagnostic tests for a bearded dragon to get to the bottom of complex medical issues.

If advanced treatment is even available in the local community, these procedures are expensive and may require more care afterward than a family can manage (Mithers, 2021). Pet insurance may cover some of these expenses, but less than 2% of American pet households have it (Spots.com, 2022). Many rural and low-income communities are not served by even basic veterinary care, much less treatments for cancer and cataracts. If the animal's caretaker does not have a car or access to public transportation, a veterinary clinic that is more than walking distance may as well not exist.

Lack of resources propel families into moral dilemmas. Accidents force the family to come up with large amounts of money with no preparation, no chance to buy time or save up. Even routine care and illnesses can cost money the family cannot afford. Guilt can be excruciating, leaving clients left to think, "If only I were a millionaire, my loved one would live".

## Reasons for Pet Loss and Their Effects on the Veterinary Team

When death comes to an animal that is in pain, cannot breathe, eliminate, or walk effectively, the sorrow of loss is tempered by a relief of suffering. Medical conditions that could be treated but are not cause significant unhappiness to those in animal health and welfare workers. A situation that is manageable to some families is not to others. For example, chronic illnesses like diabetes require frequent blood or urine tests at home, money for insulin, and injections done by the animal's caretakers. If there is no one at home to do these tasks or if that person cannot physically or emotionally provide necessary care, the pet is in danger of dying or being euthanized. Medical teams train to treat these conditions, but their hands are tied because the animal in question belongs to someone else and under the law, is considered property (Francione, 1996).

Being unable to help an animal because clients do not have the financial and emotional resources to treat it causes the veterinary or animal welfare team moral distress. (Rollin, 1988). This is a condition where the choices and conflicts one faces causes anguish, often because the person involved is forced to go against deeply held beliefs. For example, clients may demand a "convenience euthanasia" for non-medical reasons. While many veterinarians refuse, others may feel humane euthanasia is better for the pet than being dumped on a highway.

Animal health and welfare teams become attached to clients and patients over time. Veterinarians, who often make modest incomes compared to physicians, may be tempted to lower fees to enable their client to provide at least some of the recommended care. This sacrifice further damages the finances of the practice and may leave the team working longer hours to make up the difference, while they feel anger at themselves for agreeing to yet another low-cost procedure.

Other interactions with clients about loss also perturb the veterinary team. As pet lovers attempt to reach important decisions, they often turn to members of the clinic staff, from receptionist to surgeon. "What would you do if this were your pet?" These days, veterinarians and veterinary technicians have received at least minimal education on how to help clients make choices they can live with. Many veterinarians and others are skillful at having these conversations, able to show compassion and help clients consider their options.

Nevertheless, performing a euthanasia, staying calm as others sob or rage takes a toll on staff. Many vet team members thought they were picking a profession to help animals, not to get involved in difficult human emotions. In fact, they may be experiencing difficult emotions of their own. The entire staff may know the pet and its human caregivers well, know their struggles and joys. They have their own attachment and sense of loss. The situation may also be a life-threatening emergency, where everyone's adrenaline is on full blast. Whether the team has worked with the pet for five minutes or 15 years, many professionals say that they go home at night emotionally drained from human interaction.

Susan P. Cohen

# Other Kinds of Animal Loss—Service and Therapy, Farming, Research, Wildlife, Zoo, Roadkill

So far, this chapter has described grief from the loss of animals that live with or near people as companions, ones humans feel a personal connection to, ones with a name even when they do not belong to an individual. There are other, more complicated relationships that cause people grief. Readers can learn the definitions of service and support animals in other chapters. In this section, consider all that the person stands to lose when the interaction with these animals ends.

## *Service and Therapy Animals*

Working animals, including trained service dogs and mini horses, are often beloved by their human partners, but have specific tasks to perform to help people manage work or everyday activities. Without them, a blind person may have difficulty navigating city streets. A person with a seizure disorder could suffer head injuries from falls. A veteran with PTSD might live with the lingering fear that someone is sneaking up behind them.

When these service animals become unable to do their job, human caretakers have a difficult decision to make: Keep the animal as a pet, find another place for it to live, or end its life (Vogel, 2014). Keeping that dog as a pet while also maintaining a working dog can be too expensive for a person living on disability benefits. Unless the service animal training program will take it back or someone will adopt this old, possibly sick dog, euthanasia of a beloved friend and partner may be the only choice.

## *Therapy Animals*

True animal-assisted therapy partners are trained to help a person with physical, mental, or emotional challenges recover or at least improve. Facility or comfort dogs have specialized training to support people going through stressful events, such as children who are testifying in court against an abuser. Patients/clients, handlers, and staff at facilities often develop a warm feeling for these animals. Losing this contact because the animal retires or dies or because the human leaves the program can be painful, even devastating (Cohen, S. P., 2015; Kogan, 2021).

## *Emotional Support Animals*

A third, controversial group of helping companions are Emotional Support Animals (ESAs). These dogs, cats, bunnies, pigs, parrots, and peacocks often have no formal training or legal certification to provide comfort to an anxious human, but for many people, ESAs serve a real purpose. They give people who suffer from conditions like anxiety or agoraphobia the ability to leave the house, drive a car, or get on an airplane. In addition, they may also need to identify their companions as ESA or service animal to get housing or other services that accept working animals only.

## *Wildlife*

Up till now, this chapter has explored relationships with animals with whom one lives or has frequent contact. Humans also become attached to animals that live on their own whether in a backyard or thousands of miles away. "I've raised many wild animals and turned them loose when they could survive on their own. I truly cared about each one of them and felt sad when they were released, but also glad they were free". (Crooks, K., 2022, personal communication). A veterinary technician wrote:

> Creatures absolutely don't have to be family for me to grieve them. I am heartsick for all the animals that must have been terrified and/or died as a result of wildfires in our Province last year… I feel terrible that some of my mason bees didn't survive the recent cold nights, feel I should have protected them.
>
> *(Kristen @BCVetTech, personal communication)*

Thanks to the internet, certain wild animals become famous and have their own fan base. When these animals die, either because of natural causes or hunting, the sorrow is enormous and the rage at someone who deliberately took that life has real consequences for the shooter. In one famous case, Cecil, a lion with a magnificent mane who posed for photographs, seems to have been lured out of his protected sanctuary so that he could be shot with a bow and arrow by an American dentist. The backlash was so strong, the dentist was forced to close his practice (Macdonald, 2016).

Recently, a 50-year-old elephant with enormous tusks, a favorite of guides and tourists, was shot by a man who paid a hefty sum for the privilege. The hunter was dumbfounded by the response to the killing.

> [Leon] Kachelhoffer laughed and said: '[Controversy] wasn't one of the things I was thinking about. To be in a position to hunt a bull like that, it's an incredible privilege. When you take a bull like that, there's a lot of remorse, there's a lot of sadness, you think about the great life that this elephant has led'.
>
> *(Solomons, 2022)*

Apparently, for this man, remorse is not a strong enough reason not to shoot the animal in the first place.

## *Animals in Zoos*

Wild animals that are in captivity in zoos and parks also have humans who feel strongly about them. While visitors may go back occasionally to see certain animals, caretakers are with them every day. They get to know the behavior of each one, monitor its health, and celebrate its birthday with special food. Even if there is a policy against naming them to reinforce that they are not pets, attachment is inevitable. Euthanasias are done with respect for both the animal and those who have cared for it. Keepers are often given a chance for a final visit before the medical staff induces death.

In addition to regular residents, some zoos are part of special breeding and conservation programs to protect rare species at risk in the wild. The loss of one of these animals, often hand-raised by specialists, is a loss to the world. In contrast, zoos can find themselves with a surplus of certain animals, such as lions and giraffes. If they are genetically like other animals in their zoo, authorities quietly "cull" the excess that cannot be placed in another zoo. Although it is a regular practice, these deaths can be deeply distressing to zoo staff and the public (Birke, 2019; Goldman, 2014; Powell, 2016; Szita & Kotch-Jantzer, 1988).

Sadly, some zoo deaths are caused by the misbehavior of human beings. In one well-reported situation, a gorilla was shot by keepers who loved him because a child entered the gorilla's enclosure. Harambe was a 17-year-old lowland gorilla, the product of careful breeding and intensive care at two different zoos. It was hoped that he would help replenish this vanishing member of the great apes. When a three-year-old boy entered the enclosure, staff felt compelled to shoot their prize gorilla in front of horrified onlookers. The staff of both zoos required support (McPhate, 2016), and the man who had hand raised him said, "An old man can cry, too.. Harambe was my heart. It's like losing a member of the family" (Bult, 2016).

## Farming

For many reasons, the killing and eating of animals has declined in recent years. A 2018 Gallup poll reported that about 5% of Americans consider themselves vegetarian (Hyrnowski, 2022). The number who identify as vegetarian or vegan increases in younger respondents (Bourassa, 2021). For those who avoid animal products, these deaths cause both sorrow and anger.

## Research

Another controversial arena for loss involves animals used in research and education. Many life-saving drugs and procedures were pioneered on everything from Zebra fish to chimpanzees. Even if patients are aware of the connection between the injury and death, these developments cause lab animals, they may feel the benefit to human beings outweighs harm to the animals who made treatments possible. Lab animal workers who must sometimes make healthy animals sick and then euthanize them after the experiment is over often suffer significant moral stress, particularly when the research animals involved are mammals (Rollin, 1988). Rollin tells of a psychologist who suggests that workers who have to perform mass euthanasia just imagine they are in some lovely vacation spot. This tone-deaf advice ignores the underlying issues of how animals are used and discarded, as well as the anguish such tasks cause those who must perform them.

## Climate Change

Another growing trend in grief stems from climate change. Unlike the sorrow for specific wild animals, this is a more global fear that entire species may be wiped out by changes in temperature, lack of water and food supply, and the destruction of living space through competition with human needs. This fear, "eco anxiety", has become a serious source of stress, especially to teenagers and young adults who expect to see natural disasters wreak havoc on the globe (Thompson, 2022).

## Roadkill

A last category of animals whose deaths spark sorrow is one most people would never consider: Animals that die outdoors, either through accident or natural causes. One woman has decided to memorialize each body she encounters with a ceremonial display. Dr. Amanda Stronza, anthropologist, professor, and conservationist surrounds the animal with flowers, leaves, stones, and other found objects. She explains, "I cannot bear to leave them... My intention is to notice and give attention, to honor and see the animals as individuals ... 'I'm sorry' is what I say to each animal" (Stronza, A., March 17, 2022).

## Research Contributions to Understanding Pet Loss

Despite the number of programs, books, and other supports for pet loss, there has been limited rigorous scholarly research. Cindy Adams surveyed 177 people who had lost a pet within the last six weeks and found 30% suffered severe grief (Adams, 2000). A study of 409 people whose pets had been euthanized in the last year found attachment to a pet was strongly related to sorrow, anger, and guilt. Sudden death was related to anger, but death from cancer was not (Barnard-Nguyen, 2016). A survey of callers to a pet loss hotline determined that negative interactions with veterinarians and a lack of social support complicated the grief from loss of an animal companion (Remillard, 2017).

Attachment theory informs several recent studies. Grief is made worse by anxious attachment, stemming from parental neglect or inability to meet the child's needs (Field, 2009). This history fits with many who say their relationships with pets and other animals are the most unconditional, most healing that they have ever experienced. As one client put it, "Sometimes, people need their space. My dogs never need their space" (Cohen, 1998, p. 154). A survey of 429 people whose pets had recently died found two attachment styles predicted severe grief, depression, and anxiety—anxious attachment and avoidant attachment (King, 2011–2012). Avoidant attachment is believed to stem from parents who are emotionally unavailable. Their children learn to suppress emotions unwelcome to caregivers. Animals often seem more forgiving of human emotion and behavior.

Finally, some research has explored what leads to more positive outcomes. In the King study above, social support mitigated grief. A 2017 survey reported that many bereaved pet lovers experience post- traumatic growth, such as an enhanced appreciation of life and personal strength (Packman, W., 2017). A later study found that while post traumatic growth after pet loss did occur, disenfranchised grief inhibited it (Spain, O'Dwyer, & Moston, 2019).

## Help for Pet Loss

If the loss of special animals is this painful, what services can a grieving person tap into? This next section will address how to help a grieving person, what help is available in the community, and when a referral to a pet loss counselor is appropriate.

### *Pet Loss Services*

Formalized aftercare for animal companions in the United States began with the first dedicated pet cemetery. Founded in 1896, Hartsdale Pet Cemetery has more than 80,000 burials of pets belonging to ordinary citizens and the famous. In addition to dogs and cats, Hartsdale is the final resting place of, among other things, birds, ferrets, rabbits, horses, monkeys, and Goldfleck the lion, who lived with a Hungarian princess in the Plaza Hotel until his death in 1912. Pet cemeteries and crematories provide a variety of services, such as funerals including final viewings and memorial ceremonies. They sell various containers for the remains and such accessories as cremation jewelry, so that mourners can keep a tiny amount of ashes in a locket or keychain.

In addition to the support of friends and family, there are professional bereavement services for clients coping with pet illness and death. In the late 1970s, the Dean of the School of Social Work at the University of Pennsylvania, Louise Shoemaker, and later, Eleanor Ryder and PhD student, Jamie Quackenbush, began working with the School of Veterinary Medicine to help clients. (Ryder, E., 1981). Since then, most schools of veterinary medicine and some large private practices have hired a mental health practitioner to help clients make tough decisions and support them during medical procedures, including euthanasia. They often provide ongoing support after the death of the pet.

For example, Pet Loss Groups, pioneered at The Schwartzman Animal Medical Center in New York City, allow pet lovers to tell the story of their pet's illness and death, with all the accompanying self-doubt, and pain, to an audience that understands and will not derail the conversation with toxic cheer. Groups allow people to recognize that other people feel this way too. Having someone to share the experience is important because losing a pet ends their connection to others, like friends at the dog park. Finally, listening to other people berate themselves for things they did or did not do can be a revelation. "Ms. X tried everything. If she doesn't need to blame herself, maybe I can ease up on myself too".

## When to Seek Mental Health or Other Professional Care

After any loss, it is normal for a grieving person to cry, withdraw from social events, and dwell on the loved one. Deciding whether to seek professional care depends on how painful the feelings are for the person and how long this distress has lasted. If the mourner has suicidal thoughts even on Day 1, seek help immediately from a counselor who understands grief and who respects the human-animal bond.

Another time to seek help is when the person involved is stuck for a long time. It can help to speak to their veterinarian, to review the medical record, to remember how sick the pet was and all the treatments, surgeries, and medications the medical team and the family did to keep the patient alive.

Coming to terms with loss is not a straight uphill climb. There will be many ups and downs. Over time, most people make their way to a place where they have peace about what happened, knowing that life will always be different now. Grief can make a return visit even years later if an anniversary, a photo, or a fresh loss reminds someone of their loved one.

The Diagnostic and Statistical Manual of Mental Disorders (DSM-5-TR) that guides mental health practitioners has added "prolonged grief disorder", also called "complicated grief", to its list of mental health issues. Some of the criteria for this diagnosis are inconsolable grief lasting longer than a year, disbelief about the death and avoiding reminders of it, and feeling that part of oneself has died (American Psychiatric Association, 2022). In general, any feeling that causes distress, that persists and that interferes with the rest of one's life would benefit from consultation with a mental health practitioner.

One last thought about the human connection to other animals and the pain of losing them: As with any great change, one can emerge with new insights and personal growth. Losing a meaningful relationship can inspire people to value the ones they have left. Caring for a sick loved one teaches patience, compassion, and perhaps a bit of science. Advocating for a loved one in need not only secures better care for the patient, but can help the person feel strong in the moment and proud at the end. As Elizabeth Kubler-Ross wrote in her final book,

> You will not 'get over' the loss of a loved one; you will learn to live with it. You will heal and you will rebuild yourself around the loss you have suffered. You will be whole again, but you will never be the same. Nor should you be the same nor would you want to.

*(Kubler-Ross, 2005, p. 230)*

## Questions for Discussion

1. If selling a permit to shoot a large and popular wild animal brings needed revenue to a community, is it wrong?
2. Is attachment to an animal ever problematic?
3. Discuss the differences in grief response to pets vs. animals used for work or consumption.
4. What grief support is missing in the world that needs creation?
5. If pets are members of the family, should they be on our health insurance?

## References

Adams, C. L. (2000). Predictors of owner response to companion animal death in 177 clients from 14 practices in Ontario. *Journal of the American Veterinary Medical Association, 217*, 1303–1309.

American Psychiatric Association. (2022). *Prolonged grief disorder.* Retrieved from Psychiatry.org: https://www.psychiatry.org/patients-families/prolonged-grief-disorder

Barnard-Nguyen, S. B. (2016). Pet loss and grief: Identifying at-risk pet owners during the euthanasia process. *Anthrozoos, 29*(3), 421–430. https://doi.org/10.1080/08927936.1181362

Beck, A. (2014, July). The biology of the human-animal bond. *Animal Frontiers, 4*(3), 32–36.

Birke, L. H. (2019). You can't really hug a tiger: Zookeepers and their bonds with animals. *Anthrozoos, 32*(5), 597–612. https://doi.org/10.1080/08927936.2019

Blum, D. (2013, March 11). *Your neighborhood pet poisoner.* Retrieved March 11, 2022, from Wired: https://www.wired.com/2013/03/your-neighborhood-pet-poisoner/

Bourassa, L. (2021, January). *Vegan and plant-based diet statistics for 2022.* Retrieved March 2022, from Plantproteins.co: https://www.plantproteins.co/vegan-plant-based-diet-statistics/#:~:text=%20Vegan%20and%20Plant-Based%20Diet%20Statistics%20for%202022,One%20of%20the%20core%20reasons%20why...%20More%20

Bulcroft, K. (1987). Similarities and differences between the role of pets and children in the American family. *Abstracts of Presentations, Delta Society 6th Annual Conference.*

Buller, K. (2020). Living with and loving a pet with behavioral problems: Pet owners' experiences. *Journal of Veterinary Behavior, 37*, 41–47. Retrieved March 2022, from https://www.sciencedirect.com/science/article/pii/S1558787820300356

Bult, L. (2016, May 29). *Former caretaker mourns death of Harambe, gorilla killed after toddler fell into his cage; family thanks Cincinnati Zoo for 'quick action'.* Retrieved 2019, from http://www.nydailynews.com/news/national/caretaker-gorilla-killed-cincinnati-zoo-mourns-loss-article-1.2653971

Cohen, S. P. (1998). *The role of pets in some urban American families.* New York: Columbia University through UMI Dissertation Services.

Cohen, S. P. (2002). Can pets function as family members. *Western Journal of Nursing Research, 24*, 621–38.

Cohen, S. P. (2015). Loss of a therapy animal. In A. H. (Ed.), *Handbook on animal-assisted therapy: Foundations and guidelines for animal-assisted interventions* (4th ed., pp. 341–356). London: Academic Press.

Cohen, S. P., & Clark, A. (2019). The importance of human-animal relationships in the context of loss, grief, and bereavement. In P. T. (Ed.), *Transforming trauma: Resilience and healing through our Connections with animals* (pp. 395–422). West Lafayette, IN: Purdue University Press.

Dillon, N. (2022, March 24). *Lady Gaga's dog walker gives harrowing grand jury testimony, unsealed transcript reveals.* New York. Retrieved March 26, 2022, from https://www.rollingstone.com/music/music-news/lady-gaga-dog-walker-grand-jury-transcript-1326528/

Field, N. P. (2009). Role of attachment in response to pet loss. *Death Studies, 33*, 334–355. https://doi.org/10.1080/07481180802705783

Francione, G. L. (1996). *Animals as property.* Retrieved March 27, 2022, from Animal Legal & Historical Center: https://www.animallaw.info/article/animals-property

Friedmann, E. K. A. (1983). Social interaction and blood pressure: Influence of animal companions. *The Journal of Nervous and Mental Disease, 171*, 461–464.

Geggel, L. (2018, February 9). *Humans cared for sick puppies long ago, ancient burial shows.* Retrieved January 23, 2022, from Live Science: https://www.livescience.com/61717-oldest-dog-burial.html

Goldman, J. G. (2014, February 14). *Ethics at the zoo: The case of Marius the giraffe.* Retrieved 2022, from Scientific American: https://blogs.scientificamerican.com/thoughtful-animal/ethics-at-the-zoo-the-case-of-marius-the-giraffe/?msclkid=3604a8c8c5fc11ec9fc6a37d76c04dad

Gray, E. (2022, August 2). *US pet ownership statistics to know in 2022.* Retrieved 2022, from Pet Keen: https://petkeen.com/pet-ownership-statistics/

Hyrnowski, Z. (2022). *What percentage of Americans are vegetarian?* Retrieved 2022, from Gallop Polls: https://news.gallup.com/poll/267074/percentage-americans-vegetarian.aspx?msclkid=456acfa6c6c711ec83f40be4139c848f

King, L. C. (2011–2012). Attachment, social support, and responses following the death of a companion animal. *Omega, 64*(2), 119–141. Retrieved 2022, from https://journals.sagepub.com/doi/10.1177/10541373211054168

Kogan, L. R.-M. (2021, November 18). The loss of a service dog through death or retirement: Experiences and impact on partners. *Illness, Crisis, & Loss. 31*(2), 244–270. https://doi.org/10.1177/10541373211054168

Kubler-Ross, E. K. (2005). *On grief and grieving: Finding the meaning of grief through the five stages of loss.* New York: Simon & Schuster.

Macdonald, D. W. (2016, April 25). Cecil: A moment or a movement? Analysis of media coverage of the death of a lion, Panthera leo. *Animals, 6*(5), 26. Retrieved November 19, 2016, from https://doi.org/10.3390/ani6050026

McPhate, M. (2016, May 30). Zoo's killing of gorilla holding a boy prompts outrage. *The New York Times.* Retrieved from http://www.nytimes.com/2016/05/31/us/zoos-killing-of-gorilla-holding-a-boy-prompts-outrage.html

Mithers, C. (2021, November 12). *Most Americans have pets, almost one third can't afford their vet care.* Retrieved March 28, 2022, from Talk Poverty: https://talkpoverty.org/2021/11/12/low-income-veterinary-care-affordability/

Nagasawa, M. K. (2009, March). Dog's gaze at its owner increases owner's urinary oxytocin during social interaction. *Hormones and Behavior, 55*(3), 434–441. https://doi.org/10.1016/j.yhbeh.2008.12.002

Osypinska, M. (2016). *Pet cats at the early Roman red sea port of Berenike.* Egypt. Antiquity, *90* (354), ES. https://doi.org/10.15184/aqy.2016.181

Packman W, B. C. (2017, September). Posttraumatic growth following the loss of a pet. *Omega, 75*(4), 337–359. https://doi.org/10.1177/0030222816663411

Powell, D. M. (2016, March 2). Survey of U.S. zoo and aquarium animal care staff attitudes regarding humane euthanasia for population management. *Zoo Biology, 35*(3), 187–200. Retrieved March 2, 2022.

Radeska, T. (2016, December 24). *Gelert: The brave dog from Welsh folklore who saved the life of his master's son.* Retrieved March 2, 2022, from The Vintage News: https://www.thevintagenews.com/2016/12/24/gelert-the-brave-dog-from-welsh-folklore-who-saved-the-life-of-his-masters-son/?edg-c=1

Rémillard, L. M. (2017). Exploring the grief experience among callers to a pet loss support hotline. *Anthrozoos, 30*(1), 149–161. https://doi.org/10.1080/08927936.2017.1270600

Rollin, B. (1988). Animal euthanasia and moral stress. In W. J. Kay, S. P. Cohen, C. E. Fudin, A. H. Kutscher, H. A. Nieburg, R. E. Grey & M. M. Osman (Eds.), *Euthanasia of the companion animal: The impact on pet owners, veterinarians, and society* (pp. 31–41). Philadelphia, PA: The Charles Press, Publishers.

Ryder, E. (1981). Establishing a social work service in a veterinary hospital. In A. B. Fogle & Edney (Eds.), *Interrelations between people and pets* (pp. 209–220). Springfield, IL: Charles C. Thomas, Publisher. Retrieved 2022.

Solomons, A. (2022, April 19). *Trophy hunter kills Botswana's biggest 'tusker' elephant, causing fury.* Retrieved 2022, from MailOnline Daily Mail: https://www.msn.com/en-us/news/world/trophy-hunter-kills-botswana-s-biggest-tusker-elephant-causing-fury/ar-AAWnwUB?ocid=BingHp01&cvid=f2729fac1b8f4a5bf3e9f8669cd357e2

Spain, B. O. (2019). Pet loss: Understanding disenfranchised grief, memorial use, and posttraumatic growth. *Anthrozoos, 32*(4), 555–568. https://doi.org/10.1080/08927936.2019.1621545

Spots. (2022, March 25). *Pet ownership statistics [2022]: U.S pet population.* Retrieved 2022, from Spots.com: https://spots.com/pet-ownership-statistics/

Stronza, A. (2022). *Dr. Amanda Stronza (@amandastronza) • Instagram*. Retrieved 2022, from Instagram photos and videos: https://www.instagram.com/amandastronza/

Szita, B., & Kotch-Jantzer, C. (1988). Euthanasia in zoos: Issues of attachment and separation. In S. C. W. Kay (Ed.), *Euthanasia of the companion animal: The impact on pet owners, veterinarians, and society* (pp. 148–163). Philadelphia, PA: The Charles Press.

Thompson R., Fisher, H. L., Dewa, L. H., Hussain, T., Kabba, Z., & Toledano, M. B. (2022). Adolescents' thoughts and feelings about the local and global environment: A qualitative interview study. *Child and Adolescent Mental Health, 27*(1), 4–13. https://doi.org/10.1111/camh.12520

Vogel, B. (2014, November 1). *Grieving the loss of a service dog*. Retrieved 2022, from New Mobility: Life beyond wheels: http://www.newmobility.com/2014/11/grieving-service-dog/

Webb, J. (1973). All I Know. Retrieved from https://www.jimmywebb.com/lyrics

# 22
# ANIMAL HOARDING
## The Human-Animal Bond Gone Awry

*Phil Arkow*

## Introduction and Definitions

The recognition of animal hoarding as a distinct psychological concern distinguished from traditional animal ownership and necessitating multidisciplinary interventions was first described by Worth & Beck (1981) as "multiple animal ownership" and later as "animal collectors" by Lockwood and Cassidy (1988) who called the keeping of excessive numbers of animals, often in poorly managed conditions of animal husbandry, "killing with kindness." Patronek (1999) introduced the term "animal hoarding" and brought what was often thought to be purely an animal welfare concern into the public health arena by recognizing the phenomenon as a potential sentinel for mental health problems, which merit serious assessment and prompt intervention.

The specific number of animals hoarded is not as determinative as how this number overwhelms the ability of the owner to provide acceptable care (Patronek, 1999). The phenomenon occurs worldwide (e.g., Arnold et al., 2018; Calvo et al., 2014; Joffe et al., 2014; Ockenden et al., 2014; Ramos et al., 2013; Reinisch, 2009). Humane and animal control investigators frequently find hoarders' emotional attachments to animals to be so strong that the bodies of deceased pets are kept in as many as 59.3% of cases, either in the residence or a freezer (Patronek, 1999). Patronek & Nathanson (2009) speculated that hoarders' obsession over maintaining control over their possessions creates even greater anxiety when the animals' mortality represents impermanence, leading to extreme control and the failure to acknowledge death or dispose of the dead bodies.

Patronek, Loar & Nathanson (2006) of the Hoarding of Animals Research Consortium (HARC) defined four characteristics of the animal hoarding phenomenon:

- **Failure** to provide minimal standards of sanitation, space, nutrition, and veterinary care for the animals;
- **Inability** to recognize the effects of this failure on the welfare of the animals, the human members of the household and the environment;
- **Obsessive** attempts to accumulate or maintain a collection of animals in the face of progressively deteriorating conditions; and
- **Denial** or minimization of problems and living conditions for people and animals.

The terms "hoarders" and "collectors" are sometimes confused interchangeably. Hoarding is a diagnostic condition so assessment and treatment may be billable with healthcare insurances, whereas collecting is not. Distinctions must also be made between hobbyists and breeders who maintain large colonies of companion animals in stable, healthy environments, and animal hoarders who represent the extreme end of a pet-keeping continuum which can range from mild to severe to pathological (Ramos et al., 2013).

It was not until the 5th edition of the Diagnostic & Statistical Manual of Mental Disorders (DSM-5) (2013) that hoarding disorder became an official diagnosis. While DSM-5 criteria do not specify the types of possessions that are necessary for this diagnosis, because animals are legally considered property the hoarding of animals would appear to qualify (Frost et al., 2015). However, this question is not resolved in the psychiatric literature.

There may be commonalities among animal hoarders and object hoarders. Object hoarding has been described as the acquisition of and failure to discard a large number of possessions which appear to be useless or of limited value, but which provide the hoarder with a sense of security, identity, emotional comfort, and an extension of self, and the cluttering of living spaces which significantly interferes with functioning and activities of daily living (Frost & Hartl, 1996). Similar patterns of intense, compulsive acquisition, difficulty or distress in discarding, and disorganization of living spaces are commonly found among animal hoarders (Steketee et al., 2011). Both object and animal hoarders exhibit hypersentimentality over their possessions and may attribute human-like qualities to them, valuing them like, or even more than, family members or friends. These characteristics can be magnified when applied to a living creature with the capacity for reaction (Patronek & Nathanson, 2009). While case reports indicate that many animal hoarders also accumulate inanimate objects (Steketee et al., 2011), this may reflect the debris associated with keeping a large number of animals without adequate cleaning (Patronek, 1999; Worth & Beck, 1981). Both object and animal hoarders may exhibit multiple psychological co-morbidities (Arluke & Patronek, 2016).

There are, however, notable differences between object and animal hoarding. Animal hoarding is said to be more severe (Arluke & Patronek, 2016). The potential for animals to return affection and reflect back emotions creates a profoundly different and reciprocal interaction that is both engaging and free of conflict (Patronek & Nathanson, 2009). Rasmussen et al. (2014) reported that most animal hoarding cases involved squalid living conditions, whereas only a minority of object hoarding cases did. Men hoard objects about as equally as women, but animal hoarders are predominantly female (Steketee & Frost, 2014). These differences are significant enough to warrant considering animal hoarding as a distinct subtype of hoarding disorder (Ferreira et al., 2017; Frost, Patronek & Rosenfield, 2011).

## Who Are Animal Hoarders?

Studies consistently report a disproportionate over-representation of women, many of whom appear to be older, socially isolated and/or estranged from family. Patronek (1999) reported 76% of animal hoarders studied were female; 46.3% were aged 60 or older; 72.2% were single, divorced or widowed, with 55.6% living in single-person households. Other studies have reported 70% to be unmarried women with cats (Worth & Beck, 1981) and 83% to be women, with a median age of 55, three-quarters of whom were single, widowed or divorced (HARC, 2002). An Australian review of animal hoarders (Joffe et al., 2014) reported 72.4% were female and 79% were 40–64 years of age. A Brazilian study reported 73% were female, with an average age of 61 (Ferreira et al., 2017). The reasons for this marked gender disparity are unknown (Patronek & Nathanson, 2009).

Animal hoarding cuts across all socioeconomic strata. Many animal hoarders live with dependents and others who are similarly at risk of neglect and deleterious environmental health conditions. Many are successful professionals; others represent a broad spectrum of socioeconomically disadvantaged individuals (Arluke & Patronek, 2016).

The Hoarding of Animals Research Consortium (Patronek, Loar & Nathanson, 2006) suggested three types of animal hoarders:

- **Overwhelmed** caregivers own large numbers of animals that were reasonably well cared for until a change in circumstances, such as death of a spouse, loss of a job or failing health, impairs the ability to provide proper care and living conditions deteriorate. Acquisition of animals tends to be passive, with new animals resulting from breeding among the existing population. The situation is exacerbated if they develop a community reputation where people can drop off unwanted pets. They may have some awareness of their animal care problems and minimize rather than deny them. Isolation may factor into their reluctance to seek help. When confronted by authorities, they are more likely to comply with intervention than do rescuers or exploiters.
- **Rescue** hoarders are a result, in part, of the emergence of a "no-kill" movement in animal sheltering in which humane euthanasia is no longer seen as a viable alternative for unwanted, problematic or behaviorally aggressive animals. This has instigated many self-styled "rescue" groups who harbor unwanted animals indefinitely rather than adopt them into new homes with standards normally associated with more traditional "open-admission" shelters. Accordingly, some animal hoarders have a messianic sense of mission to save animals from a presumed threat of euthanasia and believe they are the only ones capable of caring for them, a condition that has been described as "Noah syndrome," a variant of Diogenes syndrome with presenting factors of psychosocial stress and loneliness, although medical conditions may also underlie the psychopathology (Saldarriaga-Cantillo & Nieto, 2015). Their acquisition tends to be more active, often searching for animals perceived to be in need of rescue. They tend to avoid authorities and outside influence. They may masquerade as an organized institution or legitimate sanctuary for hundreds of animals, even soliciting charitable donations.
- **Exploiters** exhibit sociopathic characteristics. They acquire animals to serve their own needs with little true attachment to them, appear indifferent to the animals' suffering, and lack empathy for humans and animals. They display extreme denial of any problems and flatly reject authority or outsiders' legitimate concerns. They believe their knowledge is superior to anyone else's and display an extreme need to exert control over their animals. Exploiters display many characteristics of antisocial personality disorder, including superficial charm and charisma; they are manipulative and cunning, narcissistic, and appear to lack guilt or remorse. These cases are the most serious and difficult to resolve. Prosecution, rather than educational or behavioral interventions, appears to be the most viable solution.

In addition to these three types, we might suggest a fourth category, broadly described as individuals with various mental health/psychopathology issues. An acronym for these four types – Mental health, Overwhelmed, Rescue, Exploiter – reveals their commonality, namely the desire to continually accumulate "MORE" animals.

The Hoarding of Animals Research Consortium (Patronek, Loar & Nathanson, 2006) identified factors, ranging from mild to severe, among types of animal hoarding, including:

- Presence of medical and/or psychological problems.
- Co-morbid conditions.
- Social integration and social skills.
- Jeopardy of the animals.
- Level of compulsion.
- Active or passive acquisition of animals.
- Degree of empathy.
- Attachment to humans and/or animals.
- Denial.
- Control.
- Response to authority figures.

## What Animals Are Hoarded?

While studies consistently describe the overwhelming majority of animals to be cats and dogs, other species are also represented. Patronek (1999) reported cats were involved in 65% of cases, dogs in 60%, farm animals in 11%, and birds in 11%; many cases involve multiple species. Steketee & Gibson et al. (2011) reported cats to be the most frequently collected animals, although dogs, horses, sheep, goats, reptiles, birds, rabbits, rodents, and wildlife were also represented. Berry et al. (2005) noted that 46% of prosecuted cases involved dogs and 34% involved cats; they speculated that dogs are more likely to attract prosecutorial attention, especially if dead animals are found.

Hoarded animals may be acquired through unplanned breeding, advertising, or picking up strays, rather than from intentional breeding. (Patronek, 1999). The typical number of animals range from ten to more than 900, with women more likely than men to have more than 100 animals (HARC, 2002).

## Animal Hoarding's Impact on Animal Welfare

In terms of the number of animals affected and the degree and duration of their suffering, hoarding has been called "one of the most egregious forms of animal cruelty" and "the number one animal cruelty crisis facing companion animals." An estimated 250,000 animals are victimized annually and suffer horribly, with misery that can endure for years. While often dismissed as a harmless eccentricity, many hoarded animals experience severe malnutrition, starvation, untreated medical conditions, open sores, advanced dental and eye diseases, severe psychological distress, lack of grooming, neglect, exposure to adverse weather and housing conditions, and possibly death (Animal Legal Defense Fund, 2021; Arluke & Patronek, 2016). They may be deprived of socialization and live covered in their own waste, often living among (and sometimes feeding on) the remains of dead animals (Hayes, 2010). Due to poor health, poor prognoses for recovery, widespread contagious diseases, and the large numbers of animals involved, euthanasia is often the only solution for many animals. Hoarding cases transcend incidents of individual animal neglect due to the multiplicative consequences of collective neglect in which severe crowding, increased noise, aggressive interactions among animals, restrictions of movement, entrenched behavioral problems, and the risk of disease compound the suffering. The whole is indeed much worse than the sum of its parts (Arluke & Patronek, 2016). Former prosecutor Diane Balkin has called animal hoarding "an animal cruelty case on steroids" (Schroeder, 2018).

A Canadian study examining 371 cats from "institutional" (i.e., designated rescues or shelters) and non-institutional hoarders masquerading as legitimate shelter operations reported widespread upper respiratory infections and other diseases associated with malnutrition, stress, overcrowding, and general neglect (Jacobson et al., 2020).

One of the most perplexing and contradictory aspects of animal hoarding is the paradox that while hoarders profess love and a desire to rescue and care for animals, tremendous animal suffering and neglect often result and are invariably ignored, minimized or denied. This indifference to the squalor and suffering may indicate a profound lack of insight and even dissociative behavior, which threatens the hoarder's self-image (Patronek & Nathanson, 2009).

## Animal Hoarding's Impact on People and Society

In addition to the adverse effects on animals, animal hoarding is now also seen as a community problem that affects human welfare as well. Animals are frequently kept in squalid conditions, environments often called deplorable (Hayes, 2010). 88% of hoarders allow their animals free access in the home to urinate or defecate wherever they want (Steketee & Gibson et al., 2011). Often associated with adult self-neglect, animal hoarding can also place dependent elders and children at serious risk of deteriorating health and living standards and be an economic burden to taxpayers (Patronek, Loar & Nathanson, 2006). Neighborhood conditions frequently deteriorate with many cases exposed by neighbors complaining of excessive noise, odors and sanitation issues (HARC, 2002), or through welfare checks initiated by landlords or family members.

Sanitary conditions often deteriorate to the point where residences are considered toxic biohazards, investigators must wear hazardous material garb and oxygen masks to enter the property, and dwellings are condemned by public health authorities as unfit for human habitation. Feces and urine often cover floors and/or walls, and even the hoarder's bed, creating high concentrations of urine-based ammonia and aerosolized organic contaminants and increased likelihood of zoonotic diseases, such as external parasites, rabies, ringworm, and cat-scratch disease, that pose serious health risks to occupants and investigators (Castrodale et al., 2010; Hayes 2010; Patronek, 1999). General clutter of newspapers and trash commonly pose fire hazards. Plumbing, toilet facilities and major appliances are often inoperative, making food preparation and basic sanitation impossible (HARC, 2002). Buildings may be structurally unsound due to waste accumulation and animals' tunneling attempts at escape. Infestation with insects and rodents is common (Avery, 2005). These conditions conspire to decrease the ability of hoarders to maintain basic personal hygiene and can create a self-fulfilling cycle of isolation: hoarders surround themselves with animals to avoid unpleasant experiences associated with people, and other individuals avoid hoarders because they are repelled by their odors, fur, clutter, and unsanitary hygiene. By the time conditions have deteriorated to this point, expenses for veterinary services, housing of surviving animals, criminal investigation and forensics, prosecution, and remediation or demolition of the premises can be enormously excessive (Patronek, 1999).

Social and cultural factors may affect animal hoarding. Ramos et al. (2013) hypothesized that some Brazilians harbor large numbers of cats due to a shortage of animal shelters and rehoming organizations as compared to other countries. With fewer options for organized rehoming, those with compassion for cat welfare may be more likely to take stray animals home. As more people engage in this behavior, it becomes a cultural norm and not considered inappropriate animal care.

## Practical Perspectives: Investigation and Enforcement

Arluke et al. (2017) identified five major objectives in responding to animal hoarding incidents:

1. Remove animals from harm and begin rehabilitation and placement when possible.
2. Remove children or dependent adults affected by the adverse living conditions.
3. Address hoarders' physical and psychological needs.
4. Respond to environmental issues.
5. Institute provisions for monitoring and preventing recurrence.

Traditional fragmented responses have failed to limit recidivism. Investigations and response invariably require multiple government agencies, nonprofit organizations and private practitioners, including animal control officers, humane investigators, police, sheriffs, veterinarians, legal aid, adult and child protective services, behavioral health agencies, and fire, public health, code enforcement, zoning, and sanitation departments. But differing agency priorities, problematic interagency communication, confidentiality constraints, and the absence of intergovernmental agreements can prevent the seizure of animals and impede the detection of hoarders, the exchange of information and the ability to arrive at a non-adversarial resolution (Arluke & Patronek, 2016; Hayes, 2010; Patronek, Loar & Nathanson, 2006). Conflicting perspectives as to whether animal hoarding is a law enforcement or mental health issue or merely an acceptable lifestyle choice can deter agencies from intervening and infringing on hoarders' civil liberties and private property rights.

It is not uncommon for co-occurring child maltreatment, elder abuse or assault to be found in an animal hoarding investigation. The following represent examples:

- A Hampstead, Md. couple charged with 109 counts of aggravated animal cruelty were each sentenced to 33 years in prison after authorities found 27 dead dogs and 27 living dogs in a home that the judge called a "chamber of horrors." They also had to pay $10,900 in restitution to the humane society that cared for the animals and $91,400 to the owner of the house, which had to be demolished. A psychiatrist said the woman, who was living in her car, suffered from battered spouse syndrome (Keller, 2020).
- Oxford, Mass. police removed 28 dogs and charged a woman and her ex-boyfriend with multiple counts of child sexual abuse after prosecutors alleged years of sexual, mental and physical abuse of foster children. The man, a registered sex offender, was accused of allegedly raping a boy twice, threatening his life and forcing him to eat dog feces. The woman, a nurse, was accused of forcing a former foster child to walk around naked, sleep in a dog cage and wear diapers until the child was ten years old. Police officers noticed an overwhelming odor of dog feces and urine (Moulton, 2019).
- A New York City woman's conviction on 108 counts of animal abuse for housing 55 cats, 12 dogs and two turtles in her home included a ten-year ban on owning animals. She was subsequently charged with parole violation and assaulting a peace officer after NYPD, probation and ASPCA officials reportedly recovered seven dogs, two turtles, two rabbits, two guinea pigs, two fish, and a cat crammed in an ammonia-filled, fly-infested home in 90-degree heat (Parnell & Blau, 2019).
- San Bernardino County, Calif. authorities arrested a couple on charges of cruelty to children after three malnourished youths were found living in a box for the past four years. Deputies also found 30–40 cats, human feces and mounds of trash on the

property, which had no running water or electricity. The children, aged 11, 13, and 14, were placed in Children & Family Services custody (Newkirk, 2018).
- San Diego County, Calif., sheriff's deputies and adult protective officials removed an 89-year-old woman living without adequate care after Animal Services officers removed 140 semi-feral cats from a filthy, feces-ridden apartment. Her love of cats apparently started out with good intentions, but successive breeding got out of hand and she could not find homes for all the kittens. Neighbors had an idea that there was a cat problem but could not imagine the scope of it (Hernandez & Winkley, 2017).

Because animal hoarding's implications upon human health and welfare are only now beginning to be appreciated, and because animals' suffering is more obvious than risks to humans, these cases are traditionally left to animal care and control agencies to resolve. These gaps in service are exacerbated in rural areas where such agencies and their capacity are limited; in vast regions of the United States, no viable animal welfare or control agency is empowered to investigate animal abuse, so cases default to local law enforcement for whom animal issues are not a priority (National Link Coalition, 2021a). Ignorance about the multi-faceted nature of animal hoarding, the absence of any heretofore defined effective preventive or treatment strategies, and a lack of systematic, community-mobilized public health and social services procedures for response, prevention and intervention result in failures to remediate situations and almost complete recidivism (Patronek, Loar & Nathanson, 2006).

Resolution of animal hoarding cases is procedurally cumbersome, time consuming, and costly. Seized animals are frequently treated as evidence rather than victims of a crime, held in legal limbo for what may be several years until the case is adjudicated. The excessive costs of long-term caring for animals – veterinary care, veterinary forensics, evidence collection and storage, psychological assessments, competency evaluations, eviction proceedings, and limited shelter space to house seized animals – can run into many tens of thousands of dollars. These costs, invariably paid by the shelter (sometimes with assistance from national animal protection organizations) rather than the hoarder, can be a huge disincentive for underfunded agencies (Animal Legal Defense Fund, 2021). Consequently, humane agents may strive to avoid re-victimizing animals with long-term sheltering by negotiating dropped criminal charges in exchange for gaining immediate custody of the animals and restrictions on hoarders' future ownership of animals (Berry et al., 2005).

Laws focused on intentional cruelty may fail to protect animals subjected to unintended, chronic neglect. Until veterinary forensics become more widely available, animal care and control agencies' focus on immediate removal of animals may impede gaining sufficient evidence for prosecution. The client-focused missions of child and adult protective services may create conflicts between clients' rights and animals' welfare (Patronek, Loar & Nathanson, 2006).

The emotional costs of seizing hoarded animals can be high. Investigators going into filthy, garbage-ridden, chaotic homes encounter: hostile individuals in varying states of sadness, anger and physical and psychological distress; animals in deplorable conditions; a toxic environment; and conditions where humane euthanasia may be the most viable option. Steps must be taken to monitor hoarders who feel their "children" are being taken away from them. Hoarders have reportedly committed suicide following investigations (Arluke & Patronek, 2016).

Further confounding case resolution is an overarching public sentiment which tends to trivialize animal hoarding as a harmless eccentricity. The unconventional ownership of large numbers of animals is often seen in American folklore as tolerable neighborhood "crazy cat ladies" (Arluke & Patronek, 2016). One can shop on Amazon (2021) for the Accoutrements "Crazy Cat Lady Action Figure" ("Every town has a crazy cat lady. She's the one who lives in a tiny house full of feral felines. Comes with 6 cats, includes wild-eyed look.") or at McPhee.com (2021) for the "Crazy Cat Lady Board Game" ("Collect the most cats and be the craziest cat lady. You can never have too many cats! Or can you? Fantastic feline fun for the whole family! Dressing up like a crazy cat lady while playing is optional, but encouraged.")

Many animal hoarders are reclusive and resist investigations or follow-up. Many board up windows, erect tall fences, rarely appear outside, and ignore phone calls (Arluke & Patronek, 2016). This isolation makes it difficult for investigators to gain a clear picture of conditions without court-ordered search warrants.

Barring hoarders from future ownership of pets, necessitating long-term monitoring by animal welfare or probation officials, is time- and labor-intensive and often impractical practical (Arluke & Patronek, 2016) Even after removal of the animals, all a hoarder need do is relocate to another jurisdiction where the accumulation process can start anew.

Lockwood (2018) observed that including hoarding disorder in the DSM-5 added another potential obstacle; by being classified as having a disability, hoarders who live in public housing may legally be required to have reasonable accommodation made, as long as the situation is remediated and does not violate safety, health, and welfare rules indefinitely.

## Practical Perspectives: Prosecuting Animal Hoarding Cases

There are advantages to prosecution. Adjudication offers opportunities for court-ordered evaluations of hoarders and their animals, long-term case monitoring, restitution for incurred costs, and limits on current and future animal ownership and employment involving animals. Prosecution holds offenders accountable, creates a public record and educates judges, prosecutors, legislators, and the public about the seriousness of the issue. Aggressive prosecution may be the only effective way to deal with hoarders who are irrational, uncooperative, or who are engaged in fraudulent activities by posing as reputable charitable organizations (Vaca-Guzman & Arluke, 2005).

However, there are numerous challenges to criminalizing animal hoarding cases which can serve as disincentives:

- The sheer number of animals – each of which must be captured, processed as evidence, triaged, documented, cared for, and euthanized as necessary.
- The complexity of crime scene investigations, involving multiple agencies, many of which are not skilled in search-and-seizure and forensic protocols, preserving crime scene evidence or performing necropsies.
- Costs of time, manpower, and money.
- Defendants, often elderly, who elicit sympathy from juries as rescuers with lack of criminal intent who offer a better alternative to animal shelter euthanasia.
- Lack of consensus on whether these are criminal matters or mental health concerns.
- Courts overburdened with crimes against humans, which are seen as more serious.
- Hoarding construed as neglect, a less egregious offense than deliberate abuse.

There is inconsistency as to whether hoarders should be prosecuted with one overarching charge representing all the animals, which could streamline the adjudication process but undermine the severity of the offense; or with separate charges for each animal, which risks clogging and delaying the criminal justice system and requiring each animal's forensic information to be presented separately (Hayes, 2010). Animals which are expected to recover with proper care may be the basis for misdemeanor charges; felony charges might be warranted if animals have died or suffered permanent injury or debilitation, if multiple animals are involved, or the hoarder is a repeat offender (Lockwood, 2018).

Prosecution can impede timely delivery of needed behavioral health services into what may be a contributing mental health factor. Animal hoarding often develops over a protracted time period, but intervention is frequently delayed until environmental conditions have deteriorated so substantially that neighbors report their concerns. Even under the best circumstances, the prognosis is guarded, with relapses likely. Animal hoarding has not been shown to be mitigated by customary criminal justice sentences of fines, forfeiture of some or all animals, prohibitions on future ownership, and/or incarceration (Patronek, Loar & Nathanson, 2006).

A further challenge to prosecution is that in contrast to deliberate, single acts of physical animal cruelty, animal fighting or animal sexual abuse, animal hoarders claim they do not intend to harm the animals and typically express great love and concern for them, a paradox described as pathological altruism (Nathanson & Patronek, 2011). The suffering incurred in animal hoarding accrues and increases over time and its cumulative effects can be difficult to prove in court without extensive veterinary forensics.

Rather than endure the rigors of prosecution, many cases are resolved through civil remedies, using forfeiture laws to seize animals without bringing criminal charges or by issuing an injunction against animal ownership (Lockwood, 2018). Other options include placing the hoarder under guardianship, institutional care or supervised living, or allowing the hoarder to keep a limited number of animals under continual monitoring. Because many hoarders are on limited income and public assistance, making them pay the costs of animal care and rehabilitation is difficult to enforce, and recovery of such costs has been extremely rare (Berry et al., 2005).

## Practical Perspectives: Assessment, Intervention, and Treatment

Numerous psychological characteristics and antecedents have been suggested as being prevalent among animal hoarders. Early interest suggested the Obsessive-Compulsive Disorder (OCD) model frequently applied to object hoarding might be appropriate for understanding animal hoarding (Patronek & Nathanson, 2009), but this model seems less viable now. Themes found significantly more frequently among animal hoarders include: problems with early attachment; chaotic childhood environments; significant mental health concerns; attribution of human characteristics to animals; and the presence of more dysfunctional current relationships (Steketee et al., 2011). Other suggested psychological models include delusional disorder, dementia, lack of impulse control, attachment disorders, grief, early deprivation of parental love and stability, and trauma surrounding early relationships (Ramos et al., 2013). Schizophrenia, attention deficit disorder, anxiety, PTSD, generalized social phobia, and major depressive disorder have been documented in cases of object hoarding but it remains to be seen to what extent these contribute to animal hoarding (Patronek, Loar & Nathanson, 2006).

Animal hoarding has been likened to an addiction (Frost, 2000). Hoarders may suffer from delusional thinking and feel that they have a special bond that allows them to communicate with animals (Frost & Steketee, 2010). Patronek & Nathanson (2009) suggested excessive attachment to animals becomes a substitute to replace inadequate human relationships. Animal caregiving attachments may stem from an attachment disorder in which the "unconditional love" from the human-animal bond replaces abuse, trauma, damage, or neglect experienced in childhood (Steketee et al., 2011). Animal hoarders may have grown up with parents who were insufficiently affectionate or emotionally cold (HARC, 2002). Nathanson (2009) suggested animal hoarding may be less about "love" than about animals providing a conflict-free relationship marked by control and constancy.

Brown (2011) described animal hoarding as the human-animal bond gone awry, with rigid and extreme attachments to the animals, even at the expense of their well-being. She suggested that companion animals provide hoarders with critical self-object functions, and reliance on the animals becomes intense and crucial to the hoarder's sense of well-being; loss of the animals creates a sense of fragmentation, disintegration, and depression.

Patronek & Nathanson (2009) proposed that animals may be a perfect solution for individuals seeking to repair their sense of self. Collecting unwanted, sick, homeless, feral, and otherwise disadvantaged animals who cannot refuse care and can be easily held captive and controlled is an option for individuals who are longing for intimacy in the face of a paralyzing fear of rejection and abandonment by humans. This gives hoarders what they believe to be a socially acceptable persona that deflects criticism of the failed efforts to provide proper animal care.

A common theme among animal hoarders is a state of denial of their obvious failure to provide adequate animal care and the often squalid conditions of their home; they persist in collecting animals despite the evident problems (Steketee, Gibson et al., 2011). Lockwood & Cassidy (1988) suggested a failure to accept death as a potential causal factor.

Another common feature appears to be the predilection to personify and anthropomorphize their animals. Steketee, Gibson et al. (2011) reported that 81% of animal hoarders believed the animals to have the same intelligence and characteristics as humans. One participant told interviewers, "They are not my pets....I am their pet."

Arluke & Patronek (2016) cautioned that while some animal hoarders experience mental disorders in which psychological intervention is warranted, many do not. The latter may represent a subgroup where hoarding is a result of social breakdown or highly compromised living situations stemming from personal, familial, neighborhood, or community problems. Animal hoarding may be enabled by large populations of abandoned and stray animals, the emergence of a supportive community of "no-kill" animal advocates opposed to humane euthanasia, and neighbors, family and friends who enable the hoarder or simply ignore the problem.

The unfortunate reality is that no behavioral health interventions have been identified which significantly reduce rates of recidivism, reported as ranging from 60% to 100% (Ockenden et al., 2014). Hoarders begin to collect more animals almost immediately after being charged or having their animals removed (Avery, 2005). Although at least 30 states have enacted laws mandating or permitting courts to order psychological and/or psychiatric evaluation and treatment for animal cruelty offenders, legislatures have provided little guidance on what the goals of such evaluations might be (e.g., detection of danger to public safety, detection of mental illness, identification of danger to self or others, or determination of competency and capacity to care for animals) (Lockwood, 2018).

Because no validated therapy for animal hoarding disorder is available (Frost & Patronek, 2015), Steketee & Frost (2014) recommended that clinicians be skilled in evaluating a broad spectrum of behavioral disorders and be familiar with guidelines for assessing and treating object hoarding. However, few therapists are trained to work with object hoarders, let alone animal hoarders, and few hoarders – who are generally resistant to treatment anyway – have the financial resources or insurance necessary for payment to mental health providers. Consequently, long-term monitoring of attendance to counseling sessions may fall to a beleaguered probation officer who is already overwhelmed with a caseload in a system not designed for long-term monitoring of chronic offenders (Arluke & Patronek, 2016).

In the absence of any pharmacological or cognitive behavioral therapy demonstrated to be successful with object hoarders, and no published data on successful intervention and treatment of animal hoarders, few therapists appear to have knowledge of animal hoarding or feel prepared to deal with requests to assess these very challenging clients. Furthermore, therapists should appreciate that:

- Removal of animals may not resolve the filth and compromised living spaces that ensued from the obsessive need to accumulate animals;
- Animal hoarders often have an exaggerated threat appraisal and hypervigilance which make them wary of outside intrusion into their sphere of control;
- For hoarders whose primary relationships have been with their animals, relating to or trusting another person may be exceedingly difficult; and

Cognitive impairments and difficulties in organizing and executing tasks may complicate any treatment plan follow-through, necessitating long-term interventions (Patronek & Nathanson, 2009).

Improving the assessment and treatment of animal hoarders requires attention not only to contributing psychosocial conditions, but also an awareness of the centrality of the animals to the hoarder's identity, self-esteem and sense of control, and the provision of basic social support that meets the need for bonding previously filled by animals (Patronek & Nathanson, 2009). Emphasizing that no quick fixes are likely, that no validated therapy is available, and that most individuals who have been adjudicated are reluctant to participate in therapy, Steketee & Frost (2014) recommended multidisciplinary community task forces. Representatives from housing, police, fire, public health, mental health, and legal could collaborate to engage offenders in identifying goals that matter to them while requiring compliance with non-hoarding behaviors. Of critical importance is determining whether to limit the number of animals (as a result of neglect) or require abstinence (as a result of overt cruelty).

Several communities have established task forces to coordinate early detection and a coordinated, team-based response to the diverse problems associated with these cases. Fort Wayne, Ind., Animal Care and Control has recognized that animal hoarding is more likely a mental health concern than a criminal offense. While investigating two or three cases a year – which could involve 30 dogs or 708 hamsters – the agency recognized hoarders' homes as "pits of toxic squalor" and brought together a community agency collaboration to treat hoarders' mental illnesses (Caylor, 2003).

The Fairfax County, Va., Hoarding Committee, the first of its kind in the United States when it was formed in 1998 following a fire in a heavily cluttered building that claimed four lives, created a website (https://www.fairfaxcounty.gov/code/hoarding) that offers detailed interdisciplinary responses that include animal hoarding. The committee includes animal, adult and child protective services; code compliance; mental health; fire and rescue; housing and community development; police; and other agencies.

A proactive plan in Wake County, N.C., utilizes a team approach engaging animal control officers, crisis intervention counselors, law enforcement, veterinary, and animal rescue partners (Strong et al., 2019). In New York City the ASPCA instituted a program through the Mayor's Alliance for Animals to combine the expertise of social workers, veterinarians, animal behaviorists, humane law enforcement, health and homeless services departments, community affairs, the Salvation Army, and others to respond at the earliest possible stage and monitor cases regularly (Lockwood, 2018).

## Public Policy and Legislative Perspectives

All states have statutes prohibiting various forms of animal cruelty, abuse and neglect requiring providing animals with sufficient and wholesome food, water, shelter, and necessary veterinary care (Arkow & Lockwood, 2016), and instances of animal hoarding can be charged under these generalized statutes. However, the first statutory reference to animal hoarding as a specific crime (as distinguished from general animal neglect or cruelty) was Illinois Public Act 92–0454, enacted in 2001. Sec. 510 ILCS 70/2.10 defines a "companion animal hoarder" as someone who possesses a large number of companion animals, fails to or is unable to provide adequate care, keeps them in a severely overcrowded environment, and displays an inability to recognize or understand the nature of or has a reckless disregard for their conditions and the deleterious impact on the animals' and owners' health and well-being. Hawai'i is believed to have the only other statute specifically addressing hoarding with Laws 2008, Chapter 128, defining the misdemeanor crime of animal hoarding as possessing more than 20 dogs and/or cats, failing to provide necessary sustenance for each one, and failing to correct the conditions under which they are living if the failure to provide necessary sustenance results in conditions injurious to the animal's or owner's health and well-being.

Rhode Island attempted to enact a law in 2017 which would have added "animal hoarding" to the list of animal cruelty offenses. After opposition from mental health associations, the American Civil Liberties Union and National Association of Social Workers' state chapter that this would criminalize a mental health disorder, "animal hoarding" was stricken and replaced with "hazardous accumulation of animals" and a failure to provide "adequate living conditions." Livestock are specifically excluded (Lockwood, 2018). GenLaws § 4-1-36 allows courts to order evaluations to determine a convicted offender's need for psychiatric or psychological counseling, at the offender's expense.

Hayes (2010) offered several reasons why states have not specifically addressed animal hoarding in criminal codes. Legislators may be wary because the public may view hoarding laws as infringing on rights to pet ownership. They may believe that hoarding is not a severe form of animal abuse, but merely having too much love for animals. More pragmatically, legislators may simply believe such legislation is unnecessarily redundant because hoarding can be prosecuted under animal cruelty laws. It is not clear whether isolating animal hoarding as a distinct criminal offense – as has been done in many states with dogfighting and animal sexual abuse – would offer greater public, judicial, and law enforcement awareness of the prevalence, significance and unique nature of animal hoarding and its psychological underpinnings.

Many communities have local ordinances limiting the number of companion animals any residence may harbor, and some states regulate legitimate kennel and cattery operations. These code and zoning provisions may enable local authorities to order the removal of large numbers of animals which violate the numerical standards without necessarily engaging in more complex cruelty investigations. However, these ordinances are often challenged by the pet industry, breeders and animal welfare groups as infringing on individual rights and unfairly penalizing responsible pet owners. Furthermore, numerical limits may not exist in unincorporated county jurisdictions and are found primarily in incorporated urban communities.

Incidents which attract news media are frequently more extreme and are consequently sensationalized. Reporting tends to focus on the animals' welfare, which is seen as the problem rather than a symptom of a problem, and does not always trigger a human welfare response. An analysis of more than 100 press reports suggested that media elicit a range of emotions among the general public such as revulsion, sympathy, indignation, and amusement regarding people who hoard animals, contributing confusion rather than a clinical and compassionate understanding of animal hoarding (Arluke et al., 2002).

## What We Don't Know: Directions for Future Research

Animal hoarding is one of the most widespread, severe, and multi-faceted forms of animal cruelty, a complex disorder that spotlights many issues raised at the intersection of behavioral sciences and the law, with no easy solutions (Lockwood, 2018). Considerable additional research is needed to shed further light on this dark side of the human-animal bond and to identify more effective prevention and intervention strategies.

The prevalence of animal hoarding is unknown, although estimates suggest approximately 3,000 reportable cases annually in the United States (Patronek, 2006). As the FBI's National Incident-Based Reporting System (NIBRS) becomes more systematically developed among law enforcement agencies, statistics about what NIBRS classifies as "gross neglect" will further quantify what heretofore have been primarily anecdotal reports.

Lockwood (2018) reviewed potential connections between object hoarding behaviors and genetic components, neurological disruption of brain functioning, and infection with the protozoan parasite *Toxoplasma gondii,* commonly associated with cat feces. He noted that many factors associated with animal hoarding are not mutually exclusive, especially when coupled with societal reinforcement and an individual's reliance on animal interactions to create a positive self-image. Given that the DSM-5 now recognizes hoarding as a specific disorder, there should be increased interest into more advanced study of the animal hoarding phenomenon, documenting the extent and nature of psychopathology in animal hoarders and how it is differentiated or similar to object hoarding.

Much has been written about how early trauma can alter neurobiology and how a dysfunctional or abusive childhood can lead to a disordered attachment style resulting in self-destructive addictive behaviors in later life. There is relatively little information about the potential role of antecedent traumatic events triggering the onset of animal hoarding. Longitudinal studies of animal hoarders are needed to elucidate the impact of trauma, the hoarder's ability to process the event, and the psychological, social, and environmental triggers that contribute to the phenomenon.

Additional research is needed to include the commission or witnessing of acts of animal cruelty in childhood within the constellation of Adverse Childhood Experiences (ACES) linked with greater adoption of risky health behaviors and earlier mortality in adulthood (Boat, 2014) and possible onset of animal hoarding behaviors. Similarly, it is unclear whether there is a link between histories of childhood attachments to and caring for animals and the development of compulsive caregiving and animal hoarding in adulthood (Patronek & Nathanson, 2009).

Animal hoarding often co-occurs with residential squalor. The association between premorbid personality problems and frontal lobe dysfunction and individuals living in squalor has previously been reported (Patronek & Nathanson, 2009), but it is unclear whether such associations have been demonstrated among animal hoarders. In-depth clinical assessments of severely affected populations of animal hoarders could shed additional light on possible comorbid psychological conditions.

While a multidisciplinary community agency response is invariably needed, a serious gap exists in the training of social workers who are largely unprepared to recognize the significance of the human-animal bond in the lives of their clients, let alone a response to clients' animal hoarding. The National Link Coalition (2021b) has identified only 27 U.S. schools of social work, out of 889 accredited BSW and MSW programs, in which human-animal interactions are included, even peripherally, in the curriculum. Outreach to this profession about the human-animal bond, as well as to therapists and social services sectors who respond to animal hoarding, is a critical need.

Given the lack of effective success in prosecution and therapeutic interventions, the best means to prevent recidivism has been described as a combination of constant oversight and monitoring through a coordinated effort of probation, social services, animal care and control, housing authorities, and others (Phillips & Lockwood, 2013). Research is needed to identify if these components prevent recidivism, and to develop new therapeutic and judicial approaches which may prove more effective than those heretofore employed.

Earlier recognition of incipient hoarding conditions should theoretically be advanced by a more systematic structure of statutorily-mandated cross-reporting among child, adult, and animal protective agencies. To date, mandatory or permissive cross-reporting between child and animal protection agencies is sporadic, and even more limited between adult and animal protective agencies (National Link Coalition, 2021c). This is especially problematic, given that animal hoarders tend to be older and co-occurring elder neglect or self-neglect is common. The effectiveness of such cross-reporting is unknown. Considerable research is needed to document the number and resolution of cases of child, elder and animal abuse that are cross-referred by these agencies, and to identify mechanisms which would systematize such processes.

The Hoarding of Animals Research Consortium (Patronek et al., 2006) described a five-step process to reduce fragmentation of responses and create more effective, integrated community approaches to animal hoarding:

1  Identifying relevant stakeholders, the resources and skill sets they can best provide, and areas of interagency overlap and conflict.
2  Gaining inter-agency cooperation to create practical short-term and strategic long-term interdisciplinary responses.
3  Gaining greater understanding of the types of animal hoarders and their differing motivations, behaviors, and responses.
4  Matching legal, statutory, regulatory, and psychological intervention strategies to the types of hoarders.
5  Decreasing the likelihood of recidivism through long-term approaches.

Greater awareness, acceptance and implementation of this process, and recognizing animal hoarding as a community and human health problem in addition to one of animal welfare, are needed, particularly in smaller and rural regions where agencies are especially under-resourced and where animal hoarders can more easily operate away from widespread public scrutiny.

## References

Amazon.com. (2021). *Accoutrements crazy cat lady action figure multicolored*, 8. Retrieved August 2, 2021, from https://www.amazon.com/Accoutrements-12470-Action-Figure-Multicolor/dp/B00HZSMWMY

American Psychiatric Association. (2013). *Diagnostic and statistical manual of mental disorders*, 5 (5th ed.). https://doi.org/10.1176/appi.books.9780890425596

Animal Legal Defense Fund. (2021). *Animal hoarding facts*. Retrieved August 3, 2021, from https://aldf.org/article/animal-hoarding-facts/

Arkow, P., & Lockwood, R. (2016). Definitions of animal cruelty, abuse and neglect. In M. P. Brewster & C. L. Reyes (Eds.), *Animal cruelty: A multidisciplinary approach to understanding* (2nd ed., pp. 3–23.) Carolina Academic Press.

Arluke, A., Frost, R., Steketee, G., Patronek, G., & Luke, C., et al. (2002). Press reports of animal hoarding. *Society & Animals, 10*(2), 113–135.

Arluke, A., & Patronek, G. (2016). Animal hoarding. In M. P. Brewster & C. L. Reyes (Eds.), *Animal cruelty: A multidisciplinary approach to understanding* (2nd ed., pp. 199–216). Carolina Academic Press.

Arluke, A., Patronek, G., Lockwood, R., & Cardona, A. (2017). Animal hoarding. In J. Maher, H. Pierpoint & P. Beirne (Eds.), *The Palgrave international handbook of animal abuse studies* (pp. 107–129). Palgrave Macmillan.

Arnold, A., Mackensen, H., Ofensberger, E., & Rusche, B. (2018). Assessment of recent cases of animal hoarding in Germany: The challenge for animal shelters and public authorities. *People and Animals: The International Journal of Research and Practice, 1*(1), 7.

Avery, L. (2005). From helping to hoarding to hurting: When the acts of "good Samaritans" become felony animal cruelty. *Valparaiso University Law Review, 39*(4), 815.

Berry, C., Patronek, G., & Lockwood, R. (2005). Long-term outcomes in animal hoarding cases. *Animal Law Review, 11*, 167–194.

Boat, B. W. (2014). Connections among adverse childhood experiences, exposure to animal cruelty and toxic stress: What do professionals need to consider? *National Center for Prosecution of Child Abuse Update, 24*(4), 1–3.

Brown, S.-E. (2011). Theoretical concepts from self psychology applied to animal hoarding. *Society & Animals, 19*, 175–193.

Calvo, P., Duarte, C., & Bowen, J., et al. (2014). Characteristics of 24 cases of animal hoarding in Spain. *Animal Welfare, 23*(2), 199–208.

Castrodale, L., Bellay, Y. M., Brown, C. M., Cantor, F. L., & Gibbins, J. D., et al. (2010). General public health considerations for responding to animal hoarding cases. *Journal of Environmental Health, 72*(7), 14–19.

Caylor, B. (2003, December 5). Helping the animal hoarders: City agencies, departments treat them as mentally ill. *Fort Wayne News-Sentinel*, p. 8A.

Ferreira, E. A., Paloski, L. H., Costa, D. B., Fiametti, V. S., & de Oliveira, C. R., et al. (2017). Animal hoarding disorder: A new psychopathology? *Psychiatry Research, 258*, 221–225.

Frost, R. O. (2000). People who hoard animals. *Psychiatric Times, 17*(4), 25–29.

Frost, R. O., & Hartl, T. (1996). A cognitive-behavioral model of compulsive hoarding. *Behaviour Research and Therapy, 36*, 657–664.

Frost, R. O., Patronek, G., & Rosenfield, E. (2011). Comparison of object and animal hoarding. *Depression and Anxiety, 28*(10), 885–891.

Frost, R. O., Patronek, G., Arluke, A., & Steketee, G. (2015, April 30). The hoarding of animals: An update. *Psychiatric Times, 32*(4). Retrieved June 8, 2023, from https://www.psychiatrictimes.com/view/hoarding-animals-update

Frost, R. O., & Steketee, G. (2010). *Stuff: Compulsive hoarding and the meaning of things*. Houghton Mifflin Harcourt.

Hayes, V. (2010). *Detailed discussion of animal hoarding*. East Lansing: Michigan State University College of Law Animal Legal & Historical Center. Retrieved August 3, 2021, from https://www.animallaw.info/article/detailed-discussion-animal-hoarding

Hernandez, D., & Winkley, L. (2017, November 2). *About 100 cats found in hoarding case at lakeside apartment*. Retrieved August 3, 2021, from https://www.sandiegouniontribune.com/news/public-safety/sd-me-lakeside-cats-20171102-story.html

Hoarding of Animals Research Consortium (2002). Health implications of animal hoarding. *Health & Social Work, 27*(2), 125–136.

Jacobson, L. S., Giacintl, J. A., & Robertson, J. (2020). Medical conditions and outcomes in 371 hoarded cats from 14 sources: a retrospective study (2011–2014). *Journal of Feline Medicine and Surgery, 22*(6), 484-491. https://doi.org/10.1177/1098612X19854808

Joffe, M., O'Shannessy, D., Dhand, N. K., Westman, M., & Fawcett, A. (2014). Characteristics of persons convicted for offences relating to animal hoarding in New South Wales. *Australian Veterinary Journal, 92*(10), 369–375.

Keller, M. G. (2020, July 28). *Judge sentences Hampstead woman to 7 years after dogs found dead in "chamber of horrors."* Retrieved August 3, 2021, from https://www.baltimoresun.com/maryland/carroll/news/crime/cc-filler-sentencing-blackrockdogs-20200728-xl5rb7qstbayjoxvnzrzeglntq-story.html

Lockwood, R. (2018). Animal hoarding: The challenge for mental health, law enforcement, and animal welfare professionals. *Behavioral Sciences & the Law, 36,* 698–716.

Lockwood, R., & Cassidy, B. (1988, Summer). Killing with kindness? *The Humane Society News,* 14–18.

McPhee.com. (2021). *Crazy cat lady board game.* Retrieved August 3, 2021, from https://mcphee.com/collections/cats/products/crazy-cat-lady-board-game

Moulton, C. (2019, December 10). *28 dogs removed from Oxford home where foster children allegedly were abused.* Retrieved August 3, 2021, from https://www.telegram.com/news/20191210/28-dogs-removed-from-oxford-home-where-foster-children-allegedly-were-abused

Nathanson, J. N. (2009). Animal hoarding: Slipping into the darkness of comorbid animal and self-neglect. *Journal of Elder Abuse and Neglect, 21*(4), 307–324.

Nathanson, J. N., & Patronek, G. J. (2011). Animal hoarding: How the semblance of a benevolent mission becomes actualized as egoism and animal cruelty. In B. Oakley et al. (Eds.), *Pathological altruism* (pp. 107–115). Oxford University Press.

National Link Coalition. (2021a). *National directory of abuse investigation agencies.* Retrieved August 3, 2021, from https://nationallinkcoalition.org/how-do-i-report-suspected-abuse

National Link Coalition. (2021b). *Schools of social work with human-animal interactions in curriculum.* Etowah, N. C. Author. Retrieved August 3, 2021, from https://nationallinkcoalition.org/wp-content/uploads/2021/08/Schools-of-Social-Work-with-HAI-5-2021.pdf

National Link Coalition. (2021c). *Cross reporting by reporter.* Retrieved August 3, 2021, from https://nationallinkcoalition.org/wp-content/uploads/2021/06/Cross-Reporting-by-Reporter-2021-06-30.pdf

Newkirk, B. (2018, March 1). *Three Joshua Tree children were living in a shack for years, parents arrested, police say.* Palm Springs Desert Sun. Retrieved August 3, 2021, from https://www.desertsun.com/story/news/crime_courts/2018/03/01/three-joshua-tree-children-found-living-boxparents-arrested-keeping-their-three-children-box-past-fo/387523002/

Ockenden, E. M., De Groef, B., & Marston, L. (2014). Animal hoarding in Victoria, Australia: An exploratory study. *Anthrozoos, 27*(1), 33–47.

Parnell, W., & Blau, R. (2019, January 7). *Convicted animal hoarder violates probation by owning and abusing more pets, Queens DA charges.* Retrieved August 3, 2021, from https://www.nydailynews.com/new-york/ny-metro-animal-hoarder-chargers-20190107-story.html

Patronek, G. (1999). Hoarding of animals: An under-recognized public health problem in a difficult-to-study population. *Public Health Reports, 114,* 81–87.

Patronek, G. (2006). Animal hoarding: Its roots and recognition. *Veterinary Medicine, 101,* 520–530.

Patronek, G. J., Loar, L., & Nathanson, J. N. (Eds.) (2006). *Animal hoarding: Structuring interdisciplinary responses to help people, animals and communities at risk.* Hoarding of Animals Research Consortium. Retrieved June 8, 2023, from https://nationallinkcoalition.org/wp-content/uploads/2023/05/Hoarding-HARC-Report-2006.pdf

Patronek, G. J., & Nathanson, J. N. (2009). A theoretical perspective to inform assessment and treatment strategies for animal hoarders. *Clinical Psychology Review, 29*(3), 274–281.

Patronek, G., & the Hoarding of Animals Research Consortium (2001, May/June). The problem of animal hoarding. *Municipal Lawyer, 42*(3), 6–19.

Phillips, A., & Lockwood, R. (2013). *Investigating and prosecuting animal abuse: A guidebook on safer communities, safer families & being an effective voice for animal victims.* National District Attorneys Association.

Ramos, D., da Cruz, N. O., Ellis, S. L. H., Hernandez, J. A. E., & Reche-Junior, A. (2013). Early stage animal hoarders: Are these owners of large numbers of adequately cared for cats? *Bulletin of Human-Animal Interaction, 1*(1), 55–69.

Rasmussen, J. L., Steketee, G., & Frost, R. O., et al. (2014). Assessing squalor in hoarding: the Home Environment Index. *Community Mental Health Journal, 50,* 591–596.

Reinisch, A. (2009). Characteristics of six recent animal hoarding cases in Manitoba. *Canadian Veterinary Journal, 50*(10), 1069–1073.

Saldarriaga-Cantillo, A., & Nieto, C. (2015). Noah Syndrome: A variant of diogenes syndrome accompanied by animal hoarding practices. *Journal of Elder Abuse & Neglect, 27*, 270–275.

Schroeder, L. M. (2018, June 7). *Derbe Exkhart raid: Is it a case of animal hoarding?* Retrieved August 3, 2021, from https://www.mcall.com/news/police/mc-nws-derbe-eckhart-animal-hoarding-again-20180607-story.html

Steketee, G., & Frost, R. O. (2014). Phenomenology of hoarding. In R. O. Frost & G. Steketee (Eds.), *Handbook of hoarding and acquiring* (pp. 19–32). Oxford University Press.

Steketee, G., Gibson, A., Frost, R. O., Alabiso, J., Arluke, A., & Patronek. G. (2011). Characteristics and antecedents of people who hoard animals: An exploratory comparative interview study. *Review of General Psychology, 15*(2), 114–124.

Strong, S., Federico, J., & Banks, R., et al. (2019). A collaborative model for managing animal hoarding cases. *Journal of Applied Animal Welfare Science, 22*, 267–278.

Vaca-Guzman, M., & Arluke, A. (2005). Normalizing passive cruelty: The excuses and justifications of animal hoarders. *Anthrozoös, 18*(4), 338–357.

Worth, D., & Beck, A. M. (1981). Multiple ownership of animals in New York City. *Transactions & Studies of the College of Physicians of Philadelphia, 3*(4), 280–300.

# 23
# CO-OCCURRENCE OF ANIMAL ABUSE AND INTIMATE PARTNER VIOLENCE

*Maya Gupta and Shelby McDonald*

## Introduction

Intimate partner violence (IPV) is a public health concern that impacts nearly one in three women worldwide (WHO, 2021). IPV—also termed domestic violence (DV)—is broadly defined as any act of coercive control within the context of a relationship with a current/former intimate partner (Buchanan & Humphreys, 2021; Cares, Reckdenwald, & Fernandez, 2021). Coercive control describes behaviors used to influence, intimidate, threaten, frighten, and/or manipulate an intimate partner, such as physical violence, emotional/psychological abuse, sexual violence, and/or stalking (Stark, 2012). Within the context of IPV, emotional/psychological abuse often includes non-physical forms of violence such as verbal abuse (yelling, belittling, verbal attacks), enforced isolation, micromanaging the partner's behavior, and monopolizing decision-making. Physical violence includes abuse that may range in severity such as being pushed, slapped, punched, kicked, strangled or burned, or harmed with a weapon.

Coercive control can also manifest via relationships and interactions with other family members, including children and animals. Understanding associations between IPV and animal abuse (AA) has important implications, both for family violence prevention/intervention efforts and for animal welfare.

## *Overlap of IPV and AA*

While an exhaustive review of the literature on this topic is beyond the scope of this chapter and we have therefore focused on more recent publications, useful historical reviews are provided by Ascione and Lockwood (1999), Ascione and Arkow (1999), and Ascione (2008).

Pet ownership is common in the United States, with nationally representative studies indicating that most families have at least one pet (APPA, 2021; Hoffman, Thibault, & Hong, 2021; Mueller, 2021). Nearly 95% of U.S. households consider their pets to be family members (Harris Poll, 2015; McConnell, Lloyd, & Buchanan, 2017), suggesting that pets play a central role in contemporary family systems, including households where IPV occurs. Research with IPV survivors and DV service providers has found that threats and violence against family pets are well-documented tactics of emotional/psychological abuse against intimate partners, functioning as methods of control, intimidation, and retaliation (Ascione et al., 2007; Collins et al., 2018; Onyskiw, 2007; Volant, Johnson, Gullone, & Coleman, 2008). For example, when a victim tries to leave the relationship, the abuser may threaten or actually harm/kill pets to retaliate or prevent the victim from leaving. Alternately or in addition, a partner may use threats/harm against animals as a means to isolate their victim from their family. Moreover, acts of violence experienced by animals in the home often mirror the abuse of the survivor, involving a range of behaviors including yelling, hitting, kicking, choking, burning, and injury with a weapon (Collins et al., 2018; McDonald et al., 2015, 2019).

Some evidence indicates that victims' bonds with their pet(s) may be an important factor in perpetrators' motivation to harm animals. For example, in a survey of women receiving DV shelter services, Faver and Cavazos (2007) found that 88% of participants who reported AA by their partner identified the maltreated pet as a "very important" source of emotional support; in contrast, fewer (though still 51% of) women who did not report AA said their pet was an important source of support. For this reason, many victims of IPV live within a duality of finding support in their bond with a pet while also experiencing chronic fear about having that bond exploited by their partner (Collins et al., 2018; McDonald et al., 2015).

Several studies have examined the prevalence of intersecting AA and IPV, and rates have ranged extensively across research in this area (Ascione et al., 2007; Carlisle-Frank, Frank, & Nielsen, 2004; Faver & Strand, 2003; Flynn, 2000; Volant et al., 2008). Cleary et al.'s (2021) recent systematic review of research on AA in the context of adult IPV found that the proportion of pet-owning female survivors of IPV who reported their partner threatened harm to animals ranged from 12% (Hartman et al., 2018) to 75% (Collins et al., 2018), and the prevalence of actual AA reported by female survivors varied between 23% (Travers et al., 2009) and 77% (Tiplady et al., 2012). Some scholars have suggested that there may be ethnocultural variations in the overlap between IPV and AA. For example, Hartman et al. (2018) examined rates of AA exposure among survivors of IPV and found that women whose partner was non-Hispanic and U.S.-born displayed higher rates of their partner harming pets (41%) than women with Hispanic U.S.- or Mexico-born partners (27% and 12.5%, respectively).

Another factor contributing to the wide range of co-occurrence estimates is that it is difficult to clearly define AA because relationships between humans and animals are complex. Haden, D'Emilia, and McDonald (2022) argue that speciesism impacts how we understand and determine what is cruel toward another species and emphasize that humans tend to privilege certain non-humans (e.g., companion animals) over others that are used for our consumption, experimentation, sport, equipment, and entertainment (e.g., cows, rats; Caviola, Everett, & Faber, 2019). Relatedly, there are multiple competing definitions of AA and a continuum of acts considered to constitute AA, which have contributed to methodological inconsistencies across studies. For example, Ascione (1993) described AA as "socially unacceptable behavior that intentionally causes unnecessary pain, suffering, or distress to and/or the death of an animal" (p. 51); this definition excludes animals used in hunting practices, those subjected to research, and the "humane" killing of farm animals, since these behaviors are considered acceptable in most cultures. Rowan (1999) differentiates between abuse and cruelty, defining cruelty as characterized by the perpetrator's enjoyment of the suffering whereas abuse is characterized by using an inappropriate level of force. Gullone (2014) includes repetition of the behavior in the definition of AA. Nevertheless, what legally constitutes animal cruelty depends on laws in the jurisdiction where the act occurs. In this chapter, we use the term "animal abuse" to encompass the range of harmful actions toward animals that may occur in the IPV context, but retain "animal cruelty" where specifically discussing cruelty laws and criminal cases.

Methodological inconsistencies aside, studies of IPV survivors suggest that those who experience AA by abusers may be at greater risk for more frequent and severe forms of IPV victimization when compared to survivors who do not experience AA. For example, a study of IPV situations involving law enforcement indicated that women reporting IPV by an individual with a history of AA were significantly more likely to report being forced to have sex (26%) and being strangled (76%) when compared to survivors whose abuser did not engage in AA (Campbell, Thompson, Harris, & Wiehe, 2021). Higher rates of other lethality risks were also found in households where AA was present, such as easy access to a firearm(s) and threats to harm children. Similarly, a recent nationally representative study in Canada found that threatened and actual abuse to pets by an intimate partner was a robust predictor of reporting partner-perpetrated IPV, especially IPV involving physical abuse and severe abuse (Fitzgerald et al., 2022).

## *Other Ways AA Intersects with IPV*

Although AA is often used as a mechanism of coercive control toward intimate partners, it also manifests in other ways in households where IPV occurs. It is well established that children in households where IPV occurs endure more frequent harsh physical punishment (Berger, 2005, Lee et al., 2004: Tajima, 2000). The social and emotional relationships adults have with pets are often comparable to parental relationships with children (Laurent-Simpson, 2017ab; Volsche, 2019); therefore, it is not surprising that pets in these households experience comparable disciplinary tactics (McDonald et al., 2015). Carlisle-Frank et al. (2004) found that 75% of IPV perpetrators who abused pets used harsh physical punishment to discipline the animal. McDonald et al. (2015) found that 24% of children living in households where IPV occurred had been exposed to maltreatment of a pet that was perpetrated by an adult with the goal of disciplining or punishing the animal for unwanted behaviors (chewing, elimination).

Another way that AA manifests in households impacted by IPV is as coercive control of children. Coercive parenting among fathers who engage in aggressive behaviors toward their intimate partners is well documented (Fox & Benson, 2004). Like adult victims of IPV, children often have valued and intimate relationships with family pets (McDonald et al., 2015). Although research in this area has been limited to a few qualitative investigations, studies with IPV survivors indicate that perpetrators of IPV often use the same coercive approaches to achieve child compliance and punish children in the household, including the exploitation of a child's bond with a pet (Collins et al., 2018; McDonald et al., 2019). The role of pets in coercive parenting is an area of research with notable implications for clinical practice with children and families, and warrants greater attention given that exposure to coercive parenting during childhood is a family-of-origin factor related to future IPV perpetration (Schwartz, Hage, Bush, & Burns, 2006).

Generally, children living in households where IPV occurs are at a significantly increased risk for witnessing violent AA when compared to children in non-violent families (e.g., Ascione et al., 2007; Faver & Strand, 2003, 2007; Volant et al., 2008). Like estimates of the co-occurrence of IPV and AA, the prevalence of children's exposure to AA in the context of IPV ranges extensively due to methodological inconsistencies across studies. Haden et al. (2020) report prevalence rates of child and youth exposure to pet abuse ranging from nearly a quarter to more than one-half of children across retrospective studies. One study found that 61.5% of mothers residing at a DV shelter reported that their child(ren) had heard or seen AA in the home; in contrast, the rate of children's exposure to AA in a comparison group of mothers who did not report IPV victimization was only 2.9% (Ascione et al., 2007).

Children who have been exposed to IPV, experienced direct maltreatment, and/or have witnessed AA have been consistently identified as at elevated risk for perpetrating AA in childhood and adulthood (Gullone & Robertson, 2008; Henry, 2004; Hensley & Tallichet, 2005ab; Thompson & Gullone, 2006). Consequently, children's engagement in AA behaviors is another way that AA manifests in households where IPV occurs. A recent study of families recruited from community IPV services found that approximately 16% of children who had a pet and lived in a household with IPV had engaged in some form of maltreatment of the family pet. Relatedly, Thompson and Gullone (2006) identified that adolescents who had witnessed someone close to them (family member, close friend) perpetrate abuse toward an animal reported higher levels of childhood AA behaviors. Regarding links between exposure to IPV and childhood AA, a retrospective study of Italian children between 9 and 12 years of age found that children who witnessed IPV were three times more likely to perpetrate AA (Baldry, 2005). A similar study found that, among a community sample of U.S. children aged 5–17, those who reported exposure to IPV were 2.95 times more likely to report having engaged in childhood AA than children who had not been exposed to IPV (Currie, 2006).

## *Impacts on Human Survivors' Psychological Wellbeing and Safety*

In the context of IPV, concomitant exposure to AA may exacerbate the negative effects of victimization on adult and child survivors' psychological health, physical health, and safety (Barrett et al., 2020; Collins et al., 2017; McDonald et al., 2015, 2018). Although research in this area is limited, it is reasonable to assume based on theory and preliminary evidence that the impact of experiencing threats and/or harm to pets is similar or comparable to the harmful impacts and outcomes associated with exposure to other forms of family violence (e.g., anxiety, anger, helplessness, and posttraumatic stress symptoms; Haden, McDonald, & Murphy, 2020). Some research indicates that adult female IPV survivors' emotional response to witnessing AA involves complicated, compounded emotions (Collins et al., 2018; Faver & Strand, 2007; Riggs et al., 2021). For example, qualitative studies document that survivors sometimes feel relieved that their pet was targeted instead of them or their child, yet they may be simultaneously upset by violence to the animal and feel particularly guilty in scenarios when a pet is targeted because of their bond with the animal (Collins et al., 2018; Faver & Strand, 2007; Riggs et al., 2021). Or, when an animal is abused following efforts to protect the victimized family member, the survivor may experience heightened emotional distress in response to IPV (Collins et al., 2018). For survivors who are parents and dually witness their child being controlled or intimidated by their partner's AA or related threats, knowing about this form of child coercion could exacerbate the socioemotional impacts of IPV, leading to increased self-blame and/or diminished confidence in their role as a parent (McDonald et al., 2015). A recent study by Riggs et al. (2021) examined relations between IPV and AA in people of diverse genders and sexualities and found that female participants who had experienced an animal being abused reported much higher levels of psychological distress than did male or nonbinary participants. Such findings suggest that the psychological impact of exposure to AA may relate to other aspects of the survivor's identity as well.

Regarding child survivors of IPV, the negative consequences of exposure to IPV on children's socioemotional development and health are well documented, yet children's dual exposure to IPV and AA only recently became a focus of scientific inquiry. Emerging evidence indicates that exposure to AA is associated with short- and long-term child behavioral health outcomes and that exposure to AA accounts for variance in children's psychopathology symptoms over and above co-occurring childhood victimizations, such as exposure to IPV (Girardi & Pozullo, 2015; Hawkins et al., 2019; McDonald et al., 2019). One study found that child survivors of IPV who were also exposed to AA were over five times more likely than those not exposed to AA to have severe behavioral maladjustment (i.e., clinical levels of internalizing and externalizing behavior, callous/unemotional traits, attention symptomatology; McDonald et al., 2018).

Children who witness AA in households where IPV occurs may also be more likely to intervene in violent incidents when they feel their pet is threatened, which in turn increases their risk for injury by a caregiver (McDonald et al., 2015). Prior studies suggest that this direct victimization as a result of protecting pets may increase children's risk for physical injury and psychological maladjustment (McDonald et al., 2015, McDonald et al., 2018). Given that prior research suggests AA may be associated with more severe family violence, children's attempts to protect abused pets are particularly concerning. Although AA may put children at heightened risk for adverse experiences and outcomes, this is not to suggest that children's bonds with pets solely operate as a risk factor in the context of IPV. There is evidence that bonds and social relationships with pets are a protective factor amid childhood adversity and violent victimization. For example, a recent study found that high levels of children's positive engagement with a family pet buffered the effect of exposure to IPV on children's internalizing and posttraumatic stress symptoms (Hawkins et al., 2019).

Despite empirical and anecdotal evidence that adult and child survivors of IPV have strong bonds with pets that impact survivors' decision-making when leaving abusive relationships, assessments and services related to pets have, historically, not been well integrated in DV and allied social services (e.g., child welfare). This is especially true in the context of providing safe, pet-friendly shelter and housing (Randour et al., 2021). Consequently, concerns for pets and related planning often operate as obstacles that delay and/or prevent survivors' safety-planning efforts. One study found that 48% of female IPV survivors reported that they had delayed leaving an abusive relationship due to concerns about their pets' safety (Faver & Strand, 2003). A later study found that concern for pets prevented 38% of IPV survivors utilizing community-based DV services in Colorado from being able to engage in effective safety planning (Collins et al., 2018). This qualitative study indicated that concerns for pets in the absence of pet-sheltering services and/or financial security resulted in IPV survivors utilizing alternate resources to find safe options for pets (e.g., Craigslist), which led to delays in leaving an abusive partner and seeking shelter services for their human families. In addition, safety planning may be further complicated *after* a family enters shelter if they have left an animal behind. For example, there is evidence that children's emotional responses to being separated from a beloved pet can influence parenting IPV survivors to return to the abusive environment to reunite the child and pet (Collins et al., 2018).

## Impact on the Animal's Welfare

In light of the extensive use of AA by abusers to coerce and control human family members as described above, it may seem surprising that comparatively less scholarship has been devoted to documenting the impacts on the animals in these situations. In the following section, we describe the knowledge available on this topic to date and provide recommendations on directions for further research to remedy these gaps.

## Physical Harm and Death

The most obvious impact to animals from the intersection of AA and IPV is injury, including death. Little empirical data exist on the nature and extent of injuries sustained by animals in these situations, due in part to lack of attention to this topic in the veterinary literature, the tendency of some survey-based studies to ask only whether an abuser "threatened, harmed, or killed" animals without differentiation, and the number of cases that go unreported. In a three-year study of animal cruelty cases handled by the New York Police Department, Reisman (2019) observed that the mortality rate among animal victims in cases involving alleged IPV (33%) was over three times higher than among animals in cruelty cases not involving alleged IPV. Cats had a higher mortality rate than dogs overall, but among animals weighing less than 15 pounds, mortality rates were similar between cats and dogs. Blunt force trauma was the cause of injury in 77% of all cases and the cause of death in 93% of known cases. The most common cause of injury or death was throwing an animal (against a wall or floor, or from a balcony).

While these graphic acts capture newspaper headlines and public attention, it is important to note that neglect—omission of care, such as depriving an animal of food, water, grooming, appropriate shelter, or veterinary services—can also form part of the dynamic of deliberate AA in the context of IPV. In a small study in Ireland (Gallagher et al., 2008), 22% of families who experienced IPV reported that the abuser had neglected animals. In a qualitative analysis by Johnson (2019), several survivors of IPV reported that the abuser restricted access to pet food and veterinary care, including both routine care and care for illnesses/injuries. Omissions of care may also include depriving animals of companionship or other enrichment, such as locking a dog in a small room alone. Among animals served by the PetSafe DV safe haven program at Purdue University across a 15-year timespan (Valiyamattam et al., 2022), the most common veterinary issues were dental disease and dermatological issues, though it was not reported whether these were the result of deliberate omissions of care in the context of the abuse. Further attention to this topic by researchers and practitioners is warranted, particularly given that acts of omission may be less likely to be reported to, or receive recognition from, law enforcement or animal welfare professionals.

## Psychological and Behavioral Impacts

In a recent survey by the Urban Resource Institute and the National Domestic Violence Hotline (2021), 76% of pet-owning individuals accessing the hotline reported that their pets' behavior had changed because of the abuse in the home. However, little scholarly attention has been devoted to examining psychological/behavioral effects of animals' exposure to (direct and/or indirect) abuse, either in the IPV context or in general. Given the burgeoning literature on sequelae of trauma in humans, taken together with the use of laboratory animal models of posttraumatic stress (Flandreau & Toth, 2017) and behavioral effects of traumatic brain injury (Shultz et al., 2020), it seems likely that animals in families experiencing IPV are also at risk for adverse psychological and behavioral consequences. It has been hypothesized that pets living in households where IPV occurs may be at increased risk for aggression as a result of being maltreated, neglected, and/or receiving negative training methods that reinforce aggressive behavior, contributing to a cycle of increased use of reactive methods for controlling and punishing the pet's behavior (DeGue, 2011; Hardesty et al., 2013). At the same time, resilience is an important factor in determining response to trauma in both humans and animals (Zheng et al., 2022), so it should not be assumed that all animals exposed to abuse experience adverse effects.

## *Relinquishment*

In addition to the potential for physical and/or psychological/behavioral effects, animals at the intersection of IPV and AA may be at increased risk for relinquishment: surrender to an animal shelter, rehoming, euthanasia, or abandonment. They may be relinquished by an abuser as a form of emotional abuse, or by a victim: either due to perceived inability to provide care for the animals (including care for injuries/illnesses/behavior issues caused by the abuser), or as a way to protect them from further abuse. This latter phenomenon may be especially likely in circumstances where individuals are attempting to escape the abusive relationship and encounter barriers to bringing their animals with them, such as a lack of animal-friendly housing.

There are no statistics on the number of animals relinquished due to violence in the family. However, research on overall reasons for (companion) animal relinquishment reveals housing and "caretakers' personal issues, such as relationship problems" as two commonly studied and reported categories of relinquishment reasons (Coe et al., 2014, p. 262); these likely include cases of IPV-related relinquishment.

## Practical and Policy Perspectives

The impacts—actual and potential—of the intersection of IPV and AA on both humans and animals illustrate several avenues for prevention and intervention. The following section addresses initiatives within human- and animal-focused service systems, as well as those by policymakers and the many other stakeholders in this topic, to protect and improve the safety, health, and well-being of both humans and animals through attention to these connections. For purposes of this chapter, we limit coverage of laws and programs to the United States.

## *Co-Housing and Safe Haven Programs*

Growing recognition of the nature, extent, and impacts of the overlap between IPV and AA has led to programmatic responses within both human services and animal welfare. Greatest attention has been given to emergency housing for animals whose owners/caretakers are seeking safety, since most DV and homeless shelters do not allow animals. Both onsite ("co-housing"; "co-located"; "co-sheltering") and offsite ("safe haven"; "animal safekeeping") models have emerged, with some (though only a minority of) programs offering both. Each model itself subsumes a variety of configurations. Co-housing may consist of dedicated animal kennel areas at the shelter, allowing residents to keep their animals in their rooms, or housing individuals with pets at an alternative location such as a hotel/motel, while offsite housing may comprise partnerships between DV organizations and animal sheltering organizations, veterinarians, animal boarding facilities, volunteer foster homes, and/or correctional institutions.

To date, no research has directly compared the effectiveness—in terms of improving safety and well-being for animals and people—of the different housing models described above. However, the aforementioned survey of national DV hotline callers indicated that 50% would not consider entering a shelter without their pets, while only 10% stated that they would place their pets in temporary boarding or foster care if the shelter did not accommodate them; the remaining 40% stated that they would place them with family or friends (Urban Resource Institute & National Domestic Violence Hotline, 2021). In contrast, clients of an offsite program reported trusting that their animals were safe, and several stated that it was a relief to have their animals temporarily cared for by others because this allowed them to focus on their own recovery (Johnson, 2019). It is plausible that different program models are ideally suited to individual circumstances and human needs; moreover, offering choice is consistent with the principles of trauma-informed care. The individual needs of animals also warrant alternative options in housing models: for example, animals displaying generalized fear or aggression may not be candidates for onsite housing, whereas those displaying severe separation anxiety may need to remain with their caretakers wherever possible. Situations involving large numbers of animals, or species that cannot be accommodated onsite (e.g., horses or livestock), may require creative placement options. Additional considerations in placement include situations in which abusers attempt to locate the animals, communities where the presence of animals outside a DV shelter (e.g., in outdoor runs) or in foster homes (e.g., walking an unusual dog in a small town) may threaten confidentiality, and laws in some states (e.g., North Carolina) that treat relinquishment to animal shelters (including temporary housing programs) as public record. Agencies developing new programs, or expanding existing ones, should consider the needs of both humans and animals in their communities when allocating resources, and should seek to offer as much flexibility as practicable.

Agencies offering services of this type should also consider issues of accessibility, starting with awareness: if the existence of the program is not mentioned in public-facing materials (such as the host agency's website), or by staff upon contact with the agency (including on crisis line calls and upon shelter intake), prospective clients may assume that no such service exists and consequently leave animals behind or delay/cease seeking safety. Directories of co-housing and safe haven programs, such as the Safe Havens Mapping Project (Animal Welfare Institute, 2022) and Safe Place For Pets (RedRover, 2022), assist in remedying this gap, though only for programs that consent to having their information listed. Notably, by allowing those seeking help to search either by state or by current location, these directories make it possible to locate the nearest program geographically, even if it is across state lines or in a different agency's service area. For individuals facing barriers to transportation or living in remote or other underserved areas, this radius search capability may be a valuable tool in reaching assistance.

## *Additional Housing (and Housing-Related) Considerations*

Some victims of IPV may choose not to leave their homes, yet still seek temporary safekeeping for their animals. Not all who do leave their homes enter shelters; nevertheless, animals may also be unwelcome in other temporary accommodations, such as hotels, short-term rentals, or staying with family/friends. Even when animals are allowed, financial barriers such as pet deposits, pet "rent", or higher rental rates for pet-friendly accommodations may constrain options. Similar considerations apply for individuals seeking transitional or permanent housing following a stay in emergency/temporary accommodation: there is a nationwide lack of affordable pet-friendly housing due to restrictions and fees imposed by landlords, property managers, and insurance companies (Pet-Inclusive Housing Initiative, 2021), and even those housing programs operated or subsidized by DV agencies may not allow animals. Additionally, the growth within the DV field of the Housing First approach (Washington State Coalition Against Domestic Violence, 2022), which prioritizes ensuring stable housing for individuals as quickly as possible, implies a need for equal (if not greater) emphasis on longer-term housing solutions for people and their animals compared with the current focus on emergency housing. Funding streams such as California Assembly Bill 415, which went into effect on 1/1/2020 to allow Victim Compensation Board funds to be used both for temporary pet boarding and for pet deposits and "pet rent" in rental housing, offer glimpses of a more flexible and comprehensive approach.

Finally, in the context of a focus on housing overall (see Chapters 3 and 8), it is important to keep sight of other needs that underpin survivors' restoration of safety and well-being, both for themselves and for their animals. Consideration of animals in safety planning; transportation of animals to safety alongside their caretakers, both locally and long-distance; veterinary care, both routine/preventive and related to injuries/illnesses due to abuse; animal food and supplies; and legal advocacy to protect animals and their caretakers from abusers via the court system may each be crucial components in the process of reaching and maintaining safety and well-being. These needs are often not without costs of their own, whether in terms of financial or human resources, and should be considered as part of a comprehensive approach to this issue.

## *Animals in Protection Orders*

As of this writing, 36 states plus Puerto Rico and D.C. have laws explicitly authorizing inclusion of at least some animals (more commonly companion animals) in DV protective orders (R. Wisch, 2022). Typically, these laws authorize judges to direct the care, custody, or control of animal(s) and/or to order abusers (respondents) to refrain from harming them. In some states with no such laws, judges may grant provisions regarding animals in sections pertaining to personal property or "other relief". In the professional experience of one of the authors of this chapter, such provisions have been useful in (for example) authorizing a law-enforcement escort to accompany a survivor to a residence to retrieve animals left behind during a hurried escape, or when the survivor had not known of the availability of co-housing/safe-haven programs as described above. It is important for judges, attorneys, legal advocates, and others working with human survivors of IPV to be aware of statutory provisions regarding inclusion of animals in protection orders, and to provide relevant information (and guidance, as appropriate) to survivors on requesting assistance with their animals in their petitions to the courts for such orders.

## *AA in Statutory Definitions of IPV and Related Acts*

As of March 2022, ten states have enacted laws in which acts of animal cruelty designed to intimidate, coerce, or control an intimate partner meet statutory definitions of DV. In Colorado, such acts also meet the definition of abuse of an elder or at-risk adult. In Florida, which recently became one of the states that explicitly authorize inclusion of animals in protective orders, a preexisting statutory stipulation directs judges to consider intentional injury/killing of a family pet by the respondent as one factor in determining whether the person petitioning for the protective order is currently, or is in imminent danger of becoming, a victim of DV.

Texas, Missouri, and the Navajo Nation include acts (plus, for the latter two, threats) toward animals in their definitions of stalking (National Link Coalition, 2020). Similarly, the Pet and Women Safety (PAWS) Act, signed into law as part of the 2018 Farm Bill, expands federal stalking definitions to include threats/harm toward pets, horses, and service or emotional support animals; it also creates a grant program that includes funding for animal housing and expenses (Agriculture Improvement Act, 2018).

Other states have included threats and/or harm to animals in their definitions of emotional abuse (Arkansas), coercion (Oregon), or "DV animal cruelty" (Indiana). Statutes typically reference the deliberate nature of the act to control another person's behavior and/or instill fear, and in some cases (e.g., Minnesota) these actions are accompanied by enhanced penalties.

New bills in this vein are being introduced with increasing frequency (National Link Coalition, 2022). To inform best practices and guide future legislative efforts, it is vital to examine the effectiveness of these measures (when enacted) in increasing the safety and well-being of humans and animals. We encourage the development of research questions along these lines, along with innovative approaches to answering them. Unfortunately, evaluation efforts are hampered by lack of systematic tracking of information, and by lack of access to information when it exists. We cannot ascertain whether these policies increase safety and well-being if we cannot determine the extent to which they are being applied: for example, how often are animals being included in protection orders? How often is this inclusion being requested by victims? Are victims and victim-serving agencies even aware the option exists?

## *Reporting and Cross-Reporting*

In 47 states plus D.C., Guam, Puerto Rico, American Samoa, the Northern Mariana Islands, and the U.S. Virgin Islands, at least some professions are mandated by law to report child maltreatment. Further, in 18 states, everyone is a mandated reporter of child maltreatment, and in eight states, everyone is a mandated reporter of elder abuse; this means that animal care professionals, by definition, must report such acts. No reciprocal statutes designate all individuals as reporters of animal cruelty. However, in 20 states and Guam, veterinarians are mandated reporters of at least some types of animal cruelty. In 20 additional states, they are permitted reporters. Veterinary technicians are permitted reporters of cruelty in four states, and mandated reporters in Rhode Island. Many states, regardless of whether they mandate reporting, have enacted laws removing civil and criminal liability for veterinary professionals in making good faith reports. However, ten states have no provisions regarding veterinary reporting of cruelty, leaving open questions regarding professional sanctions and liability (Wisch, 2020). Of concern is a finding from a recent national survey of veterinary professionals (Patterson-Kane et al., 2022), in which only 71% of respondents were aware of the laws pertaining to their status as a reporter, and an additional 18% had an incorrect understanding of their reporter status.

Based on recognition of IPV/AA intersections, several states have introduced "cross-reporting" of forms of abuse between agencies. In 11 states plus D.C., animal care/protection professionals (veterinarians and/or animal control officers) are required to report suspected child and/or elder abuse (National Link Coalition, 2021). Reciprocally though not in fully overlapping fashion, child protection professionals are mandated reporters of suspected animal cruelty in eight states plus D.C.; in five of these states, adult/disability protection professionals are also required to report animal cruelty. Child protection personnel are permitted reporters of suspected animal cruelty in an additional five states plus Guam; three of these states also include permitted reporting by adult/disability protection personnel. It is noteworthy that human service professionals outside these roles—for example, a social worker providing independent psychotherapy services, or a counselor teaching at a university—are typically neither mandated nor permitted reporters of animal cruelty, and may be subject to professional sanctions and/or personal liability for disclosing such acts.

## Ethical Considerations

At the juncture of so many risks to both humans and animals, and in a prevention/response framework in which the foci of diverse stakeholders (e.g., DV agencies and animal welfare organizations) may not always align, it is foreseeable that ethical dilemmas emerge. While

## Reporting AA

The intersection of AA and IPV raises both ethical and practical considerations that further complicate the already-complex topic of abuse reporting. For example, if a human victim of IPV has reason to believe that reporting the abuser's maltreatment of the animal to law enforcement will further compromise the safety of the animal and/or human family members (e.g., by triggering reprisal by the abuser) and entreats the veterinarian not to make the report, what should the veterinarian do? Does this answer change depending on whether the veterinarian is in a state that mandates reporting of animal cruelty?

A related set of questions arises concerning the implications of removing animals from the home following reports of maltreatment. Is the welfare of these animals ultimately, on balance, improved? Further, abusers are known to portray their victims as abusers, potentially accusing them of harming (or being unfit to care for) their animals. If an animal is removed from the custody of the human victim following a false accusation, this could constitute another attack on that person without benefit to the animal. Similar dilemmas have occurred in child welfare, resulting in a goal of integrating safety, permanency, and well-being for children (Wilson, 2014). In a critique of the use of AA as a "sentinel" for human violence, Patterson-Kane and Piper (2009) point to the potential for AA to be used as a reason for removing children from homes, out of proportion to its utility as a possible indicator of child maltreatment or other family violence.

## Children and AA Behaviors

A second set of considerations concerns the perpetration of AA by children. Is such an act a valid marker of IPV and/or child maltreatment, and is it a useful predictor of further violence toward animals and/or humans? In our interpretation of and response to the behavior, does it matter whether the child has been exposed to violence? How should juvenile justice, child protection, mental health, and animal protection agencies—not to mention parents/caretakers—address such situations: punitively, therapeutically, or some combination of both? What are the implications for other members of the family? As in the previous section: when the well-being of humans and animals appears to be in conflict—for example, when preserving human safety and/or the human-animal bond comes at the expense of the animal, or conversely achieving good welfare for the animal comes at the expense of human interests—which should prevail? Such questions constitute one of the core challenges of the inherently interdisciplinary work at the heart of "link" issues, which bring together passionate advocates representing what can be starkly different stakeholder perspectives.

## Unanswered Questions and Future Research Demands

Throughout this chapter, we have endeavored to draw attention to areas in which open questions remain, whether from research, practical, or ethical perspectives (in many cases all at once and interconnected). In this final section, we address a need that relates to all the work covered in this chapter: greater understanding of how this topic may or may not manifest differently across populations, including underserved populations. We conclude with a general review of methodological considerations for strengthening future work.

## *Methodological Rigor, Sample Diversity, and the Need for Longitudinal Data*

Though research on the topic of intersections between AA and IPV has grown exponentially in recent years, the methodological quality of much of this work remains lacking compared with other areas of interpersonal violence and public health. Research in this area has also heavily relied on convenience samples, which has contributed to the overrepresentation of White survivors and underrepresentation of diverse racial, ethnic, sexual, gender, and cultural backgrounds/identities. It is critical that future research adequately capture the experiences of all survivors and their animals so that programs/policies are inclusive and maximally impactful. Reliance on convenience samples (e.g., survivors in DV shelters—in other words, only those who have escaped abuse), on self-report in the absence of corroborating sources of data, and on small samples imposes limitations on our understanding of these phenomena, as does the frequent absence of a comparison group (individuals not exposed to IPV, or individuals exposed to IPV whose animals were not threatened or harmed).

Further, while numerous innovative programmatic and policy measures have been implemented in response to this issue as described above, little if any data have been collected on the effectiveness of such measures. What works best—and for whom—in housing humans and animals in IPV situations? Have the hard-fought battles to include animals in protection orders made people and/or animals measurably safer? Does the inclusion of harm to animals in statutory definitions of IPV and related acts in fact deter abusers and/or afford greater protections to human and animal victims? When reporting or cross-reporting incidents of AA and IPV is authorized or even mandated, how often does it occur—and what if any are the effects? Despite the challenges of grappling with "messy" real-world data to ask and answer such questions, research and program evaluation efforts in these areas are urgently needed.

### Potential Questions for Discussion

1. Where does available knowledge on AA in IPV illustrate similarities and differences in effects on humans vs. animals? How might both service systems for humans and those for animals need to adapt to reflect this knowledge?
2. Where are the greatest remaining areas of need in programmatic and/or policy responses to the overlap between AA and IPV—both overall and in your own community?
3. Should the development and triaging of resources in response to this topic be predicated solely on safety risk to humans and/or animals, or also on the human-animal bond? For example, should space in co-located housing programs or safe havens be extended to animals who have not been directly harmed or threatened?
4. Given the gaps in data on this topic, what should be the priorities for research that advances theoretical understanding and/or practical response?

# References

Agriculture Improvement Act of 2018. (2018). https://www.congress.gov/bill/115th-congress/house-bill/2/text

Animal Welfare Institute. (2022). *Safe havens mapping project*. https://safehavensforpets.org/

Ascione, F. R. (1993). Children who are cruel to animals: A review of research and implications for developmental psychopathology. *Anthrozoös, 6*(4), 226–247.

Ascione, F. R. (Ed.). (2008). *The international handbook of animal abuse and cruelty: Theory, research, and application*. Purdue University Press.

Ascione, F. R., & Arkow, P. (Eds.). (1999). *Child abuse, domestic violence, and animal abuse: Linking the circles of compassion for prevention and intervention*. Purdue University Press.

Ascione, F. R., Weber, C. V., Thompson, T. M., Heath, J., Maruyama, M., & Hayashi, K. (2007). Battered pets and domestic violence: Animal abuse reported by women experiencing intimate violence and by nonabused women. *Violence Against Women, 13*(4), 354–373.

Baldry, A. C. (2005). Animal abuse among preadolescents directly and indirectly victimized at school and at home. *Criminal Behaviour and Mental Health, 15*(2), 97–110. https://doi.org/10.1002/cbm.42

Barrett, B. J., Fitzgerald, A., Stevenson, R., & Cheung, C. H. (2020). Animal maltreatment as a risk marker of more frequent and severe forms of intimate partner violence. *Journal of Interpersonal Violence, 35*(23–24), 5131–5156.

Berger, L. M. (2005). Income, family characteristics, and physical violence toward children. *Child Abuse & Neglect, 29*(2), 107–133.

Buchanan, F., & Humphreys, C. (2021). Coercive control during pregnancy, birthing and postpartum: Women's experiences and perspectives on health practitioners' responses. *Journal of Family Violence, 36*, 325–335. https://doi.org/10.1007/s10896-020-00161-5

Campbell, A. M., Thompson, S. L., Harris, T. L., & Wiehe, S. E. (2021). Intimate partner violence and pet abuse: Responding law enforcement officers' observations and victim reports from the scene. *Journal of Interpersonal Violence, 36*(5–6), 2353–2372.

Cares, A. C., Reckdenwald, A., & Fernandez, K. (2021). Domestic violence and abuse through a sociological lens. In J. Devaney, C. Bradbury-Jones, R. J. Macy, C. Øverlien & S. Holt (Eds.), *The Routledge international handbook of domestic violence and abuse* (pp. 40–54). Routledge.

Carlisle-Frank, P., Frank, J. M., & Nielsen, L. (2004). Selective battering of the family pet. *Anthrozoös, 17*(1), 26–42.

Caviola, L., Everett, J. A., & Faber, N. S. (2019). The moral standing of animals: Towards a psychology of speciesism. *Journal of Personality and Social Psychology, 116*(6), 1011.

Cleary, M., Thapa, D. K., West, S., Westman, M., & Kornhaber, R. (2021). Animal abuse in the context of adult intimate partner violence: A systematic review. *Aggression and Violent Behavior, 61*, 101676.

Coe, J. B., Young, I., Lambert, K., Dysart, L., Nogueira Borden, L., & Rajić, A. (2014). A scoping review of published research on the relinquishment of companion animals. *Journal of Applied Animal Welfare Science, 17*(3), 253–273. https://doi.org/10.1080/10888705.2014.899910

Collins, E. A., Cody, A. M., McDonald, S. E., Nicotera, N., Ascione, F. R., & Williams, J. H. (2018). A template analysis of intimate partner violence survivors' experiences of animal maltreatment: Implications for safety planning and intervention. *Violence Against Women, 24*(4), 452–476.

Currie, C. L. (2006). Animal cruelty by children exposed to domestic violence. *Child Abuse & Neglect, 30*(4), 425–435.

DeGue, S. (2011). A triad of family violence: Examining overlap in the abuse of children, partners, and pets. In C. Blazina, G. Boyraz & D. Shen-Miller (Eds.), *The psychology of the human-animal bond* (pp. 245–262). New York: Springer.

Faver, C. A., & Cavazos, A. M. (2007). Animal abuse and domestic violence: A view from the border. *Journal of Emotional Abuse, 7*(3), 59–81.

Faver, C. A., & Strand, E. B. (2003). To leave or to stay? Battered women's concern for vulnerable pets. *Journal of Interpersonal Violence, 18*(12), 1367–1377.

Faver, C. A., & Strand, E. B. (2007). Fear, guilt, and grief: Harm to pets and the emotional abuse of women. *Journal of Emotional Abuse, 7*(1), 51–70.

Fitzgerald, A. J., Barrett, B. J., & Gray, A. (2022). The co-occurrence of animal abuse and intimate partner violence among a nationally representative sample: Evidence of "the link" in the general population. *Violence and Victims, 36*(6), 770–792.

Flandreau, E. I., & Toth, M. (2017). Animal models of PTSD: A critical review. *Current Topics in Behavioral Neurosciences, 38,* 47–68. https://doi.org/10.1007/7854_2016_65

Flynn, C. P. (2000). Woman's best friend: Pet abuse and the role of companion animals in the lives of battered women. *Violence Against Women, 6*(2), 162–177.

Fox, G. L., & Benson, M. L. (2004). Violent men, bad dads? Fathering profiles of men involved in intimate partner violence. In R. D. Day & M. E. Lamb (Eds.), *Conceptualizing and measuring father involvement,* 359–384.

Gallagher, B., Allen, M., & Jones, B. (2008). Animal abuse and intimate partner violence: Researching the link and its significance in Ireland - A veterinary perspective. *Irish Veterinary Journal, 61*(10), 658–667. https://doi.org/10.1186/2046-0481-61-10-658/TABLES/2

Girardi, A., & Pozzulo, J. D. (2015). Childhood experiences with family pets and internalizing symptoms in early adulthood. *Anthrozoös, 28*(3), 421–436. https://doi.org/10.1080/08927936.2015.1052274

Gullone, E. (2014). Risk factors for the development of animal cruelty. *Journal of Animal Ethics, 4*(2), 61–79. https://doi.org/10.5406/janimalethics.4.2.0061

Gullone, E., & Robertson, N. (2008). The relationship between bullying and animal abuse behaviors in adolescents: The importance of witnessing animal abuse. *Journal of Applied Developmental Psychology, 29*(5), 371–379.

Haden, S. C., McDonald, S. E., & D'Emilia, W. (2022). Psychopathy and animal cruelty offenders. In P. B. Marques, M. Paulino, & L. Alho (Eds.), *Psychopathy and criminal behavior* (pp. 445–468). Academic Press.

Haden, S., McDonald, S. E., & Murphy, J. (2020). Violence against family pets. In T. Shackelford (Ed.) *SAGE handbook of domestic violence.* Los Angeles, CA: SAGE Publications.

Hardesty, J. L., Khaw, L., Ridgway, M. D., Weber, C., & Miles, T. (2013). Coercive control and abused women's decisions about their pets when seeking shelter. *Journal of Interpersonal Violence, 28*(13), 2617–2639.

Harris Poll. (2015, July 6). *More than ever, pets are members of the family.* https://www.prnewswire.com/news-releases/more-than-ever-pets-are-members-of-the-family-300114501.html

Hartman, C. A., Hageman, T., Williams, J. H., & Ascione, F. R. (2018). Intimate partner violence and animal abuse in an immigrant-rich sample of mother–child dyads recruited from domestic violence programs. *Journal of Interpersonal Violence, 33*(6), 1030–1047.

Hawkins, R. D., McDonald, S. E., O'Connor, K., Matijczak, A., Ascione, F. R., & Williams, J. H. (2019). Exposure to intimate partner violence and internalizing symptoms: The moderating effects of positive relationships with pets and animal cruelty exposure. *Child Abuse & Neglect, 98,* 104166.

Henry, B. C. (2004). The relationship between animal cruelty, delinquency, and attitudes toward the treatment of nonhuman animals. *Society and Animals, 12*(3), 185–208. https://doi.org/10.1163/1568530042880677

Hensley, C., & Tallichet, S. E. (2005a). Animal cruelty motivations: Assessing demographic and situational influences. *Journal of Interpersonal Violence, 20*(11), 1429–1443. https://doi.org/10.1177/0886260505278714

Hensley, C., & Tallichet, S. E. (2005b). Learning to be cruel?: Exploring the onset and frequency of animal cruelty. *International Journal of Offender Therapy and Comparative Criminology, 49*(1), 37–47. https://doi.org/10.1177/0306624X04266680

Hoffman, C. L., Thibault, M., & Hong, J. (2021). Characterizing pet acquisition and retention during the COVID-19 pandemic. *Frontiers in Veterinary Science, 8,* 781403.

Johnson, T. C. (2019). *Survivors' experiences of pet abuse within the cycle of domestic violence.* (Publication No. 10933222) [Doctoral dissertation, Walden University]. ProQuest Dissertations and Theses Global.

Laurent-Simpson, A. (2017a). Considering alternate sources of role identity: Childless parents and their animal "kids". *Sociological Forum, 32*(3), 610–634.

Laurent-Simpson, A. (2017b). "They make me not wanna have a child": Effects of companion animals on fertility intentions of the childfree. *Sociological Inquiry, 87*(4), 586–607.

Lee, L. C., Kotch, J. B., & Cox, C. E. (2004). Child maltreatment in families experiencing domestic violence. *Violence and Victims, 19*(5), 573–591.

Lockwood, R., & Ascione, F. R. *Cruelty to animals and interpersonal violence: Readings in research and application*. Purdue University Press.

McConnell, A. R., Lloyd, E. P., & Buchanan, T. M. (2017). Animals as friends: The social psychological implications of human–pet relationships. In M. Hojjat & A. Moyer (Eds.), *The psychology of friendship* (pp. 157–174). Oxford University Press.

McDonald, S. E., Cody, A. M., Collins, E. A., Stim, H. T., Nicotera, N., Ascione, F. R., & Williams, J. H. (2018). Concomitant exposure to animal maltreatment and socioemotional adjustment among children exposed to intimate partner violence: A mixed methods study. *Journal of Child & Adolescent Trauma, 11*(3), 353–365.

McDonald, S. E., Collins, E. A., Nicotera, N., Hageman, T. O., Ascione, F. R., Williams, J. H., & Graham-Bermann, S. A. (2015). Children's experiences of companion animal maltreatment in households characterized by intimate partner violence. *Child Abuse & Neglect, 50*, 116–127.

McDonald, S. E., Collins, E. A., Maternick, A., Nicotera, N., Graham-Bermann, S., Ascione, F. R., & Williams, J. H. (2019). Intimate partner violence survivors' reports of their children's exposure to companion animal maltreatment: A qualitative study. *Journal of Interpersonal Violence, 34*(13), 2627–2652.

Mueller, M. K., King, E. K., Callina, K., Dowling-Guyer, S., & McCobb, E. (2021). Demographic and contextual factors as moderators of the relationship between pet ownership and health. *Health Psychology and Behavioral Medicine, 9*(1), 701–723.

National Link Coalition. (2020). *State statutes in which acts of animal abuse constitute acts of domestic violence or elder abuse*. https://nationallinkcoalition.org/wp-content/uploads/2020/11/DV-CTA-is-definition-of-DV-EA-2020.pdf

National Link Coalition. (2022). The LINK...in the legislatures. *The LINK-Letter, 15*(3), 13–23. https://nationallinkcoalition.org/wp-content/uploads/2022/03/LINK-Letter-2022-March.pdf

Onyskiw, J. E. (2007). The link between family violence and cruelty to family pets. *Journal of Emotional Abuse, 7*(3), 7–30.

Patterson-Kane, E. G., Kogan, L. R., Gupta, M. E., Touroo, R., Niestat, L. N., & Kennedy-Benson, A. (2022). Veterinary needs for animal cruelty recognition and response in the United States center on training and workplace policies. *Journal of the American Veterinary Medical Association, 260*(14), 1853–1861.

Patterson-Kane, E. G., & Piper, H. (2009). Animal abuse as a sentinel for human violence: A critique. *Journal of Social Issues, 65*(3), 589–614.

Pet-Inclusive Housing Initiative. (2021). *2021 Pet-inclusive housing report: Research and resources for pet-inclusive rental housing*. https://www.foundanimals.org/pets-and-housing/2021-pet-inclusive-housing-report/

Randour, M. L., Smith-Blackmore, M., Blaney, N., DeSousa, D., & Guyony, A. A. (2021). Animal abuse as a type of trauma: Lessons for human and animal service professionals. *Trauma, Violence, & Abuse, 22*(2), 277–288.

RedRover. (2022). *Safe place for pets*. https://safeplaceforpets.org/

Reisman, R. (2019). *Survey of New York City animal cruelty cases (2014–2016) with associated human interpersonal violence* [Unpublished master's thesis]. University of Florida.

Riggs, D. W., Taylor, N., Fraser, H., Donovan, C., & Signal, T. (2021). The link between domestic violence and abuse and animal cruelty in the intimate relationships of people of diverse genders and/or sexualities: A binational study. *Journal of Interpersonal Violence, 36*(5–6), NP3169–NP3195. https://doi.org/10.1177/0886260518771681

Rowan, A. N. (1999). Cruelty and abuse to animals: A typology. In F. R. Ascione & P. Arkow (Eds.), *Child abuse, domestic violence, and animal abuse: Linking the circles of compassion for prevention and intervention* (pp. 328–334). West Lafayette, IN: Purdue University Press.

Schwartz, J. P., Hage, S. M., Bush, I., & Burns, L. K. (2006). Unhealthy parenting and potential mediators as contributing factors to future intimate violence: A review of the literature. *Trauma, Violence, & Abuse, 7*(3), 206–221.

Shultz, S. R., McDonald, S. J., Corrigan, F., Semple, B. D., Salberg, S., Zamani, A., Jones, N. C., & Mychasiuk, R. (2020). Clinical relevance of behavior testing in animal models of traumatic brain injury. *Journal of Neurotrauma, 37*(22), 2381–2400. https://doi.org/10.1089/neu.2018.6149

Stark, E. (2012). Looking beyond domestic violence: Policing coercive control. *Journal of Police Crisis Negotiations, 12*(2), 199–217.

Tajima, E. A. (2000). The relative importance of wife abuse as a risk factor for violence against children. *Child Abuse & Neglect, 24*(11), 1383–1398.

Thompson, K., & Gullone, E. (2006). An investigation into the association between the witnessing of animal abuse and adolescents' behavior toward animals. *Society & Animals, 14*(3), 221–243.

Tiplady, C. M., Walsh, D. B., & Phillips, C. J. (2012). Intimate partner violence and companion animal welfare. *Australian Veterinary Journal, 90*(1–2), 48–53.

Travers, C., Dixon, A., Thorne, K., & Spicer, K. (2009). Cruelty towards the family pet: A survey of women experiencing domestic violence on the Central Coast, New South Wales. *Medical Journal of Australia, 191*(7), 409.

Urban Resource Institute, & National Domestic Violence Hotline. (2021). *Domestic violence and pets: PALS report and survey: Breaking barriers to safety and healing.* https://urinyc.org/wp-content/uploads/2021/05/URI-PALS-Report.pdf

Valiyamattam, G. J., Kritchevsky, J., & Beck, A. M. (2022). Demographics and outcome of dogs and cats enrolled in the PetSafe program at the Purdue University College of Veterinary Medicine: 2004–2019. *Journal of the American Veterinary Medical Association, 260*(2), 228–233.

Volant, A. M., Johnson, J. A., Gullone, E., & Coleman, G. J. (2008). The relationship between domestic violence and animal abuse: An Australian study. *Journal of Interpersonal Violence, 23*(9), 1277–1295.

Volsche, S. (2019). Understanding cross-species parenting: A case for pets as children. In L. R. Kogan & C. Blazina (Eds.), *Clinician's guide to treating companion animal issues* (pp. 129–141). London: Academic Press.

Washington State Coalition Against Domestic Violence. (2022). *Domestic violence housing first.* https://wscadv.org/projects/domestic-violence-housing-first/

Wilson, C. (2014). *Integrating safety, permanency and well-being: A view from the field.* https://www.acf.hhs.gov/sites/default/files/documents/cb/wp_overview.pdf

Wisch, R. F. (2020). *Table of veterinary reporting requirement and immunity laws.* https://www.animallaw.info/topic/table-veterinary-reporting-requirement-and-immunity-laws

Wisch, R. (2022). *Domestic violence and pets: List of states that include pets in protection orders.* Animal Legal and Historical Center at Michigan State University Website. https://www.animallaw.info/article/domestic-violence-and-pets-list-states-include-pets-protection-orders

World Health Organization. (2021). *Violence against women.* Geneva, Switzerland (2021). https://www.who.int/news-room/fact-sheets/detail/violence-against-women, Accessed 31st Jan 2022

Zheng, K., Chu, J., Zhang, X., Ding, Z., Song, Q., Liu, Z., Peng, W., Cao, W., Zou, T., & Yi, J. (2022). Psychological resilience and daily stress mediate the effect of childhood trauma on depression. *Child Abuse & Neglect, 125*, 105485.

# 24
# INTRODUCING DOGS AND CATS INTO FAMILIES

*Zarah Hedge, Amanda Kowalski, and Madelaine Steevens*

### Before You Bring Your Pet Home

#### *Canine*

##### *Breed and Age Considerations*

Before adopting a companion animal, it is best that potential adopters consider which animal, breed, and species fits their lifestyle. Age can also be relevant, and the decision on which age is appropriate for adoption should be dependent on the potential adopter's lifestyle and time availability. There are five stages of development in the dog; neonatal, transitional, socialization, juvenile, and adult. There may be specific time periods associated with these stages, and it is important to understand that the transition from one stage to another is not abrupt. While this section will discuss common needs for each age, it is important to keep in mind that they can vary by individual.

The neonatal period extends from birth to 14 days, and pups primarily nurse and sleep during this stage. They have very little motor ability until about five days, are unable to hear or see and lack the neurologic reflexes to defecate and urinate on their own (Landsberg et al., 2003; Serpell et al., 2016). Exposing puppies to mild stressors like early handling may help the dog better cope with stress later in life and create a more emotionally stable dog (Landsberg et al., 2003).

In between weeks 2–3, the pup enters the second stage of development, known as the transitional stage. During this stage, their eyes and ears begin to open, they become more coordinated, and begin to walk around (Peterson, & Kutzler, 2011). Behaviors that help facilitate social interactions, like tail wagging, will also appear during the latter part of this stage (Landsberg et al., 2003). During the transitional stage, it is good to gently expose the pup to various stimuli, such as different ground textures, sounds, and visual cues, for short periods each day.

The next period of development is the socialization period, which is approximately 3–12 weeks. At about 8–9 weeks of age, the puppies are attracted to the odors of urine and feces. They begin elimination in specific areas to avoid soiling their den. They can respond to sight or sound and have investigative behavior and recognize novel objects (Pluijmakers et al., 2010). This is also a critical stage because this is when puppies can begin to form social relationships or attachments (Serpell et al., 2016). Their social behavior and relationships are formed during this time due to interactions with littermates, their mother, or the environment (Hepper, 1985). Development of positive relationships and associations is beneficial (Peterson, & Kutzler, 2011). It is important to avoid psychological stress because puppies can develop a fear response during this developmental stage (Landsberg et al., 2003).

During the last two stages of a dog's life, from juvenile to adulthood, the dog's basic learning capacities and environmental exploration increase. Individual dogs mature at different ages; however, most breeds mature around 18 months and are then considered an adult. While many have concerns about the shorter life span of adopting an older pet and having a larger veterinary bill, there are many benefits to adopting a senior pet. Some senior pets have experience living in previous homes and may settle in quicker than a young pet who has never been in a home before. Older, more mature pets are also already full grown and have developed their personalities. This enables the adopter to understand if that pet has a personality that works with their lifestyle rather than a younger developing animal.

Before adopting a dog, it is important to consider which breed is best for the potential owner. Selecting a breed that best fits the owner and their individual family dynamic is important. Adopting a dog from a shelter saves a pet's life and reduces the burden on shelter facilities. There may be purebred dogs available for adoption in the shelter. Despite that, some owners may prefer to adopt from a breeder. If you choose to adopt from a breeder, it is important to take responsibility for the purchase of that pet and confirm that the breeder is practicing ethical breeding standards. Adopting a pet from commercial breeders, brokers, or retail stores is not recommended as those typically do not practice safe standards of care (ASPCA, 2021).

## *Preparing the Home Environment*

Before bringing home a new dog, the home environment should be assessed to meet the dog's individual needs including considerations for age, activity level, and personality. Any transition into a new environment can potentially lead to stress for both the dog and the owner.

Essential items necessary for dog ownership include a leash, collar, a front attaching harness for those dogs that like to pull, a bait bag to hold your treats and poop bags, soft bedding, food, bowls, pee pads for puppies, and a crate. All dogs will need a designated place to sleep, eat, and eliminate. Incomplete house training is one of the most reported incompatible behaviors in dogs (Learn et al., 2020; Martinez et al., 2011; Voith & Borchelt, 1985; Yeon et al., 1999) and it is a most common reason for relinquishment to a shelter (Beaver, 1994; Learn et al., 2020; Patronek et al., 1996; Salman et al., 2000). Dog owners should assume that, regardless of age, a new dog may not be sufficiently housetrained and accidents are expected. Additionally, there is some evidence to suggest that smaller dogs have a higher likelihood of incomplete house training than larger dogs, which could suggest that they may take longer to housetrain (Learn et al., 2020).

To set any dog up for success, it is essential that the owner is sure to establish a feeding and elimination routine. This will help to better predict when the dog may need to eliminate. Other activities that may predict the need to eliminate are when the dog has woken up from sleeping, even after a nap, as well as after play sessions. When outdoors, always take the dog out on a leash and go to the nearest elimination area. Once the dog has eliminated, use positive reinforcement such as praising or providing a treat immediately after to increase the chances the dog continues eliminating in the designated spot. If the dog has an accident in the home, punishment should be avoided as this will contribute to the development of fear of the owner. Instead, clean the area well with an enzymatic cleanser.

In addition to having a plan for housetraining, many dogs exhibit natural chewing behaviors. Identifying areas that could result in potential damage from the dog or to the dog is a great way for owners to be proactive. Puppies and adolescent dogs reportedly show high levels of chewing activity, especially those that are teething (Arhant et al., 2021). Providing enriching toys or treats that the dog can chew on is important to reduce boredom and prevent home destruction. Changes in daily routines, canceling activities such as walks, or participating in highly arousing events such as visits to the vet or having visitors to the home are activities related to dogs chewing on objects and suggest that dogs might use chewing to cope with negative emotions (Arhant et al., 2021). These findings support the recommendations for pet owners to provide as much consistency as possible, including appropriate outlets for chewing behaviors.

Items such as crates, exercise pens, and baby gates are helpful tools to create designated safe spaces for the dog and can also be useful in managing their behavior. It is important that owners understand that while some dogs may find crates instantly comforting, most dogs will need help learning how to enjoy spending time in the crate, especially if it is to be used for confinement. Ideally, the crate will have soft bedding inside and will be located in a quiet place. For dogs that display more anxious or fearful behaviors when confined to a crate, using an exercise pen or baby gates to block off access to certain parts of the home can provide a less stressful experience for the dog.

When bringing a dog home, it is recommended to enroll in a basic manners class or working with a private certified professional trainer in order to help the new dog and owner learn skills that allow them to communicate clearly and live together harmoniously. Choosing a class that uses reward-based techniques to teach common skill has been known to have a positive impact on short and long-term welfare, training effectiveness, and dog-owner relationships, including increased attentiveness to the owner (AVSAB, 2021). These skills will also allow for positive strategies to manage situations such as moving them off of furniture or navigating greetings with humans and other animals.

## *Introducing Your New Dog to Other Humans*

To ensure that the transition for a new pet goes as smoothly as possible, there is much to consider, including how to set up successful introductions to humans and other pets in the family. Both dogs and cats can feel overwhelmed if they are expected to meet too many humans and pets at once. When you bring your pet home, be sure to have a plan for introducing them to their immediate family, humans and animals they might meet on a walk, and guests that visit your home.

Proper introductions can set the tone for relationships in the future. It is important that dog owners are familiar with and can interpret canine body language. Deficiencies in this area can lead to miscommunications and misunderstandings about what message the dog might be trying to convey. Fear Free Happy Homes (an online service) provides free education to pet owners on a variety of behavioral topics including canine body language.

As you prepare for introducing your new dog to other members of the family, be sure you and any other person meeting the dog have plenty of small, soft, and squishy treats available to reward desirable behavior during the introduction. No matter the age, it is important that the dog feels as if they have choice and control during the introduction.

For safety, have the dog on a loose leash while you remain standing holding the leash. Instruct the person to remain standing or sitting relaxed, with their body slightly angled to the side. Allow the dog to approach and sniff the person on their own and when they do, this is a great opportunity for the person to reward their approach. If the dog is excited and starting to jump up, using a treat, you can lure them away and into a sitting position. Once they are sitting, you can try the approach again. The person greeting the dog should wait until the dog is sitting or standing before any further interactions. For dogs that may display more fear-based behaviors, have the person toss treats away and allow the dog to approach as they feel comfortable. It is important not to use the dog's interest in the treats to force an interaction by luring them closer to the person they are meant to greet. This can have the opposite effect, causing the dog to have an increased fear response once they finish eating the treat and realize they are too close for comfort. Other things to consider is the person's body language. Looming and leaning over the dog can be quite overwhelming and is best avoided.

When introducing your new dog to a child, these tips are equally, if not more, important. Children make up a large percentage of dog bite victims (CDC, 2021; Dixon et al., 2012). Family Paws® Parent Education program provides parents and dog owners with resources to ensure that both the dog and child have safe interactions. If the dog chooses to stay away, it is important that the child understands that they should leave them alone. Accidents can happen quickly, so parents should teach children to avoid chasing, wrestling, tugging, putting their face right in the dog's face and that most dogs don't like hugs (Maddie's Fund, 2019). Dogs and children should never be left unattended. Parents should practice active supervision when the dog is in the same room as the child, carefully observing both of their body languages and behaviors.

## *Introducing Your New Dog to Other Dogs*

Introductions to an existing dog in the home start before the new dog arrives. It is essential that the owner prepares the environment by picking up toys or chews, adds another location for food and water dishes, as well as beds or crates. Dog-directed aggression is commonly reported by owners, with 22% of cases happening outside of the home on walks and 8% directed toward dogs living together in the home (Casey et al., 2013). Following these steps will help to minimize any potential for conflict among the dogs once they are in the house together.

The next step involves two handlers, one for each dog. With both dogs on leash and starting in a neutral area such as a nearby park or on a walk in the neighborhood, begin a walking parallel to each other (Horwitz, 2018; San Diego Humane Society, 2021). As you are walking in the same direction, you may gradually close the distance between the dogs until eventually they are close enough to meet. It is important that both dogs have calm and relaxed body language. This technique can be used to introduce unfamiliar dogs your dog may meet on a walk, or when they are visiting another dog's home.

Once introduced outside, the dogs should continue to be monitored over the next several days in the home to ensure both dogs are acclimating to each other well. The dogs should be completely separated when eating to avoid the potential for resource guarding behaviors and should be kept physically separated when they are left alone and unattended (Horwitz, 2018).

## *Introducing Your New Dog to Cats*

Whether you are introducing a new dog to a cat or a new cat to a dog, introductions between dogs and cats should be taken slowly and with great care. For the first week, it is best to keep the dog and cat separated in different rooms. They will both be able to smell, and hear each other, and this will allow for them to begin to get accustomed to each other without visual exposure.

When the day comes to introduce the dog to your cat, be sure that you have all the safety measures in place. Keeping the dog on leash will provide a level of safety, in case either one appears to be feeling unsafe. Having your bait bag filled with tasty treats on your person can help in this case to redirect their attention and keep the interaction as calm as possible. Another helpful tool during this initial introduction is the exercise pen or baby gate. In cases where the dog is having a difficult time refocusing on the treats or the new owner, then these can be used in addition to the dog being on leash to allow for a safer introduction. Separating the dog and cat at night or when there are no humans around to actively supervise is best to prevent accidents.

If problems arise as the dog transitions and during the introductions, it is recommended to reach out to a qualified behavior professional that uses reward-based methods for assistance. The Certification Council for Professional Dog Trainers (CCPDT), International Association of Animal Behavior Consultants (IAABC), Animal Behavior Society (ABS), and the American Veterinary Society of Animal Behavior all have professional searches to find the right professional to work with.

## *Feline*

### *Breed and Age Considerations*

Despite having a reputation for being independent and aloof, the domestic cat is quite the opposite. Cats make wonderful companions, and just like other animals, individual cats have individual personalities. Deciding which cat is right for you depends on a variety of factors, from breed and age considerations to the personality of the individual cat to your lifestyle and expectations for the cat in the home. When choosing a personality, consider whether you want a cat that is social and active, one that will sit on your lap and is calm, or one somewhere in between. It's important that you find a cat that matches what you want for your family.

The American Veterinary Medical Association (2021) lists six points to consider when deciding if a cat companion will fit into your lifestyle:

1. Do you have time to devote to providing for the cat's needs for care and attention?
2. Is a cat allowed to be kept in your home if you rent?
3. Can you adapt your home environment to allow a cat to express their normal behaviors?
4. How long is your workday and do you have obligations after work that would interfere with caring for a cat?
5. Will your new cat get along with other pets in the home and vice versa? Are you willing to properly introduce a new cat into the home and provide the time and environment to allow for them to adapt?
6. Does anyone in the household have allergies to pet dander or are likely to be intolerant of normal cat behavior?

It is important to research the unique needs and care required for cats and discuss the responsibilities and medical care requirements with your veterinarian before making the decision to bring one into your home. Like other companion animals, they require a proper living environment that enables them to display their normal behaviors to ensure their health and well-being. Some long-haired cats may require daily brushing and grooming to prevent matting of their coats. Cats may develop various medical conditions throughout their life, especially as they get older, and those medical costs and care should be factored into the decision as to if it's appropriate to bring a cat into the home.

Deciding whether to get a kitten or adult cat depends on the owner's expectations for the cat and lifestyle. Kittens may require more work to litter-box train, more frequent feedings and more supervision, as kittens are more prone to get into things and injure themselves. Kittens' personalities are still forming, and it will take time before you know how affectionate, social or active they will be. Adult cats may be less maintenance and their personality is already formed and clear from the start. If there are children in the home, it is important to teach them how to behave around cats, as cats (and especially kittens) can be frightened by loud noises and require more delicate handling. Cats that are shy or fearful, or ones that tend to be more reactive or overstimulated may not be best for families with young children. If there are dogs in the home, it is important to select a cat that has a history of coexisting peacefully with dogs and be prepared to take the time to slowly introduce your new cat to other pets in the home. If there are other cats in the home, it can be helpful to match their temperaments and age groups. For example, if the cat in the home is playful and active, finding another cat that may enjoy that energy level can be helpful. Integrating a new cat into the home with other cats takes time, so owners should have realistic expectations and be willing to make the time and put in the work that is required to do so (San Diego Humane Society, 2021).

Choosing whether to adopt a cat from an animal shelter or rescue versus purchasing one from a breeder is an individual choice. However, in the United States alone, an estimated 3.2 million cats enter animal shelters each year (ASPCA, 2021). Millions of cats worldwide may benefit from being adopted each year, and there are a myriad of cat breeds, personalities, colors, and ages to choose from when adopting from an animal shelter. By adopting a cat from a local animal shelter, you are helping to save a life and reduce the unnecessary euthanasia of hundreds and thousands of cats. If you choose to purchase a cat from a breeder, it is imperative that you research the breed, their specific requirements for care, common medical conditions associated with the breed and be prepared for the financial costs associated with their care. It is also vital that you research the breeder and only purchase from reputable and ethical breeders.

## *Preparing the Home Environment*

It is crucial to understand a cat's environmental needs, which include not only their physical surroundings (both indoors and outdoors) but also how it impacts and affects social interactions with other pets and humans (Ellis et al., 2013). Several studies have found that many cats do not receive adequate environmental enrichment, which can have long-term impacts on their overall well-being (Alho & Pomba, 2016; Grigg & Kogan, 2019). An inadequate environment is associated with increased behavioral and medical conditions (Grigg & Kogan, 2019). An owner's lack of knowledge of feline behavior and unrealistic expectations of how a cat should interact in the home are factors associated with a poor human-cat bond, increased relinquishment of cats to animal shelters and increased euthanasia (Stella & Croney, 2016).

Cats prefer to have a sense of control over their environment. They thrive on routine, predictability and familiarity. When in a fearful situation, they prefer to escape and hide, and will typically fight as a last resort. Cats are considered to be non-obligate in their social behavior, opposite to dogs and humans which are considered to be socially obligate (Heath, 2020). Their capability to be flexible with their social systems means they can live alone or in social groups with other cats and other species (Heath, 2020). Cats form attachment bonds with their owners (Shreve & Utdell, 2015), similar to other domesticated species. The critical socialization period in cats is from two to seven weeks of age. It is during this time that positive handling experiences and interactions with humans result in cats that tend to display less fear, learn quicker and cope with stress better. It also helps to create a stronger human-cat bond and improves future interactions with humans (McMillin, 2002). Kittens are typically adopted after this period of time so to ensure proper socialization and building connections with humans, it will be important to provide positive experiences to a variety of people, places, objects, and handling.

An inadequate environment and poor interactions with other pets or humans in the home can lead to displays of negative feline emotions, including frustration, fear and anxiety, and pain (Heath, 2018). Cats often avoid showing obvious signs of pain, illness, stress, and anxiety, making it vitally important that owners understand common signs, subtle as they may be, so they can address them in a timely manner. Behavioral signs of these negative emotions include an increase in hiding or withdraw behavior, decrease in general grooming behaviors or overgrooming a specific area, decrease in activity level, change in eating or drinking habits, changes in mood, confrontational behavior toward humans or other animals in the home, and even purring (Heath, 2018; Merola & Mills, 2016). If these or other abnormal behaviors are noted, it is important to consult with a veterinarian and behaviorist to determine if there are underlying medical conditions and whether there are stressors or other triggers in the home environment leading to these behaviors and/or medical issues.

The Association of Feline Practitioners (AAFP) and the International Society of Feline Medicine (ISFM) describe five pillars of a healthy feline environment. The first pillar is to provide a safe place for each cat, where they can retreat and feel protected. This can include options for hiding as well as perching (Ellis et al., 2013). Cats prefer elevated surfaces so they can monitor their surroundings. Having a place to hide reduces behavioral stress, which in turn will reduce stress-related illnesses (van der Leij et al., 2019). There are a variety of cat trees, furniture, and other products to provide hiding and elevated spaces for cats. Using the cat carrier as a hiding space is an easy option, and it can help reduce stress during travel to veterinary visits or other places since the cat views the carrier as a safe location.

The second pillar to provide a healthy environment for cats is to allow for multiple and separated resources, which include feeding and drinking stations, scratching, toileting, play, and resting areas (Ellis et al., 2013). This is important for both single and multi-cat households and allows for cats to make choices as well as access these resources without competition from other cats in the home. In multi-cat homes, competition between cats, such as blocking access to resources like the litter box or food bowl, are significant causes of negative emotions, and subsequent frustration, stress, and anxiety behaviors in cats (Heath, 2020). Owners should provide one litter box per cat plus one more. Litter boxes should be large enough for the cat to comfortably fit and be able to maneuver around in. Larger cats will require more space and bigger is better for all cats in general. Like food and water, litter boxes should be placed apart from one another in the home and in different locations, including spots away from loud noises or foot traffic in the home (Herron, 2010). Cats prefer clumping, unscented, sand-like litter. Perhaps the most important aspect of providing an accessible toileting area for cats is management of the litter box. Litter boxes should be cleaned daily, new litter added as needed and boxes washed at least monthly (Carney et al., 2014). Failure to provide the appropriate litter box and failure to clean them frequently can lead to inappropriate urination issues in cats (Lees, 2021). This can also be caused by a medical condition, so seeking veterinary care when this occurs is important.

As with litter boxes, cats display different preferences for water dishes. Wild cats get a large portion of their water intake from prey they eat. Many domestic cats today are fed dry commercial pet food, and many do not drink as much water as they should. This can lead to various medical conditions, so it is important to find ways to increase water intake for cats. One method is to include canned or wet food into their diet, which has a higher water content than dry food. Cats can be selective with what food types and shapes they prefer, so not all cats will eat canned food. Other options include adding flavor (like tuna juice) to their water to entice them to drink more. Dynamic water sources include water bowls that provide circulating or free-falling water and may encourage some cats to drink more, although this has not been found to be true for all cats (Robbins et al., 2019).

Pillar three is providing opportunity for play and predatory behavior. Cats in the wild eat multiple small meals per day. Domestic cats in a home tend to be fed twice daily or ad libitum due to human convenience (Delgado & Dantas, 2020). Providing opportunities for cats to display natural predatory and foraging behaviors through feeding modalities is a great way to provide for pillar number three. This includes having cat-appropriate toys and times for play interactions with humans and compatible cats or other pets in the home (Ellis et al., 2013). Food puzzles are a good option to mimic predatory behavior and provide enrichment and there are a variety of options available commercially (Dantas et al., 2016). Object play is another method that can be used to allow cats to elicit these behaviors. Toy movement is an important factor in engaging cats in this type of play, as is ensuring the toy used allows the cat to complete the predatory sequence by physically capturing the toy (Heath, 2020).

Providing cats with outdoor access can also allow them to express some of their natural play and other behaviors. Providing uncontrolled outdoor access is controversial, especially in some countries, and is not without risks. These risks include increased disease and increased risk of trauma, predation and poisoning. There are also risks to wildlife with having cats outdoors that should be considered as well. However, there are numerous benefits to having outdoor access, including promoting natural behaviors such as hunting, exploring and climbing, preventing behavioral problems, and providing exercise opportunities. A less risky option that can still provide some of the benefits is to provide controlled outdoor access for cats. This can include access to secure enclosed outdoor spaces (like catios or yards with cat-specific fencing) or training your cat to go outside on a leash and harness. (Tan et al., 2020).

Providing positive and consistent human interaction is the fourth pillar of an appropriate feline environment. Predictable and routine interactions strengthen the human-cat bond, but it is important that the cat initiate the interaction and not be forced into it. Different cats prefer different degrees of human contact and can change over the course of a cats' lifetime (Ellis et al., 2013). The majority of pet and shelter cats in one study preferred social interaction with humans over other forms of enrichment (Shreve, Mehrkam & Udell, 2017). Regular human interaction is important as cats tend to be more social with humans who are more attentive toward them. Cats also prefer high-frequency and low-intensity interactions with humans (Heath, 2020). Something as casual as a brief pet, a verbal interaction, or spending a few minutes engaging them with a toy is important to provide multiple times daily for your cat.

The fifth and final pillar of an appropriate environment is one that accounts for the importance of a cat's sense of smell. Cats use both olfactory and chemical information (i.e., pheromones) to evaluate their environment. It is important that they be allowed to display marking behaviors, by scratching and rubbing. Cats should also be provided with surfaces for scratching, as this is an important and normal behavior with several benefits. Scratching can have functional benefits, like removing the outer sheath of the nail and providing exercise for specific muscle groups, as well as is an important form of feline communication. Providing a variety of scratching surfaces, including horizontal and vertical surfaces, allows cats to display this behavior and can prevent destructive scratching (DePorter & Elzerman, 2019). As with everything, cats can have preferences for different scratching modalities, so you may need to try different types until you find one your cat prefers. Avoiding using scented or irritating cleaning products that can interfere with a cat's sensory perception or remove their scent from the home is also important in creating a healthy environment (Ellis et al., 2013). The use of synthetic feline pheromones may provide some benefit in reducing anxiety and other behavioral issues, including aggression between cats in a multi-cat home (DePorter et al., 2019; Vitale 2018).

## *Introducing Your New Cat to Other Humans*

Proper introductions can set the tone for relationships in the future. It is important that cat owners and other family members or guests are familiar with and can interpret feline body language. Deficiencies in this area can lead to miscommunications and misunderstandings about what message the cat might be trying to convey. Fear Free Happy Homes (an online service) provides free education to pet owners on a variety of behavioral topics including feline body language.

As you prepare for introducing your new cat to other humans, be sure you and any other person meeting the cat have plenty of treats available to reward desirable behavior during the introduction. No matter the age, it is important that the cat feels as if they have choice and control during the introduction.

When you bring your cat home, provide them with a quiet and low traffic area that they can feel safe and secure in if they need to retreat. Carriers with the doors removed or crates with blankets over the top can provide security and a place to hide if they feel overwhelmed. Tall cat trees, and baby gates or exercise pens to separate rooms are also great tools to have to provide your cat security when transitioning into the home. During initial interactions with people, regardless of the human's age, it is important that the cat is allowed to come out and interact on their own. Guests and family members can drop treats as they walk by the cat to encourage the cat to come out to interact; however, it is important not to force any additional interactions after they engage.

## *Introducing Your New Cat to Other Cats*

Similarly to introducing your new cat to humans, introducing them to other cats should be done with the care to avoid negative experiences that can impact the success of a relationship. Among multi-cat households introducing a new cat, half reported fighting (Levine et al., 2005). This process should be done slowly as it can take weeks to months to introduce a new cat into the home. To set up proper introductions, first start with the cats separated. The new cat should have its own space, complete with a comfortable place to sleep, food and water, and a litter box. An open carrier is helpful to provide the cat with a safe space to hide if they would like.

After a few days, you may notice that the cats show interest in each other by sniffing under the door. If this seems to be going well and there is an absence of stress-related behaviors such as growling, hissing, or swatting under the door, then at this point, you can begin to swap bedding and other items to share between the cats. Monitor their responses over a few days and if there are no serious issues, then you can begin to allow each cat access to each other. It is best to keep these interactions short and the cats should be actively supervised.

If problems arise as the cat transitions, it is recommended to reach out to a qualified behavior professional that uses reward-based methods for assistance. International Association of Animal Behavior Consultants (IAABC), Animal Behavior Society (ABS), and the American Veterinary Society of Animal Behavior all have professional searches to find the right professional to work with.

## Adjustment Period and What to Expect

Problems with new pets while introducing them to a home can commonly lead to frustration, relinquishment, euthanasia, or abandonment (Serpell et al., 2016). The decision to relinquish an animal is typically multifactorial (Kwan & Bain, 2013). This could be due to the owner's lack of knowledge and expectations when introducing a pet to a home (Sharkin & Ruff, 2011). Many animals are relinquished after the adopter spent significant time tolerating objectionable pet behavior; specifically 66% of animals were relinquished after one week and 46% were relinquished after two weeks (Shore, 2005) The adjustment period for an individual animal varies depending on the species behavior and individual disposition or temperament. Failed adopters commonly feel they should have devoted more time, thought, and planning when considering adoption and the process (Shore, 2005). Potential adopters need to consider their lifestyle and what animal is right for them. For example, different breeds typically have different behaviors that will be beneficial to certain lifestyles.

Puppies and kittens are more receptive to socialization, adapting to new environments, and habituating to a variety of new stimuli (Landsberg et al., 2003). Therefore, young animals may be trained to the owners liking, and may have an easier adjustment period in a new home. Adopting a young pet is a great responsibility and requires an intensive understanding of the necessary stimuli and environment appropriate for the pet. Adult pets may also have a difficult time adapting to the environment, animals, or people around them. However, some adult pets may already have basic training and may be comfortable staying at home alone. Adult animals may not exhibit behaviors such as nipping, destroying the house or items by chewing, scratching, urinating, or defecating inside the home. Temperament is another important behavior consideration when adopting cats and dogs. If an animal is anxious, shy, reactive, slow, and neophobic, it may take a significant period for them to adapt to a new environment (De Palma et al., 2005). On the other hand, if the animal is bold, proactive, fast, and neophilic they may have a smoother adjustment in a new home (De Palma et al., 2005). Understanding an animal's basic needs and the time necessary for an animal to adjust to a new home may set up the pet and owner for a successful adoption.

## Keeping Them Happy and Healthy in the Home

Annual to tri-annual vaccinations and other preventive care, such as internal and external parasite control, may be recommended depending on the individual animal's species, lifestyle, and risk factors. Dental care is an important aspect of keeping the pet healthy long term. Thoughts on spaying and neutering vary globally; however, there are several benefits, like the prevention of certain reproductive cancers or the reduction of specific behaviors, such as roaming and urine marking in dogs and cats. Sterilization, especially of dogs, cats, and rabbits, has been pivotal in controlling animal overpopulation, especially in the United

States. The decision to spay or neuter and at what age should be made in conjunction with a veterinarian.

Providing continued enrichment and social interactions in and outside of the home will provide for a pet's ongoing behavioral needs and ensure they remain behaviorally healthy. Human interaction can include play, spending time outdoors, going on a walk, and training sessions. Training sessions should focus on reward-based methods and can be a great way to build and maintain the human-animal bond for both dogs and cats as well as provide mental stimulation. Training can aid in helping cats and dogs become comfortable with going in a carrier, traveling and going into new environments. Providing for a pet's physical and behavioral well-being will ensure they remain happy and healthy throughout their life.

## Conclusion

There are several steps and decisions to consider before introducing a new pet into a home. This chapter covered how to prepare a proper environment in the home for a new pet and more importantly, how to maintain a happy and healthy pet in the home. Other considerations were discussed such as bringing a pet into the home, introducing a pet to family members and existing pets, and what to expect when you bring home a new pet. This chapter is meant to help adopters consider their lifestyle and what pet would best fit into their home and routine.

### Discussion Questions

1. What are three factors to consider when deciding which dog or cat to bring into your home?
2. List three tips to successfully introduce your dog or cat to another dog or cat.
3. What are the Association of Feline Practitioners (AAFP) and the International Society of Feline Medicine (ISFM) five pillars of a healthy feline environment?

## References

Alho, M. A., Pontes, J., & Pomba, C. (2016). Guardians' knowledge and husbandry practices of feline environmental enrichment. *Journal of Applied Animal Welfare Science,* 19(2), 115–125.

American Society for the Prevention of Cruelty to Animals. (2021). *Pet Statistics.* https://www.aspca.org/helping-people-pets/shelter-intake-and-surrender/pet-statistics

American Veterinary Medical Association. (2021). *Selecting a Pet Cat.* https://www.avma.org/resources/pet-owners/petcare/selecting-pet-cat

American Veterinary Society of Animal Behavior. (2021). *Position Statement on Humane Dog Training.* https://avsab.ftlbcdn.net/wp-content/uploads/2021/08/AVSAB-Humane-Dog-Training-Position-Statement-2021.pdf

Arhant, C., Winkelmann, R., & Troxler, J. (2021). Chewing behaviour in dogs–a survey-based exploratory study. *Applied Animal Behaviour Science,* 241, 105372.

Beaver, B. V. (1994). Owner complaints about canine behavior. *Journal of the American Veterinary Medical Association,* 204(12), 1953–1955.

Carney, H. C., Sadek, T. P., Curtis, T. M., Halls, V., Heath, S., Hutchison, P., Mundschenk, K., & Westropp, J. L. (2014). AAFP and ISFM guidelines for diagnosing and solving house-soiling behaviors in cats. *Journal of Feline Medicine Surgery,* 20(6), 579–598.

Casey, R. A., Loftus, B., Bolster, C., Richards, G. J., & Blackwell, E. J. (2013). Inter-dog aggression in a UK owner survey: Prevalence, co-occurrence in different contexts and risk factors. *Veterinary Record,* 172(5), 127–127.

Centers for Disease Control and Prevention. (2021). *This Week Is Dog Bite Prevention Week.* https://www.cdc.gov/healthypets/connect/newsletter/dog-bite-prevention-week.html

Dantas, L. M., Delgado, M. M., Johnson, I., & Buffington, C. T. (2016). Food puzzles for cats: Feeding for physical and emotional wellbeing. *Journal of Feline Medicine Surgery*, September, 18(9), 723–732.

Delgado, M., & Dantas, L. (2020). Feeding cats for optimal mental and behavioral well-being. *Veterinary Clinics Small Animals*, 5, 939–953.

De Palma, C., Viggiano, E., Barillari, E., Palme R., Dufour, A. B., Fantini, C., & Natoli, E. (2005). Evaluating the temperament in shelter dogs. *Behavior*, 142, 1307–1328.

DePorter, T. L., Bledsoe, D. L., Beck, A., & Ollivier, E. (2019). Evaluation of the efficacy of an appeasing pheromone diffuser product vs placebo for management of feline aggression in multi-cat households: A pilot study. *Journal of Feline Medicine Surgery*, April, 21(4), 293–305.

DePorter, T. L., & Elzerman, A. (2019). Common feline problem behaviors: Destructive scratching. *Journal of Feline Medicine Surgery*, 21(3), 235–243.

Dixon, C. A., Mahabee-Gittens, E. M., Hart, K. W., & Lindsell, C. J. (2012). Dog bite prevention: an assessment of child knowledge. *The Journal of Pediatrics*, 160(2), 337–341.

Ellis, S. L., Rodan, I., Carney, H. C., Heath, S., Rochlitz, I., Shearburn, L. D., Sundahl, E., Westropp, J. L. (2013). AAFP and ISFM environmental needs guidelines. *Journal of Feline Medicine Surgery*, March, 15(3), 219–230.

Grigg, E. K., & Kogan, L. R. (2019). Owner's attitudes, knowledge, and care practices: Exploring the implications for domestic cat behavior and welfare in the home. *Animals*, 15(9), 978.

Heath, S. (2018). Understanding feline emotions and their role in problem behaviours. *Journal of Feline Medicine Surgery*, 20(5), 437–444.

Heath, S. (2020). Environment and feline health. *Veterinary Clinics Small Animals*, 50, 663–693.

Hepper, P. G. (1985). Sibling recognition in the domestic dog. *Animal Behavior*, 34, 288–289.

Herron, M. E. (2010). Advances in understanding and treatment of feline inappropriate elimination. *Topics Companion Animal Medicine*, 25(4), 195–202.

Horwitz, D. F. (Ed.) (2018). *Blackwell's Five-Minute Veterinary Consult Clinical Companion: Canine and Feline Behavior.* John Wiley & Sons.

Kwan, J. Y., & Bain, M. J. (2013). Owner attachment and problem behaviors related to relinquishment and training techniques of dogs. *Journal of Applied Animal Welfare Science*, 16(2), 168–183. Doi:10.1080/10888705.2013.768923

Landsberg, G. M., Hunthausen, W., & Ackerman, L. (2003). *Handbook of Behavior Problems of the Dog and Cat.* Saunders.

Learn, A., Radosta, L., & Pike, A. (2020). Preliminary assessment of differences in completeness of house-training between dogs based on size. *Journal of Veterinary Behavior*, 35, 19–26.

Lees, R. (2021). *Addressing Feline Behavioral Issues.* https://todaysveterinarynurse.com/wp-content/uploads/sites/3/2020/11/Lees-TVNWinter21_FelineBehavioralIssues.pdf

Levine, E., Perry, P., Scarlett, J., & Houpt, K. A. (2005). Intercat aggression in households following the introduction of a new cat. *Applied Animal Behaviour Science*, 90(3–4), 325–336.

Maddie's Fund. (2019). *Children and Dogs-How to Keep Interactions Safe.* https://www.maddiesfund.org/children-and-dogs-how-to-keep-interactions-safe.htm

Maddie's Fund. (2019). *Introducing Your New Dog to Cats.* https://www.maddiesfund.org/introducing-your-new-dog-to-cats.htm

Martínez, Á. G., Pernas, G. S., Casalta, F. J. D., Rey, M. L. S., & De la Cruz Palomino, L. F. (2011). Risk factors associated with behavioral problems in dogs. *Journal of Veterinary Behavior*, 6(4), 225–231.

McMillin, F. D. (2002). Development of a mental wellness program for animals. *Journal of the American Veterinary Association*, 220, 965–972.

Merola, I., & Mills, D. S. (2016). Behavioural signs of pain in cats: An expert consensus. *PLoS One*, 2016 February 24, 11(2), e0150040.

Patronek, G. J., Glickman, L. T., Beck, A. M., McCabe, G. P., & Ecker, C. (1996). Risk factors for relinquishment of dogs to an animal shelter. *Journal of the American Veterinary Medical Association*, 209(3), 572–581.

Peterson, M. E., & Kutzler, M. A. (2011). *Small Animal Pediatrics.* W.B. Saunders.

Pluijmakers, J. J. T. M, Appleby, D. L., & Bradshaw, J. W. S. (2010). Exposure to video images between 3 and 5 weeks of age decreases neophobia in domestic dogs. *Applied Animal Behavior Science*, 126, 51–58.

Robbins, M. T., Cline, M. G., Bartges, J. W., Felty, E., Saker, K. E., Bastian, R., & Witzel, A. L. (2019). Quantified water intake in laboratory cats from still, free-falling and circulating water bowls, and its effect on selected urinary parameters. *Journal Feline Medicine Surgery*, 21(8), 682–690.

San Diego Humane Society. (2021). *Adopting a Cat: Things to Consider*. https://www.sdhumane.org/behavior-and-training/resources/

San Diego Humane Society. (2021). *Introducing Dogs at Home*. https://www.sdhumane.org/resources/pdf/dog-intro-v1.pdf

Salman, M. D., Hutchison, J., Ruch-Gallie, R., Kogan, L., New Jr, J. C., Kass, P. H., & Scarlett, J. M. (2000). Behavioral reasons for relinquishment of dogs and cats to 12 shelters. *Journal of Applied Animal Welfare Science*, 3(2), 93–106.

Serpell, J., Duffy, D. L., & Jagoe, J. A. (2016). Becoming a dog: Early experience and the development of behavior. In Serpell, J. (ed.), *The domestic dog: Its evolution, behavior and interactions with people* (2nd ed., pp. 93–109). essay, Cambridge University Press.

Sharkin, B. S., & Ruff, L. A. (2011). Broken bonds: Understanding the experience of pet relinquishment. In Blazina, C., Boyraz, G., & Shen-Miller, D. (eds.), *The Psychology of the Human-Animal Bond*, 275–287. Doi:10.1007/978-1-4419-9761-6_16

Shore, E. R. (2005). Returning a recently adopted companion animal: Adopters' reasons for and reactions to the failed adoption experience. *Journal of Applied Animal Welfare Science*, 8(3), 187–198. Doi:10.1207/s15327604jaws080

Shreve, K. R., Mehrkam, L. R., & Udell, M. A. R. (2017). Social interaction, food scent or toys? A formal assessment of domestic pet and shelter cat (Felis silvestris catus) preferences. *Behavioural Processes*, 141, 322–328.

Shreve, K.R., & Udell, M.A. (2015). What's inside your cat's head? A review of cat (Felis silvestris catus) cognition research past, present and future. *Animal Cognition* 18(6), 1195–1206.

Stella, J. L., & Croney, C. C. (2016). Environmental aspects of domestic cat care and management: Implications for cat welfare. *Scientific World Journal* 2016, 6296315. Doi:10.1155/2016/6296315

Tan, S. M., Stellato, A. C., & Niel, L. (2020). Uncontrolled outdoor access for cats: An assessment of risks and benefits. *Animals,* 10(2), 258. Doi:10.3390/ani10020258

van der Leij, W. J. R., Selman, L. D. A. M., Vernooij, J. C. M., & Vinke, C. M. (2019). The effect of a hiding box on stress levels and body weight in Dutch shelter cats; a randomized controlled trial. *PLoS One*, October 14, 14(10), e0223492.

Vitale, K. R. (2018). Tools for managing feline problem behaviors: Pheromone therapy. *Journal of Feline Medicine Surgery*, November, 20(11), 1024–1032.

Voith, V. L., & Borchelt, P. L. (1985). Elimination behavior and related problems in dogs. *The Compendium on continuing education for the practicing veterinarian*. 7(7), 537–544, 546, 549.

Yeon, S. C., Erb, H. N., & Houpt, K. A. (1999). A retrospective study of canine house soiling: diagnosis and treatment. *Journal of the American Animal Hospital Association*, 35(2), 101–106.

# 25
# VETERINARY SOCIAL WORK
*Pamela Linden*

**Introduction: What Is Social Work?**

Social work is "the professional activity of helping individuals, families, groups, or communities enhance or restore their capacity for social functioning or creating societal conditions favorable to that goal" (NASW, 1973). In the United States, state laws limit the title "social worker" to individuals who have earned a bachelor's, master's, or doctoral degree in social work, or who have obtained a social work license. Social workers are unique in the way that they look at many different aspects of a problem, from the individual to the societal, from the psychological to the political (Shulman, 2012). Social welfare is an integral aspect of the practice of social work and refers to a variety of coordinated societal initiatives that promote social wellbeing. Social workers are committed to the empowerment of people and communities. They provide services to individuals and families throughout the life span, addressing the full range of biopsychosocial-spiritual and environmental issues that affect wellbeing (NASW, 2021). Common ways social workers serve clients include supportive counseling, psychotherapy, psychoeducation, advocacy, and links to public or private resources. Social workers advance social progress through policy analysis, research, and legislative advocacy.

Social workers are in a position of public trust because people often share with them personal, private, and sensitive information and they expect this to be treated with the highest level of confidentiality. The National Association of Social Workers' Code of Ethics (2021) identifies a set of core values of the social work profession: service, social justice, dignity and worth of the person, importance of human relationships, integrity, and competence. The Code provides guidance on ethical practice to social workers as they perform their daily work.

Social work with client systems is broadly organized into three distinct levels of practice: micro, mezzo, and macro. At each level of practice, social workers are trained to serve client systems, which may be individuals, families, groups, or communities. Micro social work is the most common type of practice and occurs directly with an individual client or family. They conduct biopsychosocial assessments to identify client's strengths and needs and they use evidence-supported interventions to enhance or restore the client's functioning. Mezzo social work happens on an intermediate scale, involving neighborhoods, institutions, or other smaller groups. Social workers conduct needs assessments and facilitate community organizing to bring about community level change. Macro social work refers to large-scale interventions that affect entire communities and systems of care. Social workers empower members of a community by focusing on issues that are important to them and the environments in which they work, learn, live, and play.

Many social workers pursue specialty training to enhance their skills in a particular area. Some examples of specialty social work practice include school, medical, psychiatric, military, geriatric, and hospice and palliative care social work. Social workers who are interested in developing expertise in helping people at the intersection of humans and animals may pursue training in veterinary social work.

## What Is Veterinary Social Work?

Veterinary social work is a specialty area of social work practice that attends to the human needs at the intersection of veterinary and social work practice (UTK VSW, 2022). The term "veterinary social work space" describes any setting where humans and animals interact. Many helping disciplines work within the veterinary social work space and contribute to the health and wellbeing of people who work in animal-related settings. The term "veterinary social work" was coined in 1981 by Dr. Barbara Crocken when she described a novel social work internship that provided support to owners of pets that were diagnosed with cancer at the University of Pennsylvania's College of Veterinary Medicine. They found that "in 100% of the cases where the veterinarian informed the client that a social worker would be contacting him or her, the client was not only receptive to the social worker, but eager to discuss the situation" (Crocken, 1981, p. 93). Veterinary social work was professionalized by Dr. Elizabeth Strand in 2002 when she founded the Veterinary Social Work (VSW) program at the University of Tennessee – Knoxville (UTK). In 2010, a VSW Certificate program opened enrollment to concurrent UTK Master of Science in Social Work (MSSW) students and to post-graduate licensed social workers. As of this writing, several US and international social work training programs are offering training in veterinary social work based on the UTK VSW curriculum. Graduates of the VSW training program take a Veterinary Social Work Oath, which is drawn from the veterinary oath and social work principles:

> Specializing in veterinary social work, I pledge my service to society by tending to the human needs that arise in the relationship between humans and animals. From a strengths perspective and using evidence-based practice, I will uphold the ethical code of my profession, respect and promote the dignity and worth of all species, and diligently strive to maintain mindful balance in all of my professional endeavors.
> *(UTK VSW program, 2022)*

The UTK VSW certificate curriculum is organized into four primary areas of veterinary social work activity: animal-related grief & bereavement; the link between human and animal violence; intentional wellbeing; and animal-assisted interventions (AAIs) (Table 25.1). Each area is explored in the following sections.

### *Animal-Related Grief & Bereavement*

Animals have a variety of roles in the lives of people who care for them. Some animals may be pets or companion animals, or they may serve as therapy, service, or emotional support

Table 25.1 Veterinary social work topic areas and associated activities

| VSW Topic Area | Veterinary Social Work Activities |
| --- | --- |
| Intentional Wellbeing | • Provide health and wellness resources<br>• Offer brief counseling and support<br>• Facilitate a wellness hotline<br>• Facilitate conflict meditation<br>• Provide compassion fatigue assessments<br>• Communication skill development<br>• Advocacy |
| Animal-related Grief and Bereavement | • End of life decision making<br>• Pet loss support<br>• Pet loss support group<br>• Link to grief specialists<br>• Psychoeducation |
| The Link between Human and Animal Violence | • Provide training to veterinarians and hospital staff<br>• Partner with IPV shelters<br>• Confer with courts, animal abuse task force and legislature<br>• Conduct community abuse prevention presentations |
| Animal-Assisted Interventions | • Support owner/handler teams<br>• Advocate for standards and competencies<br>• Teach, conduct, research, and evaluate animal-assisted interventions |

animals. Other animal roles may include veterinary patients, or "critters" temporarily cared for within animal welfare organizations. To farmers, animals may be stock, breeders, and sources of food production.

The American Pet Product Association's 2019–2020 National Pet Owners Survey found that approximately 67% of US homes had at least one pet, an estimated 84.9 million homes, equaling about 63 million homes with at least one dog and 42 million homes with at least one cat (APPA, 2019). Over 90% of pet owners feel their companion animal is a family member and 'family bondedness' is defined as the condition in which a person feels a positive valence emotional bond to a pet in a manner approaching, if not equivalent to, their positive valence emotional bond to a human family member. This positive emotional bond is characterized by love and affection and an emotional sense that the pet is a member of their immediate family" (Nugent & Daugherty, 2022).

For many people, the loss of an animal can be traumatic and lead to feelings of sadness and inconsolable grief (Sife, 2014). Families may have difficulty making the decision to euthanize a pet and at times pets may be in pain, suffer or die while families struggle to decide to continue treatment or work with their veterinarian to be humanely euthanized. Humane euthanasia is performed by licensed veterinary professionals (Quain, 2021); pet owners may need information about the euthanasia procedure and may not feel emotionally prepared to say goodbye to their pet. Veterinarians provide medical guidance regarding care for the pet and assess patient quality of life with the client. Although most veterinarians routinely provide comfort to clients before, during, and after an animal-related loss, they have little time and training to tend to pet owners' emotional state.

Veterinary social workers can support the veterinary-client-patient relationship by offering additional emotional support and guide families through end-of-life decision-making.

Anticipatory grief is the form of grief that occurs when confronted with a chronic/life-threatening illness or the anticipation of the death of a loved one (or oneself). In cases where pet owners have difficulty making end-of-life decisions, sick or injured animals are helped when pet owners work with veterinary social workers trained in hospice and palliative care. Brackenridge (2019) states "because of their unique training and capabilities, social workers are indispensable in this highly emotional and intimate form of veterinary practice" (p. 573). Some people may experience unremitting feelings of sadness, anger, guilt, or depression and their daily functioning may be impaired by thoughts and accompanying emotions regarding the loss.

Today, there are few socially recognized rituals or ceremonies to mark the passing of a beloved pet. It is rare to see pet funerals or sympathy cards for the loss of an animal, and paid time off to grieve is not the norm in most industries. Absent rituals to memorialize the deceased, mourners may have difficulty resolving grief. Pet-related grief is considered a disenfranchised grief, "one that is experienced when a loss cannot be openly acknowledged, socially sanctioned, or publicly mourned" (Doka, 2002). Bereaved pet owners may find help by attending a pet loss support group. Sharing their thoughts and feelings about the meaning of the loss may provide comfort and relief as they mourn. Group leaders can make referrals to licensed mental health professionals for group participants with complicated grief and whose daily functioning is impacted by the thoughts and feelings about the loss.

The decision to humanely euthanize a pet is generally made collaboratively by the veterinarian and pet owner. At times, poor communication between the veterinarian and the client may result in feelings of anger toward the veterinarian for not communicating the extent of a pet's illness (Dunn, 2005). Taking care of an ill or impaired animal may, at times, feel stressful for the caregiver, especially when pets have an acute or chronic health condition that requires caregivers to dispense medications at regular intervals, provide help to pets so that they can relieve themselves, or cope with unpleasant aspects of the pet's condition, like sour breath or gastrointestinal messes or odors.

People who are grieving may have a hard time making decisions about cremation, burial, and memorialization of the beloved pet. Veterinary social workers can help people grieving over the loss of a pet by offering emotional support, facilitating poet loss groups, and offering guidance on ways to memorialize their relationship with the deceased pet. This can facilitate the grieving process for many people, particularly those who are bonded to their non-human family member.

## *The Link between Human and Animal Violence*

Researchers have long established significant correlations between animal abuse, child abuse and neglect, domestic violence, elder abuse, and other forms of violence, referred to as "the Link" (Arkow, 2022). It has been reported that people experiencing interpersonal violence who live with companion animals remain in abusive relationships or significantly delay leaving due to concern for their animals' wellbeing (Collins et al., 2018; Taylor et al., 2019). Youth who engage in animal cruelty are known to be at increased risk of perpetrating violence on other people in their lives including peers, loved ones, and elder family members. These youths have often been exposed to family violence, including animal cruelty perpetrated on their beloved pets by violent adults (Bright et al., 2018). Communities are unlikely to be able to best protect humans from abuse and harm unless they are working to ensure the safety of animals who reside there as well (Campbell, 2022). When professional helpers are trained in the issues that arise in the intersection of people and animals, they are able to compassionately engage in problem-solving activities that ensure that people and animals reach safety.

The Pet and Women Safety (PAWS) Act became law on December 20, 2018. It establishes a grant program for entities that provide shelter and housing assistance for domestic violence survivors to enable them to better meet the housing needs of survivors with pets. The law also takes the important step of including pets, horses, service animals, and emotional support animals in federal law pertaining to interstate stalking, protection order violations, and restitution. These provisions provide law enforcement with additional tools for protecting victims from their abusers (PAWS Act (H.R. 909, S.322). The National Link Coalition (https://nationallinkcoalition.org/) raises awareness and provides education about the link between human and animal violence and developed systems of cross-reporting human and animal violence in order to prevent future violence.

Veterinary social workers can raise awareness of the link between intimate partner violence (IPV) and animal abuse by conducting presentations in libraries, hospitals, schools, and veterinary and other animal-related setting presentations to inform local communities about prevention, identification and treatment for animal abuse and violence in interpersonal relationships. Veterinary social workers can link people who may be at risk of harm with resources to protect them. Licensed mental health providers interested in using evidence-based approaches to treating children and adults who abuse animals may be interested in exploring resources developed by the Animals & Society Institute, such as the AniCare© Model of Treatment of Animal Abuse (Shapiro & Henderson, 2016). Improved cross-reporting and stronger partnerships between human and animal welfare agencies may provide an opportunity for earlier intervention and are likely to better many human and animal lives (Campbell, 2022). Veterinary social workers can advocate for domestic violence survivors by creating partnerships between shelters for people seeking safety and offer onsite or offsite boarding to ensure that the entire family is protected from abuse. While attending to the human family members, veterinary social workers advocate for the adoption and enforcement of laws and policies that protect animals living in situations where violence, neglect, and abuse are present. Veterinary social workers provide consultation and support when veterinary and other animal health-related workers suspect clients or animals in their care may be victims of neglect, abuse or other forms of violence. Veterinary social workers can provide guidance to veterinary teams on relevant laws regarding neglect, abuse, and protection of potential victims through mandatory cross-reporting to appropriate authorities. People who witness animal abuse and neglect may be emotionally affected and working closely with a veterinary social worker can help to process thoughts and emotions to avoid compassion fatigue and burnout.

## *Intentional Wellbeing*

Workplace stressors affect many types of animal-related workers, some who work directly with animals and others whose work is mainly with people responsible for the welfare of one or more animals. Some examples include:

- Veterinary medical team
- Veterinary reception team
- Clients/animal owners
- Zookeepers
- Animal rescue workers

- Animal-assisted therapy (AAT) volunteers and recipients
- Service and guide dog trainers, volunteers, handlers
- Farmers
- Animal laboratory workers
- Animal shelter workers
- Food production workers
- First responders (firefighters, law enforcement)
- Municipal animal control officers
- Crematory and pet cemetery workers

Workplace stressors that may be associated with poor mental health are poor family/work-life balance (Tomasi et al., 2019) and sizable student loan debt (Britt-Lutter & Heckman, 2020).

In September 2014, 48-year-old veterinary behaviorist and best-selling author Dr. Sophia Yin died from suicide. Dr. Yin was a trailblazer in the dog training community. She wrote books, created instructional videos, and developed tools for positive reinforcement training. Her suicide was tragic and shocking, and raised awareness of mental health and suicide in the veterinary community. In a recent study of suicides and deaths of uncertain intent among US veterinary professionals, Witte et al. (2019) found that for 34 of 73 (47%) veterinarians, the mechanism of death was classified as poisoning, with 18 of those 34 deaths (or 25% of the total) attributed to pentobarbital, the active ingredient in euthanasia solutions. A recent study of Norwegian veterinarians found the level of suicidal behavior among veterinarians is relatively high, and both individual and work-related factors contribute to serious suicidal thoughts. Findings suggest that individual factors, and particularly mental distress, played a more important role than the work-related factors, while veterinarians themselves regarded work problems as the most contributing factor to their suicidal thoughts (Dalum, Tyssen & Hem, 2022).

Compassion fatigue (CF), comprising secondary traumatic stress (STS) and burnout (BO), can adversely affect caring professions and job demands are an important risk factor for compassion fatigue in animal-related workers (Monaghan, Rohlf, Scotney, & Bennett, 2020), including laboratory animal workers (Van Hooser et al., 2021). A 2019 Merck Animal Health survey of 2,871 veterinarians found higher levels of burnout than (human) physicians, veterinarians are much more likely to think about suicide than non-veterinarians, veterinarians are 2.7 times more likely to attempt suicide, and they found an increase in serious psychological distress among women veterinarians. They found that serious psychological distress is higher in younger veterinarians, and suicidal thoughts and suicide attempts continue to remain higher in veterinarians than is found in the general public. Over half of veterinarians in their study were classified as having current serious psychological distress and just as many were not currently receiving mental health treatment (Merck Animal Health Veterinarian Wellbeing Study, 2020). A 2021 study of faculty, staff, residents, and interns at a Canadian veterinary college found the COVID-19 pandemic made worse several psychosocial demands, including home-based work challenges, increased workload and pace, reduced health and wellbeing, worsened interpersonal relations and higher work demands and stressors, which may be associated with suicide in the veterinary profession (McKee et al., 2021). Hill, LaLonde & Reese (2020) found a high level of compassion fatigue for both veterinarians and non-veterinary staff. The strongest risk factors for compassion fatigue were degree of exposure to cruelty and neglect cases and degree of stress from performing euthanasia. Work-related stressors may be found in a variety of settings where humans and animals interact, such as:

Veterinary practices
Colleges of Veterinary Medicine
Veterinary Nurse/Technician Training programs
Animal Control Agencies
Animal Welfare Organizations
Zoos & Aquariums
Wildlife Sanctuaries
Food animal/Agricultural settings
Humane Societies
Animal Behavior Training Centers
Grassroots Animal Foster, Rescue & TNR groups

AAI programs (hospitals, schools, therapeutic horseback riding facilities, jails/prisons, courts, law enforcement, mental health agencies, nursing homes, and rehabilitation facilities, addiction treatment centers, veterans' services)

Compassion fatigue affects those people who provide caretaking activities for others, including animals. A person experiencing compassion fatigue may show signs of poor self-care, have legal problems, incur large debt, experience nightmares and flashbacks to a traumatic event, have chronic physical ailments, appear apathetic or sad, complain of difficulty concentrating, be mentally and physically tired, and they may find themselves no longer finding pleasure in activities that they used to enjoy.

There is agreement that a comprehensive approach to suicide prevention among all veterinary professionals is important, which includes reducing existing barriers to seeking and accessing a continuum of mental health services, from short-term emotional support to psychiatric treatment. Veterinary social workers can help animal-related professionals recognize signs of compassion fatigue, burnout and poor mental health. They can teach them to use techniques to improve mental health and wellbeing. Veterinary social workers can work with animal-related professionals to build cohesive work teams, learn coping strategies for common stressors, and work toward occupational and compassion satisfaction.

## *Animal-Assisted Interventions*

AAI is an intervention that intentionally includes an animal as part of the therapeutic process and can be utilized in a wide variety of settings, including health, education, and human services (IAHAIO, 2014). Kruger & Serpell (2006) provide definitions that distinguish AAT and Animal-Assisted Activities (AAAs). AAT is a goal-directed intervention in which an animal that meets specific criteria is an integral part of the treatment process. AAT is directed and/or delivered by a health and human service professional with specialized expertise and within the scope of practice of his or her profession. The key features of AAT include specified goals and objectives for each individual and measured progress. In AAA, opportunities are offered for motivational, educational, recreational, and/or therapeutic benefits to enhance quality of life. AAAs are delivered in a variety of environments by specially trained professionals, paraprofessionals, and/or volunteers in association with animals that meet specific criteria. The key features of AAA include the absence of specific treatment goals; volunteers and treatment providers are not required to take detailed notes, and the visit content is spontaneous (Kruger & Serpell, 2006).

Assistance Dogs International (ADI) and International Association of Human-Animal Interaction Organizations (IAHAIO) define and disseminate AAI competencies, ethics, and standards of practice to ensure the efficacy of AAI, as well as attend to safety of people and animals involved in AAI activities. The Americans with Disabilities Act of 1990 (rev. 12/2020) defined a service animal as a dog, regardless of breed or type, that is individually trained to do work or perform tasks for the benefit of a qualified individual with a disability, including a physical, sensory, psychiatric, intellectual, or other mental disability. The US Department of Transportation (DOT) issued a final rule to amend the Air Carrier Access Act (ACAA) regulation on the transport of service animals by air to ensure that the air transportation system is safe for the traveling public and accessible to individuals with disabilities. The ACAA allows airlines to recognize emotional support animals (ESAs) as pets, rather than service animals, and permits airlines to limit the number of service animals that one passenger can bring onboard an aircraft to two service animals. ESA's are not required to have formal training and are not permitted in public spaces which do not allow animals. The US Department of Housing and Urban Development (HUD) Fair Housing Act (FHA) allows ESAs in housing when a licensed mental health provider attests that a person with a disability is helped by a companion animal. The ESA is considered a pet, not a service animal, and is not allowed in public areas where animals are not allowed. Veterinary social workers can have many roles in the planning, implementation, and evaluation of AAIs. At the most basic level, they can help the lay public understand the fundamental differences in the definitions and legal status between therapy animals, emotional support animals, companion animals, and service animals. The role of the animal determines the training requirements of the handler and the animal and the types of places that the animal and handler is allowed to enter. Serpell, Kruger, Freeman, Griffin & Ng (2020) investigated the extent to which AAI organizations attended to animal welfare, human safety, and public health. They found "many such organizations have their own policies and procedures for screening, evaluating, and instructing dogs and their owners/handlers" (p. 1). Given the lack of uniform guidelines for AAI, veterinary social workers can consult with individuals and organizations planning an AAI program by providing guidance in selecting appropriate animals and human and animal training needed to implement the program, while prioritizing the welfare of the animals. Veterinary social workers advocate for policies and standards that protect the health and welfare of animals and people providing AAI services, as well as the recipients of AAIs. Veterinary social workers may integrate AAT into their counseling practice when deemed clinically appropriate for the particular client and their treatment needs. AAIs involve ensuring and advocating for the welfare of the animals involved in helping people in their living, learning, working, and social environments.

## Veterinary Social Work and Animal Welfare

Many published studies find that owning a companion animal has physical and mental health benefits for people (Friedman & Krause-Parello, 2018). Some studies have found that the human-animal bond (HAB) may provide a buffering effect for stress and adversity (Hill, Winefield & Bennett, 2020), and attachment to a beloved pet (Julius et al., 2013) and bondedness to a pet (Nugent & Daugherty, 2022) affects the ways in which people grieve when their pet dies. Pet ownership can also bring challenges and stress that are not fully understood. Scoresby et al. (2021) reviewed 54 published studies examining the impact of pet ownership on mental health. They found that more than half of the studies had either negative, mixed positive and negative, or no such association with owner's mental health. More recent studies suggest that sometimes bondedness to pets is associated with increased depression and loneliness and attachment to pets can predict vulnerability (Peacock, Chur-Hansen & Winefield, 2012) or increased levels of emotional distress in owners (Chur-Hansen, Stern, Winefield, 2010). "Ensuring animal welfare is a human responsibility that includes consideration for all aspects of animal wellbeing, including proper housing, management, nutrition, disease prevention and treatment, responsible care, humane handling, and, when necessary, humane euthanasia" (AVMA, 2022).

The welfare of animals may be enhanced when veterinary social workers provide services to veterinary professionals, pet owners, and animal-related organizations. The humane euthanasia experiences for pet owners and veterinarians may be improved when a veterinary social worker is available to offer emotional support and resources. The veterinary professional's mental health and wellbeing may be improved by working with a licensed social worker who is prepared to integrate issues related to the human-animal relationship into assessments and interventions. Addressing individual and work-related stressors may improve retention rates for veterinarians and other animal-related workers. Social workers are beginning to collect information about non-human family members to assess an individual or family's strengths and needs (Risley-Curtiss, 2010).

A Biopsychosocial-Veterinary assessment (BPS-V) is being piloted in a national access to veterinary care initiative that explores factors that affect the health and welfare of the entire family, including non-human family members. Pet owners are expected to provide food, shelter, veterinary care, and exercise, for sustenance, wellbeing and safety. They should provide an environment that is free from fear and anxiety, provide a haven for safety and rest, and protection from harm from people, the weather, and from other animals. Veterinary social workers can assist those owners of pets whose demographic, geographic or economic characteristics impede or prevent access to veterinary care services (AVCC, 2018).

The most common barriers to access to veterinary care are lack of financial resources, lack of reliable transportation to get to and from veterinary appointments and lack of safety supplies and equipment, such as collars, leashes, portable carriers, and crates, to keep their non-human family members safe in transit (AVCC, 2018). Physical, psychological, and emotional challenges may lead to difficulty accessing veterinary care by affecting how, where and what veterinary care pets receive to get and stay healthy. The veterinary-client-patient relationship can be affected by miscommunication, limited comprehension of exam findings and treatment plans, and emotions such as fear and anxiety. When companion animals have a treatable medical condition, but the family is unable to pay for care, pets may be humanely euthanized, abandoned, or surrendered. Veterinarians and veterinary staff report that discussing cost of care with pet owners with economic limitations affected their ability to provide the desired care for those patients on a daily basis, and many cited "client economic limitations as an important contributing factor to burnout" (Kipperman, Kass, & Rishniw, 2017). The veterinary safety net should be normalized when the cost of care is out of reach of families. Many communities offer low or no cost veterinary care, veterinary stipends, and wellness and prevention clinics to increase access to sterilization and vaccinations to protect public health (e.g., rabies vaccination clinics).

Access to veterinary care is a human problem and it is also a social justice issue. Veterinary social workers can increase awareness of resources to help people and pets, support families in providing for the health and safety of pets, help secure stability in housing, food/nutrition, and healthcare. Their work can improve the veterinary-client-patient relationship by teaching and modeling effective communication skills and mediating conflicts that occur in the veterinary setting, such as disagreements between pet owners and veterinarians regarding treatment decisions.

## The Evolving Specialty of Veterinary Social Work

Licensed mental health providers who are interested in enhancing the lives of people who care for animals in a wide variety of settings may find veterinary social work appealing.

It can be rewarding for social workers to collaborate with veterinary and other animal-related professionals to cultivate a culture of health and wellbeing that ultimately improves the lives of people and animals. Arkow (2020) advocates for social workers, regardless of formal veterinary social work training, to include human-animal relationships in their practices. Veterinarians and other animal-related workers experience burnout, compassion fatigue, work-related stress, and conflict. Recent efforts to raise awareness of these issues have led to the development of both institutional and grassroots support for veterinary and other animal-related workers. Some examples include:

- Shanti's Veterinary Mental Health Initiative offers free support groups and one-on-one help from licensed mental health providers to veterinarians, veterinary technicians, and other animal-related workers (https://www.shanti.org/programs-services/veterinary-mental-health-initiative/)
- Not One More Vet is an online suicide prevention, peer support resource for veterinarians (https://www.nomv.org/)
- The Veterinary Hope Foundation offers small group sessions for veterinary professionals that are moderated by licensed mental health providers (https://veterinaryhope.org/support-groups/)
- AVMA Workplace Wellbeing Certificate Program offers veterinary team members knowledge and skills to create a culture of wellbeing (https://axon.avma.org/local/catalog/view/product.php?productid=22)

## Diversity, Equity, Inclusion, and Belonging

Social justice is a core value of the social work profession. Veterinary medicine is one of the least ethnically diverse professions in both the United States of America and the United Kingdom (Greenhill, Nelson & Elmore, 2007). There is agreement that organizations are strengthened when the voices of people with diverse social identities are valued. Social identities are the invisible and visible attributes that people use to define themselves. Examples of social identities are race, ethnicity, age, gender, gender identity, sexual preference, ability, religion, and socio-economic status (SES). Groups of veterinary professionals have formed not-for-profit organizations to enhance workplace diversity, equity, inclusivity, and belonging (DEIB). The mission of the Multicultural Veterinary Medical Association (MCVMA) is to foster a community of veterinary professionals from diverse backgrounds and cultures to create a mutual support network (MCVMA, n.d.). Pride VMC's mission and purpose is to create a better world for the LGBTQ+ veterinary community. Their vision is an empowered LGBTQ+ veterinary community that embraces wellbeing by being their authentic selves. The ethical principle underlying the social work core value of *social justice* is "Challenge social injustice and work for social change on behalf of vulnerable and oppressed people" (NASW, 2021). Veterinary social workers use their knowledge, skills, and values to ameliorate injustice by bringing attention to policies and practices that sustain discrimination and oppression for people with diverse social identities (i.e., race, ethnicity, age, ability, socio-economic status, sexual preference, gender identity). Veterinary social workers assess the degree to which animal-related settings support diversity, equity, inclusion and belonging, and they work with others to create equitable social structures.

## Conclusion

Veterinary social workers provide social work services to a wide variety of client systems in diverse animal-related settings. Their training prepares them to help families cope with pet loss, enhance AAIs to benefit people, and protect public health as they engage in prevention, education and treatment of children and adults who neglect and abuse animals. They develop and teach curricula for colleges of social work and veterinary medicine and teach humane education to increase empathy and compassion for all living creatures. There is a need for empirical research to better understand the human and animal outcomes of veterinary social work interventions. Future research should measure the impact of veterinary social work services on the mental health of veterinarians and other animal-related workers, as well as on pet owners and caretakers of unowned animals, such as animal welfare and municipal animal control workers.

There is growing interest in the specialty social work practice area of veterinary social work. The expansion of certificate level training programs in higher education and the creation of the interdisciplinary professional member organization the International Association of Veterinary Social Work (IAVSW) (www.veterinarysocialwork.org) provides opportunities to grow professional networks in the veterinary social work space, encourage empirical research on the impact of veterinary social work interventions on individual, community, and organizational outcomes, and increase educational opportunities to raise awareness about veterinary social work. The development of standards and competencies and an exam-based credential in veterinary social work education will ensure that future veterinary social workers are prepared to address the human needs at the intersection of humans and animals well into the 21st century.

### Discussion Questions

1. Discuss how the development of veterinary social work has affected the veterinary-client-patient relationship.
2. Identify how diversity might affect the interaction between a pet owner and a veterinarian. Consider if the client is a person of color, a member of the LGBTQ+ community, and a person who is physically or mentally disabled?
3. Discuss the role of animal welfare standards in the protection of animals and people in AAI.
4. What do you consider to be the needs for a person to cope with the death of a pet?
5. How does access to veterinary social workers affect the mental health and wellbeing of veterinary and other animal health-related workers?
6. What values do you hold about people and their relationships with animals?

## References

Access to Veterinary Care Coalition. (2018). *Access to Veterinary care: Barriers, current practices, and current policy*. https://pphe.utk.edu/access-to-veterinary-care-report/. Accessed August 4, 2022.

American Pet Products Association. (2019). *2019–2020 APPA national pet owners survey*. https://www.americanpetproducts.org/pubs_survey.asp. Accessed August 4, 2022.

American Veterinary Medical Association (AVMA). *Workplace wellbeing certificate program*. https://axon.avma.org/local/catalog/view/product.php?productid=22. Accessed August 4, 2022.

American Veterinary Medical Association (AVMA). *2017–2018 U.S. pet ownership statistics*. (2018). https://www.avma.org/resources-tools/reports-statistics/us-pet-ownership-statistics. Accessed August 4, 2022.

Arkow, P. (2022). The national link coalition. *The Link*. https://nationallinkcoalition.org/. Accessed August 4, 2022.

Arkow, P. (2020). Human-animal relationships and social work: Opportunities beyond the veterinary environment. *Child and Adolescent Social Work Journal, 37*, 573–588.

Assistance Dogs International. (2022). *Summary of standards*. https://assistancedogsinternational.org/standards/summary-of-standards/. Accessed August 4, 2022.

Brackenridge, S. (2019). The social worker: An essential hospice and palliative team member, *Veterinary Clinics of North America: Small Animal Practice, 49*(3), 565–574.

Bright, M. A., Huq, M. S., Spencer, T., Applebaum, J. W., & Hardt, N. (2018). Animal cruelty as an indicator of family trauma: Using adverse childhood experiences to look beyond child abuse and domestic violence. *Child Abuse & Neglect, 76*, 287–296.

Britt-Lutter, S., & Heckman, S. J. (2020). The financial life of aspiring veterinarians, *Journal of Veterinary Medical Education, 47*(1), 117–124.

Campbell, A. (2022). The intertwined well-being of children and non-human animals: An analysis of animal control reports involving children. *Social Sciences, 11*, 46. https://doi.org/10.3390/socsci11020046

Chur-Hansen, A., Stern, C., & Winefield, H. (2010). Commentary: Gaps in the evidence about companion animals and human health: Some suggestions for progress. *International Journal of Evidence-Based Healthcare, 8*, 140–146.

Collins, E., Cody, A., MacDonald, S., Nicotera, N., Ascione, F., & Williams, J. (2018). A template analysis of intimate partner violence survivors' experiences of animal maltreatment: Implications for safety planning and intervention. *Violence Against Women, 24*(4), 452–476.

Crocken B. (1981). Veterinary medicine and social work: a new avenue of access to mental health care. *Social Work Health Care, 6*(3), 91–4. http://dx.doi.org/10.1300/ J010v06n03_09. Medline:7292248

Dalum, H. S., Tyssen, R., & Hem, E. (2022). Prevalence and individual and work-related factors associated with suicidal thoughts and behaviours among veterinarians in Norway: A cross-sectional, nationwide survey-based study (the NORVET study). *BMJ Open, 12*(1–9). https://doi.org/10.1136/bmjopen-2021-055827.

Doka, K. J. (2002). *Disenfranchised grief: New directions, challenges, and strategies for practice*. Research PressPub.

Dunn, J. (2005). Naturalistic observations of children and their families. In S. G. Hogan (Ed.), *Researching children's experience* (pp. 87–101). Sage.

Friedman, E., & Krause-Parello, C. A. (2018). Companion animals and human health: Benefits, challenges, and the road ahead for human-animal interaction. *Revue Scientifique et Technique (International Office of Epizootics), 37*(1), 71–82. https://doi.org/10.20506/rst.37.1.2741. PMID: 30209428.

Greenhill, L. M., Nelson, P. D., & Elmore, R. G. (2007). Racial, cultural, and ethnic diversity within US veterinary colleges. *Journal of Veterinary Medical Education, 34*(2), 74–78.

Hill, E. M., LaLonde, C. M., & Reese, L. A. (2020). Compassion fatigue in animal care workers. *Traumatology, 26*(1), 96–108. https://doi.org/10.1037/trm0000218

Hill, L., Winefield, H., & Bennett, P. (2020). Are stronger bonds better? Examining the relationship between the human-animal bond and human social support, and its impact on resilience, *Australian Psychologist, 55*(6), 729–738.

International Association of Human-Animal Interaction Organizations. (2014). *The IAHAIO definitions for animal assisted interventions and guidelines for wellness of animals involved*. https://iahaio.org/best-practice/white-paper-on-animal-assisted-interventions/. Accessed August 4, 2022.

Julius, H., Beetz, A., Kotrschal, K., Turner, D., & Uvnas-Moberg, K. (2013). *Attachment to pets*. Hogrefe.

Kipperman B. S., Kass, P. H., & Rishniw, M. (2017). Factors that influence small animal veterinarians' opinions and actions regarding cost of care and effects of economic limitations on patient care and outcome and professional career satisfaction and burnout. *Journal of the American Veterinary Medical Association, 250*(7), 785–794. https://doi.org/10.2460/javma.250.7.785. PMID: 28306486.

Kruger, K. A., & Serpell, J. A. (2006). Animal-assisted interventions in mental health: Definitions and theoretical foundations. In A. H. Fine (Ed.), *Handbook on animal-assisted therapy: Theoretical foundations and guidelines for practice* (pp. 21–38). Academic Press.

McKee, H., Gohar, B., Appleby, R., Nowrouzi-Kia, B., Hagen, B. N. M., & Jones-Bitton, A. (2021). High psychosocial work demands, decreased well-being, and perceived well-being needs within veterinary academia during the COVID-19 pandemic. *Frontiers in Veterinary Science* 8:7467161–20. doi: 10.3389/fvets.2021.746716

Merck Animal Health Veterinarian Wellbeing Study. (2020). https://www.merck-animal-health-usa.com/about-us/veterinary-wellbeing-study/veterinary-wellbeing-study-2020. Accessed August 4, 2022.

Monaghan, H., Rohlf, V., Scotney, R., & Bennett, P. (2020). Compassion fatigue in people who care for animals: An investigation of risk and protective factors. *Traumatology*. Advance online publication. https://doi.org/10.1037/trm0000246

National Association of Social Workers. (2021). *Code of ethics of the national association of social workers*. NASW Press. https://naswpress.org/product/53535/code-of-ethics. Accessed August 4, 2022.

National Association of Social Workers (NASW). (1973). *Standards for social service manpower: Policy statement 4*. The Association, Washington, DC.

Nugent, W.R., Daugherty, L. (2022). A measurement equivalence study of the family bondedness scale: Measurement equivalence between cat and dog owners. *Frontiers in Veterinary Science* 8, 812922, 1–11. doi: 10.3389/fvets.2021.812922. PMID: 35087893; PMCID: PMC8787268.

Pet and Women Safety (PAWS) Act (H.R. 909, S.322) (2018). https://www.congress.gov/bill/115th-congress/house-bill/909

Peacock, J., Chur-Hansen, A., & Winefield, H. (2012). Mental health implications of human attachment to companion animals, *Journal of Clinical Psychology*, 68(3), 292–303. https://doi.org/10.1002/jclp.20866. Epub 2012 February 3. PMID: 22307948.

Risley-Curtiss, C. (2010). Social work practitioners and the human-companion animal bond: A national study. *Social Work*, 55(1), 38–46.

Serpell, J. A., Kruger, K. A., Freeman, L. M., Griffin, J. A., & Ng, Z. Y. (2020). Current standards and practices within the therapy dog industry: Results of a representative survey of United States therapy dog organizations. *Frontiers in Veterinary Science*, 7(35), 1–12. https://doi.org/10.3389/fvets.2020.00035.

Quain, A. (2021). The gift: Ethically indicated euthanasia in companion animal practice. *Veterinary Sciences*, 8(141), 1–11. https://doi.org/10.3390/vetsci8080141.

Scoresby, K. J., Strand, E. B., Ng, Z. Y., Brown, K. C., Stilz, C. R., Strobel, K., Barroso, C. S., & Souza, M. (2021). Pet ownership and quality of life: A systematic review of the literature. *Veterinary Sciences*, 8(332), 1–23. https://doi.org/10.3390/vetsci8120332

Shapiro, K., & Henderson, A. J. (2016). *The identification, assessment, and treatment of adults who abuse animals*. Springer. https://doi.org/10.1007/978-3-319-27362-4_2.

Shulman, L. (2012). *The skills of helping individuals, families, groups and communities* (7th ed.). Brooks/Cole.

Sife, W. (2014). *The loss of a pet* (4th ed.). Howell Book House.

Taylor, N., Riggs, D. W., Donovan, C., Signal, T., & Fraser, H. (2019). People of diverse genders and/or sexualities caring for and protecting animal companions in the context of domestic violence, *Violence Against Women*, 25(9), 1096–1115

Tomasi, S. E., Fechter-Leggett, E. D., & Edwards, N. T., et al. (2019). Suicide among veterinarians in the United States from 1979 through 2015. *Journal of the American Veterinary Medical Association*, 254, 104–112.

University of Tennessee Knoxville Veterinary Social Work (UTK VSW). (2022). *Veterinary social work certificate program*. www.vetsocialwork.utk.edu. Accessed August 4, 2022.

Van Hooser, J. P., Pekow, C., Nguyen, H. M., D'Urso, D. M., Kerner, S. E., & Thompson-Iritani, S. (2021). Caring for the animal caregiver-occupational health, human-animal bond and compassion fatigue. *Frontiers in Veterinary Science*, 8.1–10. https://doi.org/10.3389/fvets.2021.731003.

Witte, T. K., Spitzer, E. G., Edwards, N., Fowler, K. A., & Nett, R. J. (2019). Suicides and deaths of undetermined intent among veterinary professionals from 2003 through 2014. *Journal of the American Veterinary Medical Association*, 255(5), 595–608. https://doi.org/10.2460/javma.255.5.595.

# 26
# ZOONOSES
*Jason W. Stull*

## Introduction

As described elsewhere in this book, there are many direct and indirect health benefits from human-animal interactions. These benefits are often experienced by both the humans, as well as animals involved. However, in addition to these benefits, the health risks of such interactions must also be considered. This chapter addresses the infectious disease risks that are present with human-animal interactions as well as mechanisms to best reduce these risks without negatively impacting the positive benefits.

Zoonoses are infectious diseases that are naturally transmitted from living animals to humans (Hubalek, 2003). Diseases such as rabies, salmonellosis, and ringworm are perhaps some of the most well-known zoonoses. Of the more than 1,400 human infectious diseases identified, over 60% are known to be zoonotic (Taylor, Latham, & Woolhouse, 2001). Zoonotic organisms are twice as likely to be associated with emerging diseases (i.e., infections that have recently appeared within a population or whose incidence/geographic range is rapidly increasing) than non-zoonotic organisms. These diseases have the potential to cause significant morbidity and mortality in humans and animals, with resulting implications for international trade, travel, economies, and national security. The number and list of zoonotic diseases continues to increase as new organisms are identified or emerge, such as SARS-CoV-2 (the virus responsible for COVID-19). Livestock (e.g., sheep, goats, swine, poultry), horses, exotics (e.g., reptiles, amphibians, exotic mammals such as hedgehogs), and companion animals (e.g., dogs, cats) each have established and emerging zoonotic organisms that they are known to carry and transmit to people (Lappin et al., 2019; Sack, Oladunni, Gonchigoo, Chambers, & Gray, 2020; Stull, Brophy, & Weese, 2015). It is safe to assume that every animal species can carry and transmit multiple zoonotic organisms and every encountered animal is likely shedding an infectious, possibly zoonotic, organism.

Animals are frequently encountered by people (human-animal interactions). This includes the home environment (e.g., pets), workplace (e.g., food production, veterinary profession, animal sales), schools, wilderness, and even human healthcare facilities. In fact, there are few locations where people go where animals do not. In one national survey, over 60% of Canadian respondents reported contact with animals or animal food within the previous 7 days (Raschkowan et al., 2018). Additionally, there appears to be increasing popularity in the incorporation of animals into non-home environments, such as schools and human healthcare facilities. Given the high frequency with which members of society have contact with animals and the disease risks associated with these contacts, management of zoonotic disease health risks should be incorporated into all activities for which animal interactions are likely. Such health management includes education on zoonotic diseases and preventive measures to reduce disease risk, as well as steps to incorporate practices that emphasize such preventive measures.

## Epidemiology of Zoonoses

### Routes of Transmission

People can acquire zoonoses through numerous routes. Organisms can be acquired through direct or indirect contact with the skin, open wounds, and mucous membranes (i.e., eyes, mouth, nose). Direct contact includes animal bites, scratches, or contact with animal saliva, urine, and other animal body fluids. Indirect contact includes coming into contact with areas or items that have become contaminated with animal-derived organisms. This would include objects or surfaces (also called fomites), such as pet food/water bowls, pitchforks used to move animal manure, bedding, cages, toys, and aquaria. Ingestion of organisms from contaminated food, water or animal fecal material, urine, or body fluids. Organisms can be inhaled as aerosols or droplets. Some organisms can be transmitted from animals to humans when vectors, such as fleas, ticks, and mosquitoes, feed on the animal to acquire the organism and then later feed on people, transmitting the organism. In some cases, animals such as dogs and cats, bring vectors that are already infected with an organism closer to people. Many zoonoses have multiple sources and routes of infection, including direct human-to-human, and even human-to-animal transmission, so identification of a zoonotic organism in a person with recent human-animal interactions (such as pet ownership) does not necessarily indicate animal-to-human zoonotic transmission.

It is important to distinguish true zoonoses, infectious diseases naturally transmitted from living animals to humans, from other infectious disease processes. Some organisms are maintained in nature in soil, water, on vegetation, or in decaying carcasses or animal feces and can infect both humans and animals (*saproneses*). Examples include *Mycobacterium avium*–complex infection, blastomycosis, histoplasmosis, cryptococcosis, and coccidioidomycosis. For these organisms, both humans and animals can acquire infections in a similar manner but independently of each other; in most cases, infected animals do not represent a direct infection risk for people. Instead, when animals get these infections, they act as sentinels (indicators) for the risk of human infection in the region. Further, some organisms are maintained in humans (*anthroponoses*). Some of these infections can affect animals that may then transmit the infection to other animals or people, but this is not the most likely route of human infection. Examples include, Group A streptococcal infections, methicillin-resistant *Staphylococcus aureus* (MRSA) infections, *Clostridioides difficile* (often called *C diff*), and SARS-CoV-2 (COVID-19).

### Risk of Infection

Whether or not a person is infected with a zoonotic organism, and if so if they develop severe disease, is dependent on a number of human, animal, and organism factors, as well as aspects of human-animal interaction itself.

## Human Factors Influencing Infection

Although zoonoses can occur in any individual, those at the extremes of age (< 5 years; ≥ 65 years of age), pregnant, or with immunocompromised conditions are at the greatest risk of disease (referred to as "high-risk" hereafter). People at high-risk for zoonoses often suffer more severe illness, experience symptoms for a longer duration, or experience more severe or unexpected complications than others. The increased risk for these individuals occurs due to an immune system that is not functioning at the same capacity as other individuals. Children and some individuals with development disabilities often have suboptimal hygiene practices that further increase infection risk. Numerous causes can result in an immunocompromised state, such as congenital immunodeficiencies (i.e., from genetic causes), transplants (bone marrow, solid organ), infectious diseases (e.g., HIV infection), metabolic diseases (e.g., diabetes mellitus, chronic kidney failure), and cancers. It is estimated that up to 20% of the North American population has some degree of immunosuppression, with an estimated 60% of households having one or more members being high-risk (Mani & Maguire, 2009; Stull, Peregrine, Sargeant, & Weese, 2012).

In addition to the above-mentioned groups, certain occupations appear to be at increased risk for zoonoses. Veterinarians, animal health workers, and those working or training in settings with high animal contact (e.g., pet shops, shelters) may be at greater risk for zoonoses (Haagsma, Tariq, Heederik, & Havelaar, 2012; Halsby, Walsh, Campbell, Hewitt, & Morgan, 2014). As an example, during their careers, two-thirds of veterinarians reported a major animal-related injury that results in lost work or hospitalization and 47% of veterinarians in the United States reported a zoonotic infection during their work (Baker & Gray, 2009; Jackson & Villarroel, 2012).

Unlike being in a high-risk grouping, other human factors influencing zoonotic disease occurrence are often modifiable. These modifiable factors include, prompt and consistent hand hygiene (washing with soap and water or using alcohol-based hand sanitizer), wearing protective clothing in specific circumstances to reduce contamination of skin or clothing, avoiding contact with higher-risk animals (or their environment) if you are high-risk, and appropriate cleaning and disinfection of animal contact surfaces.

## Animal Factors Influencing Infection

The likelihood of a specific animal carrying (and spreading) zoonotic organisms is dependent on several animal factors. Younger animals are generally at increased risk for acquiring and transmitting zoonotic organisms. This increased risk is attributable to various factors, including inadequate vaccination, waning maternal antibodies, exposure to novel organisms, and behaviors (e.g., chewing of fomites, close play with other animals, and mouthing activity). Animals that are ill, immunosuppressed, or have suboptimal preventive care, including prevention/treatment for parasites (e.g., ticks, fleas, gastrointestinal worms) may be predisposed to infections. Feeding food/treat items that are uncooked meat- or egg-based, or unpasteurized milk, increases zoonotic organism infection risk. Husbandry, such as dense (crowded) animal living conditions, or lack of cleaning and disinfection of living or environmental areas, increases risk. Finally, the opportunity for exposure to zoonotic organisms, such as through national/international travel, involvement in unique settings such as human healthcare settings, and level of confinement (e.g., dog allowed to roam free vs. walked on a leash) impact the chance for an animal to be infected with a zoonotic organism (Robinson, 2001).

Animal species vary in their respective role and risk of transmitting zoonotic organisms. Some organisms can be transmitted to humans via numerous animal species (e.g., *Campylobacter* spp.), others can be transmitted via numerous species, but are much more common in particular animal species (e.g., the high proportion of reptiles and amphibians shedding *Salmonella* spp.), and some can be transmitted to humans by a single or small number of animal species (e.g., bites or other forms of saliva contact from cats and dogs are responsible for human cases of *Capnocytophaga canimorsus*, frequently resulting in severe illness/death) (Trevejo, Barr, & Robinson, 2005).

## *Organism Factors Influencing Infection*

Each of the main infectious organism classifications (e.g., virus, bacteria, fungus, protozoa, parasite, prion) have members that are zoonotic. Among classifications, there is variability in the environmental stability of organisms (ranging from minutes to years), impacting the potential for long-lasting environmental contamination and thus long-term transmission opportunities. An organism with short environmental stability (e.g., influenza viruses) will need to be shed from the animal and quickly transmitted to the person (close, direct human-animal contact), whereas an organism with long environmental stability (e.g., *Salmonella*) can remain infectious on a contaminated object, such that the person and animal may have indirect contact, with weeks in-between each contacting the surface.

Organisms vary with route and ease of zoonotic transmission. For instance, an organism that requires a tick to be spread (e.g., *Borrelia* bacteria the cause of Lyme disease transmitted by tick bite) will differ in terms of infection risk (and prevention strategies) from an organism transmitted by aerosols or feces (e.g., avian influenza requiring close contact between the person and infectious bird). Finally, organisms vary in their infectivity (ability to penetrate and reproduce), pathogenicity (ability to cause disease), and virulence (ability to cause severe disease or death). As an example, rabies virus has a moderate infectivity (a person bitten by a rabid animal has a moderate chance of being infected) and high virulence (if infected, very high chance of death). This can be compared to *Salmonella* bacteria that have a low infectivity and low virulence (e.g., a person whose hands are contaminated with *Salmonella* after handling a snake and inadvertently ingests the bacteria is overall less likely to be infected, and if infected, the chance for severe illness or death is limited especially if the person is not high-risk).

With over 250 zoonotic organisms, and this number continuing to increase as new organisms emerge, existing organisms adapt to be able to infect new species, and advances in technology allow recognition of zoonotic potential of existing organisms, the above represents a brief overview of the complexities of zoonotic organisms. The reader is encouraged to use one of the excellent print or on-line sources for detailed information about a specific animal species or organism (Table 26.1).

## *Human-Animal Interaction Factors Influencing Infection*

Various human-animal interaction factors play a key role in determining zoonotic transmission. These factors can be categorized as human-animal interaction frequency, intensity, and unique attributes of the interaction.

Animal contact by the public is extremely common. Up to 60% of households in North America and elsewhere have pets. These numbers are similar regardless of if high-risk individuals are living in the household (Abarca, Lopez Del, Pena, & Lopez, 2011; Stull et al.,

Table 26.1 Resources on zoonotic diseases

| Resource | Comments on Audience/Content |
|---|---|
| Center for Food Security and Public Health, Iowa State University (https://www.cfsph.iastate.edu/) | Technical and layperson factsheets on numerous zoonotic diseases and prevention measures; online education and training modules |
| Centers for Disease Control and Prevention Healthy Pets, Healthy People (www.cdc.gov/healthypets) | Resource highlighting zoonotic diseases in various animal species; risks and recommendations for high-risk individuals. Numerous brochures on pet-associated diseases in high-risk individuals. Extensive, hyperlinked list of zoonotic diseases. |
| Centers for Disease Control and Prevention Clean Hands Save Lives (www.cdc.gov/handwashing) | Handwashing guidelines for key groups, including pet owners and those involved in public settings. |
| Companion Animal Parasite Council (capcvet.org) | Information and recommendations for parasitic diseases; United States and Canada prevalence maps of select zoonotic organisms in dogs and cats. |
| Companion Animal Zoonoses. (Weese & Fulford, 2011) | Book detailing pet-associated zoonoses, with details on organisms, disease (in animals and people), and prevention. |
| National Association of State Public Health Veterinarians (www.nasphv.org) | Information and guidelines for zoonotic disease prevention. Compendium of measures to prevent disease associated with animals in public settings. |
| Pet Partners (petpartners.org) | Information on service dogs; on-line educational module on infection control/zoonoses training for animal-assisted therapy handlers and healthcare members. |
| World Health Organization (www.who.int) | Association responsible for improving human health worldwide, including zoonoses. |
| Worms and Germs Blog (www.wormsandgermsblog.com) | Numerous factsheets on pet-associated diseases designed for pet-owners, veterinarians, and human healthcare providers; additional resources on the topic. |

2012; Stull et al., 2014). In fact, in one study, 20% of households with children diagnosed with cancer acquired a new pet after their child was diagnosed (Stull et al., 2014). Even if a household does not own an animal, regular animal contact outside the home is still common (e.g., in one study 37% of non-pet owning households had a member with at least weekly animal contact) (Stull et al., 2012). Dogs and cats are the most common pets residing in households with high-risk people, but other species considered higher risk for zoonotic disease transmission (e.g., reptiles, birds, exotics, rodents) are also frequently reported (Stull et al., 2014). This combination of high-risk person with higher risk animal is concerning from a zoonotic disease standpoint.

The level of animal contact (frequency and intensity) in pet-owning households is generally very high. In fact, many pet owners consider household animals to be members of their families, to which the animals are afforded the same benefits as human household members (e.g., animals sleep in bed with household members, lick the faces of household members, household members kiss the animals). As an example, in one Canadian study, respondents reported their pet usually slept in their child's bed (dogs: 13%; cats: 30%), dogs often licked their child's face (24%) and children often cleaned up the pet's fecal material (24%). Similar findings have been reported in other studies and locations, including 30% of dog owners and 61% of cat owners in the United States and Australia reported sleeping with their pets (Chomel & Sun, 2011; Laflamme et al., 2008; Overgaauw et al., 2009).

The unique nature of the human-animal interaction can be an important determinant of zoonotic disease spread. For instance, pet ownership and contact with specific animal species has been linked to a number of national zoonotic disease outbreaks. This includes reptiles/amphibians (small turtles, bearded dragons, crested geckos, water frogs), small/exotic mammals (hedgehogs, guinea pigs, rats), and backyard poultry (Centers for Disease Control and Prevention, 2022a). Contact (direct or indirect) with wildlife poses zoonotic risks. Examples include fatal rabies infections in people bitten by rabid wildlife and gastrointestinal salmonellosis in individuals linked to wild songbirds and bird feeders (Centers for Disease Control and Prevention, 2021).

Animals are often present at various public (non-home) settings. These include schools, human healthcare facilities, animal exhibits and fairs, and petting zoos (Daly, House, Stanek, Stobierski, & National Association of State Public Health Veterinarians Animal Contact Compendium Committee, 2017). In each of these settings, there may be specific elements that increase the risk for zoonotic disease occurrence, such as very close, prolonged human-animal interactions, high-risk people involved (e.g., young children or immunocompromised), or higher risk animal activities (e.g., animals observed while giving birth). The public health impact of these human-animal interactions can be substantial. For instance, in the United States between 2010 and 2015, approximately 100 human infectious disease outbreaks involving animals in public settings were reported with substantial medical, public health, legal, and economic effects (Daly et al., 2017).

Animals are frequently permitted to enter human healthcare and long-term care facilities and interact with human patients and residents (Murthy et al., 2015; Schuurmans, Enders-Slegers, Verheggen, & Schols, 2016; Stull, Hoffman, & Landers, 2018). Zoonotic disease outbreaks in patients, residents, and staff associated with animals in these settings have been described (Moffatt et al., 2014; Murthy et al., 2015). This is perhaps unsurprising as individuals in these settings are often high-risk for zoonotic disease infection and the wide array of animals permitted to enter these facilities in conjunction with the frequent and close interactions permitted. In one US-based study, animals were permitted in 99% of long-term care facilities in the state, with dogs, cats, birds, fish, farm animals (e.g., pigs, goats), rodents, reptiles, and amphibians reported (Stull, Hoffman, & Landers, 2018). In another study, most dog owners that participated in canine human healthcare visitation programs reported a high intensity of interaction between their dog and healthcare patients; 73% allowed their dogs on patients' beds and 79% let their dogs lick patients. This close level of interactions with healthcare patients increases risk for zoonotic disease transmission. In fact, in one study, therapy dogs visiting human hospitals were shown to begin carrying two predominately human-derived zoonotic bacteria (*Clostridioides difficile* (C diff) and methicillin-resistant *Staphylococcus aureus* (MRSA) during visits to the human hospital. The risk of acquiring these zoonotic organisms was higher in dogs that were permitted to lick patients or accept treats during visits, highlighting the connection between zoonotic disease risk and intensity of human-animal interactions (Lefebvre, Reid-Smith, Waltner-Toews, & Weese, 2009).

## Drivers of Zoonoses

The science of the drivers, or why zoonotic diseases emerge or expand to new areas, has been an area of interest for some time, with multiple investigations, analyses, and theories (Institute of Medicine and National Research Council, 2009). Some of the key theorized drivers are (by domain) and interface of domains:

- Human Domain: human health, behaviors, attitudes, preferences, culture, economics, transport, trade.
- Environmental Domain: climate change, temperature, humidity, soil, vegetation type.
- Animal domain: animal health issues, behavior, geographic range, biodiversity loss, predator-prey balance, habitat, and feeding preferences.
- Human-animal interface: animal ownership, husbandry practices, wildlife management practices, habitat encroachment, animals as food.
- Human-environment interface: pollution, urban/periurban development, non-animal farming practices.
- Animal-environment interface: expansion/loss of range, invasive species, environmental effect on immunity, environmental impact on lifespan and reproduction (esp. vectors such as ticks, mosquitoes, and fleas).
- Organism domain: mutations (shift, drift).

In summary, drivers of zoonotic disease emergence and spread are a very complex issue, likely with multiple factors occurring simultaneously and interacting. Changes within one domain may lead to zoonotic organism emergence or greatly amplify and spread an ongoing emerging organism. Several examples are provided below to illustrate these concepts.

Nipah virus (a paramyxovirus) was first discovered in 1999 following an outbreak of disease in pigs and people in Malaysia and Singapore. The virus predominately lives in fruit bats. It is theorized that due to local deforestation, fruit bats were drawn to planted mango trees that were adjacent to an intensively managed, commercial pig farm. The virus was spread from the infected fruit bats by urine or saliva-contaminated partially eaten fruit on which the pigs then fed. Once infected, the pigs spread the virus to people working on the pig farm. During the 1999 outbreak, there were 105 fatalities (40% of those determined to be infected), with almost all having had close contact with infected pigs. Additional outbreaks have occurred since. Nipah virus is considered a growing threat due to its broad host range, wide geographical distribution, high case fatality (death), reports of human-to-human transmission, and the lack of vaccines or effective therapies (Skowron et al., 2021).

First identified in 2019, the SARS-CoV-2 virus (responsible for COVID-19) is believed to have begun through zoonotic transmission from one or more contained wildlife species at a wet market in Wuhan in Hubei Province, China (Mohapatra & Menon, 2022). The virus began spreading person-to-person and shortly thereafter the virus spread globally (through air travel of infected people) resulting in a pandemic with millions of deaths. Notable genetic variations in the SARS-CoV-2 virus from other coronaviruses (including the virus that resulted in the SARS outbreak of 2003) point toward genetic mutations resulting in the COVID-19 virus, and notably its wide host range (Mohapatra & Menon, 2022). Although SARS-CoV-2 is predominately transmitted from person-to-person, numerous animal species (wildlife, farmed animals, companion animals, exotic) have been shown to be naturally infected with the virus (likely from human interactions), with some resulting in illness or death. Furthermore, some species can transmit the virus to other individuals in their species (domestic cats, wild cats, hamsters, mink, ferret, beaver, gorillas, white-tailed deer) and even back to humans (e.g., mink) (Government of Canada, 2022).

A large, prolonged (2016–2021) national outbreak of multidrug-resistant *Campylobacter jejuni* infections was identified in people in the United States (Francois Watkins et al., 2021; Montgomery et al., 2018). The source of the outbreak was determined to be in close contact with puppies sold by specific pet stores and believed to be at least in-part driven by wide-scale use of antibiotics in the puppies. Numerous infected individuals developed severe illness requiring hospitalization. Dogs are known to sometimes be infected with or carry *Campylobacter jejuni*, however the close contact between puppies and household members and multidrug resistance of the bacteria (being more difficult to treat than bacteria without this drug resistance) likely contributed to this zoonotic disease event.

## *Frequency of Zoonoses*

It is estimated that zoonoses are responsible for more than 2.5 billion cases of human illness and 2.7 million human deaths worldwide each year (Christou, 2011; Gebreyes et al., 2014). Many of these disease occurrences relate to specific regions of the world and/or are transmitted by tick or mosquito vectors (e.g., Chikungunya virus) and as such may not be as important from an overall human-animal interaction perspective. Nonetheless, the occurrence and importance of zoonoses from a human-animal interaction standpoint is undeniable. For instance, it is estimated that 14% of all human illness caused by enteric organisms is attributable to direct or indirect animal contact (Hale et al., 2012). In a separate study, cats or dogs were estimated to be responsible for 26% of all reported zoonotic human enteric infections for which contact with animals or their environment was the likely mode of transmission (Whitfield et al., 2017). Despite these impressive numbers, quantifying the occurrence and particulars (e.g., animal species involved) of specific zoonoses is often challenging.

As many zoonotic organisms, especially those transmitted by pets, are not reportable to government agencies, the incidence of disease is largely unknown. Additionally, for those human zoonotic infections that are reportable to public health in most jurisdictions (e.g., salmonellosis), the vast majority may also be acquired through non-animal contact routes (e.g., food, water, direct/indirect human contact) making it difficult to determine the proportion of disease for which animals are responsible. In many cases, apparently healthy animals can carry and shed zoonotic organisms, thereby serving as an important, and often unidentified, source for human infection. Thus, for many zoonotic diseases, it is often a challenge to determine the true source of an individual's infection.

In most instances, although the proportion of potential human zoonotic infections acquired from companion animals is not known, researchers argue it is small (Angulo, Glaser, Juranek, Lappin, & Regnery, 1995; Glaser, Angulo, & Rooney, 1994; Grant & Olsen, 1999). However, this should not negate awareness and use of risk reduction/mitigation measures especially for important zoonoses, in public/healthcare settings, or when high-risk individuals are involved.

## **Prevention of Zoonoses**

Zoonotic diseases are largely preventable, especially in locations/settings that are not greatly limited by finances and resources. Recommendations for animal ownership and human-animal interactions have been established for high-risk individuals (Mani & Maguire, 2009;

Pickering, Marano, Bocchini, Angulo, & Comm Infect, 2008; Yokoe et al., 2009). Guidelines are also available for human-animal interactions in healthcare facilities and animals housed or visiting schools and other public settings (Daly et al., 2017; Murthy et al., 2015). When addressing zoonotic disease risks, it is important to also consider health benefits from human-animal interactions (discussed in other chapters in this book). Importantly, it is rarely necessary for an individual to part with household pets. However, simple measures are indicated to reduce the risk of zoonoses following human-animal interactions. Many factors that increase the chance of animals to carry zoonotic organisms and chance to transmit these organisms to people are easily modifiable.

## *Animal Selection*

This is perhaps one of the simplest, yet most often overlooked modifiable factors in human-animal interactions that impacts zoonotic disease occurrence. Because zoonotic organisms are more prevalent in young animals, the use of these animals (e.g., puppies, kittens) should be carefully considered and ideally a mature animal used instead. When acquiring a new pet, households with high-risk individuals should seek mature, healthy lower risk species (e.g., dog, cat). Contact with, or purchase of higher risk species, such as reptiles, amphibians, rodents, exotic species (e.g., hedgehogs), and young poultry (e.g., chicks), should be avoided if high-risk people will be involved. In some cases, individuals may wish to wait to acquire a new pet or take part in human-animal interactions, such as if an individual's immunocompromised state is transient or variable, until immune suppression is stable.

## *Personal Hygiene*

All individuals, and especially high-risk people, should perform hand hygiene (wash their hands with soap and water or use alcohol-based hand sanitizer) frequently, and always after handling animals or their fecal material. Hand hygiene is especially important before touching one's mucous membranes or other vulnerable sites – this includes before eating, smoking, and having contact with medical devices. Any bites or scratches from animals should be promptly and thoroughly washed with soap and water. Given the increased risk for infections, high-risk individuals should promptly consult with their healthcare provider regarding the need for antibiotics following bites or scratches. All animals should be strictly prohibited from licking open wounds, broken skin, mucous membranes (i.e., face licking), or medical devices such as intravascular catheters. Bite wounds and severe scratches are probably the most common health risk faced by high-risk individuals, resulting in direct damage from the bite, but also the transmission of zoonotic organisms. Pets that are aggressive should not be kept by households, or included in settings (e.g., public venues, healthcare) with high-risk people.

## *Care Around Animal Feces*

Contact with animal feces is a common route of zoonotic disease transmission. High-risk people should reduce contact with animal feces and animal-derived (uncooked) food treats by use of gloves or a plastic bag, or another immunocompetent individual should perform these tasks. Daily cleaning of cat litter boxes can prevent various zoonotic organisms, including *Toxoplasma*. Feline litter boxes should not be placed in food preparation areas and should be cleaned by a non-pregnant, immunocompetent adult. Animals should be prevented from jumping on kitchen counters or dining tables, where they can contaminate eating surfaces with zoonotic organisms.

Animals should be restricted from defecating in child play areas, including playgrounds. Parks should have well-marked, designated areas for pets, playground sandboxes should be kept covered when not in use, and owners should promptly pick up and dispose of pet feces. Children, and those with developmental disabilities, should be closely observed in human-animal interactions or likely contaminated environments to ensure hands are not placed into mouths and observed to effectively wash hands after.

## Care of Animals

Animals in households or included in public/healthcare settings with high-risk individuals should receive routine vaccinations and prevention for zoonotic parasites, including gastrointestinal worms, fleas, and ticks. Regular (e.g., yearly) wellness veterinary appointments are important to maintain the health of the animal. Routine tick and flea checks are indicated based on the area and lifestyle of the animal, with prompt removal of any ticks. Only cooked (commercial or home-prepared) food should be fed to animals in households or included in public/healthcare settings with high-risk individuals.

## Additional Recommendations for Individuals and Animals in Higher Risk Animal-Contact Settings

Individuals working in higher-risk animal settings, such as veterinary clinics, animal-based agriculture, zoos, and other settings where there is close human-animal interactions and animals are at increased risk of carrying a zoonotic organism (e.g., sick animal, higher risk species, dense living conditions) should use additional precautions to reduce zoonotic disease risk. High-risk workers in these settings should discuss any necessary work restrictions and precautions with their physician. This may include, avoiding handling animals with suspected or known zoonotic diseases and strictly following infection control principles, including hand hygiene and use of personal protective equipment (e.g., single use gloves, gowns), or taking additional precautions to prevent animal bites and scratches.

Animals, often dogs and cats but also others, are frequently permitted to enter human healthcare facilities and interact with human patients (Murthy et al., 2015; Schuurmans et al., 2016). Zoonotic disease protocols at facilities are often lacking or those responsible for the animals may have a limited understanding of zoonotic diseases (Lefebvre et al., 2006). Guidelines are available to reduce the risk of zoonotic disease transmission in healthcare environments and should be carefully reviewed and followed (Hardin, Brown, & Wright, 2016; Murthy et al., 2015). These recommendations build off recommendations for households with high-risk members, including further guidance on species and age involved (i.e., only mature dogs; no other species or ages) and health status (i.e., no history of vomiting or diarrhea within the past week and no known or suspected communicable diseases). Hand hygiene for both handler and participant should occur and be stressed before and after every animal contact.

## Current Barriers to Preventing Zoonoses

Given the relative simplicity of largely preventing zoonoses outlined above, it may be surprising that zoonoses continue to be a relevant human health issue. There are a number of factors that contribute to this, including (1) limited knowledge and awareness of zoonotic disease risks by the public and even health professionals, (2) limited communication between human and animal/veterinary professionals, (3) complexity of transmission, where the specific roles of given animals or settings is not easily disentangled for most zoonotic diseases, (4) limited resources dedicated to tracking zoonoses, especially those involving companion animals, and (5) emergence of new zoonoses or re-emergence of existing zoonoses, for which society may be medically unprepared to addresses (prevent, treat) and may be unpredictable. Here we will focus on the first two points as the others have been discussed elsewhere in this chapter.

One of the key barriers to zoonotic disease prevention, especially for those involving companion animals, is the low knowledge of these disease risks by the public and other healthcare professionals (Grant & Olsen, 1999; Stull et al., 2012). This limited knowledge includes high-risk individuals (Stull et al., 2014). As an example, when given a list of 11 infectious organisms, in one study, members of the public were only able to correctly classify just over half on their potential to be transmitted from pets to people. Perhaps this is unsurprising as in these studies most respondents (64%) in the general public and parents of children with cancer (>58%) indicated that they had never received information regarding pet-associated disease risks from a healthcare provider (Stull et al., 2012). In another study, 94% of surveyed physicians in the United States stated they never or rarely initiated discussions about zoonoses with patients with HIV infection or AIDS, a group that is notoriously at increased risk for infectious diseases including those that are zoonotic (Hill, Petty, Erwin, & Souza, 2012). There are several possible reasons for the limited zoonotic disease discussions occurring between human healthcare providers and their patients, including limited comfort in serving this education role, belief veterinarians should instead, or equally, advise clients about zoonotic disease, and limited overall knowledge in this area (Grant & Olsen, 1999; Hill, Petty, Erwin, & Souza, 2012). In one Australian study, veterinarians were found to have greater concern about zoonoses, be more confident in managing and giving advice about the prevention of zoonoses, and more likely to give advice about managing the risk of zoonoses compared to human general medical practitioners (Steele, Booy, Manocha, Mor, & Toribio, 2021). Further, few veterinary and human medical practices make educational materials on zoonotic diseases available to clients or patients (Hill et al., 2012; Lipton, Hopkins, Koehler, & DiGiacomo, 2008).

Perhaps one of the largest challenges is that the subject of zoonoses, especially those resulting from human-animal interactions, falls at the intersection of veterinary and human medicine. In order to effectively address this area, multiple disciplines (e.g., human healthcare, veterinary healthcare, other animal professionals, public health) need to communicate and resources need to be dedicated to the area. Unfortunately, evaluations of this area to-date document poor interdisciplinary communication (Grant & Olsen, 1999; Hill et al., 2012). In one study, 100% of physicians and 97% of veterinarians in the United States claimed to never or rarely contact individuals of the other profession when looking for advice on zoonotic disease risks or cases (Hill et al., 2012). Without a strong collaborative (One Health) approach among healthcare disciplines, human-animal interaction recommendations coming from a single healthcare group may be lacking due to limited knowledge of the area.

## Current Policy and Legislative Perspectives

Given the COVID-19 pandemic and resulting increased public awareness on zoonoses, there has been discussion on policies and legislation aimed at controlling and preventing zoonoses. However, currently in the United States and elsewhere, there are relatively few policies and legislation with the goal of reducing zoonotic disease. A few of the more notable ones are highlighted below (note: policy and legislation are highly variable between countries).

In the United States, no federal laws address the risk for transmission of zoonotic organisms at venues where animals and the public come into contact. However, local jurisdictions may have regulations surrounding animals in these settings (e.g., petting zoos, fairs, schools), that include requirements around education/signage, availability of handwashing facilities, facility design, and animal health (Daly et al., 2017; National Association of State Public Health Veterinarians, n.d). Regardless of legislation, it is strongly recommended to incorporate the zoonotic disease prevention concepts described in this chapter at these public settings when animals are involved. Posters, factsheets, signs, and specific guidance are available for these venues.

Most countries have regulations surrounding animal importation. The scope of these regulations (e.g., to which animal species they apply) will vary by country; some predominantly focus on food-production animals due to the large economic concerns with importing and transmitting diseases impacting food products and export. As an example, in the United States the Centers for Disease Control and Prevention (CDC) regulates the importation of dogs, cats, turtles, non-human primates, African rodents, bats, and civets (Centers for Disease Control and Prevention, 2022b). These regulations are in-place to address zoonotic disease concerns. Other groups, such as the United States Fish and Wildlife Services and United States Department of Agriculture may regulate other species.

In the United States, under the Americans with Disabilities Act (ADA) service dogs used by people with disabilities are permitted to enter locations animals are otherwise not permitted. Specifically, "State and local governments, businesses, and non-profit organizations that serve the public generally must allow service animals to accompany people with disabilities in all areas of the facility where the public is allowed to go" (US Department of Justice, 2020). Although, service animals may be excluded from locations where the animal's presence may be clearly detrimental, such as human healthcare operating rooms or burn units due to compromised sterility. Although not created to reduce zoonotic disease transmission, this legislation becomes important when considering human-animal interactions and implications for zoonotic disease precautions and practices.

## Unanswered Questions and Future Research Demands

Despite the high impact zoonotic diseases have on our society, there are many unanswered questions and areas in drastic need of future research.

The COVID-19 pandemic has highlighted several of these areas including,

- Are there ways to accurately predict zoonotic disease outbreaks or pandemics so that immediate strategies can be used to prevent the occurrence?
- Can we forecast, prevent, or control the development of mutations by organisms (especially those greatly impacting zoonotic disease potential or impact)? Are there ways to accurately predict organisms likely to undergo zoonotically important mutations or prone to switch animal species (e.g., human to animal, animal to human)?
- Which drivers of zoonotic disease emergence are most important? For the drivers that are most likely to be modifiable, what modifications can and should be made to reduce zoonotic disease emergence and spread?

The true scope of companion animal-associated zoonotic disease remains for the most part unknown. This results in challenges relating to risk communication and decision-making, especially when attempting to weigh disease risks against benefits of pet ownership and human-animal interactions. Answers to the following are desperately needed:

- What proportion of human disease is attributable to contact with companion animals?
- Which animals, organisms, behaviors have the greatest impact on this disease burden and what strategies are most successful in reducing disease occurrence?

## Potential Questions for Discussion

1. What are the greatest barriers to effectively preventing and controlling zoonotic infections?
2. How can society better address the knowledge gaps surrounding zoonotic diseases? Knowledge gaps in the public? Knowledge gaps in specific healthcare groups?
3. How can we address the apparent lack of collaboration and communication between healthcare disciplines (veterinary medicine, human medicine, public health)?
4. Is increased knowledge of zoonotic diseases (and prevention) likely to lead to increased uptake in prevention efforts? Why or why not? If not, what are some strategies to improve compliance?

## References

Abarca, V. K., Lopez Del, P. J., Pena, D. A., & Lopez, G. J. C. (2011). Pet ownership and health status of pets from immunocompromised children, with emphasis in zoonotic diseases. *Rev Chilena Infectol*, 28(3), 205–210.

Angulo, F. J., Glaser, C. A., Juranek, D. D., Lappin, M. R., & Regnery, R. L. (1995). Caring for pets of immunocompromised persons. *Can Vet J*, 36(4), 217–222.

Baker, W. S., & Gray, G. C. (2009). A review of published reports regarding zoonotic pathogen infection in veterinarians. *J Am Vet Med Assoc*, 234(10), 1271–1278.

Centers for Disease Control and Prevention. (2021, May 28). *Salmonella outbreak linked to wild songbirds*. Retrieved June 8, 2022, from https://www.cdc.gov/salmonella/typhimurium-04-21/index.html.

Centers for Disease Control and Prevention. (2022a, January 14). *US outbreaks of zoonotic diseases spread between animals & people*. Retrieved June 8, 2022, from https://www.cdc.gov/healthypets/outbreaks.html.

Centers for Disease Control and Prevention. (2022b, Jan 11). *Bringing an animal into the United States*. Retrieved June 8, 2022, from https://www.cdc.gov/salmonella/typhimurium-04-21/index.html.

Chomel, B. B., & Sun, B. (2011). Zoonoses in the bedroom. *Emerg Infect Dis*, 17(2), 167–172.

Christou, L. (2011). The global burden of bacterial and viral zoonotic infections. *Clin Microbiol Infect*, 17(3), 326–330.

Daly, R. F., House, J., Stanek, D., Stobierski, M. G., & National Association of State Public Health Veterinarians Animal Contact Compendium Committee. (2017). Compendium of measures to prevent disease associated with animals in public settings, 2017. *J Am Vet Med Assoc*, 251(11), 1268–1292.

Francois Watkins, L. K., Laughlin, M. E., Joseph, L. A., Chen, J. C., Nichols, M., Basler, C., . . . Friedman, C. R. (2021). Ongoing outbreak of extensively drug-resistant *Campylobacter jejuni* infections associated with US pet store puppies, 2016–2020. *JAMA Netw Open*, 4(9): 1–13.

Gebreyes, W. A., Dupouy-Camet, J., Newport, M. J., Oliveira, C. J., Schlesinger, L. S., Saif, Y. M., . . . King, L. J. (2014). The global one health paradigm: challenges and opportunities for tackling infectious diseases at the human, animal, and environment interface in low-resource settings. *PLoS Negl Trop Dis*, 8(11): 1–6.

Glaser, C. A., Angulo, F. J., & Rooney, J. A. (1994). Animal-associated opportunistic infections among persons infected with the human immunodeficiency virus. *Clin Infect Dis, 18*(1), 14–24.

Government of Canada. (2022, June 2). *Animals and COVID-19*. Retrieved June 8, 2022, from https://www.canada.ca/en/public-health/services/diseases/2019-novel-coronavirus-infection/prevention-risks/animals-covid-19.html.

Grant, S., & Olsen, C. W. (1999). Preventing zoonotic diseases in immunocompromised persons: the role of physicians and veterinarians. *Emerg Infect Dis, 5*(1), 159–163.

Haagsma, J. A., Tariq, L., Heederik, D. J., & Havelaar, A. H. (2012). Infectious disease risks associated with occupational exposure: a systematic review of the literature. *Occup Environ Med, 69*(2), 140–146.

Hale, C. R., Scallan, E., Cronquist, A. B., Dunn, J., Smith, K., Robinson, T., . . . Clogher, P. (2012). Estimates of enteric illness attributable to contact with animals and their environments in the United States. *Clin Infect Dis, 54 Suppl 5*, S472–479.

Halsby, K. D., Walsh, A. L., Campbell, C., Hewitt, K., & Morgan, D. (2014). Healthy animals, healthy people: zoonosis risk from animal contact in pet shops, a systematic review of the literature. *PLoS One, 9*(2): 1–13.

Hardin, P., Brown, J., & Wright, M. E. (2016). Prevention of transmitted infections in a pet therapy program: an exemplar. *Am J Infect Control, 44*(7), 846–850.

Hill, W. A., Petty, G. C., Erwin, P. C., & Souza, M. J. (2012). A survey of Tennessee veterinarian and physician attitudes, knowledge, and practices regarding zoonoses prevention among animal owners with HIV infection or AIDS. *J Am Vet Med Assoc, 240*(12), 1432–1440.

Hubalek, Z. (2003). Emerging human infectious diseases: anthroponoses, zoonoses, and sapronoses. *Emerg Infect Dis, 9*(3), 403–404.

Institute of Medicine and National Research Council. (2009). *Sustaining global surveillance and response to emerging zoonotic diseases*. Washington, DC: The National Academies Press.

Jackson, J., & Villarroel, A. (2012). A survey of the risk of zoonoses for veterinarians. *Zoonoses Public Health, 59*(3), 193–201.

Laflamme, D. P., Abood, S. K., Fascetti, A. J., Fleeman, L. M., Freeman, L. M., Michel, K. E., . . . Willoughby, K. N. (2008). Pet feeding practices of dog and cat owners in the United States and Australia. *J Am Vet Med Assoc, 232*(5), 687–694.

Lappin, M. R., Elston, T., Evans, L., Glaser, C., Jarboe, L., Karczmar, P., . . . Ray, M. (2019). 2019 AAFP feline zoonoses guidelines. *J Feline Med Surg, 21*(11), 1008–1021.

Lefebvre, S. L., Reid-Smith, R. J., Waltner-Toews, D., & Weese, J. S. (2009). Incidence of acquisition of methicillin-resistant *Staphylococcus aureus*, *Clostridium difficile*, and other health-care-associated pathogens by dogs that participate in animal-assisted interventions. *J Am Vet Med Assoc, 234*(11), 1404–1417.

Lefebvre, S. L., Waltner-Toews, D., Peregrine, A., Reid-Smith, R., Hodge, L., & Weese, J. S. (2006). Characteristics of programs involving canine visitation of hospitalized people in Ontario. *Infect Control Hosp Epidemiol, 27*(7), 754–758.

Lipton, B. A., Hopkins, S. G., Koehler, J. E., & DiGiacomo, R. F. (2008). A survey of veterinarian involvement in zoonotic disease prevention practices. *J Am Vet Med Assoc, 233*(8), 1242–1249.

Mani, I., & Maguire, J. H. (2009). Small animal zoonoses and immuncompromised pet owners. *Top Companion Anim Med, 24*(4), 164–174.

Moffatt, C., Appuhamy, R., Andrew, W., Wynn, S., Roberts, J., & Kennedy, K. (2014). An assessment of risk posed by a *Campylobacter*-positive puppy living in an Australian residential aged-care facility. *Western Pac Surveill Response J, 5*, pp. 1–6.

Mohapatra, S., & Menon, N. G. (2022). Factors responsible for emergence of novel viruses: an emphasis on SARS-CoV-2. *Curr Opin Environ Sci Health, 27*, 100358.

Montgomery, M. P., Robertson, S., Koski, L., Salehi, E., Stevenson, L. M., Silver, R., . . . Laughlin, M. E. (2018). Multidrug-resistant *Campylobacter jejuni* outbreak linked to puppy exposure - United States, 2016–2018. *MMWR Morb Mortal Wkly Rep, 67*(37), 1032–1035.

Murthy, R., Bearman, G., Brown, S., Bryant, K., Chinn, R., Hewlett, A., . . . Weber, D. J. (2015). Animals in healthcare facilities: recommendations to minimize potential risks. *Infect Control Hosp Epidemiol, 36*(5), 495–516.

National Association of State Public Health Veterinarians. *Animal contact compendium and resources*. Retrieved June 8, 2022, from http://www.nasphv.org/documentsCompendiumAnimals.html.

Overgaauw, P. A. M., van Zutphen, L., Hoek, D., Yaya, F. O., Roelfsema, J., Pinelli, E., . . . Kortbeek, L. M. (2009). Zoonotic parasites in fecal samples and fur from dogs and cats in The Netherlands. *Vet Parasitol, 163*(1–2), 115–122.

Pickering, L. K., Marano, N., Bocchini, J. A., Angulo, F. J., & Comm Infect, D. (2008). Exposure to nontraditional pets at home and to animals in public settings: risks to children. *Pediatrics, 122*(4), 876–886.

Raschkowan, A., Thomas, M. K., Tataryn, J., Finley, R., Pintar, K., & Nesbitt, A. (2018). Measuring animal exposure in Canada: foodbook study, 2014–2015. *Zoonoses Public Health, 65*(7), 859–872.

Robinson, R. A. (2001). Zoonoses and immunosuppressed populations. In C. N. L. Macpherson (Ed.), *Dogs, zoonoses, and public health* (pp. 273–297). New York: CABI.

Sack, A., Oladunni, F. S., Gonchigoo, B., Chambers, T. M., & Gray, G. C. (2020). Zoonotic diseases from horses: a systematic review. *Vector Borne Zoonotic Dis, 20*(7), 484–495.

Schuurmans, L., Enders-Slegers, M. J., Verheggen, T., & Schols, J. (2016). Animal-assisted interventions in Dutch nursing homes: a survey. *J Am Med Dir Assoc, 17*(7), 647–653.

Skowron, K., Bauza-Kaszewska, J., Grudlewska-Buda, K., Wiktorczyk-Kapischke, N., Zacharski, M., Bernaciak, Z., & Gospodarek-Komkowska, E. (2021). Nipah virus-another threat from the world of zoonotic viruses. *Front Microbiol, 12*, 811157.

Steele, S. G., Booy, R., Manocha, R., Mor, S. M., & Toribio, J. L. M. L. (2021). Towards one health clinical management of zoonoses: a parallel survey of Australian general medical practitioners and veterinarians. *Zoonoses Public Health, 68*(2), 88–102.

Stull, J. W., Hoffman, C. & Landers, T. (2018). Health benefits and risks of pets in nursing homes: a survey of facilities in Ohio. *J Gerontol Nurs, 44*(5):39–45.

Stull, J. W., Peregrine, A. S, Sargeant, J., M, & Weese, J. Scott. (2012). Household knowledge, attitudes and practices related to pet contact and associated zoonoses in Ontario, Canada. *BMC Public Health, 12*, 553.

Stull, J. W., Brophy, J., Sargeant, J. M., Peregrine, A. S., Lawson, M. L., Ramphal, R., . . . Scott Weese, J. (2014). Knowledge, attitudes, and practices related to pet contact by immunocompromised children with cancer and immunocompetent children with diabetes. *J Pediatr, 165*(2), 348–355.

Stull, J. W., Brophy, J., & Weese, J. S. (2015). Reducing the risk of pet-associated zoonotic infections. *CMAJ, 187*(10), 736–743.

Taylor, L. H., Latham, S. M., & Woolhouse, M. E. (2001). Risk factors for human disease emergence. *Philos Trans R Soc Lond B Biol Sci, 356*(1411), 983–989.

Trevejo, R. T., Barr, M. C., & Robinson, R. A. (2005). Important emerging bacterial zoonotic infections affecting the immunocompromised. *Vet Res, 36*(3), 493–506.

US Department of Justice. (2020, Feb 24). *Service animals.* Retrieved June 8, 2022, from https://www.ada.gov/service_animals_2010.htm.

Weese, J. S., & Fulford, M. B. (2011). *Companion animal zoonoses.* Iowa: Wiley-Blackwell.

Whitfield, Y., Johnson, K., Hobbs, L., Middleton, D., Dhar, B., & Vrbova, L. (2017). Descriptive study of enteric zoonoses in Ontario, Canada, from 2010–2012. *BMC Public Health, 17*(1), 217.

Yokoe, D., Casper, C., Dubberke, E., Lee, G., Munoz, P., Palmore, T., . . . Centers for Disease Control and Prevention. (2009). Safe living after hematopoietic cell transplantation. *Bone Marrow Transplant, 44*(8), 509–519.

# 27
# WORKING WITH EACH OTHER
## Animal Training to Coexist

*Seana Dowling-Guyer*

### Introduction

The non-human animals with whom we live and work depend on us for their care, welfare, and very survival. To properly care for them, we manage their environment, their activities, their health, and their behavior. The companion and working animals who share our lives learn to adapt to this management just as we adapt to the individual preferences and needs of each animal, with the goal being a successful human-animal relationship that enhances the lives of all species as well as accomplishing any desired tasks and activities. To support this relationship, it will inevitably be necessary to manage and direct the behavior of our companion and working animals. One way to do this is training based on learning theory and principles. Training our companion and working animals provides multiple benefits beyond just learning a behavioral task, for both the animal and the human. Our animal companions can live more harmoniously with us, perform the tasks we ask of them, stay safe and healthy, and be active physically and mentally. We ourselves can benefit from training our animal companions, strengthening our relationships with them through enhanced interaction. In fact, the more interaction we have with our animals, the better we get to know them, enabling us to provide them more optimal care.

This chapter will provide an overview of the fundamentals of learning theory commonly utilized in animal training along with research into training and outcomes. A brief review of the history of modern animal training is presented, leading to current trends in animal training. Finally, we will discuss guidelines for effective, humane, and respectful training practices and explore influences on the trainer that may impact the training experience. This chapter examines both training and behavior modification; generally, training is focused on teaching new behaviors, often related to obedience, good manners, and specialty performance training (e.g., agility, scentwork) whereas behavior modification targets existing behaviors, usually problematic ones, that need modifying or reducing. For the purposes of this chapter, we will refer to both of these goals as training. The focus of this chapter will be on companion animals such as dogs and cats who are pets or working animals; these non-human animals collectively will be called either companion animals, animals who are pets and provide companionship to the humans in their lives, or working animals, who are trained to perform a specific function or role such as a guide dog or riding therapy horse. Sometimes we will call these animals "animal learners" when discussing specific training topics. The human animals who care for and train these animals will be called owners or trainers. Keep in mind, the principles discussed here apply to other animal species such as ferrets, iguanas, horses, and goats. Note: the information presented in this chapter is meant for educational purposes and should not be used as a guide for training your own animal; please consult a professional trainer as needed. Now, let us turn to the essential points of learning theory and how animals learn.

## Learning Theory

Animals are learning all the time. From the simplest amoeba to more complex animals like mammals and reptiles, organisms have evolved the capacity to learn from their environment and their behaviors (Mackintosh, 2022). The capacity to learn is adaptive; it helps an animal survive and so has been passed on to successive generations until it is genetically coded in a species' DNA. All animals have the capacity to learn, but what they specifically learn and through what processes differ by species, by developmental period, and even by motivation or inclination. We can leverage an animal's capacity to learn when training, but it is important to remember that animals learn all the time, even when not training and especially when we least expect it!

Two types of learning most often utilized when training animals are associative and non-associative learning. Associative learning occurs when two events, or stimuli, happen close in time to each other and become linked, or associated, together. Non-associative learning does not rely on such a pairing and instead depends on exposure to a stimulus.

### *Associative Learning*

The classic story of Ivan Pavlov's dogs salivating when they heard the tick of a metronome before being given food is an example of associative learning called respondent conditioning. It is also called classical or Pavlovian conditioning. In contrast, operant (or instrumental) conditioning occurs when the animal learns from the consequences of their behavior. The animal acts and then something follows that is likely to increase or decrease the behavior due to the association between the behavior and the consequence. In both cases, learning is more likely when events and behaviors are temporally proximate to each other; the more time that passes between stimulus, behavior, or consequence, the less likely it is that these events will be associated together. See Figure 27.1 for an overview of how respondent and operant conditioning work.

### *Classical Conditioning*

Initially working with dogs to study their digestive system, Ivan Pavlov noticed his study dogs regularly salivating in response to the experimental preparation which included meat powder the dogs later consumed (Mackintosh, 2022; Todes, 2014). He began a systematic study of this phenomenon using a metronome as a neutral stimulus and the meat powder, which he described as the unconditioned stimulus, because it already naturally provoked salivation, the unconditioned response. After multiple trials, the ticking of the metronome became the conditioned stimulus, a previously neutral stimulus that now provoked salivation, which was now the conditioned response. Salivation now occurred in response to the metronome, without the presence of the meat powder. The meat powder then became a reinforcer

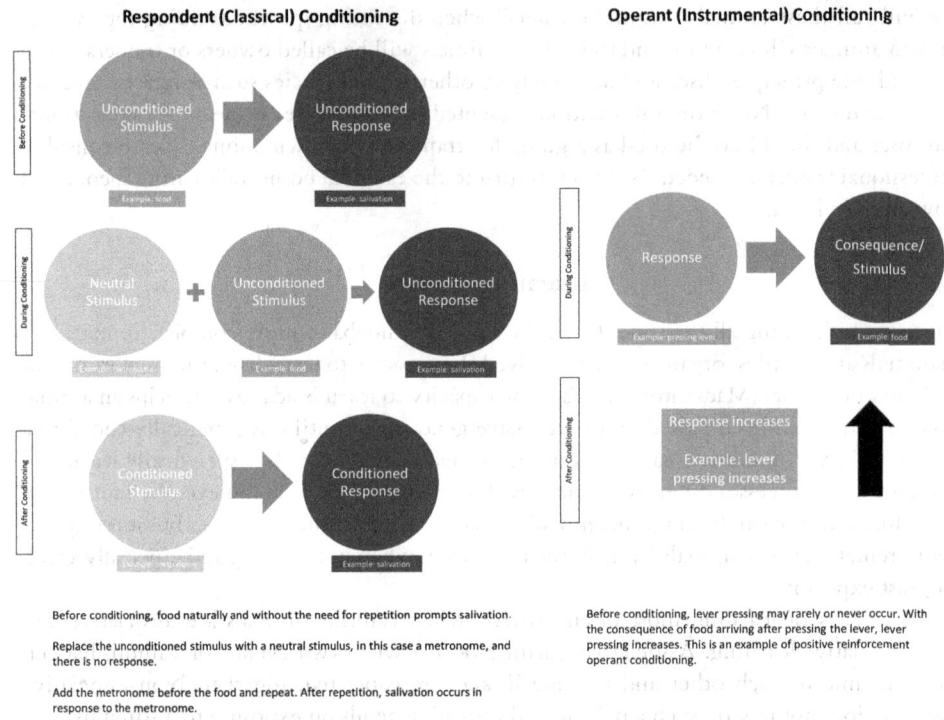

*Figure 27.1* Respondent (classical) conditioning and operant (instrumental) conditioning

of the conditioned stimulus, the metronome. At its simplest, respondent conditioning uses a stimulus-stimulus pairing to produce a desired response.

Techniques based on respondent conditioning can be powerful and effective training methods, but they must be used strategically and appropriately. Nowadays, respondent conditioning is often used to change an animal's emotional state and behavior in a situation that is uncomfortable or distressing to them (more on this below).

## Operant Conditioning

Animal training now relies heavily on operant conditioning. An important difference between respondent and operant conditioning is the role of the animal learner. Identified by Edward L. Thorndike and B.F. Skinner (Mackintosh, 2022; Skinner, 1938, 1953), operant conditioning is based on the idea that the learner's behavior prompts a consequence that then influences the occurrence of that behavior in the future. Skinner described this type of learning as, "active behavior that operates upon the environment to generate consequences" (Skinner, 1953). In Skinner's experiments with rats, rats learned over time to press a lever (the response) which then delivered food (the consequence). An association between the lever pressing and food developed because food was a reinforcement that increased the behavior of lever pressing. Operant conditioning relies on a response-consequence (also called response-stimulus) pairing to produce a desired response.

Operant conditioning can be used to increase or decrease a behavioral response through the timely delivery of consequences. Reinforcements increase or strengthen behaviors,

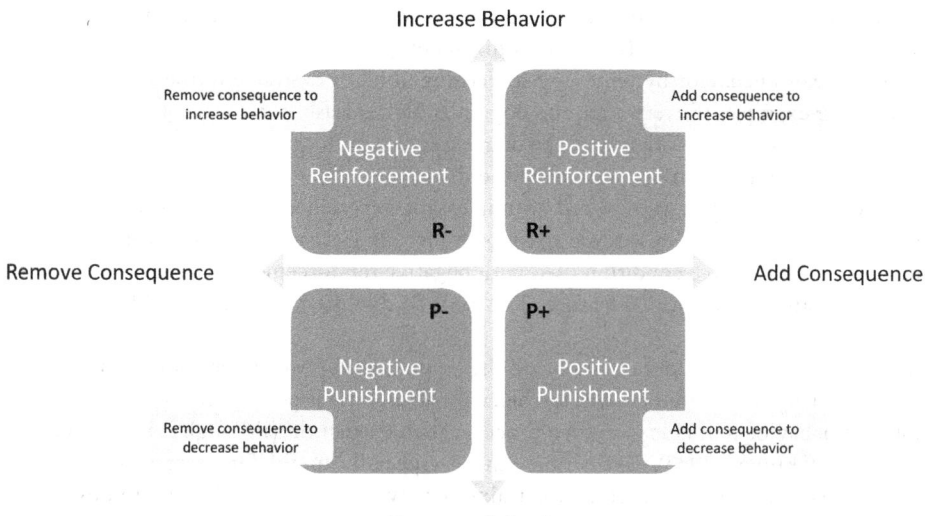

*Figure 27.2* Four quadrants of operant (instrumental) conditioning

punishments decrease or weaken behaviors. This is accomplished by adding a consequence (called a positive consequence because something is added) or removing or avoiding a consequence (called a negative consequence because something is removed or avoided). Figure 27.2 presents the four quadrants of operant conditioning, with the grid divided by the training goal (to increase or decrease behavior) and the training consequence (positive or negative). It is important to understand that positive consequences are not always good; they can be undesired or unpleasant events like scolding or "leash pops". Similarly, negative consequences are not always bad; they can be the removal or cessation of unpleasant events like the release of pressure from a taut leash or avoiding the withdrawal of an owner's attention. If this sounds confusing, it can be! Let's work through some examples to clarify these terms (see also Casey, 2013).

Imagine an owner wants to train her cat to jump in her lap and lay down. Using positive reinforcement, the owner can use petting (assuming the cat enjoys it!) to increase the behavior of lap jumping because the petting is a positive reinforcer (something added to increase the desired behavior). In another scenario, if the owner wants to reduce the cat's jumping on the table, the owner may use positive punishment to decrease that behavior. The owner can scold the cat or clang pots together to make a loud noise, both of which are positive punishers because they are added to the situation to decrease behavior. Note, it is critical that any unpleasant consequences are not too severe or distressing for the animal learner as that is not humane, may risk injury to animal learner or trainer, and may damage the bond between animal and human. We will discuss this in more detail in the next section on consequences.

Negative consequences (removing or avoiding experiences) are harder to envision and, honestly, use. Negative reinforcement aims to increase a behavior by removing or avoiding an unpleasant or undesired event. For example, an owner who wishes to increase his dog's loose-leash walking can use a collar that applies pressure on the dog's neck/head when the dog pulls, something most dogs find unpleasant. When the dog stops pulling and walks with a loose leash, the pressure is eliminated. So the dog's loose-leash walking increases because the unpleasant pressure experienced when pulling is removed (one could argue that the initial pressure introduced by the pulling was a positive punishment, but that's a discussion for another time). Negative punishment (sometimes called punishment by removal) decreases behavior by removing something good or desirable to the animal. In our dog walking example, if every time the dog pulls on the leash, the owner stops walking or turns around and goes the other way, the dog learns that pulling means the fun walk stops so eventually stops pulling in order to avoid the disruption of the walk. Notice with negative reinforcement, the owner introduced an unpleasant experience in order to remove it as a consequence. That should raise some alarm bells for you. Care must be taken when selecting and creating the negative condition used with negative reinforcement; it should be mild and not too stressful for the animal learner. This will be discussed in more detail below in the section on consequences.

Finally, trainers can intentionally break the connection between behavior and reinforcer by *not* delivering the reinforcer after the behavioral response is performed. After some repetition, the behavior will fade away in a process called extinction (although it takes time and persistence) (Skinner, 1933). Extinction can be especially useful for decreasing behaviors that were inadvertently rewarded by the trainer or owner without having to use positive or negative punishment.

## *Choosing Operant Conditioning Types and Consequences*

When determining what type of operant conditioning to use, trainers need to take into account the goal of training, their animal learner and relevant behavior, emotional state, past experience, and the training environment, among other factors. Since operant conditioning relies on consequences to change behavior, trainers need to consider what is meaningful to the animal learner. For example, reinforcers used with positive reinforcement should be seen as favorable or desirable by the learner, not necessarily the human. Using regular kibble as a food reinforcer may not be effective if the cat is full or more interested in the birds outside the window so a higher value food item might be necessary. While yelling at the dog for barking seems unpleasant to us, the dog may, in fact, interpret that consequence as a good thing because it is attention from the owner. The delivery schedule of the consequence matters as well (Morgan, 2010). In a practical training setting, when using positive reinforcement, delivering a positive reinforcer like food or praise every time the behavior is performed has few downsides although the same cannot be said for unpleasant, aversive consequences. The animal's behavior should be changed by the consequence. If the consequence does not change behavior, it is not, in fact, a consequence.

If using negative reinforcement or positive punishment, consequences are aversive events, actions, or states, either removed or added, that the animal finds uncomfortable or unpleasant. They may be mild like the withdrawal of attention, an unpleasant sound like a "tsk" from the owner, or pressure from a rider's leg. They could be moderately unpleasant like the sound of coins shaken in a can ("shake can") or even more intense like an owner yelling, an electronic collar delivering a shock (even a mild one can be painful), or a rider hitting a horse with a whip. These more intense aversive consequences should be avoided; they are mentioned here only as examples of a range of intensity. Even moderate consequences like a shake can may be distinctly unpleasant for the learner just as removing a pleasant activity as part of negative punishment may be stressful to the animal. So care must be taken when using any kind of aversive consequence so that it does not induce stress, fear, aggression, helplessness, or other negative reactions. The bottom line is that the right consequence is important; it should take into account the animal learner's preferences as well as the stage of learning, behavior being learned, learning environment, and other situational and learner-specific factors as well as being humane and respectful of the animal. It is not as easy as grabbing a piece of hot dog and training!

In addition, the risk of misuse and negative repercussions is higher when training with aversive consequences. Positive reinforcement is relatively forgiving; mistiming when treats are delivered or giving too many or not enough may be mildly confusing or frustrating for the animal learner (and often the human!) but there are few other downsides. Most times, these missteps can be corrected quickly. However, it is easier to misjudge the experience and effect of aversive consequences making the risks of negative outcomes greater. Misuse and misapplication often happen due to an incomplete understanding of the principles of learning theory, too strong of an aversive consequence, poor timing, or when the animal learner is in a fear or stressed state. A horse trainer may plan to use negative reinforcement to encourage a horse to step off to relieve the pressure from the rider's leg; but if the horse steps off and the pressure is not released on time, that pressure may become a positive punishment to the horse that will actually discourage the behavior of stepping off.

In fact, numerous studies surveying dog owners found that those who reported a higher use of aversive methods also reported higher incidences of fear and aggression in their dogs (Ziv, 2017). Confrontational methods to "correct" behavior such as hitting or kicking the dog, performing an "alpha roll" (forcing the dog to roll on to their back and keeping them there, based on the discredited idea of wolf pack dominance), forcing release of something in the dog's mouth, and grabbing the dog by neck or jowls were all related to aggressive responses in 25% or more cases of dogs presenting to a behavioral clinic (Herron et al., 2009). In several observational studies, researchers observing training sessions in different settings found more stress- and fear-related behaviors in dogs trained using aversive consequences (Deldalle & Gaunet, 2014; Haverbeke et al., 2008). In addition, these dogs were less playful with owners (Rooney & Cowan, 2011), less interactive with strangers (Rooney & Cowan, 2011), more distracted (Haverbeke et al., 2008), and took longer to learn a new task (Rooney & Cowan, 2011) or performed worse (Haverbeke et al., 2008) than dogs trained using positive reinforcement, or even less severe aversive consequences. Not only did the dogs trained with positive reinforcement methods learn the new task faster in some cases (Rooney & Cowan, 2011), they displayed fewer stress-related behaviors (Deldalle & Gaunet, 2014) and were more playful and interactive (Rooney & Cowan, 2011). They also gazed more at their owners, which is usually a behavior encouraged by trainers as it means the trainer has the dog's attention (Deldalle & Gaunet, 2014).

Although many of these studies are correlational, meaning we cannot conclude that the training methods definitively caused these negative behavioral and emotional responses, the bottom line is that there is mounting evidence, from multiple studies with different populations in different countries using different methodologies, that points to a link between aversive training methods and higher levels of fear, stress, and aggression. Ziv (2017), in a comprehensive review of 17 studies examining different dog training methods, concluded that "using aversive training methods (e.g., positive punishment and negative reinforcement) can jeopardize both the physical and mental health of dogs" (p. 51). Further, Ziv recommends, "those working with or handling dogs should rely on positive reinforcement methods and avoid using positive punishment and negative reinforcement as much as possible" (p. 51). In summary, the use of aversive consequences carries risks. While training may incorporate conditioning from all four quadrants, particularly at the start of training to decrease an undesirable behavior while increasing an incompatible desirable one, it is important to consider approaches and consequences carefully in order to best support the animal learner in a safe, comfortable, and humane learning environment. Positive reinforcement is the best approach to start with and sometimes the only one that is needed.

## Non-Associative Learning

Unlike associative learning, non-associative learning does not rely on a connection between behavior and a specific consequence; rather, it depends on exposure to a stimulus, usually repeated exposure. Two forms of this learning are sensitization and habituation. These two complementary but opposite forms of learning occur through repeated exposure to a stimulus which results in behavior change.

### Sensitization

Sensitization results in an increasing behavioral response to a stimulus after repeated exposure (Mackintosh, 2022). This process is not a form of associative learning because there is no consequence or outcome linked to the stimulus; no pleasant or unpleasant consequence occurs after the stimulus. Rather, the repeated exposure to the stimulus itself induces the behavioral change. Animals have evolved the capacity to sensitize to stimuli, particularly select stimuli relevant to each species. Sensitization is neither a positive nor negative process in and of itself but the outcome can be positive or negative for the animal, depending on the situation. Although sensitization can be utilized in animal training, more often we are interested in the opposite process, one of habituation and desensitization.

### Habituation

In contrast to sensitization, habituation occurs when an animal adapts to a stimulus until there is little effect of the stimulus on behavior; that is, there is a decrease in the animal's behavioral response to the stimulus after repeated exposure (Mackintosh, 2022). The sudden booming of thunder during a storm can be frightening at first and an animal may exhibit fearful or stress-related behavior, but over time, with repeated exposure with no negative outcomes, the thunder becomes less frightening and the animal does not exhibit those behaviors anymore in response to the thunder. The animal has habituated to the thunder. Similar to sensitization, there is an adaptive benefit to habituation. Initially, an animal may react to a novel stimulus by startling and fleeing or hiding, behaviors meant to avoid the potentially risky stimulus. But these behaviors are costly in terms of energy and time. Through habituation, an animal becomes used to some stimuli, growing to ignore their presence and conserving resources used for survival. However, dishabituation can occur, which is when an animal previously habituated to a stimulus reacts to it again. In addition, a type of overexposure can happen, where an animal exposed to a stimulus becomes so completely overwhelmed, or flooded, that he panics or shuts down, the latter being a state called learned helplessness (Seligman & Maier, 1967). It is critical to avoid flooding an animal when using habituation in training, which is why a form of habituation called desensitization is often more effective than unmodified habituation.

### Desensitization

While habituation depends on exposure to a "full" stimulus, desensitization is a strategic process used in training to induce habituation through exposure to small, less intense versions of the stimulus (Rachman, 1967). When addressing fearful or stress-related behaviors, a trainer plans exposures to small, graduated proxies of the stimulus that the animal can comfortably tolerate (McConnell, 2005a); the goal is to stay "under threshold" so the animal is never triggered into a fearful, or overly stressed state. At any sign of these behaviors, the trainer backs off to a more comfortable exposure and begins again at a less intense level. It is a process of baby steps, always working at the comfort level of the animal learner. Going forward, when discussing a habituation plan, we will focus on the use of desensitization as it is the more common technique used to address fearful or stress-related behaviors.

## *Desensitization and Counter-Conditioning*

Desensitization can be highly effective at reducing fear or stress-related behaviors, but it is not an appropriate or effective method for teaching new behaviors. The goal of desensitization is to *reduce* a behavioral response to a provoking stimulus; however, we often want to *replace* that behavioral response with another, more appropriate or desired response. This is why a technique called counter-conditioning is often paired with desensitization so that as the frequency of an undesired behavior decreases through desensitization, the frequency of a desired behavior is increased. Counter-conditioning uses respondent conditioning to pair a provoking stimulus like thunder with the desired response like relaxation through the use of a desirable stimulus like food. The combination of desensitization and counter-conditioning results in a clever two-step strategy that shifts the provoking stimulus from a negative signal to a neutral one through careful desensitization and then from a neutral signal to a positive one through counter-conditioning. Desensitization and counter-conditioning address the animal's emotional state which can influence an animal's ability to learn.

## *Behavior and Emotion in Training*

Research has amply demonstrated that animals feel emotion and that emotion influences behavior (de Waal & Andrews, 2022). When training, it is important that a trainer assess where their animal learner is emotionally, in order to develop an effective and humane plan for the desired training outcomes. No organism learns well in a fearful, anxious, or stressed state, not dogs nor cats nor horses nor iguanas – nor even humans. Or rather, to be more precise, learning can occur in these negative emotional states, but it is often suboptimal and poorly retained. If a negative emotional state such as fear or anxiety exists, the training plan needs to address those emotions first. As discussed, desensitization and counter-conditioning are often used with animals experiencing fear or stress in reaction to certain stimuli (McConnell, 2005a, 2005b). We operationalize these emotional states by assessing behavior as that is the only way for us to glean insight into an animal's emotional state, at least on a day-to-day basis. Therefore, while we focus on behavioral responses, we must always remember that these may reflect an emotional state and those emotional responses may need to be addressed first before learning a new behavior.

There are many more aspects to learning than could be presented here, but we have covered important foundational concepts. Also, there are other types of learning such as observational, social, and cultural learning, some of which can be leveraged when training animals, however, they are outside the scope of this chapter. Now that we have reviewed the fundamentals of animal learning, we will discuss current trends and best practices in training based on scientific research.

# Current Trends and Best Practices for Effective and Humane Animal Training

## A Brief History of Modern Animal Training

While a comprehensive review of the history of non-human animal training is beyond the scope of this chapter, it is helpful to learn about some of the influential recent philosophies from which modern animal training emerged. No doubt, humans have been training animals for thousands of years, much of that using dominance methods. Many early trainers in the 19th and 20th centuries came from police or military training programs where the use of stronger aversive consequences was common at that time. This style of training dominated the field for many decades, supported by a false belief that animals feel limited pain and have a limited emotional repertoire, beliefs which scientific research does not support. Unfortunately, this style of training exerted a strong influence well into the current era.

Often, animal training philosophies are based on an inaccurate understanding of an animal's natural behaviors. A good example of this is the idea of the "alpha dog" which was based on the "alpha wolf" characterization (Mech, 1999). The "alpha wolf" was described as the dominant wolf in a pack, usually male, who ruled through aggression and fear. Unfortunately, these conclusions came from observations of human-made packs where unrelated wolves were placed together in contained areas. No doubt these wolves felt fear, stress, and anxiety after capture and transport, and jockeyed for positions of power in order to gain access to resources. These behaviors are very different from the social behaviors of naturally-occurring wolf packs living as family units, with the breeding pair in the leadership positions as parents (Mech, 1999). The alpha male and alpha female lead through their natural parental authority, which rarely requires truly aggressive displays or interactions to maintain order. This fundamental misunderstanding of how actual wolf packs work has done great damage to our relationships with dogs and even compromised safety as humans used a variety of odd and sometimes unpleasant, even painful, techniques (e.g., the "alpha roll") to teach our dogs that we are the alpha being in the relationship (Ziv, 2017). Many popular training philosophies and trainers espoused this theory of "alpha dog" without realizing how wrong they were. Unfortunately, this myth is still kicking around, decades after it has been debunked. This kind of misinformation exists with other species as well, including horses.

## Current Trends in Animal Training

As our knowledge of animal behavior, cognition, and emotion has increased, so has our understanding of their needs and what constitutes good welfare. The development of the Five Freedoms framework and its successors (Mellor, 2016) has helped drive better welfare assessments and provisioning for animals in a variety of environments, from labs to farms to shelters. Research has discovered how complex animal cognition is and how deeply animals feel emotion (Safina, 2015). Some species like dogs are particularly attuned to humans, able to interpret pointing signals from humans better than primates, our closest relatives on the taxonomic tree (Reid, 2009). In addition, the role of animals in our lives has changed. More households have pets who are often seen as family members (American Veterinary Medical Association, 2022). Animals now assist humans in a variety of ways, providing help as guide and life skills support, medical detection and assistance, animal-assisted therapeutic interventions, and emotional support as well as in police and military roles. Animal training has evolved as well, moving from compulsory training methods that involved harsh aversive consequences to greater use of positive reinforcement techniques, mild aversives, and respondent conditioning to help with fear and anxiety. Different cultural and societal perspectives have expanded our understanding of human–animal relationships and informed our training practices. This paradigm shift has prompted a change in our views of the training relationship, from the idea that the human is dominant and must be obeyed to seeing training as a partnership, with a focus on the welfare, physical and mental, of the animal learner. Obviously, the goal is to achieve the desired training outcomes, but *how* we do that has changed. The animal learner is an active participant in the training relationship who is given a choice to participate. Generally, training, especially when using positive reinforcement, should be enjoyable for the animal with the trainer monitoring the animal learner's behavior for any signs of disinterest, fatigue, frustration, stress, or fear. Respect for the animal learner means allowing choices and supporting welfare, by using humane methods, going at the animal's pace, and reducing the chances of any negative emotional reactions during training.

A good example of positive reinforcement methods is Karen Pryor's Clicker Training program (Pryor, 2002). A small click or other sound appropriate to the species and setting (no need for a loud click or piercing whistle when training guinea pigs in a quiet room) is used to "mark" correct behaviors and signal the arrival of a positive reinforcer like a treat. The click is the secondary reinforcer that serves as a bridge to the primary reinforcer, the food (or other positive reinforcer). As discussed in the section on choosing consequences, it is important to choose the appropriate positive reinforcer in terms of desirability and value to the animal learner – while also considering ease of handling and deliverability for the trainer. It is difficult to juggle cans of wet cat food if you've chosen that as your positive reinforcer! While the clicker is not strictly necessary (a clicker pen, a cluck with the tongue, a simple "good job" will all work as the marker), the clicker is often appealing to people new to training because of its distinctive sound and definitive action. Regardless of whether a clicker is actually used or not, the principles that underpin clicker training are positive reinforcement.

That said, it is not unusual to dip into the other quadrants of operant conditioning when training, particularly when the goals are to reduce undesirable behaviors or to train multiple behaviors that may be interrelated. However, this should be done thoughtfully, carefully, and only when needed. Thankfully, the approach used nowadays differs dramatically from what was done in the past. A concept called "LIMA" teaches using the Least Intrusive, Minimally Aversive approaches in training and behavior modification in order to reduce the risks of negative side effects such as fear, stress, or injury (Friedman, 2022; The Association of Pet Dog Trainers (APDT), n.d.a) (see Figure 27.3). Intrusive refers to the animal learner's control over the situation; as previously discussed, respectful and humane animal training provides the animal learner with choices and the free will to choose, making the animal learner a partner in training and encouraging a relationship between learner and trainer. As already described, aversive refers to the unpleasant or undesired consequences to the learner, both those added and removed in the training scenario. The goal is to increase the use of positive reinforcement methods while reducing those that rely on negative reinforcement or punishments. Following LIMA guidelines, a trainer would always first choose positive reinforcement techniques over other approaches. Often, positive reinforcement training may be all that is needed, sometimes with some creative problem-solving looking for incompatible behaviors to teach (e.g., reducing jumping by using positive reinforcement to teach sit). Remember, punishment methods only teach the learner what *not* to do; ideally, if used, they are accompanied by positive reinforcement methods to teach the learner what *to do*. As the Association of Professional Dog Trainers (APDT) and the International Association of Animal Behavior Consultants (IAABC) state in their joint LIMA position statement, "In the

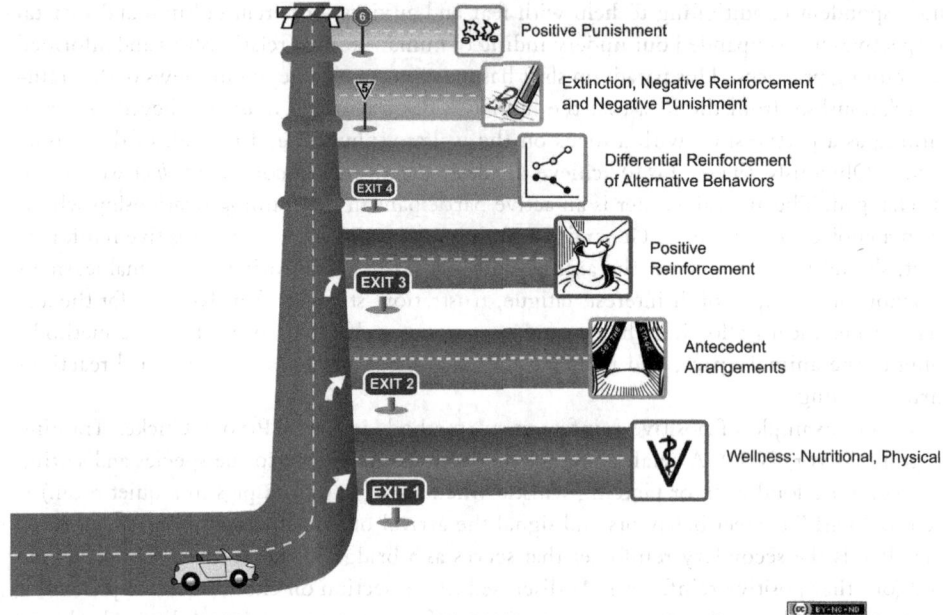

*Figure 27.3* Hierarchy of behavior change procedures according to the least intrusive, minimally aversive (LIMA) principle with Level 1 (exit 1) the most recommended to Level 6 (exit 6) the least recommended (Friedman, 2022)

vast majority of cases, desired behavior change can be affected by focusing on the animal's environment, physical well-being, and operant and respondent interventions such as differential reinforcement of an alternative behavior, desensitization, and counter-conditioning" (APDT, n.d.b)

Additionally, equipment that delivers strong aversive consequences is absolutely not recommended or endorsed. This equipment delivers an unpleasant or even painful positive punishment (often called a "correction") or creates an unpleasant or painful state that the animal is released from (negative reinforcement). Equipment such as pinch, prong, and electronic collars ("shock collars"), even bark collars, has been found to be associated with fear, stress, and aggressive reactions in dogs (Ziv, 2017). Even the electronic collars used with electric fences that beep and deliver a shock when the dog is too close to the buried fence line were associated with fearful and stress-related behaviors in some studies. Given the evidence that aversive training equipment, especially electronic collars, can be related to higher levels of fear, stress, and aggression in dogs and since it is too easy to make a mistake in timing, duration, and intensity with electronic collars, even in trained hands let alone the average owner, these types of strong aversive equipment should not be used. Several countries in the European Union have outright banned these devices. As the APDT/IAABC (n.d.b) position statement very clearly indicates, "These LIMA guidelines do not justify the use of aversive methods and tools including, but not limited to, the use of electronic, choke or prong collars in lieu of other effective positive reinforcement interventions and strategies".

While there is a general consensus amongst many professional behavior associations and their members that positive reinforcement training and a LIMA approach are effective while best positioned to support the welfare of the animal learner, in truth, there is no single standard approach amongst trainers. Positive reinforcement-only methods are criticized as ineffective at modifying problematic behavior with poor generalization to other environments and being too reliant on the delivery of food treats. Additionally, many pro-positive reinforcement trainers acknowledge that it is difficult to actually train an animal with only positive reinforcement; some trainers may ethically use natural aversive consequences, mild but delivered through negative reinforcement or positive punishment, that accompany positive reinforcement and negative punishment training. On the other hand, training methods using techniques from all four quadrants ("balanced" trainers) may not use enough positive reinforcement, moving too quickly to aversive consequences in order to achieve training goals more swiftly. The critiques of punishment methods, that animals are taught only what not to do and not what to do still apply. In addition, some anti-aversive training advocates suggest that training is less generalized and not maintained as long when aversive consequences are used. Given the mounting evidence that positive reinforcement methods are as effective as aversive ones with few risks to learner or trainer, a training program based on positive reinforcement is a compelling choice.

## *Best Practices for Effective, Humane, and Respectful Training*

Having established the benefits and uses of positive reinforcement training, let us briefly review practical considerations for developing and implementing effective, humane, and respectful animal training.

### *Developing and Implementing a Training Plan*

Formal training, such as we have discussed in this chapter, requires a plan. A formal plan provides a framework for the trainer to think through the desired goals and outcomes, which approaches and techniques will best support meeting those goals while ensuring optimal welfare for the learner, and helps identify opportunities and challenges before training. Trainers consider a host of factors when developing a training plan, from the training goals and outcomes, the animal learner and any relevant species, developmental, or individual characteristics, the learner's role (companion animal/pet, working animal, etc.), the living and learning environments, the learner's past history, the trainer's experience, the owner if involved, and cultural aspects of the learner's role, keeping, and behavior, among others.

Regular assessment of the training progress is critical to determining not only what has been achieved but also how the animal learner is doing. First and foremost, the animal learner should be observed at every session for signs of stress, anxiety, fear, aggression, or even fatigue or disinterest, as well as for any indications of potential health-related problems. These behavioral signs will be species-specific. If signs of stress or other concerning emotional states are observed, training should be paused – a break in the session for a fun game of catch or a few minutes of quiet pets can do wonders and then the animal should be observed when training re-commences. If stress or other concerning signs continue, training should end for the day and the training goals and plan re-evaluated as there may be other complicating factors that need to be addressed first. In addition to monitoring the animal learner's emotional state and reactions, training progress should be regularly assessed. Is the learner progressing as expected and achieving the behavioral goals? If not, the trainer needs to review the plan and evaluate the goal, approaches, and techniques to determine where the difficulty might lie.

## Animal Needs in Training

As is hopefully apparent at this stage in the chapter, the animal's needs in training are of utmost concern. Training new behavioral tasks using mainly positive reinforcement operant conditioning can be a fun, challenging, and rewarding activity for both animal and trainer. More intensive training to address fear or anxiety or significant problem behaviors will require a sensitive and compassionate approach that places the animal's welfare at the center of the training. The animal learner must be monitored before, during, and after training sessions for signs of stress, anxiety, fear, and aggression. Signs of disinterest and fatigue are also valuable clues to how the animal is feeling. Motivation is another factor to consider; ideally, the animal is a partner in the training, engaged and enjoying it. Even when using respondent conditioning to ease fear or anxiety, at a certain stage, positive reinforcement may be employed to teach new behaviors where the animal can be rewarded. However, there are times the animal may display signs of low motivation so the trainer should look at internal and external factors that may be contributing to this state and adjust the training plan accordingly. Perhaps training in the evening after dinner and a long exercise session is not the best time for the animal even if it works better for the owner. Perhaps training before breakfast is less effective since the animal is hungry and distracted by the food rewards. Finding the right time, as well as an environment conducive to training and positive reinforcers that are desired by the animal, can help improve motivation. If motivation is still low, perhaps the target behaviors are too challenging and need to be broken down into smaller steps or the stimulus may be too difficult to discriminate. A negative emotional state like fear may be hindering engagement in training. It is also possible that an underlying health condition may be contributing to low motivation so a visit to the veterinarian may be needed.

## The Human Factor

Throughout this chapter, we have talked frequently about the trainer and their role in the training. It is obvious that the trainer, who may be the owner, is an important part of the training relationship and key to its success. However, the human in the training partnership may be influenced by a variety of subtle and often unconscious processes.

Human knowledge of animal behavior and emotion ranges considerably, by species but also by specific behavior and emotional states, among other things. While some research has found that friendly behavior in dogs was one of the more accurately identified behaviors, research participants regularly confuse other behaviors such as play, aggressive, and fearful behaviors (Lakestani et al., 2014; Tami & Gallagher, 2009). Fearful behavior in dogs, in particular, was correctly identified in some studies (Tami & Gallagher, 2009) but not in others (Demirbas et al., 2016; Lakestani et al., 2014). People have even more difficulty identifying behavior and emotional states in cats. Dawson et al. (2019) reported that participants identified the correct positive or negative emotional state from photos of cat faces at a rate only slightly better than chance, doing better with positive states than negative. Owners are often only aware of more obvious signs of stress in their cats and dogs and miss more subtle indicators (Mariti et al., 2012; Mariti et al., 2017). In fact, owners sitting with their dogs in a veterinary waiting office rated their dogs' stress levels relatively low, which did not agree with expert ratings (Mariti et al., 2015). Knowledge and accuracy varies by the person's experience with the species and formal behavior training (Demirbas et al., 2016; Tami & Gallagher,

2009; Wan et al., 2012). Moreover, it appears that humans are prone to subtle influences by external factors such as an animal's breed (Gunter et al., 2016), coat color (Delgado et al., 2012; Fratkin & Baker, 2013), and environmental cues (Kogan & Schoenfeld, 2017) that may contribute to inaccurate interpretations of animal behavior. While results sometimes differ by study, the takeaway here is that humans, even professionals, are not always all that accurate when identifying animal behavior and emotion.

As trainers, handlers, and owners, we must be wary of quick conclusions about the behavioral and emotional health of our animals. Our understanding and interpretations of animal behavior and emotion can be influenced by our own knowledge and education, experience, beliefs, demographic characteristics, and history, as well as external factors in the environment. Regular checks on the animal learner to assess their emotional state are important, as is ongoing education, training, and peer networking for any person involved in training to help keep the animal's welfare top of mind.

## Future Directions

While this chapter has presented foundational information about animal learning and training, there is a great deal more to this topic than could be covered in a single chapter. Not to mention, there are still many areas to investigate to help refine our understanding and practices in animal training. As discussed earlier, research has generally established that there are risks to aversive training techniques with few benefits over positive reinforcement training. However, more research with a wider variety of species and training environments using observational techniques and experimental designs could provide more conclusive findings as well as clarify the comparative impacts of positive reinforcement and aversive training techniques. In addition, the benefits of engaging in training activities for both the animal learner and owner should be explored further with a wider variety of species and owners. Given the increased use of animals in service and assistance roles, research investigating effective and humane training techniques specific to these roles is needed. Finally, research examining the continued hold of dominance/compulsory training theories and tactics would help identify ways to help us move away from these outdated methods. These are just a few suggestions for future research, but hopefully they highlight the continued evolution of animal training toward a humane, respectful partnership.

## Conclusion: Enriching Lives through Humane and Respectful Training

Training the animals in our lives, whether companion animals living as family pets or working animals providing important services to the humans in their lives, has multiple and ongoing benefits to both animal and human. Training can teach animals to live more harmoniously with their human families, reducing problem behaviors and perhaps owners' frustration with challenging behaviors. Training can teach animals new behaviors, whether fun tricks or assistance tasks to help their humans. Training can also improve animals' welfare by providing stimulating enrichment and exercise as well as reduce the strain of fear and anxiety. Training often opens an owner's eyes to the intelligence, persistence, and motivation of their animal as well as help owners know their pets better, supporting the human-animal bond. It is possible to train animals of all species for a variety of tasks in a variety of settings using effective methods based on humane and respectful approaches.

## Discussion Questions

Consider the following questions as prompts for discussion:

1. How do associative and non-associative learning differ? How do desensitization and counter-conditioning work together to change an animal's emotional state and response to a stimulus? What could go wrong when using these techniques?
2. Compare and contrast respondent and operant conditioning. What is the role of the animal in each type of learning? What types of training are they best suited for?
3. Think of an example for each operant conditioning training method: positive reinforcement, negative reinforcement, positive punishment, negative punishment. How might the training go wrong and compromise the animal learner's welfare in each of those examples?
4. What are important elements to consider when developing a training plan? What might influence a person's abilities as a trainer?
5. Why do you think dominance-based, compulsory, and/or aversive training theories and practices became so popular? Why are they still so popular even with all the research and guidance against them? What can we do to help evolve the field toward more humane, respectful training?

## References

American Veterinary Medical Association. (2022). *Pet demographics and ownership sourcebook.* https://www.avma.org/news/new-report-takes-deep-dive-pet-ownership

Casey, R. (2013, October 12). *'Learning theory' terminology: Classical and operant conditioning.* Raining Cats and Dogs. https://behaviourvet.wordpress.com/2013/10/12/learning-theory-terminology-classical-and-operant-conditioning/

Dawson, L., Niel, L., Cheal, J., & Mason, G. (2019). Humans can identify cats' affective states from subtle facial expressions. *Animal Welfare, 28*(4), 519–531. https://doi.org/10.7120/09627286.28.4.519

Deldalle, S., & Gaunet, F. (2014). Effects of 2 training methods on stress-related behaviors of the dog (*Canis familiaris*) and on the dog–owner relationship. *Journal of Veterinary Behavior, 9*(2), 58–65. https://doi.org/10.1016/j.jveb.2013.11.004

Delgado, M. M., Munera, J. D., & Reevy, G. M. (2012). Human perceptions of coat color as an indicator of domestic cat personality. *Anthrozoös, 25*(4), 427–440. https://doi.org/10.2752/175303712X13479798785779

Demirbas, Y. S., Ozturk, H., Emre, B., Kockaya, M., Ozvardar, T., & Scott, A. (2016). Adults' ability to interpret canine body language during a dog–child interaction. *Anthrozoös, 29*(4), 581–596. https://doi.org/10.1080/08927936.2016.1228750

de Waal, F. B., & Andrews, K. (2022). The question of animal emotions. *Science, 375*(6587), 1351–1352. https://www.science.org/doi/full/10.1126/science.abo2378

Fratkin, J. L., & Baker, S. C. (2013). The role of coat color and ear shape on the perception of personality in dogs. *Anthrozoös, 26*(1), 125–133. https://doi.org/10.2752/175303713X13534238631632

Friedman, S. G. (2022). *Why animals need trainers who adhere to the least intrusive principle: Improving animal welfare and honing trainers' skills.* Retrieved September 8, 2022, from https://bit.ly/3BpvLeM

Gunter, L. M., Barber, R. T., & Wynne, C. D. (2016). What's in a name? Effect of breed perceptions & labeling on attractiveness, adoptions & length of stay for pit-bull-type dogs. *PLoS One, 11*(3), e0146857. https://doi.org/10.1371/journal.pone.0146857

Haverbeke, A., Laporte, B., Depiereux, E., Giffroy, J. M., & Diederich, C. (2008). Training methods of military dog handlers and their effects on the team's performances. *Applied Animal Behaviour Science, 113*(1–3), 110–122. https://doi.org/10.1016/j.applanim.2007.11.010

Herron, M. E., Shofer, F. S., & Reisner, I. R. (2009). Survey of the use and outcome of confrontational and non-confrontational training methods in client-owned dogs showing undesired behaviors. *Applied Animal Behaviour Science, 117*(1–2), 47–54. https://doi.org/10.1016/j.applanim.2008.12.011

Kogan, L. (2017, November). Black cats: Does manipulation of photographs impact perceptions? A pilot study. Paper presented at the SAWA 2017 Annual Conference & National Council on Pet Population Research Symposium, Miami, FL.

Lakestani, N. N., Donaldson, M. L., & Waran, N. (2014). Interpretation of dog behavior by children and young adults. *Anthrozoös, 27*(1), 65–80. https://doi.org/10.2752/175303714X13837396326413

Mackintosh, N. John (2022, July 29). *Animal learning.* Encyclopedia Britannica. https://www.britannica.com/science/animal-learning

Mariti, C., Gazzano, A., Moore, J. L., Baragli, P., Chelli, L., & Sighieri, C. (2012). Perception of dogs' stress by their owners. *Journal of Veterinary Behavior, 7*(4), 213–219. https://doi.org/10.1016/j.jveb.2011.09.004

Mariti, C., Raspanti, E., Zilocchi, M., Carlone, B., & Gazzano, A. (2015). The assessment of dog welfare in the waiting room of a veterinary clinic. *Animal Welfare, 24*(3), 299–305.

Mariti, C., Pierantoni, L., Sighieri, C., & Gazzano, A. (2017). Guardians' perceptions of dogs' welfare and behaviors related to visiting the veterinary clinic. *Journal of Applied Animal Welfare Science, 20*(1), 24–33. https://doi.org/10.1080/10888705.2016.1216432

McConnell, P. B. (2005a). *The cautious canine: How to help dogs conquer their fears.* McConnell Publishing Ltd.

McConnell, P. B. (2005b). *For the love of a dog: Understanding emotion in you and your best friend.* Ballantine Books.

Mech, D. L. (1999). Alpha status, dominance, and division of labor in wolf packs. *Canadian Journal of Zoology, 77,* 1196–1203.

Mellor, D. J. (2016). Updating animal welfare thinking: Moving beyond the "Five Freedoms" towards "a Life Worth Living". *Animals, 6*(3), 21. https://doi.org/10.3390/ani6030021

Morgan, D. L. (2010). Schedules of reinforcement at 50: A retrospective appreciation. *The Psychological Record, 60,* 151–172.

Pryor, K. (2002). *Don't shoot the dog!: The new art of teaching and training.* Ringpress Books Ltd.

Rachman, S. (1967). Systematic desensitization. *Psychological Bulletin, 67*(2), 93–103.

Reid, P. J. (2009). Adapting to the human world: Dogs' responsiveness to our social cues. *Behavioural processes, 80*(3), 325–333. https://doi.org/10.1016/j.beproc.2008.11.002

Rooney, N. J., & Cowan, S. (2011). Training methods and owner–dog interactions: Links with dog behaviour and learning ability. *Applied Animal Behaviour Science, 132*(3–4), 169–177. https://doi.org/10.1016/j.applanim.2011.03.007

Safina, C. (2015). *Beyond words: What animals think and feel.* Picador.

Seligman, M. E., & Maier, S. F. (1967). Failure to escape traumatic shock. *Journal of Experimental Psychology, 74*(1), 1–9. https://doi.org/10.1037/h0024514

Skinner, B. F. (1933). On the rate of extinction of a conditioned reflex. *The Journal of General Psychology, 8*(1), 114–129.

Skinner, B. F. (1938). *The behavior of organisms: An experimental analysis.* Appleton-Century-Crofts.

Skinner, B. F. (1953). *Science and human behavior.* Macmillan.

Tami, G., & Gallagher, A. (2009). Description of the behaviour of domestic dog (*Canis familiaris*) by experienced and inexperienced people. *Applied Animal Behaviour Science, 120*(3–4), 159–169. https://doi.org/10.1016/j.applanim.2009.06.009

The Association of Pet Dog Trainers (APDT). (n.d.a). *What is the least intrusive, minimally aversive (LIMA) training technique?* Retrieved August 20, 2022, from https://apdt.com/about/about-lima/

The Association of Pet Dog Trainers (APDT). (n.d.b). *Position statements* [see Least intrusive, minimally aversive (LIMA) approach]. Retrieved August 20, 2022, from https://apdt.com/about/position-statements/

Todes, D. P. (2014). *Ivan Pavlov: A Russian life in science.* Oxford University Press.

Wan, M., Bolger, N., & Champagne, F. A. (2012). Human perception of fear in dogs varies according to experience with dogs. *PLoS One, 7*(12), e51775. https://doi.org/10.1371/journal.pone.0051775

Ziv, G. (2017). The effects of using aversive training methods in dogs—A review. *Journal of Veterinary Behavior, 19,* 50–60. https://doi.org/10.1016/j.jveb.2017.02.004

# 28
# SAFE INTERACTIONS WITH ANIMALS

*Carri Westgarth, Sara Owczarczak-Garstecka, and John Tulloch*

### Safety Concerns: About Bites, Scratches, Kicks and Falls

We tend to focus on the positive aspects of interacting with animals; however, the risk of an animal-related injury is considerable. Although bites, scratches and kicks are a normal part of an animal's behavioural repertoire, they can be problematic to animal guardians and therapy participants. These behaviours also indicate that an animal is struggling to cope with the interactions or their environment, suggesting poor animal welfare.

### *Dogs*

A UK community-based study suggests that up to a quarter of people have been bitten by a dog at some point (Westgarth et al., 2018). Whilst bites by pets are common, many are bitten in the context of their occupation, for example veterinary professionals, animal rescue workers, delivery workers and social/care workers (Owczarczak-Garstecka et al., 2019). Dog bites also appear to be increasing, for example the proportion of people admitted to hospital in England for a 'dog bite or strike' tripled in a 20-year period (Tulloch et al., 2021), with an estimated incidence of 14.99 per 100,000 in 2018. Studies have estimated an annual incidence of around 25 bites per 1,000 people in India (Sharma et al., 2016), and 258 bites per 100,000 people in Brazil (Benavides et al., 2019). In the United States, emergency departments visits are 300,000–325,000 patients per year for injuries caused by dogs (CDC). In 2019 alone, these visits had a combined cost of $2.85B; however, the actual cost may be greater due to costs associated with loss of work and quality of life. These costs also exclude those associated with the burden of zoonotic diseases, such as rabies, an almost always fatal viral zoonosis typically spread through bites and scratches from dogs (see Chapter 26).

### *Cats*

The annual incidence of cat bites and scratches leading to hospital attendance is within a range of 6.36–17.9 per 100,000 population per year (Ostanello et al., 2005; Palacio et al., 2007) Cat bites are consistently the second most common bite to dogs. A cross-sectional study from Lyon, France, suggests however that the true incidence of cat bites is much higher, as 26% of cat owners reported that their cat bit someone in the past 12 months (Chomel & Trotignon, 1992). Whilst the incidence rates of feline aggression and cat-related injuries is lower than the incidence rates of dog-related injuries, cat bites and scratches are associated with higher risk of zoonotic diseases such as Bartonellosis or infection with *Pasteurella multocida* (Lloret et al., 2013; Mazur-Melewska et al., 2015) and higher rates of wound infection (Jaindl et al., 2016).

## Horses

Hospital admission rates for equestrian injuries are over three times higher than admissions due to motorcycle accidents (Mutore et al., 2021; Sorli, 2000). In the United States, over 68,000 people a year (an incidence of 21 people per 100,000 population) are presented to emergency departments for a horse-related injury (Acton et al., 2020). Almost 4% of patients admitted to hospital have severe neurological impairment and around 1% die due to their injuries (Mutore et al., 2021).

## Cattle

Cattle-related injuries are a significant and common cause of trauma, particular to those working in the agricultural sector (Fox et al., 2015; Rhind et al., 2021). Fatalities resultant of injury have been reported as high as 5%. Traumas often result in a fracture, and over half of individuals attending hospital are admitted (Lucas et al., 2013; Rhind et al., 2021). In the United Kingdom, cattle kill on average six agricultural workers each year, and they are the second most common cause of an agricultural workplace death behind accidents involving vehicles (HSE, 2021).

## Snake Bites

Of all animal species, the group that leads to the largest number of injuries and deaths are snakes. Each year, an estimated 5.4 million snake bites occur globally resulting in between 1.8 and 2.7 million envenomations, and sadly between 81,000 and 138,000 deaths (WHO, 2021). This leads to an estimated 400,000 individuals with subsequent physical and psychological disabilities, including amputations, loss of vision and extensive scarring (Gutiérrez et al., 2017). The scale of this issue is so large that the WHO has created a road map to reduce snake bite mortality and disability by 50% by 2030 (Williams et al., 2019). Snake bites have been described as an occupational disease requiring a One Health approach as they disproportionately impact the agricultural workforce, particularly in poor, rural communities (Babo Martins et al., 2019).

## Challenges around Animal Injury Statistics

Providing robust data around animal-related injuries is challenging. Essentially, three data groups that can be explored to provide estimates of injury burden and information regarding demographics and context:

a **Community data** has been explored in a range of settings predominately using surveys led primarily by public health professionals. These are usually surveys, often target specific population groups, and can provide baseline data, but are unlikely to be representative at a national level.

b **Health records** from primary care can be analysed. However, only a handful of higher income countries have enough data available for research, thus resulting in a western bias. Subsequently, cat bites can appear more problematic than snake bites for example. Primary care health data are also non-standardised (including how to record species) with a variety of coding systems. As a result, studies are not truly comparable and for many injuries, we do not fully understand the geographical and demographic differences. Further, any studies based on medical health records only capture those individuals that could get to a clinic, could afford to go to a clinic and believed travelling to that clinic would be beneficial for their health. Therefore, the inequity in access to healthcare further limits generalisability of this data.

c **Workplace accident recording data** are often reported at a national level by nation's competent authority, such as the Health and Safety Executive in the United Kingdom. However, these data often represent only the most severe of injuries that legally were required to be recorded, and the details supplied are limited.

A key issue with all animal-related injury data is selecting a denominator population. By choosing the human population, we can identify how big a burden that specific injury is to them. Also important is to know the level of exposure to the animal population causing the injury. However, data regarding the total number of animals and their geographic distribution are not always accurate and often unavailable. Companion animal national databases are relatively rare globally, based on trust and compliance of owners registering, and proxy measures such as feed sales may be poor estimates. Dog breeds and perceptions of these also complicate collection and analysis of dog bite statistics. For instance, bull-breeds and terriers are perceived as more dangerous (Clarke et al., 2016; Gunter et al., 2016) and public (and even experts) struggle to identify breeds accurately (Gunter et al., 2018; Hoffman et al., 2014). Moreover, most studies do not report information about the popularity of the breed within the study area, therefore breeds that emerge as common in bite records may simply be the most common in the area.

## Physical Safety Issues Related to the Types of Interactions We Have with Animals

### *Dog Bites*

Children appear to be more likely than adults to receive medical attention for dog bite injuries (Gilchrist et al., 2008). Children age <5 are most likely to present with injuries to the head and neck (Loder, 2019). This is thought to be a result of the child's height and that children aged <2 tend to place their face in close proximity to new or moving objects, which may trigger a reaction from a dog (Meints et al., 2010). In contrast, older children and adults are more likely to sustain injuries to their extremities, presumably due to their increased height and the limbs being the closest point to the animal. In most studies, male victims are reported to be at greater risk than females (Messam et al., 2018; Westgarth et al., 2018) but this may vary by age of the victim (Tulloch et al., 2021). The dog is typically known to the victim, owned by a family member and the bite occurs in the home (Loder, 2019; Oxley et al., 2018).

The most common context of a dog bite is related to interacting or attempting to interact with the dog (e.g. stroking, playing, handling and restraining) (Oxley et al., 2018). A study of dog bite videos on YouTube suggested that humans made more tactile contacts with dogs and changes in dog body posture and some displacement and appeasement behaviours increased approximately 20 seconds before the bite (Owczarczak-Garstecka et al., 2018). This suggests that there is at least a small window of opportunity immediately before some dog bites to avert the situation. Research into the perception of various dog-human interactions suggested that older respondents were less likely to identify several of the hazards, and those living with children were less likely to identify cuddling a dog as a hazard (Christley et al., 2021). Active supervision of children and dogs is recommended, but worryingly, in a study of dog-child supervision over half of parents reported leaving their young child alone with the dog and were much more risk tolerant than experts across a number of scenarios (Arhant et al., 2016).

## *Cat Bites/Scratches*

The majority of injuries are caused by owned cats and most victims are cat owners (Chomel & Trotignon, 1992; Palacio et al., 2007). However, some studies show that stray cats are responsible for a substantial portion (Wright, 1990). Most cat-related injuries occur in the context of play and petting (Chomel & Trotignon, 1992). Women are more commonly victims than men (Jaindl et al., 2016; Palacio et al., 2007); this may reflect patterns of ownership and also who is more likely to seek medical attention. Patterns of cat-related injuries are similar to dogs in that they mostly occur on the extremities.

## *Horses*

Around 68–75% of equine-related injuries occur when an individual is riding a horse (e.g. falls) and typically, these lead to thorax or spinal injuries (Lang et al., 2014; Van Balen et al., 2019). This has led to a push for riders to wear back protectors and helmets, though their use is highly variable ranging from 9% in Canadian hospital patients (Ball et al., 2007) to 74% in US hospital patients (Mayberry et al., 2007), illustrating how equestrian culture shapes injury pattern. In a UK survey exploring road safety and horses, 95% riders said they wore helmets, 91% wore high visibility clothing, but only 2% wore a back protector (Pollard & Grewar, 2020). Unmounted injuries account for around 25% of all horse-related injuries (Lang et al., 2014; Van Balen et al., 2019) and may be a particular risk in the context of animal-assisted interventions, which may utilise interacting with a horse on the ground. Unmounted injuries tend to be kicks and lead to head and facial injuries (Van Balen et al., 2019). In children, unmounted injuries are more severe than mounted injuries with a two times higher risk of being in an intensive care unit and eight times more likely to have a head injury (Wolyncewicz et al., 2018). Head and neck injuries produce the highest mortality rates of equine related injuries (Mutore et al., 2021). The most at risk of horse-related injuries are young female riders and adult males working with horses (Lang et al., 2014). Considering that one in five equestrians are injured during their riding career (Mayberry et al., 2007), it is critical that other solutions, in addition to wearing protective equipment, are developed to minimise the risk of injury and death from horses (Chapman & Thompson, 2016).

## *Livestock*

Farmers and vets working with cattle are most commonly injured, but members of the public can also be injured whilst walking, especially those with dogs (Fraser-Williams et al., 2016).

> ### Case Studies at Particular Risk – Animal Professionals/Vets
>
> Veterinarians are deemed to have one of the most dangerous civilian professions (Parkin et al., 2018). Over 60% of UK vets were injured by animals during their work in 2018 (Woodmansey, 2019). Cattle and horses cause the greatest number of injuries (Kabuusu et al., 2010; Lucas et al., 2009) but other hazards include dog and cat bites, cat scratches, scalpel blade cuts and back injuries from lifting animals (Jeyaretnam et al., 2000). A large UK study found that 33% of injured equine vets led to hospital admissions and 22% felt their mental health was impacted (Parkin et al., 2018). In the United States, 10% of reported injuries resulted in time off work (Cima & Larkin, 2018).
>
> ### Case studies at Particular Risk – Delivery Workers
>
> In two separate studies carried out in Brazil and Taiwan, approximately 70% of surveyed postal workers reported being bitten by a dog at some point during their career (Chen et al., 2000; Oliveira et al., 2013). In the United Kingdom on average, 6 postal workers are bitten each day (Langley, 2012). Seventy percent of these bites occur on private property, while posting a letter through a letter box, entering the front garden or handling a parcel to the dog owner.

Bulls are commonly implicated and common injury mechanisms reported are kicks and crushes, trampling and indirect trauma (such a being struck by a metal gate) (Rhind et al., 2021). Despite the potential severity of injuries, there is little data exploring in detail the context in which these injuries occurred, needed to develop injury prevention strategies. In a US study, 30% of injuries were caused when the individual was pushed into a farm structure by the cow (Fox et al., 2015). Therefore, in addition to safer cattle handling techniques, designing safer farm structures that can absorb the energy created by a striking cow is important.

### Precautions That Can Be Taken to Minimise Risk

Animal-related injuries tend to be addressed by two main methods: education and policy/legislation. An example of the latter is 'Breed-specific legislation' (BSL). Although studies investigating the prevalence of dog bites concluded that BSL or similar legislations was successful in Winnipeg (Canada) (Raghavan et al., 2013), it was not linked with a reduction in a number of dog-related hospital admissions in The Netherlands (Cornelissen & Hopster, 2010), Ireland (Creedon & Súilleabháin, 2017; Ó Súilleabháin & Doherty, 2015), Spain (Rosado et al., 2007), Denmark (Nilson et al., 2018) or England (Klaassen et al., 1996).

The most revered approach to bite prevention is based on informal educational campaigns, usually delivered by charities. This approach aims to educate the public about dog body language, under the assumption that a poor understanding of dog behaviour leads to greater risk-taking around dogs. Therefore, educational interventions have been presented as the means of reducing the number of dog bites (e.g. Dixon et al., 2012; Duperrex et al., 2009; Lakestani & Donaldson, 2015; Meints & de Keuster, 2009; Meints et al., 2010; Shen et al., 2016; Shen et al., 2017), or at least reducing bite severity (Kienesberger et al., 2022). However, education-based approaches to injury prevention can be criticised for failing to recognise a range of factors which contribute to human behaviour and portraying dog behaviour as easily foreseeable. It is unclear if dog behaviour can be reliably predicted and if it in reality follows the progressive pattern (often called the ladder of aggression, where behaviours like yawning, licking, head turns, tucked tail, body stiffening, growls and snaps are thought to occur before the bite) that educational approaches are often based on. In fact, an observational study suggested that only about 20% of dogs showed identifiable aggressive behaviours pre-bite (Owczarczak-Garstecka et al., 2018). In addition, the intensity of the educational interventions needed to change behaviour around dogs sufficiently to prevent bites mean that this approach may be difficult to scale up (Kienesberger et al., 2022). Therefore, 'Education' alone is unlikely the most effective solution to dog bite prevention. Studies point towards genetics as the most consistent predictor of dog aggression, with evidence that tendencies towards aggressive behaviour are inherited, that is particular breeding lines within a breed may show higher rates of aggression (Newman et al., 2017), and breeding that is focused temperament reduces the prevalence of fear and aggression in dogs (van der Borg et al., 2017).

Knowledge is just one of many components of behaviour change; skill development and a social and physical environment conducive of change are important as well (e.g. a car fitted with airbags and audio signalling seatbelts not being fastened) (Dahlgren & Whitehead, 1991; Hemenway, 2009; Michie & West, 2013; Reason, 1997). Public health strategies that rely on individual behaviour change (such as education campaigns) are usually less effective in improving the health of the population than top-down structural strategies (Bambra et al., 2010; Blankenship et al., 2000; Frieden, 2010; Morrison et al., 2003; Reason, 1997). For example, policies that target food reformulation, prices of fresh as well as heavily processed produce, and changes in environment that encourage active transport, are more likely to impact obesity levels than education about the importance of a healthy lifestyle.

Beliefs regarding risk affect compliance as well (Davison et al., 1992; Straughan & Seow, 1998). For example, if a person believes that a dog barking and jumping up at a postal worker is normal, low-risk behaviour, they may be less likely to modify these behaviours. Fatalistic views about risk, that is lack of control, can take a form of a belief that: (1) there are no effective ways of preventing the incident (response efficacy); (2) that an individual has no control over the incident (self-efficacy); and (3) re-evaluating the threat as smaller than it is (Peters et al., 2013). Similar common beliefs have been identified regarding the perception of risk of a dog bite (Westgarth & Watkins, 2015). For instance, those working with dogs have been observed dismissing a bite as just 'mouthing' or 'nips' in order to minimise the perceived threat (Owczarczak-Garstecka et al., 2021).

Research into perceptions of dog bites shows that victims generally do not blame the dog and instead blame themselves, and pre-incident felt that 'it wouldn't happen to me' (Westgarth & Watkins, 2015). These views are extremely similar to other recognised perceptual barriers to injury prevention including the tendency for 'victim blaming' and over-optimism (Hemenway, 2013). The view that an individual (victim, owner or parent/caregiver) is to blame for a dog bite hinders injury prevention detracts from implementing other changes. The overconfidence bias (Tversky & Kahneman, 1973) refers to placing excessive confidence in ones' own knowledge and estimation of risk, despite new information being given or having insufficient information to make a judgment. Indeed, prior experience with dogs improves confidence but not necessarily understanding of dog behaviour (Tami & Gallagher, 2009). Similarly, in human-horse interactions, greater equestrian experience and use for income generation are associated with unsafe behaviours (Chapman et al., 2020).

The wider injury prevention literature consistently shows that education – the most popular strategy in dog bite prevention – is rarely the best solution to preventing injuries. In the

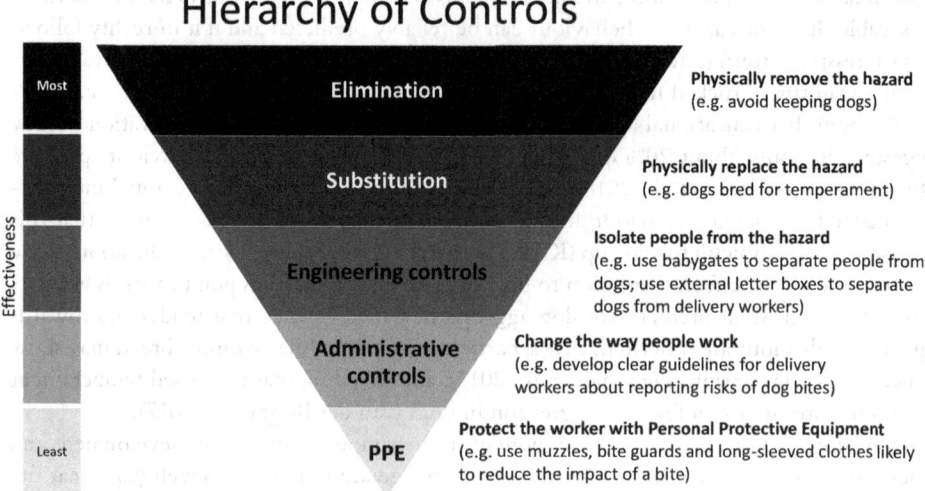

*Figure 28.1* Hierarchy of controls (CDC, 2015) as used regularly in health and safety, applied to dog bite prevention. (Image by Sara Owczarczak-Garstecka)

system-focused model, errors are seen as normal part of human behaviour and accidents are seen as consequences of poor system design. Within this model, accidents are prevented by designing multiple barriers and safeguards to prevent the consequences of a human or system error (Reason, 1990, 2000).

A commonly used framework in health and safety is the Hierarchy of Controls (Figure 28.1), where the most effective way to reduce risk is removal of the hazard completely down to least effective to use personal protective equipment to try to defend the worker from the hazard. Somewhere in between are the more realistic ideas of engineering a safer environment away from the hazard, and changing the format of the interaction with the hazard through administrative controls.

It can also be useful to view controls from an explanatory framework of whether they are managing risk before, during and after the event (Table 28.1).

Considering the lifecourse of dog ownership highlights many points of intervention that could reduce the chances of a negative interaction, way before the moment of risk interaction between the animal and the victim (Figure 28.2).

The importance of environmental controls – how can spaces be better designed to reduce the risk of an injury – is often overlooked. Examples applied to dog bite prevention include: ensuring spaces used for HAI are sufficiently large and do not make the dogs feel trapped; the use of babygates to separate dogs and people at key times; installing mirrors in passageways so that incoming dogs can be seen and avoided (e.g. to reduce dog-dog reactivity) or creating one-way systems down narrow passageways where dogs may feel intimidated.

Studies of how people interact with dogs and their perceptions of safety in the workplace (delivery workers, animal professionals) and at home, elucidated the importance of procedures in place that encourage constant surveillance (observation) for potential risk and then ways to communicate that observed risk effectively to others (Owczarczak-Garstecka et al., 2021). This aligns with other areas of safety, which recognises safety as a social process that relies on the actions of others, and the importance of procedures to formalise and support these. For example, reduction of incidents in aviation and medicine has been attributed to

*Table 28.1* Potential human-animal interaction counter-measures based on Haddon's classification of before, during and after the hazardous event (Haddon, 1973; Runyan, 2003)

*Counter-measures – measures used to prevent and reduce the impact of an injury before, during and after an incident*

| | | |
|---|---|---|
| The pre-event counter-measures | 1 | Preventing creation of the hazard in the first place (e.g. not keeping dangerous exotic animals) |
| The pre-event counter-measures | 2. | Reduction in the amount of hazard brought into being (e.g. limiting the number of dogs allowed to enter a building) |
| The pre-event counter-measures | 3. | Preventing the release of the hazard (e.g. installing locks to stop animals escaping) |
| The pre-event counter-measures | 4. | Modifying the rate at which the hazard is released or its spatial distribution (e.g. modifying the environment to change the spatial distribution that snakes will inhabit, so that they stay away from humans) |
| The during-event counter-measures | 5. | Separating the hazard from the person by time and space (e.g. avoiding building riding yards near busy roads) |
| The during-event counter-measures | 6. | Separating the hazard from the person by a physical barrier (e.g. using helmets when interacting with large animals); |
| The during-event counter-measures | 7. | Modifying relevant qualities of the hazard to make it less dangerous on impact (e.g. training dogs to wear muzzles during painful procedures, making nervous or aggressive dogs more visible and thus avoidable e.g. by wearing colourful leads or informative bandanas) |
| The during-event counter-measures | 8. | Making people or anything that is to be protected more resilient to the hazard (e.g. teaching children how to interact with dogs safely) |
| The post-event counter-measures | 9. | Countering the continuation or extension of the damage done by the hazard (e.g. CCTV that enables rapid response should an incident happen) |
| The post-event counter-measures | 10. | Stabilising, repairing and rehabilitating the damage to the injured person (e.g. physiotherapy or psychological support after injury by animals) |

an introduction of checklists (Clay-Williams & Colligan, 2015). Similarly, in human-horse interactions, training (formal or informal) in safety is also associated with positive safety behaviours, possibly by changing social norms regarding safety practices (Chapman et al., 2020). Social processes in HAI also involve emotional relationships between people and animals, which can complicate matters. Trust in an animal such as a dog, based mainly on its behaviour during past interactions (or animals that look similar to it), can influence the perception of risk to be lower even if the animal is currently showing behavioural signs incongruent with that past behaviour (Owczarczak-Garstecka, 2020). In this way, an established relationship with an animal can paradoxically increase the risk during interactions.

The before, during and after, approach to risk counter-measures (Table 28.1) also highlights an often neglected aspect of animal injury prevention strategies – that of minimising not just the risk of an injury, but the intensity of injury when it does (somewhat inevitably) occur. For example, car seat belts do not stop cars from crashing, but they reduce the extent

**Before bringing dog home**

- **Regulating dog breeding:** breeding for temperament; strict penalties for back-street breeding, unlicensed breeding and puppy-mill breeding.
- **Regulating sales of dogs:** enforcing ban on 3rd party sales of puppies; strict regulation on import of puppies from abroad; steep and enforced penalties for puppy smuggling; owner education, Puppy Contract & owner buys/ adopts a dog for their lifestyle.

**Puppyhood** (0- 6months)

**Primary socialisation & training:** puppy classes etc., setting up clear rules for interactions with other dogs and people and consistent home rules that will be followed throughout dog's life.

**Adolescence** (~5-9 months, depending on the breed)

**Preventative treatment :** vaccinations, worming & flea treatment, consider neutering)

**Socialisation & training:** managing deterioration of behaviour during adolescence and secondary socialisation during the second phase of heightened sensitivity to fear

**Adult dogs** (~12+ months, depending on the breed)

**Maintaining the established routine & clear, predictable rules**

**Ongoing provision of healthcare**

**Socialisation & training:** ongoing socialisation with dogs, people and in response to dog's current needs and experiences (e.g. introducing dog to other pets/ children); practicing known commands.

**Professional behavioural help** (if needed)

**Senior dogs** (~7-10+ years, depending on the breed)

**Geriatric healthcare:** recognising & managing pain, degenerative and long-term conditions

**Impact of ageing:** awareness and management of potential cognitive decline, sensory loss, mobility issues etc.

**Maintaining the established routine & clear, predictable rules**

**Socialisation & training**

*Ongoing socialisation & training*

*Figure 28.2* Schematic representation of matching dog–bite prevention practices with the key stages of dog's life. (Image by Sara Owczarczak-Garstecka)

## Case Study – Reducing dog bites to delivery workers (Westgarth, unpublished)

In order to reduce dog bites to postal workers in the United Kingdom, a new strategy, AVOID (Figure 28.3), was designed by postal workers and dog bite prevention experts in order to address common attitudes to dog bites and incorporate known injury prevention strategies.

*Figure 28.3* Royal Mail AVOID campaign to reduce dog attacks. (Images used with permission from Royal Mail)

This messaging is accompanied by supplying workers with a posting peg (Figure 28.4); a plastic device used to poke mail through the letterbox in the door (a technical control). Royal Mail request that all employees use a posting peg at addresses where they know a dog is present.

*Figure 28.4* Use of a posting peg. (Images used with permission from Royal Mail)

> Further, a new hazard reporting system – Animal Threat Level Reporting System (ATLAS) – has been developed so that near-miss incidents with dogs can be followed up either by the organisation or reporting to local services such as Police or Local Authority services for intervention.

of injury received during a crash. Similarly, a small scale, high-intensity educational intervention did not lead to a decrease in number of bites to children, but it led to a substantial reduction in severe dog-related injuries (from 26% of all dog-related injuries pre-interventions to 8% of all injuries post-intervention) (Kienesberger et al., 2022). Thus, application and policies around first aid for animal injuries are important, which includes the need for immediate cleaning and seeking medical aid for antibiotics given the high infection risks (NICE, 2020). Arguably, breed-specific legislation is also theoretically taking this approach so that when dog bites do occur, they are less likely to be from large powerful dogs that cause serious injury.

Finally, injury prevention research demonstrates the importance of investigating not just the potential causes of actual incidents, but also near-misses. Studies from construction and aviation industries indicate that fewer near-miss reports lead to higher chances of a serious incident (Dekker, 2014). However, evidence of investigation of dog-bite incidents and resultant changes to practice in the workplace does not appear to be common (Owczarczak-Garstecka et al., 2019).

## The Effects Safety Issues Have on HAIs

Foremost, animal-related injuries are an indicator of an animal welfare problem, but their consequences often further negatively impact animal welfare. Regarding the former, dogs that are anxious, fearful or in pain are more likely to bite (Blackshaw, 1991; Guy et al., 2001; Mills et al., 2012). Regarding the latter, dog bites can impact the owner-dog bond causing it to completely breakdown. A survey of dog bite victims suggested that as a consequence of biting, 8% were euthanised, 4% rehomed and 2% seized by police (Oxley et al., 2018). Worryingly, in more than half of incidents nothing was reported to have happened to the dog involved, meaning no changes in management or training was even undertaken. These data also suggest that over half of dogs that bite may have a previous history of aggression, and at least a quarter were known to have previously bitten a person, raising the point of whether this risk was being sufficiently addressed if they went on to bite again.

In addition to the physical impact, animal-related injuries often carry psychological costs to the victim and their owners (Peters et al., 2004; Westgarth & Watkins, 2017). Individuals are seen as responsible for managing their health and wellbeing by regulating their lifestyles (Petersen, 1997); this contributes to a perception of illness or injury as a moral failure, leading to stigma and psychological distress. The emphasis on the prevention of animal-related injury through 'education' and 'responsibility' is imbued with historical and cultural representations of risky behaviour, rather than an actual solution to the problem, and may even hinder help-seeking.

## Conclusion

There are useful frameworks that can be used to think through how to increase safety during human-animal interactions. The emotional aspect of our interactions with animals contribute to the complexity of assessing risk and how we behave around them. Multiple layers of measures are required in order provide a combination of controls that are reducing risk of a negative interaction to a minimal level – if one fails, the others will hopefully be enough to still prevent the incident. Rather than focusing what happens directly pre-incident as is the prevalent approach to injuries from animals, victim-blaming should be avoided, animal behavioural signs of stress closely monitored, and any near-miss occasions identified and investigated so that lessons can be learned and modifications made.

### Discussion questions

1. Have you had any near miss animal-related injuries and how did your behaviour change concerning that animal species?
2. Imagine you manage a riding school: what interventions would you design in your policies to minimise injuries from horses?
3. Imagine you work in a dog visitation programme in a hospital. What questions would you include on a near-miss and incident reporting form for the therapy dog programme?
4. Much focus has been placed on dog-related injuries, but less so on cats. What preventative measures can be adopted to minimise the risk of a cat in a care home facility biting or scratching residents?
5. You have been asked to design a dog visitation programme for a school to encourage children to read to animals. How would you screen the dogs for suitability for the programme in order to reduce the risk of a dog bite?

## Useful Resources

The **Merseyside Dog Safety Partnership** website contains useful resources on preventing and reducing dog bite injuries, including resources for the public to help them with their dog, and professionals to help manage risk of injuries by dogs, for example in the workplace www.merseydogsafe.co.uk.

The Health and Safety Executive (UK) website contains useful resources on how to conceptualise and manage risk within workplaces, which can apply to almost all human-animal interaction settings. https://www.hse.gov.uk/.

The UK Royal College of Nursing has developed helpful guidelines '**Working with Dogs in Health Care Settings**': A protocol to support organisations considering working with dogs in health care settings and allied health environments (2019). https://www.rcn.org.uk/Professional-Development/publications/pub-007925.

The book '**While We Were Sleeping: Success Stories in Injury and Violence Prevention**' by David Hemenway (2009, University of California Press) powerfully illuminates how public health works with more than 60 success stories drawn from the area of injury and violence prevention, that can inspire your own safety practices.

## References

Acton, A. S., Gaw, C. E., Chounthirath, T., & Smith, G. A. (2020, Jun). Nonfatal horse-related injuries treated in emergency departments in the United States, 1990–2017. *American Journal of Emergency Medicine, 38*(6), 1062–1068. https://doi.org/10.1016/j.ajem.2019.158366

Arhant, C., Landenberger, R., Beetz, A., & Troxler, J. (2016, July 01). Attitudes of caregivers to supervision of child–family dog interactions in children up to 6 years—An exploratory study. *Journal of Veterinary Behavior, 14*, 10–16. https://doi.org/10.1016/j.jveb.2016.06.007

Babo Martins, S., Bolon, I., Chappuis, F., Ray, N., Alcoba, G., Ochoa, C., Kumar Sharma, S., Nkwescheu, A. S., Wanda, F., Durso, A. M., & Ruiz de Castañeda, R. (2019). Snakebite and its impact in rural communities: The need for a One Health approach. *PLoS Neglected Tropical Diseases, 13*(9), e0007608. https://doi.org/10.1371/journal.pntd.0007608

Ball, C. G., Ball, J. E., Kirkpatrick, A. W., & Mulloy, R. H. (2007, May). Equestrian injuries: Incidence, injury patterns, and risk factors for 10 years of major traumatic injuries. *American Journal of Surgery, 193*(5), 636–640. https://doi.org/10.1016/j.amjsurg.2007.01.016

Bambra, C., Gibson, M., Sowden, A., Wright, K., Whitehead, M., & Petticrew, M. (2010). Tackling the wider social determinants of health and health inequalities: Evidence from systematic reviews. *Journal of Epidemiology & Community Health, 64*(4), 284–291.

Benavides, J. A., Megid, J., Campos, A., Rocha, S., Vigilato, M. A. N., & Hampson, K. (2019). An evaluation of Brazil's surveillance and prophylaxis of canine rabies between 2008 and 2017. *PLoS Neglected Tropical Diseases, 13*(8), e0007564–e0007564. https://doi.org/10.1371/journal.pntd.0007564

Blackshaw, J. K. (1991). An overview of types of aggressive behaviour in dogs and methods of treatment. *Applied Animal Behaviour Science, 30*(3–4), 351–361.

Blankenship, K. M., Bray, S. J., & Merson, M. H. (2000). Structural interventions in public health. *Aids, 14*, S11–S21.

CDC. *WISQUARS cost of injury: Number of injuries and associated costs.* Centers for Disease Control and Prevention. Retrieved 21 June from https://wisqars.cdc.gov/cost/

CDC. (2015, January 13). *Heirarchy of controls.* Retrieved 15 June from https://www.cdc.gov/niosh/topics/hierarchy/default.html

Chapman, M., Thomas, M., & Thompson, K. (2020). What people really think about safety around horses: The relationship between risk perception, values and safety behaviours. *Animals, 10*(12), 2222. https://www.mdpi.com/2076-2615/10/12/2222

Chapman, M., & Thompson, K. (2016). Preventing and investigating horse-related human injury and fatality in work and non-work Equestrian environments: A consideration of the workplace health and safety framework. *Animals, 6*(5), 33. https://www.mdpi.com/2076-2615/6/5/33

Chen, S.-C., Tang, F., Lee, H., Lee, C., Yen, C., & Lee, M. (2000). An epidemiologic study of dog bites among postmen in central Taiwan. *Chang Gung Medical Journal, 23*(5), 277–283.

Chomel, B. B., & Trotignon, J. (1992). Epidemiologic surveys of dog and cat bites in the Lyon area, France. *European Journal of Epidemiology, 8*(4), 619–624. <Go to ISI>://WOS:A1992JK28200022

Christley, R., Nelson, G., Millman, C., & Westgarth, C. (2021). Assessment of detection of potential dog-bite risks in the home using a real-time hazard perception test. *Anthrozoös, 34*(6), 767–786.. https://doi.org/10.1080/08927936.2021.1926710

Cima, G., & Larkin, M. (2018, October 10). *Hurt at work: Injuries common in clinics, often from animals, and usually preventable.* AVMA. Retrieved 15 June from https://www.avma.org/javma-news/2018-11-01/hurt-work

Clarke, T., Mills, D., & Cooper, J. (2016, October 11). "Type" as central to perceptions of breed differences in behavior of domestic dog. *Society & Animals, 24*(5), 467–485. https://doi.org/10.1163/15685306-12341422

Clay-Williams, R., & Colligan, L. (2015, July). Back to basics: Checklists in aviation and healthcare. *BMJ Quality and Safety, 24*(7), 428–431. https://doi.org/10.1136/bmjqs-2015-003957

Cornelissen, J. M., & Hopster, H. (2010). Dog bites in The Netherlands: A study of victims, injuries, circumstances and aggressors to support evaluation of breed specific legislation. *The Veterinary Journal, 186*(3), 292–298.

Creedon, N., & Súilleabháin, P. S. Ó. (2017). Dog bite injuries to humans and the use of breed-specific legislation: A comparison of bites from legislated and non-legislated dog breeds. *Irish Veterinary Journal, 70*(1), 23.

Dahlgren, G., & Whitehead, M. (1991). *Policies and strategies to promote social equity in health.* Stockholm: Institute for Future Studies.

Davison, C., Frankel, S., & Smith, G. D. (1992). The limits of lifestyle: Re-assessing 'fatalism' in the popular culture of illness prevention. *Social Science & Medicine, 34*(6), 675–685.

Dekker, S. (2014). The bureaucratization of safety. *Safety Science, 70*, 348–357.

Dixon, C. A., Mahabee-Gittens, E. M., Hart, K. W., & Lindsell, C. J. (2012). Dog bite prevention: An assessment of child knowledge. *The Journal of Pediatrics, 160*(2), 337–341. e332.

Duperrex, O., Blackhall, K., Burri, M., & Jeannot, E. (2009). Education of children and adolescents for the prevention of dog bite injuries. *The Cochrane Library, 2*, CD004726.

Fox, S., Ricketts, M., & Minton, J. E. (2015, January). Worker injuries involving the interaction of cattle, cattle handlers, and farm structures or equipment. *Journal of Agricultural Safety and Health, 21*(1), 3–12. https://doi.org/10.13031/jash.21.10221

Fraser-Williams, A. P., McIntyre, K. M., & Westgarth, C. (2016, January 12). Are cattle dangerous to walkers? A scoping review. *Injury Prevention*. https://doi.org/10.1136/injuryprev-2015-041784

Frieden, T. R. (2010). A framework for public health action: The health impact pyramid. *American Journal of Public Health, 100*(4), 590–595.

Gilchrist, J., Sacks, J. J., White, D., & Kresnow, M.-J. (2008). Dog bites: Still a problem? *Injury Prevention, 14*(5), 296–301. https://doi.org/10.1136/ip.2007.016220

Gunter, L. M., Barber, R. T., & Wynne, C. D. L. (2016). What's in a name? Effect of breed perceptions & labeling on attractiveness, adoptions & length of stay for pit-bull-type dogs. *PLoS One, 11*(3), e0146857. https://doi.org/10.1371/journal.pone.0146857

Gunter, L. M., Barber, R. T., & Wynne, C. D. L. (2018). A canine identity crisis: Genetic breed heritage testing of shelter dogs. *PLoS One, 13*(8), e0202633. https://doi.org/10.1371/journal.pone.0202633

Gutiérrez, J. M., Calvete, J. J., Habib, A. G., Harrison, R. A., Williams, D. J., & Warrell, D. A. (2017, September 14). Snakebite envenoming. *Nature Reviews Disease Primers, 3*(1), 17063. https://doi.org/10.1038/nrdp.2017.63

Guy, N. C., Luescher, U., Dohoo, S. E., Spangler, E., Miller, J. B., Dohoo, I. R., & Bate, L. A. (2001). Risk factors for dog bites to owners in a general veterinary caseload. *Applied Animal Behaviour Science, 74*(1), 29–42.

Haddon, W. (1973). Energy damage and the ten countermeasure strategies. *Human Factors, 15*(4), 355–366. https://doi.org/10.1177/001872087301500407

Hemenway, D. (2009). *While we were sleeping: Success stories in injury and violence prevention*. University of California Press.

Hemenway, D. (2013, September 28). Three common beliefs that are impediments to injury prevention. *Injury Prevention, 19*(4). https://doi.org/10.1136/injuryprev-2012-040507

Hoffman, C. L., Harrison, N., Wolff, L., & Westgarth, C. (2014, October–December). Is that dog a pit bull? A cross-country comparison of perceptions of shelter workers regarding breed identification. *Journal of Applied Animal Welfare Science, 17*(4), 322–339. https://doi.org/10.1080/10888705.2014.895904

HSE. (2021). *Fatal injuries in agriculture, forestry and fishing in Great Britain 2020/21*. https://www.hse.gov.uk/agriculture/resources/fatal.htm

Jaindl, M., Oberleitner, G., Endler, G., Thallinger, C., & Kovar, F. M. (2016, May). Management of bite wounds in children and adults-an analysis of over 5000 cases at a level I trauma centre. *Wien Klin Wochenschr, 128*(9–10), 367–375. https://doi.org/10.1007/s00508-015-0900-x

Jeyaretnam, J., Jones, H., & Phillips, M. (2000). Disease and injury among veterinarians. *Australian Veterinary Journal, 78*(9), 625–629. https://doi.org/10.1111/j.1751-0813.2000.tb11939.x

Kabuusu, R. M., Keku, E. O., Kiyini, R., & McCann, T. J. (2010, December). Prevalence and patterns of self-reported animal-related injury among veterinarians in metropolitan Kampala. *Journal of Veterinary Science, 11*(4), 363–365. https://doi.org/10.4142/jvs.2010.11.4.363

Kienesberger, B., Arneitz, C., Wolfschluckner, V., Flucher, C., Spitzer, P., Singer, G., Castellani, C., Till, H., & Schalamon, J. (2022, February). Child safety programs for primary school children decrease the injury severity of dog bites. *European Journal of Pediatrics, 181*(2), 709–714. https://doi.org/10.1007/s00431-021-04256-z

Klaassen, B., Buckley, J., & Esmail, A. (1996). Does the dangerous dogs act protect against animal attacks: A prospective study of mammalian bites in the accident and emergency department. *Injury, 27*(2), 89–91.

Lakestani, N., & Donaldson, M. L. (2015). Dog bite prevention: Effect of a short educational intervention for preschool children. *PloS One, 10*(8), e0134319.

Lang, J., Sathivelu, M., Tetsworth, K., Pollard, C., Harvey, K., & Bellamy, N. (2014, January). The epidemiology of horse-related injuries for different horse exposures, activities, and age groups in Queensland, Australia. *Journal of Trauma and Acute Care Surgery, 76*(1), 205–212. https://doi.org/10.1097/TA.0b013e3182a9007e

Langley, G. (2012). *Inquiry into dog attacks on postal workers*. http://www.royalmailgroup.com/sites/default/files/Langley_Report.pdf

Lloret, A., Egberink, H., Addie, D., Belák, S., Boucraut-Baralon, C., Frymus, T., Gruffydd-Jones, T., Hartmann, K., Hosie, M. J., Lutz, H., Marsilio, F., Möstl, K., Pennisi, M. G., Radford, A. D., Thiry, E., Truyen, U., & Horzinek, M. C. (2013). *Pasteurella multocida* infection in cats: ABCD guidelines on prevention and management. *Journal of Feline Medicine and Surgery, 15*(7), 570–572. https://doi.org/10.1177/1098612x13489215

Loder, R. T. (2019, March). The demographics of dog bites in the United States. *Heliyon, 5*(3), e01360. https://doi.org/10.1016/j.heliyon.2019.e01360

Lucas, M., Day, L., & Fritschi, L. (2013, Janury–February). Serious injuries to Australian veterinarians working with cattle. *Australian Veterinary Journal, 91*(1–2), 57–60. https://doi.org/10.1111/j.1751-0813.2012.01014.x

Lucas, M., Day, L., Shirangi, A., & Fritschi, L. (2009, August). Significant injuries in Australian veterinarians and use of safety precautions. *Occupational Medicine (London), 59*(5), 327–333. https://doi.org/10.1093/occmed/kqp070

Mayberry, J. C., Pearson, T. E., Wiger, K. J., Diggs, B. S., & Mullins, R. J. (2007, March). Equestrian injury prevention efforts need more attention to novice riders. *Journal of Trauma, 62*(3), 735–739. https://doi.org/10.1097/ta.0b013e318031b5d4

Mazur-Melewska, K., Mania, A., Kemnitz, P., Figlerowicz, M., & Służewski, W. (2015). Cat-scratch disease: A wide spectrum of clinical pictures. *Postepy dermatologii i alergologii, 32*(3), 216–220. https://doi.org/10.5114/pdia.2014.44014

Meints, K., & de Keuster, T. (2009). Brief report: Don't kiss a sleeping dog: The first assessment of "The Blue Dog" bite prevention program. *Journal of Pediatric Psychology, 34*(10), 1084–1090. https://doi.org/10.1093/jpepsy/jsp053

Meints, K., Syrnyk, C., & De Keuster, T. (2010). Why do children get bitten in the face? *Injury Prevention, 16*(Suppl. 1), A172–A173.

Messam, L. L. M., Kass, P. H., Chomel, B. B., & Hart, L. A. (2018, May 04). Factors associated with bites to a child from a dog living in the same home: A bi-national comparison [Original Research]. *Frontiers in Veterinary Science, 5.* https://doi.org/10.3389/fvets.2018.00066

Michie, S., & West, R. (2013, March 1). Behaviour change theory and evidence: A presentation to government. *Health Psychology Review, 7*(1), 1–22. https://doi.org/10.1080/17437199.2011.649445

Mills, D. S., Dube, M. B., & Zulch, H. (2012). *Stress and pheromonatherapy in small animal clinical behaviour.* John Wiley & Sons.

Morrison, D. S., Petticrew, M., & Thomson, H. (2003). What are the most effective ways of improving population health through transport interventions? Evidence from systematic reviews. *Journal of Epidemiology & Community Health, 57*(5), 327–333.

Mutore, K., Lim, J., Fofana, D., Torres-Reveron, A., & Skubic, J. J. (2021). Hearing hoofbeats? Think head and neck trauma: A 10-year NTDB analysis of equestrian-related trauma in the USA. *Trauma Surgery & Acute Care Open, 6*(1), e000728–e000728. https://doi.org/10.1136/tsaco-2021-000728

Newman, J., Christley, R. M., Westgarth, C., & Morgan, K. M. (2017). Chapter 10: Risk factors for dog bites - An epidemiological perspective. In D. S. Mills & C. Westgarth (Eds.), *Dog bites: A multidisciplinary perspective* (pp. 133–158). 5M Publishing.

NICE. (2020). *NICE guideline [NG184]: Human and animal bites: Antimicrobial prescribing.* https://www.nice.org.uk/guidance/ng184

Nilson, F., Damsager, J., Lauritsen, J., & Bonander, C. (2018). The effect of breed-specific dog legislation on hospital treated dog bites in Odense, Denmark—A time series intervention study. *PloS One, 13*(12), e0208393.

Ó Súilleabháin, P., & Doherty, N. (2015, August 01). Epidemiology of dog bite injuries: Dog-breed identification and dog-owner interaction [Letter]. *Journal of Plastic, Reconstructive and Aesthetic Surgery, 68*(8), 1157–1158. https://doi.org/10.1016/j.bjps.2015.03.025

Oliveira, E. A. d., Manosso, R. M., Braune, G., Marcenovicz, P. C., Kuritza, L. N., Ventura, H. L. B., Paploski, I. A. D., Kikuti, M., & Biondo, A. W. (2013). Neighborhood and postal worker characteristics associated with dog bites in postal workers of the Brazilian National Postal Service in Curitiba. *Ciencia & Saude Coletiva, 18,* 1367–1374.

Ostanello, F., Gherardi, A., Caprioli, A., La Placa, L., Passini, A., & Prosperi, S. (2005). Incidence of injuries caused by dogs and cats treated in emergency departments in a major Italian city. *Emergency Medicine Journal, 22*(4), 260–262. https://doi.org/10.1136/emj.2004.014886

Owczarczak-Garstecka, S. (2020). *Dog bites: Perception and prevention.* University of Liverpool.

Owczarczak-Garstecka, S. C., Christley, R., Watkins, F., Yang, H., Bishop, B., & Westgarth, C. (2019, October 01). Dog bite safety at work: An injury prevention perspective on reported occupational dog bites in the UK. *Safety Science, 118*, 595–606. https://doi.org/10.1016/j.ssci.2019.05.034

Owczarczak-Garstecka, S. C., Christley, R. M., Watkins, F., Yang, H., & Westgarth, C. (2021). "If You Don't See the Dog, What Can You Do?" Using procedures to negotiate the risk of dog bites in occupational contexts. *International Journal of Environmental Research and Public Health, 18*(14), 7377. https://www.mdpi.com/1660-4601/18/14/7377

Owczarczak-Garstecka, S. C., Watkins, F., Christley, R., & Westgarth, C. (2018, May 08). Online videos indicate human and dog behaviour preceding dog bites and the context in which bites occur. *Scientific Reports, 8*(1), 7147. https://doi.org/10.1038/s41598-018-25671-7

Oxley, J. A., Christley, R., & Westgarth, C. (2018, January 01). Contexts and consequences of dog bite incidents. *Journal of Veterinary Behavior: Clinical Applications and Research, 23*(Suppl. C), 33–39. https://doi.org/10.1016/j.jveb.2017.10.005

Palacio, J., León-Artozqui, M., Pastor-Villalba, E., Carrera-Martín, F., & García-Belenguer, S. (2007). Incidence of and risk factors for cat bites: A first step in prevention and treatment of feline aggression. *Journal of Feline Medicine and Surgery, 9*(3), 188–195. https://doi.org/10.1016/j.jfms.2006.11.001

Parkin, T. D. H., Brown, J., & Macdonald, E. B. (2018). Occupational risks of working with horses: A questionnaire survey of equine veterinary surgeons. *Equine Veterinary Education, 30*(4), 200–205. https://doi.org/10.1111/eve.12891

Peters, G.-J. Y., Ruiter, R. A., & Kok, G. (2013). Threatening communication: A critical re-analysis and a revised meta-analytic test of fear appeal theory. *Health Psychology Review, 7*(suppl. 1), S8–S31.

Peters, V., Sottiaux, M., Appelboom, J., & Kahn, A. (2004). Posttraumatic stress disorder after dog bites in children. *The Journal of Pediatrics, 144*(1), 121–122. https://doi.org/10.1016/j.jpeds.2003.10.024

Petersen, A. (1997). Risk, governance and the new public health. In *Foucault, Health and Medicine* (pp. 189–206). Routledge.

Pollard, D., & Grewar, J. D. (2020). Equestrian road safety in the United Kingdom: Factors associated with collisions and horse fatalities. *Animals, 10*(12), 2403. https://www.mdpi.com/2076-2615/10/12/2403

Raghavan, M., Martens, P. J., Chateau, D., & Burchill, C. (2013). Effectiveness of breed-specific legislation in decreasing the incidence of dog-bite injury hospitalisations in people in the Canadian province of Manitoba. *Injury Prevention, 19*(3), 177–183.

Reason, J. (1990). The contribution of latent human failures to the breakdown of complex systems. *Philosophical Transactions Royal Society B, 327*(1241), 475–484.

Reason, J. (1997). Engineering a safety culture. In *Managing the risks of organizational accidents*. Ashgate. https://www.taylorfrancis.com/books/mono/10.4324/9781315543543/managing-risks-organizational-accidents-james-reason

Reason, J. (2000). Human error: Models and management. *BMJ, 320*(7237), 768–770. https://doi.org/10.1136/bmj.320.7237.768

Rhind, J.-H., Quinn, D., Cosbey, L., Mobley, D., Britton, I., & Lim, J. (2021, April–June). Cattle-related trauma: A 5-year retrospective review in a adult major trauma center. *Journal of Emergencies, Trauma, and Shock, 14*(2), 86–91. https://doi.org/10.4103/JETS.JETS_92_20

Rosado, B., García-Belenguer, S., León, M., & Palacio, J. (2007). Spanish dangerous animals act: Effect on the epidemiology of dog bites. *Journal of Veterinary Behavior: Clinical Applications and Research, 2*(5), 166–174.

Runyan, C. W. (2003). Introduction: Back to the future—Revisiting Haddon's conceptualization of injury epidemiology and prevention. *Epidemiologic Reviews, 25*(1), 60–64. https://doi.org/10.1093/epirev/mxg005

Sharma, S., Agarwal, A., Khan, A. M., & Ingle, G. K. (2016, March–April). Prevalence of dog bites in rural and urban slums of Delhi: A community-based study. *Annals of Military and Health Sciences Research, 6*(2), 115–119. https://doi.org/10.4103/2141-9248.181836

Shen, J., Rouse, J., Godbole, M., Wells, H. L., Boppana, S., & Schwebel, D. C. (2016). Systematic review: Interventions to educate children about dog safety and prevent pediatric dog-bite injuries: A meta-analytic review. *Journal of Pediatric Psychology, 42*(7), 779–791.

Shen, J., Rouse, J., Godbole, M., Wells, H. L., Boppana, S., & Schwebel, D. C. (2017). Systematic review: Interventions to educate children about dog safety and prevent pediatric dog-bite injuries: A meta-analytic review. *Journal of Pediatric Psychology, 42*(7), 779–791. https://doi.org/10.1093/jpepsy/jsv164

Sorli, J. M. (2000). Equestrian injuries: A five year review of hospital admissions in British Columbia, Canada. *Injury Prevention, 6*(1), 59–61. https://doi.org/10.1136/ip.6.1.59

Straughan, P. T., & Seow, A. (1998). Fatalism reconceptualized: A concept to predict health screening behavior. *Journal of Gender, Culture and Health, 3*(2), 85–100.

Tami, G., & Gallagher, A. (2009). Description of the behaviour of domestic dog (*Canis familiaris*) by experienced and inexperienced people. *Applied Animal Behaviour Science, 120*(3), 159–169.

Tulloch, J. S. P., Owczarczak-Garstecka, S. C., Fleming, K. M., Vivancos, R., & Westgarth, C. (2021, January 19). English hospital episode data analysis (1998–2018) reveal that the rise in dog bite hospital admissions is driven by adult cases. *Scientific Reports, 11*(1), 1767. https://doi.org/10.1038/s41598-021-81527-7

Tversky, A., & Kahneman, D. (1973). Availability: A heuristic for judging frequency and probability. *Cognitive Psychology 5*(2), 207–232.

Van Balen, P. J., Barten, D. G., Janssen, L., Fiddelers, A. A. A., Brink, P. R., & Janzing, H. M. J. (2019, April). Beware of the force of the horse: Mechanisms and severity of equestrian-related injuries. *European Journal of Emergency Medicine, 26*(2), 133–138. https://doi.org/10.1097/mej.0000000000000511

van der Borg, J. A. M., Graat, E. A. M., & Beerda, B. (2017, October 01). Behavioural testing based breeding policy reduces the prevalence of fear and aggression related behaviour in Rottweilers. *Applied Animal Behaviour Science, 195*, 80–86. https://doi.org/10.1016/j.applanim.2017.06.004

Westgarth, C., Brooke, M., & Christley, R. M. (2018). How many people have been bitten by dogs? A cross-sectional survey of prevalence, incidence and factors associated with dog bites in a UK community. *Journal of Epidemiology and Community Health, 72*, 331–336. https://doi.org/10.1136/jech-2017-209330

Westgarth, C., & Watkins, F. (2015, November–December). A qualitative investigation of the perceptions of female dog-bite victims and implications for the prevention of dog bites. *Journal of Veterinary Behavior-Clinical Applications and Research, 10*(6), 479–488. https://doi.org/10.1016/j.jveb.2015.07.035

Westgarth, C., & Watkins, F. (2017). Chapter 23: Impact of dog aggression on victims. In D. S. Mills & C. Westgarth (Eds.), *Dog bites: A multidisciplinary perspective* (1st ed., pp. 309–320). 5M Publishing.

WHO. (2021). *Snakebite envenoming*. Retrieved 14 June from https://www.who.int/news-room/fact-sheets/detail/snakebite-envenoming

Williams, D. J., Faiz, M. A., Abela-Ridder, B., Ainsworth, S., Bulfone, T. C., Nickerson, A. D., Habib, A. G., Junghanss, T., Fan, H. W., Turner, M., Harrison, R. A., & Warrell, D. A. (2019, February). Strategy for a globally coordinated response to a priority neglected tropical disease: Snakebite envenoming. *PLoS Neglected Tropical Diseases, 13*(2), e0007059. https://doi.org/10.1371/journal.pntd.0007059

Wolyncewicz, G. E. L., Palmer, C. S., Jowett, H. E., Hutson, J. M., King, S. K., & Teague, W. J. (2018, May). Horse-related injuries in children – Unmounted injuries are more severe: A retrospective review. *Injury, 49*(5), 933–938. https://doi.org/10.1016/j.injury.2017.12.003

Woodmansey, D. (2019). *BVA survey: More than 6 in 10 farm vets injured at work*. Vet Times. Retrieved 15 June from https://www.vettimes.co.uk/news/bva-survey-more-than-6-in-10-farm-vets-injured-at-work/

Wright, J. C. (1990, July–August). Reported cat bites in Dallas: Characteristics of the cats, the victims, and the attack events. *Public Health Reports, 105*(4), 420–424.

# 29
# THE CASE FOR CONSERVATION IN THE AGE OF THE ANTHROPOCENE

*Eric G. Strauss and Carolyn A. Ristau*

## Introduction

The unprecedented decline in biodiversity across all scales on the planet has spurred concern and action across broad swathes of government and science. The epoch that we live in, considered by many to be the *Anthropocene*, has been witness to the highest rate of extinction at any time in the Earth's history. This is coupled to massive recent global land changes such as urbanization of developing nations and large-scale agricultural conversion of critical ecosystems such as palm oil plantations in rain forests. The emergence of conservation biology, described by Soulé as a mission-oriented crisis discipline (1985), has developed and matured under this paradigm. It serves as a lens to interpret the observed changes and offers a suite of intellectual guidelines to implement solutions.

The goals of conservation biology as stated by the Society for Conservation Biology reflect the complex multifaceted nature of the challenge: the Protection and Restoration of Biological Diversity, Ecological Integrity and Ecological Health (Trombulak et al., 2004). These goals are supported by fundamental biological concepts such as taxonomic hierarchies, the modern species concept, ecological hierarchies, genetic diversity, population growth models and species distribution. These ideas provide the underpinnings for understanding the threats to global biological diversity from human economics, patterns of human colonization and land fragmentation, emerging and proximate causes of extinction, plus the impacts of climate change (alternatively termed "global warming"). Collectively, these forces have shaped a global ecosystem that is perilous to many native species and habitats.

Although natural shifts in climate characteristics and the background rate of extinction contribute to this phenomenon, the vast majority of the cases of decline are related directly to anthropogenic influences. Habitat fragmentation, fossil fuel-driven climate change, domestication of plants and animals, plus the accelerated expansion of human cities and agricultural activities place extraordinary strain on the Earth's resilience. The result has been the collapse of fragile ecosystems, the homogenization of ecosystems by a few wildly successful domesticated or synanthropic species, such as American Crows, striped skunks and Eastern Coyotes in North America, and the creation of a heterogeneous patchwork of ecosystems urbanized through human development.

In this chapter, we will build the case for conservation through the filter of ecosystem services and interpreted through a conservation *virtuous cycle*. We will focus on the suite of values that healthy nature provides and consider best management practices towards those goals. This chapter aims to bring into focus the dynamics of the conservation challenge by investigating two significant case studies that serve as contemporary examples, elephants in the developing world and piping plover shorebirds in the urbanized habitats of North America. The similarities and differences serve as examples of the conservation plight of wild animals, the influence of domestic animals and the global scale of the problem.

The context of conservation efforts emerges from historical traditions of natural history around the idea of cataloging the organisms that one might find in any given area. This preoccupation with the results of creation, evolution and anthropogenic forces, such as urbanization and modern agricultural development, has been a driver for the emerging fields of biodiversity, conservation biology and restoration ecology. The cataloging of natural phenomena has both scientific and religious foundations that emerged in the ideas of natural philosophy.

The study of biological diversity began to accelerate in the 1980s with the work of pioneers such as Thomas Lovejoy and Terri Erwin. Investigations into species-rich ecosystems such as the rainforest tree canopies, resulted in the discovery of extraordinary species diversity resulting from hyper-specialization and endemism. For example, one species of tree in a New World tropical rain forest (*Luchea seemanii*) yielded 163 species of beetles that were found only in that tree species (Erwin, 1982). Extrapolations of these data led to estimates of nearly 30 million species of arthropods, although only 1.5 million had been cataloged to date. As field techniques have expanded and more data from other parts of the world have been gathered, these numbers have been estimated down to between 3.5 and 3.7 million and remain debatable (Alexander et al., 2010). The overarching effects of these studies on the biological community is to foster additional research and a new level of awareness among lay and scientific stakeholders.

Two case studies provide instructive insight into the complexity of conservation efforts and their impacts on local communities. Both are vertebrates that vary by orders of magnitude in size and impact their ecosystems and human inhabitants quite differently. Each provides unique conservation and management challenges with implications for future distribution and population resilience. First, we shall examine the global plight of African and Asian elephants, and second, a migratory North American shorebird, the Piping Plover.

## *Elephants*

Elephants are well-loved by many and reviled by some, those whose very lives and property are threatened by these creatures. In this section, we shall consider the elephant, its capacities and its role in various ecological and cultural environments.

The elephant looms as the largest land animal, reaching weights for an African male elephant, up to 10.4 tons. Most scientists consider there to be three species of elephants; the largest, the African savannah elephant (*Loxodonta africana*), the smallest, the African Forest elephant (*Loxodonta cyclotis*) which may interbreed with the savannah elephant when the species have overlapping ranges, and, finally, the Asian elephant (*Elephas maximus*).

## Social Elephants

There are general social differences between these elephants with the forest elephants typically considered to be much more shy than the open country savannah species, Though the species are considered to differ somewhat in social behavior, with the forest elephant the

*Figure 29.1* Savannah elephant adults in a protective group with young. Extended families meet at least occasionally over decades while multi-generations of females live together, even for entire lifetimes. In their matriarchal society, they have strong, supportive relationships and are led and guided by an old female with extensive knowledge of the environment and survival options. (Photo by Michael J. Gaade)

most shy, human contact greatly influences their shyness and aggressiveness. All three species have families with the female matriarch living for long years, sometimes to 60 + years, remaining with her offspring daughters and granddaughters. The males in all species tend to be more solitary and leave the family group as teenagers, between 10 and 20 years of age. At least for forest elephants, that family can contain up to 20 members, often including sub-adult males. To varying degrees, both during mating season and, apparently, to be "social," extended family groups meet in larger congregations (Figure 29.1).

The family members are enormously caring and supportive of each other. The tenacity with which family members try to save another, particularly a young one, can lead to their own injury and even death. At least twice in a Thai national park, adult elephants died as they attempted to save a youngster who had fallen into the torrential river below a cliff. The elephants' use of a trail close to the cliff's edge was necessitated by the park's constructing a tourist path and shops atop the elephants' former, safer route. To protest this monetization of the park to the detriment of the animals and research, one Thai conservationist engaged in a hunger strike. Nevertheless, a park director claimed people's "need to enjoy nature" had to be taken into account (Beech and Suhartono, 2020).

The old bulls are another significant aspect of the elephant story, less known because the males roam across vast distances and so are difficult to study. Recent research indicates that "solitary" old males are not quite so solitary, playing significant leadership roles particularly among young males.

Previously, old bulls had been observed to exert control over misbehaving young males. An example occurred in Pilanesberg National Park, South Africa, where introduced young orphaned males killed more than 40 white rhinos during the young males' musth period with heightened sexuality and aggressiveness. Females had rejected their sexual advances, preferring older males. The killing stopped when researchers introduced a half dozen older male elephants into the reserve (Nuwer, 2020).

In more recent work, entailing thousands of hours of videotape, adolescent males were observed to move with other males. The older bulls were usually at the head of the line, presumably either actively leading the group or merely tolerating the young "hangers-on" (Allen et al., 2020; reported in Nuwer, 2020).

Older animals, male and female, have a vast knowledge of the terrain, which the younger can learn by example. Absent the older group members, the young could probably reach waterholes on their own, following well-traversed paths that connect critical resources. However, finding water in times of prolonged drought or safer places to avoid poachers might well entail the expertise of older animals (Allen et al., 2020; Thouless, in Nuwer, 2020).

Older savannah males are also the prime breeders. On average, female savannah elephants' life expectancy is 41 years, while males', including human-induced mortality, is only 24 years. Other factors contribute to male mortality: males' more frequent involvement in wildlife conflict and the risks entailed in male dispersal to new environments. It behooves young males to associate with knowledgeable older males. Yet older males are at great risk as the prime target for both legal and illegal hunting; their tusks are the largest and most prized. Most valuable of all are the pale pink tusks of forest elephants, which grow more slowly than those of other species and are harder and preferred for carving. Elephants keep growing their tusks throughout life, though, a female forest elephant appears to stop at sexual maturity. Male Asian elephants sometimes have tusks, while the females do not; the increasing number of tuskless Asian males is believed to be due to evolutionary pressure from poaching. Conservationists who sometimes perceive a need to cull elephant herds, typically select the old bulls, thinking these males are contributing little to the group's well-being. We are beginning to know differently.

## Mating and Rumbling

Despite the difficulties, males and females get together to mate. Often, they are separated by long distances and the female is receptive for only about two days, approximately every four months. Males are most sexually active, i.e., in a state of "musth," only once a year. Male elephants have been observed to travel up to 80 miles to a female in estrous. It can be a long time between mating opportunities. Pregnancy lasts 22 months and females typically have one calf every five years. Males don't reach sexual maturity until 15–25 years, females sooner, at age 10–12 years, first reproducing a few years later. Somehow, both are communicating readiness to mate.

It is not the loud trumpeting that seems to be the primary means of communication. Pertinent research began with a rumble and continued with rumbles at its core. As the wildlife biologist Katherine B. Payne was standing near an elephant in a zoo, she felt a trembling inside her, reminding her of her youth, when "feeling" the deep sounds of her church's organ, Might the elephant be producing that sound and using it to communicate? Few humans could sense that rumbling. (Payne, 1998). After extensive experiments in Etosha Park, Namibia, her team recorded these low rumblings and obtained some measures of the distance the infrasounds could travel. Such sounds, transmitted through the ground, can travel with less disruption, attenuation and modification than the usual elephant vocalizing sent through the air (Langbauer et al., 1991). But how were these vibrations sensed? The soft fleshy bottom part of an elephant's foot was the receptive portal, the same part allowing these massive creatures to walk so silently along the ground. Internally, the rumbling is transmitted through the elephant's bones to its reception in the ear and subsequent analysis by the brain. Both male and female are communicating their readiness to mate via rumblings.

*Figure 29.2* As do all elephant species, these Asian elephants are reassuring each other with their entwined trunks. Trunks are essential and sensitive appendages, exploring the environment by touch and olfaction, gathering water and vegetation, and used in both vocal and tactile communication as shown. (Photo by Nachiketh Sharma)

## Elephant Relationships

Many observations in the field attest to the strong emotional bonds elephants have with each other within the family and their reinforcement of wider attachments through the occasional much larger assemblies.

For all elephant species, communication plays a critical role in maintaining their complex social relationships. Joyce Poole and her team have determined that African savannah elephants use over 70 sorts of vocalizations and at least 160 different visual, and tactile signals and gestures in day-to-day interactions (Elephant Voices, 2021) (Figure 29.2).

Elephants also form relationships with humans, the only predators of adult elephants. Different categories of humans pose different levels of danger to the animals. In Amboseli National Park, Kenya, researchers determined that elephants could identify, via recorded human voices, the ethnicity, gender and age of categories of humans who were most threatening and those not so. In this environment, the pastoralist Maasai came into conflict with the elephants, frequently spearing them, over the Massai cattle's access to grazing and water resources. Kamba men were agrarian without farms in the park. Elephants had learned to identify the groups (McComb et al., 2014). Elephants can even remember facial expressions from specific humans, another aid in recognizing potential friends or foes (Proops et al., 2018).

Elephants can also have strong positive relations with humans. Elephants appear to grieve for a deceased family member, standing vigil over them, touching them and engaging in other grief-like behaviors, including the mother sometimes carrying a dead infant (Pokharel et al., 2022). Elephants have been observed to stand vigil even for a very familiar human. Once a group of elephants remained two days by the home of a deceased man who had developed a close, protective relationship with them (Khanna, 2020).

## Forest Elephants

The forest elephant, though smaller than the African savannah elephant is still large, weighing up to 13,000 lb. In the Dzanga-Sangha Protected area within the Central African Republic (CAR), there are 4,000 such elephants). In daily movements from the fruiting trees of the forest to a mineral-rich watering hole, the elephants trampled the ground and toppled trees, engineering paths through the dense forest. Besides the elephants and researchers, these paths were used by many, from rodents to forest buffalos, tourists and the indigenous peoples (Remis and Jost Robinson, 2020; Helen Santoro, 2020)

And it is not just in this preserve that such walkways and other elephant modifications help establish a balanced forest ecosystem. Seeds from the fruits eaten by elephants and minerals, such as nitrogen obtained from both plants and water, are dispersed widely via the elephants' dung. That nitrogen is essential for good plant growth. Having passed through an elephant's gut, many seeds germinate faster. The elephants are also eating leaves, tree bark, and, preferentially, soft, rapidly growing woody plants, rather than hardwoods. Trees less than 12 inches tall tend to be stomped down. The elephants thereby create spaces for other plants to grow; the result is greater plant diversity. Some species, such as the wild mango (*Irvingia gabonensis*), would not be able to survive without the elephants. These various activities combine to enable slow-growing hardwoods to grow large, storing far more carbon than do softwoods. A positive effect on slowing climate change is the result.

An inadvertent experimental situation sharply reveals the elephants' impact. Due to poaching, elephants had disappeared 30 years ago from the Democratic Republic of Congo (DRC). Until recently, the other Congo, the Republic of Congo, held a large population of elephants. Researchers measured the trunk size of *every* tree within a study area and identified all the species. In only 30 years, the forests have already become different. On a much larger scale, the differences between the rainforests of Africa and those of the South American Amazon have been attributed, at least in part, to the loss of large herbivores, about 12,000 years ago, in South America. The Amazon has a higher density of smaller trees.

Should forest elephants be lost, scientists estimate that 7% of Africa's vegetation would be lost. This estimate was derived from calculations from the Ecosystem Demography model and verified from the field inventory data noted above. From the habitat loss would be a further loss of 3 billion tons of carbon dioxide absorption. Such storage is valued at $43 billion, or the equivalent of France's total $CO_2$ emissions for 27 years (Nuwer, 2019; Berzaghi et al., 2019).

On a much more local level, in the Dzanga-Sangha, the resident BaAka people traditionally used the elephants' paths to search for fruit, and medicinal plants and herbs within the forest. The paths also led them to social engagements with other communities and were used for hunting, even elephants. Only a few persons were the traditional, specialized elephant hunters; they recognized their formidable adversary, its strength, its sentience. The community's bond with elephants was strong. Said one member, "I am the elephant. The elephant is me—you are the elephant." The hunters' take was small, and the elephant's bounty provided "majestic packages of meat to sustain communities" as one hunter phrased it. That treasure was shared among many (Santoro, 2020).

But the situation is currently different and dire ... for both the elephants and the human communities. Poaching has caused a serious, continuing decline in the forest elephant populations, so much so that the CAR government has outlawed elephant hunting and restricted forest use. At present, some poachers still manage to obtain ivory tusks, but the BaAka can also no longer hunt elephants for the meat that is shared, nor can they even hunt the duikers, porcupines and other small animals. Efforts by researchers and conservation groups to modify the laws to permit traditional forest use by the BaAka are underway but is an "uphill

*Figure 29.3*  An Adult Savannah elephant, the largest pachyderm. Their valuable ivory tusks, long sought for carving and even as aphrodisiacs, are not dead like nails, but are nourished by blood vessels and extend high into the elephant's head. They are used as tools for digging, work with the trunk to pull down vegetation for food, and, if needed, as weapons in combat. (Photo by William D. Knight)

battle." (Santoro, 2020). Collaboration is needed between biologists and anthropologists along with related political and other entities (Remis and Robinson, 2020).

Some hold very different views. The noted primatologist, John Oates, has written persuasively of the continuing *failed* attempts to combine community development with conservation measures; conservation always loses (Oates, 1999).

## African Savannah Elephants

These grand pachyderms are likewise ecosystem managers (Figure 29.3). However, in environments where their population has grown extensively, and their habitat has been reduced through farming and other human incursions, they sometimes denude the environment of trees in their quest for vegetable matter. Their positive contributions include some of those similar to the forest elephants', e.g., dispersal of seeds and nutrients and fecal fertilizing. Through their use of termite mounds, savannah elephants create another significant environmental reshaping, benefiting many species. A mound can be huge, extending both high, up to 17 + feet (5 meters), and deeply below the ground. To create these homes, the termites carry earth from below ground, bringing up nutrients often not so readily available at ground level. In time, these mounds are abandoned, or sometimes elephants simply knock them over. The elephants typically dig out the lower mound, seeking minerals more abundant below. The holes enlarge and hold water since the mounds are more clay-like than nearby soil. Many species use the waterhole, the elephants even lounging in it, picking up mud and distributing it and its minerals as the mud falls off during the elephants' perambulations. (PBS, 2021).

## Asian Elephants

And finally, we consider the Asian elephant. That pachyderm is likely the most intelligent of the elephant species, with the highest Encephalization Quotient (EQ). Among terrestrial mammals, it is surpassed in EQ only by chimpanzees (*Pan troglodytes*) and humans, the highest. (EQ is a measure of relative brain size, compared to body size.) At least one manifestation of its intellect is indicated by its ability to both make and use tools. Specifically, wild and captive Asian elephants have been observed to modify tree branches, producing appropriately sized and configured branches, suitable for swishing away bothersome flies. They use different methods to do so, typically employing their dexterous trunks in the process. In the reported research, the two young watched and attempted to imitate (Hart et al., 2001).

Particularly provocative for our discussion of conservation goals and methods is the situation of these elephants in Thailand (Baker and Winkler, 2020). Presently, fewer than 10,000 Asian elephants live in Thailand, with more captive than wild. Over all the countries in their range, 30–50,000 wild Asian elephants exist, with another 10–15,000 held in some form of captivity.

For long times, they had served as "beasts of burden," hauling logs for the timber industry through the Thai forest over mountainous and other difficult terrains. After a series of flash floods destroyed villages and killed over 250 people, logging was banned in Thailand in 1989. The flooding was attributed, possibly erroneously, to unsustainable logging practices, but was more likely due to the replacement of teak forests by rubber plantations (Delang, 2002). The impact of the law was enormous, on the government's income and on the elephants and their traditional owners/trainers, the mahouts. No longer could the elephants browse in the forests where they worked, the mahouts had to grow feed or pay for it. An elephant could eat up to 150 kg of food/day.

The results were extremely deleterious for both. Elephants were "monetized" by their use in the tourist industry. They begged in the cities, causing accidents, malnutrition and disease. Next, "eco-friendly" practices were adopted whereby tourists paid to ride, feed or pose with elephants. But not all mahouts could be so employed and it was cheaper to employ outsiders. Soon, not the mahouts, but the outsiders, became the elephants' owners. To be a mahout was no longer a respected profession, learned through generations. Elephants were captured from the wild to participate in tourism.

Presumably alarmed by the reported abuses of the tourism industry, in 1997, the Queen of Thailand initiated an elephant release project. Captive-raised elephants, about 100 as of 2020, were released into national parks and wildlife sanctuaries. Most reports, but not all, indicated good integration into wild herds.

The authors, Baker and Winkler, espouse a Re-Wilding program, the "3 R" Model: "Rescue/Rehabilitate/ Rewild." They recognize that long-term "worker" elephants and those raised in captivity will have developed behavioral traits very different from those needed to integrate successfully into wild elephant herds. With help from experienced mahouts, rescued elephants would undergo rehabilitation, appropriate to each elephant. Elephants could then be released onto state land with rights held by indigenous peoples. Mahouts would patrol the land, reducing elephant-human conflict and serve as forest guides. Tourism would again finance the program, but a different kind of tourism. Visitors would pay to trek into the forest, led by mahouts, to reach the places the elephants have chosen to be. (This is similar to the highly successful gorilla visitation program in Uganda.) This is also a most ambitious program; will it ever come to be?

We have presented only a superficial view of the complexities involved in protecting the elephants and the means to forestay poaching and other corruptions. Among them are techniques such as DNA identification of each captive elephant. Ethical and philosophical issues are inherent in the captivity of any wild species: What rights do/should animals have? Which animals? On what basis?

There is, for example, a growing movement to grant "personhood" to elephants, among other species (Lepore, 2021). Two theologians, the Director and Deputy Director of the Oxford Centre for Animal Ethics, have filed an amici brief in defense of "Happy," a long-caged elephant at the center of a case being heard (2022) in New York State, USA. They cite biblical references to humans' God-given dominion over animals, which they interpret as the duty of caring for the world. Animals are not commodities; we are opposing God's wishes if we treat non-human animals' suffering as immaterial. State the theologians, "we have a moral duty rooted in Christian theology to legally recognize "Happy" as a person…" and release her to an animal sanctuary (Oxford Centre for Animal Ethics, 2022).

## *Piping Plovers*

Piping plovers (*Charadrius melodus*) are small stocky shorebirds in the Order *Charadriiformes* approximately seven inches long. During their breeding season (March–August), they inhabit coastal margins of ocean beaches, lakes and riverine systems of Eastern and Central North America. Depending on nesting location, they are Federally protected as either Endangered (Great Lakes population) or Threatened (Northern Great Plains and Atlantic Coast populations). Two subspecies are recognized, *C. melodus melodus* along the coastal margins and *C. melodus circumcinctus* with a total current population estimate of 8,000 adults (BirdLife Intl, 2020), which is an increase from the first international census attempt in 1991 (Haig and Plissner, 1993). As a result of intensive conservation efforts that began in the 1980s, populations of piping plovers are mostly increasing across their breeding range (Figure 29.4).

Their conservation saga is instructive, especially in comparison to that of elephants, which have significant economic value as draft animals, present a potential danger to local communities and are also highly intelligent and social animals, with strong family and group bonds. Although historically, piping plovers contributed feathers to the millinery trade and were victimized by hunting, their relationship to conservation has a stronger link to aesthetics, biophilia and habitat protection. The landforms they inhabit are typically ephemeral but highly attractive to humans. Often the beaches and wetlands that support their nesting and foraging or overwintering are high-value tourist locations. The local economies, including

*Figure 29.4* Adult piping plover male in full breeding plumage. (Photo by Eric G. Strauss)

restaurants, hotels and other venues rely upon human access to these locations in order to continue their businesses. This relationship is at the core of the conservation conflict.

After the passing of the Federal Migratory Bird Act of 1918, plover and other shorebird populations were on the rise, recovering from earlier declines due to hunting, and didn't start to decline again until the 1950s expansion of coastal development. The anthropogenic impacts on coastal and riparian ecosystems, such as urbanization, recreational use by pedestrians and travel by off-road vehicles, the introduction and range expansion of synanthropic predators such as coyotes (*Canis latrans*) and American crows (*Corvus brachyrhynchos*), both of whom eat eggs and chicks, and even hydrologic changes due to climate change have been identified as causes in severe indirect reduction in breeding success and subsequent population declines. Nests are also vulnerable to direct destruction and the behaviors of adult plovers and chicks disrupted by domestic animals such as dogs that are allowed to run unleashed on coastal habitats. Intervention by the Federal Agencies in the 1980s such as the US Fish & Wildlife Service, in collaboration with state and local agencies, has once again reversed the trend through targeted management changes in these affected habitats. The typical result of protecting piping plovers through beach closures has been to challenge the economic and cultural foundations of those communities and force residents to reimagine their traditional activities in these ecosystems (Figure 29.5).

The behavioral ecology of piping plovers gives unique insight into both their plight as a threatened species and pathways to their management in human-dominated landscapes. First described by ornithologist George Ord in 1824, the plaintive *peep-lo* call of the species became a signature characteristic of New England and especially Cape Cod coastal beaches and dunes. As expressed by Henry Beston in the Outermost House (1925, p. 177), "Its note is a whistled syllable, the loveliest musical note, I think, sounded by any North Atlantic bird."

Unlike least terns (*Sternula antillarum*), which utilize the same coastal habitats and nest in dense colonies as a protection from predation, piping plovers are less social and often prefer to place their nests in inconspicuous locations, thus avoiding predators through stealth. Two aspects of their behavioral ecology emerge as a result of this breeding preference. First, locating and protecting breeding piping plovers is more complicated as they're dispersed over a larger physical area. Protecting nesting habitat often requires the location of individual territories and nests. Second, their behavioral response to disturbance is a novel suite of

*Figure 29.5* A clutch of piping plover eggs. (Photo by Eric G. Strauss)

*Figure 29.6* Adult piping plover injury feigning display. (Photo by Eric G. Strauss)

behaviors associated with this genus. When they sense danger from impending predators or human disturbance, they begin an injury feigning display that appears as though they have one or two broken wings, often flopping from side to side as though they are seriously injured, squawking loudly and moving away from the nest or chicks. These behaviors are also exhibited by the more common Killdeer (*Charadrius vociferous*). This typically has the desired effect of leading potential predators away from the nest location. When the perceived threat is relieved, the bird suddenly appears uninjured and flies back to the nest. These two characteristics shape the management conundrums with this particular species (Figure 29.6).

These defensive behaviors also reveal far more sophisticated aspects than was once believed. The displaying bird does not simply reflexively produce a broken wing display upon sighting an intruder. Rather, it monitors the intruder's activities both before and while displaying. If that intruder is not following the injury-feigning bird, the plover re-approaches the menace, uttering a peep that usually attracts attention. The bird then again begins a display, always in a direction that would lead an intruder further *away* from the nest (Ristau, 1991).

In Ristau's other field experiments, piping plovers quickly learned which individuals constituted a potential threat to its nest and which likely not. The birds were even sensitive to the direction of the gaze of a human intruder. They reacted with more "anxious" behaviors to a human staring in the direction of its nest as it walked by at a goodly distance as compared to the birds' behavior toward a human who was looking in the opposite direction from its nest.

In short, piping plovers are very sensitive to humans. Significant work over the past 30 years suggested there are sensitive periods in the reproductive cycle of piping plovers that place them at greater risk to human disturbance. Typically, males return to the breeding grounds quite early in the season. On Cape Cod, this can be as early as the first week of March (Strauss, 1990) when conditions are still quite harsh. The males are sensitive to disturbance during this window of time as they're still searching for and establishing territories to defend; they engage in courtship and territorial displays on their territories when the females arrive. Males have generally high year-to-year site fidelity, and when shifted off of their historical territories suffer lower reproductive success. Once the pair bond is formed, typically by April, their fidelity to the site is relatively high, unless they're pushed away by a storm surge, flooding or other conditions that disrupt the pair bond (Haig and Oring, 1988). Once the eggs are laid and incubation begins in May, they are tenacious to the site and both males and females continue incubation despite potential human disturbance (Patterson et al., 1990).

A second highly vulnerable window of time in the reproductive cycle is after the precocial chicks have hatched and begin to feed up the water's edge (DeRose-Wilson et al., 2018). Young piping clever chicks feed themselves on marine invertebrates and insects immediately post-hatching. They're able to stand up, run and even swim just hours after they emerge from their eggs. The conservation challenge for the species revolves around the spatial characteristics of foraging use by the young chicks. They seek refuge and safety along the edges of the dunes, but must feed at the water's edge as the wet sand presents the most accessible community of marine invertebrates. As a result, the chicks must routinely traverse the beachfront between dunes and water's edge on a regular basis, - often a dozen or more times each day depending on tidal and weather conditions (Strauss, 1990). This places them in direct conflict with human beach users, their pets and any vehicular traffic that may be moving along the beachfront (Flemming et al., 1988). Unfortunately, since it is an existent path, the birds also tend to run along the same vehicular tracks that other vehicles also re-use, placing them in peril of being run over by off-road vehicles. Recent investigations suggest climate change resulting in higher sand temperatures will negatively impact life history patterns for these birds (Andes et al., 2020)

If disturbed during these windows of time, the chicks simply stop feeding and hide in the grass. This delays their development and puts them at risk of starvation. Alternatively, the parents will attempt to lead the chicks to a new location which presents a new set of affiliated risks associated with traversing unfamiliar territory. Therefore, the 40-day window between the hatching of the chicks and their ultimate fledging becomes the single largest conservation challenge in protecting piping plover populations (Strauss, 1990).

Typical management interventions, guided by the US Fish and Wildlife Service Recovery Plan (USF&WS, 1996) included fencing off recreational beach areas to reduce or remove human access, especially from unrestrained domestic dogs. On beaches that permit off-highway vehicle travel, passage along the beach in these areas is often restricted or completely closed. This prevents recreational use by people or at least forces them to find alternative access to the beachfront. The disbursed and widely distributed use of the habitat by piping plovers means that closures of beaches are often in a patchwork across the landscape or very large sections of recreational beach are closed. Typically, these conflicts occur at the height of the summer recreational season and represent an economic and management hardship for the local communities. On Sandy Neck Beach in Barnstable on Cape Cod, Massachusetts, the recreational barrier beach may decrease from 6.5 miles of beachfront access to less than 0.5 miles. Across their range, piping plovers have benefitted from these management interventions and population numbers are mostly increasing. However, the conflicts over land uses continue and have significant political implications for communities with viable plover habitat.

## Ecosystem Services as a Perspective to Understand and Derive Strategies for Existing and Future Conservation Efforts

The anthropogenic changes to our global ecosystem have long historical roots, but according to the UN Millennium Ecosystem Assessment (2005), these forces have reached unprecedented levels in the past half-century. Rapid demise of ecosystem services is considered one of the critical negative consequences of unchecked technological development and human expansion. However, the concept of these services provides a novel interpretation as a tool for extending conservation efforts among human stakeholders who may not resonate to other arguments that use a strictly biological basis for the protection of wildlife.

The concept of ecosystem services has tendrils reaching back to the mid-20th-century economic and hedonic analysis of the values we place on natural habitats. The report of The United Nations Millennium Assessment (2005) describes ecosystem services as the benefits humans derive from healthy ecological systems. Realizing that many of the activities of natural systems go unrecognized in traditional economic analytics, use of the ecosystem services model provides a holistic understanding of these values and divides them into four distinct categories.

The first is described as *provisioning services* which include the products such as food, fiber, freshwater, fuels and other products directly consumed to support human well-being. *Regulating services* reduce extreme ecosystem variation and include the climate, pollination of plants, disease regulation and the role of water as a climate modulator. *Supporting services* supplement and maintain global provisions and include such processes as nutrient cycling and soil formation. Finally, *cultural services* address the spiritual aesthetic and heritage elements of our relationship to nature and are at the core of our understanding of biophilia and human attachment to place.

By applying ecosystem services models to the conservation challenge, intrinsic values of nature can be more fully explored and systemically codified. For example, pioneering research by Kuo and Sullivan (2001) studying urban greenspaces discovered that residents with enhanced access to nature report lower levels of fear, fewer incivilities, and less aggressive and violent behavior. Expanding studies to the national scale, a team led by Sullivan and colleagues (Lu et al., 2021) studying 135 cities found significantly higher Covid infection rates among minority populations compared to their white counterparts. However, that disparity was reduced in communities where minority residents had increased access to green space, leading to the concept of nature as preventative medicine. As such, these values of green infrastructure have feedback loops on some of the costliest aspects of human society – namely crime and disease. The study carefully controlled factors of SES, demographics, pre-existing chronic disease and built-up areas. Underlying mechanisms of the disparity were hypothesized to include decreased spread of virus aerosols in open space, increased physical activities and improved air quality, but also the mental and social well-being afforded by access to nature and the ensuing stress reduction that affords better functioning of the immune system, and so offers protection against disease, including Covid.

Similar analytics have been placed on the trees themselves, along with the subsequent canopy cover that healthy tree populations provide. The benefits of the global estimate of the 73,000 tree species (Gatti et al., 2022) include reduced heat island effects, increased air quality, particulate pollution removal and associated reduction in respiratory diseases in human populations. These benefits are robust across the broad climatic range of communities in which assessments have been done (Turner-Skoff and Cavender, 2019).

## Preserving Wilderness and its Inhabitants

Besides our concern with potential beneficial environmental effects on humans, we care about environmental impacts on animals' well-being. Consider how E. O. Wilson, the famous naturalist scientist, evolved in his attitudes about such matters. From an anthropocentric and deeply pessimistic view regarding the present world and our continuing destructive activities, he developed a far more ambitious, aspirational and ethically based view. He understood that our present activities would lead to an earth more and more inhospitable to our own selves as well as to almost all other species.

We are now in the midst of the sixth great extinction event in the earth's long existence. We are the culprits, with the destruction of others' habitats, depleting resources, urban sprawl, climate change and overpopulation, which is itself the driver of so many of the other devastations. Wilson proposed that we set aside half the earth as wilderness or as "rewilded corridors" to preserve the biodiversity that still exists (Wilson, 2015). He calculates that 85% of species will be saved from extinction by saving half the land and sea. The Half-Earth Project is currently mapping the worldwide distribution of species to assist this endeavor. It's possible. Varieties of this concept have existed long before Wilson. Quite what we actually do remains intrinsically bound to our ethical values: Do other creatures on this earth merit their existence, just as we believe we do?

## Sentience and Personhood as Tools for Conservation

The successful protection of wild species is often predicated on ideas of our affiliative relationship with other living creatures. Central to our human identity is that as a species, we exhibit a high level of consciousness. Research in animal behavior has raised important questions about the degree to which other species exhibit similar characteristics. Prime among our consideration of animal capacities is determining whether they are sentient beings. If so, which? In 2021, the United Kingdom passed an Animal Welfare (Sentience) Bill, which extended the official UK government recognition that all vertebrates are sentient. With the new legislation, decapod crustaceans and cephalopod mollusks, including crabs, octopuses and lobsters, are, likewise, recognized as sentient. *Future* welfare activities will now require that sentience be considered for these invertebrates. However, *existing* industry practices will not be impacted and thus there is no direct effect on shellfish catching or in restaurant kitchens (UK Government, 2021). In the United States, animals are possessions, and animal welfare is legislated in terms of the prevention of "unnecessary suffering." Thus, animals must be rendered unconscious before slaughter. Other than that provision, sentience is considered only minimally, as in mandated provisions for animal care in agricultural and scientific enterprises.

The move towards Personhood for at least certain animals is motivated by concern for their rights, for well-being beyond that of "unnecessary suffering." Within the United States, proposed or actual court cases have focused on a chimpanzee and an elephant, both species considered to be highly cognitive, sentient beings by most scientists, with significant emotional and social relationships. A horse was also proposed. The opposition centers on such issues as the "slippery slope:" once a species is granted Personhood, where does it stop? Will horses and cats be deemed "Persons?" In the United States, corporations have achieved that status, legally, with the legal rights ensuant. Personhood, even sentience, has an impact on the use of animals in scientific research. The potential consequences are extraordinarily far-reaching, including permitted scientific methodologies.

## Final Thoughts on Conservation

Conservation essentially depends on our human value systems. Some consider ourselves "the exceptional creatures at the apex of evolution" ... "the sole species that has intrinsic value"

... " and thus we are to be privileged above all other species (Lynn, 2015). With such an anthropocentric view of nature, our relationship is defined solely in terms of its usefulness to us. Some might extend "usefulness" to include our pleasure: "biophilia." Others consider us to be one among many valued species on this planet and our values and goals must, therefore, be to co-exist in harmony with the planet: its environment, and all the other creatures.

These latter views are held in many cultures and religions. As an example, the Bible holds that humankind either (1) has dominion over all, implying that the "beasts" exist to serve us, or (2) has a role as stewards, implying we are the caretakers of the animals and must assure their well-being. Though kindly, that is patronizing. Neither case advances an attitude of respect for the other beings, appreciating their diversity, capabilities and needs. It also ignores the modern interpretation of nature through the lens of ecosystem services, which encompasses a broad spectrum of values that intact nature provides.

Recent approaches to conservation have elevated the efforts into a multi-billion-dollar global initiative involving public, private and governmental organizations. Even consumer goods are marketed with conservation considerations or direct investment in nature protection. However, the challenges are often presented as a choice between human economic health versus the protection of native ecosystems. This dichotomy introduces additional conflict into the solution process and suggests a need for an institutional ecology (Child, 2012) to address the multifaceted challenges presented. Transdisciplinary efforts to bridge the political sciences and biological conservation have yielded an approach to conservation organizations called the *virtuous cycle* (Jimenez and Basurto, 2022). Here, the synergistic relationships among five key elements create an effective and sustainable conservation effort: (1) Anchoring by a senior leadership with a shared vision, coupled to (2) well-communicated outcomes, (3) contextualized practice-based learning, (4) long-term financial viability and (5) linkages across social and geographic scales in order to access critical resources. Community involvement in decision-making and action was also considered a critical element of successful conservation plans. This approach provides additional hope for global conservation efforts that are locally impactful but embedded in the global consciousness. But what we do, if anything, depends first on our recognizing that our present activities causing global climate chaos are unsustainable for a future earth and secondly on our commitment to preserving the earth and its creatures, including humankind. We have to act.

## References

Allen, C. R., Brent, L. J., Motsentwa, T., Weiss, M. N., & Croft, D. P. (2020). Importance of old bulls: leaders and followers in collective movements of all-male groups in African savannah elephants (*Loxodonta africana*). *Scientific Reports*, *10*(1), 1–9. https://doi.org/10.1038/s41598-020-70682-y

Andes, A. K., Sherfy, M. H., Shaffer, T. L., & Ellis-Felege, S. N. (2020). Plasticity of least tern and piping plover nesting behaviors in response to sand temperature. *Journal of Thermal Biology*, 91, 102579. https://doi.org/10.1016/j.jtherbio.2020.102579

Baker, L., & Winkler, R. (2020). Asian elephant rescue, rehabilitation and rewilding. *Animal Sentience*, *5*(28), 1. https://doi.org/10.51291/2377-7478.1506

Beech, H & Suhartono, M., & Dean A. (June 5, 2020). Elephants, long endangered by Thai crowds, reclaim a national park. *New York Times. Print Edition.* (June 6, 2020). Lockdown allows Thai wildlife to reclaim a popular national park. A9.

Berzaghi, F., Longo, M., Ciais, P., Blake, S., Bretagnolle, F., Vieira, S., ... Doughty, C. E. (2019). Carbon stocks in central African forests enhanced by elephant disturbance. *Nature Geoscience*, *12*(9), 725–729. https://doi.org/10.1038/s41561-019-0395-6

Beston, H. (1928). *The outermost house: a year of life on the great beach of Cape Cod*. Henry Holt & Company.

BirdLife International. (2020). "*Charadrius melodus.*" *IUCN red list of threatened species.* 2020: e.T22693811A182083944. https://doi.org/10.2305/IUCN.UK.2020-3.RLTS.T22693811 A182083944.en

Castillo-Huitrón, N. M., Naranjo, E. J., Santos-Fita, D., & Estrada-Lugo, E. (2020). The importance of human emotions for wildlife conservation. *Frontiers in Psychology, 11*, 1277. https://doi.org/10.3389/fpsyg.2020.01277

Cazzolla Gatti, R., Reich, P. B., Gamarra, J. G., Crowther, T., Hui, C., Morera, A., ... Liang, J. (2022). The number of tree species on Earth. *Proceedings of the National Academy of Sciences, 119*(6), e2115329119. https://doi.org/10.1073/pnas.2115329119

Child, B. (2012). Innovations in state, private and communal conservation. In H. Suich, B. Child, & A. Spenceley (Eds.), *Evolution and innovation in wildlife conservation: parks and game ranches to transfrontier conservation areas* (pp. 427–440). Earthscan.

Delang, C. O. (2002). Deforestation in northern Thailand: the result of Hmong farming practices or Thai development strategies? *Society and Natural Resources, 15*(6), 483–501.

DeRose-Wilson, A. L., Hunt, K. L., Monk, J. D., Catlin, D. H., Karpanty, S. M., & Fraser, J. D. (2018). Piping plover chick survival negatively correlated with beach recreation. *The Journal of Wildlife Management, 82*(8), 1608–1616. https://doi.org/10.1002/jwmg.21552

Elephant Voices. (2021). *Elephant ethogram.* https://www.elephantvoices.org/elephant-ethogram.html

Erwin, T. L. (1982). Tropical forests: their richness in Coleoptera and other arthropod species. *The Coleopterists Bulletin, 36*(1), 74–75.

Flemming, S. P., Chiasson, R. D., Smith, P. C., Austin-Smith, P. J., & Bancroft, R. P. (1988). Piping plover status in Nova Scotia related to its reproductive and behavioral responses to human disturbance. (Estatus de charadrius melodus en Nueva Escocia, relacionado a su reproducción y respuestas de conducta a la perturbación humana). *Journal of Field Ornithology 59*(4), 321–30. http://www.jstor.org/stable/4513359.

Haig, S. M., & Oring, L. W. (1988). Mate, site, and territory fidelity in piping plovers, *The Auk, 105*(2), 268–277. https://doi.org/10.2307/4087489

Haig, S. M., & Plissner, J. H. (1993). Distribution and abundance of piping plovers: results and implications of the 1991 international census. *The Condor, 95*(1), 145–156. https://doi.org/10.2307/1369396

Hamilton, A. J., Basset, Y., Benke, K. K., Grimbacher, P. S., Miller, S. E., Novotný, V., ... Yen, J. D. (2010). Quantifying uncertainty in estimation of tropical arthropod species richness. *The American Naturalist, 176*(1), 90–95.

Hart, B. L., Hart, L. A., McCoy, M., & Sarath, C. R. (2001). Cognitive behaviour in Asian elephants: use and modification of branches for fly switching. *Animal Behaviour, 62*(5), 839–847.

Jiménez, I., & Basurto, X. (2022). An organizational framework for effective conservation organizations. *Biological Conservation, 267*, 109471. ISSN 0006-3207; https://doi.org/10.1016/j.biocon.2022.109471

Khanna, M. (2020, August 24). Guy saved elephants from poachers. They mourned his death outside his home. *India Times.* www.indiatimes.com/technology/science-and-future/elephant-conservation-emotional-quotient-521101.ht

Langbauer Jr, W. R., Payne, K. B., Charif, R. A., Rapaport, L., & Osborn, F. (1991). African elephants respond to distant playbacks of low-frequency conspecific calls. *Journal of Experimental Biology, 157*(1), 35–46.

Lepore, J. (2021, November 16). The elephant who could be a person. *Atlantic Monthly.* https://www.theatlantic.com/ideas/archive/2021/11/happy-elephant-bronx-zoo-nhrp-lawsuit/620672/?utm_source=email&utm_medium=social&utm_campaign=share

Lynn, W. (August 26, 2015). Setting aside half the Earth for 'rewilding': the ethical dimension. *The Conversation.* https://theconversation.com/setting-aside-half-the-earth-for-rewilding-the-ethical-dimension-46121

McComb, K., Shannon, G., Sayialel, K. N., & Moss, C. (2014). Elephants can determine ethnicity, gender, and age from acoustic cues in human voices. *Proceedings of the National Academy of Sciences, 111*(14), 5433–5438. https://doi.org/10.1073/pnas.1321543111

Millennium Ecosystem Assessment (Program). (2005). *Ecosystems and human well-being.* Island Press.

Nuwer, R. (2019, August 19). The thick gray line: forest elephants defend against climate change. *New York Times.* https://www.nytimes.com/2019/08/19/science/elephants-climate-change.html

Nuwer, R. (2020, September 4). Old male elephants: don't count them out. *New York Times.* https://www.nytimes.com/2020/09/04/science/male-elephants-bulls.html?searchresultposition=3

Oates, J. F. (1999). *Myth and reality in the rain forest: how conservation strategies are failing in West Africa*. University of California Press.

Oxford Centre for Animal Ethics. (2022, May 20). Press Release: Captive elephant contrary to Christian ethics, say theologians. https://www.oxfordanimalethics.com/2022/05/press-release-captive-elephant-contrary-to-christian-ethics-say-theologians

Patterson, M. E., Fraser, J. D., & Roggenbuck, J. W. (1990). Piping plover ecology, management and research needs. *Virginia Journal of Science*, 41(4A), 419–426.

Payne, K. (1998). *Silent thunder: in the presence of elephants*. Simon and Schuster.

PBS. (2021, November 3). *The elephant and the termite*. https://www.pbs.org/wnet/nature/elephant-and-termite-about-depsre/26434/

Pokharel, S. S., Sharma, N., & Sukumar, R. (2022). Viewing the rare through public lenses: insights into dead calf carrying and other thanatological responses in Asian elephants using YouTube videos. *Royal Society Open Science*, 9, 211740. https://doi.org/10.1098/rsos.211740

Proops, L., Grounds, K., Smith, A. V., & McComb, K. (2018). Animals remember previous facial expressions that specific humans have exhibited. *Current Biology*, 28(9), 1428–1432.

Remis, M. J., & Jost Robinson, C. A. J. (2020). Elephants, hunters, and others: integrating biological anthropology and multispecies ethnography in a conservation zone. *American Anthropologist*, 122(3), 459–472.

Ristau, C. A. (1991). Aspects of the cognitive ethology of an injury-feigning bird, the piping plover. In C. A. Ristau (Ed.), *Cognitive ethology: the minds of other animals* (pp. 91–126). Hillsdale, NJ: Lawrence Erlbaum Associates.

Santoro, H. (2020, October 15). How humans benefit from a highway of trails created by African forest elephants. Smithsonianmag.com

Soulé, M. E. (1985). What is conservation biology? *BioScience*, 35(11), 727–734.

Strauss, E. G. (1990). Reproductive success, life history patterns, and behavioral variation in a population of Piping Plovers subjected to human disturbance. Doctoral Dissertation Tufts University.

Trombulak, S. C., Omland, K. S., Robinson, J. A., Lusk, J. J., Fleischner, T. L., Brown, G., & Domroese, M. (2004). Principles of conservation biology: recommended guidelines for conservation literacy from the Education Committee of the Society for Conservation Biology. *Conservation Biology*, 18(5), 1180–1190.

Turner-Skoff, J. B., & Cavender, N. (2019). The benefits of trees for livable and sustainable communities. *Plants, People, Planet*, 1, 323–335. https://doi.org/10.1002/ppp3.39

U.S. Fish and Wildlife Service. (1996). Piping plover (*Charadrius melodus*). *Atlantic coast population, revised recovery plan* (258 pp.). Hadley, MA.

Wilson, E. O. (2016). *Half-earth: our planet's fight for life*. WW Norton & Company.

# 30
# ETHICS, WELLBEING AND WILD LIVES

*William S. Lynn, Liv Baker, Francisco Santiago-Ávila, and Kristin L. Stewart*

The animal turn in scholarship directs our attention to the empirical and theoretical dimensions of humanity's complex relationship with other animals.[1] Composed of multiple turns in distinct fields of scholarship, the study of human-animal interactions is richly interdisciplinary. This helps us to understand how ideas about—and practices toward—animals have developed over time and in space, as well as how humans and other animals co-create their social and ecological environments. Two turns are of direct interest to this chapter, the turn to animal wellbeing in science and a companion turn to animal ethics in the humanities and social sciences.

## The Turn to Welfare and Wellbeing

While ghosts of behaviorism still haunt the halls of the academy, the failure of behaviorism in the social sciences freed us from dogmas that sought to eviscerate the individual and collective agency of human beings. The belief that human agency can be reduced to conditioning or social determinants of behavior ran into a buzz saw of contrary evidence and a failure of analytical power to produce deterministic, positivist explanations for human beings and society (Toulmin, 1992).

Studies in cognitive ethology, animal psychology, and animal welfare science do much of the same work to rescue the agency of animals from reductionism and mechanism. The knowledge from these scientific disciplines—alongside insight from the humanities—is co-producing a paradigm shift in how we understand the worlds of other animals. We know that animals have deep capacities to experience physical and psychological suffering, to manifest positive emotions, an ability to exercise their own agency in their lives, to develop and pass along cultural knowledge, and even exercise moral and social sensibilities that are at times not so different from our own. And yet, we still suffer a failure of imagination in how we navigate and constitute our shared world.

Taken together, this work gradually opened a window to investigate ontic and ontological aspects of sentience, sapience, and sociality in wild and domesticated animals within and across diverse ecological and social communities. Sentience, sapience, and sociality are concepts to elucidate varying degrees of evolved awareness, from the phenomenal to subjective and objective awareness. However, this investigation remains frustrated by the residue of Cartesian beliefs and behaviorism where animal minds are regarded as beyond the scope of scientific apprehension. Such claims perpetuate the theological and philosophical rationalizations for human exceptionalism, but do not accord with empirical findings about the complexity, individuality and agency of the animal world.

While an emergent paradigm shift is palpable, its effects remain modest. The gap in knowledge, understanding, and importantly of synthesis of the animal world has not been properly filled by existing animal-related fields. It may be the case that a new (sub)field of applied study is needed—one that has as its first principle, that nonhuman animals—like us—have an interest in the outcomes of their lives; and asks as we do in research on human health, what makes life worth living for other animals. To ask such a question demands of us greater humility, creativity, and openness than we have thus far made manifest. Two distinct, yet interdependent concepts central to this effort are welfare and wellbeing, dovetailing what the science is telling us while underwriting ethical thinking about why animals matter.

Concepts such as welfare, wellbeing, and quality of life are often deemed fungible by laypeople and professional practitioners alike, and many would argue that defining these terms for humans is itself difficult. So, what then for other animals? To address what may lead to a higher quality of life for humans, the World Health Organization (1946) defines the concept of health as, a state of complete physical, mental and social wellbeing and not merely the absence of disease or infirmity." Within this definition is a profound distinction *with* a difference: that wellbeing is *something other than* welfare, as alluded to by "the absence of disease or infirmity." Despite this notion of health being pertinent to other animals, animal wellbeing as such has not been given just consideration. Instead, welfare as a concept has dominated and circumscribed our scientific and ethical approach to our understanding and treatment of animals.

The animal welfare movement developed from concerns over the quality of life of domesticated and captive animals, focusing on reducing pain and suffering. Rooted in these interests, animal welfare science emerged in the 1960s. As David Fraser, a preeminent thinker in the field writes, "...animal welfare science occurred as a response to ethical concerns about the treatment of animals and debate about the kind of life they should be allowed to live" (Fraser, 2008, p.6). This said, it is important to keep in mind that this concern primarily extended to those animals under our direct care and management. This ultimately served to perpetuate a culture of animal use, and to exclude certain animals from the sphere of concern (e.g., free-living wild animals). Consequently, the historical position of animal welfare science has been to focus its scientific efforts on mitigating negative states, and conditions in which animals are afforded little agency (Brambell, 1965).

By far, the Five Freedoms model of animal welfare is the most widely recognized outside the field, and perhaps the most extensively implemented within it (Webster, 2008). In more recent years, models of animal welfare have made strides to understand and integrate mental states of animals, including full emotional valence. The Five Domains (Mellor & Reid, 1994), which updates the Freedoms to include mental states, has possibly supplanted the Five Freedoms as guiding principles for animal welfare science. Lesser known, but an increasingly used model is the Qualitative Behavior Assessment (Wemelsfelder et al., 2001), which tasks human observers with assessing the behavior of the "whole animal" to determine if an animal is in a positive or negative emotional state. Lesser known still is the two-dimensional model of emotion development, which relies on cognitive and affective biases as key measures (Mendl et al., 2010).

In the last few years, there has been a partial turn to "positive welfare." By way of example, the "Welfare through Competence" framework (Webber et al., 2022) privileges the possibility of competence through opportunities for choice and control, in environments that cultivate complexity and variety. In general, positive welfare—and similar concepts—has occurred to redress the focus that historically has been placed on the alleviation from suffering, by widening the field of view to include the benefits of positive experiences (Rault et al., 2020). While the improvements are undeniable, each of these approaches remains constrained, to varying degrees, by the conventional practice and precedent of animal welfare that were born of animal use and mitigation of harms.

On the one hand, the scientific position has been to acknowledge animals' physiological and biological needs, alongside their cognitive capacities; and on the other hand, to deny the bearing of this acknowledgment on animals and situations where the "science is not yet in." Instead, for non-human animals, we should be advancing beyond a welfarist approach that has come to focus almost exclusively on mitigating conditions in which animals are afforded little agency. We could be practicing instead a "science of animal wellbeing" that also sees animal health as "a state of complete physical, mental and social wellbeing and not merely the absence of disease or infirmity" (Baker, 2016).

Thus, wellbeing is conceptually and operationally beyond welfare. A model of animal wellbeing, in contrast to one of welfare, is not relativistic. It actively privileges autonomy and agency. And in the service of understanding and achieving high degrees of autonomy and agency for the animals, attending to an array of meaningful intrinsic and extrinsic factors is paramount. Such factors include the animal's eco-social environment; environmental engagement; behavioral repertoire; expression of privileged behaviors (e.g., play); self-expression; culture; opportunities for reasonable challenges; access to shelter, safety, and proper nutrition; physical health; knowledge acquisition via learning and exploration; expression of emotional valence/positive affect; and lived experience (see Figure 30.1). Importantly, the constellation of these factors is situated within an understanding of the evolutionary legacy of the animals and their species.

## The Ethical Turn

The roots of the ethical turn are ancient and complex. In one sense, there have always been moral norms about the use of animals for food, labor, and fiber. This is to be expected by virtue of humanity's inescapable interdependence with animals throughout our history. This interdependence has always been practical as humans have collaborated with and exploited animals for our benefit. Moral primates that we are, we will try to find meaning and purpose in this interdependence, some of which may be illuminating and some not. There exists a wealth of examples in the world of normative ethics that guide how we might consider and care for animals.

The Chippewa, a first nation of North America, is one example. For the Chippewa, animals are kin and interlaced in complex relationships of respect and use. Wolves, for instance are regarded as relatives and therefore as extended members of the tribe, exemplars of social values like courage and loyalty. They may not be killed for trivial reasons but only in the case of vital needs (Gilbert et al., 2022). The difference between blood sports and hunting for survival makes this point clear.

For the purposes of this volume, the ethological turn underwrites a formal ethical consideration of animals, both wild and domesticated. As we come to know better animals of many kinds, we recognize that they have manifold desires and perspectives of their own, which human interference (intentional or not) may support or suppress utterly. Mary Midgley's concept of the mixed community notes that we have always implicitly understood this as it is the basis for diverse interspecies communication between humans and domesticated and wild animals (Midgley, 1998). So too, George Perkins Marsh's concept of "geographical agency"—humanity's capacity to alter the planet for good or ill—equally applies to animals as it does to people and nature as a whole (Marsh, 1965).

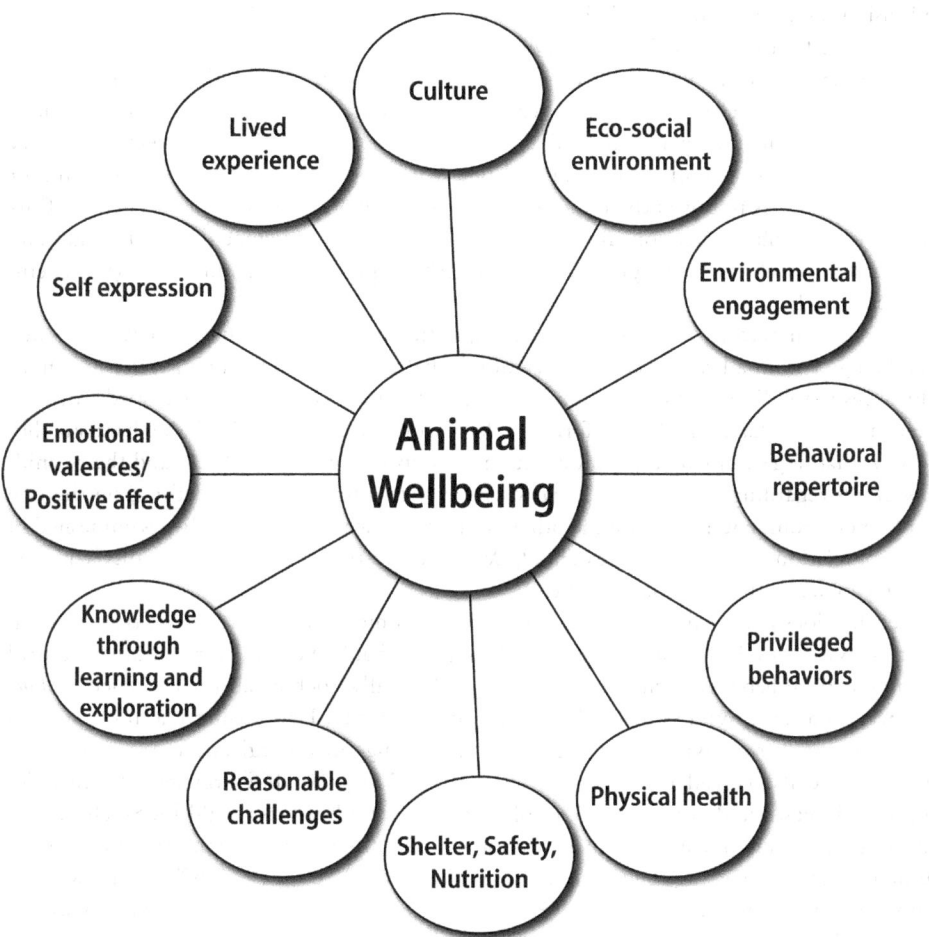

*Figure 30.1* Spheres of wellbeing. (Original by Liv Baker, 2020)

Seen from this light, our depredations on wild lives are particularly severe. Global heating and the climate emergency, the decimation of biodiversity, habitat degradation through agricultural conversion and urban sprawl, pollution of our shared environments, and market and trophy hunting are some of the direct threats to wildlife (Gregory et al., 2021; Kolbert, 2014). That only 4% of the terrestrial biomass of the globe is constituted by wild animals—with 96% composed of humans and domesticated animals used for food—is a devastating illustration of this depredation (Bar-On et al., 2018). It is a situation desperately in need of ethical interpretation and action.

## Ethics

What then is ethics? In the words of Socrates, it is both the study of and deliberation about "how we ought to live" (Plato, 1987). We study ethics to understand what is good, right, and just (to use but three ethical concepts), and we deliberate with others to best discern how to apply ethics in our individual and collective lives. This involves a two-fold project of critique and vision, as we identify moral challenges in our ways of life today while we envision a better future for tomorrow (Midgley, 1993).

Like science, law, medicine, and other forms of practical thinking, ethical arguments and claims are justified or not according to reason and evidence. Thus, ethics cannot be reduced to religious commands, personal preferences, or cultural norms. These aspects of society may inform or reflect ethical thought and practice to be sure, but they do not define or circumscribe it. As importantly, ethics is not for angels but for fallible human beings. Ethics cannot prescribe perfection. Rather, it serves as practical guidance for our thought and conduct as we seek to be in right relationship with people, animals, and nature (Bernstein, 1991; Singer, 1993).

When it comes to wildlife, ethics raise the question of how we ought to live with nonhuman beings within a broader community of life that includes human beings. We may have ethical responsibilities to the bear and her cubs in the forest behind our backyard, while at the same time we have an ethics-informed relationship with the neighbor's dog harassing her cubs, or that dog's human companion whom we know loves both wildlife and their canid. Any ethics regarding wildlife is not narrowly restricted to only animal ethics, on the one hand, or environmental ethics on the other because—as we are virtually always enmeshed in a web of relationships—both are essential. We reference this complex relationality with the phrase people, animals and nature (PAN).

Yet this does not mean we cannot bracket for a moment the full weight of that web and focus on wild lives. For us, *wild* animals include those that live outside direct human control in more natural habitats, such as bears and deer. Typically, such creatures are called *wildlife*; however, as a term, wildlife is fluid in its use and meaning. Free-roaming animals such as coyotes (*Canis latrans*) who live alongside human beings in ecologically novel ecosystems like cities are also considered wildlife (Gehrt et al., 2009). Wildlife arguably also includes formerly domesticated animals who acculturated or were born to a wild life, such as feral cats, dogs, pigs, and horses. And there are some wild animals who live mutualistically with human beings, such as the Ruddy mongoose (*Urva smithii*) (Lodrick, 1982). While wildlife is not a perfect term, we mean it to reference a range of wild animal lives, each of whom—individually and as part of their complex relationships with each other, humanity, and the natural world—deserve careful ethical consideration.

Ethics asks the deceptively simple question, "how ought we live," for which there are many answers. It may be tempting to think the answers are found only in books written by experts. That is not the case. Moral norms are manifest in a wide variety of beliefs and behaviors. These may or may not be formalized in scholarly writing. Thus, we understand ethical theory to be ubiquitous. It is latent in ethical presuppositions and tacit sensibilities. It becomes manifest when latent ideas about ethics are formalized and innovated upon. The development of theory, practical applications of moral concepts, and a praxis between theory and concrete realities are all at play. Nothing is so powerful as an insightful theory put to good use, as it helps us orient our thoughts, actions, and indeed ourselves in the world, and in our instance, with respect to the wellbeing of animals.

## Animals and Nature in Ethical Theory

Animals and nature appear in ethical theory in a couple of categories—animal ethics and environmental ethics—alongside the social ethics of humanity. As the names suggest, the focus of these categories differs in emphasis. In the first, animals are the primary subject and point of departure for ethical reasoning. In the other, animals may play a supporting role, but if they do, it is in a larger tapestry of environmental and social concern. The divisions between these three approaches are a result of analytic choices, which serve to highlight certain topics, and such divisions belie how beings of all kinds engage with the world.

In some cases, analytic distinctions have ossified into dichotomies. This ossification is especially acute in conservation, which categorically asserts that individual animals and their wellbeing are peripheral concerns. In conservation biology, invasion ecology, and cognate fields, individual animals are viewed primarily as functional units of ecosystems and providers of ecological services, not the proper focus of its ethical or scientific concerns (Callicott, 1997).

This is matched in the dominant discourses of sustainability and development ethics. Here, animals are primarily regarded as resources for supporting human development and addressing the inequities of capitalist and colonialist society. In political ecology—a Marxist-inspired approach to conservation and development—nonhuman animals and the rest of nature are reduced to the social constructions of political-economy. In all these cases, questions that are putatively of social ethics—what do we owe to human beings—come to displace questions of what we owe to animals and the rest of nature (Crist, 2004; Lynn, 2020).

We ask readers to consider a different approach and take inspiration from an earlier generation of ethicists who fully appreciated the interdependencies and interconnections between animal, environmental, and social ethics. As an early theorist of green politics, John Rodman grappled with the implications of ecology for society, while at the same time addressing the rights of dolphins (Rodman, 1974, 1977). Mary Midgley's writings on ethics, science, animals, and Gaia always sought to avoid reductionism and highlight the resonances between ethical theories and issues (Midgley, 1995, 1996, 2007).

Ecofeminists are sharply critical of the separation of animal and environmental ethics, especially the exclusion of animals from environmental concern. They see this exclusion as an artifact of patriarchal hierarchies that fail to appreciate ecological inter-relationalities. Some ecofeminists therefore speak in terms of social and nature ethics, where nature ethics combines both animal and environmental ethics (Peterson, 2013). Others emphasize interlaced "spheres" of ethical concern for animals, the environment, and society (Plumwood, 2000). We take the ecofeminist critique to heart. We see people, animals, and nature as three distinct yet interacting and interdependent spheres of concern. Humans are animals too, embodied and embedded in a wider natural world. We participate with many other beings in mixed ecological and social communities. We subsequently share a moral ecology, so to speak, among people, other animals, and the nature we all share.

To take an example, house cats are often loved as individual family members. The extraordinary risks some Ukrainians have taken during the Russian invasion to save their companion felines exemplifies this. Yet cats are simultaneously a carnivorous species of mammal (*Felis (sylvestris) catus*), obligate predators who hunt other species of diverse kinds. This has prompted a heated dispute over their ecological effects and our ethical responsibilities in that light. There is even a dispute over which cats are domesticated and which are mutualistic wild beings. The ecological, ethological, and ethical aspects of house cats are quite complex (Bradshaw, 2013; Lynn et al., 2019; Wald & Peterson, 2020)! That we have ethically significant relationships and responsibilities to people, animals, and nature illustrates how all three spheres of ethics are inextricably interlaced. This is true to varying degree for most human-animal interactions.

As authors, we are quite skeptical of discussing environmental ethics as if it is separate from questions of ethics that touch on animals or society. Therefore, we think in terms of pan-ethics—ethics for people, animals, and nature. How then shall we discuss ethics and how it applies to wild lives? One way is to sidestep unproductive theoretical disputes, and instead focus on the major families of ethical reasoning and the cardinal concepts used in ethical thinking about animals and nature. This helps us cut through the noise of narrow moral theories, while identifying the building blocks most ethical theories have in common. In technical terms, we are taking a step back from the epistemological details of discrete theories, and instead focusing on the meta-ethical frameworks and conceptual tools that tie the theories together.

## Families of Ethics

The philosopher Ludwig Wittgenstein (1889–1951) developed the idea of a "family resemblance" between various kinds of philosophical thought. Here, family is a metaphor that references the physical features of biological families. While these resemblances are real, they also admit of great variety, are constantly shifting, and yet persist over time. Facial resemblances, recessive traits and other heritable features may ebb and flow but are nonetheless traceable through a family over time. Families do not have a shared essence *per se* (i.e., look or think exactly alike), nor do they conform to a typology (i.e., have a stable list of traits and subtraits). Instead, they have a collection of overlapping traits, the sum total of which is shared through shifting combinations (Eco, 1995; Wittgenstein, 2001).

A similar reasoning is applied by Ernst Mayr (1904–2005) with respect to species. As a taxon of biological organisms, species do not have a single essence (as thought by Aristotle) nor a common set of traits and subtraits that fit into a neat typology. Instead, species are polytypic, sharing in a bundle of features even while they may differ, within as individuals and between as populations. According to Mayr, it is the relative differences among the bundle of genes and morphologies that distinguish one species and subspecies from the next (Mayr, 1982).

Ethical theories are much the same. Within the bundles of concepts, distinctions, values, and worldviews are a set of family relationships that can help us make sense of and identify their key features. There are two main family lineages of ethical theory, the axiomatic and the interpretive. This is a distinction originally outlined by Albert Jonsen and Stephen Toulmin who along with other interpretive ethicists guide our analysis (Bernstein, 1991; Jonsen & Toulmin, 1988; Midgley, 1993; Weston, 2006).

As the name suggests, *axiomatic theories* of ethics treat their principles as axioms—universal moral truths derived from reason or revelation. These theories tend toward monolithic understandings of our ethical relationships and responsibilities and is applied as a rule for reasoning to deduce the correct moral position to take or course of action to follow. These theories also tend to be siloed from one another. One applies the same theory repeatedly in all cases, irrespective of the circumstances. An ethical theory becomes a hammer, and all the world is its nail. Examples include utilitarian (consequence-focused) and deontological (right-based) animal ethics (e.g., Singer, 1975; Regan, 1983).

*Interpretive theories* undertake ethics in a different way. They conceive of ethical principles not as universal truths, but as rules of thumb. Instead of grasping tightly to a single absolute truth, this tradition is pluralistic and welcomes the insights of diverse moral principles. No one principle is taken to have the final word. Rather than applying an *a priori* truth heedless of context, interpretive ethics seeks to match real-world cases with those principles that best reveal the moral issues at stake and provide guidance on what to do about them. The ethics of care and multispecies justice are two examples of interpretive ethical theories (Peterson, 2013; Santiago-Ávila & Lynn, 2020).

To be sure, the axiomatic approach emphasizes a rigorous use of deductive logic. If we think of ethics and other philosophical issues as conceptual puzzles, then much can be learned about the assumptions, argumentative rigor, and implications of our theorizing. For this reason, the interpretive family honors the insights of axiomatic theories and uses their principles when warranted. Yet the axiomatic family is insensitive to context, and this undermines its benefits when thinking about real-world issues regarding wild lives.

For the interpretive family, context is indispensable to ethical reasoning. This context may be situational or relational, featuring differing sites and situations with various ecological or social features. Interpretive ethics therefore adjusts its reasoning to the requirements of different cases. Moreover, in examining specific cases it does not generalize from specific cases to the universe of moral problems. This is the cardinal error of overgeneralizing in case-based research. Rather, the moral insights applicable to or generated from specific cases are used to sharpen interpretive ethical theory so it is better prepared for the next case of a moral problem.

## Cardinal Concepts

Despite their differences, both the axiomatic and interpretive share cardinal concepts. By *cardinal* we are referencing the root meaning of the Latin *cardo* or "hinge." Cardinal concepts are especially important concepts upon which the meaning and application of ethics pivot. Cardinal concepts are used as tools to create the principles, distinctions, and arguments of ethical theory. The concepts below are not the only cardinal concepts of ethics as they apply to animals, but they are some of the most common and important.

*Absolute and Relative Dismissal.* One commonplace concept in discussions about wildlife is the distinction between absolute and relative dismissal (Midgley, 1998). *Absolute dismissal* is an outright rejection of moral consideration for non-human beings. *Relative dismissal* is less absolutist. Moral consideration is partially acknowledged but the interests of humans are firmly placed before that of animals. Midgley critiques these dismissals by noting that they frequently presume non-existent conflicts between animal and human wellbeing or miss how a moral consideration of animals helps protect the wellbeing of people, animals, and nature. The roots of pandemic disease in human depredations on wildlife and wildlands are a case in point.

*Equal Consideration.* Giving equal consideration to the interests, rights, or wellbeing of nonhuman animals is one way to counter the problem of absolute and relative dismissal. This concept serves as a principle in Peter Singer's utilitarian animal ethics but is more broadly applicable in other moral theories (Singer, 1993). Significantly, giving equal consideration does not imply equal treatment. While we may give due regard to the interest, rights, or wellbeing of other animals, this does not mean we treat humans and other animals in exactly the same way. Rather, we treat them in ways appropriate to their individuality, their circumstances and their species' capacities. While dogs cannot vote like humans, they can lead lives of agency and fulfillment in a way that is recognizable in terms of human understandings of agency and flourishing. Wild lives have a wellbeing that can be helped or harmed, and we have an ethical responsibility to give their wellbeing equal consideration with our own.

*Moral Value.* Another point of departure is that of moral value. Moral value is the keystone concept for all ethics, as entities (e.g., persons, beings, communities, systems) without moral value do not require moral consideration. Within this concept is the distinction between *intrinsic* and *extrinsic value*. Those persons, beings, communities, and systems with intrinsic value matter in and of themselves, irrespective of their use or the interests of others. In contrast, to have only extrinsic value (a.k.a. instrumental value) means something does not matter ethically in and of itself, and its value lies only in its use or interest to others (Rolston, III, 1988).

For example, from an extrinsic value point of view, a pine tree may be of value to the bear scratching her back or the human logging the tree for saw boards. Still, the tree does not have value in and of itself. Contrast that with how we humans regard ourselves. We are intrinsically valuable while the bear and tree are mere resources for our use and thus of extrinsic value. Such a view is contestable but illustrates the contrast between extrinsic and intrinsic value.

Whether or not one has intrinsic or extrinsic value determines whether one is entitled to moral consideration (a.k.a., moral considerability). If wildlife stands within a circle of moral concern, they matter in and of themselves and are part of a more-than-human moral community. If they are outside the circle, they are not part of the moral community and will be thought of only in terms of extrinsic values. What this means in practice is that wildlife excluded from moral considerability is exploited, used as natural resources, and harmed for human sport with only an eye for human benefits.

*Value Paradigms.* Worldviews of who or what has intrinsic value are known as *value paradigms* (Lynn, 1998). In more formal language, they are often called axiologies—theories of value — that complement epistemologies (theories of knowledge) and ontologies (theories of existence). The dominant value paradigm the world over is *anthropocentrism*. It centers intrinsic value only or mostly in human beings, whether individually or collectively. Routinely invoking absolute or relative dismissal of animals (and nature), it renders wildlife as valuable insofar as they serve human wellbeing. Recent discussions within critical animal studies and critical race theory have argued for a conceptualization of anthropocentrism as historically constituted of interlocking prejudices and oppressions against nonhumans and humans alike. Instead of privileging the entire human species, anthropocentrism privileges a particular class of beings that share in qualities historically considered quintessentially human. Being white, male, able-bodied and rational is one cogent example (Calarco, 2011).

There are a variety of alternative paradigms to anthropocentrism, collectively termed *non-anthropocentrism*. One that focuses on individual living beings beyond humanity is called *biocentrism*. All living beings or only highly cognitive ones may be its focus. The reverence for life manifest in the writings of Albert Schweitzer is an example of a biocentric ethic (Schweitzer, 1987). *Ecocentrism* turns biocentrism on its head and locates intrinsic value only in ecological wholes such as populations, species and ecosystems. When it comes to animals, a telling and contradictory exception is made by ecocentrism for individual human beings. This again reflects a variation on the concept of absolute and relative dismissal. In this case the dismissal is for wildlife as individuals, whether complete or partial. The ecocentric wholism of J. Baird Callicot is an example of ecocentrism (Callicott, 1997). *Geocentrism* differs from all the foregoing value paradigms. In this case, intrinsic value is understood as manifesting in diverse ways through people, animals, and nature. This includes individuals and the ecological and social communities of which they are a part. A geocentric approach to intrinsic value also scales from individuals through ecological and social wholes (Lynn, 1998).

*Consequences.* A common claim of many ethicists is that the consequences of one's individual or collective action are the standard by which to judge what is ethical. Here the idea is that the best thing to do is what does the most good. It features prominently when discussing the effects of human actions (e.g., hunting, habitat destruction) on the wellbeing of animals as both individuals and groups. Ethical theories that primarily look to outcomes are termed *consequentialist*. There is a wide variety of outcomes featured in different consequentialist theories. Hedonistic versions like Jeremy Bentham's look to happiness, and how pain or pleasure affects that happiness. In contrast, rule-based consequentialism focuses not on actions *per se*, but on normative rules which guide those actions toward producing the good (Hallgarth, 1998). These are but two of the options in a very popular way to think about human interactions with wildlife.

*Rights and Duties*. The idea that moral rights and correlative duties are the standard of ethics is one alternative to consequentialism. These are called deontological theories (Hallgarth, 1998). Instead of focusing on the outcomes of an action, deontological ethicists believe that it is the character of the conduct itself—doing the right thing irrespective of the consequences—that tells us whether an action is right or wrong. Like consequentialism, what makes for ideal conduct varies. Divine commands are a commonly cited source of one's duties, as when the Judeo-Christian God commanded Moses to "go forth and subdue the earth"—a justification invoked for human supremacy over all animals and nature. In contrast, Tom Regan claimed that animals who are "subjects of a life" have inherent value and thus right (Regan, 1983). As a result, humans have explicit duties to respect them.

*Virtue*. In virtue-oriented ethics, what is good does not depend on the outcomes or the character of one's conduct. It depends on the goodness of one's intentions. This does not mean that virtue ethics is entirely insensitive to outcomes or conduct, only that intentions remain the focus of ethical analysis (Louden, 1998). Further, virtues are dispositions to act well in specific circumstances, and virtue ethics emphasizes the building of character to make the right ethical choices when faced with moral problems. The seven virtues of wisdom, courage, temperance, justice, charity, love, and faith are an amalgam of classical Stoic and medieval Christian virtues. When it comes to the wellbeing of wildlife, the virtue of compassion for other beings is strongly emphasized as a counterweight to consequentialist reasoning that sacrifices animal lives for human ends (Santiago-Ávila & Lynn, 2020).

*Care*. Care became a cardinal concept of ethics in the 20th century and is particularly associated with the ethics of wildlife and wildlands through ecofeminism. Caring in this sense is founded on empathy and concern for the wellbeing of others. In feminist circles where the ethics of care is a central development, the impartiality of ethics assumed by consequentialist and deontological ethics is criticized as violative of the context-dependent relationality that characterizes our relationship to other people, animals, and nature (Donovan & Adams, 2007). This does not restrict care to individuals per se, but rather features an enriched sense of how personal, social, and ecological relationships open our eyes to the wellbeing of others and our responsibilities in that light. Ecofeminist approaches to wildlife are often founded on an ethic of care.

*Sentientism*. In contemporary debates over animals, *sentientism* is an increasingly important variation on biocentrism. It locates all or most intrinsic value in conscious individual beings. Sentientism often allies itself with hedonistic consequentialism to emphasize animals' experience of pain, pleasure and associated emotions. As a result, the alleviation of animal suffering is a primary goal of sentientists. To paraphrase Jeremy Bentham, with respect to animals, it doesn't matter if they can speak or reason, only that they can suffer (Bentham, 1789).

What this means for wildlife is that only those individual creatures who are deemed sentient have intrinsic value and are members of our moral community, while ecological relationships and social communities lack intrinsic value. Combined with the belief that the lives of sentient individuals are of no greater or lesser value; all sentient life is fungible. There is no ethical consideration given to whether an individual or species is native or non-native, rare or common, wild or domesticated. This is a profound challenge to the established modes of conservation, wildlife management and sustainability (Lynn, 2020).

*Personhood.* While attending to sentience is a necessary condition for ethics, many argue it is neither sufficient on its own nor always appropriate when determining our ethical responsibilities to wild lives. Ethologically, many animals have a complex consciousness that is not only sentient (aware), but sapient (self-aware), and social (relational). This applies to many taxa (Bekoff et al., 2002). The wellbeing of sapient and social creatures thus requires a distinct ethical reasoning that incorporates yet transcends sentience alone. Personhood is one concept that captures the gestalt of the many capacities of nonhuman animals that are morally relevant to our consideration of their wellbeing (Andrews et al., 2018).

## Animal Wellbeing in Practice

Strategies of conservation and animal protection that address the wellbeing of individuals have emerged in mixed-community situations. This has occurred partly in response to a recognized need to overcome the domination of other animals by humans, and as an appreciation of the integrated social lives of human and other animals, wherein manipulative power structures affect vulnerable populations of humans and nonhumans alike (Baker & Winkler, 2020). A case in point is that of Asian elephant and the human lives in countries where elephants geographically range.

Captive elephants are largely ignored in conservation strategies. Instead, most efforts on improving their lives focus on their welfare in captivity (Baker & Winkler, 2020). Improving the lives of captive elephants is undeniably important in and of itself, as well as considering how deeply embedded these elephants are in the economic and cultural traditions of many rangeland countries, and the unlikelihood that all elephants will live outside of human ownership and management (Baker & Winkler, 2020).

In Thailand alone, the population of legally designated captive elephants is estimated at 3,000–4,500, which is more than the number of legally designated wild elephants of 1,600–3,700. Importantly, these two legally distinct groups are governed by different laws, with real-world consequences for treatment and protection (Baker & Winkler, 2020). The Thai Wild Animal Reservation and Protection Act (WARPA, 1992) defines wild elephants as an endangered species and bans their killing, capture, and trade of their parts; whereas owned (captive) elephants fall under the 1939 Draught Animal Act (or Beast of Burden Act), which regulates the sale and trade of elephants as livestock. A telling loophole is that any elephant poached from the wild or an elephant fathered by a wild bull but born to a captive female, falls under the control of the Draught Animal Act (Baker & Winkler, 2020).

In Thailand, the valuation of captive elephants has reached an all-time high and recent sales indicate that the price may be as high as 60,000 USD. Yet the fine for poaching a wild elephant is only 4% at most of the current market price (Baker & Winkler, 2020; Nijman, 2014). This creates a system of perverse incentives for elephant abuse and trafficking that, through the explosion of the elephant tourism industry, has exacerbated severe welfare and conservation problems.

That there are captive and wild elephant populations is principally a product of legal and societal drivers, rather than anything biological. We know captive elephants are one and the same species as their wild counterparts. They retain quintessential capacities to forage native foods in native habitat, to form new social groups, navigate new terrain, and even integrate into wild populations at large (Baker & Winkler, 2020). Asian elephants are liminal creatures in this sense. This raises possibilities about the rewilding of Asian elephants to improve their wellbeing, and one such project is underway amongst certain Karen communities of northern Thailand, in collaboration with Mahouts Elephant Foundation and their Director of Research, Liv Baker.

Rewilding has conventionally been approached as a way to restore ecological function to an ecosystem. Increasingly utilized as a conservation practice across the globe, it generates both praise for its contributions to long-term sustainability as well as opposition for its feared impacts on human development. When framed in terms of the 3Cs—carnivores, cores, and corridors—it has sought to use apex predators over spatially extensive landscapes to drive trophic processes that promote biodiversity (Soulé & Noss, 1998). Yet the animals involved are treated as a biological means to an ecological end, and their wellbeing is given little direct attention. The mass starvation of animals at the Oostvaardersplassen (OVP) rewilding experiment in The Netherlands was a horrific example. OVP underscored calls for a rewilding with compassion that cares for the wellbeing of all animals affected by the practice, something framed in terms of the 4Cs—carnivores, cores, corridors, and compassion (Kopnina & Cryer, 2019; Lynn, 2020).

The rewilding of Asian elephants in collaboration with Karen people is a more evolved example of this. Here rewilding is viewed as a psycho-ecological and sociocultural event with the power to benefit the wellbeing of Asian elephants by increasing their accessibility—both ecological and psychological—to "wild" states (Baker & Winkler, 2020; Baker, 2020). Preliminary rewilding efforts are showing that elephants,

> including those who had been working in intensively managed tourist camps, can learn to navigate space and terrain, to forage for themselves on a host of native plant species, and to form natural social groupings[...], the behavioral repertoire and activity budgets of rewilded elephants appear to be similar to those of their wild counterparts and stereotypic behaviors born of captivity, restricted space and activity begin to abate
> 
> *(Baker & Winkler, 2020)*

Moreover, increased autonomy for the elephants can be facilitated by positive guardianship of experienced, knowledgeable Karen (traditional) mahouts (Baker et al., unpublished data). The conservation efforts that are most likely to be successful for all, are those that recognize that the wellbeing of humans and of other animals are inextricably intertwined (Baker & Winkler, 2020).

## Questions

- How would shifting from welfare to wellbeing change how you think about human-animal interactions? Your personal and professional relationships with animals?
- What did ethics mean to you before you read this chapter? How might the distinction between axiomatic and interpretive families of ethical theory reinforce or change how you think about your ethics?
- Which of the cardinal concepts of ethics do you use most when you think about or interact with animals? Do you use other cardinal concepts when you think about people and nature?
- Which cardinal concepts do you see reflected in or deployed to understand the relationship between Asian elephants and the Karen community?
- Some cats, Asian elephants, and mongooses my inhabit a liminal space in human-animal interaction, being neither domesticated nor entirely wild in the binary way these terms are typically used. What does that mean for the ethics of wild lives?

## Acknowledgements

A special note of appreciation to our handling editor, Jose M. Peralta DVM, PhD. His patience and editorial guidance were greatly appreciated.

## Note

1 Since humans are biological and social animals too, we use term like "animals," "other animals," and the like to refer to nonhuman animals. We use the terms "wildlife," wild animals," and "wild lives" for those animals not domesticated by human beings.

## References

Andrews, K., Comstock, G. L., Crozier, G. K. D., Donaldson, S., Fenton, A., John, T. M., Johnson, L. S. M., Jones, R. C., Kymlicka, W., & Meynell, L. (2018). *Chimpanzee Rights: The Philosophers' Brief*. Routledge.
Baker, L. (2016). What's the common sense of just some improvement of some welfare for some animals. *Animal Sentience*, 1(7), 3. https://www.wellbeingintlstudiesrepository.org/cgi/viewcontent.cgi?article=1079&context=animsent
Baker, L. (2020). Rewilding and mixed-community collaboration in conservation. *Animal Sentience*, 28(21). https://www.wellbeingintlstudiesrepository.org/animsent/vol5/iss28/21/
Baker, L., & Winkler, B. (2020). Asian Elephants rescue, rehabilitation and rewilding. *Animal Sentience*, 28(1). https://doi.org/10.3389/fevo.2022.782840
Bar-On, Y. M., Phillips, R., & Milo, R. (2018). The biomass distribution on earth. *Proceedings of the National Academy of Science of the United States of America*, 115(25), 2–6. https://doi.org/10.1073/pnas.1711842115
Bekoff, M., Allen, C., & Burghardt, G. (Eds.). (2002). *The Cognitive Animal: Empirical and Theoretical Perspectives on Animal Cognition*. MIT Press.
Bentham, J. (1789). *Introduction to the Principles of Morals and Legislation*. T. Payne and Sons.
Bernstein, R. J. (1991). *Beyond Objectivism and Relativism: Science, Hermeneutics and Praxis*. University of Pennsylvania Press.
Bradshaw, J. (2013). *Cat Sense: How the New Feline Science Can Make You a Better Friend to Your Pet*. Basic Books.
Brambell, F. W. R. (1965). *Report of the Technical Committee to Enquire into the Welfare of Animals Kept Under Intensive Livestock Husbandry Systems (The Brambell Report)*. Her Majesty's Stationary Office.
Calarco, M. (2011). Identity, difference, indistinction. *The New Centennial Review*, 11(2), 41–60. https://www.jstor.org/stable/41949742
Callicott, J. B. (1997). Conservation values and ethics. In G. K. Meffe & C. R. Carroll (Eds.), *Principles of Conservation Biology* (Second ed., pp. 29–55). Sinauer Associates.
Crist, E. (2004). Against the social construction of nature and wilderness. *Environmental Ethics*, 26, 5–24.
DeMello, M. (Ed.). (2010). *Teaching the Animal: Human-Animal Studies Across the Disciplines*. Lantern Books.
Donovan, J., & Adams, C. J. (Eds.). (2007). *The Feminist Care Tradition in Animal Ethics*. Columbia University Press.
Eco, U. (1995). Ur-fascism. *The New York Review of Books*. https://www.nybooks.com/articles/1995/06/22/ur-fascism/
Fraser, D. (2008). *Understanding Animal Welfare: The Science in its Cultural Context*. Blackwell.
Gehrt, S. D., Anchor, C., & White, L. A. (2009). Home range and landscape use of coyotes in a metropolitan landscape: Conflict or coexistence. *Journal of Mammalogy*, 90(5), 1045–1057. https://doi.org/10.1644/08-MAMM-A-277.1
Gilbert, J. H., David, P., Price, M. W., & Oren, J. (2022). Ojibwe perspectives toward proper Wolf Stewardship and Wisconsin's February 2021 Wolf Hunting Season. *Frontiers in Ecology and Evolution*, 257. https://doi.org/10.3389/fevo.2022.782840

Gregory, J., Stouffer, R., Molina, M., Chidthaisong, A., Solomon, S., Raga, G., Friedlingstein, P., Bindoff, N., Treut, H., Rusticucci, M., Lohmann, U., Mote, P., Randall, D., Christensen, J., Hoskins, B., Stocker, T., Manning, M., Denman, K., Lemke, P., … Stone, D. (2021). Climate change 2021—The physical science basis. *Chemistry International*. https://doi.org/10.1515/ci-2021-0407

Hallgarth, M. (1998). Consequentialism and deontology. In Ruth Chadwick et al. (Ed.), *Encyclopedia of Applied Ethics* (Vol. 1, pp. 609–621). Academic Press.

Jonsen, A. R., & Toulmin, S. (1988). *The Abuse of Casuistry: A History of Moral Reasoning*. University of California Press.

Kolbert, E. (2014). *The Sixth Extinction*. Henry Holt and Company.

Kopnina, H., & Cryer, P. (2019). The golden rules of rewilding: Examining the case of Oostvaardersplassen. *Ecos*, *40*(6), 1–19. https://www.ecos.org.uk/ecos-406-the-golden-rules-of-rewilding-examining-the-case-of-oostvaardersplassen/

Lodrick, D. O. (1982). Man and mongoose in Indian culture. *Anthropos*, *77*(1/2), 191–214. https://www.jstor.org/stable/40460438

Louden, R. (1998). Virtue ethics. In Ruth Chadwick et al. (Ed.), *Encyclopedia of Applied Ethics* (Vol. 4, pp. 491–498). Academic Press.

Lynn, W. S. (1998). Contested moralities: Animals and moral value in the dear/Symanski debate. *Ethics, Place and Environment*, *1*(2), 223–242.

Lynn, W. S. (2020). Rewilding the covenant of life with compassion: A future for global and sustainability ethics. In P. Burdon, K. Bosselmann, & K. Engel (Eds.), *The Crisis in Global Ethics and the Future of Global Governance: Fulfilling the Promise of the Earth Charter Debate* (pp. 225–245). Edward Elgar Publishing.

Lynn, W. S., Santiago-Ávila, F. J., Lindenmayer, J., Hadidian, J., Wallach, A., & King, B. J. (2019). A moral panic over cats. *Conservation Biology*, *33*(4), 768–776. https://doi.org/10.1111/cobi.13346

Marsh, G. P. (1965). *Man and Nature, or, Physical Geography as Modified by Human Action*. Belknap Press. Originally published in 1864.

Mayr, E. (1982). *The Growth of Biological Thought: Diversity, Evolution, and Inheritance*. Harvard University Press.

Mellor, D. J., & Reid, C. S. W. (1994). Concepts of animal well-being and predicting the impact of procedures on experimental animals. https://www.wellbeingintlstudiesrepository.org/cgi/viewcontent.cgi?article=1006&context=exprawel

Mendl, M., Burman, O. H. P., & Paul, E. S. (2010). An integrative and functional framework for the study of animal emotion and mood. *Proceedings of the Royal Society B: Biological Sciences*, *277*(1696), 2895–2904. https://doi.org/10.1098/rspb.2010.0303

Midgley, M. (1993). *Can't We Make Moral Judgements?* St. Martin's Press.

Midgley, M. (1995). *Beast and Man: The Roots of Human Nature*. Routledge.

Midgley, M. (1996). Is a dolphin a person? In *Utopias, Dolphins and Computers: Problems of Philosophical Plumbing* (pp. 107–117). Routledge.

Midgley, M. (1998). *Animals and Why They Matter* (Reissue ed.). University of Georgia Press.

Midgley, M. (2007). *Earthy Realism: The Meaning of Gaia*. Imprint Academic.

Nijman, V. (2014). An assessment of the live elephant trade in Thailand. *Traffic International*, 1–38.

Peterson, A. L. (2013). *Being Animal: Beasts and Boundaries in Nature Ethics*. Columbia University Press.

Plato. (1987). *The Republic* (D. Lee, Trans. Second revised ed.). Penguin.

Plumwood, V. (2000). Ecological ethics from rights to recognition: Multiple spheres of justice for humans, animals and nature. In N. Low (Ed.), *Global Ethics and the Environment* (pp. 188–212). Routledge.

Rault, J.-L., Hintze, S., Camerlink, I., & Yee, J. R. (2020). Positive welfare and the like: Distinct views and a proposed framework. *Frontiers in Veterinary Science*, 370. https://doi.org/10.3389/fvets.2020.00370

Regan, T. (1983). *The Case for Animal Rights*. University of California Press.

Rodman, J. (1974). The Dolphin papers. *North American Review*, Spring, 13–26.

Rodman, J. (1977). The liberation of nature? *Inquiry*, *20*(1), 83–145.

Rolston, H., III. (1988). *Environmental Ethics: Duties to and Values in the Natural World*. Temple University Press.

Santiago-Ávila, F. J., & Lynn, W. S. (2020). Bridging compassion and justice in conservation ethics. *Biological Conservation*, *248*, 108648. https://doi.org/10.1016/j.biocon.2020.108648

Schweitzer, A. (1987). The ethics of reverence for life. In *The Philosophy of Civilization* (pp. 307–329). Prometheus Books.
Singer, P. (1975). *Animal Liberation: A New Ethics for Our Treatment of Animals.* Avon Books.
Singer, P. (1993). *Practical Ethics* (Second ed.). Cambridge University Press.
Soulé, M. E., & Noss, R. (1998). Rewilding and biodiversity: Complementary goals for continental conservation. *Wild Earth, 8*(3), 19–28.
Toulmin, S. (1992). *Cosmopolis: The Hidden Agenda of Modernity.* University of Chicago Press.
Wald, D. M., & Peterson, A. L. (2020). *Cats and Conservationists: The Debate over Who Owns the Outdoors.* Purdue University Press.
WARPA. (1992). *Wild Animal Reservation and Protection Act B.E. 2535.*
Webber, S., Cobb, M. L., & Coe, J. (2022). Welfare through competence: A framework for animal-centric technology design. *Frontiers in Veterinary Science.* https://doi.org/10.3389/fvets.2022.885973
Webster, J. (2008). *Animal Welfare: Limping towards Eden: A Practical Approach to Redressing the Problem of our Dominion over the Animals.* Wiley-Blackwell.
Wemelsfelder, F., Hunter, T. E. A., Mendl, M. T., & Lawrence, A. B. (2001). Assessing the 'whole animal': A free choice profiling approach. *Animal Behaviour, 62*(2), 209–220. https://www.wellbeingintlstudiesrepository.org/cgi/viewcontent.cgi?article=1096&context=acwp_asie
Weston, A. (2006). *A Practical Companion to Ethics* (Third ed.). Oxford University Press.
Wittgenstein, L. (2001). *Philosophical Investigations.* Wiley-Blackwell.
World Health Organization. (1946). *Constitution of the World Health Organization.* https://www.who.int/about/governance/constitution

# 31
# TOURISM AND ITS IMPACTS ON WILDLIFE

*Neil Carr and Paul Tully*

## Introduction

Tourism is a human-built, human-driven, traditionally solely human-benefitting phenomenon that affects non-human animals. Before the disruption caused by the COVID-19 pandemic, tourism was a US$9.2 trillion global industry (World Travel & Tourism Council, 2021). It directly sustains approximately 330 million jobs worldwide and overlaps with many different industries, creating additional indirect employment opportunities in the process (Tsionas, 2021). However, to talk only of tourism as a commercial powerhouse ignores the point that, at its core, tourism is about human desires and experiences (MacCannell, 2011). The overwhelming majority of people use the tourism industry for a holiday, which is an oversimplified word for millions of diverse trips that can be about beaches, rurality, diving, cities, disasters, mountaineering, roads, sex, backpacking, nature, cruises, death, music, space, and so much more. What is common to all those consumers engaging with the tourism industry is a wish for their desires to be met and the activity to be enjoyable (Carr & Broom, 2018). In short, tourism is about the fulfilment of human pleasures and business profit-making. This tourism takes place in a time and space away from the everyday mundane, as a liminal experience as people travel, temporarily, outside their home environment to engage in, partially or wholly, leisure-oriented activities.

Due to the desires that people want fulfilling and the income that businesses can gain from facilitation of these, tourism activity pulls into its powerful influence social, cultural, environmental, and economic aspects of life (Sharpley, 2009). The tourism industry incorporates whatever is necessary to create and facilitate pleasures and profits. The Rwandan genocide, the Beetles legacy, Route 66, the waterfalls of Iguazú, waxwork figures, Italian cuisine, the Pyramids of Giza, the Petronas Towers, and the fictional world of Harry Potter all provide examples of how the tourism industry manipulates facets of society and the environment for its own benefit and those of its consumers. Given the omni-presence of animals in everyday society, they are predictably affected in this human-centred creation in multiple different ways (Markwell, 2015). It should be of no surprise that tourists and the tourism industry have seen them as attractive, useable, and marketable resources. Less transparently, tourists and the tourism industry often also see animals as unattractive, unusable pests. As well as affecting animals directly, the scale and power of tourism mean it can affect them indirectly as both seek space in the world. Wild animals tangle with tourism activity in multifarious ways (Bertella, 2021), and this chapter considers the impacts this has on their lives. This discussion is situated within the knowledge that animals are sentient beings with associated rights and welfare needs (Carr & Broom, 2018).

The definition of wild animals in tourism is fuzzy, with a lack of a clear distinction between them and domesticated ones. This fuzziness suggests a continuum between the two concepts rather than seeing them as dichotomous (Carr & Young, 2018). Carr and Young (2018) talk of the problematic idea of whether animals need to be in the wild (which leads to the equally problematic question of what is the wild) to be defined as wild animals, and equally how domesticated an animal has to be before it ceases to be wild. The tiger exemplifies this definitional problem, as they can be encountered in a variety of spaces, including the uncontrolled wilderness, managed reserves, and zoos. The degree to which these spaces, and the animals within them, are under human control differs across locations. Consequently, human definitions of the wild also differ.

Linking to the continuum noted by Carr and Young (2018), Cohen (2009) describes tourist encounters with wild animals across four types of settings (fully-natural, semi-natural, semi-contrived, fully-contrived). Each setting has differing links to the idea, or ideal, of the wild. Within this context, the human ideal of wildness, and hence wild animals, is fully natural, beyond human control. It is these spaces and animals that are most venerated and desired by tourists. This highlights the point that the attractiveness of a wild species to tourists is not homogenous. It is, rather, at least partially, linked to where the individual animal is located. The situation is further complicated by recognising that not all species are equally appealing to humans in general and tourists in particular. There are animals that humans find desirable, love, and venerate, some they are ambivalent to, and others they hate. The most desirable tend to be mammals, exotic, intelligent, and have cute and cuddly babies. In contrast, reptiles and insects tend to be hated by a majority (Carr, 2016). Wild animals, and the spaces they live in, are culturally and spatially distinguished, so a British tourist's perception of a wild animal may not be the same view as a Vietnamese one, for example.

Based on the recognition that tourists and, therefore, the tourism industry view different species of wild animal and wild animals in different spaces differently, this chapter explores the negative and positive impacts that tourism has and can have on wildlife. In order to do this, the discussion begins by exploring the multifaceted position of wild animals in tourism before discussing the negative and positive impacts of tourism on wildlife. The chapter then considers the implications of the changing societal perceptions of animal sentience, rights, and welfare for the position of wild animals in tourism. It ends by examining potential futures and poses a variety of questions to challenge readers. It is impossible to detail all the ways wild animals are involved in and impacted by tourism in a single book chapter. Rather than attempt to do so, what is presented here are a series of examples that illustrate the diversity of the intersection between wild animals and tourism. The chapter then deliberately targets the same intersections to highlight not just negative and positive impacts, but also the complexities associated with trying to assess consequences for animal welfare.

## The Positions of Wild Animals in Tourism

Wild animals are drawn into humans' tourism activities in numerous different ways. All of these are managed from the human perspective to bring people the pleasures and profits of tourism. Whilst animals play an active part in these activities, they do so with no understanding of tourism and little, or in most cases no, knowledge of their role requirement (Bertella et al., 2019). Animals may learn behaviours or be trained/encouraged to perform in specific ways, but they never act because it will create touristic benefits. Rather, their participation in tourism always happens due to human action. In this section, the principal ways humans draw animals into tourism activity are explained, and the discussion considers how these scenarios correspond to the creation of benefits.

The largest placement of animals into tourism activity comes via captivity. Captive animal attractions engage hundreds of millions of tourists annually, with the most well-known form being the zoo and its marine-based equivalent, the aquarium. Specific animals like lions, tigers, giraffes, elephants, penguins, giant pandas, and dolphins are considered as the 'box office' attractions that draw in visitors. Yet zoos and aquaria also keep a variety of other animals, such as frogs, snails, ducks, squirrels, and otters, whose more common presence, and mundane appearance in society pushes them into the background of operations (Carr, 2016). Zoos and aquaria exist in a state of tension between their contemporary roles as tourist attractions and conservation orientated institutions (Carr & Cohen, 2011). As attractions, they position animals for the 'convenience-viewing' of humans who benefit from decreased complexities and reduced dangers in structured animal gazing (Knight, 2010: 748). The zoo space removes for people who desire 'wild' animal encounters the need to travel far from their home environment and comfort zone into time-consuming, uncontrolled, and potentially dangerous wild environments with a heightened risk of animal attacks. Instead, they create an accessible, organised, relatively cheap, and sanitised entertainment stage for tourists to enjoy animals on. In this context, the animals are always on display (Wilkinson, 2018), despite the nature of how they are displayed changing in response to evolving views regarding animal welfare. The justification behind this captive animal staging in the 21st century, amid heightened awareness of the history of humans' poor animal treatment, has become about protection and learning. At the heart of zoo strategies, or at least their promotional campaigns, are conservation programs, animal research, and human education (Table 31.1). If this protection and learning is the rationale behind zoo operations today, the visitor entertainment aspect is the means of making it happen. As Lee Durrell, Honorary Director of Durrell Wildlife Conservation Trust, writes regarding operations at Jersey Zoo, "our single main source of income, as for most zoos, arises from our visitors – gate receipts and secondary spend" (Durrell, 2018: 82). Captive animals are positioned in two roles for creating benefits at zoo and aquaria attractions, as an entertaining revenue generator and an inspirational educator.

*Table 31.1* Missions of zoo and aquaria associations

| Association | Mission |
|---|---|
| World Association of Zoos and Aquariums (WAZA, 2021) | WAZA is the voice of a global community of high standard, conservation-based zoos and aquariums and a catalyst for their joint conservation action. |
| British and Irish Association of Zoos and Aquariums (BIAZA, 2021) | Our members are dedicated to achieving the highest standards of animal care, conserving the natural world through research and conservation, and educating and inspiring their visitors. |
| European Association of Zoos and Aquariums (EAZA, 2021) | To facilitate cooperation within the European zoo and aquarium community towards the goals of education, research, and conservation. |

Activities that can see tourists ride elephants, watch dancing monkeys, hold sloths, witness snake charming, and visit crocodile farms exemplify other captive wild animal tourist attractions. This use of captive animals benefits from a contemporary tourist desire for physical animal encounters (Cui, 2020). As such, animals once captured are tamed to make them 'safe' for tourist encounters. For example, Cohen (2012) details how tigers in a Thai temple were taken from their mothers as cubs and harshly treated, thus allowing tourists to stroke and pose with them for a selfie.

The desire to see and interact with animals has also seen a burgeoning wish for encounters in animals' natural habitats, creating a billion-dollar niche within the wildlife tourism sector (Rizzolo, 2021). This niche aims to capitalise on the popularity of charismatic animals amongst the general public, which are predominantly the same as the 'box office' zoo attractions. As such, activities like jeep safaris, cetacean viewing boat trips, and jungle treks see people encroach into animal habitats, potentially prioritising the seeking of tourism gains over animals' lives. Food provisioning to draw wild animals into convenient locations for tourist interaction, exemplified by swim with dolphin (Mann et al., 2018) attractions, is an integral part of the wildlife tourism industry. Such strategies are nothing new. Indeed, Yellowstone National Park was providing food to attract bears to viewing areas from the 1920s until the practice ended in 1942 (Biel, 2006).

While many wild animal encounters occur in a controlled and manipulated manner, they can also happen in completely unstructured/spontaneous ways when tourism activity sees animals and humans sharing space without design or control. Sometimes these unstructured encounters involve tourists trespassing into animal habitats, such as when sharks come across a swimmer. But the majority involve those animals who live wild in anthropogenic environments, as exemplified by the seagulls who congregate in seaside resorts, the feral hogs who roam urban parks and the cockroaches who scuttle around hotel rooms. In such instances, the wild animal may be transformed from desirable into a pest in the eyes of humans. Thus, encounters with them are viewed as being detrimental to tourism activity. Animals like these are dealt with to protect the tourist experience and destination reputation. Carr and Broom (2018) discuss an airbrushing process that often takes place to maintain pristine and sanitised spaces. Tourists have no wish to deal with insects and other perceived nuisances that innocently share their tourism environments. Consequently, human-viewed pests are erased by the industry.

Dead animals can also be of benefit to tourism. Hunting is a worldwide recreational activity that occurs in diverse ways and generates millions of dollars in tourism revenue (Lovelock, 2015). Tourist hunters benefit from continuing traditions, social bonding, time outdoors, a trophy, and the gaining of meat (Lovelock, 2018). Many hunters take part in the growing and lucrative phenomena of trophy hunting. Here, management stakeholders have commoditised recreational hunting, placing a price on heads of individual species (Cohen, 2014), as illustrated by the $5,000 leopard trophy available in Namibia (Namibia Hunting Safaris, 2021). International trophy hunting facilitates tourist thrills and the accumulation of social kudos whilst management stakeholders gain income (Lovelock, 2018), which is, ironically, not dissimilar to the outcomes associated with photographing wild animals during wildlife tourism.

Killing animals, including wild ones, also contributes to another vast aspect of tourism, namely food and drink. The liminal setting of the tourism experience arguably gives the desire and opportunity to consume the unusual, including wild rather than domesticated animals, to create memorable experiences (Stone et al., 2018). What is more, many destinations have world-renowned, iconic dishes, the thought of which can excite the tourist (Jenkins,

2018). As a result, tourists seek out local treats in places, such as kangaroo burgers in Australia, civet coffee in Indonesia, lobster in Nova Scotia, and whitebait in New Zealand. These culinary experiences give tourists a sense of unique locality, authenticity, and novelty (Stone et al., 2018). These are also experiences that can bring significant economic benefits back into areas and foster local pride in traditional food offerings (Zhang et al., 2019), helping to ensure the continuation of the cultures associated with them in the process (Chitrakar et al., 2020). Therefore, it is no surprise that restaurants promote the consumption of exotic meats.

## Tourism and Negative Animal Welfare

The welfare of each animal changes when used by humans for tourism purposes. If an animal copes well in its environments, such as through appropriate social interaction, nutrition, and physical activity, welfare impacts can be positive, whereas adverse conditions will produce negative effects (Broom, 2014). As animals are positioned in tourism impacts on their welfare are guaranteed but what is less sure is whether the outcomes are positive or negative. These potentials are dependent on human choices. Carr and Broom (2018) detail how the welfare of animals revolves around a collection of physical, psychological, and social needs. How the human-centric nature of tourism ignores animals' needs and, thus, causes potential negative impacts on welfare is now explored.

Since the 1970s, people have sought to better the welfare of animals held in zoos and aquariums, largely via improvements in animal enclosures (Cole & Fraser, 2018) and calls for their closure (Safina, 2018). However, many questions remain about the welfare quality experienced by animals in these attractions. Animals' placement in zoos and aquariums can lead to, amongst other consequences, increases in human interaction and decreases in appropriate sociality. As a result, signs of compromised welfare when visitor numbers increase around modern-day 'interactive' enclosures are visible, for example, with Kangaroos (Sherwen et al., 2015). Animals, such as elephants, show behavioural indicators of a similar outcome due to social separation and isolation in zoos (Greco et al., 2016). In addition, placing animals into the zoo industry regularly involves transporting them over long distances (Linhart et al., 2008), which can stress animals during and for multiple days after movement (Dembiec et al., 2004). Even arguments that zoos and aquaria can aid the protection of species ignore the difference between species welfare and the welfare of individual animals, failing to see animals as individuals with unique needs.

Non-zoo attractions based on captivity also have questionable animal welfare conditions. A World Animal Protection (2018) investigation in Indonesia found that none of the 26 attractions studied that provide tourists with the chance to encounter animals like elephants, monkeys, orangutans, turtles, and dolphins, achieved even basic animal needs. Such attractions occur globally, offering encounters that place tourist pleasure and operator profit ahead of animal needs. For instance, riding, patting, and washing elephants have become extremely popular activities and are big business for Thailand's tourism sector, but, behind the tourism stage, captive elephants are found to be held in appalling conditions, subjected to cruel training techniques, and recipients of violent control mechanisms from their mahout (National Geographic, 2019). In Latin America, the growing practice of tourists handling jungle species like snakes and sloths is made possible by abusive techniques to make animals submissive (World Animal Protection, n.d.). Despite the questionable animal welfare issues, the use of captive animals as attractions continues to be a significant part of global tourism.

In theory, living in the wild enables animals to freely perform behaviours that follow personal instincts to meet their needs and wants. This is put at risk, however, thanks to tourism encroachment. Take the example of whale and dolphin watching, which has grown into a billion-dollar industry (Mann et al., 2018). This activity can lead to changes in natural whale behaviour, exemplified by decreases in foraging and resting plus increases in avoidance tactics (Machernis et al., 2018). Furthermore, when this tourism activity goes unregulated (or regulations are flouted), and vessels get too close to enable tourists to get the *ideal* picture, animals are at risk of injury and death (Sitar et al., 2016). Similar problems occur when food provisioning is utilised to benefit the tourists' ability to encounter wild animals as it risks interfering with natural behaviours. For instance, provisioning can negatively impact interactions between adult dolphins and their calves, inhibiting the ability of calves to rest and survive (Foroughirad & Mann, 2013).

The presence of so-called pest animals potentially has undesirable effects on the tourist experience and industry operations, thus risking the pleasure and profit outcomes (Carr & Broom, 2018). Therefore, for a successful (as defined by humans) experience to potentially occur, unwelcome animals are continually removed from tourism environments. Unappealing animals like rats, mice, and cockroaches are eradicated in huge numbers via strategies that risk the pain and suffering of the target animal, interference with adults caring for their young, and non-targeted animals suffering secondary poisoning (Mason & Littin, 2003). This removal not only applies to insects and other pests. For instance, when tourists swim in waters populated by sharks, authorities often employ techniques for human safety, arguably, demonstrating a misunderstanding of shark behaviour. Attempts to remove the shark from the tourism space can be lethal or non-lethal (Niella et al., 2021), yet all impact the wellbeing not only of sharks but also other species who, for example, get caught in nets unintentionally. With the focus on maximising human pleasures and profits, animals we love, desire to see, and interact with can be transformed into pests depending on how humans view their actions and positions within the tourism experience. Examples include bears who are viewed as a majestic attraction in a National Park but also a pest when they seek out humans' food in the park campsites.

Other experiences that involve intentional death are equally problematic for animals' welfare. Trophy hunting, as discussed earlier, is a growing niche tourism sector that predominantly involves tourists shooting an animal with a gun or bow and arrow. In such a scenario, death may not be instantaneous, thus leaving an animal in pain before finally succumbing hours later to either the injuries or a second shot, as was reported as happening with one of the most iconic trophy hunting victims of recent times, Cecil the Lion (Loveridge, 2018). The death of a wild animal via hunting can also have implications for the welfare of other animals and the ecosystem in which they are situated. Cecil, for example, at the time of his death, was both a key habitat predator and head of two lion prides.

Humans' hunting for fun has detrimental effects on animals' physical, psychological, and social lives. Jones and Draper (2018) detail how the development of trophy hunting has seen captive breeding become big business, with animals bred in large numbers and used as a financial resource throughout their development before eventual use in a 'staged' or 'canned' hunt. The captivity experiences of these animals are akin to the worst type of zoo, where the emphasis is on human benefit and there is no concern for the welfare of animals beyond ensuring they live long enough to be profitable. In these scenarios, animals suffer inappropriate living conditions that inhibit natural behaviour, enforced breeding, separation of mothers and cubs, and intense human interaction (Jones & Draper, 2018). The commercial activity of hunting uses animal life as a resource rather than having concern about a sentient being with inherent needs.

Tourists' food and drink experiences also take precedence over animal needs. Take the civet coffee, also known as kopi luwak, experience of Indonesia, for instance, which meets touristic desire for a sense of the local and authenticity (Cohen, 1988). The authentic version comes from coffee beans collected in the wild from the droppings of palm civet, or luwak, who eat the coffee cherries but cannot digest the beans (Wild, 2013). Carder et al. (2016) found that growth in demand for this experience had led to the capturing and caging of wild palm civets, potentially imperilling the authenticity of the coffee (Cohen, 1988) though not the demand. They uncovered, via investigation of 16 tourism-focused coffee plantations, that every civet held experienced compromised welfare. The situation is similar for many traditionally wild animal species who are now increasingly farmed to provide human protein. For example, the tourism sector benefits from the compromised welfare of salmon, farmed in increasing numbers over the last 50 years (Olesen, Myhr & Rosendal, 2011). Such tourism situations are inseparable from the issues between animal welfare and farming that are present in society.

In all the touristic entanglements with animals discussed in this section, the human outcomes, be it the tourists' desire and experience or business profits or, in most cases, both, take primacy over animal needs. Hence, the potential for negative animal welfare proves inseparable from human-centric tourism activity.

## Tourism and Positive Animal Welfare

As Carr and Broom (2018) have said, there are many bad zoos whose concern for animal welfare, as noted above, can be minimal. However, there are also good zoos that emphasise animal welfare. This is not to say that those good zoos do not experience pressure to provide visitors with an attractive experience and the ability to get up close to animals. Rather, while labouring under the economic reality that these zoos, just like others, must generate income through tourist visitations to remain viable, they do so primarily to ensure animal wellbeing rather than for human benefit. Rather than thinking in binary terms, we need to recognise that zoos exist somewhere along a continuum between good and bad regarding animal welfare. In all zoos, it is necessary to look beyond preconceived ideas to understand fully what is going on. One example of this is the orangutan enclosure in Jersey Zoo. On first examination, the enclosure is set up to enable the viewing of these animals by the paying public, albeit in a setting that allows the orangutans to express their natural behaviour, as depicted in Figure 31.1. However, Neil was lucky to talk with Lee Durrell, the widow of the founder of the Zoo, Gerald Durrell, who pointed out that the children's play area in the zoo was deliberately situated to enable the orangutans to see it and be entertained by what went on in it. In this action, we see a focus on the stimulation of the animals rather than a presentation of entertainment for humans, albeit in an innovative way that meets dual needs. Thus, hidden under the surface is an emphasis on animal welfare.

As well as being concerned with individual animal wellbeing, zoos are also increasingly concerned with species survival. Good zoos, of which it is possible to identify Jersey Zoo as one while recognising that *good* is not the same as *perfect*, have been responsible for helping to increase animal populations in the world, including species that have little public appeal (Carr & Cohen, 2011). Zoos have also sold themselves and been identified as vehicles to cultivate an appreciation of wild animals and hence a commitment to help ensure their survival among the general population. The success of such a process is, however, still a matter of strong debate (Carr & Cohen, 2011).

Finally, there is the emotive debate surrounding whether animals should be caged in zoos as a form of Ark to ensure species survival while those in the wild and their habitats are destroyed. The most extreme animal rights position sees this as being contrary to the rights of

*Figure 31.1* Orangutan enclosure in Jersey Zoo, UK. (Photograph from Neil Carr)

animals and argues against it. Does such an argument imperil the survival of species? This requires consideration from an animal rather than a human perspective. So, to an animal, what is a cage? What is freedom? What is the wild? Here we must recognise that how humans view enclosures or cages may be very different to how animals do. There is little argument that placing animals that would normally range across large areas, such as elephants, in small, enclosed spaces is not beneficial to their welfare. However, can the same be said of a spider, for example? By removing the opportunity to hunt, do we adversely affect the welfare of a carnivore such as a tiger? Yet, at the same time, do we negatively impact the welfare of a species, such as a meerkat, by removing it from the sight of those who naturally prey on it?

Looking beyond zoos and aquaria, elephant attractions in countries such as Thailand have rightly been criticised for their often-negative impacts on animal welfare. However, it is necessary to question if such practices are uniformly found throughout the tourism industry. The answer is that they are not. Instead, as Rizzolo and Bradshaw (2018) have noted, elephant sanctuaries exist where tourists can fulfil their desires to meet these pachyderms while the focus is on the welfare of the animals. The emphasis in these sanctuaries is on developing "a sense of independence, authority, and control over their lives [which] help promote elephant self-determination" (Rizzolo & Bradshaw, 2018: 125). These sanctuaries are not primarily about providing an ethical, animal-centric tourism experience. Rather, they are primarily about animal welfare, rescuing elephants who have suffered abuse at the hands of humans working in tourism or other industries (see The Mahouts Elephants Foundation – https://www.mahouts.org/).

Can animal welfare in the wild be improved or ensured by tourism? A traditional argument surrounding whale-watching tourism is that it places an economic value on whales that discourages their hunting (Hoyt & Hvenegaard, 2002). The implicit argument is that without this value whales would be killed instead. Yet, the validity of this claim is questionable. Whale hunting has not declined because hunters see more value in taking people to view the whales. Instead, declining demand for whale meat and a long-running campaign to change social opinions has driven most countries to abandon whale hunting. However, whale watching experiences may have played a part in this change. This relates to the point that, as in the case of zoos, viewing wildlife in their natural habitats has been identified as an important tool to help educate people about the importance of these animals, in their own right and as integral components of a healthy ecosystem (Packer & Ballantyne, 2012). As in the case of zoos, reports that support and contradict this idea currently exist.

Do all unwanted wild animals suffer as a consequence of tourism? It is important to remember that while most people are attracted to the types of animals identified earlier in this chapter, some find exoticness and attractiveness in animals such as reptiles and insects. In doing so, they can help to turn snakes and crocodiles, both seen as pests by many local inhabitants, into valued tourism attractions. Thus, we begin to see an economic reason for the preservation of these animals, potentially helping to transform them from pests to valued animals. The same process is occurring when local farmers are encouraged to revaluate as tourist attractions carnivores, such as lions and tigers, normally seen as pests due to predation of livestock. These arguments have echoes of the economics of whales noted earlier in this section. While transforming animals from pests to valued components of the tourism industry may ultimately ensure their survival, it does not recognise them as sentient beings with unique welfare needs. Instead, their wellbeing is intimately tied to their value to the tourism industry. In this way, their sentience is irrelevant, and the sustainability of their welfare is dependent on the continuation of tourists being willing to pay to see them. This is, undeniably, a precarious position though it is better than being killed after being labelled a pest.

Can there ever be such a thing as 'ethical hunting'? In other words, can hunting ever be said to consider animal rights and welfare? Lord and Winter (2021) explored this complex question in the context of duck hunting, examining the different perspectives of the hunting community, management bodies, and animal activists. The pro-hunting lobby talks of the ability of hunters to enact a 'clean kill', where animals do not suffer from the pain of a lingering death (Lord & Winter, 2021). Lovelock (2015) also talks of the potential of hunting tourism to be ethical concerning its ability to control introduced species (aka animals defined by humans as pests). Such hunting may also help prevent overpopulation of particular species within an ecosystem or weed out weak, old, and injured animals who may be unlikely to survive the winter.

The argument about protecting the ecosystem through the hunting of introduced animals or where there are too many of a particular species can still be said to hold some validity. This is despite the reality that such practices are generally not aimed at the eradication of these species. After all, some benefit to the ecosystem is, surely, better than none. However, it is important to recognise that, in both instances, we are talking about species-level wellbeing rather than individual rights and welfare. It is also the case that it is humans as those in power deciding which and how many animals should be killed. Humans also choose which animals are transformed from merely being pests to be eradicated by whatever means necessary into worthy adversaries of the hunters, who are to be venerated and hunted using *ethical* techniques. Here we can see the distinction in New Zealand, for example, between possums that are indiscriminately killed using organofluorine toxicants, like the poison 1080, and the thar

Lovelock (2018) talks of killing with a single shot. If we consider hunting through the eyes of the individual animals affected and their rights and welfare, then it becomes infinitely harder to justify tourism hunting of wildlife.

The consumption of meat from wild animals in the tourism experience cannot be said to have any welfare benefit for individual animals. Whether it can even aid species survival through the association of an economic value is questionable. In unregulated markets, the development of such a value is more likely to increase the number of animals killed (e.g., the increasing demand for bush meat). Alternatively, where demand increases, the domestication of a species will possibly occur (e.g., the development of salmon farming). Similar to the cases above, could the witnessing of animals in the wild be a tool to persuade tourists that instead of seeking to eat wild animals they should let them live? Mkono (2015) supports this idea, but notes there is much more work needed to assess its validity.

## Changing Perceptions of Animal Sentience, Rights, and Welfare: Implications for Wildlife Tourism

We are witnessing an increase in people's awareness of animal sentience and, consequently, increasing recognition of their rights and welfare needs, including those identified as 'wild' (Carr & Broom, 2018). This is arguably partially fuelled by and fuelling interest in wildlife tourism. Consequently, the pressures on animal wellbeing noted earlier in this chapter are, ironically, related to our increasing concern for them. At the same time, this increasing concern is driving people to demand wildlife tourism experiences that increasingly take into account the welfare of individual animals alongside the welfare of species.

Growing consumer demand to consider individual animal wellbeing is visible in how providers of wildlife tourism encounters are altering practices, exemplified by the changing way many, though not all, zoos house their animals. These changes increase the focus on animals' welfare as sentient beings as opposed to objects of the human gaze. Likewise, there has been a rise of guidelines surrounding cetacean encounters that have been designed to minimise impacts of tourism. Within National Parks, multiple attempts have also been made to encourage humans to keep their distance from wildlife. The emphasis in such messages is that tourists, whether they be in zoos or the wild, should respect the animals they wish to see, thus witnessing them without adversely affecting them. In addition, tourism companies are increasingly refusing to send tourists to attractions that do not meet their standards of animal welfare. As part of this, the Federation for Tour Operators (FTO), now a part of ABTA (the Association of British Travel Agents), published its *Travelife Animal Attraction Handbook* in 2008. This was followed in 2013 by ABTA's *Global Animal Welfare Guidelines*. In the same year, the tourism operator TUI Group, by its own admittance the world-leading tourism group, with an annual turnover of €19bn (TUI Group, 2022), began requiring all animal attractions it dealt with to conform to the ABTA guidelines (Jenkinson & Felton, 2018). That ABTA's guidelines were the product of cooperation with The Born Free Foundation since 2004 shows what can be achieved by the tourism industry working together with activist organisations (Turner, 2018).

Given a rise in awareness of animal sentience and associated welfare needs, it can be suggested that governments and businesses have reacted to improve the welfare of wild animals, both at the species and individual level. Yet it is important to look at this shift critically, just like studies of environmentally friendly and anti-gender discrimination policies and practices have raised concerns of fem and green-washing (Khoo-Lattimore et al., 2019; Lyon &

Montgomery, 2015). In this way, we must ask whether animal welfare washing is present in the marketing of the wildlife tourism sector. Certainly, zoos have become wise to the need to adapt their image, from one of human entertainment to one of education and conservation (Carr & Cohen, 2011). Likewise, operators offering animal encounters in the wild now value their ability to display their animal welfare credentials to potential customers. The distinction between those providing a truly animal-welfare focused operation and those selling themselves at least partially through such a label while remaining focused on profit and human entertainment (those guilty of animal welfare washing) is important, though difficult to identify.

It is important to recognise that it is not only operators and governments that engage in animal welfare washing. Tourists also do the same. This links to the social kudos to be gained from seeing wild animals close-up and showing evidence of this to the world whilst also publicly displaying support for animal welfare. It is this mix that operators and governments attempt to react to but also drive through an ongoing feedback loop. The result is a constant battle where tourists want to see wild animals up close while at the same time having and wishing to display increasing concerns for their welfare. Herein lies the problem of disentangling authentic concerns and actions for animal wellbeing versus human attempts to profit from displaying such concerns while in reality blithely ignoring them.

All of this speaks to the continued disempowered position of wild animals. Animal welfare is, even in the face of growing concern about it, still only within the powers of humans to bestow on animals. As long as this situation persists, the potential for abuse of wild animals, indeed all animals, will exist.

## Looking to the Future

Animal research in general and, in particular, within tourism studies needs to exist at the boundary between research and activism. Research for research's sake, whereby we push the boundaries of knowledge is interesting in an intellectual sense, but it is only through the transmutation of this knowledge into action, which is where activism comes in, that animal welfare can be improved and ensured. As Carr and Berdychevsky (2021) have called for, it is time, actually beyond time, for scholars to get out of their ivory towers and engage in a meaningful and helpful way with the *real* world. This does not mean academics need to abandon their research to become activists necessarily, but that they need to ensure what they are learning and saying about animal welfare reaches the ears of others to help make things change for the benefit of animals.

Researchers are ideally situated to help drive change by doing what they are good at – research. As identified in this chapter, there are many questions to answer about the wellbeing of wild animals within tourism encounters and how such meetings can be developed to enhance wellbeing. Such work feeds nicely into the multidisciplinary nature of tourism studies. However, there is a need for this multidisciplinarity to expand. Probably not surprisingly, the attention of tourism studies has traditionally been focused within the social sciences and business studies disciplines. Where it has stepped into the natural sciences, its engagement with zoology and biology has tended to be relatively limited. To understand animal wellbeing in relation to tourism, this needs to change. We also need an approach to research and activism that blends human and animal-centric views and thinking in a way that sees beyond the duopoly of welfarism and animal rights activism where human ideals and feelings are dominant (Carr, 2018). The ultimate aim must be a state of co-existence where the rights and welfare of all are respected (Fennell, 2018).

## Potential Questions for Discussion

This chapter has hopefully raised many questions in the minds of the reader.

- How do readers feel about animal sentience, rights, and wellbeing?
- Within this context, readers are encouraged to think about what level of wellbeing they feel is appropriate and its applicability to all wild animals in all contexts. Are we willing to sacrifice our close-up selfie with a quokka for its wellbeing? Will we only visit restaurants that provide vegan dishes? Will we only stay in hotels that pledge to not kill any wild animals within their property, including rats, mice, spiders, and cockroaches?
- Should zoos and aquaria be allowed to continue to exist in the 21st century?
- Can recreational hunting ever be anything other than an unethical blood sport?
- Is the tourism industry guilty of animal welfare washing?

## References

Bertella, G. (2021). Introduction: Welcome to the futures of wildlife tourism. In G. Bertella (Ed.), *Wildlife tourism futures: Encounters with wild, captive and artificial animals* (pp. 1–6). Channel View Publications.

Bertella, G., Fumagalli, M. & Williams-Grey, V. (2019). Wildlife tourism through the co-creation lens. *Tourism Recreation Research*, 44(3), 300–310.

BIAZA. (2021). *Home*. British & Irish Association of Zoos & Aquariums. https://biaza.org.uk/.

Biel, A. W. (2006). *Do (not)feed the bears: The fitful history of wildlife and tourists in Yellowstone*. University Press of Kansas.

Broom, D. (2014). *Sentience and animal welfare*. CABI.

Carder, G., Proctor, H., Schmidt-Burbach, J. & D'Cruze, N. (2016). The animal welfare implications of civet coffee tourism in Bali. *Animal Welfare*, 25(2), 199–205.

Carr, N. (2016). An analysis of zoo visitors' favourite and least favourite animals. *Tourism Management Perspectives*, 20, 70–76.

Carr, N. (2018). Tourist desires, and animal rights and welfare within tourism: A question of obligations. In B. Grimwood, K. Caton, L. Cooke & D. Fennell (Eds.), *New moral natures in tourism* (pp. 116–1130). Routledge.

Carr, N. & Berdychevsky, L. (2021). Delving into the uncharted terrains of sex in tourism: Conclusions and ways forward. In N. Carr & L. Berdychevsky (Eds.), *Sex in tourism: Exploring the light and the dark* (pp. 264–280). Channel View Publications.

Carr, N. & Broom, D. (2018). *Tourism and animal welfare*. CABI.

Carr, N. & Cohen, S. (2011). The public face of zoos: Balancing entertainment, education, and conservation. *Anthrozoos*, 24(2), 175–189.

Carr, N. & Young, J. (2018). Wild animals and leisure: An introduction. In N. Carr & J. Young (Eds.), *Wild animals and leisure: Rights and welfare* (pp. 1–11). Routledge.

Chitrakar, K., Carr, N. & Albrecht, J. (2020). Community-based homestay tourism and its influence on indigenous gastronomic heritage. *Journal of Gastronomy and Tourism*, 4(2), 81–96.

Cohen, E. (1988). Authenticity and commoditization in tourism. *Annals of Tourism Research*, 15, 371–386.

Cohen, E. (2009). The wild and the humanized: Animals in Thai tourism. *Anatolia*, 20(1), 100–118.

Cohen, E. (2012). Tiger tourism: From shooting to petting. *Tourism Recreation Research*, 37(3), 193–204.

Cohen, E. (2014). Recreational hunting: Ethics, experiences and commoditization. *Tourism Recreation Research*, 39(1), 3–17.

Cole, J. & Fraser, D. (2018). Zoo animal welfare: The human dimension. *Journal of Applied Animal Welfare Science*, 21, 49–58.

Cui, Q. (2020). Wildlife tourism in (un)sustainable futures. In G. Bertella (Ed.), *Wildlife tourism futures: Encounters with wild, captive and artificial animals* (pp. 9–23). Channel View Publication.

Dembiec, D. P., Snider, R. J. & Zanella, A. J. (2004). The effects of transport stress on tiger physiology and behaviour. *Zoo Biology, 23*(4), 335–346.

Durrell, L. (2018). A tale of two zoos: Tourism and zoos in the 21st century. In N. Carr & D. Broom (Eds.), *Tourism and animal welfare* (pp. 81–84). CABI.

EAZA. (2021). *About us.* https://www.eaza.net/about-us/.

Fennell, D. (2018). Afterword: On tourism, animals, and suffering – lesson from Aeschylus' *Oresteia*. In C. Kline & J. Rickly (Eds.), *Exploring non-human work in tourism: From beasts of burden to animal ambassadors* (pp. 255–262). De Gruyter Publishers.

Foroughirad, V. & Mann, J. (2013). Long-term impacts of fish provisioning on the behaviour and survival of wild bottlenose dolphins. *Biological Conservation, 160*, 242–249.

Greco, B. J., Meehan, C. L., Hogan, J. N., Leighty, K. A., Mellen, J., Mason, G. J. & Mench, J. A. (2016). The days and nights of zoo elephants: Using epidemiology to better understand stereotypic behaviour of African elephants (*Loxodonta Africana*) and Asian elephants (*Elephas Maximus*) in North American zoos. *PLoS One, 11*(7). https://doi.org/10.1371/journal.pone.0144276

Hoyt, E. & Hvenegaard, G. (2002). A review of whale-watching and whaling with applications for the Caribbean. *Coastal Management, 30*, 381–399.

Jenkins, B. (2018). Who pays for our cheap meat? The impact of modern meat production on slaughterhouse workers: Considerations for tourists. In C. Kline (Ed.), *Tourism experiences and animal consumption* (pp. 42–57). Routledge.

Jenkinson, C. & Felton, H. (2018). Animal welfare – driving improvements in tourism attractions. In N. Carr & D. Broom (Eds.), *Tourism and animal welfare* (pp. 135–137). CABI.

Jones, M. & Draper, C. (2018). Trophy hunting and animal welfare. In A. Butterworth (Ed.), *Animal welfare in a changing world* (pp. 46–56). CABI.

Khoo-Lattimore, C., Yang, E. & Je, J. (2019). Assessing gender representation in knowledge production: A critical analysis of UNWTO's planned events. *Journal of Sustainable Tourism, 27*(7), 920–938.

Knight, J. (2010). The ready-to-view wild monkey: The convenience principle in Japanese wildlife tourism. *Annals of Tourism Research, 37*(3), 744–762.

Linhart, P., Adams, D. B. & Voracek, T. (2008). The international transportation of zoo animals: Conserving biological diversity and protecting animal welfare. *Veterinaria Italiana, 44*(1), 49–57.

Lord, D. & Winter, C. (2021). The contradictory ethics of native duck shooting: Recreation, protection and management. *Annals of Leisure Research.* https://doi.org/10.1080/11745398.2021.1974904

Lovelock, B. (2015). Troubled-shooting: The ethics of helicopter-assisted guided trophy hunting by tourists. In K. Markwell (Ed.), *Animals and tourism: Understanding diverse relationships* (pp. 91–105). Channel View Publications.

Lovelock, B. (2018). Ethical hunting. In N. Carr & D. Broom (Eds.), *Tourism and animal welfare* (pp 147–150). CABI.

Loveridge, A. (2018). *Lion hearted: The life and death of Cecil & the future of Africa's iconic cats.* Simon and Schuster.

Lyon, T. P. & Montgomery, A. W. (2015). Means and end of greenwash. *Organization & Environment, 28*(2), 223–249.

MacCannell, D. (2011). *The ethics of sightseeing.* University of California Press.

Machernis, A. F., Powell, J. R., Engleby, L. & Spradlin, T. R. (2018). An updated literature review examining the impacts of tourism on marine mammals over the last fifteen years (2000–2015) to inform research and management programs. *NOAA Technical Memorandum NMFS SER, 7.* https://doi.org/10.7289/V5/TM-NMFS-SER-7

Mann, J., Senigaglia, V., Jacoby, A. & Bejder, L. (2018). A comparison of tourism and food-provisioning among wild bottlenose dolphins at Monkey Mia and Bunbury, Australia. In N. Carr & D. Broom (Eds.), *Tourism and animal welfare* (pp. 85–96). CABI.

Markwell, K. (2015). Birds, beasts and tourists: Human–animal relationships in tourism. In K Markwell (Ed.), *Animals and tourism: Understanding diverse relationships* (pp. 1–24). Channel View Publications.

Mason, G. J. & Littin, K. E. (2003). The humaneness of rodent pest control. *Animal Welfare, 12*, 1–37.

Mkono, M. (2015). 'Eating the animals you come to see': Tourists' meat-eating discourses in online communicative texts. In K. Markwell (Ed.), *Animals and tourism: Understanding diverse relationships* (pp. 211–226). Channel View Publications.

Namibia Hunting Safaris. (2021). *Hunting leopards.* http://www.namibiahuntingsafaris.com/hunting-namibia/hunting-leopard/#leopard_hunting_package

National Geographic. (2019). *Inside the dark world of captive wildlife tourism* [video]. https://www.youtube.com/watch?v=ITlo2ZBJOWU

Niella, Y., Peddemors, V. M., Green, M., Smoothey, A. F. & Harcourt, R. (2021). A "wicked problem" reconciling human-shark conflict, shark bite mitigation, and threatened species. *Frontiers in Conservation Science, 78*. https://doi.org/10.3389/fcosc.2021.720741

Olesen, I., Myhr, A. I. & Rosendal, G. K. (2011). Sustainable aquaculture: Are we getting there? Ethical perspectives on salmon farming. *Journal of Agricultural and Environmental Ethics, 24*(4), 381–408.

Packer, J. & Ballantyne, R. (2012). Comparing captive and non-captive wildlife tourism. *Annals of Tourism Research, 39*(2), 1242–1245.

Rizzolo, J. B. (2021). Wildlife tourism and consumption. *Journal of Sustainable Tourism*. https://doi.org/10.1080/09669582.2021.1957903

Rizzolo, J. B. & Bradshaw, G. A. (2018). Human leisure/elephant breakdown: Impacts of tourism on Asian elephants. In N. Carr & J. Young (Eds.), *Wild animals and leisure: Rights and welfare* (pp.113–131). Routledge.

Safina, C. (2018). Where are zoos going—or are they gone? *Journal of Applied Animal Welfare Science, 21*, 4–11.

Sharpley, R. (2009). *Tourism development and the environment: Beyond sustainability?* Earthscan.

Sherwen, S. L., Hemsworth, P. H., Butler, K. L., Fanson, K. V. & Magrath, M. J. (2015). Impacts of visitor number on Kangaroos housed in free-range exhibits. *Zoo Biology, 34*(4), 287–295.

Sitar, A., May-Collado, L. J. Wright, A. J., Peters-Burton, E., Rockwood, L. & Parsons, E. C. M. (2016). Boat operators in Bocas del Toro, Panama display low levels of compliance with national whale-watching regulations. *Marine Policy, 68*, 221–228.

Stone, M. J., Soulard, J., Migacz, S. & Wolf, E. (2018). Elements of memorable food, drink, and culinary tourism experiences. *Journal of Travel Research, 57*(8), 1121–1132.

Tsionas, M. G. (2021). COVID-19 and gradual adjustment in the tourism, hospitality, and related industries. *Tourism Economics, 27*(8), 1828–1832.

TUI Group (2022). *About TUI Group*. https://www.tuigroup.com/en-en/about-us/about-tui-group

Turner, D. (2018). Managing tourism's animal footprint. In N. Carr, N. & D. Broom (Eds.), *Tourism and animal welfare* (pp. 106–111). CABI.

WAZA. (2021). About us. World Association of Zoos and Aquariums. https://www.waza.org/about-waza/

Wild, T. (2013, September 13). Civet coffee: Why it's time to cut the crap. *The Guardian*. https://www.theguardian.com/lifeandstyle/wordofmouth/2013/sep/13/civet-coffee-cut-the-crap

Wilkinson, S. (2018). Being Camilla: The 'leisure' life of a captive chameleon. In N. Carr & J. Young (Eds.), *Wild animals and leisure: Rights and welfare* (pp. 96–112). Routledge.

World Animal Protection. (2018). *Bali horror: Wildlife tourist attractions are living hell for animals*. https://www.worldanimalprotection.org.uk/news/bali-horror-wildlife-tourist-attractions-are-living-hell-animals

World Animal Protection. (n.d.). *A close up on cruelty: The harmful impact of wildlife selfies in the Amazon*. https://www.worldanimalprotection.org.uk/sites/default/files/media/wap_wildlifenotentertainers_report_final_uk_092617hr_2.pdf

World Travel & Tourism Council. (2021). *Economic impact reports*. https://wttc.org/Research/Economic-Impact

Zhang, T., Chen, J. & Hu, B. (2019). Authenticity, quality, and loyalty: Local food and sustainable tourism experience. *Sustainability, 11*(12), 3437. https://doi.org/10.3390/su11123437

# 32
# HUMAN-WILDLIFE CONFLICT AND INTERACTION

*Brian J. MacGowan*

## Introduction to the Issue (Including Definitions)

Types of wildlife conflicts vary widely in their relative amount of damage or level of threat to humans. These factors play a role in how people react (Reiter et al. 1999). Threats to personal safety is a high priority conflict that may require use of lethal force. To a lesser extent, attacks to pets may be viewed similarly. The risk of human fatalities from wildlife airplane collisions often requires airports to institute hazing or lethal control of some wildlife species on airport grounds. Wildlife serving as a vector for disease transmission could require lethal control. Bites involving rabid or potentially rabid animals would lead to the dispatching of that animal. Rabies is just one of many zoonoses, or diseases that can be spread from wildlife to humans. However, since rabies is a lethal disease, the action threshold is very low. Wildlife also may play a role in transmitting diseases to things humans value such as pets (e.g., distemper) or livestock (e.g., swine fever).

For most people, wildlife conflicts they experience center around some type of damage occurring in many forms. Woodpeckers can cause structural damage to wood siding through drumming activities. Moles digging and burrowing behavior can damage lawns. White-tailed deer can damage trees and or other plants by feeding on new growth or bucks rubbing to remove the velvet from hardened antlers. Similarly, farmers can have a variety of agronomic crops damaged by wildlife. Voles can girdle fruit trees by feeding on the cambium of the trunk at or below ground level. Raccoons can damage large patches of field corn during the milk stage of plant development.

Finally, conflicts with wildlife may also be categorized as a nuisance. Tree squirrels consuming bird seed at a feeder is a nuisance to many of the millions of people who feed birds. There is certainly an economic cost of having to refill the feeder more often, but there is no damage or safety threat. A chorus of calling frogs on a warm spring evening may be viewed as a nuisance to those who had their sleep impacted. These and other encounters may be viewed positively by some individuals rather than a nuisance to be eliminated.

The particular wildlife species involved in a conflict can influence how one perceives a conflict and how they respond. Some people value various types or species of animals differently. An encounter with a rat snake will likely invoke a different response than an encounter with a moose, regardless of differences in actual danger to the person(s) involved in each scenario. Killing the snake might be acceptable to many while killing the moose may be acceptable to very few. This variation in response is due to regulatory differences for each species, but also how people perceive the value of each species.

## Cultural/Religious and/or Historical Perspectives, as Relevant

People and wildlife have a long history of interactions that have had negative consequences for the former. Once human civilizations began farming agricultural food crops or domesticating livestock, wild animals have surely damaged crops and food stores, or depredated animals (Conover 2002:19). Civilizations that cohabitated with animals capable of causing personal harm or death faced additional challenges. Despite an everchanging evolution of technology and advancements, we actually face similar problems with wildlife now.

How societies have dealt with wildlife damage management has evolved. During the early 20th century, federal agencies provided financial incentives and bounties to control wolf and coyote depredation of livestock (Miller 2007). Beginning in the 1960s, wildlife damage control began to shift away from a strictly control-oriented approach to one that included wildlife protection and an integrated approach used today (Conover 2002, Miller 2007). Our evolution in wildlife damage management philosophy has coinciding with a changing population.

In the United States, urban population growth (12.1%) outpaced national growth (9.7%) from 2000 to 2010, resulting in 80.7% of the U.S. population residing in urban areas (U.S. Census Bureau 2012). Urbanization is only a relatively recent trend in the United States. The U.S. Census of 1920 was the first census where more than half of the population lived in urban areas (Auch et al. 2004). Today, this urban population lives on only 3.1% of the conterminous United States (Nowak et al. 2005). Urbanization leads to questions of disconnect between people and natural resources. This disconnection with rural lands by the majority of citizens likely plays a role in how the public approaches wildlife damage management.

## Impact on People

Wildlife conflicts can impact people in many ways, but most of the published literature focus on economic costs, human injuries, and deaths (Figure 32.1). National crash databases have facilitated the impact of vehicle and airplane collisions with wildlife. The number of wildlife-vehicle collisions in the United States has been estimated between 1 and 2 million per year (Romin and Bissonette 1996, Clevenger et al. 2008). Collisions annually result in about 26,000 injuries (Conn et al. 2004) and 200 deaths (Clevenger et al. 2008). Dolbeer et al. (2021) summarized civilian aircraft wildlife strike events from multiple datasets from 1990 to 2020. Birds were involved in 96% of the 243,064 strikes reported during the 31-year period (Dolbeer et al. 2021). They estimated the total economic cost of all civilian aircraft wildlife strikes during this period to exceed $870 million.

Economic losses from wildlife damage are estimated based on direct assessments or perceptions of loss. Current published data on economic losses, frequency, and wildlife species involved are sparse. More recent data has been summarized from insurance claims (e.g., crop insurance, wildlife vehicle collisions) or required reporting (e.g., bird airplane collisions). McKee et al. (2021) estimated wildlife damage annually losses of $592.6 million to corn, soybeans, wheat, and cotton in the U.S. (Figure 32.2). Conover (2002) estimated all wildlife damage in the United States costs an estimated $22 billion annually. This estimate included damage to property as well as money and time spent addressing problems. Conover (2002)

*Figure 32.1* Impacts of wildlife vehicle collisions are summarized and reported from impacts to people. Data on wildlife impacts are lacking although collisions likely result in direct wildlife mortality and indirectly if only adults with dependent young or eggs are killed in the collision. (Photo credit: Brian MacGowan)

considered this to be a conservative estimate and would be over $33 billion in 2021 dollars using an average inflation rate of 2.18%.

## Impact on the Animal's Welfare

Although few studies have empirically estimated the impact of wildlife conflicts and mitigation strategies on wildlife, there are some examples other than direct mortality. Translocation is the movement of an animal outside of its home range. Researchers have investigated the fate of translocated animals by capturing nuisance animals, anesthetizing them, affixing radio transmitters, releasing them to a suitable "natural" habitat, and then tracking their movements and fate. Movements and survival are sometimes compared to resident animals within the release areas. Data from these studies indicate that translocated animals rarely remain where you release them, even if the habitat is relatively large and of perceived good quality. Nuisance wildlife is often relatively abundant. Moving these animals to other locations likely places them in direct competition for resources with resident animals of the same species. Newly-translocated wildlife may be moving to find available resources or attempting to return to their natal area.

While releasing animals may seem more humane, post-release mortality is often high. In a study of translocated gray squirrels, 17 of 38 animals died (or were probably dead) within

*Figure 32.2* Many wildlife species including white-tailed deer, raccoon (pictured), and blackbirds damage corn. Protecting agricultural crop damage from wildlife is challenging because of the size of crop fields and the total area in cropland. (Photo credit: Mike Kerper)

54 days after release (Adams et al. 2004). Predation is the most common cause of mortality, although other causes likely exist that have yet to be determined. Since researchers are experienced and use anesthesia to reduce the risk of released animals dying from capture myopathy (Paterson 2014:177), the rate of mortality is likely higher than indicated by research studies.

The translocation of wildlife can also contribute to the spread of diseases. Wild animals may carry a particular disease or parasite, but may not show any clinical signs. By moving animals to areas not affected, you can unknowingly spread disease. Legal restrictions on where or how far trapped animals may be released from the point of capture in part address this issue.

Euthanasia is derived from the Greek words *eu* meaning good and *thanatos* meaning death. It is generally interpreted as a death that minimizes or eliminates pain, although the rapidness of death, the degree of stress involved, and the justification may also be a consideration. Some published literature provides guidance on acceptable methods, but information on free-ranging wildlife is limited.

The American Veterinary Medical Association produced and regularly updates the AVMA Guidelines for the Euthanasia of Animals. These recommendations are intended for veterinarians "who must use their professional judgment in applying them to the various settings where animals are to be euthanized" (AVMA Panel on Euthanasia 2020). While these guidelines have a section on free-ranging wildlife, it recognizes the difficulty in meeting standards in these situations. "Free-ranging animals may need to be killed quickly and efficiently in ways that may not fulfill the criteria for euthanasia established by the [AVMA] (AVMA Panel on Euthanasia 2020)." Since these guidelines are written for veterinarians, some of the acceptable methods are not possible for most people due to a lack of expertise or the availability of drugs and specialized equipment. The American Society of Mammalogists published "Guidelines of the American Society of Mammalogists for the Use of Wild Mammals in Research" (Sikes, & the Animal Care and Use Committee of the American Society of Mammalogists 2016). The ASM wrote these guidelines for individuals conducting research involving wild mammals so some of their recommendations are not applicable or possible for the general public.

## Practical Perspectives

The four steps in addressing wildlife conflicts are problem definition, ecology of the problem species, control method, and evaluation of control (Dolbeer et al. 1994). Defining the problem includes determining if a problem exists and the wildlife species that is causing the problem. This may involve direct observation such as seeing white-tailed deer browsing tree seedlings. However, in many cases indirect observation of signs, including scat, tracks, digging or burrowing, and the physical characteristics of the damage is only possible. To identify a plan of action, the ecology of the problem species must be considered. Are they capable of climbing or digging? Do they have sharp incisors capable of penetrating some materials? Are they active during the daytime or nighttime? What is their diet and how does it change seasonally?

Using this information, one can develop a plan to address the problem. Several general approaches are used to address wildlife conflicts. Behavioral modification involves the use of devices or repellents to alter how an animal behaves or moves. Taste repellents can prevent or minimize feeding on certain plants. Electronic sound devices that mimic raptor calls or avian distress calls can minimize bird use of areas (e.g., garden) within close proximity. Exclusion is the process of preventing access to some area or plant. Fencing can protect larger areas, while plastic tubes and wire cages can protect individual plants (Figure 32.3). People

*Figure 32.3* Fencing and other exclusion devices can effectively deter wildlife damage. Polypropylene fencing is a relatively cost-effective option for protecting large areas from deer. (Photo credit: Brian MacGowan)

can modify cultural practices to minimize attractiveness to wildlife, which are attracted to areas because they provide food, water, and/or cover. Manipulating one or more of these components can reduce the attractiveness of an area or even eliminate its use altogether. The final strategy for dealing with wildlife conflicts is to eliminate the problem animal(s) through either trap and relocation or euthanasia. While this approach is often required to immediately alleviate damage, it will rarely eliminate the problem since most wildlife species that cause damage are relatively abundant. New animals will eventually reoccupy the vacated space if no other action is taken.

Evaluation of methods used to address wildlife problems is the last step defined by Dolbeer et al. (1994). This process may include observing a reduction or elimination of damage. It may also include observations of wildlife species or even measured changes in the relative abundance of animals through time. A homeowner who is experiencing woodpecker damage to cedar siding may hang mylar tape or other reflective deterrents on the side of the house after repairs are made. Periodic visual inspections of the house reveal no additional damage so the repair and visual deterrents are deemed successful. Evaluating other scenarios such as reducing deer damage to agricultural crops are more challenging. A farmer may anecdotally observe less deer browsing to corn after implementing deer reduction hunts. However, assessing deer populations over time or quantifying the actual reduction in damage (or corresponding increase in yield) is much more challenging. In most cases, this level of evaluation is unnecessary.

Many of these approaches can be deployed seasonally when damage peaks. In some cases, they may be required for several years until plants are mature and can withstand occasional damage. Often two or more methods are used in conjunction with each other to increase efficacy. As a general approach, preventing wildlife conflicts is always preferrable to reacting and solving damage and conflict after it occurs. However, there is no clear set of guidelines that work in all situations. Research is lacking on why people choose different methods and under what circumstances they view them to be acceptable. Logically, those experiencing wildlife conflicts that are completely intolerable to them (e.g., threats to human safety, high value of damage) will choose the most effective or extreme methods regardless of cost or perception of others. Similarly, those with more economic resources or practical skills may choose different methods, install practices themselves, or hire professionals to implement wildlife control practices.

## Current Policy and Legislative Perspectives

Wildlife in the United States is managed by state and federal agencies under the public trust doctrine, which establishes a trustee relationship between government and wildlife (The Wildlife Society 2010). Wildlife is owned by the public, entrusted to the government agencies charged with their care. This Public Trust Doctrine serves as the cornerstone of the North American Model of wildlife management and sets it apart from other models where the general public receive little or no benefit from wildlife resources (The Wildlife Society 2010).

Migratory birds native to the United States are regulated under federal law under the Migratory Bird Treaty Act (MBTA) of 1918. "The MBTA prohibits the take (including killing, capturing, selling, trading, and transport) of protected migratory bird species without prior authorization by the Department of Interior U.S. Fish and Wildlife Service (USFWS 2022)." Wildlife conflicts involving migratory birds (e.g., woodpeckers, Canada goose) that require "take" are limited under what is allowable under the MBTA and specific actions by

MBTA permits. Some activities associated with migratory birds also require permits from state or other federal agencies.

State fish and wildlife agencies have authority over native birds, mammals, fish, reptiles and amphibians. Methods of take and other allowable actions for dealing with wildlife conflicts within these taxa are included within this authority. For managing game species that cause wildlife conflicts with people, state agencies can adjust annually hunting and trapping season lengths and timing, bag limits, and method of take. Longer season lengths, higher bag limits, and allowing more efficient methods can increase the annual harvest, and thus, lower the population. States apply strategies across the state or large management units, but wildlife population responses will be unequal at the local level since vegetation type, habitat quality, and hunting pressure vary across the landscape.

State hunting and trapping regulations are designed in part to maintain populations at a socially and biologically acceptable level. These rules also help ensure in part that wildlife hunting and trapping is done so in an ethical manner which is a cornerstone of fair chase and the North American Model (The Wildlife Society 2010). The Boone and Crocket Club define fair chase as "*the ethical, sportsmanlike, and lawful pursuit and taking of any free-ranging wild game animal in a manner that does not give the hunter an improper or unfair advantage over the game animals* (Boone and Crocket Club 2022)."

Regulations designed for hunting and trapping wild game animals may not be applicable for some human wildlife conflicts. For example, fair chase may not be equally applied for human wildlife conflicts where the goal may be the immediate reduction of wild animals at the local scale. Technology such as drones or remote cameras that provide real-time images wirelessly to a smart phone, or hunting at night using thermal or night-vision scopes may be unlawful to use for sport hunting but may be allowable for wildlife damage control. Wildlife damage control may require taking or trapping of animals outside of the regulated hunting or trapping season. In these situations, state agencies can issue individuals a permit to take regulated wildlife out of season. These permits will typically outline the allowable timeframe, the species of wildlife and the sex, number that may be killed, and the geographic location. States will also set guidelines on what can be done with live trapped nuisance wild animals. In Indiana, for instance, animals captured live must be euthanized or released in the county of capture, cannot be possessed more than 24 hours, and must have prior consent of the landowner or property manager (312 IAC 9–10–11).

## Ethical Considerations

It is reasonable to assume that we will never eliminate all conflicts with wildlife. Some of our experiences with wildlife will be negative and some will be positive. The question becomes how to strike a harmonious balance when both must coexist (König et al. 2020). Regulatory agencies provide us some ethical standards for dealing with wildlife conflicts when they allow certain types of traps and banning other devices. They also limit lethal take of some species of wildlife. However, in many instances people are left with their own moral compass on how to address conflicts with wildlife.

One could make the case that preventing conflicts from occurring is a more ethical approach to dealing with them. For example, feeding bears is discouraged (or unlawful) because bears associate people with food, which leads to increased likelihood of attacks on humans. Bears that cannot break this connection are sometimes euthanized. Avoidance of the simple act of feeding bears eliminates the need for more extreme measures. While this is a specialized example, the same principle can be applied to many common situations. A homeowner can apply repellents to vegetation or erect a fence around a garden prior to damage occurring. A failure to do so in areas with abundant wildlife may be viewed as unethical by some. Wild animals will likely consume unprotected food crops leading some to take more extreme measures such as shooting or trapping. Initiating low-impact interventions prior to damage has the potential to eliminate more higher-impact interventions after damage occurs.

Solutions to wildlife damage often include trapping and relocation of wild animals or even euthanasia. Once a trap is set, the operator has a responsibility to any animal caught in the trap. Trapping wild animals leads to many ethical questions without clear answers. Should a trapped animal be translocated or euthanized? What methods of euthanasia are acceptable? How can non-target animals be avoided in traps? Does the trapped animal have dependent young? The answers to these questions likely vary among individuals and depend on the severity of the conflict. Timing of wildlife damage often overlaps during the breeding season. However, whether or not an adult has dependent young is not always clear. While one can observe a sow raccoon with her kits in an attic, the reproductive status of a sow raccoon in a box trap is unknown. The trap operator can only use the time of year as a guide.

Animal welfare is a top public concern in regards to trapping wildlife (White et al. 2021). Trapping is acceptable to the majority of the public, but acceptance is higher when the purpose is damage control or population management (Responsive Management 2016). Laws prevent wanton waste and the killing of wildlife for no reason. Wildlife damage control efforts may require the use of a lethal trap or require a live-trapped animal to be euthanized. Some wildlife species suffer high rates of mortality when relocated after trapping (Adams et al. 2004). To reduce the spread of wildlife diseases, some states require certain species of wildlife to be euthanized when trapped by a professional wildlife control operator. Under these various circumstances, some may view the killing of wildlife unethical and not in the interest of animal welfare. Others may view relocating a trapped animal to be unethical when its probability of survival is low. Additionally, many of the species involved in human wildlife conflicts are relatively abundant. Relocating these species may be viewed as unethical or irresponsible by moving the problem to someone else.

## Unanswered Questions and Future Research Demands

As Benjamin Franklin once said, "An ounce of prevention is worth a pound of cure." Preventing human wildlife conflicts is better than solving them. How can we do a better job of achieving this? Local and regional planning accounts for green space, which provides wildlife habitat and viewing opportunities. Incorporating strategies such as wildlife crossings or barriers in regional roadway plans could reduce vehicle strikes. Proper culvert design and placement near areas of amphibian "hot spots" can reduce mortality where animals cross the road (Patrick et al. 2010). Many species of wildlife damage or inhabit structures both old and new. Research that empirically assesses new construction design flaws that lead to interior access or unwanted animal use (e.g., perching) and test different solutions would be a first step in reducing conflicts associated with new construction. Implementation of improved methods would require codifying them in appropriate building construction codes.

Despite our best efforts in preventing them, human wildlife conflicts will always occur. There are resources available to help the public identify wildlife damage and methods to prevent or control it. While these resources are a great benefit, they leave it to the user—most have little knowledge of wildlife biology and behavior—to determine which strategy is best for them. A better understanding of the economic costs and benefits of various options to prevent or control damage has the potential to facilitate selection of the best practices.

A variety of methods are used in dealing with wildlife conflicts including repellents, hazing, barriers. Most of these methods have been researched for their efficacy in preventing or reducing damage. Outside of testing the efficacy of appropriately labeled toxins or a limited number of studies on the survival and behavior of translocated wildlife, we have little knowledge of what impact various control measures have on wildlife. Research that addresses the impacts of common control practices on animal welfare, survival, reproduction, and behavior would provide a more complete picture of control efforts and help inform selection of proper methods.

**Potential Questions for Discussion**

*To act or not to act*—Under what circumstances do people tolerate wildlife conflicts? Each situation varies based on the type of conflict, animal involved, and characteristics of the person involved (age, income, etc). What factors influence an individual to initiate wildlife damage prevention and control practices?

*Lethal control*—Under what conditions is lethal control of wildlife allowable? What methods of euthanasia are acceptable and how might these differ among wildlife species? Should the age of the animal be considered?

*Impacts to animals*—Outside of some translocation studies, there is little to no information on how other control measures impact the survival of wild animals. Should this be a concern since most damage is from relatively common species?

## References

Adams, L. W., Hadidian, J., & Flyger, V. (2004). Movement and mortality of translocated urban-suburban grey squirrels. *Animal Welfare, 13,* 45–50.

Auch, R., Taylor, J., & Acevedo, W. (2004). *Urban growth in American cities, circular 1252,* U.S. Department of the Interior and U.S. Geological Survey. https://pubs.usgs.gov/circ/2004/circ1252/pdf/circ1252.pdf

AVMA Panel on Euthanasia. 2020. *AVMA guidelines for the euthanasia of animals.* American Veterinary Medical Association. https://www.avma.org/sites/default/files/2020-02/Guidelines-on-Euthanasia-2020.pdf

Boone and Crocket Club. (2022, April 2). *Fair chase statement.* https://www.boone-crockett.org/fair-chase-statement

Clevenger, T., Cypher, B. L., Ford, A., Huijser, M., Leeson, B. F., Walder, B., & Walters, C. (2008). *Wildlife-vehicle collision reduction study report to Congress, FHWA-HRT-08-034.* U.S. Department of Transportation Federal Highway Commission. https://www.fhwa.dot.gov/publications/research/safety/08034/08034.pdf

Conn, J. M., Annest, J. L., & Dellinger, A. (2004). Nonfatal motor-vehicle animal crash-related injuries—United States, 2001–2002. *Journal of Safety Research, 35,* 571–574. https://doi.org/10.1016/j.jsr.2004.10.002

Conover, M. R. (2002). *Resolving human-wildlife conflicts.* Lewis Publishers.

Dolbeer, R. A., Begier, M. J., Miller, P. R., Weller, J. R., & Anderson, A. L. (2021). *Wildlife strikes to civil aircraft in the United States, 1990–2020.* Federal Aviation Administration and U. S. Department of Agriculture, Wildlife Services. https://www.faa.gov/airports/airport_safety/wildlife/media/Wildlife-Strike-Report-1990-2020.pdf

Dolbeer, R. A., Holler, N. R., & Hawthorne, D. W. (1994). Identification and assessment of wildlife damage: an overview. In S. E. Hygnstrom, R. M. Timm, & G. E. Larson (Eds.), *Prevention and control of wildlife damage* (4th ed., pp. 1–18). University of Nebraska Cooperative Extension.

König, H. J., Kiffner, C., Kramer-Schadt, S., Fürst, C., Keuling, O. & Ford, A. T. (2020). Human–wildlife coexistence in a changing world. *Conservation Biology, 34*, 786–794. https://doi.org/10.1111/cobi.13513

McKee, S. C., Shwiff, S. A. & Anderson, A. M. (2021). Estimation of wildlife damage from federal crop insurance data. *Pest Management Science, 77*, 406–416.

Miller, J. E. (2007). Evolution of the field of wildlife damage management in the United States and future challenges. *Human-Wildlife Conflicts, 1*, 13–20.

Nowak, D. J., Walton, J. T., Dwyer, J. F., Myeong, S., & Kaya, L. G. (2005). The increasing influence of urban environments on U.S. forest management. *Journal of Forestry, 103*, 377–382.

Paterson, J. (2014). Capture myopathy. In G. West, D. Heard, & N. Caulkett (Eds.), *Zoo animal and wildlife immobilization and anesthesia* (2nd ed., pp. 171–179). John Wiley & Sons, Inc.

Patrick, D. A., Schalk, C. M., Gibbs, J. P., & Woltz, H. W. (2010). Effective culvert placement and design to facilitate passage of amphibians across roads. *Journal of Herpetology, 44*, 616–626.

Reiter, D. K., Brunson, M. W., & Schmidt, R. H. (1999). Public attitudes toward wildlife damage management and policy. *Wildlife Society Bulletin, 27*(3), 746–758.

Responsive Management. (2016). Awareness of and attitudes toward trapping in Connecticut, Indiana, and Wisconsin. Association of Fish and Wildlife Agencies. http://www.fishwildlife.org/files/AFWA_Public_Attitudes_Toward_Trapping_2016_final.pdf

Romin, L. A., & Bissonette, J. A. (1996). Deer vehicle collisions: Status of state monitoring activities and mitigation efforts. *Wildlife Society Bulletin, 24*, 276–283.

Sikes, R. S., & the Animal Care and Use Committee of the American Society of Mammalogists. (2016). Guidelines of the American Society of Mammalogists for the use of wild mammals in research and education. *Journal of Mammalogy, 97*(3), 663–688. https://doi.org/10.1093/jmammal/gyw078

The Wildlife Society. (2010). *The public trust doctrine: Implications for wildlife management and conservation in the United States and Canada, Technical Review 10–01.* https://wildlife.org/wp-content/uploads/2014/05/ptd_10-1.pdf

U.S. Census Bureau. (2012 March 26). *Growth in urban population outpaces rest of nation.* https://www.census.gov/newsroom/releases/archives/2010_census/cb12-50.html

USFWS. (2022 June 2). *Migratory Bird Treaty Act of 1918.* Department of Interior U.S. Fish and Wildlife Service. https://www.fws.gov/law/migratory-bird-treaty-act-1918

White, H. B., Batcheller, G. R., Boggess, E. K., Brown, C. L., Butfiloski, J. W., Decker, T. A., Erb, J. D., Fall, M. W., Hamilton, D. A., Hiller, T. L., Hubert, G. F., Jr., Lovallo, M. J., Olson, J. F. & Roberts, N. M. (2021). Best management practices for trapping furbearers in the United States. *Wildlife Monographs, 207*, 3–59. https://doi.org/10.1002/wmon.1057

# 33
# OH, THE PLACES I'VE SEEN AND THE PLACES WE WILL GO

## A Commentary on the Evolution of AAI

*Aubrey H. Fine*

### Introduction

"Oh, the places I've seen and the places we will go." Who would have imagined that the words of Dr. Seuss would become my inspiration for beginning this commentary on the evolution and growth of animal-assisted interventions (AAIs)? Professionally, I have been involved in the field since the early 1970s. My engagement was quite serendipitous when I accidentally discovered the therapeutic impact of my small gerbil on a group of children enrolled in a social skills training program developed for youngsters who were diagnosed with learning disabilities. Her impact on the youngsters was remarkable and was the impetus for my next five decades of engagement in the practice of what we call today AAIs. For sure, so much has changed in the field since my early introduction, but the psychological and emotional significance of the therapeutic interactions with animals has not. I personally have watched the field go from a misunderstood form of therapy that was supported by numerous anecdotal testimonials as evidence. Today, AAI has blossomed to a more respected complementary therapy built around a growing body of scientific evidence-based findings.

The field of animal-assisted therapy has witnessed a tremendous evolution over the past six decades since its modern-day inception. In one of his earliest published writings entitled *The Dog as a Co-Therapist*, Boris Levinson (1962) explains his early discovery and his impressions of the importance of house pets in providing psychological support to various family members. He noted in this seminal article, how he made his first discovery of how he integrated a dog as "an accessory in the treatment of children with disturbances." Within this pioneering article, he highlights his now famous account about how his dog Jingles unexpectedly met and interacted with one of his young patients. He was amazed at their first introduction and the tremendous response it had. Intuitively, he believed that the experience provided a catalyst for his relationship with the child and enhanced their therapeutic alliance. It was in the early 60s that he began to appreciate the psychological benefits of providing what he called pet therapy (now considered AAIs) to his clients. He once noted that "through relationships with animals and in using them as transitional objects to gain meaningful and altruistic relationships human beings, man in the year 2000, will we discover the meaning of life... pets will become a very important safety valve" (Levinson, 1975, p. 159)

Over time, Levinson became aware that his dog brought a sense of comfort and calmness to the therapeutic environment. He believed that the animal's presence could help break the ice in fostering a relationship with the therapist. In a paper written a few years after his first publication, Levinson began to question his initial insights and argued for the need for further research on a variety of questions regarding his discovery of Pet Therapy. He advocated that there needs to be a better understanding of the therapeutic mechanisms that were impacted. Specifically, he emphasized the need to better understand what the children and the dogs needed to form and support an effective therapeutic relationship. He also was very interested in understanding how to prescribe the pet for certain kinds of disorders. He concluded the paper with a challenge for the future (Levinson, 1975). He stated that it did not seem that far-fetched to conceive that in the future we may witness a systematic integration of dogs in therapy. He suggested that in the future, the field of psychotherapy could envision the creation of a national well-trained canine counseling core in the United States (and perhaps globally) for children. His assumptions were reinforced by the fact that a dog core was already in existence serving this country in military, rescue, and police tasks. So, he believed it wouldn't be that unfathomable to consider the psychotherapeutic aid of dogs as being that unrealistic. Today, his initial viewpoint has become a global reality but in the early years was treated with speculation and uncertainties. His early impressions and remarks elicited tremendous snickers and sneers by many, including delegates at an American Psychological Association convention held in New York in the early 60s. My good friend and colleague, Dr. Stanley Coren, was at that meeting. Spotlight 1 highlights his early interactions with Dr. Levinson and his memories of that session at the American Psychological Association.

## Spotlight 1 on Boris Levinson

### How Therapy Dogs Almost Didn't Come to Be

*Stanley Coren*

The formal study of the use of the human-animal bond in psychotherapy in North America can probably be traced to Boris Levinson and his 1969 book, *Pet-Oriented Child Psychotherapy*. The book resulted from observations Levinson made while treating children who had severe emotional and communicative problems, such as autism. Levinson noted an improvement in his ability to get through to these children when his Golden Retriever, Jingles, was present during the therapy sessions.

In the 1960s, at the time Levinson was gathering data for his book, I was teaching at the Graduate Faculty of the New School for Social Research, located in a building at the corner of Fifth Avenue and Fourteenth Street in New York City. Yeshiva University, where Levinson held a position as a professor, was located only a few blocks south on Fifth Avenue, and occasionally he would join me and other colleagues for coffee and conversation.

Although Levinson was a specialist in child psychology, he never struck me as a person who would get along with children. His angular face, deep-set eyes, and severe goatee made him look more like the stereotype of an evil sorcerer. Nonetheless, quite

> incongruously, he had a merry note in his voice, and I later learned that most of his young patients referred to him as "Uncle Boris."
>
> It just so happened that the meeting of the American Psychological Association was being held in New York and Levinson decided to give a report on his data showing how a dog might assist in therapy sessions there. This would turn out to be the first formal presentation of animal-assisted therapy given before a national audience. Because of my interest in dogs and Levinson's work, I decided to attend his talk.
>
> The reception of his report was not positive, and the tone in the room did not do credit to the psychological profession. Many of the psychologists attending treated his work as a laughing matter. One even catcalled from the floor, "What percentage of the therapy fees do you pay the dog?" This did not bode well for the future of such research and therapy, and I thought that it was likely that I would never hear about such use of animals in therapeutic interactions again.
>
> I might have been correct; however, a person whose voice could not be ignored by the psychological community essentially argued in favor of animal-assisted therapy from his grave. At the time, it was only around 15 years after Sigmund Freud's death. Just by chance, translations of many of his letters and journals were just appearing in English. From these, we learned that Freud often had his Chow, Jofi, in his office with him during psychotherapy sessions. Freud noticed that the presence of the dog seemed to relax his patients. This effect was most marked when dealing with children or adolescents. It seemed to him that the patients seemed more willing to talk openly when the dog was in the room. Thus, it became clear that Freud had observed very much the same phenomena that Levinson described.
>
> With evidence that Sigmund Freud was willing to entertain the usefulness of animal helpers in psychotherapy, the climate warmed, the laughter stopped, and some serious consideration of animal-assisted therapy began.

Fine (2016) reflected on Levinson's early pioneer work and debated if his early insights of the field are relevant today? His famous text entitled *Pet Oriented Child Psychotherapy* (Levinson, 1969) was published in 1969 and was the seminal textbook in the field. It presented an early foundation and perceptions of the practice of AAI. His writings left his readers with a call for action to enrich the field with more scientific support to help the field gain more respectability. Levinson challenged the field to become more diligent in its efforts to redefine itself and enhance its image. He continued this argument at a lecture in 1981 at the University of Pennsylvania (Levinson, 1982). Levinson urged the delegates to continue investigating their research on several prominent areas confronting the new field of Anthrozoology. He also advocated for scientists to better clarify and investigate the roles of human-animal interactions (HAIs) in various human cultures and ethnic groups, so that the scientific community could better understand how animals fit in to the global context. This challenge continues to be a priority today and the world has become a much smaller globe for all of us to interact within. His major challenge he argued at the meeting was directed to AAIs. He encouraged researchers in the scientific community to better equip themselves to understand the direct relevance of the therapeutic value of animals in formal psychotherapy. Again, this proposed challenge continues to be strongly relevant to researchers and practitioners today.

In his chapter Forecast for the Year 2000, Levinson made a couple of predictions of where he believed the field would go in the future (Levinson, 1975). His prediction seems very apropos today. Levinson noted that

> suffering from even greater feelings of alienation than those which are already attacking our emotional health, future man will be compelled to turn to nature and the animal world to recapture some sense of unity with the world that otherwise will seem chaotic and meaningless.

Levinson concluded in his remarks that animals will become junior partners and friends, effecting a revolutionary transformation of man's attitudes. Katz (2005), an American journalist and author, remarked that Levinson's commentary written over 45 years ago may be becoming the new perceived job description for the new roles we are witnessing some of our companion animals fulfilling. Dogs are becoming fully enmeshed in our lives, including becoming therapy animals.

In the mid-1990s, Dr. Gerald Mallon updated the first edition of the textbook with the permission of Dr. Levinson's wife, AIda. Mallon decided to keep the initial manuscript intact and integrated his personal comments and suggestions at the end of each of the chapters. He began the new edition with an introduction to who Boris Levinson was. The following spotlight highlights Mallon's interactions with Aida and his thoughts about updating the seminal book *Pet Oriented Child Psychotherapy*.

---

**Spotlight 2**

**An Interview with Dr. Gerald Mallon**

**Updating Pet-Oriented Child Psychotherapy**

The following is from an interview with Dr. Mallon on February 15, 2022

FINE: How did you first come upon the idea of updating Levenson's book? Do you have any personal memories of your interactions with her, and did she share any memories of her late husband as it related to AAI?

MALLON: I knew that Mrs. Levinson lived in Queens, NY, which was where I lived at the time. I reached out to her to say hi and to introduce myself – At that time, I worked at Green Chimneys and had a very keen interest in animal-assisted therapy. Initially, when we met, I didn't say anything about updating the book. We just talked. After several of these meetings, she asked if I would be interested in updating his book. The book had been out of print for many years. We both knew that if it was updated, the publisher (Charles C. Thomas), might be interested in re-issuing the book. We agreed and got a contract with the publishers, and I began to work on it.

When I finished the second edition, I shared it with Mrs. Levinson. Initially, she was not at all happy with what I had done. She felt I changed too much of the original material. We agreed that I would modify my work and simply annotate the updated version with updated references. I would also write a new introduction, which was essentially a re-write of an article I had previously written entitled – *A Generous Spirit*. We were both happy with the final version when the book was printed. Many may not realize that one of the residents at Green Chimneys, Alberto, illustrated that cover of the updated version.

I never saw Mrs. Levinson again after the book was published (she has now passed away). She did introduce me to their sons who were very happy with the final product. Over the time we interacted, I enjoyed hearing about Dr. Levinson. She talked warmly about her husband and told me that he was a very serious clinician. She made a point to tell me that he kept a small notebook by his bedside and when he woke in the morning, he jotted down notes from ideas that came to him during the night. She also emphasized that Dr. Levinson was a very kind and generous man and loved the children he worked with. She suggested I speak to a few other people who knew him, both personally and professionally. They all confirmed that he was a dedicated clinician and a wonderful man.

FINE: Was there a special quote in the book that you remember writing?
MALLON: "My only regret is that you could not have met him yourself. If he were alive, he would have been more than happy to spend time with you, to talk to you about his work. He was a generous, generous man. He was a good friend to many people, a kind, genuine, generous, honest, and gentle man (A. Levinson, interview, February 17, 1994-- vii)
FINE: In looking back at the book, how relevant do you think some of the insights generated are today?
MALLON: I think it was relevant. He was the Father of novel interventions in working with children, and he was not afraid to try new things. This is something that clinicians today should do more of. Dr. Levinson was a very serious clinician but was ridiculed by some colleagues when he introduced this intervention. Nevertheless, he was undeterred. A brave man and as I have said before - a generous spirit.

I guess what was most relevant in the book was the way that Dr. Levinson naturally allowed the child to interact with Jingles – it was initially what he called "a happy accident." He usually would not permit Jingles to be in the consultation room, but one of his young patients and his mother arrived early, and Dr. Levinson did not have the time to put Jingles in another room. When the child saw Jingles, he immediately began to interact with him, whereas the mother immediately cautioned against it. Dr. Levinson motioned for her to allow them to interact and thus, the intervention was born. I love how he was brave enough to permit this to happen. He also had great success utilizing companion animals with children who were non-verbal. I was also impressed with how he utilized the companion animal to talk to the child – noting how children naturally calmed themselves with an animal present.

FINE: The book brought so much attention to what Levinson called pet therapy. What is the strongest message that you believe comes from the book?
MALLON: That is simple! Don't be afraid to try things, even out of the box things especially when working with children. Animals are wonderful partners in working with them and should be introduced whenever possible and whenever their collaboration is warranted.

Although Levinson has been given the accolades for being the founder and pioneer of AAIs, the work of Bossard (1944) may be less known but has significance for the growing recognition of the field. In his paper published in Mental Hygiene, Bossard (1944) discussed the important roles that domestic animals played in family life. He emphasized their role in fostering and enriching the mental health of all its members. He specifically highlighted the psychological importance of companion animals on children and argued the need to better understand their roles. A few years later in 1950, Bossard wrote an additional article called I Wrote about Dogs. To his astonishment, after writing his first article in 1944, he received over 1,000 letters asking for reprints (which was quite an accomplishment). He concluded in this follow-up that the outcomes left him with no doubt that the love of animals by humans is incredible. He concluded that household pets should be considered as an integral part of family life and should be thought of as a viable element in optimal mental hygiene. The article, which is now over seven decades old, enlightens readers to appreciate how animals are a source of unconditional love, and serve as an outlet for us to express our love. These early insights, as well as the findings of Levinson, appear to be two of the earliest testimonials of the value of HAI/AAI. In essence, relationships with pets can be a lens through which a child can approach life and reality.

Both individuals demonstrated in the early years that pets could have an impact on decreasing alienation of humans by providing an avenue for communication with natural contact with others. When looking at the historical contributors to the field, one must not also forget the initial contributions of both Sam and Elizabeth Corson and their scholarly contributions to the field. Both were considered early researchers of the value of canine-assisted interventions. They inadvertently discovered that some of their patients with psychiatric disorders were keenly interested in the dogs and observed that the patients appeared to communicate more easily with each other and the staff when the dogs were present (Corson et al., 1975; Corson et al., 1977). The early researchers and practitioners in the field were instrumental in opening doors to advocate those relationships with animals were not only beneficial for humans but also valuable in therapeutic environments.

Finally, as I look back at our initial history, one additional milestone within the field that should be highlighted was the establishment of the Delta Foundation, which was founded in Portland Oregon in 1977. The initial founders were Dr. Leo Bustad, Dr. R K Andersen, Dr. Stanley Diesch, Dr. Joe Quigley, Dr. Alton Hopkins, as well as both Doctors Michael and Bill McCulloch. Dr. Michael McCulloch, who was a practicing psychiatrist, was the first president of the organization. The organization initially got the name by looking at the Delta triangle that emphasized the animal, the patient as well as the environment. In 1981, the Foundation changed its name to the Delta Society, because of the greater interest in the practice and research in HAI/AAI.

One of the major contributions in the early years of the Delta Society was the establishment of the journal Anthrozoos, which was initially the organization's peer-reviewed Journal. Many reading this chapter may be familiar with the journal, as it has become an independent journal hosted by The international Society of Anthrozoology.

Over the years, Delta Society has developed numerous initiatives, including the Pet Partner program, which began in 1991. The program was the first registration system in the field for volunteers and healthcare professionals to register their therapy animal/handler teams. It highlighted the urgency of evaluating whether a team had the prerequisite skills to engage in therapy work. Anne Howie and her dog Falstaff were the first registered team. They are both highlighted in the following spotlight that discusses the early years of Pet Partners. For more information on Howie's work in the field, readers are referred to Howie, 2015.

## Spotlight 3

## Do You Want to be the First Pet Partner?

*Ann R. Howie, CCA, CCFT, LICSW, ACSW*

"Would you be willing to be the first Pet Partner?" asked Maureen MacNamara, Vice President of Delta Society (now known as Pet Partners). She was referring to a brand-new national program she was developing to register dogs and other domesticated animals as therapy animals. What might you say in response to that question? My response was an unqualified and eager "yes!" (Experience has taught me that asking a few clarifying questions could have been useful.)

Becoming the first Pet Partner involved many steps and tasks: I took multiple lengthy written exams as Delta staff decided which test questions were most likely to give them the information they needed about the handler's knowledge. We discussed (at length!) procedures in various facilities in which I had worked, both with and without a dog at my side. (I started in 1988 including a dog in my work as a psychotherapist in a psychiatric hospital and as a volunteer in a physical rehabilitation hospital.) My dogs underwent several behavioral evaluations given by veterinarians in clinics and parking lots as the animal testing was formalized. (At the beginning, it was a test solely for the animal.) I was honored (and a bit tired) to be at the forefront of this vanguard program. Finally, Falstaff (our Great Dane/Boxer cross) received the designation of Pet Partner #0001 on his tag. Qui (pronounced "key," our oversized Shetland Sheepdog) was #2.

From the time I learned of a thing then called "pet therapy" to now, there have been extensive changes in the field. In 1987, my early networking with other programs was typing a letter, mailing it through the post, and receiving personalized responses, sometimes handwritten and always happy to help someone starting out. There were no generally accepted definitions of the kinds of work being done with animals until the mid-1990s when Delta Society brought a group of people from all over the United States to Seattle to put our minds together and develop initial definitions. The energy in that group was powerful, dynamically fueling many programs and careers. Those early meetings gave this emerging field definitions for animal-assisted activities (AAAs), animal-assisted therapy (AAT), animal-assisted education (AAE), human-animal support services (HASS), and the first Standards of Practice.

The field has expanded dramatically since then. Now AAIs are worldwide. Today evaluations of animals go beyond skills (does the dog sit when asked?) to aptitude (does the animal *enjoy* interacting with strangers?). The animal and handler are considered a *team*, with greater emphasis on the handler's role to support and advocate for the therapy animal. Some colleges and universities offer individual classes and AAI certificate programs at introductory to graduate levels. Therapy animal welfare is more consistently considered by handlers, programs, and researchers. The Therapy Animal's Bill of Rights is being used internationally. Some volunteer therapy animal organizations have identified AAI *competencies* for both handlers and animals, training for those competencies and assessing them before allowing teams to begin working

> in AAI. Both the American Counseling Association and the American Psychological Association have published and support competencies for professionals utilizing AAI.
>
> We have come a long way and we have a long way to go. It has taken an unfortunately long time to recognize the value and necessity of considering the welfare of our therapy animals, not just the potential benefits to humans. As we recognize and celebrate our many milestones, may we act with respect for *all* the lives involved, automatically holding therapy animal welfare as paramount rather than an afterthought.

In February 2012, the organization changed its name from the Delta Society to Pet Partners. Today, we still see many handlers across the United States and in several countries affiliating with Pet Partners who practice AAI with various species of animals. Since 2012, the organization has instituted a couple of other specialty programs, including Read with Me, and an Animal Assisted Crisis Response Program.

## Trends in Research in the Field of AAI

Over the past several decades, there have been numerous studies that have been initiated to investigate the impact of animals in supporting optimal mental health in humans. One of the early studies that stands out was conducted by Mugford and M'Comisky (1975) in the United Kingdom. Their study found that older people who lived independently with a budgie significantly improved in their social attitudes and appeared happier in their daily lives than those members in a control group after a five-month period. Their research also pointed out that animals living with the terminally ill or visiting those with similar constraints seemed to lessen the residents' fears and sense of loneliness.

Fine, Beck, and Ng (2019) and Fine and Andersen (2021) provided a concise overview of some of the changes that the field is presently witnessing regarding research and scholarship. It is evident that the research has become more sophisticated over the past 15 years than those reported in various meta-analyses that highlighted numerous flaws in the research (Nimer and Lundahl, 2007; Maujean et al., 2015). Some of the weaknesses that have been reported in the various meta-analyses over the decades include comments about the small sample sizes and poorly designed research questions. Furthermore, many have also commented on the few studies in the literature that employed control groups, applied randomized assignments to conditions, and integrated long-term follow-ups on the subjects involved.

Today with more opportunities for funding, research studies are now applying more randomized controlled trials, the gold standard in intervention research. Griffin et al. (2019) suggested that researchers that are preparing meta-analysis papers in the future should consider using their findings from their reviews to form recommendations regarding future research directions, including topics such as design and measures. These points could be valuable in contributing to future research in the area.

Although funding is still a challenge in HAI research, there are increasingly more options for researchers to apply for funding for their research (e.g., joint funding from NICHD/Waltham, Human Animal Bond Research Institute (HABRI) and the Horses and Humans Research Foundation (HHRF)). Fine and Andersen (2021) provide a brief overview of some of the resources presently in place. Specifically, in 2008, a public-private partnership was established between the Eunice Kennedy Shriver National Institute of Child Health and Human Development (NICHD) at the US National Institutes of Health (NIH) and the

WALTHAM Foundation. Over the past 15 years, the partnership has provided resources and funding for quality research as well as forums presenting findings and future research directions. On the other hand, HABRI is a non-profit organization that also funds innovative research studies, and the Horses and Humans Research Foundation (HHRF) supports research related to equine-assisted interventions.

A significant change in research is the explosion of a new generation of scientists and practitioners who are initiating a diverse agenda of research, including attempts to analyze the mechanisms influenced by AAIs. Their research is becoming more laser pointed to also investigate the efficacy of the various approaches to develop more evidence-based protocols. The research platform of today is much more robust than in the past, and there are more scientific investigations assessing the impact of AAI on a variety of populations (Serpell et al., 2017; O'Haire, 2012, 2017; Gabriels et al., 2015; Schuck et al., 2015, 2018; Fine and Friedman, 2018). Additionally, one significant shift in the scientific approach is that researchers are applying more physiological measures as methods assessing the efficacy of AAI, in combination with more traditional measures such as surveys and observational research. These measurements include the collection of oxytocin and cortisol as a source of change as well as the collection of heart rate variability and blood pressure utilizing various types of wristband monitors (Powell et al., 2019; Rodriguez et al., 2018; Crossman et al., 2018; MacLean et al., 2021).

## Contemporary Concerns for the Field

I believe that one of the most significant continued misunderstandings or misrepresentations of the field of AAI is the potential exaggeration of the impact it has on humans served. In the formative years, AAI's reputation of an evolving therapy was primarily supported by more anecdotal comments regarding its significance than actual scientific evidence-based findings. According to Linda Hines (2003) in the early years of the field's evolution, many mainstreamed people were reached through well-planned media campaigns in regards to the value of HAI an AAI. Hines points out, that although media attempt to bridge the pool of knowledge of science to the public, many of the media stories accompanied by heartwarming photographs of animals in the laps of the elderly and the children helped instigate significant changes on how people viewed and valued companion animals. In many ways, media brought to the public attention the wonderful work that was occurring at that time, including the efforts of service animals. In looking back at how HAI has attracted the attention of media, some of the campaigns may have been somewhat misleading and either attracted or distracted possible followers. These misrepresentations could have been attributed to the excessive unscientific focus on the relationship of humans and animals to attract more popular press play. MacLean et al. (2021) stress that scholars in HAIs are highly fortunate that the public is intrinsically interested in this area of research and service. However, they caution investigators to do their due diligence to assure that their findings are accurately communicated to the public. Researchers must not overstate the implications of their findings in press releases or conversations because these distortions can mislead the public into misrepresenting the actual findings of the research.

One thing that is for certain in all fields of inquiry, scientific research has always been used to understand better the world around us. The more robust the science adds greater credibility to the findings, it can help validate or demystify our preconceived notions and theories. Society needs the scientific findings to accurately document the positive and negative benefits of HAIs to make changes to present day policies. There is a need to translate much of this research to bridge science to practice. Eventually, as the field continues to grow, it is our hope that these findings will assist in identifying and developing best practices and more effective evidence-based treatment protocols. It is evident that the future of AAI must align itself to its new pathway toward the future.

One of the strongest contributions derived from reliable and valid scientific findings is that science can help advocate for more resources to be allocated to the general field of HAIs and more specifically animal-assistant interactions. I have always believed that there is a strong need by all our constituents (researchers, practitioners, and stakeholders) to promote policy changes at various political levels, social and healthcare resources to help the field become more entrenched and more accessible. Arkow (2017) uses the analogy "feed the monster" to explain that many public policy figures seem more interested in the outcomes of HAI/AAI research demonstrating tangible benefits for humans. It is incumbent for both practitioners and researchers in this field to better understand the landscape of key influencers who will support sustainable change and acceptance of the field. These policymakers can open new frontiers and enrich the pathways that are unchartered territories at this point (Fine & Andersen, 2021). The efforts to influence and support policymakers will also ensure that there is a less misrepresentation of the science of AAI. Forums that act as consortiums promoting collaboration and problem solving between researchers, stakeholders, and policymakers to act as change agents for the future are valuable assets that may influence the future trajectory of AAI. Hughes et al. (2021) strongly endorsed this option of nurturing innovative partnerships. They conclude by urging others to consider using successful platforms such as think tanks and global leadership summits to advance the field and foster policy changes.

## Animal Welfare

As the field continues to evolve, there are two areas of significance that have dramatically changed since the inception of AAI. AAI was initially established with the primary focus on human well-being. Thankfully, today's practice of AAI gives greater attention to the welfare of the animals who are involved. Logically, that makes sense since good welfare has a direct positive impact on the overall efficacy of the treatment provided. It was unfortunate that during the early years of the field, little attention was given in the literature assessing the implications for animals working in the field. The practice and research in the field have continued to flourish and much more attention is now being given to the impact of therapy on the animals themselves (Ng et al., 2018; Hatch, 2007; Zamir, 2006). Research and best practice protocols are now being instituted globally to not only to investigate welfare concerns, but to institute models that preserve the integrity, respect, and safety of all involved. Today, much more has been written on identifying the stressors that the animals may experience in this process, and guidelines and suggestions are being written to be sure we offset these challenges (Haubenhofer and Kirchengast, 2006, 2007; King et al., 2011; Glenk et al., 2013; McCullough et al., 2018; van Houtert et al., 2022; Uccheddu et al., 2022; Ng et al., 2014). Even in his early years, Levinson elegantly pointed out that provisions had to be made for the animals' well-being and safety, an insight that the field is continuing to refine and embrace

## The Professionalization of the Field

Looking back at our history and toward the future, one can truly recognize that the field has come to a crossroad. This professionalization crossroad is the final area that I want to discuss. AAI today is a multidisciplinary field with scholars and practitioners from diverse fields and academic backgrounds. Clear professional standards for practice have not been in place, and there is a strong void in identifying clear professional competencies that must be in place as well as the need for training of those individuals providing AAIs both professionally as well as in volunteer capacities. Today, although there are several professional organizations within the field, that void continues to grow. Many professionals who apply AAI may have a strong background in their field of practice but may not be as well trained in animal welfare, ethology, and behavior. One of the significant voids that is present is the lack of graduate level educational opportunities, including more advanced continuing education options. Today, we are witnessing a growth in various AAI/HAI centers housed in universities that are attempting to fill the void by providing continuing education opportunities and academic courses meeting the training needs of the growing multidisciplinary group of professionals interested in the area.

On the other hand, over the past several decades, there has been a growth in numerous organizations that are attempting to enrich the practice by implementing standards of practice and providing educational forums. Some of these organizations include Pet Partners, Animal Assisted Intervention International (AAII), the Norwegian Organization for Animal Assisted Therapy, the Alliance of Therapy Dogs, Animal Therapies Inc., and the International Society for Animal Assisted Therapy which is based out of Switzerland.

In a white paper instituted by Pet Partners in 2020, the organization focused on surveying numerous professionals and identifying the challenges they were experiencing in providing AAIs. It seems evident that, although many practitioners are cognizant of the research on HAI/AAI, many seem to need support in translating the body of research to practice. As noted earlier, there is the need for advanced continuing education opportunities enhancing the critical information needed that is necessary in providing safe and effective therapy.

The field continues to need leadership and stewardship in how it can meet the needs of the new multidisciplinary generation of professionals providing services as well as those engaged in translational research. This guidance must also include enhancing the clinicians' knowledge base not only in providing services for their patients/clients but in their awareness of becoming a more fluid and proactive partner with a therapy animal(s). Regarding researchers, the field would also benefit from research that translates outcomes that can also be put into practice. Fine and Griffin (2022) argue that translational research promotes a stronger connection from basic science to the actual practice of AAI. They emphasized that there is a stronger need in the field today for more research to be conducted formulating possible solutions that will make the experiences more effective, reliable, and safer for the humans and therapy animals involved.

One of the challenges that was highlighted in the white paper is the difficulties expressed by some AAI professionals in getting a buy in from their colleagues in respecting the value of HAIs in clinical settings. Although we have come a great way since the pioneer years of Boris Levinson, there continues to be some resistance by some colleagues who need more clarity on the scope of AAI. Some of their concerns pertain to infection control and zoonosis along with some of the liability and legal issues relating to having a therapy animal in a clinical setting. For these reasons and many others, some of the organizations are investigating into options for providing liability insurance to cover professional visits that incorporate AAIs.

The following two snapshots highlight two organizations that are moving the professionalization of the practice of AAI forward. AAII was established in 2013. The organization's goal is to foster a global community that would support practitioners in providing professional services in the field and to establish a level of standard practice for that service. Additionally, in March of 2022, the organization Animal Assisted Intervention Professionals was established. The purpose of the organization is to provide professionals with resources to implement safe and effective AAI. AAIP will also institute a national certification exam to register those professionals who meet the criteria and to designate those professionals as meeting a standard of competency in the field of AAI. Let's look at the two organizations with a brief snapshot of each.

---

**Spotlight 4**

**The Formation of Animal Assisted Intervention International**

*Melissa Winkle, OTR/L, FAOTA, CPDT-KA*

AAII is a European-based non-profit organization that planted seeds long before its official business status in 2014. Historically, assistance dog training organizations also placed dogs in AAI situations, and were aware of the need for different standards, competencies, preparation and training methods, and evaluation for humans and animals for AAIs. International representatives from assistance dog training organization executives, dog trainers and handlers, human healthcare and education professionals, and other experts in HAIs came together to create AAII.

While AAII recognized that there was a lot to be developed in the realm of AAIs, they also appreciated that there were other individuals and organizations with already established significant contributions that provided the root system in the flourishing field of AAIs. AAII has always been a member-driven organization with the goal to create friendly, professional collaborations with other related organizations, so we can nurture and grow the vast potential of AAIs. Internationally, we share similar theories, models, approaches, techniques, and literature, so it made sense to have international teams of members and experts collaborate to establish standards of practice and competencies that embraced and included what had already been established and add new concepts to meet the needs of today's AAI needs.

The group drafted its first version for Standards of Practice in 2012 in the areas concerning handlers (volunteer and professional), participants (patients, students, clients), animal assisted activities, animal-assisted therapy, animal assisted education, and animal support (for anyone who trains or handles the dogs for or within practice). The group launched as an official European-based entity, named AAII, in 2014. By August of 2015, AAII sponsored a member visionary and development meeting in which the standards were updated, and an accreditation process was drafted. Over the past few years, AAII has developed collaborations with several other international organizations, and is working together to professionalize the practice of AAIs in several areas with both long- and short-term goals, that will positively impact the international milieu.

The 2020–2022 AAII revisions to membership include:

- Adding collaborative models to membership
- Further development of the membership category for AAI educational institutions
- Providing high quality educational opportunities for both members and the general public with consistent conferences, webinars, certification specialties, and hand-on workshops for practitioners and dogs trainers involved in AAIs.

The standards and competencies have been revised and updated for:

1 Standards of Practice for the Administration of Programs
2 Standards of Practice for the Ethical Treatment and Welfare of Participants
3 Standards of Practice for Dog Handlers and the Support of Dogs
4 Standards of Practice for the Health, Welfare, Wellbeing and Training of Dogs
5 Standards of Practice for Animal Assisted Activity (AAA), Animal Assisted Therapy (AAT), and Animal Assisted Education (AAE)
6 Standards of Practice Collaborative Animal Assisted Therapy (C-AAT) and Education (C-AAE)

## Spotlight 5

## The Establishment of AAIP

*Comments from Taylor Chastain Griffin*

"How hard could it be?" I remember asking myself as I prepared to bring Rex, my registered therapy animal, to work with me for the first time. With the experience of having been a professional dog trainer for years combined with the expertise in mental health counseling I was learning through my master's program, it seemed to me at the time that I surely had all the skills I needed to practice animal-assisted therapy. I had even spent years working with Rex in volunteer therapy animal settings, and I had taught a training course preparing therapy animal handlers for their work. However, on that first day sitting across from a client with Rex, I suddenly felt unprepared and anxious about determining the most meaningful way to incorporate him into my formal treatment plans. I still felt that I needed more education tailored to the services I would be offering, and I needed ways to build my confidence and competencies specific to my work with therapy animals.

Over the years since, I've had conversations with hundreds of other professionals who can relate to my story. I've also been able to collaborate with many leaders and innovators in this space who successfully "figured it out" and now lead the field in realizing the best ways to partner with therapy animals.

In my role as National Director of AAI Advancement at Pet Partners, I've been able to spend the last two years diving in to truly understand the needs of AAI professionals at all points of their careers. This has been an experience that has not only affirmed by own journey, but it has informed the creation resources made available through the newly launched Association for Animal-Assisted Intervention Professionals (AAAIP). This is a Pet Partners-affiliated association, and it answers a long-awaited call for education, connection, research, and community tailored specifically for professionals who aim to integrate therapy animals into vocational practice. The field's first true certification has also been launched (AAI-S), providing a way for professionals to demonstrate an awareness of best practices that protect client safety and animal welfare. Through AAAIP, professionals from various fields share best practices, develop working groups, create best practices, and ultimately advocate for the intervention by bringing these standards of practice into their larger vocational identities and associations. AAAIP not only provides a roadmap for those getting started in the field, but it will also pave the way to professionalizing AAI long into the future.

As our team has worked together with countless experts from the field to create this association, I've often ended my workdays by taking a moment to look over at Rex, – now retired and gray – and I imagine just how far we've come since those early days together in practice. I recall the ways that we were able to intuitively bring about positive change for our clients, and I imagine just how much more can be accomplished as professionals come together to advance this intervention. It is my belief that we've only scratched the surface in realizing the ways HAI can inspire wellness for the people who we serve, and I cannot wait to witness the growth of AAI through efforts like the launch of AAAIP.

## Concluding Remarks

As I conclude this commentary, I reflect on where the field started and where the field is headed. I am excited about the trajectory of the field's growth, but I also believe we must honor our past to make sense of where our future will take us. Dr. Levinson and many of our early pioneers were trailblazers in making a difference in working with children and adults. Specifically, Levinson believed in innovation and thinking outside the conventional box of therapeutic interventions in providing services for children in need. In the early years, many were unenlightened to the possibilities of employing AAI in diverse settings. They found themselves using their gut instincts in believing in this modality. Many, including myself, used their professional knowledge in which they were clinically trained in, but may have turned to others to enrich their understanding about animal ethology, behavior, and welfare. Collaboration between human health care and animal professionals has enriched our body of knowledge to learn how to work more effectively in tandem with animals and to appreciate and respect their needs in staying safe and healthy. I urge practitioners to be open minded to new alternatives and to enrich their repertoire of therapeutic alternatives. Even in my early years of practice (I can only imagine Levinson's), there were several professionals and lay people who were skeptical of AAI and questioned why one would bring animals into various therapeutic environments. Some were even misinformed in thinking that the approach was trivial with little value. They couldn't fathom and conceive all the therapeutic benefits and values. They also didn't perceive how hard this work was for the animals who were engaged. In essence, many of the early pioneers were able to see the trees within the forest and acknowledged how animals' presence altered a milieu, made it more comfortable for the clients and impacted therapeutic alliance and engagement. In essence, a major transition from Boris Levinson's pet therapy to what we are now witnessing today, resulting in a new generation of scientists and practitioners that are more informed about AAI. Collectively, they are determined to enrich the body of knowledge within the field to ensure that it becomes a more respected complementary form of therapy.

A comment that I've often applied over the years, is the idea "that what you think you know, sometimes blinds you from what you could know." This statement accurately portrays a myriad of challenges that the field has faced for decades. In the first place, although there are many followers and believers, the field continues to be confronted with disbelievers and skeptics who are unclear about the efficacy of HAI. On the other hand, there are those who are so entrenched in their own practices that they don't recognize how the field is continuing to evolve. In some ways, they are stagnate in their forward thinking about the future. It is incumbent for all professionals in, or about to enter the field, to appreciate that the practice and scholarship of AAI are much different today and to embrace these changes and become part of this change. The professional community must be open to this growth. There continues to be a strong need to fortify the bridge between science and practice that will strengthen the efficacy and reputation of the field. Scientist and practitioners must consider working in harmony to help the field move to the next threshold of acceptance. Each will bring a different dimension to the credibility of AAI. These collaborative efforts will also assist policy makers in their understanding and potentially infuse change.

According to Leonard Sweet, a Professor Emeritus at Drew University, *the future is not something we enter. The future is something we create.* Our future in AAI will depend on our collective efforts to move the field with more credible science and with evidence-based approaches that include consideration for animal welfare and professionalism. Following this premise, I anticipate new directions becoming achievable. I'm quite certain that Boris Levinson would be smiling (from above) acknowledging many of the changes we are witnessing. Who would have believed that when he remarked over 70 years ago that one day, we may see a canine brigade of a therapy animals supporting mental health professionals in their work that his remarks would become a reality! Today, that perception has turned into a reality. We are globally witnessing a brigade of numerous species of animals, working alongside a multidisciplinary group of professionals and handlers in making a difference in the lives of many.

## References

Arkow, P. (2017). A species-spanning approach to animal welfare advancing. In *Proceedings of the Mid-Year Pet Partners Board Meeting*, Phoenix, AZ, USA.

Bossard, J. H. S. (1944). Mental hygiene of owning a dog. *Mental Hygiene*, 28, 408–413.

Corson, S. A., Arnold, L. E., Gwynne, P. H., & Corson, E. O. L. (1977). Pet dogs as nonverbal communication links in hospital psychiatry. *Comprehensive Psychiatry*, 18(1), 61–72. https://doi.org/10.1016/s0010-440x(77)80008-4

Corson, S., O'leary Corson, E., & Gwynne, P. (1975). Pet-facilitated psychotherapy. In R. S. Anderson (Ed.), *Pet Animals and Society* (pp. 19–35). London: Baillière Tindal.

Crossman, M. K., Kazdin, A. E., Matijczak, A., Kitt, E. R., & Santos, L. R. (2018). The influence of interactions with dogs on affect, anxiety, and arousal in children. *Journal of Clinical Child & Adolescent Psychology*, 49(4), 535–548. https://doi.org/10.1080/15374416.2018.1520119

Fine, A. H. (2016). Standing the test of time: Reflecting on the relevance today of Levinson's pet-oriented child psychotherapy. *Clinical Child Psychology and Psychiatry*, 22(1), 9–15. https://doi.org/10.1177/1359104515589638

Fine, A. H., & Andersen, S. J. (2021). A commentary on the contemporary issues confronting animal assisted and equine assisted interactions. *Journal of Equine Veterinary Science*, 100, 103436. https://doi.org/10.1016/j.jevs.2021.103436

Fine, A. H., Beck, A., & Ng, Z. (2019). The state of animal-assisted interventions: Addressing the contemporary issues that will shape the future. *International Journal of Environmental Research and Public Health*, 16(20), 3997. https://doi.org/10.3390/ijerph16203997

Fine, A. H., & Friedmann, E. (2018). Involving our pets in relationship building—pets and elder well-being. *Social Isolation of Older Adults*. https://doi.org/10.1891/9780826146991.0012

Fine, A. H., & Griffin, T. C. (2022). Protecting animal welfare in animal-assisted intervention: Our ethical obligation. *Seminars in Speech and Language*, 43(01), 008–023. https://doi.org/10.1055/s-0041-1742099

Gabriels, R. L., Pan, Z., Dechant, B., Agnew, J. A., Brim, N., & Mesibov, G. (2015). Randomized controlled trial of therapeutic horseback riding in children and adolescents with autism spectrum disorder. *Journal of the American Academy of Child & Adolescent Psychiatry, 54*(7), 541–549. https://doi.org/10.1016/j.jaac.2015.04.007

Glenk, L. M., Kothgassner, O. D., Stetina, B. U., Palme, R., Kepplinger, B., & Baran, H. (2013). Therapy dogs' salivary cortisol levels vary during animal-assisted interventions. *Animal Welfare, 22*(3), 369–378. https://doi.org/10.7120/09627286.22.3.369

Griffin, J. A., Hurley, K., & McCune, S. (2019). Human-animal interaction research: Progress and possibilities. *Frontiers in Psychology, 10.* https://doi.org/10.3389/fpsyg.2019.02803

Hatch, A. (2007). The view from all fours: A look at an animal-assisted activity program from the animals' perspective. *Anthrozoös, 20*(1), 37–50. https://doi.org/10.2752/089279307780216632

Haubenhofer, D. K., & Kirchengast, S. (2006). Physiological arousal for companion dogs working with their owners in animal-assisted activities and animal-assisted therapy. *Journal of Applied Animal Welfare Science, 9*(2), 165–172. https://doi.org/10.1207/s15327604jaws0902_5

Haubenhofer, D. K., & Kirchengast, S. (2007). 'Dog handlers' and dogs' emotional and cortisol secretion responses associated with animal-assisted therapy sessions. *Society & Animals, 15*(2), 127–150. https://doi.org/10.1163/156853007x187090

Hines, L. M. (2003). Historical perspectives on the human-animal bond. *American Behavioral Scientist, 47*(1), 7–15. https://doi.org/10.1177/0002764203255206

Howie, A. R. (2015). *Teaming with Your Therapy Dog.* West Lafayette, IN. Purdue University Press.

Hughes, A. M., Braun, L., Putnam, A., Martinez, D., & Fine, A. (2021). Advancing human–animal interaction to counter social isolation and loneliness in the time of Covid-19: A model for an interdisciplinary public health consortium. *Animals, 11*(8), 2325. https://doi.org/10.3390/ani11082325

Katz, J. (2005). *Katz on Dogs: A Commonsense Guide to Training and Living with Dogs.* Villard.

King, C., Watters, J., & Mungre, S. (2011). Effect of a time-out session with working animal-assisted therapy dogs. *Journal of Veterinary Behavior, 6*(4), 232–238. https://doi.org/10.1016/j.jveb.2011.01.007

Levinson, B. M. (1962). The dog as a "co-therapist." *Mental Hygiene, 46,* 59–65

Levinson, B. M. (1969). *Pet-oriented Child Psychotherapy.* C.C. Thomas.

Levinson, B. M. (1982). The future of research into relationships between people and their animal companions. *International Journal for the Study of Animal Problems, 3*(4), 283–294.

Levinson, B. M. (1975). Forecast for the year 2000. In R. S. Anderson (Ed.), *Pet Animals and Society* (pp. 155–159). London: Bailliere Tindall.

MacLean, E. L., Fine, A., Herzog, H., Strauss, E., & Cobb, M. L. (2021). The new era of canine science: Reshaping our relationships with dogs. *Frontiers in Veterinary Science, 8.* https://doi.org/10.3389/fvets.2021.675782

Maujean, A., Pepping, C. A., & Kendall, E. (2015). A systematic review of randomized controlled trials of animal-assisted therapy on psychosocial outcomes. *Anthrozoös, 28*(1), 23–36. https://doi.org/10.2752/089279315x14129350721812

Mugford, R. A., & M'Comisky, J. G. (1975). Some recent work on the psychotherapeutic value of caged birds with old people. In R. S. Anderson (Ed.), *Pet Animals and Society* (pp. 54–65), London: Bailliere Tindall.

Nimer, J., & Lundahl, B. (2007). Animal-assisted therapy: A meta-analysis. *Anthrozoös, 20*(3), 225–238. https://doi.org/10.2752/089279307x224773

MacLean, E. L., Gesquiere, L. R., Gee, N., Levy, K., Martin, W. L., & Carter, C. S. (2018). Validation of salivary oxytocin and vasopressin as biomarkers in domestic dogs. *Journal of Neuroscience Methods, 293,* 67–76. https://doi.org/10.1016/j.jneumeth.2017.08.033

McCullough, A., Jenkins, M. A., Ruehrdanz, A., Gilmer, M. J., Olson, J., Pawar, A., Holley, L., Sierra-Rivera, S., Linder, D. E., Pichette, D., Grossman, N. J., Hellman, C., Guérin, N. A., & O'Haire, M. E. (2018). Physiological and behavioral effects of animal-assisted interventions on therapy dogs in pediatric oncology settings. *Applied Animal Behaviour Science, 200,* 86–95. https://doi.org/10.1016/j.applanim.2017.11.014

Ng, Z., Morse, L., Albright, J., Viera, A., & Souza, M. (2018). Describing the use of animals in animal-assisted intervention research. *Journal of Applied Animal Welfare Science, 22*(4), 364–376. https://doi.org/10.1080/10888705.2018.1524765

Ng, Z. Y., Pierce, B. J., Otto, C. M., Buechner-Maxwell, V. A., Siracusa, C., & Werre, S. R. (2014). The effect of dog–human interaction on cortisol and behavior in registered animal-assisted activity dogs. *Applied Animal Behaviour Science, 159,* 69–81. https://doi.org/10.1016/j.applanim.2014.07.009

O'Haire, M. E. (2012). Animal-assisted intervention for autism spectrum disorder: A systematic literature review. *Journal of Autism and Developmental Disorders*, *43*(7), 1606–1622. https://doi.org/10.1007/s10803-012-1707-5

O'Haire, M. E. (2017). Research on animal-assisted intervention and autism spectrum disorder, 2012–2015. *Applied Developmental Science*, *21*(3), 200–216. https://doi.org/10.1080/10888691.2016.1243988

Powell, L., Guastella, A. J., McGreevy, P., Bauman, A., Edwards, K. M., & Stamatakis, E. (2019). The physiological function of oxytocin in humans and its acute response to human-dog interactions: A review of the literature. *Journal of Veterinary Behavior*, *30*, 25–32. https://doi.org/10.1016/j.jveb.2018.10.008

Rodriguez, K. E., Bryce, C. I., Granger, D. A., & O'Haire, M. E. (2018). The effect of a service dog on salivary cortisol awakening response in a military population with posttraumatic stress disorder (PTSD). *Psychoneuroendocrinology*, *98*, 202–210. https://doi.org/10.1016/j.psyneuen.2018.04.026

Schuck, S. E. B., Emmerson, N., Abdullah, M. M., Fine, A. H., Stehli, A., & Lakes, K. D. (2018). A randomized controlled trial of traditional psychosocial and canine-assisted interventions for ADHD. *Human-Animal Interaction Bulletin*, *6*, 64–80.

Schuck, S. E. B., Emmerson, N. A., Fine, A. H., & Lakes, K. D. (2015). Canine-assisted therapy for children with ADHD: Preliminary findings from the positive assertive cooperative kids study. *Journal of Attention Disorders*, *19*(2), 125–137. https://doi.org/10.1177/1087054713502080

Serpell, J., McCune, S., Gee, N., & Griffin, J. A. (2017). Current challenges to research on animal-assisted interventions. *Applied Developmental Science*, *21*(3), 223–233. https://doi.org/10.1080/10888691.2016.1262775

Uccheddu, S., Ronconi, L., Albertini, M., Coren, S., Da Graça Pereira, G., De Cataldo, L., ... Pirrone, F. (2022). Domestic dogs (*Canis familiaris*) grieve over the loss of a conspecific. *Scientific Reports*, *12*(1), 1–9.

van Houtert, E. A., Endenburg, N., Vermetten, E., & Rodenburg, T. B. (2022). Hair cortisol in service dogs for veterans with post-traumatic stress disorder compared to companion dogs (*Canis Familiaris*). *Journal of Applied Animal Welfare Science*, 1–11.

Zamir, T. (2006). The moral basis of animal-assisted therapy. *Society & Animals*, *14*(2), 179–199. https://doi.org/10.1163/156853006776778770

# 34
# THE PRACTICE OF ANIMAL-ASSISTED INTERVENTIONS

Exploring the Value of AAT/AAA

*Taylor Chastain Griffin, Jen VonLintel, Erin Trella, and Annie Kate Hudson*

## Introduction

It is widely known that connection with animals can have positive effects on humans in a multitude of ways. From aiding in mental health and stress, to improving behaviors such as communication and social skills, the positive impact of animals in intervention settings is increasingly supported by the literature (Maujean et al., 2015; Jones et al., 2019; Hillen & Kirnan, 2020). As our appreciation for the human-animal bond has developed over time, the field of animal-assisted interventions (AAIs) has evolved. AAI covers a wide range of interventions with therapy animals, such as animal-assisted therapy (AAT), animal-assisted education (AAE), and animal-assisted activities (AAA). Through continued research in the field, AAI and its subcategories have become more precisely defined, and the health and wellbeing of the animals and their counterparts has been examined and made a priority. The importance of relationship and collaborative practice with therapy animals have been realized, and training to understand both animal and human cues has emerged as a vital part of the process (Standards of Practice, 2021). As an emerging practice area, continued research and investigation into AAI is vital. Professionals seeking to incorporate animals into practice must have sound reasoning that is clinically backed to aid in legitimizing the field and propelling the practice forward. This chapter delves into the multifaceted field that is AAI and the value it brings to both humans and animals alike.

## The Value of AAI

Interaction with therapy animals has been found to promote wellbeing for people across many different contexts, proving to be an intervention capable of bringing about positive change whether a person is young or old, sick or well, presenting for a treatment concern or simply combatting daily stressors (Maujean et al., 2015; Jones et al., 2019; Hillen & Kirnan, 2020). Multiple literature reviews on AAI have noted its association with decreased psychological stress, improved mental health (impacting depression, anxiety, and trauma), and benefits to psychosocial behaviors like communication and social skills (Morrison, 2007; Maujean et al., 2015; O'Haire et al., 2015; Jones et al., 2019). Many of the benefits of AAI have been measured with outcomes such as heart rate, blood pressure, and salivary cortisol

(Morrison, 2007), while other impacts of therapy animal interaction remain more difficult to define and quantify. Some researchers have utilized self-report or parental report measures to determine the intervention's effect on outcomes such as quality of life and self-perceived health status (Johnson et al., 2008; McCullough et al., 2018). Findings from these kinds of studies have been promising, ultimately indicating that participants note significant changes in their wellness as related to their interactions with therapy animals (Johnson et al., 2008; McCullough et al., 2018).

Those who receive AAI are not the only ones with good reports of the intervention's efficacy. Professionals whose job it is to bring about wellness for their clients are also often attracted to partnering with therapy animals. In a study of 300 mental health practitioners, animal-assisted counseling was reported to be viewed as a legitimate modality resulting in positive outcomes for patients, most commonly impacting issues such as anxiety, depression, trauma, grief, and abuse (Hartwig & Smelser, 2018). Professionals from various disciplines have recognized how therapy animals can help build rapport, motivate client participation, and allow for safe and appropriate comforting touch during treatment (Chandler, 2012). Having noted just some of the ways in which therapy animals have been reported to assist people throughout the lifespan, there is no wonder so many people are interested in learning more about how to become involved in the field. However, before one attempts to unleash the power of therapy animals, there is foundational knowledge about the intervention that needs to be obtained so that AAI is as safe and effective as possible. The state of the current AAI literature, which includes mixed findings and a need for further investigation must also be realized so that AAI advocates can accurately speak to the state of the field.

## Summary of AAI Literature

Research on therapy animals has increased considerably as the intervention has become more sought after by clients and professionals alike. There is a foundational need for the anecdotal experiences about the influence of animals to be bolstered by empirical investigation, especially as the intervention is applied as a professional modality in formal treatment settings. Scientists have studied the influence of therapy animals in a wide range of settings: mental health treatment centers, pain clinics, hospitals, residential care facilities, schools, and others. Therapy animals have also been considered in terms of how they impact certain populations. For example, there is research on the intervention's specific impact on children, people with autism, college students, older adults, people with dementia, and more. With so many faucets of AAI having been explored, the literature reviews that have considered subsections within the research are extremely helpful in establishing an appreciation for the state of the intervention. Overall, these reviews have pointed to the fact that interaction with therapy animals may be beneficial across many diverse applications. In a review that examined studies involving canine-assisted psychotherapy, the presence of a therapy animal was commonly reported to have a positive impact on a person's diagnosis and symptomatology, impacting variables such as anxiety, anger, engagement, and social interaction (Jones et al., 2019). A literature review conducted by O'Haire et al. (2015) focused on AAI as it is applied in treatment for trauma. In this publication, the studies that were considered identified decreases in depression, PTSD, and anxiety associated with AAT interventions (O'Haire et al., 2015). Hoagwood et al. (2017) reviewed studies on AAT for children and adolescents who were at a risk for mental health concerns, with the most significant positive findings being for children with autism who engaged in equine-assisted therapy. Similarly, reviews of the research on

AAT for older adults have provided evidence of positive impacts across psychosocial, cognitive, and behavioral domains (Chang et al., 2021). While the positive outcomes noted across the literature have helped the intervention grow to be one that is empirically supported, other themes have emerged throughout collective reviews of the literature, ultimately pointing to gaps in our empirical understanding and the need for further investigation on therapy animals. Some research projects have found AAI to have a limited impact on participants, yielding suggestions about how to strengthen future research with measures such as larger sample sizes and longer AAI interventions (Barker et al., 2015; McCullough et al., 2018; Mueller et al., 2021). It's also worth recognizing that the designs that are implemented across AAI studies tend to be tremendously varied, and there is often little information provided within the literature about the therapy animal intervention itself (Jones et al., 2019). Many of the existing AAI studies lack methodological rigor, calling for future studies that utilize more sophisticated designs such as randomized control groups (O'Haire et al., 2015; Hoagwood et al., 2017). Funding to support this research is becoming more readily available through grants such as ones offered by the Human-Animal Bond Research Institute and Pet Partners, and many universities are now partnering with practitioners to address the gaps in our empirical appreciation of AAI.

## Defining Terminology

Although many people are attracted to the field of AAI, there is tremendous confusion within the public at large about the key terminology that is used to describe this work. Professionals and laypeople alike often conflate the roles of various animals such as therapy animals, assistance/service animals, facility animals, and emotional support animals (ESAs). Beyond understanding the roles and subsequent accommodation rights of these animals, it can be challenging to appreciate the different kinds of AAI's such as AAT, AAE, and AAA. Clarity on each of these terms is essential for anyone who desires to practice with a therapy animal, as it enables the practitioner to best research, represent, and advocate for the work they do. Consider an educator who is attempting to get a district policy passed that allows AAI in the school setting. The superintendent and school board members may have a perspective that teachers will start bringing their pet dogs into classrooms. By informing and clarifying key terminology to decision makers, as well as explaining the expected behaviors of a therapy animal, a new perspective can be revealed. A basic understanding of the professionalism and work that is required to establish an AAI program may help encourage the approval of new policies.

## Different Kinds of Working Animals

The first step in developing a foundational appreciation for the field of AAI is to understand the title of the animal itself. Therapy animals are the animals who are involved in AAI, as they can provide physical, psychological, and emotional benefits to those they interact with, typically in facility settings such as healthcare, assisted living, and schools (Pet Partners, n.d.). Therapy animals do not have any special rights of access to public locations and must be invited in by facility management for the purposes of AAI. While most therapy animals are dogs, there are many other species who might work in this capacity. Pet Partners, one of the world's largest international therapy animal organizations, currently registers nine different species of therapy animals including dogs, cats, horses/donkeys, birds, rabbits, guinea pigs, rats, miniature pigs, and llamas/alpacas. These animals register with their human handlers by completing an evaluation to test for their propensity for this kind of work, and they are then able to visit in settings that match their team's preferences and capabilities. For example, a dog may demonstrate relaxed body language when visiting a retirement center but display stress signals in a school. A handler may feel comfortable around young children but have discomfort communicating with teenagers. These feelings and behaviors are key to determining the appropriate setting for a team's interactions.

While this chapter will focus on the work of therapy animals, it's worth noting the key features of other working animals so that handlers can help provide education on the differences that exist within these roles. Facility animals are closely related to therapy animals, as they are specially trained for extended interactions with clients or residents of a given facility where they work, which may include AAA, AAE, or AAT (Pet Partners, n.d.). Assistance animals, commonly called service animals, are defined as dogs and in some cases miniature horses that are individually trained to do work or perform tasks for people with disabilities (Pet Partners, n.d.). Emotional support animals are prescribed by a licensed mental health professional for a person with a mental illness (Pet Partners, n.d.). Refer to Figure 34.1 to learn more about the differences between these kinds of animals.

## Types of AAI

Having considered the differences between the roles of animals, it's also important to note that there are various interventions included within the umbrella term of AAI. AAT describes a goal-oriented, planned, structured, and documented therapeutic intervention directed by a health and/or human service provider that is being delivered as part of their professional role (Pet Partners, n.d.). AAE is also a structured intervention, but it is directed by a general education or special education professional with the objective of enhancing academic goals, prosocial skills, and cognitive functioning for students with progress being both measured and documented (Pet Partners, n.d.). AAA differs from the other interventions under the AAI umbrella in that they are more informal in nature while still providing meaningful opportunities for motivational, educational, and/or recreational benefits to enhance the quality of life of the people who visit with the therapy animal team (Pet Partners, n.d.). Furthermore, AAA are conducted by someone who is volunteering their time rather than working with the therapy animal in a professional capacity.

Classifying these different kinds of interventions requires a thorough assessment of the type of programming that is occurring with a therapy animal. For example, an assumption may be made that all interventions in a school setting would be included in AAE. However, due to the diversity of individuals delivering the interventions, the diversity of individuals receiving the intervention, and the intervention itself; AAT, AAA or AAE may be used in schools. An occupational therapist working with a volunteer therapy dog team and a student with gross motor goals would be conducting AAT. An art teacher greeting students in the morning with their therapy dog to aid in developing a positive school environment would be conducting AAA. A reading interventionist working with their own therapy dog and a group of students with reading comprehension goals would be implementing AAE, as would a social-emotional learning specialist who is teaching a humane education class to improve students' knowledge of human and dog body language with a volunteer therapy dog team.

When determining which kind of intervention a therapy animal is involved in, it's essential to determine the role and competencies of the person who is directing and/or delivering the intervention. It is also helpful to evaluate the goal of the intervention to determine which kind of programming is the best fit. If an organizational goal can be met by an informal visit with therapy animals, AAA is called for. Examples would be a librarian wanting to increase use of the library by the community, or a retirement center hoping to inspire social interactions among residents. These goals can be met by informal visits conducted by volunteer therapy animal teams, which account for much of the work that therapy animal teams do. However, even professionals may find they utilize AAA in some situations. A school counselor may want to increase positive relationships with families at their school, so they may be present with their therapy animal at pick-up/drop-off times or school events attended by parents and guardians.

If the goal of the intervention is specific to an individual or small group and it is directed and/or delivered by a professional in that field, then an evaluation of AAT or AAE is determined. Note that just because a volunteer therapy animal is involved in an intervention, it does not necessarily mean that AAA is taking place. Many professionals partner with volunteer therapy animal teams to offer AAT or AAE services when a therapy animal of their own is either not available or it has been determined that having a designated handler present is the best modality given the intervention at hand. If it is not clear which term to use, then stating the intervention is AAI may be most appropriate. Figure 34.2 further outlines the different kinds of interventions that fall within the AAI umbrella.

## Standards of Practice

As should be the case with any formal intervention, there are minimum standards of practice that need to be in place any time a person is working with a therapy animal. These standards address practices related to handlers, therapy animals, assessment, animal welfare, and risk management. There are several points of inconsistency within the field in terms of the practices associated with various therapy animal organizations, and it is therefore essential that each individual handler develop an appreciation of these guidelines to ensure they are following best practices (Serpell et al., 2020).

While professional competencies related to AAI will be further discussed later in this chapter, there is a basic level of knowledge, skills, and aptitude that all people who handle therapy animals must have to uphold the ethical integrity of the work (Kerulo et al., 2020). By registering with a reputable therapy animal organization, a handler is assessed to ensure they have training specific to AAI and are a responsible pet owner with an in-depth knowledge of their animal (Standards of Practice, 2021). The registration process that a therapy animal team undergoes also includes considerations of standards related to the therapy animal. These animals should have core training skills, a suitable temperament, appropriate grooming and health status, and a developed bond with their handler (Standards of Practice, 2021).

Both animals and people change over time, which is why another crucial standard of practice in AAI relates to the need for thorough and regular assessment of all therapy animal teams (Lefebvre et al., 2008; Standards of Practice, 2021). Continued assessment of therapy animals helps to protect animal welfare in AAI, which is a standard of practice that directly impacts safety for the intervention's participants. Therapy animals should not simply tolerate their role: they should enjoy it. They need to demonstrate active consent and should be protected by time limitations, routine veterinary care, positive training techniques, and the ability to end an intervention at any time the animal cues the handler to do so (Standards of Practice, 2021). Although these measures are widely considered to be best practices by AAI experts, not all therapy animal organizations build these standards into their guidelines. In fact, in a 2020 study on therapy animal organizations, only half of the included organizations imposed time limits for AAA visitation, and only a small group of the organizations prohibit potentially harmful diets such as raw feeding (Serpell et al., 2020).

## What Makes a Therapy Animal Different?

| | Therapy Animal | Assistance Animal | Emotional Support Animal |
|---|---|---|---|
| Primary role is to benefit many people. | ✓ | ✗ | ✗ |
| Primary role is to support one individual. | ✗ | ✓ | ✓ |
| Okay to approach and pet in public places. | ✓ | ✗ | ✗ |
| Has been evaluated to be tolerant of a wide variety of environments. | ✓ | ✓ | ✗ |
| Able to live in housing with "No Pets" policies. | ✗ | ✓ | ✓ |
| Special rights of access in public such as stores, restaurants, and airplanes. | ✗ | ✓ | ✗ |

*Assistance animals include service, hearing, and guide dogs.

www.petpartners.org

*Figure 34.1* Pet partners' chart: what makes a therapy animal different

Finally, there are standards for risk management that are essential in protecting all involved in AAI. Therapy animal teams should be insured either by a volunteer therapy animal organization or through professional liability insurance (Hartwig & Smelser, 2018; Standards of Practice, 2021). They should adhere to vaccine and dietary requirements that minimize the risk of infection and zoonosis and implement preventative measures such as hand hygiene and barrier use (Hardin et al., 2016; Standards of Practice, 2021). There should also be steps in place to ensure a therapy animal handler knows

**Pet Partners**

**Human-Animal Bond**
(HAB)
The human-animal bond is a mutually beneficial and dynamic relationship between people and animals that positively influences the health and well-being of both.

**Human-Animal Interactions**
(HAI)
Includes, but is not limited to, emotional, psychological, and physical interactions of people, animals, and the environment.

**Animal-Assisted Intervention**
(AAI)
Animal-assisted interventions are goal oriented and structured interventions that intentionally incorporate animals in health, education and human service for the purpose of therapeutic gains and improved health and wellness.

**Animal-Assisted Activities**
(AAA)
Includes:
- Hospital Visits
- Nursing Home Visits
- Memory Care
- Stress Reduction Visits at Universities, Airports, Conferences
- Hospice
- At-Risk Youth

Also Includes:
- Animal-Assisted Crisis Response
- Animal-Assisted Workplace Well-being

**Animal-Assisted Therapy**
(AAT)
Includes:
- Animal-Assisted Occupational Therapy
- Animal-Assisted Physical Therapy
- Animal-Assisted Counseling
- Animal-Assisted Social Work
- Animal-Assisted Speech Therapy
- Paraprofessional AAT Service Model

**Animal-Assisted Education**
(AAE)
Includes:
- Reading/Literacy Program
- Humane Education

*Figure 34.2* Pet partners' chart: AAI terminology

what to do should an incident with the therapy animal occur (Howie, 2021). Incidents are often complex in nature and may encompass everything from an injury to the participant or animal to a perceived incident, such as a participant experiencing a negative emotional state and/or fear of the therapy animal. Incidents should be well-documented and responded to in a timely manner that empowers the therapy animal team to make changes or reconsider their work depending on what has occurred (Howie, 2021; Standards of Practice, 2021).

## Roadmap to Partnering with a Therapy Animal

Given the state of confusion that is so commonly associated with AAI terminology and standards of practice, it's not surprising that so many handlers aren't sure where to get started when attempting to begin work as a therapy animal team (Chastain-Griffin et al., 2021). Professionals interested in this work may decide to partner with an experienced volunteer team to learn more about this modality and slowly build their program before establishing a comprehensive model for their own practice (VonLintel & Bruneau, 2021). Having established what a therapy animal does and realizing the structures that should be in place to protect the intervention at a basic level, a handler is ready to consider how they can best partner with an animal to bring the positive impacts of AAI to their community.

**Preparing with a Therapy Animal** As mentioned in the standards of practice considerations above, a therapy animal is one that doesn't merely tolerate interactions with a wide variety of people in a wide range of settings. Animals who are a good fit for this kind of work should have an affiliative nature (Crowley-Robinson & Blackshaw. 1998), meaning that they seek to connect with people outside of the members of their family. Their enjoyment for their role can be appreciated by viewing body language signals that are more often relaxed as compared to indicators of being aroused or avoidant (Howie, 2021). While there are certain characteristics to look for in determining which animals will be suited for AAI, therapy animals are ultimately beloved pets who have shown a propensity for working in this realm (Ng & Fine, 2019). Because animals are sentient beings, the characteristics of potential therapy animals need to be considered at an individual level when selecting candidates for AAI (Santaniello et al., 2021).

Therapy animals are best set up for success when they have undergone some level of training with their handlers. By working through a training program, the pair not only becomes prepared to face the novel stimuli they'll encounter in their work, but they also begin to establish a common language and working relationship (Standards of Practice, 2021). All training that a therapy animal participates in should be based on relationship-building and positive rewards. Aversive training methods should never be used to motivate therapy animal behavior. Not only have positive training methods been found to enhance the relationship that a handler has with their animal, but they've also been demonstrated to be more effective than other methods (Blackwell et al., 2008; Makowska, 2018; China et al., 2020). Working together through trained cues is a great way for a handler to establish a deep bond with their animal, and this kind of developed relationship is key to safe and effective AAI (Stewart et al., 2013). If a handler is working with a professional trainer to prepare for AAI, it can be very helpful that the trainer is familiar with the role of a therapy animal and can help the pair prepare accordingly (Howie, 2021).

**Encouraging a Therapy Animal to Thrive** At a minimum, we must protect the welfare of all animals who are involved in AAI. First proposed by the Farm Animal Welfare Council in 1965 (Brambell, 1965), the Five Freedoms have often been cited as a conceptual starting place for ensuring welfare of therapy animals. This framework highlights the importance of keeping all animals free from thirst, hunger, malnutrition, discomfort, pain, injury, disease, and distress while also allowing them to express normal behavior, such as the ability to stay or leave in response to any given stimuli. Many of the standards of practice and expected competencies outlined in this chapter have been created with the intention of protecting animal welfare in these ways. However, empirical investigation on the impact that AAI has on therapy animals is limited, and more research is needed to best determine the ways that welfare can be promoted. A 2017 literature review prepared by Glenk (2017) found that therapy animal welfare is influenced by considerations such as the frequency and duration of AAI sessions, novelty of the environment, controllability, age of therapy animal, and familiarity of recipients. There have also been studies that measured the stress of therapy dogs engaged in AAI, finding that measures such as salivary cortisol and stress-associated behaviors did not increase for dogs during their sessions (Glenk et al., 2013; Ng et al., 2014).

However, we owe it to our therapy animals to go beyond the assurance of basic welfare, moving into the space of encouraging a sense of true thriving. Having selected a therapy animal based upon their propensity for the role, and working from a foundation of essential standards, there is much that a handler can do to make the role of a therapy animal as enjoyable as possible. Therapy animals should be monitored to determine their preferences according to what is objectively rewarding for them, and they should be matched to a population accordingly (Standards of Practice, 2021). A professional who is partnering with a volunteer handler may have a discussion with available teams to determine what their animal enjoys. As they are considering which team to match with each client, they may consider how they can meet a client's goals by utilizing an animal's favorite ways to interact. A dog who enjoys retrieving balls may enjoy working with an occupational therapist and a client who is working on upper body gross motor skills. A dog who enjoys going for slow walks may also enjoy a mindfulness session with a counselor and client. A client working on executive functioning or emotional regulation may benefit from learning agility with a dog passionate about that sport.

Positive reinforcements such as treats, play time, and praise are commonly utilized tools to motivate and reward animal behavior (Blackwell et al., 2008), and these rewards can be incorporated into a therapy animal's work. However, handlers must also be aware of the fact that many therapy animals have a high desire to please, and we can thus limit our authentic communication with an animal and their subsequent ability to choose their interactions by asking for too many cues or over-rewarding the behavior we desire (Ng, 2019). Therapy animal handlers need to be prepared for the fact that an animal's preferences and abilities change throughout the lifetime, and a therapy animal can only thrive when their inclinations determine whether or not the intervention will continue. Careful assessment of therapy animals, often in conjunction with repeated evaluations and veterinary and/or behaviorist consultation, must inform when an animal should either take a break from their role or retire completely (Ng & Fine, 2019).

## Developing Professional Competencies

There are competencies specific to practicing AAI across all applications of the intervention. Everyone who partners with a therapy animal, from volunteer handlers to vocation-specific professionals, need to be able to demonstrate that they've established core capabilities that have prepared them for their work. Professionals who bring therapy animals into practice need to develop specialized skills (Stewart et al., 2016). Not only do they need to meet the minimum requirement for licensure within their respective field, but they should also have background of additional education to support their specialization as an AAI practitioner (Trevathan-Minnis et al., 2021). Many of the professional competencies that currently exist in the field are based on the dissertation research of Leslie Stewart who created a theoretical framework based on qualitative investigation with AAT practitioners (2014). Through these interviews, it was determined that professionals need to possess specific knowledge, skills, and attitudes before they work with a therapy animal. This includes training, assessment, and supervision as well as in-depth knowledge about the animal they partner with so that they can practice integrative ethics (Stewart, 2014). These professionals also must demonstrate a mastery of their vocation prior to intentionally incorporating an animal into their treatment plans (Stewart, 2014). They must be able to objectively assess the animal's behavior while realizing that handlers are prone to experiencing bias in assessing the therapy animals with whom they partner (Stewart, 2014). AAI professionals must respect animals and understand that animal welfare is directly linked to client safety (Stewart, 2014). These foundational competencies have since been built upon by many professionals and have even been adopted by various vocational professional associations, moving the field forward in the goal to increase standardization for the betterment of AAI. Figure 34.3 outlines these

| Handler Competency Level | Knowledge | Skills | Attitudes |
|---|---|---|---|
| Core Competencies: Required of animal-handler teams operating all levels and, in all capacities, (volunteer, paraprofessional, and professional). | • In depth knowledge of their therapy animal<br>• Knowledge of appropriate training techniques<br>• Ability to create and maintain a strong working relationship<br>• Understanding of relevant cultural facts<br>• Ability to promote safe interactions | • Ability to address the animal stress/fatigue and prevent burnout<br>• Ability to identify an animal's suitability, strengths and limitations | • Prioritization of their animal's welfare and safety<br>• Striving towards AAI professional values including enthusiasm for AAI, flexibility and creativity, and empathy |
| | Demonstrated through belonging to a therapy animal organization that adheres to the field's Standards of Practice (www.therapyanimalstandards.org) | | |
| Intermediate Competencies: Required of animal handler teams in paraprofessional and professional capacities. | • Discipline specialty training Appropriate risk management | • Skillful and intentional selection of appropriate interventions<br>• Ability to demonstrate effective judgement when assessing the session's impact on the therapy animal and client | |
| | Demonstrated through participating in training specific to the human-animal bond and to the discipline in which the services are being provided, as well as appropriate documentation, liability knowledge and insurance. | | |
| Professional Competencies: Discipline-specific knowledge, skills, and attitudes required of animal-handler teams in a professional capacity | • Knowledge of AAI specific techniques appropriate to their discipline<br>• Participation in supervised practice<br>• Aware of AAI specific ethical consideration and can integrate those within their ethical professional practice | • Demonstrate a mastery of discipline specific professional skills prior to integrating AAI<br>• Able to incorporate the animal's responses in a therapeutically meaningful way | • Professional advocate for AAI |
| | Demonstrated through AAI Specialist Certification, active involvement in continuing education and belonging to network of AAI professionals that allows for collaboration and unbiased feedback | | |

*While this chart outlines some of the best practices associated with handler competency levels, it is not exhaustive.

*Figure 34.3* Pet partners' chart: demonstration of handler competencies

competencies, their associated standards of practices, and some avenues by which a handler might demonstrate these competencies at each level.

## Preparing Clients for AAI

Not all people desire to interact with therapy animals, and the intervention should never be forced upon any participants – human and animal alike. Volunteer handlers who are adequately prepared for AAI have been trained on how to approach a person and give them the ability to choose their degree of engagement with an animal (Standards of Practice, 2021). Professionals who work with therapy animals should have a means by which they inform clients about the intervention, its risks, and the expectations of all involved (Chandler, 2012; Ng, 2019). This preparatory process helps set the stage for proactive handling in which the client knows to rely on the handler to direct interactions while also allowing for all involved to play active roles as advocates for the animal (Standards of Practice, 2021). The person who is overseeing AAI should always consider whether or not each client is a good fit for the interaction. By and large, participants should be accepting and approachable (Ng, 2019). In some cases, people with fears and phobias of animals, those with a history of animal abuse, and participants with a diminished capacity to understand the intervention might not be a good fit for working with a therapy animal (Chandler, 2012). The cultural implications of the human-animal relationship for each individual client must also be considered and respected prior to the introduction of AAI (Valiyamattam et al., 2018). Individuals with allergies and a strong desire to engage in AAI should not be automatically restricted from this work. Professionals may consider a discussion regarding the nature of the allergy and medications, alternative therapies/interventions, potential changes to the proposed animal-assisted model to include switching intervention locations and significantly limiting any physical contact. This may allow clients to safely participate with their physician's approval.

## Investing in Continued Education and Professional Community

An AAI handler is on a never-ending journey of learning more about their animals and about the intervention as a whole. By taking the position of a lifelong learner, handlers can remain aware of advances in theory, evidence-based research, animal behavior, and specific therapy animal techniques (Trevathan-Minnis et al., 2021). Handlers who are working within a sub-specialty in AAI, such as crisis-response, should also undergo training reflective of the niche area in which they are working (Chandler, 2012). It is equally important that those working in the field of AAI have a network of professionals with whom they can share information and receive consultation from those who are outside of a handler's intimate relationship with their animal. This allows for more unbiased feedback to be generated, as all handlers are prone to blind spots when it comes to judging the actions of their own animal (Howie, 2021; Trevathan-Minnis et al., 2021).

More and more avenues for continued education and professional networking are becoming available within the field of AAI. There are many universities that have centers of human-animal interaction which offer in-depth course work on subjects related to working with therapy animals. There are also organizations such as the International Association of Human-Animal Interaction Organizations, Animal Assisted Interventions International, the International Society for Anthrozoology, and the Association for Animal-Assisted Intervention Professionals (AAAIP), which offer education and connection aimed at growing the field. With its launch in 2022, AAAIP made available the field's first true certification exam for Animal-Assisted Intervention Specialists. Through this offering, professionals can demonstrate they have learned the crucial points of knowledge associated with working with a therapy animal, and they are required to show evidence of continued learning for their certification to stay active. The availability of this certification exam is a key step for the field in terms of generating a way for professionals to not only demonstrate their preparedness for working in AAI, but for clients and consumers of the intervention to check for competence in the handlers who introduce them to therapy animals.

## Advocating for the Intervention

Having outlined some of the areas of growth that are still called for within AAI, tremendous development has been achieved by the field as more and more professionals come together to advance the intervention. This progress wouldn't have been possible if it weren't for the core commitment to advocacy that AAI professionals have – a value so important that it has been identified as a key competency for therapy animal professionals (Stewart et al., 2013). To effectively advocate for the work that therapy animal teams do, we must continue to come together in utilizing shared language while also educating others on the terminology that is frequently referenced in this space (Parish-Plass, 2014; Schlote, 2009). The use of this common language will not only help to decrease conflation, but it will also assist those who are newer to the field in being able to get started while following best practices and standards in a meaningful way (Parish-Plass, 2014).

There is also a need to further advocate for the impact of AAI by contributing to empirical research on the intervention. While anecdotal experiences of the impact of therapy animals are powerful, scientific data related to the intervention's outcomes help to increase the merit of the intervention. Though there have been many promising research findings on AAI over the past several decades, future studies with increased rigor, random clinical control groups, clear internal validity, and solid theoretical basis are called for (Stewart et al., 2016; Fine & Beck, 2019; López-Cepero, 2020). As the field expands upon its commitment to promote therapy animal wellbeing, there is a need for continued investigation on the impact that AAI has on therapy animals (Fine & Beck, 2019; Glenk 2017; Ng, 2019). Moving towards increased understanding of specific outcomes such as true pleasure indicators of therapy animals will assist handlers in moving beyond the provision of welfare and into the encouragement of therapy animal thriving (Santaniello et al., 2021).

## Next Steps for the Field

It is no surprise that the field of AAI is burgeoning across so many different disciplines as more people witness the potential that therapy animals have in inspiring wellness. Many crucial steps are already underway to help make this intervention not only more accessible but safer and more standardized for all who wish to access it. Increased research has enabled translational findings to inform the development of best practices in AAI (Fine et al., 2019). Clarity around terminology has allowed for more effective advocacy, both within the field and externally. The creation of professional associations and a certification exam have moved the needle in terms of professionalizing the practice of partnering with therapy animals. Many professionals have become familiar with the power of human-animal interactions by working together with volunteer teams. As this modality advances, we may see volunteer teams that specialize in certain types of AAI. This is especially true if the volunteer handler has experience in that particular treatment area. For example, a retired teacher may find that specializing in reading interventions with their dog is rewarding for themselves, their dog, the students, and a classroom teacher that is directing the intervention.

As the field continues to grow, room is being made for the various activities and vocations represented by AAI professionals to be differentiated, documented, and individually studied so that specific standards related to qualities such as dosage and the ideal format for AAI within each unique application can be determined (Jones et al., 2019). Finally, there is a call within the field to enhance public policy efforts and to inspire buy-in for the intervention outside the community of advocates who have already witnessed the power of the human-animal bond and AAI (Fine et al., 2019). As these efforts come together, AAI will become more recognized and respected as the scientific intervention that it is.

## Conclusion

In many ways, the field of AAI is just getting started. The pioneers in practice and research have made it possible for us to realize that intentional interaction with therapy animals is extremely versatile in its ability to reach not only a wide range of diagnoses and client goals, but to meet the needs of diverse individuals along the entire lifespan. While AAI is already present across many vocational domains, with continued research and increasingly universal terminology, it is unknown how many other fields could also benefit by partnering with therapy animals. To keep growing this field in an ethical direction, it is vital to understand the work that goes into being an active participant of AAI as a volunteer, health and human service provider, or other professional. As mentioned throughout the chapter, handlers and individuals facilitating AAI need to be adaptable and committed to being lifelong learners. Working in this field means constantly being aware of evidence-based research, investing in continued education, advocating for your animal and the use of this modality, acquiring professional competencies, adhering to standards, and evaluating and reevaluating that your animal is enjoying their role during each interaction. Creating a strong foundation will undoubtedly allow AAI to ethically expand, becoming a highly recognized intervention capable of making major impacts in the lives of both the people and the animals who partner together for this mission.

## Discussion Questions

1. What are the differences between therapy animals, emotional support animals, and services animals?
2. How does animal-assisted therapy differ from animal-assisted activities and animal-assisted education?
3. How can a practitioner protect the wellbeing of the therapy animals involved in their practice? How can they prepare clients to take an active role in this critical duty?
4. What steps can be taken to improve future empirical investigation on AAI?
5. Why is it important for an AAI practitioner to assume the position of a lifelong learner?

# References

Barker, S. B., Knisely, J. S., Schubert, C. M., Green, J. D., & Ameringer, S. (2015). The effect of an animal-assisted intervention on anxiety and pain in hospitalized children. *Anthrozoös, 28*(1), 101–112.

Blackwell, E. J., Twells, C., Seawright, A., & Casey, R. A. (2008). The relationship between training methods and the occurrence of behavior problems, as reported by owners, in a population of domestic dogs. *Journal of Veterinary Behavior, 3*(5), 207–217.

Brambell, F. W. R. (1965). *Report of the technical committee to enquire into the welfare of animals kept under intensive livestock husbandry systems: Command paper 2836*. London: Her Majesty's Stationary Office.

Chandler, C. K. (2012). *Animal assisted therapy in counseling*. New York: Routledge

Chang, S. J., Lee, J., An, H., Hong, W. H., & Lee, J. Y. (2021). Animal-assisted therapy as an intervention for older adults: A systematic review and meta-analysis to guide evidence-based practice. *Worldviews on Evidence-Based Nursing, 18*(1), 60–67.

Chastain-Griffin, T., Hudson, A., & Trella E. Pet Partners. (2021). The state of animal-assisted interventions in professionalized settings [White paper]. Retrieved from https://petpartners.org/learn/aat-professionals/aai-in-professionalized-settings/

China, L., Mills, D. S., & Cooper, J. J. (2020). Efficacy of dog training with and without remote electronic collars vs. a focus on positive reinforcement. *Frontiers in Veterinary Science, 7*, 508.

Crowley-Robinson, P., & Blackshaw, J. K. (1998). Nursing home staffs' empathy for a missing therapy dog, their attitudes to animal-assisted therapy programs and suitable dog breeds. *Anthrozoös, 11*(2), 101–104.

Fine, A.H., & Beck, A. (2019). Understanding our kinship with animals: Input for healthcare professionals interested in the human/animal bond. In Fine A.H. (Ed.), *Handbook on animal-assisted therapy: Theoretical foundations for guidelines and practice* (3rd ed., pp. 3–12). San Diego, CA: Elsevier Inc.

Fine, A. H., Beck, A. M., & Ng, Z. (2019). The state of animal-assisted interventions: Addressing the contemporary issues that will shape the future. *International Journal of Environmental Research and Public Health, 16*(20), 3997.

Glenk, L. M., Kothgassner, O. D., Stetina, B. U., Palme, R., Kepplinger, B., & Baran, H. (2013). Therapy dogs' salivary cortisol levels vary during animal-assisted interventions. *Animal Welfare, 22*(3), 369–378.

Glenk, L. M. (2017). Current perspectives on therapy dog welfare in animal-assisted interventions. *Animals, 7*(2), 7.

Hardin, P., Brown, J., & Wright, M. E. (2016). Prevention of transmitted infections in a pet therapy program: An exemplar. *American Journal of Infection Control, 44*(7), 846–850.

Hartwig, E. K., & Smelser, Q. K. (2018). Practitioner perspectives on animal-assisted counseling. *Journal of Mental Health Counseling, 40*(1), 43–57.

Hillen, N., & Kirnan, J. (2020). Animal-assisted intervention on college campuses. *TCNJ Journal of Student Scholarship, 23*, 1–11.

Hoagwood, K. E., Acri, M., Morrissey, M., & Peth-Pierce, R. (2017). Animal-assisted therapies for youth with or at risk for mental health problems: A systematic review. *Applied Developmental Science, 21*(1), 1–13.

Howie, A. R. (2021). *Assessing handlers for competence in animal-assisted interventions*. Purdue University Press.

Johnson, R. A., Meadows, R. L., Haubner, J. S., & Sevedge, K. (2008, March). Animal-assisted activity among patients with cancer: Effects on mood, fatigue, self-perceived health, and sense of coherence. *Oncology Nursing Forum, 35*(2), 225–232.

Jones, M. G., Rice, S. M., & Cotton, S. M. (2019). Incorporating animal-assisted therapy in mental health treatments for adolescents: A systematic review of canine assisted psychotherapy. *PloS One, 14*(1), e0210761.

Kerulo, G., Kargas, N., Mills, D. S., Law, G., VanFleet, R., Faa-Thompson, T., & Winkle, M. Y. (2020). Animal-assisted interventions: Relationship between standards and qualifications. *People and Animals: The International Journal of Research and Practice, 3*(1), 4.

Lefebvre, S. L., Golab, G. C., Castrodale, L., Aureden, K., Bialachowski, A., Gumley, N., ... Writing Panel of the Working Group. (2008). Guidelines for animal-assisted interventions in health care facilities. *American Journal of Infection Control, 36*(2), 78–85.

López-Cepero, J. (2020). Current status of animal-assisted interventions in scientific literature: A critical comment on their internal validity. *Animals, 10*(6), 985.

Makowska, I. J. (2018). Review of dog training methods: Welfare, learning ability, and current standards. In *Dog training standards, version 2 prepared by the BC society for the prevention of cruelty to animals* (pp. 1–47). Vancouver, BC.

Maujean, A., Pepping, C. A., & Kendall, E. (2015). A systematic review of randomized controlled trials of animal-assisted therapy on psychosocial outcomes. *Anthrozoös, 28*(1), 23–36.

McCullough, A., Ruehrdanz, A., Jenkins, M. A., Gilmer, M. J., Olson, J., Pawar, A.,... O'Haire, M. E. (2018). Measuring the effects of an animal-assisted intervention for pediatric oncology patients and their parents: A multisite randomized controlled trial. *Journal of Pediatric Oncology Nursing, 35*(3), 159–177.

Morrison, M. L. (2007). Health benefits of animal-assisted interventions. *Complementary Health Practice Review, 12*(1), 51–62.

Mueller, M. K., Anderson, E. C., King, E. K., & Urry, H. L. (2021). Null effects of therapy dog interaction on adolescent anxiety during a laboratory-based social evaluative stressor. *Anxiety, Stress, & Coping, 34*(4), 365–380.

Ng, Z. (2019). Advocacy and rethinking our relationships with animals: Ethical responsibilities and competencies in animal-assisted interventions. In Tedeschi, P., & Jenkins, M. A. (Eds.), *Transforming trauma: Resilience and healing through our connection with animals* (pp. 55–90). West Lafayette, IN: Purdue University Press.

Ng, Z. Y., & Fine, A. H. (2019). Considerations for the retirement of therapy animals. *Animals, 9*(12), 1100.

Ng, Z. Y., Pierce, B. J., Otto, C. M., Buechner-Maxwell, V. A., Siracusa, C., & Werre, S. R. (2014). The effect of dog–human interaction on cortisol and behavior in registered animal-assisted activity dogs. *Applied Animal Behaviour Science, 159*, 69–81.

O'Haire, M. E., Guérin, N. A., & Kirkham, A. C. (2015). Animal-assisted intervention for trauma: A systematic literature review. *Frontiers in Psychology, 6*, 1121.

Parish-Plass, N. (2014). Order out of chaos revised: A call for clear and agreed-upon definitions differentiating between animal-assisted interventions. *Animals and Society, 50*, 71–79. Retrieved April 2021

Pet Partners. (n.d.). *Terminology*. Retrieved December 13, 2021, from https://petpartners.org/learn/terminology/.

Santaniello, A., Garzillo, S., Cristiano, S., Fioretti, A., & Menna, L. F. (2021). The research of standardized protocols for dog involvement in animal-assisted therapy: A systematic review. *Animals, 11*(9), 2576.

Schlote, S. M. (2009). *Animal-assisted therapy and equine-assisted therapy/learning in Canada: Surveying the current state of the field, its practitioners, and its practices* (Doctoral dissertation).

Serpell, J. A., Kruger, K. A., Freeman, L. M., Griffin, J. A., & Ng, Z. Y. (2020). Current standards and practices within the therapy dog industry: Results of a representative survey of United States therapy dog organizations. *Frontiers in Veterinary Science, 7*, 35.

*Standards of practice in animal-assisted interventions*. therapyanimalstandards.org. (n.d.). Retrieved December 13, 2021, from https://therapyanimalstandards.org/.

Stewart, L. A. (2014). Competencies in animal assisted therapy in counseling: A qualitative investigation of the knowledge, skills and attitudes required of competent animal assisted therapy practitioners, Ph.D. dissertation. Atlanta, GA: Georgia State University

Stewart, L. A., Chang, C. Y., & Rice, R. (2013). Emergent theory and model of practice in animal-assisted therapy in counseling. *Journal of Creativity in Mental Health, 8*(4), 329–348.

Stewart, L. A., Chang, C. Y., Parker, L. K., & Grubbs, N. (2016). *Animal-assisted therapy in counseling competencies*. Alexandria, VA: American Counseling Association, Animal-Assisted Therapy in Mental Health Interest Network.

Trevathan-Minnis, M., Johnson, A., & Howie, A. R. (2021). Recommendations for transdisciplinary professional competencies and ethics for animal-assisted therapies and interventions. *Veterinary Sciences, 8*(12), 303.

Valiyamattam, G., Yamamoto, M., Fanucchi, L., & Wang, F. (2018). Multicultural considerations in animal-assisted intervention. *Human-Animal Interaction Bulletin, 6*, 82–104.

VonLintel, J., & Bruneau, L. (2021). Pathways for implementing a school therapy dog program: Steps for success and best practice considerations. *Journal of School Counseling, 19*(14), 1–34.

# 35
# ANIMALS IN EDUCATION

*Nancy R. Gee*

## Background

### What Is Animal-Assisted Education?

Broadly speaking, animal-assisted interactions encompass all possible types of interactions between humans and animals. As you can see in Figure 35.1 these interactions can be separated into interventions (AAI) and activities (AAA). An intervention is planned, goal-oriented, often structured, and assessed for efficacy by a trained professional, such as an educator or a therapist. Activities are frequently unstructured, may or may not include broadly stated goals (e.g., mood elevation, stress reduction, reading improvement) and are not typically assessed for efficacy. Further, activities do not require the presence of a trained professional, such as a therapist or teacher, to be present during their delivery.

Animal-Assisted Education (AAE) according the International Association of Human Animal Interaction Organizations (IAHAIO) white paper is "…a goal oriented, planned and structured intervention directed and/or delivered by educational and related service professionals" (IAHAIO, 2018, p. 5). It is delivered by a qualified educator, either individually or to groups of students, who determine their own assessment of student learning outcomes. The focus of AAE is typically on relevant goals related to specific educational content, or on the development of educationally relevant skills such as social skills or cognitive functioning.

Educationally relevant, but adjacent to AAE, is Animal-Assisted Therapy (AAT), which likewise is "…a goal oriented, planned and structured intervention" (IAHAIO, 2018, p. 5) but in this case is delivered most often by a trained, and usually licensed, health services professional, such as a psychologist, social worker, or occupational or speech therapist. This, individual, trained to deliver a therapeutic service, documents progress in the enhancement of physical, emotional, cognitive, behavioral, and/or social functioning. AAT is considered adjacent to AAE because enhancement of functioning can have direct and indirect impacts on educational outcomes and classroom adjustment.

Understanding these terms can be complicated because the same type of interaction between humans and animals can be classified differently depending on how it is delivered. For example, consider the very popular reading to dogs programs (Hall et al., 2016). Depending upon how these programs are structured and delivered they can be classified as an

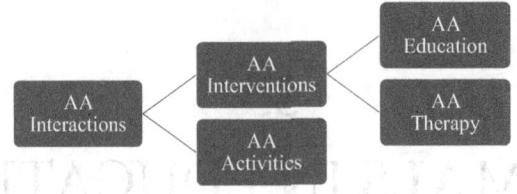

*Figure 35.1* Structure of Animal-Assisted (AA) terms

AAA, an AAE, or an AAT. If the program were designed and implemented by an educator or researcher who set up specific reading goals, used standardized measures of reading, determined delivery of the details of the reading program and then assessed the efficacy of the program, that would constitute an AAE. If the program were set up at a local library with reading to dogs times advertised and any child could take advantage of the opportunity to read any book to a dog (and their handler) without assessment or any other formal structure, this would be an AAA. If a speech therapist, worked with a child to practice reading to a dog as part of their ongoing speech therapy (involves therapeutic goals and assessment) that would be an AAT.

## *Why Teachers and Students Are Drawn to AAE*

Educators have long recognized the value of incorporating animals into educational experiences for students because the practice helps teachers to meet the developmental and educational goals they set for students (Gee & Fine, 2019). The inclusion of animals provides a way for students to deepen their knowledge, pique their interests and increase in their motivation for learning. Humane education, or teaching about the animal's side of the equation, helps people to understand the value of all living beings, to learn about how the animal experiences their world, to foster critical thinking with regards to the needs of the animals and society, and to encourage the development of empathy, compassion, integrity, and wisdom in the process.

## *Historical Perspective*

Though not well documented, the role of animals in educational curricula in the United States can be consistently identified throughout different eras (Smilie, 2022). Pet ownership was ubiquitous around the time of the Civil War and as such pets are commonly represented in instructional texts and the literature of the time, although there is no evidence supporting the presence of animals in the classroom. Since, even if they did not own pets, most Americans lived in rural areas and interacted with farm animals routinely in the 19th century, this is not surprising (Smilie, 2022).

Smilie (2022) uses the analogy of "beasts of burden" to describe the role assumed by animals in American K-12 classrooms beginning in the 20th century: they shouldered the responsibility for teaching children about the uncomfortable topics of sex education and death. Additionally, in the early 1900s, educators advocated for benefits stemming from children having direct contact with animals and nature through class related projects. Classroom pets also began serving as a source of comfort or a way to alleviate alienation for students who were processing heavy topics or stressful situations. By the early 1980s, threats of nuclear war served as a background to what was happening in the classroom and pets helped students process stress and uncertainty.

In short, schools have historically used animals in a way that mirrored society's use of animals as exploitative, though also providing educational value. As AAE continues to advance, it is important to recognize this exploitative history and purposefully move toward an approach that recognizes the value and dignity of the animals, grants them agency, and teaches mutual respect and shared benefits.

## *Settings*

AAE takes place in a wide variety of settings including typical and special needs classrooms (Meints et al., In Press), classrooms across the developmental spectrum from preschool classrooms (for a summary, see Gee et al., 2015) to university settings (Pendry et al., 2021) and in school and public libraries (Hall et al., 2016). But AAE isn't limited to standard school-based settings, it also takes place in farm settings (Flynn et al., 2020), at aquariums and zoos (Ballantyne et al., 2021), in natural settings (Russell, 2019), and even on the poles of the planet (Choi et al., 2021). It is hard to imagine a setting in which one cannot teach or learn about, or with, the inclusion of animals.

## *Populations*

The literature indicates that AAIs have been successfully deployed with a wide variety of educational populations including typical preschool children and those with language impairments (for a summary, see Gee et al., 2015), typical and special needs grade-school children (Meints et al., In Press), children with reading difficulties (Le Roux et al., 2014), children with Autism Spectrum Disorder (Dimolareva & Dunn, 2021) or Attention Deficit Hyperactivity Disorder (Schuck et al., 2015), and typical and at-risk college students (Pendry et al., 2021).

## *Animals Involved*

It is entirely likely that AAE has included a larger variety of animals than any other type of AAI because educational opportunities can take place in such wide variety settings (as described above), allowing for the inclusion of an equally wide variety of animals. For example, classroom pets commonly include fish, turtles, rabbits, insects, snakes, birds, rodents, and other small mammals. Dogs and other therapy animals, such as horses and miniature horses, cats, miniature pigs, alpacas, and donkeys, are frequently brought to into school settings for a variety of educational experiences for students. Field trips to farms, zoos and aquariums provide opportunities for students of all ages to learn about the wide variety of animals. In biology and ecology classes, students often capture and learn about insects and microscopic life forms from local waterways, or spend time bird watching and observing a wide variety of local wild bird species. There are even learning-based excursions such as whale watching and safaris designed to help people of all ages experience and learn about creatures that we would otherwise only see in video recordings.

Finally, there are controversial wild animal experiences, aimed at educating people about those animals. For example, dolphin encounters allow people to swim with dolphins and give them food to encourage the dolphins to swim close enough for the human to touch them. One such program charges from $80 to about $650 for these encounters, depending on how much interaction the person wants to have with the dolphin, such as belly rides, kisses and handshakes, getting towed on the dorsal fin or be a "Dolphin Trainer for a Day" which includes more educational opportunities (Costhelper.com, n.d.). The controversy in these programs surrounds the idea that the wild animals are being exploited by humans. The dolphins are required to live in comparatively tiny spaces and interact directly with humans (often requiring physical contact) in order to eat. All of which creates an impoverished environment and stressful life experience for the animal, but a profitable venture or fun experience for humans.

## Evidence for Efficacy

### Children

Evidence indicates that interacting with a dog has a stress reducing/anti-anxiety effect in children, which may prevent the evolution of emotional problems into full-fledged emotional, mental or behavioral disorders (Purewal et al., 2017). Animals may provide a source of comfort and safety for children, potentially alleviating distress and reducing behavioral problems. They appear to enhance social support directly, as a source of comfort, and indirectly, through facilitation of social interactions. For example, O'Haire and colleagues have demonstrated improved social functioning in children with Autism Spectrum Disorder following an 8-week animal-assisted activity involving guinea pigs (O'Haire et al., 2014) and in another study they demonstrated an increase in positive social behaviors following just three 10-minute interactions with a guinea pig (O'Haire et al., 2013).

Other positive outcomes linked to AAE include increased attention and concentration (Hediger & Turner, 2014), improvements in feelings of student connectedness (Fedor, 2018), attitudes toward learning and school attendance, development of empathy and positive emotions, task performance, classroom and emotional cohesion, prosocial behaviors, and aggression (Scandurra et al., 2021). It is interesting to note that most, but not all, of the studies showing these effects have taken place with special needs populations. It is possible that due to the additional challenges these children face, the presence of the animal may be particularly beneficial in an educational setting due to their ability to reduce stress and facilitate communication.

The major cognitive deficit in children with ADHD is impaired executive function including reduced self-regulation and failure to attend to social cues (Schuck & Crinella, 2005). Standard interventions include school-based behavior modification. Schuck and colleagues (Schuck et al., 2015) implemented a canine-assisted intervention over 12-weeks and included a comparison to the standard intervention. They found reduced ADHD symptoms in the canine group when compared to the control condition. Later, they reported improvements in inattention and social skills, including social initiation in the dog group (Schuck et al., 2018). These findings indicate the beneficial effect of the presence of therapy dogs in a school-based behavior modification plan for children with ADHD, suggesting unique mechanisms of action which appear to differentially enhance standard approaches to treatment.

Particularly relevant to AAE is the evidence showing educationally related benefits derived from involving animals in educational settings. I conducted a series of studies involving registered therapy dogs in a number of educationally relevant tasks (for a summary, see Gee et al., 2015). Most of these studies compared the performance of preschool children executing a task with a dog and their handler, with a stuffed toy dog (similar in appearance to the real dog) and their handler, and with the handler alone. My team and I found that in the dog condition relative to the other conditions, the children adhered to instructions better (Gee et al., 2009), required fewer instructional prompts to complete the task (Gee et al., 2010), had better memory performance (Gee, Belcher, et al., 2012), made fewer errors on a categorization task (Gee et al., 2010), categorized animate objects better (Gee, Gould, et al.,

2012), and completed motor skills tasks faster without sacrificing accuracy (Gee et al., 2007). Although this work involved small sample sizes, the methodological rigor of the studies and the effect sizes noted both suggest that under the right conditions involving animals in the education of children can be beneficial for the performance of tasks that are beneficial to academic performance and success.

## *Young Adults*

University-based animal visitation programs are often deployed around final exam time, with the goal of alleviating student stress and promoting well-being (Crossman, 2019). These programs usually involve therapy dogs and their handlers visiting with students in informal and unstructured interactions. They tend to be widespread and wildly popular with students, often requiring crowd control. According to Crossman (2019) who summarized the relevant literature, there has been tremendous growth in research evaluating these programs. Studies show, for example, that these programs are associated with improved mood and life satisfaction, reduced anxiety and stress, reduced depression, homesickness, and loneliness. We must be cautious in our interpretation of these results because a number of studies have failed to show effects on physiological measures as blood pressure, heart rate, heart rate variability, and electrodermal activity. Crossman also points out that there are studies showing effects that are opposite of predicted outcomes.

In one randomized control study, university students at-risk of academic failure experienced improved executive function after repeated opportunities to interact with a therapy dog compared to engaging in formalized university approved stress management programming (Pendry et al., 2021). This is strong evidence that interacting with dogs, rather than devoting resources to plan and deliver formalized stress management programming can be cost-effective approach to improving executive functioning while also reducing stress-related indices in students.

## *Reading and Literacy Programs*

Reading to dogs programs have been growing in popularity since the early 2000s (Henderson et al., 2020). Generally, these programs take place in a school setting or public library and involve a child or small group of children reading for a set amount of time (20–30 minutes) to a registered therapy dog and their handler. There is variability in the amount of instructional support provided by the handler, but most programs have the handler provide periodic encouragement or assist with a word or words only occasionally. A systematic review of the evidence indicated that reading to a dog may have a beneficial effect on a number of behavioral processes linked to the reading environment contributing to improved reading performance (Hall et al., 2016). For example, there is some evidence that reading to dogs may improve children's motivation (Rousseau & Tardif-Williams, 2019), well-being, perceptions of themselves as readers, attitudes toward reading and pro-social classroom behaviors (Henderson et al., 2020). Students with special needs may be particularly poised to benefit from these programs as they may serve to increase motivation, ability to stay on task, and elevate confidence and self-esteem (Kirnan et al., 2020). In a randomized control trial, poor readers (Le Roux et al., 2014) had significantly improved reading accuracy and comprehension in the reading to dog condition compared to the control conditions. Importantly, teachers also perceived benefits to children's social, emotional, and behavioral outcomes as well as their motivation for, and confidence in, reading (Steel et al., 2021).

## Summary/Evaluation of evidence

In a recent meta-analysis of the effects of dogs on learning, 20 studies were examined which included 66 independent effect sizes (Reilly et al., 2020). For reading outcomes, the overall effect size for 12 studies conducted in the United States and in Africa was moderate in size. For social outcomes, the effect sizes for studies conducted in the United States were moderately large compared to studies in Europe or Asia. For emotional/behavioral outcomes, studies conducted in the United States had small but significant effect sizes when compared to those conducted in Europe. Overall, this analysis of studies conducted across multiple countries revealed the benefit of dogs on student learning. It also highlights that the majority of evidence in AAE tends to involve dogs, which indicates that other types of animals frequently involved in AAE (classroom pets, farm, zoo, aquarium settings) have not been studied as extensively.

Another systematic review of the evidence related to AAE included 25 studies, most of which reported significant positive effects of animals in educational settings (Brelsford et al., 2017). The studies included varied widely in methodologies and design, intervention types, sample sizes, measures used, and duration of exposure to the animal. Furthermore, this review indicated that there is a glaring paucity of risk assessment and attention to animal welfare. As you will see in the next section, there is reason to be concerned about the welfare of animals housed in classrooms or more generally involved in AAE.

## Impact on Animals – Animal Welfare Concerns

The roles of animals in classrooms have generally matched the ways they have been treated, often exploited, by humans in other settings (Smilie, 2022). They have been required to reproduce to demonstrate the life cycle and served as the subject of dissection for anatomy lessons, to cite a few examples. Pets kept in classrooms full time may experience many stressors. They may be housed in small or inadequate enclosures from which they are regularly removed to be handled by many different unfamiliar and unskilled hands. Often inappropriate choices are made about which animals to include in a classroom setting, such as animals that are nocturnal (e.g., mice, hedgehogs). Those animals are stressed by constantly being awakened during their natural sleeping time. Many classroom pets suffer abuse from children, even well-meaning children who do not understand how to treat animals. They may also suffer from neglect when they are left alone for long periods of time over weekends or school breaks. Some teachers may allow students to take the animal home over these breaks as a way of learning responsibility, but a household full of other pets or children may be a dangerous and stressful place for the animal.

Animals that are brought into classrooms for "show and tell" experiences, or "bring your pet to school" day are also likely to find the experience extremely uncomfortable. For those animals, being taken to school and handled by large groups of children and teachers, potentially kept at school all day and subject to frequent and abrupt handling is likely to be stressful, overwhelming and potentially dangerous to the animal or the children in the classroom if the animal feels that it must defend itself.

Some teachers bring their own pets to school and include them as classroom pets during the course of the school day. There are many potential concerns in these situations, most significantly, that the teacher's attention is divided between all the children in the room and the animal. The teacher may be focused on a child or group of children, while another child or group of children begins to interact with the animal. The teacher cannot be in all places at once and the animal may suffer harm or stress or may need to defend itself. One simple way to avoid this potential problem is to invite therapy animal teams into the classroom. Each team includes an animal and their human handler. The handler's primary concern is the welfare of their own animal. That leaves the classroom teacher more freedom to focus on the needs of the students in the room.

Teachers may or may not seek the approval of their school administrator to bring their pet into the classroom. Even if they do, approval from a person in a position of authority at the school is generally meaningless in terms of assessing the suitability of their animal to the environment and the student population. In many cases these animals have no formal training, no habituation to the environment, or no independent evaluation of their suitability to be present in a classroom. This situation is risky for everyone and represents a potentially serious animal welfare concern. To avoid over-taxing the animals, trained and registered therapy animals are typically given a two-hour maximum working time in a given day. By contrast, these animals work the same hours the teachers work, in the same Monday through Friday schedule.

Without a doubt, the best-case scenario from an animal welfare perspective is one in which the animal has their own dedicated advocate in the room. This is a person who can recognize behavioral signs of stress and enjoyment in the animal, and is dedicated to meeting their needs without the distraction of also having to meet someone else's needs. When a trained and registered therapy animal and their handler visit a classroom, the handler can focus on the needs of their animal and act immediately in the best interest of that animal.

## Practical Issues

### Policy Needs

Despite the growing popularity of AAE, there are no unified standardized policies or requirements for how it should be delivered, and there is a glaring absence of oversight in place to ensure the protection of the students and animals. There are existing guidelines for animal-assisted interventions (see Table 35.1), but often teachers and school administrators are unaware of them and there is no formal requirement for adherence. An overarching policy at the state or federal level is required to ensure that these guidelines are followed. This could be accomplished initially with a very simple policy stating minimal requirements that could be refined and modified over time.

In a recent review of current standards and practices in the therapy dog industry across the United States (Serpell et al., 2020) described a number of areas as departures from what is consider best practices for AAI. These include: lack of limits on working times for the dogs, lack of agreement on reward-based training approaches, lack of ban on raw-fed diets for dogs representing a serious zoonotic infection concern, lack of re-evaluation of animals for ongoing suitability to therapy dog work, inconsistency with regard to vaccination requirements, nail clipping, or continuous flea/tick treatment for the dogs and lack of information provided to handlers about hand hygiene, handling of animal waste, and avoiding visitation when a member of their household has a communicable disease. Unfortunately, there is a glaring lack of standards and best practices noted for other species.

### Risk Assessment

All of this indicates that AAE currently takes place in an unregulated manner with sparse recognition and implementation of best practice standards. In order to safely implement an

Table 35.1 Examples of organizational guidelines relevant to AAE

| Source | Type of Guidelines | How to Access |
|---|---|---|
| Animal-Assisted Intervention International (AAII) | Comprehensive AAT, AAA, AAE, etc… | https://otaus.com.au/publicassets/80cfd523-2030-ea11-9403-005056be13b5/AAII-Standards-of-Practice.pdf |
| American Veterinary Medical Association (AVMA) | General AAI | https://www.avma.org/resources-tools/avma-policies/animal-assisted-interventions-guidelines |
| Pet Partners | General AAI | https://petpartners.org/standards/ |
| International Organization of Human-Animal Interaction Organizations (IAHAIO) | White Paper – definitions and guidelines for animal wellness | https://iahaio.org/best-practice/white-paper-on-animal-assisted-interventions/ |
| IAHAIO | Training and welfare for small animals | https://iahaio.org/iahaio-international-guidelines-on-care-training-and-welfare-requirements-for-small-animals-in-animal-assisted-interventions/ |
| Centers for Disease Control (CDC) | Zoonotic Disease Prevention in AAE settings | https://www.cdc.gov/healthypets/specific-groups/schools.html |

AAE program, a complete risk assessment with a corresponding risk mitigation plan is necessary to ensure the wellbeing of all involved: humans and animals. The *LEAD* Risk Assessment Tool was developed for this purpose (Brelsford et al., 2020). The tool is freely available to all, and was designed to be broadly applicable and adaptable to AAE and other settings. It includes a structure for establishing ongoing communication with key stakeholders (administration, parents, teachers), hazard identification and control measures, the development of an action plan, and, importantly, a dog care plan tailored to the individual dogs involved.

A risk assessment/mitigation approach to AAE should be a universal requirement for the implementation of AAEs. Additionally, ongoing evaluation of all aspects of a risk assessment plan should be required. Because so few health and safety incidents are reported in AAEs, teachers and administrators are not motivated to take this relatively time-consuming approach. Realistically, however, the lack of reporting does not mean that those incidents are not happening. Adding oversight to the process would also provide a clearer picture of the frequency and severity of the incidents that do occur.

## *Educational Goals and Measures of Success*

Returning to the IAHAIO definition of AAE, it is important that they are "…goal oriented, planned and structured…" (IAHAIO, 2018, p. 5). Essentially, this means that there should be an educational purpose involving specific student learning outcomes that can be assessed to determine the degree to which the AAE is successfully meeting the stated objectives. Table 35.2 provides a few examples, but it should be noted that student learning outcomes and potential AAE activities may vary widely. AAE provides a rich learning opportunity for those who are serious about using it to educate rather than entertain students.

*Table 35.2* Examples of student learning outcomes, relevant AAE activities, and methods of assessment

| Student Learning Outcome | AAE Activity | Method of Assessment |
|---|---|---|
| *Preschool Age* | | |
| Students will engage actively in turn taking and apply the use of "please" and "thank you". | Students raise their hand and wait their turn. When called on they use "please" to ask if they can throw a soft toy to a dog. When the dog brings it to them, they say "thank you" and then toss the toy back to the dog. | Behavioral observation – does each child follow the "rules" of the game? Do they generalize the use of "please" and "thank you" to other situations? |
| *Grade School Age* | | |
| Students will communicate effectively about dog care and wellbeing. | Humane educational material will be presented to the students along with specific demonstrations (involving a therapy dog) about appropriate handling techniques and basic care such as brushing, teeth and ear cleaning, and petting consent testing. | Written or oral assessment through specific questions or scenarios. Presentation to the group about petting consent, or brushing, or teeth/ear cleaning using a stuffed dog. |
| *University Students and Other Adults* | | |
| Students will appreciate the complexities involved in keeping animals for various reasons; economic, environmental, and personal (as pets). | Field trips are arranged that include visits to aquariums, zoos, large- and small-scale farms and dairies, and different therapy animal specifies are invited to the classroom where students meet the animals and engage in conversation with the handlers. | Written or oral assessment through specific questions or scenarios. Classroom presentation or debate on animal welfare and how it is juxtaposed against benefits to humans. |

# Ethical Considerations

## Attitudes Toward Animals

Our relationships with animals are complicated and rife with contradiction, as Hal Herzog points out in his book "Some we Love, Some we Hate, Some we Eat; Why it's so Hard to Think Straight about Animals" (Herzog, 2021). We classify some animals as pets (e.g., dogs and cats), which often means that we give them names, bring them into our homes, feed them, collect their feces, and let them sleep in our beds. Other animals we consider food animals (e.g., cows and turkeys); these are kept outside of the home, usually in barns or fenced in fields, not given names and "processed" (slaughtered and butchered) for food. Still, other animals are feared or despised (e.g., poisonous snakes or coyotes) and are avoided or destroyed. This complicated relationship combined with religious texts supporting the human domain over animals has set us on a path of "using" animals to our benefit.

## Animal Protections

Only recently are governments beginning to enact laws providing for the sentient status of animals. For example, Spain recently enacted legislation in the Spanish Civil Code, stating that pets and wild species are now considered to be sentient (Mercer, 2021). Likewise, as a field, our perspective on the role of the animals in AAE is evolving. Rawlings (2021) has written a notable, thoughtful consideration of the ethical issues involved in AAI with specific applications to AAE. He proposes more questions than solutions, but ultimately his discussion centers around one simple question: do the benefits outweigh the risks? As we balance these issues, it is important to include the risk to animal welfare and their quality of life in our deliberations. (For a more detailed discussion of these issues, see Peralta & Fine, 2021.)

## Animal Training

Consistent with our evolving attitudes toward animals, there has been a movement toward using reward-based training techniques with animals in AAIs (for a summary, see *Provoost* 2021). Aversive techniques have been shown to have a number of important drawbacks including degrading the human-animal bond, inhibition of spontaneous problem solving, and a higher incident of stress behaviors and negative emotions. Positive reinforcement techniques not only improve the animal's quality of life, but these techniques have also been documented repeated as being effective (e.g., Shapiro et al., 2003). For these reasons, it is suggested by leading authorities in the field that only positive reinforcement techniques are recommended for use in AAE settings.

## Animal Agency

It is critical to weighing risks and benefit that we conceptualize animals as our partners, deserving of a good life, and entitled to agency in their roles. Teaching students about the consent to pet test (for a description and example, see: https://somuchpetential.com/petting-a-dog-the-consent-to-pet-test/) is one example of how we can educate others about giving animals agency in their AAE activities. We also need to consider carefully the language we use when speaking about animals in AAE. A first step would be to stop "using" therapy dogs (they are not tools) and start "involving" them in our work (as partners). In this chapter, the word "use" has been purposefully deployed to indicate when animals have been exploited for human gain, or when they have not been given choices regarding their roles in AAIs. We can and should lead by example, treating all animals with the respect and dignity they deserve and it all flows from the language we use to communicate about the animals involved in AAE.

## The Future of Animal-Assisted Education

Several repeated themes regarding the current status of AAE are threaded through this chapter. Some are positive:

- AAE is popular with students and educators. It is used in a variety of settings with diverse types of animals and assorted objectives. Related to this theme, is the idea that getting "buy in" from school administrators is key to safe and effective implementation of AAE.

- Existing evidence supports the benefits of AAE in multiple populations; the greatest effect seems to be among children with special needs or adults enduring stressful circumstances.

But others represent challenges:

- There are few legal or regulatory policies controlling the conditions under which AAEs are implemented, frequently resulting in a "freewheeling" approach that lacks safety controls.
- Risk assessment and mitigation processes are largely absent from AAE programs, leaving open the potential for harm to students or animals uncontrolled and incidents undocumented. Using a risk assessment tool like the LEAD tool described above can effectively address, and in most cases, eliminate these concerns.
- The animal participants in AAE must rely on the humans who initiate their involvement to protect their welfare.

To overcome the challenges, AAE practitioners must enhance their efforts to balance the welfare of the animals involved in AAE against the benefits accruing to students. It is incumbent upon those teachers, researchers and other individuals who initiate ways to integrate animals into educational settings to consider whether the circumstances of the proposed engagement are fair to the animal. Animals participating in AAE should not be our "beasts of burden" but our partners.

## Discussion Questions

1. What paths exist to implement animal protections at the local, regional, and national levels?
2. What would a tool to balance animal welfare against student benefits look like?
3. What approach would be best to communicate to classroom teachers the need to provide animal protections and carefully consider their overall quality of life?

## References

Ballantyne, R., Hughes, K., Lee, J., Packer, J., & Sneddon, J. (2021). Facilitating zoo/aquarium visitors' adoption of environmentally sustainable behavior: Developing a values-based interpretation. *Tourism Management, 84*, 104243: DOI: 10.1016/j.tourman.2020.104243

Brelsford, V., Dimolareva, M., Gee, N. R., & Meints, K. (2020). Best practice standards in animal-assisted interventions: How the LEAD risk assessment tool can help. *Animals, 10,* 974. DOI: 10.3390/ani10060974

Brelsford, V. L., Meints, K., Gee, N. R., & Pfeffer, K. (2107). Animal-assisted interventions in the classroom – A systematic review. *International Journal of Environmental Research and Public Health, 14,* 669. doi:10.3390/ijerph14070669

Choi, H., Chung, S., Choi, Y., Kang, H., Jeon, J., & Shin, D. (2021). Analysis of polar education programs. *Journal of Korean, Earth Science Society, 42*(1), 102–117. DOI: 10.5467/JKESS.2021.42.1.102

Costhelper.com. (n.d.). *Swimming with dolphins cost.* Retrieved February 9, 2022 from https://travel.costhelper.com/swimming-with-dolphins.html#:~:text=Swimming%20with%20Dolphin%20opportunities%20are,location%20and%20the%20dolphins'%20behavior

Crossman, M. K., (2019). Animal visitation programs in colleges and universities: An efficient model for reducing student stress. In A.H Fine *Handbook on animal-assisted therapy: Foundations and guidelines for animal-assisted interventions* (pp. 343–350). Academic Press, San Diego, CA.

Dimolareva, M., & Dunn, T. J. (2021). Animal-assisted interventions for school-aged children with Autism Spectrum Disorder: A meta-analysis. *Journal of Autism and Developmental Disorders, 51*, 2436–2449. DOI: 10.1007/s10803-020-04715-w

Fedor, J. (2018). Animal-assisted therapy supports student connectedness. *NASN School Nurse, November.* DOI: 10.1177/1942602X18776424.

Flynn, E., Mueller, M. K., Luft, D., Geldhof, G. J., Klee, S., Tedeschi, P., & Morris, K. N. (2020). Human-animal-environment interactions and self-regulation in youth with psychosocial challenges: Initial assessment of the Green Chimneys model. *Human-Animal Interaction Bulletin, 8*(2), 55–67.

Gee, N. R., Belcher, J., Grabski, J., DeJesus, M., & Riley, W. (2012). The presence of a therapy dog results in improved object recognition performance in preschool children. *Anthrozoös, 25*, 289–300.

Gee, N. R., Church, M. T., & Altobelli, C. L. (2010). Preschoolers make fewer errors on an object categorization task in the presence of a dog. *Anthrozoös, 23*, 223–230.

Gee, N. R., Crist, E. N., & Carr, D. N. (2010). Preschool children require fewer instructional prompts to perform a memory task in the presence of a dog. *Anthrozoös, 23*, 178–184.

Gee, N. R., & Fine, A. (2019). Animals in educational settings: Research and application. In A. Fine (Ed.), *Animal assisted therapy: Theoretical foundations and guidelines for practice* (5th ed.). New York: Academic Press.

Gee, N. R., Fine, A., & Shuck, S. (2015). Animals in educational settings: Research and application. In A. Fine (Ed.), *Animal assisted therapy: Theoretical foundations and guidelines for practice* (4th ed.). New York: Academic Press.

Gee, N. R., Gould, J. K., Swanson, C. C., & Wagner, A. K. (2012). Preschoolers categorize animate objects better in the presence of a dog. *Anthrozoös, 25*, 187–198.

Gee, N. R., Harris, S. L., & Johnson, K. L. (2007). The role of therapy dogs in speed and accuracy to complete motor skills tasks for preschool children. *Anthrozoös, 20*, 375–386.

Gee, N. R., Sherlock, T. R., Bennett, E. A., & Harris, S. L. (2009). Preschoolers' adherence to instructions as a function of the presence of a dog, and motor skills task. *Anthrozoös, 22*, 267–276.

Hall, S. S., Gee, N. R., & Mills. D. S. (2016). Children reading with dogs: A systematic review of the literature. *PloS ONE, 11*(2), e0149759. DOI: 10.371/journal.pone.0149759

Hediger, K.& Turner, D. C. (2014). Can dogs increase children's attention and concentration performance? A randomized controlled trial. *Human-Animal Interaction Bulletin, 2*(2), 21–39.

Henderson, L., Grove, C., Lee, F., Trainer, L., Schena, H., & Prentice, M. (2020). An evaluation of a dog-assisted reading program to support student wellbeing in primary school. *Children and Youth Services Review, 118*, 105449.

Herzog, H. (2021). *Some we love, some we hate, some we eat; Why it's so hard to think straight about animals.* Harper Perennial.

International Association of Human Animal Interaction Organizations (IAHAIO). (2018). *The IAHAIO definitions for animal assisted interventions and guidelines for wellness of animals involved in AAI.* Retrieved February 9, 2022 from https://iahaio.org/best-practice/white-paper-on-animal-assisted-interventions/

Kirnan, J., Shar, S., & Lauletti, C. (2020). A dog-assisted reading programme's unanticipated impact in a special education classroom. *Educational Review, 72*(2), 196–219. DOI: 10.1080/00131911.2018.1495181

Le Roux, M. C., Swartz, L., & Swart, E. (2014). The effect of an animal-assisted reading program on reading rate, accuracy and comprehension of grade 3 students: A randomized control study. *Child Youth Care Forum, 43*, 655–673. DOI: 10.1007/s10566-014-9262-1

Mercer, T. (12/6/2021). *Upworthy.* Retrieved on February 9, 2022 from https://www.upworthy.com/new-law-spain-classifies-animals-sentient-beings?rebelltitem=3#rebelltitem3

O'Haire, M. E., McKenzie, S. J., Beck, A. M., & Slaughter, V. (2013). Social behaviors increase in children with autism in the presence of animals compared to toys. *PLOS One, 8*(2), e57010. DOI: 10.1371/journal.pone.0057010.

O'Haire, M. E., McKenzie, S. J., McCune, S., & Slaughter, V. (2014). Effects of classroom animal-assisted activities on social functioning in children with autism spectrum disorder. *The Journal of Alternative and Complementary Medicine, 20*(3), 162–168. DOI: 10.1089/acm.2013.0165.

Pendry, P., Carr, A. M., Vandagriff, J. L., & Gee, N. R. (2021). Incorporating human–animal interaction into academic stress management programs: Effects on typical and at-risk college students' executive function. *AERA Open, 7*, DOI: 10.1177/23328584211011612

Peralta, J. M., & Fine, A. H. (Eds.) (2021). *The welfare of animals in animal-assisted interventions: Foundations and best practice methods*. New York: Springer. DOI: 10.1007/978-3-030-69587-3

Provoost, L. (2021). Behavior and training for optimal welfare in therapy settings. In J. M. Peralta, & A. H. Fine (Eds.), *The welfare of animals in animal-assisted interventions: Foundations and best practice methods*. New York: Springer. DOI: 10.1007/978-3-030-69587-3

Purewal, R., Christley, R., Kordas, K., Joinson, C., Meints, K., Gee, N. R., & Westgarth, C. (2017). Companion animals and child/adolescent development: A systematic review of the evidence. *International Journal of Environmental Research and Public Health, 14*, 234; DOI: 10.3390/ijerph14030234

Rawlings, J. M. (2021). Ethics of animal-assisted interventions. In J. M. Peralta & A. H. Fine (Eds.), *The welfare of animals in animal-assisted interventions* (pp. 43–57). Cham: Springer.

Reilly, K. M., Adesope, O. O., & Erdman, P. (2020). The effects of dogs on learning: A meta-analysis. *Anthrozoös, 33*(3), 339–360, DOI: 10.1080/08927936.2020.1746523

Rousseau, C. X., & Tardif-Williams, C. Y. (2019). Turning the page for spot: The potential of therapy dogs to support reading motivation among young children. *Anthrozoös, 32*(5), 665–677, DOI: 10.1080/08927936.2019.1645511.

Russell C. (2019). An intersectional approach to teaching and learning about humans and other animals in educational contexts. In T. Lloro-Bidart & V. Banschbach (Eds.), *Animals in environmental education*. Palgrave Studies in Education and the Environment. Cham: Palgrave Macmillan. DOI: 10.1007/978-3-319-98479-7_3

Scandurra, C., Santaniello, A., Cristiano, S., Mezza, F., Garzillo, S., Pizzo, R., Menna, L. F., & Bochicchio, V. (2021). An animal-assisted education intervention with dogs to promote emotion comprehension in primary school children—The Federico II Model of Healthcare Zooanthropology. *Animals, 11*, 1504. DOI: 10.3390/ ani11061504

Schuck, S. E., & Crinella, F. M. (2005). Why children with ADHD do not have low IQs. *Journal of Learning Disabilities, 38*(3), 262–280.

Schuck, S. E. B., Emmerson, N., Abdullah, M. M., Fine, A. H., Stehli, A., & Lakes, K. D. (2018). A randomized controlled trial of traditional psychosocial and canine-assisted interventions for ADHD. *Human-Animal Interaction Bulletin, 6*, 64–80.

Schuck, S. E. B., Emmerson, N. A., Fine, A. H., & Lakes, K. D. (2015). Canine-assisted therapy for children with ADHD: Preliminary findings from the positive assertive cooperative kids study. *Journal of Attention Disorders, 19*(2), 125–137. DOI: 10.1177/108705471302080

Serpell, J. A., Kruger, K. A., Freeman, L. M., Griffin, J. A., & Ng, Z. Y. (2020). Current standards and practices within the therapy dog industry: Results of a representative survey of United States therapy dog organizations. *Frontiers in Veterinary Science, 7*, 35. doi: 10.3389/fvets.2020.00035

Shapiro, S. J., Bloomsmith, M., A., & Laule, G. E. (2003). Positive reinforcement training as a technique to alter nonhuman primate behavior: Quantitative assessments of effectiveness. *Journal of Applied Animal Welfare, 6*(3), 175–187.

Smilie, K. D. (2022). Sex, death, and alienation: The burdened history of classroom pets in the American curriculum, *Paedagogica Historica, 58*(2), 233–251. DOI: 10.1080/00309230.2020.1822889

Steel, J., Williams, J. M., & McGeown, S. (2021). Reading to dogs in schools: An exploratory study of teacher perspectives, *Educational Research, 63*(3), 279–301, DOI: 10.1080/00131881.2021.1956989

# 36
# AROUND THE ARENA
## Equine-Assisted Services' Foundation in Understanding Human-Horse Interactions

*Aviva Vincent and Robin Peth-Pierce*

### A Unique Therapeutic Space

What benefits does the horse offer that other animals who collaborate in therapeutic capacities do not? Many characteristics make horses unique, including their size, capacity to "mirror" human emotions, their domestic environment, and the purposeful effort that must be made to partner with them. Counterintuitive to their large presence, horses are naturally sensitive. They use their senses to assess their environment and mirror the energy and characteristics facing them. In the barn environment, human senses are on high alert, with sensory input interpreted through the nervous and proprioceptive systems, flooding all five senses. The horses' unique smell or odor, array of colors and sizes, and sounds of munching on hay and nickering and whinnying to each other create a unique opportunity to engage all the senses simultaneously. The barn environment and characteristics of the horse engage all of the senses, providing humans the opportunity to create an emotional connection and a feeling of being grounded. Horse-human communication relies on touch, emotional connection, and becoming attuned to physical movements.

This unique therapeutic space is why practitioners of equine-assisted services (EAS) tend to lean on the evidence-based practices of Mindfulness Based Stress Reduction (Vincent et al., 2021). Counter to human predatory nature, pausing these natural predator instincts and shifting to be in harmony with a large prey animal, rewards the internal biopsychosocial system, important to holistically addressing the physical, psychological and social factors that comprise human mental health and well-being.

Horses hover between domesticated and wild animals, offering a vastly different type of opportunity for interaction than any other domesticated animal. Though domesticated, horses do not live in the home or share the home as do many other types of pets. To work with these animals, clients have to go to the horse(s) in a purposeful, planned way. Clients schedule time and pay a fee, as most EAS are not covered by insurance. This represents a significant investment of personal time, money, and effort, which alludes to their perceived value of the EAS. Perception of value is critical, especially for youth participants) because parents'/adults' perceptions of treatment have been shown to predict whether they continue participation (Horwitz et al., 2012).

## History of Humans and Horses

Paleolithic artists drew horses' facial expressions and body language, documenting the early human understanding of the horse (Leste-Lasserre, 2019). Archeologists unearthed burial sites of Greek and Celtic warriors to find their trusted steeds buried alongside them (Daley, 2019). In this formative relationship, horses were recognized as providing physical benefits to humans, specifically when injured in wartime. How strange it is that the same horses that brought humans into war and battle—served as facilitators in the healing process as well.

Hippocrates, the "Father of Medicine," wrote of the benefits of horseback riding around 500 B.C. (Jacobson, 2019). Wounded warriors sat astride horses for healing from war and trauma. The horses' natural walking gait, which mirrors that of a typical human walking gait, stimulated the soldier's nerves and muscles. Humans physically healed through the partnership with their trustworthy steeds, but the mental health implications were not lost on all who served (Jacobson, 2019). Florence Nightingale and others of the medical mind understood that healing with non-human animals provided more than physical therapy (Fine, 2019). The horse's stature, speed, carriage, and innate ability to connect with humans have provided a robust history that arcs from warfare through to healing. Worldwide, horses and humans have a long history entrenched in warfare and recovery of trauma from war, though the positive benefits of working collaboratively with horses is not limited to war.

## *Relationships and Communication with Humans*

Working with horses, including riding horses (physically getting on their back) goes against the very nature of horses' prey instinct. Like humans, horses have a range of fight-or-flight reaction responses. As a means of protection, horses tend to choose flight because they do not have a strong ability to defend against predators based on their natural body mechanics (e.g., flat surfaced teeth, hooves instead of fingers/claws). Their greatest defense is to run away, fast. It is a psychological challenge for horses to perceive humans as anything other than a threat (Thiel, 2013); which is why humans must approach horses in an open and neutral manner, making the horse-human bond unique.

The core attributes of a successful, bidirectional, interspecies relationship are trust, utility, respect, and acceptance (Wilsie & Vogel, 2016). When humans partner with horses, they report feelings of trust, non-judgmental interaction, calm, and love (Hallberg, 2017; Lac, 2017). As a result, the behavior between horses and humans does not need to be one of oppression, power, and mastery. People who work with horses often use the language of empowerment and mindfulness to explain their experiences. This is why the language of "partner" and "collaborate" is preferred over "use" when referring to sessions with horses.

Working with horses engages cognitive and physical capacities of horses and humans. Horses demand complete attention and focus (de Kunffy, 1993). As sensitive beings that require minimal pressure to move and respond to cues, horses are extremely responsive to body language and subconscious cues (Scopa et al., 2019). This quick reaction to human body language, emotion, and energy, is commonly referred to as mirroring (Wilsie & Vogel, 2016).

Humans can learn to identify and name horse behavior, akin to learning their language (Bachi, 2013; Gabriels et al., 2015). In partnership, the horse and human work together to complete activities, or just simply share space and breathe. However, even if a rider is not outwardly distraught, horses perceive and internalize the discordance between emotion and behavior (Vincent & Farkas, 2017). When in harmony with the horse, a single human exhale can bring a horse to a stop.

## Theoretical Foundations of Equine-Assisted Services

The intention of theory is to provide a framework to *predict, describe,* and *explain* an experience or intervention. While there is no unifying theory related to EAS, there are a number of working theories that can be helpful to researchers, mental health clinicians, equine professionals, consumers, and advocates. Theories provide a framework for understanding EAS in relation to human functioning as identified by the National Institute of Mental Health's (NIMH) Research Domain Criteria (RDoC) which includes: negative valence, positive valence, cognitive processing, social experience, emotional regulation/arousal, and the sensorimotor system. The goal of the RDoC framework is to provide information about basic biological and cognitive processes that lead broadly to mental health and illness (National Institute of Mental Health, 2021). In Table 36.1, dominant theories emergent in the literature and research are highlighted (though not an exhaustive list), with each theory tied to a domain(s) of human functioning.

Some use theories to support the use of EAS with specific populations. At an environmental level, the biophilia hypothesis is a common reference point; the framework theorizes that in general, mammals are drawn to each other for co-regulation (Harper et al., 2019). Kemp and colleagues (Kemp et al., 2013) applied Rogerian theory to "at-risk" youth to suggest that they flourish within an environment of unconditional positive support. The unconditional love and support is necessary to allow youth to feel that they can explore, express, and discover their true selves safely, without fear of reprimand, expectation, or disappointment.

*Table 36.1* Common theoretical foundations for EAS with historical and modern references

| Theory and Context | | Foundational Theorists | Citations for the Application of the Theory to EAS |
|---|---|---|---|
| Environmental, person-in-environment | Ecology theory Biophilia Hypothesis Attention restoration theory | Bronfenbrenner Wilson Kaplan | Dunlop and Tsantefski (2018) |
| Emotional Regulation | Emotional Transfer Hypothesis Rogerian Theory | Schachter and Singer Rogers | Kemp et al. (2013) |
| Cognitive Systems | Cognitive Behavior Therapy | Beck | Acri et al. (2021) |
| Social Processes | Social Learning Theory Empowerment Theory | Bandura Rappaport, Zimmerman | McCullough et al. (2015) |
| Arousal/Regulatory | Attachment Theory | Bowlby, Ainsworth, Winnicott | Vincent and Farkas (2017) |
| Sensorimotor | Psychoanalytic Theory | Freud, Erikson | Lentini and Knox (2009) |
| Aggregate | Comprehensive overview of theory | N/A | Vincent et al., (2023) Kruger and Serpell (2010) |

## Therapeutic Modalities

Over the last several decades, equine and mental health professionals worldwide were using a wide range of terminology in an effort to most clearly describe the services provided. However, the inconsistency of language became a challenge for all; in particular, consumers did not have a common phrase to search for support, and equine professionals were using different language for services delivered (Wood et al., 2021). In 2018, a consensus-building process, led by the Professional Association of Therapeutic Horsemanship, International (PATH Intl), brought together a working group to address the problem of "ambiguous, imprecise, and confusing terminology" (Wood et al., 2021). The resulting document proposed a recommended nomenclature to be used to describe the range of EAS delivered worldwide.

By more clearly defining what is considered "therapy," "learning," and "horsemanship," Wood and colleagues (2021) suggest that the therapeutic horsemanship field is moving toward narrowing the scope of equine professionals' practice. The new nomenclature demarcates that equine professionals cannot provide services that they call "therapy" unless they are "licensed" therapy professionals. However, they may provide "adaptive riding" or "therapeutic services," which are "services that focus on skillfully adapting riding and making natural healthful benefits of riding and horses accessible to individuals and groups with diverse needs," with the focus on developing treatment plans and meeting identified goals.

According to this new nomenclature, therapy may be rendered in five distinct service categories: counseling, psychotherapy, occupational therapy, physical therapy, and speech language pathology. Each of these services is to be provided by a licensed therapy professional and use *therapy-first* language, which looks like "physical therapy with equines" or "psychotherapy with equines." This *therapy-first* language allows for the professional to be specific about their scope of practice, first and foremost, and also acknowledge the horses for their contribution. This language tends to become confusing because of the different iterations that have historically been used to explain the same experience: equine-assisted therapy, equine-facilitated therapy, equine-assisted/facilitated psychotherapy—or as those who *use* the service refer to it: "horse therapy" (Table 36.2).

Mental health clinicians are trained from a variety of fields including, but not limited to social work, counseling, and psychology (Chandler, 2012; International Association of Human Animal Interaction Organizations, 2018). Session content can vary greatly from groundwork to riding, and from individual sessions to group-based work. As a mental health intervention, a licensed mental health clinician and client work collaboratively with a horse; best practices include a third person in the therapeutic triad who is solely responsible for the horse throughout the session; this person is referred to as a volunteer or horse wrangler (PATH Intl, 2021). All participants have a clear mental health challenge and objective(s) for each session. The inclusion of the horse is intended to achieve outcomes that otherwise would be more difficult to achieve or have been found ineffective in other environments (Nimer & Lundahl, 2007). The horse is an active collaborator in the therapeutic process (Chandler et al., 2010; Matuszek, 2010), who aids in building social rapport, confidence, and creates an opportunity for introspective reflection for the client.

Participants in these interventions demonstrate decreases in stress across populations, including at-risk youth, veterans, and older adults (e.g., Wilkie et al., 2016; Kinney et al., 2019; Stergiou et al., 2017). Overall, studies cite the positive impact of the horse-human relationship. Calls for increased rigor in equine interventions, for example, as outlined by Anestis and colleagues (2014), include normalizing professional training requirements, coding for

Table 36.2 Categories of equine-assisted services

| Therapy | Learning | Horsemanship |
|---|---|---|
| **Counseling, Occupational Therapy, Physical Therapy, Psychotherapy, Speech-Language Pathology** | **Equine-Assisted Learning in Education, Equine-Assisted learning in Organizations, Equine-Assisted Learning in personal Development** | **Adaptive Equestrian Sport, Adaptive riding or therapeutic riding, Driving, Interactive Vaulting** |
| Therapy is inclusive of mental and physical health and well-being. Thus, in the practical application of **Therapy** an individual may be engaged in mental health services. While counseling and psychotherapy and the specific modalities referenced (Wood et al., 2021), social work and other professionals are actively engaged in providing mental health services. In contrast, or in addition to mental health, individuals who may be engaged in therapy for their physical well-being such as occupational, physical, or speech and language pathology. A point of contention in the new nomenclature is the use of *"**Hippotherapy**"*. The use of horses as a therapeutic tool for humans to regain physical mobility (i.e., physical therapy or occupational therapy) is referred to as Hippotherapy or physical therapy with horses. This is the only instance where the horse is referred to as a tool—and not a collaborator or partner. While the American Hippotherapy Association holds fast to their name, other providers are aligning with the therapy first language for clarity in the services provided and to move the field toward insurance coverage (Ballard et al., 2020). | Services offered under this category are non-therapeutic, and are offered in three primary environments: education, organizations and personal development. The focus of these services is experiential learning, rather than therapy, and professionals have training, experience and skills in equine-assisted learning.<br><br>Two examples of structured and formalized equine-assisted learning programs include: (1) Brook Hill Farm equine-assisted learning model (Virginia), which when combined with a traditional high school education program and tutoring, improved high school graduation rates to 100% (Kruger & Serpell, 2010). | This final category of services, provided by equine professionals with *specialized training or certifications*, emerges from what are known as the traditional equine disciplines-- horseback riding, driving and vaulting-- and now includes: adaptive equestrian sport, adaptive riding or therapeutic riding, driving and interactive vaulting, with the word adaptive specifically conveying that the riding has been adapted for the *diverse* needs of individuals. Activities include "horse handling, grooming, longing, driving, riding, and vaulting" (McCardle, 2011). Regardless of diagnosis, the cognitive aim of activities is typically to increase a sense of curiosity, build self-esteem/confidence and trust in a safe and non-judgmental environment. An example is the services provided by many PATH-accredited centers to youth and adults (e.g., Bachi et al., 2012; Ewing et al., 2007). |

EAS treatment delivery, and adherence to progress notes are required to move the field forward. Rigorous research aims to clarify the benefits as well as the mechanisms of change. In totality, a wide range of populations are served by equine professionals—from preschoolers to geriatric populations—and tackle a heterogenous set of mental, emotional, and behavioral challenges.

## *Industry Adaptations to New Nomenclature*

Equine organizations have begun to adapt their organization's standards to meet these new definitions. For example, in 2021, PATH International Service Standards were revised to separate the delivery of (1) *equestrian* services (i.e., the goal/outcome of a session for the participant to gain equestrian skills); (2) *medical* services (i.e., the goal/outcome achieved by licensed/credentialed therapist); and (3) *mental health* services (i.e., the goal/outcome achieved by licensed/credentialed mental health professionals) (Anestis et al., 2014; PATH, 2021). However, the majority of equine professionals are not also *licensed mental health* professionals (i.e., PATH- or EAGALA-certified and licensed marriage and family therapist [LMFT], licensed clinical social worker [LCSW], or similar). This nomenclature narrows what services can be provided, which may impact potential government and/or commercial insurance coverage, and with it, potentially increased rates to providers of EAS.

## *Impact on Participants in Services*

For those taking part in services, one effect of the new nomenclature is that it both clarifies the qualifications of the individual delivering the service, as well as the nature of EAS services being delivered. For recipients of what has historically been called "therapeutic riding," this has been mostly an adjunct service, with many receiving other services from licensed and credentialed professionals (i.e., LCSWs, LMFT). Importantly, with this change in nomenclature, the three-tiered service delivery model may also lead to clearer research designs and manualized interventions, which could in part provide better information on what works and for whom, potentially opening the door to eventual insurance coverage (Ballard et al., 2020).

## *Case Example: Psychotherapy with Equines*

Services typically provided to veterans diagnosed with post-traumatic stress (PTS) feature small group sessions with a treatment team, which includes a licensed mental health professional and a trained equine professional. For a group of five clients, three horses are brought into the session; having fewer horses than the number of people allows for natural opportunities to collaborate and step-back. In one recent study with veterans diagnosed with PTS, a mental health professional, an equine specialist, and a horse wrangler delivered a manualized intervention, which focused on boundary setting, anxiety and frustration tolerance, and promoting self-efficacy (Arnon et al., 2020). In an evaluation of the program's services, 50% of the veterans had reductions in their PTS symptoms which were correlated with changes in brain structure (Fisher et al., 2021; Zhu et al., 2021).

## The Environment of the Horse

Horses inherently need more space than domesticated animals, like cats and dogs. To participate in EAS, individuals often travel to barns housing the horses. The act of committing time and traveling aids in creating a separation from typical routines and can aid in priming the mind and body for the therapeutic work to come. Once at the facility, participants arrive in an environment created by equine professionals to meet the needs of the client(s). The equine professional does this through creating intentional physical and collaborative space in which to work, selecting attributes of physical and collaborative spaces that meet each client's individual needs. Sessions can occur in a variety of places on the property including, but not limited to cross-ties, arena/ring, trails, and even in the horse's stall. Each space has a unique and intentional purpose to the therapeutic process.

## Professionalization of Equine-Assisted Services

Central to the creation of the EAS industry was Lis Hartel, an equestrian competitor who medaled in the 1952 Helsinki Olympics despite being impaired by polio, and who publicly shared her success as an equine professional. Through her, healthcare providers began to understand the power of the horse-human bond. Therapeutic riding emerged from barns in the mid-1900s, with the early years of EAS narrowly focused on riders with physical disabilities. Certifications for providers were conducted through the North American Riding for the Handicapped Association, Inc. (NARHA), established in 1969, the first therapeutic riding centers opened in Toronto and Michigan; The Community Association of Riding for the Disabled, and the Cheff Center for the Handicapped, respectively. Organizations sprang up worldwide to support these professionals, including the Canadian Therapeutic Riding Association (CANTRA), founded in 1980, and The Federation of Horses in Education and Therapy International (HETI), founded in 1972 to "facilitate the worldwide collaboration between organizations and individuals whose objectives are philanthropic, scientific and educational in the field of equine-assisted activities and therapies" (HETI, 2021).

In the United States, the Equine Assisted Growth and Learning Association (EAGALA), founded in 1999, was one of the first international organizations to present a model for EAS activities focused exclusively on groundwork. The American Hippotherapy Association separated from NARHA in 2003 and continues to focus on the physical needs of individuals. In 2011, as EAS began to focus on the mental and psychological challenges faced by clients, NARHA was renamed as the Professional Association of Therapeutic Horsemanship International (PATH Intl).

Over the last 50 years since horses have been employed therapeutically to improve both physical and mental health outcomes, the field has become increasingly professionalized, with multiple certifications available, and even more importantly, more professionals involved in research who are examining the impact of therapeutic riding on an array of outcomes (see case study examples, below). For example, one certification available through EAGALA incorporates a "licensed mental health professional and a qualified equine specialist" (EAGALA, 2021), wherein the team focuses on groundwork to meet therapeutic goals. Recently, EAGALA's co-founder established Arenas for Change, with the intention of broadening the scope of EAS services "beyond the barn and equines" to include a "community of facilitators in mental health, organizational development, coaching, military, and education services" (Arenas for Change, 2021). Along with Arenas for Change, other models, such as HERD Institute and Natural Lifemanship, offering training, education, and certifications that aim to advance the EAS field while also carving out niche specialties for practitioners.

## Research on Horse-Human Interactions

The growing professionalization of EAS calls for a parallel increased and focused research effort to assess whether these services are making a difference, how (i.e., mechanisms of change), for whom, and in what contexts. Of equal importance, research efforts need to assess equine welfare, as the horses are equal collaborators and partners in the delivery of EAS.

### *Methods of Research to Explore and Evaluate Horse-Human Bond*

In the last decade, new methodologies have been developed to assess the physiological impact of EAS interventions, and include measuring salivary biomarkers to assess stress or bonding, taking fMRI and MRI scans of brain structures to assess brain changes, and measuring heart rate variability (HRV) to assess an individual's ability to recover from stress. In 2015, the first *randomized* trial assessing the impact of equine-assisted psychotherapy on youth with ASD collected and assessed salivary biomarkers of stress (i.e., cortisol) (Gabriels et al., 2015, 2018). Since then, the recommendation of Beetz and colleagues (2012) has come to fruition, with more EAS research now including the collection of oxytocin (Hoagwood et al., 2022). This is an important move forward for the field because researchers are able to move past measuring only impacts on negative stress, to measuring the positive impact of an intervention. The findings from such research may help answer the "why horses?" question.

### *Measuring Equine Welfare in Equine-Assisted Services*

With the ease and efficacy of human physiological data collection well-established, researchers are now able to heed the call-to-action to address equine welfare by collecting physiological data from equines in tandem with the participants (Ekholm Fry, 2021; Griffin et al., 2019; Ng, 2021). Currently, it is estimated that 1,500 facilities in the United States offer EAS services by certified equine professionals, and this estimate includes only two of the largest certifying bodies (i.e., PATH and EAGALA) that provide a range of EAS to youth and adults. In regard to equine welfare, PATH centers assess welfare according to the following criteria: "age appropriate to the activity and workload, soundness appropriate to carry out the work; temperament; height, build, conformation and movement appropriate for the activity and participants; gender and herd dynamics…" (PATH, 2021). Assessing equine welfare is *foundational* to EAS service delivery, for without horses there would be no EAS programs.

The importance of assessing the welfare broadly before, during, and after their HAI "career" has been well-documented (Fine, 2019). Horses who become partners in EAS join these programs later in life, and often have physical limitations or behavioral problems. Studies of equine welfare in EAS primarily use observable behaviors to assess welfare (Rankins et al., 2021). While there are observable signs of distress (i.e., ears pinned back, pawing ground), no studies to date have been conducted to determine whether humans in EAS can accurately assess equine distress using only these behavioral signs (Ekholm Fry, 2021), and in a way that correlates their behavior to physiologic biomarkers of stress (e.g., cortisol).

Hoagwood and colleagues 2022 are the first known team in the United States to assess the impact of interventions in EAS on the trio of rider, horse, and volunteer. Globally, novel ways to assess equine welfare include the development of a new type of equine bit to more easily collect saliva to assess biomarkers of stress and/or well-being. Saliva collection, via equine bit, is hypothesized to be less-invasive than collection by standard cotton swab (Contreras-Aguilar et al., 2019). Vincent and colleagues (2021) replicated the modified equine bit developed by Contreras-Aguilar and colleagues (2019) to gauge equine stress *prior* to a planned intervention and *after* the intervention. In the replication, the bit was used to collect and analyze cortisol and oxytocin toward understanding the potential impact EAS.

## Future Research

The breadth of EAS has broadened during the last five decades, from addressing primarily physical disabilities to more broadly addressing mental health and social-emotional challenges. Research on the latter has been further focused on impacts of EAS, largely *by mental health disorder* (i.e., ASD, PTS, trauma, and conduct disorder). More recently, Latella and Abrams (2019) characterized the last decade of research on EAS and therapies as outcomes-driven (e.g., assessing physical, psychosocial, or behavioral changes). This last decade of EAS research growth was fueled, in part, by an influx of funding to the field, with ad-hoc funding driven largely by organizational missions. The pools of funding reflect the wide breadth of research efforts, including human-centric (e.g. National Institute of Child Health and Human Development (NICHD) and human-animal bond (e.g., Horses and Humans Research Foundation, Human-Animal Bond Research Institute).

This decade-long investment provided a boost to EAS research, producing several seminal studies, including one of the only randomized controlled trials (RCTs) to date assessing the impact of EAS on youth with ASD (Gabriels et al., 2015; Gabriels et al., 2018; Peters, et. al., 2022), as well as new tools to measure human-animal interaction (HAI) (Guérin et al., 2018) and efforts to assess equine welfare in study outcomes (Vincent et al., 2021). Building on these infrastructure investments, researchers in HAI are now drawing on resources from other fields like psychology that have successfully implemented RCTs to achieve status as an evidence-based practice (EBP) (e.g., PracticeWise). HAI researchers broadly (González-Ramírez et al., 2013), and those in the field of EAS (Acri et al., 2021), are working toward melding evidence-based interventions into EAS practice.

Through collaboration, opportunities for advancing EAS research are limitless. For example, researchers from Columbia University applied knowledge about the impact of post-traumatic stress (PTS) on brain structure by using fMRI measures in a new protocol delivered in the Man-O-War program (Arnon et al., 2020). Others are considering the effect of EAS on heart rate variability (HRV) and are integrating interventions, such as those developed by the HeartMath© Institute (Baldwin, 2017), into their funded EAS research efforts. While nearly all of the EAS research field agrees that rigorous research is needed (Esposito et al., 2011; Elkholm Fry, 2013; Griffin et al., 2019), there is however, conversation about the *degree* to which the EAS field needs to meet the traditional "gold standard" of research: the double-blind randomized control trial (RCT). The main issue is that participants in EAS will inherently know if they are in the treatment group or control group by virtue of being paired with a horse or not.

The closest design is a waitlist control group, and while considered nearly as rigorous, is still not an RCT. Thus, the idea that an RCT is needed to move toward evidence-based practice is too narrow of a framework for the research needed to move the EAS field ahead. Instead of focusing only on design, researchers are advancing the field by integrating novel measures (e.g., Gabriels et al., 2015, Acri et al., 2021) and consolidating findings through systematic reviews and meta-analyses (e.g., Collado-Mateo et al., 2020; Kinney et al., 2019; Stergiou et al., 2017; Wilkie et al., 2016).

> ### Case Study—Mental Health: Veterans Experiencing Post-Traumatic Stress
>
> Historically, the trauma experienced by military persons was referred to as "soldier's heart," "shell shock," or "combat fatigue." Presently, the National Institute of Mental Health and Veterans Affairs (VA) refers to this trauma clinically as post-traumatic stress (Department of Veterans Affairs, 2021; National Institute of Mental Health, 2021). The VA offers a range of evidence-based treatments, predominantly cognitive processing therapy (CPT), prolonged exposure therapy, and eye movement desensitization and reprocessing (EMDR) therapy. Other treatments include dialectical behavior therapy, present-centered therapy, transcranial magnetic stimulation, and pharmacological interventions.

While the PTS diagnosis has been helpful in researching and refining intervention treatments, the negative connotation of labeling veterans as "broken" steers them away from services. EAS offers a way forward; veterans' partner with horses, in addition to the team of equine professional and mental health clinician, to address their stress. The horse as the primary partner provides a much different medium for connection and power different than traditional talk therapy.

Veterans' typical behavior of hyperarousal and defensive stance is matched by the horse's natural behaviors; this is an important point of connection. Many horses have experienced stress; some are resilient, and others develop maladaptive coping mechanisms. Additionally, horses utilize their prey instinct to mirror and attune, to meet their counterpart where they are at. Thus, sessions for individuals with PTS to work in collaboration with horses allows space for veterans to be present, grounded, and explore their cognitive state—and that of their horse.

While some have advocated that clients *not* to engage in EAS until the research inexplicably validates its positive impact (Anestis et al., 2014), EAS professionals have been supporting the veteran community for decades. Researchers have advanced research (e.g., Marchard et al., 2021), while the practice community has also manualized training and education. PATH Intl has developed specialized training to support certified instructors in facilitating structured programs for veterans (e.g., Equine Specialist in Mental Health and Learning and Equine Services for Heroes, respectively). One example is the Man-O-War project (Arnon et al., 2020), described briefly above; in this protocol, teams that include a certified therapeutic instructor, a licensed mental health professional, and an equine specialist complete a three-session training to learn about research, practice, and implementation of the curriculum.

### *Recommendations for Engagement in Research and Future Directions*

Data is essential to moving the field of EAS forward. The collection and storage of de-identified micro (i.e., client), mezzo (i.e., horse, equine professional, and volunteers), and macro (i.e., program and/or center) level data would be an asset to the field and would allow for multilevel and longitudinal research efforts to come to fruition. The collection of this multilevel data would support the identification of the essential components of EAS. In 2015, Hybel and colleagues undertook the TRAIN initiative (**T**herapeutic **R**iding **A**ssessment of **I**mpact **N**etwork) in order to garner support for and momentum in measuring outcomes to assess the impact of therapeutic riding. The focus was on collecting multilevel data (i.e. participant, program, center, instructor, industry) to facilitate outcomes research, largely based around the use of Goal Attainment Scaling (GAS), a measure commonly used to assess outcomes in social work. However, the effort did not expand past five PATH International-accredited centers in New England. A unified outcomes measure, utilized by equine professionals, and standardizing data collection around the six domains of human functioning, would certainly help build a database.

By linking data to individual and aggregate outcomes, the EAS industry could establish a set of evidenced-based EAS practices, identifying what—and importantly, *for whom*—works best. At a macro level, this could contribute the necessary evidence to garner insurance parity for EAS alongside traditionally-delivered mental health services (Ballard et al., 2020). These essential data elements could allow for a holistic view of EAS services globally, for what is now a quagmire of largely case-study or small trial heterogenous data points of services, delivered to a broad spectrum of the population.

A new initiative has emerged to make sense of the current heterogeneous data: two nonprofit equine-focused organizations, Humans and Horses Research Foundation (HHRF) and Federation of Horses in Education and Therapy International (HETI), are jointly studying the feasibility of creating "a single research platform that will act as a repository for all information pertinent to the field of Equine-assisted Services research" (HETI Federation, 2021). This is the first time the EAS industry has proposed a single database, international in scope, to allow for the organized collection of data and facilitation of analysis across the various populations and disorders. The availability of a global repository, in combination with a coalescence of a common EAS nomenclature (Wood et al., 2021) has significant potential to increase the rigor and reproducibility of EAS research.

The EAS field must find unity beyond nomenclature; the EAS profession is at a remarkably interesting juxtaposition. On the one hand, it is a business and industry where centers and professionals need to make a living wage and ego dictates a desire to excel. On the other hand, the majority of the industry is not-for-profit and in-service to marginalized and underrepresented persons. Between these two points, there exists competition of mission and intention between professional groups. In recognition that there is substantial demand for all organizations and associations to have their needs met, and providers to render services to the community, those in the industry have a responsibility to come together for the common good of those we collectively aim to serve. This is where research and data, in addition to a common nomenclature, serve to bring the field together.

For example, understanding the outcomes of a therapeutic riding intervention compared to a ground-based intervention does not (and should not) be perceived as putting the PATH Intl model against the EAGALA model, and especially should not be leveraged as such by funders. Instead, this would be an opportunity to understand if specific groups of individuals benefit less, same, or more from being in the barn—or on the back of a horse. For those who have walked into a barn, we hear "just pulling into the parking lot, I feel better," or just "smelling the horses, hearing the horses, or being near them." But is that so? Research is the only way to empirically know if what we collectively hear in anecdotes is valid and reliable. Furthermore, research is the way that the EAS industry can come together in unity to support all the persons we aim to serve.

## A Final Thought

The need to bridge research and practice is evident. Research allows for a collective understanding of impact. Importantly, researchers and practitioners must consistently consider the participant's perspective, desire, and understanding of services offered. Even with research, education, and best practices, after all is said and done—the most important perspective is that of those we serve. A participant shared:

> I started with the grief therapy program that I saw on the news. When I came, I learned about the veteran groups. When I joined the Friday night group I met the social worker and bonded with Smoke. In the round pen Smoke kept coming up to me; he picked me. When I don't see the horses during the week my depression sinks in and my PTS gets worse. The horses make me feel like I can face the world.

And that is why we do what we do; and *that* is why horses.

## References

Acri, M., Morrissey, M., Peth-Pierce, R., Seibel, L., Seag, D., Hamovitch, E. K., Guo, F., Horwitz, S., & Hoagwood, K. E. (2021). An Equine-Assisted Therapy for Youth with Mild to Moderate Anxiety: Manual Development and Fidelity. *Journal of Child and Family Studies*. 30(10), 2461–2467. https://doi.org/10.1007/s10826-021-02011-4

Anestis, Mi., Anestis, J. C., Zawlinski, L., Hopkins, T. M., & Lilienfeld, S. (2014). Equine-Related Treatments for Mental Disorders Lack Empirical Support: A Systematic Review of Empirical Investigations. *Journal of Clinical Psychology*, 70(12), 1115–1132.

Arenas for Change. (2021). *Arenas for Change*. Arenas for Change. https://arenasforchange.com/

Arnon, S., Fisher, P. W., Pickover, A., Lowell, A., Turner, J. B., Hilburn, A., Jacob-McVey, J., Malajian, B. E., Farber, D. G., Hamilton, J. F., Hamilton, A., Markowitz, J. C., & Neria, Y. (2020). Equine-Assisted Therapy for Veterans with PTSD: Manual Development and Preliminary Findings. *Military Medicine*, 185(5–6), e557–e564. https://doi.org/10.1093/milmed/usz444

Bachi, K. (2013). Application of Attachment Theory to Equine-Facilitated Psychotherapy. *Journal of Contemporary Psychotherapy*, 43(3), 187–196. https://doi.org/10.1007/s10879-013-9232-1

Bachi, K., Terkel, J., & Teichman, M. (2012). Equine-Facilitated Psychotherapy for At-Risk Adolescents: The Influence on Self-Image, Self-Control and Trust. *Clinical Child Psychology and Psychiatry*, 17(2), 298–312. https://doi.org/10.1177/1359104511404177

Baldwin, A. (2017, January 17). *Research: Heart Math*. https://ridingbeyond.org/research-heart-math/

Ballard, I., Vincent, A., & Collins, C. (2020). Equine Facilitated Psychotherapy with Young People: Why Insurance Coverage Matters. *Child and Adolescent Social Work Journal*, 37(6), 657–663. https://doi.org/10.1007/s10560-020-00712-1

Beetz, A., Uvnäs-Moberg, K., Julius, H., & Kotrschal, K. (2012). Psychosocial and Psychophysiological Effects of Human-Animal Interactions: The Possible Role of Oxytocin. *Frontiers in Psychology*, 3(234), 1–15. https://www.frontiersin.org/article/10.3389/fpsyg.2012.00234

Chandler, C. K. (2012). Equine Assisted Counseling. In *Animal Assisted Therapy in Counseling (Second Edition)* (pp. 205–228). Routledge. http://www.webofscience.com/wos/woscc/full-record/WOS:000314875200010

Chandler, C., Portrie-Bethke, T. L., Minton, C. A., Fernando, D. M., & O'Callaghan, D. (2010). *Matching Animal-Assisted Therapy Techniques and Intentions with Counseling Guiding Theories*. https://doi.org/10.17744/MEHC.32.4.U72LT21740103538

Collado-Mateo, D., Lavín-Pérez, A. M., Fuentes García, J. P., García-Gordillo, M. Á., & Villafaina, S. (2020). Effects of Equine-Assisted Therapies or Horse-Riding Simulators on Chronic Pain: A Systematic Review and Meta-Analysis. *Medicina (Kaunas, Lithuania)*, 56(9), E444. https://doi.org/10.3390/medicina56090444

Contreras-Aguilar, M. D., Henry, S., Coste, C., Tecles, F., Escribano, D., Cerón, J. J., & Hausberger, M. (2019). Changes in Saliva Analytes Correlate with Horses' Behavioural Reactions to An Acute Stressor: A Pilot Study. *Animals : An Open Access Journal from MDPI*, 9(11). https://doi.org/10.3390/ani9110993

Daley, J. (2019, December 10). Archaeologists Unearth Celtic Warrior Grave Complete With Chariot, Elaborate Shield. *Smithsonian Magazine*. https://www.smithsonianmag.com/smart-news/celtic-chariot-grave-found-england-includes-horses-and-elaborate-shield-180973730/

de Kunffy, C. (1993). *The Ethics and Passions of Dressage.* Half Halt Press.
Department of Veterans Affairs. (2021). *Posttraumatic Stress Disorder (PTSD).* Post Traumatic Stress Disorder. https://www.research.va.gov/topics/ptsd.cfm
Dunlop, K., & Tsantefski, M. (2018). A Space of Safety: Children's Experience of Equine-Assisted Group Therapy. *Child & Family Social Work, 23*(1), 16–24. https://doi.org/10.1111/cfs.12378
EAGALA. (2021). *EAGALA- A Global Standard in Equine-Assisted Psychotherapy and Personal Development.* https://www.eagala.org/org
Elkholm Fry, N. (2013). Equine-Assisted Therapy: An Overview. In *Biotherapy—History, Principles and Practice: A Practical Guide to the Diagnosis and Treatment of Disease Using Living Organisms* (pp. 255–284). Springer Science + Business Media. https://doi.org/10.1007/978-94-007-6585-6_10
Ekholm Fry, N. (2021). Welfare Considerations for Horses in Therapy and Education Services. In J. M. Peralta & A. H. Fine (Eds.), *The Welfare of Animals in Animal-Assisted Interventions: Foundations and Best Practice Methods* (pp. 219–242). Springer International Publishing. https://doi.org/10.1007/978-3-030-69587-3_9
Esposito, L., Mccune, S., Griffin, J., & Maholmes, V. (2011). Directions in Human–Animal Interaction Research: Child Development, Health, and Therapeutic Interventions. *Child Development Perspectives, 5,* 205–211. https://doi.org/10.1111/j.1750-8606.2011.00175.x
Ewing, C., MacDonald, P., Taylor, M., & Bowers, M. (2007). Equine-Facilitated Learning for Youths with Severe Emotional Disorders: A Quantitative and Qualitative Study. *Child and Youth Care Forum, 36,* 59–72. https://doi.org/10.1007/s10566-006-9031-x
Fine, A. H. (2019). *Handbook on Animal-Assisted Therapy: Foundations and Guidelines for Animal-Assisted Interventions.* Cambridge, MA: Academic Press.
Fisher, P. W., Lazarov, A., Lowell, A., Arnon, S., Turner, J. B., Bergman, M., Ryba, M., Such, S., Marohasy, C., Zhu, X., Suarez-Jimenez, B., Markowitz, J. C., & Neria, Y. (2021). Equine-Assisted Therapy for Posttraumatic Stress Disorder Among Military Veterans: An Open Trial. *The Journal of Clinical Psychiatry, 82*(5). https://doi.org/10.4088/JCP.21m14005
Gabriels, R. L., Pan, Z., Dechant, B., Agnew, J. A., Brim, N., & Mesibov, G. (2015). Randomized Controlled Trial of Therapeutic Horseback Riding in Children and Adolescents With Autism Spectrum Disorder. *Journal of the American Academy of Child & Adolescent Psychiatry, 54*(7), 541–549. https://doi.org/10.1016/j.jaac.2015.04.007
Gabriels, R. L., Pan, Z., Guérin, N. A., Dechant, B., & Mesibov, G. (2018). Long-Term Effect of Therapeutic Horseback Riding in Youth with Autism Spectrum Disorder: A Randomized Trial. *Frontiers in Veterinary Science, 5,* 156. https://doi.org/10.3389/fvets.2018.00156
González-Ramírez, M. T., Ortiz-Jiménez, X. A., & Landero-Hernández, R. (2013). Cognitive–Behavioral Therapy and Animal-Assisted Therapy: Stress Management for Adults. *Alternative and Complementary Therapies, 19*(5), 270–275. https://doi.org/10.1089/act.2013.19505
Griffin, J. A., Hurley, K., & McCune, S. (2019). Human-Animal Interaction Research: Progress and Possibilities. *Frontiers in Psychology, 10,* 2803. https://doi.org/10.3389/fpsyg.2019.02803
Guérin, N. A., Gabriels, R. L., Germone, M. M., Schuck, S. E. B., Traynor, A., Thomas, K. M., McKenzie, S. J., Slaughter, V., & O'Haire, M. E. (2018). Reliability and Validity Assessment of the Observation of Human-Animal Interaction for Research (OHAIRE) Behavior Coding Tool. *Frontiers in Veterinary Science, 5,* 268. https://doi.org/10.3389/fvets.2018.00268
Hallberg, L. (2017). *The Clinical Practice of Equine-Assisted Therapy: Including Horses in Human Healthcare (First Edition).* Routledge.
Harper, D. N. J., Rose, K., & Segal, D. (2019). *Nature-Based Therapy: A Practitioner's Guide to Working Outdoors with Children, Youth, and Families.* New Society Publishers.
HETI. (2021). *The Federation of Horses in Education and Therapy International AISBL/HETI.* About us. Retrieved from: https://hetifederation.org/
Hoagwood, K., Vincent, A., Acri, M., Morrissey, M., Seibel, L., Guo, F., ... & Horwitz, S. (2022). Reducing Anxiety and Stress among Youth in a CBT-Based Equine-Assisted Adaptive Riding Program. *Animals, 12*(19), 2491.
Horwitz, S., Demeter, C., Hayden, M., Storfer-Isser, A., Frazier, T. W., Fristad, M. A.,... & Findling, R. L. (2012). Parents' Perceptions of the Benefit of Children's Mental Health Treatment and Continued Use of Services. *Psychiatric Services, 63*(8), 793–801.
International Association of Human Animal Interaction Organizations. (2018). *IAHAIO Definitions for Animal Assisted Intervention and Guidelines for Wellness of Animals Involved in AAI.* https://iahaio.org/wp/wp-content/uploads/2021/01/iahaio-white-paper-2018-english.pdf

Jacobson, T. (2019, July 18). *The History of Equine Therapy*. Of Horse. https://www.ofhorse.com/view-post/The-History-of-Equine-Therapy

Kemp, K., Tania Signal, Helena Botros, Taylor, N., & Prentice, K. (2013). Equine Facilitated Therapy with Children and Adolescents Who Have Been Sexually Abused: A Program Evaluation Study. *Journal of Child and Family Studies, 23*, 558–566. https://doi.org/10.1007/s10826-013-9718-1

Kinney, A. R., Eakman, A. M., Lassell, R., & Wood, W. (2019). Equine-Assisted Interventions for Veterans with Service-Related Health Conditions: A Systematic Mapping Review. *Military Medical Research, 6*(1), 28. https://doi.org/10.1186/s40779-019-0217-6

Kruger, K. A., & Serpell, J. A. (2010). Animal-Assisted Interventions in Mental Health: Definitions and Theoretical Foundations. In A. H. Fine (Ed.), *Handbook on Animal-Assisted Therapy (Third Edition)* (pp. 33–48). Academic Press. https://doi.org/10.1016/B978-0-12-381453-1.10003-0

Lac, V. (2017). *Equine-Facilitated Psychotherapy and Learning: The Human-Equine Relational Development (HERD) Approach (First Edition)*. Academic Press.

Latella, D., & Abrams, B. (2019). Chapter 10—The Role of the Equine in Animal-Assisted Interactions. In A. H. Fine (Ed.), *Handbook on Animal-Assisted Therapy (Fifth Edition)* (pp. 133–162). Academic Press. https://doi.org/10.1016/B978-0-12-815395-6.00010-9

Lentini, J. A., & Knox, M. (2009). A Qualitative and Quantitative Review of Equine Facilitated Psychotherapy (EFP) with Children and Adolescents. *The Open Complementary Medicine Journal, 1*(1). https://benthamopen.com/ABSTRACT/TOALTMEDJ-1-51

Leste-Lasserre, C. (2019, September 30). *The Horse-Human Relationship: From Prehistory to Today*. https://thehorse.com/179351/the-horse-human-relationship-from-prehistory-to-today/

Matuszek, S. (2010). Animal-Facilitated Therapy in Various Patient Populations: Systematic Literature Review. *Holistic Nursing Practice, 24*(4), 187–203. https://doi.org/10.1097/HNP.0b013e3181e90197

McCardle, P. (2011). *How Animals Affect Us: Examining the Influences of Human–Animal Interaction on Child Development and Human Health*, Ed. S. McCune, J. A. Griffin, & V. Maholmes (pp. xvi, 228). American Psychological Association. https://doi.org/10.1037/12301-000

McCullough, L., Risley-Curtiss, C., & Rorke, J. (2015). Equine Facilitated Psychotherapy: A Pilot Study of Effect on Posttraumatic Stress Symptoms in Maltreated Youth. *Journal of Infant, Child, and Adolescent Psychotherapy, 14*(2), 158–173. https://doi.org/10.1080/15289168.2015.1021658

National Institute of Mental Health. (2021). *Post-Traumatic Stress Disorder*. Post Traumatic Stress Disorder. https://www.nimh.nih.gov/health/publications/post-traumatic-stress-disorder-ptsd

Ng, Z. (2021). Strategies to Assessing and Enhancing Animal Welfare in Animal-Assisted Interventions. In *The Welfare of Animals in Animal-Assisted Interventions: Foundations and Best Practice Methods* (pp. 123–154). Cham: Springer International Publishing.

Nimer, J., & Lundahl, B. (2007). Animal-Assisted Therapy: A Meta-Analysis. *Anthrozoös, 20*(3), 225–238. https://doi.org/10.2752/089279307X224773

Peters, B. C., Pan, Z., Christensen, H., & Gabriels, R. L. (2022). Self-Regulation Mediates Therapeutic Horseback Riding Social Functioning Outcomes in Youth With Autism Spectrum Disorder. *Frontiers in Pediatrics, 10*.

Professional Association for Therapeutic Horsemanship. (2021). *PATH International Standards for Certification and Accreditation*. https://fontevacustomer-15cf09b5446.force.com/CPBase__item?id=a135G000004TLOLQA4

Rankins, E. M., Wickens, C. L., McKeever, K. H., & Malinowski, K. (2021). A Survey of Horse Selection, Longevity, and Retirement in Equine-Assisted Services in the United States. *Animals, 11*(8), 2333. https://doi.org/10.3390/ani11082333

Scopa, C., Contalbrigo, L., Greco, A., Lanatà, A., Scilingo, E. P., & Baragli, P. (2019). Emotional Transfer in Human-Horse Interaction: New Perspectives on Equine Assisted Interventions. *Animals: An Open Access Journal from MDPI, 9*(12), E1030. https://doi.org/10.3390/ani9121030

Stergiou, A., Tzoufi, M., Ntzani, E., Varvarousis, D., Beris, A., & Ploumis, A. (2017). Therapeutic Effects of Horseback Riding Interventions: A Systematic Review and Meta-analysis. *American Journal of Physical Medicine & Rehabilitation, 96*(10), 717–725. https://doi.org/10.1097/PHM.0000000000000726

Thiel, U. (2013). *Ridden: Dressage from the Horse's Point of View* (C. Hughes, Trans., 2013th ed.). Trafalgar Square Books.

Vincent, A., & Farkas, K. J. (2017). Application of Attachment Theory to Equine-Facilitated Therapy. *Society Register*, *1*(1), 7–22.

Vincent, A., O'Reily, A., McKissock, B. (2023). Animal Assisted Interventions. In S. Loue & P. Linden (Eds.), *The Comprehensive Guide to Interdisciplinary Veterinary Social Work*. Springer Nature.

Vincent, A., Peth-Pierce, R. M., Morrissey, M. A., Acri, M. C., Guo, F., Seibel, L., & Hoagwood, K. E. (2021). Evaluation of a Modified Bit Device to Obtain Saliva Samples from Horses. *Veterinary Sciences*, *8*(10), 232. https://doi.org/10.1080/08927936.2016.1189747

Wilkie, K. D., Germain, S., & Theule, J. (2016). Evaluating the Efficacy of Equine Therapy Among At-risk Youth: A Meta-analysis. *Anthrozoös*, *29*(3), 377–393. https://doi.org/10.1080/

Wilsie, S., & Vogel, G. (2016). *Horse Speak: An Equine-Human Translation*. https://www.thriftbooks.com/w/horse-speak-an-equine-human-translation-guide-conversations-with-horses-in-their-language_gretchen-vogel_sharon-wilsie/11645910/

Wood, W., Alm, K., Benjamin, J., Thomas, L., Anderson, D., Pohl, L., & Kane, M. (2021). Optimal Terminology for Services in the United States That Incorporate Horses to Benefit People: A Consensus Document. *The Journal of Alternative and Complementary Medicine*, *27*(1), 88–95. https://doi.org/10.1089/acm.2020.0415

Zhu, X., Suarez-Jimenez, B., Zilcha-Mano, S., Lazarov, A., Arnon, S., Lowell, A. L., Bergman, M., Ryba, M., Hamilton, A. J., Hamilton, J. F., Turner, J. B., Markowitz, J. C., Fisher, P. W., & Neria, Y. (2021). Neural Changes Following Equine-Assisted Therapy for Posttraumatic Stress Disorder: A Longitudinal Multimodal Imaging Study. *Human Brain Mapping*, *42*(6), 1930–1939. https://doi.org/10.1002/hbm.25360

# 37
# ASSISTANCE AND EMOTIONAL SUPPORT ANIMALS

*Marguerite E. O'Haire, Leanne O. Nieforth, Clare L. Jensen, and Sarah C. Leighton*

## Introduction to the Issue

Animals have filled numerous roles to assist humans over many centuries, and recent decades have included substantial additions to the specialized roles that animals may play to promote human health. As assistance roles have expanded in applicability to meet physical, mental, and emotional health needs, so too has the demand for these animals (Walther et al., 2017). With popular interest on the rise, a need has emerged for consistent and accessible terminology, definitions, and information.

### *Assistance Animals*

An **assistance animal** is trained to complete tasks specifically intended to mitigate a disability for their handler. There are several distinct categories, including guide, hearing, mobility, medical alert, and psychiatric assistance dogs. Though miniature horses can also be assistance animals in the United States, there is no research on their impact; thus, this chapter will focus on assistance dogs. Although the term service dog is also common in the United States and is sometimes used interchangeably with assistance dog, the assistance dog terminology is favored as being more globally consistent.

**Guide dogs** are trained to lead their handler (who is partially sighted or blind) through their physical environment, taking directions and navigating accordingly.

**Hearing dogs** alert their handler who is d/Deaf or hard of hearing to important sounds, leading to increased environmental awareness.

**Mobility assistance dogs** are trained to provide physical assistance specific to the handler's disability and needs, such as retrieving dropped items or helping to propel a wheelchair.

**Medical alert assistance dogs** monitor for signs of a medical event and either alert their handler at the onset or respond to the event. Examples may include alerting to a change in blood sugar for an individual with diabetes, or the onset of a seizure for an individual with epilepsy.

**Psychiatric assistance dogs** intervene to mitigate symptoms of a psychological disability or neurodevelopmental disorder. Examples may include waking an individual with posttraumatic stress disorder (PTSD) from a nightmare or providing deep calming pressure to an individual with autism spectrum disorder (ASD).

## Emotional Support Animals

In the United States, another role for animals that has gained in popularity over recent decades is that of the ESA. An **emotional support animal** is an animal providing *comfort and generalized aid* to someone with a psychological disability or mental health condition. In contrast to therapy animals intended to provide comfort and support to multiple people, ESAs aid just one primary handler. The critical component distinguishing an ESA from an assistance animal is that ESAs do not require specific training to benefit their handlers in their intended roles. Rather, ESAs are intended to benefit their handlers through their presence alone, similar to a companion animal. The key difference between a companion animal (pet) and an ESA is that the ESA is medically recommended, while the companion animal is not. As ESAs are not required to have any training and are not regulated by any types of behavioral standards, they are also not permitted in public spaces where pets are otherwise prohibited. Although ESAs are restricted from public spaces, handlers of ESAs are entitled to access certain types of housing with their animals regardless of pet restrictions. This ensures that handlers can receive and maintain support from their ESAs while still having equal access to the same housing opportunities for individuals without psychological disabilities or mental health diagnoses. Given that ESAs are only recognized in the United States (not globally) and that research on their use is nearly non-existent, the remainder of the chapter will focus largely on assistance animals only.

See Table 37.1 for a comparison of characteristics among assistance, emotional support, therapy, companion, and facility animals.

*Table 37.1* Comparison of characteristics

| | Assistance Animal | Emotional Support Animal | Therapy Animal | Companion Animal | Facility Animal |
|---|---|---|---|---|---|
| **Specially trained** to work with an individual handler | ✓ | | | | |
| **Specially trained** to work with a group of people other than the handler | | | | | ✓ |
| **Evaluated** for appropriateness in role | ✓ | | ✓ | | ✓ |
| Granted special **legal considerations**, which may vary by country | ✓ | ✓ | | | ✓ |
| Examples | Guide dog, psychiatric assistance animal | Dog who provides comfort to a person with anxiety | Therapy dog in an after-school reading program | Pet dog, pet cat, pet hamster | Facility dog in a hospital |

*Note:* Shading and a check mark indicates that the characteristic applies to that type of working animal role.

## Impact on People

There is almost no research on ESAs, thus understanding their impact is best gleaned from companion animal literature (see Section I of this Handbook). For assistance dogs, each team is unique, and individual impacts may vary; nevertheless, research has demonstrated some commonalities according to the dog's trained role.

### *Impact on Handlers*

**Guide dogs** allow their handlers to be more independent, often creating an improved quality of life and overall psychological wellbeing (Glenk et al., 2019). Guide dogs broaden their handler's world, increasing travel by improving orientation and wayfinding, and increasing safety around traffic. Partnership with a guide dog can decrease fatigue, and promote health benefits through increased exercise (Whitmarsh, 2005).

**Hearing dogs** create an environment where handlers feel less fear, less anxiety, and greater safety (Guest et al., 2006). Safety is enhanced by ensuring they are aware of potential threats such as fire, an ambulance siren, or the whistle of a tea kettle (Rintala et al., 2008). The dog also creates an opportunity for social interaction and community engagement (Guest et al., 2006; Sachs-Ericsson et al., 2002). Handlers report a decrease in reliance on friends and family or on assistive devices, and improvements in sleep (Guest et al., 2006; Rintala et al., 2008).

**Mobility assistance dogs** can improve their handler's wellbeing. This includes psychological wellbeing, emotional functioning, self-esteem, and vitality (Rodriguez et al., 2020; Winkle et al., 2012). Attachment theory suggests that assistance dogs and owners may have a reciprocal relationship with positive and secure attachment (Kwong & Bartholomew, 2011), although one review found that most psychosocial health outcomes (68%) were not significant (Rodriguez et al., 2020). Handlers also report increased independence and decreased reliance on caretakers/family, which can mean significant financial savings in some cases. Mobility assistance dogs enable their handler to reallocate their energy away from physical tasks and toward other activities. Research has demonstrated improvements in physical fitness, strength, and motor control through partnership and interaction with the assistance dog (Rintala et al., 2008).

**Medical alert assistance dogs**, similar to mobility assistance dogs, can also improve the psychosocial wellbeing (Rodriguez et al., 2020; Winkle et al., 2012). Additionally, for handlers with medical alert assistance dogs, the potential impacts of accurate and reliable alerts could have enormous health and safety implications; research on dogs' ability to accurately and reliably detect medical events appears promising, albeit inconclusive at this time (Catala et al., 2019; Gonder-Frederick et al., 2017).

**Psychiatric assistance dog** research has focused primarily on placements with veterans with military-connected PTSD, with some limited research also examining impacts for individuals with ASD. Veterans with PTSD who are paired with psychiatric assistance dogs demonstrate clinically significant reductions in PTSD symptoms, are less depressed, and are less anxious (O'Haire & Rodriguez, 2018). They are also more satisfied with their life, feel a greater sense of wellbeing, and in some cases have better relationships and social lives (Crowe et al., 2018). Biological impacts have also been identified via cortisol, a hormone that is important in the stress response system. Veterans with assistance dogs have awakening cortisol levels closer to healthy adults when compared to veterans on the waitlist for an assistance dog (Rodriguez et al., 2018).

For individuals with ASD, possible benefits include improved social responsiveness, language, and decreased parent-reported problem behaviors (Nieforth, Schwichtenberg et al., 2021). Psychiatric assistance dogs may serve as a catalyst for social interaction and may improve overall wellbeing and mental health of both children with ASD and their families (Tseng, 2022). Improvements have also been noted in initial physiological measures such as the cortisol awakening response (Viau et al., 2010).

## *Impact on Caregivers of Individuals with Disabilities*

Psychiatric assistance dogs for ASD may reduce caregiver stress, create a sense of safety for caregivers and be a friend to the caregiver (Burrows et al., 2016). They may also strengthen the bond between family members and increase social interaction with the community (Burgoyne et al., 2014). Caregivers of individuals with mobility and medical alert assistance dogs often experience similar benefits including emotional support, stress reduction and bond formed with the animal (Nieforth, Rodriguez et al., 2021). Spouses of military veterans with psychiatric assistance dogs for PTSD report experiencing both benefits and challenges. These impacts may be part of a resilience process where interaction with the assistance dog helps to build emotional reserves and facilitates relational maintenance (Nieforth, Craig et al., 2021).

## *Drawbacks and Challenges*

Prior to partnership, challenges may occur during acquisition. While some organizations provide assistance animals free of charge, in other cases costs can exceed tens of thousands of dollars. Applicants often face multi-year waiting lists, during which time symptoms may worsen. Then, the pairing process itself can be demanding and fatiguing. Many organizations employ onsite classes as part of the pairing process, requiring weeks away from home and travel costs. Once home, significant lifestyle adjustments are required to incorporate the animal into the day-to-day living – particularly for first-time animal owners. Handlers must account for ongoing costs of care including food, health, and supplies. For some, these costs can preclude partnership, even if they might otherwise benefit (Lessard et al., 2018). However, research suggests that financial savings through increased independence, decreased need for hired caretakers, and health savings may somewhat balance costs, and possibly lead to net savings (Sachs-Ericsson et al., 2002).

Family members of assistance dog handlers may also experience challenges. While caregiver burden may decrease through enhanced independence (Rintala et al., 2002), it may increase in other areas, most commonly as a result of the care and upkeep involved and the shared financial burden (Nieforth, Rodriguez et al., 2021). Additionally, if the assistance animal takes on a task that was previously the responsibility of the caregiver, it can be difficult for the caregiver to accept that they are not needed anymore in that situation (Whitworth et al., 2020). Assistance dogs may also strain some relationships through interrupting intimacy or other interactions (Nieforth, Craig et al., 2021).

In some cases, assistance animals may have negative impacts on human health, similar to companion animals. These could include allergies, injuries, or infection (Davis et al., 2004). Given that assistance dogs receive temperament screening and training prior to placement, presumably the risk of injuries is lower than for companion or ESAs; however, this has not been confirmed by research.

Assistance animals require ongoing training and behavioral maintenance. The emergence of problematic behavior can be a significant burden for handlers and their families, demanding time and emotional investment (Davis et al., 2004). Nevertheless, these issues are part of the reality of partnership with an animal, and the time investment is important to ensure not only efficacy in the working role but also the animal's well-being (Sachs-Ericsson et al., 2002). Placement failure (e.g., needing to retire the animal from its working role or return the animal to the organization) can occur if behavior or training issues become too extreme, lead to safety concerns, or due to personal circumstances. This can be a very difficult experience for the handler, particularly if they have already bonded with the animal.

While thankfully representing only a small fraction of the industry, of significant concern are organizations that do not maintain acceptable standards, and as a result directly harm the individuals with disabilities that they claim to serve. Reports exist of clients never receiving the animal following payment or being matched with an animal with a severe lack of training, behavioral problems, or health issues. These cases are damaging to the individual, the animal, and the public's perception of the industry as a whole (Breed, 2019).

The use of an assistance animal in public can also come with challenges. Public access denials can be stressful and disruptive (Zier, 2020). Increased attention due to the presence of a dog can be negative, due to the public's lack of knowledge of proper etiquette. This can result in rude questioning, inconvenient delays, injuries from aggressive pet dogs, or dangerous distractions from an assistance dog's work. Moreover, for individuals with invisible disabilities, the mere presence of the assistance animal is effectively a disclosure of disability status, which can lead to stigma and invasive questioning among other problems (Mills, 2017).

Finally, a significant consideration for any animal is grief surrounding loss, given animals' shorter lifespans. For assistance dog handlers, this is compounded by the additional layer of the working relationship, and can occur not only with death but also at the time of retirement from the dog's working role (Kogan et al., 2021). Indeed, anticipation of this grief can be a source of stress even during the partnership (McLaughlin & Hamilton, 2019). Nevertheless, research indicates that a majority of these handlers go on to seek out additional assistance dogs, suggesting that the value and benefits of the partnership outweigh the emotional distress upon loss (Kogan et al., 2021).

## Impact on the Animal's Welfare

Compared to the body of research on how assistance animals may affect humans, information is sparse regarding the effects of these roles on the animals themselves. Potential welfare concerns among assistance animals have been raised in early qualitative studies, including greater prevalence of physical, emotional, and social stressors (e.g., lack of time to relieve themselves, lack of consistent daily routine, limited recreational time) (Burrows et al., 2008). Although evidence for the enduring well-being of therapy dogs (Clark et al., 2020) has been used to make speculations about similar well-being in assistance dogs, fundamental differences in therapy and assistance animal roles indicate a need for targeted research on welfare of assistance animals specifically. Emerging research has begun to do exactly that. Two publications have found no evidence of welfare detriments measured using cortisol, a hormone indicative of physiological stress levels. Assistance dogs maintained relatively low levels of salivary cortisol during an assistance dog training session, suggesting the dogs did not experience increased short-term stress (van Houtert et al., 2021). Further, when assessing levels of hair cortisol indicative of long-term stress (i.e., over the prior two months), no significant differences were found between assistance dogs and pet dogs (van Houtert et al., 2022).

To meet behavioral standards required for public access and to perform specific tasks for an individual with a disability, assistance animals spend a greater than average proportion of their lives in training. Training and learning activities that engage an animal's mind can provide cognitive enrichment or mental exercise, which are commonly considered to be important strategies for promoting and maintaining an animal's welfare. Indeed, research in canine neuroscience has found that dogs receiving more lifelong training maintain better cognitive health as they age (Chapagain et al., 2017). Further, research has found stable or lowered levels of the stress hormone cortisol in response to training sessions among assistance dogs for veterans with PTSD (van Houtert et al., 2021).

The type of training methods used can play an important role in well-being and overall welfare. Within the assistance dog community, accredited organizations have moved to embrace the Least Intrusive, Minimally Aversive (LIMA) (Assistance Dogs International, 2022b) approach to training in alignment with numerous other major dog-training bodies. The LIMA approach calls for trainers and handlers to work consistently within a hierarchy of procedures which prioritizes constructive environmental changes and positive reinforcement over punishment. In this way, trainers and handlers ensure use of the least intrusive techniques necessary to accomplish the desired behavior change, while limiting risk of adverse effects (International Association of Animal Behavior Consultants, 2022). Methods such as this may help to protect and maintain assistance dog welfare, as use of excessively forceful training methods have been linked to higher levels of the stress hormone cortisol, and greater instances of aggression and fear in dogs (Ziv, 2017). In contrast, dogs receiving more reward-based training have been found to demonstrate fewer behaviors indicative of stress (Ziv, 2017).

Throughout training and in subsequent partnership, trainers and handlers should recognize and respond appropriately to signals of stress, fear, or discomfort from the animal (Wenthold & Savage, 2007). Even if certain behaviors may be modified, underlying factors leading to these behaviors could also be signs that the specific temperament of an individual animal is not well-suited to the role. For assistance dog partnerships to be healthy and sustainable, it is essential to recognize subtle changes in dog behaviors and monitor for signs of distress (Wenthold & Savage, 2007). These signs may indicate that the dog needs a break, or time to play freely without focusing on the handler's needs. Over time, small or consistent welfare concerns in an animal not well-suited for the role may compound into a major event that compromises the health and safety of the animal, the handler, or others.

Although research has begun to predict which dogs will graduate from assistance dog training (Bray et al., 2019; MacLean & Hare, 2018) or benefit their handlers, we are missing research to predict which dogs will have the best welfare in the role. There are also unique welfare concerns that may emerge at the time of the animal's retirement. Retirement usually takes place when the animal can no longer comfortably or effectively meet the needs of their handler. If the animal's health or physical condition is deteriorating, retirement may be necessary to maintain the animal's welfare by reducing or preventing any pain and fatigue. Once retired, the subsequent changes to an animal's life and routine may also influence their cognitive and emotional well-being. After spending years in constant proximity to their handler, a newly retired assistance animal may experience distress when their handler starts to go places without them. To limit any detrimental effects of retirement, many assistance dog organizations work closely with the handler to help prepare for this transition. Providing long-term support and proper education of trainers and handlers can have a substantial influence over assistance animal welfare.

## Current Policy and Legislative Perspectives

### *Legal Rights and Restrictions*

#### *United States*

Multiple laws govern assistance animals (referred to in law as service animals) in the United States (Zier, 2020). For brevity, this section focuses on federal law; a link to state laws is provided below.

The **Air Carrier Access Act of 1986 (ACAA)** provides legal guidance for commercial airlines in accommodating passengers with disabilities. Under the ACAA's recent revision, service animals (defined as "a dog, regardless of breed or type, that is individually trained to do work or perform tasks for the benefit of a qualified individual with a disability") may accompany their handler in the cabin. This is contingent on the handler attesting to the dog's appropriate behavior and training, ability to refrain from relieving itself in flight, and health status (U.S. Department of Transportation, 2020). As of 2020, ESAs are considered equivalent to pets and no longer have the right to travel in the cabin. This change was made in response to growing problems with ESAs during flights (U.S. Department of Transportation, 2020).

The **Americans with Disabilities Act of 1990 (ADA)** protects the rights of individuals with disabilities in public spaces and in the workplace and consists of Titles I, II, and III. Under Title I, employers are required to consider both assistance animals and ESAs as potential accommodations for employees with disabilities (Carroll et al., 2020). Titles II and III cover state and local government buildings and private entities serving the general public. In these venues, service animals (defined as "dogs [or miniature horses] that are individually trained to do work or perform tasks for people with disabilities") may accompany their handler (U.S. Department of Justice, 2010), but ESAs do not have public access rights.

Under Titles II and III, businesses or venues may ask two questions (U.S. Department of Justice, 2010):

1. Is the dog a service animal required because of a disability?
2. What work or task has the dog been trained to perform?

The animal must also behave appropriately and remain under control. If they aren't, or if both questions are not answered satisfactorily, businesses may ask for the animal to be removed (U.S. Department of Justice, 2010). Additionally, if the presence of an animal would "fundamentally alter the nature of the business," they may be excluded (e.g., operating rooms, parts of zoos, public pools) (Kisor & Goodier, 2019).

The **Fair Housing Act of 1968 (FHA)** requires landlords to accommodate assistance animals (defined as "animals that do work, perform tasks, assist, and/or provide therapeutic emotional support for individuals with disabilities") in certain types of housing, regardless of pet restrictions. Landlords may ask for additional information (such as a doctor's letter) from the tenant if the nature of their need for accommodation is not readily apparent (U.S. Department of Housing and Urban Development, 2020).

#### *Global*

Outside of the United States, ESAs are not legally acknowledged, and legislation on assistance animals is inconsistent. In the rest of North America, assistance dogs are well-recognized in

Canada, but considered legally equivalent to pets in Mexico, without special public access rights (International Association of Assistance Dog Partners, 2022; Kisor & Goodier, 2019). At the time of writing, no laws are listed for countries in Africa or Asia on the Assistance Dogs International resources page (Assistance Dogs International, 2022a). Australia defines assistance animals with verbiage similar to the ADA (Bremhorst et al., 2018). Assistance dogs are also well-recognized in Europe, but legislation is managed by country and is highly variable (Bremhorst et al., 2018). Finally, in South America, few countries formally recognize assistance animals within law (Assistance Dogs International, 2022a).

## Resources

- Public Access Laws. https://assistancedogsinternational.org/resources/public-access-laws/. Assistance Dogs International.
- Table of State Service Animal Laws. https://www.animallaw.info/topic/table-state-assistance-animal-laws Michigan State University Animal Legal & Historical Center.

## Policy

Within the assistance dog community, umbrella organizations establish standards and manage accreditations. These include Assistance Dogs International and the International Guide Dog Federation, between them overseeing over 200 organizations and 40,000 assistance dog teams (Assistance Dogs International, 2019; International Guide Dog Federation, 2020). These organizations do not oversee non-accredited facilities, nor individuals who obtain their dog outside of an organization. No similar organizations or standards exist for ESAs.

Despite the legal equivalence between an assistance animal and any other type of assistive technology, at the of the time of writing, no private insurance companies cover related costs (Sachs-Ericsson et al., 2002). For military veterans in particular, the VA provides financial benefits for the costs of acquisition, care, and equipment; however, they have been hesitant to extend these benefits to include psychiatric assistance dogs for PTSD, citing a lack of empirical evidence as to their effectiveness (Richerson et al., 2020).

## Cultural Perspectives

### Individual Perceptions

Cultural, social, and demographic diversity influence how one experiences and relates to their environment (Randy J. Virden, 1999), how many and which opportunities are accessible in their lifetime (Butler, 2018), and how one may be treated by people and structures in power (Vassar & Howard, 2021). Religious beliefs may also influence perceptions about animals in assistance roles (Smith, 2019). Although research on diversity as it relates to assistance animals is largely nonexistent, some inferences may be made from research on companion animal or pet relationships.

Meaningful bonding with pets is common among pet owners across the United States, regardless of race or ethnicity (Risley-Curtiss, Holley, Cruickshank et al., 2006), although research indicates that white people may be more likely to have pets (Brown, 2002; Risley-Curtiss, Holley, & Wolf, 2006). For example, results of a survey conducted among 133 veterinary students demonstrated that, compared to Black or African American students, white students were more likely to have pets (100% vs. 86%), had a greater number of pets (4 vs. 2), and were more likely to allow their pet(s) to sleep in bed with them (70% vs. 53%) (Brown, 2002). A later survey of 587 adults in the United States found that white and Indigenous respondents had the highest likelihood of having pets, followed Hispanic/Latinx respondents and Black respondents, whereas Asian respondents had the lowest likelihood (Risley-Curtiss, Holley, & Wolf, 2006). An additional publication in the same year went on to interpret qualitative interviews with 15 women of color about pets in their lives. Among interviewees, 87% considered their pets to be family and 53% reported this to be a common view among other members of their ethnic communities (Risley-Curtiss, Holley, Cruickshank et al., 2006).

Despite differences in ownership, findings that individuals from culturally diverse communities form familial bonds with companion animals might also suggest comparable benefits from companionship, support, and aid offered by assistance animals or ESAs. However, diversity among handlers remains largely unexamined. Further, the additional barriers faced by historically minoritized individuals may influence accessing resources for and about assistance animals and ESAs. Among these barriers are financial barriers, as well as the greater inequities faced by underrepresented and disenfranchised minority groups in the healthcare system (Rosenkranz et al., 2021). For example, studies have shown that patients who are Black, Indigenous, or people of color are less likely to be admitted for more intensive treatment after an emergency hospital visit and are less likely to be referred for specialized interventions (Blanchard et al., 2003; Zhang et al., 2020). As assistance animals are often considered specialized or complementary health interventions, it seems plausible that information, resources, and recommendations for assistance animals may also be offered less frequently to individuals who are Black, Indigenous, or people of color.

Although disparities in healthcare overall could include greater barriers to interventions with animals, researchers have yet to examine how inequities may apply to assistance animal recommendation and provision. Stakeholders in the field of human-animal interaction have noted the importance of considering contexts, cultures, and geographical locations in human-animal research and an increasing number of field members have called for greater considerations of diversity, equity, inclusion, and belonging (DEIB) (Smith, 2019).

## *Public Perceptions*

Research in the United States and Australia indicates that the public generally views assistance dogs favorably and perceives them as happier, better cared for, and more loved than pet dogs (Gibson & Oliva, 2021; Schoenfeld-Tacher et al., 2017). In contrast, research in Japan found generally low levels of support and demand for assistance animal use, concluding that cultural, historical, and environmental influences had an important role in acceptance of working dogs (Yamamoto et al., 2014). Public opinion toward ESAs in the United States is mixed; it is unclear whether this is driven by lack of understanding of their role, or by perceptions of illegitimacy (Schoenfeld-Tacher et al., 2017).

Despite overall positive perceptions in the United States, misconceptions around assistance dogs are widespread (Zier, 2020). Confusion around inconsistent terminology and the lack of standardized certification has led to a growing problem with fraud (Schoenfeld-Tacher et al., 2017). When members of the public misrepresent ill-behaved pets as assistance animals, they can create dangerous distractions and leave negative impressions resulting in increased stigma against legitimate teams (Zier, 2020). Research indicates that up to two-thirds of assistance dog handlers have experienced discrimination as a result of their use of an assistance dog, and this is significantly higher for those with invisible disabilities (Mills, 2017; Zier, 2020).

Supporting the disabled community, including assistance animal handlers, is an important part of any conversation around DEIB. Assistance animal handlers experience both positive and negative effects with regards to inclusion and discrimination, and members of the public are broadly unaware of appropriate etiquette when encountering a team (Mills, 2017). In public venues, workplaces, and schools, the steps that leaders do or do not implement to prepare for and welcome an assistance animal can either promote a positive environment or lead to challenges. Further education, research, and clarity is needed to drive true inclusion.

This confusion around regulations and rights for assistance animal teams remains one of the biggest barriers to widespread understanding and support. And yet, it is important to acknowledge that any solution to these issues must also consider the rights and dignity of the disabled community. The idea of a standardized credentialing process may appear to be an appealingly straightforward solution, but if such a system were not implemented appropriately it would at best risk increased discrimination or privacy violations, and at worst could be considered an abuse of fundamental civil rights.

## Practical Perspectives

### *Interacting in Public Spaces*

When assistance animals are working, they should not be distracted. It is important to be respectful of the handler and recognize that they have no obligation to talk about their experiences. In multiple studies, a drawback to having an assistance dog for both handlers and family members is negative or probing interaction from individuals in public spaces (Nieforth, Rodriguez et al., 2021). For additional information on interacting with assistance dog handler teams in public spaces visit https://canine.org/about/resources/for-the-community/.

### *Finding and Partnering with an Assistance Animal*

There is substantial variability among assistance animal provider processes, programs, and lifetime support. Given the growing popularity of assistance dogs, waitlists are increasingly long. Since the beginning of the COVID-19 pandemic, waitlists have increased in length and currently range from six months to five years. Once approved for an assistance dog, the training process varies dependent upon the organization. Some organizations train the dog prior to pairing followed by a two- to three-week intensive onsite program for handlers and their dogs. Other organizations train the handler to train the assistance dog themselves. The origin of assistance dogs ranges from purpose-bred to shelter rescues. The costs can be upward of $30,000, but most are provided free of cost with some organizations providing financial support throughout the dog's lifetime.

### *Clinical Recommendations: When Is an Assistance Animal an Appropriate Clinical Recommendation?*

The recommendation of an assistance animal or ESA should be made after reviewing an individual's current symptoms, treatment goals, and history (Porter et al., 2021). Assistance animals are trained to do specific tasks that mitigate an individual's symptoms. Clear treatment goals and motivation to improve are important, given the previously described drawbacks and challenges (Porter et al., 2021). Identifying expectations and educating the individual about real-life experiences may be critical to creating realistic understanding and a willingness to persevere and address issues that may arise. If specific tasks are not needed to mitigate symptoms, an ESA (in the United States) or a pet (globally) may be the better choice. Regardless of choosing an assistance animal, ESA, or pet, it is critical that the individual be aware of the care and commitment associated with adding an animal to the household. Additionally, animal welfare must be considered. Individuals with a history of violence or animal abuse are not well-equipped to take on assistance animals or ESAs.

Practical resources include general guidelines for clinicians (Hoy-Gerlach et al., 2019; Younggren et al., 2020), veterinarian specific information (*Service, Emotional Support, and Therapy Animals*, 2022), legal implications (Carroll et al., 2020), and PTSD specific recommendations (Porter et al., 2021).

## Ethical Considerations

As a working animal, consent to do the work is important for a healthy, mutually beneficial partnership. Consent in an assistance animal partnership is still being explored; however, it is important to consider the assistance animal as an individual being as part of a *team* rather than an object that is being *used* by the human (Horowitz, 2021). Each assistance animal is a sentient being with emotions, personality, and preferences, and this sentience should be accounted for in interactions with the animal. Handlers must account for the animal's preferences and needs in all areas including offering the animal agency (i.e., the ability to act independently and make its own choices) (Spinka, 2019). On an individual level, some animals are better suited for different roles, perhaps as pets, rather than as assistance animals. An assistance animal should ideally thrive in its role, finding the work itself rewarding. Consent and agency are critical as these choices may influence the quality of the human-animal bond (Iannuzzi & Rowan, 1991).

Considering the agency of the assistance animal and the human involved, it is important to have realistic expectations for the partnership. Pairing an animal with a human is not a standalone treatment, and should not replace other interventions (e.g., medications, physical therapy, psychotherapy, etc.). Instead, it is an opportunity for an adjunct integrative health option to be used alongside other mainstream, evidence-based care. For example, a service dog should not be the *only* treatment a veteran receives. Instead, the service dog should be *added* to other treatments, such as cognitive behavioral therapy with a licensed mental health practitioner *plus* a service dog in the home. Assistance dog organizations generally expect individuals to work alongside their physician or mental health professional to incorporate the canine into other ongoing care to achieve treatment goals. Having appropriate expectations of the assistance animal may help to keep the animal's welfare at the forefront of the interaction, thus potentially mitigating challenges that may arise.

## Unanswered Questions and Future Research Demands

### *Diversity, Equity, Inclusion, and Belonging (DEIB)*

Given that assistance animals and ESAs work with individuals with disabilities and/or mental health diagnoses, research on the subject should be performed through the lens of DEIB. And yet, such research is all but nonexistent. Of particular concern are indications that the mere presence of an assistance dog can lead to *increased* discrimination against the handler (Mills, 2017). Therefore, an important priority for future research is to explore DEIB in the assistance animal and ESA world, including measurement of differentiating access and impacts across diverse groups, exploring intersectional discrimination, and understanding costs, scalability, and sustainability.

## Toward a Rounded and Nuanced Understanding

Reviews of existing literature find assistance animal interventions to be promising, while emphasizing the need for further rigorous research (O'Haire, 2013; Whitmarsh, 2005); and evidence is entirely lacking with regards to ESAs (Carroll et al., 2020). A great deal of assistance animal research is limited by over-reliance on self-reported information and therefore impacted by expectancy bias (Rodriguez et al., 2020). Research designs frequently set out to prove perceived benefits and do not always account for potential drawbacks and challenges. Meanwhile, provider organizations themselves often do not use valid and reliable measures as part of their assessments, and may not repeat measures when tracking outcomes (Butterly et al., 2013). And yet, medical professionals, government agencies, and medical insurance companies rely on strong empirical evidence to support their decision-making in recommending specific medical interventions, including an assistance animal (Richerson et al., 2020; Vincent et al., 2016). The effort to collect useable, objective, nuanced outcome information will ultimately require collaboration and coordination between researchers and these provider organizations.

The potential impact of robust evidence is important: increased funding for assistance animals (including through insurance) would result in shorter wait times, better matches, and more resources invested in follow-up support (Rintala et al., 2008). A better understanding of the factors involved in successful partnerships could enhance services, experiences, and outcomes for both humans and animals. Ultimately, achieving a rounded and nuanced understanding of the effects of assistance animals and ESAs that satisfy these conditions will move the field beyond the binary of whether or not these interventions "work" to how, why, for whom, and under what circumstances.

**Proposed Questions for Discussion**

- What did you know about assistance or emotional support animals before reading this chapter? Have you ever encountered a team before? How has your perspective and understanding changed?
- Describe two different scenarios where an assistance animal may benefit the mental and/or physical health of their handler, including a nuanced view of potential challenges.
- Discuss the importance of combining research outcomes and practical perspectives when pairing an assistance animal with an individual.
- How might greater consistency in legislation, policy, and public education improve the experience and well-being of assistance animal users?
- How can major ethical concerns be mitigated by the assistance animal provider, the clinician, and/or the handler?

## References

Assistance Dogs International. (2019). *Member Program Statistics*. Assistance Dogs International. https://assistancedogsinternational.org/members/member-program-statistics/

Assistance Dogs International. (2022a). *Public Access Laws*. Assistance Dogs International. https://assistancedogsinternational.org/resources/public-access-laws/

Assistance Dogs International. (2022b). *Summary of Standards*. Assistance Dogs International. https://assistancedogsinternational.org/standards/summary-of-standards/

Blanchard, J. C., Haywood, Y. C., & Scott, C. (2003). Racial and Ethnic Disparities in Health: An Emergency Medicine Perspective. *Academic Emergency Medicine*, *10*(11), 1289–1293. https://doi.org/10.1197/S1069-6563(03)00501-3

Bray, E. E., Levy, K. M., Kennedy, B. S., Duffy, D. L., Serpell, J. A., & MacLean, E. L. (2019). Predictive Models of Assistance Dog Training Outcomes Using the Canine Behavioral Assessment and Research Questionnaire and a Standardized Temperament Evaluation. *Frontiers in Veterinary Science*, *6*, 49. https://doi.org/10.3389/fvets.2019.00049

Breed, A. G. (2019, May 5). Service Dog Trainers Barely Regulated, Leading to Incompetence, Fraud. *Associated Press*. https://www.usatoday.com/story/money/2019/05/05/service-dog-trainers-barely-regulated-leading-incompetence-fraud/1111805001/

Bremhorst, A., Mongillo, P., Howell, T., & Marinelli, L. (2018). Spotlight on Assistance Dogs—Legislation, Welfare and Research. *Animals: An Open Access Journal from MDPI*, *8*(8). https://doi.org/10.3390/ani8080129

Brown, S.-E. (2002). Ethnic Variations in Pet Attachment among Students at an American School of Veterinary Medicine. *Society & Animals*, *10*(3), 249–266.

Burgoyne, L., Dowling, L., Fitzgerald, A., Connolly, M., P Browne, J., & Perry, I. J. (2014). Parents' Perspectives on the Value of Assistance Dogs for Children with Autism Spectrum Disorder: A Cross-sectional Study. *BMJ Open*, *4*(6), e004786–e004786. cmedm. https://doi.org/10.1136/bmjopen-2014-004786

Burrows, K. E., Adams, C. L., & Millman, S. T. (2008). Factors Affecting Behavior and Welfare of Service Dogs for Children with Autism Spectrum Disorder. *Journal of Applied Animal Welfare Science: JAAWS*, *11*(1), 42–62. https://doi.org/10.1080/10888700701555550

Burrows, K. E., Adams, C. L., & Spiers, J. (2016). Sentinels of Safety: Service Dogs Ensure Safety and Enhance Freedom and Well-Being for Families with Autistic Children. *Qualitative Health Research*, *18*(12), 1642–1649. https://doi.org/10.1177/1049732308327088

Butler, T. T. (2018). Black Girl Cartography: Black Girlhood and Place-Making in Education Research. *Review of Research in Education*, *42*(1), 28–45. https://doi.org/10.3102/0091732X18762114

Butterly, F., Percy, C., & Ward, G. (2013). Brief Report: Do Service Dog Providers Placing Dogs with Children with Developmental Disabilities Use Outcome Measures and, If So, What Are They? *Journal of Autism and Developmental Disorders*, *43*, 2720–2725. https://doi.org/10.1007/s10803-013-1803-1

Carroll, J. D., Mohlenhoff, B. S., Kersten, C. M., McNiel, D. E., & Binder, R. L. (2020). Laws and Ethics Related to Emotional Support Animals. *The Journal of the American Academy of Psychiatry and the Law*, *48*(4), 509–518. https://doi.org/10.29158/jaapl.200047-20

Catala, A., Grandgeorge, M., Schaff, J.-L., Cousillas, H., Hausberger, M., & Cattet, J. (2019). Dogs Demonstrate the Existence of an Epileptic Seizure Odour in Humans. *Scientific Reports*, *9*(1), 4103. https://doi.org/10.1038/s41598-019-40721-4

Chapagain, D., Virányi, Z., Wallis, L. J., Huber, L., Serra, J., & Range, F. (2017). Aging of Attentiveness in Border Collies and Other Pet Dog Breeds: The Protective Benefits of Lifelong Training. *Frontiers in Aging Neuroscience*, *9*. https://www.frontiersin.org/article/10.3389/fnagi.2017.00100

Clark, S. D., Martin, F., McGowan, R. T. S., Smidt, J. M., Anderson, R., Wang, L., Turpin, T., Langenfeld-McCoy, N., Bauer, B. A., & Mohabbat, A. B. (2020). Physiological State of Therapy Dogs during Animal-Assisted Activities in an Outpatient Setting. *Animals*, *10*(5), 819. https://doi.org/10.3390/ani10050819

Crowe, T. K., Sánchez, V., Howard, A., Western, B., & Barger, S. (2018). Veterans Transitioning from Isolation to Integration: A Look at Veteran/Service Dog Partnerships. *Disability and Rehabilitation*, *40*(24), 2953–2961. https://doi.org/10.1080/09638288.2017.1363301

Davis, B. W., Nattrass, K., O'Brien, S., Patronek, G., & MacCollin, M. (2004). Assistance Dog Placement in the Pediatric Population: Benefits, Risks, and Recommendations for Future Application. *Anthrozoös*, *17*(2), 130–145.

Gibson, P. E., & Oliva, J. L. (2021). Public Perceptions of Australian Assistance Dogs: Happier and Better Used Than Companion Dogs. *Journal of Applied Animal Welfare Science*, 1–13. https://doi.org/10.1080/10888705.2021.1931869

Glenk, L. M., Přibylová, L., Stetina, B. U., Demirel, S., & Weissenbacher, K. (2019). Perceptions on Health Benefits of Guide Dog Ownership in an Austrian Population of Blind People with and without a Guide Dog. *Animals : An Open Access Journal from MDPI*, *9*(7), 428. https://doi.org/10.3390/ani9070428

Gonder-Frederick, L. A., Grabman, J. H., & Shepard, J. A. (2017). Diabetes Alert Dogs (DADs): An Assessment of Accuracy and Implications. *Diabetes Research and Clinical Practice*, *134*, 121–130. https://doi.org/10.1016/j.diabres.2017.09.009

Guest, C. M., Collis, G. M., & McNicholas, J. (2006). Hearing Dogs: A Longitudinal Study of Social and Psychological Effects on Deaf and Hard-of-Hearing Recipients. *Journal of Deaf Studies and Deaf Education*, *11*(2), 252–261. https://doi.org/10.1093/deafed/enj028

Horowitz, A. (2021). Considering the "Dog" in Dog–Human Interaction. *Frontiers in Veterinary Science*, *8*. https://www.frontiersin.org/article/10.3389/fvets.2021.642821

Hoy-Gerlach, J., Vincent, A., & Lory Hector, B. (2019). Emotional Support Animals in the United States: Emergent Guidelines for Mental Health Clinicians. *Journal of Psychosocial Rehabilitation and Mental Health*, *6*(2), 199–208. https://doi.org/10.1007/s40737-019-00146-8

Iannuzzi, D., & Rowan, A. (1991). Ethical Issues in Animal-Assisted Therapy Programs. *Anthrozoos*, *4*, 154–163. https://doi.org/10.2752/089279391787057116

International Association of Animal Behavior Consultants. (2022). *IAABC Statement on a Least Intrusive, Minimally Aversive (LIMA) Approach to Behavior Modification and Training*. International Association of Animal Behavior Consultants. https://m.iaabc.org/about/lima/

International Association of Assistance Dog Partners. (2022, January). *Assistance Dog Laws and Legal Resources*. https://www.iaadp.org/canadianlaws.html

International Guide Dog Federation. (2020, December 31). *Facts and Figures*. International Guide Dog Federation. https://www.igdf.org.uk/about-us/facts-and-figures/

Kisor, H., & Goodier, C. (2019). *Traveling with Service Animals: By Air, Road, Rail, and Ship across North America* (Illustrated edition). University of Illinois Press.

Kogan, L. R., Packman, W., Currin-McCulloch, J., Bussolari, C., & Erdman, P. (2021). The Loss of a Service Dog through Death or Retirement: Experiences and Impact on Partners. *Illness, Crisis & Loss*, 10541373211054168. https://doi.org/10.1177/10541373211054168

Kwong, M. J., & Bartholomew, K. (2011). "Not just a dog": An Attachment Perspective on Relationships with Assistance Dogs. *Attachment & Human Development*, *13*(5), 421–436. https://doi.org/10.1080/14616734.2011.584410

Lessard, G., Vincent, C., Gagnon, D. H., Belleville, G., Auger, É., Lavoie, V., Besemann, M., Champagne, N., Dumont, F., & Béland, E. (2018). Psychiatric Service Dogs as a Tertiary Prevention Modality for Veterans Living with Post-Traumatic Stress Disorder. *Mental Health and Prevention*, *10*, 42–49. https://doi.org/10.1016/j.mhp.2018.01.002

MacLean, E. L., & Hare, B. (2018). Enhanced Selection on Assistance and Explosive Detection Dogs Using Cognitive Measures. *Frontiers in Veterinary Science*, *5*, 236. https://doi.org/10.3389/fvets.2018.00236

McLaughlin, K., & Hamilton, A. L. (2019). Exploring the Influence of Service Dogs on Participation in Daily Occupations by Veterans with PTSD: A Pilot Study. *Australian Occupational Therapy Journal*, *66*(5), 648–655. https://doi.org/10.1111/1440-1630.12606

Mills, M. L. (2017). Invisible disabilities, visible Service Dogs: The Discrimination of Service Dog Handlers. *Disability & Society*, *32*(5), 635–656. https://doi.org/10.1080/09687599.2017.1307718

Nieforth, L. O., Craig, E. A., Behmer, V. A., MacDermid Wadsworth, S., & O'Haire, M. E. (2021). PTSD Service Dogs Foster Resilience among Veterans and Military Families. *Current Psychology: A Journal for Diverse Perspectives on Diverse Psychological Issues*. https://doi.org/10.1007/s12144-021-01990-3

Nieforth, L. O., Rodriguez, K. E., & O'Haire, M. E. (2021). Benefits and Challenges of Mobility and Medical Alert Service Dogs for Caregivers of Service Dog Recipients. *Disability and Rehabilitation: Assistive Technology*, 1–9. https://doi.org/10.1080/17483107.2021.1916630

Nieforth, L. O., Schwichtenberg, A. J., & O'Haire, M. E. (2021). Animal-Assisted Interventions for Autism Spectrum Disorder: A Systematic Review of the Literature from 2016 to 2020. *Review Journal of Autism and Developmental Disorders*, 1–26. https://doi.org/10.1007/s40489-021-00291-6

O'Haire, M. E. (2013). Animal-Assisted Intervention for Autism Spectrum Disorder: A Systematic Literature Review. *Journal of Autism and Developmental Disorders*, *43*(7), 1606–1622. https://doi.org/10.1007/s10803-012-1707-5

O'Haire, M. E., & Rodriguez, K. E. (2018). Preliminary Efficacy of Service Dogs as a Complementary Treatment for Posttraumatic Stress Disorder in Military Members and Veterans. *Journal of Consulting and Clinical Psychology*, *86*(2), 179–188. https://doi.org/10.1037/ccp0000267

Porter, M., Winkle, M., & Herlache-Pretzer, E. (2021). Considerations for Recommending Service Dogs versus Emotional Support Animals for Veterans with Post-Traumatic Stress Disorder. *People and Animals: The International Journal of Research and Practice, 4*(1). https://docs.lib.purdue.edu/paij/vol4/iss1/4

Randy J. Virden, G. J. W. (1999). Ethnic/Racial and Gender Variations Among Meanings Given to, and Preferences for, the Natural Environment. *Leisure Sciences, 21*(3), 219–239. https://doi.org/10.1080/014904099273110

Richerson, J. T., Saunders, G. H., Skelton, K., Abrams, T., Storzbach, D., Fallon, M. T., Biswas, K., Wagner, T. H., Magruder, K. M., Stock, E. M., Mi, Z., Beaver, B. V., McGovern, S. R., Dorn, P. A., Huang, G. D., Frakt, A., Pizer, S., Leighton, S. B., Kennedy, B. S., ... Groer, S. (2020). *A Randomized Trial of Differential Effectiveness of Service Dog Pairing to Improve Quality of Life for Veterans with PTSD* (p. 186). Department of Veterans Affairs. https://www.research.va.gov/REPORT-Study-of-Costs-and-Benefits-Associated-with-the-Use-of-Service-Dogs-Monograph1.pdf

Rintala, D. H., Matamoros, R., & Seitz, L. L. (2008). Effects of Assistance Dogs on Persons with Mobility or Hearing Impairments: A Pilot Study. *The Journal of Rehabilitation Research and Development, 45*(4), 489–504. https://doi.org/10.1682/JRRD.2007.06.0094

Rintala, D. H., Sachs-Ericsson, N., & Hart, K. A. (2002). The Effects of Service Dogs on the Lives of Persons with Mobility Impairments: A Pre-Post Study Design. *SCI Psychosocial Process, 15*(2), 70–82.

Risley-Curtiss, C., Holley, L. C., Cruickshank, T., Porcelli, J., Rhoads, C., Bacchus, D. N. A., Nyakoe, S., & Murphy, S. B. (2006). "She Was Family": Women of Color and Animal-Human Connections. *Affilia, 21*(4), 433–447. https://doi.org/10.1177/0886109906292314

Risley-Curtiss, C., Holley, L. C., & Wolf, S. (2006). The Animal-Human Bond and Ethnic Diversity. *Social Work, 51*(3), 257–268. https://doi.org/10.1093/sw/51.3.257

Rodriguez, K., Bibbo, J., & O'Haire, M. E. (2020). The Effects of Service Dogs on Psychosocial Health and Wellbeing for Individuals with Physical Disabilities or Chronic Conditions. *Disability and Rehabilitation, 42*(10), 1350–1358. https://doi.org/10.1080/09638288.2018.1524520

Rodriguez, K., Bryce, C., Granger, D., & O'Haire, M. (2018). The Effect of a Service Dog on Salivary Cortisol Awakening Response in a Military Population with Posttraumatic Stress Disorder (PTSD). *Psychoneuroendocrinology, 98*, 202–210. https://doi.org/10.1016/j.psyneuen.2018.04.026

Rosenkranz, K. M., Arora, T. K., Termuhlen, P. M., Stain, S. C., Misra, S., Dent, D., & Nfonsam, V. (2021). Diversity, Equity and Inclusion in Medicine: Why It Matters and How Do We Achieve It? *Journal of Surgical Education, 78*(4), 1058–1065. https://doi.org/10.1016/j.jsurg.2020.11.013

Sachs-Ericsson, N., Hansen, N. K., & Fitzgerald, S. (2002). Benefits of Assistance Dogs: A Review. *Rehabilitation Psychology, 47*(3), 251–277.

Schoenfeld-Tacher, R., Hellyer, P., Cheung, L., & Kogan, L. (2017). Public Perceptions of Service Dogs, Emotional Support Dogs, and Therapy Dogs. *International Journal of Environmental Research and Public Health, 14*(6), 642.

*Service, Emotional Support, and Therapy Animals.* (2022). American Veterinary Medical Association. https://www.avma.org/resources-tools/animal-health-and-welfare/service-emotional-support-and-therapy-animals

Smith, Y. (2019). Pets and Human Diversity: Toward Culturally Competent, Culturally Humble Psychotherapy. In L. Kogan & C. Blazina (Eds.), *Clinician's Guide to Treating Companion Animal Issues* (pp. 477–496). Academic Press. https://doi.org/10.1016/B978-0-12-812962-3.00025-3

Spinka, M. (2019). Animal Agency, Animal Awareness and Animal Welfare. *Animal Welfare, 28*, 11–20. https://doi.org/10.7120/09627286.28.1.011

Tseng, A. (2022). Brief Report: Above and Beyond Safety: Psychosocial and Biobehavioral Impact of Autism-Assistance Dogs on Autistic Children and their Families. *Journal of Autism and Developmental Disorders.* https://doi.org/10.1007/s10803-021-05410-0

U.S. Department of Housing and Urban Development. (2020). *Assistance Animals Notice.* Office of Fair Housing and Equal Opportunity. https://www.hud.gov/sites/dfiles/PA/documents/HUDAsstAnimalNC1-28-2020.pdf

U.S. Department of Justice. (2010). *Americans with Disabilities Act 2010 Revised Requirements: Service Animals.* https://www.ada.gov/service_animals_2010.htm

U.S. Department of Transportation. (2020). *14 CFR Part 382 Nondiscrimination on the Basis of Disability in Air Travel.* Office of the Secretary, U.S. Department of Transportation. https://www.transportation.gov/sites/dot.gov/files/2020-12/Service%20Animal%20Final%20Rule.pdf van Houtert, E. A. E., Endenburg, N., Rodenburg, T. B., & Vermetten, E. (2021). Do Service Dogs for Veterans with PTSD Mount a Cortisol Response in Response to Training? *Animals*, *11*(3), 650. https://doi.org/10.3390/ani11030650

van Houtert, E. A. E., Endenburg, N., Vermetten, E., & Rodenburg, T. B. (2022). Hair Cortisol in Service Dogs for Veterans with Post-traumatic Stress Disorder Compared to Companion Dogs (*Canis Familiaris*). *Journal of Applied Animal Welfare Science*, 1–11. https://doi.org/10.1080/10888705.2022.2033119

Vassar, R. R., & Howard, T. C. (2021). Examining Black Male Identity Through a Prismed Lens: Critical Race Theory, Intersectionality, and the Complexities of Black Males' Experiences. In *Handbook of Critical Race Theory in Education* (2nd ed.). Routledge.

Viau, R., Arsenault-Lapierre, G., Fecteau, S., Champagne, N., Walker, C.-D., & Lupien, S. (2010). Effect of Service Dogs on Salivary Cortisol Secretion in Autistic Children. *Psychoneuroendocrinology*, *35*(8), 1187–1193. https://doi.org/10.1016/j.psyneuen.2010.02.004

Vincent, C., Poissant, L., Gagnon, D. H., Corriveau, H., & The, M. (2016). Consensus Building for the Development of Guidelines for Recommending Mobility Service Dogs for People with Motor impairments. *Technology and Disability*, *28*(3), 67–77. https://doi.org/10.3233/TAD-160445

Walther, S., Yamamoto, M., Thigpen, A. P., Garcia, A., Willits, N. H., & Hart, L. A. (2017). Assistance Dogs: Historic Patterns and Roles of Dogs Placed by ADI or IGDF Accredited Facilities and by Non-Accredited U.S. Facilities. *Frontiers in Veterinary Science*, *4*, 1. https://doi.org/10.3389/fvets.2017.00001

Wenthold, N., & Savage, T. A. (2007). Ethical Issues with Service Animals. *Topics in Stroke Rehabilitation*, *14*(2), 68–74. https://doi.org/10.1310/tsr1402-68

Whitmarsh, L. (2005). The Benefits of Guide Dog Ownership. *Visual Impairment Research*, *7*(1), 27–42. https://doi.org/10.1080/13882350590956439

Whitworth, J., O'Brien, C., Wharton, T., & Scotland-Coogan, D. (2020). Understanding Partner Perceptions of a Service Dog Training Program for Veterans with PTSD: Building a Bridge to Trauma Resiliency. *Social Work in Mental Health*, *18*(6), 604–622. https://doi.org/10.1080/15332985.2020.1806181

Winkle, M., Crowe, T. K., & Hendrix, I. (2012). Service Dogs and People with Physical Disabilities Partnerships: A Systematic Review. *Occupational Therapy International*, *19*(1), 54–66. https://doi.org/10.1002/oti.323

Yamamoto, M., Hart, L. A., Ohta, M., Matsumoto, K., & Ohtani, N. (2014). Obstacles and Anticipated Problems Associated with Acquiring Assistance Dogs, as Expressed by Japanese People with Physical Disabilities. *Human-Animal Interaction Bulletin*, *2*(1), 59–79.

Younggren, J. N., Boness, C. L., Bryant, L. M., & Koocher, G. P. (2020). Emotional Support Animal Assessments: Toward a Standard and Comprehensive Model for Mental Health Professionals. *Professional Psychology: Research and Practice*, *51*(2), 156–162. https://doi.org/10.1037/pro0000260

Zhang, X., Carabello, M., Hill, T., Bell, S. A., Stephenson, R., & Mahajan, P. (2020). Trends of Racial/Ethnic Differences in Emergency Department Care Outcomes among Adults in the United States from 2005 to 2016. *Frontiers in Medicine*, *7*. https://www.frontiersin.org/article/10.3389/fmed.2020.00300

Zier, E. R. (2020). Which One to Follow? Service Animal Policy in the United States. *Disability and Health Journal*, *13*(3), 100907. https://doi.org/10.1016/j.dhjo.2020.100907

Ziv, G. (2017). The Effects of Using Aversive Training Methods in Dogs—A Review. *Journal of Veterinary Behavior*, *19*, 50–60. https://doi.org/10.1016/j.jveb.2017.02.004

# 38
# HANDLER-DOG INTERACTION
## Role of Operational Working Dogs from Past to Present

*Melissa A. Singletary, Sarah Krichbaum, and Lucia Lazarowski*

### Introduction and History of Working Dogs

In modern times, there is no shortage of books, movies, documentaries, and media headlines that recount the personal stories of hander-working dog teams and reflect on the powerful bonds that exist within these partnerships. The field of working dogs is a dynamic, challenging, and rewarding one that demands an extensive cooperation and bond which characterizes this human-animal interaction (HAI). This unique relationship, more descriptively the handler-dog interaction, has deep-rooted historical origins.

The association of canids to humans for interspecies cooperation and the earliest indications of domestication can be seen in ancient artifacts dating thousands of years ago and suggest canids initially played a role assisting humans in hunting and herding. By some estimates, this relationship started at least within the last 11,000 years (Li et al., 2014; Wang et al., 2019) with genetic information suggesting domestication of dogs may have begun over 135,000 years ago (Olson, 2002; Suplee, 1997). The domestication of canids resulted in a companion that would evolve and change to fit the increasing needs and interests of humans within a changing environment. As various traits become priority for function and breeding, these disparate selection pressures led to a significant neuroanatomical variation across the resulting breeds present in modern time (Hecht et al., 2019).

Historians believe that variations in dog breeds began with a selection of traits advantageous for hunting. For example, dogs may have benefited human hunters by chasing down animals that were too fast for humans or by detecting prey that human senses could not. Evidence suggests ancient Assyrians employed dogs during military warfare (Turbak, 2000) and the ancient Greeks employed dogs in defense and warfare during the Peloponnesian War (431–404 BCE) (Rujoiu & Rujoiu, 2018). By the 1300s, there were several distinct dog "varieties" based on the type of game being hunted (e.g., wolfhound vs deerhound). Shortly after, the number of breeds began to expand. In addition to hunting dogs, early working dogs included livestock guarding dogs, herding dogs, and sled dogs. During the American Revolutionary War (1775–1783), information suggests dogs were involved in the war efforts as well (Rujoiu & Rujoiu, 2018). More recently (approximately the 1800s), breeding selection changed its emphasis and moved toward the practice of line-breeding to accentuate desirable physical traits rather than working ability. This has resulted in many of the breeds that make up the modern dog (Parker, 2012). The first breeds associated with military operations were the Dutch shepherd, Belgian Malinois, and German shepherd during World War I (WWI) for physical-based work but were originally used for herding sheep and cattle in Europe. During WWI, the intelligence of these dogs enabled over 25,000 of them to be deployed to the battlefield primarily as runners and draft animals (Turbak, 2000). In the few reports from this early physical-based work, the intangible element of animal companionship and the role of emotional support were also apparent in accounts, such as those shared by Josepf Weber in his 1939 book "The Dog in Training." Weber shares the following excerpt from a military report referenced in World War I;

> Seven days ago we occupied Height No. X, which lies a two and a half hours' climb from the base below. Our dog, "Girl," covers this distance five and six times a day without appreciable effort, because she takes routes and short cuts which we could never venture to negotiate successfully. During observation duty at night, "Girl," with her superior alertness and hearing, is also our most dependable listener and observer. Each one of us, when on night guard, relies on her. For when we begin to dream and the overstrained tension of our ears causes imagination to play tricks on us so that we begin to see non-existent spirits, "Girl" is the great influence that calms us down, and her mood is the best basis for judging the circumstances that actually prevail. As long as she is quiet, we can rest assured that we have only had a case of nerves. In addition to this economy she represents in the small amount of material she consumes and in the man power she equals, the latter by grace of the fact that she alone does the work of three or four men, she is an excellent friend and comrade.
>
> (Weber, 1939)

More modern applications of working dogs include their use in tracking and detection during World War II. On March 13th 1942, the United States Army formed the U.S. K-9 Corps and widely adopting the use of military working dogs during the Vietnam War and Korean War, where dogs assisted in detecting explosives or tracking enemy soldiers (Rujoiu & Rujoiu, 2018). Advancements in training methodology and recognition for the spectrum of application continue to place the working dog in an invaluable position within our society. However, the modern use of working dogs, especially operational working dogs, carries inherent occupational risks and hazards. The occupational conditions shared between the handler and the dog can create a context conducive to establishing a strong, highly valued and often lifelong bond. Though these dogs do not volunteer for their service, they have demonstrated for thousands of years to be some of the most faithful, effective, and dependable partners in the field.

This chapter attempts to amalgamate the experience and knowledge of subject matter experts in the field of working dogs through public reports of first-hand experience and it highlights relevant published studies within the field. Though HAIs are the focus of many studies in the companion animal (i.e. pet) sector, few focus on working dogs and more specifically in operational canine teams. Though the focus of this discussion will be heavily weighted on military and some law enforcement canines as examples, we will also touch on the vast array of working dog specialties.

## Modern Working Dogs: Operational Team Specialties

Dogs are used across domains for their utility as aids and tools for humans to complete designated tasks, such as detection or patrol work. One of the primary functions that working dogs perform in olfactory detection and currently represents a unique niche for which no equivalent alternative exists. Analytical methods for detection are available with specialized equipment, such as gas chromatography mass spectroscopy (GC-MS), and techniques are improving in the field but the mobile, real-time, and rapidly programmable nature of the dog as an intelligent sensor is unrivaled (Walper et al., 2018).

The terminology and classification of working dogs are not well standardized with noted inconsistencies across the various fields. However, we will provide some generalized terms (Figure 38.1) and examples for working dogs with a more detailed focus on operational canines (OpsK9s/Operational K9s) in detection and protection specialties.

A working dog is a dog that is trained to accomplish a specific, defined task(s) (AAFS Standards Board, 2017). The term working dog includes a wide array of trained canines with some examples highlighted in Figure 38.1 (Olson, 2002; Worth & Cave, 2018).

A subcategory of working dogs includes the specialties of detection and protection dogs, which can be placed under the group operational canines. This subcategory will be the main focus of this chapter. Operational canines perform in tactical and high-risk roles serving

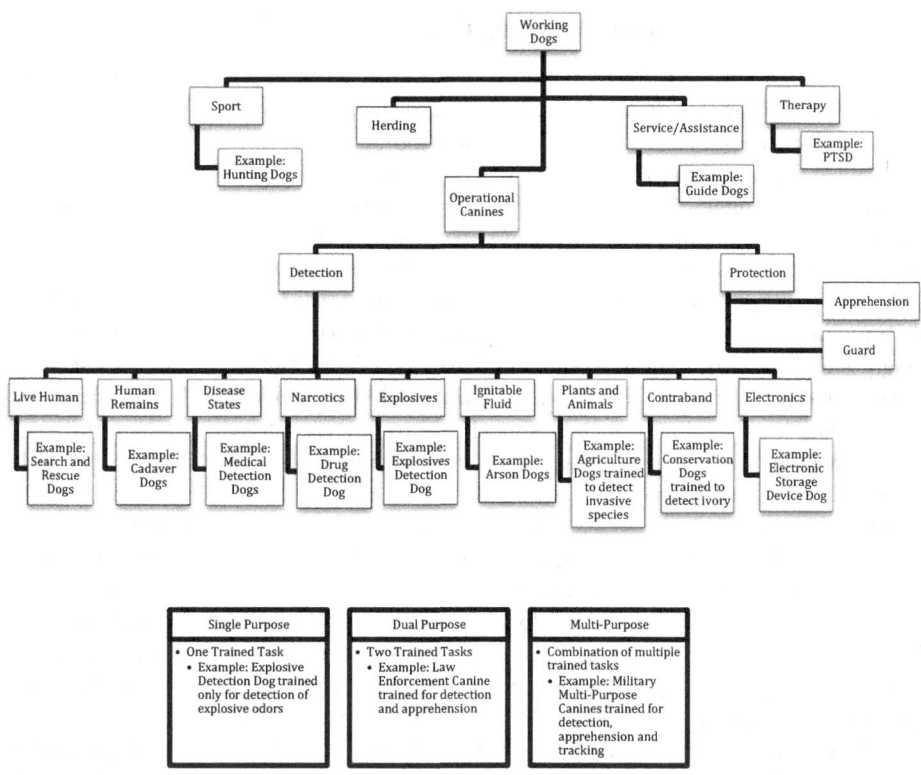

*Figure 38.1* An example of working dog classification and categorization scheme (top) and highlighted categorization of three types of operational canines; single purpose, dual purpose, and multi-purpose (bottom). Figure provided by author

in various disciplines throughout local, state, federal, and military positions across various governmental and non-governmental organizations (Olson, 2002; Palmer et al., 2015; Parr & Otto, 2013). Current estimates suggest 15,000 operational working dogs are in service within the United States, though exact figures are unknown (Force, 2022).

## Select Definitions

*Detection Dog* (detection K9/detection canine/detector canine/detector dog) is defined as "a canine trained to detect and alert to the presence of certain scents or odors for which it has been trained" (AAFS Standards Board, 2017).

*Protection Dog* (protection/security/patrol canine or K9) is a term that includes dogs trained to "alert to and deter human or animal threats" (Otto et al., 2021).

*Canine Team* (K9 team) is a dyad that consists of "a human and working canine that train and work together as an operational unit" (AAFS Standards Board, 2017).

*Handler* (dog handler, canine or K9 handler) is a "person who has successfully completed a recognized course of canine handling in a specific discipline and maintains those abilities through field applications, maintenance training, certification, recertification, and department-, agency-, or organization-required continuing canine education" (AAFS Standards Board, 2017).

## Important Working Dog Characteristics

Working dogs require a specific set of traits to successfully perform their tasks. Dogs must be structurally sound and healthy to meet the physical demands of a working career. However, behavioral characteristics are considered the most significant determinant of a dog's ability to be trained for a working role. While some characteristics vary by specific career, some are common to the majority of working dogs. For example, trainability (e.g. responsiveness to commands and consequences) is critical to a dogs' ability to learn to perform the tasks required for its role and work effectively with a handler (Bray et al., 2021).

The exception to requiring a high degree of trainability are working dogs that perform tasks for which they were originally bred (e.g. Livestock Guarding Dogs). Their behavior is strongly genetically controlled, and they work largely independently. However, most modern working dog tasks require learning to perform a task that is not naturally occurring (e.g., detection dogs learning to alert to synthetic chemicals that are otherwise biologically irrelevant), responding to handler commands, and some degree of obedience to work effectively in anthropocentric environments.

Another behavioral characteristic common to most working dogs is confidence or a lack of fearfulness, which may vary in degree depending on the dogs' operational environment (Bray et al., 2021). For example, detection dogs working in airports or large event venues, must be able to work effectively despite loud noises, crowds of people, traffic, and a range of other novel and unpredictable stimuli. Exceptions to requiring a high degree of confidence might include dogs working in more static, low-stimulus environments, such as detection dogs screening samples in medical laboratories or cargo in warehouses.

Characteristics critical to working dog training and operational success are the product of both genetic and environmental factors. The effectiveness of dogs as working partners with humans stems from a genetically-based heightened sociability as a result of domestication (Vonholdt et al., 2010). However, a second wave of selection led to phenotypic variability among breeds with specialized characteristics for different types of working tasks. For example, dogs originally bred for working cooperatively with a human partner, such as herding dogs and gundogs, tend to exhibit high levels of trainability due to a selection for the ability to work at a distance from a human partner while maintaining visual or verbal contact. By contrast, "independent workers" such as sled dogs, scent hounds, and livestock guarding dogs are less responsive to human social cues and are considered to be less trainable (Gácsi et al., 2009). Within these breed groups, certain breeds are more effective than others for specific jobs. For example, Labrador Retrievers are a popular breed for many types of work including assistance and detection tasks due to high trainability and sociability. Still, they lack the defensive tendencies of breeds favored for patrol and protection work like the German Shepherd Dog and Belgian Malinois (Bray et al., 2021). Finally, working dog breeding program organizations utilize selective breeding to further enhance desirable traits for a specific task.

## Operational Dog/Handler Interaction

The occupational environment in which humans and animals interact can drastically impact the interaction and relationships that develop. Under conditions that operational working dog teams frequently encounter, a unique bond and partnership can develop. A balance exists between appreciating the job that a working dog needs to complete as an asset and the need to build a relationship that contributes to the success of that asset's performance.

Dogs can be viewed as passive sensors for field application, but due to our close social relationship with dogs, it is relatively difficult to accept this reductionistic view. In fact, findings from ethnographic research suggest that detection dog handlers view their dogs as individuals with feelings and emotions that require specific training and handling (Holland, 2022). It is this attunement that allows the team to grow together, and undoubtedly affects their performance (Sanders, 2006).

The demands of field work place the dog and handler at the front lines in high-stress high-stakes conditions which establish a co-dependence between the dog and handler. In some fields, such as the U.S. military, it is a tradition that the canine outrank their handler, highlighting the value placed on their role (Linda Crippen, 2017).

The trust that is needed under this context has to be built over months to years of training and practice as a team. Acting as a singular unit with the ability for both the dog and the handler to effectively read the other's behavior and respond accordingly is critical to the team's success. The interview with an operations superintendent at the Military Working Dog School by a U.S. Army Training and Doctrine Command TRADOC stated:

> Instructors say that handlers are there to make sure the dog gets food, water and rest. The handlers are there to motivate the dog when it's tired and ready to quit. The dog is the worker bee of the team. Perhaps the most important part of the bond is that handlers are there to translate what the dog is saying.
>
> *(Linda Crippen, 2017)*

There is increasing awareness of the effects of war on the soldier's health and wellbeing. However, both ends of the leash are important to consider in their physical, mental, and emotional status as both suffer from injury in the line of duty. In one estimate, up to 10%

of U.S. military working dogs in 2014 suffered from post-traumatic stress disorder (PTSD) post-war (Brait, 2015).

Because working dogs function as a dyad with a handler, the relationship between dog and handler is imperative to the team's success. The remainder of this chapter will focus on the interaction between operational dogs and their handlers, specifically detection dog/handler teams. Detection dogs benefit humans by providing protection and security from threats, but because they work as a team with a handler and can only be as successful as the team's success, the dog-handler relationship is an important aspect of their utilization.

## *The Bond and Influential Factors*

External factors can also influence the bond between detection dog and handler. Studies have suggested that handler stress can affect detection work. Jezierski et al. (2014) found that detection teams had longer search times and more false alerts during formal compared to informal certification exams, possibly due to increased handler stress from the pressure of the formal certification (Jezierski et al., 2014). However, a small amount of manageable stress might be beneficial. Zubedat et al. (2014) found that when the handler's anxiety levels increased under specific stressors associated with a search task, teams had shorter detection times (Zubedat et al., 2014). This effect was attributed to increased handlers' anxiety leading to less control over their dogs. Without a strong influence over their dog's behavior, dogs may have been more efficient in their search. Together, these findings may be explained by the Yerkes-Dodson phenomenon, which suggests that the influence of stress on performance follows a bell-shaped curve such that a large amount of stress negatively influences performance, while a certain degree of manageable stress creates enough attention to detail for optimal performance (Calabrese, 2008).

The social interactions of handlers and operational canines may have influence based on context. The as the human side of the "leash" may exhibit behaviors associated with stress and anxiety that evoke a response and behavioral change in the dog (Jones & Josephs, 2006). While the human side of the "leash" may also exert a calming effect on a dog during a stressful event by providing gentle stroking (Hennessy et al., 1998; Tuber et al., 1996). As the "leash" represents a tethered bond between the handler and dog, it suggests that emotional states of either end may easily affect the other. A quote from Farr et al. (2021b):

> The dogs don't seem to do as bad as the people do, but I think the people affect the dogs. The dog's hooked to the leash with the guy that's basically not happy to be there, and that has an effect more so on the dog than the actual weather itself. I've been in situations where it was raining, and it was cold, and the dogs appeared to be cold as well, but a lot of times it's that behavior that runs down the leash to the dog.
>
> *(Farr et al., 2021b)*

## *The Team Pairing*

The working dog is a mission asset that requires a skilled and attentive operator. There is a need to regulate a balanced relationship that ensures mission readiness and peak performance of duties while maintaining the team's physical and psychological health and welfare. This also relies on a level of interoperability for canine teams to adapt to changes on either end of the leash, where a handler may be out of operation and another handler assigned to a particular dog or vice versa. The flexibility of this team for some organizations, such as the U.S. military, can be critical to ensuring the mission's success. This leads to organization specific variations between rotational canine teams versus dedicated canine teams. Rotational teams involve handlers and dog pairing to rotate based on mission, cycle or duty location as needed. Dedicated teams are partnered teams that remain paired for the duration of the dog's operational career. Dog exposure to multiple personnel can have both positive and negative potential outcomes for performance and welfare (Haverbeke et al., 2008; Horváth et al., 2008; Lefebvre et al., 2007). Rotation provides cross-training and operational redundancy; however, it may result in minimization or severance of the human-animal bond as one study rotational handling resulted in lower percentage of correct detections (Nolan & Gravitte, 1977). There are departments that require a familiar reserve handler to be designated for emergency situations where the control and care of the canine would occur separate from the primary handler. An interview with a Military Police Sergeant aired on National Public Radio captured the adaptability of rotational team handling and limitations of separation based on dog personality;

> One of the most challenging and important aspects of being a dog handler is learning each dog's personality. He's worked with eight different dogs in five years…I would say as a young handler, it probably had more of an effect. As soon as I got more experience, I kind of got over the fact that I was going to be changing dogs quite often… [He] views it as an asset that he's worked with a range of dogs; it gives him more experience with different behaviors and personalities. But, [he] says, the dogs don't always adjust well to changing handlers…Some get separation anxiety if they develop a long bond with someone.
>
> *(Hancock, 2013)*

The interview by the TRADOC writer referenced earlier also captured the emotional difficulty associated with rotational handling and separation:

> When she was leaving for a new duty station and saying her goodbyes, [she] said one of her Soldiers became emotional. [She] patted her on the back and told her that it wasn't a big deal. The next day, [she] went to the kennel to tell her dog goodbye…'It was the dog I deployed with in Iraq. I walk out, and I'm bawling - bawling like a baby! My husband was like, 'What's wrong with you' Your Soldier was crying because you're leaving, and you showed little emotion. You tell your dog goodbye, and you're a basket case.' She said other dog lovers understand. 'That bond between you and the dog, there's nothing else like it. I would totally trust my life with a dog.
>
> *(Linda Crippen, 2017)*

Another organization specific variation occurs between the handler's choice of canine vs. assignment of a canine partner. The latter of which corresponds more closely to organizational structures that also support rotational teams vs. dedicated teams.

Quote from Master Sergeant Valentin Rosiaoara:

> … The course lasted until August after which I meet Max who become my responsibility, and later my teammate. On the first day, the canine instructors there talked with us to see what dog would suit each of us. They wanted to find out more about who we really are like as individuals. We are talking about drawing a profile, because the canine

> instructors already knew the temperament and behavior of the dogs and what kind of energy they have. The instructors asked me even what my zodiac sign was, what I like to do, what I know about dogs. Observing that I am an active person, they decided to show me a dog called Igor considering that he is fit for me. However, I, later, chose to change his name into Max.
>
> *(Rujoiu & Rujoiu, 2018)*

In the U.S. Army Regulation 190-12, the basic needs of the canine are discussed as basic responsibilities and duties that a military working dog handler should have or perform. Amongst the standard (i.e. grooming, daily care, kennel cleaning, physical fitness) it acknowledges that a handler should, "exhibit a high degree of affection for the military working dog" (Luckwaldt, 2017). While a great deal of emphasis has been placed on identifying the ideal characteristics of a successful working dog and the methods to improve canine selection, less emphasis has been placed on the individual characteristics important to a handler's success. However, studies on pets have demonstrated various relationships of owner personality and the health and cognitive abilities of their pets. For example, Kotrschal et al. (2009) found that owners who scored high in neuroticism had dogs with lower stress but were less successful in an evaluation of task-oriented activities (Kotrschal et al., 2009). In line with this finding, the only study to examine personality in operational dog handlers showed that they typically score low in neuroticism and high in agreeableness. Additional research is needed to determine the utility of personality assessments as screening tools for the selection of operational dog handlers and its variations across disciplines and operational assignments (e.g. military, law enforcement, search, and rescue).

Numerous personal accounts of handler-dog heroism and life-saving events on behalf of both the dog and human have characterized the strong emotional attachments that can result from this career pairing (Frankel, 2014; Rujoiu & Rujoiu, 2018; Smith & Thompson, 2014; Turbak, 2000). To quote Frankel (2014),

> Inside the world of handlers and their working dogs is a culture of dedication and sacrifice, even grief, all the things ones might expect to find. But there was something else I encountered, something surprising. Resistance to the idea of love.
>
> *(Frankel, 2014)*

There is a suggestion that the handler and dog connection must be aimed toward gaining trust rather than a feeling of love as more of "function rather than feeling" (Frankel, 2014). However, the stories and personal accounts of working dog teams indicate that the bonds developed can be lifelong and deep. The loss of a handler, for example the death of a Marine handler from the Marines Expeditionary Force (II-MEF) impacted his partner, military working dog Sirius, causing indications that, *"Sirius was in distress"* and appeared to be searching for his partner during the ramp ceremony and agitated during the event (Frankel, 2014). Alternatively, the impact of the loss of a dog on the handler and unit can be significant.

> I was overwhelmed and sad, even before Max died, especially from a physical point of view. The fact that I could not help my best friend made me feel helpless. ... I was upset, nervous about what happened to him. Why him? Why us? Why couldn't his condition be treated? I was desperate. I really did not know what to do. What can I

tell you. ... Everybody has feelings for a family member, but I had special feelings for Max who, if necessary, would have saved my life. We were always in the line of fire, on the lookout to see if the road is mined or not. He was the first in line and I was behind him at 10–12 meters. ...First of all, Max was a friend. Secondly, he was my partner. When I was upset, and I missed my human family the most, he felt this and came to me and stayed close to me. He was there for me. He made me forget about any kind of problems. He was special. I think every person who has an animal companion at home can say about his pet that he is special if he really cares for him. As a military working dog, Max executed the orders I asked him to, even when they were very difficult. He understood the fact that we were here to do the job. He was next to me, regardless of the weather conditions at hand such as: the heat, the sandstorms, the rain or the snow, the cold. ...For me, he was my best friend!!! ... Max was part of my family!!!

*(Rujoiu & Rujoiu, 2018)*

In the U.S. military, a common tradition prevails that when a military working dog has died, a ceremony is held where the poem, 'Guardians of the Night', is read or displayed and the dog's food bowl is ceremonially turned upside down:

Trust in me, my friend, for I am your comrade. I will protect you with my last breath.
When all others have left you, and the loneliness of the night closes in, I will be at your side.
My eyes are your eyes to watch you and to protect you.
My ears are your ears to hear and detect evil minds in the dark.
My nose is your nose to scent the invader of your domain.
And so you may live, my life is also yours.
Together we will conquer all obstacles, and search out those who might wish to harm others.
It is for you that I will unselfishly give my life, and fill my nights without rest.
Although our days together may be marked by the passing of the seasons,
Know that each day at your side is my reward.
My days are measured by the coming and going of your footsteps.
I am your right arm, the sword at your side, your defender and protector.
I attempt to do what you bid of me. I seek only to please you and remain in your favor.
Together you and I shall experience a bond only others like us will understand
I will listen to you without question, nor will your spoken words ever be repeated.
I will remain ever silent, ever vigilant, always faithful and loyal.
When our time together is done, and you move on in the world,
Remember me with kind thoughts and tales of a time we were unbeatable.
If needed at another time and place, I would gladly take up your fight.
I am a military working dog, and together we are guardians of the night.

-author unknown

This tradition is also commonly conducted in law enforcement canine units, with replacement of "military working dog" with "police dog" in the last line of the poem (Unknown, Unknown). This ceremony and poem highlight the relationship and bond of a handler and working dog team, referring to each other as friends and comrades that have a trust and bond that, "only others like us will understand." Though this is a poem written by an unknown author, the recognition and use for ceremonial purposes place these words in high regard as a shared sentiment amongst operational canine teams and are a reminder that the handler-dog interaction can represent more than "work."

Though handlers may bond and work closely with their canine counterparts, they do not "belong" to the handlers. Decisions for their pairing, housing, duty assignment and disposition are mostly commonly guided by the owning organization or entity. This dynamic of ownership may create conflict under certain circumstances.

The disposition of military working dogs has been an area of advancement in recent decades. Following the Vietnam War, the disposition of thousands of U.S. military working dogs was primarily euthanasia as they were left in country with the Army of the Republic of Vietnam and only a handful of the overall dogs were returned to the United States (Clark, 1991). This event sparked handlers to rally efforts toward alternative options for disposition, which led to the passing of Public Law 106–446 "Promotion of Adoption" commonly known as "Robby's Law" by the U.S. Congress in 2000. This law enables military working dogs to return to service in the country or be adoptable by former handlers, law enforcement agencies, or qualified civilians (Congress, 2000). There are risks associated with working dog retirement and adoption given the nature of their training, particularly protection training, which is an important factor in the disposition of working dogs. However, with careful evaluation of temperament and adoptee selection, the field has seen high success rates with adoption at retirement. The development of a strong bond is reflected in the retirement adoption rates for operational dogs, with estimates representing a large majority (exceeding 90%) of retiring dogs in the United States are adopted by their handlers or prior handlers (McIntyre, 2017).

> The dogs aren't just U.S. government property...these dogs are our partners
> *(Linda Crippen, 2017).*

## *Training and Housing*

As early as World War II, it was reported that dog/handler teams worked best if they trained together and then were assigned to duty as a pair. To maintain and protect this bond, handlers were assigned as the sole providers of the dogs' care and were the primary source of their human interactions. In a recent scientific approach to understanding the influence of the bond on performance, detection dogs worked a search with a familiar and unfamiliar handler. Dogs' detection accuracy was higher when handled by their familiar handler and showed increased distractibility when handled by an unfamiliar handler (Jamieson et al., 2018). Though it is clear that familiarity of the team influences performance, it is unknown how long a team should train together to maintain optimal performance in the field.

The training methods used in working dogs have evolved over the decades and still vary by trainer and organization. However, the predominant training approach selected and used across working dog organizations is through operant conditioning with primarily positive reinforcement (Adams & Johnson, 1994; Rooney et al., 2016). Approaches using aversive or negative reinforcement techniques such as shock collars, pinch collars, etc. are fading within the field.

Studies in companion animals show negative effects such as increased aggression, increased physiological indicators of stress, and/or undesired behaviors associated with the use of aversive training techniques (Arhant et al., 2010; Blackwell et al., 2008; Casey et al., 2014; Cooper et al., 2014; Herron et al., 2009; Hiby et al., 2004). For example, dogs trained with aversive methods showed more stress markers and negative expectations of an ambiguous situation compared to those trained with more balanced methods including positive reinforcement (de Castro et al., 2020). Reward-based training has also been associated with a more consistent secure attachment (de Castro et al., 2019). The specific use of shock collars in working dogs has been associated with reduced training outcomes (Arnott et al., 2014). Numerous countries have banned or restricted the use of electronic collars (Masson et al., 2018).

The handler, in many cases, whether the dog is domestically-housed or kennel-housed, plays in part the role of the caregiver. This role often, in and of itself, contributes to the development of a bond (Beck & Katcher, 1996). As the animal's basic needs are met, such as food, water, shelter and healthcare, a dependency likened to that of a parental kind develops (Beck & Katcher, 1996).

The time dedicated to this varies across the field and can also be impacted by housing arrangements. For dogs housed domestically with a handler there are studies to suggest that the performance of the working dog team is increased (Lefebvre et al., 2007). Additionally, co-sleeping is thought to potentially increase the perceived closeness of the human and dog (Smith & Thompson, 2014). For dogs kennel-housed, studies suggest that an additional investment of time outside of regular working and training hours to practice has a similar impact on improving performance. Extra time investment can consist of various activities including agility, obedience, athletics, and apprehension (Lefebvre et al., 2007).

## *How the Bond Creates Biases: for Better or Worse*

The use of dogs harnesses their trainability and sociability paired with superior senses, especially those of smell. Dogs are well adept at socializing and communicating with humans and demonstrate a remarkable sensitivity to human emotional states (Albuquerque et al., 2016), and respond to social gestures (Hare et al., 2002), and through complex cues such as gaze alternation (Miklósi et al., 2003). This responsiveness to human signaling and communication makes dogs great candidates for cooperative work but also increases the ability for inadvertent human communication to result in undesired outcomes. For example, dogs tend to look at their human partner when faced with a challenge and may perceive unintentional cues from the human that can influence the dog's behavior (Zubedat et al., 2014). Indeed, the dogs 'trust' in human cues has shown to be overridden both olfactory and visual indications for a food-based task (Szetei et al., 2003).

Working dog training should consist of a balance between obedience and independence, where the dog can follow directions but does not rely upon the handler for decisions, such as an odor-driven alert response when encountering their trained target odor (e.g., an explosive odor).

> They can be high drive and fast, but I like to see a dog that will take direction easily and work away from the handler very well, but still doesn't rely on the handler. So, either on or off lead, it doesn't change the way they work, you shouldn't see a change. They shouldn't be relying on the handler at all is one of the things that I look for. They shouldn't check back in too much with the handler. So, if I'm watching a dog work, I don't particularly like when they keep checking back in with the handler, or back to people looking for those cues.
>
> *(Farr et al., 2021a)*

The bond between a handler and a dog inevitably creates biases that can influence performance. It is common for handlers to be lenient when reporting on their dogs' work. For example, in a study comparing handler ratings of performance to those of an impartial observer, handlers ranked their dogs higher on performance characteristics such as "Strength of Indication" (Clark et al., 2020). However, a critical view may also suggest that a seasoned handler-dog team may be attentive to subtle behavioral changes that increase the handler's perceptive value of a "strong" vs "weak" indication. As self-reporting is very common in the dog training industry, it is important to consider all the potential biases to provide an objective performance measure.

## Future Directions

There is a valuable foundation of meaningful studies being performed that hold merit toward further evaluations of the handler-dog bond in the working dog field. A standardized measure or series of measures that could be developed to evaluate the human-dog interaction and bond would go a long way in establishing specific contexts for individual working dog specialties and generalized contexts for optimal performance of this dyad. There are areas within this field that expand across a range of organizational structures, which lead to variations in breeding, development, selection, housing, training, and general animal husbandry practices. Further exploration of the handler-dog bond, its implications for welfare, and how canine welfare may influence the development of the bond are needed. Future research efforts are needed to evaluate the dynamic of training time versus housing conditions to evaluate the optimal time investment needed for kennel-housed dogs compared to those housed domestically with handlers to establish similar bonds and the resulting effects on performance. An evaluation of the efficacy and efficiency of a primarily aversive training approach, reward-based training only approach, and those with a mixed approach has not been systematically conducted in operational canines. However, it is an important area for future work. A focus on the handler, from selection, training and pairing to temperament and personality profiles, is needed to establish a 'best-fit' for optimal bonding. A deeper dive is needed into the effects of handler rotation on the bond vs on the mission capability (adaptability for interoperability) to consider the tradeoffs that would be assumed as the system shifts across that spectrum. Continuing the high standards for animal health and welfare within the working dog industry and promoting an educational framework to advance the practice and knowledge of positive bond development are key for industry sustainability.

The impact of the working dog on human evolution and modern society is expansive and will continue to play a critical role in the future advancement of our world. The handler-dog bond is at the heart of this impact.

## Discussion Questions

1. The selective breeding practices for phenotypic selection in working dogs have had a profound impact on the population. What future implications of propagating this practice are there for the human-animal bond?
2. Consider the paradox of a strong rapport and the requirements for task-focused operability of a working dog. What implications could this have for the human-animal bond?

# References

AAFS Standards Board, L. (2017). Crime scene/death investigation – Dogs and sensors terms and definitions. *ASB Technical Report 025*. Retrieved February 23, 2022, from

Adams, G. J., & Johnson, K. G. (1994). Sleep, work, and the effects of shift work in drug detector dogs Canis familiaris. *Applied Animal Behaviour Science, 41*(1), 115–126. https://doi.org/10.1016/0168-1591(94)90056-6

Albuquerque, N., Guo, K., Wilkinson, A., Savalli, C., Otta, E., & Mills, D. (2016). Dogs recognize dog and human emotions. *Biology Letters, 12*(1), 20150883.

Arhant, C., Bubna-Littitz, H., Bartels, A., Futschik, A., & Troxler, J. (2010). Behaviour of smaller and larger dogs: Effects of training methods, inconsistency of owner behaviour and level of engagement in activities with the dog. *Applied Animal Behaviour Science, 123*(3–4), 131–142.

Arnott, E. R., Early, J. B., Wade, C. M., & McGreevy, P. D. (2014). Environmental factors associated with success rates of Australian stock herding dogs. *PLoS One, 9*(8), e104457.

Beck, A. M., & Katcher, A. H. (1996). *Between Pets and People: The Importance of Animal Companionship*. Purdue University Press.

Blackwell, E. J., Twells, C., Seawright, A., & Casey, R. A. (2008). The relationship between training methods and the occurrence of behavior problems, as reported by owners, in a population of domestic dogs. *Journal of Veterinary Behavior, 3*(5), 207–217.

Brait, E. (2015). Canine PTSD: How the US military's use of dogs affects their mental wellbeing. Retrieved February 23, 2022, from https://www.theguardian.com/society/2015/nov/11/canine-ptsd-us-military-working-dogs

Bray, E. E., Gruen, M. E., Gnanadesikan, G. E., Horschler, D. J., Levy, K. M., Kennedy, B. S., Hare, B. A., & MacLean, E. L. (2021). Dog cognitive development: A longitudinal study across the first 2 years of life. *Animal Cognition, 24*(2), 311–328. https://doi.org/10.1007/s10071-020-01443-7

Calabrese, E. J. (2008). Converging concepts: Adaptive response, preconditioning, and the Yerkes–Dodson Law are manifestations of hormesis. *Ageing Research Reviews, 7*(1), 8–20.

Casey, R. A., Loftus, B., Bolster, C., Richards, G. J., & Blackwell, E. J. (2014). Human directed aggression in domestic dogs (Canis familiaris): Occurrence in different contexts and risk factors. *Applied Animal Behaviour Science, 152*, 52–63.

Clark, C. C. A., Sibbald, N. J., & Rooney, N. J. (2020). Search dog handlers show positive bias when scoring their own dog's performance. *Frontiers in Veterinary Science, 7*, 612. https://doi.org/10.3389/fvets.2020.00612

Clark, W. H. H. (1991). *The History of the United States Army Veterinary Corps in Vietnam, 1962–1973*. W.H. Wolfe Associates. https://books.google.com/books?id=-WrxAAAAMAAJ

Congress. (2000). *Text - H.R.5314-106th Congress (1999–2000): To require the immediate termination of the Department of Defense practice of euthanizing military working dogs at the end of their useful working life and to facilitate the adoption of retired military working dogs by law enforcement agencies, former handlers of these dogs, and other persons capable of caring for these dogs.*: Library of Congress Retrieved from Congress.gov

Cooper, J. J., Cracknell, N., Hardiman, J., Wright, H., & Mills, D. (2014). The welfare consequences and efficacy of training pet dogs with remote electronic training collars in comparison to reward based training. *PLoS One, 9*(9), e102722.

de Castro, A. C. V., Barrett, J., de Sousa, L., & Olsson, I. A. S. (2019). Carrots versus sticks: The relationship between training methods and dog-owner attachment. *Applied Animal Behaviour Science, 219*, 104831.

de Castro, A. C. V., Fuchs, D., Morello, G. M., Pastur, S., de Sousa, L., & Olsson, I. A. S. (2020). Does training method matter? Evidence for the negative impact of aversive-based methods on companion dog welfare. *PLoS One, 15*(12), e0225023.

Farr, B. D., Otto, C. M., & Szymczak, J. E. (2021a). Expert perspectives on the performance of explosive detection Canines: Operational requirements. *Animals (Basel), 11*(7). https://doi.org/10.3390/ani11071976

Farr, B. D., Otto, C. M., & Szymczak, J. E. (2021b). Expert perspectives on the performance of explosive detection Canines: Performance degrading factors. *Animals (Basel), 11*(7). https://doi.org/10.3390/ani11071978

Force, A. D. D. T. (2022). *FAQ*. Retrieved July 9, 2022 from https://www.akc.org/akc-detection-dog-task-force/faqs/

Frankel, R. (2014). *War Dogs: Tales of Canine Heroism, History, and Love*. St. Martin's Press.

Gácsi, M., Gyoöri, B., Virányi, Z., Kubinyi, E., Range, F., Belényi, B., & Miklósi, Á. (2009). Explaining dog wolf differences in utilizing human pointing gestures: Selection for synergistic shifts in the development of some social skills. *PLoS One, 4*(8), e6584.

Hancock, S. J. (2013). *Sniffing Out Bombs in Afghanistan: A Job That's Gone to the Dogs* [Interview]. https://www.npr.org/2013/03/10/173815691/sniffing-out-bombs-in-afghanistan-a-job-thats-gone-to-the-dogs

Hare, B., Brown, M., Williamson, C., & Tomasello, M. (2002). The domestication of social cognition in dogs. *Science, 298*(5598), 1634–1636.

Haverbeke, A., Laporte, B., Depiereux, E., Giffroy, J. M., & Diederich, C. (2008). Training methods of military dog handlers and their effects on the team's performances. *Applied Animal Behaviour Science, 113*(1–3), 110–122. https://doi.org/10.1016/j.applanim.2007.11.010

Hecht, E. E., Smaers, J. B., Dunn, W. D., Kent, M., Preuss, T. M., & Gutman, D. A. (2019). Significant neuroanatomical variation among domestic dog breeds. *Journal of Neuroscience, 39*(39), 7748–7758. https://doi.org/10.1523/JNEUROSCI.0303-19.2019

Hennessy, M. B., Williams, M. T., Miller, D. D., Douglas, C. W., & Voith, V. L. (1998). Influence of male and female petters on plasma cortisol and behaviour: Can human interaction reduce the stress of dogs in a public animal shelter? *Applied Animal Behaviour Science, 61*(1), 63–77.

Herron, M. E., Shofer, F. S., & Reisner, I. R. (2009). Survey of the use and outcome of confrontational and non-confrontational training methods in client-owned dogs showing undesired behaviors. *Applied Animal Behaviour Science, 117*(1–2), 47–54.

Hiby, E., Rooney, N., & Bradshaw, J. (2004). Dog training methods: Their use, effectiveness and interaction with behaviour and welfare. *Animal Welfare-Potters Bar then Wheathampstead, 13*(1), 63–70.

Holland, K. E. (2022). Ambiguity, ambivalence, and affective encounters: An ethnographic account of medical detection dog–trainer relationships. *Anthrozoös, 35*(2), 259–271.

Horváth, Z., Dóka, A., & Miklósi, Á. (2008). Affiliative and disciplinary behavior of human handlers during play with their dog affects cortisol concentrations in opposite directions. *Hormones and Behavior, 54*(1), 107–114.

Jamieson, L. T. J., Baxter, G. S., & Murray, P. J. (2018). You are not my handler! Impact of changing handlers on dogs' behaviours and detection performance. *Animals, 8*(10), 176.

Jezierski, T., Adamkiewicz, E., Walczak, M., Sobczyńska, M., Gorecka-Bruzda, A., Ensminger, J., & Papet, E. (2014). Efficacy of drug detection by fully-trained police dogs varies by breed, training level, type of drug and search environment. *Forensic Science International, 237*, 112–118.

Jones, A. C., & Josephs, R. A. (2006). Interspecies hormonal interactions between man and the domestic dog (*Canis familiaris*). *Hormones and Behavior, 50*(3), 393–400.

Kotrschal, K., Schöberl, I., Bauer, B., Thibeaut, A.-M., & Wedl, M. (2009). Dyadic relationships and operational performance of male and female owners and their male dogs. *Behavioural Processes, 81*(3), 383–391. https://doi.org/https://doi.org/10.1016/j.beproc.2009.04.001

Lefebvre, D., Diederich, C., Delcourt, M., & Giffroy, J. M. (2007). The quality of the relation between handler and military dogs influences efficiency and welfare of dogs. *Applied Animal Behaviour Science, 104*(1–2), 49–60. https://doi.org/10.1016/j.applanim.2006.05.004

Li, Y., Wang, G. D., Wang, M. S., Irwin, D. M., Wu, D. D., & Zhang, Y. P. (2014). Domestication of the dog from the wolf was promoted by enhanced excitatory synaptic plasticity: A hypothesis. *Genome Biol Evol, 6*(11), 3115–3121. https://doi.org/10.1093/gbe/evu245

Linda Crippen, T. (2017). *Military Working Dogs: Guardians of the Night.* Retrieved February 23, 2022, from https://www.army.mil/article/56965/military_working_dogs_guardians_of_the_night

Luckwaldt, A. (2017, 2019). *Career Profile: Military Working Dog Handler.* Retrieved February 23 2022 from https://www.thebalance.com/career-profile-army-military-working-dog-handler-2356458

Masson, S., de la Vega, S., Gazzano, A., Mariti, C., Pereira, G. D. G., Halsberghe, C., Leyvraz, A. M., McPeake, K., & Schoening, B. (2018). Electronic training devices: Discussion on the pros and cons of their use in dogs as a basis for the position statement of the European Society of Veterinary Clinical Ethology. *Journal of Veterinary Behavior, 25*, 71–75.

McIntyre, C. (2017). *Army K-9 Corps: Man's Best Friend Raises Paw, Serves Too.* Retrieved February 23 2022, from https://www.army.mil/article/189131/army_k_9_corps_mans_best_friend_raises_paw_serves_too

Miklósi, Á., Kubinyi, E., Topál, J., Gácsi, M., Virányi, Z., & Csányi, V. (2003). A simple reason for a big difference: Wolves do not look back at humans, but dogs do. *Current Biology, 13*(9), 763–766.

Nolan, R., & Gravitte, D. (1977). *Mine-detecting Canines.* https://apps.dtic.mil/sti/pdfs/ADA048748.pdf

Olson, P. N. (2002). The modern working dog--a call for interdisciplinary collaboration. *Journal of the American Veterinary Medical Association*, *221*(3), 352–355. https://doi.org/10.2460/javma.2002.221.352

Otto, C. M., Darling, T., Murphy, L., Ng, Z., Pierce, B., Singletary, M., & Zoran, D. (2021). 2021 AAHA working, assistance, and therapy dog guidelines. *Journal of the American Animal Hospital Association*, *57*(6), 253–277. https://doi.org/10.5326/JAAHA-MS-7250

Palmer, L. E., Maricle, R., & Brenner, J. A. (2015). The Operational Canine and K9 Tactical Emergency Casualty Care Initiative. *Journal of Special Operations Medicine*, *15*(3), 32–38. https://doi.org/10.55460/RMVA-7381

Parker, H. G. (2012). Genomic analyses of modern dog breeds. *Mammalian Genome*, *23*(1–2), 19–27. https://doi.org/10.1007/s00335-011-9387-6

Parr, J. R., & Otto, C. M. (2013). Emergency visits and occupational hazards in German Shepherd police dogs (2008–2010). *Journal of Veterinary Emergency and Critical Care*, *23*(6), 591–597. https://doi.org/10.1111/vec.12098

Rooney, N. J., Clark, C. C. A., & Casey, R. A. (2016). Minimizing fear and anxiety in working dogs: A review. *Journal of Veterinary Behavior: Clinical Applications and Research*, *16*, 53–64. https://doi.org/10.1016/j.jveb.2016.11.001

Rujoiu, O., & Rujoiu, V. (2018). Losing max, the first Romanian army's military working dog hero. *Journal of Loss & Trauma*, *23*(7), 608–621. https://doi.org/10.1080/15325024.2018.1530182

Sanders, C. R. (2006). "The Dog You Deserve" ambivalence in the K-9 officer/patrol dog relationship. *Journal of Contemporary Ethnography*, *35*(2), 148–172.

Smith, B., & Thompson, K. (2014). Should we let sleeping dogs lie... with us? Synthesizing the literature and setting the agenda for research on human-animal co-sleeping practices. *Humanimalia*, *6*, 114–127. https://doi.org/10.52537/humanimalia.9930

Suplee, C. (1997). Barking up the evolutionary tree. *Denver Post*.

Szetei, V., Miklósi, Á., Topál, J., & Csányi, V. (2003). When dogs seem to lose their nose: An investigation on the use of visual and olfactory cues in communicative context between dog and owner. *Applied Animal Behaviour Science*, *83*(2), 141–152.

Tuber, D. S., Hennessy, M. B., Sanders, S., & Miller, J. A. (1996). Behavioral and glucocorticoid responses of adult domestic dogs (Canis familiaris) to companionship and social separation. *Journal of Comparative Psychology*, *110*(1), 103.

Turbak, G. (2000). Saluting canine courage. *VFW, Veterans of Foreign Wars Magazine*, *87(6)*, 12–15.

Unknown. (Unknown). *Guardians of the Night*. Retrieved February 23 2022 from https://www.pawsofhonor.org/guardians-of-the-night/

Vonholdt, B. M., Pollinger, J. P., Lohmueller, K. E., Han, E., Parker, H. G., Quignon, P., Degenhardt, J. D., Boyko, A. R., Earl, D. A., Auton, A., Reynolds, A., Bryc, K., Brisbin, A., Knowles, J. C., Mosher, D. S., Spady, T. C., Elkahloun, A., Geffen, E., Pilot, M., ... Wayne, R. K. (2010). Genome-wide SNP and haplotype analyses reveal a rich history underlying dog domestication. *Nature*, *464*(7290), 898–902. https://doi.org/10.1038/nature08837

Walper, S. A., Lasarte Aragonés, G., Sapsford, K. E., Brown, C. W., Rowland, C. E., Breger, J. C., & Medintz, I. L. (2018). Detecting biothreat agents: From current diagnostics to developing sensor technologies. *ACS Sensors*, *3*(10), 1894–2024. https://doi.org/10.1021/acssensors.8b00420

Wang, G.-D., Larson, G., Kidd, J. M., vonHoldt, B. M., Ostrander, E. A., & Zhang, Y.-P. (2019). Dog10K: The International Consortium of Canine genome sequencing. *National Science Review*, *6*(4), 611–613. https://doi.org/10.1093/nsr/nwz068

Weber, J. (1939). *The Dog in Training*. McGraw-Hill Book Company, Incorporated.

Worth, A. J., & Cave, N. J. (2018). A veterinary perspective on preventing injuries and other problems that shorten the life of working dogs. *Revue Scientifique et Technique*, *37*(1), 161–169. https://doi.org/10.20506/rst.37.1.2749

Zubedat, S., Aga-Mizrachi, S., Cymerblit-Sabba, A., Shwartz, J., Leon, J. F., Rozen, S., Varkovitzky, I., Eshed, Y., Grinstein, D., & Avital, A. (2014). Human–animal interface: The effects of handler's stress on the performance of canines in an explosive detection task. *Applied Animal Behaviour Science*, *158*, 69–75.

# 39
# WORKING ANIMALS AND HUMAN-ANIMAL INTERACTIONS

*Tamara Tadich and João B. Rodrigues*

## Introduction

A working animal is one that provides a resource that is essential for their caretakers, who depend largely on these animals for their subsistence. Animals are used in a diversity of activities that require different capacities from the animal. For example, in some cases the work performed is done by traction, but in other cases by carrying loads, or by learning to perform a specific task. Globally, there are different species used as working animals, with the most frequent ones being bovids (cows, oxen, and buffalo) and equids (horses, donkeys, and mules) (FAO, 2014). Depending on the geoclimatic region, we can also find camelids, goats, elephants, and dogs performing activities that are of great importance for the livelihood of their communities. The use of animals for work is an ancient activity (Guthiga et al., 2007) and originated when people realized that some species, in addition to being a source of food, had muscle power that could be harnessed as a source of force for different activities (Davis, 1987). Working animals have played an important role in people's life throughout history and have been linked to important technological innovations such as the invention of the wheel, different types of vehicles, and harnessing equipment (Nóbrega, 2018).

While working, these animals need to be in continuous communication with their handler, thus human-animal interaction (HAI) is a central aspect to consider. The term HAI can be used to describe a simple contact between a human and a nonhuman animal, but it has also been defined as a dynamic process where previous interactions between the animal and humans form the foundation of a relationship that then exerts a feedback effect on the nature and perception of future interactions (Waiblinger et al., 2006). Handlers of working animals interact with them through tactile cues such as pressure and release, voice cues, visual cues, or a mixture of them. These interactions can be neutral, positive, or negative in nature and will define the relationship with the handler and the effects on animal welfare.

Currently, there are hundreds of millions of animals working globally, in particular in Low and Middle Income Countries (LMIC) (FAOSTAT, 2020). These animals increase the resilience of their communities and are involved in many aspects of their caregivers' life; many times being considered as part of the family (Clancy et al., 2021; Luna et al., 2017). Nevertheless, in some cases HAIs can be negative in nature, due to a lack of knowledge of the animal's behavior, use of inappropriate instruments that are in contact with the animal (harnesses), economic constraints, or psychosocial constructs (empathy, attitudes, perceived control, and perceptions) (Luna et al., 2019). In these cases, both animal welfare and human well-being are at risk.

In this chapter, the impact of the human-working animals' interactions on the livelihoods of people and the welfare of the animal will be explored. Special consideration will be given to the current challenges and opportunities that working animals provide for achieving sustainability.

## Impact of Working Animals on Livelihoods and Sustainability

Working animals are essential for direct and indirect income generation of millions of families, although they continue to be invisible for part of the society. This results in working animals not being considered by local governments in their policies, leaving them without any legal protection and not being considered in animal health strategies in many cases. A person's livelihood comprises their capabilities and their means of living, including food security, income, and tangible and intangible assets (Chambers & Conway, 1991). According to FAO (2009), these assets can be divided into five categories:

1. Human capital includes the workforce, people's ability and health to work, and the knowledge and skills they have acquired over generations of experience and observation.
2. Natural capital, including the access to land, water, forest resources, and livestock.
3. Financial capital, which may come from the conversion of production into cash or other incomes.
4. Physical capital may include tools and equipment, as well as infrastructure that influence people's ability to earn an adequate livelihood.
5. Social capital refers to the way in which people work together, both within the household and in the wider community. In many communities, social obligations exist that link people allowing reciprocal exchange, trust, and mutual support.

A working animal could be considered in more than one of these categories, for example Pritchard (2014) described that working horses could be considered within the natural, financial, physical, and social capital of their caregivers.

Working animals are multipurpose animals, providing draft power for agriculture while being used for transport, milk, wool, fuel, and meat at the end of their lives. Camels are an excellent example of this multipurpose function and have been essential to the livelihoods of people in Ethiopia where they account for 34.5% of Africa's camel population (Zeleke, 2017). The use of bovids as draft animals is still common in Asia, Latin America, and Africa, in particular when work activities require forces greater than those that people can exert and no motorized equipment is available. These animals are also an important source of transport for people and at the end of their working life, a source of food. The use of cattle has decreased due to the motorization of agriculture and forestry tasks; however, they are still widely used by small-scale farmers, production units located on terrains that do not favor motorization, organic systems, and in LMIC (Mota-Rojas et al., 2021, Okello et al., 2015).

It is important to understand these multiple roles that working animals can have within a family, since it allows them to use a capital asset substitution strategy, as a coping style, in times of need (Manlosa et al., 2019), which explains why working animals are so important for resilience capacity. For example, a cow or an ox can be considered a natural capital asset based on its characteristic as an animal. They can also be considered a physical capital asset since their primary function in the same farm can be draft work. In times of need of more financial capital, a family can rent their working animal to another farmer; in this case, the substitution mechanism is increasing their financial capital by decreasing their physical capital. These substitution mechanisms allow to interchange and reorder the capital assets, but without an exhaustive use of the capital (Manlosa et al., 2019). This provides the household the opportunity to continue functioning in challenging situations (resilience) without losing their working animal. Conversely, when communities sell their working animals to increase their financial capital, they may not be able to replace it in the future. This is the case of the donkey skin trade for the Chinese ejiao market, where the skin of donkeys is used to obtain collagen, which is used to manufacture a traditional medicine called ejiao. The global trade of donkeys for this purpose has shown ethical issues associated with poor welfare and biosecurity problems. This, added to the difficulties in donkey farming systems, makes it an unsustainable market leading to a decline of donkey populations in many parts of the world and a decrease in resilience of the communities of origin (The Donkey Sanctuary, 2019; Tatemoto et al., 2021; Fernándes Lima et al., 2022).

Religion and other cultural aspects can also affect the way in which humans interact with working animals. In India, zebu cattle, mithun, and buffalos are the main source of draft power, but only males are used for work purpose due to the social and religious considerations associated with cows. This fact has affected the number of draft animals available in the country since the male: female ratio in cattle has decreased over time (Singh, 2000). On the other hand, camel, donkey, mule, horse, and yak are also used in India, independent of sex, mainly as packed animals and in carts for transportation (Singh, 2000). Social status can also affect the way in which working animals are perceived. In the case of equids, historically there has been a bias against mules and donkeys and in favor of horses. In colonial times in America, the hierarchy between the Spanish elite, *mulatto* laborers and indigenous workforce was rendered as an analogy of the hierarchy between horses, mules, and donkeys (Schuerch, 2020). Today, there is still an association of donkeys with poorer communities, although the value of mules has increased due to their higher work capacity (Galindo et al., 2018; Lagos et al., 2021). These cultural perceptions influence the monetary value of the animals. For example, in Ethiopia, donkeys are also the cheapest option for a family; nevertheless, they have enormous advantages in terms of working capacity, rusticity, suitability for being used by women and children, and ease of training (Zeleke, 2017).

Animal traction, the use of animal power to pull farm equipment, vehicles or other types of load, is one of the important reasons why people keep livestock, since it is a cost-effective labor-saving technology, in particular for small scale and subsistence farmers. Studies have shown that households that have oxen, buffalo, or cows for draft work had better crop yields, lower labor requirements and higher incomes, making them more efficient and sustainable (Teweldmehidin & Conroy, 2010; Adenisa, 1992; Jaeger & Matlon, 1990). When work is more efficient, it positively impacts the well-being of people since they can spend more time in other activities with their family or community and physical labor is decreased. Working animals are essential to the livelihoods of communities in rural and urban areas around the globe, as they allow relieving families, and in particular women, from energy consuming tasks, supporting farming activities, transport, and other traditional and ceremonial events (Zeleke, 2017).

This diversity of roles that working animals perform increases resilience capacity, but at the same time supports the achievement of several of the sustainable development goals (SDGs). These SDGs, also known as Global Goals, are 17 goals adopted by the United Nations and to which many countries have committed. They are meant to be a call to action to end poverty, protect the planet and to facilitate that by 2030 people can enjoy peace and prosperity (United Nations, 2015). They support SDG 1 no poverty through income generation, SDG 5 gender equality by alleviating women's labor and developing empowerment of women, SDG 6 clean water and sanitation with equids being used as a means of water transport, SDG 8 decent work and economic growth by facilitating work and transport of goods and SDG 13 climate action by reducing soil compaction and decreasing the use of fossil fuels. Although working animals and animal welfare are not directly mentioned in any of the 17 SDGs, working to achieve these goals is compatible with working to improve animal welfare with the presence of mutual benefits for people and animals (Keeling et al., 2019).

## Impact of HAI on the Welfare of Working Animals

HAIs can affect both the well-being of humans as well as the welfare of animals, making them an essential component of the One Welfare framework (Luna & Tadich, 2019; Waiblinger et al., 2006). The nature of the interactions can be modulated by different factors. From the human perspective, the intrinsic psychological attributes (empathy, attitudes, and perception), and cognitive attributes (knowledge, experience) will affect the way in which they interact with the working animal combined with external factors such as the equipment chosen to communicate with the animal. The animal will respond according to the species, temperament/personality, health, and cognitive abilities (ease of training, previous experiences). Nevertheless, for many working animals' environmental factors should also be considered, such as the geoclimatic conditions under which they are asked to work (Figure 39.1). Each of these factors should be considered when trying to understand the welfare problems that can arise in working animals.

When interactions are negative in nature, they will stimulate a fear response in the animal, which can become a source of stress. In turn, the handling of a fearful animal will become more difficult and can place the handler and the animal in a dangerous situation

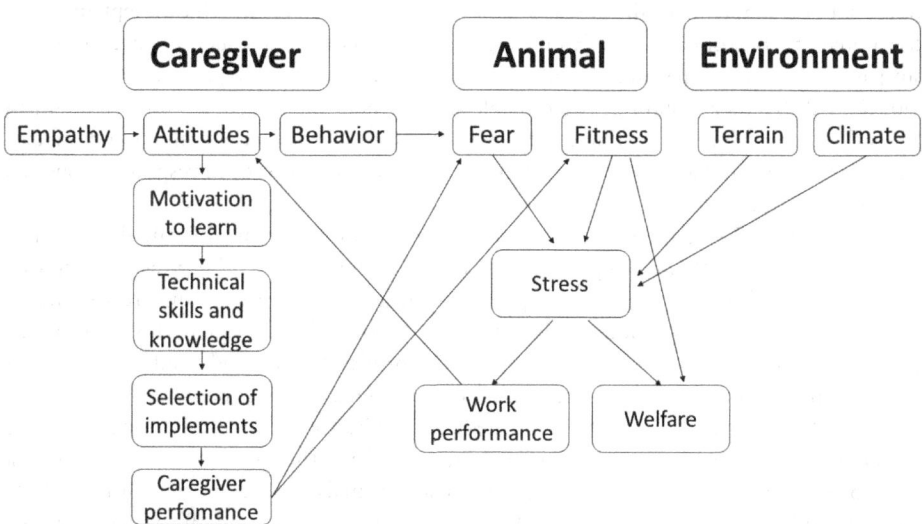

*Figure 39.1* Representation of the interactions between human, animal, and environmental factors that modulate human-working animal interactions. Adapted from Hemsworth and Coleman (2011)

(Breuer et al., 2000, 2003; Hemsworth, 2003). Pritchard et al. (2005) have shed light on the importance that the assessment of the HAI has for the establishment of appropriate strategies to improve the welfare of working equids and their owners. The authors propose that "without a degree of social bonding between the owner or user and his animal (...), there is little motivation to improve welfare". We will focus on elephants and equids, species for which the effects of HAIs have been studied in more detail.

## *Elephants*

Asian elephants *(Elephas maximus)* are still used for different work activities such as riding for religious processions or tourism, anti-poaching activities and ridden by park rangers when carrying out wildlife census and logging, which requires them to spend many hours daily in close contact with humans. The training of elephants in Asia is an ancient tradition with its roots in the Indus Valley (Fleischer et al., 2001) and as in other species the training techniques used are part of the folklore and have been passed through generations (McLean, 2021). The way in which the elephant handlers or mahouts have traditionally interacted with elephants is through a fixed hierarchy where rank is maintained by ritualized aggression (McLean, 2021). This form of training and relationship can lead to dangerous risks for mahouts and animals alike. More recently, these training methods are being modified by introducing the application of more humane ones such as learning theory (pressure-release techniques). An example is the Working Elephant Program Asia (WEPA), an organization based in Finland that has started the first elephant training system based on learning theory.

The lack of a proper human interaction between mahouts and elephants can lead to high death rates, and, although there is not much data, some studies suggest that 10–20 mahouts are killed annually in work accidents (Lair, 1997). If elephants are not properly trained these risks are even higher when they are asked to interact with tourists, which may not have any knowledge about elephant's behavior.

The use of inadequate training methods in elephants can lead to different types of welfare problems. For example, if a painful stimulus is used as punishment, which the elephant is not able to predict, control or escape from, they can show behavioral responses that can range from passive coping mechanisms (trunk movements, inactivity) to hyperreaction (Moberg, 2000; Keay & Bandler, 2008). Learned helplessness, where the animal becomes tolerant to pain, is also a possible response that can be very dangerous for both the mahout and the elephant since the affected animal can attack under stressful situations or in response to sudden noises (Bradshaw, 2009).

Handler familiarity can also affect the elephants' success at performing the desired tasks (Liehrmann et al., 2021). The elephant and his mahout develop a working relationship that starts at the time of taming, when the elephants are around four years old and are assigned to a mahout. This relationship traditionally lasted for a lifetime, but nowadays this has changed since many mahouts leave this profession for more profitable ones and results in elephants having continued changes of handlers, at times for younger and less-experienced ones, which can negatively affect the HAIs (Liehrmann et al., 2021, Crawley et al., 2021). Longer positive HAIs allow to build a stronger bond between the animal and the handler and increased trust in each other, which is important in particular when animals are confronted with novel situations. Neophobia reflects the animals' fear of unfamiliar stimuli and affects the way they respond to them; these responses can be modified through experience (Hemsworth et al., 2018). Liehrmann et al. (2021) found that when timber elephants in Myanmar were confronted to walk on a novel surface, those guided by an unknown mahout or with a shorter relationship were more reluctant to walk on it compared to elephants guided by mahouts they had known for over a year. These results are interesting because the prior experience of the mahout did not affect the elephant's response, rather the time they had worked as a dyad. The same group of researchers found that timber elephants responded faster to the commands given by their own mahouts than to those given by unknown ones. This was accompanied by lower levels of serum enzymes associated with muscle damage in animals engaged in longer relationships, likely due to greater trust, which in turn reduced the need to use force and punishments to control the elephant (Crawley et al., 2021).

Due to social pressure and animal rights campaigns, riding elephants is perceived as a negative activity and has pushed the industry to develop new ways of interacting with elephants. Some countries such as Thailand have modified the traditional elephant camps and now tourists get involved in other types of interactions with them, like feeding and bathing them (Bansiddhi et al., 2018; McLean, 2021; Baker & Winkler, 2020). This new type of interaction raised other elephant welfare concerns, such as those associated with increased obesity levels and metabolic problems. Norkaew et al. (2019) compared elephant camps in Thailand with different types of activities and work intensity. The results showed significant differences across camps where supplemental feeding and lack of exercise resulted in negative consequences for the elephants' welfare, as determined by higher fecal glucocorticoid metabolites in elephants in camps where they are not ridden. Increased obesity and higher insulin responses were also a concern for these elephants, probably associated with the high glycemic index foods (bananas, watermelon, sugarcane, or pumpkins) that tourists are allowed to feed the animals with no restriction (Norkaew et al., 2018, 2019). These results are important since they support that riding elephants per se may not necessarily give way to poor welfare, while limiting the interactions between tourists and elephants to feeding and bathing may not be the ideal type of HAI. There is no evidence that the application of humane training techniques is detrimental for elephants (McLean, 2021). Conversely, a well-trained elephant could expand HAIs to positive situations such as engaging in more physical activities and natural social behaviors beneficial for their health and welfare.

## *Equids*

Donkeys, horses, and mules are used for diverse tasks worldwide. They are involved in urban contexts as a means of transport, in forestry to haul logs and equipment, in agricultural activities such as plowing, and supporting many other activities in rural settings. Equids can be used as pack, ridden, or draft animals, making them particularly versatile working animals.

In all activities where equids are used, they interact with their handlers through a harness system, with four main functions: move the load, carry the load, break the load and communication, with the last one based on pressure/release cues. This system can be supported with voice commands, and in some cases only voice cues are used. The harness system also allows for a distribution of the force exerted by the animal to pull a load on a cart or to carry the load on its back safely and minimizing the risk of lesions. This interaction between the equid and the handler requires knowledge about training methods and the selection of the appropriate harness. A lack of knowledge can result in improper decision making and subsequent health and welfare problems (Rushen & de Passillé, 2010). Frequent welfare problems in working equids are associated with a lack of knowledge about feeding practices, incorrect shoeing, and poor harnessing, which result in poor body condition scores, lameness, and lesions (De Aluja, 1998; Pritchard et al., 2005, 2008; Tadich et al., 2008; Ali et al., 2015). The lack of knowledge can be associated with a high percentage of illiteracy among working equids handlers (Tadich & Stuardo-Escobar, 2014) and a scarce availability of training courses suited for them, together with low accessibility due to their multidimensional vulnerability (Burn et al., 2009; Lanas et al., 2018). Contrary to what one would expect, the multidimensional vulnerability of working equid owners, which includes their socioeconomic status, does not explain the welfare state of their animals (Lanas et al., 2018). This is relevant since most working equids are kept by the poorest communities in LMICs and implies that there are other factors that modulate HAIs that need to be considered.

One factor that can modulate HAIs is empathy. A study by Luna et al. (2018) found that human-animal empathy was a good predictor of the welfare state of working equids. Owners with higher human-animal empathy did not only keep their horses in a better welfare state, but also had higher perception of pain, this independent of their socio-economic status. Some implications are that more empathic owners could interact in a more positive way with their equids since they are able to put themselves in the place of the animal, and as a result make better decisions about their management or avoid situations that may put the animal at risk.

The way in which handlers perceive their working equid can also affect the quality of the HAIs. The handlers' perception can vary from an affective to an instrumental one. For example, in a study with urban draft horses, the affective perception was predominant, with most owners considering their horses as part of their family or as a friend, while owners in more rural settings had a more instrumental perception, where horses were considered as an income source (Luna et al., 2017, 2019). In the case of mules used by the army, soldiers' perception has also been described as more instrumental, as they describe mules as strong and noble animals that are valuable for work under difficult field conditions and as an essential component of the army's logistics (Lagos et al., 2021). Particularly, in the case of mules, they have been historically considered superior to horses and donkeys for work due to their higher physical capacity, but at the same time they have been considered as more aggressive, making them a difficult animal to work with (Sasimowski, 1984; Hameed et al., 2016; Ali et al., 2019). These perceptions and negative attitudes toward mules can predispose handlers to interact with mules in a more aggressive way and to make decisions that can negatively affect the mules' welfare.

Many of the welfare problems identified in working equids are connected to human welfare problems and to misconceptions about equids. Thus, improving human welfare is essential if improvements to animal welfare are to be achieved (Rodrigues et al., 2020). The implementation of animal welfare strategies that include a One Welfare approach is vital to ensure long-term improvement (Tadich, 2020). The development of human-animal empathy needs to be incorporated at a young age in communities where working animals are an essential part of their livelihoods, since children are involved in many of the equid care routines. For example, poor handling skills, including aggressive interactions, were observed between owners and kiln donkeys in Nepal, and these negative interactions were replicated by young children who probably learned these bad practices by observing adults (Rodrigues et al., 2020). Similarly, in Mexico, children were able to identify some basic needs of the donkeys they cared for, such as water and food, but had more difficulty in identifying other needs such as veterinary care (Tadich et al., 2016).

## Current Policy and Legislative Perspectives

Animals can be categorized differently based on human preferences or the activities they are involved in, even though they share biological and behavioral characteristics, such as intelligence, sociality, or sentience (Clement, 2011). For example, two common classifications of domestic animals used by countries to develop guidelines and regulations are "companion animals" and "production animals" (Short-McKendree et al., 2014). This results in one set of standards on how companion animals, such as dog, cats or other species kept for companionship, should be treated and a different standard for farm animals, those kept for production purposes, despite their inherent similarities. In the case of working animals, most countries do not usually consider them in either of these two categories creating a challenge when trying to regulate working practices and the animals' welfare.

Most working animals are livestock, but, unlike farm animals in production systems, they are managed individually, interact closely and can form strong bonds with their handlers (Luna & Tadich, 2019). The quality of HAIs has major implications for working animals' welfare and productivity, and thus they should be regulated together with their basic needs. In 2016, the World Organization for Animal Health (WOAH) published Chapter 7.12 on the Welfare of Working Equids within the Terrestrial Animal Health Code. This chapter applies to horses, donkeys, and mules that are used for traction, transport, and income generation and provides guidelines for defining responsibilities, criteria for measuring their welfare, and recommendations for measures applied to working equids (WOAH, 2016). These recommendations can be adapted by member countries into local regulations, in particular by countries where working equids are still common. Although there are no international guidelines for other working animal species, the WOAH chapter could be used as a basis for their development.

To date, most countries have some general anti-cruelty or animal welfare acts, but when it comes to working animals not many countries have developed specific regulations. For instance, in India, the Prevention of Cruelty to Animals Act (1960) has some relevant prohibitions that apply to working animals. These include prohibitions of beating, over-riding, over-driving, or over-loading. The same Act also includes a specific section on the Prevention of Cruelty to Draft and Pack Animals Rules, which sets out maximum loads for draft animals by species. The same document set the maximum passengers that an animal can transport, recommend regular rests, and prohibit the use of spiked bits or harnesses. In Pakistan, the Prevention of Cruelty to Animals Act (1890) prohibits to overload draught animals and to employ a sick or injured animal for such purposes. This Act also allows provincial governments to develop regulations regarding maximum loads and other parameters, but this has not occurred. In Thailand, the Prevention of Cruelty to Animals and Provision of Animal Welfare Act (2014) states that any animal owner or person involved with transporting or using an animal for work shall provide appropriate animal welfare, in accordance with the criteria, procedures and conditions prescribed by the Minister, but these conditions have not been developed. This is similar to the case of Myanmar where the Animal Health and Development Law (1993) does contain a specific reference to animals used for draft purposes. Article 18 specifies that the government will make provisions to limit the weight carried by certain species and breach of any such prescribed limits is a punishable offense under Article 25, but there are no specific norms. Ethiopia has a large number of working animals, nevertheless they only have some by-laws in the Halaba region that allow euthanasia of working horses when they suffer from extreme injuries or disease.

Colombia is probably the country that has developed the most specific legislation where Law 84 of 1989 includes specific provisions by which animals should not be overloaded and in general prohibits animals not fit for purpose (such as blind or wounded animals) from being used for work.

It is a challenge to develop regulations for working animals because of differences between species, but also because of the diversity of working contexts. For example, there is no universal rule for estimating work capacity, since it will depend on environmental factors (terrain, temperature, humidity), the workload, the frequency with which the animal is asked to work, the quality of the harnesses, and individual variables such as fitness and training.

## Ethical Considerations

Lately, there has been an increase in interest in our relationship and interactions with animals and the moral issues that arise from them. It is widely accepted that we owe moral obligations to animals (Garner, 2010), but what these obligations are will depend on different ethical views making it difficult to establish in practice a unique moral code toward animals (Cui & Xu, 2019).

Animal welfare ethics are considered a weak anthropocentrism or a framework for protecting nature from an environmental ethics point of view, where the use of natural resources, including animals, should be done only after careful and rational consideration (Norton, 1984). In this sense, animal welfare recognizes the moral status of animals but does not reject the use of animals for human benefit (Cui & Xu, 2019), as it happens, for instance, with working animals. In particular, the moral importance given to non-human animals has been associated with sentiency, or the capacity to experience pain and pleasure (Garner, 2010).

According to Price (2002), domestication of animals can be viewed as a biological event that benefits both humans and animals, creating a symbiosis. For example, a working animal could have all its biological needs provided by its owner. The benefits for each species can vary widely depending on the use given to the animal and most times the human-animal relationship is unequal since it involves a high level of control of the animal by humans (Hemsworth, 2007). This relationship between the working animal and its handler will be shaped by the interactions that occur, which are more frequent than those occurring with other types of livestock. For many people it is acceptable that these animals are used for human benefit or profit provided that such use is humane. For others, the use of animals for any purpose is unethical and considered a form of exploitation. Zamir (2017) makes a clear distinction between "use" and "exploitation": A person "uses" an animal when the person perceives the animal as a means of furthering its own well-being (financial or other), but it turns into "exploitation" when the person is willing to act in a way that is substantially detrimental to the animals' welfare. According to this, a responsible and ethical use of working animals could be achieved when HAIs are positive, no harm is done to the animal and a mutually beneficial relationship is formed.

Nevertheless, modern animal ethics do not only include our obligations toward animals but also our duties to animal users and society in general (Hemsworth, 2007; Levy, 2004). This again raises difficult questions, such as whether it is ethical to ban the use of working animals and leave a family without their means of transport or income, or if it is ethical to maintain an activity that is considered cruel by society.

A ban of some activities that involve working animals has been established in many countries, in particular in big cities and other urban areas. For example, in Chile, horse-drawn carriages used for tourism have been banned in the city of Viña del Mar since 2020, after pressure from animal rights advocate groups and a public consultation carried out by the municipality. These carriages, known as Victorias, were a traditional tourism attraction for the city, and the ban left almost a hundred families without income and many of these horses ended in slaughterhouses due to the incapacity of the owners to maintain them.

On the other hand, countries such as Israel have approved a ban on horse- and donkey-drawn carts in cities but still allow carts for tourism purposes. In Colombia, horse-drawn carts have been banned in cities such as Medellín, but they are allowed in smaller cities and rural areas, while Romania has banned horse carts from national roads (Starkey, 2011). It is more frequent to find petitions to ban the use of working animals that belong to species considered more charismatic or closer to humans, as in the case of elephants and horses, than for cattle, which raises other ethical concerns on how decisions about public policy are made and the effect that these decisions could have on the welfare of species considered less charismatic.

## Future Perspectives and Challenges in the Use of Working Animals

The use of working animals is still present globally, although many times invisible for those living in highly urbanized areas. These animals represent not only a source of income for their caregivers, but they are also strongly rooted in their culture and are many times considered part of their family. The substantial differences that exist between the level of understanding or attitudes toward working animals and animal welfare issues between agrarian and urban societies depend largely on people's degree of contact with animals (Bonafos et al., 2010). These differences frequently end in regulations that ban the use of working animals, since societal pressure in urban areas can be very strong.

Climate change is a current concern worldwide and it especially affects mountain areas with small and rural systems in communities where working animals are an indispensable power and resilience source. Without working animals, the vulnerability of many communities would be even more exposed and thus future research should consider their role in alleviating the impact of disasters and emergencies (Clancy et al., 2021). There is a need for establishing sustainable practices to mitigate the effects of climate change and working animals should be considered as a suitable way to promote sustainable practices, in particular in LMIC. The use of working animals could promote social, environmental, and economic aspects of sustainable production systems since they promote gender equality by enabling women to be economically active, increasing their status within the community and personal resilience and access to essential services, replace the use of fossil fuels, reduces the dependency of external factors of production, and generate an income for their owners. Nevertheless, their use requires prioritizing animal welfare and a change in socio-cultural values. One of the biggest challenges that societies face is to be able to design appropriate animal power strategies that ensure animal welfare and allow a sustainable development instead of banning it and providing incentives to mechanized technologies that depend on fossil fuels.

There are working practices with animals that must be rethought, in particular where working animals are used only for the entertainment of people. For example, some activities with elephants designed for tourists are disappearing, but these elephants can be involved in other activities together with their handlers as in anti-poaching and wildlife census in National Parks. Unlike other types of vehicles, elephants have the advantage of not disturbing wildlife, providing a tall vantage point and being able to traverse difficult terrains (McLean, 2021). As in any other work activity, the animals involved need to be properly trained and their welfare taken care for.

The segregation of communities in which working animals are still of great importance, the loss of trades related to the training of these animals and the construction of adequate harnesses generate a great challenge in terms of developing human capacities. Efforts should be put in the development of training programs for working animals' handlers since the promotion of positive HAIs is the only way to ensure the welfare of these animals.

## Discussion Questions

1 What are the ethical and welfare implications of using animals in working activities?
2 How can working animals impact sustainability?
3 Why are human-animal interactions an essential aspect to consider when using working animals?

## References

Adenisa, A.A. (1992). Oxen cultivation in semi-arid West Africa: Profitability analysis in Mali. *Agriculture Systems*, 38, 131–47. DOI: 10.1016/0308-521X(92)90037-O

Ali, A.B.A., El Sayed, M.A., McLean, A.K., & Heleski, C.R. (2019). Aggression in working mules and subsequent aggressive treatment by their handlers in Egyptian brick kilns—Cause or effect? *Journal of Veterinary Behavior*, 29, 95–101. DOI: 10.1016/j.jveb.2018.05.008

Ali, A.B.A., Matoock, M.Y., Fouad, M.A., & Heleski, C.R. (2015). Are mules or donkeys better adapted for Egyptian brick kiln work? (Until we change the kilns). *Journal of Veterinary Behavior*, 10, 158–165. DOI: 10.1016/j.jveb.2014.12.003

Baker, L. & Winkler, R. (2020). Elephant rewilding. *Animal Sentience*, 28(1). DOI: 10.51291/2377-7478.1506

Bansiddhi, P., Brown, J.L., Thitaram, C., Punyapornwithaya, V., Somgird, C-, Edwards, K.L., & Nganvongpanit, K. (2018). Changing trends in elephant camp management in northern Thailand and implications for welfare. *PeerJ*, 23(6), e5996. DOI: 10.7717/peerj.5996

Bonafos, L., Simonin, D., & Gavinelli, A. (2010). Animal welfare: European legislation and future perspectives. *Journal of Veterinary Medical Education*, 37, 26–29. DOI: 10.3138/jvme.37.1.26.

Bradshaw, G. (2009). Inside looking out: Neuroethological compromise effects of elephants in captivity. In Forthman, D.L., Kane, L.F., Hancocks, D. & Waldau, P.F. (eds) *An Elephant in the Room*. North Grafton, MA: Tufts Center for Animal Policy, p. 58.

Breuer, K., Hemsworth, P.H., Barnett, J.L., Matthews, L.R., & Coleman, G.J. (2000). Behavioural response to humans and the productivity of commercial dairy cows. *Applied Animal Behaviour Science*, 66, 273–288. DOI: 10.1016/s0168–1591(99)00097-0

Breuer, K., Hemsworth, P.H., & Coleman, G.J. (2003). The effect of positive or negative handling on the behavioural and physiological responses of nonlactating heifers. *Applied Animal Behaviour Science*, 84, 3–22.DOI: 10.1016/S0168-1591(03)00146-1

Burn, C.C., Pritchard, J.C., Whay, H.R. (2009). Observer reliability for working equine welfare assessment: problems with high prevalences of certain results. *Animal Welfare*, 18, 177–187.

Chambers, R., & Conway, G.R. (1991). Sustainable rural livelihoods: Practical concepts for the 21st century. Institute of Development, Discussion Paper 296. http://www.ids.ac.uk/publication/sustainable-rural-livelihoods-practical-concepts-for-the-21st-century

Clancy, C., Watson, T., & Raw, Z. (2021). Resilience and the role of equids in humanitarian crises. *Disasters*. Ahead of print DOI: 10.1111/disa.12501

Clement, G. (2011). "Pets or Meat"? Ethics and domestic animals. *Journal of Animal Ethics*, 1, 46–57. DOI: 10.5406/janimalethics.1.1.0046

Crawley, J.A.H., Liehrmann, O., Franco dos Santos, D.J., Brown, J., Nyein, U.K., Aung, H.H. Htut, W., Min Oo, Z., Seltmann, M.W., Webb, J.L., Lahdenperä, M., & Lummaa, V. (2021). Influence of handler relationships and experience on health parameters, glucocorticoid responses and behavior of semi-captive Asian elephants. *Conservation Physiology*, 9, coaa116. DOI: 10.1093/conphys/coaa116

Cui, Q., & Xu, H. (2019). Situating animal ethics in Thai elephant tourism. *Asia Pacific Viewpoint*, 60, 3. DOI: 10.1111/apv.12221

Davis, S.J.M. (1987). *The Archaeology of Animals*. London: Routledge.

De Aluja, A.S. (1998). The welfare of working equids in Mexico. *Applied Animal Behaviour Science* 59, 19–29. DOI: 10.1016/S0168-1591(98)00117-8

Fernándes Lima, Y., Tatemoto, P., Reeves, E., Burden, F.A. & Santurtun, E. (2022). Donkey skin trade and its non-compliance with legislative framework. *Frontiers in Veterinary Science*, 9, 849193. DOI: 10.3389/fvets.2022.849193

Fleischer, R.C., Perry, E.A., Muralidharan, K., Stevens, E.E. and Wemmer, C.M. (2001). Phylogeography of the Asian elephant (*Elephas maximus*) based on mitochondrial DNA. *Evolution*, 55(9), 1882–1892. DOI: 10.1111/j.0014-3820.2001.tb00837.x.

Food and Agriculture Organization (FAO). (2009). *The Livelihood Assessment Tool-Kit, Analysing and Responding to the Impact of Disasters on the Livelihoods of People* (1st ed.). Geneva, Switzerland: Food and Agriculture Organization of the United Nations (FAO), Rome and International Labour Organization (ILO).

Food and Agriculture Organization (FAO). (2014). The role, impact and welfare of working (traction and transport) animals. *Report of the FAO-The Brooke Expert Meeting*. Rome: FAO, pp. 53.

Food and Agriculture Organization Database (FAOSTAT). (2020). Rome, Italy: Food and Agriculture Organization, United Nations, p. 1. Available: http://www.fao.org/faostat/en/

Galindo, F., de Aluja, A., Cagigas, R., Huerta, L.A., & Tadich, T.A. (2018). Application of the hands-on donkey tool for assessing the welfare of working equids at Tuliman, Mexico. *Journal of Applied Animal Welfare Science*, 21, 93–100. DOI: 10.1080/10888705.2017.1351365

Garner, R. (2010). Animal, ethics and public policy. *The Political Quarterly*, 81, 1.

Guthiga, P.M., Karugia, J.T., & Nyikal, R.A. (2007). Does use of draft animal power increase economic efficiency of smallholder farms in Kenya? *Renewable Agriculture and Food Systems*, 22, 290–296.

Hameed, A., Tariq, M., & Yasin, M.A. (2016). Assessment of welfare of working donkeys and mules using health and behavior parameters. *Journal of Agriculture Science and Food Technology*, 2, 69–74.

Hemsworth, P.H. (2003). Human–animal interactions in livestock production. *Applied Animal Behaviour Science*, 81, 185–198. DOI: 10.1016/S0168-1591(02)00280-0

Hemsworth, P.H. (2007). Ethical stockmanship. Education, ethics and welfare. *Australian Veterinary Journal*, 85, 194–200. DOI: 10.1111/j.1751-0813.2007.00112.x

Hemsworth, P.H., & Coleman, G.J. (2011). Human-livestock interactions. In Chapter 6: *A Model of Stockperson-Animal Interactions and their Implications for Livestock*. Oxfordshire: CAB International, p. 132.

Hemsworth, P.H., Sherwen, S.L., & Coleman, G.J. (2018). Human contact. In Appleby, M.C., Olsson, I.A.S., & Galindo, F. (eds) *Animal Welfare* (3rd ed.). Wallingford: CAB International, pp. 294–314.

Jaeger, W., & Matlon, P. (1990). Utilization, profitability and adoption of animal draft power in West Africa. *American Journal of Agricultural Economics*, 72, 35–48. DOI: 10.2307/1243143

Keay, K., & Bandler, R. (2008). Emotional and behavioural significance of the pain signal and the role of the midbrain periaqueductal gray (PAG). In Basbaum, A.I., Kaneko, A., Shepherd, G.M., & Westheimer, G. (eds) *The Senses: A Comprehensive Reference. Volume 5, Pain*. Cambridge, MA: Academic Press, pp. 627–634.

Keeling, L., Tunón, H., Olmos Antillón, G., Berg, C., Jones, M., Stuardo, L., Swanson, J., Wallenbeck, A., Winckler, C., & Blokhuis, H. (2019). Animal welfare and the United Nations sustainable development goals. *Frontiers in Veterinary Science*, 6, 336. DOI: 10.3389/fvets.2019.00336

Lagos, J., Rojas, M., Rodrigues, J.B., & Tadich, T. (2021). Perceptions and attitudes towards mules in a group of soldiers. *Animals*, 11, 1009. DOI: 10.3390/ani11041009

Lair, R.C. (1997). *Gone Astray: The Care and Management of the Asian Elephant in Domesticity* (1st ed.). FAO Regional Office for Asia and the Pacific. Bangkok: Dharmasarn Co., Ltd.

Lanas, R., Luna, D., & Tadich, T. (2018). The relationship between working horse welfare and their owners' socio-economic status. *Animal Welfare*, 18, 27–54. DOI: 10.7120/09627286.27.1.047

Levy, N. (2004). *What Makes Us Moral? Crossing the Boundaries of Biology*. Oxford: OneWorld.

Liehrmann, O., Crawley, J.A.H., Seltmann, M.W., Feillet, S., Nyein, U.K., Aung, H.H., Htut, W., Lahdenperä, M., Lamsade, L., & Lummaa, V. (2021). Handler familiarity helps to improve working performance during novel situations in semi-captive Asian elephants. *Scientific Reports*, 11, 15480. DOI: 10.1038/s41598-021-95048-w

Luna, D., & Tadich, T.A. (2019). Why should human-animal interactions be included in research of working equids' welfare? *Animals*, 9, 42. DOI: 10.3390/ani9020042

Luna, D., Vásquez, R.A., Rojas, M., & Tadich, T.A. (2017). Welfare status of working horses and owners' perceptions of their animals. *Animals*, 7, 56. DOI: 10.3390/ani7080056

Luna, D., Vásquez, R.A., & Tadich, T. (2019). Exploring the relationship between socio-demographic background and empathy toward nonhuman animals in working horse caretakers. *Society & Animals*, 1–20. DOI: 10.1163/15685306-12341607

Luna, D., Vásquez, R.A., Yáñez, J.M., & Tadich, T. (2018). The relationship between working horse welfare state and their owners' empathy level and perception of equine pain. *Animal Welfare*, 27, 115–123. DOI: 10.7120/09627286.27.2.115

Manlosa, A., Schultner, J., Dorresteijn, I., & Fischer, J. (2019). Capital asset substitution as a coping strategy: Practices and implications for food security and resilience in southwestern Ethiopia. *Geoforum*, 106, 13–23. DOI: 10.1016/j.geoforum.2019.07.022

McLean, A. (2021). Modernizing human-elephant interactions. In Laws, E., Scott, N., Font, X., & Koldowski, J. (eds) *The Elephant Tourism Business*. Wallingford: CAB International, pp. 219–231.

Moberg, G.P. (2000). Biological response to stress. In Moberg, G.P., & Mench, J.A. (eds) *The Biology of Animal Stress. Basic Principles and Implications for Animal Welfare*. Wallingford: CAB International, pp. 1–21.

Mota-Rojas, D., Braghieri, A., Álvarez-Macías, A., Serrapica, F., Ramírez-Bribiesca, E., Cruz-Monterrosa, R., Masucci, F., Mora-Medina, P., & Napolitano, F. (2021). The use of draught animals in rural labour. *Animals*, 11, 2683. DOI: 10.3390/ani11092683

Nóbrega, R.R. (2018). The ethnozoological role of working animals in traction and transport. In Nóbrega, R.R., & Albuquerque, U.P. (eds) *Ethnozoology: Animals in Our Lives* (pp. 339–349). DOI: 10.1016/b978-0-12-809913-1.00018-1

Norkaew, T., Brown, J.L., Bansiddhi, P., Somgird, C., Thitaram, C., Punyapornwithaya, V., Punturee, K., Vongchan, P., Somboon, N., & Khonmee, J. (2018). Body condition and adrenal glucocorticoid activity affects metabolic marker and lipid profiles in captive female elephants in Thailand. *PLOS ONE*, 13(10), e0204965. Epub 2018/10/03. pmid:30278087.

Norkaew, T., Brown, J.L., Thitaram, C., Bansiddhi, P., Somgird, C., Punyapornwithaya, V., Punturee, K., Vongchan, P., Somboon, N., & Khonmee, J. (2019). Associations among tourist camp management, high and low tourist seasons, and welfare factors in female Asian elephants in Thailand. *PLoS ONE*, 14(6), e0218579. DOI: 10.1371/journal.pone.0218579

Norton, B.G. (1984). Environmental ethics and weak anthropocentrism. *Environmental Ethics*, 6(2), 131–148.

Okello, W.O., Muhanguzi, D., MacLeod, E.T., Welburn, S.C., Waiswa, C., & Shaw, A.P. (2015). Contribution of draft cattle to rural livelihoods in a district of southeastern Uganda endemic for bovine parasitic diseases: An economic evaluation. *Parasites and Vectors*, 8, 571. DOI: 10.1186/s13071-015-1191-9

Price, E.O. (2002). *Animal Domestication and Behavior*. Oxon: CABI Publishing.

Pritchard, J.C., Lindberg, A.C., Main, D.C.J., & Whay, H.R. (2005). Assessment of the welfare of working horses, mules and donkeys, using health and behaviour parameters. *Preventive Veterinary Medicine*, 69, 265–283. DOI: 10.1016/j.prevetmed.2005.02.002

Pritchard, J. (2014). What role do working equids play in human livelihoods, and how well is this currently recognised? *Proceedings of the 7th International Colloquium on Working Equids*, 1–3 July 2014, Royal Holloway, University of London, UK.

Pritchard, J.C., Burn, C.C., Barr, A.R.S., & Whay, H.R. (2008). Validity of indicators of dehydration in working horses: A longitudinal study of changes in skin tent duration, mucous membrane dryness and drinking behaviour. *Equine Veterinary Journal*, 40, 558–564. DOI: 10.2746/042516408X297462

Rodrigues, J.B., Sullivan, R.J.E., Judge, A., Norris, S.L., & Burden, F.A. (2020). Quantifying poor working equid welfare in Nepalese brick kilns using a welfare assessment tool. *Veterinary Journal*, 187, 445. DOI: 10.1136/vr.106135

Rushen, J., & de Passillé, A.M. (2010). The importance of good stockmanship and its benefits for the animals. In Grandin, T. (ed.) *Improving Animal Welfare a Practical Approach*. Cambridge: CAB International, Cambridge University Press.

Sasimowski, E. (1984). Management and utilization of equine animals for work. In *Animal Energy in Agriculture in Africa and Asia; FAO Series, Publication No. 42*. Rome, Italy: FAO.

Schuerch, I. (2020). Of horses and men, mules and labourers. Human animal semantics, practices and cultures in the early modern Caribbean. *Iberoamericana*, 73, 13–35.

Short-McKendree, M.G., Croney, C.C., & Olynk-Widmar, N.J. (2014). BIOETHICS SYMPOSIUM II: Current factors influencing perceptions of animals and their welfare. *Journal of Animal Science*, 92, 1821–1831. DOI: 10.2527/jas.2014–7586

Singh, G. (2000). Empowering farmers through animal traction in India. In Kaumbutho, P.G., Pearson, R. A., & Simalenga, T. E. (eds) *Empowering Farmers with Animal Traction. Proceedings of the Workshop of the Animal Traction Network for Eastern and Southern Africa*, September 1999, South Africa.

Starkey, P. (2011). Livestock for traction and transport: World trends, key issues and policy implications. *Animal Production and Health Working Paper*. Animal Production and Health Division, Food and Agricultural Organisation (FAO) of the United Nations. DOI: 10.13140/RG.2.2.29674.34245

Tadich, T. (2020). Working equids: Linking human and animal welfare. *Veterinary Record*, 187, 442–444. DOI: 10.1136/vr.m4572

Tadich, T., de Aluja, A., Cagigas, R., Huerta, L.A., & Galindo, F. (2016). Children's recognition of working donkeys' needs in Tuliman, Mexico: Preliminary observations. *Veterinaria México OA*, 3(4). DOI: 10.21753/vmoa.3.3.404

Tadich, T., Escobar, A., & Pearson, R.A. (2008). Husbandry and welfare aspects of urban draught horses in the south of Chile. *Archivos de Medicina Veterinaria*, 40, 267–273. http://dx.doi.org/10.4067/S0301-732X2008000300007

Tadich, T.A., & Stuardo-Escobar, L.H. (2014). Strategies for improving the welfare of working equids in the Americas: A Chilean example. *Revue Scientifique et Technique*, 33, 203–211. DOI: 10.20506/rst.33.1.2271

Tatemoto, P., Fernandes Lima, Y., Santurtun, E., Reeves, E., & Raw, Z. (2021). Donkey skin trade: Is it sustainable to slaughter donkeys for their skin? *Brazilian Journal of Veterinary Research and Animal Science*, 58(special issue), e174252. DOI: 10.11606/issn.1678-4456.bjvras.2021.174252

Teweldmehidin, M.Y., & Conroy, A.B. (2010). The economic importance of draught oxen on small farms in Namibia's Eastern Caprivi Region. *African Journal of Agricultural Research*, 5(9), 928–34.

The Donkey Sanctuary. (2029). Under the skin: Update on the global crisis for donkeys and the people who depend on them [Internet]. United Kingdom; 2019 Available from: https://www.thedonkeysanctuary.org.uk/sites/uk/files/2019-12/ under-the-skin-report-english-revised-2019.pdf

United Nations. (2015). What are the sustainable development goals? Available from: https://www.undp.org/sustainable-development-goals

Waiblinger, S., Boivin, X., Pedersen, V., Tosi, M.V., Janczak, A.M., Visser, E.K., & Jones, R.B. (2006). Assessing the human-animal relationship in farmed species: A critical review. *Applied Animal Behaviour Science*, 101, 185–242. DOI: 10.1016/j.applanim.2006.02.001

WOAH. (2016). Chapter 7.12. Welfare of Working Equids. Terrestrial Animal Health Code. World Organisation for Animal Health.

Zamir, T. (2017). The moral basis of animal-assisted therapy. In Armstrong, S.J., & Botzler, R.G. (eds) *The Animal Ethics Reader* (3rd ed.). New York: Routledge, Taylor and Francis Group, pp 650–663.

Zeleke, B. (2017). Status and growth trend of draught animals population in Ethiopia. *Journal of Dairy, Veterinary and Animal Research*, 6(1), 00167. DOI: 10.15406/jdvar.2017.06.00167

# 40
# HUMAN-NONHUMAN INTERACTIONS IN THE ANIMAL LABORATORY

*Lesley Sharp, Garet Lahvis, Amy Robinson-Junker, and Larry Carbone*

**Introduction**

Laboratory animals – like the humans in charge of their lives – have subjective experiences and thus can experience pain and pleasure, suffering and joy, boredom and play, and loneliness and sociality (Mellor, 2019). Common animal housing practices and experimental methods that cause animal suffering or constrain their opportunities to live full lives can decrease animal welfare, diminish the validity of scientific studies, and lower the quality of life of human staff. As we call attention to the ways in which current practices work against human and nonhuman thriving and valid, reproducible science, we acknowledge that the ultimate goal of ending animal research is well in the future, and that advances – such as cures for various cancers or vaccines for Covid or other infections – currently depend on animal research.

The modern vivarium, or laboratory animal facility, contains a wide array of human and nonhuman animals. You will find dogs, monkeys, swine, fish, and tens of thousands of genetically modified rodents, plus many other species. The animals reside in whatever cages the humans choose for them, cages that meet the letter of regulations but rarely allow the animals the rich lives these unwilling conscripts of science deserve. The people in the vivarium are similarly varied from students and technicians who directly interact with the animals – feeding them, cleaning cages, conducting experiments, and killing them – to senior scientists directing those activities, with or without ever visiting the vivarium. Frontline staff with direct animal contact are often lower-level staff with minimal authority to control the work they are assigned to perform. Meanwhile, ethics committee members and inspectors may pass through only briefly or periodically to oversee the animals' health and welfare; veterinarians too may enter animal rooms at irregular intervals.

Human-animal interactions in the lab range broadly. Humans often build relationships with individual animals with whom they work or play, especially dogs, cats, and nonhuman primates. They may advocate vigorously that some animals should be found post-research lives in homes or sanctuaries. On a day-to-day basis, these direct interspecies interactions are an essential component of laboratory work, including frequent handling, restraint, as well as killing or euthanizing animals. How humans perform husbandry tasks and research procedures profoundly affects animal welfare, as well as the data derived from the animals.

Managers and lead scientists who interact less directly with animals make far-reaching decisions about what cages to buy, which experiments to run, how many animals a vivarium will house, the number of humans needed to staff it. Scientists and administrators thus exert near-total control over animals whom they may never meet.

We describe here some common practices, flag essential literature, and draw on our own combined work and observations in animal laboratories. We do so as professionals from diverse yet complementary fields, including veterinary medicine and ethics, ethology and neuroscience, animal welfare and behavior, and the anthropology of science. Our shared goal is the effort to promote practices that can improve the experiences of both human and nonhuman animals in laboratories. We begin with some of the most easily implemented interventions that staff can implement right away. We then move to proposals at the institutional and regulatory levels, especially those that may require reconfiguring people's work, redesigning experiments, or investing in animal housing. Many reforms will require institutional commitments which are unlikely without changes in the regulations under which scientists operate. Our proposals rest on the belief that we must always remain alert to the subjective lives of laboratory animals and the constraints that science imposes on them.

Our ideal is that though confined, animals should live in environments that allow them a range of natural behaviors they are intrinsically motivated to perform. They should be able to explore their surroundings, seeking out favored foods, meeting their species-mates, and checking that no dangers are lurking. They should be able to dig or climb or build nests as they choose. In an environment that fosters normal physical and mental development, they should have some control over their lives. If they are relatively content, the data humans acquire from studying them can be of high-quality, not tainted by effects of impoverished housing or stressful handling.

In this laboratory ideal, people understand their animals' behavior and biology. They see them as thinking, feeling individuals with rich subjective inner lives. Staff have the time to work patiently, acclimating the animals to handling and restraint. Research scientists and animal caregivers communicate and respect each other's expertise, collaborating in a shared respect for their animals' needs. Their animals come to know them and do not fear them. When scientists must perform painful procedures or cause disease in their research animals, they inflict the least harm possible, with pain medications and nursing care. If animal suffering exceeds the human ability to ameliorate it, staff will euthanize animals in the gentlest manner. When euthanasia is not necessary, a robust adoption or rehoming program places animals in homes or sanctuaries to live out their post-research lives. Staff at all levels look to the day when replacements for all harmful animal experiments are in hand, and animal research is obsolete. Until then, they take pride that they use animals only in truly important high-quality experiments and that they afford these unconsenting subjects the very best possible care.

## Challenges to Human and Animal Welfare in the Laboratory

Efficiency, economy and hygiene are human-centric values that presently take precedence over animal comfort mental well-being and human-animal interactions in choices of how to house animals (Gaskill & Gouveia, 2020). Regulations set minimum housing standards that we believe are inadequate. Pigs cannot root, mice cannot burrow, and monkeys cannot run and climb and have normal group interactions in enclosures that just barely meet regulations.

Efficiency deprives staff of the time they need to watch their animals and learn their behaviors. A single room may house a thousand mice or more, near identical to the human observer. Hygiene and safety concerns create physical barriers such as sealed individually ventilated cages and layers of protective clothing, even for handling dogs and cats, that discourage the closeness humans and companion animals enjoy together in a home.

Scientific ideology pushes people to regard animals as disposable tools (Rollin, 1989). Narrowly specialized scientists may see their animals for their neurons or immune cells, without appreciating the animal holistically. They may undervalue how the animal's psychology and emotions will affect the biology that they are studying. They may fear that welfare efforts, such as painkillers, enriched environments, or earlier interventions, may hurt their experiments when in reality the opposite is often true (Committee on Recognition and Alleviation of Distress in Laboratory Animals, 2008; Committee on Recognition and Alleviation of Pain in Laboratory Animals, 2009; Elkhoraibi et al., 2019; Wurbel, 2007a).

Workplace culture sometimes discourages showing concern or attachment for the animals or questioning whether a painful experiment is justified. Researchers and animal care staff may find themselves in adversarial relationships when they could be mutually supportive, each informing the other of the needs of the experiment and the needs of the animals. Too often, staff come to believe that it is unprofessional to care too much about the animals. Even to name an animal instead of using an impersonal number seems somehow unscientific (Arluke, 1994; Sharp, 2018).

For over a century, outside activists have campaigned against animal research, resulting in increased regulation of animal welfare in labs. Inside the lab, workers know that not just activists, but other outsiders disapprove of the work they do. Laboratory work can thus be isolating, as workers shy from telling outsiders about their experiences, fearing public disapproval of animal research. Even apart from animal activists' targeted campaigns, workers at all levels may carry a daily sense that they should reveal as little as possible about their work, with likely effects on their emotional health. Scientists may avoid writing of animals' mental lives, knowing that such an acknowledgment raises the ethical cost of their experiments. Or they may shy from conducting studies for treatments of animal pain or distress, as publishing such work invites criticism that their attempts to treat animal pain actually cause pain in the process.

Not all of these barriers to human and animal welfare are inevitable, even given the context in which laboratory animals suffer illness to model human disease. Rather, they are the choices of human actors who can choose instead to devote the resources and change their practices to foster human-animal thriving.

## HAI Impacts on Animal welfare and Animal Data in Laboratories

The human-controlled animal environment affects animals' welfare but also causes physiological stresses that negatively affect the quality of data in animal experiments (Gaskill & Gouveia, 2020). For example, comfortable temperatures for a clothed human feel cold to a mouse given insufficient material to build a cozy nest (Gaskill et al., 2013). Cold stress is both a welfare concern and a physiological stressor that can affect tumor growth in cancer studies and glucose metabolism in obesity studies (Hylander et al., 2022). Humans, comfortable in these environments, may not even notice their animals' discomfort.

Animals' environments affect their cognitive and emotional development. A standard cage severely restricts opportunities to explore, to have control of one's social approach or avoidance, to segregate functions like nest-building and toileting, to thermoregulate, to solve cognitive challenges, or even to stand upright on hind limbs (Makowska et al., 2019;

Makowska & Weary, 2020). Mice in "conventional housing" versus larger, more diverse and socially rich environments have shorter life spans and more severe disease in a range of medical research fields (Cait et al., 2022).

The data/welfare overlap is particularly strong in experiments where scientists are observing animals' behaviors to understand how their brains work. To study animal behaviors of pain, anxiety, or fearfulness, scientists unwittingly put animals into conventional lab cages that provoke anxiety, or they handle the animals in ways that frighten or stress them (Lahvis, 2017; Wurbel, 2001). These situations affect cognitive development as well as emotions. Thus, constricted, impoverished environments that people choose for their animals not only diminish their welfare but make them less normal and useful as research subjects.

Animal welfare and data quality do not always align. Unfortunately, many experiments cause significant suffering, but efforts to improve welfare could interfere with the experiment. Some 7% of experiments reported annually in the United States are classed as causing significant, unalleviated pain or distress (Carbone, 2021). Examples include placebo-treated animals in pain studies, animals in experiments on advanced cancer, or animals used to model psychiatric conditions such as depression or learned helplessness.

From a laboratory animal's perspective, every experience is mediated by humans. From birth to death, humans dictate their standard of living, and their physical and social environments. Human presence is rarely neutral for caged animals (Davis & Balfour, 1992). One person may bring food, or the chance for play and exercise. Another person may bring pain, restraint, distress, or fear. Animals remain vigilant about human approach, even listening for human activity outside their room. Standard practice includes daily checks on the animals, whether the animals want that well-meaning intrusion into their lives, or actually even benefit from it. Occasionally, animals have the control and are given the time to initiate contact with people if they choose; far more often, people initiate the contact, forcing their own schedule on the animals. At the minimum, people should take the time to gently acclimate animals to whatever restraint or manipulation is necessary.

Laboratory animals benefit from patient, gentle handling, even if rough restraint seems the quicker and surer way to move an animal or collect a research sample. Mice handled by coaxing into a tube rather than the traditional tail grab method demonstrate less anxious behavior (Gouveia & Hurst, 2019). Similarly, rats who have been "tickled" (which mimics play with another rat) are quicker to approach a human hand, even after an aversive procedure (Cloutier et al., 2018). With proper training, nonhuman primates, dogs, and pigs will willingly approach the researcher and submit to a mildly painful blood sample, or present for their person to hook a leash to their collar (Allen et al., 2020). Training monkeys to cooperate with blood sample collection (rather than squeezing them against the cage wall for an anesthetic injection) yields more accurate blood glucose measurements, a crucial necessity in diabetes research (Graham et al., 2012). Research on the effects of acclimation, training, and other human-animal interactions with other laboratory species such as birds, cephalopods, amphibians, or fish is presently lacking.

"Do laboratory animals know their individual people?" is a serious and important welfare question. Most animal species seem capable of recognizing individual humans and have preferences to associate with some and avoid others (Anderson et al., 2010; Davis, 2002; Davis & Balfour, 1992). Anecdotally, one of us managed a colony of marmoset monkeys, who experienced a month-long colony-wide drop in body weight, and no evident medical conditions. Once their familiar daily caregiver returned from a long absence, their appetites and weights rebounded. Individual monkeys display profoundly different behaviors toward different humans; one human may be consistently greeted with a threat display while another is solicited for a grooming session. Likewise with dogs and cats, signs of stress are lower when their familiar human is present (Behnke et al., 2021; Stellato et al., 2020).

Not all animals can have the same familiar humans every day, as evenings, weekends, and holidays disrupt schedules. To partially make up for this, lab members and caregivers can coordinate to use the same handling and training practices. Even so, animals may characterize people into groups, with greater affinity for – or fear of – one sex or another, people in lab coats versus scrubs, or other factors. Male scientists (and women scientists wearing a male colleague's shirt) appear to cause a stress-induced analgesia in mice; essentially, the animals are scared enough in the presence of male pheromones that they temporarily feel less pain (Sorge et al., 2014). This phenomenon has both animal welfare implications and bears evidence of skewing data outcomes when male versus female researchers conduct their studies. Imagine how the quest for new painkillers suffers if a researcher's mere presence can disrupt the pain under study.

Likewise, humans come to know and care about individual animals, treating some individuals as laboratory mascots, or working to find post-laboratory homes for favorite animals, even adopting a familiar animal themselves. Adoption programs can be synergistic for human and animal welfare, and some states in the US-mandate adoption programs for healthy post-research dogs and cats who meet certain criteria. Adoption programs require institutional resources, including time and veterinary care, and with no legal pressures on administrations to provide them, they are often bottom-up staff-driven programs (Carbone et al., 2003; Martin, 2017). With this reality in mind, an important reform could entail including the cost of rehoming – and ongoing support of care costs for involved animals – as an integral part of any research proposal.

The "Three Rs" is a useful framework for pursuing alternatives to harmful animal experiments (Russell & Burch, 1959). Scientists should seek to Replace sentient animals with cells or other non-sentient material and Reduce the numbers of animals that they use to the minimum necessary. Scientists should Refine animal care and use to minimize pain and distress and, we argue, to maximize positive experiences for the animals (Makowska & Weary, 2020). The Three Rs focus on harm reduction, not on the justifications of using animals in experiments. That requires an ethical consideration of the potential benefits of research, not just the harms (Beauchamp & DeGrazia, 2019). A Three Rs analysis should recognize that the minimum requirements of current regulations may cause outright suffering and certainly limit the positive experiences of their animals.

Ethologists study how wild animals make complex decisions throughout the day – averting dangers, finding foods, raising their young, choosing mates. Their environments are dynamic and responsive; animals are driven to seek a safe, new nesting site, a source of food, social interactions, and in the right environment, they can find some success (Markowitz & Timmel, 2005; Watters et al., 2019). Laboratory scientists confine animals to cages that filter out the variation, the exploration, the satisfactions, and then all too often, veterinarians and other laboratory personnel come to see these impoverished environments as normal for the animals. Attempts to refine animal care with small "enrichments" such as extra nesting material or chew toys do not approach the rich lives of animals in their natural state (Burghardt, 1999). We suggest that a full understanding of animals' mental capacities and needs, shared among scientists, veterinarians, and animal caregivers would lead all to reject the current standards and strive for animal environments that maximize animals' opportunities in their confined lives (Makowska & Weary, 2020).

## Animal-Animal Interactions and their Effects on Welfare and Data Outcome

Animals' interactions with people are episodic, seconds or hours per day. By contrast, animals' social lives – whether living in groups or in solitary confinement – are full time. In nature, animals may come together to feed or sleep, disperse to forage, and control their distance from threatening conspecifics. Animal-animal interactions are important to them, but in most labs, they have no control over their social environment. All too often, the choices are either enforced togetherness with minimal individual flight zones; or enforced isolation, far from the natural situation in which social groups' size and composition fluctuates.

In a large and complex environment, animals can escape hostile or agonistic encounters. In a standard lab cage, they cannot. Outright fights are sometimes dramatic and obvious, requiring separating rodents, rabbits, and others to solitary enclosures, suturing dogs after dog fights or digit amputations after monkey fights. Animal-on-animal hostilities can be clandestine, without obvious wounds, seen only with round-the-clock video monitoring, but stressful to the subordinate animal nonetheless (Giles et al., 2018). Hostilities can even occur among singly caged animals within a room, in settings where a subordinate monkey, for example, cannot escape the watchful glare of a dominant animal.

Placing animals into solitary caging is an unsatisfactory response to animal aggression, with loneliness, frustration, and boredom as the potential welfare costs of solitary caging. Behavior specialists have strategies for gradual increases in contact with conspecifics and for arranging cage space to allow animals to escape hostile or stressful relationships. Unfortunately, few research institutions have trained experts on hand to guide behavioral management of animals. Instead, vivaria may rely on veterinarians or technicians with neither the training nor the time for in-depth behavior evaluations. We encourage creating and funding such a role in research vivaria, investing authority in it, and having the behaviorist report independently to the animal ethics committee rather than to cost-conscious facility management.

Interaction with known, trusted people may partially compensate for animals' social isolation, but often, human approach can be just one more stressor in animals' lives. With time and patience, many animals acclimate to human handling and may even welcome it. Individual laboratory animals of many species actively choose to approach and engage with their people. For animals living in social isolation, human interaction can be beneficial, but in institutions that house hundreds of individually caged animals, giving those animals more than even a few seconds a day of such attention would not fit with most current workplace practices and caregiver-to-animal ratios.

Though animal-animal interactions affect research outcomes, scientists are not always careful to maintain standardized stable social environments even within a single experiment, or to report social conditions in their scientific manuscripts. Stressful social isolation, for instance, can be a confounding factor that leads to different cancer rates in rodents compared to animals housed in compatible groups (Hermes et al., 2009). Scientists should factor in these potentially confounding differences as they design their experiments, including mid-experiment changes in social grouping if animals are removed from the experiment. Reporting housing conditions in scientific manuscripts allows for more accurate attempts to replicate published results (Percie du Sert et al., 2020; Wurbel, 2007b).

Animals can serve as social buffers, or as support animals, for other animals undergoing painful or stressful experiments. The general benefits of emotional resiliency in having another animal present are enhanced when that other individual is a familiar cage mate with whom one has bonded (Denommé & Mason, 2022). Perhaps known and trusted humans could similarly be sources of emotional support for their animals rather than sources of fear.

## HAI Impacts on Human Welfare in Laboratories

Interspecies interactions shape the well-being of lab personnel, not just the animals with whom they work. Staff will often develop strong attachments to individual animals, or more generally, to whole species (Sharp, 2018). Efforts to exceed minimum requirements for animal care affect not only animal well-being, but also the individual person's sense of self-worth. That sense of self-worth is, in turn, shaped by a sense that a person can make a positive difference.

Humans perform a variety of jobs in the laboratory, and their labor is inevitably hierarchical. These hierarchies are evident in the flow of authority and associated knowledge, research decisions and practices, and how different personnel interact with the laboratory animals. The tasks performed, time spent, and responsibilities toward animals of a lab supervisor, postdoctoral student, animal caregiver, veterinarian, or undergraduate intern will vary dramatically. Each reasonably claims animal expertise; yet how, and by whom, decisions are made about animal care reflects how their work and knowledge is valued. A caregiver may possess deep knowledge of individual animals' daily behaviors, preferences, and quirks, but their exclusion from research team discussions flags a devaluing of their expertise within a lab's labor hierarchy. Animal caregivers report higher job satisfaction when scientists show they value the caregivers' expertise in animal care. Supervisors are crucial to ensuring that the needs – and knowledge – of all personnel are integrated into laboratory work (Kelly, 2015).

Less well studied are the job satisfactions and stresses of the researcher scientists, including junior scientists in training. They must live with their decisions to conduct painful animal experiments, and those with the most direct animal contact and least authority to decide animals' fates strike us as particularly susceptible to animal-related distress. Veterinarians also face workplace stressors, which can vary with their seniority, experiences, and place in the veterinary team's hierarchy. They may experience moral stress, caught between staff pushing them to advocate for animals and researchers expecting them to do their all to keep the animal projects moving forward (Carbone, 2022; Engel et al., 2020).

Killing animals can be a stressful task, often relegated to non-research staff, such as caregivers or entry-level interns. Euthanasia, or killing animals, is often one of the first tasks a newly hired person must master and may even function as an initiation of sorts (Arluke, 1994; Friese & Clarke, 2012; Smith, 1995). An elaborate set of euphemisms for killing proliferates in animal labs, such as describing animal death as the "sacrifice" (or, in abbreviated form, the "sac[k]ing") of animals (Scotney et al., 2015; Sharp, 2018). "Euthanasia" itself is a euphemism insofar as its use implies mercy killing for suffering animals, when the animals might be perfectly healthy and capable of a long post-laboratory life (AVMA Panel on Euthanasia, 2020; Carbone, 2013). Language matters: euphemisms are powerful tools for discouraging talk among staff about the associated hardships of interspecies work. Killing is the "animal caregiver's burden." Staff need to be able to speak openly of their hardships in ending animals' lives (Birke et al., 2007).

Compassion fatigue, first recognized in overburdened nurses as the loss of an ability to nurture is a concern among workers in lab animal care (Joinson, 1992; LaFollette, 2020; LaFollette et al., 2019; Rank et al., 2009). We urge anticipating and minimizing the causes of emotional burnout and associated ethical regret (Gluck, 2016; Randall et al., 2021). Mental health professionals should be available for animal workers, and should develop preventive programs for workers' emotional health as occupational health physicians guide physical injury and illness prevention programs.

One tool to promote human welfare, ironically perhaps, is to put animals first as a form of self-conscious practice, as opposed to merely following standardized procedures, internalizing a moral commitment to the animals in their charge (Bertelsen & Hawkins, 2020). People self-report feeling better if they put time and effort into naming their animals, talking to them, petting them, or generally having frequent positive human-animal interactions, even if those animals are destined for euthanasia (LaFollette et al., 2020).

Scientists at all levels of their training might improve their animal care and their own welfare by trying to understand the experiences of their animals. Watching videos of "their" species in free-ranging settings, and quietly watching their animals in the lab, can improve their empathy and understanding of the animals they study, even if their studies are primarily at the microscope. Watching rats in videos play, climb, and forage supplies a dramatic counterpoint to laboratory housing and experiences, and might motivate scientists to push for richer lives for the animals. One of the present authors taught writing classes for science students, encouraging journaling as a tool for students to reflect on their role as animal experimenter. Animal behaviorists can guide learning to think like an animal as another tool to foster empathetic animal care and human welfare.

Mutual respect and support are essential. Lab directors could invite staff across the full spectrum of roles, daily animal caregivers included, to staff meetings and journal clubs. This approach instills among all staff a greater investment in the science and commitment to the team, while also helping researchers to see animal care staff as important partners and animal lives as integral to every experiment they discuss (Sharp, 2018). A separation between the work of science and the work of animal care is unhelpful.

Management must be alert to, and compassionate toward, the emotional burdens of animal work among all members of a lab community (Pekow, 1994). Shaming or discrediting human expressions of hardship, or egging co-workers to "toughen up," erodes the integrity of ethical research and detrimentally affects humans and animals in a lab. Animal death and animal killing are frequent in laboratories and vivaria and can be hard on humans performing or observing. Some labs approach this through memorial practices that recognize the work performed by deceased animals: these range from bulletin boards that display portraits of favorite animals, to gatherings reminiscent of wakes, to annual ceremonies that commemorate animal lives of all species (Judy MacArthur Clark, 2019; Sharp, 2018). It is important that all staff are invited as participants in these efforts and events.

## Current Regulation and Policy in the United States

Regulation of laboratory animal welfare in the United States is a confusing patchwork. Two federal laws set standards, though each has exclusions. No agency tracks how many animals are in US laboratories nor how many are outside of federal welfare protections (Carbone, 2021). The *Guide for the Care and Use of Laboratory Animals* sets welfare standards, though only for federally funded research, while the Animal Welfare Act (AWA) covers all facilities, but excludes mice and rats, overwhelmingly the most common research mammals (Institute for Laboratory Animal Research, 2011; USDA Animal Care, 2019). Despite the many animals excluded, the AWA and the *Guide*'s standards for animal and human welfare deserve attention.

For nonhuman primates, but only for nonhuman primates, the Animal Welfare Act requires provisions for psychological well-being (USDA Animal Care, 2019). The requirements lack specificity but do allow that "interaction with the care giver or other familiar and knowledgeable person" could have a place, especially if they are being caged singly. Otherwise, this law is mostly silent on human-animal interactions.

The 2011 edition of the *Guide* reflects greater and more up-to-date awareness than the AWA that human interactions can affect animals of various species and it encourages training and acclimation rather than forced handling for experimental procedures. It notes that "Dogs, cats, rabbits, and many other animals benefit from positive human interaction," though it does not require such interactions, and that, for singly housed animals, "enrichment should be offered such as positive interaction with the animal care staff" (Institute for Laboratory Animal Research, 2011). These are suggested practices that are not requirements in the *Guide*.

Neither the *Guide* nor the AWA discuss effects of HAI on people. Requirements for occupational health programs apply only to physical conditions, such as ergonomics and exposure to infections, chemicals, and allergens. The *Guide* makes no mention of human mental health and well-being, with the sole exception of its mention of the psychological distress of euthanizing animals, and the need for supervisor sensitivity to this issue (Institute for Laboratory Animal Research, 2011). It mandates no protections for workers such as those who wish to avoid killing or hurting animals with whom they have bonded. Current US law thus puts minimal pressure on institutions to promote positive HAI for animals, and even less to promote positive HAI for workers.

These regulations set minimum standards for animal care, which institutions are free to exceed. Staff may push for HAI initiatives, but if those cost money or staff time, they will face administrator resistance. Two of the present authors who worked together encountered this resistance when promoting social housing for colony sentinel mice and cozy nesting material for all the mice, with veterinarian administrators placing cost above animal welfare, even to the point of requesting an ethics committee exemption from standards. The failure of our "middle up" efforts (one veterinarian/ethics committee coordinator and one behavior and enrichment specialist) illustrates yet another barrier to human and animal thriving in the laboratory: the conflict of roles of institutional veterinarians. Whereas vets have an obligation to promote animal health and welfare, often they are the directors of the vivarium as well, with an institutional responsibility to contain costs (Carbone, 2022). When innovations that go beyond the minimal standards of the regulations push up against the financial spreadsheet, they may not be able to move past veterinary campus-wide leadership. If and when an update to the AWA or the *Guide* dictates programs for employee mental health support or for more animal-centric welfare standards, institutions will need to comply.

## Recommendations

We have described some of the complexities of animal-human interactions in labs, with suggestions to improve the welfare of both human and nonhuman animals. Greater attention to human-animal interactions can improve not just animal and human welfare, but also the quality of experimental data. We have emphasized that human-animal interactions include direct encounters, but of utmost importance are the systemic decisions humans make on how to care for animals, perform experiments, and structure the human workplace.

Many of our suggestions could be implemented immediately, with institutional buy-in. Everyone who handles mice can switch right now from grabbing them by the tail to gentler restraint methods. Research teams can begin today to bring their own lab techs and animal care team together into lab meetings, making animal care an integral part of all animal research and a shared responsibility of all personnel.

Animals and their carers will be in laboratories for decades to come. Covid vaccines, alongside many other medical advances, are testimony to the value of animal testing though many critics, including many scientists, point out disappointing rates of translational success in bringing drugs from animal models to the human population (Fernandez-Moure, 2016; Garner et al., 2017). We believe that more holistic attention to animals as subjects could improve science outcomes and that better attention to animal welfare improves the harm-benefit balance of animal experiments. Finally, we argue that the quality of life for animals in labs bears profound implications for the humans who work with, manage, care for, and euthanize them.

Scientists should take a leadership role, individually and in their disciplines' journals and governing bodies. When they see their animals holistically, they will want higher standards of care than what the regulations mandate. They will abandon passive-voice narratives, taking responsibility as "I" or "We" agents when describing their animal methods in their journal articles. They will thank their animals in their Acknowledgments, not just listing them in their Materials and Methods with cells and chemicals. Journals will require descriptions of steps taken to enrich lab animals' lives. Scientists will train their students, no matter how reductionist their experiments on cells or chemicals, to sit and watch and learn about their animals as complex, thinking, feeling, and social creatures.

Research institutions will not lead the way to better HAI if improvements require money or staff time. College campuses and private companies seem to see no huge pay-off in aggressively improving, and publicizing, their animal welfare. Rather, they rely on public statements that they are accredited for their adherence to regulatory standards, or to general claims that they endorse the "Three Rs" framework of alternatives. Thus, regulatory standards must change if institutions are to change, and both the AWA and the *Guide* are in need of updates.

We begin with animals' needs. The *Guide* should update its professional guidance, at the very least, on animal housing. "Recommended" cages sizes for many species remain unchanged since 1968 and no institution will invest in – and no cage manufacturer will build – bigger or better enclosures, at least for mice and rats (Institute for Laboratory Animal Research, 2011; Institute of Laboratory Animal Resources, 1968). Current "recommended" minimum cage sizes do not meet the *Guide*'s own performance criteria of being large enough to avoid aggression or for animals to rest apart from their feces and urine. Performance standards for evaluation should be clear and measurable and should reflect what ethologists find to be important behaviors for animals to perform. An updated version of the *Guide* should address what types of experiment, if any, warrant solitary housing (Denommé & Mason, 2022; Gaskill & Gouveia, 2020). It should be explicit that picking animals up by the tail is inappropriate (Gouveia & Hurst, 2019).

Veterinarians have a mandated role as advisers on animal health and as voting members of the ethics committee. It is time to include animal welfare scientists as required members as well. We would avoid placing them under facility directors' oversight and budgets, given the conflicting interests we have seen when the same veterinarian is in charge of animal welfare and in charge of cost containment as vivarium director. Staff training programs should be under behaviorists' guidance. Accreditation of zoos and aquaria requires a program for annual welfare assessment of every animal (Association of Zoos and Aquariums, 2022; Watters et al., 2019). Laboratories should follow their lead, with a behaviorist monitoring animals for abnormal behaviors, adequate environments, and reaction to human approach, as starters, with documented audits and improvement plans submitted to the ethics committee.

We also flag human needs. The *Guide* has always mandated occupational health and safety programs for staff, with an exclusive focus on physical health, not mental health. The *Guide* should require a mental health program for all researchers and care staff. Staff training programs should include education on how to prevent, recognize, alleviate, and seek help for distress and incipient burnout.

## Conclusion

Finally, we call for a reset of our human interactions with animals in laboratories, at the individual and the institutional level. Animal research is a powerful tool and scientists need not fear that society will take that tool away any time soon. But scientists can and should take steps now to improve their stewardship of vulnerable animals, and institutions should reenforce this. Right now, we ask that every person in a lab or vivarium to pause, take some time to simply *be* in their animals' presence. Watch them. Appreciate their and your own subjectivity, the activities that you are each motivated to pursue, as you watch each other through their cages' walls. Reflect on what you require of them, giving them no chance for them to opt out of the environments and experiments you impose on them.

## Questions

- Do you think that activities like "rat tickling," providing enrichment to animals, or naming them make it easier or harder for staff to work with them, knowing that they may be euthanized at the end of a study?
- Whether you are for or against animal research, what information or changes in policy would change your perspective? Why?
- What do you think is the most important aspect of human-animal interaction in the lab, from both the human and animal perspectives? Why?
- If you work in science, what barriers limit the extent to which you see your study animals as thinking, feeling individuals?

## Acknowledgments

Authors Sharp, Lahvis, and Carbone received support from the Brooks McCormick Jr. Animal Law & Policy Program, Harvard Law School.

## References

Allen, C., Sórensen, D. B., & Ottesen, J. L. (2020). The Dog. In B. Gaskill & S. Cloutier (Eds.), *Animal-centric Care and Management* (pp. 149–162). CRC Press.

Anderson, R. C., Mather, J. A., Monette, M. Q., & Zimsen, S. R. (2010). Octopuses (*Enteroctopus dofleini*) recognize individual humans. *J Appl Anim Welf Sci, 13*(3), 261–272. https://doi.org/10.1080/10888705.2010.483892

Arluke, A. (1994). The ethical socialization of animal researchers. *Lab Anim, 23*(6), 30–35.

Association of Zoos and Aquariums. (2022). *Accreditation Basics*. Retrieved March 8 from https://www.aza.org/becoming-accredited?locale=en

AVMA Panel on Euthanasia. (2020). *AVMA Guidelines for the Euthanasia of Animals: 2020 Edition*. American Veterinary Medical Association. www.avma.org/KB/Policies/Documents/euthanasia.pdf

Beauchamp, T., & DeGrazia, D. (2019). *Principles of Animal Research Ethics*. Oxford University Press.

Behnke, A. C., Vitale, K. R., & Udell, M. A. R. (2021). The effect of owner presence and scent on stress resilience in cats. *Applied Animal Behaviour Science, 243*, 105444. https://doi.org/10.1016/j.applanim.2021.105444

Bertelsen, T., & Hawkins, P. (2020). A Culture of Care. In B. Gaskill & S. Cloutier (Eds.), *Animal-centric Care and Management* (pp. 15–29). CRC Press.

Birke, L., Arluke, A., & Michael, M. (Eds.). (2007). *The Sacrifice: How Scientific Experiments Transform Animals and People*. Purdue University Press.

Burghardt, G. M. (1999). Deprivation and enrichment in laboratory animal environments. *J Appl Anim Welf Sci, 2*(4), 263–266. https://doi.org/10.1207/s15327604jaws0204_1

Cait, J., Cait, A., Scott, R. W., Winder, C. B., & Mason, G. J. (2022, Jan 13). Conventional laboratory housing increases morbidity and mortality in research rodents: Results of a meta-analysis. *BMC Biol, 20*(1), 15. https://doi.org/10.1186/s12915-021-01184-0

Carbone, L. (2013). Euthanasia and Laboratory Animal Welfare. In K. Bayne & P. Turner (Eds.), *Laboratory Animal Welfare* (pp. 157–169). Academic Press, Elsevier.

Carbone, L. (2021). Estimating mouse and rat use in American laboratories by extrapolation from Animal Welfare Act-regulated species. *Scientific Reports, 11*(493). https://doi.org/https://doi.org/10.1038/s41598-020-79961-0

Carbone, L. (2022). Laboratory Animals. In B. Kipperman & B. E. Rollin (Eds.), *Ethics in Veterinary Practice: Balancing Conflicting Interests*. Wiley Blackwell.

Carbone, L., Guanzini, L., & McDonald, C. (2003, October). Adoption options for laboratory animals. *Lab Anim, 32*(9), 37–41.

Cloutier, S., LaFollette, M. R., Gaskill, B. N., Panksepp, J., & Newberry, R. C. (2018, May 8). Tickling, a technique for inducing positive affect when handling rats. *J Vis Exp*(135). https://doi.org/10.3791/57190

Committee on Recognition and Alleviation of Distress in Laboratory Animals, I. f. L. A. R., National Research Council. (2008). *Recognition and Alleviation of Distress in Laboratory Animals*. The National Academies Press.

Committee on Recognition and Alleviation of Pain in Laboratory Animals, N. R. C. (2009). *Recognition and Alleviation of Pain in Laboratory Animals*. The National Academies Press

Davis, H. (2002). Prediction and preparation: Pavlovian implications of research animals discriminating among humans. *ILAR J, 43*(1), 19–26. https://doi.org/10.1093/ilar.43.1.19

Davis, H., & Balfour, D. (1992). The Inevitable Bond. In H. Davis & D. Balfour (Eds.), *The Inevitable Bond* (pp. 1–6). Cambridge University Press.

Denommé, M. R., & Mason, G. J. (2022, Jan 1). Social buffering as a tool for improving rodent welfare. *J Am Assoc Lab Anim Sci, 61*(1), 5–14. https://doi.org/10.30802/AALAS-JAALAS-21-000006

Elkhoraibi, C., Robinson-Junker, A., Alvino, G., & Carbone, L. (2019). The Mental Health of Laboratory Animals. In F. D. McMillan (Ed.), *Mental Health and Well-Being in Animals, 2nd edition* (pp. 277–290). CABI. https://www.google.com/books/edition/Mental_Health_and_Well_being_in_Animals/Lge9DwAAQBAJ?hl=en&gbpv=1

Engel, R. M., Silver, C. C., Veeder, C. L., & Banks, R. E. (2020, Mar 1). Cognitive dissonance in laboratory animal medicine and implications for animal welfare. *J Am Assoc Lab Anim Sci, 59*(2), 132–138. https://doi.org/10.30802/AALAS-JAALAS-19-000073

Fernandez-Moure, J. S. (2016). Lost in translation: The gap in scientific advancements and clinical application. *Front Bioeng Biotechnol, 4*, 43. https://doi.org/10.3389/fbioe.2016.00043

Friese, C., & Clarke, A. E. (2012). Transposing bodies of knowledge and technique: Animal models at work in reproductive sciences. *Soc Stud Sci, 42*(1), 31–52. https://journals.sagepub.com/doi/pdf/10.1177/0306312711429995

Garner, J. P., Gaskill, B. N., Weber, E. M., Ahloy-Dallaire, J., & Pritchett-Corning, K. R. (2017, Mar 22). Introducing therioepistemology: The study of how knowledge is gained from animal research. *Lab Anim, 46*(4), 103–113. https://doi.org/10.1038/laban.1224

Gaskill, B. N., Gordon, C. J., Pajor, E. A., Lucas, J. R., Davis, J. K., & Garner, J. P. (2013, Feb 17). Impact of nesting material on mouse body temperature and physiology. *Physiol Behav, 110–111*, 87–95. https://doi.org/10.1016/j.physbeh.2012.12.018

Gaskill, B. N., & Gouveia, K. (2020). The Mouse. In B. Gaskill & S. Cloutier (Eds.), *Animal-centric Care and Management* (pp. 103–119). CRC Press.

Giles, J. M., Whitaker, J. W., Moy, S. S., & Fletcher, C. A. (2018, Apr 18). Effect of environmental enrichment on aggression in BALB/cJ and BALB/cByJ mice monitored by using an automated system. *J Am Assoc Lab Anim Sci*. https://doi.org/10.30802/AALAS-JAALAS-17-000122

Gluck, J. P. (2016). *Voracious Science and Vulnerable Animals. A Primate Scientist's Ethical Journey*. University of Chicago Press.

Gouveia, K., & Hurst, J. L. (2019, Dec 30). Improving the practicality of using non-aversive handling methods to reduce background stress and anxiety in laboratory mice. *Sci Rep, 9*(1), 20305. https://doi.org/10.1038/s41598-019-56860-7

Graham, M. L., Rieke, E. F., Mutch, L. A., Zolondek, E. K., Faig, A. W., Dufour, T. A., Munson, J. W., Kittredge, J. A., & Schuurman, H. J. (2012, Apr). Successful implementation of cooperative handling eliminates the need for restraint in a complex non-human primate disease model. *J Med Primatol, 41*(2), 89–106. https://doi.org/10.1111/j.1600-0684.2011.00525.x

Hermes, G. L., Delgado, B., Tretiakova, M., Cavigelli, S. A., Krausz, T., Conzen, S. D., & McClintock, M. K. (2009, Dec 29). Social isolation dysregulates endocrine and behavioral stress while increasing malignant burden of spontaneous mammary tumors. *Proc Natl Acad Sci U S A, 106*(52), 22393–22398. https://doi.org/10.1073/pnas.0910753106

Hylander, B. L., Repasky, E. A., & Sexton, S. (2022). Using mice to model human disease: Understanding the roles of baseline housing-induced and experimentally imposed stresses in animal welfare and experimental reproducibility. *Animals, 12*(3), 371. https://www.mdpi.com/2076-2615/12/3/371

Institute for Laboratory Animal Research, N. R. C. (2011). *Guide for the Care and Use of Laboratory Animals* (8th ed.). National Academies Press.

Institute of Laboratory Animal Resources, N. R. C. (1968). *Guide for Laboratory Animal Facilities and Care* (3rd ed.). National Institutes of Health.

Joinson, C. (1992, Apr). Coping with compassion fatigue. *Nursing, 22*(4), 116, 118–119, 120. https://www.ncbi.nlm.nih.gov/pubmed/1570090

Judy MacArthur Clark, P. C., Wendy Jarrett, Cynthia Pekow. (2019). Communicating about animal research with the public. *ILAR Journal, 60*(1), 34–42. https://doi.org/10.1093/ilar/ilz007

Kelly, H. (2015). Overcoming compassion fatigue in the biomedical lab. *ALN*, 12–15. https://helenkellywriting.files.wordpress.com/2015/08/overcoming-compassion-fatigue-in-the-biomedical-lab.pdf

LaFollette, M. (2020). Human-Animal Interaction. In B. Gaskill & S. Cloutier (Eds.), *Animal-centric Care and Management*. CRC Press.

LaFollette, M. R., Cloutier, S., Brady, C., Gaskill, B. N., & O'Haire, M. E. (2019). Laboratory animal welfare and human attitudes: A cross-sectional survey on heterospecific play or "rat tickling." *PLoS One, 14*(8), e0220580. https://doi.org/10.1371/journal.pone.0220580

LaFollette, M. R., Riley, M. C., Cloutier, S., Brady, C. M., O'Haire, M. E., & Gaskill, B. N. (2020). Laboratory animal welfare meets human welfare: A cross-sectional study of professional quality of life, including compassion fatigue in laboratory animal personnel. *Front Vet Sci, 7*, 114. https://doi.org/10.3389/fvets.2020.00114

Lahvis, G. P. (2017, Jun 29). Unbridle biomedical research from the laboratory cage. *Elife, 6*. https://doi.org/10.7554/eLife.27438

Makowska, I. J., Franks, B., El-Hinn, C., Jorgensen, T., & Weary, D. M. (2019, Apr 16). Standard laboratory housing for mice restricts their ability to segregate space into clean and dirty areas. *Sci Rep, 9*(1), 6179. https://doi.org/10.1038/s41598-019-42512-3

Makowska, I. J., & Weary, D. M. (2020). A good life for laboratory rodents? *ILAR J*. https://pubmed.ncbi.nlm.nih.gov/11657070/

Markowitz, H., & Timmel, G. B. (2005). Animal Well-Being and Research Outcomes. In F. D. McMillan (Ed.), *Mental Health and Well-Being in Animals* (pp. 277–283). Blackwell Publishing.

Martin, L. (2017). Beagle freedom laws: Finding homes for cats and dogs after life in the laboratory. *Anim Law, 13*–16.

Mellor, D. J. (2019, Jul 13). Welfare-aligned sentience: Enhanced capacities to experience, interact, anticipate, choose and survive. *Animals (Basel), 9*(7). https://doi.org/10.3390/ani9070440

Pekow, C. (1994). Suggestions from research workers for coping with research animal death. *Lab Anim, 23*(10), 28–29.

Percie du Sert, N., Hurst, V., Ahluwalia, A., Alam, S., Avey, M. T., Baker, M., Browne, W. J., Clark, A., Cuthill, I. C., Dirnagl, U., Emerson, M., Garner, P., Holgate, S. T., Howells, D. W., Karp, N. A., Lazic, S. E., Lidster, K., MacCallum, C. J., Macleod, M., Pearl, E. J., Petersen, O. H., Rawle, F., Reynolds, P., Rooney, K., Sena, E. S., Silberberg, S. D., Steckler, T., & Wurbel, H. (2020, Jul). The ARRIVE guidelines 2.0: Updated guidelines for reporting animal research. *PLoS Biol, 18*(7), e3000410. https://doi.org/10.1371/journal.pbio.3000410

Randall, M. S., Moody, C. M., & Turner, P. V. (2021, Jan 1). Mental wellbeing in laboratory animal professionals: A cross-sectional study of compassion fatigue, contributing factors, and coping mechanisms. *J Am Assoc Lab Anim Sci, 60*(1), 54–63. https://doi.org/10.30802/AALAS-JAALAS-20-000039

Rank, M. G., Zaparanick, T. L., & Gentry, J. E. (2009). Nonhuman-animal care compassion fatigue. *Best Pract Ment Health, 5*(2), 40–61.

Rollin, B. E. (1989). *The Unheeded Cry: Animal Consciousness, Animal Pain and Science*. Oxford University Press.

Russell, W. M. S., & Burch, R. L. (1959). *The Principles of Humane Experimental Technique*. Methuen & Co. Ltd.

Scotney, R. L., McLaughlin, D., & Keates, H. L. (2015, Nov 15). A systematic review of the effects of euthanasia and occupational stress in personnel working with animals in animal shelters, veterinary clinics, and biomedical research facilities. *J Am Vet Med Assoc, 247*(10), 1121–1130. https://doi.org/10.2460/javma.247.10.1121

Sharp, L. A. (2018). *Animal Ethos: The Morality of Human-Animal Encounters in Experimental Lab Science*. Univserity of California Press.

Smith, M. (1995). Rats: The road to her PhD was paved with dead rodents. *The Utne Reader*, (May–June), 85–87.

Sorge, R. E., Martin, L. J., Isbester, K. A., Sotocinal, S. G., Rosen, S., Tuttle, A. H., Wieskopf, J. S., Acland, E. L., Dokova, A., Kadoura, B., Leger, P., Mapplebeck, J. C., McPhail, M., Delaney, A., Wigerblad, G., Schumann, A. P., Quinn, T., Frasnelli, J., Svensson, C. I., Sternberg, W. F., & Mogil, J. S. (2014, Jun). Olfactory exposure to males, including men, causes stress and related analgesia in rodents. *Nat Methods, 11*(6), 629–632. https://doi.org/10.1038/nmeth.2935

Stellato, A. C., Dewey, C. E., Widowski, T. M., & Niel, L. (2020, Nov 15). Evaluation of associations between owner presence and indicators of fear in dogs during routine veterinary examinations. *J Am Vet Med Assoc, 257*(10), 1031–1040. https://doi.org/10.2460/javma.2020.257.10.1031

USDA Animal Care, A. a. P. H. I. S. (2019). *Animal Welfare Act and Animal Welfare Regulations*. https://www.aphis.usda.gov/animal_welfare/downloads/bluebook-ac-awa.pdf

Watters, J. V., Krebs, B. L., & Pacheco, E. (2019). Measuring Welfare through Behavioral Observation and Adjusting it with Dynamic Environments. In A. B. Kaufman, M. J. Bashaw, & T. L. Maple (Eds.), *Scientific Foundations of Zoos and Aquariums: Their Role in Conservation and Research* (pp. 212–240). Cambridge University Press. https://doi.org/doi:10.1017/9781108183147.009

Wurbel, H. (2001, Apr). Ideal homes? Housing effects on rodent brain and behaviour. *Trends Neurosci, 24*(4), 207–211. https://www.ncbi.nlm.nih.gov/pubmed/11250003

Wurbel, H. (2007a). Environmental enrichment does not disrupt standardisation of animal experiments. *ALTEX, 24 Spec No*, 70–73. https://www.ncbi.nlm.nih.gov/pubmed/19835063

Wurbel, H. (2007b, Mar 15). Publications should include an animal-welfare section. *Nature, 446*(7133), 257. https://doi.org/10.1038/446257a

# 41
# COMPANION ANIMALS IN TIMES OF CRISIS
## Contemporary Evidence from the COVID-19 Pandemic

*Jaume Fatjó and Jonathan Bowen*

### Introduction

The impact of distressing or challenging life events results from the interplay between individual characteristics and the person's physical and social environment. Extensive research into the neurobiology of stress has led to an over-emphasis on the individual's personality and resilience as the main stress-factors, which has undermined the role of the stressful events themselves and environment-related risk and protective factors (Fava et al., 2012).

The role of stressful life episodes as risk factors in the development of physical and psychiatric conditions such as depression and anxiety is well established. The causal relationship between acute or chronic periods of stress and the advent of depression and anxiety-related conditions has been confirmed by prospective studies (Brown & Harris, 1989; McLaughlin & Hatzenbuehler, 2009).

Environment-related factors include socioeconomic and cultural determinants, as well as the composition and accessibility of the individual's social network (Dahlgren & Whitehead, 2021).

In this chapter, we will examine what a crisis consists of, what social support is and how it might be affected by a crisis and what role companion animals might have in providing different forms of social support that could help people through difficult times. We will use experience from the recent COVID-19 pandemic as a contemporary lens through which to look at the role of companion animals during crises.

### What Is a Crisis?

Within the study of crisis intervention, there are numerous definitions of a crisis. Some examples include:

- "People are in a state of crisis when they face an obstacle to important life goals – an obstacle that is, for a time, insurmountable by the use of customary methods of problem solving. A period of disorganisation ensues, a period of upset, during which many abortive attempts at a solution are made" (Caplan, 1964).
- "Crisis is a crisis because the individual knows no response to deal with a situation" (Carkhuff & Berenson, 1977).

# Companion Animals in Times of Crisis

- "A temporary state of upset and disorganisation, characterised chiefly by an individual's inability to cope with a particular situation using customary methods of problem-solving, and by the potential for a radically positive or negative outcome" (Slaikeu, 1990).
- "Crisis is a perception or experience of an event or situation as an intolerable difficulty that exceeds the person's current resources and coping mechanisms" (James & Gilliland, 2005).

There has been considerable development in the understanding of what a crisis is, but the common factors between the definitions are that the crisis is not determined by the event itself but by the effects it has on people, and that the pathognomonic characteristic is that the affected people are somehow overwhelmed. The event is only the trigger, and could include illness, domestic abuse, bereavement, natural disaster, terrorism or war.

Immediately after World War 2, there was interest in understanding how people responded to a crisis, and how they might be better supported through one. The most enduring model, called the ABC-X Model of Crisis, was developed by Hill (1958). This saw personal and family crisis as an interaction between the triggering negative event or situation (A), the perception or significance of that event or situation (C) and the resources available to cope with it (B), creating the resulting acute disequilibrium and immobilisation of the family or personal system (X). This model was further developed by McCubbin and Patterson (1982) into the Double ABC-X Model, which included a consideration of the pre-crisis condition; the pre-existing stressors, perceptions and resources that were present before the crisis and could contribute to family or personal resilience or vulnerability to the coming crisis event (see Figure 41.1).

Whether, and how, an individual will be overwhelmed depends on their conditions, the support resources available to them, and their coping abilities before and at the onset of the crisis.

Resources may be tangible (e.g., food & shelter) or intangible (e.g., social support). Different kinds of resources exist at the individual and group (e.g., family, community) level. For example, at an individual level, personal finance, education, health and personality characteristics could be important resources, whereas at the family unit level, resources could include cohesion, communication and shared interests. On a wider community or extended family level,

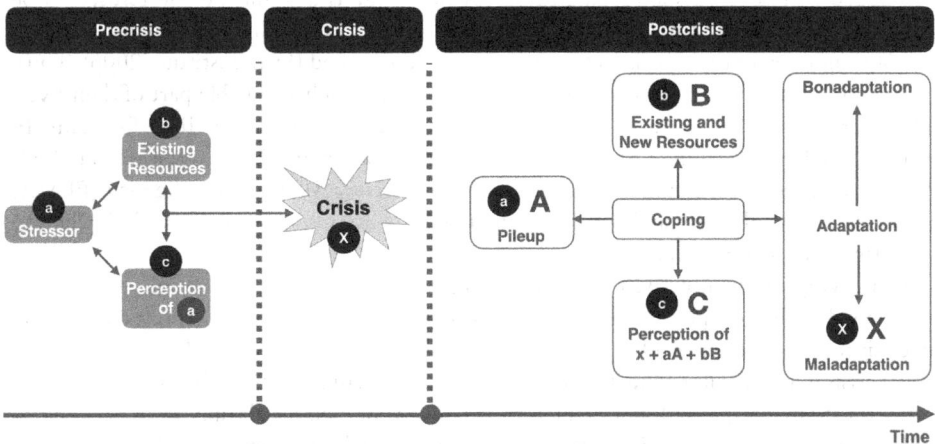

*Figure 41.1* The Double ABC-X model to describe the impact of crises on individuals and families. Adapted from McCubbin and Patterson (1982)

resources include social networks and the services provided by state institutions (e.g., social care and welfare, the healthcare system and emergency services). The more that available resources match and meet the demands of a situation, the less it is likely to be perceived as a crisis.

So, although crisis-triggering events can occur at an individual, group, community or national level, the chief differences are the number of individuals affected and the degree to which the availability of support resources is affected. For example, the partner of a homicide victim may still have access to a full range of support resources from family, friends, the workplace, the community, and regional and national institutions (e.g., health and social care services). However, in the situation of a mass shooting, certain kinds of support resources may be sapped at up to community level, and with an even more widespread threat such as an invasion, hurricane or pandemic even regional and national resources may be collapsed.

It is important to identify all the potential resources that are available to the individual and the community, so that we can understand how they may be used to help people. Whilst most resources are well understood and are included in considerations of crisis management, some are overlooked. The support that people obtain from their companion animals is one of these commonly overlooked resources.

## The Role of Companion Animals in a Crisis

In 2020, our research group conducted a series of studies in different countries to estimate the role of companion animals as a source of social support during the COVID-19 initial outbreak and its associated lockdown periods. As we predicted, in a moment of social isolation and stress, people turned to their pets as sources of support to mitigate the profound effects of the lifestyle and emotional impact of the confinement. Compared to the pre-pandemic situation, the support provided by dogs and cats dynamically adapted to compensate for the negative impact of the lockdown on the availability and accessibility of other members of social networks (Bowen et al., 2021; Riggio et al., 2022).

However, despite their benefits as sources of support, companion animals can become a problem of their own in times of crisis. For example, when Hurricane Katrina threatened large parts of the United States in 2005 many people were forced to abandon their homes (Fritz Institute, 2006). At the time, emergency services such as the US Coast Guard were not obliged to rescue pets along with their owners, and as a result, people refused to be evacuated. It is estimated that about 50% of people who stayed behind did so because they did not want to leave their pets, and many of those people died (Fritz Institute, 2006). Thus, at a time of life-threatening crisis, people saw their pets as such a valuable part of their lives, and a source of emotional support, that they were willing to risk their lives for them. In response, the US government introduced the Pets Evacuation and Transportation Standards (PETS) Act 2006, which authorises the Federal Emergency Management Agency (FEMA) to include household pets and service animals within its response framework (including evacuating and caring for them).

By applying the Double ABC-X framework, we could predict that people who have a strong bond with their pets and have concerns that those pets will not be protected during the crisis might experience additional stress during the pre-crisis period. Not only do they enter a crisis with a greater expectation that the situation will be overwhelming, but also that during the crisis, the pet will not be present as a source of emotional support.

In a smaller scenario of crisis, such as a family or personal crisis, people could also struggle to keep their pets with them due to external constraints. Common examples include victims of domestic violence who are forced to remain in the abusive environment to protect the animal if they are offered access to a shelter that has a no-pet policy (Newberry, 2017). The same situation is observed in people experiencing homelessness who are prevented from accessing social services where animals are not allowed (Anderson et al., 2021).

Public policies should take into account the protection of the human-animal bond and should promote the recognition of pets as a source of social support.

## What Is Social Support and Why Is It so Important?

The range of coping strategies individuals activate when faced with challenging events can be divided into those focused on dealing with the source of stress (problem-focused) and those aimed to minimize the psychological and emotional impact of the situation (emotion-focused). Problem-solving strategies include active coping and planning. Emotion-focused strategies include positive reinterpretation, acceptance, venting emotions and denial (Litman, 2006). Uncontrollable events demand mainly emotional support, whereas controllable events require more instrumental and informational kinds of support focused on evaluation and problem solving.

Within the mental health literature, social support is commonly described as the availability of other individuals to provide assistance or comfort to an individual when they are facing stressful events. So, the social support available to a person is determined by the size and quality of their social network, the availability of the people within it (especially during times of stress) and that person's competence in communicating their needs and soliciting the support that they need.

There is overwhelming evidence that social support is one of the most important predictors of good health and wellbeing (Alegría et al., 2018).

According to ecological approaches to health, social and community networks play a broad and long-lasting role in the four mechanisms influencing the determinants of health: availability and accessibility to resources, degree of exposure to risk factors, individual resilience and disease consequences (Figure 41.2) (Dahlgren & Whitehead, 2021). In other words, social networks not only have a proximal positive impact by providing direct support in times of need, but also a distal effect on the individual's resilience and access to protective factors for good health and wellbeing.

It is also important to remember that social support is not only recognised as a protective factor for good physical and mental health, but also as a therapeutic goal in itself. For instance, interventions to strengthen the individual's social network have been proposed in the short- and long-term management of severe mental disorders (Tenhula et al., 2009).

## What Are the Main Sources of Social Support?

People tend to obtain more social support – from closer relationships, such as partners, family members, relatives and friends. Without the ability to form, diversify and expand attachments, social support – could not exist. Attachment theory, which was constructed after studies of the relationship between the infant and the mother, provides the best biological framework to understand how emotional ties with the parental figure are formed, maintained and spread to others. Attachment is also proposed as the biological basis of the human-animal bond, particularly between people and their dogs (Zeng et al., 2021; Payne et al., 2015; Simpson & Belsky, 2008).

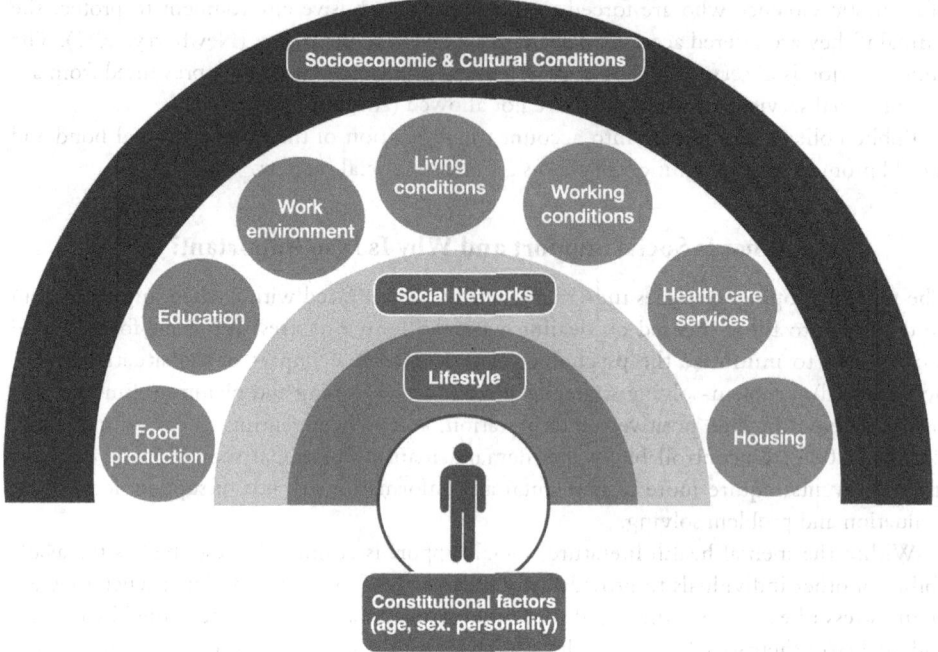

*Figure 41.2* The socio-ecological theory of health describes a multi-factorial, multi-layered model to understand the interaction and interdependence between individual strengths and vulnerabilities, life-style factors, social networks and socioeconomic and cultural influences. Adapted from Dahlgren and Whitehead (1991)

An attachment figure is an individual whose presence supports positive emotional balance by providing both a safe-haven during stressful events and a secure base from which to explore the environment (Zilcha-Mano et al., 2011). During development, and particularly during puberty, individuals gradually expand the range of attachment figures from their primary caregivers to other individuals, including peer, romantic-ties and eventually companion animals (Gillath et al., 2016).

Each person develops an attachment style, which relates to their self-perception of being worthy of care and support, and their perception of their parental attachment figures as effective, dependable and available social supports. Effective and reliable parental attachment figures foster a secure attachment style, whilst those who are ineffective and unreliable foster an insecure attachment style.

Attachment styles established early in life tend to permeate the way people build up and maintain social relationships during the rest of their lives, and an insecure attachment style may be one of the main internal psychological barriers to the individual securing access to available sources of social support. Attachment style is relatively persistent if the main elements of the social network remain the same, but it is open to a degree of transient or more enduring change in response to life events (Fraley et al., 2021).

## Companion Animals as Attachment Figures

Evidence from research on the human-animal bond indicates that people can treat their pets as attachment figures (Allen et al., 2002; Zilcha-Mano et al., 2012). The ability to do this originates in the natural tendency to attribute human characteristics to non-human animals, including behaviours, intentions and emotions. The more similar one animal is to humans, both morphologically and behaviourally, the more intense will be this process of humanisation (or anthropomorphism) (Urquiza-Haas & Kotrschal, 2015).

The 'matching' and 'compensation' hypotheses are opposing models of how people transfer attachment-related functions from their primary attachment figures to other social ties (Zilcha-Mano et al., 2011), including peers and companion animals. According to the matching hypothesis, people who show an insecure attachment style towards humans should also show the same attachment style towards their dogs and cats, leading to attachment-related anxiety or avoidance. The compensation hypothesis predicts that people with insecure attachment styles will use animals as a compensatory source of 'secure' attachment and support.

In a series of related studies, Zilcha-Mano and collaborators have confirmed the matching hypothesis. However, the moderate correlations that were found suggest that there could be still individual differences in the way people feel about human vs animal attachment figures (Zilcha-Mano et al., 2011).

## What Are the Specific Psychological Benefits of Social Support?

Social support is an overarching construct that operates through different psychological and behavioural interdependent mechanisms, which include social influence and social comparison, social control, behavioural guidance and mattering, self-esteem, belonging and companionship, and stress-buffering (Thoits, 2011).

- Social influence and social comparison: We adjust our behaviour and habits to match those displayed by reference individuals and/or groups. A functional social network will tend to promote healthy habits and normative behaviour in the individual, which would result in better health and adaptation to the environment.
- Social control: Social network members can actively encourage and monitor positive health and wellbeing practices.

However, social influence and social control can be perceived as intrusive and judgemental, which can result in a backfire effect.

Although a pet dog might not be considered as a reference individual itself, owning a dog compels owners to engage in activities that are positive for physical and mental wellbeing, such as physical exercise and access to green spaces. So, it provides social control through *gentle obligations,* as opposed to impositions, which should promote engagement and increase adherence. Unlike the influence of human social network members, which can be seen as intrusive and judgemental, the active encouragement provided by a pet dog is unlikely to produce a backfire effect as the dog is not perceived as being critical or having any agenda other than satisfying its own needs. In addition, whilst meeting the animal's needs, these actions can also bring people into contact with human reference individuals and groups that do act as social influencers in favour of healthy lifestyle. For example, joining a group of dog owners that regularly meet or go on walks together. It is important that health professionals are aware of these advantages so they can help society to fully accept and recognise the value of living with companion animals to improve health and quality of life.

- Behavioural guidance, purpose and mattering: Assuming obligations linked to the roles played within the social networks establishes barriers for behaviours that could be risky for the individual. For example, the parent who avoids smoking in front of children. Role-identities provide purpose and meaning and help to establish ties with other individuals or groups sharing those same identities. Mattering can be defined as the belief that one is important to another individual, who depends on one for having a good quality of life.

People often talk about themselves as valuable and even irreplaceable when it comes to their relationship with their pets. They feel that they matter to their pet. People also seem to derive meaning and purpose from the relationship they have with their pets. In our own research, we found that the majority of Spanish dog owners (77.8%) agreed or strongly agreed with the statement "My dog gives me a reason to get up in the morning".

- Self-esteem: According to the relational-self theory, our self-perception is strongly influenced by the way our significant and similar others perceive us. People describe themselves differently depending on what other relationship they have been asked to think about (Kring & Johnson, 2013).
- Belonging and companionship: One very important aspect of a functional social network is that the individual feels accepted by others. Companionship specifically refers to having individuals with whom we can share activities. Companionship provides positive affect, which increases psychological wellbeing. Conversely, lack of companionship is related to negative indicators of wellbeing, including depression and anxiety.

The highly dependent and non-judgemental relationship that companion animals establish with their families increases self-esteem and a feeling of accomplishment. Being a good caregiver for a companion animal increases the person's perception of control and mastery. One of the unique characteristics of pets is that they provide acceptance without judgement, and thus a unique sense of belonging for the owner.

- Stress-buffering: A person acts as a stress-buffer when their presence counteracts the real and perceived effects of a stressful situation, which may be brief or prolonged. For example, being accompanied by a friend to a medical appointment or living with a partner during a period of workplace harassment. To be effective, the stress-buffer must be readily available and accessible, as availability is one of the main predictors of perceived social support (Rush, 2002).

Support in stressful situations includes being attentive to the person, showing concern (sympathy/empathy), listening, spending time with the person in silence (simply "being there"), confirming the sense of mattering, sustaining the sense of self-worth and bolstering the perception of belonging to the group.

Extensive research on the human-animal bond indicates that people perceive their animals as very consistent and reliable stress-buffering sources of support that enable them to get through difficult times.

## What Are the Practical and Functional Dimensions of Social Support Obtained from Companion Animals?

The main challenge of understanding and quantifying the social support people get from their pets is that the kind of support that is available is so different from that which we get from people.

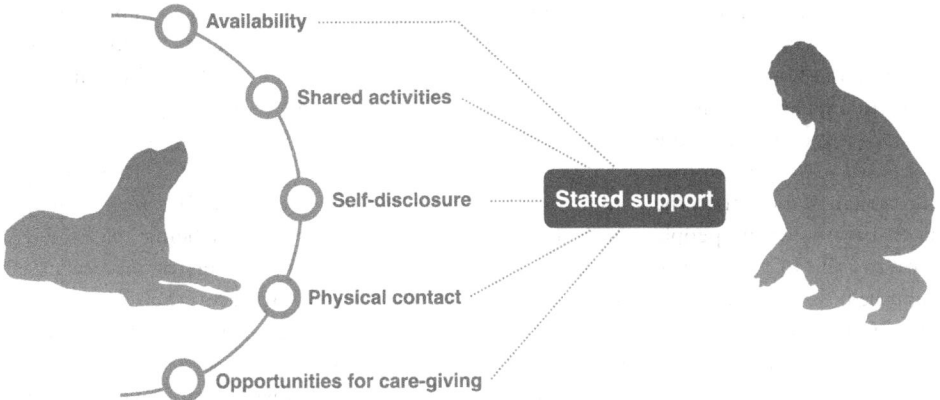

*Figure 41.3* Components of social support provided by companion animals

Whilst we accept that pets do provide their owners with some kind of social support, we have to characterise the phenomenon from a practical perspective. We need to determine how widespread it is, how flexible it is in adapting to circumstances and whether people genuinely turn to it in times of need.

Availability and closeness are two of the main features of any source of social support (Rush, 2002); availability relates to how easily accessible a source of social support is, and closeness includes interdependence, shared activities and emotional support (Berscheid et al., 1989). These are quite general and flexible categories that can be readily applied to non-human sources of social support.

Within this framework, we can identify five features of the social support that people could get from dogs or other companion animals: availability, shared activities, physical contact, opportunity for care giving and being a confidant (Figure 41.3) (Bowen et al., 2021).

## *Availability*

Pet ownership is quite commonplace, with 69% of US homes having a dog and 45% having a cat in 2021–2022 (Insurance Information Institute). Pet ownership in Europe is somewhat lower, with 25% of homes having a dog and 26% having a cat, with dog ownership being higher in some countries (e.g., Poland & the Czech Republic) and cat ownership higher in others (like Romania and Latvia) (FEDIAF, 2021). This makes pets more available to those individuals and families than most friends and relatives.

During the COVID-19 pandemic, pet ownership increased in many countries, and reached an all-time high in the United States (PFMA. APPA). In the United Kingdom, the percentage of dog-owning households may have increased from 22% to 33% during the pandemic (Braemar, UK). So, it appears that even non-pet-owning households were attracted to pet ownership during this crisis, which suggests that people were turning to pets for social support.

Availability in the socially supportive relationship could be considered in two ways: mere presence and pro-activity. By merely being present within a particular setting, the animal is available for the person to interact with, to provide them with company or physical contact when they require it. This is important to people, and in a survey of a pre-pandemic representative sample of Spanish dog owners we found that 90.6% of owners agreed or strongly agreed with the statement "If everyone else left me, my dog would still be there for me", 90.2% with the statement "My dog provides me with constant companionship" and 87% with the statement "My dog is there whenever I need to be comforted" (Bowen et al., 2021). In a separate dataset, collected from a convenience sample of dog owners during the COVID-19 pandemic, those feelings were found to be intensified. In that sample, 24.3%, 52.2% and 23.7% of dog owners, respectively, reported that they experienced those feelings more or much more than they did before the pandemic (Bowen et al., 2021). We can imagine that for the many people who feel this way about their pets, simply being able to have access to them in a wider range of situations (such as at work or when receiving medical care) might help them significantly. On the other hand, not being able to access their pets due to, for example, temporarily moving into crisis accommodation that restricts pet ownership or being hospitalised, could be very detrimental to the person's wellbeing.

However, pets are not merely present, they are proactive. Dogs have been found to be able to detect and respond to human emotional responses, including negative emotions (Huber et al., 2017; Sanford et al., 2018), and dogs approach people more often when they are crying (Custance and Mayer, 2012). Although responding to emotional states is probably part of the repertoire of sociable mammalian species, the way that dogs respond to people is probably the product of selection pressure, and selective breeding, during the process of domestication (Hare et al., 2002). Dogs can also respond to emotional contagion and are able to express what owners perceive to be prosocial helping behaviours (Sanford et al., 2018). This is supported by our own work (Bowen et al., 2021); we found that 73.6% of owners in the pre-pandemic representative sample agreed or strongly agreed with the statement "My dog is constantly attentive to me". Sixty-two point two of respondents in the pandemic convenience sample reported that they agreed with that sentiment more or much more than before the pandemic.

So, it seems that pets are not only commonplace, but their owners do see them as highly available sources of support to which they can readily turn during a crisis.

## Shared Activities

The degree to which a companion animal can participate in shared activities will depend on the species as well as the temperament and behaviour of the individual animal. Pet dogs can accompany people on trips, such as going shopping, or sit with a person while they are relaxing watching TV or reading a book, which may be quite a passive role but still provides company. They can join people to visit family and friends, as well as on walks or other forms of outdoor physical activity, or engage in play, during which there may be more interactive. Dog ownership appears to encourage walking and outdoor activity (Westgarth et al., 2014), and play with dogs is particularly good for the social-emotional development of pre-school children (Wenden et al., 2021).

More than 90% of participants in the Spanish representative sample population we studied said that they played with their dogs at least once a day or once every few days. Just under 90% of respondents stated that they had their dog with them while they were relaxing at least once a day or once every few days (Bowen et al., 2021). The frequency of these two shared activities increased during the pandemic, with 61.9% and 41.8% of respondents, respectively, reporting that they did them more or much more often. Shared activity was therefore the feature of support that showed the greatest overall increase during the pandemic, probably because the restrictions on social mixing meant that people were unable to go on walks with friends or family members, and they could not have visitors in their homes. As a result, the presence of the pet became more valuable as a substitute.

Again, this indicates that human wellbeing is likely to be improved by dogs being able to join their owners in the widest possible range of situations, simply as companions. By implication, anything that prevents the person from being with the dog, such as restrictions on access to public transport or to open spaces (due to dog bans in parks, for example) is probably bad for the person.

## *The Pet as Confidant*

The main characteristic of a confidant relationship is self-disclosure (Isaacs et al., 2015). Some well-intentioned acts of support can be perceived as challenging and distressing by the recipient. This helps to explain the added value of animals as non-judgemental confidants.

It would be easy to be dismissive of this confidant relationship between owner and pet because the pet is incapable of understanding what is being disclosed. However, evidence from the use of text-based chatbots and artificial intelligence (AI)-based counselling systems indicates that people benefit from disclosure even when they know that the counsellor is a robot.

The first evidence of computerised chatbots as convincing conversational partners capable of eliciting self-disclosure comes from the ELIZA studies at MIT (Weizenbaum, 1966). One of the response scripts used in those studies was a simulation of Rogerian psychotherapy (Bassett, 2019). Although people were only conversing with the system using a keyboard, and receiving written replies on-screen, they treated the dialogue as if it was with a counsellor and appeared to benefit from the interaction.

However, that primitive system was unable to respond to more complex human responses and emotions. As a result of the ubiquity of personal computing systems, and in particular mobile devices, there has been a drive to develop AI-driven chatbots that can analyse user inputs and respond convincingly to human emotions in clinical applications within psychology and psychotherapy (Bendig et al., 2019). However, this is something that dogs do naturally. So, although they may not be able to understand the content of what we are saying, they can respond to the emotion. In fact, not being able to understand the content of what is being disclosed may be the unique advantage of the pet as confidant. Not wanting to burden others with your troubles, particularly if those troubles reflect badly on you or could have implications for the future of the confidant, are reasons why people fail to disclose. The pet cannot be judgemental, and it cannot be concerned by the implications of the disclosure for it.

The only evidence we have of the extent to which people confide in their pets comes from our data from the Spanish representative sample, in which 69.6% of people said that they told their dog things they don't tell anyone else once a day or once a week (Bowen et al., 2021). In the pandemic convenience sample population, 21.4% of respondents reported that they confided with their dogs in this way more or much more often (Bowen et al., 2021). We suspect that this modest increase was because, unlike other forms of social support, the ability to confide in people was less impacted by pandemic restrictions because many people could still use telecommunications to maintain contact with the family and friends they used as confidants.

## *Physical Contact*

Physical contact and touch are important for socio-emotional, physical and psychological well-being (Field, 2010; Gallace & Spence, 2010; Jakubiak & Feeney, 2017). Touch contact has been found to reduce the perception of loneliness, at least in a laboratory setting, especially among single people (Heatley et al., 2020). In a study looking at touch contact between married partners, whilst they were discussing personal stressors, disclosers who received more touch contact perceived that they were better able to overcome the causes for their stress. They reported greater decreases in self-reported stress, greater increases in self-esteem and had a more positive view of their partners (Jakubiak & Feeney, 2019).

Pets are a readily available source of physical contact, with 86.8% of owners in our pre-pandemic representative sample indicating that they hugged their dogs once a day or once a week, and 69.1% kissing their dog with the same frequency. In the pandemic sample population that was collected separately, 48.5% and 33.9% of people reported that they did these things more or much more often during the pandemic than before. Physical contact was the second most increased aspect of social support after shared activity in this time of crisis.

Hugging dogs appears to be a more common physical interaction than kissing them, which may be related to sensible concerns about hygiene and safety. However, there is evidence that kissing dogs is specifically associated with elevated levels of the hormone oxytocin, which is associated with bonding and positive feelings (Payne et al., 2015). Animals are not just recipients of physical contact, but they also initiate it.

For marginalised people living alone during a period of forced isolation, such as during the pandemic, companion animals can be valuable sources of physical contact. In terms of the dimensions of human social support, physical contact appears to contribute to stress-buffering, feelings of self-esteem and mattering.

## *Opportunities for Caregiving*

Caregiving has been associated with psychological benefits for the caregiver, particularly when caring behaviour is not perceived to be a burden (Schulz & Sherwood, 2008). It keys into the human social support dimensions of purpose and mattering to another, self-esteem, belonging and companionship. In addition, because forms of care-giving such as grooming involve physical contact, it may also have a stress-buffering effect.

In our representative pre-pandemic sample population, 56.8% of people gave their dog food treats at least once a day or every few days, and 54.7% groomed their dog at least once a day or every few days. In the pandemic sample population, respondents reported that the frequency of these interactions increased more or much more by 33.7% and 24.2% of people, respectively, during the pandemic. So, it appears that although caring for the pet is important to people, it may be a much less powerful part of social support than other aspects. This needs to be investigated, but it may hint that the social support people get from animals is quite egocentric and unidirectional. So, being a source of social support to a person may not protect the animal from being exploited or inadequately provided for. Indeed, being a source of something as positive as social support may make the animal vulnerable to a false consensus effect in which its interests are subsumed into those of the owner (Fatjó et al., 2021).

## *Stated Support*

Having discussed the five main areas of social support that appear to be relevant to human-animal interactions, we can say that these are quite valuable and ubiquitous within pet keeping. In one of these areas, as the confidant, the dog has unique benefits that give it an advantage over humans.

It should probably be no surprise that when we look at two key statements that summarise the social support benefit people obtained from their dogs "My dog helps me get through tough times" and "My dog gives me a reason to get up in the morning", we found that 80%

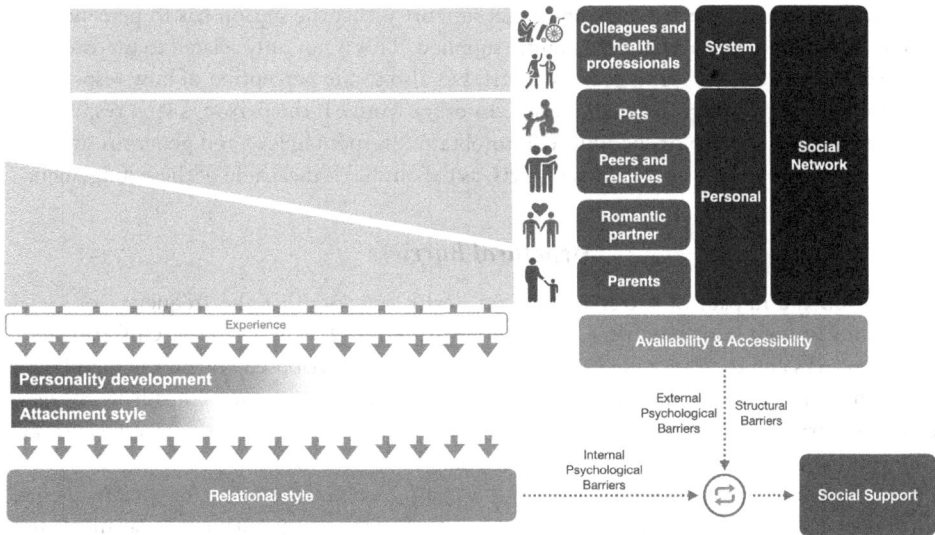

*Figure 41.4* The dynamics of social support. The level of received social support depends on the interplay between the availability of the different sources of support and the ability of the individual to negotiate and maintain a functional social network

of people agreed or strongly agreed with the first statement, and 77.8% did the same with the second statement.

## Barriers to Effective Social Support and the Effect of Crises

Barriers to social support can be divided into two main types: psychological and structural (Figure 41.4).

### *Psychological Barriers*

Psychological barriers could be internal or external, depending on whether they refer to the receiver or the provider of social support.

Obtaining effective social support partially relies on the individual's ability to build up a social network and to mobilise it in times of need (Offer, 2012). Internal psychological barriers include self-stigmatisation, an insecure attachment style (which leads to lack of competence in managing interpersonal relationships), or what could be termed *empathetic rejection of support,* where the person refuses to seek support from one particular individual that could be directly or indirectly affected by the situation. The potential provider of social support is often exposed to the same stressors and is therefore not considered as a valid provider of help. This is particularly relevant to close social ties, which also happen to be the best providers of support.

External psychological barriers include stigmatisation and prejudice. Stigmatised individuals are perceived by at least certain members of society as deviating from normative behaviour and standards, which then justifies providing them with less access to social support. Stigmatisation and prejudice constitute a very important barrier to negotiate a functional social network (Li et al., 2020). Furthermore, stigmatisation and prejudice not only prevent forming social ties, but also result in negative social interactions based on rejection and conflict, which become a source of stress by themselves.

A very important aspect of efficient social support is that the person has to perceive that support is spontaneously and not reluctantly supplied. This is not only related to the effects of cost-related factors on the provider of support, but also to the perception of how responsible the person is when facing a particular crisis. In other words, if the person is seen responsible for what is happening, he or she will tend to obtain less spontaneous and generous support. These negative effects do not occur with pets, as the animal is incapable of these judgements.

## *Structural Barriers*

Research on the impact of social support on health suggests that the frequency of social interactions is more important than their amount (Rihmer & Angst, 2009). Unfortunately, social networks and their provided support are not evenly distributed within the population (Diderichsen et al., 2019) and many of these barriers will be exaggerated during a crisis. Structural barriers refer to all those factors that limit the opportunities of an otherwise socially competent person to benefit from a functional social network.

The most obvious structural barrier is the lack of potential sources of social support, which is a very common scenario for elderly people. Sources of support may be available but not accessible due to physical or social barriers. A clear example of this is the lack of accessibility to relatives by people who have been forced to move to a different location or a different country. Companion animals are often prevented from accessing certain places, such as retirement homes or shelters for people experiencing homelessness or from travelling abroad, and this could have a very profound impact on an already compromised social network.

In our study (Bowen et al., 2021), we found that many of the social support benefits of a dog were significantly negatively associated with how difficult it was to look after the dog, including the two general statements about the dog helping the person to get through tough times and giving the person a reason to get up in the morning. The harder it was to look after the dog, the lower the person rated the support the dog gave them. The same was true with the perceived costs of owning the dog in general, which included the financial burden the dog poses as well as the negative effect it has on the owner's lifestyle (e.g., interfering with work and leisure activities).

An overlooked source of structural barriers is the socio-political environment. The Western neoliberal model of society promotes the ideology that individualism and self-reliance are the only ways to achieve happiness and success. This has led to *psychocentrism*; the assumption that all human problems reside within the individual mind/body rather than being a product and expression of social, political, historical and economic problems. This has created atomised citizens and an industry of self-help that profits from turning people in on themselves and away from sources of social support (Rimke, 2020). It has exploited psychological barriers such as self-stigmatisation, insecure attachment style and empathetic rejection of support that we have previously mentioned, which cause some individuals to hesitate or even to feel ashamed to obtain support from interpersonal relationships. In this context, pets could become unhealthy substitutes for human social support, rather than the healthy elements of a person's broad and flexible social support network that they ought to be.

## Conclusions

Although more research is required to fully understand the role of companion animals as sources of social support, the evidence that we already have points to some of the unique, advantageous characteristics of the support that people get from companion animals that make it more valuable, particularly in certain situations or for certain people or groups. It may be more than merely a second-class substitute for other sources of social support. Our data from the COVID-19 pandemic showed that people turned to their dogs for support during a time of crisis, and in particular as a source of physical contact during social isolation. This indicates that companion animals should be factored into public health decisions about how to manage crises and interventions to help people who are in distress. This could include taking advantage of the presence of a dog in the household as part of psychological interventions or finding ways to reduce the difficulties of ownership that can impair perceived support from the pet.

### Discussion Questions

- Which aspects of the social support that people can get from either people or animals are equivalent?
- What are the most obvious barriers to social support?
- Should the relationship between people and their pets be protected under law? For example, the ability of people to keep their pets with them wherever they go.

## References

Alegría, M., NeMoyer, A., Falgàs Bagué, I., Wang, Y., & Alvarez, K. (2018). Social Determinants of vMental Health: Where We Are and Where We Need to Go. *Current Psychiatry Reports, 20*(11), 95. https://doi.org/10.1007/s11920-018-0969-9

Allen, K., Blascovich, J., & Mendes, W. B. (2002). Cardiovascular Reactivity and the Presence of Pets, Friends, and Spouses: The Truth about Cats and Dogs. *Psychosomatic medicine, 64*(5), 727–739. https://doi.org/10.1097/01.psy.0000024236.11538.41

Anderson, M. C., Hazel, A., Perkins, J. M., & Almquist, Z. W. (2021). The Ecology of Unsheltered Homelessness: Environmental and Social-Network Predictors of Well-Being among an Unsheltered Homeless Population. *International Journal of Environmental Research and Public Health, 18*(14), 7328. https://doi.org/10.3390/ijerph18147328

APPA: https://www.americanpetproducts.org. Reported in https://www.petfoodindustry.com/articles/10325-new-us-pet-ownership-study-confirms-pandemic-led-growth

Bassett, C. (2019). The Computational Therapeutic: Exploring Weizenbaum's ELIZA as a History of the Present. *AI & Society, 34*, 803–812. https://doi.org/10.1007/s00146-018-0825-9

Bendig, E., Erb, B., Schulze-Thuesing, L., Baumeister, H. (2019). The Next Generation: Chatbots in Clinical Psychology and Psychotherapy to Foster Mental Health – A Scoping Review. *Verhaltenstherapie*. 1–13. https://doi.org/10.1159/000501812

Berscheid, E., Snyder, M., & Omoto, A. M. (1989). The Relationship Closeness Inventory: Assessing the Closeness of Interpersonal Relationships. *Journal of Personality and Social Psychology, 57*(5), 792–807. https://doi.org/10.1037/0022-3514.57.5.792

Bowen, J., Bulbena, A., & Fatjó, J. (2021). The Value of Companion Dogs as a Source of Social Support for their Owners: Findings From a Pre-pandemic Representative Sample and a Convenience Sample Obtained During the COVID-19 Lockdown in Spain. *Frontiers in Psychiatry, 12*, 622060.

Braemar: https://www.braemarfinance.co.uk/news-and-insights/facts-figures-pet-ownership

Brown, G. W., & Harris, T. O. (1989). Depression. In T. O. Harris & G. W. Brown (Eds.) *Life Events and Illness* (pp. 49–93). Guildford.

Caplan, G. (1964). *Principles of Preventive Psychiatry*. Basic Books.

Carkhuff, B. G., & Berenson, R. R. (1977). *Beyond Counseling and Therapy* (p. 165). Holt, Rinehart and Winston.

Custance, D., & Mayer, J. (2012). Empathic-Like Responding by Domestic Dogs (*Canis familiaris*) to Distress In Humans: An Exploratory Study. *Animal Cognition*, *15*(5), 851–859. https://doi.org/10.1007/s10071-012-0510-1

Dahlgren, G., & Whitehead, M. (1991). *Policies and Strategies to Promote Social Equity in Health*. Institute for Futures Studies.

Dahlgren, G., & Whitehead, M. (2021). The Dahlgren-Whitehead Model of Health Determinants: 30 Years on and Still Chasing Rainbows. *Public Health*, *199*, 20–24. https://doi.org/10.1016/j.puhe.2021.08.009

Diderichsen, F., Hallqvist, J., & Whitehead, M. (2019). Differential Vulnerability and Susceptibility: How to Make Use of recent Development in Our Understanding of Mediation and Interaction to Tackle Health Inequalities. *International Journal of Epidemiology*, *48*(1), 268–274. https://doi.org/10.1093/ije/dyy167

Fatjó, J., Bowen, J., & Calvo, P. (2021). Stress in Therapy Animals. In J. M. Peralta & A. H. Fine (Eds.) *The Welfare of Animals in Animal-Assisted Interventions*. Springer. https://doi.org/10.1007/978-3-030-69587-3_5

Fava, G. A., Sonino, N., & Wise, T. N. (2012). The Psychosomatic Assessment. Strategies to Improve Clinical Practice. *Advances in Psychosomatic Medicine*, *32*, 58–71. https://doi.org/10.1159/000330004

FEDIAF. (2021). Facts and Figures. https://europeanpetfood.org/about/statistics/ (last accessed: June 2022).

Field, T. (2010). Touch for Socioemotional and Physical Well-Being: A Review. *Developmental Review*, *30*(4), 367–383. https://doi.org/10.1016/j.dr.2011.01.001

Fraley, R. C., Gillath, O., & Deboeck, P. R. (2021). Do Life Events Lead to Enduring Changes in Adult Attachment Styles? A Naturalistic Longitudinal Investigation. *Journal of Personality and Social Psychology*, *120*(6), 1567–1606. https://doi.org/10.1037/pspi0000326

Fritz Institute. (2006). *Hurricane Katrina: Perceptions of the Affected*. Fritz Institute. http://beta.fritzinstitute.org/PPTs/FI-Harris_Katrina_0506.pdf

Gallace, A., & Spence, C. (2010). The Science of Interpersonal Touch: An Overview. *Neuroscience and Biobehavioral Reviews*, *34*(2), 246–259. https://doi.org/10.1016/j.neubiorev.2008.10.004

Gillath, O., Karantzas, G. C., & Fraley, R. C. (2016). How Stable Are Attachment Styles in Adulthood? In O. Gillath, G. C. Karantzas & R. C. Fraley (Eds.) *Adult Attachment* (pp. 129–157), Academic Press.

Hare, B., Brown, M., Williamson, C., & Tomasello, M. (2002). The Domestication of Social Cognition in Dogs. *Science (New York, N.Y.)*, *298*(5598), 1634–1636. https://doi.org/10.1126/science.1072702

Heatley Tejada, A., Dunbar, R., & Montero, M. (2020). Physical Contact and Loneliness: Being Touched Reduces Perceptions of Loneliness. *Adaptive Human Behavior and Physiology*, *6*(3), 292–306. https://doi.org/10.1007/s40750-020-00138-0

Hill, R. (1958). Generic Features of Families under Stress. *Social Casework*, *39*(2–3), 139–150. https://doi.org/10.1177/1044389458039002-318

Huber, A., Barber, A., Faragó, T., Müller, C. A., & Huber, L. (2017). Investigating Emotional Contagion in Dogs (*Canis familiaris*) to Emotional Sounds of Humans and Conspecifics. *Animal Cognition*, *20*(4), 703–715. https://doi.org/10.1007/s10071-017-1092-8

Insurance Information Institute. https://www.iii.org/fact-statistic/facts-statistics-pet-ownership-and-insurance

Isaacs, J., Soglian, F., & Hoffman, E. (2015). Confidant Relations in Italy. *Europe's Journal of Psychology*, *11*(1), 50–62. https://doi.org/10.23668/psycharchives.1120

Jakubiak, B. K., & Feeney, B. C. (2017). Affectionate Touch to Promote Relational, Psychological, and Physical Well-Being in Adulthood: A Theoretical Model and Review of the Research. *Personality and Social Psychology Review*, *21*(3), 228–252. https://doi.org/10.1177/1088868316650307

Jakubiak, B. K., & Feeney, B. C. (2019). Interpersonal Touch as a Resource to Facilitate Positive Personal and Relational Outcomes during Stress Discussions. *Journal of Social and Personal Relationships*, *36*(9), 2918–2936. https://doi.org/10.1177/0265407518804666

James, R. K., & Gilliland, B. E. (2005). *Crisis Intervention Strategies* (5th ed.). Thomson Brooks/Cole Publishing Co.

Kring, A. M., & Johnson, S. L. (2013). Current Paradigms in Psychopathology. In *Abnormal Psychology* (12th ed., pp. 56). John Wiley & Sons.

Li, J., Liang, W., Yuan, B., & Zeng, G. (2020). Internalized Stigmatization, Social Support, and Individual Mental Health Problems in the Public Health Crisis. *International Journal of Environmental Research and Public Health*, *17*(12), 4507. https://doi.org/10.3390/ijerph17124507

Litman, J. A. (2006). The COPE Inventory: Dimensionality and Relationships with Approach- and Avoidance-Motives and Positive and Negative Traits. *Personality and Individual Differences, 41*(2), 273–284. https://doi.org/10.1016/j.paid.2005.11.032

McLaughlin, K. A., & Hatzenbuehler, M. L. (2009). Stressful Life Events, Anxiety Sensitivity, and Internalizing Symptoms in Adolescents. *Journal of Abnormal Psychology, 118*(3), 659–669. https://doi.org/10.1037/a0016499

McCubbin, H., & Patterson, J. (1982). Family Adaptation to Crises. In H. McCubbin, A. Cauble, & J. Patterson (Eds.) *Family Stress, Coping and Social Support* (pp. 26–47). Thomas.

Newberry, M. (2017). Pets in Danger: Exploring the Link Between Domestic Violence and Animal Abuse. *Aggression and Violent Behavior, 34*, 273–281. https://doi.org/10.1016/j.avb.2016.11.007

Offer, S. (2012). Barriers to Social Support among Low-Income Mothers. *International Journal of Sociology and Social Policy, 32*, 120–133. https://doi.org/10.1108/01443331211214712

Payne, E., Bennett, P. C., & McGreevy, P. D. (2015). Current Perspectives on Attachment and Bonding in the Dog-Human Dyad. *Psychology Research and Behavior Management, 8*, 71–79. https://doi.org/10.2147/PRBM.S74972

PFMA: https://www.pfma.org.uk/news/pfma-confirms-dramatic-rise-in-pet-acquisition-among-millennials-

Riggio, G., Borrelli, C., Piotti, P., Grondona, A., Gazzano, A., Di Iacovo, F. P., Fatjó, J., Bowen, J. E., Mota-Rojas, D., Pirrone, F., & Mariti, C. (2022). Cat-Owner Relationship and Cat Behaviour: Effects of the COVID-19 Confinement and Implications for Feline Management. *Veterinary Sciences, 9*(7), 369.

Rihmer, Z., & Angst, J. (2009). Mood Disorders: Epidemiology. In B. J. Sadock, V. A. Sadock & P. Ruiz (Eds.) *Kaplan & Sadock's Comprehensive Textbook of Psychiatry* (9th ed., pp. 1651). Lippincott Williams & Wilkins.

Rimke, H. (2020). Self-help, Therapeutic Industries, and Neoliberalism. In D. Nehring, O. J. Madsen, E. Cabanas, C. Mills & D. Kerrigan (Eds.) *The Routledge International Handbook of Global Therapeutic Cultures* (pp. 37–50). Routledge. https://doi.org/10.4324/9780429024764-5

Rush, M. M. (2002). Perceived Social Support: Dimensions of Social Interaction Among Sober Female Participants in Alcoholics Anonymous. *Journal of the American Psychiatric Nurses Association, 8*(4), 114–119. https://doi.org/10.1067/mpn.2002.126673

Sanford, E. M., Burt, E. R., & Meyers-Manor, J. E. (2018). Timmy's in the Well: Empathy and Prosocial Helping in Dogs. *Learning & Behavior, 46*(4), 374–386. https://doi.org/10.3758/s13420-018-0332-3

Simpson, J. A., & Belsky, J. (2008). Attachment Theory within a Modern Evolutionary Framework. In J. Cassidy & P. R. Shaver (Eds.) *Handbook of Attachment: Theory, Research, and Clinical Applications* (pp. 131–157). The Guildford Press.

Schulz, R., & Sherwood, P. R. (2008). Physical and Mental Health Effects of Family Caregiving. *The American Journal of Nursing, 108*(9 Suppl), 23–27. https://doi.org/10.1097/01.NAJ.0000336406.45248.4c

Slaikeu, K. A. (1990). *Crisis Intervention: A Handbook for Practice and Research* (2nd ed., pp. 15). Allyn and Bacon.

Tenhula, W. N., Bellack, A. S., & Drake, R. E. (2009). Schizophrenia: Psychosocial Approaches. In B. J. Sadock, V. A. Sadock & P. Ruiz (Eds.) *Kaplan & Sadock's Comprehensive Textbook of Psychiatry* (9th ed., pp. 1570). Lippincott Williams & Wilkins.

Thoits, P. A. (2011). Mechanisms Linking Social Ties and Support to Physical and Mental Health. *Journal of Health and Social Behavior, 52*(2), 145–161. https://doi.org/10.1177/0022146510395592

Urquiza-Haas, E. G., & Kotrschal, K. (2015). The Mind Behind Anthropomorphic Thinking: Attribution of Mental States to Other Species. *Animal Behaviour, 109*, 167–117.

Weizenbaum, J. (1966). ELIZA—A Computer Program for the Study of Natural Language Communication between Man and Machine. *Communications of the ACM, 9*, 36–45.

Wenden, E. J., Lester, L., Zubrick, S. R., Ng, M., & Christian, H. E. (2021). The Relationship Between Dog Ownership, Dog Play, Family Dog Walking, and Pre-Schooler Social-Emotional Development: Findings from the PLAYCE observational Study. *Pediatric Research, 89*(4), 1013–1019. https://doi.org/10.1038/s41390-020-1007-2

Westgarth, C., Christley, R. M., & Christian, H. E. (2014). How Might We Increase Physical Activity Through Dog Walking?: A Comprehensive Review of Dog Walking Correlates. *International Journal of Behavioral Nutrition and Physical, 11*(83), 1–14. https://doi.org/10.1186/1479-5868-11-83

Zeng, H., Yu, Z., Huang, Q., & Xu, H. (2021). Attachment Insecurity in Rats Subjected to Maternal Separation and Early Weaning: Sex Differences. *Frontiers in Behavioral Neuroscience, 15*, 637678. https://doi.org/10.3389/fnbeh.2021.637678

Zilcha-Mano, S., Mikulincer, M., & Shaver, P. R. (2011). An Attachment Perspective on Human–Pet Relationships: Conceptualization and Assessment of Pet Attachment Orientations. *Journal of Research in Personality, 45*(4), 345–357. https://doi.org/10.1016/j.jrp.2011.04.001

Zilcha-Mano, S., Mikulincer, M., & Shaver, P. R. (2012). Pets as Safe Havens and Secure Bases: The Moderating Role of Pet Attachment Orientations. *Journal of Research in Personality, 46*(5), 571–580. https://doi.org/10.1016/j.jrp.2012.06.005

# 42
# HUMAN-ANIMAL INTERACTIONS IN AN AGRICULTURAL AND PRODUCTION SETTING

*Jose M. Peralta and Taylor Haefs*

### The Beginnings of the Interaction: A Brief History of Farm Animal Domestication

Recent genetic evidence comparing ancient dog and wolf fossils suggests that dogs may have been domesticated from wolves some 30,000 years ago (Thalmann et al., 2013). That was a time during the Mesolithic stone age when humans were still hunter-gatherers. There is no evidence of any farm animal domestication dating that far back in time. The lifestyle of the hunter-gatherer, mostly nomadic, required constant movement and while basecamp sites may have existed, any settlement may have been temporary and short lived (Crombé & Robinson, 2019).

The end of the last Glacial Period around 15,000 years ago was characterized by an abrupt climate change that resulted in more temperate weather around the world (Severinghaus & Brook, 1999). With the more benevolent conditions, humans started to transition from nomadic hunter-gatherers to living in more stable sedentary settlements. About 12,000 years ago, at the beginning of the Neolithic period, the improved environmental conditions and the settled human communities facilitated the evolution of agriculture (Lewin, 2009). As humans settled in this warmer climate, some ideal circumstances occurred that facilitated the domestication of different plant species. The addition of farm animals only strengthened those settlements and gave those humans broader opportunities to thrive. Domesticated farm animals provided a dependable source of digestible energy and nutritious protein. This only resulted in improved human health and an enhanced quality of life.

Domestication is a process that starts with an initial taming stage in which wild animals habituate to the presence of humans and adapt to the conditions in which humans keep animals of a given species. A second and equally necessary step is accomplished with selective breeding that leads to changes in gene expression (Albert et al., 2012). This results in the intentional modification of the genetic material in favor of certain traits that are considered beneficial primarily to humans (Sandøe et al., 1996). In the case of farm animals, some of these traits were behavioral, like the selection for less aggressive animals by culling those that were more difficult to handle, while other selected traits helped provide a more steady and consistent supply of food and nutrients. Humans also benefited from some of the animal by-products, like their skin for clothing or their manure that nourished the fields. This provided opportunities to take advantage of all available elements from the animals to maximize human benefit (Mora et al., 2019).

Domestication of farm animals started approximately 10,000 years ago. Interestingly, archeological findings suggest that this happened at around the same time period in different regions around the world. For example, cattle, sheep, goats, pigs and buffalo were domesticated in East and Southwest Asia, while llamas and alpacas were domesticated in South America (Bruford et al., 2003). This supports the idea that domestication was not a random event that started at one point and spread from there. Instead, it happened globally, likely supported by external factors like the sudden modification in climate patterns or the demographic, cultural and behavioral changes occurring to humans at that time (Lewin, 2009; Vigne, 2015).

It is possible that at least at first the process of domestication was somewhat serendipitous, with some animals getting closer to existing settlements, possibly because they were less fearful or maybe younger individuals who were more accepting of the presence of humans (Hemsworth, 2007). This could have been due to their recognition of some advantage to that proximity; perhaps their association with humans made them feel protected from other predators, or maybe they gained access to feed and other resources that were scarcer in the wild. As a result, a form of symbiotic relationship developed that facilitated increasing interactions. With that accepted proximity, humans gained greater control over these animals and started benefitting from their presence in the settlements.

The harmonious integration of plant and animal agriculture was improved over the centuries and gave way to a mutually beneficial and sustainable agricultural cycle that favored all, especially humans. Farm animals were cared for and protected from predators while they supplied humans with food and their waste fertilized agricultural land to produce more crops. An abundance of crops provided additional food sources for humans and the leftovers were fed to livestock that turned these indigestible parts into additional protein to feed humans (McIntyre et al., 2009).

This agricultural cycle based on a pastoral model changed only minimally for thousands of years and resulted in a relatively efficient food production system. Familiarity with the animals' needs and wants resulted in proper care being provided, and in return, the pantries were always well supplied. That was the case until approximately the middle of the 20th century, when agriculture underwent a process of industrialization of food production (Fraser, 2001). In principle, that seemed like an improvement in the system and was initially embraced and encouraged. Animal and crop production became specialized, genetic selection centered on improving quality, our understanding of animal nutritional needs was maximized, and as a result, productivity and efficiency increased (Thompson, 2001). The availability of an abundant and cheap food supply allowed the feeding of an ever-growing human population. Unfortunately, what became a success story for producers and consumers also changed the quality of life of farm animals and affected the relationship and the interactions between humans and animals (Rollin, 2004).

This industrial setting has the animals in confinement and their management and care has been automated. Caretakers have in many cases been replaced by equipment and when the pastoral model had one person caring for tens of animals in small herds under constant supervision, now in the industrial model, one person is responsible for hundreds if not thousands of animals at once in large farms, with more distant monitoring often delegated to electronic devices. That change obviously modified the nature and type of interactions and affected the bonding with individual animals, transitioning from an individualized relationship to a generalized relationship. Even though the training and skills of the caretakers influences the quality of the interactions and the care the animals receive, individualized relationships require individual recognition and that is not as easy when animals count into the hundreds or the thousands as is common in many of these large operations (Waiblinger & Menke, 1999).

## Relationships between Human and Animals in Agricultural Settings

An important aspect of domestication includes selective breeding, a process that introduces permanent changes in gene expression and function, making the domesticated animals different from their wild ancestors (Albert et al., 2012). The intentional selection for animals has resulted in changes to their physiology and their behavior. As a result, thousands of years of genetic modification have affected how animals interact with humans. Modern-day farm animals are less aggressive toward humans and other animals, and this is the result of a planned strategy to facilitate interactions (Cesarani & Pulina, 2021). There are limitations to how genetic alterations can support human-animal interactions (HAIs), and the relationship between animals and humans is also influenced by the attitude of the caretakers. That must also be taken into consideration, since, for example, negative behavioral attitudes toward cows make the animals wary of their relationship with humans, which results in increased fearful responses and a decreased interest in engaging human contact (de Boyer des Roches et al., 2016).

This likely occurs because, despite the changes to their genotype associated with domestication, farm animals still see themselves as prey and perceive humans as predators. In a study that looked at the electrical activation of neurons in the temporal cortex of sheep, the same brain neuron groups were active when the sheep were looking at a person or at a dog, while different areas of the brain fired up when they were looking at other sheep (Kendrick, 1991). They have unquestionably become more comfortable having us around, but that innate prey behavior has not been completely erased from their genetic makeup and still determines many of their behavioral reactions when they interact with us.

Individual variation still exists in how animals express their comfort level in our presence, and that is in part influenced by how they interact with humans (Rushen et al., 1999). Farm animals react differently to different types of human behaviors, but they avoid aversive contact so consistently that measuring the animals' response when interacting with humans has been proposed as an effective strategy to determine the quality of the handling. There is such an alignment that, conversely, an assessment of the way the caretakers handle the animals has equally been suggested as a reliable estimator of the animals' response (Pajor et al., 2000; Ellingsen et al., 2014).

Farm animals can discriminate between different handlers and the type and quality of the interaction determines the response that the animals display. This needs to be kept in mind and monitored when designing training for caretakers, because both calves and cows can recognize individuals and become fearful of those who may have engaged in aversive interactions with them in the past (de Passillé et al., 1996; Munksgaard et al., 1997). This ability to recognize individual caretakers has been demonstrated in other farm animal species as well, with pigs and sheep also being able to tell people apart by differences in body size and facial characteristics (Koba & Tanida, 2001; Knolle et al., 2017).

Furthermore, they can also differentiate people by the quality of their interactions, favoring those who are kind and gentle to them. Farm animals can remember those who engage in negative interactions and can also identify those who are neutral in their behavior when compared to those who engage in positive behaviors with the animals. Animals can generalize and, if cattle and pigs have been treated positively by one person, they can be more comfortable around other people even if they do not know them (Hemsworth et al., 1996). That said, most commonly, both species demonstrated longer contact time with people they already knew when compared with those who were unfamiliar (de Passillé et al., 1996; Tanida & Nagano, 1998).

Positive interactions should be promoted to take advantage of these abilities that farm animals have to recognize us and respond according to the nature of our behavior toward them. Aversive handling causes fear, and this has negative consequences to the quality of life and productivity of farm animals (Rushen et al., 1999). In a study conducted in multiple dairy farms, stockperson attitude, determined by the nature of their interactions with the cows, was compared with some production parameters. They found that cows that were fearful of their caretakers showed a decrease in milk production when compared to those in farms where the staff displayed positive behaviors toward the cows. These positive interactions were as simple as patting and talking to the cows (Breuer et al., 2000). In another study, farms where employees called cows by name reported higher levels of milk production than those with cows identified by numbers (Bertenshaw & Rowlinson, 2009). Of course, there could be other factors in these different farms that also helped increase the productivity, but the constant individualization of the animals reflected by them being named could be an indication of a general attitude and improved quality of care that incidentally impacted their productivity positively.

Most farm animals are social and gregarious (Nawroth et al., 2019). This is a trait that is common among prey species that find strength in numbers. They thrive in the presence of compatible conspecifics and seek their company. When we bring them into captivity, we may alter the normal grouping to meet our needs. If we house social animals in isolation, this may result in stressful conditions that get in the way of what they consider normal behaviors and interactions, challenging one of the Five Freedoms of animal welfare and having a negative impact on their mental state. When beef heifers were housed in isolation, their heart rate went up and they were more active, both signs of stress and anxiety. Even the presence of a mirror where they could see their reflection was effective at lowering both stress parameters (Piller et al., 1999).

Dairy heifers are separated from their mothers at birth and often housed in individual hutches. Most of them are in visual contact with other calves, but they cannot physically interact with one another. Single-housed calves are still quite common, even though recent recommendations favor group housing (Costa et al., 2016). In these conditions, where they cannot rely on the herd for protection, heifers are particularly vulnerable and benefit from always having positive interactions with their caretakers. Similarly, pigs, another social species, became less fearful and more comfortable when caretakers displayed positive attitudes in the barn (Coleman et al., 2000).

Farm animals, despite having been domesticated for thousands of generations, still perceive us as predators. Even though they grew accustomed to our close-up presence as they became tamed, there is still an element of fear that cannot be disregarded, especially if we behave unkindly toward them. Our efforts at engaging them with gentle interactions will help them become more comfortable having us around (Lürzel et al., 2016).

## Human and Farm Animal Interactions

Companion animals spend a considerable amount of their time around us and are influenced by our presence in their lives. One can say that we have meaningful relationships with animals inside our homes, equivalent to the relationships we may have with other humans, if not on quality, at least on intent. In a production setting, our interactions with farm animals are more limited. Farm staff head to the barn for three main reasons. First, we take care of the animals, feed them, make sure they have access to water and clean after them. Second, we manage their environment and social interactions, sort them into compatible groups and make efforts to reduce the presence of agonistic interactions. Finally, we reach out as appropriate to collect farm animal by-products, and as such we take them to the milking parlor or shear their wool or transport them to slaughter when they reach market weight. Farm animals also interact with other people, like veterinarians, which show up at the farm exclusively to take care of those in need of medical attention. In general terms, it can be said that the interactions that farm animals have with humans always have a purpose and while there may be an affective component, in a commercial setting, this likely does not meet the level of emotional connection of the social relationship that more often develops with our companion animals.

The lives of farm animals are by and large under the control of humans. We provide them with a place to live and food to eat, and also control their social interactions by deciding how and when they are grouped. We also impose ourselves in their daily lives as it fits us without giving them much of a choice. Domestication may have reduced the diameter of their flight zone when compared to their wild conspecifics, but farm animals still have some of that primordial prey behavior permeating in our interactions with them. They may have become comfortable enough to accept our presence, but the predator-prey perception still drives our interactions and defines the relationship we have with them. It is important to keep that in mind as we design their environment and define how we plan to interact with them (Grandin, 1997).

Because of the size and strength of farm animals, it is often argued that in order for our interactions to be safe and to facilitate their handling, a certain component of dominance is needed. However, this does not mean they need to be continuously subdued by brute physical force. That would be counterproductive, as those negative interactions would make the animals anxious and harder to handle. They would also have an impact on their productivity. For example, farm staff displaying negative behaviors toward pigs caused a decrease in production parameters, including a drop in reproductive performance (Coleman et al., 2000). Negative handling also resulted in a behavioral and emotional response, as it happens to heifers that, after being slapped by their caretakers, displayed fearful signs and increased their flight distance in future interactions (Breuer et al., 2003).

It has been known for quite some time that stressful events, like loading cattle into a truck, increase heart rate more dramatically than when they were calmly interacting with other cattle in a pen. After stressful events of this kind, it is best to let them calm down for as long as 20–30 minutes (Stermer et al., 1982), which indicates a long-lasting impact on their well-being. Conversely, positive interactions resulted in improved care and relationship with the animals. For example, stroking the chin or allowing calves to lick the caretakers' fingers after being fed, resulted in calves that were healthier and less agitated and displayed more comfortable behaviors when approaching both familiar and unfamiliar people (Lensink et al., 2000a, 2000b). Incidentally, as already stated, these studies also demonstrated that positive interactions also affected productivity, resulting in calves with improved growth rates and meat quality, parameters that are of economic interest to the producer.

For an ideal interaction with farm animals, management practices and routine handling must be respectful of the animals' nature and designed keeping in mind the animals' needs. It is important to recognize the uniqueness of each animal and how they may respond differently to the same given situation. Much like it happens with other domesticated species, this individualized response in farm animals is not only genotypic, but also has a phenotypic component and it is heavily influenced by previous experiences that each animal has had when interacting with humans (de Passillé et al., 1996). This individuality of each animal also needs to be taken into consideration and used to everybody's advantage as we plan our interactions with farm animals.

Farm animals, like all other animals including humans, use their senses to perceive the world around them. In their interactions with us, they may use primarily vision and hearing and, to a lesser extent, some tactile and olfactory perception. Farm animals see us as soon as we enter their field of vision, and as prey animals, that is a wide field. They pay attention to our movement and our gestures as we approach them and how we behave is taken into consideration and influences their response to our presence. If we walk fast or move our arms or body quickly in their presence, the animals will become agitated, flinch, step and kick more often, all signs of discomfort and stress (Breuer et al., 2000). Keeping in mind that farm animals have a memory of previous interactions that influence future contacts, it is additionally beneficial to always be quiet and calm when approaching farm animals to avoid unnecessary agitation and to facilitate an optimal interaction.

Farm animals also have a good sense of hearing and use it to detect our presence even before they see us. Their ears move in the direction of the sound, and they focus their attention as they attempt to locate its source. Once we get closer to them, they listen to any noise that we may make and use that information to educate their response. Interestingly, farm animals identify different tones of human voice and find loud voices particularly alarming. In a study with beef cattle, they reported that human shouting was more stressful than metal clanging and resulted in a higher heart rate and increased movements, both indicators of discomfort and stress (Waynert et al., 1999). This was further corroborated in preference studies that demonstrated that cattle chose people talking in a gentle voice over those who were shouting, which suggests that loud voices can be found to be aversive by cows (Pajor et al., 2003). This is important to be kept in mind as we interact with farm animals, as the tone and volume of our voice may be calming or upsetting and influence the quality of the relationship (Waiblinger et al., 2006).

Regarding our tactile interaction with farm animals, this is often limited due to the nature of our relationship with them and to the limited opportunities for close contact. When we engage with them in physical contact, it is often some specific manipulations that we impose on the animals and that the animals rarely seek. Perhaps because our direct contact is often forced on farm animals, they may not find it rewarding, no matter how gentle our petting or brushing might be, and they do not necessarily look for this contact (Pajor et al., 2003). That said, they accept kind interactions and prefer them to more aversive ones. Our only manner of contact with farm animals should be kind at all times, because there is extensive evidence that rough and aversive handling is detrimental to the animals and to the human-animal relationship. Calves handled adversely with nose tongs or with an electric prod by a handler while another petted them kindly learned to discriminate between the two handlers and the calves preferred the company of the kinder handler while they avoided the other (de Passillé et al., 1996). Similar studies in adult animals produced the same results, with cows accepting the presence of a kind handler that offered food and petted them over that of a handler who slapped them on the muzzle (Munksgaard et al., 1997).

## Human Role in the Interactions

There is a general mythology in Western society that views our relationship with farm animals through a pastoralist lens under which animals are well respected and receive ideal care. However, recent decades of agricultural industrialization resulted in a distortion of that perception. With the specialization of animal agriculture came its intensification and confinement, which produced notable economic benefits for the industry, but resulted in an alteration of the pastoral reality that may still prevail in the minds of those in urban settings (Fraser, 2001). The industrialization of agriculture also gave way to the automatization of several tasks that only a few decades ago were done by hand. This resulted in more limited contact between humans and animals, making the development of personal relationships with animals a rarer event. With decreased interactions and farm animals being seen as a group, it is easier for the caretakers to start considering them a commodity rather than an individual, resulting in reduced connectivity and less concern about their quality of life (Losada-Espinosa et al., 2020).

When farm animals were originally domesticated, the relationship they established with humans could have been described as mutualistic, with all species involved benefitting from the closeness and the interaction with one another (Salonen, 2019). Over time, that relationship has evolved. More recently, the symbiosis may no longer be truly mutually beneficial and now has an anthropocentric tinge (Hemsworth, 2007). After all, we may not have absolute control over the lives of animals; some of them die unexpectedly under our care despite our supervision, but we still undoubtedly have a great deal of control over their lives (Buller & Morris, 2003). Farm animals benefit from the balanced diet we provide or the veterinary care they receive or the fact that we actively protect them against predators with a certain degree of success. However, we do all that while we decide where they live, what and how much they eat, with whom they live, socialize and breed, and how their offspring are raised. All this and, perhaps most importantly, the fact that we control when and how they die gives us an upper hand in our relationship with farm animals that makes it tilted in our favor. Under these conditions, it is hard to justify that we call it truly a mutualistic symbiosis.

The fact that the relationship between farm animals and humans is unbalanced and that not all involved benefit equally does not mean that anything goes and that we have absolute power over their lives. We have an obligation, for both personal and societal reasons, to treat farm animals with respect and to honestly try to satisfy their needs and provide what they want (Averós et al., 2013; Mellor & Stafford, 2008). We do this by educating ourselves about the physiological, behavioral and natural characteristics of the different species we work with and use this knowledge to guide our decisions and our training. That by itself may not be enough, as characteristics such as being empathetic and having a positive attitude when working with animals are also important to ensure a good quality of interaction with farm animals (Leon et al., 2020). This concern has been described for two subpopulations who frequently interact with farm animals, as both farmers and large animal veterinarians have demonstrated that those with higher levels of empathy toward animals have greater concerns about animal welfare issues (Norring et al., 2014; Kielland et al., 2010). A challenge in increasing the level of empathy displayed by farm staff may be associated with the industrialization of agriculture mentioned earlier. With the mechanization of tasks came a decrease in opportunities to interact, a weaker emotional connection and reduced empathy (Losada-Espinosa et al., 2020). Despite that, it behooves farm managers and owners to provide timely training about their animals' needs and wants as well as to select empathetic and kind people who are given time to connect with the animals. These caretakers will be prepared to make the best of the HAIs in the farm.

It is not only the personality and training of the staff working with animals in a farm that will influence the interactions. The caretakers' attitudes during those interactions are also going to have an impact. Certain production systems, like the dairy industry, may involve larger amounts of contact time between stockpeople and animals than other systems. For example, farm staff move dairy cows to the milking parlor and feed the young calves daily. These interactions shape how an animal perceives that person and their surroundings. The attitude farm personnel display in the presence of the animals has a direct and predictable effect on their behavior and actions, determining how they will react toward humans. This response happens in general or on an individual basis, since animals can discriminate between people based on the differences in their attitude (Welp et al., 2004). Displaying positive attitudes around cattle results in cows that avoid humans less, which is considered a sign of reduced fear. Conversely, a poor attitude toward an animal increases vigilance behavior and other expressions of fear toward humans (Leon et al., 2020; Waiblinger et al., 2002).

Caretakers who are naturally nervous or that display aggressive or dominant behaviors may make the animals uncomfortable (Ellingsen et al., 2014). In this regard, it is important to keep in mind that the aggression or dominance does not have to be deliberate. What matters is that a given movement or behavior can be perceived as aggressive by the animal and that might be enough to trigger fear or apprehension. A frightened or anxious animal will often be more difficult to handle, which may lead to caretaker frustration and poorer handling, resulting in an unfortunate repetitive cycle that does not benefit neither animals nor humans. A rushed hand movement, a fast approach or the kicking of a bucket in frustration may have a negative impact on the animal, even if it is not done with that intent. This is particularly true for farm animals that consider us predators and are instinctually wired to interpret fast movements and loud noises as signs of imminent danger (Grandin, 1998). Calm and slow, steady movements by caretakers should be encouraged and expected as they make the animals more comfortable and lead to more ideal and productive interactions (Mota-Rojas et al., 2020).

## Interactions with Farm Animals and Human Welfare

The HAI is an involved reciprocal relationship where the exchange that occurs results in a variety of effects on each other. We often assume that the impact of the interaction is mutually beneficial, but those assumptions deserve to be tested using the scientific method. As a result, in recent times, a variety of projects have begun to focus on the interactions of farm animals and humans and whether they are beneficial and how they result in improvements to the quality of life of the humans and animals involved. Before we explore some examples, it is important to acknowledge that there is a physical and a mental component to an ideal quality of life and both need to be enhanced for optimal well-being.

Our interactions with farm animals may result in injuries and those responsible for their care are at risk. The physical size and strength of farm animals make them dangerous to people, even if they do not mean to cause harm or if they are just responding naturally to our presence. Most farm animals are docile and accustomed to our presence, but they can be startled or frightened and, in those circumstances, they become fearful and may respond differently. For example, when caretakers yell and raise their voices when handling both beef and dairy cattle, the cows become agitated and there are more kicking and trampling incidents reported (Waynert et al., 1999; Lindahl et al., 2016). On the other hand, less injuries and improved worker safety have been described when animals are handled gently and in a manner that leads to a reduction in fearful responses (Waiblinger et al., 2004). Human health and physical integrity can be positively impacted by training farm staff and encouraging them to be kind to animals.

Caretaker interactions with farm animals can also have an impact on human well-being and this is affected by the nature of the interactions. There are indirect effects associated with the impact of inappropriate handling and the response it instigates. Fearful animals are more difficult to handle, which makes the caretakers frustrated and angered because they have to work with animals that refuse to cooperate and arguably have to spend more time working with them while putting themselves in physical danger. Even though the animals respond this way as a consequence of the caretaker's actions, having to work with animals that are difficult to manage and handle can have a negative effect on the workers, resulting in decreased job satisfaction and motivation (Waiblinger et al., 2002). Furthermore, difficult interactions give way to more injuries and increased concerns about workplace safety. Employees become dissatisfied, which results in decreased motivation to perform well and negatively impacts their mental well-being (McCaughey et al., 2013). Conversely, when some workers at a swine farm were trained on better ways to handle the pigs, their retention rate increased compared to other workers who did not participate in the training, and this happened only after a few months of implementing these skills (Coleman et al., 2000). Not only did the pigs benefit from the kinder interactions, but the job satisfaction of the trained workers also improved, which suggests a more direct impact on their satisfaction and well-being.

## Interactions with Farm Animals and Animal Welfare

Farm animals have been selected over thousands of generations to have a kinder temperament and to accept our presence in closer proximity than that demonstrated by feral or wild species (Cesarani & Pulina, 2021). Farm animals are less fearful of humans and more comfortable around us, but the interactions are still primarily driven by the humans and a lot of the innate prey behaviors still prevail in how the animals respond. A good deal of behaviors displayed by farm animals may have an innate basis, but they are modified by learned associations. It is in this way that farm animals learn to be trusting or fearful of us (Nicol, 1996). If our interactions are consistently positive, the animal will be relaxed and respond in kind; on the other hand, if the animals are handled inappropriately or unpredictably, or suffer during their interaction with us, they become stressed when anticipating future interactions.

The influence of our presence in the animals' environment and the quality of our interactions can have a great impact on their response to different situations. This obviously can influence their welfare, as has been suggested in some examples already mentioned. Kind handling has resulted in more comfortable animals, while yelling makes the animals distrust the caretakers (Waiblinger et al., 2002; Coleman et al., 2000). There is always a memory of our previous interactions with farm animals, and this memory influences future interactions. If we treat animals with kindness, a positive affective state will dominate our interactions and they will approach us with increasing interest and comfort in the future. They will even remember the location of those interactions and will approach us faster in an area where positive interactions occurred in the past (Rushen et al., 1999). Even our posture can impact the animals' response, with both pigs and goats among other species having demonstrated that they approach a person who is standing or squatting faster than one who is approaching them (Hemsworth et al., 1986; Lyons et al., 1988). Clearly, the quality and nature of the interactions that we have with farm animals affect their emotional state and as such have an impact on their welfare.

Our interactions with animals not only affect their mental state. As previously noted, productivity has also been influenced by the quality of our interactions (Hemsworth, 2009; Tallet et al., 2018). Stressful and aversive handling of pigs has a negative impact on their weight gain (Gonyou et al., 1986; Hemsworth & Barnett, 1991). In some cases, even unintended behaviors can result in a negative impact on productivity; having nervous caretakers around cows decreases milk yield and increases veterinary costs in dairy farms (Arias & Špinka, 2005). Negative interactions also increase stress, as demonstrated by higher milk cortisol levels, and even affect milk quality, resulting in lower protein and fat concentration (Hemsworth et al., 2000).

Improved interactions between farm animals and humans can benefit the quality of life of both animals and humans alike. That should be a good incentive for the implementation of programs that encourage them. Improved interactions can also improve the productivity of farm animals, which should serve as an additional incentive for those responsible for these operations. These are strong reasons to encourage those involved in animal agriculture to be proactive in the implementation of steps to ensure and encourage the establishment of positive interactions.

## Conclusion

Our interaction with farm animals, much like the interactions we have with other species, is predominantly dominated by humans. As such, we have a greater responsibility in ensuring that the interactions are positive and that both parties benefit. A combination of positive and negative experiences affects the relationship between caretakers and farm animals. A good relationship needs to be established and any interactions must have an overall positive valence. Any potential negative experience needs to be counterbalanced with an excess of positive interactions. Proper socialization and habituation to humans, something that needs to start when the animals are young and most receptive, as well as positive handling throughout all our interactions is necessary if we want to make the best of our interactions with farm animals. It is important to keep in mind that, in our interactions with animals, it is the impact that we have on them what matters in the end.

### Discussion questions

- What can those living in urban settings do to facilitate improved interactions between caretakers and animals in farms?
- What practical steps can be taken to improve the interactions between farm animals and their caretakers?
- How can production levels be maintained at present levels while respect for animals and improved interactions are increased?

## References

Albert FW, Somel M, Carneiro M, Aximu-Petri A, Halbwax M, Thalmann O, Blanco-Aguiar JA, Plyusnina IZ, Trut L, Villafuerte R, Ferrand N, Kaiser S, Jensen P, Pääbo S (2012) A Comparison of Brain Gene Expression Levels in Domesticated and Wild Animals. PLOS Genetics 8(9): e1002962. https://doi.org/10.1371/journal.pgen.1002962

Arias JLP, Špinka M (2005) Associations of Stockpersons' Personalities and Attitudes with Performance of Dairy Cattle Herds. Czech Journal of Animal Science 50: 226–234.

Averós X, Aparicio MA, Ferrari P, Guy JH, Hubbard C, Schmid O, Ilieski V, Spoolder HAM (2013) The Effect of Steps to Promote Higher Levels of Farm Animal Welfare across the EU. Societal versus Animal Scientists' Perceptions of Animal Welfare. *Animals* 3: 786–807. https://doi.org/10.3390/ani3030786

Bertenshaw C, Rowlinson P (2009) Exploring Stock Managers' Perceptions of the Human-Animal Relationship on Dairy Farms and an Association with Milk Production. *Anthrozoös* 22(1): 59–69. https://doi.org/10.2752/175303708X390473

Breuer K, Hemsworth PH, Barnett JL, Matthews LR, Coleman GJ (2000) Behavioural Response to Humans and the Productivity of Commercial Dairy Cows. *Applied Animal Behaviour Science* 66(4): 273–288. https://doi.org/10.1016/S0168-1591(99)00097-0

Breuer K, Hemsworth PH, Coleman GJ (2003) The Effect of Positive or Negative Handling on the Behavioural and Physiological Responses of Nonlactating Heifers. *Applied Animal Behaviour Science* 84(1): 3–22. https://doi.org/10.1016/S0168-1591(03)00146-1

Bruford MW, Bradley DG, Luikart G (2003) DNA Markers Reveal the Complexity of Livestock Domestication. *Nature Reviews Genetics* 4(11): 900–910. https://doi.org/10.1038/nrg1203

Buller H, Morris C (2003) Farm Animal Welfare: A New Repertoire of Nature-Society Relations or Modernism Re-embedded? *Sociol Ruralis* 43(3): 216–237 https://doi.org/10.1111/1467-9523.00242

Cesarani A, Pulina G (2021) Farm Animals Are Long Away from Natural Behavior: Open Questions and Operative Consequences on Animal Welfare. *Animals* 11: 724–738. https://doi.org/10.3390/ani11030724

Coleman GJ, Hemsworth PH, Hay M, Cox M (2000) Modifying Stockperson Attitudes and Behaviour Towards Pigs at a Large Commercial Farm. *Applied Animal Behaviour Science* 66(1–2): 11–20. https://doi.org/10.1016/S0168-1591(99)00073-8

Costa JHC, von Keyserlingk MAG, Weary DM (2016) Effects of Group Housing of Dairy Calves on Behavior, Cognition, Performance, and Health. *Journal of Dairy Science* 99: 1–15. https://doi.org/10.3168/jds.2015-10144

Crombé P, Robinson E (2019) *European Mesolithic: Geography and Culture, in Encyclopedia of Global Archaeology*. Living Edition, Ed. Claire Smith, Springer Nature, Switzerland. ISBN: 9781441904652

de Boyer des Roches A, Veissier I, Boivin X, Gilot-Fromont E, Mounier L (2016) A Prospective Exploration of Farm, Farmer, and Animal Characteristics in Human-Animal Relationships: An Epidemiological Survey. *Journal of Dairy Science* 99(7): 5573–5585. https://doi.org/10.3168/jds.2015-10633

de Passillé AM, Rushen J, Ladewig J, Petherick C (1996) Dairy Calves' Discrimination of People Based on Previous Handling. *Journal of Animal Science* 74(5): 969–974. https://doi.org/10.2527/1996.745969x

Ellingsen K, Coleman GJ, Lund V, Mejdell CM (2014) Using Qualitative Behavior Assessment to Explore the Link Between Stockperson Behavior and Dairy Calf Behavior. *Applied Animal Behavior Science* 153: 10–17. https://doi.org/10.1016/j.applanim.2014.01.011

Fraser D (2001) Farm Animal Production: Changing Agriculture in a Changing Culture. *Journal of Applied Animal Welfare Science* 4(3): 175–190. https://doi.org/10.1207/S15327604JAWS0403_02

Gonyou HW, Hemsworth PH, Barnett JL (1986) Effects of Frequent Interactions with Humans on Growing Pigs. *Applied Animal Behaviour Science* 16: 269–78.

Grandin T (1997) The Design and Construction of Facilities for Handling Cattle. *Livestock Production Science* 49(2): 103–119. https://doi.org/10.1016/S0301-6226(97)00008-0

Grandin T (1998) Reducing Handling Stress Improves Both Productivity and Welfare. *The Professional Animal Scientist* 14: 1–10. https://doi.org/10.15232/S1080-7446(15)31783-6

Hemsworth PH, Gonyou HW, Dziuk PJ (1986) Human Communication with Pigs: The Behavioral Response of Pigs to Specific Human Signals. *Applied Animal Behaviour Science* 15: 45–54.

Hemsworth PH, Barnett JL (1991) The Effects of Aversively Handling Pigs, Either Individually or in Groups, on their Behavior, Growth and Corticosteroids. *Applied Animal Behaviour Science* 30: 61–72.

Hemsworth PH, Price EO, Borgwardt R (1996) Behavioural Responses of Domestic Pigs and Cattle to Humans and Novel Stimuli. *Applied Animal Behaviour Science* 50(1): 43–56.

Hemsworth PH, Coleman GJ, Barnett JL, Borg S (2000) Relationships between Human-Animal Interactions and Productivity of Commercial Dairy Cows. *Journal of Animal Science* 78: 2821–2831.

Hemsworth PH (2007) Ethical Stockmanship. *Australian Veterinary Journal* 85(5): 194–200. https://doi.org/10.1111/j.1751-0813.2007.00112.x

Hemsworth PH (2009) Impact of Human-Animal Interactions on the Health, Productivity and Welfare of Farm Animals. In Aland A, Madec F (Eds.), *Sustainable Animal Production: The Challenges and Potential Developments for Professional Farming* (pp. 57–68). Wageningen: Academic Publishers.

Kendrick KM (1991) How the Sheep's Brain Controls the Visual Recognition of Animals and Humans. *Journal of Animal Science* 69: 5008–5016.

Kielland C, Skjerve E, Østerås O, Zanella AJ (2010) Dairy Farmer Attitudes and Empathy Toward Animals Are Associated with Animal Welfare Indicators. *Journal of Dairy Science* 93(7): 2998–3006. https://doi.org/10.3168/jds.2009-2899

Koba Y, Tanida H (2001) How Do Miniature Pigs Discriminate Between People?: Discrimination Between People Wearing Coveralls of the Same Colour. *Applied Animal Behaviour Science* 73(1): 45–58. https://doi.org/10.1016/S0168-1591(01)00106-X

Knolle F, Goncalves RP, Morton AJ (2017) Sheep Recognize Familiar and Unfamiliar Human Faces From Two-Dimensional Images. *Royal Society Open Science* 4: 171228. https://doi.org/10.1098/rsos.171228

Leon AF, Sanchez JA, Romero MH (2020) Association between Attitude and Empathy with the Quality of Human-Livestock Interactions. *Animals (Basel)* 10(8): 1304. https://doi.org/10.3390/ani10081304

Lensink BJ, Boivin X, Pradel P, Le Neindre P, Veissier I (2000a) Reducing Veal Calves' Reactivity to People by Providing Additional Human Contact. *Journal of Animal Science* 78(5): 1213–1218.

Lensink BJ, Fernandez X, Boivin X, Pradel P, Le Neindre P, Veissier I (2000b) The Impact of Gentle Contacts on Ease of Handling, Welfare and Growth of Calves and on Quality of Veal Meat. *Journal of Animal Science* 78(5): 1219–1226.

Lewin R (2009) The Origin of Agriculture and the First Villagers. In *Human Evolution: An Illustrated Introduction* (5th ed.). Malden, MA: John Wiley & Sons, p. 250. ISBN 978-1-4051-5614-1

Lindahl C, Pinzke S, Herlin A, Keeling LJ (2016) Human-Animal Interactions and Safety During Dairy Cattle Handling—Comparing Moving Cows to Milking and Hoof Trimming. *Journal of Dairy Science* 99(3): 2131–2141. https://doi.org/10.3168/jds.2014-9210

Losada-Espinosa N, Miranda-De la Lama GC, Estévez-Moreno LX (2020) Stockpeople and Animal Welfare: Compatibilities, Contradictions, and Unresolved Ethical Dilemmas. *Journal of Agricultural and Environmental Ethics* 33: 71–92. https://doi.org/10.1007/s10806-019-09813-z

Lürzel S, Windschnurer I, Futschik A, Waiblinger S (2016) Gentle Interactions Decrease the Fear of Humans in Dairy Heifers Independently of Early Experience of Stroking. *Applied Animal Behaviour Science* 178: 16–22. https://doi.org/10.1016/j.applanim.2016.02.012

Lyons DM, Price EO, Moberg GP (1988) Individual Differences in Temperament of Domestic Dairy Goats: Constancy and Change. *Animal Behaviour* 36: 1323–1333.

McCaughey D, DelliFraine J, McGhan G, Bruning N (2013) The Negative Effects of Workplace Injury and Illness on Workplace Safety Climate Perceptions and Health Care Worker Outcomes. *Safety Science* 51: 138–147. https://doi.org/10.1016/j.ssci.2012.06.004

McIntyre BD, Herren HR, Wakhungu J, Watson RT (2009) *Agriculture at a Crossroads: The Global Report*. The International Assessment of Agricultural Knowledge, Science and Technology for Development (IAASTD), Washington, DC.

Mellor DJ, Stafford KJ (2008) Integrating Practical, Regulatory and Ethical Strategies for Enhancing Farm Animal Welfare. *Australian Veterinary Journal* 79(11): 762–768. https://doi.org/10.1111/j.1751-0813.2001.tb10895.x

Mora L, Toldra-Reig F, Reig M, Toldra F (2019) Possible Uses of Processed Slaughter Byproducts. In Galanakis CM (Ed.), *Sustainable Meat Production and Processing*. Academic Press, Cambridge MA. https://doi.org/10.1016/B978-0-12-814874-7.00008-0

Mota-Rojas, D, Broom DM, Orihuela A, Velarde A, Napolitano F, Alonso-Spilsbury M (2020) Effects of Human-Animal Relationship on Animal Productivity and Welfare. *Journal of Animal Behaviour and Biometeorology* 8: 196–205. https://doi.org/10.31893/jabb.20026

Munksgaard L, de Passille' AM, Rushen J, Thodberg K, Jensen MB (1997) Discrimination of People by Dairy Cows Based on Handling. *Journal of Dairy Science* 80(6): 1106–1112. https://doi.org/10.3168/jds.S0022-0302(97)76036-3

Nawroth C, Langbein J, Coulon M, Gabor V, Oesterwind S, Benz-Schwarzburg J, von Borell E (2019) Farm Animal Cognition—Linking Behavior, Welfare and Ethics. *Frontiers in Veterinary Science* 6: 24. https://doi.org/10.3389/fvets.2019.00024

Nicol CJ (1996) Farm Animal Cognition. *Animal Science* 62(3): 375–391. https://doi.org/10.1017/S1357729800014934

Norring M, Wikman I, Hokkanen AH, Kujala MV, Hänninen L (2014) Empathic Veterinarians Score Cattle Pain Higher. *Veterinary Journal* 200(1): 186–190. https://doi.org/10.1016/j.tvjl.2014.02.005

Pajor EA, Rushen J, de Passillé AMB (2000) Aversion Learning Techniques to Evaluate Dairy Cattle Handling Practices. *Applied Animal Behaviour Science* 69(2): 89–102. https://doi.org/10.1016/S0168-1591(00)00119-2

Pajor EA, Rushen J, de Passillé AMB (2003) Dairy Cattle's Choice of Handling Treatments in a Y-Maze. *Applied Animal Behaviour Science* 80(2): 93–107. https://doi.org/10.1016/S0168-1591(02)00119-3

Piller CAK, Stookey JM, Watts JM (1999) Effects of Mirror-Exposure on Heart Rate and Movement of Isolated Heifers. *Applied Animal Behaviour Science* 63(2): 93–102. https://doi.org/10.1016/S0168-1591(99)00010-6

Rollin BE (2004) Animal Agriculture and Emerging Social Ethics for Animals. *Journal of Animal Science* 82: 955–964

Rushen J, Taylor AA, de Passillé AM (1999) Domestic Animals' Fear of Humans and its Effect on their Welfare. *Applied Animal Behaviour Science* 65(3): 285–303. https://doi.org/10.1016/S0168-1591(99)00089-1

Salonen AS (2019) Dominion, Stewardship and Reconciliation in the Accounts of Ordinary People Eating Animals. *Religions* 10(12): 669. https://doi.org/10.3390/rel10120669

Sandøe P, Holtug N, Simonsen HB (1996) Ethical Limits to Domestication. *Journal of Agricultural and Environmental Ethics* 9: 114–122. https://doi.org/10.1007/BF03055297

Severinghaus JP, Brook EJ (1999) Abrupt Climate Change at the End of the Last Glacial Period Inferred from Trapped Air in Polar Ice. *Science* 286(5441): 930–934. https://doi.org/10.1126/science.286.5441.930

Stermer R, Camp TH, Stevens DC (1982) Feeder Cattle Stress During Transportation. *Transactions of the ASAE* 25(1): 246–249. https://doi.org/10.1303/2013.33513

Tallet C, Brajon S, Devillers N, Lensink J (2018) Pig-human Interactions: Creating a Positive Perception of Humans to Ensure Pig Welfare. In Špinka M (Ed.), *Advances in Pig Welfare* (pp. 381–398). Amsterdam: Elsevier Science.

Tanida H, Nagano Y (1998) The Ability of Miniature Pigs to Discriminate between a Stranger and their Familiar Handler. *Applied Animal Behaviour Science* 56(2–4): 149–159. https://doi.org/10.1016/S0168-1591(97)00095-6

Thalmann O, Shapiro B, Cui P, Schuenemann VJ, Sawyer SK, Greenfield DL, Germonpré MB, Sablin MV, López-Giráldez F, Domingo-Roura X, Napierala H, Uerpmann HP, Loponte DM, Acosta AA, Giemsch L, Schmitz RW, Worthington B, Buikstra JE, Druzhkova A, Graphodatsky AS, Ovodov ND, Wahlberg N, Freedman AH, Schweizer RM, Koepfli KP, Leonard JA, Meyer M, Krause J, Pääbo S, Green RE, Wayne RK (2013) Complete Mitochondrial Genomes of Ancient Canids Suggest a European Origin of Domestic Dogs. *Science* 342(6160): 871–874. https://doi.org/10.1126/science.1243650

Thompson PB (2001) Animal Welfare and Livestock Production in a Postindustrial Milieu. *Journal of Applied Animal Welfare Science* 4(3): 191–205. https://doi.org/10.1207/S15327604JAWS0403_03

Vigne JD (2015) Early Domestication and Farming: What Should We Know or Do for a Better Understanding? *Anthropozoologica* 50(2): 123–150. https://doi.org/10.5252/az2015n2a5

Waiblinger S, Menke C (1999) Influence of Herd Size on Human-Cow Relationships. *Anthrozoös* 12(4): 240–247

Waiblinger S, Menke C, Coleman G (2002) The Relationship Between Attitudes, Personal Characteristics and Behaviour of Stockpeople and Subsequent Behaviour and Production of Dairy Cows. *Applied Animal Behaviour Science* 79(3): 195–219. https://doi.org/10.1016/S0168-1591(02)00155-7

Waiblinger S, Menke C, Korff J, Bucher A (2004) Previous Handling and Gentle Interactions Affect Behaviour and Heart Rate of Dairy Cows During a Veterinary Procedure. *Applied Animal Behaviour Science* 85: 31–42.

Waiblinger S, Boivin X, Pederson V, Tosi M, Janczak A, Kathalijne Visser E, Jones R (2006) Assessing the Human-Animal Relationship in Farmed Species: A Critical Review. *Applied Animal Behaviour Science* 101(3–4): 185–242.

Waynert DF, Stookey JM, Schwartzkopf-Genswein KS, Waltz CS (1999) The Response of Beef Cattle to Noise During Handling. *Applied Animal Behaviour Science* 62(1): 27–42. https://doi.org/10.1016/S0168-1591(98)00211-1

Welp T, Rushen J, Kramer DL, Festa-Bianchet M, de Passillé AM (2004) Vigilance as a Measure of Fear in Dairy Cattle. *Applied Animal Behaviour Science* 87(1–2): 1–13. https://doi.org/10.1016/j.applanim.2003.12.013

# 43
# FARM-BASED INTERVENTIONS CONSIDERING ONE HEALTH – ONE WELFARE

*Lena Lidfors, Bente Berget, and Karen Thodberg*

## Introduction

Farm animals can be very suitable for Animal-assisted Interventions (AAIs), and farms usually offer more than one species and many other activities for the clients. There are many different names for this type of intervention, which will be explained. Research has been conducted over several decades on how farm-based interventions can support different client groups and ages, which will be presented shortly in this chapter. However, when taking help from farm animals to support humans, one needs to consider that these animals may not be as familiar with close interactions with humans as dogs or horses. We will therefore explain what research has shown considering the effects of early positive contact with humans for farm animals. Animal welfare friendly housing of cattle, sheep, swine and poultry that is suitable for farm-based interventions will be presented. The chapter will end with a discussion of ethical considerations.

When considering farm-based interventions involving farm animals, one needs always to think about the One Health – One Welfare aspects (Jegatheesan et al., 2018). It is important to consider the health of both humans and animals to avoid diseases and injuries in both, and safeguard that both have a good welfare. We hope this chapter will not only raise the awareness of farm animals as possible species for AAI, but also raise the awareness about what is required for this to be a good experience for both parties.

## Care Farming, Green Care and Farm-Based Interventions

Care farming is a service that has developed within the agricultural sector in Europe (Hassink & Van Dijk, 2006). It aims to promote mental and physical health and is increasingly used in mental health rehabilitation. This service has many different names, i.e. 'Green Care', 'Care farming', 'Social farming', 'Farming for health', 'Inn på tunet' (Norwegian) and 'Grön arena' (Swedish) (Berget & Braastad, 2008). The main idea of care farming is that people in need of health-related support may, on a regular basis, participate in farm-related activities alongside a farmer. Most commonly, these are group activities, but they should always be adapted to each individual participant's mental and physical needs.

Green Care is a well-established international concept using animals, plants and nature in an active process to offer health-promoting activities for people (Gallis, 2013). The idea about nature's potentially healthy effects on human health is not new. Within prisons, hospitals, monasteries and churches, this tradition goes back to before the Middle Ages (Gerlach-Spriggs et al., 1998). These concepts have recently gained a renewed interest in a quite different society with strong confidence in modern science.

Studies on care farming highlight group participation, the social setting and the farmer's supportive supervision as important (Ellingsen-Dalskau et al., 2015). Care farming programs have been described as a suitable transition between marginalization related to illness and inclusion in society. This kind of service comprises practical and varied work experiences that may include livestock farming; forest management; the cultivation of grains, fruits or vegetables; or other businesses on the farm, such as working in a farm shop or café (Ellingsen-Dalskau et al., 2016). It is important that the participants can take part in relevant and useful work tasks, independent of their mental and physical day-to-day functioning (Berget & Braastad, 2018).

Care farming services include a variety of different client groups within the broad agricultural context. The Netherlands, a pioneer country in the development of care farms, exceeded 1,000 care farms (Hassink et al., 2017). In Norway, care farming is used in many municipalities in collaboration with the health- and social sector and educational institutions. The care farms are subjected to a qualification system to be approved. Care farming has been established in several countries within and outside of Europe, even if individual programs are conceptualized differently (Berget et al., 2021).

In a review of Green Care for people out of work or school in the Nordic countries, Steigen et al. (2016) found the following main outcomes of the participants:

1. Contact with animals, giving an inner peace and the possibility of physical contact with another living being.
2. Supportive natural environments, by, e.g. getting the possibility to see plants grow and then harvest them in the fall, and as such experience the different seasons.
3. The service leader as a significant important other by giving advice, encouragements and supporting the participants in their working processes.
4. Social acceptance and fellowship with the other participants. Doing work tasks together strengthening their solidarity and self-efficacy.
5. Meaningful and individually adapted activities in which mastery can be experienced.

Similarly, other studies have shown that care farms offer real and meaningful work activities that provide an opportunity to learn new skills and build confidence (Elings & Hassink, 2010). In addition, it has been emphasized that care farms offer a flexible work environment, where clients have an opportunity to experience ordinary working life, work at their own pace and have freedom to switch between activities according to their interests and levels of functioning (Pedersen et al., 2012).

The social community on the farm includes other clients that can increase feelings of security and acceptance in a group (Hauge et al., 2014). The farmer who provides practical and emotional support, guidance and feedback, has also been described as an important factor in prevocational training on care farms. Many farmers have taken courses in Green Care in the Nordic countries, and several farmers have professional backgrounds in health, welfare services or education (Ellingsen-Dalskau et al., 2016). Farmers often use the farm environment actively to facilitate contact with animals and nature experiences for the clients (Pedersen et al., 2016). Working with animals or being in nature also usually includes physical activities like hiking trips, horseback riding or fishing. In addition, farmers emphasize the importance of creating a structured daily routine for the clients (Pedersen et al., 2016).

## How Farm Animals Can Support Humans, Especially Vulnerable People

In Green Care, a big proportion of the interactions with the farm animals is to provide care for the animals. Being able to provide care for another living being, the possibility to contribute to society, to achieve something, being able to help and feel needed gives a positive self-perception and enhances feelings of efficacy. Furthermore, animals and working with them are very effective distractions from personal problems and/or illnesses. Due to the prevailing attraction and positive attitude to the animals, the Green Care activities provide a focus of attention for longer time spans than many other distractions, such as reading or watching TV, since it also involves physical activity and demands a certain degree of personal presence.

Nonverbal feedback, e.g. hugging and brushing an animal is specific for AAIs, even with farm animals, and is often a very potent way of showing empathy, social support and thereby assists in calming a person. In a relatively calm and relaxed state, and with a feeling of safety, persons are able to work effectively with problematic topics such as a trauma. The animals give prompt and honest feedback. This is particularly important for persons who experience difficulties and stress in interpersonal interactions and tend to misinterpret interpersonal communications (Beetz & Enders-Slegers, 2013).

Although Animal-assisted Therapy (AAT) for humans with mental disorders has been well documented with pets, there are few controlled studies of farm animals as therapeutic agents for psychiatric patients. However, in a randomized controlled trial on adults with different psychiatric disorders, a six-month intervention resulted in increased self-efficacy and coping and decreased anxiety and depression compared to the control group (Berget et al., 2008, 2011). In a longitudinal study, a higher intensity and exactness in work tasks were observed at the end of the intervention compared to the beginning, which correlated with the increase in self-efficacy and decreased symptoms of anxiety (Berget et al., 2007).

Hassink et al. (2017) showed that farm animals are important to many client groups. Potential benefits included: providing a meaningful day and occupation, generating valued relationships, mastering tasks, providing opportunities for reciprocity, distracting from problems, providing relaxation, facilitating relationships with other people, and stimulating healthy behavior. Furthermore, Hassink et al. (2017) revealed that the participants were involved in productive activities as well as more recreational activities. For youngsters with behavioral problems, the focus was mostly on productive work. For people with mental illness, participants responded well to farm animals as well as pets, like cats and dogs. Flexibility in work tasks and activities with the animals, e.g. having a bad or good day, was important (Pedersen et al., 2016). This shows the multi-facetted outcomes of interacting with animals on care farms.

## Human-Animal Relationships and How It Affects Farm Animal Welfare

Animals' relationships with individual humans depend on their previous experiences. Farm animals are all handled to a greater or lesser extent and are subjected to varying degrees of contact with humans. If the animal's experiences with certain people are positive and pleasant, for example during feeding, and if the contact is predictable, the animal will associate these humans with something positive and show reduced fearfulness toward them. The animal may even begin to seek out these persons itself. The opposite situation, where the animal is handled in a negative way, e.g. involving painful procedures, fixation and unpredictability, will cause the animal to associate persons handling them this way with something negative, and their fearfulness toward these persons will increase. It is important for the animal to be able to predict whether a given person is going to handle it positively or negatively, as unpredictability is a stress factor for the animal (Waiblinger et al., 2006).

A positive animal-human relationship is of great importance for the welfare of animals that are handled by humans, mainly due to a reduced feeling of fearfulness toward the persons they are in contact with (Figure 43.1a and 43.1b), but also due to the presence of positive affective states. A positive relationship will also be beneficial for the human part, as a relationship based on previous and present positive interactions with humans will enhance safety when handling the animals during daily caretaking, and when using the animals for therapeutic purposes (e.g. Lindahl et al., 2016).

*Figure 43.1a*   Positive human-beef cattle interactions on a pasture in Sweden (Photo: Lena Lidfors)

*Figure 43.1b*   Positive human-goat interaction on a farm in Australia (Photo: Lena Lidfors)

Many farm animals are able to distinguish between, and react differently, to different humans, according to whether or not they have had positive or negative experiences with them. Cows, for example, are able to distinguish between different people based on the color of their clothes. When cows were handled positively or negatively by two humans, each wearing clothes of a specific color, the cows subsequently kept the greatest distance from the person who had handled them negatively (Munksgaard et al., 1997). This was also the case if the cows were tested in a different environment (home environment) than where the treatments took place (Rushen et al., 1998). Similar studies in pigs have shown that young pigs are able to distinguish people who handled them negatively from others, even several days after the handling took place (Sommavilla et al., 2016).

In addition, many animals will, based on positive or negative experiences with certain individuals, generalize their perception of these specific humans to unknown individuals, and this can be reflected in their behavior. Studies of sub-adult pigs handled negatively by a specific person, show that the pigs generalized and reacted fearfully to strangers (Hemsworth et al., 1996). Chickens and hens handled positively, or exposed to visual contact with humans became less fearful of humans (Silvera et al., 2017). Even just seeing other birds being handled positively reduced hens' fearfulness of humans (Zulkifli & Azah, 2004).

The way humans interact with animals thus has consequences for how these animals subsequently respond to contact with humans, both to these specific individuals, but also for how they react to humans in general. Animals kept by humans, like farm animals, are handled in many different ways, and with different frequency. Daily positive contact with caretakers will reduce animals' fear of humans.

Handling animals early in their life to reduce their fearfulness toward humans is effective. Shortly after mammals are born or birds are hatched, they are especially sensitive to environmental input. At this stage in life, animals develop and learn about their social preferences with conspecifics under natural conditions. The effect of positive handling of calves can be measured until at least 12 weeks (Schütz et al., 2012) and often many months after the handling (Silva et al., 2017). However, in studies where such individuals were tested one year later, it was not possible to measure significant differences in the animals' reaction to humans as a response to early positive handling outside ordinary routines (Silva et al., 2017). This deterioration of the human-animal relationship could be due to negative experiences with humans, e.g. in connection to management procedures in the intermediate period, counteracting the effect of the early positive experiences with humans.

Almost all farm animals undergo management procedures where they are restrained and may be exposed to pain. Examples of some management procedures are: tail docking, dehorning, hoof trimming, branding, transport and veterinarian treatments. Several studies of animals' responses to such procedures find increased measures of physiological stress and fearful behavior, e.g. during tail docking and castration of piglets (Prunier et al., 2005; Sutherland et al., 2012), and during dehorning of cattle (Stewart et al., 2013). Several studies have shown that the human-animal relationship worsened after such management procedures. Lürzel et al. (2015) found that calves showed more avoidance behavior toward humans after being dehorned. Chickens became more fearful of humans after being lifted by their legs with their head down, compared to being handled more carefully with support under the belly (de Lima et al., 2019). Even human behavior that does not involve physical contact can be perceived negatively by animals; for example cattle find it aversive when people shout at them (Waynert et al., 1999).

The sum of both the positive and negative experiences affects the individual animal's relationships with humans. This means that even though animals and humans have developed a good relationship, this relationship must be maintained, in order to counterbalance potential negative experiences with positive interactions. This is particularly relevant when using farm animals for AAI, where people often have limited experience interacting with animals and special attention should be given to preparing the animals for this type of interaction. This would include proper socialization to humans, positive handling, and habituation to being in near proximity of different humans.

## Animal Welfare Friendly Housing and Management of Some Farm Animal Species

Farm animals are mainly kept for production of meat, milk, eggs or fur, which has led to intensive housing conditions of large numbers of animals. Here, the focus is on how to house and manage farm animals involved in AAI so that their natural behavior, health and positive affective state are promoted. In farm-based interventions, it is important that the clients meet animals that have good welfare; otherwise, the purpose of the intervention may not be fulfilled. The species presented here are chickens, sheep, pigs and cattle, but other animals such as goats, llamas, alpacas, camels, ducks and other fowl may also be involved in farm-based interventions.

### *Chickens*

#### *Chicken Origin and Natural Behavior*

Chickens originate from the red jungle fowl (*Gallus gallus*) in South-East Asia, where domestication started around 8,000 years ago (Mench, 2017). The domesticated chicken (*Gallus gallus domesticus*) has been bred into two main lines: laying hens that produce many eggs and fast-growing boiler chickens that produce meat. Mench (2017) describes natural behaviors of chickens, a brief summary is presented below.

Jungle fowl live in groups with one male and several females, and other males live solitary or in small all-male groups. Chickens have good color vision, hearing (and individual body odors), which are involved in communication. Chickens have a pecking order that is separate for the two sexes. When aggressive, they peck at the head region of other birds who then demonstrate submissive behaviors.

Jungle fowl or chickens spend a large part of the day ground pecking, ground scratching and grazing. They eat grass, shrubs, roots, leaves, berries, invertebrates and even vertebrates such as lizards and mice. Chickens drink water several times a day by scooping up water in their beak and then raise their head so gravity helps them swallow the water.

Chickens preen frequently during the day by using their beak and face to spread oil from the uropygial gland on the base of the tail through its feathers, and removing mites and ticks. They dust-bathe to remove excess oil from the feathers by lying down and pulling loose substrates toward the body, after which they rub themselves on the substrate and shake the wings and body.

Chickens have a distinct daily rhythm and during nighttime, they sleep sitting high up on branches of trees, which protects them from ground predators. In the morning, the hen searches for a nest to lay her egg. Jungle fowl lay eggs and brood them for 21 days until the chicks hatch. Within a day of hatching, chicks start pecking at food items.

## Chicken Housing and Management

A hobby farm with laying hens usually consists of a small flock of hens with a cock. The animals are kept in a henhouse with an opening so that they can go outside during daytime. Animal welfare friendly housing of laying hens should include a nest where they can hide and lay their eggs, perches where they can sit during the dark hours, and sand or sawdust where they can perform dust bathing (Figure 43.2a and 43.2b). Possibilities to go outside is important for good chicken welfare.

*Figure 43.2a*  Housing of chickens at Kiipula vocational college in Finland. Students feed and take care of the chickens as part of their practical training in the school. (Photo: Lena Lidfors)

*Figure 43.2b*  Housing of chickens at the school Green Chimney in Brewster, New York. Students feed and take care of the chickens as part of their practical training in the school. (Photo: Lena Lidfors)

Commercial grain is provided ad libitum in troughs placed on the floor. A separate container with ground mollusk shells is available to provide extra calcium for improved eggshell quality. Water is also freely available in a trough or from a nipple container.

In organic poultry production, the hens should be allowed to go outdoors all year around and during the summer have access to pasture areas (KRAV Standards, 2022). A movable hen house can be used to give the hens access to fresh grass during the growing season.

Hobby farmers may allow the hens to brood their eggs. However, as the modern laying breeds are not good brooding hens, even hobby farmers may have to buy chickens from breeders. Chickens for commercial production are usually hatched in hatchery machines in specialized farms.

Broiler chickens bred for meat production, grow very fast and are usually slaughtered when they are 6–8 weeks old. There are broiler breeds with slower growth that suit farm-based interventions better. For more information about the most animal welfare friendly housing of broilers, see report by Bracke et al. (2020).

## How to Involve Chickens in Farm-Based Interventions

As chickens are small animals (on average 70 cm (27.6 inches) tall and 2.6 kg (5.7 pounds)), they may be perceived as less challenging for persons with no previous experience with farm animals. Meeting the hens when they are outdoors gives participants opportunities to see the hens' natural behavior and provides a nice opportunity to interact with them. However, some of the participants may have an innate fear of the birds (ornithophobia), and as such it is important to observe how the participants react to the birds before interacting with them.

Picking up eggs, giving concentrate and water and cleaning out the henhouse are very good activities. As cocks may sometimes become aggressive toward humans, they should be kept separate when clients are visiting.

Clients may appreciate seeing hens brooding their eggs and to interact with the newly hatched chickens. However, the hens may protect their newly hatched chickens. Chickens hatched in machines imprint on humans very easily when being handled and fed during their first days after hatching.

Lifting and holding a chicken or an adult hen may be a very positive interaction for clients. It is important to lift the hens in an animal welfare friendly way, by taking a firm grip around their body and wings and holding them against the body. Chickens can be trained to approach humans if they are given treats, as for example mealworms. It is also possible to clicker train them (Hazel et al., 2015).

# Sheep

## Sheep Origin and Natural Behavior

Sheep originate from the Asiatic mouflon (*Ovis orientalis*) and were domesticated about 9,000–12,000 years ago (Clutton-Brock, 1999). Sheep have been bred for production of either fur/wool, meat or milk. Dwyer (2017) describes natural behaviors of sheep, a brief summary is presented below.

Sheep live in small family groups consisting of females, their lambs and up to two-year-old males. Rams are only present in these family groups during the rutting season. Ewes have low aggression level and seek close body contact with members of the group. The size of a group depends on natural environmental conditions, e.g. on the availability of food sources, and breed. As sheep are small prey animals, they flock together and flee in case of danger. If a single animal is separated from the herd, it vocalizes to keep in contact with its group. Sheep can remember the faces of more than 50 individual sheep for up to two years (Kendrick et al., 2001).

Sheep graze on average 9–11 hours a day over the course of 4–7 feeding periods (Figure 43.3). They eat both grass and leaves and prefer clover to grass. Sheep select their food depending on the nutritional value and often pick the most nutritious part of the plant. Sheep ruminate about eight hours per diurnal cycle during eight rumination periods. Water intake varies with breed, climate and reproductive stage in sheep.

Ewes try to withdraw from the family group when giving birth to 1–2 lambs, but it depends on breed and opportunities to show this behavior. Immediately after the first lamb is born, the ewe stands up, licks it, and performs low-pitched bleats to activate and bond with the lamb. The same behavioral elements are shown for the next lamb. Ewes nurse their lambs in very short bouts, and in natural conditions lambs are usually weaned at six months. Lambs perform locomotor and social play with a peak at 10–14 days.

## Sheep Housing and Management

Indoor housing of sheep during wintertime is practiced in countries with a cold climate. The stable should be draught-free, dry and clean, and have an adequate resting area with straw. Sheep can be kept outdoors on pasture all year round but need access to protection from the sun during summertime. In wintertime, their dense wool protects them from precipitation and low temperatures in most cases. However, they need access to shelters with bedding, and the ground must be non-slippery and dry to avoid injuries.

Sheep need to change pasture regularly to protect them from internal parasites. Lambs learn from their mother what to eat, and different types of concentrate or other feed should be provided to the ewe when she is still nursing her lambs. To limit the potential for disease, sudden changes of feed quality or quantity should be avoided. When given fresh, nutrient-rich feed, a sufficient amount of roughage must be offered. Sheep should always have free access to fresh water and minerals, as they have capacity to compensate for mineral deficiencies when they are offered.

*Figure 43.3* Grazing in growing lambs on pasture at SLU Götala Beef and Lamb Research Centre in Sweden (Photo: Lena Lidfors)

Ewes should be kept in small family groups with their lambs until weaning, and should never be kept alone, as that causes stress. Rams can be kept in small groups, but should be limited to small spaces when mixed; fighting rams back off from each other and then run in high speed to clash their heads, which can cause serious injuries or even death.

Domesticated ewes can birth one-to-five lambs. To avoid disturbance from other sheep during the ewe-lamb bonding, the animals are usually moved to a separate pen. As the ewes only have two teats, some lambs may have problems accessing milk during the short nursing. Farmers can give the lambs extra milk in a bottle while they are in the family group or remove the lamb from the group and provide milk in a teat-bucket.

Lambs can be separated from the ewes and weaned from ten weeks of age, but preferably later. Lambs should be kept in groups of separate sex after weaning, as males are sexually mature from four months of age if not castrated. Learn more about sheep welfare in Ferguson et al. (2017).

## *How to Involve Sheep in Farm-Based Interventions*

Sheep are, due to their smaller size and low aggression level, suitable for farm-based interventions. Meeting sheep outdoors when they approach and make contact can be a positive experience. Watching lambs play in the fields and feeding sheep and lambs can be a fun activity and a positive human-animal interaction. Sheep remember humans, which can be important when interacting with those who visit the farm repeatedly.

Rams should not be used in AAI as they may suddenly attack humans.

## *Swine*

### SWINE ORIGIN AND NATURAL BEHAVIOR

Domesticated swine originate from the wild boar (*Sus scrofa*) and domestication began around 9,000–10,000 years ago (Clutton-Brock, 1999). The appearance and size of domesticated swine have changed considerably compared to their ancestors, but not their behavioral repertoire. The main behavioral difference is a reduced threshold for eliciting fear in domesticated swine, and that they reproduce all year round (D'Eath and Turner, 2009). Špinka (2017) describes natural behaviors of swine. A brief summary is presented below.

In nature, swine live in groups of 2–4 adult sows and their offspring. The social hierarchy is based on dominance relationships where the older and larger individuals dominate the smaller and younger animals. Young boars leave the group when mature (aged 6–10 months) and join bachelor groups, whereas older boars live solitarily and only visit the groups of adult females for mating.

Swine spend most of their hours awake searching for food and exploring their surroundings. They are omnivores, meaning that they eat both vegetation and animals, such as earthworms, insects, rodents and carrions. Rooting and digging are part of their feed seeking and exploratory behavior and they plough up the ground with their snout to find roots and bulbs.

Pregnant sows isolate from the group 2–3 days before farrowing and start preparing a nest approximately 24 hours prior. The nest consists of a hollow with a pile of grass and branches arranged on top, and the sow lies down quietly inside the nest before farrowing starts. The sow and her piglets remain around the nest for 7–10 days before returning to the group. A teat order is established during the first 24 hours, and thereafter suckling follows a cyclic pattern, with nursing every 40–80 minutes. In natural conditions piglets are fully weaned when they are 14–17 weeks old.

## Pig Housing and Management

Sows and growing pigs are kept in groups on deep bedding or slatted floors, whereas adult boars are kept individually. Young pigs are housed in groups with individuals of identical age and size. The animals are frequently regrouped, which leads to unacquainted pigs fighting to restore a strict social hierarchy. Swine need space for establishing their hierarchy, to feed, lie down and perform exploratory behavior. Swine prefer to eat together, thus feeding them in meals, with sufficient feeder space for all animals, is the most animal welfare friendly option. Swine should be provided with suitable substrate, such as straw or other material, to fulfill their need for exploration.

Keeping swine outdoors will give them more space and possibilities to explore. They should then have access to shelters, to keep warm or to avoid sunburn, depending on the climate. In hot weather, a wallow is necessary, as swine cannot sweat and can get easily overheated, especially large or pregnant sows. They wallow in mud or water to increase evaporative cooling (Figure 43.4a). As swine turn over the grass cover, leading to problems with percolation of manure to the groundwater, they should be moved to new areas regularly.

Pregnant sows kept outdoors need access to a farrowing hut large enough for them to turn around and where they can isolate themselves 2–3 days before farrowing (Figure 43.4b). All sows should be provided with proper materials to build a farrowing nest. Newborn piglets are in danger of hypothermia right after birth, and before they start nursing, and thus a heat lamp is often placed in the farrowing area. Sows may give birth to more piglets than they have teats, and farmers often move excess piglets to sows with fewer piglets. This should be dome before the establishment of the teat order, i.e. within the forst 24 hours after birth. Pigs are usually weaned at 4–7 weeks after farrowing.

## How to Involve Swine in Farm-Based Interventions

When swine are involved in farm-based interventions, they should be kept in stable social groups to increase safety for the animals and humans involved. Moving swine to new outdoor areas or pens, cleaning and feeding, but also training them are suitable activities.

*Figure 43.4a* Outdoor kept pigs wallowing in a mud bath on a hot day (Photo: Trine Sund Kammersgaard)

*Figure 43.4b* Sows kept outdoors with huts where they can build a nest before farrowing (Photo: Trine Sund Kammersgaard)

Swine are very easy to train (similar to dogs) which, together with their curiosity, make them an obvious choice for farm-based interventions if caretakers keep in mind their needs. Most pigs can be trained with positive reinforcement, such as clicker training, or by luring them (Jønholt et al., 2021).

Newly farrowed sows and their piglets should not be involved in farm-based interventions, as some sows will react aggressively toward intruders to protect their offspring. However, observing suckling piglets from a distance may be an interesting activity as long as the human presence do not disturb the sow.

## *Cattle*

### *Cattle Origin and Natural Behavior*

Cattle originate from the extinct aurochs (*Bos primigenius*) and domestication began around 9,000 years ago (Clutton-Brock, 1999). There are two main types of cattle: those kept for milk production where the calf commonly is removed from the cow shortly after birth, and those kept for meat production where the calf stays with its mother. Hall (2002) describes natural behaviors of cattle. A brief summary is presented below.

In nature, cattle live in maternal groups with cows of different ages along with their calves. The bull calves leave the group before they are two years old. Young bulls live in bachelor groups and bulls older than 4–5 years are solitary. Cows develop a linear rank order, and the stable hierarchy keeps aggression levels down.

Cattle live on open fields with access to trees for protection against the sun and precipitation. They eat grass and herbs in 4–5 bouts over the day as they ambulate. In between grazing bouts, they ruminate, mostly while lying down, for 6–8 hours, with longer times when the grass contains more fiber. They may also eat leaves from bushes and trees both in the spring and autumn. Cattle get water from the wet grass in the morning and drink water from streams.

Cows separate themselves from the herd shortly before giving birth to bond with the calf without disturbance from other cows. The calf suckles the dam several times per day until natural weaning at 8-11 months. During growth, calves play with each other to train muscles and social skills.

## Cattle Housing and Management

Cattle are kept indoors during wintertime in countries with colder climate, and outdoors all year round in countries with warmer climates (Figure 43.5a). The best welfare for cattle is if they can be loose housed in straw-bedded pens where they have a clean and soft lying surface, plenty of space to stand up and lie down, and where they can rest in their preferred lying position (Ekesbo & Gunnarsson, 2018). During summertime, cattle could be kept on pasture in small groups (i.e. 10–20 cows) of varying ages to keep aggression level low.

Dairy cows are usually fed a mixture of concentrate, silage and hay/straw, called Total Mixed Ration (TMR). The energy content of the TMR is very important for high producing dairy cows as they can produce up to 50 kg of milk daily. Individual cows can be fed extra concentrate on special feeding stations in the loose housing or during milking. Beef cows nurse one calf for 5–8 months, and their energy demands are not as high. They are usually fed silage and/or hay. Growing beef cattle can be raised on good-quality silage. Cattle should always have free access to water of good quality; water intake is influenced by animal-related factors, environmental factors, feed content and water quality (Golher et al., 2021).

Dairy cows should be placed in individual pens some days before expected calving, but with the possibility to see and hear their herd mates. Beef cows may give birth outdoors but need supervision in case something goes wrong. Dairy calves may also stay with the mother several weeks after birth to allow them to suckle parallel with the mother being milked by automatic milking systems (Figure 43.5b).

*Figure 43.5a* Dairy calves being raised by their mother in a loose-housing barn with automatic milking system at SLU Swedish Livestock Research Centre Lövsta (Photo: Lena Lidfors)

*Figure 43.5b* Group-housing in pens with deep litter and a scraped feed alley for growing dairy steers at SLU Götala Beef and Lamb Research Centre in Sweden (Photo: Lena Lidfors)

## How to Involve Cattle in Farm-Based Interactions

Domestication has led to cattle being less aggressive and fearful toward humans. However, they are large animals and need to be handled with caution when used in AAIs.

Interactions with dairy calves during the milk-feeding process can reduce fearfulness of humans and build the human-animal bond by gently stroking them around their chin and neck. Around the weaning process, positive interactions with calves may build positive relationships with humans, which prepare them for AAI.

Cattle involved in AAI should be kept in small stable groups. Adding and removing animals from the group should be avoided as this may put humans at risk. Feeding cattle provides good opportunities for positive human-animal interactions and clients can then feel that they are doing something important for the animals.

Cows with their calf during the first weeks after calving are not recommended to be used in AAI, as the cow may protect the calf. Bulls should also not be involved in AAI, as they may become aggressive.

## Animal Welfare and Ethical Considerations

All animals kept by man should be housed, managed and fed, so that their health and welfare are kept as good as possible, supporting the One Health – One Welfare concept. Brambell (1965) presented the Five Freedoms, which state that animals should be offered Good feeding, Good housing, Good health, Preventing fear and distress and Opportunities for natural behavior. The Welfare Quality project has expanded these and developed protocols for dairy cows, pigs, chickens, etc., which are used to decide if a farm has a good general welfare or not (Welfare Quality Network, 2018).

There are laws and regulations in the EU and in individual countries around the world that aim to improve the welfare of farm animals. Some countries do not have such detailed laws and regulations, and extra care needs to be taken to ensure that the animals involved in farm-based interventions are well treated. This is important not only for the animals' welfare, but also for the humans' welfare, as the benefits of the intervention may be compromised if clients observe suffering in the animals. The International Association of Human-Animal Interaction Organizations (IAHAIO) has developed best practice guidance (IAHAIO international guidelines on care, training, and welfare requirements for farm animals involved in animal-assisted interventions, 2021).

Farm animals are usually raised to be slaughtered or to produce milk or eggs for human consumption. This may raise ethical questions, such as whether it is acceptable to kill them, or if they should be allowed to live as retired animals until they die naturally or are euthanized. The human perception of slaughtering farm animals may be important when involving them in farm-based interventions. Vulnerable people can be more affected by the notion that farm animals are mainly used for production, and that they will be slaughtered before the natural end of their lives. However, this can also be turned to something therapeutic, where one can discuss life and death. The therapist should prepare the client for the possibility of euthanasia or slaughter early in the process of engaging with the animal. This will give the client a choice as to whether he or she should build a relationship with that animal or not.

Many farms involved in farm-based interventions have several different species, which give the clients possibilities to interact with the species they prefer. Small farms can offer more individualized interactions with the farm animals, as for example at Green Chimneys (https://www.greenchimneys.org/). Young persons can learn about life through the daily activities on the farm. It is also pedagogic, as youngsters and clients learn about where the food comes from, and about different species, their biology and their needs, which may increase their interest in animals as such. Knowledge about animals and their needs may also increase the interest in and awareness about animal as sentient beings, and the importance of safeguarding their welfare.

## Conclusion

Farm-based intervention is an alternative to the more traditional involvement of a dog or horse in AAI. On a farm, participants can choose which animal species they want to feed, care for and interact with in different ways. They also have the possibility to grow plants, harvest, cook, make handicraft, take walks in nature or the forest and interact socially with other participants and the farmer. Taken together, this gives participants more freedom and possibilities to perform meaningful tasks, including interactions with the animals. It is important that the animals are kept and managed so that their welfare is good.

### Discussion Questions

1. Why are farm animals suitable for therapy, activity, education, and coaching within AAI?
2. How can farm animals be included in AAI?
3. When is it not suitable to include farm animals in AAI?
4. What are the natural behaviors of chickens, sheep, pigs, and cattle?
5. In what way should farm animals be housed and managed to be most suitable for inclusion in AAI?

## References

Beetz, A. & Enders-Slegers, M.J. (2013). Theoretical background of AAIs in a Green Care context. In C. Gallis (ed.), *Green Care. For human Therapy, Social Innovation, Rural Economy, and Education* (pp. 65–91). Nova Science Publishers. Inc., New York.

Berget, B. & Braastad, B.O. (2008). *Kunnskapsstatus og forskningsbehov for Inn på tunet* (*Knowledge Status for IPT*) (p. 69). Universitetet for miljø- og biovitenskap, Ås.

Berget, B. & Braastad, B.O. (2018). Teoretisk grunnlag for dyreassisterte intervensjoner. In B. Berget, E. Krøger, & A.B. Thorød (eds), *Antrozoologi. Samspill mellom dyr og menneske* (pp. 46–60). Universitetsforlaget, Oslo.

Berget, B., Ekeberg, Ø. & Braastad, B.O. (2008). Animal-assisted therapy with farm animals for persons with psychiatric disorders: Effects on self-efficacy, coping ability and quality of life, a randomized controlled trial. *Clinical Practice and Epidemiology in Mental Health*, 4(1), 1–7. https://doi.org/10.1186/1745-0179-4-9

Berget, B., Ekeberg, Ø., Pedersen, I. & Braastad, B.O. (2011). Animal-assisted therapy with farm animals for persons with psychiatric disorders: Effects on anxiety and depression, a randomized controlled trial. *Occupational Therapy in Mental Health*, 27(1), 50–64. https://doi.org/10.1080/0164212X.2011.543641

Berget, B., Höglund, J., Yrjas, C. & Lidfors, L. (2021). *The farm as a resource for society and the organization of its welfare service* (In Swedish with English summary). Report 11, SLU Future Animals Nature and Health. https://pub.epsilon.slu.se/26289/

Berget, B., Skarsaunet, I., Ekeberg, Ø. & Braastad, B.O. (2007). Humans with mental disorders working with farm animals. A behavioral study. *Occupational Therapy in Mental Health*, 23(2), 101–117. https://doi.org/10.1300/J004v23n02_05

Bracke, M., de Jong, I., Gerritzen, M., Jacobs, L., Nalon, E., Nicol, C., O'Connell, N. & Porta, F. (2020). The welfare of broiler chickens in the EU – From science to action. *Report from Eurogroup for Animals*, Brussels.

Brambell, F.W.R. (Chairman) (1965). *Report of the Technical Committee to Enquiry into the Welfare of Animals Kept Under Intensive Livestock Husbandry Systems*, HMSO, London.

Clutton-Brock, J. (1999). *A Natural History of Domesticated Mammals*, 2nd edn. Cambridge University Press, Cambridge.

D'Eath, R.B. & Turner, S.P. (2009). The natural behaviour of the pig. In J.N. Marchant-Forde (ed.), *The Welfare of Pigs* (pp. 13–45). Springer, Dordrecht.

de Lima, V.A., Ceballos, M.C., Gregory, N.G. & Da Costa, M.J.P. (2019). Effect of different catching practices during manual upright handling on broiler welfare and behavior. *Poultry Science*, 98(10), 4282–4289. https://doi.org/10.3382/ps/pez284

Dwyer, C. (2017). The behaviour of sheep and goats. In P. Jensen (ed.), *The Ethology of Domestic Animals: An Introductory Text* (pp. 199–213). CAB International, Wallingford.

Ekesbo, I. & Gunnarsson, S. (2018). *Farm Animal Behaviour – Characteristics for Assessment of Health and Welfare*. CAB International, Glasgow.

Elings, M. & Hassink, J. (2010). The added value of care farms and effects on clients. In *Building Sustainable Rural Futures*. Proceedings of the 9th European IFSA Symposium, 4–7 July 2010, Vienna (Austria), pp. 239–247.

Ellingsen-Dalskau, L.H., Berget, B., Pedersen, I., Tellnes, G. & Ihlebæk, C. (2015). Understanding how prevocational training on care farms can lead to functioning, motivation and well-being. *Disability and Rehabilitation*, 38(25), 2504–2513. http://doi.org/10.3109/09638288.2015.1130177

Ellingsen-Dalskau, L.H., Morken, M., Berget, B. & Pedersen, I. (2016). Autonomy support and need satisfaction in prevocational programs on care farms: The self-determination theory perspective. *Work*, 53(1), 73–85. http://doi.org/10.3233/WOR-152217

Ferguson, D., Lee, C. & Fisher, A. (eds) (2017). *Advances in Sheep Welfare*. Woodhead Publishing Cambridge.

Gallis, C. (2013). *Green Care: For Human Therapy, Social Innovation, Rural Economy, and Education* (Vol. 12). Nova Science Publishers, New York.

Gerlach-Spriggs, N., Kaufmann, R.E., Warner, S.B. (1998). *Restorative Gardens: The Healing Landscape*. New Haven, CT: Yale University Press.

Golher, D.M., Patel, B.H.M., Bhoite, S.H., Syed, M.I., Panchbhai, G.J. & Thirumurugan, P. (2021). Factors influencing water intake in dairy cows: A review. *International Journal of Biometeorology*, 65(4), 617–625. https://doi.org/10.1007/s00484-020-02038-0

Hall, S.J. (2002). Behaviour of cattle. In P. Jensen (ed.), *The Ethology of Domestic Animals: An Introductory Text* (pp. 131–143). CAB International, Wallingford.

Hassink, J. & Van Dijk, M. (eds) (2006). *Farming for Health: Green-care Farming across Europe and the United States of America* (Vol. 13). Springer Science & Business Media, Dordrecht.

Hassink, J., De Bruin, S.R., Berget, B. & Elings, M. (2017). Exploring the role of farm animals in providing care at care farms. *Animals*, 7(6), 45. https://doi.org/10.3390/ani7060045

Hauge, H., Kvalem, I.L., Berget, B., Enders-Slegers, M.J. & Braastad, B.O. (2014). Equine-assisted activities and the impact on perceived social support, self-esteem and self-efficacy among adolescents–an intervention study. *International Journal of Adolescence and Youth*, 19(1), 1–21. https://doi.org/10.1080/02673843.2013.779587

Hazel, S.J., O'Dwyer, L. & Ryan, T. (2015). "Chickens are a lot smarter than I originally thought:" Changes in student attitudes to chickens following a chicken training class. *Animals*, 5(3), 821–837. https://doi.org/10.3390/ani5030386

Hemsworth, P.H., Verge, J. & Coleman, G.J. (1996). Conditioned approach-avoidance responses to humans: The ability of pigs to associate feeding and aversive social experiences in the presence of humans with humans. *Applied Animal Behaviour Science*, 50(1), 71–82. https://doi.org/10.1016/0168-1591(96)01065-9

*IAHAIO International Guidelines on Care, Training and Welfare Requirements for Farm Animals Involved in Animal-Assisted Interventions* (2021). version 1.1. https://iahaio.org/iahaio-international-guidelines-on-care-training-and-welfare-requirements-for-farm-animals-involved-in-animal-assisted-interventions/

Jegatheesan, B., Beetz, A., Ormerod, E., Johnson, R., Fine, A., Yamazaki, K.,... & Choi, G. (2014). IAHAIO Whitepaper 2014 (updated for 2018). The IAHAIO definitions for animal assisted intervention and guidelines for wellnes of animals involved in AAI. http://iahaio.org/best-practice/white-paper-on-animal-assisted-interventions

Jønholt, L., Bundgaard, C. J., Carlsen, M. & Sørensen, D.B. (2021). A case study on the behavioural effect of positive reinforcement training in a novel task participation test in Göttingen mini pigs. *Animals*, 11(6), 1610. https://doi.org/10.3390/ani11061610

Kendrick, K.M., da Costa, A.P., Leigh, A.E., Hinton, M.R. & Peirce, J.W. (2001). Sheep don't forget a face. *Nature*, 414(6860), 165–166. https://doi.org/10.1038/35102669

Lidfors, L., Loberg, J., Nielsen, P.P. & Jensen, M.B. (2009). Influence of milk feeding methods on the welfare of dairy calves. In A. Aland & F. Madec (eds), *Sustainable Animal Production - The Challenges and Potential Developments for Professional Farming* (pp. 161–168). Wageningen Academic Publishers, Wageningen. https://doi.org/10.3920/978-90-8686-685-4

Lindahl, C., Pinzke, S., Herlin, A. & Keeling, L.J. (2016). Human-animal interactions and safety during dairy cattle handling—Comparing moving cows to milking and hoof trimming. *Journal of Dairy Science*, 99(3), 2131–2141. https://doi.org/10.3168/jds.2014-9210

Lürzel, S., Münsch, C., Windschnurer, I., Futschik, A., Palme, R. & Waiblinger, S. (2015). The influence of gentle interactions on avoidance distance towards humans, weight gain and physiological parameters in group-housed dairy calves. *Applied Animal Behaviour Science*, 172, 9–16. https://doi.org/10.1016/j.applanim.2015.09.004

Mench, J.A. (2017). Behaviour of domesticated birds: Chickens, turkeys and ducks. In P. Jensen (ed.), *The Ethology of Domestic Animals – An Introductory Text*. CABI, Glasgow.

Munksgaard, L., De Passillé, A.M., Rushen, J., Thodberg, K. & Jensen, M.B. (1997). Discrimination of people by dairy cows based on handling. *Journal of Dairy Science*, 80(6), 1106–1112. https://doi.org/10.3168/jds.S0022-0302(97)76036-3

Pedersen, I., Ihlebæk, C. & Kirkevold, M. (2012). Important elements in farm animal-assisted interventions for persons with clinical depression: A qualitative interview study. *Disability and Rehabilitation*, 34(18), 1526–1534. https://doi.org/10.3109/09638288.2011.650309

Pedersen, I., Patil, G., Berget, B., Ihlebaek, C. & Gonzalez, M.T. (2016). Mental health rehabilitation in a care farm context: A descriptive review of Norwegian intervention studies. *Work*, 53(1), 31–43. https://doi.org/10.3233/WOR-152213

Pedersen, I., Patil, G., Berget, B., Ihlebæk, C., Gonzalez, M.T (2016). Mental health rehabilitation in a care farm context: A descriptive review of Norwegian intervention studies. *Work*, 53, 31–43.

Prunier, A., Mounier, A.M. & Hay, M. (2005). Effects of castration, tooth resection, or tail docking on plasma metabolites and stress hormones in young pigs. *Journal of Animal Science*, 83(1), 216–222. https://doi.org/10.2527/2005.831216x

Rushen, J., Munksgaard, L., de Passille, A.M.B., Jensen, M.B. & Thodberg, K. (1998). Location of handling and dairy cows' responses to people. *Applied Animal Behaviour Science*, 55(3-4), 259-267. doi: 10.1016/s0168-1591(97)00053-1

Schütz, K.E., Hawke, M., Waas, J.R., McLeay, L.M., Bokkers, E.A.M. Van Reenen, C.G.,... & Stewart, M. (2012). Effects of human handling during early rearing on the behaviour of dairy calves. *Animal Welfare-The UFAW Journal*, 21(1), 19–26. https://doi.org/10.7120/096272812799129411

Silva, L.P., Sant'Anna, A.C., Silva, L.C.M. & Paranhos da Costa, M.J.R. (2017). Long-term effects of good handling practices during the pre-weaning period of crossbred dairy heifer calves. *Tropical Animal Health and Production*, 49(1), 153–162. https://doi.org/10.1007/s11250-016-1174-7

Silvera, A.M., Wallenbeck, A., Butterworth, A. & Blokhuis, H.J. (2017). Modification of the human–broiler relationship and its potential effects on production. *Acta Agriculturae Scandinavica, Section A—Animal Science*, 66(3), 161–167. https://doi.org/10.1080/09064702.2017.1286379

Sommavilla, R., Titto, E.A.L., Titto, C.G. & Hötzel, M.J. (2016). Ninety one-days-old piglets recognize and remember a previous aversive handler. *Livestock Science*, 194, 7–9. https://doi.org/10.1016/j.livsci.2016.10.008

Špinka, M. (2017). Behaviour of pigs. In P. Jensen (ed.), *The Ethology of Domestic Animals – An Introductory Text*. CABI, Glasgow.

Standards for KRAV-certified production (2022). https://www.krav.se/en/standards/download-krav-standards/ (Accessed 2022–03–30)

Steigen, A.M., Kogstad, R. & Hummelvoll, J.K. (2016). Green Care services in the Nordic countries: An integrative literature review. *European Journal of Social Work*, 19(5), 692–715. https://doi.org/10.1080/13691457.2015.1082983

Stewart, M., Shepherd, H.M., Webster, J.R., Waas, J.R., McLeay, L.M. & Schütz, K.E. (2013). Effect of previous handling experiences on responses of dairy calves to routine husbandry procedures. *Animal*, 7(5), 828–833. https://doi.org/10.1017/s175173111200225x

Sutherland, M.A., Davis, B.L., Brooks, T.A. & Coetzee, J.F. (2012). The physiological and behavioral response of pigs castrated with and without anesthesia or analgesia. *Journal of Animal Science*, 90(7), 2211–2221. https://doi.org/10.2527/jas.2011-4260

Waiblinger, S., Boivin, X., Pedersen, V., Tosi, M.V., Janczak, A.M., Visser, E.K. & Jones, R.B. (2006). Assessing the human–animal relationship in farmed species: A critical review. *Applied Animal Behaviour Science*, 101(3–4), 185–242. https://doi.org/10.1016/j.applanim.2006.02.001

Waynert, D.F., Stookey, J.M., Schwartzkopf-Gensewein, K.S., Watts, J.M. & Waltz, C.S. (1999). The response of beef cattle to noise during handling. *Applied Animal Behaviour Science*, 62(1), 27–42. https://doi.org/10.1016/S0168-1591(98)00211-1

Welfare Quality Network (2018). http://www.welfarequality.net/en-us/home/ (Accessed 2022–04-04)

Zulkifli, I. & Azah, A.S.N. (2004). Fear and stress reactions, and the performance of commercial broiler chickens subjected to regular pleasant and unpleasant contacts with human being. *Applied Animal Behaviour Science*, 88(1–2), 77–87. https://doi.org/10.1016/j.applanim.2004.02.014

# 44
# PUBLIC PERCEPTIONS OF FARM ANIMAL WELFARE
## The Case of Cow-Calf Separation

*Lara V. Sirovica and Marina A.G. von Keyserlingk*

## Introduction

Human-animal interactions (HAIs) in animal production and management settings are complex, involving not only different species with different needs, but different people and interested parties with different perceptions and concerns. Understanding these perceptions is essential for identifying and addressing concerns around animal welfare, including those practices that fail to resonate with societal values. In order to maintain public trust in farming, there is a need to accept that some traditional animal husbandry practices may no longer be deemed acceptable by society. For such practices, there is urgent need to identify what aspects must be modified or changed to allow the industry to retain its social license. What follows is a discussion of the consideration of animal welfare, including the importance of public perceptions and how they can be incorporated to aid in identifying practices that are socially acceptable. The contentious issue of early cow-calf separation, a long-established management practice of the dairy industry, is used to illustrate how consideration of the public as an important and credible partner early on in the discussions can help identify a path forward when addressing contentious animal management issues, especially involving animal welfare concerns, that will hopefully improve public trust in dairy farming.

## In Consideration of Animal Welfare

### Definitions and Perceptions of Animal Welfare

Animal welfare refers to the quality of life that animals have and is often broken down into the intersecting concepts of biological functioning, affective state, and naturalness (Fraser et al., 1997). Biological functioning refers to the health of an animal and their ability to achieve normal functioning through physiological processes such as growth and development; affective state refers to how an animal is feeling, including experiences of emotions of positive or negative valence; and naturalness refers to an animal's ability to use their natural capabilities to perform behaviors that satisfy motivational needs. Naturalness is also sometimes used to refer to an animal's ability to live a 'natural life,' although in captive environments, especially those characteristic of intensive production systems, living a 'natural life' may not be possible, let alone always an ideal model for good welfare (Rushen et al., 2011; Beaver et al., 2019a).

Fraser et al.'s (1997) framework highlights the complex, multidimensional nature of animal welfare and enables a discussion that takes into consideration the values of different groups. All groups are arguably important at some point along the value chain, but vary in degrees of knowledge. Many of them are at the interface of HAIs, such as farmers, and veterinarians, whilst others are more removed, such as the public and regulatory bodies (e.g., World Organization for Animal Health (WOAH)).

Different interest groups often ascribe different levels of importance to the intersecting concepts that make up this framework of animal welfare. For example, Te Velde et al. (2002) and Vanhonacker et al. (2008) both reported that lay people (those with little to no experience of farm animal agriculture) tended to focus most on how animals were feeling and their abilities to live a natural life. In contrast, farmers tended to focus more on issues directly relating to animal health and productivity. Interestingly, these authors also noted a discrepancy in how lay people and farmers evaluated animal welfare, with farmers having a more positive perception of farm animal housing conditions compared to lay people.

It follows that while welfare can be viewed as intrinsic to the animal, animal welfare as a concept is not just about animals. It exists in practice (i.e., the field of animal welfare science has evolved to measure and to improve or protect animal welfare) because of the weight and importance that humans give it and in relation to our ethical framing of animal issues (Fraser, 1999; Broom, 2011). How we perceive animals and their treatment is integral to the practice of animal welfare. Therefore, understanding these differences in perceptions is essential to understanding and promoting animal welfare.

At the nexus of human and animal interests inherent within the field of farm animal production, animal welfare is affected and largely determined by human behaviors (Fraser & MacRae, 2011). It follows that animal welfare as a practice must also be defined in terms of human involvement. Fraser et al. (1997) highlighted this acknowledgment of the human dimension of animal welfare when defining their framework, explaining that "Scientific research on 'animal welfare' began because of ethical concerns over the quality of life of animals … [and] The conception of animal welfare used by scientists must relate closely to these ethical concerns" (p. 187). Thus, in addition to the aspects (biological functioning, affective state, naturalness) identified as key areas of ethical concern, all of which can also be measured on an animal level, there also exists a concept of a 'standard of care.' This concept is not new and is consistently brought up by scientists (e.g., Weary & Robbins, 2019) and the public (e.g., Spooner et al., 2014a) alike. It stems from our ancient and ongoing relationship with animals and the idea that by profiting from and engaging in these relationships, we owe the animals a duty of care (Engster, 2006). It follows that there is a minimum quality of care, or 'humane care,' that animals must receive from humans to ensure the animals' welfare. This implies an acknowledgment of responsibility placed on human actions toward animals as an inherent part of animal welfare. This concept highlights the importance of our role in affecting animal welfare, i.e., the idea that we are responsible for animal welfare and that good animal welfare cannot exist without a conscious effort on our part (whether as producers, veterinarians, scientists, or the public – as citizens and consumers) to care for animals and their needs.

Particularly relevant to animal agriculture is the discussion of the human relationship with animals and the standard of care concept discussed in terms of the 'ancient contract' (Rollin, 2008). The late Bernie Rollin argued that in many cases industrial agriculture has placed us in a position that we are in breach of this contract, and that advances in technology have allowed us to increase production even at the expense of sustainable animal care practices. He also stated that when "we fail to create sustainable systems, we will eventually be unable to raise animals or cultivate the land, in the absence of affordable (in all senses) input" (p. 19). Rollin further argued that in addition to this "prudential dimension" to needing to restore "husbandry," with "the essence of husbandry [being] care," there is an "ethical dimension [which] incorporates our obligations to the future of humanity and to renewing our ancient contract with the animals" (Rollin, 2008, p. 19). It is important to note that others have argued that the effects of the industrialization of animal agriculture do not necessitate poor animal welfare; for example, the use of certain technologies, access to more resources and the standardization of good animal care practices across larger, industrialized farms have been shown in some instances to benefit animal welfare (see review and discussion by Robbins et al., 2016). However, as discussed by Rollin (2008), it is clear that the industrialization of agriculture challenges our way of thinking about our relationship with animals and in turn how we care for them.

## *Importance of Understanding Perceptions of Farm Animal Welfare and Management Practices*

With the role that human actions play in determining animal care on farms, it follows that the goal of improving farm animal welfare is inherently human action focused. As human behavior is influenced by attitudes, or people's evaluations and perceptions of a particular concept, object, or situation (Vaske, 2008), it is important to study human attitudes and underlying perceptions as they relate to animal welfare, including specific animal management practices. Understanding these, especially from the public, is essential both in the evaluation of existing management practices and the development of more socially sustainable and welfare-friendly practices (e.g., in the case of egg production: Swanson et al., 2011).

The public is increasingly interested in the social sustainability of food products (Vermeir & Verbeke, 2006), including dairy (e.g., de Graaf et al., 2016). Animal welfare is a critical component of the social sustainability pillar (von Keyserlingk et al., 2013), which also includes the socio-cultural concept that reflects people's concerns about a practice being consistent with their values and able to preserve itself and its components for the future (i.e., not self-limiting in its exploitation of resources) (Boogaard et al., 2008). This further ties into the social license of animal industries (as discussed in terms of social ethics for animals by Rollin, 2004), which is the idea that while society (in large part, the public) acknowledges that it does not understand these industries enough to regulate them, society trusts these industries to self-regulate so long as their practices resonate with societal values. However, if this trust is breached through a failure of industries to self-regulate in this way, then society will step in and regulate, despite acknowledging that as outsiders and lay persons, their understanding of farming may not be complete. It follows that societal values must be identified and arguably better understood and taken into consideration as farm animal industries contemplate animal care practices; failure to do so could undermine their social license to practice. There is increasing recognition and emphasis being placed on the importance of animal welfare, including public perceptions of animal welfare, to social license as an essential issue for farm animal industries (e.g., review by Alonso et al., 2020).

The value placed on animal welfare can be seen in part through it being expressed as an area of concern by many consumers, with links having been discussed between perceptions of animal welfare and consumption intent and behavior toward animal products (Clark et al., 2017). For a dairy example, 93% of public respondents to a UK survey investigating perceptions of milk marketing and animal welfare reported that they would pay more for 'good dairy welfare' (Ellis et al., 2009). Similarly, there are reports of higher willingness to pay for products from 'welfare-friendly' animal production systems compared to standard systems or those perceived as less 'welfare-friendly' (e.g., Miranda-de la Lama et al., 2017).

Beyond concerns over consumer willingness to pay for animal welfare (with care needed in not over interpreting links between reported willingness to pay and actual consumer behavior which is often influenced by a complex interaction of situational factors and different underlying values (e.g., Liebe et al., 2011)), there is increasing recognition of the role of social pressure and movement outside of what is expressed by consumers with their wallets. For example, a survey of the Australian public found that 95% of people reported farm animal welfare to be an issue and that 91% wanted at least some reform to address it (Futureye, 2018). A recent European Citizen's Initiative ('End the Cage Age') was also successful in inspiring an EU-wide resolution to investigate and propose policy options toward ending the use of cages in animal agriculture, with a potential ban being considered to take effect as early as 2027 (European Commission, 2021).

There is little doubt that the public is increasingly having an influence on government and corporate approaches toward farm animal care practices and policies, both through their purchasing habits and by expressing their perceptions of such practices. Concerns for animal welfare and the sustainability of animal production practices will likely become especially important as the public becomes more aware of production practices and also more involved in expressing their concerns on what it means to employ welfare-friendly livestock practices (e.g., through pressure on governmental and non-governmental policy and regulatory bodies) (Buller et al., 2018).

Examples of specific farm animal management practices of concern include intensive housing of pigs (e.g., Ryan et al., 2015) and cages for laying hens (e.g., Savory, 2004). Considerable research has already gone into assessing alternative management options to the contentious practices of gestation stalls for sows (as reviewed by Barnett et al., 2001) and to conventional cages (often referred to as 'battery cages') for hens (as reviewed by Schuck-Paim et al., 2021).

Importantly, work investigating public and industry perceptions has also been performed in the case of specific alternative practices for sows (e.g., industry, Spooner et al., 2014b; public, Ryan et al., 2015; Vandresen & Hötzel, 2021) and for hens (e.g., industry, Stadig et al., 2015; public, Heng et al., 2013). Participants across these studies expressed a range of perceptions; however in general, industry perceptions of animal welfare effects of practices tended to focus more on perceived biological functioning benefits of conventional practices (i.e., gestation and farrowing crates for sows and conventional cages for laying hens, respectively), while public perceptions were more negative toward these practices that were perceived as providing poor animal welfare and being restrictive in nature. If industry or scientist led alternative management systems do not resonate with societal values, there is a risk that these alternatives may be rejected once public awareness of them grows (Weary et al., 2016).

A key example of this is the failure of the public to accept enriched cages as an adequate alternative to conventional cages for laying hens, despite widespread support for and investment in developing this specific alternative by the scientific community (Weary et al., 2016). While there now exists some research on public perceptions toward alternatives to conventional cages for laying hens, there was little knowledge in this area prior to the investment in enrichment cages by scientists and the poultry industry. Industry scientists working to develop enriched cages focused on factors associated with hen health (such as reduced risks of feather pecking and cannibalism) and behavioral and affective state benefits (such as the ability to perform certain motivated behaviors including nesting and dust bathing), but likely did not fully consider and address societal values around naturalness and rejection of the behavioral restriction resulting from cage housing (Weary et al., 2016). This disconnect resulted in the development of an alternative management system that failed to win public support and was ultimately doomed to market failure (Savory, 2004).

The example of modified cages for laying hens illustrates the need to consider public perceptions in addition to other interested parties' values and to animal outcomes (including animal welfare) when assessing and developing production practices thought to be mutually sustainable and more welfare friendly. To help avoid a similar situation to that of modified cages for laying hens, research on public perceptions toward alternative management practices to other conventional and contentious animal management issues is necessary. What follows is a case example of a contentious issue in the dairy industry.

## A Case Example of a Contentious Issue: Immediate and Permanent Separation of the Cow and Calf at Birth

To produce milk, cows must first give birth to calves, the majority of which are removed from their mother cows within a few hours of birth and then individually housed (e.g., Brazil, Hötzel et al., 2014; Canada, Winder et al., 2018; Europe, Marcé et al., 2010; US, USDA, 2016). Animal welfare concerns surrounding these practices and research into public perceptions suggest that these are controversial aspects of cow-calf management that will have to be addressed by the dairy industry sooner rather than later to maintain social license.

### *Early Cow-Calf Separation: Concerns and Perceptions*

Common reasons reported in support of separating cows and calves early on farm include improved cow and calf health and minimizing loss of saleable milk; however, there is little scientific evidence supporting this (reviewed by Beaver et al., 2019b; Meagher et al., 2019). There is also considerable public concern over early separation (North America, Ventura et al., 2013, 2016, Sirovica et al., 2022; Germany and US, Busch et al., 2017; Brazil, Hötzel et al., 2017; the Netherlands and Norway, Boogaard et al., 2010). Concerns with early separation versus cow-calf contact include 'naturalness,' cow and calf emotions, milk production, and health (including calf nutrition and disease risk).

In contrast to poor public perceptions of early cow-calf separation, prolonged cow-calf contact is often viewed positively and is associated by the public with 'naturalness'; indeed, many opponents of early separation cite 'unnaturalness' as a key concern surrounding early separation (e.g., Boogaard et al., 2011; Ventura et al., 2013, 2016). This result is consistent with the more general perception that 'naturalness,' or the ability of animals to have a more natural existence, leads to better welfare (e.g., Ventura et al., 2016).

Cow and calf emotions, specifically distress associated with separation, are also an area of concern for many people (e.g., mix of people with and without industry involvement or knowledge, Ventura et al., 2013; farmers, Hötzel et al., 2014; public, Busch et al., 2017, Sirovica et al., 2022). Proponents of early separation (often – but not exclusively – primarily include those involved with the dairy industry, including many veterinarians and farmers) often cite the reduction of acute distress responses following early compared to later separation (Ventura et al., 2013). There is some evidence indicating that separating cows and calves, regardless of age of the calves, will involve some distress responses and limit the ability of cows to perform and of calves to receive motivated and beneficial maternal behaviors (reviewed by Meagher et al., 2019). Considering the method of separation (i.e., with step-down levels of cow-calf contact versus abrupt separation) may help mitigate cows' and calves' acute distress responses to separation (reviewed by Meagher et al., 2019). However, opponents of early separation often bring up concerns around the emotional distress that any separation of mother and offspring potentially causes (Ventura et al., 2013; Busch et al., 2017; Sirovica et al., 2022), suggesting that separation will likely be an area of contention regardless of how it is approached in practice.

Such concerns that the emotional distress of cows and calves may extend beyond the acute effects of separation commonly addressed in the literature (e.g., reviewed Meagher et al., 2019) and cited by proponents of early separation (Ventura et al., 2013) are in line with the potential for long-term stress resulting from separation in both cows and calves (reviewed by Meagher et al., 2019). For example, calves separated early display more stereotypic behaviors than calves allowed extended cow-calf contact (Roth et al., 2009), and maternal deprivation is associated with stress and the development of stereotypic behavior in many species (Latham & Mason, 2008). That dairy cows displayed a reduction in brush use during the postpartum period and following early separation suggests these are challenging events for cows that may lead to negative affective states (Lecorps et al., 2021) and there is even some speculation that cows may suffer from something like postpartum depression (e.g., Meagher et al., 2019); especially considering the high incidences of both disease and stress in cows during the transition period (three weeks pre- and three weeks post-calving) (reviewed by Sundrum, 2015) and links between depression and reduced immunity and health in humans (e.g., Kiecolt-Glaser & Glaser, 2002). Potential similarities in effects of separation and separation distress postpartum across species have also been brought up by some survey participants presented with cow-calf management systems involving early separation who discussed the cow-calf relationship in terms of, and in relation to, the human parent-child relationship and how separating any parent from a child regardless of species is distressing (Sirovica et al., 2022).

Separate from concerns regarding how cows and calves feel, is the concern that milk production may be reduced by nursing (brought up by proponents of early separation) and the contrasting belief that it may be increased when the calf is able to suckle (brought up by opponents of early separation) (Ventura et al., 2013; Busch et al., 2017; Sirovica et al., 2022). In their recent systematic review summarizing this body of research, Meagher et al. (2019) reported that the available evidence suggests that while cows will temporarily produce less-saleable milk when nursing calves, this temporary reduction in saleable milk yield may balance out over the entire lactation cycle, depending on when suckling calves are weaned. It is also important to note that farmers will not have the extra expense and labor of feeding nursing calves with milk replacer. Ultimately, concerns surrounding milk production, production costs, and profitability are important to all people, not just producers. For example, these concerns are frequently brought up by members of the public, whether or not they actually support or oppose the practice of early separation (Ventura et al., 2013; Busch et al., 2017; Sirovica et al., 2022). This suggests that some members of the public are sympathetic to challenges faced by farmers in balancing cow-calf care and productivity. However, other public members have explicitly expressed 'dissatisfaction with industry motivations' in prioritizing productivity and profit over cow-calf care through the practice of early separation (Ventura et al., 2013; Busch et al., 2017; Sirovica et al., 2022).

Calf nutrition in early separation systems is another area of concern to many people regardless of industry involvement (Ventura et al., 2013). Members of the public have suggested that calves would have better access to good quality milk and colostrum if suckling from their mothers as opposed to being fed artificially by the farmer (Ventura et al., 2013; Sirovica et al., 2022). Much work has been done to assess the production, behavioral, and health benefits of increased milk allowance to calves (e.g., Ballou, 2012; Rosenberger et al., 2017). However, milk allowance can be increased with or without access to cows, so this concern is not specific to early separation.

There are additional health effects thought to be associated both with cow-calf contact and lack of contact. While cow udder health has been brought up as a reason to separate cows and calves, recent reviews suggest that udder health, in particular the risk of mastitis, is improved in cows suckled by calves (Johnsen et al., 2016; Beaver et al., 2019b). Disease risk to calves from prolonged cow-calf contact is of particular concern to those supporting early cow-calf separation, often brought up by those involved in the dairy industry (Ventura et al., 2013). Separating cows and calves to reduce disease risk has also often been advocated by veterinarians and researchers (e.g., McGuirk, 2008). However, despite these long-held beliefs, there is no conclusive evidence to support the argument that keeping calves in contact with cows increases disease in calves (as reviewed by Beaver et al., 2019b). In contrast, when brought up by lay person participants in a recent North American survey, disease risk was more often discussed as a consequence of calves being denied access to their mother's milk (Sirovica et al., 2022). Many people involved with the dairy industry also express concerns over the physical safety of calves kept with cows (e.g., risk of getting crushed), although this appears less of a perceived concern to the public (Ventura et al., 2013). Clearly, any changes to the amount of contact between cows and calves will have to be made with considerations to the physical environment that cows and calves are kept in. As with the large role of environmental hygiene in disease risk, there is a role of housing environment in physical safety risk that must be accounted for and disentangled from risks of actual cow-calf contact (reviewed by Beaver et al., 2019b). In fact, as demonstrated by many cow-calf contact studies (e.g., Fröberg et al., 2011; Wenker et al., 2020), it is possible to house cows and calves together in production settings (although some housing adjustments may have to be made depending on individual farm design). The public is likely not unaware of challenges farmers may face in having to make changes to implement cow-calf contact, with some people acknowledging or even empathizing with the changeability of dairy farming structures as a barrier to accepting cow-calf contact (Ventura et al., 2013; Busch et al., 2017; Sirovica et al., 2022). However, keeping cows and calves together (as opposed to separating them) was also discussed by some survey participants as a beneficial change for farmers and milk production, and was ultimately perceived more positively than any system involving early cow-calf separation (Sirovica et al., 2022).

As awareness has grown of concerns over the conventional system of removing calves from their mothers immediately after birth and then housing calves individually, so too has an interest in alternative practices. For example, Johnsen et al. (2016) discussed the feasibility of different levels of cow-calf contact, suggesting that while more research is needed into the full effects and implementation of systems differing in level of cow-calf contact, early work indicates that these systems have potential. In addition to considering the effects of different levels of social contact on cows and calves, it is also important to consider what forms of social contact (i.e., calf-calf, any cow-calf, or only mother cow-calf) are of value to the public in cow-calf management systems. What follows is a discussion of concerns and perceptions surrounding two examples of alternative practices involving different forms of social contact, with social housing of calves in contrast to individually housing calves but not addressing cow-calf contact, and keeping calves with foster cows in contrast to restricting all cow-calf contact but not addressing the immediate separation of calves from their own mothers.

## Social Housing of Calves

Housing calves with other calves (social or group housing) is an alternative practice seeing increasing use and interest, in part as a result of concerns over negative social and physical effects of isolating calves (Cantor et al., 2019). Benefits of housing calves with other calves are well documented and include improved solid feed intake and calf weight gains, improved social skills and coping abilities, and improved behavioral development (reviewed by Costa et al., 2016). Calves also appear highly motivated to engage in full social contact with other calves (Holm et al., 2002).

There is some evidence of public concerns regarding individual housing (Boogaard et al., 2010; Ventura et al., 2013). Furthermore, American participants in one survey perceived the practice of allowing dairy cattle to interact with other cattle (i.e., as provided by social housing) as a positive management practice (Widmar et al., 2017), and participants in another survey preferred social housing over individual housing when asked to choose solely between these two practices (Perttu et al., 2020). Some opponents of early cow-calf separation in the study by Ventura et al. (2013) also suggested that they may view cow-calf management systems involving early separation more favorably if these systems provided what they perceived would be 'better' care for the calf following separation, such as through social housing of calves with conspecifics.

However, social housing of calves (in the absence of any maternal figure) does not appear likely to be perceived by the public as an adequate alternative solution to concerns surrounding the standard cow-calf management system of early separation from the mother cow followed by individually housing calves. For example, North American participants expressed negative perceptions of animal welfare and were negative in terms of their overall attitude toward any system involving early separation. In contrast, participants presented with a scenario that maintained contact between the mother cow and her calf had positive perceptions of animal welfare within system and positive attitudes toward the system (Sirovica et al., 2022).

## Foster Cows

The use of foster cows (also sometimes referred to as 'nurse cows') is relatively uncommon and less discussed in the dairy industry; however, it has been suggested as a potential alternative cow-calf management practice (see review by Kälber & Barth, 2014). Foster cows are often further along in lactation (and thus already producing lower quantities of saleable milk) and each cow may be used to nurse multiple calves, reducing the loss of saleable milk compared to each cow nursing her own calf. However, while foster cows are arguably able to perform some natural maternal behaviors, potentially benefiting cow and calf, there is concern for the welfare of cows and calves forced into these foster relationships, as cows form specific and differential bonds with their own calves (von Keyserlingk & Weary, 2007; Kent, 2020). For example, potential complications of such specific and differential cow-calf bonds in foster cow systems are ensuring the acceptability of alien calves and even distribution of maternal care to multiple calves at once by each foster cow (reviewed by Johnsen et al., 2016). In addition, cows and calves in foster cow systems still have to undergo the stress of early separation (from their biological mother cow/calf) in addition to later separation stress (from their foster cow/calf at weaning). Indeed, foster cows and calves have been observed to display similar distress responses to separation (including increased vocalizations) as do biological cow-calf pairs (Loberg et al., 2007, 2008).

However, whilst there may be some potential benefits of foster cow systems such as the calves being able to suckle from a cow, this system is predicted to not be accepted given that it still requires the mother cow-calf bond to be severed at birth; a fact that resulted in this system being rejected by North American participants unfamiliar with dairy farming (Sirovica et al., 2022).

## Implications for the Future of Socially Sustainable Cow-Calf Management Systems

There are concerns surrounding the quality of life and care that both cows and calves receive in cow-calf management systems, with value also placed on the mother cow-calf bond. This suggests that alternative practices to the conventional system (i.e., early separation followed by individually housing calves) that do not address concerns for both cows and calves and this bond (i.e., early separation followed by social housing of calves and early separation followed by keeping calves with foster cows) will likely fail in the long run. This does not dismiss the animal welfare benefits of socially housing calves over individually housing calves after early cow-calf separation. Nor does it preclude in the interim socially housing separated calves. In fact, the potential ban on cages being discussed in the EU will include a ban on individual calf pens/cages (European Commission, 2021).

There are also important concerns around the social sustainability of socially housing calves that should be contemplated given the public concerns over any system that relies on immediate separation of the cow and her calf. Thus, it may be advantageous to invest now in developing management solutions that will allow for highly valued mother cow-calf contact, rather than relying on interim practices that are likely to be socially unacceptable in the long term. It is important that animal agriculture consider whenever possible the underlying values of all groups involved and affected, including the public, when contemplating the future of their industry. By doing so they may avoid missteps such as that made by the egg industry when they embraced the furnished cage without consulting the public. By engaging the public early on it may be possible to better identify practices that will be socially acceptable from the outset rather than over-investing resources on practices that the industry thinks will be acceptable but then are rejected (i.e., as seen in the previously discussed case of enriched cages for laying hens (Weary et al., 2016)). Figure 44.1 illustrates similarities in these two case examples.

## Conclusion

Concerns around farm animal welfare, including the standard of care provided to dairy cows and calves, are underlying values of importance to the public surrounding farm animal management and it is incumbent on the farm animal industries to incorporate animal care practices that resonate with societal values; failure to do so risks their social license to practice. In the case of the common practice of immediately separating the cow from her offspring at birth, the dairy industry is encouraged to investigate practices that show that they have heard the public's concerns and that they are attempting to identify alternative practices

*Figure 44.1* A laying hen next to an enriched cage. The enriched cage was promoted as an alternative to the much smaller barren cages used by the egg industry that were heavily criticized by the public. However, there is growing evidence that this industry led proposed solution will ultimately fail given that it still restricts movement of the birds. The cow and her calf are next to a group of calves housed together after being separated from their cows. Although awareness of cow-calf separation is to date limited amongst the public, based on lessons learned from the egg industry, alternative housing systems that promote social housing but still include immediate separation of the calf from their mother will likely fail given that they do not address underlying concerns raised by the public regarding the breaking of the maternal bond. Illustration by Ann Sanderson

that resonate with societal values but also maintain high standards of animal welfare. Future research on this topic is necessary as it can provide insights into the path forward on how to best manage cow-calf contact systems that are mutually sustainable for all involved and affected parties, including the animals living within the systems. Ideally, research on attitudes should occur in parallel with further natural science research on practices thought to improve cow-calf care, with the results informing the discussion, as well as identifying science-based approaches that resonate with public values.

## Discussion Questions

1. How do public values and perceptions contribute to social sustainability of farm animal management practices?
2. What are some of the greatest risks to farm animal agriculture losing its social license to practice?
3. What barriers to change may some farmers face when implementing proven welfare improvements that include transitioning to a system and practices that they have little experience in managing? How might these barriers influence their perceptions of management practices compared to public perceptions of management practices?
4. How might someone's perception of animal welfare influence their approach to interacting with animals? Does this change, based on the type of interaction? For example, how might someone's perception of animal welfare influence their approach to animal agricultural practices as a farmer? How might it influence their approach as a consumer of animal products?

# References

Alonso, M. E., González-Montaña, J. R., & Lomillos, J. M. (2020). Consumers' concerns and perceptions of farm animal welfare. *Animals 10*(3), 385. https://doi.org/10.3390/ani10030385

Ballou, M. A. (2012). Immune responses of Holstein and Jersey calves during the preweaning and immediate postweaned periods when fed varying planes of milk replacer. *Journal of Dairy Science, 95*(12), 7319–7330. https://doi.org/10.3168/jds.2012-5970

Barnett, J. L., Hemsworth, P. H., Cronin, G. M., Jongman, E. C., & Hutson, G. D. (2001). A review of the welfare issues for sows and piglets in relation to housing. *Australian Journal of Agricultural Research, 52*(1), 1–28. https://doi.org/10.1071/AR00057

Beaver, A., Ritter, C., & von Keyserlingk, M. A. G. (2019a). The dairy cattle housing dilemma: Natural behavior versus animal care. *Veterinary Clinics of North America – Food Animal Practice, 35*(1), 11–27. https://doi.org/10.1016/j.cvfa.2018.11.001

Beaver, A., Meagher, R. K., von Keyserlingk, M. A. G., & Weary, D. M. (2019b). Invited review: A systematic review of the effects of early separation on dairy cow and calf health. *Journal of Dairy Science, 102*, 5784–5810. https://doi.org/10.3168/JDS.2018-15603

Broom, D. M. (2011). A history of animal welfare science. *Acta Biotheoretica, 59*, 121–137. https://doi.org/10.1007/s10441-011-9123-3

Boogaard, B. K., Bock, B. B., Oosting, S. J., Krogh, E., & Boogaard, B. K. (2010). Visiting a farm: An exploratory study of the social construction of animal farming in Norway and the Netherlands based on sensory perception. *The International Journal of Sociology of Agriculture and Food, 17*(1), 24–50. https://doi.org/10.48416/ijsaf.v17i1.266

Boogaard, B. K., Bock, B. B., Oosting, S. J., Wiskerke, J. S. C., & van der Zijpp, A. J. (2011). Social acceptance of dairy farming: The ambivalence between the two faces of modernity. *Journal of Agricultural and Environmental Ethics, 24*, 259–282. https://doi.org/10.1007/s10806-010-9256-4

Boogaard, B. K., Oosting, S. J., & Bock, B. B. (2008). Defining sustainability as a socio-cultural concept: Citizen panels visiting dairy farms in the Netherlands. *Livestock Science, 117*(1), 24–33. https://doi.org/10.1016/j.livsci.2007.11.004

Buller, H., Blokhuis, H., Jensen, P., & Keeling, L. (2018). Towards farm animal welfare and sustainability. *Animals, 8*(6), 81. https://doi.org/10.3390/ani8060081

Busch, G., Weary, D. M., Spiller, A., & Von Keyserlingk, M. A. G. (2017). American and German attitudes towards cowcalf separation on dairy farms. *PLoS One, 12*(3). https://doi.org/10.1371/journal.pone.0174013

Cantor, M. C., Neave, H. W., & Costa, J. H. C. (2019). Current perspectives on the short- and long-term effects of conventional dairy calf raising systems: A comparison with the natural environment. *Translational Animal Science, 3*(1), 549–563. https://doi.org/10.1093/TAS/TXY144

Clark, B., Stewart, G. B., Panzone, L. A., Kyriazakis, I., & Frewer, L. J. (2017). Citizens, consumers and farm animal welfare: A meta-analysis of willingness-to-pay studies. *Food Policy 68*, 112–127. https://doi.org/10.1016/j.foodpol.2017.01.006

Costa, J. H. C., von Keyserlingk, M. A. G., & Weary, D. M. (2016).. Invited review: Effects of group housing of dairy calves on behavior, cognition, performance, and health. *Journal of Dairy Science, 99*(4), 2453–2467. https://doi.org/10.3168/jds.2015-10144

Ellis, K. A., Billington, K., McNeil, B., & McKeegan, D. E. F. (2009). Public opinion on UK milk marketing and dairy cow welfare. *Animal Welfare, 18*(3), 267–282.

Engster, D. (2006). Care ethics and animal welfare. *Journal of Social Philosophy, 37*(4), 521–536. https://doi.org/10.1111/j.1467-9833.2006.00355.x

European Commission. (2021). Communication from the Commission on the European Citizens' Initiative (ECI) 'End the Cage Age' (2021/C274/01). *Official Journal of the European Union, 274*(9.7.2021), 1–13. https://eur-lex.europa.eu/legal-content/EN/TXT/?uri=uriserv%3AOJ.C_.2021.274.01.0001.01.ENG&toc=OJ%3AC%3A2021%3A274%3ATOC#document1

Fraser, D. (1999). Animal ethics and animal welfare science: Bridging the two cultures. *Applied Animal Behaviour Science, 65*, 171–189.

Fraser, D., & MacRae, A. M. (2011). Four types of activities that affect animals: Implications for animal welfare science and animal ethics philosophy. *Animal Welfare, 20*(4), 581–590.

Fraser, D., Weary, D. M., Pajor, E. A., & Milligan, B. N. (1997). A scientific conception of animal welfare that reflects ethical concerns. *Animal Welfare, 6*(3), 187–205.

Fröberg, S., Lidfors, L., Svennersten-Sjaunja, K., & Olsson, I. (2011). Performance of free suckling dairy calves in an automatic milking system and their behaviour at weaning. *Acta Agriculturae Scandinavica, Section A – Animal Science*, *61*(3), 145–156. https://doi.org/10.1080/09064702.2011.632433

Futureye. (2018). *Australia's Shifting Mindset on Farm Animal Welfare*. https://www.outbreak.gov.au/sites/default/files/documents/farm-animal-welfare.pdf

de Graaf, S., Van Loo, E. J., Bijttebier, J., Vanhonacker, F., Lauwers, L., Tuyttens, F. A. M., & Verbeke, W. (2016). Determinants of consumer intention to purchase animal-friendly milk. *Journal of Dairy Science*, *99*(10), 8304–8313. https://doi.org/10.3168/jds.2016-10886

Heng, Y., Peterson, H. H., & Li, X. (2013). Consumer attitudes toward farm-animal welfare: The case of laying hens. *Journal of Agricultural Economics*, *38*, 418–434.

Holm, L., Jensen, M. B., & Jeppesen, L. L. (2002). Calves' motivation for access to two different types of social contact measured by operant conditioning. *Applied Animal Behaviour Science*, *79*(3), 175–194. https://doi.org/10.1016/S0168-1591(02)00137-5

Hötzel, M. J., Cardoso, C. S., Roslindo, A., & von Keyserlingk, M. A. G. (2017). Citizens' views on the practices of zero-grazing and cow-calf separation in the dairy industry: Does providing information increase acceptability? *Journal of Dairy Science*, *100*(5), 4150–4160. https://doi.org/10.3168/JDS.2016-11933

Hötzel, M. J., Longo, C., Balcão, L. F., Cardoso, C. S., & Costa, J. H. C. (2014). A survey of management practices that influence performance and welfare of dairy calves reared in Southern Brazil. *PLoS One*, *9*(12), e114995. https://doi.org/10.1371/journal.pone.0114995

Johnsen, J. F., Zipp, K. A., Kälber, T., Passillé, A. M. de, Knierim, U., Barth, K., & Mejdell, C. M. (2016). Is rearing calves with the dam a feasible option for dairy farms?—Current and future research. *Applied Animal Behaviour Science*, *181*, 1–11. https://doi.org/10.1016/j.applanim.2015.11.011

Kälber, T., & Barth, K. (2014). Practical implications of suckling systems for dairy calves in organic production systems—A review. *Landbauforschung*, *64*(1), 45–58. https://doi.org/10.3220/LBF_2014_45-58

Kent, J. P. (2020). The cow-calf relationship: From maternal responsiveness to the maternal bond and the possibilities for fostering. *Journal of Dairy Research*, *87*(S1), 101–107. https://doi.org/10.1017/S0022029920000436

von Keyserlingk, M. A. G., & Weary, D. M. (2007). Maternal behavior in cattle. *Hormones and Behavior*, *52*(1), 106–113. https://doi.org/10.1016/j.yhbeh.2007.03.015

von Keyserlingk, M. A. G., Martin, N. P., Kebreab, E., Knowlton, K. F., Grant, R. J., Stephenson, M., Sniffen, C. J., Harner, J. P., Wright, A. D., & Smith, S. I. (2013). Invited review: Sustainability of the US dairy industry. *Journal of Dairy Science*, *96*(9), 5405–5425. https://doi.org/10.3168/JDS.2012-6354

Kiecolt-Glaser, J. K., & Glaser, R. (2002). Depression and immune function central pathways to morbidity and mortality. *Journal of Psychosomatic Research*, *53*(4), 873–876. https://doi.org/10.1016/S0022-3999(02)00309-4

Latham, N. R., & Mason, G. J. (2008). Maternal deprivation and the development of stereotypic behaviour. *Applied Animal Behaviour Science*, *110*(1–2), 84–108. https://doi.org/10.1016/j.applanim.2007.03.026

Lecorps, B., Welk, A., Weary, D. M., & von Keyserlingk, M. A. G. (2021). Postpartum stressors cause a reduction in mechanical brush use in dairy cows. *Animals*, *11*(11), 3031. https://doi.org/10.3390/ANI11113031

Liebe, U., Preisendörfer, P., & Meyerhoff, J. (2011). To pay or not to pay: Competing theories to explain individuals' willingness to pay for public environmental goods. *Environment and Behavior*, *43*(1), 106–130. https://doi.org/10.1177/0013916509346229

Loberg, J. M., Hernandez, C. E., Thierfelder, T., Jensen, M. B., Berg, C., & Lidfors, L. (2007). Reaction of foster cows to prevention of suckling from and separation from four calves simultaneously or in two steps. *Journal of Animal Science*, *85*(6), 1522–1529. https://doi.org/10.2527/jas.2006-813

Loberg, J. M., Hernandez, C. E., Thierfelder, T., Jensen, M. B., Berg, C., & Lidfors, L. (2008). Weaning and separation in two steps-A way to decrease stress in dairy calves suckled by foster cows. *Applied Animal Behaviour Science*, *111*(3–4), 222–234. https://doi.org/10.1016/j.applanim.2007.06.011

Marcé, C., Guatteo, R., Bareille, N., & Fourichon, C. (2010). Dairy calf housing systems across Europe and risk for calf infectious diseases. *Animal*, *4*(9), 1588–1596. https://doi.org/10.1017/S1751731110000650

McGuirk, S. M. (2008). Disease management of dairy calves and heifers. *Veterinary Clinics of North America – Food Animal Practice, 24*(1), 139–153. https://doi.org/10.1016/j.cvfa.2007.10.003

Meagher, R. K., Beaver, A., Weary, D. M., & von Keyserlingk, M. A. G. (2019). Invited review: A systematic review of the effects of prolonged cow-calf contact on behavior, welfare, and productivity. *Journal of Dairy Science, 102*(7), 5765–5783. https://doi.org/10.3168/jds.2018-16021

Miranda-de la Lama, G. C., Estévez-Moreno, L. X., Sepúlveda, W. S., Estrada-Chavero, M. C., Rayas-Amor, A. A., Villarroel, M., & María, G. A. (2017). Mexican consumers' perceptions and attitudes towards farm animal welfare and willingness to pay for welfare friendly meat products. *Meat Science, 125*, 106–113. https://doi.org/10.1016/j.meatsci.2016.12.001

Perttu, R. K., Ventura, B. A., & Endres, M. I. (2020). Youth and adult public views of dairy calf housing options. *Journal of Dairy Science, 103*(9), 8507–8517. https://doi.org/10.3168/jds.2019-17727

Robbins, J. A., Von Keyserlingk, M. A. G., Fraser, D., & Weary, D. M. (2016). Invited review: Farm size and animal welfare. *Journal of Animal Science, 94*(12). https://doi.org/10.2527/jas2016-0805

Rollin, B. E. (2004). Annual Meeting Keynote Address: Animal agriculture and emerging social ethics for animals. *Journal of Animal Science, 82*(3), 955–964. https://doi.org/10.1093/ANSCI/82.3.955

Rollin, B. E. (2008). The ethics of agriculture: The end of true husbandry. In M. S. Dawkins & B. Roland (Eds.), *The Future of Animal Farming: Renewing the Ancient Contract* (pp. 7–19). Blackwell Publishing Ltd.

Rosenberger, K., Costa, J. H. C., Neave, H. W., von Keyserlingk, M. A. G., & Weary, D. M. (2017). The effect of milk allowance on behavior and weight gains in dairy calves. *Journal of Dairy Science, 100*(1), 504–512. https://doi.org/10.3168/jds.2016-11195

Roth, B. A., Barth, K., Gygax, L., & Hillmann, E. (2009). Influence of artificial vs. mother-bonded rearing on sucking behaviour, health and weight gain in calves. *Applied Animal Behaviour Science, 119*(3–4), 143–150. https://doi.org/10.1016/j.applanim.2009.03.004

Rushen, J., Butterworth, A., & Swanson, J. C. (2011). Animal behavior and well-being symposium: Farm animal welfare assurance: Science and application. *Journal of Animal Science, 89*(4), 1219–1228. https://doi.org/10.2527/jas.2010-3589

Ryan, E. B., Fraser, D., & Weary, D. M. (2015). Public attitudes to housing systems for pregnant pigs. *PLoS One, 10*(11), 1–14. https://doi.org/10.1371/journal.pone.0141878

Savory, C. J. (2004). Laying hen welfare standards: A classic case of "power to the people". *Animal Welfare, 13*(S), 153–158.

Schuck-Paim, C., Negro-Calduch, E., & Alonso, W. J. (2021). Laying hen mortality in different indoor housing systems: A meta-analysis of data from commercial farms in 16 countries. *Scientific Reports, 11*(1), 3052. https://doi.org/10.1038/s41598-021-81868-3

Sirovica, L. V., Ritter, C., Hendricks, J., Weary, D. M., Gulati, S., & von Keyserlingk, M. A. G. (2022). Public attitude toward and perceptions of dairy cattle welfare in cow-calf management systems differing in type of social and maternal contact. *Journal of Dairy Science, 105*(4), 3248–3268. https://doi.org/10.3168/JDS.2021-21344

Spooner, J. M., Schuppli, C. A., & Fraser, D. (2014a). Attitudes of Canadian citizens toward farm animal welfare: A qualitative study. *Livestock Science, 163*(1), 150–158. https://doi.org/10.1016/j.livsci.2014.02.011

Spooner, J. M., Schuppli, C. A., & Fraser, D. (2014b). Attitudes of Canadian pig producers toward animal welfare. *Journal of Agricultural and Environmental Ethics, 27*(4), 569–589. https://doi.org/10.1007/s10806-013-9477-4

Stadig, L. M., Ampe, B. A., Van Gansbeke, S., Van Den Bogaert, T., D'Haenens, E., Heerkens, J. L. T., & Tuyttens, F. A. M. (2015). Opinion of Belgian egg farmers on hen welfare and its relationship with housing type. *Animals, 6*(1). https://doi.org/10.3390/ANI6010001

Sundrum, A. (2015). Metabolic disorders in the transition period indicate that the dairy cows' ability to adapt is overstressed. *Animals, 5*(4), 978–1020. https://doi.org/10.3390/ani5040395

Swanson, J. C., Lee, Y., Thompson, P. B., Bawden, R., & Mench, J. A. (2011). Integration: Valuing stakeholder input in setting priorities for socially sustainable egg production 1. *Poultry Science, 90*(9), 2110–2121. https://doi.org/10.3382/ps.2011-01340

Te Velde, H., Aarts, N., & Van Woerkum, C. (2002). Dealing with ambivalence: Farmers and consumers' perceptions of animal welfare in livestock breeding. *Journal of Agricultural and Environmental Ethics, 15*(2), 203–219. https://doi.org/10.1023/A:1015012403331

USDA. (2016). *Dairy 2014: Dairy cattle management practices in the United States, 2014*. https://www.aphis.usda.gov/animal_health/nahms/dairy/downloads/dairy14/Dairy14_dr_PartI_1.pdf

Vandresen, B., & Hötzel, M. J. (2021). Pets as family and pigs in crates: Public attitudes towards farrowing crates. *Applied Animal Behaviour Science, 236*, 105254. https://doi.org/10.1016/J.APPLANIM.2021.105254

Vanhonacker, F., Verbeke, W., Van Poucke, E., & Tuyttens, F. A. M. (2008). Do citizens and farmers interpret the concept of farm animal welfare differently? *Livestock Science, 116*(1–3), 126–136. https://doi.org/10.1016/j.livsci.2007.09.017

Vaske, J. J. (2008). Linking theory and concepts to survey research. In *Survey Research and Analysis: Applications in Parks, Recreation and Human Dimensions* (pp. 17–34). Venture Publishing, Inc.

Ventura, B. A., von Keyserlingk, M. A. G., Schuppli, C. A., & Weary, D. M. (2013). Views on contentious practices in dairy farming: The case of early cow-calf separation. *Journal of Dairy Science, 96*(9), 6105–6116. https://doi.org/10.3168/jds.2012-6040

Ventura, B. A., von Keyserlingk, M. A. G., Wittman, H., & Weary, D. M. (2016). What difference does a visit make? Changes in animal welfare perceptions after interested citizens tour a dairy farm. *PLoS One, 11*(5), 1–19. https://doi.org/10.1371/journal.pone.0154733

Vermeir, I., & Verbeke, W. (2006). Sustainable food consumption: Exploring the consumer "attitude – behavioral intention" gap. *Journal of Agricultural and Environmental Ethics, 19*(2), 169–194. https://doi.org/10.1007/s10806-005-5485-3

Weary, D., & Robbins, J. (2019). Understanding the multiple conceptions of animal welfare. *Animal Welfare, 28*(1), 33–40. https://doi.org/10.7120/09627286.28.1.033

Weary, D. M., Ventura, B. A., & von Keyserlingk, M. A. G. (2016). Societal views and animal welfare science: Understanding why the modified cage may fail and other stories. *Animal, 10*(2), 309–317. https://doi.org/10.1017/S1751731115001160

Wenker, M. L., Bokkers, E. A. M., Lecorps, B., von Keyserlingk, M. A. G., van Reenen, C. G., Verwer, C. M., & Weary, D. M. (2020). Effect of cow-calf contact on cow motivation to reunite with their calf. *Scientific Reports, 10*(1), 14233. https://doi.org/10.1038/s41598-020-70927-w

Widmar, N. O., Morgan, C. J., Wolf, C. A., Yeager, E. A., Dominick, S. R., & Croney, C. C. (2017). US resident perceptions of dairy cattle management practices. *Agricultural Sciences, 8*, 645–656. http://doi.org/10.4236/as.2017.87049

Winder, C. B., Bauman, C. A., Duffield, T. F., Barkema, H. W., Keefe, G. P., Dubuc, J., Uehlinger, F., & Kelton, D. F. (2018). Canadian National Dairy Study: Heifer calf management. *Journal of Dairy Science, 101*(11), 10565–10579. https://doi.org/10.3168/jds.2018-14680

# 45
# HUMAN-ANIMAL INTERACTIONS IN WESTERN ART

*Laura D. Gelfand*

## Introduction to the Issue

We can tell a great deal about the history of human-animal interactions from how humans have represented animals over time and how these representations have changed. Images of animals vary from naturalistic to abstract, and animals have been shown engaged with humans in all sorts of activities and locations including as targets of and participants in hunts, accompanying humans into war and the afterlife, as food, pets, and as symbolic stand-ins for abstract concepts (Klingender, 1971). The close connection humans have with nonhuman animals has been represented in every possible medium, and these representations have appeared in caves and castles, churches, tchotchkes, and everywhere in between. This chapter focuses on the history of representations of human and nonhuman interaction in Western Art, and it begins in the caves of France where we find some of the oldest depictions of animals and humans together. Reading human-animal interaction through art helps illuminate Steve Baker's important question, "why is it that our ideas of the animal—perhaps more than any other set of ideas—are the ones which enable us to frame and express ideas about *human* identity?" (Baker, 2001, p. 6).

## Cultural, Religious, and Historical Perspectives

Humans have always been driven to create visual records of the world around them, and the function of these images has varied as have the materials from which they have been made and the representations themselves. Images depicting interactions between humans and animals have been produced consistently throughout history, and they provide irrefutable evidence that animals are not only good to think with, they are also good to see with (Lévi-Strauss, 1964). In what follows, I trace changes in human-animal interaction through the history of art. The trajectory of this change may be seen to follow an arc that begins with humans appearing to consider themselves part of a shared world in which nonhuman animals are of primary importance, to humans assuming and asserting a dominant role over animals, an attitude that eventually evolves into a more equal balance between humans and animals. Contemporary art showcases a range of attitudes toward human-animal relations that are not easily categorized but indicate a renewed interest in human and nonhuman interaction and the recognition that animals and humans share this planet.

## *Paleolithic Art*

Prehistoric paintings in European caves include a wide variety of animals together with human figures, and, although it is not possible to know precisely why these images were made or what purpose they served, it has been suggested that Paleolithic images reflect rituals that were intended to establish a dialog with animal prey which had the ontological status of kin (Ingold, 1994; McNiven, 2010). It is also possible that these representations commemorated specific hunts or other events (Mithen, 1988). Although we do not know how Paleolithic representations were interpreted by those who made them, the extraordinary naturalism of these images indicates that their creators were not simply skilled observers of the world around them but were also interested enough in animals to render their appearance with precision and care (Antl-Weiser, 2018, p. 64). Perhaps, this was because, as John Berger noted, "Animals first entered the imagination as messengers and promises" (2009, p. 12).

The remarkable paintings found in 1940 in the caves of Lascaux, France by four local boys and their dog, were made between 18,900 and 18,600 BCE (Aujoulat, 2004; Antl-Weiser, 2018, p. 66). The cave includes over 6,000 images of animals, but only one human figure, a man who appears to have been wounded or killed by a bison. In *Man and Bird, Panel of the Wounded Man* (Figure 45.1), the man seems to be wearing the headdress of a bird and he extends his arm toward a bird-like totem. Although the figure is recognizable as human, it is essentially a stick figure, and the rendering of the form is notably primitive when compared to the remarkable naturalism of the animals in the cave. The artists who worked at Lascaux exploited the natural irregularities of the cave walls to heighten the verisimilitude and

*Figure 45.1* *Man and Bird, Panel of the Wounded Man*, 18,900 and 18,600 BCE, Lascaux, France; HIP/Art Resource, NY

dynamism of the animals they depicted, and they achieved remarkable three-dimensionality thanks to careful shading and color. The liveliness of the illusion would have been even more striking when the caves were lit by flickering light from carved stone lamps—over 100 of which have been found—that would have made the animals appear to move (Wachtel, 1993). Intriguingly, bison dominate the cave's decoration, but reindeer, which make up less than 4% of the figures in European prehistoric caves, were the main food source for Paleolithic humans (Leroi-Gourhan, 1968, p. 48). We don't know why the cave's decorative scheme includes only a single human figure, or what this man's violent interaction with the bison next to him signaled to the earliest viewers of the image. However, Yusoff taps into its elemental nature in describing the human as "there to see the death that he has inflicted, to bow before the horns of the bison, head covered, sex erect, vulnerable and yet inexplicably placed in the space of death that is opening up in the wound of the bison" (2015, pp. 393–394). The paintings in the caves of Lascaux convey both the intimacy of the human-nonhuman connection and the danger that our interactions entail for both partners.

## *Ancient Egypt and the Aegean*

Thanks to their ubiquity, familiarity, and importance, animals appear frequently with humans throughout the history of art. After the Paleolithic period animals were often integrated into symbolic systems in which they represent diverse concepts including, but in no way limited to, fertility, divinity, and political power (Klingender, 1971). Artwork from nearly every time and place uses animals symbolically, and when humans interact with them, it is often a reflection of religious beliefs, symbolic meaning, or ritual practices. Images showing human-animal interaction reveal some of the many ways humans have worshipped and revered animals. In Egyptian tombs scenes of human-animal interaction are usually restricted to small sculptures that show humans worshipping animal gods or engaging in daily activities with domesticated animals that were meant to provide sustenance for the dead in the afterlife (Colonna, 2021, p. 3).

The dangerous practice of bull-leaping is closely associated with the Minoan culture of Knossos in the Aegean. Surviving images reveal that athletic young male athletes would leap up and over the top of bulls, flipping their bodies through the air as they somersaulted across the bull's back, in a sport that must have been extraordinarily exciting to watch and terrifying to participate in. Ceramic vessels that show small leaping human figures attached to the horns of enormous bulls are found as early as around 2000 BCE, about half a century before the famous Toreador wall painting from Knossos (Figure 45.2). According to Morgan, "The participants of the bull-sports are always depicted as young and athletic, and the social function of the sport was undoubtedly as an expression of youthful vigour and daring in acts involving physical challenge and human dominance over animal power" (1998, p. 18). Morgan associates Egyptian and Aegean representations of humans suppressing and/or killing animals with an assertion that man is superior to animals, and that through such victories man assumes the power the animal once wielded (p. 31). This interpretation explains the innumerable images of hunting and killing animals that dominate depictions of human-animal interactions.

## *Ancient Greece*

Ancient Greek art frequently includes scenes of human-animal interaction that, like those of Minoan culture, may reflect actual practices as well as functioning symbolically. Humans

*Figure 45.2* Bull-leaping (Toreador fresco), c. 1400 BCE, Palace of Knossos, Crete; Scala/Art Resource, NY

are often shown in combat with real and imaginary animals on Greek vases (Robertson, 1951). According to established visual conventions for vase decoration, the combatant on the viewer's left should be understood to be the winner (Shapiro, 1990), and it is significant that humans are not always victorious over their animal foes.

Human-animal interactions are frequently shown on Greek funeral stele, where we find the first examples of humans memorializing beloved pets, a practice that continues to the present day. Memorial representations on stele reveal how some ancient Greeks felt about their animal companions and allow us to see that for some it was much like the relationships we enjoy with our pets today. According to Louise Calder, in Greek art, "the holding and caressing of animals certainly does feature in ways suggesting that many derived reassurance and pleasure [from it]," (2017, p. 63). Calder also notes that, "Dogs so frequently accompany male owners publicly in Greek art that they seem to have been almost essential accessories of respectable leisure" (p. 66). A grave stele from Ilissos shows the deceased as an idealized young man who is mourned by his father, a young boy, and a dog with its head lowered and ears drawn back in an unmistakable expression of canine distress (Figure 45.3).

Hunt scenes showing dogs chasing hares on vases were frequently juxtaposed with pederastic scenes (Barringer, 2001, p. 83). When used in the drinking parties known as symposia, vases from the late-sixth and early fifth century BCE were meant to charm their elite male users while providing instruction on the strictly ritualized codes of conduct that governed the complex cultural conventions of pederastic practice (Haworth, 2018). In these erotic scenes, dogs are used to highlight inappropriate behaviors within pederastic courtship and provide humorous as well as cautionary information about suitable conduct. According to Haworth, these dogs are "not merely attendants in the scene, but full participants, humorously doubling the actions of the men" (p. 10). But dogs in Greek art were not only associated with men, and they were often associated with women, however, such associations were, according to Cristiana Franco, "a precision instrument for reflecting on women as a race, distinct from male humans yet intrusively situated in the heart of humanity" (2014, p.

*Figure 45.3* Ilissos Sculptor, Grave Stele, c. 340–330 BCE, National Archaeological Museum, Athens; Album/Art Resource, NY

158). Like women, dogs exist in reality and symbolically. By associating women with dogs, Greek men reinforced the idea that women were inferior and justified their subordinate position in society.

## *Roman Art*

Human-animal interactions in Roman art are found primarily in scenes of hunting, an aristocratic activity in ancient Rome and one that was associated with the nobility for centuries afterward. Roman handbooks published as early as the second century BCE, such as that by Arian, focus on hunting with hounds (Anderson, 1985; Toynbee, 1973). Images that correlated lion and boar hunting with royal activity were particularly prevalent in the late fourth century BCE, during the reigns of Philip II and Alexander the Great (Cohen, 2018, p. 91). Thanks to the ubiquity of these representations, according to Cohen, "Alexander was remembered in Hellenistic, Roman and later European art as a relentless hunter of felines" (p. 92). Hunting scenes promote the concept of human superiority over animals, an idea that was ritually reenacted in the slaughter of thousands of wild animals for bloodthirsty crowds in the Roman arena. These propagandistic performances buttressed the military state during the *Pax Romana* and the Late Republic (Foss, 2012, p. 102).

Less-violent examples of Roman art include the many sculptures of Eros that were used as funerary memorials for children. In the guise of the god Eros, these winged children sleep surrounded by attributes associated with immortal gods, and they were intended to comfort parents and other mourners. These sculptures often include a lizard that, according to Sorabella,

> is integral to this comforting message. It is present as both playmate and plaything, amusement, and protector. [The lizard] refers to the childhood of Eros rather than his divinity; it humanizes him and harmonizes his image with the viewer's memory of a particular human child.
>
> *(2007, p. 367)*

Artwork produced during the Roman Empire used images of wild and domestic animals to convey symbolic messages of power and comfort, and it codified the concept of human superiority over animals, making it an assumption rather than just an assertion (Toynbee, 1973, pp. 21–23). This had a significant impact on how humans interacted with animals in Western culture from classical antiquity to the present, and it is clearly reflected in the history of art.

## *Medieval Art*

Up to this point, it has been possible to characterize human-animal interactions in relatively general terms, but in the medieval period, representations of these interactions took on more nuanced and changeable aspects that complicate easy categorization. Although my emphasis is on the meaning these representations would have had for those who created and consumed them, then as now, medieval people would surely have found looking at representations of animals pleasurable. Thus, in religious contexts, images of human-animal interaction have a visual appeal that can supersede their sacred meaning and artists may have included them to attract viewers to the image.

Following Roman authors and the Old Testament, the New Testament promoted the idea that God made man superior to all animals, and every living creature was created to aid mankind. Thus, throughout the Medieval and Early Modern periods, animals were consistently represented as inferior to humans and appropriately subject to human control. This concept is illustrated in the writings of the apostle Paul, who decried the sinfulness of Egyptian animal worship, claiming that the Egyptians, "exchanged the truth of God for a lie and revered and worshipped the creation instead of the Creator" (Rom. 1:23–28; Grant, 1999, p. 7). However, images of animals were also extraordinarily common in Early Christian art, appearing in almost every possible context as symbols and as representations of the natural world.

Heavily illustrated, the bestiary reflected and shaped conceptions of natural history throughout the medieval period (Heck and Cordonnier, 2012). Depictions of Adam naming the animals illustrate mankind's supposed superiority and God-given dominion over every nonhuman creature. A charming illustration from the Peterborough Bestiary (Figure 45.4), shows Adam authoritatively holding up a long finger and gesturing toward a variety of animals as he names them, included among them is a lion who looks somewhat skeptically toward the viewer. A large brown and white bird nestles in a tree behind Adam, leaning toward his ear as if to whisper suggestions. Three birds perch in the tree canopy behind Adam where they look as if they are ready to sing their approval of the entire enterprise. My reading anthropomorphizes these animals, but it is in keeping with medieval ideas about

*Figure 45.4* Adam Naming the Animals, Peterborough Bestiary, 14th c., Cambridge, Corpus Christi College, MS 53, f. 195v; © Creative Commons

them. As the foremost resource for information on animals in the Middle Ages, the bestiary "encompasses behavioral, social, moral, and spiritual meanings within its classificatory project." Crane continues, "A taxonomic impulse to find relationships as well as to differentiate presses the bestiary to situate humans among the other animals, not only as their master but also as their similar. The category confusion latent in the scholastic premise that *homo est animal* enriches the depictions of other animals" (2013, p. 6). As a result, animals are not simply used as moral exempla or symbols, rather, they take on human characteristics and attributes to convey increasingly complex meaning through the medieval period.

In Early Christian artworks, animals served as symbolic representations or attributes of Christian holy figures: the lamb represented Christ, and three of the four evangelists were associated with specific animals (Mark: lion, Luke: ox; John: eagle). The earliest hagiographic texts connect saints with animals to highlight the saint's purity through their connection to nature, their miraculous ability to tame wild animals and alter the world around them, and to showcase their sacred separation from the sinfulness of the human world, a state of being so lonely that the holy individual turns to animals for companionship (Alexander,

2008). St. Jerome was shown with a lion because he was said to have removed a thorn from a lion's paw and afterward it followed him through the desert like a housecat (Ring, 1945). St. Margaret was supposed to have been swallowed by a dragon, and her prayers for salvation were answered when she burst forth from the beast's belly, making her an ideal patron saint of childbirth (Parker, 2020). In addition to dragons, imaginary animals like unicorns were also popular during the Middle Ages. Real or fictive, images of saints with animals helped medieval preachers and their devotees visualize complex theological concepts about salvation (Resl, 2007, p. 179).

The connection between hunting and aristocracy that began with the ancient Romans was further strengthened and codified throughout the early medieval period, reaching something of an acme with the hunt à force, a day-long pursuit of a single animal that involved an enormous number of retainers, trained dogs, and other resources (Judkins, 2013). This inherently inefficient activity was not about killing the animal but, according to Crane, "structures contact with the hunted animal and the pack of hounds to perform and reinforce the rightness of aristocratic superiority. Animal death, ritualized to evoke the powers of sacrifice, is only a final expression of this superiority" (p. 7). In the prologue to his popular treatise on hunting, written between 1387 and 1389, Gaston Phoebus speaks of the activity's many positive aspects (for humans) including avoiding the seven deadly sins, enjoying fresh morning air, a celebratory meal at the successful conclusion, and excellent sleep (Klemettilä, 2015).

## *Renaissance Art*

Traditions established during the Medieval and Renaissance periods were deeply entrenched in Western art for centuries. However, naturalistic rendering of human and animal forms became the subject of intense interest during the Renaissance. Representations of animals during this period typically serve symbolic ends, and they also broadcast information about the patron's wealth and social status. The tension that results when symbolic content is fused with visual naturalism is seen in Leonardo da Vinci's painting, the *Lady with an Ermine* (Figure 45.5). The painting shows Cecilia Gallerani, the young mistress of Lodovico Sforza, the duke of Milan, holding a white animal that is far too large to be an ermine. Despite the painting's traditional title, the animal has been identified as a member of the weasel family, most likely a ferret (Niemelä and Örmä, 2016). Medieval conceptions of animal behaviors and the symbolic meanings attached to them continued throughout the fifteenth and sixteenth centuries, and ermines were imbued with a wide variety of diverse meanings including purity, chastity, and moderation. It was believed that the animals would kill themselves if their immaculate white winter coats were stained, which is why their fur was often used to line the clothing of noble rulers including, perhaps most famously, the virgin Queen Elizabeth I (Mazzola, 2006). Gallerani may have been pregnant with Sforza's illegitimate son Cesare when Leonardo painted her portrait, and ermine were also associated with childbirth (Musacchio, 2001). The animal in Leonardo's painting has also been connected to Lodovico Sforza's investiture in the Order of the Ermine, which occurred shortly before the painting was created. Following his investiture, Lodovico was known as *l'Ermellino*, or little ermine, and thus, in Leonardo's painting, Sforza's mistress is shown caressing a cleverly concealed representation of her lover (Pedretti, 1990, p. 167).

Any consideration of human–animal interaction in Western art would be incomplete without considering the inclusion of animals in pornographic scenes that also show humans engaged in sexual activity. As with all such imagery, this was intended to stimulate a sexual

*Figure 45.5* Leonardo da Vinci, *Lady with an Ermine*, c. 1490, Czartoryski Museum, Kraków; Erich Lessing/Art Resource, NY

response in human viewers, and numerous authors have considered animals in pornography including Derrida, Fudge, and Haraway (Derrida, 2002; Fudge, 2000a; Haraway, 2008). There seems to have been something of an explosion of such imagery in the Early Modern period, and it is not clear if our impression comes from the quantity of surviving imagery, or if it truly was more common and popular from the sixteenth through the eighteenth centuries. One of the more striking examples of this kind of imagery shows a copulating heterosexual couple atop an ejaculating horse, a work discussed by Mary Catherine Kinniburgh, who illustrates it and situates it within a framework of queer visuality noting that it "demonstrated the limits of any well-defined boundary between species and their sexualities" (2016, p. 75).

## Early Modern Art

Centuries of symbolic representations of animals in combination with greater naturalism meant that after the Renaissance, animals continued to be imbued with religious and political significance. While artwork produced in Catholic countries, like Italy and Spain, showed human-animal interactions in more traditional ways, Protestant countries, including the Dutch Republic, endowed animal representations with new kinds of political meaning, and introduced animals into artwork in ways that were not possible previously. During the Thirty Years War (1618–1648), Gerard Ter Borch, painted several versions of the *Horse and Rider* in response to the Netherland's battle for independence from Spain (Figure 45.6) (Kunzle, 2000). Here, the exhausted human rider and horse appear to have become one unified form. Neither the horse nor the soldier is given a specific identity, their faces are not shown so that they represent any, or every, human and animal that fights for freedom. The soldier's metal cuirass glints in the sun, as do the horse's glossy flanks, and their forms merge into a single body in which the horse's hind legs seem to belong to the man and the soldier's torso could be that of the horse. Ter Borch seems to show, among other things, that man and animal, individually and together, share emotional states including exhaustion, despair, and triumph.

In Paulus Potter's remarkable painting, *The Bull* (Figure 45.7), painted on the eve of the Treaty of Westphalia, a farmer observes his small herd of livestock while situated behind a short fence and a tree with new spring growth bursting from its sturdy, pollarded trunk. The man is surrounded by animals that, like the tree, represent the fecundity of Dutch lands. The life-size bull dominates the composition and looks toward the viewer, meeting our gaze and drawing us into the image. The bull's pose is powerful and protective, its genitalia on emphatic display, the bull seems prepared to protect the cow and other livestock relaxing nearby. The painting was extraordinarily popular when it was first shown, in part because the young bull was understood to represent the power and pride of the new Dutch republic and its people. It was thus seen as a particularly egregious offence when the painting was confiscated by Napoléon's troops and displayed in Paris at the Musée

*Figure 45.6* Gerard ter Borch, *Man on Horseback*, 1634, Museum of Fine Arts, Boston. Photograph © 2022, Museum of Fine Arts, Boston

*Figure 45.7* Paulus Potter, *The Bull*, 1647, Mauritshuis, The Hague; © DeA Picture Library/Art Resource, NY

Napoléon from 1795 to 1815. When it was finally returned to The Hague it was accompanied by a military escort and met with patriotic enthusiasm by the Dutch (Walsh et al., 1994–1995, pp. 74–77). Potter's painting of the young bull marks a rupture in the trajectory of human–animal interaction in art. As Nathaniel Wolloch writes, "never before seventeenth-century Netherlandish painting, were images of animals given such prominent, insightful, and consistent attention in art, let alone with such talent and artistry. This novel perception opened up new avenues of meaning" (Wolloch, 2019, p. 193). The artists that follow produce artwork that vividly documents the changing relationships humans have had with animals since the seventeenth century.

One of the clearest illustrations of these changes can be seen in *The Four Stages of Cruelty*, a series of engravings produced by the British artist William Hogarth in 1750–1751. Hogarth is most famous for his satirical prints which were intended to educate and inform, but he was also a well-known animal lover who had a cemetery for his pets, including Dick the drake and Pompey the dog at his home in Chiswick (Uglow, 1997). In one of his self-portraits (Figure 45.8), Hogarth included his beloved pug, Trump, for whom he apparently "had conceived a greater share of attachment than is usually bestowed on these domestic animals" (Einberg and Egerton, 1988). *The Four Stages of Cruelty* was printed on inexpensive paper so the images would be affordable for members of the lower classes, their intended audience (Steintrager, 2001). The prints are an early representation of the idea that those who are cruel toward animals go on to become violent toward humans (Beirne, 2013). Hogarth conveyed his moralizing message using text and image to make it as clear as possible. The series follows the villain, Tom Nero, from his childhood, when he is seen torturing dogs and cats. In the second plate, Nero is seen viciously beating his coach horse which has broken its leg. In the third plate, we find Nero having murdered his lover after convincing her to steal from her mistress. And in the final plate, we see his body, having been taken from the gallows, and subjected to public dissection as a dog speculatively sniffs his entrails. The series shows how ideas about animals and their treatment shifted dramatically during the long eighteenth

*Figure 45.8* William Hogarth, *The Painter and his Pug*, 1745, Tate Gallery, Britain; © Tate, London/Art Resource, NY

century. While Hogarth's argument rests primarily on how cruelty toward animals leads to violence against humans, the work also reflects a growing consensus that animal cruelty was morally abhorrent in and of itself (Thomas, 1983, pp. 150–165). Please see Chapter 23 for additional information regarding violence toward animals.

## Nineteenth-Century Art

In the nineteenth century, it becomes difficult, if not impossible, to generalize about human-animal interaction in art because so much more artwork has survived, and the subjects and approaches used by artists are so diverse. During the nineteenth century, art academies shifted to an even greater focus on the human form and male artists were able to attain higher status than women in part because only men were allowed to study nude models (Havice, 1981). Rendering the human form from life was essential to artistic success in most academic genres, so women artists like Rosa Bonheur specialized in painting animals because they had greater access to animal bodies. Bonheur's enormous painting, *The Horse Fair* (Figure 45.9) was produced from a series of sketches the artist made at a Paris horse market she visited twice a week. In the painting, Bonheur includes a portrait of herself dressed as a man and looking out toward viewers (Saslow, 1992; Garb, 1993, pp. 233, 239). The power and the

*Figure 45.9* Rosa Bonheur, *The Horse Fair*, 1852–1855, Metropolitan Museum of Art, New York; © The Metropolitan Museum of Art. Image source: Art Resource, NY

*Figure 45.10* Edwin Landseer, *Man Proposes, God Disposes*, 1864, Royal Holloway, University of London; Photo credit, Royal Holloway, University of London

muscular athleticism of the horses is the artist's primary focus, and the men charged with managing them are of decidedly secondary interest. During the Victorian era, many women recognized that they shared second-class status with animals and marginalized groups of people, and some became leaders in anti-vivisection and abolition campaigns (Elston, 1987).

Edwin Landseer specialized in animal painting, and attained tremendous success, although his work was viewed by academic artists as lower in status than paintings that featured naturalistic renderings of the human form (Ormond, 1981). One of Landseer's late paintings illustrates a situation in which humans no longer dominate animals and are instead overpowered by natural forces. The painting neatly intertwines the Victorian fascination with science and the sublime, an esthetic that combines beauty and terror to provoke a strong emotional response (Guerlac, 1991). In *Man Proposes, God Disposes* (Figure 45.10), Landseer creates an enormous, nightmarish scene that imagines the fate of the Franklin expedition, which set out to find the Northwest Passage in 1845 and promptly vanished without a trace for decades (Moore, 2009, pp. 32–37; Nugent, 2003, pp. 132–135). The fate of Franklin and his crew of 128 men stirred the Victorian imagination, and Landseer capitalized on this in his large-scale depiction of the wooden boat that was found in 1859 on King William Island by Francis Leopold McClintock (Lüders & Nanna, 2020). The small boat was dragged across the rocky island by members of Franklin's crew after their ships, the *Erebus* and *Terror*, became inextricably locked in the polar ice. McClintock's report included a description of two human skeletons found in the ruined boat, both apparently dismembered and eaten by wild animals. McClintock suggested these were wolves, but Landseer transformed them into polar bears, animals that are far larger and more terrifying, and that he could study from life in the London Zoo. The polar bear on the viewer's right is tearing into a human ribcage with apparent relish, squeezing a rib bone between its powerful jaws. Landseer included several objects that were brought back to England by McClintock, and these serve as relics and *memento mori* devices. Between the fearsome animals and the reminders of humankind's earthly transience, Landseer's painting drives home the message that humans are not always dominant over animals and nature, nor are they meant to be.

## *Modern and Contemporary Art*

From the nineteenth century on, artwork showing human-animal interaction has split into at least two dominant camps, one populated by artists who produce work that is stylistically akin to illustration; Fredrick Remington and Winslow Homer typify the American approach to this kind of artwork, which remains popular to the present day. A second group is made up of avant-garde artists who are less interested in creating naturalistic imagery and more concerned with exploring the expressive possibilities of formal elements. Their artwork frequently serves as means of self-expression and/or social critique, and experimental twentieth-century artists have explored human-animal interaction in many ways.

One particularly interesting example is the work of Franz Marc, a German artist who was a founding member of the *Blaue Reiter* group. Marc's painting, *Fate of the Animals* (Figure 45.11), was originally titled *The Trees Show Their Rings, The Animals Their Veins* (Levine, 1976, pp. 269–277). Marc identified so closely with the animals he painted that his works represent a very different kind of human-animal interaction than any addressed thus far. For Marc, human-animal interaction is conceptual rather than objective. Marc's ideas about human interactions with animals are as radical a rupture with the past as his formal experimentations with abstraction. On the reverse of the *Fate of the Animals*, he painted a subtitle for the painting: *Und alles Sein ist flammend Leid,* "And All Being is Flaming Suffering." In 1915, when Marc, who was then serving in the German Army saw a reproduction of the painting, he wrote to his wife that "At first glance I was completely shaken. It is like a premonition of this war, horrible and gripping. I can hardly believe I painted it!" (Marc, 12 April, 1915, trans. Chipp, 1968, p. 182). Tragically, Marc was killed near Verdun a short time later, and the painting was damaged in a fire while being transported to a memorial exhibition for the artist. *Fate of the Animals* depicts numerous animals in a forest; as fire rains down on them, the trees burn, and the animals die in agony. The composition is ordered along a series of diagonal lines and the animals include deer, horses, and boars. In writing about his artistic expressionism, Marc compared his own experiments and those of artists in his circle to the Impressionists who came before them, noting, "Today we seek behind the veil of appearance the hidden things in nature that seem to us more important than the discoveries of the Impressionists" (Marc, 1912, p. 469). Levine argued that in this painting, Marc recalled the Nordic Eddas and expressed a wish for a destructive cleansing that would usher in a new

*Figure 45.11* Franz Marc, *Fate of the Animals*, 1913, Basel, Kunstmuseum; Art Resource, NY

world of hope and happiness (Levine, p. 276). In his work, Marc embraces and destroys representations of animals as an extension and embodiment of himself and all humanity.

Contemporary artists have explored a diverse range of media and methods, and it is impossible to present an overview that captures the depth and breadth of how Modern and Contemporary art explores human-animal interaction. The examples I include here are meant to give the reader an understanding for how, from Lascaux to the present, representations of the connection between humans and animals have some fundamental elements in common. Contemporary artists have transitioned from using animal bodies to make art, to using art to draw attention to ecological destruction, social injustice, and other profoundly challenging topics. In doing so, some artists have interacted with animals, using them as collaborators, as with the German artist Joseph Beuys whose "action," piece, *Coyote: I Like America and America Likes Me,* took place during three days in 1974 in the René Block Gallery in New York. Arriving at JFK airport, Beuys was transferred by ambulance to the gallery where he remained locked in the space with a live coyote, and several other objects and materials including felt, a shepherd's crook, and newspapers that were delivered daily. Upon the work's completion, Beuys stated that, "I believe I made contact with the psychological trauma point of the United States's energy constellation: the whole American trauma with the Indian, the Red Man. You could say that a reckoning had to be made with the coyote, and only then can the trauma be lifted" (Kuoni 1990, p. 142). The work was initially met with nearly universal enthusiasm by critics and others, and aspects of Beuys's ecological discourse still resonate. However, his appropriation of Native-American culture, together with concern about the coyote's experience during the event, are now viewed less rapturously (Gandy, 1997, p. 644).

The final work discussed here, Nicholas Galanin's *Inert Wolf* (2009, Figure 45.12), connects current events with a history of violent human-animal interactions and returns us to

*Figure 45.12* Nicholas Galanin, *Inert Wolf*, 2009. Material republished with the express permission of: Vancouver Sun, a division of Postmedia Network, Inc

the conceptual territory of the cave paintings at Lascaux. In this work, Galanin combined two taxidermied wolves into a single, monstrous creature. The back half of the wolf is a flattened pelt, a trophy incapable of motion. The front half of the wolf, however, has been mounted by the taxidermist to appear as if it is still struggling to move, to survive. It looks at the viewer with a pleading expression as it attempts to crawl forward, away from its own destroyed body. Galanin, a member of the Tlingit Tribe, created the work for an exhibition about society's impact on the environment. The sculpture engages with historical trauma, the inertia of current policies, and hope for survival. At Lascaux, the dead or dying man reached his arm toward a bird in what may have been a plea for help, in Galanin's work, the wolf looks to us. Here, an animal that has been, and continues to be decimated by weapons, snares, and other human tools pleads for our intercession. Transformed from an apex predator into a pathetic trophy, Galanin's wolf looks to us, the human viewer, asking us to see ourselves as part of a shared world in which human–animal interaction is not based on domination but on compassion.

## Ethical Considerations

The section above focused on representations of human–animal interactions in Western art but did not consider the ways in which animal parts are commonly and historically used in the creation of artwork. This has been and continues to be common practice. Animal skin glue, egg tempera, hair for brushes, the list of animal parts is long, and many are still fundamental elements in a diverse range of artistic processes. As discussed above, this includes the use of living animals, as was the case with Joseph Beuys's coyote, or dead animals, as in Nicholas Galanin's work. Numerous contemporary artists are aware of the problematic aspects of this and are working to establish more ethical, vegan practices in their own work.

## Unanswered Questions and Future Research Demands

The "animal turn" in art history is entering its third decade, and art historians are looking ever more closely at representations of human–animal interactions. As they do so, the need for a broad theoretical framework to assess these representations remains. Such a framework is essential to help current and future scholars understand how and why these changes have occurred, what they mean, and what they tell us about our relationships with animals and ourselves. This need is especially urgent at a time when concern about climate change and mass extinction is growing.

## Important Authors Working on This Interdisciplinary Topic Include

Giovanni Aloi, Steve Baker, Erica Fudge, Donna Haraway, Linda Kalof, Susan McHugh, Annie Potts, Harriet Ritvo, Nigel Rothfels, Boria Sax, Karl Steel, Naomi Sykes, and Cary Wolfe.

### Potential Questions for Discussion

- How do visual representations of human-animal interactions affect our understanding of our relationship with animals across time?
- Do the broad trends we see in depictions of human-animal interaction reflect reality or ideals?
- How have we used animals to represent ourselves, and do these uses change when animals and humans are shown interacting with one another?
- Do representations of human-animal interaction that show domesticated animals differ from those that are wild? What do these representations tell us about human-animal interactions with diverse species?
- Do the various media used over time reflect different ideas about human-animal interaction? Beyond reflecting changes in technology, what do changes in the materials used to make art tell us about our relationships with animals?

## References

Alexander, D. (2008). *Saints and Animals in the Middle Ages.* Woodbridge: Boydell Press.
Anderson, J. K. (1985). *Hunting in the Ancient World.* Los Angeles: University of California Press, 83–153.
Antl-Weiser, W. (2018). Beyond hides and bones – Animals, animal representations and therianthropic figurines in palaeolithic art. *Annalen Des Naturhistorischen Museums in Wien. Serie A Für Mineralogie Und Petrographie, Geologie Und Paläontologie, Anthropologie Und Prähistorie, 120*: 51–70.
Aujoulat, N. (2004). *Lascaux, le geste, l'espace et le temps,* Paris: Le Seuil.
Baker, S. (2001). *Picturing the Beast: Animals, Identity, and Representation.* Champaign, IL: University of Illinois Press.
Barringer, J. M. (2001). *The Hunt in Ancient Greece,* Baltimore: The Johns Hopkins University Press.
Beirne, P. (2013). Hogarth's Animals. *Journal of Animal Ethics, 3*(2): 133–162.
Berger, J. (2009). *Why Look at Animals?* London: Penguin.
Calder, L. (2017). Pet and Image in the Greek World: The Use of Domesticated Animals in Human Interaction. In Fö, T. and Thomas, E. (eds.). *Interactions Between Animals and Humans in Graeco-Roman Antiquity.* Berlin: Walter de Gruyter Gmbh.
Chipp, H. B. (1968). *Theories of Modern Art.* Berkeley, LA, and London: University of California Press.
Cohen, A. (2018). Turning Heads: Alexander and the Animals. *Illinois Classical Studies, 43*(1): 88–136.
Colonna, A. (2021). *Religious Practice and Cultural Construction of Animal Worship in Egypt from the Early Dynastic to the New Kingdom: Ritual Forms, Material Display, Historical Development.* Oxford: Archaeopress, 1–27.

Crane, S. (2013). *Animal Encounters, Contacts and Concepts in Medieval Britain*. Philadelphia: Univ. of Pennsylvania Press.

Derrida, J. (2002). The Animal That Therefore I Am (More to Follow). Wills, D. (trans). *Critical Inquiry, 28*(2): 369–418

Einberg, E. and Egerton, J. (1988). *The Age of Hogarth: British Painters Born 1675–1709*, II. London: Tate Gallery Collections.

Elston, M. A. (1987). Women and Anti-Vivisection in Victorian England, 1870–1900. In Rupke, M.A. (ed.). *Vivisection in Historical Perspective*. New York: Croon Helm, 259–94.

Franco, C. (2014). *Shameless: The Canine and the Feminine in Ancient Greece*. Fox, M. (trans.). Berkeley, CA: University of California Press.

Fudge, E. (2000a). *Perceiving Animals: Humans and Beasts in Early Modern English Culture*. Champaign, IL: University of Illinois Press.

Fudge, E. (2000b). Monstrous Acts: Bestiality in Early Modern England. *History Today, 50*(8): 20–25

Foss. T. F. (2012). Roman Ideas in the Late Republic about Animals: Pervasive Cruelty as Indicated and Propagated in the 'Bellum Cantilinae" of Sallust and Interrelating Narrative. *I Quaderni Urbinati di Cultura Classica, 102*(3): 95–121.

Gandy, M. (1997). Contradictory Modernities: Conceptions of Nature in the Art of Joseph Beuys and Gerhard Richter. *Annals of the Association of American Geographers, 87*(4): 636–659.

Garb, T. (1993). Gender and Representation. *Modernity and Modernism: French Painting in the Nineteenth Century*. New Haven, CT: Yale University Press.

Grant, R. M. (1999). *Early Christians and Animals*. London and New York: Routledge.

Guerlac, S. (1991). The Sublime in Theory. *MLN, 106*(5): 895–909.

Haraway, D. (2008). Foreword: Comparing Companion Species, Mis-recognition, and Queer Worlding. In Giffney, N. and Hird, M. (eds.). *Queering the Non/Human*. Farnham, UK: Ashgate.

Havice, C. (1981). In a Class by Herself: 19th Century Images of the Woman Artist as Student. *Woman's Art Journal, 2*(1): 35–40.

Haworth, M. (2018). The Wolfish Lover: The Dog as a Comic Metaphor in Homoerotic Symposium Pottery. *Archimède: Archéologie et histoire ancienne, 5*: 7–23.

Heck, C. and Cordonnier, R. (2012). *The Grand Medieval Bestiary, Animals in Illuminated Manuscripts*. New York: Abbeville Press.

Ingold, T. (1994). From trust to domination: an alternative history of human-animal relations. In Manning, A. and Serpell, J. (eds.). *Animals and Human Society: Changing Perspectives*. London and New York: Routledge, 1–22.

Judkins, R. R. (2013). The Game of the Courtly Hunt: Chasing and Breaking Deer in Late Medieval English Literature. *The Journal of English and Germanic Philology, 112*(1): 70–92.

Kinniburgh, M. C. (2016). Equine Erotics, Possible Pleasures: Early Modern Bestiality and Interspecies Queerness in Plate Five of L'Academie des dames. *Journal for Early Modern Cultural Studies* (16/4): 72–95

Klemettilä, H. (2015). *Animals and Hunters in the Late Middle Ages: Evidence from the BnF MS fr. 616 of the Livre de chasse by Gaston Fébus*, Museum Studies, New York: Routledge.

Klingender, F. (1971). *Animals in Art and Thought to the end of the Middle Ages*. Antal, E. and Harthan, J. (eds.). New York: Routledge.

Kunzle, D. (2000). The Soldier Redeemed. Art and Reality in a Dutch Province at War 1650–1672: Gerard Ter Borch in Deventer. *Marburger Jahrbuch Für Kunstwissenschaft, 27*: 269–298.

Kuoni, C. (1990). *Energy Plan for the Western Man: Joseph Beuys in America. Writings by and Interviews with the Artist*. New York: Four Walls Eight Windows.

Leroi-Gourhan, A. (1968). *The Art of Prehistoric Man in Western Europe*. London: Thames and Hudson.

Lévi-Strauss, C. (1964). *Totemism*. Needham, R. (trans). Boston: Beacon Press.

Levine, F. S. (1976). The Iconography of Franz Marc's Fate of the Animals. *The Art Bulletin, 58*(2): 269–277.

Lüders, K. and Nanna, L. (2020). What Happened to John Franklin? Danish and British Perspectives from Francis McClintock's Arctic Expedition, 1857–59. *Journal of Victorian Culture, 25*(2): 300–314.

Marc, F. (1912). Die neue Malerei. *Pan, 11*: 469.

Mazzola, E. (2006). Something Old, Something New, Something Borrowed, Something Ermine: Elizabeth I's Coronation Robes and Mother's Legacies in Early Modern England. *Early Modern Women, 1*: 115–136.

McNiven, I. J. (2010). Navigating the Human-Animal Divide: Marine Mammal Hunters and Rituals of Sensory Allurement. *World Archaeology, 42*(2): 215–230.

Mithen, S. J. (1988). To Hunt or to Paint: Animals and Art in the Upper Palaeolithic. *Man, 23*(4): 671–695.

Moore, A. (2009). Sir Edwin Landseer's *Man Proposes, God Disposes*: And the Fate of Franklin. *British Art Journal, 9*(3): 32–37.

Morgan, L. (1998). Power of the Beast: Human-Animal Symbolism in Egyptian and Aegean Art. *Ägypten Und Levante/Egypt and the Levant,* 7: 17–31.

Musacchio, J. M. (2001). Weasels and Pregnancy in Renaissance Italy. *Renaissance Studies, 15*(2): 172–187.

Niemelä, P. and Örmä, S. (2016). Lady with an 'Ermine. *Source: Notes in the History of Art, 35*(4): 302–310.

Nugent, F. (2003). *Seek the Frozen Lands: Irish Polar Explorers 1740–1922*. Cork: Collins Press.

Ormond, R. (1981). *Sir Edwin Landseer*. New York: Rizzoli International Pub.

Parker, L. P. (2020). The Life of St. Margaret of Antioch (11th c.). In McNabb, C. H. (ed.). *Medieval Disability Sourcebook: Western Europe*. New York: Punctum Books, 210–219.

Pedretti, C. (1990). La dama con l'ermellino come allegoria politica. In Baldini, A.E. and Barcia, F. (eds.). *Studi Politici in Onore di Luigi Firpo*. Milan, I, 161–181.

Phoebus, G. (1387–9). *Livre de chasse*. MS fr. 616, Bibliothèque nationale de France, Paris.

Resl, B. (2007). Beyond the Ark: Animals in Medieval Art. In Resl, B. (ed.). *A Cultural History of Animals in the Medieval Age*. London: Bloomsbury Pub.

Ring, G. (1945). St. Jerome Extracting the Thorn from the Lion's Foot. *The Art Bulletin, 27*(3): 188–194.

Robertson, M. (1951). The Place of Vase-Painting in Greek Art. *The Annual of the British School at Athens, 46*: 151–159.

Saslow, J. M. (1992). 'Disagreeably Hidden': Construction and Constriction of the Lesbian Body in Rosa Bonheur's 'Horse Fair'. In Broude, N. and Garrard, M.D. (eds.). *The Expanding Discourse: Feminism and Art History*. New York: Routledge.

Shapiro, H. A. (1990). Old and New Heroes: Narrative, Composition, and Subject in Attic Black-Figure. *Classical Antiquity, 9*(1): 114–148.

Sorabella, J. (2007). Eros and the Lizard: Children, Animals and Roman Funerary Sculpture. *Hesperia Supplements, 41*: 353–370.

Steintrager, J.A. (2001). Monstrous Appearances: Hogarth's "Four Stages of Cruelty" and the Paradox of Inhumanity. *The Eighteenth Century, 42*(1): 59–82.

Thomas, K. (1983). *Man and the Natural World: Changing Attitudes in England, 1500–1800*. Oxford: Oxford University Press.

Toynbee, J.M.C. (1973). *Animals in Roman Life and Art*. London: Thames and Hudson.

Uglow, J. (1997). *Hogarth: A Life and a World*. New York: Farrar, Straus and Giroux.

Wachtel, E. (1993). The First Picture Show: Cinematic Aspects of Cave Art. *Leonardo, 26*(2): 135–140.

Walsh, A., Buijsen, E., and Broos, B. (1994–1995). *Paulus Potter: Paintings, Drawings and Etchings*. The Hague: Mauritshuis.

Wolloch, N. (2019). *The Enlightenment's Animals*. Amsterdam: Amsterdam University Press.

Yusoff, K. (2015). Geologic Subjects: Nonhuman Origins, Geomorphic Aesthetics and the Art of Becoming Inhuman. *Cultural Geographies, 22*(3): 383–407.

# 46
# SINISTER CANINES IN THE MEDIA
## Animals in Popular Culture

*Margo DeMello*

## Introduction

When most people think about human-animal relationships, they think of our interactions and relationships with the animals whose lives best intersect with those of our own – companion animals – or perhaps with working animals like service dogs, therapy horses, and the like. But one of the most significant places where animals come into human lives is through the media we consume. Even as we stumble through a growing ecological crisis and the largest animal extinction event in 250 million years, with hundreds (if not thousands) of species vanishing from this planet each year, animals are more ubiquitous in culture than ever.

Animals occupy an important place in our art, our poetry, our folktales, and our literature. They play a significant role in our religious beliefs and practices, they are sprinkled throughout our languages, and, as symbols, they help us to make sense of our own lives and our place in the world.

We have been writing, painting, and thinking with animals since we have been human – if not before. The oldest known cave paintings, those found at Maltravieso cave in Cáceres, Spain, feature animals like bison, mammoths, and horses, and have been dated to 66,000 years ago (Hoffmann et al 2018). If this date is correct (and not everyone agrees that it is; see Pons-Branchu, et al 2020 for the most recent discussion) then this find is significant because it dates to before the arrival of anatomically modern humans (*Homo sapiens sapiens*) in Europe, which means the paintings must have been created by Neanderthals.

With changes in media technology come changes in how we tell stories. Animals have traveled with us as our stories move off the cave walls and onto, eventually, the written page, and, in the last two centuries, onto television and film and onto the Internet. But while an increasingly animal welfare-savvy public, and a growing community of scholars in anthrozoology, is concerned about the lives of other animals – especially companion animals and other animals who live in captivity – there is very little thought or attention given to the ways in which animals are represented in popular media.

This chapter takes seriously the presence of other animals in the media and attempts to lay out a path that might help us to better reckon with this presence. I hope to demonstrate that how we think about, and represent, the lives of other species matters, in that it can tell us not only how we think about ourselves but can shape our treatment of others.

## Understanding Popular Culture

Popular culture refers to media forms like television shows, podcasts, or YouTube videos, created primarily by the culture industry for mass, mainstream consumption (Fiske 2010; Strinati 2014). While art, folktales, and myths predate mass media, popular culture is now typically thought of in terms of those forms of media which are found in industrial and postindustrial capitalist societies and which are marked by mass production and consumption of goods as well as significant leisure time.

Popular culture is designed to inform and entertain audiences, but it does much more than that. Part of what media of any kind does is to *mediate* reality. It tells us what to look at, and how to look at it. YouTube videos, to pick one example, make our worlds larger: they bring people, animals, ideas, and things from around the world (and increasingly from outside of this world), things that we might otherwise never see, into our homes and our lives, and, importantly, into our spheres of influence. Without ever traveling outside of my country, or indeed without ever leaving my bed where I spend hours scrolling on my tablet, I can see people and animals with whom I would otherwise never come in contact. In this sense, popular culture can expand our world in vast and incalculable ways.

But viewing the world through the mediating lens of film and a television screen, for example, shrinks it as well. This is because of the ways that technology and the intent of the producer of the media form combine to produce a particular story, while at the same time closing off additional stories.

Stories – from *Aesop's Fables* to the latest installment of *The Planet of the Apes* film franchise – are creations of particular creators. These creators bring their own values and experiences to the media text that they create, and those values and experiences are shaped by the subject position of the producer. In this way, social norms surrounding class, ethnicity and racial position, gender and sexuality, are all replicated through the media form. And because the culture industry remains dominated by English-speaking, Caucasian, straight, middle-class men, then it is the values and experiences of those men that will become disseminated through the media, thus reinforcing a particular value system and habitus, and spreading it around the world (Lauzen 2012; Erigha 2018; Bernardi 2009). This is an example of cultural hegemony: the transmission and replication of the values and norms of the ruling class via cultural products like fashion, films or music videos, but also attitudes about family, community, animals, and more (see Hall 2006; Hebdige 2012 on popular culture and ideology). This centuries-long stranglehold over control of the media has been changing, however. We have seen, over the last two decades in particular, more women, people of color, and sexual minorities take positions of influence within the culture industry. This has resulted in new films and television shows like *Encanto*, *Us*, and *Bridgerton*, and new musical artists like Lil Nas X, that both reflect and speak to the diversity of voices that make up our culture, and that challenge long-held cultural norms.

And finally, because media companies are so highly consolidated (i.e. a tiny number of companies and individuals own the majority of all media in the world), it limits the number of voices and ideas that circulate to those that are most acceptable to the majority, and those most supportive of the status quo (Wellstone 1999; Ho & Quinn 2008). One of the reasons why popular culture largely promotes conservative values has to do with the funding model for most media: advertising. As long as media is primarily funded through advertising revenue, the messages that we see in media will remain messages that advertisers and the companies they represent can support (Jhally 2003).

In this sense, popular culture is far more than mindless entertainment. Popular culture *does* something. It makes us see and experience the world through a very narrow lens, and how we see the world then shapes what we do.

Popular culture theorists have demonstrated that who and what we see in the media, and how what we see is represented, matters deeply. For example, the 1933 Hollywood film *King Kong* (and to a slightly lesser extent, the subsequent remakes), made during a period of deep racism in the united states and much of the rest of the world, is based on a set of misunderstandings about the "violent" nature of gorillas combined with racist ideas about the "animalistic" nature of Black men, ideas which had been circulating in American popular culture since the eighteenth century (Bradley 2000; Quallen 2016). After all, gorillas had only been first "discovered" by Europeans in the mid-nineteenth century (the first mountain gorilla not until 1902), and Western primatology was still decades away when *King Kong* was made. The result is not just a dozen films about a dangerous giant ape, but a deeply disturbing history of using animals to dehumanize African and African American men (see Creed 2007; Hund 2015; Roche Fahmi 2017; Roche Carcel 2021).

## Theories of Popular Culture

Theorists of mass media approach the subject through a variety of ways. A historical analysis is one that focuses on the historical context of a media text, and how changing social, economic, and cultural forces result in changes to the text. For example, the Walt Disney company is responsible for shaping the minds and imaginations of millions of children around the world. The size and power of this company's reach make Disney an extremely important subject for analysis. One issue that a historical analysis might investigate, for example, is how Disney, through their *True Life Adventures* nature series (1946–1960), shaped generations of children's ideas about animals, and also created the template for the modern nature documentary (Lutts 1992; Robertson 1999; Cruz Jr. 2012; Mitman 2012). This template, based on a presentation of nature as peaceful, free of humans, and populated by hyper-cute animals who live in American-style nuclear families, has helped to create the idea of nature that many Americans still hold today. Bambi (Algar 1942) not only helped to establish the concept of the nature documentary, but it helped shape generations' of Americans' ideas about hunting, forest fires, and conservation in general.

A content analysis is a type of approach which focus on the media text itself, and assess the ways in which meaning was encoded into the text by the producer(s) via images, words, music, etc. Popular culture texts are polyvocalic, which means that there might be a number of different, even contradictory, meanings found in the text. Because consumers are anchored in a particular social context, different consumers will pick up different meanings from the same cultural product.

Sometimes the main message of a popular culture text conflicts with other messages inserted into the text. So a 2013 article on climate change on the website POPSCI is titled "How extreme weather links the fates of four adorable Arctic species." The article is accompanied by a picture of an adorable arctic fox (Doyle 2013). While the article's focus was on disseminating the results of a recent scientific paper on climate change, the headline and the photo focus less on the crisis of climate change with its resulting threats of extinction and more on the attractiveness of some of the threatened species. We already know that perceptions of cuteness influence our ideas about which species deserve conservation (Gunnthorsdottir 2001), welfare (Wolfensohn 2020), or even to be eaten or not (Silva 2016; Zickfeld et al 2018). This article reinforces – without much subtlety – the idea that only cute animals deserve moral consideration. This is related to the idea that, through much of popular culture, animals are, in the words of Steve Baker (2001) Disneyfied: reduced to their cuteness through the use of images, sounds, and texts that highlight this cuteness. Baker writes, "the animal is the sign of all that is taken not very seriously in contemporary culture; the sign of that which doesn't really matter" (1993: 174).

Another focus of content analysis is how stereotypes – and in particular, stereotypes about animals – are deployed in a text. We know, for example, that one of the most common uses of an animal in a media text is through anthropomorphism. Anthropomorphism refers to the projection of allegedly human traits onto nonhuman animals. Dressing animal characters in clothing, having them engage in human activities like birthday parties or weddings, and having them speaking a human language are all examples of anthropomorphism and can give audience members – especially children – an unrealistic understanding of animals (Geerdts 2016; Li et al 2019).

Another trope that we see in media texts that use animals is zoomorphism (Kordos 2014). Zoomorphism is the use of animal behavior to help us to understand human behavior and is a common feature in nature documentaries where the behavior of wild animals is used to explain contemporary human social relations. Where anthropomorphism can paint an unrealistic image of an animal, zoomorphism reduces the complexities of social behavior to an (already reduced) animal model. Both practices can have implications for animals.

In order to understand how audience members decode popular culture texts, an audience analysis is used. This type of analysis can not only tell us how different social positions produce different readings of a text, but an audience analysis can also help us to understand the impact that representation – or lack thereof – has on us.

This analytical framework has been most commonly used to show how minority groups are either underrepresented or heavily stereotyped in the media, and, crucially, how that representation impacts identity formation, mental health, the development of prejudice, unequal opportunities, and much more (see Gross 1991; Merskin 1998; Reny & Manzano 2016; Kiprotich & Chang'orok 2017; Schug et al 2017; Wigger 2019).

One of the reasons why social inequities are replicated in popular culture has to do with who creates media, and under what conditions. A production analysis is a type of analysis which looks at who produces media texts, and how the values of those producers get disseminated through the culture. As long as writers, directors, editors, and producers of media content are drawn from a narrow demographic, then we will continue to consume media content which only represents a few of us (Ho & Quinn 2008). One way that we might use a production analysis to understand the role of animals in the media is to look at animal actors. How animal actors are procured, trained, and disposed of, as well as their living conditions while not working, is an important, but largely invisible aspect of television shows, movies, and plays and can tell us much about how much we value those animals (Iacona 2016; Aldrich 2018).

## Why Animals?

Throughout history, and in every culture, animals serve as symbols. Symbols are things to which people give meaning, which stand for, or represent, something else. The meaning of symbols depends on the cultural context in which they appear, and shape how we think about things. The cross, a nation's flag, and the swastika are all symbols that are imbued with different meanings in different cultural and historical contexts. There is no inherent meaning in the symbols themselves.

Animals are used to symbolize a whole host of characteristics that we see in ourselves, or want to project onto others, but that may be dangerous or foreign to us (Rowland 1973). Thus, animals can be lustful, deceitful, murderous, or promiscuous. They can also symbolize more positive qualities like love, altruism, and sacrifice. The second-century Roman writer Aelian's book, *On the Nature of Animals,* a collection of stories about animals, tells us that the octopus is sneaky, fish are well behaved, and owls are wily, like a sorceress. This type of part-scientific, part-fanciful discussion of animals would remain popular through the classical world and into the Middle Ages. Medieval bestiaries were books filled with images of animals, a bit about their natural history, and what the animals symbolized, all accompanied by a moral lesson (McCulloch 2017).

Why are animals so commonly used as symbols? Anthropologist Claude Lévi-Strauss, in his work on totems (1963), said that animals are chosen as totems not because they are good to eat, but because they are "good to think (with)." In other words, animals have great metaphorical potential, and, for Lévi-Strauss, they are especially useful for representing social classifications like clans and other aspects of kinship systems.

Animals are both *like us*, but also *unlike us*. Because of this ambiguity, they are a perfect vehicle for expressing information about ourselves, to ourselves. As we've noted, we both bestialize people (who we call bitches, cows or foxes, or, in the case of whole groups of people, beasts, or vermin – see Diallo 2006; Arseneault 2012; Jackson 2021), and we humanize animals via anthropomorphism. And while we can use animals to highlight a person's good qualities (brave like a lion), we more commonly use animals negatively (cunning like a fox), and especially to denigrate minorities (Haslam, Loughnan, & Sun 2011; Plemenitaš 2017; Gambert & Linné 2018; Donovan 2020; Cordeiro-Rodrigues Forthcoming).

But why do we choose *some* animals to symbolize specific things? On the one hand, symbols are arbitrary – nothing in the stars and stripes of the American flag that indicate freedom or democracy. Cats, for instance, can be symbolic of bad luck, of evil, or of witchcraft in Western cultures (Lawrence 2003). Why the cat was chosen tells us something about both cats (who famously see in the dark, who are solitary, and who have a blood-curdling cry when fighting) and about the people in the cultures that use cats in this way (see Darnton 2009 for a chilling look at the reality of cats' lives in eighteenth-century France.)

## Animals in Children's Media

Perhaps the area of popular culture that most utilizes animals is children's books, clothing, toys, and decor. Animals may also play such an important role in children's literature because children seem to naturally love animals (Melson 2005; Cole & Stewart 2016). Children in all cultures are drawn to animals from a very young age, forming attachments to them and making them central in their lives. Children also anthropomorphize animals; scholars say that children have not reached the point that so many adults have where the animal and human worlds have become separate; animals are, to many children, playmates, parents, friends, and teachers. Because of this, children project allegedly "human" characteristics and traits onto animals – and while scientists are just now coming around to the idea that feelings like love, anger, or jealousy are most definitely shared by both humans *and* many animals, children seem to intuitively do this, using their own bodily experiences to relate to the experiences of animals (Endenburg & van Lith 2011; Melson & Fine 2015).

In his classic work on children's fairy tales, *The Uses of Enchantment* (2010), Bruno Bettelheim writes that the line between humans and animals is much less sharply drawn for children than adults, so the idea that animals can be children, or can turn into humans seems quite possible. In addition, children are drawn to animals which seem like little people – bears, for example, while fierce in real life, with their fat bodies, big heads, short limbs and upright gait, seem like little people – and thus was born the teddy bear (Geerdts 2016).

Children relate to animals, and, since at least the nineteenth century when children began to be understood and treated as a distinct class, adults have understood that they could use the natural affinity between children and animals to teach children valuable social skills (Grier 2010). Through reading about and caring for animals, the idea is that children learn empathy, relationship skills, kindness, and compassion (Ascione 1992). They also use their relationships with animals to help develop their own identities as people. Finally, children use animals as an emotional safety net – retreating to them when feeling sad or upset or scared. Because of children's connection to real animals, representations of animals are also popular with children, explaining why animals so heavily populate children's toys, children's books, and children's art.

If animals are "good for children to think," the corollary is that as children develop the skills, they need to be adults, that they will (or should) no longer need animals at all. As children grow up, they are expected to shrink from animals, and if they don't, they are thought to be immature, to be hanging onto childhood. Many children's books, while emphasizing the closeness between child and animal, end with the child growing up and, sometimes, the animal's death. The most obvious example of this was found in Fred Gipson's *Old Yeller* (1956), which details the love between a boy named Travis and his dog, Yeller, but after Yeller contracts rabies, the boy must shoot him.

## Animals in the News

But while we associate animals in the media with children's texts, animals are an important element of adult media as well. One of the most important places this occurs is in the news media.

Animals, in the words of Claire Parker Molloy (2011), sell papers. Animals who escaped from slaughter, cruelty cases, extraordinary pet stories, baby animal births, and animal attacks – stories like this are the bread and butter of much of local news. They are also considered "soft" news, i.e. news stories created by women, often read by women, and considered to be trivial compared to politics or crime (Packwood Freeman 2009).

Animal news stories, like popular culture in general, are conservative, in that they tend to ignore or whitewash elements of the story that might make the audience uncomfortable or might alienate advertisers. So a story about a cow who jumped off of a truck to escape slaughter is reported in a way that highlights the elements of the story that are familiar to us (like the hero's journey) over the elements that might provide needed context (like the conditions in modern agricultural settings – see Colling (2014) for more on media portrayals of escaped livestock). Another example are news pieces about the welfare problems associated with dogs living on First Nations reservations in Canada which focus on the lack of care by First Nations communities but not on the many systemic obstacles to that care (Fraser-Celin & Rock 2021).

Animal news stories are the original form of clickbait. Used by tabloids, broadsheets, and other inexpensive forms of media aimed at the poor and working classes, these stories aimed (and still do) to titillate, excite, or outrage the reader (Herzog & Galvin 1992). And while we might be forgiven for assuming that tabloid coverage of animals is a relatively new phenomenon, it is not.

In the remainder of this chapter, I want to discuss the story of the Beast of Gévaudan, a creature responsible for a series of animal attacks that terrorized the peasants of France in the eighteenth century, and, when the story went international, may have played a small role in the coming French Revolution. It is my hope that this story will allow us to better understand the ways in which media accounts can radically shape lives – both animal and human.

## The Wolf

Wolves are one of the most persecuted of all wild animals but also one of the most culturally powerful. They populate our myths, legends, and folktales, and are a powerful symbol of loyalty, freedom, and strength.

Wolves are apex predators, which means that they have no predators of their own (except humans), and they are a keystone species, which means that they play a vital role in maintaining an ecosystem. Wolves once roamed much of Europe, where their relationships with humans ranged from uneasy détente to outright warfare. In eastern Europe, they were highly regarded, especially among the Slavic countries.

The ability of the wolf to quietly stalk and viciously kill their prey (they are endurance predators rather than ambush predators) makes wolves both respected and feared hunters. After humans began domesticating livestock starting around 10,000 years ago, wolves began to be the subject of widespread persecution. After the rise of Christianity, wolves were not just an economic threat, but were becoming an *existential* threat. The portrayal of wolves started to shift from a strong, independent, and fierce predator to an amoral, greedy, and evil – even demonic – creature. *The Malleus Malleficarium*, a fifteenth-century guidebook for hunting witches, described wolves as either agents of the devil or agents sent by God to punish men. This is why Jesus is known as the Good Shepherd: he protects his flock of sheep (who represent early Christians) from wolves (who represent the devil). Ultimately, the wolf suffered greatly from these portrayals, as wolves have been subject to systematic slaughter in much of the western world (Kvangraven 2019; Kvangraven 2019; Pluskowski 2005; Beresford 2013).

Starting in the Middle Ages, Eurasian wolves were targeted by a concerted campaign which spanned much of western Europe and which resulted in the wolf population disappearing entirely in England in the sixteenth century, followed by Scotland and Ireland. Later, all the Scandinavian nations eventually wiped out their wolf populations. One by one, countries in western Europe, followed by central Europe, eliminated all their wolves; only eastern Europe, thanks to its extensive forests, did not lose all of its wolves (Fernández & Ruiz de Azua 2010). Once wolves developed this more sinister reputation, it did not take long for the fear and hatred of wolves to be expressed through tales like Charles Perrault's Little Red Riding Hood or through the creation of the werewolf.

## The Beast of Gévaudan

One of the most frightening of all wolf stories is the saga of the Beast of Gévaudan – a creature or creatures who terrorized the people living in the province of Gévaudan (now the Lozère/Haute-Loire), in the south-central region of France, during the mid-eighteenth century. Starting in the summer of 1764, both livestock and villagers began disappearing in the vicinity of the forest, sometimes turning up mutilated. The first survivor of the beast was a 14-year-old girl named Marie-Jeanne Valet who was attacked while crossing a river; the girl only survived because she stabbed the wolf with a wooden spear. (There is now a statue of Valet fighting the beast in the town of Auvers.)

The beast was described as huge, much larger than wolf, with red or gray fur, a stripe down its back, and blazing red eyes, and it could stand on four feet or just two. Some witnesses reported that it could repel bullets, disappear into thin air, or do other unnatural things.

Before the crisis was over in 1767, over 200 people were attacked, and over 100 killed, with an unknown number of animals losing their lives. The victims were maimed by their attackers, often having their faces or throats ripped open. Many of the accounts include a sexual element; victims (who were overwhelmingly women and children, who were primarily responsible for caring for flocks) were reportedly "ravaged" while being killed. Eyewitnesses and survivors of some of the attacks reported that the attackers were either dogs, wolves, or some sort of large, bipedal humanoid. Other witnesses reported seeing multiple animals: sometimes a pair, and sometime a mother and offspring.

Stories of the attacks were first reported in the local newspapers, but the story quickly gained traction, appearing in newspapers around the world. Just as modern tabloids feature sensational stories of dubious origin, so did newspapers of the past. Each newspaper story, poster, handbill, or pamphlet that covered the Beast included language and images intended to excite and titillate the readers. In fact, many of the reports simply repeated the claims found in French newspapers, whether or not they were sensationalistic or just incorrect.

One can argue that it was not the attacks, but the press coverage of the attacks, that created the Beast. François Morénas, the editor of the *Courrier d'Avignon*, was perhaps more responsible than anyone else for the more fantastical aspects of the story, as he was intentionally trying to add stories of a more exciting nature to the paper. It was Morénas' reports, which emphasized the unnatural and supernatural aspects of the story, that served as the template for other journalists to follow (Smith 2011).

As the press reports spread, the attacks spread as well. While the first attacks were located in the region of Gévaudan, as more newspapers began reporting on the story, the attacks started to spread outward from Gévaudan, as the fear spread throughout the French countryside.

France, during this time, was experiencing a number of social, political, and economic strains. France at this time was divided into three "estates": the clergy, known as the first estate, the nobility, known as the second estate, and the rest of the population, which made up the third estate. It was only the third estate who were responsible for paying taxes, so peasants were burdened by constantly increasing taxes. In addition, continuous fighting between French Catholics and French Protestants (known as Hugenots), the end of the Seven Year War between France and Britain, a rising population which put huge pressures on the production of grain, leading to an inflation crisis, all created fertile ground for the Beast to emerge.

After reports of the unexplained deaths reached King Louis XV, a number of hunters were dispatched to Gévaudan to track and kill the beast. A wanted poster printed in 1764 described the beast as follows, before offering a reward of 2,700 livres to the person who kills it:

> Picture of the monster desolating the Gévaudan, This Beast is the size of a young Bull, it likes to attack Women and Children, it drinks their Blood, cuts off their Heads, and carries them off.

Like many of the reports from this time, this poster emphasized the beast's interest in women and children as well as the vicious way that it attacks. Some of the drawings of the beast that were made at this time depict it as a large wolf, while others show a more ambiguous, human-like creature. In fact, one of the reasons why this is such a well-known case is because the eighteenth century was a time when the popular presses in Europe began to rapidly expand. European newspaper breathlessly covered the Beast of Gévaudan and heavily embellished the stories of the attacks, but also spread the story far and wide, embarrassing King Louis XV, who could not keep his people safe. It was thanks to these international stories, and the embarrassment that they brought France, that Louis sponsored an official government program to kill the beast.

After the first hunters returned to the capitol without having caught the beast, Francois Antoine, the King's Lieutenant of the Hunt, arrived in the region in June 1765. In June, Antoine shot and killed a gray wolf that was identified by some of the survivors as the culprit, and later killed that wolf's entire family; the first wolf was stuffed and sent to Versailles to be presented to the King. The scene at Versailles was reported in newspapers around the world. The *Elgin Courier* (from Elgin Texas) reported:

> the astonishment and the disappointment of all ... — the terrible wild beast of Gévaudan, whose sanguinary career had for so many months excited such dismay there and wonder elsewhere — and found that it was only a wolf after all, and not a very large one!

Even though Antoine was both hailed for his bravery and paid generously, it was less than three months later that two young boys were attacked (and survived). This attack was followed by more attacks, and more deaths, as well as additional hunting expeditions (on one such raid, over 2,000 wolves were killed) until another wolf was killed by a local hunter named Jean Chastel on June 19, 1767. Chastel used a silver bullet (which is known to be able to kill werewolves) to kill the creature. During the preparation of the body for stuffing, the last victim was found in its stomach.

Ultimately, it is unknown how many animals were responsible for the attacks in Gévaudan, although it is now assumed that the killings were done by multiple wolves, perhaps working in a pack, although the killers could have been dogs as well. The killings could also have been committed by a lion or another big cat; these animals were kept in private menageries all over Europe and were known to escape from time to time. Whether it was one wolf, a rabid wolf or dog, multiple wolves, or some other animal, well over countless wolves were killed before the attacks ended in 1767. Unfortunately for the French crown, while the troubles associated with the Beast were finally over, the rest of the troubles plaguing France were not, and just ten years later, the French Revolution erupted in an orgy of killing.

The fact that a wolf was killing peasants in the mountainous countryside of southern France should not surprise us. The fact that a wolf (or pack of wolves) could terrorize an entire nation and could weaken one of the most powerful kings in all of Europe should surprise us. But it is an illustration of the power of the media in creating reality, and in shaping lives. Louis himself recognized the power of the press, and most likely intervened to ensure that French newspapers stop covering the story. It shouldn't surprise us to learn that when the press coverage stopped, the killings stopped as well.

## Conclusion

In his influential essay, "Why Look at Animals" (1973), Jon Berger discusses our evolving relationship with other species, and, like many scholars to follow him, suggests that the original place of animals in human society was as symbol, rather than as food, feather, or fur. But industrialism transformed the human-animal relationship, leaving more animals living in captivity, under the watchful gaze of humans. Where animals and humans may once have looked at each other from a position of equality, this, to Berger, is no longer possible. Regarding that gaze, Berger notes that while other animals may look *at us,* the difference between the animal gaze and the human gaze is the power that lies beneath it. When we magnify the power of humans to see every aspect of an animal's life and death, we also magnify human control over those lives and deaths.

But the power that humans hold, and the substantial power that accrues to media content creators, can be used to shape public opinion and to help animals. Most Americans are unaware of this, but the most important piece of federal legislation protecting animals in the United States, the Animal Welfare Act, was passed in its original form in 1966 because of two news stories, one in *Life Magazine* and one in *Sports Illustrated*, about a stolen pet dalmatian named Pepper. Pepper was kidnapped from his Pennsylvania yard in 1965 and was later found dead in a New York hospital, where she had died while undergoing experimental heart surgery. The public outrage generated from the media coverage over Pepper's death (which occurred only four years after the release of Disney's *101 Dalmatians*, the story of a family of stolen dalmatians, including one named Pepper) was so intense that it resulted in the creation of what was then called the Laboratory Animal Welfare Act.

The Act contained, among other things, details regarding how and where animals used for experimentation could be obtained, as well as care requirements once at the facility. By 1966, when Pepper's story first hit the press, Americans were already primed to be concerned about the theft of pet dogs; Disney's *101 Dalmatians* was released in 1961, and its story of the kidnapping of a litter of dalmatian puppies (who were going to be killed for fur, rather than vivisection) helped to make it one of Disney's most successful films. And the fact that one of the stolen puppies in the Disney film was also named Pepper made the connection between the film and the real-life Pepper (who was most likely named after Disney's dog) all the more real.

Sixty years later, *Blackfish*, an independent documentary on the killing of dolphin trainer Dawn Brancheau (and two others) by a killer whale named Tillikum, was released and resulted in SeaWorld, where Tillikum was a performing whale, losing millions of dollars in ticket and merchandise sales as well as corporate partnerships. Ultimately, the public sentiment surrounding the film resulted in the end of all orca shows and the orca breeding program at SeaWorld. The "Blackfish effect," as it is now known, has not yet been repeated with this kind of success elsewhere, but media producers and animal rights activists clearly recognize the potential for such transformative results (Brammer 2015; Boissat et al 2021). Only time will tell whether animals will continue to be trivialized in the media, used to generate clicks, likes, and outrage, and whether animal advocates will continue to use the media in new ways in order to generate action on behalf of animals.

## Discussion Questions

1. Why are animals, in the words of Lévi-Strauss, "good to think?"
2. What do the stories of the Beast of Gévaudan and the twentieth-century story of Pepper the dalmatian tell us about how popular culture shapes our understanding of, and treatment of, nonhuman animals?
3. How can, or how should, animal advocates use the media to change the public's attitudes toward animals?

## References

Aldrich, B.C. (2018). The use of primate "actors" in feature films 1990–2013. *Anthrozoös*, *31*(1), 5–21.

Algar, J., Armstrong, S., Hand, D., Heid, G., Roberts, B., Satterfield, P., Wright, N., Davis, A., & Geronimi, C. (1942). *Bambi*. RKO Radio Pictures.

Arseneault, J. (2012). Brute violence and vulnerable animality: A reading of postcoloniality, animals, and masculinity in Damon Galgut's the beautiful screaming of pigs. *Postcolonial Text*, *7*(4), 1–23.

Ascione, F. R. (1992). Enhancing children's attitudes about the humane treatment of animals: Generalization to human-directed empathy. *Anthrozoös*, *5*(3), 176–191.

Baker, S. (2001). *Picturing the beast. Animals, identity, and representation*. Urbana and Chicago: University of Illinois Press. First published 1993.

Beresford, M. (2013). *The White Devil: The werewolf in European culture*. Reaktion Books.

Bernardi, D. (2009). *Filming difference: Actors, directors, producers, and writers on gender, race, and sexuality in film*. University of Texas Press.

Bettelheim, B. (2010). *The uses of enchantment: The meaning and importance of fairy tales*. Vintage.

Boissat, L., Thomas-Walters, L., & Veríssimo, D. (2021). Nature documentaries as catalysts for change: Mapping out the 'Blackfish Effect'. *People and Nature*, *3*(6), 1179–1192.

Bradley, K. (2000). Animalizing the slave: The truth of fiction. *The Journal of Roman Studies*, *90*, 110–125.

Brammer, R. (2015). Activism and antagonism: The 'Blackfish' effect. *Screen Education*, (76), 72–79.

Cole, M., & Stewart, K. (2016). *Our children and other animals: The cultural construction of human-animal relations in childhood*. Routledge.

Colling, S. (2014). *Animals without borders: Farmed animal resistance in New York*.

Cordeiro-Rodrigues, L. (Forthcoming). Nonhuman animal metaphors and the reinforcement of homophobia and heterosexism. In David Nibert (Ed.), *Capitalism and animal oppression*.

Creed, B. (2007). What do animals dream of? Or King Kong as Darwinian screen animal. In Simmons, L. (ed.), *Knowing animals* (pp. 57–78). Brill.

Cruz Jr, R. (2012). *The animated roots of wildlife films: Animals, people, animation and the origin of Walt Disney's' True-Life Adventures'* (Doctoral dissertation, Montana State University-Bozeman, College of Arts & Architecture).

Darnton, R. (2009). *The great cat massacre: And other episodes in French cultural history*. Basic Books.

Diallo, A. C. (2006). *"More Approximate to the Animal": Africana resistance and the scientific war against black humanity in mid-nineteenth century America*. ProQuest.

Donovan, J. (2020). Animal equality: Language and liberation (2001). *The American Journal of Semiotics*, *17*(4), 359–362.

Doyle, R. (2013). *How extreme weather links the fates of four adorable arctic species*. Popsci.

Endenburg, N., & van Lith, H. A. (2011). The influence of animals on the development of children. *The Veterinary Journal*, *190*(2), 208–214.

Erigha, M. (2018). On the margins: Black directors and the persistence of racial inequality in twenty-first century Hollywood. *Ethnic and Racial Studies*, *41*(7), 1217–1234.

Fahmi, M. E. E. (2017). Peter Jackson's King Kong (2005): A critique of postcolonial/animal horror cinema. *English Language and Literature Studies* *7*(2), 15–30.

Fernández, J. M., & Ruiz de Azua, N. (2010). Historical dynamics of a declining wolf population: Persecution vs. prey reduction. *European Journal of Wildlife Research*, *56*(2), 169–179.

Fiske, J. (2010). *Understanding popular culture*. Routledge.

Fraser-Celin, V. L., & Rock, M. J. (2021). One Health and reconciliation: Media portrayals of dogs and Indigenous communities in Canada. *Health Promotion International*, *37*(2), 1–10

Gambert, I., & Linné, T. (2018). From rice eaters to soy boys: Race, gender, and tropes of 'plant food masculinity'. *Gender, and Tropes of 'Plant Food Masculinity'* (December 9, 2018).

Geerdts, M. S. (2016). (Un) real animals: Anthropomorphism and early learning about animals. *Child Development Perspectives*, *10*(1), 10–14.

Grier, K. C. (2010). *Pets in America: A history*. UNC Press Books.

Gross, L. (1991). Out of the mainstream: Sexual minorities and the mass media. *Journal of Homosexuality*, *21*(1–2), 19–46.

Gunnthorsdottir, A. (2001). Physical attractiveness of an animal species as a decision factor for its preservation. *Anthrozoös*, *14*(4), 204–215.

Hall, S. (2006). Popular culture and the state. *The Anthropology of the State: A Reader*, *9*, 360.

Haslam, N., Loughnan, S., & Sun, P. (2011). Beastly: What makes animal metaphors offensive?. *Journal of Language and Social Psychology, 30*(3), 311–325.

Hebdige, D. (2012). 12 (i) From culture to hegemony; (ii) subculture: The unnatural break. *Media and Cultural Studies: Keyworks*, 124–136

Herzog, H. A., & Galvin, S. L. (1992). Animals, archetypes, and popular culture: Tales from the tabloid press. *Anthrozoös, 5*(2), 77–92.

Ho, D. E., & Quinn, K. M. (2008). Viewpoint diversity and media consolidation: An empirical study. *Stanford Journal of International Law, 61*, 781.

Hoffmann, D. L., Standish, C. D., García-Diez, M., Pettitt, P. B., Milton, J. A., Zilhão, J., ... & Pike, A. W. (2018). U-Th dating of carbonate crusts reveals Neandertal origin of Iberian cave art. *Science, 359*(6378), 912–915.

Hund, W. D. (2015). Racist King Kong fantasies. *Simianization: Apes, Gender, Class, and Race, 6*, 43.

Iacona, J. (2016). Behind closed curtains: The exploitation of animals in the film industry. *Journal of Animal & Natural Resource Law, 12*, 25.

Jackson, Z. I. (2021). Animality and blackness. In McHugh, S., & Aloi, G. (eds.), *Posthumanism in art and science: A Reader*, (pp. 55–58). Columbia University Press.

Jhally, S. (2003). Image-Based Culture: Advertising and popular culture. In Dines, G., (ed.), *Gender, race, and class in media: A Text-reader* (pp. 77–87). Sage.

Kiprotich, A., & Chang'orok, D. (2017). Gender communication stereotypes: A depiction of the mass media. *IOSR Journal of Humanities and Social Science, 20*(11), 69–77.

Kordos, P. (2014). Talking animalish in science-fiction creations. Some thoughts on literary zoomorphism. *Dialogue and Universalism*, (1), 219–226.

Kvangraven, E. H. (2019). *The wolf as an icon of wildness: Romanticization and demonization* (Master's thesis).

Lauzen, M. M. (2012). Where are the film directors (who happen to be women)?. *Quarterly Review of Film and Video, 29*(4), 310–319.

Lawrence, E. A. (2003). Feline fortunes: Contrasting views of cats in popular culture. *Journal of Popular Culture, 36*(3), 623.

Lévi-Strauss, C. (1963). *Totemism* (R. Needham, Trans.). Boston: Beacon.

Li, H., Eisen, S., & Lillard, A. S. (2019). Anthropomorphic media exposure and preschoolers' anthropomorphic thinking in China. *Journal of Children and Media, 13*(2), 149–162.

Lutts, R. H. (1992). The trouble with Bambi: Walt Disney's Bambi and the American vision of nature. *Forest and Conservation History, 36*(4), 160–171.

McCulloch, F. (2017). *Medieval Latin and French Bestiaries*. The University of North Carolina Press.

Melson, G. F. (2005). *Why the wild things are: Animals in the lives of children*. Harvard University Press.

Melson, G. F., & Fine, A. H. (2015). Animals in the lives of children. In *Handbook on animal-assisted therapy* (pp. 179–194). Academic Press.

Merskin, D. (1998). Sending up signals: A survey of Native American1 media use and representation in the mass media. *Howard Journal of Communication, 9*(4), 333–345.

Mitman, G. (2012). *Reel nature: America's romance with wildlife on film*. University of Washington Press.

Molloy, C. (2011). *Popular media and animals*. Springer.

Packwood Freeman, C. (2009). This little piggy went to press: The American news media's construction of animals in agriculture. *The Communication Review, 12*(1), 78–103.

Plemenitaš, K. (2017). Metaphorical elements in gendered slurs. *BAS British and American Studies*, (23), 207–217.

Pluskowski, A. (2005). The tyranny of the gingerbread house: Contextualising the fear of wolves in Medieval Northern Europe through material culture, ecology and folklore. *Current Swedish Archaeology, 13*(1), 141–160.

Pons-Branchu, E., Sanchidrián, J. L., Fontugne, M., Medina-Alcaide, M. Á., Quiles, A., Thil, F., & Valladas, H. (2020). U-series dating at Nerja cave reveal open system. Questioning the Neanderthal origin of Spanish rock art. *Journal of Archaeological Science, 117*, 105120.

Quallen, M. (2016). *Making animals, making slaves: Animalization and slavery in the Antebellum United States* (Doctoral dissertation).

Reny, T., & Manzano, S. (2016). The negative effects of mass media stereotypes of Latinos and immigrants. *Media and Minorities, 4*, 195–212.

Robertson, T. (1999). *The wonderful world of Walt Disney: Nature and nation in Bambi and the" true life" adventures* (Doctoral dissertation).

Roche Cárcel, J. A. (2021). King Kong, the Black Gorilla. *Quarterly Review of Film and Video*, 1–45.

Rowland, B. (1973). *Animals with human faces*. University of Tennessee Press.

Schug, J., Alt, N. P., Lu, P. S., Gosin, M., & Fay, J. L. (2017). Gendered race in mass media: Invisibility of Asian men and Black women in popular magazines. *Psychology of Popular Media Culture*, 6(3), 222.

Silva, C. R. P. D. (2016). *Am I too cute to eat? The effect of cuteness appeal on the promotion of a more plant-based diet* (Doctoral dissertation).

Smith, J. M. (2011). *Monsters of the Gévaudan: The making of a beast*. Harvard University Press.

Strinati, D. (2014). *An introduction to studying popular culture*. Routledge.

Wellstone, P. (1999). Growing media consolidation must be examined to preserve our democracy. *Federal Communications Law Journal*, 52, 551.

Wigger, I. (2019). Anti-Muslim racism and the racialisation of sexual violence: 'intersectional stereotyping' in mass media representations of male Muslim migrants in Germany. *Culture and Religion*, 20(3), 248–271.

Wolfensohn, S. (2020). Too cute to kill? The need for objective measurements of quality of life. *Animals*, 10(6), 1054.

Zickfeld, J. H., Kunst, J. R., & Hohle, S. M. (2018). Too sweet to eat: Exploring the effects of cuteness on meat consumption. *Appetite*, 120, 181–195.

# 47
# I'LL GET YOU... AND YOUR LITTLE DOG TOO!
## Animal Cruelty and Kindness in the Cinema over the Last Century

*Randall Lockwood*

The righteous care for the needs of their animals but the kindest acts of the wicked are cruel

*Proverbs 12:10*

## Introduction

Centuries of concern about the moral, ethical, and psychological significance of cruelty to animals has spanned many disciplines, including religion (Regenstein, 1991; Linzey, 1995; 2016), philosophy (Singer, 1975; Regan, 2004; Rollin, 1981), and law (Favre, 2014). More specifically, much of the focus has been on what we can learn about the character of an individual or a culture by looking at how it treats or mistreats animals. Lockwood & Ascione (1998) provide a compendium of essays covering centuries of writings examining this topic from the perspective of many different fields. Related to this is the converse, an interest in the development of kindness to others as a sign of moral strength (Schweitzer, 1965) and the special connection between human and non-human life (Wilson, 1984).

How can we best gauge a culture's perception of the significance of kindness and cruelty to animals? One way is to examine its *artistic depiction* of human-animal interactions. The earliest such representations come in the form of cave paintings and generally depict hunting and killing prey (Kalof, 2007). Millennia later, one of the most direct artistic portrayals of the connection between animal treatment and an individual's character was provided by William Hogarth in his 1751 series of woodcuts entitled *The Four Stages of Cruelty* (Shesgreen, 1973; Lockwood, 1999). The first of the woodcut series, *The First Stage of Cruelty*, depicts an urban street scene where several young boys are inflicting a variety of tortures on animals. A fighting dog is being turned loose on a cat, a cockfight is underway, and several boys are tormenting cats. At the center of the scene, a young Tom Nero (the protagonist of Hogarth's woodcuts) plunges an arrow into the rectum *of* a dog while another boy tries to stop him by offering him a cake. In the second scene, *The Second Stage of Cruelty*, Nero is an adult coach driver. His greed led him to overload his coach, causing his horse to break a leg. In his efforts to get the horse moving, Nero has beaten and blinded the animal. The third scene, *Cruelty in*

*Perfection,* takes us further into Nero's future. He is a highwayman, depicted in a churchyard beside the pregnant, hacked corpse of his mistress Ann Gill, who had stolen her employer's silver at Nero's command. His deeds discovered, he is captured by a group of farmers. The final woodcut entitled *The Reward of Cruelty* shows Nero taken to the gallows and hanged. His corpse is carried to a surgical suite, where it is subjected to numerous forms of mutilation. In the foreground, an old dog, perhaps the same one he tortured as a youth, is eating Nero's discarded heart!

Animal maltreatment continues to be addressed through visual art in modern times, such as through Sue Coe's sketches of her experiences visiting slaughterhouses in her book *Dead Meat* (1995) and Banksy's *Sirens of the Lambs* project (2013) designed to highlight the issue of animals being farmed for their meat by mimicking the transport of animals to slaughter using a truckload of stuffed animals.[1]

A more direct way of assessing a society's view of cruelty to animals is by looking at its laws. Our society generates our laws and our laws define what we stand for as a society. For most Americans, kindness to animals, or at least an aversion to the intentional infliction of animal suffering, is one of the core values that define the laws of a civilized society (Favre & Tsang, 1993).

In the 1980s, law-enforcement interest in animal cruelty grew dramatically, in part fueled by the proliferation of research into the association between animal cruelty and interpersonal violence (Arkow & Lockwood, 2016). This was further driven by the dramatic growth in size and political impact of animal protection and animal rights groups in the United States. This was mirrored in the dramatic strengthening of animal cruelty laws and the inclusion of felony penalties for some forms of animal cruelty in every state's law by 2016 (Ascione & Lockwood, 2001). A major indicator of growing concern about such crimes is the fact that the Federal Bureau of Investigation (FBI) added animal cruelty to the crimes tracked within the National Incident Based Reporting System (NIBRS) beginning in 2016 (DeSousa, 2016).

One of the most powerful tools for gaining insight into a modern society's views of cruelty and kindness to animals is to examine its portrayal of these ideas in movies. Like its laws, a society's films both *reflect* and often *create* attitudes that can permeate culture. Unlike law and static art, films reflect a *dynamic* connection to the viewer. Their popularity depends on striking what Schwartz (1974) termed a "responsive chord" with the beliefs and emotions of the viewer. Movies also provide a rich resource for examining attitudes across a wide range of demographic backgrounds. Prior to the Covid pandemic, 76% of the U.S./Canada population age two and over went to the movies at least once in 2017 and those in the 18–29 age group averaged seeing nine movies a year in 2007 (MPA, 2021). Increasingly, movies depend on widespread international distribution, adding to the likelihood that they will both reflect and shape worldwide ideas.

We would expect changes in the public's attitude toward animal cruelty would be reflected in how human-animal relationships are portrayed in movies. As Kean (1998) notes:

> The changes that would take place in the treatment of animals relied not merely on philosophical, religious or political stances but in the way in which animals were literally and metaphorically seen. The very act of seeing became crucial in the formation of the modern person.
>
> *(pp. 26–27)*

We will look at some examples of how human-animal interactions are depicted in movies, to identify common **tropes**. In cinema, a trope is what Rizzo (2014) defines as "a universally identified image imbued with several layers of contextual meaning creating a new visual metaphor" (p. 513). In some cases a trope may have become overused to the point that it has become a **cliché**, and may ultimately be used ironically or humorously. Looking at the recurrent human-animal-oriented themes and tropes in movies not only helps reveal deep-seated beliefs about this relationship common in a culture, but it can also provide a rich resource for generating testable hypotheses regarding the objective reality of these beliefs and can help stimulate new research into the human-animal bond.

## The Depiction of Animal Cruelty

Early films served primarily to show off the ability of this new technology to portray scenes that might be new to the viewer. The earliest film depicting animal cruelty is likely an 1895 record of a Spanish bullfight, which, according to an 1897 review "makes us sorry for the animals, and for those who can enjoy such a spectacle" (Burt, 2002, p. 119). The Lumiere brothers introduced a similar short film – *A Spanish Bullfight* in 1900. In that same year William N. Selig made *The Chicago Stockyards—From Hoof to Market* showing the full meatpacking process from cattle being unloaded at the stockyards to canning (Erish, 2012). The most famous early depiction of animal cruelty is Thomas Edison's *Electrocuting an Elephant* (1903). The 74 second film documents Edison's use of alternating current to "execute" Topsy, a 28-year-old elephant kept at Coney Island who had killed three men, including a severely abusive trainer who had attempted to feed her a lit cigarette. The ASPCA opposed plans to kill Topsy by poisoning or hanging as inhumane. Edison offered to execute her using alternating current, which was done before an audience of 1,500 people. The movie was then packaged in Edison kinetoscopes and shown to audiences across the country. Edison clearly had a commercial purpose for placing this disturbing film into wide circulation. He and his colleagues routinely demonstrated the electrocution of dogs, cats, and other animals to illustrate the hazards of alternating current that was being promoted by his archrival, George Westinghouse. This was a major tactic in the competition between Westinghouse's technology of using alternating current as a means of producing and distributing electric power and Edison's direct current generating plants that eventually lost out. Topsy's electrocution and the widespread showing of the film aimed to show that this new form of power could even kill an elephant (Essig, 2003; Daly, 2013).

Many subsequent films documenting commercial exploitation and abuse have had a significant impact on shaping public opinion and legislation. One of the earliest was *The Blood of Beasts (Le San des bêtes)* (1949), a powerful black and white film that contrasts peaceful scenes of Parisian suburbia with scenes from a slaughterhouse where horses, cattle, and sheep are processed. Seeing that film as a college freshman helped make me a lifelong vegetarian! *The Animals Film* (1981) depicted an international variety of forms of abuse in laboratories, factory farms, and the military and helped fuel the emerging animal rights movement of the 1980s. *The Cove* (2009) profiled dolphin hunting practices in Japan and received an Academy Award for the best documentary. One of the most influential animal documentaries in history is *Blackfish* (2013), focusing on the treatment of orcas in captivity. Public and political response was enormous, including major financial impact on SeaWorld and updates to the Animal Welfare Act to address concerns about captive cetaceans. Ultimately, SeaWorld announced plans to end its captive orca breeding program and phase out live orca performances.

## Animal Cruelty and Character

Although non-fiction films documenting animal cruelty can help shape public opinion about specific forms of abuse or exploitation – it is the more prevalent portrayal of human-animal relationships in fictional films that can provide insight into major themes of these relationships. Tropes related to animal cruelty as an indicator of the nature of an individual's character did not begin to emerge until well into the 20th century. The most basic theme is superficially quite simple – how an individual treats animals is a mirror of how other people are likely to be treated. Thus a cruel act helps establishes someone as a cruel person. Conversely, when character is shown being kind for no apparent gain, it helps create a character the audience is meant to care about and cheer for. Both devices are used to help the audience become emotionally invested in the story. Often the tropes will coexist as a kind animal advocate confronts an animal abuser/exploiter.

As noted earlier, the concept of a connection between the maltreatment of animals and interpersonal violence or criminal behavior has a centuries-long history (Lockwood & Ascione, 1998) and has been reinforced by hundreds of academic studies.[2] Different studies of this association tend to focus on two different approaches. Some suggest a *graduation* or *progression* approach in which early childhood or adolescent cruelty to animals is a precursor to escalation into crimes against property and people, up to and including homicides or serial or mass killing (Walters, 2013; Wright &Hensley, 2003). Others see the connection as part of an overall pattern of *general deviance* (DeGue & DiLillo, 2009; McGee & Newcomb, 1992). Evidence for the progression hypothesis is weaker, perhaps in part because reporting of and response to animal cruelty committed by younger offenders has, until recently, been rare so it has been more difficult to document early acts of cruelty.

## Progressively Evil

Several influential films have depicted the development of violence and the progression of harm. In *The Bad Seed* (1956, remade in 1985), young Rhoda wants a dog but quickly tires of it and it "falls" from a window. A rival classmate wins a medal and soon drowns on a school picnic. More deaths follow, including a maintenance man who is burned to death when Rhoda learns that he has discovered her crimes. Her adoptive mother learns that Rhoda's biological mother was an executed serial killer and attempts to kill her with sleeping pills before shooting herself. The 1956 version seeks closure by having Rhoda killed by lightning, but the 1985 version keeps the novel's ending, having Rhoda survive the overdose and be free to continue her crimes. R.D. Hare, a leading authority on psychopathic personality, praised the story as an accurate portrayal of the development of psychopathy in childhood (Hare, 1999).

In *Gummo* (1997), several disaffected boys in a decaying Ohio town spend their time hunting feral cats, drowning or shooting them. One boy, Tummler, called "downright evil" by his friends, shoots his sister's cat and eventually disconnects his grandmother from life support.

In *The Good Son* (1993), young Mark is sent to stay with his cousins Henry and Connie after the death of his mother. Mark finds that Henry has a fascination with death and witnesses him shooting a dog with a crossbow. Later, Henry takes Connie ice skating and leads her to thin ice, where she falls in and eventually winds up in a coma. Henry tries to suffocate her in her hospital bed. She recovers and grows closer to Mark. Ultimately, the boys fight at the edge of a cliff, with both going over while Connie holds on to them. Unable to hold both, she chooses to let Henry fall to his death.

## Generally Deviant

In film, the depiction of animal cruelty as part of a broader pattern of meanness and general deviance has been the most popular approach for decades. The most widely recognized early use of this trope – and the source of our title – appears in *The Wizard of Oz* (1939) in the character of Almira Gulch/the Wicked Witch of the West. In the "real" world of Kansas – the wealthy Ms. Gulch is a dog-hater who gets a sheriff's order authorizing her to seize Dorothy's dog Toto to be euthanized for allegedly biting her. Toto escapes and Dorothy runs away with him. In Oz, Dorothy's first encounter with The Wicked Witch of the West comes after the witch has failed in her attempt to get the ruby slippers off the feet of her deceased sister, killed by Dorothy's falling house. She leaves in a puff of smoke uttering the classic curse to Dorothy – "I'll get you, my pretty, and your little dog too!" Of course, she ultimately dies when, after setting fire to Dorothy's friend the Scarecrow, she is melted by the water Dorothy throws to put out the fire. Although released in 1939, The *Wizard of OZ* was first shown on television in 1956 and became an annual tradition, usually reaching more than 50% of viewers. As such, it was the first time that millions of people, young and old, saw the tale of an evil character, cruel to animals, children, and just about everyone and everything else.

Perhaps the second most memorable cinema depiction of the animal cruelty-evil character connection comes in *The Godfather* (1972). The godson of Vito Corleone, singer Johnny Fontane seeks Vito's help in securing a movie role. Vito dispatches his henchman, Tom Hagen, to persuade studio head Jack Woltz to give Johnny the part. Woltz complies only after he finds the severed head of his prized stallion in his bed as a warning. The scene is particularly shocking in that involved a real severed horse head obtained from a dog-food company.[3]

Animal cruelty and the Mafia converge again in *The Freshman* (1990) in which Marlon Brando portrays mobster Carmine Sabitini (who is said to look like Vito Corleone) who runs a restaurant where patrons are served endangered species, paying extra if they get to eat an animal that is the last of its kind, although it is eventually revealed that he is conning his patrons.

Another film classic that subtly builds on the cruelty trope is *Psycho* (1960), loosely based on the real-world serial killer Ed Gein. In the parlor scene in which Norman Bates first speaks with his eventual victim, Marion *Crane*, he is shown gently fondling a taxidermied *bird* on a shelf. The view then shifts to show him framed by two stuffed owls as he tells her "my hobby is stuffing things", which is later revealed to be the fate of his dead mother. Mondal (2017) describes Hitchcock's work as "a discourse on the preservation/destruction duality that is central to taxidermy" (p. 1).

Other classic cinema characters have continued to reinforce the animal cruelty-evil character connection. In the animated and live action versions of the 101 Dalmatians stories (*101 Dalmatians* 1961, 1996; *102 Dalmatians*, 2000; *Cruella*, 2021) the lead villain, Cruella DeVil is the epitome of animal cruelty, not only in her desire to turn puppies in coats, but also to turn a rare white tiger into a rug. As part of her overall deviance, she also smokes constantly, drives recklessly, and is abusive to her goons and other staff. Cruella's 2021 origin story links the origins of her cruel obsession to witnessing her mother being pushed off a cliff by a group of Dalmatians who are pursuing her after a jewel heist.

In *American Psycho* (2000), we are first introduced to investment banker Patrick Bateman at a dinner with other wealthy friends where he is upstaged by the superior quality of a rival's business card. Soon after, he vents his rage by stabbing a homeless man and stomping his dog to death. He eventually kills many more people, including a shooting a woman who interrupted his attempt to shoot a cat. The finale, however, suggests that Bateman's many crimes might all be in his mind.

A recent invocation of this theme is presented with some complexity in *The Power of the Dog* (2021). Phil, a coarse and volatile rancher constantly belittles his brother as well as his brother's wife and son. He is shown violently punching a horse and chooses to burn surplus cow hides rather than donate them to local Native Americans. Ironically, the abused son, a gentle medical student, secretly infects Phil with anthrax from a diseased cowhide, resulting in his death.

Several films have dealt with the well-documented general deviance seen in people associated with organized dogfighting. Dogfighting is strongly linked to other crimes beyond animal cruelty, including drug and weapons violation and assaults (Lockwood, 2011). The abusive training methods as well as the horrors of fights themselves are realistically depicted in *Eyes of an Angel* (1991), *White God* (2014), and *Amores perros* (2000). This last film features three scenarios linking the cruelty of humans toward both animals and other humans. It received considerable attention due to the convincing depiction of dogfights, which in fact involved careful editing of playing dogs overdubbed with fighting sound effects.[4] Response to the movie helped lead to dogfighting becoming illegal in Mexico in 2017.

## Animal Cruelty and Interpersonal Violence

In addition to general deviance and criminality, research has linked animal cruelty to child abuse (DeViney et al., 1983), domestic violence (Ascione, 1998; Faver & Strand, 2007), and elder abuse (Cooke-Daniels, 1999; Lockwood, 2002). All of these connections have been reflected in the portrayal of animal cruelty in movies.

## Child Abuse

The villains in *Benji* (1974) are child kidnappers whose crime is witnessed by the children's adopted stray, Benji. To show how really nasty the kidnappers are, one of them viciously kicks a Tiffany, a small white poodle who is Benji's love interest. Ultimately, the heroic dog leads the FBI to the villains.

*Radio Flyer* (1992) presents a complex tale of escape from child abuse told by Mike, an adult survivor. In the retelling, 11-year-old Mike and 8-year-old Bobby move with their German Shepherd Shane when their mother remarries. Their stepfather "The King" is an abusive alcoholic who beats Bobby, whom Shane tries to protect. Shane is also nearly killed by the King on several occasions. In the tale Bobby flies away off a cliff in a plane made from a wagon – sending postcards from all over the world. It is left unclear whether Bobby is simply a personality created by Mike who disappears when the abuse ends, or if he has actually committed suicide or died in a real escape attempt.

In *All the Little Animals* (1998), a mentally challenged boy named Bobby runs away from an abusive stepfather who has killed his pet mouse and other pets and "screamed at his mother until she died" and even told him to dig his own grave. He joins up with Mr. Summers, a kindly man who gives burials to road kills and eventually teams up to confront the stepfather, who is portrayed as pure evil..

## Domestic Violence

*The Burning Bed* (1984) is a powerful depiction of the use of animal cruelty for power and control in the context of domestic violence. It is based on the true story of Francine Hughes, who suffered 13 years of physical abuse from her husband Mickey. As depicted in the film, the final straw for her came when Mickey refused to allow her to take her pregnant dog to the vet when she was in difficult labor, instead putting her outside in the cold Michigan night, where she died. Soon after, Francine put her children in her car and poured gasoline on her sleeping husband, setting him on fire. She then drove to the police station to confess. A sympathetic jury found her not guilty by reason of temporary insanity in one of the first cases involving "battered-woman syndrome" as a defense.

Although female perpetration of interpersonal violence coupled with animal cruelty is relatively rare (Renzetti, 1988), it is seen in some prominent films, often as a prelude to a murderous jealous rage. One of the most memorable films incorporating this theme is *Fatal Attraction* (1987) in which Dan Gallagher has a brief affair with Alex Forest who quickly begins to cling to him. She cuts her wrists to force him to stay longer to look after her. She continues to call him, but is ignored. Dan moves his family to another town, but Alex continues to stalk him, eventually breaking into his home, killing his daughter's pet rabbit and putting it on the stove to boil. She later picks up the daughter from school and takes her to an amusement park, resulting in Dan's wife Beth being injured in a car accident when searching for her. Eventually, Alex tries to attack Beth at home while she is taking a bath. Dan rescues her and Beth fatally shoots Alex.

In *Single White Female* (1992), Heddy is obsessed with her roommate Allie. She erases voicemails from Allie's ex-fiancé Sam seeking to reconcile. Heddy buys Allie a puppy but is jealous when the puppy ignores her and when Allie announces plans to move in with Sam. Heddy pushes the dog out a window, claiming it was an accident. She becomes even more controlling, eventually taking on Allie's appearance. Heddy eventually kills Sam and others about to expose her, before she is finally killed by Allie.

## Elder Abuse

Although there have been many documentaries addressing elder abuse and neglect, the theme of animal cruelty connected to elder abuse has rarely been portrayed in film fiction. As noted above, the cat killer of *Gummo* (1997) eventually kills his ailing mother. We see this theme arise in *Whatever Happened to Baby Jane* (1962). Blanche, a fading Hollywood actress now paralyzed from a car accident and her sister Jane, a psychotic has-been child star, resentful of her sister's fame, are now living together in an old house. Jane, fearing Blanche will have her committed, keeps Jane cut off from the outside world, often tied to her bed. She taunts her, at one point bringing her a covered dinner tray that reveals a dead rat. She kills the housekeeper who has discovered her deeds.

## Animal Cruelty on the Job

Many of the cinematic instances of people portrayed as bad characters who harm animals come in the context of people doing jobs that exploit animals. These situations often result in conflicts with "good" people seeking to protect or rescue creatures from harm.

### *Zoos and Circuses*

In *Soul of the Beast* (1923), Ruth, a young girl, runs away from an abusive stepfather, who owns a circus, and takes the circus' trained elephant, her only friend, with her. In the animated version of *Dumbo* (1941), the Ringmaster, though not truly evil, is a greedy, and arrogant man who exploits workers and animals. His character is rehabilitated in the live-action version (2019), but Vandevere, the owner of Dreamland who acquires Dumbo, seeks to have Dumbo's mother euthanized. *Water for Elephants* (2007) pits the vet student Vincent against a ringmaster who abuses horses and the star elephant Rosie. The duplicitous Park owners in *Free Willy* (1993) plan to break the orca's water tank to kill Willy for the insurance until he is released into the wild by Jesse, a troubled 12-year-old abandoned by his estranged mother.

## "Dog Catchers"

In the early 20th century, the local dog catcher served as an easy villain, portrayed most ominously in one of the first Our Gang shorts – *Pete the Pooch* (1935) in which the mean dog catcher is determined to kill Stymie's dog in the gas chamber, but is foiled when the chamber is out of gas and a kindly lady pays the five-dollar license fee to reclaim him, whereupon he is bitten by Stymie and the gang, who also release the dogs from his truck. The dog catcher also serves as an easy villain in *Lassie Come Home* (1943). As the reputation of dog catchers has evolved into professional Animal Control Officers, their portrayal has become more balanced and positive, as in *Hotel for Dogs* (2009).

## Mad Scientists and Evil Super Villains

From the 1930s through the 1970s, mad scientists were frequently portrayed as practicing their evil procedures on both animals and people. The classic characterization of this trope came from the diabolical Dr. Moreau, seeking to transform animals into humanoids, and vice-versa in his "house of pain", first in *The Island of Lost Souls* (1932) and later in *The Island of Dr. Moreau* (1966, 1977). Likewise, *Dr. Cyclops* (1940) sought to save the world by shrinking animals and people with his "radium ray", which ultimately spelled his doom. The evil scientist trope continued into modern times in *The Day of the Dolphin* (1973) where representatives of an evil research foundation subverted beneficent studies of language learning in dolphins to training the animals to carry out political assassination.

Evil scientists are also responsible for the creation of killer dogs in a variety of films either through training (*The Breed*, 2006), genetic engineering (*Man's Best Friend*, 1993), or "military pheromones" (*Dogs*, 2006).

The James Bond universe provides many examples of evil geniuses bent on world domination or destruction who exploit animals – particularly in their disposal of their enemies. Most Bond villains clearly meet the psychiatric diagnosis of psychopaths (Kavanagh & Cavanna, 2020). The films also employ the trope of showing a villain, e.g. Ernst Blofeld, fondly stroke a white cat while plotting evil deeds. This imagery appears repeatedly in *From Russia with Love* (1963), *You Only Live Twice* (1967), *On Her Majesty's Secret Service* (1969), and *Diamonds are Forever* (1971), presumably adding a bit of sympathetic complexity to the evil character. This trope likely had its origins in the portrayal of Cardinal Richileu, historically a known cat lover, who serves as the villain in a popular early version of *The Three Musketeers* (1948) where he is portrayed by Vincent Price stroking a white cat! Blofeld also keeps a tank of piranhas. Emilio Largo keeps a shark tank (*Thunderball*, 1965), as do Karl Stromberg (*The Spy Who Loved Me*, 1977) and Hugo Drax (*Moonraker*, 1979), while Dr. Kananaga keeps a crocodile pit in *Live and Let Die* (1973). This cliché is played for laughs in *Austin Powers in Goldmember* (2002), a Bond parody in which the villain, Dr. Evil, not only holds a hairless cat in his lap, but also acquires a pool of "sharks with frickin laser beams attached to their heads".

## Hunters

The American Film Institute's list of greatest cinema villains includes "Man", as #44 representing the villain responsible for killing the mother of *Bambi* (1942). Early films generally treated big game hunters uncritically, with the earliest appearance being a silent short - *Buffalo Hunting in Indo-China* (1908). The tradition continued with *Bring 'Em Back Alive* (1932) and similar "Great White Hunter" movies. The ethics and morality of such characters began to be questioned in characterizing the exploitative adventurers of *King Kong* (1933), *Mighty Joe Young* (1949), and *Valley of Gwangi* (1969). The link to violence against people is fully realized in the arch hunter Zaroff in *The Most Dangerous Game* (1932) who has tired of hunting animals and sponsors human hunts. Such characterizations stand in contrast to positive cinematic portrayals of indigenous hunters such as the Inuit seal hunter of *Nanook of the North* (1922) or the Aboriginal boy providing for and guiding lost teens to safety in *Walkabout* (1971).

## Killing with Mercy

Sometimes the killing of an animal is depicted as a sign of strong character, particularly if it is seen as a necessary act of mercy that adds to the growth and complexity of the character's development arc. The archetype of this trope is young Travis's shooting of his beloved dog in *Old Yeller* (1957), after it develops rabies from an attack by a rabid wolf. Similarly, in *The Yearling* (1946), young Jody must put his pet deer Flag out of its misery after it has been shot by his mother for eating the family's crops. Atticus Finch, the hero of *To Kill a Mockingbird* (1962) is considered by the American Film Institute to be the #1 heroic character in American cinema.[5] However, early in the film, he kills a rabid dog with a single shot to protect his children.

## Animal Kindness and Character

Research has demonstrated that depicting people engaged with animals in a positive or even neutral way leads viewers to see them as friendlier, happier, bolder, more relaxed, and wealthier (Lockwood, 1985). It is no surprise that an influential guide to screen writing is titled *Save the Cat!* (Snyder, 2005) after a powerful plot technique for defining who a hero is and making the audience like him.

The perception that a character is kind to animals not only makes him/her likeable and humane – it can even make an *inhuman* character someone we root for. In the *Star Trek* universe of TV and films, the android Data is always striving to be closer to human. This is beautifully portrayed through his deep connection to his pet cat Spot (Neuwirth, 2018). Similarly, in *Hellboy* (2004), a demon from Hell that has been adopted by a scientist is shown to be almost human by his devotion to kittens, and becomes an ally of the FBI. In *Ghost in the Shell* (2017) Mira Killian is a cyborg fashioned to be a counter-terrorist, but she reveals her remaining human characteristics in stopping to feed stray dogs while on her missions. *The Iron Giant* (1999) asks the question "What if a gun had a soul?" The 50 foot alien robot war machine befriends a small boy and is exposed to the concept of death in witnessing a hunter shoot a deer. He then becomes a protector of the townspeople despite the efforts of the military to destroy him. In *Starman* (1984) an alien has occupied the cloned body of Jenny's deceased husband. He is regarded with fear and suspicion, but proves his kindness when he encounters a dead deer tied to a hunter's fender and uses one of his limited supply of cosmic globes to bring it back to life and set it free.

The ability to show empathy to the suffering of others, including animals, lies at the heart of distinguishing between humans and artificial replicants in *Blade Runner* (1982). The

Voigt-Kampff test used to identify replicants looks for autonomic reactions to situations such as finding a tortoise on its back or receiving a calf-skin wallet. It is particularly poignant, since real animals have almost completely disappeared in the film's future world (which is identified as 2019!). As Marder (1991) notes: "humans can only determine their difference from the species they have created (androids) by invoking their nostalgic empathy for the species that they have already destroyed (animals)" (p. 91).

Perhaps the earliest use of showing kindness to an animal as indication of good character is the story of Androcles and the Lion in which a runaway Roman slave encounters a lion with a painful thorn in its paw. He removes the thorn and later, when captured and thrown to lions in the Circus Maximus, he encounters the same lion and is spared. He is pardoned and released, keeping the lion as a companion. The earliest surviving account is found in Aulus Gellius's 2nd-century book *Attic Nights*.[6] George Bernard Shaw's play based on the story was the basis of the film *Androcles and the Lion* (1952).

The "save the cat" trope took an amphibian/alien turn in *E.T.: The Extraterrestrial* (1982) when young Elliot, his empathy increased through his psychic connection to E.T., creates havoc in his biology class by releasing the frogs facing dissection. The same theme occurs in *Disney's Freaky Friday* (1995) when adult Ellen, occupying her teenage daughter Annabelle's body, speaks out against dissection in class. The theme was revisited in the real-world case of Jenifer Graham, a California high-school student whose refusal to participate in dissection led to changes in state law allowing students to use non-animal alternatives (Lockwood, 1989). Her story was the basis of the slightly fictionalized movie *Frog Girl* (1989).

Sometimes defenders of animals are portrayed as fragile or on the fringe, like Mr. Summers in *All the Little Animals* mentioned earlier. In *The Misfits* (1961), young divorce' Roslyn moves in with aging cowboy Gaylord. She objects when he plans to kill rabbits eating his garden. She becomes upset when she finds out a flank strap is used to irritate rodeo horses into bucking. When she finds out that the wild-caught mustangs are meant for dog food, she loses it, but eventually succeeds in convincing Gaylord to free the wild stallions they have captured. In *Short Circuit* (1986) a kind but somewhat flakey animal rescuer finds a runaway military robot that has become sentient after being struck by lightning and helps him improve his language and eventually helps hide him away from his warmongering pursuers.

Other "good guys" who save animals are more mainstream. The heroes of *Project X* (1987), airman Jimmy Garret and chimp trainer Teri MacDonald, rescue a chimp named Virgil who is being used in military experiments to see how long subjects will be able to pilot a simulator after being exposed to lethal radiation. They break Virgil and other chimps out of the facility and attempt to steal a plane but are caught by military police. Their distraction allows Virgil to escape by piloting the plane. He crashes into the Everglades as the movie ends he is seen hiding in the bushes with his chimp girlfriend.

In *Okja* (2017), young Mija lives with her grandfather and Okja, one of a new breed of giant Superpig. Okja had saved Mija from falling off a cliff. As Okja is being reclaimed by the Mirando company that produced him, the truck is intercepted by the Animal Liberation Front, which wants to plant a recording device in Okja's ear to document the cruelty of the company. After many travails, Mija is able to rescue Okja from a slaughter plant and return to her life with her grandfather, Okja and another superpiglet.

## Redemption through Caring

In film and in real life, kindness is not always universal. Not all movie characters that treat animals well are completely good. A common theme is to add depth to a violent or otherwise unsavory character through their legitimate bond to an animal or animals. This theme is the key feature of *Birdman of Alcatraz* (1962), the true story of Robert Stroud, a convicted murderer held in solitary confinement in Alcatraz prison who was allowed to care for many birds, eventually becoming a renowned expert on avian diseases.

In fiction, other violent characters have been given a noble motive by acting in response to the animal cruelty of others. *John Wick* (2014) receives Daisy, a beagle puppy sent to him by his wife just before her death from a terminal illness. He runs afoul of Russian gangsters wanting to buy his vintage car, which he refuses to sell. They break into his home, kill Daisy and steal the car. He seeks revenge on New York's Russian Mafia, eventually killing many of them, including their leader. He is stabbed in his final confrontation and breaks into a veterinary clinic to treat his injuries, where he steals a pit bull puppy that is scheduled to be euthanized and walks home.

Similarly, in *The Drop* (2014), Bob Saginowski works at a bar owned and frequented by Chechen mobsters involved in many payoff schemes that he is a part of. He discovers a wounded pit bull in a trash bin and begins to bond with it and his neighbor Nadia, who helps him care for the dog he names Rocco. He later meets a local thug, Deeds, who proves he is the owner of the dog and admits to abusing him. Deeds threatens to take Rocco back and starve him unless Bob gives him $10,000. Bob later confesses to Nadia and Deeds that he had once killed a rich customer of the bar and used the money to pay off a debt to the Chechens, a crime that Deeds had claimed credit for. Bob then kills Deeds mainly because " he was going to hurt Rocco". Bob, Nadia and Rocco then walk home.

## The Last Survivors

Surviving the collapse of civilization can demand some nasty behavior if you are to stay alive. Having an animal companion makes the survivor more sympathetic and relatable, despite the violent things they may need to do. The theme of a lone survivor with an animal companion dates back at least as far as Mary Shelly's long-overlooked novel – *The Last Man* (1826), which describes the travels of an almost solitary survivor of a worldwide plague in 2092 who travels the landscape with a sheepdog as his main companion. We see such a partnership in *The Last Man on Earth* (1964) and its update *I am Legend* (2007) as well as *The Road Warrior* (1982) and *Finch* (2021). The hero of Finch is a surviving engineer when Earth's ozone layer has been destroyed. He is dying of radiation poisoning, so his main quest is to build a robot to care for his dog Goodyear after his death.

The most memorable example of this trope comes in *A Boy and His Dog* (1975) in which young Vic travels the nuclear wasteland of 2024 Topeka with his genetically engineered telepathic dog Blood who helps him locate food and women to meet his needs. Vic is captured and sent to an underground city to be used as breeding stock. He escapes to the surface with Quilla June, a girl who has fallen in love with him. The pair relocate Blood, who is now starving after having been left topside. Vic feeds her to the starving dog, because "a boy loves his dog".

## Conclusion

Themes of kindness and cruelty to animals as a reflection of human character are a powerful tool in storytelling that have shown little change over the last century. Many of the assumptions of these tropes have been empirically validated. However, some contemporary animal issues have so far received little attention in mainstream cinema. For example, there has been much recent attention given to the frequent occurrence of animal cruelty in the history of mass shooters, including school shooters (Arluke et al., 2018). None of the films representing such instances have made the connection to this powerful warning sign, including *Targets* (1968), *Heart of America* (2002), *Elephant* (2003), *Zero Day* (2003), or *Run, Hide, Fight* (2020).

One of the most common forms of animal cruelty, affecting thousands of animals annually, is animal hoarding (Lockwood, 2018). Although the issue has seen significant coverage on television, few mainstream movies have dealt with the problem other than a single documentary – *Cat Ladies* (2009) and passing reference to the cats and raccoons occupying the Beale house in *Grey Gardens* (1975). Cat ladies have occasionally appeared as peripheral characters in fictional stories such as *Big Fish* (2003) and even *A Clockwork Orange* (1971). Given the complex psychodynamics of such cases, it could provide rich material for cinematic treatment.

Another form of animal cruelty of growing concern is bestiality or animal sexual assault, which has also been linked to potential involvement in human sexual assaults (Beetz, 2005; Beirne et al., 2017). Currently, 48 states have amended their animal cruelty laws to specifically address bestiality. However, the issue has received little attention from mainstream cinema other than the documentary *Zoo* (2007), which describes a Pacific Northwest engineer's fatal sexual encounter with a horse. Although a Sundance and Cannes favorite, the film was also criticized for breaching the "last taboo".

As in the past, movies will continue to help explore some of the main questions surrounding cruelty to animals and its connection to interpersonal and societal violence. They will help frame big questions – "how does such cruelty originate?", "how can we identify those who are most likely to create problems for animals and people?", and "what can we do to better prevent and respond to such cruelty". Paying closer attention to how those themes have been addressed and are addressed in future films may provide fertile ground for the growth of new ideas and more insightful filmmaking.

## Notes

1 https://allthatsinteresting.com/banksys-sirens-of-the-lambs.
2 For updates on "link" research and a detailed up-to-date bibliography visit www.nationallinkcoalition.org.
3 *The Godfather*, DVD commentary featuring Francis Ford Coppola, 2001.
4 *Amores perros*, DVD production commentary track, 2001.
5 https://www.afi.com/afis-100-years-100-heroes-villians/.
6 *The Attic Nights of Aulus Gellius*, with An English Translation by John C. Rolfe. London 1927, Book I, section XIV, p. 255ff.

## References

Arkow, P. & Lockwood, R. (2016). Defining animal cruelty. In C.L. Reyes & M. Brewster (Eds.). *Animal Cruelty: A Multidisciplinary Approach to Understanding* (2nd ed.). Durham: Carolina Academic Press, 3–23.

Arluke, A., Lankford, A. & Madfis, E. (2018). Harming animals and massacring humans: Characteristics of public mass and active shooters who abused animals. *Behavioral Sciences & the Law*, 36(6), 739–751.

Ascione, F.R. (1998). Battered women's reports of their partners' and their children's cruelty to animals. *Journal of Emotional Abuse*, 1, 119–133.

Ascione, F.R. & Lockwood, R. (2001). Cruelty to animals: Changing psychological, social, and legislative perspectives. In D. J. Salem & A.N. Rowan (Eds.). *The State of the Animals 2001*. Washington, DC: Humane Society of the U.S., 39–54.

Beetz, A. M. (2005). Bestiality and zoophilia: Associations with violence and sex offending. In A.M. Beetz & A.L. Podberscek (Eds.). *Bestiality and Zoophilia: Sexual Relations with Animals*. West Lafayette, IN: Purdue University Press, 46–70.

Beirne, P., Maher, J., & Pierpoint, H. (2017). Animal sexual assault. In J. Maher, H. Pierpoint, & P. Beirne (Eds.). *The Palgrave International Handbook of Animal Abuse Studies*. London: Palgrave Macmillan, 59–85.

Burt, J. (2002). *Animals in Film*. London: Reaktion Books.

Coe, S. (1995). *Dead Meat*. New York: Four Walls Eight Windows.

Cooke-Daniels, L. (1999). The connection between animals and elder abuse. *Victimization of the Elderly and Disabled*, 2(3), 37–47.

Daly, M. (2013). *Topsy*. New York: Atlantic Monthly Press.

DeGue, S., & DiLillo, D. (2009). Is animal cruelty a "red flag" for family violence? Investigating co-occurring violence toward children, partners, and pets. *Journal of Interpersonal Violence*, 24(6), 1036–1056.

DeSousa, D. (2016). *NIBRS User Manual for Animal Control Officers and Humane Law Enforcement*. http://www.nacanet.org/page/NIBRS_Manual.

DeViney, L., Dickert, J. & Lockwood, R. (1983). The care of pets within child abusing families. *International Journal for the Study of Animal Problems*, 4(4), 321–336.

Erish, A. A. (2012). *Col. William N. Selig: The Man Who Invented Hollywood*. Austin, TX: University of Texas Press.

Essig, M. (2003). *Edison and the Electric Chair*. New York: Walker & Company.

Faver, C. A., & Strand, E.B. (2007). Fear, guilt, and grief: Harm to pets and the emotional abuse of women. *Journal of Emotional Abuse*, 7(1), 51–70.

Favre, D.S. (2014). *Animal Law: Welfare, Interests, and Rights*. New York: Aspen Publishers.

Favre, D., & Tsang, V. (1993). The development of anti-cruelty laws during the 1800's. *Detroit College of Law Review*, 1, 1–35.

Hare, R.D. (1999). *Without Conscience: The Disturbing World of the Psychopaths Among Us*. New York: Guilford Press.

Kean, H. (1998). *Animal Rights: Political and Social Change in Britain since 1800*. London: Reaktion Books.

Kalof, L. (2007). *Looking at Animals in Human History*. London: Reaktion Books.

Kavanagh, C., & Cavanna, A. E. (2020). James Bond villains and psychopathy: A literary analysis. *Journal of Psychopathology*. https://doi. org/10.36148/2284-0249-351.

Linzey, A. (1995). *Animal Theology*. Champaign, IL: University of Illinois press.

Linzey, A. (2016). *Christianity and the Rights of Animals*. New York: Crossroad Publishing.

Lockwood, R. (1985). The role of animals in our perception of people. *Veterinary Clinics of North America: Small Animal Practice*, 15(2), 377–385.

Lockwood, R. (1989). Following your conscience in the classroom: An interview with Jenifer Graham. *Humane Society News*, 34(4), 27–29.

Lockwood, R. (1999). Animal cruelty and violence against humans. *Animal Law*, 5, 81–87.

Lockwood, R. (2002). Making the connection between animal cruelty and abuse and neglect of vulnerable adults. *The Latham Letter*, 23(1), 10–11.

Lockwood, R. (2011). *Dogfighting toolkit for law enforcement: Addressing dogfighting in your community*. Washington, DC: Community Oriented Policing Services, U.S. Department of Justice.

Lockwood, R. (2018). Animal hoarding: The challenge for mental health, law enforcement, and animal welfare professionals. *Behavioral Science and the Law*, 36(6), 1–19.

Lockwood, R. & Ascione, F. (Eds.). (1998). *Animal Cruelty and Interpersonal Violence: Readings in Research and Application*. West Lafayette, IN: Purdue University Press.

Marder, E. (1991). Blade Runner's moving still. *Camera Obscura*, 9(3), 88–107.

McGee, L. & Newcomb, M.D. (1992). General deviance syndrome: Expanded hierarchical evaluations at four ages from early adolescence to adulthood. *Journal of Consulting and Clinical Psychology*, 60(5), 766.

Mondal, S. (2017). "Did he smile his work to see?" - Gothicism, Alfred Hitchcock's *Psycho* and the art of taxidermy. *Palgrave Communications*, 3, 17044, 1–9.

Motion Picture Association. (2021). *Theme Report 2021*. motionpictures.org, accessed July 28, 2022.

Neuwirth, M. (2018). "Absolute Alterity"? The alien animal, the human alien, and the limits of posthumanism in star trek. *European Journal of American Studies*, 13(1), 1–20.

Regan, T. (2004). *The Case for Animal Rights*. Berkeley: University of California Press.

Regenstein, L.G. (1991). *Replenish the Earth*. New York: Crossroad.

Renzetti, C. M. (1988). Violence in lesbian relationships: A preliminary analysis of causal factors. *Journal of Interpersonal Violence*, 3(4), 381–399.

Rizzo, M. (2014). *The Art Direction Handbook for Film* (2nd ed.). Waltham, MA: Focal Press.

Rollin, B. (1981). *Animal Rights and Human Morality*. Buffalo, NY: Prometheus Books.

Schwartz, T. (1974). *The Responsive Chord: How Radio and TV Manipulate You… Who You Vote For… What You Buy… and How You Think*. New York: Doubleday.

Schweitzer, A. (1965). *The Teaching of Reverence for Life*. New York: Holt, Rinehart and Winston.

Shesgreen, S. (Ed.). (1973). *Engravings by Hogarth* (p. 10020). New York: Dover.

Singer, P. (1975). *Animal Liberation*. New York: New York Review Press.

Snyder, B. (2005). *Save the cat!* Sudio City, CA: Michael Wise Productions.

Wilson, E.O. (1984). *Biophilia*. Cambridge, MA: Harvard University Press.

Walters, G. D. (2013). Testing the specificity postulate of the violence graduation hypothesis: Meta-analyses of the animal cruelty–offending relationship. *Aggression and Violent Behavior*, 18(6), 797–802.

Wright J., & Hensley, C. (2003). From animal cruelty to serial murder: Applying the graduation hypothesis. *International Journal Offender Therapy and Comparative Criminology*, 47(1), 71–88.

# 48
# ANIMALS IN OUR LEISURE AND PLEASURE
## Research for a More Species Equitable Future

*Carmel Nottle, Janette Young, Torben Nielsen, Kate Dashper, and Zoei Sutton*

### Introduction

The intimate intersection of human lives with those of other species is a global and historical phenomenon. Archeological explorations consistently reveal animals buried with care and tenderness indicating the same intimate connections with other species that over half of the world's current population live with as they share their domestic/leisure life with non-human others. While there is little consistent data collected or collated on the numbers of animals who live with humans (aka pets), it is estimated that over 382 million birds, fish, reptiles, dogs, cats, horses and small animals can be characterised as filling this role in human life and society (Animal Medicine Australia, 2019). An unknown additional number of animals play a core role in other human leisure spaces including zoos, wildlife parks, equine activities such as horse "rentals" and seaside donkeys, forming part of nature experiences for humans (Carr & Young, 2018; Young & Carr, 2018). Animals are embedded in human leisure, and we gain pleasure from our leisure engagements with animals. But the question is – what about the animals involved? When and how, if at all, does multi-species *shared* leisure-pleasure occur? This chapter explores this question from varying research fields to collectively increase our understandings of how multi-species rather than simply human-centric understandings/experiences of leisure can be encompassed with a focus on not only how humans interact with animals during their leisure, but also the reciprocal, animals with humans.

### What Is Multi-Species Leisure?

While the definition of the word "leisure" may be commonly understood to include terms such as *time when not working, time free from other duties, time for enjoyment* or *time to relax*, what represents leisure, and what an individual may identify as leisure is highly personal and far more complex than commonly understood definitions may suggest (Elkington & Stebbins, 2014; Veal, 2017). Regardless of the definition, with over half of the population globally sharing their life with a pet (GfK, 2022) it is unquestionable that for many humans, animals are an integral part of their leisure pursuits. With the most commonly accepted definition of "pet": "a domesticated animal kept for pleasure rather than utility" (Merriam-Webster,

2022), the definition clearly identifies a human-centric focus and why pets and other animals are often considered part of human leisure. The unique human-animal relationship that exists for many however, has seen not only a growing understanding of the term multi-species leisure (Nottle & Young, 2019), but also an acceptance that such leisure is moving past a purely anthropocentric focus to also consider what constitutes leisure for the animals with whom we share our lives (Dashper, 2019).

## *Leisure with Dogs*

With dogs the most commonly owned pet worldwide (Gfk, 2022), it is not surprising they feature widely in multi-species leisure literature. While the reasons for dog ownership can broadly be placed into one of three categories – companionship, therapeutic or assistance/service/working (Hicks & Kramer, 2020) – the roles that dogs can play in human-animal leisure are almost as varied as dogs breeds themselves. For many dog owners walking their dogs (Brown & Rhodes, 2006; Fletcher & Platt, 2018), or playing with their dog primarily at home, makes up the main components of their co-leisure experiences (Greenebaum, 2004). On the other hand, there are also those who spend time in the realm of serious leisure with their companion animals (Stebbins, 2011).

From canicross, lure coursing and dock jumping, to conformation, obedience and agility, the list of sports that humans and dogs can do together is ever expanding (Sundance, 2010). Yet despite high rates of dog ownership and international broadcasting of such events as the Crufts (British Veterinary Association, 2016) and Westminster dog shows (Wang, 2019), it is a world still unknown to many dog owners and their dogs (Gillespie, Leffler & Lerner, 2002). For example, Australia is home to an estimated 5.5 million dogs (Pet Keen, 2022); however, 2021 membership statistics to the Australian National Kennel Council, which administers competitions in over 13 different sports reveals only 30, 272 members (Dogs Australia, n.d.). These low membership to dog ownerships rates are mirrored in other countries (The American Kennel Club Inc, 2022; The Kennel Club, 2022).

Work by Hultsman (2012) examined the motivations of couples participating in agility dog sports and why they actually choose these outlets. The findings from their research show that the interviewees spoke of both human-human and human-animal leisure constructs. They note their own enjoyment and desires, both from an individual and a couple leisure perspective that fits well within Stebbins' (2011) definition of serious leisure. The participants also noted the development of a *strong bond* with their dogs as a key motivator for participation. However, they focused largely on their own perspectives and provided little insight on how their four-legged companions enjoy the experience.

Similar to this, research by Farrell, Hope, Hulstein and Spaulding (2015) examined the motivators for individuals to participate in a variety of dog sport covering five disciplines: conformation, obedience, rally, agility and/or field competitions. Notably, of the 85 participants in the research, 80 reported participating in more than one sport with their dogs. Participants reported enjoyment in the sports due to the social engagement and the associated physical activity, including the socialisation with others while connecting with their dogs. Interestingly, however, some were also motivated to a degree by external factors such as prizes and titles that can be gained during competitions, and the element of achieving personal goals (Farrell et al., 2015). As with Hultsman's (2012) participants, however, little insight into dog sports as a leisure experience for the dog participants is provided.

Leisure is usually understood to be uncoerced and therefore something that is done by choice (Stebbins, 2011). Dog sports, either as training or competition, would easily represent leisure for a human participant. However, can it be considered uncoerced and something engaged in through choice by the dogs? Are dog sports leisure for the dogs involved? Unlike the human participants, the dogs are not able to directly express their motivations for participation (Forry, 2016). Some authors suggest that understanding human motivations offers insight into animal motivations. In particular, many dog sports are performed off-lead and are therefore seen to represent a degree of willingness and therefore "choice" on the dogs' behalf (Forry, 2016). Willingness on behalf of the dog is something that has been discussed frequently in sled-dog racing (Seselja, 2021; Sloan, 2017).

Interestingly, Dabrowska (2018) concluded that while conformation dog shows represent casual leisure for the human participants, it is the development and maintenance of human to human, not human to animal, relationships which is the dominating determinant. Furthermore, the author concluded that considering such events as leisure is likely not a view reciprocated by the canine participants, despite such events involving an element of sensory stimulation and aerobic activity (Stebbins, 2011). This was due to the idea that some "unpleasant" aspects of conformation for the dogs, such as exhaustion and stress, would override any enjoyable elements. Similarly, Pastore et al. (2011) previously concluded that participation in agility can be stressful for the dog and may pose a threat to the dog's welfare long term. As such the degree to which participation in dog sports truly represents willingness versus a conditioned response (McGreevy, Starling, Branson, Cobb & Calnon, 2012), and thus truly represents leisure, is not yet well understood (Danby, Dashper & Finkel, 2019).

Perhaps the most relatable aspect of animal leisure for pets revolves around the notion of animal enrichment for captive animals. Animal enrichment is driven by concerns for animal welfare and the ethical need to allow animals freedom to express normal behaviour (Webster, 2001). Pet enrichment specifically is a growing area for the prevention of obsessive behaviours, boredom, and to increase both mental and physical stimulation concerns, many of which can involve both the human and the animal (Bailey, 2021; Whelan, 2010).

If leisure in humans is about work-life balance and participating willingly during nonwork time, do those animals with "jobs" also have leisure as we understand it in humans? Assistance, quarantine, detection, search and rescue, military, herding (muster) dogs, police horses, farm horses and cattle, military dolphins, animal actors, are just a few of the animal groups that have well recognised "jobs" (Animal Ethics, 2021). There has been minimal consideration of these animals' nonwork life and therefore their potential to enjoy leisure. One notable exception is Knight and Sang's (2020) study of police dogs and their handlers. The human handlers recognised that the police dogs were sometimes "off duty", and therefore able to relax and behave differently than when they are working. The handlers did not recognise this as dog leisure per se, but they did acknowledge that these dogs were more than just workers.

The concept of canine leisure is explored in a paper by Nottle and Young (2019) and seeks to demonstrate how we as humans can try to understand and describe leisure relationships with our canine pets. Our pets provide us numerous outlets that promote more optimal living, and support many humans in coping with difficult periods of time, including suicide protection (Young et al., 2020). Furthermore, many studies have reported on the importance of touch and physical proximity with our animal companions as an important dimension in supporting the relationships between humans and non-human animals (Young, Pritchard, Nottle & Banwell, 2020). It is not uncommon for humans to seek out the companionship of their companion animals in an informal way of relaxation and leisure. While this could be seen to focus on the leisure we as humans derive from our pets, many of those interviewed spoke of the touch initiated by their pets, reinforcing the concept of reciprocity in the joint leisure as something that is done "willingly" (Stebbins, 2011). As with dog sports, the idea that any captive animal, including pets, can do anything willingly remains contentious (McFarland & Hediger, 2009) and something that pet owners and the scientifically motivated researcher will continue to explore and debate.

## *Leisure with Cats*

It is not just dogs that humans frequently share their lives with. The second most popular pet worldwide is the cat, with an estimated 45 million households in the United State home to a cat (Insurance Information Institute Inc., 2022), and approximately 370 million pet cats worldwide (Bedford, 2020). Although cats are often viewed as more independent of their human companions than dogs (Crowley, Cecchetti & McDonald, 2020; Vitulli, 2006), most human-cat leisure remains within the definition of casual leisure (Stebbins, 2011) and revolves around our shared living environments and the companionship that this offers (Mertens, 1991). Despite this perceived independence, regulations governing how humans and cats should co-exist are a changing landscape (Crowley et al., 2020; Grayson, Calvar & Styles, 2002), with regulations progressively limiting cat agency based on human values and ideals of what is acceptable (Jongman, 2007; van Eeden et al., 2021; Zito, Vankan, Bennett, Paterson & Phillips, 2015).

Also similar to dogs, many human-cat relationships expand into the realm of serious leisure. Cat fancier is a commonly used term to describe individuals who are involved in breeding and conformation exhibiting (showing) cats as a hobby; as distinct from those that breed cats as a commercial venture (Stone, 2022). These individuals reportedly do so for human companionship and the social opportunities it provides, as well as development of skills and knowledge, and for competition (Stone, 2019). Participation can contribute to the individual's identity, with serious leisure participants developing their own subculture and language (Stone, 2022) the same as that seen in human (Green, 2001) and dogs sports (Gillespie et al., 2002). As with the dog world however, it is suggested that while the experience is shared by humans and cats, it likely only represents a pleasurable leisure experience for the human participant. Research by Stone (2019) concluded that, while it is not the case for all cats, cats were seemingly denied expression of agency and that human not animal interests where prioritised at cat shows.

## *Birds and Human Leisure*

Like our four-legged small companion animals, birds too have a long and varied tradition as participants and co-participants in human leisure (Birkhead & van Balen, 2008; Storer, 1933). In 2022, there were an estimated 1.6 million pet birds in the United Kingdom with this figure excluding other avian species such as chickens and other fowls (Statista, n.d.). In Indonesia, an estimated 12 million household are home to over 70 million caged birds, a portion of which are trapped and traded from wild bird populations (Marshall et al., 2020). Pigeons are reported to be one of the oldest domesticated birds with breeding for aesthetics documented as early as the 16th century (Shapiro & Domyan, 2013), although they have been used by humans for much longer (Queensland Racing Pigeon Federation Inc Australia, n.d.). A detailed mapping of species-specific experiences of pleasurable activity (Young, Nielsen, Dashper, Sutton & Nottle, 2022) seeks to assist people to become better informed of species-specific needs. For example, many pet loving people are probably not aware that their loved little budgerigar (a small, internationally known Australian parrot) may well delight as much in human companionship as their dog. As a social species of bird, humans can at least partly substitute for other parrots in the same manner as they can for dogs. Knowing this can help parrot guardians to facilitate better leisure for their loved non-human companion.

Like the cat fancier, the bird fancier also finds leisure in the breeding and showing of birds with manuals for enthusiasts dating back to the late 1800s (Browne, 1892). Research by Marhsall et al. (2020), found the most common reason for keeping birds was pleasure and entertainment, particularly of songbirds (Marshall et al., 2021), or as a hobby. While these represent internal motivators, the external motivator of keeping up with peers or family members was also commonly reported. Although as previously mentioned, pigeons were amongst the first birds bred for aesthetics, breeding for sport is also documented to have occurred with the start of pigeon racing in Belgium in the early 1800s (Marlier, 2022). The history of the sport is said to have been a symbol of masculine retreat for the working class (Baker, 2013; Johnes, 2007), and while today's racing or homing pigeons are all genetic descendants of the wild rock dove (Walcott, 1996), the motivation of competition today has seen the humble pigeon fancier evolve into part higher performance manager, part coach (Marshall, n.d.). The feathered participants' health is also sometimes put at risk by human participants, with doping of birds to improve performance known to occur not unlike that seen in other human and animal sports (Marlier, 2022).

It is not just captive birds that are a source of leisure. Garden or backyard feeding of birds is a global phenomenon (Fuller, Warren, Armsworth, Barbosa & Gaston, 2008) with up to 57 million people in the United States alone reported to feed wild birds in their backyards (U.S. Fish & Wildlife Service, 2016). Backyard feeding is also seen to have educational benefits. Beck, Melson, da Costa and Liu (2001) conducted a study focusing on elementary age children involved in a home-based ten-week program feeding wild birds. They concluded that the younger children in the sample (seven- to nine-year-olds) developed a positive outlook on the value of bird feeding. Additionally, 90% of the families were still involved in backyard feeding a year after the project was completed with family involvement a key motivator for continuing. Additionally, research by Horvath and Roelans (1991) into motivators for backyard bird feeding included not just the aesthetic value of the birds they attracted, but also the entertainment and companionship value of the birds. Interestingly, individuals also reported a belief that the birds *appreciated the food,* and of a feeling of *duty to provide for their well-being.* These sentiments are mirrored in later research by Clark, Jones and Reynolds (2019) with themes such as *pleasure* in seeing and hearing the birds, *companionship* from the potential interaction, and *survival* in terms of providing nourishment and alternative food sources emerging. Interestingly, the themes of *making amends* out of guilt about how humans in general treat the environment, and *personal atonement* to make up for environmental damage they may have personally caused also emerged.

Taking leisure with birds one step further are birdwatchers who can be further subdivided into those that participate near their homes, and the more serious "birders" who travel in pursuit of their pastime (Wilkinson, Waitt & Gibbs, 2014). While many birdwatchers are probably also bird feeders, they are more likely to note or record common and new species seen in their local area and may even have purchased equipment such as binoculars to enhance their participation experience (White, 2019). The 2016 report from the United States Fish and Wildlife Service on fishing, hunting and wildlife-association recreation estimated that around 20% of the population of the United States, or 45.1 million people, participated in birdwatching (U.S. Fish & Wildlife Service, 2016). Interestingly, regardless of the level of commitment to their birdwatching leisure pursuits, social connections do not rate highly as motivators for participation. Instead, conservation is reported to be a primary motivator

(Cottman-Fields, Brereton, Wimmer & Roe, 2014) with high levels of bird related motivators, not personal motivators, also reported (Hvenegaard, 2002). Research also suggests that the "typical" birder is slowly changing with a younger, more diverse demographic now participating, although birders are more likely to enjoy other outdoor activities than their non-birder counterparts (Cordell & Herbert, 2002). Additionally, while many birdwatchers may reside in populated cities, more serious birding pursuits tend to happen in remote and rural areas (Connell, 2009). However, the degree to which bird watching is multi-species leisure can be questioned. While both bird and birder are performing activities willingly at the same point in time, the motivations for doing such activity will vary, and the motivations of the bird remain unknown. For example, if a bird watcher is watching a bird building a nest, is that bird performing an activity of leisure or is it work (Álvarez & Barba. 2008)?

## *Horses and Leisure*

Akin to the shared leisure experiences between humans and dogs, humans and horses also have a long and varied history of both casual and serious leisure together (Monterrubio & Pérez, 2021). This is in addition to the working relationship that many humans have with horses (Coulter, 2019). Horses used for human leisure are increasing in numbers globally (Vial, Aubert & Perrier-Cornet, 2011) with the use of horses commonly placed into one of four categories: recreational/social, breeding, sporting/competition, meat production (Endenburg, 1999). Regardless of the role taken in human leisure, horses are both temporally (time) and financially demanding (Dashper, 2017b). Indeed, the demands of horse keeping for leisure can be so high that Dashper, Abbott and Wallace (2020) asked the question "Do horses cause divorces?". While their conclusion was that they do not, their research highlights the intense commitment and relationship that many horse owners have with their non-human leisure partners. This strong relationship is also highlighted by Mukherjee (2020) regarding child-horse leisure. There is a growing body of literature in the horse related leisure field that is considering animal agency and the idea of co-constitution in horse-human leisure engagements (Dashper, 2017a; Dashper & Brymer, 2019). This includes how these cross-species relationships can shape human identity (Dashper, 2016; Linghede, 2019) and the importance of understanding equine welfare in leisure animals as well as for animals used in work and sport (Dittmann, et al., 2020; Janczarek & Wilk, 2017).

The list of horse sports available to humans and their horses is extensive, with sports such as dressage, jumping, cross-country and endurance just to name a few (Ollenburg, 2005). The game of polo is thought to be one of the first ever team sports, played as early as 600 B.C. (Johnson, n.d.). Originally played as training for cavalry rather than sport (Latham, 2021) today it is played worldwide but with relatively low participation rates of just over 20,000 in 2016 (Polo +10 GMBH, 2016). As a comparison, there are approximately 33, 000 members of the United States Dressage Federation alone (USDF, n.d.). The United States horse industry, which includes the professional racing organisations, is estimated to be worth $40 billion annually (Emma, 2022).

With horse-related leisure activities typically including horse riding in some form (Gilbert & Gillett, 2014), annually an estimated 36 per 100,000 of the United States population experience horse riding in some form per year (Emma, 2022). As with previous multi-species leisure experiences, however the extent to which this represents leisure for the horse is unknown and continues to be explored (Birke & Hockenhull, 2015). From the human perspective, the motivations for horse riding, like all leisure is personal, however age and gender analyses' do show some strong trends. Overall, male riders tend to be attracted to competition sports particularly at the higher levels of competition, while female participation rates are higher in non-competitive participation and rider education pursuits (Calamatta, 2012). For the younger demographic, socialisation, thrill and skill development are key driving motivators while the older demographic are more likely to report using horse riding as a method of relaxation and escape from routine (Wu, Saxena & Jayawardhena, 2015). Other reported motivations for horse riding include self-development (Cynarski & Obodynski, 2008) with some riders reporting that their leisure has helped them learn valuable life lessons such as patience and persistence (Rashid, 2011).

Horses also play an increasing role in the tourism sector, particularly eco-tourism (Helgadóttir & Sigurðardóttir, 2008). Horse-related activities are often promoted as being part of the cultural experience (Monterrubio & Pérez, 2021) and seen as important part of national heritage in many countries (Pickel-Chevalier, 2015). As with birders, there is an increasing market for wild horse tourism like that seen for other wild animals such as dolphins (Orams, 1997), whales (Lambert, Hunter, Pierce & MacLeod, 2010) and large game animals (Akama, Maingi & Camargo, 2011; Di Minin, Fraser, Slotow & MacMillan, 2013). For horses, however the use of the term "wild" is often contentious with most wild horses considered feral domesticated animals whose population is closely managed in many countries (Oelke, 2012). Regardless of the definition, the extent to which this also represents true multi-species leisure is unknown given the leisure experience of the horse can only be assumed. Those that participate in this form of leisure however often report a connection with the horses, describing them as having charisma, and evoking feelings of a spiritual connection (Notzke, 2016) not unlike the feelings reported by "pet" horse owners (Dashper, 2017a, Dashper, 2017b).

## How Can We Better Understand Multi-Species Leisure?

As can be seen, the motivations for participation in multi-species leisure are largely human-centric and somewhat divided based on the concept of pet versus wild animal participation. This intriguing divide between conceptual frameworks of rights and positive experiences of leisure for wild and domestic animals has been documented previously in two edited books that split the animal foci into "domestic" (Young & Carr, 2018) and "wild" (Carr & Young, 2018) animals. Wild animal rights are embedded in conceptions of needing to be separated from intimate human engagement ("no touching"). This can be seen in the literature relating to birding leisure pursuits discussed previously (Hvenegaard, 2002) or in tourism adventures (the readers are encouraged to review Chapter 31 on Tourism for further information), or bears freed from inhumane cages, but unable to be returned to the wild, seen as needing to be kept separate from humans (Scutter & Young, 2018). Conversely, "domestic" animals are seen as requiring focus and attention from humans to facilitate their well-being, where personal motivations are central to participation such as in dog sports (Dabrowska, 2018). There are, however, presumptions in both. Wild animals are denied individual agentic potential of wanting to engage with humans, while the quality of domestic animals' leisure is seen as inherently tied to humans. Sitting in a liminal space are also the animals whose roles within many human societies have changed dramatically over the last 100 or so years as human-animal co-labour has been replaced with technologies (Yaxley, Joiner & Abbass, 2021; Zhou, Ma & Li, 2018). Having engineered the instincts and drives of these animals for many generations, we now often consign these animals to endless, potentially unfulfilling, lives of "leisure", connecting only with humans in the space of leisure; rather than shared purposeful labour (Young & Baker, 2018).

Outside of serious leisure pursuits, even the keeping of pets for pleasure and the information gained from this can inform us on how to care for and interact with animals. This in turn provides at least an understanding of the degree to which casual leisure may truly be shared. Different aspects of keeping and interacting with pets have been investigated and found to be affecting the animals. Stabling horses individually can lead to stereotypical behaviours (Lansade et al., 2014) and punishment of cats and dogs can lead to unwanted behaviours or stress (Grigg & Kogan, 2019; de Castro et al., 2020). One group where negative effects on the animals of pet keeping may be particularly important is small mammals, as there can be issues with housing, feeding, interactions and handling due to ignorance and misinterpreting the animals' behaviour (Harrup & Rooney, 2020). For example, some of these species can freeze when they see a perceived threat, which people might see as the pet accepting touch or waiting to be picked up. However, after learning to interact appropriately with these species, interactions can be positive, such as tickling a pet rat and getting a positive response in return (LaFollette, O'Haire, Cloutier & Gaskill, 2018). The important point is the need to be careful in how we as humans interpret pet behaviour. Play, whether it be solo play, with other species or with people, is often seen as positive and pleasurable for the animal. However, some types of play can also be indicative of poor welfare. For example, early weaned kittens might increase the time they spend playing with objects which could indicate they are training for accessing prey to make up for a nutritional shortfall (Ahloy-Dellaire et al., 2018).

## Research Challenges and the Study of Leisure with Human

Serious leisure pursuits with our pets are also not immune to complicated understandings of shared leisure that need further investigation. As previously highlighted, many serious leisure pursuits are human-centric and do not consider fully the animals' willingness to participate in the shared leisure experience, or the adverse effects of participation that can arise for the human and/or the animal. Some examples include the increased risk of the spread of infectious diseases at cat shows (Stone, 2019), stress (Pastore et al., 2011) and serious injury from participation in dog agility (Inkilä et al., 2022; Kerr, Fields & Comstock, 2014; Tomlinson & Manfredi, 2018), bird fancier's lung (Morell et al., 2008) and *Chlamydia psittaci* in racing pigeons (Ling et al., 2015). Consideration of both human and animal welfare and the disease or injury risk associated with pets (Chomel, 2015; Goldstein, 1991) allows for development of risk management tools (Young et al., 2022) and policy considerations (Damborg et al., 2016). Risk management tools in particular can scope out the risks to both humans and animals, identifying risk reduction strategies and best practise examples to ensure high quality of life for all species involved. When considering such literature, however, it must be remembered that this emerging field of awareness of animals as having rights will always be human-centric. As human beings, we have no experience of the life of being a domestic dog or wild lion, or any other creature. Even though through our animal research and human imagination we may genuinely seek to care for the non-human species that share our everyday lives (Dashper & Buchmann, 2019), understanding how animals (literally) see their world is not knowledge that is accessible to us as humans at this present time. Our best hope is to seek out connections with both other animals and humans who have studied animal species at length.

Within the research relating to human-horse shared leisure, ethnographic research in particular embraces the challenges, messiness and potential of multispecies ethnography to recognise other animals as *subjects* within leisure. The struggle in this area however is the constraint of the researcher's own perceptual human limitations, the inadequacy of human language to convey the embodied aspects of interspecies communication and, of course, the rigidities of academic research. However, despite this, researchers have remained committed to trying to give space to non-humans within leisure research by ongoing support from lifelong interactions with various individual animals. Such companions have featured in numerous publications and have proven to be invaluable "research assistants" – helping authors gain access to different social worlds, teaching the language of interspecies communication and, through developing bonds, reminding the researchers that all animals are individual "people" themselves, worthy of human care and attention (Dashper, 2021a). Such research allows for the exploration of the varied meanings leisure animal owners attach to their animal activities (Dashper, 2017a, 2017b), from conceptualising horses and other species as workers in the tourism industry (Dashper, 2020a, 2021b), to considering if animals themselves have leisure, and if so, what might a holiday for an animal look like (Dashper, 2020b). The use of ethnography to understand human-animal leisure has been used to explore the experiences of dogs (Robins, Sanders, Cahill, 1991), cats (Alger & Alger, 1999) and horses (Ford, 2019) and while complicated does allow for consideration of animals as subjects (Madden, 2014).

## Concluding Remarks: Looking to the Future of Multi-Species Leisure

It is difficult to deny that multi-species leisure does not exist, but there is much more to learn. While our understanding of other species experiences will always be human-centric, there is an increasing body of research that aims to better understand how animals may enjoy leisure. Only by constantly questioning what we humans consider leisure with other species, and showing increasing respect for animals as sentient beings, can we hope to move towards a greater understanding of multi-species leisure. In this process of self-reflection and respectful engagements there is hope that we can increasingly engage in truly *multi*-species leisure; rather than just sharing our human leisure time with other species, regardless of their willingness to participate or experience leisure for themselves in such engagements.

### Questions for Further Discussion

1. What strategies can we (humans) use to better understand the animals in our lives and their enjoyment/leisure?
2. Can we really call shared leisure with animals "*multi-species* leisure"? Is there a better term that we could or should use?
3. What questions and ideas raised in this chapter might be different when considering multi/cross-species leisure that does not involve humans?

## References

Ahloy-Dallaire, J., Espinosa, J., & Mason, G. (2018). Play and optimal welfare: does play indicate the presence of positive affective states? *Behavioural Processes*, *156*, 3–15. https://doi.org/10.1016/j.beproc.2017.11.011

Akama, J. S., Maingi, S., & Camargo, B. A. (2011). Wildlife conservation, safari tourism and the role of tourism certification in Kenya: a post colonial critique. *Tourism Recreation Research*, *36*(3), 281–291. https://doi.org/10.1080/02508281.2011.11081673

Alger, J., & Alger, S. (1999). Cat culture, human culture: an ethnographic study of a cat shelter. *Society and Animals*, 7(3), 199–218. https://doi.org/10.1163/156853099X00086

Álvarez, E., & Barba, E. (2008). Nest quality in relation to adult bird condition and its impact on reproduction in Great Tits Parus major. *Acta Ornithologica*, 43(1), 3–9. https://doi.org/10.3161/000164508X345275

Animal Medicine Australia. (2019). *Pets in Australia: A National Survey of Pets and People*. Retrieved September 2, 2022, from, https://animalmedicinesaustralia.org.au/wp-content/uploads/2019/10/ANIM001-Pet-Survey-Report19_v1.7_WEB_low-res.pdf

Animal Ethics. (2021). *Animals Used as Work*. Retrieved September 2, 2022, from, https://www.animal-ethics.org/animals-used-workers/

Bailey, B. (2021). The importance of enrichment for our pets. *Veterinary Nursing Journal*, 36(12), 340–341. https://doi.org/10.1080/17415349.2021.1999626

Baker, A, R. H. (2013). Pigeon racing clubs in Pas-de-Calais, France, 1870–1914. *Journal of Historical Geography*, 41, 1–12.

Beck, A. M., Melson, G. F., da Costa, P. L., & Liu, T. (2001). The educational benefits of a ten-week home-based wild bird feeding program for children. *Anthrozoös*, 14(1), 19–28. https://doi.org/10.2752/089279301786999599

Bedford, E. (2020). *Global Dog and Cat Pet Population 2018*. Retrieved September 23, 2022, from, https://www.statista.com/statistics/1044386/dog-and-cat-pet-population-worldwide/

Birke, L., & Hockenhull, J. (2015). Journeys together: horses and humans in partnership. *Society and Animals*, 23, 81–100. https://doi.org/10.1163/15685306-12341361

Birkhead, T. R., & van Balen, S. (2008). Bird-keeping and the development of ornithological science. *Archives of Natural History*, 35(20), 281–305. https://doi.org/10.3366/E0260954108000399

British Veterinary Association (2016). Celebrations – and controversy – at the 125th Crufts dog show. *The Veterinary Record*, 178(12), 281. https://doi.org/10.1136/vr.i1540

Brown, S. G., & Rhodes, R. E. (2006). Relationships among dog ownership and leisure-time walking in Western Canadian adults. *American Journal of Preventative Medicine*, 30(2), 131–136. https://doi.org/10.1016/j.amepre.2005.10.007

Browne, D. G. (1892). *The Bird Fancier; or, How to Breed, Rear and Car for Song and Domestic Birds; with their Diseases and Remedies*. Orange Judd Company.

Calamatta, K. (2012). A qualitative study of gender and work in a British riding school. Doctor of Philosophy, University of Sussex, Brighton, UK.

Carr, N., & Young, J. (Eds.). (2018). *Wild Animals and Leisure: Rights and Wellbeing*. Routledge.

Chomel, B. B. (2015). Diseases transmitted by less common house pets. *Microbiology Spectrum*, 3(6). https://doi.org/10.1128/microbiolspec.IOL5-0012-2015. PMID: 27337276

Clark, D. N., Jones, D. N., & Reynolds, S. J. (2019). Exploring the motivations for garden bird feeding in south-east England. *Ecology and Society*, 24(1). https://doi.org/10.5751/ES-10814-240126

Connell, J. (2009). Birdwatching, twitching and tourism: towards and Australian perspective. *Australian Geographer*, 40(2), 203–217. https://doi.org/10.1080/00049180902964942

Cordell, H. K. & Herbert, N. G. (2002). The popularity of birding is still growing. *Birding*, February, 54–61.

Cottman-Fields, M., Brereton, M., Wimmer, J., & Roe, P. (2014). *Collaborative Extension of Biodiversity Monitoring Protocols in the Bird Watching Community*. 13th Participatory Design Conference, Namibia. https://doi.org/10.1145/2662155.2662193

Coulter, K. (2019). Horses' labour and work-lives: new intellectual and ethical directions. In *Equine Cultures in Transition* (pp. 17–31). Routledge.

Crowley, S. L., Cecchetti, M., & McDonald, R. A. (2020). Our wild companions: domestic cats in the athropocene. *Trends in Ecology & Evolution*, 35(6). https://doi.org/10.1016/j.tree.2020.01.008

Cynarski, W. J., & Obodynski, K. (2008). Horse-riding in the physical education, recreation and tourism – axiological reflection. *Research Yearbook*, 14(1), 37–43.

Dabrowska, M. (2018). Dog shows as casual leisure. Asymmetry of human and animal experience. In J. Young & N Carr (Eds.), *Domestic Animals, Humans, and Leisure: Rights, Welfare and Wellbeing* (pp. 32–46). Routledge.

Damborg, P., Broens, E. M., Chomel, B. B., Guenther, S., Pasmans, F., Wagenaar, J. A., Weese, J. S., Wieler, L. H., Windahl, L. H., Vanrompay, D., & Guardabassi, L. (2016). Bacterial zoonoses transmitted by household pets: state-of-the-art- and future perspectives for targeted research and policy actions. *Journal of Comparative Pathology*, 155(1 Suppl 1), S27-S40. https://doi.org/10.1016/j.jcpa.2015.03.004

Danby, P., Dashper, K., & Finkel, R. (2019). Multispecies leisure: human-animal interactions in leisure landscapes. *Leisure Studies*, *38*(2), 291–302. https://doi.org/10.1080/02614367.2019.1628802

Dashper, K. (2016). Strong, active women: re(doing) rural femininity through equestrian sport and leisure. *Ethnography*, *17*(3), 350–368. https://doi.org/10.1177/14661381156093

Dashper, K. (2017a). Listening to horses: developing attentive interspecies relationships through sport and leisure. *Society & Animals*, *25*(3), 207–224. https://doi.org/10.1163/15685306-12341426

Dashper, K. (2017b). *Human-animal Relationships in Equestrian Sport and Leisure*. Routledge.

Dashper, K. (2019). Moving beyond anthropocentrism in leisure research: multispecies perspectives. *Annals of Leisure Research*, *22*(2), 133–139. https://doi.org/10.1080/11745398.2018.1478738

Dashper, K. (2020a). More-than-human emotions: multispecies emotional labour in the tourism industry. *Gender, Work & Organization*, *27*(1), 24–40. https://doi.org/10.1111/gwao.12344

Dashper, K. (2020b). Holidays with my horse: human-horse relationships and multispecies tourism experiences. *Tourism Management Perspectives*, *34*, 100678. https://doi.org/10.1016/j.tmp.2020.100678

Dashper, K. (2021a). Close encounters: individual interspecies relationships and research practices. In L. Birke & H. Wels (Eds.). *Dreaming of Pegasus: Equine Imaginings* (pp. 53–66). Victorina Press.

Dashper, K. (2021b). Conceptualising nonhuman animals as 'workers' within the tourism industry: theoretical, practical and ethical implications. In C. Kline & J. Rickly (Eds.). *Working Animals of Tourism: From Beasts of Burden to K-9 Security* (pp. 21–36). De Gruyter Oldenbourg.

Dashper, K., Abbott, J., & Wallace, C. (2020). 'Do horses cause divorces?' Autoethnographic insights on family, relationships and resource-intensive leisure. *Annals of Leisure Research*, *23*(2), 304–321. https://doi.org/10.1080/11745398.2019.1616573

Dashper, K., & Brymer, E. (2019) An ecological-phenomenological perspective on multispecies leisure and the horse-human relationship in events. *Leisure Studies*, *38*(3), 394–407. https://doi.org/10.1080/02614367.2019.1586981

Dashper, K., & Buchmann, A. (2019). Multispecies event experiences: introducing more-than-human perspectives to event studies. *Journal of Policy Research in Tourism, Leisure and Events*, *12*(3), 293–309. https://doi.org/10.1080/19407963.2019.1701791

de Castro, A. C. V., Fuchs, D., Morello, G. M., Pastur, S., de Sousa, L., & Olsson, I. A. S. (2020). Does training method matter? Evidence for the negative impact of aversive-based methods on companion dog welfare. *PLoS One*, *15*(12), e0225023. https://doi.org/10.1371/journal.pone.0225023

Di Minin, E., Fraser, I., Slotow, R., & MacMillan, D. C. (2013). Understanding heterogeneous preference of tourists for big game species: implications for conservation and management. *Animal Conservation*, *16*, 249–258. https://doi.org/10.1111/j.1469-1795.2012.00595.x

Dittmann, M. T., Latif, S. N., Hefti, R., Hartnack, S., Hungerbühler, V., & Weishaupt, M. A. (2020). Husbandry, use, and orthopedic health of horses owned by competitive and leisure riders in Switzerland. *Journal of Equine Veterinary Science*, *91*, 103107. https://doi.org/10.1016/j.jevs.2020.103107

Dogs Australia (n.d.). *Dogs Australia Membership Statistics 1995–2021*. Retrieved September 21, 2022, from, https://www.dogsaustralia.org.au/about-dogs-australia/statistics/

Endenburg, N. (1999). Perception and attitudes towards horses in European societies. *Equine Veterinary Journal*, *28*(Suppl), 38–41. https://doi.org/10.1111/j.2042-3306.1999.tb05154.x

Elkington, S. & Stebbins, R. (2014). *The Serious Leisure Perspective*. Routledge.

Emma (2022). *15 Surprising Horse Statistics to Know in 2022*. Retrieved October 12, 2022, from https://pawsomeadvice.com/pets/horse-statistics/

Farrell, J. M., Hope, A. E., Hulstein, R., & Spaulding, S. J. (2015). Dog-sport competitors: what motivates people to participate with their dogs in sporting events? *Anthrozoös*, *28*(1), 61–71. https://doi.org/10.2752/089279315X1412935072201

Fletcher, T., & Platt, L. (2018). (Just) a walk with the dog? Animal geographies and negotiating walking spaces. *Social & Cultural Geography*, *19*(2), 211–229. https://doi.org/10.1080/14649365.2016.1274047

Ford, A. (2019). Sport horse leisure and the phenomenology of interspecies embodiment. *Leisure Studies*, *38*(3), 329–340. https://doi.org/10.1080/02614367.2019.1584231

Forry, J. G. (2016). Why some animal sports are not sports, In S. E. Klein (Ed.), *Defining Sports: Conceptions and Borderlines* (pp. 175–192), Lexington Books.

Fuller, R. A., Warren, P. H., Armsworth, P. R., Barbosa, O., & Gaston, K. J. (2008). Garden feeding predicts the structure of urban avian assemblages. *Diversity and Distributions*, *14*, 131–137. https://doi.org/10.1111/j.1472-4642.2007.00439.x

Gfk (2022). *Mans Best Friend: Global Pet Ownership and Feeding Trends*. Retrieved September 15, 2022, from https://www.gfk.com/insights/mans-best-friend-global-pet-ownership-and-feeding-trends

Gilbert, M., & Gillett, J. (2014). Into the mountains and across the country: emergent forms of equine adventure leisure in Canada. *Society and Leisure, 37*(2), 313–325. https://doi.org/10.1080/07053436.2014.936168

Gillespie, D. L., Leffler, A., & Lerner, E. (2002). If it weren't for my hobby, I'd have a life: dog sports, serious leisure, and boundary negotiations. *Leisure Studies, 21*(3-2), 285–304. https://doi.org/10.1080/0261436022000030632

Goldstein, E. J. (1991). Household pets and human infections. *Infectious Disease Clinics of North America, 5*(1), 117–130.

Grayson, J., Calver, M., & Styles, I. (2002). Attitudes of suburban Western Australians to proposed cat control legislation. *Australian Veterinary Journal, (80),* 536–543. https://doi.org/10.1111/j.1751-0813.2002.tb11030.x

Green, B. C. (2001). Leveraging subculture and identity to promote sports events. *Sport Management Review, 4*(1), 1–19. https://doi.org/10.1016/S1441-3523(01)70067-8

Greenebaum, J. (2004). It's a dog's life: elevating status from pet to "fur baby" at yappy hour. *Society & Animals, 12*(2), 117–135. https://doi.org/10.1163/1568530041446544

Grigg, E. K., & Kogan, L. R. (2019). Owners' attitudes, knowledge, and care practices: exploring the implications for domestic cat behavior and welfare in the home. *Animals, 9*(11), 978. https://doi.org/10.3390/ani9110978.

Harrup, A. J., & Rooney, N. (2020). Current welfare state of pet guinea pigs in the UK. *Veterinary Record, 186*(9), 282–282. https://doi.org/10.1136/vr.105632

Helgadóttir, G., & Sigurðardóttir, I. (2008). Horse-based tourism: community, quality and disinterest in economic value. *Scandinavian Journal of Hospitality and Tourism, 8*(2). 105–121. https://doi.org/10.1080/15022250802088149

Hicks, J. R., & Kramer, M. (2020). Therapy dog ownership as serious leisure for members of a therapy dog volunteer group. *People and Animals: The International Journal of Research and Practice, 3*(1), 1–12. https://docs.lib.purdue.edu/paij/vol3/iss1/5

Horvath, T. & Roelans, A. M. (1991). Backyard feeders: not entirely for the birds. *Anthrozoös, 4*(4), 232–236. https://doi.org/10.2752/089279391787057080

Hultsman, W. Z. (2012). Couple involvement in serious leisure: examining participation in dog agility. *Leisure Studies, 31*(2), 231–253. https://doi.org/10.1080/02614367.2011.619010

Hvenegaard, G. T. (2002). Birder specialization differences in conservation involvement, demographics and motivations. *Human Dimensions of Wildlife, 7*(1), 21–36. https://doi.org/10.1080/108712002753574765

Inkilä, L., Hyytiäinen, H. K., Hielm-Björkman, A., Junnila, J., Bergh, A., & Boström, A. (2022). Part II of Finnish agility dog survey: agility-related injuries and risk factors for injury in competition-level agility dogs. *Animals, 12*(3), 227. https://doi.org/10.3390/ani12030227

Insurance Information Institute Inc. (2022). *Facts + Statistics: Pet Ownership and Insurance.* Retrieved September 23, 2022, from https://www.iii.org/fact-statistic/facts-statistics-pet-ownership-and-insurance

Janczarek, I., & Wilk, I. (2017). Leisure riding horses: research topics versus the needs of stakeholders. *Animal Science Journal, 88,* 953–958. https://doi.org/10.1111/asj.12800

Johnes, M. (2007). Pigeon racing and working-class culture in Britain, C. 1870–1950. *Cultural and Social History, 4*(3), 361–383. https://doi.org/10.2752/147800407X219250

Johnson, B. (n.d.). *The Origins of Polo.* Retrieved October 12, 2022, from https://www.historic-uk.com/CultureUK/The-Origins-of-Polo/

Jongman, E. C. (2007). Adaption of domestic cats to confinement. *Journal of Veterinary Behavior, 2,* 193–196. https://doi.org/10.1016/j.jveb.2007.09.003

Kerr, Z. Y., Fields, S., & Comstock, R. D. (2014). Epidemiology of injury among handlers and dogs competing in the sport of agility. *Journal of Physical Activity and Health, 11*(5), 1032–1040. https://doi.org/10.1123/jpah.2012-0236

Knight, C., & Sang, K. (2020). At home, he's a pet, at work he's a colleague and my right arm: police dogs and the emerging posthumanist agenda. *Culture and Organization, 26*(5–6), 355–371. https://doi.org/10.1080/14759551.2019.1622544

LaFollette, M. R., O'Haire, M. E., Cloutier, S., & Gaskill, B. N. (2018). A happier rat pack: the impacts of tickling pet store rats on human-animal interactions and rat welfare. *Applied Animal Behaviour Science, 203,* 92–102. https://doi.org/10.1016/j.applanim.2018.02.006

Lambert, E., Hunter, C., Pierce, G. J., & MacLeod, C. D. (2010). Sustainable whale-watching tourism and climate change: towards a framework of resilience. *Journal of Sustainable Tourism, 18*(3), 409–427. https://doi.org/10.1080/09669581003655497

Lansade, L., Valenchon, M., Foury, A., Neveux, C., Cole, S.W., Layé, S., Cardinaud, B., Lévy, F., & Moisan, M.P. (2014). Behavioral and transcriptomic fingerprints of an enriched environment in horses (equus caballus). *PLoS One, 9*(12), e114384. https://doi.org/10.1371/journal.pone.0114384

Latham, R. C. (2021, July 23). *Polo. Encyclopedia Britannica.* Retrieved October 10, 2022, from https://www.britannica.com/sports/polo

Ling, Y., Chen, H., Chen, X., Yang, X., Yang, J., Bavoli, P. M., & He, C. (2015). Epidemiology of Chlamydia psittaci infection in racing pigeons and pigeon fanciers in Beijing, China. *Zoonoses and Public Health, 62*(5), 401–406. https://doi.org/10.1111/zph.12161

Linghede, E. (2019). Becoming horseboy(s) – human-horse relations and intersectionality in equiscapes. *Leisure Studies, 38*(3), 408–421. https://doi.org/10.1080/02614367.2019.1584230

Madden, R. (2014). Animals and the limits of ethnography. *Anthrozoös, 27*(2), 279–293. https://doi.org/10.2752/175303714X13903827487683

Marshall, H., Collar, N. J., Lees, A. C., Moss, A., Yuda, P., & Marsden, S. J. (2020). Characterizing bird-keeping user-groups on Java reveals distinct behaviours profiles and potential for change. *People and Nature, 2*, 877–888. https://doi.org/10.1002/pan3.10132

Marshall, H., Glorizky, G. A., Collar, N. J., Lees, A. C., Moss, A., Yuda, P., & Marsden, S. J. (2021). Understanding motivations and attitudes among songbird-keeps to identity best approaches to demand reduction. *Conservation Science and Practice, 3*, e507. https://doi.org/10.1111/csp2.507

Marshall, R. (n.d.). *Racing Pigeons.* Retrieved September 23, 2022, from https://www.birdhealth.com.au/racing-pigeons#:~:text=In%20Australia%2C%20many%20thousands%20of, any%20other%20team%20of%20athletes

Marlier, D. (2022). Doping in racing pigeons (Columba livia domestica): a review and actual situation in Belgium, a leading country in this field. *Veterinary Science, 9*, 42. https://doi.org/10.3390/vetsci9020042

McFarland, S. E., & Hediger, R. (2009). Approaching the agency of other animals: an introduction. In S. E. McFarland & R Hediger (Eds.). *Animals and Agency* (pp. 1–20). Brill.

McGreevy, P. D., Starling, M., Branson, N. J., Cobb, M. L., & Calnon, D. (2012). An overview of the dog-human dyad and ethograms within it. *Journal of Veterinary Behavior, 7*, 103–117. https://doi.org/10.1016/j.jveb.2011.06.001

Merriam-Webster. (2022). *Pet.* Retrieved August 18, 2022, from https://www.merriam-webster.com/dictionary/pet

Mertens, C. (1991). Human-cat interactions in the home setting. *Anthrozoös, 4*(4), 214–231. https://doi.org/10.2752/089279391787057062

Monterrubio, C., & Pérez, J. (2021). Horses in leisure events: a posthumanist exploration of commercial and cultural values. *Journal of Policy Research in Tourism, Leisure and Events, 13*(2), 147–171. https://doi.org/10.1080/19407963.2020.1749063

Morell, F., Roger, A., Reyes, L., Cruz, M. J., Murio, C., & Munoz, X. (2008). Bird fancier's lung. *Medicine, 87*(2), 110–130. https://doi.org/10.1097/MD.0b013e31816d1dda

Mukherjee, U. (2020). Caring, relating and becoming: child-horse relationships in equestrian leisure. *Journal of Childhood Studies, 45*(2), 85–97. https://doi.org/10.18357/jcs452202019741

Nottle, C., & Young, J. (2019) Individuals, instinct and moralities: exploring multi-species leisure using the serious leisure perspective, *Leisure Studies, 38*(3), 303–316. https://doi.org/10.1080/02614367.2019.1572777

Notzke, C. (2016). Wild horse-based tourism and wildlife tourism: the wild horse as the other. *Current Issues in Tourism, 19*(12), 1235–1259. http://dx.doi.org/10.1080/13683500.2014.897688

Oelke, H. (2012). *Wild Horses Then and Now.* Kierdorf Verlag.

Ollenburg, C. (2005). Worldwide structure of the equestrian tourism sector. *Journal of Ecotourism, 4*(1), 47–55. https://doi.org/10.1080/14724040508668437

Orams, M. B. (1997). Historical accounts of human-dolphin interaction and recent developments in wild dolphin based tourism in Australasia. *Tourism Management, 18*(5), 317–326. https://doi.org/10.1016/S0261-5177(96)00022-2

Pastore, C., Pirrone, F., Balzarotti, F., Faustini, M., Pierantoni, L., & Albertini, M. (2011). Evaluation of physiological and behavioral stress-dependent parameters in agility dogs. *Journal of Veterinary Behavior, 6*(3), 188–194. https://doi.org/10.1016/j.jveb.2011.01.001

Pet Keen (2022). *10 Australian Pet Ownership Statistics to Know in 2022: How Many Australians Have Pets?* Retrieved September 21, 2022, from https://petkeen.com/australia-pet-ownership-statistics/#:~:-text=About%2061%25%20of%20pets%20are, over%205.5%20million%20dogs%20total

Pickel-Chevalier, S. (2015). Can equestrian tourism be a solution for sustainable tourism development in France?. *Society and Leisure, 38*(1), 110–134. https://doi.org/10.1080/07053436.2015.1007580

Polo +10 GMBH (2016). *Polo in Numbers*. Retrieved October 12, 2022, from https://poloplus10.com/polo-in-numbers-63934/

Queensland Racing Pigeon Federation Inc Australia (n.d.). *Pigeon Racing*. Retrieved September 12, 2022, from http://www.qrpf.com/pigeon-racing/

Rashid, M. (2011). *Life Lessons for a Ranch Horse*. Skyhorse Publishing.

Robins, D. M., Sanders, C. R., & Cahill, S. E. (1991). Dogs and their people: pet-facilitated interaction in a public setting. *Journal of Contemporary Ethnography, 20*(1), 3–25. https://doi.org/10.1177/08912419102000100

Scutter, & Young, J. (2018). Volunteering for bear charities…what's in it for the bears? In N. Carr & J. Young (Eds.), *Wild Animals and Leisure: Rights and Wellbeing* (pp. 132–148). Routledge.

Seselja, E. (2021, May 23). Sled dog racing the 'overlooked' winter sport that could keep you and your four-legged friend fit, *ABCNews*. Retrieved from https://www.abc.net.au/news/2021-05-23/sled-dog-racing-could-keep-you-and-your-pup-fit/100138514

Shapiro, M. D. & Domyan, E. T. (2013). Domestic pigeons. *Current Biology, 23*(8), R302. Retrieved from https://www.cell.com/current-biology/pdf/S0960-9822(13)00132-2.pdf.

Sloan, H. (2017, August 29). Are sled dogs forced to run?. *Teacher on the Trails*. Retrieved from https://iditarod.com/edu/are-sled-dogs-forced-to-run/

Statista (n.d.). *Leading Pets Ranked by Estimated Population Size in the United Kingdom (UK) in 2022*. Retrieved September 23, 2022, from https://www.statista.com/statistics/1044386/dog-and-cat-pet-population-worldwide/

Stebbins, R. A. (2011). Basic concepts. *The Serious Leisure Perspective (SLP)*. Retrieved from https://www.seriousleisure.net/concepts.html

Stone, E. (2019). What's in it for the cats?: cats shows as serious leisure from a multispecies perspective. *Leisure Studies, 38*(3), 381–393. https://doi.org/10.1080/02614367.2019.1572776

Stone, E. (2022). *Cat-people: Human-Cat Interrelatedness in the Cat Fancy*. Taylor & Francis.

Storer, T. I. (1933). Relations between man and birds in California. *The Condor 35*(2), 55–59. https://doi.org/10.2307/1363647

Sundance, K. (2010). *101 Ways To Do More With Your Dog: Make Your Dog a Superdog with Sports, Games, Exercises, Tricks, Mental Challenges, Crafts, and Bonding*. Quarry Books.

The American Kennel Club Inc. (2022). *AKC Facts and Stats*. Retrieved September 21, 2022, from https://www.akc.org/press-center/articles-resources/facts-and-stats/

The Kennel Club (2022). *About the Kennel Club*. Retrieved September 21, 2022, from https://www.thekennelclub.org.uk/about-us/about-the-kennel-club/

Tomlinson, J. E., & Manfredi, J. M. (2018). Return to sport after injury: a web-based survey of owners and handlers of agility dogs. *Veterinary and Comparative Orthopaedics and Traumatology, 31*(6), 473–478. https://doi.org/10.1055/s-0038-1670676

USDF (n.d.). *About USDF*. Retrieved October 12, 2022, from https://www.usdf.org/about/about-usdf/

U.S. Fish and Wildlife Service. (2016). Birding in the United States: a demographic and economic analysis. Addendum to the 2016 National Survey of Fishing, Hunting, and Wildlife-Associated Recreation, 20220. Report 2016-2.

Vial, C., Aubert, M., & Perrier-Cornet, P. (2011). The organizational choices of pleasure horse owners in rural spaces. Économie rurale, 321, 42–57. https://doi.org/10.4000/economierurale.2911

van Eeden, L. M., Hames, F., Faulkner, R., Geschke, A., Squires, Z. E., & McLeod, E. M. (2021). Putting the cat before the wildlife: exploring cat owners' beliefs about cat containment as predictors of own behavior. *Conservation Science and Practice, 3*, e502. https://doi.org/10.1111/csp2.502

Veal, A. J. (2017). The serious leisure perspective and the experience of leisure. *Leisure Science, 39*(3), 205–223. https://doi.org/10.1080/01490400.2016.1189367

Vitulli, W. F. (2006). Attitudes towards empathy in domestic dogs and cats. *Psychological Reports, 99*, 981–991. https://doi.org/10.2466/PR0.99.3.981-991

Walcott, C. (1996). Pigeon homing: observations, experiments and confusions. *The Journal of Experimental Biology, 199*, 21–27.

Wang, A. B. (2019). At this Westminster dog show competition, it's the humans who are judged. *Washington Post*, 11 February 2019, viewed at https://www.washingtonpost.com/science/2019/02/12/this-westminster-dog-show-competition-its-humans-who-are-judged/

Webster, A. J. F. (2001). Farm animal welfare: the five freedoms and the free market. *The Veterinary Journal, 161*(2), 229–237. https://doi.org/10.1053/tvjl.2000.0563

Whelan, F. (2010). Environmental enrichment for pets. *Veterinary Nursing Journal, 25*(3), 27–28. https://doi.org/10.1111/j.2045-0648.2010.tb00023.x

White, J. (2019). *How Popular Is Birdwatching?* Retrieved September 23, 2022, from https://birda.org/how-popular-is-birdwatching/

Wilkinson, C., Waitt, G., & Gibbs, L. (2014). Understanding place as 'home' and 'away' through practices of bird-watching. *Australian Geographer, 45*(2), 205–220. https://doi.org/10.1080/00049182.2014.899029

Wu, J., Saxena, G. & Jayawardhena, C. (2015). Does age and horse ownership affect riders' motivation? In C. Vial & R. Evans' (Eds.), *The New Equine Economy in the 21st Century* (pp. 113–122). Wageningen Academic Publishers.

Yaxley, K. J., Joiner, K. F., & Abbass, H. (2021). Drone approach parameters leading to lower stress sheep flocking and movement: sky shepherding. *Scientific Reports, 11*, 7803. https://doi.org/10.1038/s41598-021-87453-y

Young, J. & Baker, A. (2018a). From labour to leisure: the relocation of animals in modern Western society. In J. Young & N. Carr (Eds.), *Domestic Animals, Humans, and Leisure: Rights, Welfare, and Wellbeing* (pp. 125–141). Routledge.

Young, J., Bowen-Salter, H., O'Dwyer, L., Stevens, K., Nottle, C., & Baker, A. (2020) Pets as Suicide protection in older people. *Anthrozoös, 33*(2), 191–205. https://doi.org/1080/08927936.2020.1719759

Young, J. & Carr, N. (2018). *Domestic Animals, Humans, and Leisure: Rights, Welfare, and Wellbeing.* Routledge.

Young, J., Neilsen, T., Dashper, K., Sutton, Z. & Nottle, C. (2022). Pets and People: Whose Leisure? Whose Health? In H. Maxwell, R. McGrath, J. Young & N. Peel (Eds.), *Exploring the Leisure - Health nexus: pushing the boundaries globally* (pp. 164–188). CABI.

Young, J., Pritchard, R., Nottle, C., & Banwell, H. (2020). Pets, touch, and COVID-19: health benefits from non-human touch through times of stress. *Journal of Behavioural and Economic Policy, 4 (COVID-19 Special Issue 2)*, 25–33. https://sabeconomics.org/wordpress/wp-content/uploads/JBEP-4-S2-3.pdf

Zhou, X., Ma, W., & Li, G. (2018). Draft animals, farm machines and sustainable agriculture production: insights from China. *Sustainability, 10*(9), 3015. https://doi.org/10.3390/su10093015

Zito, S., Vankan, D., Bennett, P., Paterson, M., & Phillips, C. J. C. (2015). Cat ownership perception and caretaking explored in an internet survey of people associated with cats. *PLoS One, 10*(7), e0133293. https://doi.org/10.1371/journal.pone.0133293

# 49
# HUNTING AND THE CONSUMPTIVE USE OF WILDLIFE

*Jacob L. Berl and Joshua T. Wise*

## Hunting and the Consumptive Use of Wildlife: A Historical Perspective

### Hunting and Human Evolutionary History

Hunting and the killing of wild animals for consumptive use was fundamental to human evolutionary history. A consistent body of scientific evidence suggests that hunting wild animals (i.e., wildlife) for food was central to the lives and existence of early humans (Peterle, 1977). Archeological evidence shows that humans have been hunting wildlife, including with the use of sophisticated tools, for millennia (Wilkins et al., 2012). In southern Africa, stone points that were likely used by early humans for hunting and self-protection have been discovered that date back at least 500,000 years (Wilkins et al., 2012). Some evidence suggests that true subsistence hunting, or reliance on hunting as a primary means of food procurement, would have been limited until the development complex and efficient hunting tools (Carrier et al., 1984). Indeed, the earliest evidence of sophisticated hunting tools, such as those made by heat-treated stone, does not appear in the archeological record until approximately 100,000 years ago (Brown et al., 2012). However, early humans may have utilized means of hunting other than the killing of animals by hand or tool. For example, persistence hunting (also known as endurance hunting) was likely used by early humans to subsistence hunt without the use of sophisticated tools (Carrier et al., 1984). Early humans developed physiological adaptations, such as evaporative body cooling and bipedal locomotion, that allowed them to out-endure their prey and in such cases, a successful hunt would end when prey succumbed from heat exhaustion (Carrier et al., 1984; Wheeler, 1991; Bramble & Lieberman, 2004; Lieberman, 2011). Persistence hunting by modern-day hunter-gatherer tribes in central Africa provides further evidence that this form of hunting may have been an integral part of human evolution (Liebenberg, 2006).

Ancient paintings and other forms of rock art (e.g., petroglyphs) provide further archeological evidence and insight to the role of hunting and early human life. For example, cave paintings from Indonesia, dating over 40,000 years old, depict scenes of early humans hunting wild pigs and other animals (Aubert et al., 2019). Such paintings provide a visual image of early humans that hunted for food, had deep connection with their food source (prey), and likely believed such bonds were spiritual in nature by immortalized hunting scenes though artwork (Aubert et al., 2019). Similar scenes from ancient paintings have been identified that depict the importance of hunting to the livelihood of early humans throughout evolutionary history (Bellezza, 2002). Recent examples (2,000–4,000 years ago) include painted crystalline rock shelters surrounding celestial lakes in northern Tibet that display scenes of hunting and farming as a part of daily life (Bellezza, 2002). These examples present compelling evidence that humans have relied on hunting as a means of survival (subsistence hunting) throughout evolutionary history, and that hunting was an integral part of early human livelihood and culture.

Although it is clear that early humans lived in concert with their natural environments, likely deeply connected to the wild animals they relied upon for food and sustenance, early humans also overharvested natural resources and were directly responsible for several extinction events. For example, extinctions of some Pleistocene megafauna (large-bodied animals) in North America can be attributed in part to early humans crossing the land bridge from Asia and exploiting virgin ecosystems (Martin, 1973; Cooper et al., 2015; Broughton & Weitzel, 2018). North American megafauna had evolved in isolation from humans, and the sudden introduction of the human hunter likely had direct effects on the extinctions of several species through overharvest (e.g., mammoths [*Mammuthus* spp.], and mastodons [*Mammut* spp.]; (Martin, 1973)). Other examples of pre-industrial revolution extinctions as a result of human overharvest are abundant. Human settlement of New Zealand resulted in the overharvest and extinction of several endemic species, such as moa (Order: Dinornithiformes), that had evolved in the absence of large mammalian predators (Holdaway et al., 2014). Similarly, the island of Madagascar experienced the rapid extinction of much of its megafauna shortly after human colonization approximately 2,000 years ago (Burney et al., 2004). In their recent analysis, Andermann et al., (2020) explored how humans have influenced extinction rates throughout human history. Their predictive model, which included human population size, human land occupation, and other environmental predictors (e.g., global temperature), identified that human population size alone predicted extinctions with 96% accuracy.

## *Hunting and Early Conservation*

In modern, post-industrial revolution history, hunting continued to be a common form of human interaction with wildlife. In the developing United States, hunting and other forms of consumptive use of wildlife (e.g., trapping) were central to livelihoods, culture, and economies during the 18th and 19th centuries (Hickey, 1974). However, this era was also characterized by unregulated harvest and overexploitation of wildlife ('era of overexploitation'), with a little management of natural resources as a whole. Examples of North American species that were overharvested to extinction or near-extinction during this era are abundant. The passenger pigeon (*Ectopistes migratorius*), which was once the most abundant bird on earth, was overhunted to extinction largely due to commercial demand for its meat (Bucher, 1992). The great auk (*Pinguinus impennis*) was similarly hunted to extinction (Thomas et al., 2019). Many early conservation movements in the United States developed in response to this period of clear overexploitation. Primarily during the early 20th century, wildlife policy in the United States shifted from unregulated harvest and overexploitation to conservation and management of wildlife ('era of conservation'; Arnett & Southwick, 2015). Indeed, many pioneers of the early conservation movement in the United States, such as Theodore Roosevelt and Aldo Leopold, were avid hunters and developed their conservation ethic in part because of their experiences while hunting (Reiger, 1975; Trefethen & Corbin,

1975). Efforts of these early conservationists ushered in the new era of wildlife conservation in the United States, of which regulated hunting was a key component.

## Hunting and the Conservation and Management of Wildlife

### Biological Basis for Regulated Hunting

Modern regulated hunting is defined here as the legal harvest of wildlife by human hunters. Presently in the United States, hunting of wildlife is highly regulated by federal, state, and local laws. Hunting is therefore distinguished from the unlawful harvest of wildlife, often referred to as poaching, which is the killing of wildlife in violation of hunting laws or regulations (Eliason, 2008). Despite the seeming dichotomy between conservation and the hunting (killing) of wildlife, there is a biological basis that provides for sustainable harvest over time. Scientifically-based regulated hunting is a pillar of wildlife management in the United States, and melds principles of species biology (e.g., reproductive capacity and behavior), population ecology (e.g., population dynamics and species interactions), and human dimensions (e.g., regulated participation) to develop sustainable harvest regulations (Campbell & Mackay, 2009). These principles are in contrast to pre-industrial revolution hunting practices, which typically occurred in the absence of regulation. Science-based regulations seek to minimize the impacts of what has been coined 'the tragedy of the commons' (Hardin, 1968), whereby shared public resources (wildlife) are overexploited in the absence of common-sense regulation (Finley & Oreskes, 2013). Sustainable harvest of wildlife is therefore achieved when regulations are guided by science. For example, complex population models are used to predict implications of different levels of hunter harvest and strategies. One contemporary example is the use of a maximum sustained yield to regulate harvest for optimal societal and economic gain. In theory, models such as this can help policy makers set harvest limits that both maximize harvest and economic yield, while also sustainably conserving wildlife populations.

### Hunting and Wildlife Conservation Funding

As a key component of the 'North American Model of Wildlife Conservation' (Organ et al., 2002), a disproportionate amount of funding for wildlife conservation in the United States is derived from hunting-related revenue, which allows state and federal agencies to fund wildlife research, habitat management, land acquisition, conservation law enforcement, and other conservation activities (Nie, 2004; Jacobson et al., 2010; Heffelfinger et al., 2013). Indeed, state wildlife conservation agencies are often majority funded by revenue generated by hunting, either through the direct purchase of hunting licenses and permits, or through federal excise taxes on firearms, ammunition, and hunting equipment (i.e., Federal Aid in Wildlife Restoration Act; Smith, 2011; Heffelfinger et al., 2013, Peterson & Nelson, 2017). However, in recent years, a steady decline in the number of people participating in hunting, in part due to an aging population of hunters and low recruitment of new recreationists, has led to funding constraints (Auger et al., 2018). With fewer hunters, revenue for state conservation agencies has progressively declined and has resulted in strained capacity of conservation program implementation (Price et al., 2018). Maintaining funding for wildlife conservation in the United States is therefore dependent on either preserving current hunting-related revenue, or developing alternative funding mechanisms for wildlife conservation (e.g., through general tax dollars), which to date have been limited in implementation (Echols et al., 2019).

In response, many states have initiated targeted programs to attract new hunters and retain current recreationists (Auger et al., 2018). In part to maximize the effectiveness of these recruitment and retention programs, state wildlife managers have sought to identify the social and ethical reasons why people chose to participate or abstain from hunting. Recent evidence suggests that hunter recruitment and retention programs are most effective when designed to encourage individuals to become hunters, instead of simply encouraging people to go hunting (Ryan & Shaw, 2011). Encouraging a person to become a hunter is to encourage them to join a subculture of society that is dedicated to wildlife conservation, a love for the outdoors, and spending time with friends and family. These points should be emphasized, as many non-hunters may associate hunting exclusively with ending a life (Ryan & Shaw, 2011). Furthermore, recruitment efforts have focused on the mentor-mentee relationship that appears requisite for new hunter recruitment. Hunting is an activity with a high initial learning curve. Numerous logistical (e.g., specialized equipment) and knowledge-based (e.g., hunting techniques) barriers can hinder new recruitment (Larson et al., 2014). Indeed, most hunters are introduced at a young age when a family member takes them hunting and most hunters speak to the importance of the bond between themselves and their family members in their decision to continue hunting later in life. Unsurprisingly, having a robust and ongoing social support network is a key retention factor for individual hunters over time (Enck et al., 2000; Martin & Miller, 2009).

## *Hunting as a Wildlife Management Tool*

In addition to the perceived benefit of acquiring food or resources on behalf of the individual hunter, hunting is used widely in the United States for the management of wildlife populations. Hunting is used as a tool to manage wildlife, such as elk (*Cervus canadensis*) and deer (*Odocoileus* spp.), which populations can exceed environmentally or socially acceptable levels in the modern-day absence of natural predators (Ripple & Beschta, 2004; Garrott, 2005; Chitwood et al., 2015). When natural predators are extirpated from an area, prey populations can increase to unsustainable levels and damage ecosystems (e.g., by over-browsing) or create human-wildlife conflicts (e.g., deer-vehicle collisions). In Yellowstone National Park, after the extirpation of large predators (i.e., wolves ([*Canis lupus*]) in the early 20th century, prey populations (elk) nearly doubled (from 5–7 to 8–12 elk per km) in the absence of predation by natural predators and harvest by human hunters (disallowed within the park). As a result, elk over-browsed young trees, particularly aspen (*Populus tremuloides*) in riparian areas, which limited natural forest regeneration and had cascading negative effects on the ecosystem (White & Garrott, 2005). Hunters can also help control nuisance wildlife populations or invasive species that are outcompeting native species for space and resources (Estévez et al., 2015). In Spain, wild boar (*Sus scrofa*) are considered a nuisance species because they are a major contributor to human-wildlife vehicle accidents and agricultural damage. Hunting has been used to regulate their populations. Quirós-Fernández et al., (2017) compared wild boar populations during years when hunting was allowed to years when wild boar hunting was banned. They found that wild boar populations were nearly seven times higher in years that hunting was banned, concluding that in their study area, hunters provide important societal, economic, and ecosystem services by regulating nuisance wildlife populations (Quirós-Fernández et al., 2017).

However, although hunting can be an important and effective tool for managing certain wildlife populations, it is not uniformly a net positive for native ecosystems. As an example, wild boar are an invasive species in the United States and were first introduced for the purposes of recreation hunting and associated consumptive use. However, despite the well-known negative ecological impacts of this invasive species, some hunting groups are resistant to wild boar eradication efforts based on cultural and recreational grounds (Nelson et al., 2005). In this context, the argument that hunting is the sole means to control wildlife populations is often misapplied and used as a blanket pro-hunting argument, even when not applicable (Nelson et al., 2005).

Lastly, because hunters and fishers frequently interact with wildlife, often in a personal and intimate capacity, they provide a unique opportunity to gather usable scientific data on wildlife populations that would otherwise be challenging or expensive to obtain. For example, collection of wings from hunter-harvested sage grouse (*Centrocercus urophasianus*) in the Gunnison basin of Colorado were used to identify that wings from grouse collected in that area were significantly smaller than those from other locations throughout the species' range. Subsequent investigations revealed significant behavioral and genetic differences and resulted in the identification of the Gunnison sage grouse (*Centrocercus minimus*) as a taxonomically distinct species (Young et al., 2000). Similarly, mandatory hunter harvest reporting programs (e.g., migratory bird harvest information program) and participation in mark-recapture studies (e.g., migratory bird band reporting program) are important sources of data and staples of wildlife population monitoring and modeling in the United States (Arnold et al., 2020).

## Recreational Hunting: Participation, Motivations, and Ethics

### Who Hunts and Why?

According to the national survey of hunting, fishing, and wildlife-associated recreation, approximately 11.5 million people participated in hunting in the United States in 2016, or roughly 3.5% of the national population (USFWS, 2016). Of the 11.5 million that participated in hunting, 90% were male, which highlights a glaring gender gap in the hunting community that has remained stable for decades (Gigliotti & Metcalf, 2016). Relative per capita participation in hunting varies by state and region, with as few as 2% of the population participating in hunting in the northeastern United States and as high as 8% in the southeastern United States. Nationally, per capita participation in hunting (i.e., proportion of United States population that hunts) has steadily declined for decades, as the total number of hunters has declined despite substantial overall population growth. Even with relatively low per capita participation in hunting as a function of the total United States population, hunting is a major recreational industry and wields a large economic footprint. In the United States, hunters spend around 26 billion each year on equipment, trip-related expenses, and license/permit fees (Arnett & Southwick, 2015). To put those numbers in context, individual hunters spent, on average, $803 on hunting trip-related expenses and this is comparable to expenditures from other popular forms of outdoor recreation (e.g., camping). Furthermore, hunting is a major economic engine for many small, rural communities in which economies are often centered wholly or partially on outdoor recreation and tourism, including hunting (Arnett & Southwick, 2015).

In addition to declining participation rates, the demographics of current hunters are skewed by older age cohorts, with 60% of hunters over the age of 45 and only 11% under the age of 25 (USFWS, 2016). These demographic trends point to continued declines in hunting participation as older hunters are phased out of the hunting community (Gigliotti & Metcalf, 2016). On average, hunters spent 16 days afield each year; a substantial time and energy investment (USFWS, 2016). This is a testament to hunter avidity, and highlights how hunters are deeply connected to their chosen pastime and are typically not casual participants. The majority (80%) of hunters in the United States pursue big game animals, such as deer, elk, and turkey, which offer a sizeable consumptive return (food) per animal harvested. Fewer hunters pursue small game (squirrel, rabbit, etc.) or migratory birds (ducks, geese, etc.), which is a trend in contrast to prior generations whereby small game (e.g., squirrel, rabbit) where the most commonly pursued game species.

## *Motivations for Hunting and Harvesting Wildlife*

While historically the primary motivation for human participation in hunting was likely the need to acquire food for self-subsistence (Peterle, 1977), throughout human history, the motivations for participating in hunting have evolved. Presently, motivations for hunting are diverse and nuanced. Numerous studies have attempted to identify the motivations for why individuals choose to participate in hunting. Results from these studies suggest that in the United States, participation in hunting has important cultural and traditional value and that hunters feeling strongly about their opportunity to harvest wildlife (Hayslette et al., 2001; Black et al., 2018). Hunters are often introduced to the activity by close family members, typically along a paternal lineage, and their environmental ethic and views toward animals and wildlife are shaped from these early life experiences (Larson et al., 2014). Rarely does someone decide to hunt by happenstance. Rather, hunters are typically introduced into a sub-culture that strongly self-identifies with certain traditions and values inherent to the pastime (e.g., clothing, language, and leisure activities; Larson et al., 2014).

The specific motivations for participation in hunting are varied. For example, surveys of registered hunters from the central United States indicated the primary reason individuals decided to hunt varied based on the type of game that they pursued (e.g., waterfowl, big game [deer and elk], small game [rabbits and quail], or fishing; Hinrichs et al., 2021). However, regardless of type of game pursued, some of the most commonly cited motivations for participating in hunting were (1) acquiring food, such as filling the freezer or knowledge of food source, (2) interaction with nature, such as viewing wildlife and spending time outdoors, (3) challenge, such as harvesting a quality animal and using skills and equipment, and (4) social interaction, such as spending quality time with friends and family (Hinrichs et al., 2021). Similarly replicated studies suggest that, although there are nuanced differences, many of the key findings indicate that motivations for participating in hunting go beyond the simple necessity or desire to acquire food (Beardmore et al., 2011; Gigliotti & Metcalf, 2016).

Spending time outdoors and immersive experiences in nature are well known to have aesthetic and psychological benefits (Hammitt et al., 1990; Jackson et al., 2021). In a time when human interactions with the natural world are rapidly declining (Jackson et al., 2021), hunting has been used as a reminder of societal reliance on the natural environment (Peterson et al., 2011). For example, hunting is often a primary means of outdoor recreation and interaction with nature. Furthermore, with recent heighted public interest in acquiring organic, locally-sourced food (i.e., 'locavore' movement), opportunities for individuals to hunt and harvest their own food are appealing and provide a unique human connection to food sources. This is particularly true for hunters targeting large game (e.g., deer and elk) for which there is a sizeable return of meat for hunters and their families. Although securing food is often a key motivation for participation in hunting, it appears the actual harvest or killing of an animal is relatively unimportant to hunters. For example, Tynon (1997) found that, when comparing different factors most important to elk hunters in Idaho, spending time outdoors, being away from people/others, and simply seeing elk were most important and the actual harvest of an elk was the least important metric in determining the perceived quality or enjoyment of a hunt (Tynon, 1997).

## *Hunting Ethics*

Although at face value hunting can provide numerous benefits to the individual (e.g., interaction with nature and acquiring food) and society (e.g., wildlife management), the practice of killing wild animals remains and will continue to be the subject of ethical controversy (Causey, 1989; Nelson et al., 2005). Primarily, there are three ethical dilemmas that are linked to the hunting of wildlife, including that (1) animals are sentient beings that experience pain and suffering when hunted and killed, (2) hunting harms ecosystems through overharvest, elimination of genetic diversity, and disruption of natural processes, and (3) in modern day, humans are too far removed from nature to fully appreciate the death of the animals that they kill (Nelson et al., 2005). These ethical challenges have been refuted by hunting advocates and scholars by highlighting that humans are not separate from nature and have evolved alongside other species; it is therefore natural that humans are in direct competition with other species and that some suffering to sustain life is inherently inevitable (e.g., both domestic and wild animals are regularly killed for human consumption; Peterle, 1977). An ethical defense to the second challenge includes that regulated hunting can help manage wildlife populations that may inadvertently harm ecosystems, and that instances of overharvest in regulated systems during modern times are indeed rare (Hewitt, 2015). Lastly, the third challenge is refuted with the notion that hunters are often more connected to nature than non-hunters, spend more time outdoors, and regularly participate in the life and death cycle (Shepard, 2011). How harvested animals are utilized makes clear distinction amongst perceptions toward hunting, with relatively high public support for regulated hunting when animals are killed for consumptive use, and little moral rational or support when hunting occurs primarily for obtaining non-edible animal parts (i.e., 'trophy hunting' for hides and horns; Horowitz, 2019).

An additional ethical paradox arises because most hunters claim to love the wildlife that they hunt and pursue. The hunting community often simultaneously expressed notions of love toward the wildlife they kill, revealing a complex narrative (Kelly & Rule, 2013). It begs the ethical question: how can someone claim to love something, and then subsequently kill it? This multifaceted discourse has been regarded by some ethicists to be an ethically irresolvable paradox (e.g., Cohen 2014). Others, like Loftin (1984), claim that the issue may be resolved by remembering that hunters in North America make up a significant portion of conservation funding for wildlife beyond the individual animals they pursue. In this regard, the harvest of individual animals secures the conservation of other wildlife that is not harvested, and benefits the collective ecosystem (Loftin, 1984). Other authors (e.g., Tynon, 1997) resolve the paradox by pointing to the relative unimportance of killing in the general motivation for hunting. Hunters often stress that 'the kill' is rarely considered the highlight of a typical hunt. Further, the modern-day practice of subsistence hunting in some indigenous cultures, or maintaining intimate husbandry of domestic animals that are destined for slaughter, are examples of genuine reasons that people could simultaneously love, appreciate, and kill (Kelly & Rule, 2013). Ultimately, whether or not hunting and the killing of wildlife can be justified or rationalized depends on personal conservation ethics that govern an individual's tolerance toward such activities.

## Discussion Questions

1. Why is it important to have hunting laws and regulations that incorporate scientifically-based management objectives?
2. In addition to harvesting and securing food, what are some other reasons and motivations that individuals choose to participate in hunting?
3. Describe the potential impacts that low hunter recruitment and reduced participation in hunting can have for the conservation and management of wildlife in the United States.
4. What are the ethical challenges that make recreational hunting a controversial practice and ethical paradox?

## References

Andermann, T., Faurby, S., Turvey, S. T., Antonelli, A., & Silvestro, D. (2020). The past and future human impact on mammalian diversity. *Science Advances*, 6(36).

Arnett, E. B., & Southwick, R. (2015). Economic and social benefits of hunting in North America. *International Journal of Environmental Studies*, 72(5), 734–745.

Arnold, T. W., Alisauskas, R. T., & Sedinger, J. S. (2020). A meta-analysis of band reporting probabilities for North American waterfowl. *The Journal of Wildlife Management*, 84(3), 534–541.

Aubert, M., Lebe, R., Oktaviana, A. A., Tang, M., Burhan, B., Jusdi, A.,... & Brumm, A. (2019). Earliest hunting scene in prehistoric art. *Nature*, 576(7787), 442–445.

Auger, D., Leclerc, V., Roult, R., & Adjizian, J. M. (2018). Review of best practices in recruitment and retention of North American hunters and anglers. *Loisir et Société/Society and Leisure*, 41(3), 472–484.

Beardmore, B., Haider, W., Hunt, L. M., & Arlinghaus, R. (2011). The importance of trip context for determining primary angler motivations: are more specialized anglers more catch-oriented than previously believed?. *North American Journal of Fisheries Management*, 31(5), 861–879.

Bellezza, J. V. (2002). Gods, hunting and society animals in the ancient cave paintings of Celestial Lake in Northern Tibet. *East and West*, 52(1/4), 347–396.

Black, K. E., Jensen, W. F., Newman, R., & Boulanger, J. (2018). Motivations and satisfaction of North Dakota deer hunters during a temporal decline in deer populations. *Human-Wildlife Interactions*, 12(3), 427.

Bramble, D. M., & Lieberman, D. E. (2004). Endurance running and the evolution of Homo. *Nature*, 432(7015), 345–352.

Broughton, J. M., & Weitzel, E. M. (2018). Population reconstructions for humans and megafauna suggest mixed causes for North American Pleistocene extinctions. *Nature Communications*, 9(1), 1–12.

Brown, K. S., Marean, C. W., Jacobs, Z., Schoville, B. J., Oestmo, S., Fisher, E. C.,... & Matthews, T. (2012). An early and enduring advanced technology originating 71,000 years ago in South Africa. *Nature*, 491(7425), 590–593.

Bucher, E. H. (1992). The causes of extinction of the passenger pigeon. In *Current ornithology* (pp. 1–36). Springer.

Burney, D. A., Burney, L. P., Godfrey, L. R., Jungers, W. L., Goodman, S. M., Wright, H. T., & Jull, A. T. (2004). A chronology for late prehistoric Madagascar. *Journal of Human Evolution*, 47(1–2), 25–63.

Campbell, M., & Mackay, K. J. (2009). Communicating the role of hunting for wildlife management. *Human Dimensions of Wildlife*, 14(1), 21–36.

Carrier, D. R., Kapoor, A. K., Kimura, T., Nickels, M. K., Satwanti, Scott, E. C., ... & Trinkaus, E. (1984). The energetic paradox of human running and hominid evolution [and comments and reply]. *Current Anthropology*, 25(4), 483–495.

Causey, A. S. (1989). On the morality of hunting. *Environmental Ethics*, *11*(4), 327–343.

Chitwood, M. C., Peterson, M. N., Bondell, H. D., Lashley, M. A., Brown, R. D., & Deperno, C. S. (2015). Perspectives of wildlife conservation professionals on intensive deer management. *Wildlife Society Bulletin*, *39*(4), 751–756.

Cohen, E. (2014). Recreational hunting: Ethics, experiences and commoditization. *Tourism Recreation Research*, *39*(1), 3–17.

Cooper, A., Turney, C., Hughen, K. A., Brook, B. W., McDonald, H. G., & Bradshaw, C. J. (2015). Abrupt warming events drove Late Pleistocene Holarctic megafaunal turnover. *Science*, *349*(6248), 602–606.

Eliason, S. L. (2008). Wildlife crime: Conservation officers' perceptions of elusive poachers. *Deviant Behavior*, *29*(2), 111–128.

Enck, J. W., Decker, D. J., & Brown, T. L. (2000). Status of hunter recruitment and retention in the United States. *Wildlife Society Bulletin*, *28*(4), 817–824.

Estévez, R. A., Anderson, C. B., Pizarro, J. C., & Burgman, M. A. (2015). Clarifying values, risk perceptions, and attitudes to resolve or avoid social conflicts in invasive species management. *Conservation Biology*, *29*(1), 19–30.

Finley, C., & Oreskes, N. (2013). Maximum sustained yield: a policy disguised as science. *ICES Journal of Marine Science*, *70*(2), 245–250.

Gigliotti, L. M., & Metcalf, E. C. (2016). Motivations of female Black Hills deer hunters. *Human Dimensions of Wildlife*, *21*(4), 371–378.

Hammitt, W. E., McDonald, C. D., & Patterson, M. E. (1990). Determinants of multiple satisfaction for deer hunting. *Wildlife Society Bulletin (1973–2006)*, *18*(3), 331–337.

Hardin, G. (1968). The tragedy of the commons: the population problem has no technical solution; it requires a fundamental extension in morality. *Science*, *162*(3859), 1243–1248.

Hayslette, S. E., Armstrong, J. B., & Mirarchi, R. E. (2001). Mourning dove hunting in Alabama: motivations, satisfactions, and sociocultural influences. *Human Dimensions of Wildlife*, *6*(2), 81–95.

Heffelfinger, J. R., Geist, V., & Wishart, W. (2013). The role of hunting in North American wildlife conservation. *International Journal of Environmental Studies*, *70*(3), 399–413.

Hewitt, D. G. (2015). Hunters and the conservation and management of white-tailed deer (Odocoileus virginianus). *International Journal of Environmental Studies*, *72*(5), 839–849.

Hickey, J. J. (1974). Some historical phases in wildlife conservation. *Wildlife Society Bulletin (1973–2006)*, *2*(4), 164–170.

Hinrichs, M. P., Vrtiska, M. P., Pegg, M. A., & Chizinski, C. J. (2021). Motivations to participate in hunting and angling: a comparison among preferred activities and state of residence. *Human Dimensions of Wildlife*, *26*(6), 576–595.

Holdaway, R. N., Allentoft, M. E., Jacomb, C., Oskam, C. L., Beavan, N. R., & Bunce, M. (2014). An extremely low-density human population exterminated New Zealand moa. *Nature Communications*, *5*(1), 1–8.

Horowitz, A. (2019). Trophy hunting: A moral imperative for bans. *Science*, *366*(6464), 435.

Jackson, S. B., Stevenson, K. T., Larson, L. R., Peterson, M. N., & Seekamp, E. (2021). Outdoor activity participation improves adolescents' mental health and well-being during the COVID-19 pandemic. *International Journal of Environmental Research and Public Health*, *18*(5), 2506.

Jacobson, C. A., Organ, J. F., Decker, D. J., Batcheller, G. R., & Carpenter, L. (2010). A conservation institution for the 21st century: implications for state wildlife agencies. *The Journal of wildlife management*, *74*(2), 203–209.

Kelly, J. R., & Rule, S. (2013). The hunt as love and kill: hunter-prey relations in the discourse of contemporary hunting magazines. *Nature and Culture*, *8*(2), 185–204.

Larson, L. R., Stedman, R. C., Decker, D. J., Siemer, W. F., & Baumer, M. S. (2014). Exploring the social habitat for hunting: Toward a comprehensive framework for understanding hunter recruitment and retention. *Human Dimensions of Wildlife*, *19*(2), 105–122.

Liebenberg, L. (2006). Persistence hunting by modern hunter-gatherers. *Current Anthropology*, *47*(6), 1017–1026.

Lieberman, D. E. (2011). Human locomotion and heat loss: an evolutionary perspective. *Comprehensive Physiology*, *5*(1), 99–117.

Loftin, R. W. (1984). The morality of hunting. *Environmental Ethics*, *6*(3), 241–250.

Martin, P. S. (1973). The Discovery of America: The first Americans may have swept the Western Hemisphere and decimated its fauna within 1000 years. *Science*, *179*(4077), 969–974.

Nelson, F., Kerasote, T., King, R. J., Bateson, P., Eves, H. E., Wolf, C. M., & Yaich, S. C. (2005). The ethics of hunting. *Frontiers in Ecology and the Environment, 3*(7), 392–397.

Nie, M. (2004). State wildlife policy and management: the scope and bias of political conflict. *Public Administration Review, 64*(2), 221–233.

Organ, J. F., Geist, V., Mahoney, S. P., Williams, S., Krausman, P. R., Batcheller, G. R., … Rentz, T. (2012). The North American model of wildlife conservation. *The Wildlife Society Technical Review, 12*(4), 1–60.

Peterle, T. J. (1977). Hunters, hunting, anti-hunting. *Wildlife Society Bulletin, 5*(4), 151–161.

Peterson, M. N., & Nelson, M. P. (2017). Why the North American Model of Wildlife Conservation is problematic for modern wildlife management. *Human Dimensions of Wildlife, 22*(1), 43–54.

Peterson, M. N., Hansen, H. P., Peterson, M. J., & Peterson, T. R. (2011). How hunting strengthens social awareness of coupled human-natural systems. *Wildlife Biology in Practice, 6*(2), 127–143.

Price Tack, J. L., McGowan, C. P., Ditchkoff, S. S., Morse, W. C., & Robinson, O. J. (2018). Managing the vanishing North American hunter: a novel framework to address declines in hunters and hunter-generated conservation funds. *Human Dimensions of Wildlife, 23*(6), 515–532.

Quirós-Fernández, F., Marcos, J., Acevedo, P., & Gortázar, C. (2017). Hunters serving the ecosystem: the contribution of recreational hunting to wild boar population control. *European Journal of Wildlife Research, 63*(3), 1–6.

Reiger, J. F. (1975). *American sportsmen and the origins of conservation*. Winchester Press.

Ripple, W. J., & Beschta, R. L. (2004). Wolves and the ecology of fear: can predation risk structure ecosystems?. *BioScience, 54*(8), 755–766.

Ryan, E. L., & Shaw, B. (2011). Improving hunter recruitment and retention. *Human Dimensions of Wildlife, 16*(5), 311–317.

Shepard, P. (2011). *The tender carnivore and the sacred game*. University of Georgia Press.

Smith, C. A. (2011). The role of state wildlife professionals under the public trust doctrine. *The Journal of Wildlife Management, 75*(7), 1539–1543.

Thomas, J. E., Carvalho, G. R., Haile, J., Rawlence, N. J., Martin, M. D., Ho, S. Y.,… & Knapp, M. (2019). Demographic reconstruction from ancient DNA supports rapid extinction of the great auk. *Elife, 8*, 1–35.

Trefethen, J. B., & Corbin, P. (1975). *American crusade for wildlife*. Winchester Press.

Tynon, J. F. (1997). Quality hunting experiences: a qualitative inquiry. *Human Dimensions of Wildlife, 2*(1), 32–46.

Wheeler, P. E. (1991). The thermoregulatory advantages of hominid bipedalism in open equatorial environments: the contribution of increased convective heat loss and cutaneous evaporative cooling. *Journal of Human Evolution, 21*(2), 107–115.

White, P. J., & Garrott, R. A. (2005). Northern Yellowstone elk after wolf restoration. *Wildlife Society Bulletin, 33*(3), 942–955.

Wilkins, J., Schoville, B. J., Brown, K. S., & Chazan, M. (2012). Evidence for early hafted hunting technology. *Science, 338*(6109), 942–946.

Young, J. R., Braun, C. E., Oyler-McCance, S. J., Hupp, J. W., & Quinn, T. W. (2000). A new species of sage-grouse (Phasianidae: Centrocercus) from southwestern Colorado. *The Wilson Bulletin, 112*(4), 445–453.

# 50
# HUMAN-ANIMAL INTERACTIONS WITH NON-DOMESTIC ANIMALS IN CAPTIVE SETTINGS

*Petty Grieve and Marcy Souza*

## Introduction

In this chapter, we will discuss human-animal interactions (HAIs) with non-domestic animals in captive settings including in homes as a pet or in a zoo or aquarium. When considering terminology used for animals discussed in this chapter, we must recognize the variety of terms including exotic pets, pocket pets, wildlife, and zoo animals that are sometimes also used to describe subsets of the same group of animals. For example, rabbits (*Oryctolagus cuniculus domesticus*) and ferrets (*Mustela furo*) are often included in the "exotic" pet category but are actually domesticated. We will focus on animals that are less commonly kept as pets when compared to dogs and cats and non-domestic animals that are kept in zoos or aquariums.

While dogs and cats are considered more "traditional" pets in some societies, many other types of animals are also kept in our homes. The AVMA periodically performs a survey to determine the number of pets kept in the United States, and in 2016, there were approximately 70 million dogs and 74 million cats (AVMA, 2018). The same survey estimated 8.3 million birds, 84.6 million specialty and exotic animals (various species), 57.8 million fish, and 5.3 million reptiles kept as pets. While culture can play a role in pet ownership, studies have shown that owners of non-domestic pets often think of their pets as part of the family, similar to what is often reported for dogs and cats (Guimaraes, 2018; Kieswetter, 2017).

Local or national laws regarding pet ownership of non-domestic animals vary. For example, many states such as Illinois have a complete ban on the ownership of "dangerous animals", while others such as Nevada allow possession of some dangerous animals such as primates or non-native felids (MSU, 2022). Similarly, animals kept in zoological settings may be subject to various inspections to ensure welfare and safety. These regulations vary with location, whether the public is allowed to view animals, and whether a zoo is accredited by a governing body such as the Association of Zoos and Aquariums. At the time of writing, there are 238 AZA-accredited zoos and aquariums globally. More specific information on regulation and accreditation of non-domestic animals in captive settings will be covered later in the chapter.

## "Non-Traditional" Animals in History and Current Times

Throughout history, non-domestic animals have been kept in captivity as pets, to provide labor, to serve in religious or cultural activities, and in zoos. Animals have been kept as pets for thousands of years, and studies of epitaphs have found evidence of insects, birds, and mammals kept as pets in Greco-Roman time (3rd/2nd centuries BC to 4th/5th centuries AD; Bodson, 2000). Dogs were a prominent pet in ancient times, but others included cheetahs, monkeys, grass snakes, and fish (Bodson, 2000). In more recent times, native people of the Amazon River basin commonly keep wildlife as pets (Erikson, 2000), and numerous examples of non-domestic pet ownership and how it can go awry are described in *Forbidden Creatures* (Laufer, 2010). As previously stated, the almost 100 million "exotic" pets in the United States provide further evidence of the ubiquitous nature of keeping less traditional species of animals as pets (AVMA, 2018).

Asian elephants have been kept in captivity for hundreds of years to provide labor for timber harvesting. The practice of capturing and keeping elephants can be dangerous and inhumane (Munster, 2016). But, a bond does exist between the elephant and handler, also known as a mahout; Munster (2016) reports that 'love' is an important part of the relationship. Additionally, captive elephants and their mahouts may play a critical role in ecosystem conservation in southern India by providing transport to wildlife officials in challenging terrain, aiding in anti-poaching efforts, and intervening in human-wildlife conflict (Munster, 2016).

Venomous snake handling began to be incorporated into Appalachian Pentecostal church practices in the early 20th century. There are many theories as to why snake handling began and continues, including direct reference to scripture. Snake handling is now practiced throughout much of Appalachia from West Virginia to Alabama, USA (Tidball and Tourney, 2007). While the number of people injured or killed through snake handling practices is not readily available, the dangerous nature of these animals poses significant risk to those involved.

Lastly, zoological gardens have likely existed for thousands of years and were described in ancient Egyptian writings, paintings, and carvings (Foster, 1999). These early zoos were associated with royalty, and caretakers for animals and plants were employed. While zoo facilities and standards have greatly changed over time, their place in ancient history is well documented. Human interactions with non-domestic animals, in a variety of settings, have been documented for thousands of years and we are now just beginning to examine those relationships.

## The Impact of Human-Animal Interactions on People

### Positive Impacts

People are impacted both positively and negatively by HAIs with non-domestic animals in captive settings. The human-animal relationship or bond (HAR/HAB) is the result of the quantity and quality of shared interactions between a human and an animal (Hosey and Melfi, 2014). Humans interact with a variety of non-domestic animals in many ways including as pets, and in zoos and aquariums, sanctuaries, circuses, and roadside zoos. Caregivers across all of these settings typically enjoy interacting with animals. A study examining the human reptile bond and its' implications for welfare of semi-aquatic turtles found that 65.8% of owners considered their turtle to be a "member of the family", 64.0% of people claimed to talk to their turtle more than five times a week, and 70.2% pet them at least once a week (Guimarães, 2018). While the study did not find that viewing turtles as family members increased their welfare, it did find that turtles were viewed similarly to dogs and cats as members of the family (Guimarães, 2018; Arahori et al., 2017). People can also observe zoo animals through web cameras or other online platforms; future research should examine how these types of interactions affect the human observers' feelings or opinions.

A survey on the HAB in a laboratory setting found that caretakers were drawn to the career by a desire to work with animals, and they found the positive animal interactions to be highly rewarding. One woman even responded that she had maximum job satisfaction and wrote,

> Working with these animals has been the most rewarding thing I have ever done in my entire life. It has caused me to put off personal goals I had thought I would be working on by now, such as raising a family. I still plan to accomplish this goal but feel that the animals need me more right now
>
> (Chang and Hard, 2002)

Zoo keepers often refer to their jobs as passion work or their life's calling instead of a career choice. Zookeepers have historically been paid low wages; although this is slowly changing. Zoo Knoxville has made significant efforts to improve keeper salary by increasing base pay each of the last five years (2016–2021). The AZA routinely performs a compensation survey and states that accredited institutions must provide compensation sufficiently competitive to recruit and retain professional, qualified staff (AZA, 2022). Within the zoo industry, the cultural norm has been that the price of having your dream job is struggling financially while having to work weekends and holidays. While these struggles are negative, the value keepers place on their relationship with the animals under their care outweighs the hardship. Keepers also have a strong sense of responsibility for the direct wellbeing of their animals and a fear that if they weren't there, the animals' care would be substandard. This belief contributes to a sense of specialness and being part of a unique community that is caring for the world's endangered animals.

In addition to being intrinsically motivated by caring for non-domestic animals, zoo-keepers are highly reinforced by the operational and affective benefits of a positive HAR/HAB. Operational benefits for zoo professionals are those that enable better, easier or more efficient management of the animals in their care (Hosey and Melfi, 2012). Affective benefits are the emotional aspects of the bond between a zoo keeper and animal; zoo keepers reported that this bond was emotionally rewarding (Hosey and Melf, 2012). In the author's experience (PG), the results of Hosey and Melfi's study are still relevant today. The primary goal of modern animal training programs is to encourage animals to participate in their own healthcare. This is primarily done through positive reinforcement training where an animal receives a desired consequence for performing a voluntary behavior. It has become best practice for keepers to train their animals for a wide variety of voluntary medical procedures such as injections, venipuncture, nail or hoof trims, mouth inspection, and subcutaneous fluids which are procedures that would otherwise require anesthesia if the behavior was not trained. From the keeper's perspective, the ability to train your animal to voluntarily participate in a healthcare procedure to avoid the use of anesthesia is one of the most rewarding aspects of the job. Similarly, training your animal to accept a voluntary anesthetic injection instead of having to be darted for a medical procedure is also highly rewarding. Training animals for complex voluntary behaviors, such as standing for radiographs (Figure 50.1), would not be possible without a positive HAR/HAB that has been created through a series of more positive than negative interactions overtime.

*Figure 50.1* African elephant at Zoo Knoxville participates in voluntary radiographs. The elephant holds the target pole in her trunk to signal that she is ready for the technicians to approach and take the radiograph. (Photo courtesy of Zoo Knoxville)

## *Negative Impacts*

While there are positive aspects of interacting with non-domestic animals, negative impacts on caretakers or the public can occur including zoonoses, physical trauma, and stress associated with care. Zoonoses are infectious diseases that can be transmitted to humans from animals. Studies estimate that 75% of emerging human infectious diseases are zoonotic, with non-domestic animals commonly acting as a reservoir species (Chomel et al., 2007). While reservoir species are typically unaffected by the zoonotic agent, they carry and shed the organism infecting humans.

Zoonotic infections include bacteria, parasites, fungi, and viruses and human infections have occurred with exotic pet ownership, zoo visitation, and in zoo animal caretakers. This chapter will not cover the long list of zoonoses transmitted by non-domestic animals, but zoonotic infections in humans can range from minor, such as ringworm (caused by a fungus) to life-threatening, such as rabies (Chomel et al., 2007; Souza, 2009; Souza, 2011; Beserra-Santos et al., 2021). Salmonellosis is one of the most common zoonoses transmitted by non-domestic species and has been associated with turtles (Back et al., 2016), Komodo dragons (Freidman et al., 1998), green iguanas (Burnham et al., 1998), water frogs (Hall et al., 2010), and African pygmy hedgehogs (Riley and Chomel, 2005). Education of potential risks and preventive measures are crucial to reduce zoonotic spread from non-domestic animals (Moorhouse et al., 2021; Bender and Shulman, 2004).

Similar to interactions with any animal, bites and scratches are possible with non-domestic animals. There are few studies examining the prevalence of injuries from non-domestic animals, but large felids and non-human primates, whether kept as pets or in zoo settings, can inflict severe damage (Shepherd et al., 2014; Ivory, 2007). Treatment of traumatic injuries from large non-domestic animals may be similar to other major traumas including stabilization, soft tissue and/or orthopedic surgery, wound care, and antibiotics to prevent infection (Shepherd et al., 2014). Reported attacks from dangerous non-domestic animals can spur changes in local, state, or federal regulations regarding ownership of these animals (Shepherd et al., 2014; Nyhus et al., 2003).

Compassion fatigue is well recognized in the veterinary profession and has also been documented in other types of animal care workers (ACWs; Hill et al., 2020). A recent study of ACWs found that almost half suffered from symptoms of depression in the last month and two-thirds had difficulty coping (Marton et al., 2020). While the close relationship between caretakers and the animals they care for can be positive, it can also lead to stress particularly when an animal is sick or dies. Professional support for ACWs is slowly increasing as the broad impact of compassion fatigue is recognized (White et al., 2021).

Research on the impacts of pet ownership has mostly focused on traditional animals such as dogs and cats, but similar bonds could be formed with animals less commonly kept as pets. While pet ownership has been shown to have positive physical and mental health effects in some cases, there are many studies that show either mixed or negative impacts for the owner (Barroso et al., 2021; Scoresby et al., 2021). While pets are often considered part of the family, pet ownership can be particularly stressful when the animal is sick or the owner is financially unable to provide appropriate care. A recent survey indicated that the inability to obtain needed veterinary care negatively impacted the owner's mental and emotional health (AVCC, 2018).

## The Impact of Human-Animal Interactions on Animals

### *Positive Impacts*

Animals' welfare can be elevated by positive HAIs. Martin and Friedman (2013) refer to a "trust bank account" in their article "The Power of Trust". They explain that to build a trusting relationship with an animal, one must have more positive than negative interactions. Deposits are made into the account every time there is a positive interaction or when the animal has the opportunity to make a choice. Inevitably, one will have to make a withdrawal for a critical procedure, and the goal is to have enough trust built up in the account that the relationship will not be damaged beyond repair (Friedman and Martin, 2013).

Some of the earliest studies of HARs and the effect on productivity, as a sign of good welfare, started in the agricultural industry. Studies have shown the quality of skill and approach/attitude a stocksperson has with their livestock decreases fearfulness of humans and increases productivity (Hemsworth, 2003). When you consider successful biological functions as one indicator of welfare, it is apparent that HARs play a role in animal welfare (Claxton, 2011; Mellor et al., 2020).

The AZA defines welfare as an animal's collective physical, mental, and emotional state over a period of time and is measured on a continuum from good to poor. The AZA requires each accredited zoo to have a process in which animal welfare is assessed annually and a method in which welfare concerns are reported and addressed (Silver, 2021). Two common animal welfare models are the 'Five Freedoms' and the 'Five Domains' (Mellor, 2016). The freedoms and domains are similar in that they both have four elements that evaluate the physiologic health of the animal with the fifth element evaluating the mental health of the animal resulting from physiological conditions. The 5 Freedoms are: freedom from hunger and thirst; freedom from discomfort; freedom from pain, injury, and disease; freedom to express normal and natural behavior, and freedom from fear and distress. The 5 Freedoms model is based on mitigating negative effects to the animal, whereas the 5 Domains model has been updated to focus on providing positive opportunities for the animal (Mellor et al., 2020). The most recent update to the 5 Domains model includes an emphasis on HAIs. The current five Domains model includes: *Nutrition, Physical Environment, Health, Behavioral Interactions* (formerly Behavior), and *Mental State* (Mellor, et al., 2020).

Even though Mellor did not include HAI as an element of animal welfare until 2020, studies in the 1990s evaluated the relationship between non-domestic animals and their caretakers. A study on factors influencing reproductive success in small captive exotic felids found a positive correlation with reproductive success and a husbandry style where the keeper was allowed adequate time to develop a positive relationship with the animal (Mellen, 1991). Similarly, a study evaluating the management of fishing cats in North American zoos to enhance breeding found that institutions that used positive reinforcement training had more successful copulations (Fazio et al., 2019). In addition to reproductive success, fecal corticoids have been used to measure stress levels as a means of evaluating welfare. White rhinos that were given a rating of "highly" friendly to keepers had significantly lowered average concentrations of fecal corticoids than those that were rated "low" in friendly to keepers (Carlstead and Brown, 2005).

Advancements in nutrition and veterinary medicine have had a positive impact on animal welfare. A study comparing longevity and senescence in captive and wild mammals found that 84% of the 59 examined species lived longer in zoos than their wild counterparts (Tidiére et al., 2016). The study went on to find that mammals with fast-paced lives, such as carnivores, benefited more from human care than those with slower-paced lives like elephants. Increased longevity is likely due to a decrease in predation risk, competition for resources, and disease for fast-paced living animals; whereas, the slower paced animals already have a lower predation rate and are long living animals. There is still work to be done to improve longer lived animals' life span in human care. For example, advancements in husbandry and facility design have been made in recent years for animals such as elephants; determination of the true impact of these advancements on life expectancy and onset of aging is still many years away (Tidiére et al., 2016).

Animal caretakers in zoos and aquariums also have an important role to play for animal welfare through implementation of best husbandry practices that include animal training and environmental enrichment. Both training and enrichment were developed through behavior analysis, an area of science that studies the relationship between the environment and the individual animal's behavior. Animal caretakers can arrange environmental antecedents to encourage animals to use their behavior to gain certain outcomes or consequences (Fernandez and Martin, 2021; Alligood et al., 2018). Both training and enrichment programs are goal driven and science based. AZA-accredited zoos' training programs are based on operant conditioning with a focus on positive reinforcement (AZA, 2022). Environmental enrichment programs are often based on a model referred to as S.P.I.D.E.R., which stands for goal **S**etting, **P**lanning and approval process, **I**mplementation, **D**ocumentation, **E**valuation, and **R**eassessment. This same principle can also be used to develop a training program.

When considering how to modify the environment to elicit an animal to change its behavior, it is important to consider ethical standards in addition to efficacy. Dr. Susan Friedman

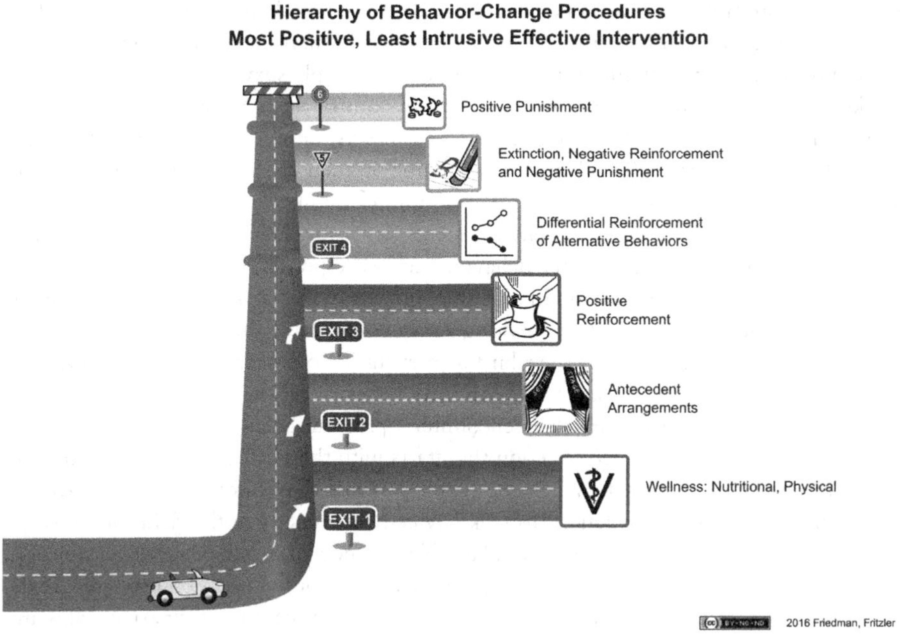

*Figure 50.2* Suggested hierarchy of behavior change procedures according to the least intrusive, effective intervention principle. (Reprinted with permission from Friedman, 2018)

has developed the "Hierarchy of Behavior Change Procedures" in which operant training procedures are arranged from the least intrusive to the most intrusive intervention (Friedman, 2008, 2020). Detailed examples at each level are described in: "What's Wrong with This Picture: Effectiveness Is Not Enough" (Friedman, 2018) (Figure 50.2).

## Negative Impacts

Just as animal welfare can be elevated by HAIs, it can be degraded by them. When considering how to manage a particular species in a captive setting, one needs to fully understand that animal's biology. In a survey of veterinarians in private practice in Ireland, their major concern with exotic pet ownership is the owners' lack of knowledge, as well as their own knowledge and having limited resources (Goins and Hanlon, 2021). Knowledge of a species' natural history is necessary to provide proper nutrition, social structure, and an appropriate environment. Malnutrition is a common problem with exotic pets and occurs when owners do not know the intricacies of how, when, and what their pet eats in the wild. For example, carnivores commonly have metabolic bone disease when fed a diet primarily of skeletal meat and/or neonatal prey such as day-old chicks. Carnivores need to consume whole, well-nourished prey animals, including the bones and organs in addition to the skeletal meat to have proper levels of calcium and vitamin D (Donoghue and Langenberg, 1994).

Social structure also plays an important role in an animal's welfare. Animals should be housed in a social setting that mimics their natural history. However, an individual's history must also be considered because some animal's individual history has impacted the animal in such a way that it may prevent the animal from living with conspecifics. In the author's experience (PG), animals housed in appropriate social groupings similar to their wild counterparts spend more time exhibiting normal desirable behaviors than animals not housed in appropriate social groups. When animals are in captivity, humans have the responsibility to house them in such a way that allows them to thrive, not simply survive.

It makes sense to consider that behavior evolved to control the environment in ways that result in animals' surviving and thriving. As Friedman says, animals are "built to behave", meaning in the simplest terms that all animals are acting on their environment to either obtain something or get away from something (personal communication). From this perspective, maximizing animals' control, often through environments rich with choices, is essential to animal welfare in zoos. Forceful tactics to alter behavior such as response blocking should be rare. Response blocking, also known as flooding, is the procedure of reducing unwanted behavior by physically preventing an animal from making a response (Friedman, 2002). An example of this is when a parrot owner restrains their bird to prevent it from biting them. The owner is focusing only on the undesired behavior of biting and not on what they want the parrot to be doing instead. When owners or caretakers encounter a problem behavior, they have an ethical responsibility to choose a course of action that starts with the most positive, least intrusive method to achieve the goal (Friedman, 2008, 2020). When trying to formulate a plan to decrease unwanted behavior, it's beneficial to seek advice from experts in the fields of ethology and applied behavior analysis. An expert would be able to guide an owner or caregiver through a functional assessment of the behavior and help design a behavior intervention program. Texts addressing behavior issues are available and can serve as a resource for veterinarians who are typically the first point of contact for owners (Friedman et al., 2006; Friedman and Haug, 2010).

## Ethical Considerations of Non-Domestic Animals in Captivity

Ethics of keeping non-domestic animals in captive settings is a divisive topic. When considering whether or not it is ethical to keep non-domestic animals as pets or in zoos, one must recognize that there are differing levels in the quality of care. Zoos that are accredited by the AZA are all similar in their missions to save animals from extinction. They work toward this goal through a variety of avenues including conservation education, research, recovery, and reintroduction programs, habitat restoration, and development of relevant technologies (Hutchins et al., 2003). In 2020, AZA member institutions contributed $208.8 million to field conservation and $20.5 million to research (AZA, 2020). Numerous species, such as the American red wolf, California condor, and black-footed ferret, would have gone extinct without the aid of accredited zoos. AZA zoos play an important role in wildlife conservation, and they are equally as focused on providing the highest standard of care and welfare to their individual animals. While non-accredited zoos typically still must comply with local or federal regulations, they have less consistency in care, welfare, and mission when compared to AZA-accredited zoos.

Exotic pet owners are rarely required to follow regulatory agency guidelines for husbandry or welfare. These animals can fall victim to owners that have good intentions but don't have the knowledge or experience to provide adequate care. Exotic animals have complex needs and can suffer malnutrition, stress from isolation or improper social grouping, and inadequate environments without the opportunity to perform natural behaviors (McLennan, 2012). The "difficulty" of care certainly varies by species, making some non-domestic animals more difficult to keep in captivity, especially with inexperienced owners.

Harvesting animals from the wild to live in a captive setting is an even more controversial subject. A recent study estimated that greater than 3.2 billion live animals were imported into the United States from 2000 to 2014 (Eskew et al., 2020); these animals end up in a variety of captive settings. Collecting animals from the wild for either pet ownership or zoos can negatively impact wild populations. For example, the increasing popularity and harvesting of ornamental fish threatens wild fisheries in some parts of the world (Livengood and Chapman, 2020). Reptiles have also historically been harvested from the wild and traded, but over the last few decades captive breeding of many species has increased and imports of animals into the United States have decreased (Collis and Fenili, 2011). Released, unwanted exotic pets, such as lion fish and Burmese pythons, can also negatively impact native wildlife (NOAA, 2022; USGS, 2022). Captive breeding and sustainable collection practices must be employed to decrease the impact of non-domestic animal ownership on wild populations.

In 2016, three AZA zoos imported 17 African elephants from Swaziland (AZA, 2016). The two fenced areas where elephants had been reintroduced in Swaziland in 1987 and 1994 could not handle the growing herds; the elephants had decimated most of the large trees, which required them to eat shrubs and grasses. If left unchecked, the elephants could destroy habitat for other animals, namely rhinos (USFWS, 2016). Habitat degradation was the same reason cited by Swaziland when they exported 11 African elephants to two AZA zoos in 2003. Swaziland had planned to cull the elephants to ensure suitable rhino habitat and exportation was seen as a way to reduce habitat degradation and save the elephant lives, both contributing to conservation. However, some conservation organizations such the Amboseli Trust, stated that there was no evidence of rhinos sharing the space with elephants in Swaziland.

Conservation organizations speculate that the 2003 shipment reinforced Swaziland's outdated conservation efforts, which included culling and exporting, and led to the 2016 export to the United States. The Amboseli Trust advocates for water point management, corridor creation, and translocation as best practices for conservation over culling (Amboseli, 2015). Each institution, whether it be the AZA or the Amboseli Trust, selects which facts to put in their press releases in order to justify their goals and actions. If you were not directly involved, it is difficult to know the entire truth of the situation.

## Legislation Regarding Non-Domestic Animals in Captivity

Regulations for keeping wildlife in captivity vary by country, territory or state, and federal legislation, such as the US Animal Welfare Act, often focuses on minimum standards of care (USDA, 2021). Accreditation of zoos and aquariums by organizations such as the Association of Zoos and Aquariums (AZA) provides assurance that the accredited institutions adhere to certain guidelines and standards when caring for captive species and contributing to conservation. These guidelines typically exceed legislation and address more than minimum standards of care for animals in captivity. For example, the USDA stipulates the type of enclosure and amount of space required for a particular sized chimpanzee (USDA, 2019), but the AZA standards also include housing specifications with consideration for locomotion of the animal, foraging opportunities, and overall exhibit design (AZA, 2010). While not necessarily required to operate, many zoos globally seek accreditation as a standard of excellence. Accreditation standards from the AZA and European Association of Zoos and Aquaria are available online (AZA, 2022; EAZA, 2021a). These accrediting organizations also work with national or local governments to impact legislation (EAZA, 2021b).

Pet ownership of non-domestic animals is generally less regulated than animals in agriculture or those kept in zoo or aquarium settings (USDA, 2021; Broom et al., 2017), and rules vary greatly usually on a local basis. For example, in the United States, the legality to own different types of Class I non-domestic animals varies by state (Born Free, 2016). Some states may have complete bans on ownership of certain species while others may allow ownership following a permitting process. Other less-dangerous animals may have restrictions at a more local level. For example, in the authors' home city, ownership of large constrictor snakes requires a permit (City of Knoxville, 2021). However, even if legislation exists regarding ownership of non-domestic animals as pets, enforcement of the rules may be lacking, or absent, rendering them useless. Owners of non-domestic animals should seek advice from veterinarians and other reputable sources to ensure their pets are provided appropriate care and enrichment, while also protecting the health and safety of the humans that come in contact with the animal.

## Practical Perspectives – Negative Impacts of Captivity on Non-Domestic Animals

The care of non-domestic animals can be challenging in captivity, especially when the animal is long-lived and the complexity of their natural environment cannot be duplicated. Abnormal behavior in the animal may result which can be self-destructive at times and may lead owners to neglect or even abandon pets. Two examples of animals with behavioral issues are provided to demonstrate abnormalities and some steps to address them.

### Example 1 – The Bald Parrot

In a survey conducted by the AVMA (2018), approximately 7.5 million birds are kept as pets in the United States. This number does not include poultry but does include many parrots, such as cockatoos, macaws, and African grays. Parrots are interactive and talkative, leading many potential owners to believe that they will make a great pet. However, due to their

*Figure 50.3* Feather picking or self-destructive behavior can occur in captive parrots as shown in this African gray parrot (*Psittacus erithacus*). (Photo courtesy of Dr. Cheryl Greenacre)

Non-Domestic Animals in Captive Settings

inherently social nature, pet owners can rarely provide the emotional stimulation and support in captivity needed by these highly intelligent birds. Parrots may also bond with their owner as they would with a mate of the same species leading to frustration due to an inability to reproduce. Feather picking or feather destructive behavior, as shown in this African gray (*Psittacus erithacus*), often results when a parrot's social and emotional needs are not met (Figure 50.3). While any parrot with feather destructive behavior needs to be examined for underlying medical problems, behavioral issues are often the cause. Medical and behavioral steps, as well as environmental enrichment, can be taken to address the feather destructive behavior but are not always successful resolving the problem. Additionally, even if the destructive behavior is discontinued, the feather follicles may be permanently damaged, and feathers will not regrow.

### Example 2 – Elephant with Self-Injurious Behavior

Edie is an African elephant (*Loxodonta africana*) at Zoo Knoxville that the author (PG) works with directly. Edie began a self-injurious behavior (SIB) in 2014 that has been labeled "tusking"; Edie lifts one of her front legs and places it onto a stationary object and presses her tusk into her carpus and/or toenails forcefully causing wounds. In 2015, the author began working with Dr. Susan Friedman, an Applied Behavior Analyst, on a behavior intervention program for Edie. The process started at the bottom of the Hierarchy of Behavior Change Procedures (Figure 50.1) with a veterinary exam and analgesics to see if the underlying cause was pain in her feet. During the course of analgesic treatment, the elephant team performed a functional assessment of the tusking behavior; there are five situations that Edie is likely to perform the tusking behavior. In each situation, Edie used the SIB for different functions such as access to a different location, food, enrichment, ending a training session, and seeking attention from keepers. When analgesics did not affect the SIB, fluoxetine was started to allow time for the staff to begin implementing the behavior intervention (Figure 50.4).

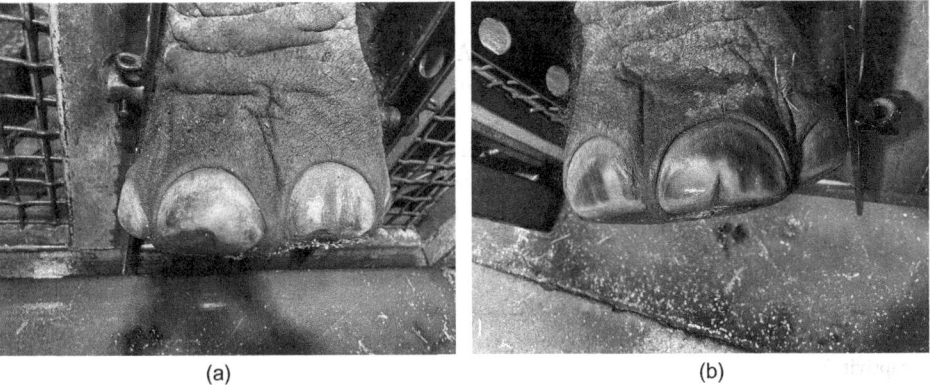

*Figure 50.4* (a) Following two years of a behavioral intervention program, nail damage due to self-inflicted behavior in an African elephant is improved. (Photos courtesy of Zoo Knoxville). (b) Following six years of a behavioral intervention program, nail damage due to self-inflicted behavior in an African elephant is resolved. (Photos courtesy of Zoo Knoxville)

The elephant team addressed the problem situations one at a time. After numerous environmental changes throughout 2016 and 2017, Edie's SIB was reduced to near zero and has been maintained at near zero for the last four years. In addition to SIB frequency, the team monitors Edie's foot health as an indicator of the behavior intervention program's success. This level of success has only been achievable because the elephant team works proactively to ensure Edie's environment is meeting her needs on a daily basis. It was critical for the team to understand that Edie's behavior was a function of her environment, and they had the power to change her behavior by changing her environment. We know her environment meets her needs when her time is spent exhibiting natural behaviors and the frequency of her SIB is near zero.

## Unanswered Questions and Future Research Directions

While humans and non-domestic animals have interacted for thousands of years, there is still limited research examining this relationship leading to many questions for research and discussion (Table 50.1).

Organizations are beginning to explore some of these questions including: Does seeing exotic animals in captivity increase peoples' concern for those same animals in the wild? Is this a way to help global conservation efforts? Measuring the impact of zoo and aquarium visits on the conservation efforts of zoo visitors is important because conservation is a shared mission across AZA accredited zoos and aquariums. Godinez and Fernandez (2019) reviewed various types of studies that assessed the impact a zoo visit had on the general public. In this literature review, the general public was separated into two categories: people that visit zoos and those that do not. Studies found several factors that influence zoo visitors to act for conservation. Having the ability to participate in conservation action while at the zoo, observing species appropriate behavior from active animals, and connecting with zoo staff at interactive stations had a positive influence on visitor conservation efforts post visit. In addition, people who visit the same zoo more than once were more likely to participate in conservation action. Most research has been conducted with zoo visitors, and more research is needed with people that do not visit zoos to better provide an analysis of the effects zoos and aquariums have on influencing the general public to act to protect wildlife (Godinez and Fernandez, 2019).

The Why Zoos and Aquariums Matter (WZAM) initiative was launched in 2001 (WZAM, 2022). Since its inception, WZAM has coordinated three separate multi-year research projects focused on understanding how zoos and aquariums contribute to American society. To date, the research projects have included: "What Are the Impacts of Zoos & Aquariums?", "The Human Dimensions of Wildlife Conservation", and the current project "STEM Matters: Investigating the Confluence of Visitor and Institutional Agendas" (WZAM, 2022).

*Table 50.1* Potential future questions for research and discussion

| | |
|---|---|
| 1 | What species do we need to have additional efforts to ensure appropriate care and welfare in captivity? |
| 2 | Are there certain animals that should be banned from being kept in captivity? In which settings? |
| 3 | How does the HAB with non-domestic pets differ from the HAB with more common pets such as dogs or cats? Or is it essentially the same? |
| 4 | Should non-AZA accredited zoos and aquariums be required to have permits to breed animals? |

*Table 50.2* Social science research questions that need to be explored as determined by the Association of Zoos and Aquariums (Kubarek et al., 2020).

| | |
|---|---|
| 1 | How can zoos and aquariums help build a more equitable society through critical reflection on their internal operations, culture, and communications? How can zoo and aquarium diversity, equity, access and inclusion (DEAI) efforts support this? |
| 2 | What is the role of zoos and aquariums in communities, including in the context of striving for environmental and social justice? |
| 3 | What is the role of zoos and aquariums in contributing to social change toward conservation? |
| 4 | What is the role of zoos and aquariums in contributing to the development of a person's intellectual, social, emotional, and physical well-being? |
| 5 | How can zoos/aquariums maximize their systemic impact on conservation? |

The inaugural project, "What Are the Impacts of Zoos and Aquariums" found that adult visitors' mindsets and perceptions on conservation were impacted in a measurable way when they went to an AZA-accredited zoo or aquarium in North America. This study evaluated the visitor's impression of what their role in conservation efforts are and what they think about protecting the environment (Falk et al., 2007). From this information, a visitor evaluation toolbox was created to enable member institutions to learn what personal context visitors bring to their zoo/aquarium, how visitors spend their time and why, and what impact the visit had on their conservation-related attitudes (AZA, 2009). With this information, zoos and aquariums can make better decisions on future habitat design, conservation messaging, and educational programs to engage visitors and increase likelihood of acting for conservation (Falk et al., 2007). The AZA has also identified numerous social science questions that focus on the interactions of zoos and human communities (Table 50.2; Kubarek et al., 2020).

## Conclusion

Non-domestic animals have been kept in captivity for thousands of years with varying quality of care, welfare, and bonds with their owners. While the care and welfare of these animals has generally improved over time, there is certainly still room for improvement. Bonds do exist with these animals, and many people report similar bonds as those described with more commonly kept animals, such as dogs and cats. Future research should examine methods to improve the bond for both animal and human while also contributing to welfare and conservation of these non-domestic animals.

### Potential Questions for Discussion in a Class Setting

1. Do you think keeping non-domestic animals as pets or in zoos is ethical?
2. What would you consider an exotic pet?
3. Describe difficulties that may occur when keeping non-domestic animals in captivity.
4. Describe potential benefits that may occur following interactions with non-domestic animals in captivity.

## References

Access to Veterinary Care Coalition (AAVC) (2018). *Access to veterinary care barriers, current practices, and public policy*. Retrieved from https://pphe.utk.edu/wp-content/uploads/2020/09/avcc-report.pdf.

Alligood, C., Skurski, M., Miller, A. (2018). Building a science-based behavioral husbandry program. *The Science of Behavioral Husbandry 45*(4), 107–110. Retrieved from https://static1.squarespace.com/static/57c607322e69cf3a289944bb/t/5acf4bde575d1f0b6edc1cf4/1523534814670/Building+a+-Science-based+Behavioral+Husbandry+Program.pdf.

Amboseli (Amboseli Trust for Elephants) (2015). *Statement on Swaziland*. Retrieved from http://www.elephanttrust.org/index.php/articles/item/statement-on-swaziland.

Arahori, M., Kuroshima, H., Hori, Y., Tagachi, S., Chijiwa, H., Fujita, K. (2017). Owners' view of their pets' emotions, intellect, and mutual relationship: cats and dogs compared. *Behavioural Processes 141*(3), 316–321.

AVMA (2018). *AVMA pet ownership and demographics sourcebook 2017–2018 edition*. Retrieved from https://www.avma.org/sites/default/files/resources/AVMA-Pet-Demographics-Executive-Summary.pdf.

AZA (2009). *Visitor evaluation toolbox*. Retrieved from https://wzam.org/wp-content/uploads/ILI.AZA_.2007_Why-Zoos-and-Aquariums-Matter-Visitor-Evaluation-Toolbox.pdf.

AZA (2010). *Chimpanzee care manual*. Retrieved from https://assets.speakcdn.com/assets/2332/chimpanzeecaremanual2010r.pdf.

AZA (2016). *Statement from AZA regarding the USFWS approval of an African elephant import permit*. Retrieved from https://www.aza.org/aza-news-releases/posts/statement-from-aza-regarding-the-usfws-approval-of-an-african-elephant-importpermit?locale=enhttps://www.fws.gov/international/pdf/questions-and-answers-swaziland-elephant-import.pdf.

AZA (2020). *2020 Annual report on conservation and science*. Retrieved from https://assets.speakcdn.com/assets/2332/aza_arcshighlights_2020_final.pdf.

AZA (2022). *Association of zoos and aquariums accreditation standards*. Retrieved from https://assets.speakcdn.com/assets/2332/aza-accreditation-standards.pdf.

Back, D.S., Shin, G.W., Wendt, M., Heo, G.J. (2016). Prevalence of *Salmonella spp.* in pet turtles and their environment. *Laboratory Animal Research 32*(3), 166–170. https://doi.org/10.5625/lar.2016.32.3.166

Barroso, C., Brown, K., Laubach, D., Souza, M., Daugherty, L., Dixson, M. (2021). Cat and/or dog ownership, cardiovascular disease, and obesity: a systematic review. *Veterinary Science 8*(12), 333. https://doi.org/10.3390/vetsci8120333

Bender, J.B., Shulman, S.A. (2004). Reports of zoonotic disease outbreaks associated with animal exhibits and availability of recommendations for preventing zoonotic disease transmission from animals to people in such settings. *Journal of the American Veterinary Medical Association 224*(7), 1105–1109. https://doi.org/10.2460/javma.2004.224.1105

Beserra-Santos, M.A., Mendoza-Roldan, J.A., Thompson, R.C.A., Dantas-Torres, F., Otranto, D. (2021). Illegal wildlife trade: a gateway to zoonotic infectious diseases. *Trends in parasitology 37*(3), 181–184. https://doi.org/10.1016/j.pt.2020.12.005

Bodson, L. (2000). Motivations for pet-keeping in Ancient Greece and Rome: a preliminary survey. In Podbersceke, A.L., Paul, E.S., Serpell, J.A. (Eds.), *Companion animals and us: exploring the relationships between people and pets* (pp. 27–41). Cambridge: Cambridge University Press.

Born Free (2016). *Summary of state laws relating to private possession of exotic animals*. Retrieved from https://www.bornfreeusa.org/campaigns/animals-in-captivity/summary-state-laws-exotic-animals/.

Broom, D.M. (2017). *Animal Welfare in the European Union*. Retrieved from https://www.researchgate.net/profile/Donald-Broom/publication/315721435_Animal_Welfare_in_the_European_Union/links/58dea012aca272059aaac6f2/Animal-Welfare-in-the-European-Union.pdf.

Burnham, B.R., Atchley, D.H., DeFusco, R.P., Ferris, K.E., Zicarelli, J.C., Lee, J.H., Angulo, F.J. (1998). Prevalence of fecal shedding of Salmonella organisms among captive green iguanas and potential public health implications. *Journal of the American Veterinary Medical Association 213*(1), 48–50.

Carlstead, K., Brown, J.L. (2005). Relationship between patterns of fecal corticoid excretion and behavior, reproduction, and environmental factors in captive black (*Diceros bicornis*) and white (*Ceratotherium simum*) rhinoceros, *Zoo Biology 24*, 215–232.

Chang, F.T., Hard, L.A. (2002). Human-animal bonds in the laboratory: How animal behavior affects the perspective of caregivers. *ILAR Journal 43*(1), 10–18.

Chomel, B.B., Belotto, A., Meslin, F.X. (2007). Wildlife, exotic pets, and emerging zoonoses. *Emerging Infectious Diseases 13*(1), 6–11.

City of Knoxville. (2021). *Knoxville Municipal Code Section 5–30*. Retrieved from https://library.municode.com/tn/knoxville/codes/code_of_ordinances?nodeId=PTIICOOR_CH5AN#fn_39%23TOPTITLE.

Claxton, A.M. (2011). The potential of the human-animal relationship as an environmental enrichment for the welfare of zoo-housed animals. *Applied Animal Behavior Science 133*(1––2); 1–10.

Collis, A.H., Fenili, R.N. (2011). *The Modern U.S. Reptile Industry*. Retrieved from https://reptifiles.com/wp-content/uploads/2020/01/The_Modern_US_Reptile_Industry_05_12_2011Final.pdf

Donoghue, S., Langenberg, J. (1994). Clinical nutrition of exotic pets. *Australian Veterinary Journal 71*, 337–341.

EAZA. (2021a). *European Association of Zoos and Aquaria Membership and Accreditation Manual*. Retrieved from https://www.eaza.net/assets/Uploads/Governing-documents/EAZA-Membership-and-Accreditation-Manual-2021.pdf.

EAZA. (2021b). *European Association of Zoos and Aquaria Policy and Legislation*. Retrieved from https://www.eaza.net/about-us/areas-of-activity/policyandlegislation/.

Erikson, P. (2000). The social significance of pet-keeping among Amazonian Indians. In Podberscke, A.L., Paul, E.S., Serpell, J.A. (Eds.), *Companion Animals and Us: Exploring the Relationships Between People and Pets* (pp. 7–26). Cambridge: Cambridge University Press.

Eskew, E.A., White, A.M., Ross, N. et al. (2020). United States wildlife and wildlife product imports from 2000–2014. *Scientific Data 7*, 22. https://doi.org/10.1038/s41597-020-0354-5

Falk, J.K., Reinhard, E.M., Vernon, C.L., Bronnenkant, K., Heimlich, J.E., Deans, N.L. (2007). *Why zoos & aquariums matter: assessing the impact of a visit to a zoo or aquarium*. Retrieved from https://www.researchgate.net/profile/Cynthia-Vernon/publication/253004933_Why_Zoos_Aquariums_Matter_Assessing_the_Impact_of_a_Visit_to_a_Zoo_or_Aquarium/links/5705627a08ae44d70ee342a8/Why-Zoos-Aquariums-Matter-Assessing-the-Impact-of-a-Visit-to-a-Zoo-or-Aquarium.pdf.

Fazio, J.M., Freeman, E.W., Bauer, E., Rockwood, L., Parsons, E.C.M. (2019). Evaluation of management in North American zoos to enhance success of the fishing cat (*Prionailurus viverrinus*) ex situ population. *Zoo Biology 38*, 189–199.

Fernandez, E.J., Martin, A.L. (2021). Animal training, environmental, and animal welfare: a history of behavior analysis in zoos. *Journal of Zoological and Botanical Gardens 2*, 531–543.

Foster, K.P. (1999). The earliest zoos and gardens. *Scientific American 281*(1), 64–71.

Friedman, C.R., Torigian, C., Shillman, P.J., Hoffman, R.E., Heltzel, D., Beebe, J.L., Malcolm, G., DeWitt, W.E., Hutwanger, L., Griffin, P.M. (1998). An outbreak of salmonellosis among children attending a reptile exhibit as a zoo. *Journal of Pediatrics 132*(5), 802–807. https://doi.org/10.1016/s0022-3476(98)70307-5

Friedman, S.G. (2002). *Alternatives to breaking parrots: Reducing aggression and fear through learning*. Retrieved from https://www.behaviorworks.org/files/articles/Alternatives%20to%20Parrot%20Breaking%202002.pdf.

Friedman, S.G. (2018). *What's wrong with this picture? Effectiveness is not enough*. Retrieved from https://www.behaviorworks.org/files/articles/What%27s%20Wrong%20With%20this%20Picture-General.pdf.

Friedman, S.G. (2020). *Why animals need trainers who adhere to the least intrusive principle: improving animal welfare and honing trainers' skills*. Retrieved from https://www.behaviorworks.org/files/articles/Why%20Animals%20Need%20Trainers%20Who%20Adhere%20to%20a%20Procedural%20Hierarchy.pdf.

Friedman, S.G., Edling, T., Cheney, C.D. (2006). The natural science of behavior. In G.J. Harrison, T.L. Lightfoot (Eds.), *Clinical Avian Medicine* (pp. 46–59). Palm Beach, FL: Spix Publishing.

Friedman, S.G., Haug, L. (2010). From parrots to pigs to pythons: Principles and procedures of learning. In V.V. Tynes (Ed.), *Behavior of Exotic Pets* (pp. 191–205). Ames, IA: Wiley-Blackwell.

Godinez, A.M., Fernandez, E.J. (2019). What is the zoo experience? How zoos impact a visitor's behaviors, perceptions, and conservation efforts. *Frontiers in Psychology* 10:1746. https://doi.org/10.3389/fpsyg.2019.01746

Goins, M., Hanlon, A.J. (2021). Exotic pets in Ireland: 2. Provision of veterinary services and perspectives of veterinary professionals' on responsible ownership. *Irish Veterinary Journal* 74, 13. https://doi.org/10.1186/s13620-021-00191-5

Guimaraes, M.L.L. (2018). *The human-reptile bond and its implications for the welfare of captive semiaquatic turtles in Portugal*. Retrieved from https://comum.rcaap.pt/handle/10400.26/24505.

Hall, J., Poulson, M., Fawcett, L., et al. (2010). Multistate outbreak of human Salmonella Typhimurium infections associated with aquatic frogs – United States, 2009. *Morbidity and Mortality Weekly Report* 58(51&52); 1433–1436.

Hemsworth, P.H. (2003). Human-animal interactions in livestock production. *Applied Animal Behaviour Science* 83(3), 185–198.

Hill, E.M., LaLonde, C.V.M., Reese, L.A. (2020). Compassion fatigue in animal care workers. *Traumatology* 26(1), 96–108. https://doi.org/10.1037/trm0000218

Hosey, G., Melfi, V. (2012). Human-animal bonds between zoo professionals and the animals in their care: Human-animal bonds in zoos. *Zoo Biology* 31(1), 13–26. https://doi.org/10.1002/zoo.20359

Hosey, G., Melfi, V. (2014). Human-animal interactions, relationships and bonds: a review and analysis of the literature. *International Journal of Comparative Psychology* 27(1), 13–26.

Hutchins, M. Smith, B., Allard, R. (2003). In defense of zoos and aquariums: the ethical basis for keeping wild animals in captivity. *Journal of the American Veterinary Medical Association* 223(7), 958–962.

Ivory, A.S. (2007). *Animal Legal and Historical Center: Incidents and attacks involving captive chimpanzees*. Retrieved from https://www.animallaw.info/article/incidents-attacks-involving-captive-chimpanzees.

Kieswetter, S. (2017). *The motivations behind obtaining exotic pets: a discussion paper*. Retrieved from https://www.zoocheck.com/wp-content/uploads/2017/10/The-Motivations-Behind-Obtaining-Exotic-Pets_September-2017.pdf.

Kubarek, J., Ogden, J., Grow, S., Rutherford, A. (2020). *AZA Social Science Research Agenda 2020*. Retrieved from https://assets.speakcdn.com/assets/2332/aza_2020_socsciresearchagenda.pdf.

Laufer, P. (2010). *Forbidden creatures*. Connecticut, Lyons Press.

Livengood, E.J., Chapman, F.A. (2020). *The ornamental fish trade: an introduction with perspectives for responsible aquarium fish ownership*. Retrieved from https://edis.ifas.ufl.edu/publication/FA124

Martin, S., Friedman, S.G. (2013). *The power of trust*. Retrieved from https://www.behaviorworks.org/files/articles/The%20Power%20of%20Trust.pdf.

Marton, B., Kilbane, T., Nelson-Becker, H. (2020). Exploring the loss and disenfranchised grief of animal care workers. *Death Studies* 44, 31–41. https://doi.org/10.1080/07481187.2018.1519610

McLennan, S. (2012). *Keeping of exotic animals: welfare concerns*. Retrieved from https://fve.org/cms/wp-content/uploads/Staci-McLennan.pdf.

Mellen, J.D. (1991). Factors influencing reproductive success in small captive exotic felids (Felis spp.): A multiple regression analysis. *Zoo Biology* 10(2), 95–110. https://doi.org/10.1002/zoo.1430100202

Mellor, D.J. (2016). Updating animal welfare thinking: moving beyond the "Five Freedoms" towards "A Life Worth Living". *Animals* 6(3), 21. https://doi.org/10.3390/ani6030021

Mellor, D.J., Beausoleil, N.J., Littlewood, K.E., McLean, A.N., McGreevy, P.D., Jones, B., Wilkins, C. (2020). The 2020 five domains model: Including human-animal interactions in assessments of animal welfare. *Animals* 10(10), 1870. https://doi.org/10.3390/ani10101870

Michigan State University (MSU) (2022). *Animal Legal & Historical Center*. Retrieved from https://www.animallaw.info/content/map-private-exotic-pet-ownership-laws.

Moorhouse, T.P., D'Cruze, N.C., Macdonald, D.W. (2021). Information about zoonotic disease risks reduces desire to own exotic pets among global consumers. *Frontiers in Ecology and Evolution* 9(609547). https://doi.org/10.3389/fevo.2021.609547

Munster, U. (2016). Working for the forest: the ambivalent intimacies of human-elephant collaboration in South Indian wildlife conservation. *Ethnos* 81(3), 425–447. https://doi.org/10.1080/00141844.2014.969292

NOAA (2022). *Impacts of invasive lionfish*. Retrieved from https://www.fisheries.noaa.gov/southeast/ecosystems/impacts-invasive-lionfish.

Nyhus, P.J., Tilson, R.L., Tomlinson, J.L. (2003). Dangerous animals in captivity: ex situ tiger conflict and implications for private ownership of exotic animals. *Zoo Biology* 22, 573–586.

Riley, P.Y., Chomel, B.B. (2005). Hedgehog zoonoses. *Emerging Infectious Diseases* 11(1), 1–5.

Scoresby, K., Strand, E., Ng, Z., Brown, K., Strobel, K., Stilz, R., Souza, M. (2021). Pet ownership and quality of life: a systematic review of the literature. *Veterinary Science* 8(12), 332. https://doi.org/10.3390/vetsci8120332

Shepherd, S.M., Mills, A., Shoff, W.H. (2014). Human attacks by large felid carnivores in captivity and in the wild. *Wilderness and Environmental Medicine 2*, 220–230. https://doi.org/10.1016/j.wem.2014.01.005

Silver, K. (2021). *Elevating animal welfare*. Retrieved from https://www.aza.org/connect-stories/stories/elevating-animal-welfare-zoos-aquariums.

Souza, M.J. (2009). Bacterial and parasitic zoonoses of exotic pets. *Veterinary Clinics: Exotic Animal Practice 12*(3), 401–415. https://doi.org/10.1016/j.cvex.2009.06.003

Souza, M.J. (2011). One Health: zoonoses in the exotic animal practice. *Veterinary Clinics: Exotic Animal Practice 14*(3), 421–426. https://doi.org/10.1016/j.cvex.2011.05.007

Tidball, K.G., Tourney, C. (2007). Serpents, sainthood, and celebrity: symbolic and ritual tension in Appalachian Pentecostal snake handling. *Journal of Religion and Popular Culture 17*, 1–17.

Tidière, M., Gaillard, J.M., Berger, V. et al. (2016). Comparative analyses of longevity and senescence reveal variable survival benefits of living in zoos across mammals. *Scientific Reports 6*, 36361.

USDA. (2019). *USDA Animal Care Animal Welfare Act and Animal Welfare Regulations*. Retrieved from https://www.aphis.usda.gov/animal_welfare/downloads/bluebook-ac-awa.pdf.

USDA. (2021). *United States Department of Agriculture Animal Welfare Act*. Retrieved from https://www.nal.usda.gov/legacy/awic/animal-welfare-act.

USFWS. (2016). *Importation of elephants from Swaziland*. Retrieved from https://www.fws.gov/international/pdf/questions-and-answers-swaziland-elephant-import.pdf.

USGS. (2022). *How have invasive pythons impacted Florida ecosystems?* Retrieved from https://www.usgs.gov/faqs/how-have-invasive-pythons-impacted-florida-ecosystems.

White, B., Yeung, P., Chilvers, B.L., O'Donoghue, K. (2021). Reducing the "cost of caring" in animal-care professionals: social work contribution in a pilot education program to address burnout and compassion fatigue. *Journal of Human Behavior in the Social Environment 31*(7), 828–847. https://doi.org/10.1080/10911359.2020.1822249

WZAM (Why Zoos & Aquariums Matter; 2022). *Outcomes from the WZAM project*. Retrieved from https://wzam.org/.

# 51
# EPILOGUE
## A Roadmap for the Future of Anthrozoology

*Aubrey H. Fine, Megan K. Mueller, Zenithson Y. Ng,
Alan M. Beck, and Jose M. Peralta*

The field of Anthrozoology, defined as the study of interactions and relationships between human and nonhuman animals, has evolved over the recent decades and has become a more in-depth scholarly endeavor. This book is a testimony to the breath that this vast field has unveiled. Our contributors have attempted to synthesize several of the contemporary issues and topics discussed by the scholarly community. Within this epilogue, the editors will provide a projected roadmap for the field and our vision for the future.

### Lessons We Learned While Working on the Text: Some Highlights

#### *The Discipline of HAI and Anthrozoology Is Here to Stay*

As was noted throughout the book, the science of anthrozoology and human-animal interactions (HAIs) has had a rich history. Globally, we are living in a world where our commitment to our interactions with nonhuman animals as well as the living environment has become much more clearly understood and appreciated. Ever since the early 1970s, there has been steadily increasing interest by a team of multidisciplinary scholars who are genuinely interested in studying our relationships with animals on various academic platforms. Our understanding of HAI today reaches beyond western belief systems and is now being studied through a multicultural lens. The field continues to expand and includes a wide range of academic disciplines including conservation and wildlife, ecotourism, art and literature, and various animal-assisted interventions for distinct populations. Additionally, there are now more contributions from many diverse disciplines including epidemiology, child development, gerontology, psychology, animal welfare, veterinary medicine, and clinical medicine, including nursing and hospice care.

From a historical point of view, we have witnessed a plethora of new scientific publications that highlight the expansion of intellectual engagement within the field. What started in the 1970s with the establishment of one University Center in the United States at the University of Pennsylvania and one primary journal, Anthroozoös, has evolved with the establishment of numerous new university centers and departments, as well as several respected peer-reviewed journals.

We all must agree that the field has indeed advanced and will continue to evolve with the ambitions and leadership of veteran scholars and a new generation of scientists. All these individuals appear committed to better conceptualizing the research and practice agenda of HAIs. We are also witnessing a stronger commitment by both government and private agencies in funding research and best practice approaches in numerous aspects of the field. Finally, it is important to note that media coverage of the findings of some of the research in the field has had an impact on society's awareness of the intricacies that HAIs face but also has had a hand in promoting the value of HAI to the public. At times, some of the findings are over sensationalized in the media, but the general attention that the field receives does nonetheless signify its relevance and the general interest that exists at the present time. Researchers must be guarded in not over-representing and misleading their findings, providing an unbiased perspective of neutral and negative findings along with the positive ones.

It does behoove the new leadership in this field to take charge and continue to advance the study of anthrozoology into the future. One does not know exactly where the field will be several decades from now, but we are certain it will continue to evolve and transform.

## *There Are Infinite Ways/Examples to Frame Human-Animal Interactions and Places Where People and Animals (and Environment) Interact*

As we have seen reflected across many of the chapters in this volume, the ways in which animals are involved in our lives are diverse and ever-changing. As the field of anthrozoology and HAI grows, we continue to learn more about the complex mental, physical, cultural, and historical processes involved in these interactions. For example, research on our interactions with companion animals in our homes has increased exponentially, with scholarship addressing human-animal relationships from youth through adulthood, in many different populations and contexts. This volume addresses many of these sub-fields of anthrozoology, touching on various points in the lifespan, key areas of focus (e.g., caregiver burden, the link between animal abuse and domestic violence, and others), as well as summarizing current perspectives on the health effects of HAIs and opportunities for improving the evidence base in the future. Crucially, the field is continually expanding in scope. As reflected in several of the chapters, understanding anthrozoology through a global, multicultural lens is critical to understanding the diversity in our interactions with animals. Exploring the animal side of the connection is also key, with research on animal cognition and emotions unlocking new ways of thinking about interactions.

Exploring the human side also requires additional attention. There is a noticeable lack of diversity in all areas of anthrozoology and HAI. While there is increasing use of dogs with veterans with PTSD, there is much evidence that minority veterans in general receive less psychotherapy with dogs than white counterparts (Spoont et al., 2017). Additionally, there is a clear lack of diversity among the animal-related professions. According to the 2021 Bureau of Labor Statistics, 93.3% of veterinarians in the United States are White, making it one of the one of the top "Whitest jobs in America." Over 62% of animal welfare organizations surveyed reported that they had no African American employees, with only 4% of the total employee population being African American (Brown, 2005). Among the general population, pet ownership and type of pet-owner relationship varies among individuals of different backgrounds and experiences. Of course, there are cultural differences on how pet attachment is expressed, which is an area that deserves more study (Brown, 2002). It is important to highlight the chapters in this textbook that have specifically addressed these diversity and inclusion topics so rarely discussed in the field of HAI. We should prioritize addressing the lack of diversity in researchers and practitioners across all levels as this area of scholarship continues to expand.

Furthermore, as we learn more about animals in our lives, it has become abundantly clear that HAIs take many different forms and have many different outcomes. There is not, and never will be, a "one size fits all" approach to understanding these relationships, which is in fact what makes the field of anthrozoology such an interesting area of scholarship. There are many opportunities to continue to learn from other intersecting fields that can help us better understand the many shapes and forms of HAI.

Finally, the impact that our interactions with nature and wildlife have on the environment and the influence that they have on conservation are aspects that deserve attention. The need to protect these ecosystems and to ensure that our interactions with wildlife are as positive and unobtrusive as possible is clearly relevant. Our presence in nature needs to leave the place in better condition than we found it, and that includes having a positive impact on the lives that we find in our visits and interactions. An understanding of the needs and wants of animals we connect with in the wild is an important first step in this process. The existence of conflicting situations needs to be addressed from an educated perspective to ensure a positive outcome. The impact, for example, that different forms of ecotourism have on the environment and on wildlife needs to be considered before we embark on those trips. An educated exploration through that prism that allows us to anticipate the impact that our presence will have is a necessary step. There may not be just one answer to these questions, but the importance of reflecting on them before and after the interactions will benefit all involved, both human and nonhuman animals alike.

## When Interacting with Any Animal in Any Way, We Must Prioritize the Welfare of the Animal

A varied assortment of situations and animal species are represented in the field of anthrozoology, as demonstrated by the wide range of interactions between humans and animals described in the different chapters of this book. Our interactions with companion animals (at home or in therapy sessions, for instance), often come first to mind as these animals are constantly close and present in our lives. This does not mean that we can ignore those interactions we have with horses, wildlife, research, working, or farm animals, just because they feel more distant for most of us. HAIs impact us, very often for the better, as has been documented in many examples found throughout this book. HAIs equally impact animals, and the valence of that interaction must also be positive.

The broadening of the field of anthrozoology to involve such a variety of experiences and species over the last several decades has occurred in parallel with an increasing societal preoccupation with the quality of life of the animals we interact with. In any exchange, the human has as much of an opportunity to influence the animal as the animal has to impact the human. However, as humans, we have a greater control of these interactions. We decide on the time, the length, the location, the extent, the quality of our relationships with animals. As such, we have a greater responsibility to ensure that the interactions are equally positive for all involved, not just for us. For the interactions to be successful, they must prove advantageous to the animals just as they are for the humans.

In this context, ensuring that animals also benefit is primarily our responsibility. In most cases, we impose the interaction on the animals and the animals are not offered a choice but to engage. As a result, we have a moral duty to ensure a satisfactory encounter. For the animals involved, it is not what we think of the interaction that matters. What is ultimately important for the animals is how they feel during and after the interactions. Of particular concern should be the impact on the emotional and mental state of the animals. In this regard, every effort must be made to examine any HAI from the perspective of the animal. It is our responsibility to monitor and adjust the conditions accordingly with an ultimate goal of ensuring that the interactions benefit the humans and the animals alike.

## *The Health and Welfare of People Depends on a Complementary Relationship Between Animals and People*

Historically, the driving force behind the study of HAIs has been the animals' impact on people. The human-animal bond is defined by the American Veterinary Medical Association's Committee on the Human-Animal Bond as "a mutually beneficial and dynamic relationship between people and animals that is influenced by behaviors essential to the health and wellbeing of both." The human-animal bond is not a one-way relationship, and we must be mindful that the impact that humans have on animals is just as important as the impact that animals have on humans. The benefits of these interactions will transpire if the needs of both human and animal parties are met and balanced. If the consequences outweigh the benefits for either party, strategies should be employed to mitigate any negative effects. Future research should aim to simultaneously investigate the outcomes for both humans and animals in their interactions. This considers the One Health One Welfare perspective, ensuring best practices to achieve meaningful connections and maximize both human and animal health and welfare. Today, we are seeing many scientists investigate the welfare of animals, both domestic and wild, to assure that as a global society we are doing our due diligence to support the well-being of all creatures that inhabit this world. This is a major effort, and we believe it will continue to be on the working agenda of numerous scholars and scientists.

## Future Directions in the Field

To help organize our thoughts, we have subdivided this section into five components. Each section provides a brief synopsis of areas that we believe are key to expanding the field in the years to come.

### *Research/Funding Opportunities*

While the inherently interdisciplinary nature of the field creates perennial challenges for the funding landscape, HAI and anthrozoology are increasingly becoming recognized areas of scientific scholarship, with funding opportunities diversifying. For example, in the United States, the National Institutes of Health has now devoted millions of dollars to large-scale research projects. Furthermore, there are many private sector funding resources that help support research in HAIs. For example, since 2014, the Human Animal Bond Research Institute has funded numerous grants investigating the importance of HAIs, as well as the health outcomes of pet ownership. By continuing to educate the scientific community about the broader field and its place in more "mainstream" scientific communities, we can enhance and diversify available funding streams. Creativity and interdisciplinary collaboration in research and funding opportunities are critical to the health and longevity of the field.

### *Educational/Training Opportunities*

As the field of anthrozoology continues to grow, opportunities for education and training will continue to expand. Numerous academic institutions deliver education in the form of formal degrees, individual classes, certificates, or continuing education. If a university does not offer specific coursework in anthrozoology or HAIs, students can typically find virtual educational opportunities globally at institutions that do offer them. Virtual coursework across the globe allows for rapid expansion and recognition of this field. This education may not be presented as anthrozoology, but rather as specific areas of interest within broader disciplines such as the social sciences and animal science. In addition, practical experience is invaluable. We encourage students to further their understanding of HAI by seeking mentorship from an individual with experience in the field. Internships or shadowing mentors provide insight into the science of anthrozoology and potential for careers.

Researchers, academics, and practitioners invite young scholars to further their understanding of HAIs. Any subject that takes into consideration the intersection between humans and animals falls within the scope of anthrozoology. This textbook touches on the diverse range of topics that can be further explored. It is important to note that this is a multi- and interdisciplinary field with valuable expertise and perspectives from each discipline. The same subject may be taught from different perspectives, such as the benefits of animal-assisted therapy through the lens of a mental health specialist versus an animal behaviorist. Each specialist and discipline have their own expertise, and it is critical that each recognizes their limitations while embracing the contributions of others, utilizing consistent terminology to garner mutual respect and understanding. What is beautiful and unique about the field is the ability to bring together multiple experts to holistically advocate for both human and animal health and welfare.

## Organizational Opportunities and the Advancement of Public Policy

As anthrozoology continues to blossom, there are now numerous professional and academic organizations representing distinct dimensions of the discipline. Collectively, the leadership of these organizations provides directions for the field as a whole. Additionally, these organizations are charged to assist professionals in networking and become more united.

With the expansion of the field comes increasing responsibilities for those involved. Participation in the existing organizations is an important first step to contribute to the growth and to remain informed on the evolution of the field. As it continues to expand, additional structure may be needed, and organizations provide a viable forum to initiate those discussions. As some of the areas of HAIs are of general societal interest, there will be an increasing need for the development of public policy to provide a framework to coordinate different relevant aspects of the interactions. The involvement of those in the field, both individually and as members of the different organizations, will be crucial to those discussions and to the development of sound policies and guidelines.

## Career Opportunities

There is no limit to careers that are centered on the interaction between humans and animals. As the field continues to grow and expand to connect with many other disciplines, there has been a corresponding increase in career opportunities for new scholars. Veterinary medicine, psychology, and human medicine have traditionally been fields that support anthrozoology, and careers within the niche of HAI within these fields are becoming more viable. It is often thought that the only practical application of the study of HAIs is therapists delivering animal-assisted therapy. In fact, professionalization of animal-assisted interventions is growing, with rigorous education and certification programs now available for those wishing to pursue careers as practitioners. However, as this textbook illustrates, careers ranging from public health experts to policy makers to animal trainers to environmentalists to veterinarians also benefit from the practical applications of anthrozoology. Of course, careers involving the research and teaching application of HAIs demonstrate the importance of this growing field.

Careers can be cultivated through appropriate mentorship. If you have not found a mentor in this area, we invite you to contact any of the authors in this textbook for further insight and guidance on career aspirations. In addition, networking at conferences in HAIs will provide opportunities for mentorship and exposure to the diverse range of careers and job openings available. This field is intentionally multidisciplinary and constantly evolving, allowing for creativity, and thinking outside the box to create jobs that may not currently exist.

## *Diversity, Equity, and Inclusion*

Including perspectives of HAI specialists of various backgrounds other than the Caucasian-dominant perspective will truly enrich our understanding of how *all* humans interact with animals. To encourage diversity in the field, we should bring awareness of the possibilities of HAI-oriented careers in youth education, with a particular focus on engaging diverse populations. This may involve animal-related professionals attending career days in schools to offer mentorship, implementing animal-assisted education or other types of animal programming, and spotlighting examples of these career opportunities on platforms of high visibility to youth such as social media. Additionally, scholarships to offset higher education costs in HAI should be widely available to support youth and early career students, particularly those from diverse backgrounds. Efforts should be made to translate research and scholarly information into other languages so those in countries that may not have resources to prioritize human-animal relationships as an academic area of interest have access to this knowledge and can foster awareness and opportunities. These action steps require advocates of HAI to develop appropriate cultural competence to recognize that every individual has a different perspective of and relationships with animals and diverse human-animal relationships must be approached with sensitivity and grace. The ultimate goal of achieving a high level of diversity to the scholarly field of HAI will not happen overnight, but we can continue to openly discuss and prioritize the issue to eventually bring about the change we want to see.

## *Final Comments*

Providing final comments about a volume that is close to a half a million words has been an arduous and challenging task. We are honored that we have had so many brilliant contributors who have synthesized the present state of their distinct topics. Collectively, the chapters provide a roadmap which is not only reflective but is forward in thinking. So much has been integrated into this book that provides our readers with a rich introduction to numerous topics that are critical to our understanding of HAIs. It is our hope that this Handbook supports further research in the field and continues to promote more serious engagement by the multidisciplinary scholars who do the work. We as editors believe that the collection of these chapters supports the conviction that HAI is now a well-recognized part of science that justifies and clarifies that humans are part of nature. The chapters promote a better understanding that our care, study, and respect for nature, including the animals that share our lives, benefit all. However, as our contributors suggest, these relationships are more complicated than we all have imagined and more needs to be done to continue advancing and refining the field of anthrozoology. Lincoln once stated that "the best way to predict the future is to create it." The field of anthrozoology is in good hands with a new generation of scientists who will do just that! They will continue to question and discover, as they help advance the importance of our unique and complex relationships with animals.

## References

Brown, S. E. (2002). Ethnic variations in pet attachment among students at an American school of veterinary medicine. *Society & Animals, 10*(3), 249–266.

Brown, S. E. (2005). The under-representation of African American employees in animal welfare organizations in the United States. *Society & Animals, 13*(2), 153–162.

Bureau of Labor Statistics. Household data annual averages. Accessed November 4, 2022. https://www.bls.gov/cps/cpsaat11.pdf

Spoont, M. R., Sayer, N. A., Kehle-Forbes, S. M., Meis, L. A., & Nelson, D. B. (2017). A prospective study of racial and ethnic variation in VA psychotherapy services for PTSD. *Psychiatric Services, 68*(3), 231–237.

# APPENDIX 1

## Glossary of Terms

*Stefanie Malzyner, Nicole Lorig, Kylie Bell, and Morgan Starkweather*

**American Hippotherapy Association (AHA)** formed in 1992 initially as a section of the North American Riding for the Handicapped Association (NARHA, now PATH Intl.) as a resource for guiding standards of practice and educating practitioners about the use of horses in skilled therapy services. AHA officially became an independent nonprofit organization in 2003 (American Hippotherapy Association, n.d.).

**Animal-Assisted Activity (AAA)** a subcategory of animal-assisted intervention in which human-animal teams participate in a planned, goal-oriented visitation with individuals or groups. Human-animal teams must undergo training and assessment to ensure adequate preparation prior to participation in activities (Fine, 2019; International Association of Human-Animal Interaction Organizations, 2018).

**Animal-Assisted Education (AAE)** a subcategory of animal-assisted intervention in which degree holding, general and special education teachers guide goal oriented, structured intervention for individuals or groups (Fine, 2019; International Association of Human-Animal Interaction Organizations, 2018).

**Animal-Assisted Intervention (AAI)** the intentional inclusion of trained animals in structured interventions to provide restorative benefits for humans within health, education, and/or human service contexts. Human-animal teams are incorporated into interventions implemented utilizing an interdisciplinary approach. Animal-assisted intervention is a more general term that often encompasses the following specific interventions – animal-assisted activity, animal-assisted education, and animal-assisted therapy (Fine, 2019; International Association of Human-Animal Interaction Organizations, 2018).

**Animal-Assisted Intervention Professionals** competent, trained professionals in mental health, allied health (occupational, physical, recreational, speech, and language therapists), and education fields who facilitate animal-assisted interventions (AAIs) within their practice or a dedicated animal handler. Professionals must demonstrate competence within their chosen profession and have received further training in the selection, training, socialization, and understanding of animal ethology and behavior. Handlers must also demonstrate competence in animal knowledge, animal training, and animal behavior while also receiving basic education regarding the populations that the animal-human team will be working with (Fine, 2019; VanFleet, 2022).

**Animal-Assisted Interventions International (AAII)** a member-driven, international, interdisciplinary entity consisting of animal trainers, health care providers, human service professionals, and animal handlers. AAII developed standards of practice, competence, and accreditation for professionals to ensure animal and human welfare. AAII is a third-party organization designed to regulate preparation, education, evaluation, practice for provision of animal-assisted interventions to ensure ethical practice (Winkle, Gorbing, Rogers, & Vancoppernolle, 2020).

**Animal-Assisted Therapy (AAT)** a subcategory of animal-assisted intervention in which licensed human service professionals facilitate planned, structured, goal-oriented therapeutic activities alongside an animal. The animal may be handled by either the human service professional or by an individual under the supervision of the human service professional (Fine, 2019; International Association of Human-Animal Interaction Organizations, 2018).

**Animal Enrichment** strategies used to better the overall welfare of captive animals. "Improvements in enclosures, feeding devices, appropriate social groupings, and sensory stimuli are some examples of animal enrichment. Enrichment is just as necessary as nutrition and veterinary care for the welfare of animals" (Claxton, 2011; "Animal Enrichment", 2021).

**Animal Personhood** the idea that supports providing animals "rights" within legal applications (Klinkenborg, 2014).

**Animal Welfare** a reflection of how the animal feels and what the animal wants in a particular setting or situation. There are various means of measuring animal welfare such as health, productivity, behavior, and physiological response. Animal welfare assessment is challenging because of our limitation in interpreting the animal's perspective, which is the one that actually matters (Hewson, 2003; "Animal welfare: What is it?", 2022).

**Anthropocene** geological era when the collective actions of humanity impact the environment and create geological change (Whitehead, 2014).

**Anthropocentrism** the philosophical idea that human beings have higher intrinsic value and are the most important beings in the universe based on their superior moral worth (Arkow, 2020)

**Anthropomorphism** ascribing of human traits and characteristics to non-human beings (Fine, 2019; Tedeschi & Jenkins, 2019).

**Anthrozoology** a multi-disciplinary field that studies human-animal interactions. Anthrozoology considers how the welfare of both animals and humans is impacted by their interactions instead of only the humans' perception of animal welfare (Siddiq & Habib, 2016).

**Assistance Animals** an animal that is specifically trained to perform specialized tasks to functionally assist an individual with a physical, sensory, psychiatric, intellectual, or other mental disability. The tasks must be directly linked to an individual's disability. In the United States, per Titles I and II of the Americans Disability Act, dogs and miniature horses are the only species of animal that can serve as an assistance animal (Assistance Dogs International, 2022; Brennan, 2014; Fine, 2019).

**Attachment** any form of behavior that results in an individual attaining or maintaining proximity to some other clearly defined individual who is perceived as better able to cope with the world. This term is used within the context of the paradigm proposed by John Bowlby in 1958 originally to describe mother-child relationships, but has now been shown to extend to other forms of relationships including those between humans and animals. Pets, or other animals in human care, have a biological attachment rooted

Appendix 1: Glossary of Terms

in protection as they rely on their owners for food, shelter, and care. On the contrary, humans tend to rely on pets for social support, which is said to be the foundation for healthy functioning and mental health (Bowlby, 1970).

**Biodiversity** also known as biological diversity, can be defined as the variety of species, or living organisms within an ecosystem (Swingland, 2013). This includes the five kingdoms of living organisms or phyla; animals, plants, molds, yeasts and mushrooms, algae, ciliates and associated microbes, and bacteria (Margulis et al., 1999).

**Biocentrism** a concept suggesting that all living things are morally considerable. In contrast with anthropocentrism, which holds the belief that only human beings are considerable, biocentrism encompasses all living organisms (Attfield, 2016).

**Biophilia** hypothesis that human beings retain an innate inclination to affiliate with nature, natural life forms, and other sentient beings (Fine, 2019; Melson, 2006; Olmert, 2009; Starkweather, 2020; Tedeschi & Jenkins, 2019).

**Conservation** can be defined as the act of protecting, preserving, and promoting sustainable use of wildlife and associated resources (Hambler & Canney, 2013).

**Cortisol** a steroid hormone synthesized from cholesterol within the adrenal cortex. It is consistently produced in the body, with peak daily production occurring at first awakening. Cortisol also serves as a component of the stress-response and is stimulated by physiological and psychosocial stress via activation of the hypothalamus-pituitary-adrenal axis. In response to acute stress, cortisol promotes processes such as glucose metabolism, immune system suppression, and heart rate moderation to ensure energy stores are directed toward survival. To aid in survival effort, cortisol production has been shown to have a cognitive effect by promoting fear and avoidance behaviors as well as anxiety. Chronic production of cortisol has been shown to negatively influence overall health and well-being via continuous suppression of vital processes and influence on mental health. It is often measured as a biomarker of negative well-being (Chu et al., 2022; Kozlowska et al., 2015; Pulopulos et al., 2020).

**Domestication** a cultural and biological process where animals are assimilated within social structures of human societies. Animals change over time during domestication to better integrate within human society (Clutton-Brock, 1992).

**EAGALA (Equine Assisted Growth and Learning Association)** a nonprofit organization that was established in 1999 by Lynn Thomas, LCSW. Using their self-titled model, EAGALA promotes the use of members of the equine species in ground-based equine facilitated psychotherapy for group or individual sessions. The model follows a team approach by incorporating a licensed and EAGALA-certified mental health professional, an EAGALA-certified equine specialist, and the equine (EAGALA, n.d.).

**E3A (Equine Experiential Education Association)** founded in 2007 as a professional membership organization offering education to EAS practitioners about implementation of EAL programs. E3A provides education in the use of an experiential approach to growth and learning while ensuring the safety and well-being of clients and horses involved (Equine Experiential Education Association, 2022).

**Ecological Resilience** the ability of an ecosystem to return to a normal, functioning state following a disruption. A measure of resilience can be used to understand the overall state of that specific ecosystem (Gunderson, 2000).

**Ecology** the study of living organisms and their ecosystems. This includes understanding the affiliation between organisms and the resulting impact on the environment (Odum & Barrett, 2009).

**Ecosystem Services** the many life-sustaining benefits we as humans gain from the surrounding ecosystems. These services range from the purification of water and air to climate regulation and pollination for food crops (Daily, 2017).

**Ecotourism** a type of tourism business for nature enthusiasts who participate in specialty tours. Ecotourism aims to advance tourism's "potential for conservation development, and minimizing the negative impact tourism has on ecology, culture, and aesthetics." This concept upholds the strong commitment for environmental, economic, and social responsibility among the tourism business. The goal of ecotourism is to "provide positive experiences for both visitors and host, build low-impact facilities and awareness, while recognizing the rights and spiritual beliefs of indigenous people within communities" (Lindberg & Hawkins, 1993; "What is Ecotourism", 2022).

**Emotional Support Animals (ESA)** a companion animal whose presence provides comfort and therapeutic support to individuals with mental health and/or emotional conditions (Assistance Dogs International, 2022). ESA are not under the legislation covered by the American Disability Act.

**Ethology** the study of how animals live, behave, and interact with one another, often in their natural environment. Understanding animal behavior from a scientific perspective includes both observation and interpretation. For example, behavior observations may include "analyzing movement patterns, vocalizations and body posture" while interpretation of that behavior focuses on understanding why the animal is displaying such behavior (Ewer & Leyhausen, 1976; Immelmann, 1980).

**Equine-assisted Services (EAS)** developed by Wood and colleagues in 2021, EAS is a broad unifying term to describe all ways in which members of the equine species are incorporated into professional services. This includes therapeutic, learning, and horsemanship contexts. Within the realm of therapy, equines can be included in physical therapy, psychotherapy, speech therapy, and occupational therapy. Within the realm of learning, equines can be included in education, professional development, and team building. Lastly, within the realm of horsemanship, traditional horsemanship skills may be adapted to individuals with physical disabilities, which may include adaptive sports, adaptive riding, adaptive driving, and interactive vaulting. The horsemanship realm may also refer to use of the horse as a means of therapeutic treatment for mental health such as therapeutic riding and driving as well as grooming, and groundwork (Wood et al., 2021).

**Equine-assisted Therapy (EAT)** the incorporation of members of the equine species into therapeutic activities, which may include counseling, physical therapy, psychotherapy, speech-language pathology, and occupational therapy. EAT is considered within the therapeutic context of Equine Assisted Services. To be considered EAT, an activity-appropriate certified professional such as a physical therapist, a mental health professional, a speech therapist, or an occupational therapist should be included in the planning and instruction of activities (Wood et al., 2021).

**Equine-facilitated psychotherapy (EFP)** the incorporation of members of the equine species into psychotherapy sessions with a licensed mental health professional. EFP promotes immersion into the non-traditional therapeutic setting of the natural living environment of the horse or other equine. Typically, EFP sessions are ground-based and involve communication, trust-building, and/or relationship growth with the horse. Personal growth through equine interactions may be extrapolated into the client's everyday life. This term may be used interchangeably with Equine-Assisted Psychotherapy (EAP) (Wood et al., 2021).

*Appendix 1: Glossary of Terms*

**Equine-assisted learning (EAL)** activities performed with members of the equine species that are designed to engage individuals of any age in learning processes, personal development, and/or growth in life skills. These activities may be individual or group sessions and may be facilitated by non-licensed persons such as coaches, teachers, or equine professionals. In contrast to EAT, EAL does not work toward treatment goals (White-Lewis, 2019; Wood et al., 2021).

**Equine Assisted Activities and Therapies (EAAT)** the many ways in which members of the equine species are incorporated into various therapeutic and non-therapeutic contexts. Used of EAAT has been replaced by EAS which more cohesively describes the realms of services in the field (Wood et al., 2021).

**Equine-assisted intervention** programs which incorporate members of the equine species to promote human health and well-being. While this term can be used in recreational and educational contexts, the term equine-assisted interventions is traditionally used most in the context of therapeutic interventions (White-Lewis, 2019).

**Equine-assisted interaction** ways in which humans may interact with one or several members of the equine species. Equine-assisted interactions may be positive or negative in nature for either or both the human and equine. This term is used in all contexts of EAS (White-Lewis, 2019).

**Facility Animals** certified animals that are specially trained and placed with a trained professional to perform specific skilled tasks within a therapeutic environment (facility) to participate in provision of animal-assisted interventions for multiple clients (individuals or groups) within a residential or clinical setting (Assistance Dogs International, 2022).

**Fight or flight** refers to the autonomic nervous system's first "fast" response to a threat or a stressor. During times of perceived stress, the sympathetic branch of the autonomic nervous system is activated, stimulating the release of epinephrine (adrenaline) and norepinephrine. Binding of these biological chemicals results in increased blood flow to the heart, lungs, muscles, and certain areas of the brain, and decreased blood flow to the gastrointestinal tract in an effort to prepare the body to survive the perceived threat. The secondary, "slow", component of the stress response is associated to the release of cortisol from the adrenal cortex, which signals the body to conserve energy by halting functions such as the immune system, and growth, and metabolizes glucose to increase energy supply (Chu et al., 2022; Kozlowska et al., 2015; Pulopulos et al., 2020; Wolter, Stefanski & Krueger, 2018).

**Five Domains** a framework to assess animal welfare. The domains include four physical factors (nutrition, environment, health, behavior), and a mental state. After the physical domains are assessed, the last domain, mental, determines the animal's perspective and is a reflection of its welfare state (Mellor & Beausoleil, 2015).

**Five Freedoms** due to the investigation of the welfare of farmed animals, the 'Five Freedoms of Animal Welfare' was developed by the UK Farm Animal Welfare Council (FAWC) in 1979. The five freedoms include: "1. Freedom from hunger and thirst; 2. Freedom from discomfort; 3. Freedom from pain, injury, or disease; 4. Freedom to express normal behavior; 5. Freedom from fear and distress". They have been used as a framework for the development of guidelines for the assessment of animal welfare in different species (McCausland, 2014; Farm Animal Welfare Council, 2012).

**Green Infrastructure** refers to city planning and infrastructure that supports solutions to challenges between urban developments, and the surrounding ecosystems specific to the protection and management of water-based resources (Grabowski et al., 2022). Examples include stormwater systems, rainwater harvesting, and land conservation (EPA, 2022).

**Habitat**   refers to the environment in which an animal, plant or organism lives, and/or develops (Yapp, 1922).

**Heart rate variability (HRV)**   a quantitative value describing the fluctuation in the beat-to-beat interval between consecutive heartbeats. Changes in this value reflect the action of the autonomic nervous system. Heart rate is intended to increase or decrease in response to physiological needs. Activation of the sympathetic branch of the autonomic nervous system results in increased heart rate during occurrences such as exercise or fight-or-flight response. In contrast, the parasympathetic branch of the autonomic nervous system is activated in response to occurrences such as sleep or digestion, resulting in reduced heart rate. This ability of the body to transition quickly between sympathetic and parasympathetic activity ensures that the animal can adequately cope with its environment. However, in an animal that has experienced chronic stress, overactivation of the sympathetic nervous system results in reduced HRV, leaving it more vulnerable to acute stressors. Thus, HRV is often used as a means to measure an individual's ability to cope with stress and may be included in assessments of well-being as a biological parameter (Kozlowska et al., 2015).

**Herd animal**   an animal belonging to a species whose natural social structure is that of a group. The size of a herd varies according to the species. Typically, these animals are highly gregarious and are often herbivores. An evolutionary benefit of the herd social structure is a decreased risk of predation in a large group. As such, members of a herd mimic the behaviors of their conspecifics to react quickly to potential threats (De Santis et al., 2017).

**Hippotherapy**   describes the use of equine movement during mounted activities as a treatment tool for individuals with diverse physical needs. Hippotherapy is often incorporated as a supplemental therapy component within an overall treatment plan. Thus, unlike therapeutic riding, hippotherapy facilitators must be licensed professionals such as occupational therapists, physical therapists, or speech-language pathologists (Wood et al., 2021).

**Human-animal interaction (HAI)**   the multitude of ways in which humans and non-human animals can associate with one another, whether positively or negatively. There is a growing body of research within this topic to uncover mechanisms driving these interactions, and the influence HAIs have on the health of both the human and non-human participant (Fine & Beck, 2019).

**Human-animal bond (HAB)**   phenomenon recognized widely throughout history as the strong connection that can occur between human and non-human animals, similar in nature to that of the bond experienced between two individuals within the same species such as family members, romantic partners, or close friends. This connection is most often noted between humans and companion animals such as dogs and cats, but may be seen in other domesticated species such as horses, or non-domesticated species including exotic animals. To be classified as a true HAB, the bond must be mutual, voluntary, and long-lasting with both parties being able to recognize one another, trust one another, and recognize one another's needs. The mutual, positively valanced relationship is said to be rooted in the attachment theory, theory of social support, and biophilia hypothesis (Beck, 1999; Fine & Beck, 2019).

**Human-animal-environment interaction (HAEI)**   a one health concept defined as intentional, mutually influential exchanges between animals, plants, and/or nature. Each component of HAEI can influence the others positively or negatively. Use of positive HAEI is incorporated into therapeutic contexts in which humans are given the opportunity to be in an environment with animals and plants. This sentiment is rooted in the biophilia hypothesis, originally proposed by E. O. Wilson (1984), which emphasizes that humans have evolved a desire to connect with living organisms and nature, suggesting that the brain is structured to pay selective attention to living things (Flynn et al., 2020; Wood et al., 2021).

**International Association of Human Animal Interaction Organizations (IAHAIO)** an international organization founded in 1990 that "advances leadership in the field of Human Animal Interactions through research, education and collaboration" (UIA,n.d).

**International Society of Anthrozoology** a global non-profit organization that focuses on "the scientific and scholarly study of human-animal interactions".

**Mirroring** innate ability of members of the equine species to perceive and mimic the emotions and behavior of those around them. This evolutionarily beneficial behavior originated as a component of the herd instinct enabling equines to react quickly to potential threats based on the behaviors of other herd members without reliance on their own awareness of the threat. Mirroring can be noted during equine-assisted interactions in which equines may mirror the emotions portrayed by humans (Berckmans, 1996; Duranton & Gaunet, 2016).

**National Audubon Society** non-profit organization within the United States of America. They aim to protect avian species through the conservation of ecosystems across the nation. Public education and ongoing research play a key role in accomplishing the mission of this organization (Audubon, 2022).

**Natural Lifemanship** founded in 2010 by husband-and-wife pair Tim Jobe and Bettina Shultz-Jobe, this nonprofit organization certifies practitioners in trauma-focused equine-assisted psychotherapy. Certified individuals undergo extensive training in the science and application of trauma-informed care using the latest neurobiology research and understanding of the influence of relationships on trauma-care (*Natural Lifemanship*, 2022).

**One Health and One Welfare** concept that the health and well-being of all natural world systems and beings are interconnected based on complex, dynamic, causal relationships. The One Health model is an interdisciplinary endeavor to address and advocate for the collective cooperation amongst nations to establish healthy human-animal, human-human, and human-nature relationships while preserving and healing nature in response to the maladaptive consequences resulting from anthropocentrism. One Health recognizes that the health of humans, animals, and the environment are interrelated and dependent on one another. Further, One Welfare considers the relationship between human wellbeing, animal welfare and the conservation of the environment. (Centre for Animal Welfare, 2022; Fine, 2019; International Association of Human-Animal Interaction Organizations, 2018).

**Oxytocin** peptide neurohormone produced by the hypothalamus primarily in response to proprioceptive stimulation, or touch. Oxytocin is most well-known for its maternal roles in producing uterine contractions during labor and promoting milk ejection, as well as its roles in social bonding. Release of oxytocin promotes prosocial behavior and has a variety of other effects including reduction of the stress response, decreased pain response, and decreased blood pressure. Oxytocin may be released following social interactions within species such as those between parent-child, friends, and romantic partners. It has also been shown that oxytocin is released in response to positive social interactions between members of different species including during human-animal interactions. Oxytocin measurement has, in recent years, been increasing as a biomarker of positive human-animal interaction (De Santis et al., 2017; Coulon et al., 2013; Insel et al., 1998).

**PATH Intl. (Professional Association of Therapeutic Horsemanship International)** nonprofit organization founded in 1969 to promote safe and effective therapeutic horseback riding by accrediting therapeutic riding facilities across North America and Internationally. Originally called the North American Riding for the Handicapped Association (NARHA), the organization now serves as one of the largest and most recognized accrediting organizations in the field of Equine Assisted Services (Ensuring excellence and changing lives, 2022).

**Pet Partners (originally known as the Delta Society)** a non-profit, national organization comprised of volunteer human-animal teams that are trained, assessed, and registered according to the organization's standards of practice. The teams participate in animal-assisted interventions across diverse populations, communities, and settings to demonstrate the physical, social, and emotional benefits of the human-animal bond (Pet Partners, 2022).

**Prey animal** any animal that is hunted by a predator animal for food. An animal may be a prey animal in some instances but a predator in others (De Santis et al., 2017).

**Sentience** capacity for beings (including nonhumans) to experience, perceive, and respond to feelings and sensations (Arkow, 2020).

**Social capital** all the resources, both tangible and intangible, that contribute to the value of social networks. Humans are naturally social animals and rely on connections with others and formation of community for physical and emotional well-being. The outcome of strong social capital is reciprocity, trust, and cooperation within a community, which promotes leadership, community participation, civic engagement, and social ties. Evidence has suggested that animals may serve as social capital. Pet ownership and presence of pets has shown to catalyze communication between neighbors or strangers, increase perception of trust, promote reciprocal exchange of favors between neighbors, and increase general sense of community (Arkow, 2019).

**Synchronization** Synchronization or synchrony is defined as the conscious or unconscious matching of behavior between two or more individuals that may manifest in matched timing of any behavior changes, performance of identical behavior, or performance of any behavior within the same location. This adaptive mechanism is associated with improved chances of survival, reduced energy expenditure, and increased social cohesion within and between species. Synchronization is more likely to occur between familiar dyads or groups including pets and their owners and may be measured as a marker of the human-animal relationship (Duranton & Gaunet, 2016; Wolter, Stefanski & Krueger, 2018).

**Synanthropic species** species associated with urban areas more than with other ecosystems (Francis & Chadwick, 2012). It is important to note that while one species may be considered synanthropic in one area, the same species may not have a concentrated presence in other urban areas.

**Therapeutic Riding** the use of mounted equine-assisted interactions to promote the natural benefits of riding and interacting with horses for individuals or groups with diverse physical or psychological needs. Sometimes referred to as adaptive riding, equipment may be adapted to accommodate a rider's individual needs. Therapeutic riding is categorized within the horsemanship context of EAS and thus does not require facilitators of these services to be licensed professionals (White-Lewis, 2019).

*Appendix 1: Glossary of Terms*

**Therapeutic Driving**  the use of equine driving activities among individuals with diverse physical and/or psychological needs. The practice of equine driving involves hitching one or more members of the equine species to a carriage, wagon, cart, or other vehicle and guiding it to pull. Natural benefits of participation in driving include improved physical fitness and increased cognitive ability. Sometimes referred to as adaptive driving, therapeutic driving promotes these natural benefits of driving by enabling individuals with diverse physical and/or psychological needs to safely participate in activities with equines without being mounted (White-Lewis, 2019).

**Trust Bank Account**  a metaphorical concept that shows the complex nature of human and animal relationships. The idea is to build enough trust with animals that it withstands "withdrawals" within the relationship. With each positive interaction, "deposits" are made to a "trust account." Withdrawals are made when negative interactions occur. For example, when we use "force, threats, or punishment", a withdrawal is made. And if the withdrawal is large where it surpasses the positive balance, we may jeopardize the relationship. Therefore, making more deposits than withdrawals will only prove beneficial within a human animal relationship.

**Watchable Wildlife**  observing wildlife without a means for human consumption such as hunting or fishing. Wildlife viewing is a recreational activity that introduces humans to an animal's habitat in person or online ("Watchable Wildlife", 2022; Clayton & Mendelsohn, 1993).

**Wildlife**  animals that have not been domesticated and continue to roam freely within their environment. This term is specific to "terrestrial vertebrae" which includes, "reptiles, amphibians, birds and mammals" (Yarrow, 2009).

**Wildlife Conservation Society**  non-profit organization that defines their mission as saving "wildlife and wild places worldwide through science, conservation action, education, and inspiring people to value nature" (WCS, 2022).

**Zoo Visitor Experience**  aim to educate visitors while cultivating an appreciation for wildlife conservation. Modern zoos influence those who visit them. Therefore, visitors who can actively engage and "interact with animals, exhibits, and programming/staff, are more likely to have positive perceptions of zoos. Repeat visitors are more likely to seek out conservation efforts." Therefore, the zoo experience impacts the future of wildlife conservation (Godinez & Fernandez, 2019).

**Zoonosis**  a disease that is transmitted from nonhuman animals to humans. Examples of Zoonosis include: "anthrax, brucellosis, influenza, hantavirus syndromes, plague, and rabies" (Keck & Lynteris, 2018).

## References

American Hippotherapy Association, Inc. (n.d.). Retrieved May 1, 2022, from https://www.americanhippotherapyassociation.org/

*Animal enrichment.* Smithsonian's National Zoo. (2021, June 22). Retrieved April 12, 2022, from https://nationalzoo.si.edu/animals/animal-enrichment

*Animal welfare: What is it?* American Veterinary Medical Association. (n.d.). Retrieved April 12, 2022, from https://www.avma.org/resources/animal-health-welfare/animal-welfare-what-it

Arkow, P. (2019). Chapter 5- The social capital of companion animals: Pets as a catalyst for social networks and support… and a barometer of community violence. In A. H. Fine (Ed.), *Handbook on Animal-Assisted Therapy (Fifth Edition)* (pp. 51–60). Academic Press. https://doi.org/10.1016/B978-0-12-815395-6.00005-5

Arkow, P. (2020). Human-animal relationships and social work: opportunities beyond the veterinary environment. *Child & Adolescent Social Work Journal: C & A, 37*(6), 573–588. https://doi.org/10.1007/s10560-020-00697-x

Assistance Dogs International. (2022). *ADI terms & definitions.* Assistance Dogs International. https://assistancedogsinternational.org/resources/adi-terms-definitions/

Attfield, R. (2016). Biocentrism. *International Encyclopedia of Ethics*, 1–10. https://doi.org/10.1002/9781444367072.wbiee670.pub2

Audubon. (2022). Retrieved April 20, 2022, from https://www.audubon.org/about#:~:text=Since%201905%2C%20Audubon's%20vision%20has, is%20a%20nonprofit%20conservation%20organization.

Beck, A. M. (1999). Companion animals and their companions: Sharing a strategy for survival. *Bulletin of Science, Technology & Society, 19*(4), 281–285. https://doi.org/10.1177/027046769901900404

Berckmans, P. R. (1996). Behavioral expression and related concepts. *Behavior and Philosophy, 24*(2), 85–98.

Bowlby, J. (1970). Disruption of affectional bonds and its effects on behavior. *Journal of Contemporary Psychotherapy, 2*(2), 75–86. https://doi.org/10.1007/BF02118173

Brennan, J. (2014). *Service animals and emotional support animals.* (V. Nguyen, Ed.). ADA: National Network. https://adata.org/guide/service-animals-and-emotional-support-animals#main-content

Centre for Animal Welfare. (2022). *One health, one welfare.* Centre for Animal Welfare. https://animalwelfaremalawi.org/one-health-one-welfare/

Chu, B., Marwaha, K., Sanvictores, T., & Ayers, D. (2022). Physiology, stress reaction. In *StatPearls.* StatPearls Publishing. http://www.ncbi.nlm.nih.gov/books/NBK541120/

Claxton, A. M. (2011). The potential of the human–animal relationship as an environmental enrichment for the welfare of zoo-housed animals. *Applied Animal Behaviour Science, 133*(1–2), 1–10. https://doi.org/10.1016/j.applanim.2011.03.002

Clayton, C., & Mendelsohn, R. (1993). The value of watchable wildlife: A case study of McNeil river. *Journal of Environmental Management, 39*(2), 101–106. https://doi.org/10.1006/jema.1993.1057

Clutton-Brock, Juliet. (1992). The process of domestication. *Mammal Review, 22*(2), 79–85. https://doi.org/10.1111/j.1365-2907.1992.tb00122.x

Coulon, M., Nowak, R., Andanson, S., Ravel, C., Marnet, P. G., Boissy, A., & Boivin, X. (2013). Human–lamb bonding: Oxytocin, cortisol and behavioural responses of lambs to human contacts and social separation. *Psychoneuroendocrinology, 38*(4), 499–508. https://doi.org/10.1016/j.psyneuen.2012.07.008

Daily, G. C. (2017). Nature's services: Societal dependence on natural ecosystems. In L. Robin (Ed.), *The Future of Nature.* New Haven, CT: Yale. https://doi.org/10.12987/9780300188479-039

De Santis, M., Contalbrigo, L., Borgi, M., Cirulli, F., Luzi, F., Redaelli, V., Stefani, A., Toson, M., Odore, R., Vercelli, C., Valle, E., & Farina, L. (2017). Equine assisted interventions (EAIs): Methodological considerations for stress assessment in horses. *Veterinary Sciences, 4*(3), 44. https://doi.org/10.3390/vetsci4030044

Duranton, C., & Gaunet, F. (2016). Behavioural synchronization from an ethological perspective: Overview of its adaptive value. *Adaptive Behavior, 24*(3), 181–191. https://doi.org/10.1177/1059712316644966

*Eagala—A global standard in equine-assisted psychotherapy and personal development.* (n.d.). Retrieved May 14, 2022, from https://www.eagala.org/index

*Ensuring excellence and changing lives.* (n.d.). PATH International. Retrieved May 14, 2022, from https://pathintl.org/

Environmental Protection Agency. (2022). *EPA.* Retrieved April 20, 2022, from https://www.epa.gov/green-infrastructure/what-green-infrastructure

*Equine Experiential Education Association E3A* (n.d.). Retrieved May 14, 2022, from https://e3assoc.org/about/vision-mission/

Ewer, R. F., & Leyhausen, P. (1976). Ethologie der säugetiere: Mit... 2 tabellen. Parey.

Farm Animal Welfare Council. (2012). FAWC updates the five freedoms. *Veterinary Record, 131*, 357.

Fine, A. (Ed.) (2019). *Handbook on animal-assisted therapy: Foundations and guidelines for animal-assisted interventions.* Elsevier.

Fine, A. H., & Beck, A. M. (2019). Chapter 1- Understanding our kinship with animals: input for health care professionals interested in the human–animal bond. In A. H. Fine (Ed.), *Handbook on*

*Animal-Assisted Therapy* (5th ed.). Academic Press. p. 3–12. Available from: https://www.sciencedirect.com/science/article/pii/B9780128153956000018

Flynn, E., Denson, E. B., Mueller, M. K., Gandenberger, J., & Morris, K. N. (2020). Human-animal-environment interactions as a context for youth social-emotional health and wellbeing: Practitioners' perspectives on processes of change, implementation, and challenges. *Complementary Therapies in Clinical Practice, 41*, 101223. http://doi.org/10.1016/j.ctcp.2020.101223

Francis, R. A., & Chadwick, M. A. (2012). What makes a species synurbic? *Applied Geography, 32*(2), 514–521. https://doi.org/10.1016/j.apgeog.2011.06.013

Godinez, A. M., & Fernandez, E. J. (2019). What is the zoo experience? How zoos impact a visitor's behaviors, perceptions, and conservation efforts. *Frontiers in Psychology, 10*. https://doi.org/10.3389/fpsyg.2019.01746

Grabowski, Z. J., McPhearson, T., Matsler, A. M., Groffman, P., & Pickett, S. T. A. (2022). What is green infrastructure? A study of definitions in US city planning. *Frontiers in Ecology and the Environment, 20*(3), 152–160. https://doi.org/10.1002/fee.2445

Gunderson, L. H. (2000). Ecological resilience-in theory and application. *Annual Reviews*. Retrieved April 20, 2022, from https://www.annualreviews.org/doi/full/10.1146/annurev.ecolsys.31.1.425

Hambler, C., & Canney, S. M. (2013). *Conservation*. Cambridge: Cambridge University Press.

Hewson, C. J. (2003). What is animal welfare? Common definitions and their practical consequences. *The Canadian Veterinary Journal = La revue veterinaire canadienne, 44*(6), 496–499.

Immelmann, K. (1980). *Introduction to ethology*. Plenum Press, div. of Plenum Publishing Corp.

Insel, T. R., Winslow, J. T., Wang, Z., & Young, L. J. (1998). Oxytocin, vasopressin, and the neuroendocrine basis of pair bond formation. In H. H. Zingg, C. W. Bourque, & D. G. Bichet (Eds.), *Vasopressin and Oxytocin: Molecular, Cellular, and Clinical Advances* (pp. 215–224). Springer US. https://doi.org/10.1007/978-1-4615-4871-3_28

International Association of Human-Animal Interaction Organizations. (2018). *IAHAIO white paper 2014: Definitions for animal assisted intervention and guidelines for wellness of animals involved.* https://iahaio.org/best-practice/white-paper-on-animal-assisted-interventions/

*International Association of Human-Animal Interaction Organizations: UIA Yearbook Profile*. International Association of Human-Animal Interaction Organizations | UIA Yearbook Profile | Union of International Associations. (n.d.). Retrieved April 12, 2022, from https://uia.org/s/or/en/1100062682

Keck, F., & Lynteris, C. (2018). Zoonosis. *Medicine Anthropology Theory, 5*(3). https://doi.org/10.17157/mat.5.3.372

Klinkenborg, V. (2014). Animal 'personhood': Muddled alternative to real protection. *Yale Environment, 360*. Retrieved May 6, 2022, from https://e360.yale.edu/features/animal_personhood_muddled_alternative_to_real_protection#:~:text=Animal%20personhood%20is%20meant%20to,those%20rights%20to%20adhere%20to.

Kozlowska, K., Walker, P., McLean, L., & Carrive, P. (2015). Fear and the defense cascade: Clinical implications and management. *Harvard Review of Psychiatry, 23*(4), 263–287. https://doi.org/10.1097/HRP.0000000000000065

Lindberg, K., & Hawkins, D. E. (1993). Defining ecotourism. In *Ecotourism: A Guide for Planners and Managers* (pp. 7–11). essay, Ecotourism Society.

Margulis, L., Schwartz, K. V., & Dolan, M. (1999). *Diversity of life: The illustrated guide to the five kingdoms*. Jones and Bartlett Publishers.

McCausland, C. (2014). The five freedoms of animal welfare are rights. *Journal of Agricultural and Environmental Ethics, 27*, 649–662. https://doi.org/10.1007/s10806-013-9483-6

Mellor, D. J., & Beausoleil, N. J. (2015). Extending the 'five domains' model for animal welfare assessment to incorporate positive welfare states. *Animal Welfare, 24*(3), 241–253. https://doi.org/10.7120/09627286.24.3.241

Melson, G. F. (2006). Companion animals and the development of children: Implications of the biophilia hypothesis. In A. Fine (Ed.), *Handbook on Animal-Assisted Therapy* (2nd ed., pp. 375–383). Elsevier. https://doi.org/10.1016/B978-012369484-3/50019-0

*Natural Lifemanship*. (n.d.). Retrieved May 14, 2022, from https://naturallifemanship.com/

Odum, E. P., & Barrett, G. W. (2009). *Fundamentals of ecology*. Boston, MA: Cengage Learning.

Olmert, M. (2009). *Made for each other: The biology of the human-animal bond*. Philadelphia, PA: Da Capo Press.

Pet Partners. (2022). *Who we are*. Pet Partners. https://petpartners.org/about-us/who-we-are/

Pulopulos, M. M., Baeken, C., & De Raedt, R. (2020). Cortisol response to stress: The role of expectancy and anticipatory stress regulation. *Hormones and Behavior, 117*, 104587. https://doi.org/10.1016/j.yhbeh.2019.104587

Siddiq, A. B., & Habib, A. (2016): Anthrozoology - an emerging robust multidisciplinary subfield of anthropological science. *Figshare. Journal Contribution.* https://doi.org/10.6084/m9.figshare.13159736.v1

Starkweather, M. (2020). Pawsitive purpose: The impact of autism assistance dogs on levels of participation and engagement in occupations of children with autism spectrum disorder. [Unpublished master's thesis]. Ithaca College.

Swingland, I. R. (2013). Biodiversity, definition of. *Encyclopedia of Biodiversity* (pp. 399–410). New York: Academic Press. https://doi.org/10.1016/b978-0-12-384719-5.00009-5

Tedeschi, P., & Jenkins, M. A. (Eds.) (2019). *Transforming trauma: Resilience and healing through our connections with animals.* Purdue University Press.

VanFleet, R. (2022). *Why competencies for professionals in animal assisted interventions?* International Institute for Animal Assisted Play Therapy. https://iiaapt.org/why-competencies-for-professionals-in-animal-assisted-interventions/

*Watchable wildlife grant program.* Washington Department of Fish & Wildlife. (n.d.). Retrieved April 12, 2022, from https://wdfw.wa.gov/get-involved/watchable-wildlife-grants#:~:text=Watchable%20wildlife%20or%20wildlife%20viewing, as%20defined%20by%20RCW%2077.32.

WCS.org. (2022). Retrieved May 6, 2022, from https://www.wcs.org/about-us

*What is ecotourism - the International Ecotourism Society.* (n.d.). Retrieved April 12, 2022, from https://ecotourism.org/what-is-ecotourism/

White-Lewis, S. (2019). Equine-assisted therapies using horses as healers: A concept analysis. *Nursing Open, 7*(1), 58–67. https://doi.org/10.1002/nop2.377

Whitehead, M. (2014). Environmental transformations. https://doi.org/10.4324/9781315832678

Wilson, E. O. (1984). *Biophilia: The human bond with other species.* Cambridge: Harvard University Press.

Winkle, M., Gorbing, P., Rogers, J., & Vancoppernolle, D. (2020). *Animal assisted interventions international frequently asked questions.* Animal Assisted Intervention International. https://aai-int.org/wp-content/uploads/2021/02/AAII-Frequently-Asked-Questions-Jan-18-2021.pdf

Wolter, R., Stefanski, V., & Krueger, K. (2018). Parameters for the analysis of social bonds in horses. *Animals: An Open Access Journal from MDPI, 8*(11), 191. https://doi.org/10.3390/ani8110191

Wood, W., Alm, K., Benjamin, J., Thomas, L., Anderson, D., Pohl, L., & Kane, M. (2021). Optimal terminology for services in the United States that incorporate horses to benefit people: A consensus document. *The Journal of Alternative and Complementary Medicine, 27*(1), 88–95. https://doi.org/10.1089/acm.2020.0415

Yapp, R. H. (1922). The concept of habitat. *The Journal of Ecology, 10*(1), 1. https://doi.org/10.2307/2255427

Yarrow, G. (2009, May 1). Wildlife and wildlife management. Digital Collections Home. Retrieved May 6, 2022, from https://dc.statelibrary.sc.gov/handle/10827/41263

# INDEX

Note: **Bold** page numbers refer to tables; *Italic* page numbers refer to figures and page numbers followed by "n" denote endnotes.

AAA *see* animal-assisted activities (AAA)
AAE *see* animal-assisted education (AAE)
AAI *see* animal-assisted interventions (AAI)
AAT *see* animal-assisted therapy (AAT)
ABC-X Model of Crisis 597
Abrams, B. 530
absolute dismissal 445
access to veterinary care (AVC) 131, 248–250, 280, 281, 368
Action Plan for Animal Welfare 31n3
adaptive coping behaviors 218
ADHD 41, 267, 512
adjustment period, feline 356
adolescent development: adaptive coping behaviors 218; attachment and identity 215–216; cognitive and language development 217; emotional and social support 216; physical activity and health 218–219; practical and policy perspectives 220; social skills 217; stress and anxiety 217–218
adoption programs 586
Adult Protective Service (APS) 236, 320
adults: animal visitation programs 513; developmental disabilities 262; human-animal interactions 70–71; intellectual disabilities 262
Adverse Childhood Experiences (ACES) 326
Aegean art 660
*Aesop's Fables* 94, 678
African elephant *732*, 737, 739, 740
African gray 738, *738*
African Savannah elephants 422, 425, 427, *427*
age friendly framework 236
ageism 124, 229, 235

aggression ladder 409
aging 228; in care facilities 70, 230, 237, 238; definitions 228; with dementia 232–233; healthy 228–231, 236, 296; hoarding 100, 233, 314–327, 701; impact on animals 233, 514; in place 229–230; practitioners implications 235; psychological and social health 231–232
aging well 228–231
agriculture 19, 20, 28, 52, 53, 93, 184, 299, 382, 384, 569, 613, 615–616, 645; harmonious integration 614; human welfare 620–621; industrialization 619
Air Carrier Access Act (ACAA) 367, 543
allegorical form 97
alpacas/llamas 496, 511, 614, 631
Amboseli Trust 737
American Heart Association 68
American Hippotherapy Association (AHA) 526, 528
American Psychological Association 2, 11, 125, 478, 479, 484
American Revolutionary War 553
Americans with Disabilities Act (ADA) 67, 265, 367, 384, 543, 544
American Veterinary Medical Association (AVMA) 2, 11, 13, 52, 53, 58, 351, 369, 470, 471, **516,** 729, 738
ancient contract 645, 646
Ancient Egypt art 50, 65, 301, 660, 730
Ancient Greece art 660–662, *661, 662*
Andermann, T. 720
Anderson, D. 5

# Index

Anderson, E. C. 218
Anderson, R. K. 3, 4
Anderson, R. S. 2
Andersen, S. J. 484
Anestis, Mi. 525
animal abuse (AA): by children 341; co-housing and safe haven programs 337–338; definitions 339–340; intersects 333–334; overlap 331–333; physical harm and death 336; relinquishment 337; reporting 341
animal actors 680, 706
animal agriculture 19, 614, 619, 622, 645, 647, 652, 653
Animal and Plant Health Inspection Service (APHIS) 52
animal-animal interactions 587
animal-assisted (AA) interventions 11, 14, 55, 70–75, 80, 110, 146, 150, 217–218, 238, 267, 306, 366, 396, 477, 479, 480, 494, 504–506, 509, *510*, 512, 515, 518, 628, 640, 750, 751
animal-assisted activities (AAAs) 71, 72, 143–145, 261, 262, 366, 383, 489, 494, 497–499, 504, 510
animal-assisted education (AAE) 497, 499, 509; animal protections 26, 124, 320, 327, 341, 448, 457, 518, 691; animals agency 518; attitudes toward animals 86, 99, 215, 220, 517–518; classrooms settings 511; current status 518–519; populations 511; risk assessment 515–516, **516**; success, goals and measures 516; teachers and students 510
Animal Assisted Intervention International (AAII) 487, 488–489, **516**
Animal-Assisted Intervention Professionals (AAAIP) 489–490, 504–505
animal-assisted interventions (AAIs) 55, 74, 81, 86, 139, 238, 477, 750–751; competencies 502–504; cultural factors and utilitarian value 144–146; education and professional community 504–505; foundational appreciation 496; manualization and treatment integrity 87–88; multicultural settings 146–150; religion 141–143; social identities 143–144; standards of practice 499–501, *500*; therapy animal 495–496, 501–502; types 497–499, *498*; value 494–495; veterinary social work 366–367
animal-assisted therapy (AAT) 11, 71, 74, 80, 306, 396, 477, 479, 480, 483, 488, 489, 494–506, 509, 628, 750
animal birth control (ABC) program 145; *see also* ABC-X Model of Crisis
animal caretakers 19, 732, 734
animal care workers (ACWs) 733
animal companionship 235, 602
animal cruelty 101, 102, 219, 317, 319, 322, 323, 325, 326, 332–333, 336, 339, 340, 341, 363, 669, 691–696, 701; and character 693; and interpersonal violence 695–696; on job 696–697; killing with mercy 698; redemption through caring 699–700
animal domain 379
animal emotional expressions 170, 171, 173, 176; automated recognition 176–177; challenges 171–172; ear movements 172; facial expressions 170, 171, 173, 174, 176, 177, 425, 523; mouth movements 172–173; perceiving and recognising 174–176; study 170–171
animal enrichment 706
animal-environment interface 379
animal extremist groups 18
animal factors influencing infection 375–376
animal farmers 19; *see also* animal agriculture
animal health 8, 30, 31n3, 51–53, 55, 59, 79, 110, 204, 221, 249, 250, 279, 305, 364, 365, 375, 379, 384, 440, 564, 569, 590, 591, 645, 749; animal-assisted interventions 55; pet ownership 9, 37, 56, 66, 68–71, 73, 80, 86, 109–111, 113–115, 118, 128, 131, 200–202, 208–210, 214, 215, 220, 229–231, 234–237, 246–251, 259–267, 276, 277, 281, 325, 331, 378, 385, 510, 603, 729; trap-neuter-release programs 55–56; wildlife conservation 55, 721, 722, 736; *see also* animal welfare
Animal Health and Development Law (1993) 575
animal hoarding 314, 320; assessment, intervention, and treatment 322–325; cases 321–322; cats and dogs 317; definitions 314–315; impact on animal welfare 317–318; impact on people and society 318; investigation 319–320; investigation and enforcement 319–321; legislation 325–326; prevalence 326; public policy 325–326; recognition 314; resolution 320; social and cultural factors 318; types 316–317
animal industry groups 31n5
animal kindness and character 698–699
*Animal Liberation: A New Ethics for Our Treatment of Animals* (Singer, 1975) 18
*Animal Machines* (Harrison, 1964) 18, 101
animal maltreatment 691, 693
*The Animal Mind* (Washburn, 1908) 154
animal models 54, 98, 336, 591, 680
animal news stories 682–683
animal-related injury 375, 404, 406, 414; cats 404–405, 407; cattle 405; challenges 405–406; dogs 404, 406–407; horses 405, 407; livestock 407–408; precautions 408–411, *410*, **411**, *412, 413*, 413–414, *414*; snake bites 405
animal-robot chimeras development 29
Animals and Provision of Animal Welfare Act (2014) 575
animals, zoonoses 382

*Index*

Animal Threat Level Reporting System (AT-LAS) 414
animal traction 570
animal training 306, 367, 388, 394, 396, 397, 399, 401, 518, 731, 734; *see also* modern animal training
animal welfare 1, 2, 17, 19, 21–31, 45, 55, 85, 98, 101–103, 105, 110–112, 115, 131, 171, 177, 219, 233, 235, 236, 238, 248, 249, 251, 266, 301, 317, 321, 325, 327, 331, 337, 340, 362, 364, 366, 370, 414, 434, 439, 440, 455, 457, 459–464, 474, 486, 487, 490, 499, 501–503, 514–515, 547, 568, 571, 575–577, 582–591, 616, 621, 631–633, 638, 644–653, 686, 711, 733–737, 747–749; assistance animal 1, 14, 42–45, 67, 71, 75, 396, 497, 537, 538, 540–548; definition 644–646; early cow-calf separation (*see* cow-calf separation); Five Freedoms model 439, 734; foster cows 651–652; laboratory animals 6, 582–588; management practices 646–648; social structure role in 735; societal values 646; veterinary social work 367–368; *see also* animal health
Animal Welfare Act (AWA) 45, 589, 590, 591, 686, 692
Animal Welfare (Sentience) Act 103
animal welfare risk 233; hoarding 233; neglect 233; relinquishment 233
animal welfare science (AWS) 17, 18, 25, 26
Anthony, D. W. 191
Anthropocene 18, 22, 23, 24, 26, 93–105, 421–435
anthropocentrism 93–105, 174, 446, 576; *see also* anthropomorphism
anthropomorphism 93; characteristics 94; defined 680; errors types 97; forms 97; legislation and litigation 101–103; rise and fall 94–96; taxonomies 96–97; types 96; value 97–98
anthrozoology 746–747; academics and practitioners 1; career opportunities 750–751; defined 1, 746; educational/training opportunities 749–750; field of 2, 5; organizational opportunities 750; research/funding opportunities 749
Anthrozoös 2, 6–7
anti-anthropomorphism 98
antibiotic resistance and hormones 54
anticipatory grief 363
antimicrobial resistance 54
anxiety 9, 66, 67, 69, 71, 72, 74, 83, 100, 204, 216–218, 250, 260, 263, 264, 288–291, 294, 304, 306, 308, 309, 314, 322, 334, 338, 353, 355, 368, 395–401, 494, 495, 513, 527, 539, 558, 559, 585, 596, 601, 602, 616, 628, 637
APHIS *see* Animal and Plant Health Inspection Service (APHIS)

applied animal behavior 68, 99
applied anthropomorphism 104
applied ethology 17, 18, 25, 26
applied form 97
aquaria 455, **455**
Aristotle 93
Arkow, P. 331, 369, 486
Arluke, A. 319, 323
'Aryan' peoples 189
Ascione, F. R. 331, 332
Asian elephants *(Elephas maximus) 425,* 427–429, 572–573
'Asiatic nations' 184
assistance animal 537; clinical recommendation 546–547; drawbacks and challenges 540–541; ethical considerations 547; finding and partnering with 546; individual perceptions 544–545; legal rights and restrictions 543–544; public perceptions 545–546
Assistance Dogs International (ADI) 366
Association for Animal-Assisted Intervention Professionals (AAAIP) 489–490, 504–505
Association for Pet Obesity Prevention 263, 266
Association of Feline Practitioners (AAFP) 353
Association of Pet Dog Trainers (APDT) 397
Association of Zoos and Aquariums (AZA) 729, 730, 736, 737, 741
associative learning 389; classical conditioning 389–390, *390*; operant conditioning 390–394, *391*
Atherton, G. 100
attachment theory 203, 215–216
attention, cognitive domains 158
attention oddity (AO) 160
Australian National Kennel Council 705
autism spectrum disorder (ASD) 71, 100, 260–261, 540
automated emotion recognition 176–177
availability 603–604
aversive techniques 518
AVMA *see* American Veterinary Medical Association (AVMA)
AVMA Workplace Wellbeing Certificate Program 369
AVOID campaign 413
AWS *see* animal welfare science (AWS)
axiomatic theories 444
AZA *see* Association of Zoos and Aquariums (AZA)

backyard feeding 708
Baker, O. 276
Beast of Gévaudan 683–685
beasts of burden 428, 510, 519
Beetz, A. 529
Beck, A. M. 2, 3, 4, 314, 708
behavioral observation 84–85, 221, **517**

behaviorism 95, 98, 153, 438, 439
Bekoff, M. 101
Belyaev, D. K. 206
Bennett, J. R. B. 200
Berdychevsky, L. 463
Berry, C. 317
Bharatharaj, J. 104
biocentrism 446, 447
biological diversity 422
biological functioning 644–645, 647
biophilia hypothesis 55, 68, 139, 149, 202–203, 524
biopsychosocial model 67–69, *68,* 204–205
Biopsychosocial-Veterinary assessment (BPS-V) 368
bird keeping 208
birdwatching 708
Blackfish effect 686, 692
Born Free Foundation 462
*Bos primigenius see* cattle
Bovenkerk, B. 31n6
Bowers, K. 59
Bradshaw, J. 1, 5, 9, 99
Brambell, R. 101
breeding animals 27
breed-specific legislation (BSL) 408, 413
Breland, K. 96, 97
Breland, M. 96, 97
Brewbaker, E. J. 277
Broom, D. 456, 457, 459
Brown, S. E. 141, 323
brucellosis 53, 57, 265
Bruni, D. 98
Buddhism 141
The Bull 667, *668*
bull-leaping 660, *661*
Burger, I. 5
Burghardt, G. 96
burnout (BO) 288, 290, 291, 364, 365, 366, 368
Bustad, L. 3, 4, 6, 7

camels 142, 569, 631
*Campylobacter jejuni* infections 380
Canadian Symposium on Pets and Society 9
Canadian Therapeutic Riding Association (CANTRA) 528
canine: breed and age considerations 347–348; home environment 348–349; new dog to cat 350–351; new dog to other dogs 350; new dog to other humans 349–350
canine team (K9 team) 554, 556, 558, 559, 562
captive animals 439, 455, 456, 457, 706
captive breeding 458, 737
captive elephants 428, 448, 457, 730
captivity, non-domestic animals in: ethical considerations 736–737; legislation 737–738; negative impacts 738–740

Carder, G. 459
cardinal concepts 444–448
care around animal feces 381–382
care farming 626, 627
caregivers/caretakers 19, 24, 25, 28, 29, 54, 99, 104, 144, 162, 203, 221, 234, 235, 248, 286–296, 305–307, 316, 335, 337–339, 341, 370, 538, 539, 540, 568, 577, 614–622, 730, 736
caregivers implications 234–235
Carlisle, G. K. 100, 260
carnivores 460, 461, 734, 735
Carr, N. 454, 457, 459, 463
Carson, R. 18
Cassels, M. T. 220
Cassidy, B. 314, 323
cats: bites/scratches 404–405, 407; categorical emotions 169; companion 351; fancier 707, 708; sensory capabilities 156
cattle: cattle-related injuries 405; domestication 180, 188, 189; farm-based interactions 638; housing and management 638, *639;* origin and natural behavior 637–638
cattle-related injuries 405
Cavazos, A. M. 332
cave paintings 673, 677, 690, 719–720
C-BARQ scale 99
CDC *see* Centers for Disease Control and Prevention (CDC)
CENSHARE (Center to Study Human-Animal Relationships and Environments) 4
Center for the Human-Animal Bond 2
Center on Interaction of Animals and Society 2, 3
Centers for Disease Control and Prevention (CDC) 52, 384
*Centrocercus minimus see* Gunnison sage grouse
*Centrocercus urophasianus see* sage grouse
cerebral cortex 154
*Cervus canadensis see* elk
Chadwick, J. 115
"chamber of horrors" 319
Chang, L. 220
*Charadrius melodus see* piping plovers
Cherokee Nation v. Georgia case 37
chickens: farm-based interventions 633; housing and management *632,* 632–633; origin and natural behavior 631
child abuse 219, 250, 363, 695
child-animal bond 214
childhood development 214; adaptive coping behaviors 218; attachment and identity 215–216; cognitive and language development 217; emotional and social support 216; graduation/progression approach 693; physical activity and health 218–219; practical and policy perspectives 220; social skills 217; stress and anxiety 217–218

child-pet relationships 214, 219–220
children: animal abuse 341; children's media 681–682; developmental disabilities 260–261; efficacy, evidence for 512–513; human-animal interactions 70
Chile 576
chimpanzees 21, 97, 103, 161, 174, 205, 308, 428, 434, 737
Chippewa 440
Christian, H. 218
Christianity 65, 141, 683
Christiansen, S. B. 288
Civil Rights Act 43
classical conditioning 154, 155, 389–390, *390*
Cleary, M. 332
clergy 302, 684
climate change 23, 24, 56–57, 308, 421, 426, 430, 432, 434, 577, 679
coercive control 331, 333
co-evolution 17, 20, 28, 29, 184, 206
cognate fields 443
cognition 153–154; brain and learning 154–155; ethical considerations 163; impact on people 161–162; impact on welfare 162; and language development 217; limitation 161; practical perspectives 162; social 160–161
cognitive domains: attention 158; delayed non-match to position 159; discrimination 159; language development 217; non-social cognition 159; pattern learning and visuospatial memory 158–159; perception 157; reversal learning 160; task switching 160; visual search/attention oddity 160
cognitive ethology 98, 103–105, 438
Cohen, A. 662
Cohen, E. 454, 456
Cohen, R. 129–130
collective imaginaries 18, 22–23
collective liberation 124, 133
Colombia 575, 577
commodity, housing and pets 37–39, 619
communal pets 237–238
community-based research methodologies 132
community data 405–406
Companion Animal Bonding Scale 100
companion animal caregiver burden 287–289, 292–295; determinants 293–294; ethical considerations 294–295; historical and diverse perspectives 287–288; *vs.* human family member 292–293; impact 288, 291–292; veterinary care provider 290–291, **291**
companion animal hoarder 323
companion animals 596–609; attachment figures 216, 600–601; benefits and risks of 113–115; crisis 598–599; housing pathways perspective 110–115; not allowing risks 114–115; social support 599–600

compassion 23, 24, 30, 38, 203, 248, 305, 310, 318, 370, 510, 682
compassion fatigue (CF) 290, 364–366, 369, 588, 733
compensation hypotheses 601
complicated grief 310, 363
conceptual/philosophical framings 18
Conover, M. R. 468
consequences 27, 28, 38, 57, 67, 72, 73, 114, 115, 219, 250, 291, 334, 390–393, 396–399, 410, 433, 446–447
consequentialist 446, 447
conservation 1, 27, 307, 422, 427, 428, 435, 455, 679, 708, 720; animal protection 448; conservation biology 421, 443; sentience and personhood 434; wildlife conservation 55, 721
contemporary art 658–659, 672
contemporary artists 672, 673
Contreras-Aguilar, M. D. 530
control/comparison group 80–82
Cook, P. 99
cortisol 67, 82, 218, 485, 502, 529, 539–542
counter-conditioning 395, 398, 402
COVID-19 pandemic 18, 49, 69, 265, 384, 603, 691
cow-calf separation: emotional distress 648, 649; foster cows 651–652; management systems 652; milk production 649; nutrition 650; production costs 649; sensory capabilities 156
criminalizing animal hoarding cases 321
crisis: ABC-X Model 597; companion animals 598–601; crisis-triggering events 598; social support 599; study 596–597
critical anthropomorphism 94, 96, 98, 105
Cross, L. 100
Crossman, M. K. 513
cultural difference 126, 150, 288, 747
cultural factors 140, 144–146, 318
cultural humility 126, 130, 133, 139, 140, 150
cultural services 433
*Curious Naturalists* (Tinbergen, 1958) 95
cuteness effect 205

Dabrowska, M. 706
da Costa, P. L. 708
dairy heifers 615
Damiano, L. 104
Darwin, C. 94, 98, 99, 101, 153, 170, 185, 186
Davis, R. 58
Dawkins, M. S. 26, 96
Dawson, L. 400
deer 195, 302, 472, 698, 722; European red deer 195; white-tailed deer 467, 470, 471
deficit-focused language 131
DEIB *see* diversity, equity, inclusion, and belonging (DEIB)
delayed non-match to position (DNMP) 159

Delta Foundation 3–6, 9, 11, 12, 13, 482–484
dementia 10, 70, 86, 162, 232–233, 237, 238, 322, 495
D'Emilia, W. 332
dental care 356
deontological theories 447
Descartes, R. 21, 93
Descola, P. 183, 184
desensitization 394–395, 398, 531
Desmet, C. 99
detection dog 556–558
developmental disabilities 71, 217, 382; adults with 262; children with 260–261
developmental disorders 71, 537
De Waal, F. B. M. 98
Dewsbury, D. A. 154
Diagnostic & Statistical Manual of Mental Disorders (DSM-5) 260, 310, 315, 321, 326
Diesch, S. 3, 4
DiFiore, J. 103
dimensional emotions 169–170
disabilities 14, 41–45, 67, 74, 131, 228, 250, 259–267, 306, 321, 340, 367, 375, 382, 405, 477, 528, 537, 545
"Discovery Doctrine" 37
discrimination 43, 45, 102, 125, 128, 135, 147, 159, 160, 229, 245, 246, 248, 249, 251, 369, 462, 545, 546–548
diversity 22, 55, 86, 128, 139, 194, 200, 237, 288, 295, 342, 369, 421, 434, 449, 454, 497, 544, 545, 570, 678, 751
diversity, equity, inclusion, and belonging (DEIB) 123, 124, 369, 547–548; considerations 126–130; defined 132, 133; importance 124–125
Dixon, C. A. 219
dizygotic (DZ) 207
dog-bite prevention practices *412*
dogs: bites 404, 406–407; domestication 186; family dogs 208; feral dogs 208; frequencies *195*; interactions 67; neighborhood dogs 208; programs 513; restricted dogs 208
Dolbeer, R. A. 468
dolphins 21, 153, 443, 455–458, 511, 686, 692, 706, 710
DOM2 192–195
domestication 17, 20, 50, 183–196, 205, 553, 576, 613–615; of canids 553; concept 184; explanations 186; narratives 188; traditional approaches to 184–193
domestication syndrome (DS) 185
*The Domestic Dog* (Serpell, 1996) 5
domestic violence (DV) 219, 247, 250, 251, 304, 331, 336, 338, 363, 364, 599, 695–696
donkeys 183, 185, 300, 496, 511, 570, 573–575, 704
Double ABC-X model 597, *597*

downright evil 693
Draper, C. 458
Draught Animal Act 448
Drucker, P. 16
Ducos, P. 189
Dumouchel, P. 104
Duncan, I. J. H. 26
Durie, M. 149

early Christian artworks 663, 664
early modern art 666–669, *667, 668, 669*
EAS *see* equine-assisted services (EAS)
ecocentrism 446
ecofeminists 443, 447
*E. coli* 53
ecosystem 28, 49, 50, 54, 57, 58, 421, 422, 426, 430, 432–433, 435, 443, 449, 461, 683, 720, 722, 730, 748
ecosystem services 422, 432–433, 435, 722
ecotourism 1, 710, 746, 748
*Ectopistes migratorius see* passenger pigeon
education *see* animal-assisted education (AAE)
efficacy, evidence for: children 512–513; evaluation of evidence 514; reading and literacy programs 513; young adults 513
Egyptian tombs scenes 660
Ein, N. 218
ejiao, traditional medicine 570
Ekman, P. 169
elder maltreatment/abuse 219, 236, 319, 340, 363, 395, 696
elephants 422–424, *423,* 448; African Savannah 427; Asian 427–429; forest 425–427; mating and rumbling 424–425; relationships 425
*Elephas maximus see* Asian elephants
*Elgin Courier* 685
elk 722, 725
emotional support animals (ESAs) 367, 538, **538**
emotions 169; behavioural changes 170; categorical 169; dimensional 169–170; expressions 170–172; identify 170; and social support 216; in training 395
empathetic rejection of support 607, 608
Encephalization Quotient (EQ) 427–428
environmental domain 379
environmental enrichment programs 734
environmental health and welfare: climate change 56–57; pollution 57; urbanization 57
epidemic villains 17
Epley, N. 96, 99
equal consideration 445
Equine Assisted Growth and Learning Association (EAGALA) 528
equine-assisted services (EAS) 522; industry adaptations to new nomenclature 527; measure equine welfare 529–530; participants 527; perception of value 522;

professionalization 528; psychotherapy 527; theoretical foundations 524; therapeutic modalities 525–527, **526**
equine-assisted therapy 3, 71, 72, 81, 495, 525
equine welfare 529–530
equity 43, 123–132, 220, 251, 369, 406, 545, 574, 751
ESAs *see* emotional support animals (ESAs)
Esteves, S. 261
estoppel 43
ethical considerations: assistance animal 547; captivity, non-domestic animals in 736–737; cognition 163; companion animal caregiver burden 294–295; farm animals 638–640, *639*; intimate partner violence 340–341; non-domestic animals 736–737; working animal 576–577
ethical inquiry *see* valuational inquiry
ethical responsibility 23
ethical theory 442–445
ethics 18–24, 26, 29, 231, 281, 366, 438–449, 503, 576, 583, 590, 646, 698, 725, 736; animals and nature 442–444; families of 444–445; human-animal relationships 18–20; right *vs.* privilege of pet ownership 236
Ethiopia 149, 570, 575
euphemisms 588
European Citizen's Initiative 647
Euthanasia 26, 55, 56, 286, 290–293, 303, 305–308, 317, 356, 365, 368, 470–472, 474, 575, 583, 588, 640
evictions 37, 40–41, *42,* 44, 111, 118, 273, 320
evidence base HAI, future directions: AAI manualization and treatment integrity 87–88; research findings synthesis 87; transparent reporting 88
ewes 633, *634*
exclusivist sentiments 22
exotic pets 56, 729, 730, 735, 737
explanatory form 97
explicit/episodic memory 158
exploiters 316, 693
extensive research 596, 602
external psychological barriers 607
extinction process 392
extrinsic value 445, 446

Fabre, J. H. 94
Facial Action Coding System (FACS) 174, 176, 177
Fair Housing Act (FHA) 43–45, 367, 543
family bondedness 362
family dogs 208
Family Paws® Parent Education program 350
FAO *see* Food and Agriculture Organization (FAO)
farm animals 640; caretaker interactions 621; cattle 637–638; chickens 631–633, *632*; cow 630; domestication 613–615; ethical considerations 638–640, *639*; farm-based interventions 626; green care 626–627; human-animal relationships 628–631, *629*; human interactions 616–620; sheep 633–635, *634*; swine 635–637, *636, 637*
farm animal welfare *see* animal welfare
Farm Animal Welfare Advisory Committee 102
farm-based interventions 626, 631, 633, 635–637
farming for health *see* care farming
Farr, B. D. 558
Fate of the Animals 671, *672*
Faver, C. A. 332
Federal Aid in Wildlife Restoration Act 721
Federal Bureau of Investigation (FBI) 101, 326, 691, 695
Federation for Tour Operators (FTO) 462
The Federation of Horses in Education and Therapy International (HETI) 532, 528
feline: adjustment period 356; breed and age considerations 351–352; home environment 352–355; home, happy and healthy 356–357; new cat to other cats 355–356; new cat to other humans 355
feral dogs 208
Ferres, K. 177
ferrets 309, 389, 729
Fine, A. H. 200, 203, 204, 479, 484, 487
Fisher, J. A. 97
Five Domains model 439, 733–734
Five Freedoms model 102, 439, 733–734; from discomfort 102; to express normal behavior 102; from fear and distress 102; from hunger and thirst 102; from pain, injury, and disease 102
Fleure, H. J. 184
flooding *see* response blocking
fluoxetine 739
Food and Agriculture Organization (FAO) 52, 569
Food Safety and Inspection Service (FSIS) 53
Food Supply Veterinary Medicine (FSVM) 53
forebrain 154
forest elephants 422–427
formal ethical standards 31n3
foster cows 651–652
Four Cs (carnivores, cores, corridors, and compassion) 449
*The Four Stages of Cruelty* 668
framings, philosophical significance of 20–22
Fraser, D. 23
Freeman, L. M. 367
Fresco, J. 16
Friedmann, E. 3, 5, 8
Frost, R. O. 323, 324
FSIS *see* Food Safety and Inspection Service (FSIS)
FSVM *see* Food Supply Veterinary Medicine (FSVM)

*Gallus gallus domesticus see* chickens
Geller, J. 280
gene editing 27
genetic engineering 27, 697
geographical agency 440
Geraldine Dodge Foundation 3
gerontological models 230
Gibson, A. 317, 323
Glenk, L. M. 502
Global Goals 570
global legislation 115–118
global precedence hypothesis 157
Goal Attainment Scaling (GAS) 532
goats 73, 173, 176, 183, 191, 317, 389, 568, 621
Goffman, E. 276
Good Shepherd 683
gorillas 380, 679
Gottlieb, M. 150
grave stele 661, *662*
Gray, P. B. 207
great auk 720
green care 55, 626–628
Griffin, D. R. 98
Griffin, J. A. 367, 487
Griffin, T. C. 484, 489–490
Grön arena *see* care farming
guide dogs 14, 148, 260, 388, 537, 539
Guide Dogs for the Blind Association 66
*Guide for the Care and Use of Laboratory Animals* 589–590
Gullone, E. 332, 334
Gunnison sage grouse 723
Gustafsson, E. 215
Guthrie, S. E. 93
Gutierrez, L. J. 126–127
gyri 154

Habib, A. 1
habitat degradation 441, 737
habituation 394–395, 515, 622, 631
Haden, S. C. 332, 333
handler-dog interaction *see* working dog
handler familiarity 572
handling skills 574
Harding, E. J. 26
harness system 573
Harris, C. R. 99
Harrison, R. 18
Hart, B. *3*, 4, 5
Hartel, L. 528
Hart, L. *3*, 4, 5
Hartman, C. A. 332
harvesting animals 737
Harvey, A. 110
Hawkins, R. D. 215
Hayes, V. 325
Headey, B. 113

health: animal social determinants of 248; biomedical model of 66; concept of 439; health promotion 232; health records 406; social determinants of 131, 134, 248–251; socio-ecological theory *600*
health in social-ecological systems (HSES) 57
healthy aging 228–231, 236, 296
hearing dogs 537, 539
heart rate variability (HRV) 485, 513, 529, 530
Helmer, D. 189
Herzog, Jr, H. A. 139, 517
Hierarchy of Behavior Change Procedures *398*, 735, 739
Hierarchy of Controls 410, *410*
Hill, E. M. 365
Hill's Pet Nutrition 4, 11
Hinduism 141, 142
Hines, L. 1, 2, 6, 14
hippotherapy 71, 72, **526,** 528
Hoagwood, K. E. 495
hoarders 100, 233, 314–316, 318–324, 326, 327
Hoarding disorder 99–100, 315, 321, 323
Hoarding of Animals Research Consortium (HARC) 314, 316, 327
hobby farmers 633
hoboes 274
Hogg, S. 279
homelessness 272; animal's welfare impact 279–280; challenges of experiencing with pets 276–279; current policy implications 280–281; ethics 281–282; gender, race and ethnicity 273; history 274–275; psychosocial challenges 275–276; risk factors 273; World War II 274
homeownership 38, 45
Hongo, H. 189
Hopper, K. 272
Horse and Rider 666–667, *667*
*The Horse Fair* 669, *670*
horse-related injury 405, 407
horses 573; domestication 191; equestrian injuries 405; sensory capabilities 155
Horses and Humans Research Foundation (HHRF) 484, 485
Horvath, T. 708
housing: cattle 638, *639*; calves 650–652; chickens *632*; commodity 37–39; companion animals within 110–112; fair housing rights 43–45; farm animals 631; global legislation 115–118; intimate partner violence 338–339; sheep *634*, 634–635; swine 636, *636*, 637
housing pathways perspective 110–112
Hughes, E. C. 275, 486
Hull, K. E. 246
Hulstein, R. 705
Hultsman, W. Z. 705

human-animal bond (HAB) 3, 12–13, 14, 202, 367; attachment theory 203; biophilia hypothesis 202–203; biopsychosocial model 204–205; cuteness effect 205; definition 11, 749; the IKEA effect 205; social supports 203–204
Human Animal Bond Research Institute (HABRI) 484, 485
human-animal empathy 574
human-animal interactions (HAIs) 8–10, 14; animal-assisted interventions 81, 86, 733–736; autism and developmental disorders 71; biopsychosocial model 67–69, *68*; child development 70; control groups 80–82; evidence-based practice 79; factors influencing infection 376–378; measurement 82–85; older adults 70–71; on people 730–733; physical and mental outcomes 69; physiological assessments 85; possible negative effects 72–73; variability 85–86
human-animal relationship or bond (HAR/HAB) 730, 731; *see also* human-animal bond (HAB)
human-dog interactions 99, 145, 187, 203, 564
human domain 176, 379, 517
human-environment interface 379
human evolution, hunting and 719–720
human factors influencing infection 375
human health and welfare: antibiotic resistance and hormones 54; medical advances 54; safeguarding food supply 53–54; zoonoses 53
human-horse interactions: environment 528; history 523; methods of research to explore 529; relationships and communication 523–524
human-machine interactions 104–105
human mental health 98–100, 99–100, 158, 522, 590
Humans and Horses Research Foundation (HHRF) 484, 485, 532
Hummel, H. I. 176
Humphrey, T. 172
hunting 456; aristocracy 665; early conservation 720–721; ethics 725; and human evolution 719–720; motivations for 725–726; recreational 723–725; regulated, biological basis for 721; and wildlife conservation funding 721–722; as wildlife management tool 722–723
husbandry 19, 20, 23, 26, 28, 29, 185, 314, 375, 379, 564, 582, 644, 646, 725, 734, 736
hybridization 190, 193, 194
hypersociability 206
hypoallergenic pets 265
hypothalamus 154, 155

identity 124, 133; cultural identity 208; gender identity 86, 150; individual's identity 150; moral identity 22, 276; racial identity 86; sexual identity 150; social identity 143, 279
the IKEA effect 205
Ilissos Sculptor 661, *662*
illegal evictions 40
imaginaries 18, 31n6
imaginative error 97
implicit memory 158
inadequate training methods 572
inclusion 21, 115, 123–134, 230, 252, 261, 339, 342, 369, 510, 751
India 570; anti-poaching efforts 730; constitution of 118; diversity of 140; Jallikattu (bull fighting), social identity 143; pet ownership in 209; upper middle-class population 144
Indonesia, cave paintings 719–720
Industrial Revolution 93, 720
Inert Wolf 672–673, *673*
infection risk: animal factors influencing 375–376; human-animal interaction factors 376–378; human factors influencing 375; organism factors influencing 376; zoonoses drivers 379–380
insurance companies 45, 128, 338, 544, 548
intellectual disabilities 84, 262, 264, 265
Intentional Wellbeing Workplace stressors 364–365
*The Interactions Bibliography* (Anderson, 1988–1999) 5
internal working model 215
International Association of Human Animal Interaction Organizations (IAHAIO) 2, 115, 366, 509, 640
International Society for Anthrozoology (ISAZ) 2, 5, 504
International Society of Feline Medicine (ISFM) 353
interpersonal violence 342, 363, 691, 695–696
interpretive theories 444, 445
inter-species families 273
intimate partner violence (IPV) 331, 364; animal abuse intersects with 333–334; co-housing and safe haven programs 337–338; definitions 339–340; ethical considerations 340–341; human survivors' psychological wellbeing and safety 334–335; overlap 331–333; protection orders 339; psychological and behavioral impacts 336; relinquishment 337; reporting and cross-reporting 340
intrinsic value 433, 434, 445–447
*Introduction to Comparative Psychology* (Morgan, 1894) 94
introgressive capture 190
invasion ecology 443
inverse Clever Hans error 98
IPV *see* intimate partner violence (IPV)

Irvine, L. 276, 277, 280, 281
Islam 118, 141–142
Italian cuisine 453

Jacobson, K. C. 220
Jakeman, M. 219
Jallikattu (bull fighting) 143
Jamieson, D. 101
Jarolmen, J. 219
Jegatheesan, B. 150
Jezierski, T. 558
Johnson, T. C. 336
Johnson v. McIntosh case 37
Joint Advisory Committee on Pets in Society 2
Jones, M. 458
*Journal of the Delta Society* (Beck, 1984-1985) 4
Judaism 141, 146
justice 37, 103, 128, 129, 132–133, 232, 281, 322, 360, 444, 672
Justice Rivera 103

Kant, I. 21, 93
Katcher, A. 2, 3
katie dobutsu concept 146
Kelly, P. E. 128, 129
Kerns, K. A. 218
Keulartz, J. 31n6
K-12 Humane Animal Care Curriculum 4
Kidd, A. H. 276, 279
Kidd, R. M. 276, 279
killing animals 456, 588, 660
Kim, C. H. 273
*King Kong* 679
*King Solomon's Ring* (Lorenz, 1952) 95
Ko, A. 124
Koch, I. J. 206
Kohler, W. 95
kopi luwak 458–459
Kotrschal, K. 560
Kramer, S. C. 104
Kruger, K. A. 367
Kubler-Ross, R. 310

laboratory animals 582; animal welfare 584–586; challenges 583–584; current regulation and policy 589–590; data 584–586; facility 582; human welfare 588–589
Laboratory Animal Welfare Act 686
Labrador Retrievers 277, 557
*Lady with an Ermine* 665, *666*
LaLonde, C. M. 365
lambs 633–635
Larson, G. 190
Latella, D. 530
Leach, E. 141
LEAD Risk Assessment Tool 516
learned helplessness 394

learning theory 154, 388, 389, 393, 572
Least Intrusive, Minimally Aversive (LIMA) 397, 542
legislation, non-domestic animals in captivity 737–738
leisure: birds and human 707–709; with cats 707; definition 704; with dogs 705–707; horses and 709–710
Leonardo's painting 665
Leopold, A. 720
*l'Ermellino* 665
Levinson, B. M. 11, 66
LGBTQIA+: human-animal interactions 246–248; identities and labels 245–246; pet-inclusive policy and practices 251; potential benefits 246–247; social determinants of health 248–251; stressors 247–248
Librado, P. 192
*Life Magazine* 686
LIMA *see* Least Intrusive, Minimally Aversive (LIMA)
*Listeria* species 53
Liu, T. 708
livestock 407–408
LMIC *see* Low and Middle Income Countries (LMIC)
Loar, L. 314
Locke, J. 65
Lockwood, R. 314, 321, 323, 326, 331
Loftin, R. W. 725
Lorenz, K. 95
Lord, D. 461
Lovelock, B. 462
Low and Middle Income Countries (LMIC) 568
*Loxodonta africana see* African elephant
Louv, R. 104

MacLean, E. L. 485
macro social work 361
Mahouts Elephant Foundation 448
Malik, R. 110
*The Malleus Malleficarium* 683
malnutrition 317, 318, 428, 501, 735
Man and Bird, Panel of the Wounded Man 659, *659*
Man-O-War project 531
Man Proposes, God Disposes 670, *670*
Maori model 149
Marines Expeditionary Force (II-MEF) 559
Markowitz, H. 96
Marshall, J. 37
Marshall-Pescini, S. 203
"Marshall Trilogy" 37
Marxist-inspired approach 443
Mary, N. 194
Matchock, R. L. 218
Matijczak, A. 248

McCarthy, M. 209
M'Comisky, J. G. 484
McCulloch, B. 3, 4, 11, 12, *12*, 13
McCulloch, M. 3, *12*
McDonald, S. E. 332, 333
McKee, S. C. 468
McKinney-Vento Homeless Assistance Act 272
McMillan, F. D. 102
Meadow, R. H. 184
measurement in HAI research 82–85; behavioral observation 84–85; physiology 85; population variability 85–86; qualitative methods 84; questionnaires 83–84
media 677, 681–682; audience analysis 680; content analysis 679, 680; historical analysis 679; production analysis 680; sinister canines in 677–686; social inequities 680
medical advances 54, 590
medical alert assistance dogs 537, 539, 540
Medieval art 663–665, *664*
Meints, K. 219
Melson, G. F. 104, 105, 708
memory: cognitive domains 158; explicit/episodic memory 158; implicit memory 158; semantic memory 158; visuospatial memory 158–159; working memory 158
Menakem, R. 125
Mendl, M. 26
Merck Animal Health survey 365
Meyer, I. H. 245
mezzo social work 360
microaggressions 125, 133, 247, 248
micro social work 360
Migratory Bird Act (1918) 430
Miklosi, Á. 99
Military Police Sergeant 559
military working dogs 554, 557, 558, 561–562
Military Working Dog School 557
Mkono, M. 462
mobility assistance dogs 537, 539
modern and contemporary art 671–673, *672, 673*
modern animal training: animal needs 400; behavior and emotion in 395; current trends 396–399; developing and implementing 399; effective, humane, and respectful 399–400; human factor 400–401
modern vivarium 582
monozygotic (MZ) 207
moral and practical approaches 23
moral individualism 21
moral relationalism 21
moral rights 447
moral value 445–446
Morgan, C. L. 94, 95
Morgan, L. 660
Morgan, L. H. 184

Mugford, R. A. 484
Morgan's Canon 94–95
Mueller, M. K. 217, 218, 220, 229
mules 570, 573–575
Multicultural Veterinary Medical Association (MCVMA) 369
multi-species leisure 704–712
Munro, N. D. 189
Munster, U. 730
Muraco, A. 246
*Mustela furo see* ferrets
Myanmar 572, 575
My Pit Bull Is Family 39, 45

Nagel, T. 101
Nathanson, J. N. 314, 322, 323
National Association of Social Workers' Code of Ethics 360
national health surveys 113
National Incident Based Reporting System (NIBRS) 101, 326, 691
National Institute of Child Health and Human Development (NICHD) 484
National Institute of Mental Health's (NIMH) 524
National Institutes of Health (NIH) 13, 214, 296, 484, 749
National Link Coalition 327, 364
2020–2021 National Pet Owners Survey 37, 259
Natterson-Horowitz, B. 59
naturalism 183, 184, 196, 659, 665
naturalness 644–645, 648
natural occurring disease models 162
Nature deficit disorder 104
Nawroth, C. 102
negative animal welfare 219, 457–459
negative consequence 391
negative reinforcement 391–393, 397–399, 562
neighborhood dogs 208
Neolithic Revolution 183
neophobia 572
nests 154, 430, 583
*New Perspectives on Our Lives with Companion Animals* (Katcher and Beck, 1983) 2, 3
news, in animals 682–683
Newton, E. K. 273
New York v. Oneida Indian Nation of New York case 37–38
New Zealand 112, 149, 457, 461, 720
Ng, Z. Y. 367, 484
NhRP *see* Nonhuman Rights Project (NhRP)
Nightingale, F. 260
nineteenth-century art 669–671, *670*
Nipah virus 379
nobility 662, 684
non-anthropocentrism 446

non-associative learning 394; counter-conditioning 395; desensitization 394–395; habituation 394; sensitization 394
non-domestic animals: ethical considerations 736–737; history and current times 730; legislation 737–738; negative impacts 738–740
Nonhuman Rights Project (NhRP) 102–103
non-social cognition 153, 159
non-zoo attractions 457
North American Model of Wildlife Conservation 721
North American Riding for the Handicapped Association, Inc. (NARHA) 528
Norwegian Organization for Animal Assisted Therapy 487
Not One More Vet 369
Nottle, C. 706
novel imaginaries 23, 25, 30
nurse cows *see* foster cows

obesity 219, 262–263, 266, 409, 573, 584
objectives, animal hoarding 319
*Odocoileus* spp. *see* deer
Office International des Epizooties (OIE) 52
Office, M. E. 495, 512
older adults 109; human-animal interactions 70–71
old homelessness 274, 275
*Old Yeller* (Gipson, 1956) 682
*On Aggression* (Lorenz, 1966) 95
*101 Dalmatians* 686
One Health 24, 110, 280; action 52, 57–58; animal health and welfare 55–56; definition 49, *51*; environmental health and welfare 56–57; global level 52; history 50–51; human health and welfare 53–54; opportunity 58–60; organizational 51–53
One Health Act (2016) 58
One Health Commission 52
One Health Day 60
One Health Initiative 60
"one size fits all" approach 748
One Welfare approach 24, 59, 110, 574
*On the Nature of Animals* (Aelian) 681
Oostvaardersplassen (OVP) 449
Open Science Framework 88
operant conditioning 155, 390–394, *391*
oppression 124, 125, 129, 130, 134, 135, 369, 523
organism factors influencing infection 376
organizational one health 51–53
*Origin of Species* (Darwin, 1872) 95
Ortyl, T. A. 246
*Oryctolagus cuniculus domesticus see* rabbit
OSA *see* osteosarcoma (OSA)
osteosarcoma (OSA) 54
overwhelmed caregivers 316
*Ovis orientalis see* sheep

Pakistan 575
Paleolithic art *659*, 659–660
Paleolithic cave art 65
Paleolithic period animals 660
Parish-Plass, N. 148, 149
parrots 300, 306, 738, *738*
passenger pigeon 720
Pastore, C. 706
Patronek, G. J. 314, 315, 317, 322, 323
pattern learning 158–159
Patterson-Kane, E. G. 341
Paul, E. S. 26
Pavlovian conditioning 389
Pavlov, I. P. 95, 154, 389
Peake, H. J. 184
Peloponnesian War (431–404 BCE) 553
people with developmental disabilities: animal welfare 266–267; benefits of pets 260–261; pet ownership 259–260
perceptual learning 154
permanent cultural enlightenment 139
person-first language 125, 133
personhood 21, 22, 26, 231, 429, 434, 448
personification form 97
Pet and Women Safety (PAWS) Act 339, 364
*Pet Animals in Society* (Anderson, 1975) 2
pet care assistance programs 235, 238
Pet Committee 116
*The Pet Connection: Its Influence on Our Health and Quality of Life* (Anderson, Hart, 1984)
Peterborough Bestiary 663, *664*
pet-inclusive housing policy 39, 114, 115, 251
pet keeping, global perspectives of 207–209
pet loss: acceptance 302; emotional support 306; farming 308; fiction 302; history 301; reasons for 305; roadkill 308; services 306, 309–310; therapy 306; understanding 309; wildlife 306–307
pet lovers: accident 303; behavior issues 303–304; lack of resources 304–305; life changes, disappearance, deliberate harm 304; unassisted death 303
*Pet-Oriented Child Psychotherapy* (Levinson, 1969) 478, 479
pet ownership 9, 37, 56, 66, 68–71, 73, 80, 86, 109–111, 113–115, 118, 128, 131, 200–202, 208–210, 214, 215, 220, 229–231, 234–237, 246–251, 259–267, 276, 277, 281, 325, 331, 378, 385, 510, 603, 729, 733, 738; burdens of 264–266; cross-cultural considerations 215; definition 200–201; and families 201–202; guardianship 229; housing 110; mental health 109; for people with developmental disabilities 259–260; right *vs.* privilege of 236
pet-owning tenants 116
Pet Partners 4, 9, 483, 487; chart *498, 500, 503*

pets: in child and adolescent development 215–219; commodity 37–39; definition 37; demographic and cultural shifts 109; evictions and 40–41, *42*; homeownership 45; as property 45–46
*Pets, Benefits & Practice* (Burger, 1990) 5
Pets Evacuation and Transportation Standards (PETS) Act 73, 112, 598
pet therapy 477, 478, 481, 483, 490
*Philosophical Dictionary* (Voltaire, 1763) 100
philosophical significance: of framings 18, 20–22; of imaginaries 22–29
physical activity and health 218–219
physical ailments 263, 366
physical and mental outcomes 69
physical contact 605–606
physical disabilities 262–264
physical health 231
physiology measurement, HAI research 85
pig *see* swine
Pinchak, N. P. 113
*Pinguinus impennis see* great auk
Piper, H. 341
piping plovers *429*, 429–432, *430*, *431*
*The Planet of the Apes* (film) 678
point-in-time methods 272
Polheber, J. P. 218
political ecology 443
pollution 52, 55, 57, 379, 433, 441
POPSCI 679
popular culture: definition 678–679; theories 679–680
population variability 85–86
positionality 84, 123, 129, 133–134
positive animal welfare 459–462, *460*
positive child-pet relationships 219–220
positive consequence 391
positive reinforcement 348, 365, 392–394, 397–401, 518, 542, 562
positive reinforcement training 365, 397, 398, 399, 401, 731, 734
positive welfare 29, 439, 440
possible negative effects 72–73
post-traumatic stress (PTS) 527, 531
post-traumatic stress disorder (PTSD) 68, 71, 263, 306, 322, 495, 537, 539, 540, 558
Potter, K. 262
power 27, 28, 37, 38, 60, 94, 117, 124, 129, 134, 146
Power, E. R. 111, 117
practical perspectives 220, 337; animal companionship 235; caregivers, implications for 234–235; cross-sector collaboration 235; need for planning 234; practitioners, implications for 235
prevalence 214, 222, 229, 250, 262, 264, 265, 272, 325–326, 332, 333, 408, 409, 541, 733

Prevention of Cruelty to Animals Act: 1890 575; 1960 575
Prevention of Cruelty to Animals and Provision of Animal Welfare Act (2014) 575
Primatt, H. 101
primitive system 605
prisoners, vulnerable populations 72
Professional Association of Therapeutic Horsemanship, International (PATH Intl) 525, 528
professionalization 486–488, 528, 529, 750
profit-planet-people trade-offs 29
prolonged grief disorder 310
Promotion of Adoption 562
protection dog 555, 556
Prouvost, C. 99
provisioning services 433
prudential dimension 646
Przewalski horses 192–194
*Psittacus erithacus see* African gray
psychiatric assistance dogs 537, 539, 540
psychiatric patients, vulnerable populations 71–72
psychocentrism 608
psychological and social health 231–232
psychotherapy 66, 82, 128, 130, 148, 150, 340, 478–480, 527
public policy 274, 325, 486, 506, 577, 750
punishment 392, 398
Purugganan, M. D. 184

Qualitative Behavior Assessment 439
qualitative methods, HAI research 84
quality of life 17, 20, 25, 71, 101, 158, 232, 288, 289, 292, 293, 366, 439, 497, 539
Quigley, J. 4
Quirós-Fernández, F. 722

rabbit 73, 205, 220, 309, 319, 496, 511, 587, 590, 696, 724, 729
Rainbow Nation 147
Ramos, D. 318
rams 633–635
randomized control trial (RCT) 513, 530
Rasmussen, J. L. 315
Rawlings, J. M. 518
reasonable accommodation 43–45, 321
recreational hunting 456, 723–725
Reese, L. A. 365
Regan, T. 21, 22, 31n4
regulated hunting 473, 721, 725
regulating services 433
Reid, P. 99
Reisman, R. 336
relational learning 155
relative dismissal 445, 446
religions 141–143, 230

Renaissance art 665–666, *666*
renting with pets 39–40
reptiles 154, 201, 317, 373, 377, 378, 389, 454, 473, 729, 737
rescue hoarders 316
Research Domain Criteria (RDoC) 524
resource deserts 131, 249
respondent conditioning *see* associative learning
respondent learning 389–390
response blocking 736
responsible pet ownership 56, 66, 209–210
responsive chord 691
restricted dogs 208
reversal learning 159–160
reverse anthropocentrism 98
reward-based training 515, 518, 542, 563
rewilding of Asian elephants 448, 449
Re-Wilding program 428
Risley-Curtiss, C. 128
"Robby's Law" 561
robotic pets 105, 237; therapeutic study 104
Roelans, A. M. 708
Rojas, L. 127–129
Rollin, B. E. 27
Roman art 662–663
Romanes, G. 94, 153
Romania 577, 603
Roosevelt, T. 720
Route 66 453
Rowan, A. N. 4, 6, 11, 332
Ruddy mongoose 442
Russow, L. -M. 22
Rütimeyer, L. 184, 185, 194
Rwandan genocide 453

safeguarding food supply 53–54
sage grouse 723
Saint Augustine 93
Saint Thomas Aquinas 93
salivary cortisol levels 70, 82, 494, 541
*Salmonella* 53, 376
salmonellosis 265, 373, 732
San Miguel, S. F. 249
SARS-CoV-2 virus 17, 379
Schmitz, R. M. 251
Schwabe, C. 3
Scoresby, K. J. 367
sculpture from Illisos 661, *662*
secondary traumatic stress (STS) 365
second estate 684
self-injurious behavior (SIB) 738–740
self-reporting 176, 564
semantic memory 158
sensitization 394–395, 398, 531
sensory capabilities 155–156, **156–157**; cats 156, **156–157**; cow 156; horse 155
sentience 99, 101–103, 169–177, 447, 462, 575

sentientism 447
Serpell, J. A. 5, 99, 188, 367
Servais, V. 96, 97
service animals 43, 67, 75, 260, 262, 306, 327, 367, 384, 485, 497, 543, 544, 598; *see also* assistance animal
Shamanism 65
Shanti's Veterinary Mental Health Initiative 369
Sharkey, A. 104
Sharkey, N. 104
Shaw, A. M. 100
sheep 172, 176, 183, 185, 195, 317, 483, 554, 614, 615; farm-based interventions 635; housing and management *634*, 634–635; origin and natural behavior 633–634
Sheets-Johnstone, M. 98
Shibata, T. 104
Shuldiner, E. 206
Siddiq, A. 1
*Silent Spring* 18
Singer, P. 18, 21, 22, 101
sinister canines, in media 677–686
situational error 97
Skinner, B. F. 95, 390
Smilie, K. D. 510
Smith, Y. 139, 140, 150
snake bites 405–406
social cognition 71, 102, 153, 154, 160–161
social community 627
social farming *see* care farming
social identities 143–144, 276, 369
social justice 128, 132, 232, 248, 281, 360, 368, 369
"social license to operate" 19
Social Robotics (SR) 104
social skills 65, 100, 217, 477, 494, 497, 512, 651, 682
Social Skills Improvement System Rating Scale 100
social support 599; availability 603–604; caregiving 606; physical contact 605–606; practical and functional dimensions 602–603; psychological barriers 607–608; psychological benefits 601–602; shared activities 604–605; sources 599–600; stated support 606–607, *607*
social support theory 14, 69
social work 360
social workers 147, 295, 325, 327, 360–364, 366–370
Society for Companion Animal Studies (SCAS) 9
socio-ecological framework 230–231
socio-ecological theory *600*
Spaulding, S. J. 705
speciesism 21, 124, 125–126, 134
S.P.I.D.E.R 734
Spitznagel, M. B. 293
*Sports Illustrated* 686

stability 200, 246, 247, 322, 368
staff training programs 591
Stebbins, R. 705
Steiner, G. 93
Steketee, G. 317, 323, 324
*Sternula antillarum* 430
Stewart, S. 147
stimulus-response learning 155
Stokes, T. 261
Stone, E. 707
Strand, E. 361
stress 66–75, 109, 125–127, 217–218, 245–250, 287–294, 334, 347, 348, 353, 355, 365, 367, 392–400, 470, 485, 494, 497, 510–518, 525, 527–531, 539–542, 558, 571, 586, 596–602, 606, 618, 725, 734
stress-buffering 601, 602, 606
Strimple, E. 4
student learning outcomes 509, 516, **517**
subjects-of-a-life lacks, definition 22
Summons & Complaint 40, 41
superficial form 97
support animals 43, 45, 260, 273, 306, 339, 367, 497, 506, 538, 541
supporting services 433
supremacy 124–126, 447
*Sus scrofa see* swine
sustainable development goals (SDGs) 570–571
swine: farm-based interventions 636–637; housing and management 636, *636, 637*; origin and natural behavior 635
symbols, in animals 680, 681
symposia 98, 661

Tamura, T. 104
task-switching paradigms 160
taurine cattle 188, 190, 194
technological advancements 29
*Telos* 27
Tenancy Fees Act (2019) 114
tenant 40–44, 111, 112, 114–118, 281, 685
Thailand 208, 209, 428, 448, 460, 573, 575
'*thali*' 140
Theory of Mind (ToM) 97, 99, 100
therapeutic modalities 525–527, **526**
therapeutic riding **526,** 527, 528, 531, 532
therapy-first language 525
third estate 684
Thomas, S. 9
Thompson, K. 334
Thompson, P. 28
Thorndike, E. 154
Three Cs (carnivores, cores, and corridors) 449
"Three Rs" 586
Timpano, K. R. 100
Timperio, A. 219
Tinbergen, N. 95

TMR *see* Total Mixed Ration (TMR)
TNR programs *see* trap-neuter-release (TNR) programs
Tolman, E. C. 161
Total Mixed Ration (TMR) 638
touch contact 605–606
tourism 55; consumers 453; implications 462–463; industry 453; negative animal welfare 457–459; positive animal welfare 459–462, *460*; wild animals 454–457, **455**
*Toxoplasma gondii* 53
tramps 274
trans-animalism development 29
trap-neuter-release (TNR) programs 55–56
"Trapping Agony" (Darwin, 1863) 101
trauma: diversity, equity, and inclusion 125–126; vulnerable populations 71
treatment integrity 87–89
Triebenbacher, S. L. 200
*True Life Adventures* 679
Trypanosomiasis 51, 53
Tsing, A. L. 185, 188, 194
Tuke, W. 65, 259
Turgot, A. R. J. 184, 188
Turner, C. H. 153
Turner, D. 5
twin registry 206, 207
Tynon, J. F. 724

Ukraine 31n2, 126, 191
United Kingdom 1, 2, 9, 67, 101, 103, 115, 116, 200, 214, 215, 229, 273, 369, 405, 406, 413, 434, 484, 707
United Nations Environment Programme (UNEP) 49, 52
United States 2; benefits of allowing pets 113; Civil Rights Act 43; housing and pet ownership 37; 2020–2021 National Pet Owners Survey 37
United States Dressage Federation (USDF) 709
university-based animal visitation programs 513
University of Tennessee – Knoxville (UTK) 361
University Press of New England (UPNE) 6
unlawful detainer/eviction action 40
urbanization 57, 274, 421, 422, 430, 468
*Urva smithii see* Ruddy mongoose
US Animal Welfare Act 737
U.S. Department of Agriculture (USDA) 53
*The Uses of Enchantment* (Bettelheim, 2010) 681
utilitarian value 144–146

vaccination 146, 266, 279, 356, 368, 375, 382, 515
valuational (ethical) inquiry 17, 18, 19, 29, 30n1; with AWS and applied ethology 26
value paradigms 446
Verdugo, M. P. 190
Veterans Affairs (VA) 266, 531

veterinarians 3, 4, 13, 18, 22, 23, 25, 28, 49, 50, 52, 53, 59, 129, 249, 265, 278–280, 290, 305, 325, 340, 341, 357, 362, 365, 369, 383, 408, 470, 547, 591
veterinary care 29, 128, 131, 201, 219, 231, 235, 248–250, 266, 279–281, 305, 320, 336, 339, 368, 574
veterinary-client-patient relationship 362, 368
Veterinary Hope Foundation 369
*Veterinary Medicine and Human Health* (Schwabe, 1984) 3
veterinary social work (VSW): animal-assisted interventions 366–367; animal welfare 367–368; areas and associated activities **362**; grief & bereavement 361–363; human and animal violence 363–366; principles 361; specialty 368–369
Victorias 576
Vietnam War 562
Vigne, J. -D. 185
virtue-oriented ethics 447
virtuous cycle 422, 435
visual search tasks 159, 160
visuospatial memory 157–159
Volsche, S. 209
von Frisch, K. 95
Von Holdt, B. M. 206
VSW *see* veterinary social work (VSW)
vulnerable populations 69, 115, 282, 448; developmental disorders 71; individuals with autism 71; prisoners 72; psychiatric patients 71–72; trauma, individuals 71

Wada, K. 104
waiver 40, 43
Walt Disney 679
WALTHAM Foundation 485
Washburn, M. F. 96, 154
Washington State University (WSU) People-Pet Partnership 3
Wasserman, E. A. 154
Wathes, C. 102
Watson, J. B. 95, 153
waxwork figures 453
Weary, D. M. 26
Welfare Quality project 639
Welfare through Competence framework 439
wellbeing 79, 81–83, 85, 86, 88, 118, 124, 149, 246, 248, 251, 334, 360, 361–369, 438–449; in practice 448–449; spheres of *441*
Wells, H. K. 95
Western art 658; Ancient Egypt and Aegean 660; Ancient Greece 660–662, *661, 662*; cultural, religious, and historical perspectives 658–659; early modern art 666–669, *667, 668, 669*; Medieval art 663–665, *664*; modern and contemporary art 671–673, *672, 673*; nineteenth-century art 669–671, *670*; Renaissance art 665–666, *666*; Roman art 662–663
Westgarth, C. 215
white supremacy 124, 134
Why Zoos and Aquariums Matter (WZAM) initiative 741
Wild Animal Reservation and Protection Act (WARPA) 448
wild animals 454–457, **455**
wildlife 442; conservation and management 721–723; and consumptive use 719–721; as management tool 722–723; motivations for hunting and harvesting 724–725; sustainable harvest 721
wildlife conservation 55, 721, 722, 736
wildlife conservation funding 721–722
wildlife problems 472
Williams–Beuren syndrome 206
Williams, D. L. 279
Winter, C. 461
Wilson, E. O. 101, 139
Winchester, B. 4
Wise, S. 102
Witte, T. K. 365
Wittgenstein, Ludwig 444
Wood, W. 525
wolves 124, 186–188, 194, 205, 393, 396, 440, 613, 673, 683–685, 722
Worcester v. Georgia case 37
working animal 568; current policy and legislative perspectives 574–576; elephants 572–573; equids 573–574; ethical considerations 576–577; handlers 568; livelihoods and sustainability 569–571; welfare 571, *571, 572*
working dogs 553–554; bond and influential factors 558; categorization 555, *555*; characteristics 556–557; classification 555, *555*; operational dog/handler interaction 557–558; team pairing 558–562; trainability and sociability 563–564; training and housing 562–563
Working Elephant Program Asia (WEPA) 572
working memory 158–160
workplace accident recording data 406
workplace culture 584
work-related stressors 365–366
World Animal Protection 457
World Health Organization (WHO) 52, 115, 208, 280, 377, 439
World Organization for Animal Health (WOAH) 52, 280, 575, 645
World War I (WWI) 67, 554
World War II 274, 554, 562
Worth, D. 314

Xenophenes 93

Yellowstone National Park 456, 722
Yerkes, R. 95
Young, J. 204, 454, 706
Young, S. M. 207
youth-pet relationships 221
YouTube videos 678

Zarit Burden Interview (ZBI) 289
Zeder, M. A. 185, 186, 191
Zinsstag, J. 57
Ziv, G. 393
Zoobiquity 59
zookeepers 731
zoomorphism 680
zoonoses 23, 49, 50–53, 58, 114, 732; animal selection 381; barriers 382–383; care around animal feces 381–382; epidemiology 374–380; frequency 380; personal hygiene 381; prevention 380–381
Zoonotic diseases 24, 50–53, 55–58, 72, 110, 146, 208, 265, 318, 374, 375, **377,** 378–385, 404, 405, 516
zoonotic infections 375, 380, 385, 515, 732
zoos 7, 10, 20, 66, 67, 148, 281, 307–308, 378, 454, 455, **455,** 457, 459–464, 463, 511, 543, 591, 696, 704, 730, 734, 736, 737, 740, **741**
Zubedat, S. 558